D0086670

Plant Physiology

Third Edition

Plant Physiology

Third Edition

Lincoln Taiz
University of California, Santa Cruz

Eduardo Zeiger
University of California, Los Angeles

Sinauer Associates, Inc., Publishers
Sunderland, Massachusetts

Front Cover

A prickly pear cactus blooms profusely in Big Bend National Park, Texas.
Photo © Willard Clay.

***Plant Physiology*, Third Edition**

For information or to order, address:

Sinauer Associates, Inc.
23 Plumtree Road/PO Box 407
Sunderland, MA 01375 U.S.A.
FAX: 413-549-1118
Email: publish@sinauer.com
www.sinauer.com

Library of Congress Cataloging-in-Publication Data
Taiz, Lincoln.
Plant physiology / Lincoln Taiz, Eduardo Zeiger.-- 3rd ed.
p. cm.
Includes bibliographical references (p.) and index.
ISBN 0-87893-823-0 (hardcover)
1. Plant physiology. I. Zeiger, Eduardo. II. Title.
QK711.2 .T35 2002
571.2--dc21 2002009242

4 3 2

Preface

IT IS WITH A SENSE OF ACCOMPLISHMENT and gratitude that we present to you the third edition of Plant Physiology. The sense of accomplishment is elicited by the outstanding quality of this new edition; our gratitude goes to the large team of players that made it possible.

The first edition, published in 1991, was 565 pages long; the second, published in 1998, had 792 pages. As we began planning the third edition, it was all too evident that providing proper treatment of new developments in the plant sciences would have increased the length of the book still further, while reducing the portion of the book dealing with topics normally covered in an undergraduate plant physiology course. Our answer to this challenge has been to decrease the length of the hard copy of the textbook, and provide users with a fully dedicated, companion web site: **www.plantphys.net**.

This web site includes Chapter 2, *Energy and Enzymes* and Chapter 14, *Gene Expression and Signal Transduction* from the second edition, which provide basic reviews of concepts needed to understand major subjects treated in the book. It also features most of the boxes from the second edition and selected advanced material, not often covered in typical undergraduate courses.

The most dramatic impact of the companion web site concept is that it adds a unique, dynamic dimension to the textbook. Freed from the severe space constraints of a hard copy book, the web site features novel web essays—short articles on cutting-edge subjects written by expert researchers. These essays will be frequently updated, and new ones will be added to accommodate emerging breakthroughs of interest. In addition, the web site adds a multimedia component to the textbook, and it will feature videos and sound clips that are beyond the scope of a conventional textbook. We hope and trust that this new, integrated textbook–web site approach will facilitate their use in both introductory and advanced courses.

Pedagogic improvement and user-friendliness have been the major goals of this new edition. The text has been extensively streamlined and revised for clarity; illustrations and photos are now presented in full color; a comprehensive glossary has been added; and study questions are now available on the web site. At the same time, we have endeavored to incorporate important new developments that reflect the progress of plant biology into the "post-genomic" era.

As in the previous two editions, the third edition would not have been possible without the participation of a large number of dedicated professionals. First and foremost, the outstanding group of contributing authors that superbly updated the chapters and ensured their accuracy and completeness. As in previous editions, we have divided responsibilities on the oversight of the chapters and their integration into a cohesive whole. E.Z. was in charge of chapters 3–12, 18, and 25, while L.T. was in charge of 1, 2, 13–17 and 19–24. We also wish to acknowledge the lasting intellectual legacy of Paul Bernasconi, Malcolm Drew, James W. Siedow, and Wendy K. Silk, contributing authors from the second edition who did not join us for the third.

We continue to benefit from the wisdom of James Funston, our developmental editor, whose association with the book goes back nearly as far as our own. His thoughtful advice has been indispensable for keeping us focused on our pedagogical goals. Most special thanks go to Kathaleen Emerson, our managing editor at Sinauer Associates, who navigated the treacherous waters of the editorial and production process with grace, kindness, and the highest standard of quality. We are also indebted to Stephanie Hiebert who performed her usual outstanding job as copy editor and grammar sleuth; to David McIntyre for resourcefully tracking down images throughout cyberspace and beyond. We were delighted to have Elizabeth Morales back on our team; she rejoined us as artist from the first edition; her beautiful renditions of the art have dramatically raised the quality of our illustration program. We also thank our Web Master, Jason Dirks, and Chris Small, Susan McGlew, Joan Gemme, Marie Scavotto, and Sydney Carroll for their valuable work

We are deeply grateful to our publisher, Andy Sinauer, for his solid good judgment on policy issues, for his patience, and for his willingness to take risks in the interest of quality. We couldn't imagine working for a more congenial and creative publisher.

Finally, we could not have devoted as much time to this project as it inevitably required without the patience and understanding of those around us: our departmental colleagues, research associates, students, and postdocs. We would especially like to thank our wives, Lee Taiz and Yael Zeiger-Fischman for their enthusiasm and support from start to finish.

Lincoln Taiz

Eduardo Zeiger

July 2002

The Authors

Lincoln Taiz is a Professor of Biology at the University of California at Santa Cruz. He received his Ph.D. in Botany from the University of California at Berkeley in 1971. Dr. Taiz worked for many years on the structure and function of vacuolar H^+-ATPases. He has also worked on plant metal tolerance and the role of flavonoids and aminopeptidases in auxin transport. His current research is on UV-B receptors and their roles in phototropism and stomatal opening.

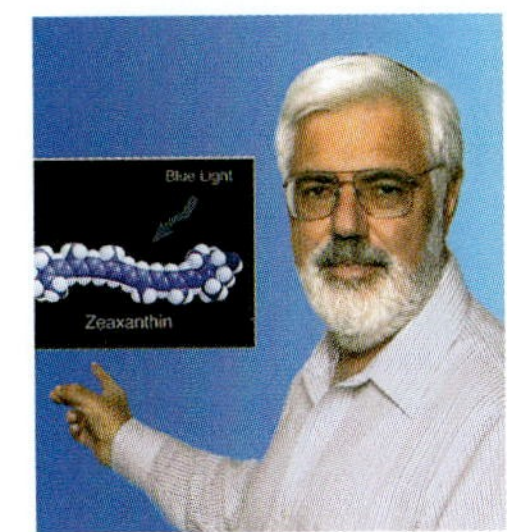

Eduardo Zeiger is a Professor of Biology at the University of California at Los Angeles. He received a Ph.D. in Plant Genetics at the University of California at Davis in 1970. His research interests include stomatal function, the sensory transduction of blue light responses, and the study of stomatal acclimations associated with increases in crop yields.

Principal Contributors

Richard Amasino is a Professor in the Department of Biochemistry at the University of Wisconsin-Madison. He received a Ph.D. in Biology from Indiana University in 1982 in the laboratory of Carlos Miller, where his interests in the induction of flowering were kindled. One of his research interests continues to be the mechanisms by which plants regulate the timing of flower initiation. (Chapter 24)

Robert E. Blankenship is a Professor of Chemistry and Biochemistry at Arizona State University in Tempe. He received his Ph. D. in Chemistry from the University of California at Berkeley in 1975. His professional interests include mechanisms of energy and electron transfer in photosynthetic organisms, and the origin and early evolution of photosynthesis. (Chapter 7)

Arnold J. Bloom is a Professor in the Department of Vegetable Crops at the University of California at Davis. He received a Ph.D. in Biological Sciences at Stanford University in 1979. His research focuses on plant-nitrogen relationships, especially the differences in plant responses to ammonium and nitrate as nitrogen sources. (Chapters 5 and 12)

Ray A. Bressan is a Professor of Plant Physiology at Purdue University. He received a Ph.D. in Plant Physiology from Colorado State University in 1976. Dr. Bressan has studied the basis of salinity and drought tolerance for several years. His recent interests have also turned toward the way plants defend themselves against insects and fungal disease. (Chapter 25)

John Browse is a Professor in the Institute of Biological Chemistry at Washington State University. He received his Ph.D. from the University of Aukland, New Zealand, in 1977. Dr. Browse's research interests include the biochemistry of lipid metabolism and the responses of plants to low temperatures. (Chapter 11)

Bob B. Buchanan is a Professor of Plant and Microbial Biology at the University of California at Berkeley. After working on photosynthesis, Dr. Buchanan turned his attention to seed germination, where his findings have given new insight into germination and led to promising technologies. (Chapter 8)

Daniel J. Cosgrove is a Professor of Biology at the Pennsylvania State University at University Park. His Ph.D. in Biological Sciences was earned at Stanford University. Dr. Cosgrove's research interest is focused on plant growth, specifically the biochemical and molecular mechanisms governing cell enlargement and cell wall expansion. His research team discovered the cell wall loosening proteins called expansins and is currently studying the structure, function, and evolution of this gene family. (Chapter 15)

Peter J. Davies is a Professor of Plant Physiology at Cornell University. He received his Ph.D. in Plant Physiology from the University of Reading in England. His present interests are using genotypes and polygene analysis to elucidate the role of hormones in potato tuberization, stem elongation, and plant senescence. He is the compiler and editor of the principal monograph on plant hormones and has also worked on the isolation of genes of gibberellin biosynthesis. (Chapter 20)

Susan Dunford is an Associate Professor of Biological Sciences at the University of Cincinnati. She received her Ph.D. from the University of Dayton in 1973 with a specialization in plant and cell physiology. Dr. Dunford's research interests include long-distance transport systems in plants, especially translocation in the phloem, and plant water relations. (Chapter 10)

Ruth Finkelstein is a Professor in the Department of Molecular, Cellular and Developmental Biology at the University of California at Santa Barbara. She received her Ph.D., also in Molecular, Cellular and Developmental Biology, from Indiana University in 1986. Her research interests include mechanisms of abscisic acid response, and their interactions with other hormonal, environmental, and nutrient signaling pathways. (Chapter 23)

Donald E. Fosket is a Professor of Developmental and Cell Biology at the University of California at Irvine. He received his Ph.D. in Biology from the University of Idaho and subsequently did postdoctoral work at Brookhaven National Laboratory and at Harvard. (Chapter 16)

Jonathan Gershenzon is a Director of the newly established Max Planck Institute for Chemical Ecology, Jena, Germany. He received his Ph.D. from the University of Texas at Austin in 1984 and did postdoctoral work at Washington State University. His research focuses on the biosynthesis of plant secondary metabolites, and in establishing the roles of these compounds in plant–herbivore interactions. (Chapter 13)

Paul M. Hasegawa is a Professor of Plant Physiology at Purdue University. He earned a Ph.D. in Plant Physiology at the University of California at Riverside. His research has focused on plant morphogenesis and the genetic transformation of plants. He has used his expertise in these areas to study many aspects of stress tolerance in plants, especially ion homeostasis. (Chapter 25)

N. Michele Holbrook is a Professor in the Department of Organismic and Evolutionary Biology at Harvard University. She received her Ph.D.from Stanford University in 1995. Dr. Holbrook's research group focuses on water relations and water transport through the xylem. (Chapters 3 & 4)

Joseph Kieber is an Associate Professor in the Biology Department at the University of North Carolina at Chapel Hill. He earned his Ph.D. in Biology from the Massachusetts Institute of Technology in 1990. Dr. Kieber's research interests include the role of hormones in plant development, with a focus on the signaling pathways for ethylene and ctyokinin, as well as circuitry regulating ethylene biosynthesis. (Chapters 21 & 22)

Robert D. Locy is a Professor of Biological Science at Auburn University in Auburn, Alabama. He received his Ph. D. in Plant Biochemistry from the Purdue University in 1974. His professional interests include biochemical and molecular mechanisms of plant tolerance to abiotic stress, and undergraduate plant science education. (Chapter 25)

Ian Max Møller is a Professor of plant biochemistry at Risø National Laboratory in Denmark. He received his Ph.D. in plant biochemistry from Imperial College, London, UK and has worked for a number of years at Lund University, Sweden. Professor Møller has investigated plant respiration throughout his career and his current interests include the respiratory NAD(P)H dehydrogenases, formation of reactive oxygen species and the functional proteomics of plant mitochondria. (Chapter 11)

Angus Murphy is an Assistant Professor in the Department of Horticulture and Landscape Architecture at the Purdue University. He earned his Ph.D. in Biology from the University of California, Santa Cruz in 1996. Dr. Murphy studies the regulation of auxin transport and the mechanisms by which transport proteins are asymmetrically distributed in plant cells. (Chapter 19)

Ronald J. Poole is a Professor of Biology at McGill University, Montreal. He received his Ph.D. from the University of Birmingham, England, in 1960. Dr. Poole's research interests are in ion transport in plant cells, including electrophysiology, biochemistry, and molecular biology of ion pumps and channels. (Chapter 6)

Allan G. Rasmusson is an Associate Professor at Lund University in Sweden. He received his Ph.D. in plant physiology at the same university in 1994. Dr. Rasmusson's current research centers on expressional regulation of respiratory chain enzymes, especially NAD(P)H dehydrogenases, and their physiological significance. (Chapter 11)

Jane Silverthorne is an Associate Professor in the Department of Biology at the University of California at Santa Cruz. She received her Ph.D. in Biology from the University of Warwick in the United Kingdom in 1980. Her research interests focus on the role of phytochrome in the regulation of molecular aspects of plant development. (Chapter 17)

Thomas C. Vogelmann is a Professor of Plant Physiology at the University of Vermont and State Agriculture. He received his Ph.D. from Syracuse University in 1980, specializing in plant development. His current research is focused on how plants interact with light. Specific research areas include plant tissue optics, leaf structure function related to photosynthesis and environmental stress, and plant adaptations to the environment. (Chapter 9)

Ricardo A. Wolosiuk is a Professor in the Instituto de Investigaciones Bioquímicas at the University of Buenos Aires. He received his Ph.D. in Chemistry from the same university in 1974. Dr. Wolosiuk's research interests concern the modulation of chloroplast metabolism and the structure and function of plant proteins. (Chapter 8)

Reviewers

Steffen Abel
University of California, Davis

Lisa Baird
University of San Diego

Wade Berry
University of California, Los Angeles

Mary Bisson
The University of Buffalo

Nick Carpita
Purdue University

James Ehleringer
University of Utah

Steven Huber
North Carolina State University, Raleigh

Tatsuo Kakimoto
Osaka University, Osaka, Japan

J. Clark Lagarias
University of California, Davis

Park S. Nobel
University of California, Los Angeles

David J. Oliver
Iowa State University

Anne Osbourn
The Sainsbury Laboratory, Norwich, U.K.

Phil Reid
Smith College

Eric Schaller
University of New Hampshire

Julian Schroeder
University of California, San Diego

Susan Singer
Carleton College

Edgar Spalding
University of Wisconsin, Madison

Tai-Ping Sun
Duke University

Heven Sze
University of Maryland

Robert Turgeon
Cornell University

Jan Zeevaart
Michigan State University

Contents in Brief

Table of Contents

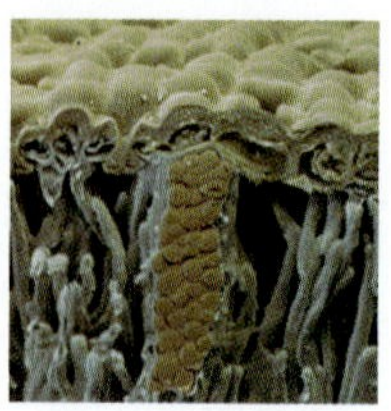

UNIT I

Transport and Translocation of Water and Solutes 31

3 Water and Plant Cells 33

4 Water Balance of Plants 47

5 *Mineral Nutrition 67*

6 *Solute Transport 87*

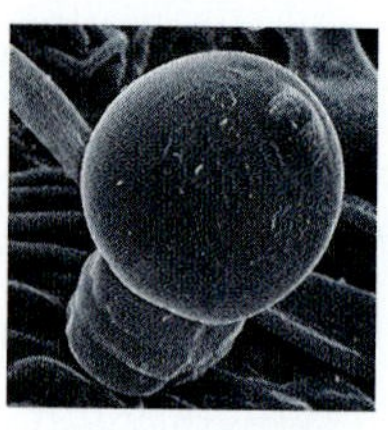

UNIT II

Biochemistry and Metabolism 109

7 Photosynthesis: The Light Reactions 111

8 Photosynthesis: Carbon Reactions 145

9 Photosynthesis: Physiological and Ecological Considerations 171

10 Translocation in the Phloem 193

11 Respiration and Lipid Metabolism 223

12 *Assimilation of Mineral Nutrients 259*

13 Secondary Metabolites and Plant Defense 283

UNIT III
Growth and Development 309

18 Blue-Light Responses: Stomatal Movements and Morphogenesis 403

19 Auxin: The Growth Hormone 423

22 *Ethylene: The Gaseous Hormone 519*

23 *Abscisic Acid: A Seed Maturation and Antistress Signal* 539

24 *The Control of Flowering* 559

Chapter

Plant Cells

THE TERM *CELL* IS DERIVED from the Latin *cella*, meaning storeroom or chamber. It was first used in biology in 1665 by the English botanist Robert Hooke to describe the individual units of the honeycomb-like structure he observed in cork under a compound microscope. The "cells" Hooke observed were actually the empty lumens of dead cells surrounded by cell walls, but the term is an apt one because cells are the basic building blocks that define plant structure.

This book will emphasize the physiological and biochemical functions of plants, but it is important to recognize that these functions depend on structures, whether the process is gas exchange in the leaf, water conduction in the xylem, photosynthesis in the chloroplast, or ion transport across the plasma membrane. At every level, structure and function represent different frames of reference of a biological unity.

This chapter provides an overview of the basic anatomy of plants, from the organ level down to the ultrastructure of cellular organelles. In subsequent chapters we will treat these structures in greater detail from the perspective of their physiological functions in the plant life cycle.

PLANT LIFE: UNIFYING PRINCIPLES

The spectacular diversity of plant size and form is familiar to everyone. Plants range in size from less than 1 cm tall to greater than 100 m. Plant morphology, or shape, is also surprisingly diverse. At first glance, the tiny plant duckweed (*Lemna*) seems to have little in common with a giant saguaro cactus or a redwood tree. Yet regardless of their specific adaptations, all plants carry out fundamentally similar processes and are based on the same architectural plan. We can summarize the major design elements of plants as follows:

- As Earth's primary producers, green plants are the ultimate solar collectors. They harvest the energy of sunlight by converting light energy to chemical energy, which they store in bonds formed when they synthesize carbohydrates from carbon dioxide and water.

- Other than certain reproductive cells, plants are nonmotile. As a substitute for motility, they have evolved the ability to grow toward essential resources, such as light, water, and mineral nutrients, throughout their life span.
- Terrestrial plants are structurally reinforced to support their mass as they grow toward sunlight against the pull of gravity.
- Terrestrial plants lose water continuously by evaporation and have evolved mechanisms for avoiding desiccation.
- Terrestrial plants have mechanisms for moving water and minerals from the soil to the sites of photosynthesis and growth, as well as mechanisms for moving the products of photosynthesis to nonphotosynthetic organs and tissues.

OVERVIEW OF PLANT STRUCTURE

Despite their apparent diversity, all seed plants (see **Web Topic 1.1**) have the same basic body plan (Figure 1.1). The vegetative body is composed of three organs: **leaf**, **stem**, and **root**. The primary function of a leaf is photosynthesis, that of the stem is support, and that of the root is anchorage and absorption of water and minerals. Leaves are attached to the stem at **nodes**, and the region of the stem between two nodes is termed the **internode**. The stem together with its leaves is commonly referred to as the **shoot**.

There are two categories of seed plants: gymnosperms (from the Greek for "naked seed") and angiosperms (based on the Greek for "vessel seed," or seeds contained in a vessel). **Gymnosperms** are the less advanced type; about 700 species are known. The largest group of gymnosperms is the conifers ("cone-bearers"), which include such commercially important forest trees as pine, fir, spruce, and redwood.

Angiosperms, the more advanced type of seed plant, first became abundant during the Cretaceous period, about 100 million years ago. Today, they dominate the landscape, easily outcompeting the gymnosperms. About 250,000 species are known, but many more remain to be characterized. The major innovation of the angiosperms is the flower; hence they are referred to as *flowering plants* (see **Web Topic 1.2**).

Plant Cells Are Surrounded by Rigid Cell Walls

A fundamental difference between plants and animals is that each plant cell is surrounded by a rigid **cell wall**. In animals, embryonic cells can migrate from one location to another, resulting in the development of tissues and organs containing cells that originated in different parts of the organism.

In plants, such cell migrations are prevented because each walled cell and its neighbor are cemented together by a **middle lamella**. As a consequence, plant development, unlike animal development, depends solely on patterns of cell division and cell enlargement.

FIGURE 1.1 Schematic representation of the body of a typical dicot. Cross sections of (A) the leaf, (B) the stem, and (C) the root are also shown. Inserts show longitudinal sections of a shoot tip and a root tip from flax (*Linum usitatissimum*), showing the apical meristems. (Photos © J. Robert Waaland/Biological Photo Service.)

Plant cells have two types of walls: primary and secondary (Figure 1.2). **Primary cell walls** are typically thin (less than 1 μm) and are characteristic of young, growing cells. **Secondary cell walls** are thicker and stronger than primary walls and are deposited when most cell enlargement has ended. Secondary cell walls owe their strength and toughness to **lignin**, a brittle, gluelike material (see Chapter 13).

The evolution of lignified secondary cell walls provided plants with the structural reinforcement necessary to grow vertically above the soil and to colonize the land. Bryophytes, which lack lignified cell walls, are unable to grow more than a few centimeters above the ground.

New Cells Are Produced by Dividing Tissues Called Meristems

Plant growth is concentrated in localized regions of cell division called **meristems**. Nearly all nuclear divisions (mitosis) and cell divisions (cytokinesis) occur in these meristematic regions. In a young plant, the most active meristems are called **apical meristems**; they are located at the tips of the stem and the root (see Figure 1.1). At the nodes, **axillary buds** contain the apical meristems for branch shoots. Lateral roots arise from the **pericycle**, an internal meristematic tissue (see Figure 1.1C). Proximal to (i.e., next to) and overlapping the meristematic regions are zones of cell elongation in which cells increase dramatically in length and width. Cells usually differentiate into specialized types after they elongate.

The phase of plant development that gives rise to new organs and to the basic plant form is called **primary growth**. Primary growth results from the activity of apical meristems, in which cell division is followed by progressive cell enlargement, typically elongation. After elongation in a given region is complete, **secondary growth** may occur. Secondary growth involves two lateral meristems: the **vascular cambium** (plural *cambia*) and the **cork cambium.** The vascular cambium gives rise to secondary xylem (wood) and secondary phloem. The cork cambium produces the periderm, consisting mainly of cork cells.

Three Major Tissue Systems Make Up the Plant Body

Three major tissue systems are found in all plant organs: dermal tissue, ground tissue, and vascular tissue. These tis-

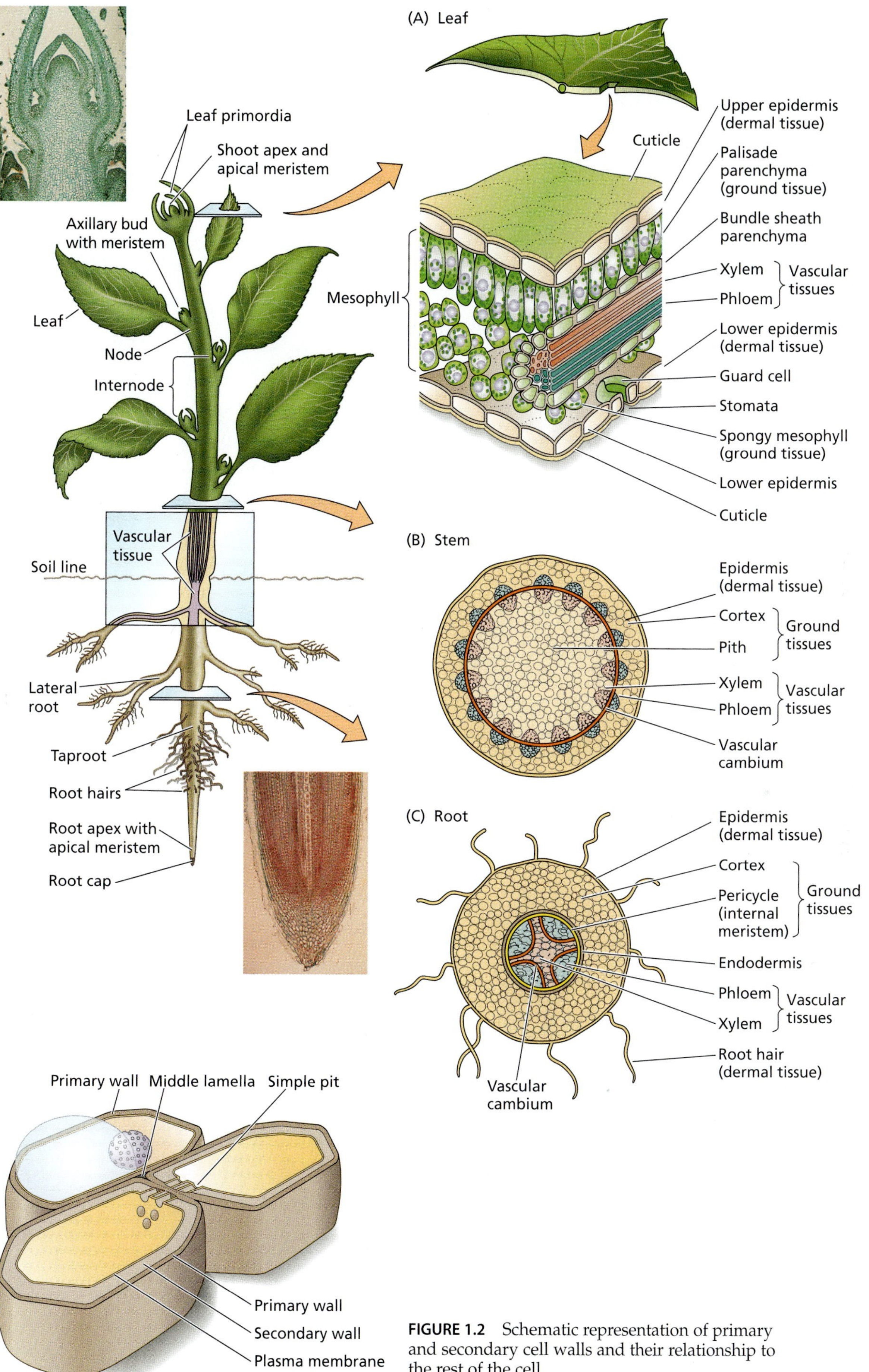

FIGURE 1.2 Schematic representation of primary and secondary cell walls and their relationship to the rest of the cell.

(A) Dermal tissue: epidermal cells

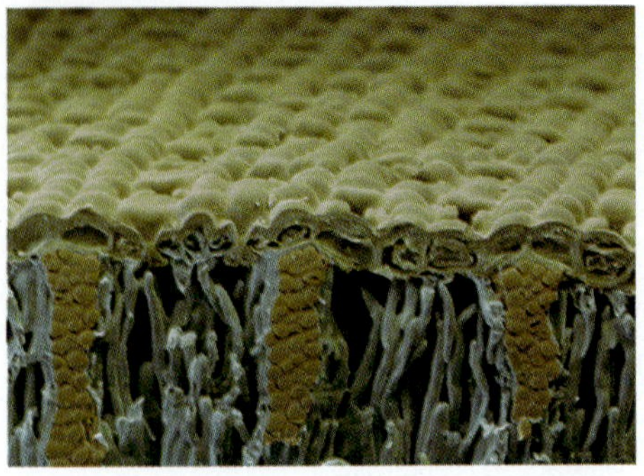

(B) Ground tissue: parenchyma cells

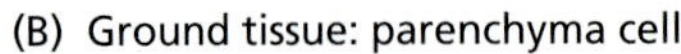

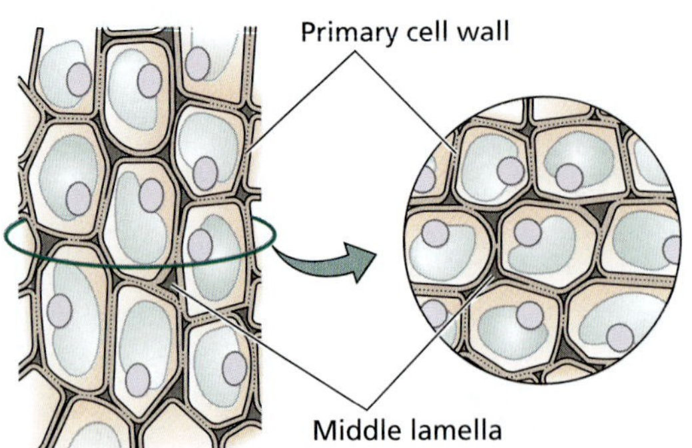

(C) Ground tissue: collenchyma cells

(D) Ground tissue: sclerenchyma cells

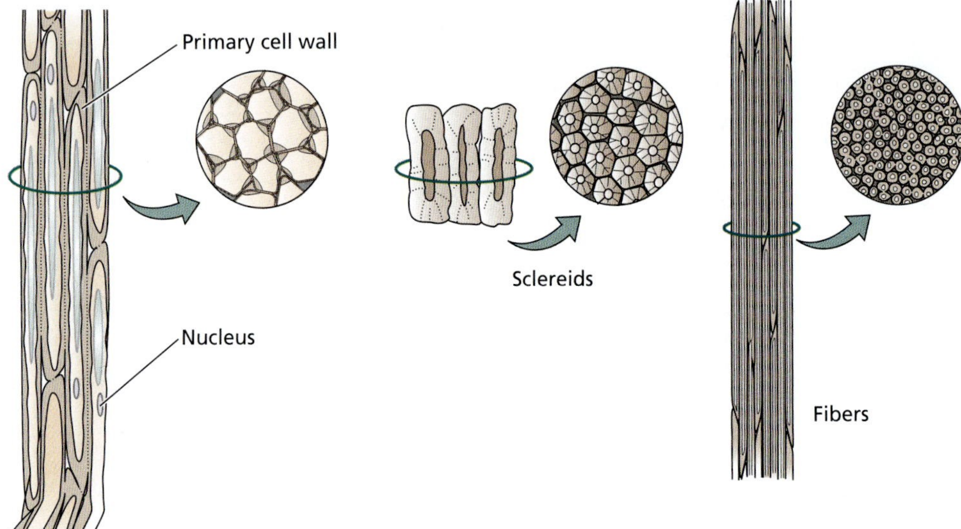

(E) Vascular tisssue: xylem and phloem

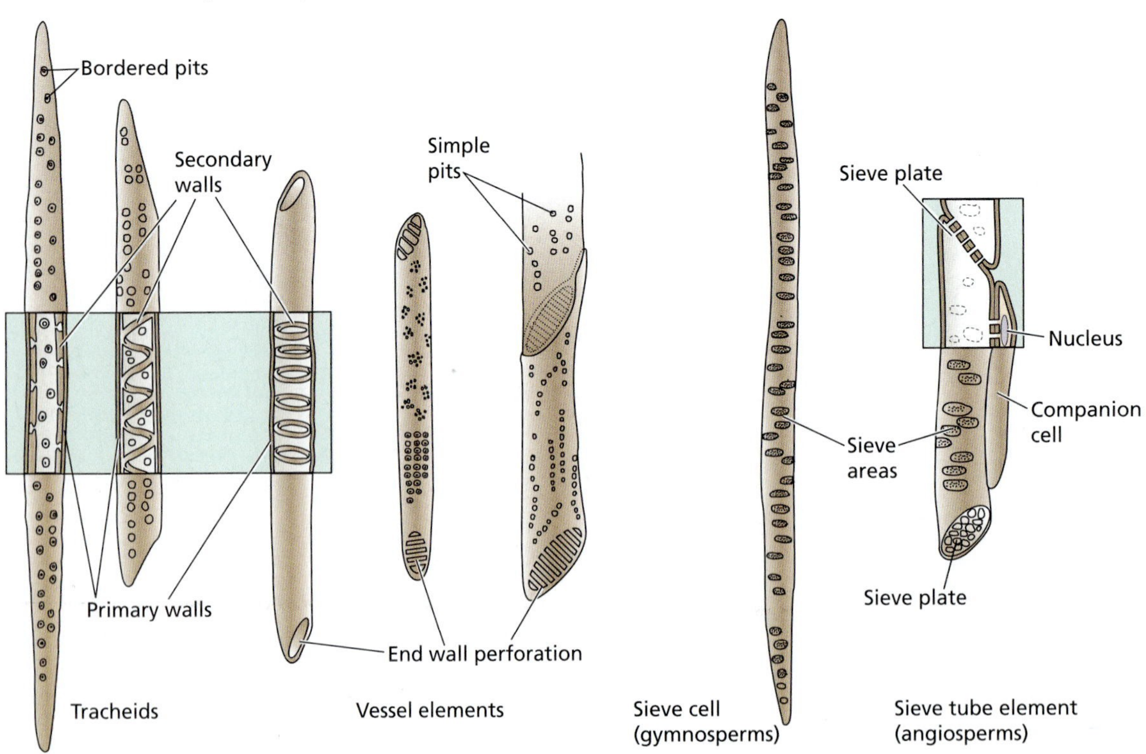

◀ **FIGURE 1.3** (A) The outer epidermis (dermal tissue) of a leaf of *Welwischia mirabilis* (120×). Diagrammatic representations of three types of ground tissue: (B) parenchyma, (C) collenchyma, (D) sclerenchyma cells, and (E) conducting cells of the xylem and phloem. (A © Meckes/Ottawa/Photo Researchers, Inc.)

sues are illustrated and briefly chacterized in Figure 1.3. For further details and characterizations of these plant tissues, see **Web Topic 1.3**.

THE PLANT CELL

Plants are multicellular organisms composed of millions of cells with specialized functions. At maturity, such specialized cells may differ greatly from one another in their structures. However, all plant cells have the same basic eukaryotic organization: They contain a nucleus, a cytoplasm, and subcellular organelles, and they are enclosed in a membrane that defines their boundaries (Figure 1.4). Certain structures, including the nucleus, can be lost during cell maturation, but all plant cells *begin* with a similar complement of organelles.

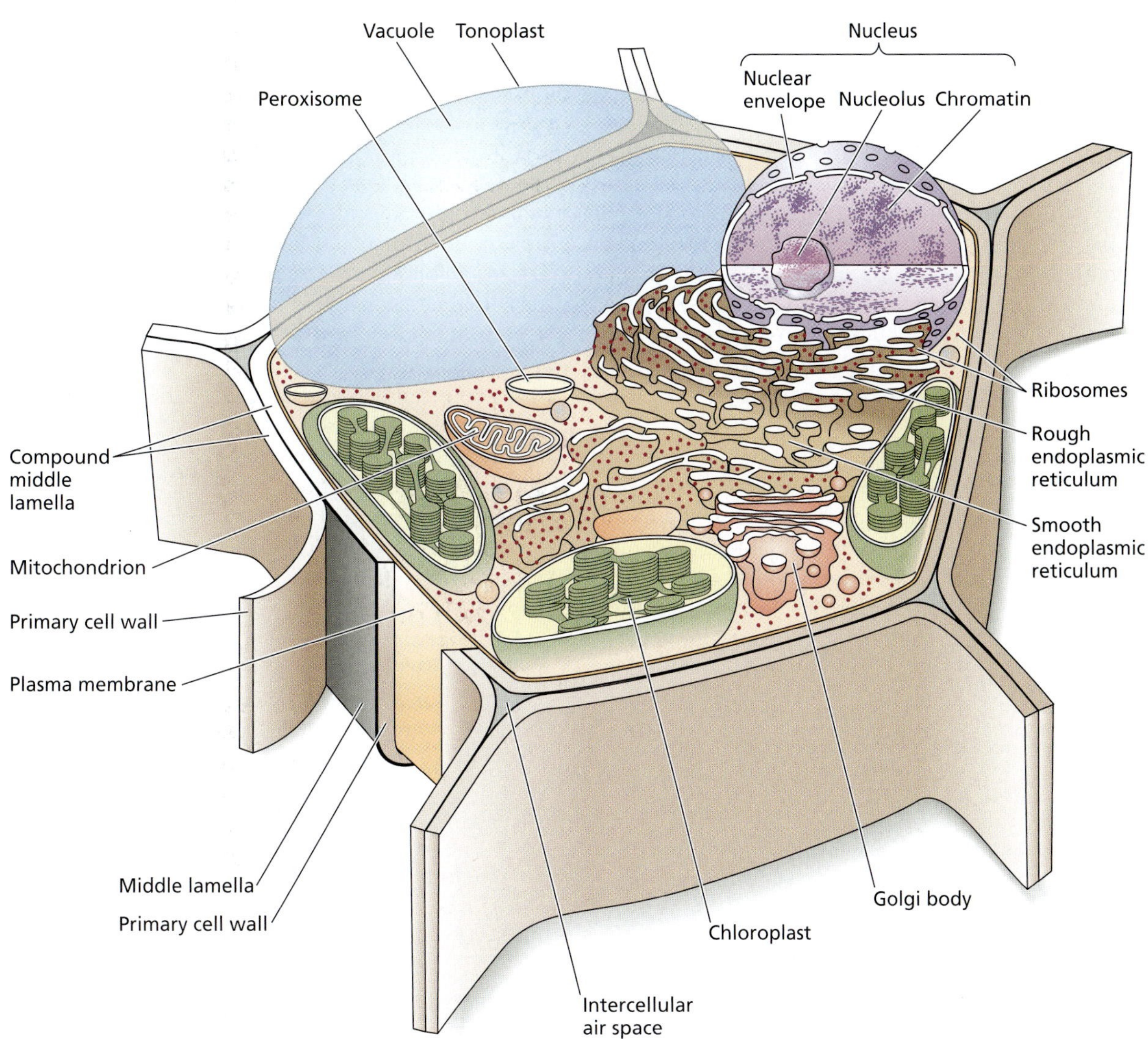

FIGURE 1.4 Diagrammatic representation of a plant cell. Various intracellular compartments are defined by their respective membranes, such as the tonoplast, the nuclear envelope, and the membranes of the other organelles. The two adjacent primary walls, along with the middle lamella, form a composite structure called the compound middle lamella.

An additional characteristic feature of plant cells is that they are surrounded by a cellulosic cell wall. The following sections provide an overview of the membranes and organelles of plant cells. The structure and function of the cell wall will be treated in detail in Chapter 15.

Biological Membranes Are Phospholipid Bilayers That Contain Proteins

All cells are enclosed in a membrane that serves as their outer boundary, separating the cytoplasm from the external environment. This **plasma membrane** (also called **plasmalemma**) allows the cell to take up and retain certain substances while excluding others. Various transport proteins embedded in the plasma membrane are responsible for this selective traffic of solutes across the membrane. The accumulation of ions or molecules in the cytosol through the action of transport proteins consumes metabolic energy. Membranes also delimit the boundaries of the specialized internal organelles of the cell and regulate the fluxes of ions and metabolites into and out of these compartments.

According to the **fluid-mosaic model**, all biological membranes have the same basic molecular organization. They consist of a double layer (*bilayer*) of either phospholipids or, in the case of chloroplasts, glycosylglycerides, in which proteins are embedded (Figure 1.5A and B). In most membranes, proteins make up about half of the membrane's mass. However, the composition of the lipid components and the properties of the proteins vary from membrane to membrane, conferring on each membrane its unique functional characteristics.

Phospholipids. Phospholipids are a class of lipids in which two fatty acids are covalently linked to glycerol, which is covalently linked to a phosphate group. Also attached to this phosphate group is a variable component, called the *head group*, such as serine, choline, glycerol, or inositol (Figure 1.5C). In contrast to the fatty acids, the head groups are highly polar; consequently, phospholipid molecules display both hydrophilic and hydrophobic properties (i.e., they are *amphipathic*). The nonpolar hydrocarbon chains of the fatty acids form a region that is exclusively hydrophobic—that is, that excludes water.

Plastid membranes are unique in that their lipid component consists almost entirely of **glycosylglycerides** rather than phospholipids. In glycosylglycerides, the polar head group consists of galactose, digalactose, or sulfated galactose, without a phosphate group (see **Web Topic 1.4**).

The fatty acid chains of phospholipids and glycosylglycerides are variable in length, but they usually consist of 14 to 24 carbons. One of the fatty acids is typically *saturated* (i.e., it contains no double bonds); the other fatty acid chain usually has one or more *cis* double bonds (i.e., it is *unsaturated*).

The presence of *cis* double bonds creates a kink in the chain that prevents tight packing of the phospholipids in the bilayer. As a result, the fluidity of the membrane is increased. The fluidity of the membrane, in turn, plays a critical role in many membrane functions. Membrane fluidity is also strongly influenced by temperature. Because plants generally cannot regulate their body temperatures, they are often faced with the problem of maintaining membrane fluidity under conditions of low temperature, which tends to decrease membrane fluidity. Thus, plant phospholipids have a high percentage of unsaturated fatty acids, such as oleic acid (one double bond), linoleic acid (two double bonds) and α-linolenic acid (three double bonds), which increase the fluidity of their membranes.

Proteins. The proteins associated with the lipid bilayer are of three types: integral, peripheral, and anchored. **Integral proteins** are embedded in the lipid bilayer. Most integral proteins span the entire width of the phospholipid bilayer, so one part of the protein interacts with the outside of the cell, another part interacts with the hydrophobic core of the membrane, and a third part interacts with the interior of the cell, the cytosol. Proteins that serve as ion channels (see Chapter 6) are always integral membrane proteins, as are certain receptors that participate in signal transduction pathways (see Chapter 14). Some receptor-like proteins on the outer surface of the plasma membrane recognize and bind tightly to cell wall consituents, effectively cross-linking the membrane to the cell wall.

Peripheral proteins are bound to the membrane surface by noncovalent bonds, such as ionic bonds or hydrogen bonds, and can be dissociated from the membrane with high salt solutions or chaotropic agents, which break ionic and hydrogen bonds, respectively. Peripheral proteins serve a variety of functions in the cell. For example, some are involved in interactions between the plasma membrane and components of the cytoskeleton, such as microtubules and actin microfilaments, which are discussed later in this chapter.

Anchored proteins are bound to the membrane surface via lipid molecules, to which they are covalently attached. These lipids include fatty acids (myristic acid and palmitic acid), prenyl groups derived from the isoprenoid pathway (farnesyl and geranylgeranyl groups), and glycosylphosphatidylinositol (GPI)-anchored proteins (Figure 1.6) (Buchanan et al. 2000).

The Nucleus Contains Most of the Genetic Material of the Cell

The **nucleus** (plural *nuclei*) is the organelle that contains the genetic information primarily responsible for regulating the metabolism, growth, and differentiation of the cell. Collectively, these genes and their intervening sequences are referred to as the **nuclear genome**. The size of the nuclear genome in plants is highly variable, ranging from about 1.2 $\times 10^8$ base pairs for the diminutive dicot *Arabidopsis thaliana* to 1×10^{11} base pairs for the lily *Fritillaria assyriaca*. The

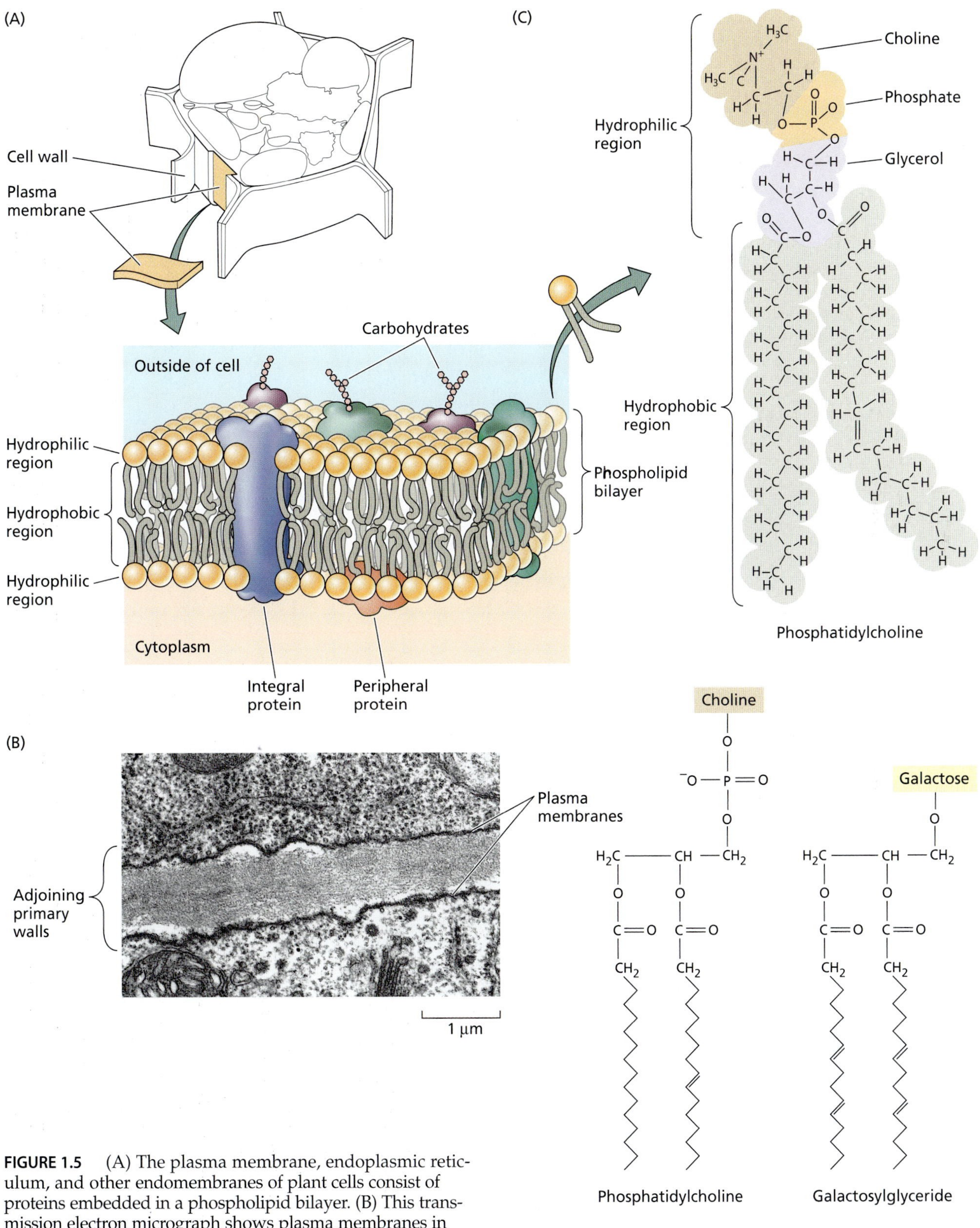

FIGURE 1.5 (A) The plasma membrane, endoplasmic reticulum, and other endomembranes of plant cells consist of proteins embedded in a phospholipid bilayer. (B) This transmission electron micrograph shows plasma membranes in cells from the meristematic region of a root tip of cress (*Lepidium sativum*). The overall thickness of the plasma membrane, viewed as two dense lines and an intervening space, is 8 nm. (C) Chemical structures and space-filling models of typical phospholipids: phosphatidylcholine and galactosylglyceride. (B from Gunning and Steer 1996.)

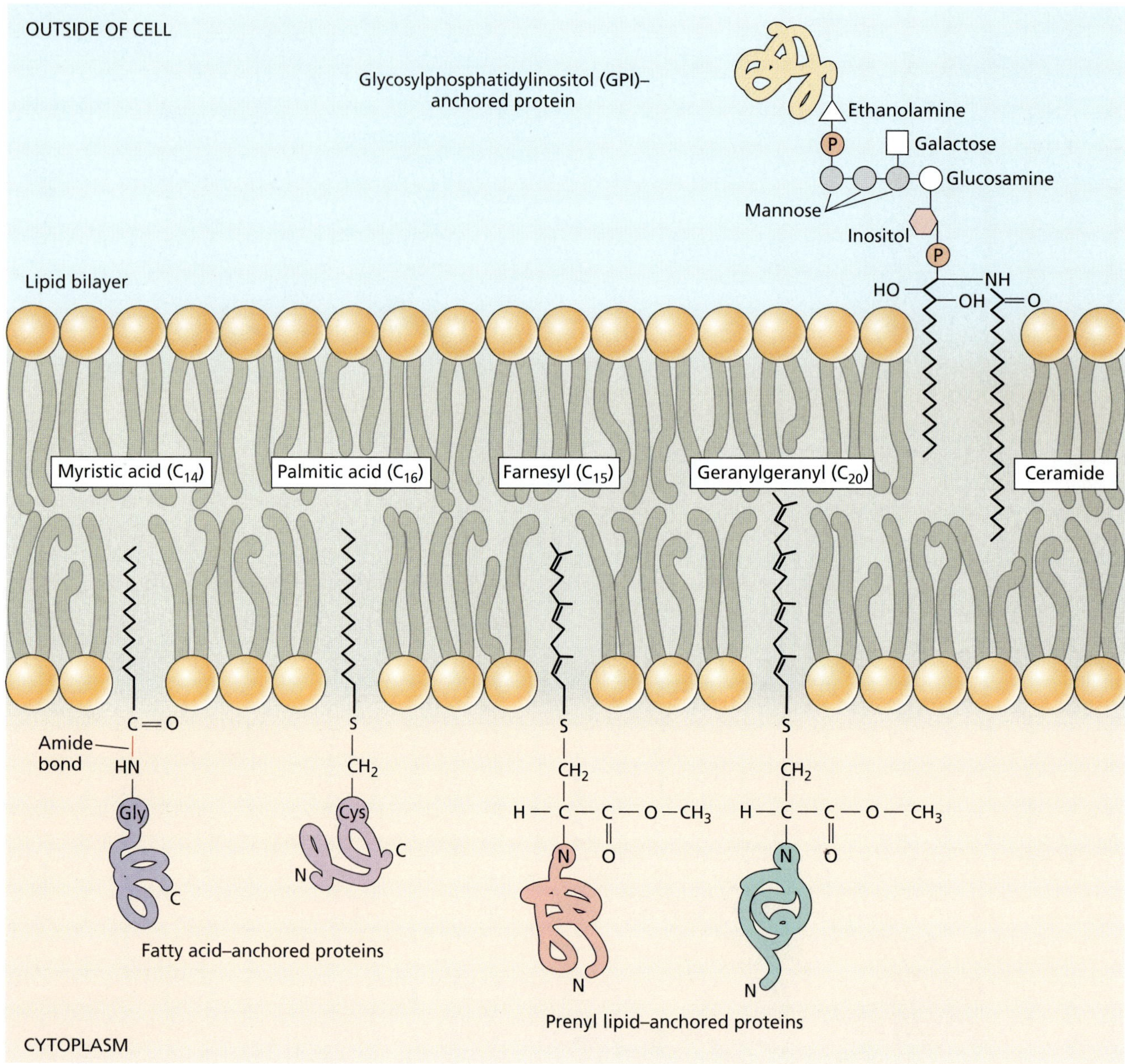

FIGURE 1.6 Different types of anchored membrane proteins that are attached to the membrane via fatty acids, prenyl groups, or phosphatidylinositol. (From Buchanan et al. 2000.)

remainder of the genetic information of the cell is contained in the two semiautonomous organelles—the chloroplasts and mitochondria—which we will discuss a little later in this chapter.

The nucleus is surrounded by a double membrane called the **nuclear envelope** (Figure 1.7A). The space between the two membranes of the nuclear envelope is called the **perinuclear space**, and the two membranes of the nuclear envelope join at sites called **nuclear pores** (Figure 1.7B). The nuclear "pore" is actually an elaborate structure composed of more than a hundred different proteins arranged octagonally to form a **nuclear pore complex** (Figure 1.8). There can be very few to many thousands of nuclear pore complexes on an individual nuclear envelope. The central "plug" of the complex acts as an active (ATP-driven) transporter that facilitates the movement of macromolecules and ribosomal subunits both into and out of the nucleus. (Active transport will be discussed in detail in Chapter 6.) A specific amino acid sequence called the **nuclear localization signal** is required for a protein to gain entry into the nucleus.

The nucleus is the site of storage and replication of the **chromosomes**, composed of DNA and its associated proteins. Collectively, this DNA–protein complex is known as

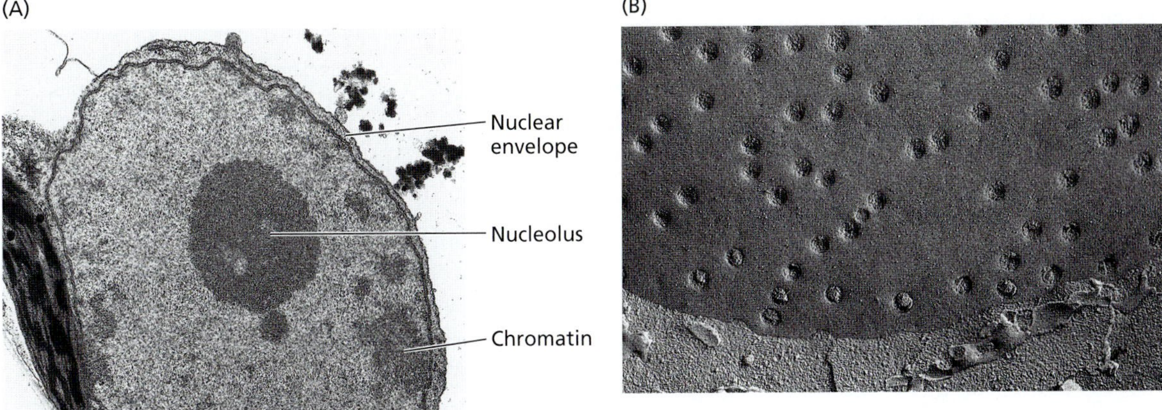

FIGURE 1.7 (A) Transmission electron micrograph of a plant cell, showing the nucleolus and the nuclear envelope. (B) Freeze-etched preparation of nuclear pores from a cell of an onion root. (A courtesy of R. Evert; B courtesy of D. Branton.)

chromatin. The linear length of all the DNA within any plant genome is usually millions of times greater than the diameter of the nucleus in which it is found. To solve the problem of packaging this chromosomal DNA within the nucleus, segments of the linear double helix of DNA are coiled twice around a solid cylinder of eight **histone** protein molecules, forming a **nucleosome**. Nucleosomes are arranged like beads on a string along the length of each chromosome.

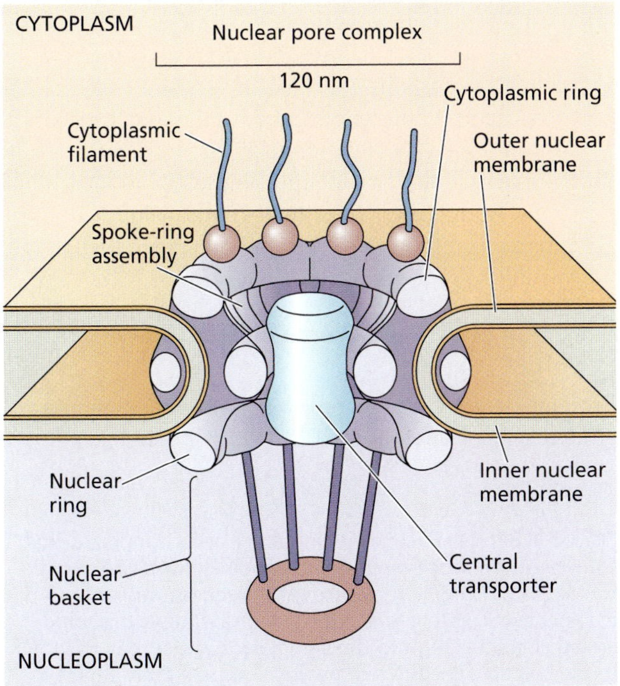

FIGURE 1.8 Schematic model of the structure of the nuclear pore complex. Parallel rings composed of eight subunits each are arranged octagonally near the inner and outer membranes of the nuclear envelope. Various proteins form the other structures, such as the nuclear ring, the spoke-ring assembly, the central transporter, the cytoplasmic filaments, and the nuclear basket.

During mitosis, the chromatin condenses, first by coiling tightly into a **30 nm chromatin fiber**, with six nucleosomes per turn, followed by further folding and packing processes that depend on interactions between proteins and nucleic acids (Figure 1.9). At interphase, two types of chromatin are visible: heterochromatin and euchromatin. About 10% of the DNA consists of **heterochromatin**, a highly compact and transcriptionally inactive form of chromatin. The rest of the DNA consists of **euchromatin**, the dispersed, transcriptionally active form. Only about 10% of the euchromatin is transcriptionally active at any given time. The remainder exists in an intermediate state of condensation, between heterochromatin and transcriptionally active euchromatin.

Nuclei contain a densely granular region, called the **nucleolus** (plural *nucleoli*), that is the site of ribosome synthesis (see Figure 1.7A). The nucleolus includes portions of one or more chromosomes where ribosomal RNA (rRNA) genes are clustered to form a structure called the **nucleolar organizer**. Typical cells have one or more nucleoli per nucleus. Each 80S ribosome is made of a large and a small subunit, and each subunit is a complex aggregate of rRNA and specific proteins. The two subunits exit the nucleus separately, through the nuclear pore, and then unite in the cytoplasm to form a complete ribosome (Figure 1.10A). **Ribosomes** are the sites of protein synthesis.

Protein Synthesis Involves Transcription and Translation

The complex process of protein synthesis starts with **transcription**—the synthesis of an RNA polymer bearing a base

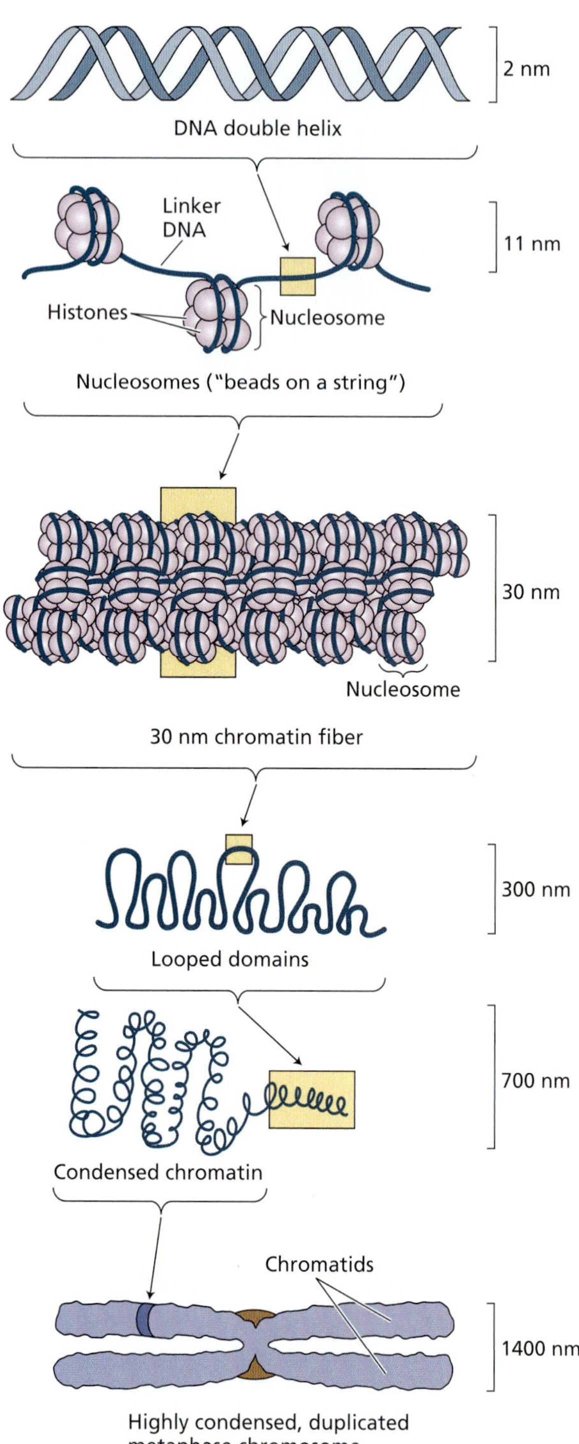

FIGURE 1.9 Packaging of DNA in a metaphase chromosome. The DNA is first aggregated into nucleosomes and then wound to form the 30 nm chromatin fibers. Further coiling leads to the condensed metaphase chromosome. (After Alberts et al. 2002.)

sequence that is complementary to a specific gene. The RNA transcript is processed to become messenger RNA (mRNA), which moves from the nucleus to the cytoplasm. The mRNA in the cytoplasm attaches first to the small ribosomal subunit and then to the large subunit to initiate translation.

Translation is the process whereby a specific protein is synthesized from amino acids, according to the sequence information encoded by the mRNA. The ribosome travels the entire length of the mRNA and serves as the site for the sequential bonding of amino acids as specified by the base sequence of the mRNA (Figure 1.10B).

The Endoplasmic Reticulum Is a Network of Internal Membranes

Cells have an elaborate network of internal membranes called the **endoplasmic reticulum** (**ER**). The membranes of the ER are typical lipid bilayers with interspersed integral and peripheral proteins. These membranes form flattened or tubular sacs known as **cisternae** (singular *cisterna*).

Ultrastructural studies have shown that the ER is continuous with the outer membrane of the nuclear envelope. There are two types of ER—smooth and rough (Figure 1.11)—and the two types are interconnected. **Rough ER** (**RER**) differs from smooth ER in that it is covered with ribosomes that are actively engaged in protein synthesis; in addition, rough ER tends to be lamellar (a flat sheet composed of two unit membranes), while smooth ER tends to be tubular, although a gradation for each type can be observed in almost any cell.

The structural differences between the two forms of ER are accompanied by functional differences. **Smooth ER** functions as a major site of lipid synthesis and membrane assembly. Rough ER is the site of synthesis of membrane proteins and proteins to be secreted outside the cell or into the vacuoles.

Secretion of Proteins from Cells Begins with the Rough ER

Proteins destined for secretion cross the RER membrane and enter the lumen of the ER. This is the first step in the

FIGURE 1.10 (A) Basic steps in gene expression, including transcription, processing, export to the cytoplasm, and translation. Proteins may be synthesized on free or bound ribosomes. Secretory proteins containing a hydrophobic signal sequence bind to the signal recognition particle (SRP) in the cytosol. The SRP–ribosome complex then moves to the endoplasmic reticulum, where it attaches to the SRP receptor. Translation proceeds, and the elongating polypeptide is inserted into the lumen of the endoplasmic reticulum. The signal peptide is cleaved off, sugars are added, and the glycoprotein is transported via vesicles to the Golgi. (B) Amino acids are polymerized on the ribosome, with the help of tRNA, to form the elongating polypeptide chain. ▶

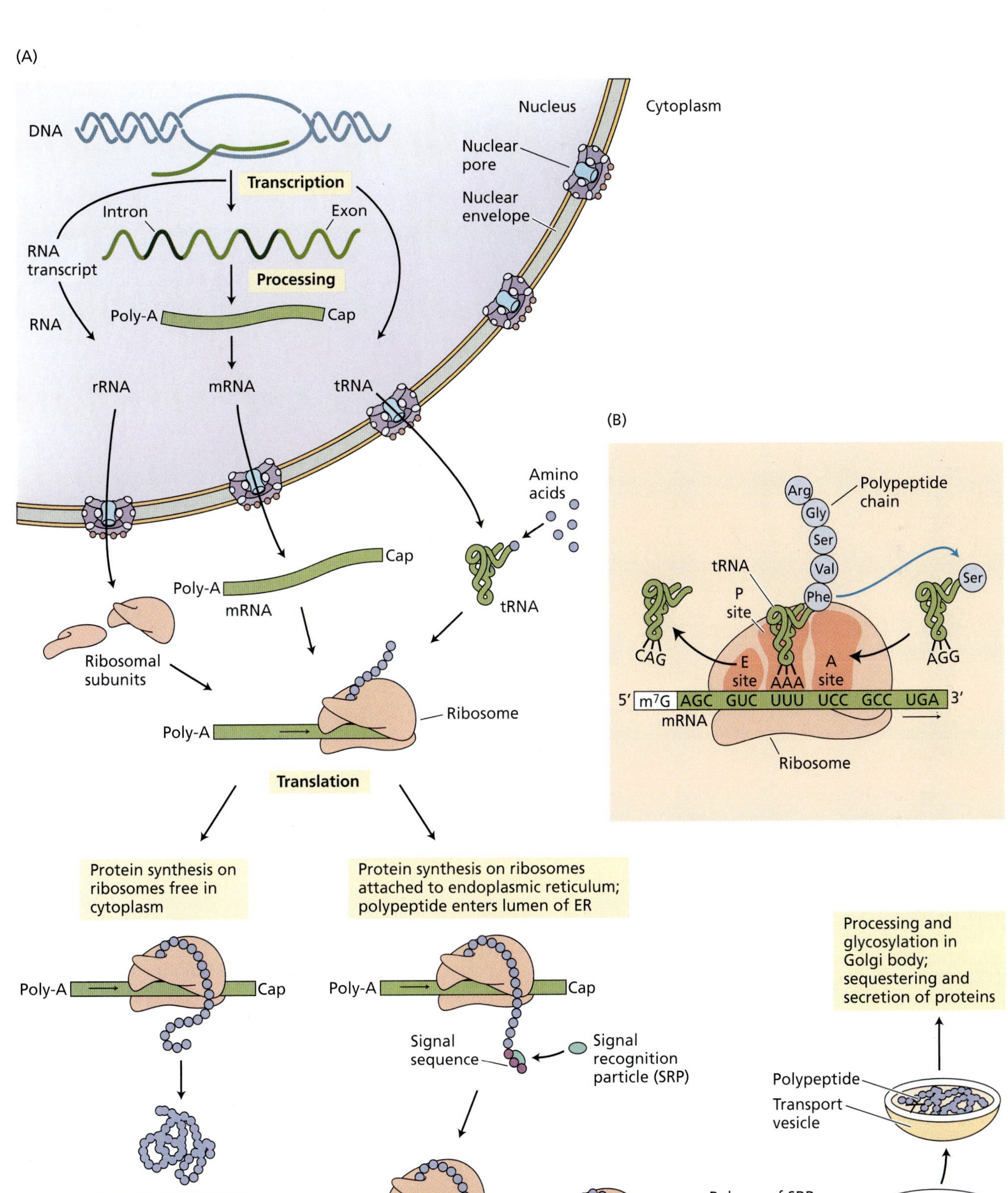
(A)
DNA
Transcription
Intron
Exon
RNA transcript
Processing
Poly-A
Cap
RNA
rRNA
mRNA
tRNA
Nucleus
Cytoplasm
Nuclear pore
Nuclear envelope
Amino acids
Poly-A
Cap
mRNA
tRNA
Ribosomal subunits
Poly-A
Ribosome
Translation
Protein synthesis on ribosomes free in cytoplasm
Protein synthesis on ribosomes attached to endoplasmic reticulum; polypeptide enters lumen of ER
Poly-A
Cap
Poly-A
Cap
Signal sequence
Signal recognition particle (SRP)
Polypeptides free in cytoplasm
Poly-A
Cap
Poly-A
Cap
Release of SRP
SRP receptor
Cleavage of signal sequence
Carbohydrate side chain
Rough endoplasmic reticulum
Polypeptide
Transport vesicle
Processing and glycosylation in Golgi body; sequestering and secretion of proteins
(B)
Arg
Gly
Ser
Val
Phe
Polypeptide chain
tRNA
P site
Ser
CAG
E site
AAA
A site
AGG
5′
m7G
AGC GUC UUU UCC GCC UGA
3′
mRNA
Ribosome

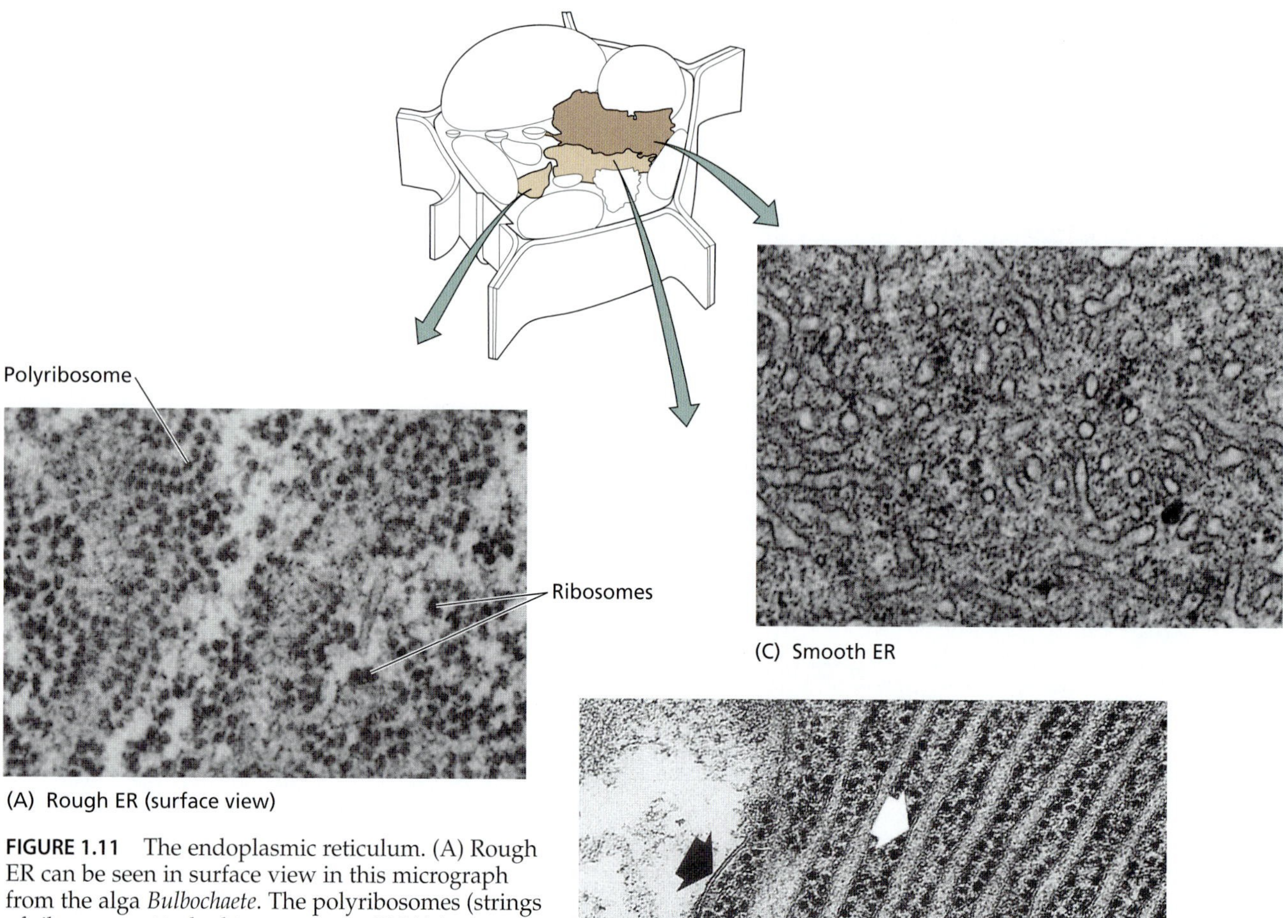

FIGURE 1.11 The endoplasmic reticulum. (A) Rough ER can be seen in surface view in this micrograph from the alga *Bulbochaete*. The polyribosomes (strings of ribosomes attached to messenger RNA) in the rough ER are clearly visible. Polyribosomes are also present on the outer surface of the nuclear envelope (N-nucleus). (75,000×) (B) Stacks of regularly arranged rough endoplasmic reticulum (white arrow) in glandular trichomes of *Coleus blumei*. The plasma membrane is indicated by the black arrow, and the material outside the plasma membrane is the cell wall. (75,000×) (C) Smooth ER often forms a tubular network, as shown in this transmission electron micrograph from a young petal of *Primula kewensis*. (45,000×) (Photos from Gunning and Steer 1996.)

secretion pathway that involves the Golgi body and vesicles that fuse with the plasma membrane.

The mechanism of transport across the membrane is complex, involving the ribosomes, the mRNA that codes for the secretory protein, and a special receptor in the ER membrane. All secretory proteins and most integral membrane proteins have been shown to have a hydrophobic sequence of 18 to 30 amino acid residues at the amino-terminal end of the chain. During translation, this hydrophobic leader, called the **signal peptide** sequence, is recognized by a **signal recognition particle** (**SRP**), made up of protein and RNA, which facilitates binding of the free ribosome to **SRP receptor** proteins (or "docking proteins") on the ER (see Figure 1.10A). The signal peptide then mediates the transfer of the elongating polypeptide across the ER membrane into the lumen. (In the case of integral membrane proteins, a portion of the completed polypeptide remains embedded in the membrane.)

Once inside the lumen of the ER, the signal sequence is cleaved off by a signal peptidase. In some cases, a branched oligosaccharide chain made up of *N*-acetylglucosamine (GlcNac), mannose (Man), and glucose (Glc), having the stoichiometry $GlcNac_2Man_9Glc_3$, is attached to the free amino group of a specific asparagine side chain. This carbohydrate assembly is called an *N-linked glycan* (Faye et al. 1992). The three terminal glucose residues are then removed by specific glucosidases, and the processed glycoprotein (i.e., a protein with covalently attached sugars) is ready for transport to the Golgi apparatus. The so-called **N-linked glycoproteins** are then transported to the Golgi apparatus via small vesicles. The vesicles move through the cytosol and fuse with cisternae on the *cis* face of the Golgi apparatus (Figure 1.12).

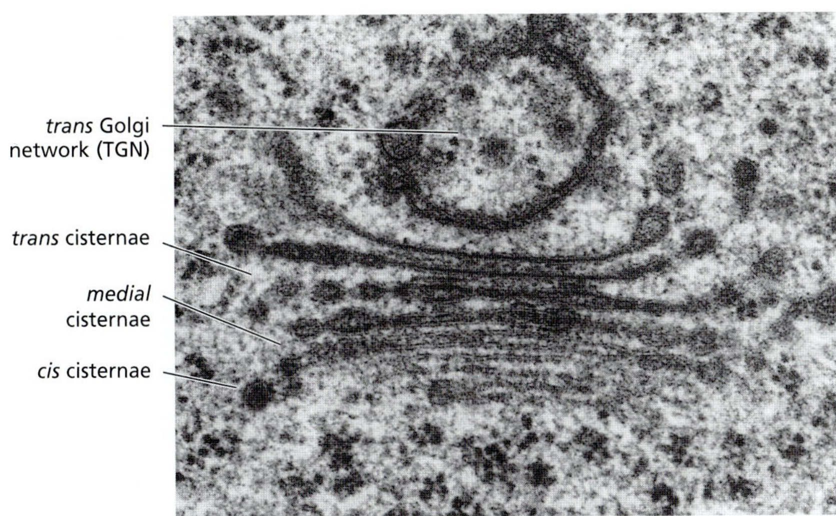

FIGURE 1.12 Electron micrograph of a Golgi apparatus in a tobacco (*Nicotiana tabacum*) root cap cell. The *cis*, *medial*, and *trans* cisternae are indicated. The *trans* Golgi network is associated with the *trans* cisterna. (60,000×) (From Gunning and Steer 1996.)

Proteins and Polysaccharides for Secretion Are Processed in the Golgi Apparatus

The **Golgi apparatus** (also called **Golgi complex**) of plant cells is a dynamic structure consisting of one or more stacks of three to ten flattened membrane sacs, or cisternae, and an irregular network of tubules and vesicles called the ***trans* Golgi network (TGN)** (see Figure 1.12). Each individual stack is called a **Golgi body** or **dictyosome**.

As Figure 1.12 shows, the Golgi body has distinct functional regions: The cisternae closest to the plasma membrane are called the *trans* face, and the cisternae closest to the center of the cell are called the *cis* face. The *medial* cisternae are between the *trans* and *cis* cisternae. The *trans* Golgi network is located on the *trans* face. The entire structure is stabilized by the presence of **intercisternal elements**, protein crosslinks that hold the cisternae together. Whereas in animal cells Golgi bodies tend to be clustered in one part of the cell and are interconnected via tubules, plant cells contain up to several hundred apparently separate Golgi bodies dispersed throughout the cytoplasm (Driouich et al. 1994).

The Golgi apparatus plays a key role in the synthesis and secretion of complex polysaccharides (polymers composed of different types of sugars) and in the assembly of the oligosaccharide side chains of glycoproteins (Driouich et al. 1994). As noted already, the polypeptide chains of future glycoproteins are first synthesized on the rough ER, then transferred across the ER membrane, and glycosylated on the —NH_2 groups of asparagine residues. Further modifications of, and additions to, the oligosaccharide side chains are carried out in the Golgi. Glycoproteins destined for secretion reach the Golgi via vesicles that bud off from the RER.

The exact pathway of glycoproteins through the plant Golgi apparatus is not yet known. Since there appears to be no direct membrane continuity between successive cisternae, the contents of one cisterna are transferred to the next cisterna via small vesicles budding off from the margins, as occurs in the Golgi apparatus of animals. In some cases, however, entire cisternae may progress through the Golgi body and emerge from the *trans* face.

Within the lumens of the Golgi cisternae, the glycoproteins are enzymatically modified. Certain sugars, such as mannose, are removed from the oligosaccharide chains, and other sugars are added. In addition to these modifications, glycosylation of the —OH groups of hydroxyproline, serine, threonine, and tyrosine residues (**O-linked oligosaccharides**) also occurs in the Golgi. After being processed within the Golgi, the glycoproteins leave the organelle in other vesicles, usually from the *trans* side of the stack. All of this processing appears to confer on each protein a specific tag or marker that specifies the ultimate destination of that protein inside or outside the cell.

In plant cells, the Golgi body plays an important role in cell wall formation (see Chapter 15). Noncellulosic cell wall polysaccharides (hemicellulose and pectin) are synthesized, and a variety of glycoproteins, including hydroxyproline-rich glycoproteins, are processed within the Golgi.

Secretory vesicles derived from the Golgi carry the polysaccharides and glycoproteins to the plasma membrane, where the vesicles fuse with the plasma membrane and empty their contents into the region of the cell wall. Secretory vesicles may either be smooth or have a protein coat. Vesicles budding from the ER are generally smooth. Most vesicles budding from the Golgi have protein coats of some type. These proteins aid in the budding process during vesicle formation. Vesicles involved in traffic from the ER to the Golgi, between Golgi compartments, and from the Golgi to the TGN have **protein coats. Clathrin-coated vesicles** (Figure 1.13) are involved in the transport of storage proteins from the Golgi to specialized protein-storing vacuoles. They also participate in **endocytosis,** the process that brings soluble and membrane-bound proteins into the cell.

The Central Vacuole Contains Water and Solutes

Mature living plant cells contain large, water-filled central vacuoles that can occupy 80 to 90% of the total volume of the cell (see Figure 1.4). Each vacuole is surrounded by a **vacuolar membrane**, or **tonoplast**. Many cells also have cytoplasmic strands that run through the vacuole, but each transvacuolar strand is surrounded by the tonoplast.

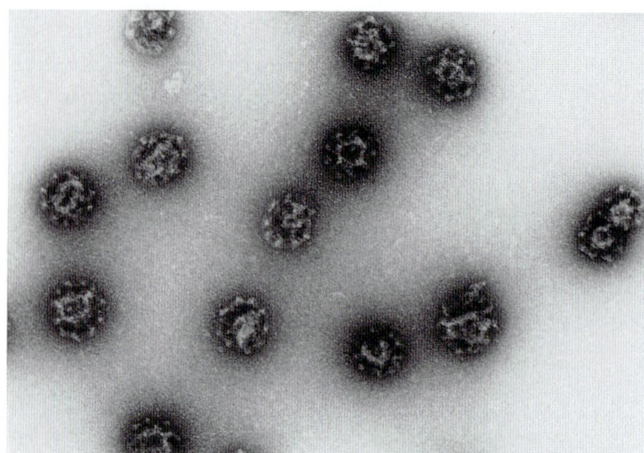

FIGURE 1.13 Preparation of clathrin-coated vesicles isolated from bean leaves. (102,000×) (Photo courtesy of D. G. Robinson.)

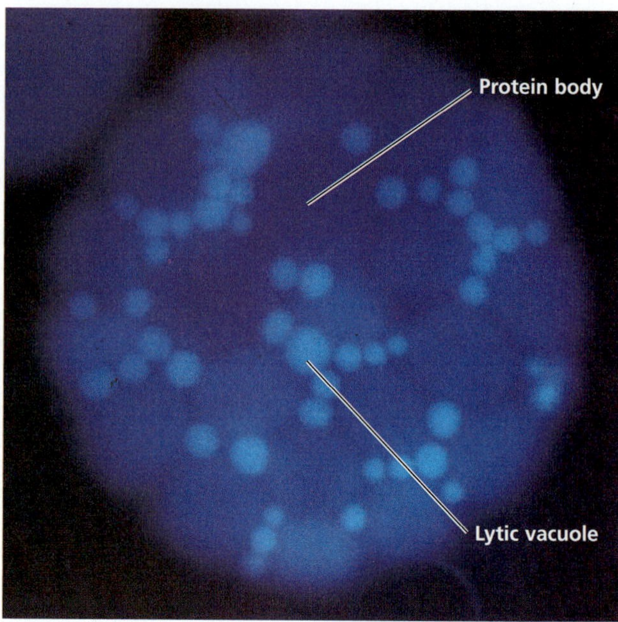

FIGURE 1.14 Light micrograph of a protoplast prepared from the aleurone layer of seeds. The fluorescent stain reveals two types of vacuoles: the larger protein bodies (V_1) and the smaller lytic vacuoles (V_2). (Photo courtesy of P. Bethke and R. L. Jones.)

In meristematic tissue, vacuoles are less prominent, though they are always present as small **provacuoles**. Provacuoles are produced by the *trans* Golgi network (see Figure 1.12). As the cell begins to mature, the provacuoles fuse to produce the large central vacuoles that are characteristic of most mature plant cells. In such cells, the cytoplasm is restricted to a thin layer surrounding the vacuole.

The vacuole contains water and dissolved inorganic ions, organic acids, sugars, enzymes, and a variety of secondary metabolites (see Chapter 13), which often play roles in plant defense. Active solute accumulation provides the osmotic driving force for water uptake by the vacuole, which is required for plant cell enlargement. The turgor pressure generated by this water uptake provides the structural rigidity needed to keep herbaceous plants upright, since they lack the lignified support tissues of woody plants.

Like animal lysosomes, plant vacuoles contain hydrolytic enzymes, including proteases, ribonucleases, and glycosidases. Unlike animal lysosomes, however, plant vacuoles do not participate in the turnover of macromolecules throughout the life of the cell. Instead, their degradative enzymes leak out into the cytosol as the cell undergoes senescence, thereby helping to recycle valuable nutrients to the living portion of the plant.

Specialized protein-storing vacuoles, called **protein bodies**, are abundant in seeds. During germination the storage proteins in the protein bodies are hydrolyzed to amino acids and exported to the cytosol for use in protein synthesis. The hydrolytic enzymes are stored in specialized **lytic vacuoles**, which fuse with the protein bodies to initiate the breakdown process (Figure 1.14).

Mitochondria and Chloroplasts Are Sites of Energy Conversion

A typical plant cell has two types of energy-producing organelles: mitochondria and chloroplasts. Both types are separated from the cytosol by a double membrane (an outer and an inner membrane). **Mitochondria** (singular *mitochondrion*) are the cellular sites of respiration, a process in which the energy released from sugar metabolism is used for the synthesis of ATP (adenosine triphosphate) from ADP (adenosine diphosphate) and inorganic phosphate (P_i) (see Chapter 11).

Mitochondria can vary in shape from spherical to tubular, but they all have a smooth outer membrane and a highly convoluted inner membrane (Figure 1.15). The infoldings of the inner membrane are called **cristae** (singular *crista*). The compartment enclosed by the inner membrane, the mitochondrial **matrix**, contains the enzymes of the pathway of intermediary metabolism called the Krebs cycle.

In contrast to the mitochondrial outer membrane and all other membranes in the cell, the inner membrane of a mitochondrion is almost 70% protein and contains some phospholipids that are unique to the organelle (e.g., cardiolipin). The proteins in and on the inner membrane have special enzymatic and transport capacities.

The inner membrane is highly impermeable to the passage of H^+; that is, it serves as a barrier to the movement of protons. This important feature allows the formation of electrochemical gradients. Dissipation of such gradients by the controlled movement of H^+ ions through the transmembrane enzyme **ATP synthase** is coupled to the phosphorylation of ADP to produce ATP. ATP can then be released to other cellular sites where energy is needed to drive specific reactions.

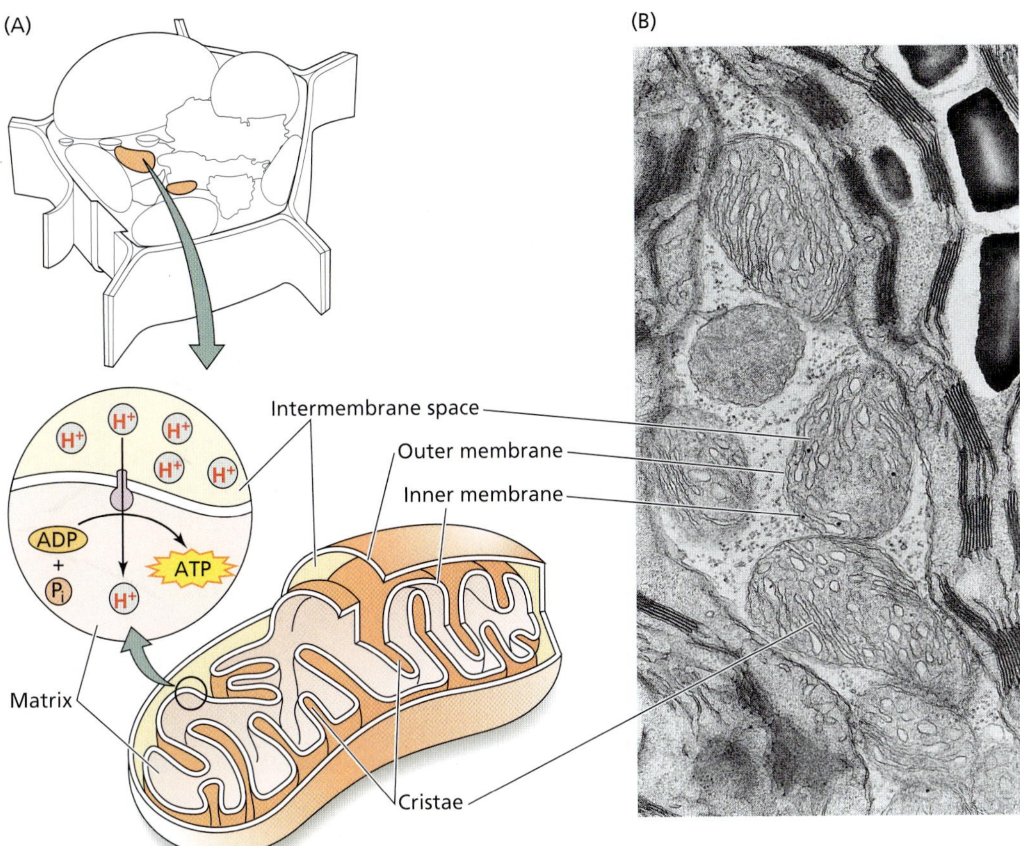

FIGURE 1.15 (A) Diagrammatic representation of a mitochondrion, including the location of the H^+-ATPases involved in ATP synthesis on the inner membrane. (B) An electron micrograph of mitochondria from a leaf cell of Bermuda grass, *Cynodon dactylon*. (26,000×) (Photo by S. E. Frederick, courtesy of E. H. Newcomb.)

Chloroplasts (Figure 1.16A) belong to another group of double membrane–enclosed organelles called **plastids**. Chloroplast membranes are rich in glycosylglycerides (see Web Topic 1.4). Chloroplast membranes contain chlorophyll and its associated proteins and are the sites of photosynthesis. In addition to their inner and outer envelope membranes, chloroplasts possess a third system of membranes called **thylakoids**. A stack of thylakoids forms a **granum** (plural *grana*) (Figure 1.16B). Proteins and pigments (chlorophylls and carotenoids) that function in the photochemical events of photosynthesis are embedded in the thylakoid membrane. The fluid compartment surrounding the thylakoids, called the **stroma**, is analogous to the matrix of the mitochondrion. Adjacent grana are connected by unstacked membranes called **stroma lamellae** (singular *lamella*).

The different components of the photosynthetic apparatus are localized in different areas of the grana and the stroma lamellae. The ATP synthases of the chloroplast are located on the thylakoid membranes (Figure 1.16C). During photosynthesis, light-driven electron transfer reactions result in a proton gradient across the thylakoid membrane. As in the mitochondria, ATP is synthesized when the proton gradient is dissipated via the ATP synthase.

Plastids that contain high concentrations of carotenoid pigments rather than chlorophyll are called **chromoplasts**. They are one of the causes of the yellow, orange, or red colors of many fruits and flowers, as well as of autumn leaves (Figure 1.17).

Nonpigmented plastids are called **leucoplasts**. The most important type of leucoplast is the **amyloplast**, a starch-storing plastid. Amyloplasts are abundant in storage tissues of the shoot and root, and in seeds. Specialized amyloplasts in the root cap also serve as gravity sensors that direct root growth downward into the soil (see Chapter 19).

Mitochondria and Chloroplasts Are Semiautonomous Organelles

Both mitochondria and chloroplasts contain their own DNA and protein-synthesizing machinery (ribosomes, transfer RNAs, and other components) and are believed to have evolved from endosymbiotic bacteria. Both plastids and mitochondria divide by fission, and mitochondria can also undergo extensive fusion to form elongated structures or networks.

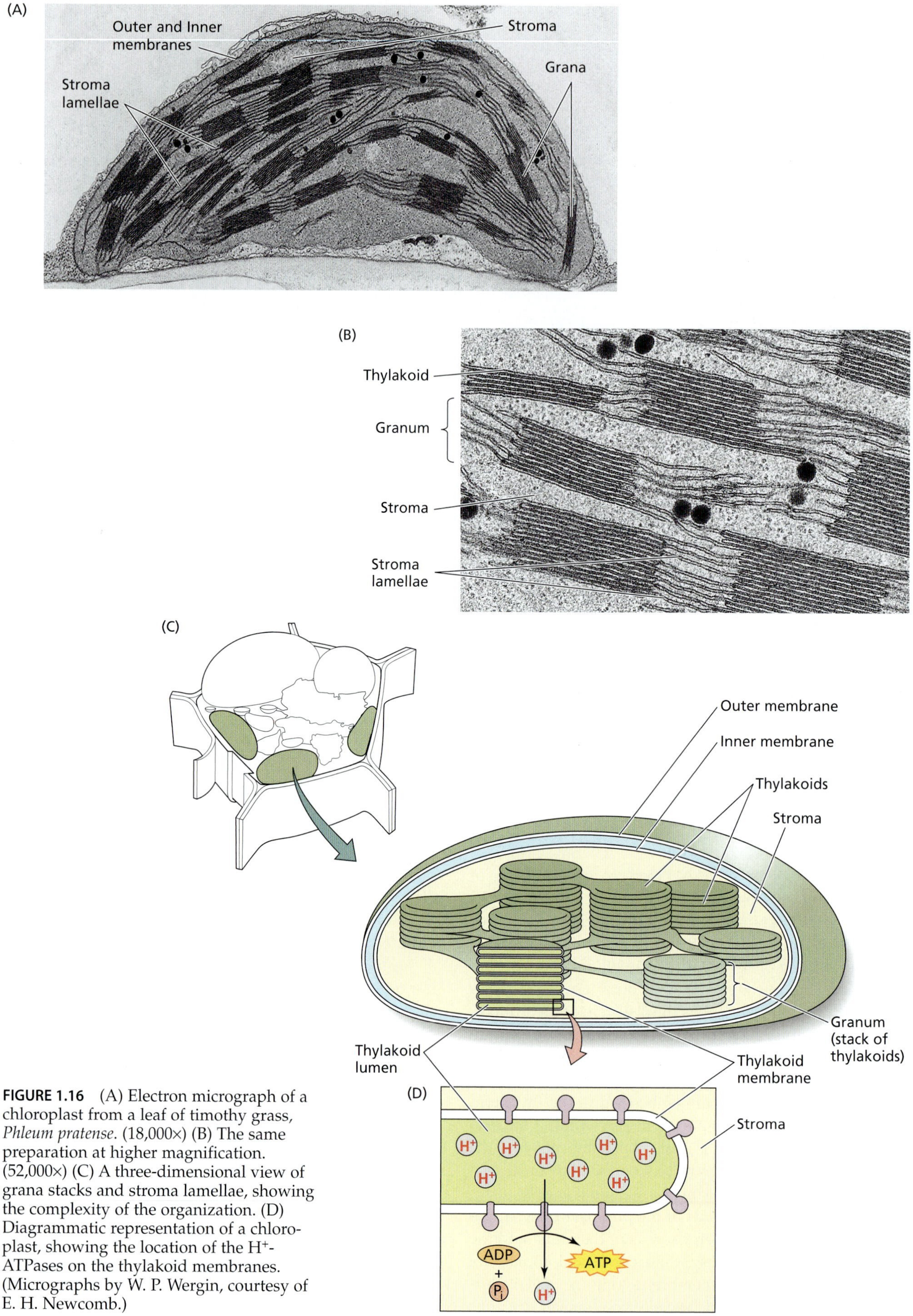

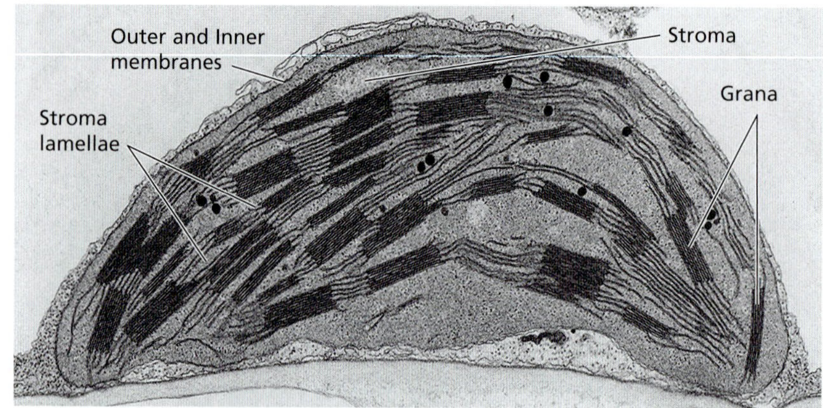

FIGURE 1.16 (A) Electron micrograph of a chloroplast from a leaf of timothy grass, *Phleum pratense*. (18,000×) (B) The same preparation at higher magnification. (52,000×) (C) A three-dimensional view of grana stacks and stroma lamellae, showing the complexity of the organization. (D) Diagrammatic representation of a chloroplast, showing the location of the H^+-ATPases on the thylakoid membranes. (Micrographs by W. P. Wergin, courtesy of E. H. Newcomb.)

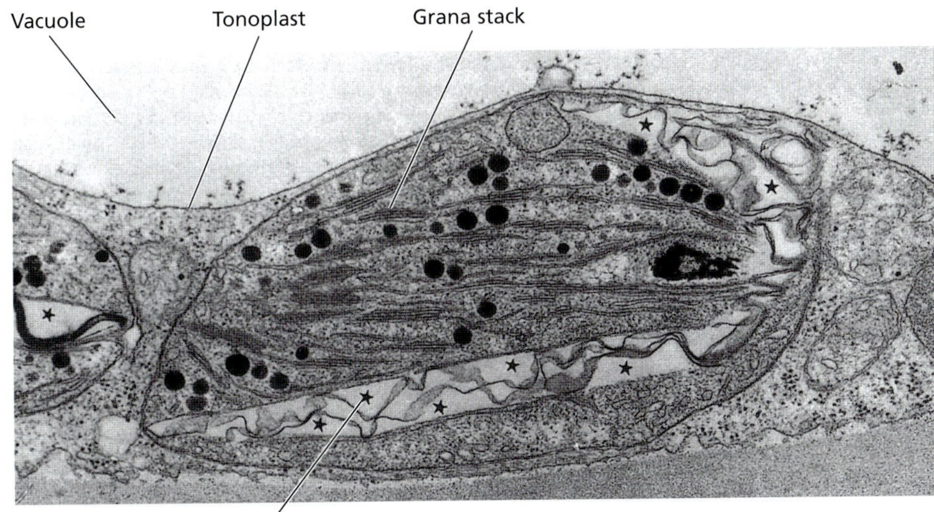

FIGURE 1.17 Electron micrograph of a chromoplast from tomato (*Lycopersicon esculentum*) fruit at an early stage in the transition from chloroplast to chromoplast. Small grana stacks are still visible. Crystals of the carotenoid lycopene are indicated by the stars. (27,000×) (From Gunning and Steer 1996.)

The DNA of these organelles is in the form of circular chromosomes, similar to those of bacteria and very different from the linear chromosomes in the nucleus. These DNA circles are localized in specific regions of the mitochondrial matrix or plastid stroma called **nucleoids**. DNA replication in both mitochondria and chloroplasts is independent of DNA replication in the nucleus. On the other hand, the numbers of these organelles within a given cell type remain approximately constant, suggesting that some aspects of organelle replication are under cellular regulation.

The mitochondrial genome of plants consists of about 200 kilobase pairs (200,000 base pairs), a size considerably larger than that of most animal mitochondria. The mitochondria of meristematic cells are typically polyploid; that is, they contain multiple copies of the circular chromosome. However, the number of copies per mitochondrion gradually decreases as cells mature because the mitochondria continue to divide in the absence of DNA synthesis.

Most of the proteins encoded by the mitochondrial genome are prokaryotic-type 70S ribosomal proteins and components of the electron transfer system. The majority of mitochondrial proteins, including Krebs cycle enzymes, are encoded by nuclear genes and are imported from the cytosol.

The chloroplast genome is smaller than the mitochondrial genome, about 145 kilobase pairs (145,000 base pairs). Whereas mitochondria are polyploid only in the meristems, chloroplasts become polyploid during cell maturation. Thus the average amount of DNA per chloroplast in the plant is much greater than that of the mitochondria. The total amount of DNA from the mitochondria and plastids combined is about one-third of the nuclear genome (Gunning and Steer 1996).

Chloroplast DNA encodes rRNA; transfer RNA (tRNA); the large subunit of the enzyme that fixes CO_2, ribulose-1,5-bisphosphate carboxylase/oxygenase (rubisco); and several of the proteins that participate in photosynthesis. Nevertheless, the majority of chloroplast proteins, like those of mitochondria, are encoded by nuclear genes, synthesized in the cytosol, and transported to the organelle. Although mitochondria and chloroplasts have their own genomes and can divide independently of the cell, they are characterized as *semiautonomous organelles* because they depend on the nucleus for the majority of their proteins.

Different Plastid Types Are Interconvertible

Meristem cells contain **proplastids**, which have few or no internal membranes, no chlorophyll, and an incomplete complement of the enzymes necessary to carry out photosynthesis (Figure 1.18A). In angiosperms and some gymnosperms, chloroplast development from proplastids is triggered by light. Upon illumination, enzymes are formed inside the proplastid or imported from the cytosol, light-absorbing pigments are produced, and membranes proliferate rapidly, giving rise to stroma lamellae and grana stacks (Figure 1.18B).

Seeds usually germinate in the soil away from light, and chloroplasts develop only when the young shoot is exposed to light. If seeds are germinated in the dark, the proplastids differentiate into **etioplasts**, which contain semicrystalline tubular arrays of membrane known as **prolamellar bodies** (Figure 1.18C). Instead of chlorophyll, the etioplast contains a pale yellow green precursor pigment, **protochlorophyllide**.

Within minutes after exposure to light, the etioplast differentiates, converting the prolamellar body into thylakoids and stroma lamellae, and the protochlorophyll into chlorophyll. The maintenance of chloroplast structure depends on the presence of light, and mature chloroplasts can revert to etioplasts during extended periods of darkness.

Chloroplasts can be converted to chromoplasts, as in the case of autumn leaves and ripening fruit, and in some cases

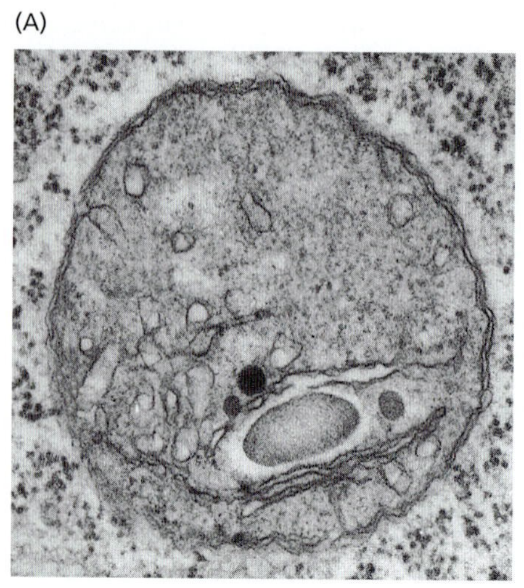

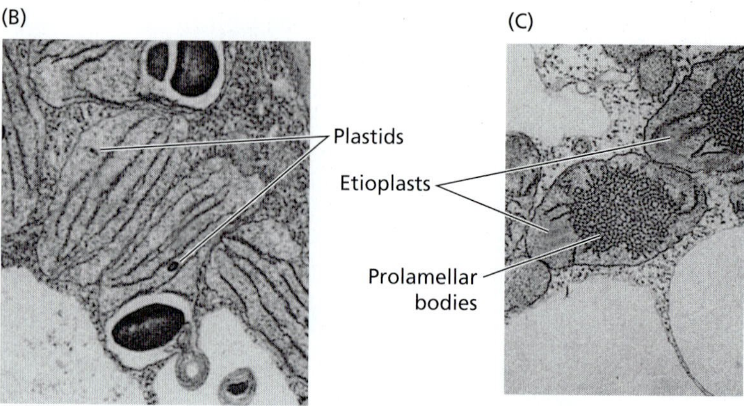

FIGURE 1.18 Electron micrographs illustrating several stages of plastid development. (A) A higher-magnification view of a proplastid from the root apical meristem of the broad bean (*Vicia faba*). The internal membrane system is rudimentary, and grana are absent. (47,000×) (B) A mesophyll cell of a young oat leaf at an early stage of differentiation in the light. The plastids are developing grana stacks. (C) A cell from a young oat leaf from a seedling grown in the dark. The plastids have developed as etioplasts, with elaborate semicrystalline lattices of membrane tubules called prolamellar bodies. When exposed to light, the etioplast can convert to a chloroplast by the disassembly of the prolamellar body and the formation of grana stacks. (7,200×) (From Gunning and Steer 1996.)

this process is reversible. And amyloplasts can be converted to chloroplasts, which explains why exposure of roots to light often results in greening of the roots.

Microbodies Play Specialized Metabolic Roles in Leaves and Seeds

Plant cells also contain **microbodies**, a class of spherical organelles surrounded by a single membrane and specialized for one of several metabolic functions. The two main types of microbodies are peroxisomes and glyoxysomes.

Peroxisomes are found in all eukaryotic organisms, and in plants they are present in photosynthetic cells (Figure 1.19). Peroxisomes function both in the removal of hydrogens from organic substrates, consuming O_2 in the process, according to the following reaction:

$$RH_2 + O_2 \rightarrow R + H_2O_2$$

where R is the organic substrate. The potentially harmful peroxide produced in these reactions is broken down in peroxisomes by the enzyme catalase, according to the following reaction:

$$H_2O_2 \rightarrow H_2O + \frac{1}{2}O_2$$

Although some oxygen is regenerated during the catalase reaction, there is a net consumption of oxygen overall.

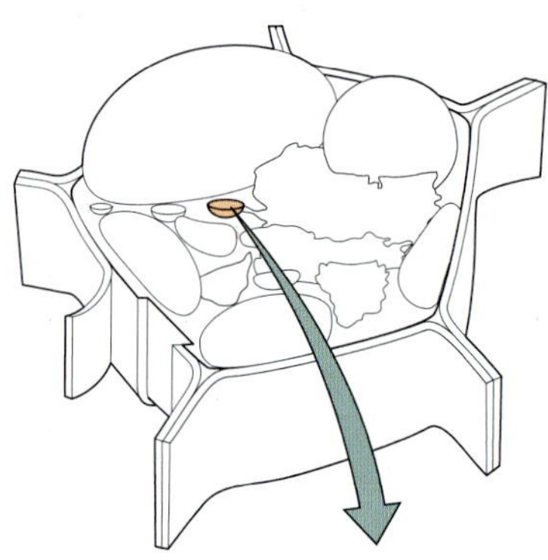

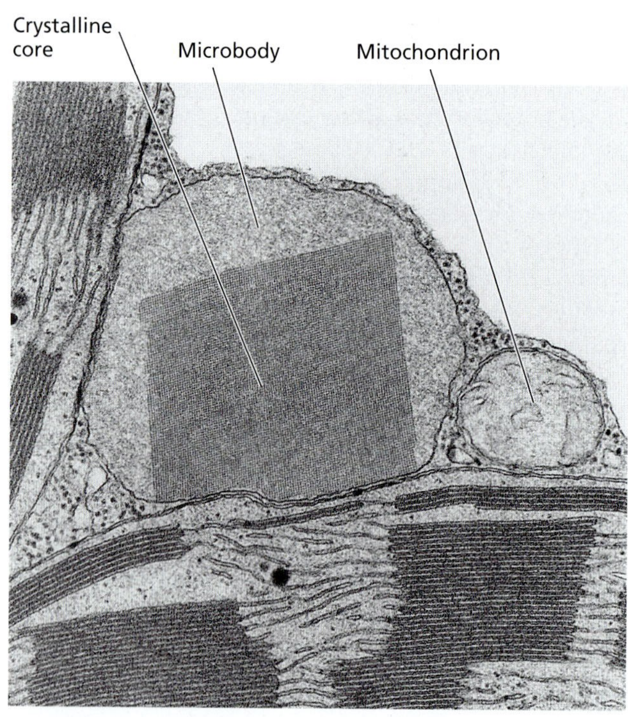

FIGURE 1.19 Electron micrograph of a peroxisome from a mesophyll cell, showing a crystalline core. (27,000×) This peroxisome is seen in close association with two chloroplasts and a mitochondrion, probably reflecting the cooperative role of these three organelles in photorespiration. (From Huang 1987.)

Another type of microbody, the **glyoxysome**, is present in oil-storing seeds. Glyoxysomes contain the *glyoxylate cycle* enzymes, which help convert stored fatty acids into sugars that can be translocated throughout the young plant to provide energy for growth (see Chapter 11). Because both types of microbodies carry out oxidative reactions, it has been suggested they may have evolved from primitive respiratory organelles that were superseded by mitochondria.

Oleosomes Are Lipid-Storing Organelles

In addition to starch and protein, many plants synthesize and store large quantities of triacylglycerol in the form of oil during seed development. These oils accumulate in organelles called **oleosomes**, also referred to as *lipid bodies* or *spherosomes* (Figure 1.20A).

Oleosomes are unique among the organelles in that they are surrounded by a "half–unit membrane"—that is, a phospholipid monolayer—derived from the ER (Harwood 1997). The phospholipids in the half–unit membrane are oriented with their polar head groups toward the aqueous phase and their hydrophobic fatty acid tails facing the lumen, dissolved in the stored lipid. Oleosomes are thought to arise from the deposition of lipids within the bilayer itself (Figure 1.20B).

Proteins called **oleosins** are present in the half–unit membrane (see Figure 1.20B). One of the functions of the oleosins may be to maintain each oleosome as a discrete organelle by preventing fusion. Oleosins may also help other proteins bind to the organelle surface. As noted earlier, during seed germination the lipids in the oleosomes are broken down and converted to sucrose with the help of the glyoxysome. The first step in the process is the hydrolysis of the fatty acid chains from the glycerol backbone by the enzyme lipase. Lipase is tightly associated with the surface of the half–unit membrane and may be attached to the oleosins.

THE CYTOSKELETON

The cytosol is organized into a three-dimensional network of filamentous proteins called the **cytoskeleton**. This network provides the spatial organization for the organelles and serves as a scaffolding for the movements of organelles and other cytoskeletal components. It also plays fundamental roles in mitosis, meiosis, cytokinesis, wall deposition, the maintenance of cell shape, and cell differentiation.

Plant Cells Contain Microtubules, Microfilaments, and Intermediate Filaments

Three types of cytoskeletal elements have been demonstrated in plant cells: microtubules, microfilaments, and intermediate filament–like structures. Each type is filamentous, having a fixed diameter and a variable length, up to many micrometers.

Microtubules and microfilaments are macromolecular assemblies of globular proteins. **Microtubules** are hollow

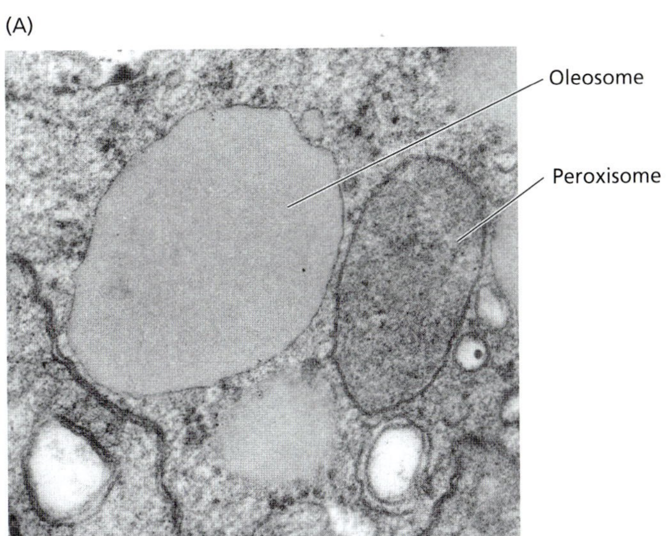

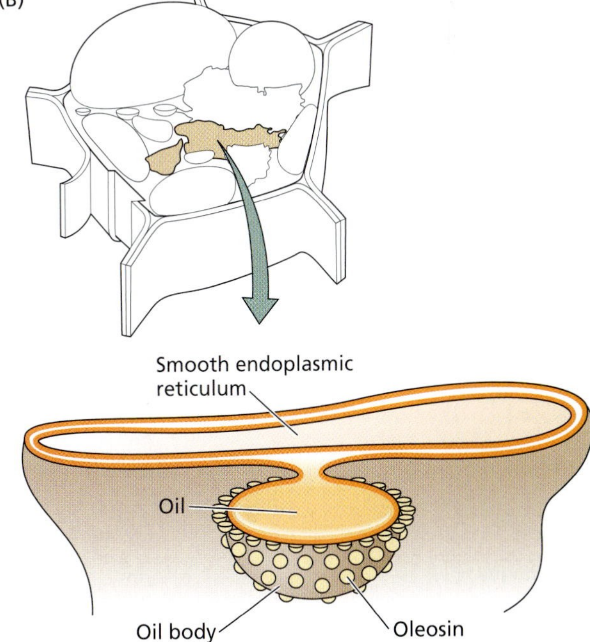

FIGURE 1.20 (A) Electron micrograph of an oleosome beside a peroxisome. (B) Diagram showing the formation of oleosomes by the synthesis and deposition of oil within the phospholipid bilayer of the ER. After budding off from the ER, the oleosome is surrounded by a phospholipid monolayer containing the protein oleosin. (A from Huang 1987; B after Buchanan et al. 2000.)

cylinders with an outer diameter of 25 nm; they are composed of polymers of the protein **tubulin**. The tubulin monomer of microtubules is a heterodimer composed of two similar polypeptide chains (α- and β-tubulin), each having an apparent molecular mass of 55,000 daltons (Figure 1.21A). A single microtubule consists of hundreds of thousands of tubulin monomers arranged in 13 columns called *protofilaments*.

Microfilaments are solid, with a diameter of 7 nm; they are composed of a special form of the protein found in muscle: globular actin, or **G-actin**. Each actin molecule is composed of a single polypeptide with a molecular mass of approximately 42,000 daltons. A microfilament consists of two chains of polymerized actin subunits that intertwine in a helical fashion (Figure 1.21B).

Intermediate filaments are a diverse group of helically wound fibrous elements, 10 nm in diameter. Intermediate filaments are composed of linear polypeptide monomers of various types. In animal cells, for example, the **nuclear lamins** are composed of a specific polypeptide monomer, while the **keratins**, a type of intermediate filament found in the cytoplasm, are composed of a different polypeptide monomer.

In animal intermediate filaments, pairs of parallel monomers (i.e., aligned with their —NH_2 groups at the same ends) are helically wound around each other in a **coiled coil.** Two coiled-coil dimers then align in an antiparallel fashion (i.e., with their —NH_2 groups at opposite ends) to form a tetrameric unit. The tetrameric units then assemble into the final intermediate filament (Figure 1.22).

Although nuclear lamins appear to be present in plant cells, there is as yet no convincing evidence for plant keratin intermediate filaments in the cytosol. As noted earlier, integral proteins cross-link the plasma membrane of plant cells to the rigid cell wall. Such connections to the wall undoubtedly stabilize the protoplast and help maintain cell shape. The plant cell wall thus serves as a kind of cellular exoskeleton, perhaps obviating the need for keratin-type intermediate filaments for structural support.

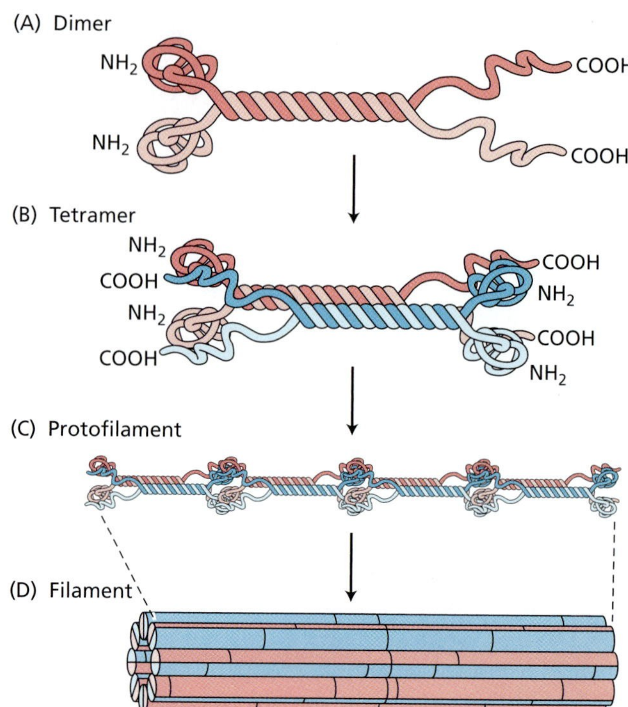

FIGURE 1.22 The current model for the assembly of intermediate filaments from protein monomers. (A) Coiled-coil dimer in parallel orientation (i.e., with amino and carboxyl termini at the same ends). (B) A tetramer of two dimers. Note that the dimers are arranged in an antiparallel fashion, and that one is slightly offset from the other. (C) Two tetramers. (D) Tetramers packed together to form the 10 nm intermediate filament. (After Alberts et al. 2002.)

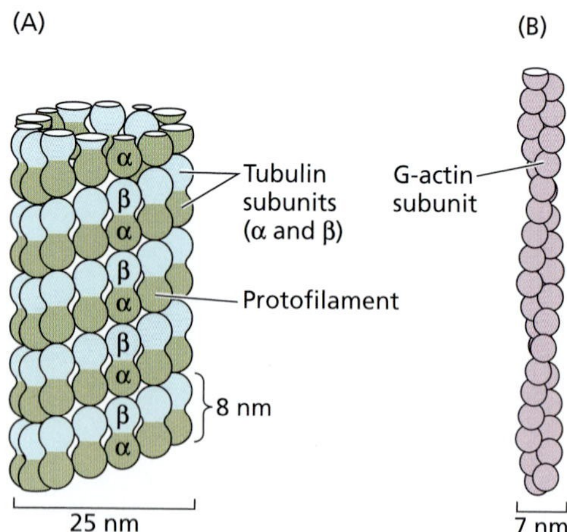

FIGURE 1.21 (A) Drawing of a microtubule in longitudinal view. Each microtubule is composed of 13 protofilaments. The organization of the α and β subunits is shown. (B) Diagrammatic representation of a microfilament, showing two strands of G-actin subunits.

Microtubules and Microfilaments Can Assemble and Disassemble

In the cell, actin and tubulin monomers exist as pools of free proteins that are in dynamic equilibrium with the polymerized forms. Polymerization requires energy: ATP is required for microfilament polymerization, GTP (guanosine triphosphate) for microtubule polymerization. The attachments between subunits in the polymer are noncovalent, but they are strong enough to render the structure stable under cellular conditions.

Both microtubules and microfilaments are polarized; that is, the two ends are different. In microtubules, the polarity arises from the polarity of the α- and β-tubulin heterodimer; in microfilaments, the polarity arises from the polarity of the actin monomer itself. The opposite ends of microtubules and microfilaments are termed *plus* and *minus*, and polymerization is more rapid at the positive end.

Once formed, microtubules and microfilaments can disassemble. The overall *rate* of assembly and disassembly of these structures is affected by the relative concentrations of free or assembled subunits. In general, microtubules are more unstable than microfilaments. In animal cells, the half-life of an individual microtubule is about 10 minutes. Thus microtubules are said to exist in a state of *dynamic instability*.

In contrast to microtubules and microfilaments, intermediate filaments lack polarity because of the antiparallel orientation of the dimers that make up the tetramers. In addition, intermediate filaments appear to be much more stable than either microtubules or microfilaments. Although very little is known about intermediate filament–like structures in plant cells, in animal cells nearly all of the intermediate-filament protein exists in the polymerized state.

Microtubules Function in Mitosis and Cytokinesis

Mitosis is the process by which previously replicated chromosomes are aligned, separated, and distributed in an orderly fashion to daughter cells (Figure 1.23). Microtubules are an integral part of mitosis. Before mitosis begins, microtubules in the cortical (outer) cytoplasm depolymerize, breaking down into their constituent subunits. The subunits then repolymerize before the start of prophase to form the **preprophase band** (**PPB**), a ring of microtubules encircling the nucleus (see Figure 1.23C–F). The PPB appears in the region where the future cell wall will form after the completion of mitosis, and it is thought to be involved in regulating the plane of cell division.

During prophase, microtubules begin to assemble at two foci on opposite sides of the nucleus, forming the **prophase spindle** (Figure 1.24). Although not associated with any specific structure, these foci serve the same function as animal cell centrosomes in organizing and assembling microtubules.

In early metaphase the nuclear envelope breaks down, the PPB disassembles, and new microtubules polymerize to form the mitotic spindle. In animal cells the spindle microtubules radiate toward each other from two discrete foci at the poles (the centrosomes), resulting in an ellipsoidal, or football-shaped, array of microtubules. The mitotic spindle of plant cells, which lack centrosomes, is more boxlike in shape because the spindle microtubules arise from a diffuse zone consisting of multiple foci at opposite ends of the cell and extend toward the middle in nearly parallel arrays (see Figure 1.24).

Some of the microtubules of the spindle apparatus become attached to the chromosomes at their **kinetochores**, while others remain unattached. The kinetochores are located in the **centromeric** regions of the chromosomes. Some of the unattached microtubules overlap with microtubules from the opposite polar region in the spindle midzone.

Cytokinesis is the process whereby a cell is partitioned into two progeny cells. Cytokinesis usually begins late in mitosis. The precursor of the new wall, the **cell plate** that

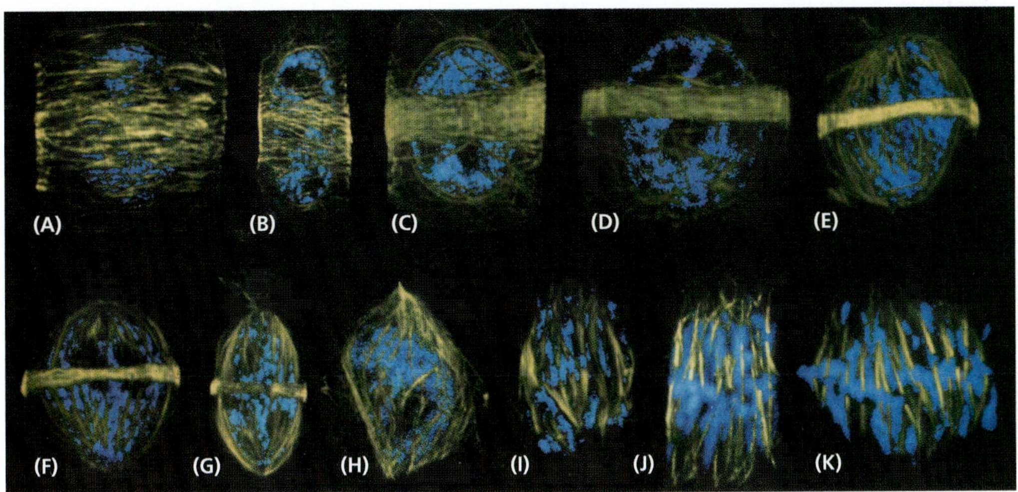

FIGURE 1.23 Fluorescence micrograph taken with a confocal microscope showing changes in microtubule arrangements at different stages in the cell cycle of wheat root meristem cells. Microtubules stain green and yellow; DNA is blue. (A–D) Cortical microtubules disappear and the preprophase band is formed around the nucleus at the site of the future cell plate. (E–H) The prophase spindle forms from foci of microtubules at the poles. (G, H) The preprophase band disappears in late prophase. (I–K) The nuclear membrane breaks down, and the two poles become more diffuse. The mitotic spindle forms in parallel arrays and the kinetochores bind to spindle microtubules. (From Gunning and Steer 1996.)

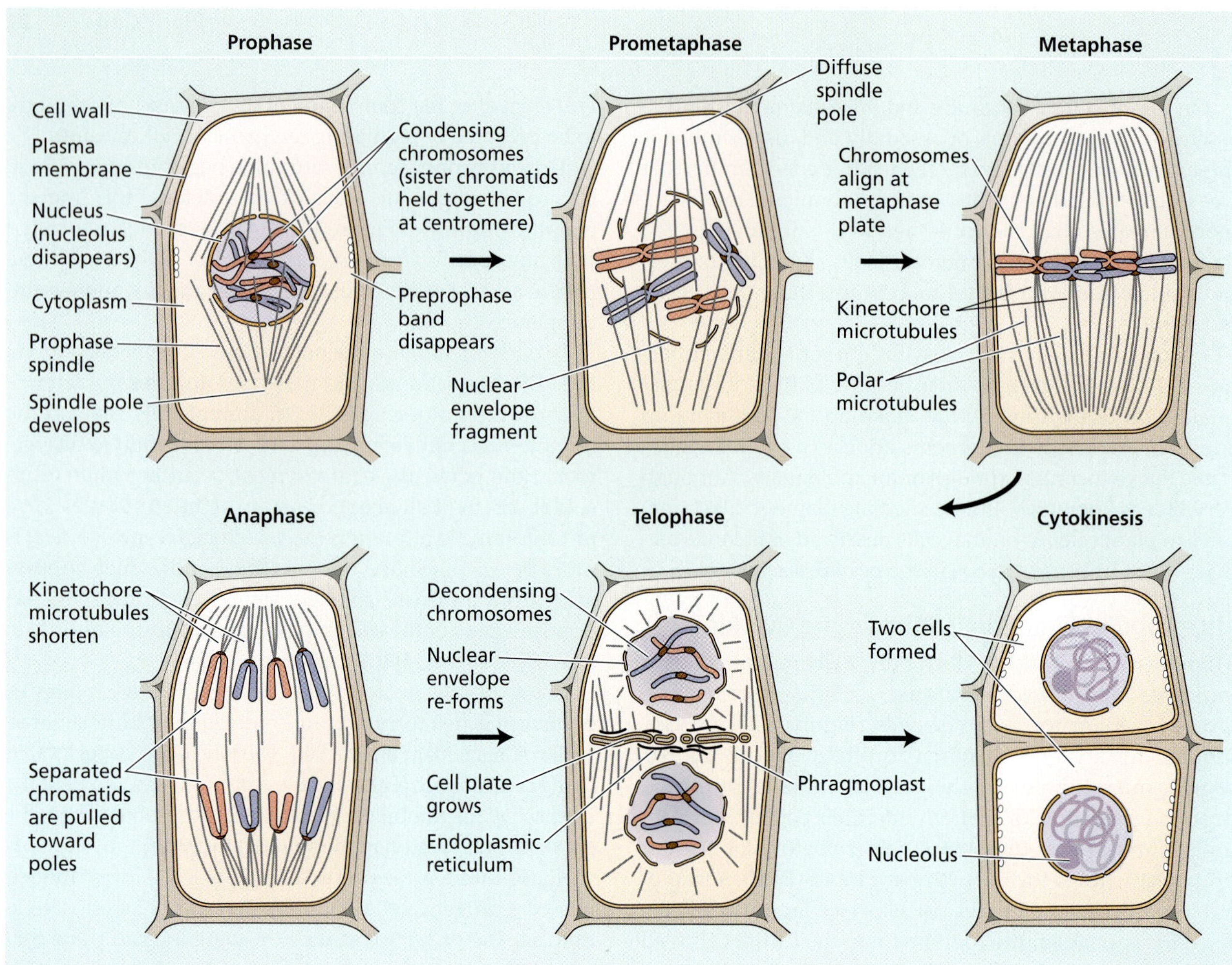

FIGURE 1.24 Diagram of mitosis in plants.

forms between incipient daughter cells, is rich in pectins (Figure 1.25). Cell plate formation in higher plants is a multistep process (see **Web Topic 1.5**). Vesicle aggregation in the spindle midzone is organized by the **phragmoplast**, a complex of microtubules and ER that forms during late anaphase or early telophase from dissociated spindle subunits.

Microfilaments Are Involved in Cytoplasmic Streaming and in Tip Growth

Cytoplasmic streaming is the coordinated movement of particles and organelles through the cytosol in a helical path down one side of a cell and up the other side. Cytoplasmic streaming occurs in most plant cells and has been studied extensively in the giant cells of the green algae *Chara* and *Nitella*, in which speeds up to 75 μm s^{-1} have been measured.

The mechanism of cytoplasmic streaming involves bundles of microfilaments that are arranged parallel to the longitudinal direction of particle movement. The forces necessary for movement may be generated by an interaction of the microfilament protein actin with the protein myosin in a fashion comparable to that of the protein interaction that occurs during muscle contraction in animals.

Myosins are proteins that have the ability to hydrolyze ATP to ADP and P_i when activated by binding to an actin microfilament. The energy released by ATP hydrolysis propels myosin molecules along the actin microfilament from the minus end to the plus end. Thus, myosins belong to the general class of **motor proteins** that drive cytoplasmic streaming and the movements of organelles within the cell. Examples of other motor proteins include the **kinesins** and **dyneins**, which drive movements of organelles and other cytoskeletal components along the surfaces of microtubules.

Actin microfilaments also participate in the growth of the pollen tube. Upon germination, a pollen grain forms a tubular extension that grows down the style toward the embryo sac. As the tip of the pollen tube extends, new cell wall material is continually deposited to maintain the integrity of the wall.

A network of microfilaments appears to guide vesicles containing wall precursors from their site of formation in the Golgi through the cytosol to the site of new wall formation at the tip. Fusion of these vesicles with the plasma membrane deposits wall precursors outside the cell, where they are assembled into wall material.

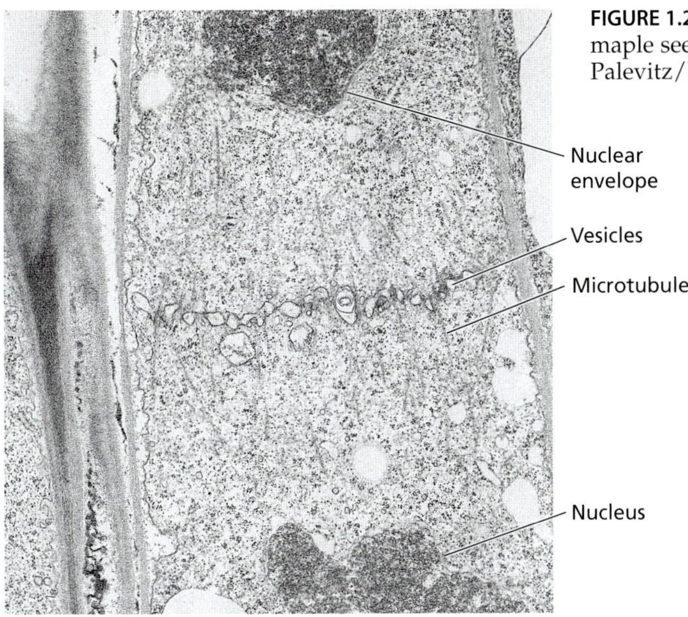

FIGURE 1.25 Electron micrograph of a cell plate forming in a maple seedling (10,000×). (© E. H. Newcomb and B. A. Palevitz/Biological Photo Service.)

Intermediate Filaments Occur in the Cytosol and Nucleus of Plant Cells

Relatively little is known about plant intermediate filaments. Intermediate filament–like structures have been identified in the cytoplasm of plant cells (Yang et al. 1995), but these may not be based on keratin, as in animal cells, since as yet no plant keratin genes have been found. Nuclear lamins, intermediate filaments of another type that form a dense network on the inner surface of the nuclear membrane, have also been identified in plant cells (Frederick et al. 1992), and genes encoding laminlike proteins are present in the *Arabidopsis* genome. Presumably, plant lamins perform functions similar to those in animal cells as a structural component of the nuclear envelope.

CELL CYCLE REGULATION

The cell division cycle, or cell cycle, is the process by which cells reproduce themselves and their genetic material, the nuclear DNA. The four phases of the cell cycle are designated G_1, S, G_2, and M (Figure 1.26A).

Each Phase of the Cell Cycle Has a Specific Set of Biochemical and Cellular Activities

Nuclear DNA is prepared for replication in G_1 by the assembly of a prereplication complex at the origins of replication along the chromatin. DNA is replicated during the S phase, and G_2 cells prepare for mitosis.

The whole architecture of the cell is altered as cells enter mitosis: The nuclear envelope breaks down, chromatin condenses to form recognizable chromosomes, the mitotic spindle forms, and the replicated chromosomes attach to the spindle fibers. The transition from metaphase to anaphase of mitosis marks a major transition point when the two chromatids of each replicated chromosome, which were held together at their kinetochores, are separated and the daughter chromosomes are pulled to opposite poles by spindle fibers.

At a key regulatory point early in G_1 of the cell cycle, the cell becomes committed to the initiation of DNA synthesis. In yeasts, this point is called START. Once a cell has passed START, it is irreversibly committed to initiating DNA synthesis and completing the cell cycle through mitosis and cytokinesis. After the cell has completed mitosis, it may initiate another complete cycle (G_1 through mitosis), or it may leave the cell cycle and differentiate. This choice is made at the critical G_1 point, before the cell begins to replicate its DNA.

DNA replication and mitosis are linked in mammalian cells. Often mammalian cells that have stopped dividing can be stimulated to reenter the cell cycle by a variety of hormones and growth factors. When they do so, they reenter the cell cycle at the critical point in early G_1. In contrast, plant cells can leave the cell division cycle either before or after replicating their DNA (i.e., during G_1 or G_2). As a consequence, whereas most animal cells are diploid (having two sets of chromosomes), plant cells frequently are tetraploid (having four sets of chromosomes), or even polyploid (having many sets of chromosomes), after going through additional cycles of nuclear DNA replication without mitosis.

The Cell Cycle Is Regulated by Protein Kinases

The mechanism regulating the progression of cells through their division cycle is highly conserved in evolution, and plants have retained the basic components of this mechanism (Renaudin et al. 1996). The key enzymes that control the transitions between the different states of the cell cycle, and the entry of nondividing cells into the cell cycle, are the **cyclin-dependent protein kinases,** or **CDKs** (Figure 1.26B). Protein kinases are enzymes that phosphorylate proteins using ATP. Most multicellular eukaryotes use several protein kinases that are active in different phases of the cell cycle. All depend on regulatory subunits called cyclins for their activities. The regulated activity of CDKs is essential for the transitions from G_1 to S and from G_2 to M, and for the entry of nondividing cells into the cell cycle.

CDK activity can be regulated in various ways, but two of the most important mechanisms are (1) cyclin synthesis and destruction and (2) the phosphorylation and dephosphorylation of key amino acid residues within the CDK protein. CDKs are inactive unless they are associated

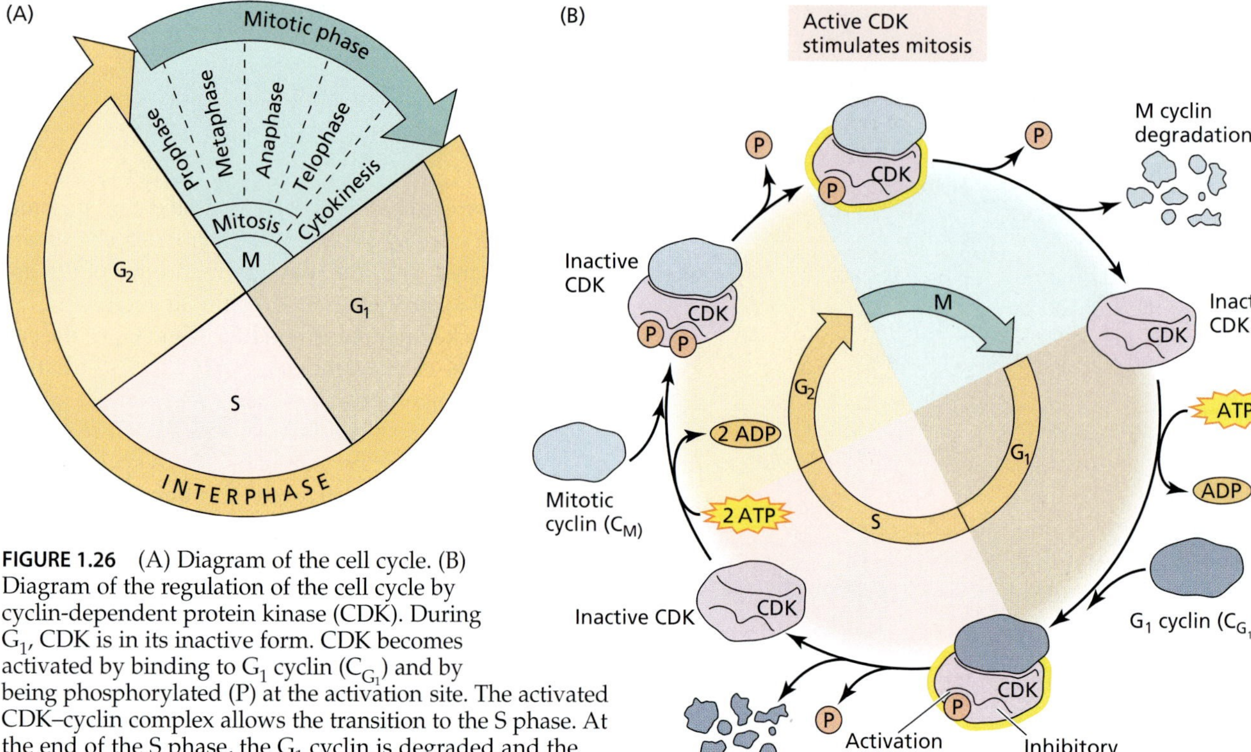

FIGURE 1.26 (A) Diagram of the cell cycle. (B) Diagram of the regulation of the cell cycle by cyclin-dependent protein kinase (CDK). During G_1, CDK is in its inactive form. CDK becomes activated by binding to G_1 cyclin (C_{G_1}) and by being phosphorylated (P) at the activation site. The activated CDK–cyclin complex allows the transition to the S phase. At the end of the S phase, the G_1 cyclin is degraded and the CDK is dephosphorylated, resulting in an inactive CDK. The cell enters G_2. During G_2, the inactive CDK binds to the mitotic cyclin (C_M), or M cyclin. At the same time, the CDK–cyclin complex becomes phosphorylated at both its activation and its inhibitory sites. The CDK–cyclin complex is still inactive because the inhibitory site is phosphorylated. The inactive complex becomes activated when the phosphate is removed from the inhibitory site by a protein phosphatase. The activated CDK then stimulates the transition from G_2 to mitosis. At the end of mitosis, the mitotic cyclin is degraded and the remaining phosphate at the activation site is removed by the phosphatase, and the cell enters G_1 again.

with a cyclin. Most cyclins turn over rapidly. They are synthesized and then actively degraded (using ATP) at specific points in the cell cycle. Cyclins are degraded in the cytosol by a large proteolytic complex called the **proteasome**. Before being degraded by the proteasome, the cyclins are marked for destruction by the attachment of a small protein called *ubiquitin*, a process that requires ATP. Ubiquitination is a general mechanism for tagging cellular proteins destined for turnover (see Chapter 14).

The transition from G_1 to S requires a set of cyclins (known as **G_1 cyclins**) different from those required in the transition from G_2 to mitosis, where **mitotic cyclins** activate the CDKs (see Figure 1.26B). CDKs possess two tyrosine phosphorylation sites: One causes activation of the enzyme; the other causes inactivation. Specific kinases carry out both the stimulatory and the inhibitory phosphorylations.

Similarly, protein phosphatases can remove phosphate from CDKs, either stimulating or inhibiting their activity, depending on the position of the phosphate. The addition or removal of phosphate groups from CDKs is highly regulated and an important mechanism for the control of cell cycle progression (see Figure 1.26B). Cyclin inhibitors play an important role in regulating the cell cycle in animals, and probably in plants as well, although little is known about plant cyclin inhibitors.

Finally, as we will see later in the book, certain plant hormones are able to regulate the cell cycle by regulating the synthesis of key enzymes in the regulatory pathway.

PLASMODESMATA

Plasmodesmata (singular *plasmodesma*) are tubular extensions of the plasma membrane, 40 to 50 nm in diameter, that traverse the cell wall and connect the cytoplasms of adjacent cells. Because most plant cells are interconnected in this way, their cytoplasms form a continuum referred to as the **symplast**. Intercellular transport of solutes through plasmodesmata is thus called **symplastic transport** (see Chapters 4 and 6).

There Are Two Types of Plasmodesmata: Primary and Secondary

Primary plasmodesmata form during cytokinesis when Golgi-derived vesicles containing cell wall precursors fuse to form the cell plate (the future middle lamella). Rather than forming a continuous uninterrupted sheet, the newly deposited cell plate is penetrated by numerous pores (Figure 1.27A), where remnants of the spindle apparatus, consisting of ER and microtubules, disrupt vesicle fusion. Further deposition of wall polymers increases the thickness of the two primary cell walls on either side of the middle lamella, generating linear membrane-lined channels (Figure 1.27B). Development of primary plasmodesmata thus provides direct continuity and communication between cells that are clonally related (i.e., derived from the same mother cell).

Secondary plasmodesmata form between cells after their cell walls have been deposited. They arise either by evagination of the plasma membrane at the cell surface, or by branching from a primary plasmodesma (Lucas and Wolf 1993). In addition to increasing the communication between cells that are clonally related, secondary plasmodesmata allow symplastic continuity between cells that are not clonally related.

Plasmodesmata Have a Complex Internal Structure

Like nuclear pores, plasmodesmata have a complex internal structure that functions in regulating macromolecular traffic from cell to cell. Each plasmodesma contains a narrow tubule of ER called a **desmotubule** (see Figure 1.27). The desmotubule is continuous with the ER of the adjacent cells. Thus the symplast joins not only the cytosol of neighboring cells, but the contents of the ER lumens as well. However, it is not clear that the desmotubule actually represents a passage, since there does not appear to be a space between the membranes, which are tightly appressed.

Globular proteins are associated with both the desmotubule membrane and the plasma membrane within the pore (see Figure 1.27B). These globular proteins appear to be interconnected by spokelike extensions, dividing the pore into eight to ten microchannels (Ding et al. 1992). Some molecules can pass from cell to cell through plasmodesmata, probably by flowing through the microchannels, although the exact pathway of communication has not been established.

By following the movement of fluorescent dye molecules of different sizes through plasmodesmata connecting leaf epidermal cells, Robards and Lucas (1990) determined

(A)
ER
Plasma membrane
Neck
Desmotubule
Cell wall
ER

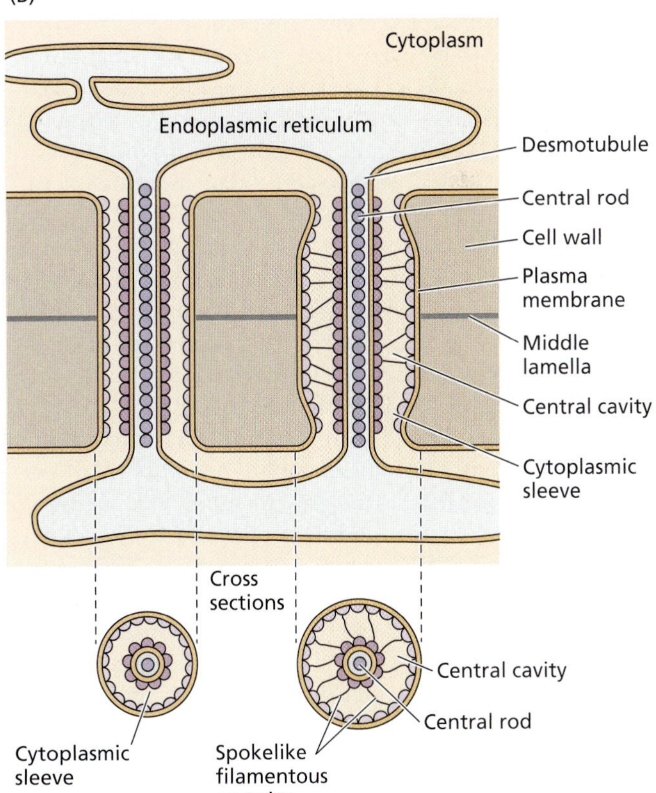

FIGURE 1.27 Plasmodesmata between cells. (A) Electron micrograph of a wall separating two adjacent cells, showing the plasmodesmata. (B) Schematic view of a cell wall with two plasmodesmata with different shapes. The desmotubule is continuous with the ER of the adjoining cells. Proteins line the outer surface of the desmotubule and the inner surface of the plasma membrane; the two surfaces are thought to be connected by filamentous proteins. The gap between the proteins lining the two membranes apparently controls the molecular sieving properties of plasmodesmata. (A from Tilney et al. 1991; B after Buchanan et al. 2000.)

the limiting molecular mass for transport to be about 700 to 1000 daltons, equivalent to a molecular size of about 1.5 to 2.0 nm. This is the **size exclusion limit**, or **SEL**, of plasmodesmata.

If the width of the cytoplasmic sleeve is approximately 5 to 6 nm, how are molecules larger than 2.0 nm excluded? The proteins attached to the plasma membrane and the ER within the plasmodesmata appear to act to restrict the size of molecules that can pass through the pore. As we'll see in Chapter 16, the SELs of plasmodesmata can be regulated. The mechanism for regulating the SEL is poorly understood, but the localization of both actin and myosin within plasmodesmata, possibly forming the "spoke" extensions (see Figure 1.27B), suggests that they may participate in the process (White et al. 1994; Radford and White 1996). Recent studies have also implicated calcium-dependent protein kinases in the regulation of plasmodesmatal SEL.

SUMMARY

Despite their great diversity in form and size, all plants carry out similar physiological processes. As primary producers, plants convert solar energy to chemical energy. Being nonmotile, plants must grow toward light, and they must have efficient vascular systems for movement of water, mineral nutrients, and photosynthetic products throughout the plant body. Green land plants must also have mechanisms for avoiding desiccation.

The major vegetative organ systems of seed plants are the shoot and the root. The shoot consists of two types of organs: stems and leaves. Unlike animal development, plant growth is indeterminate because of the presence of permanent meristem tissue at the shoot and root apices, which gives rise to new tissues and organs during the entire vegetative phase of the life cycle. Lateral meristems (the vascular cambium and the cork cambium) produce growth in girth, or secondary growth.

Three major tissue systems are recognized: dermal, ground, and vascular. Each of these tissues contains a variety of cell types specialized for different functions.

Plants are eukaryotes and have the typical eukaryotic cell organization, consisting of nucleus and cytoplasm. The nuclear genome directs the growth and development of the organism. The cytoplasm is enclosed by a plasma membrane and contains numerous membrane-enclosed organelles, including plastids, mitochondria, microbodies, oleosomes, and a large central vacuole. Chloroplasts and mitochondria are semiautonomous organelles that contain their own DNA. Nevertheless, most of their proteins are encoded by nuclear DNA and are imported from the cytosol.

The cytoskeletal components—microtubules, microfilaments, and intermediate filaments—participate in a variety of processes involving intracellular movements, such as mitosis, cytoplasmic streaming, secretory vesicle transport, cell plate formation, and cellulose microfibril deposition. The process by which cells reproduce is called the cell cycle. The cell cycle consists of the G_1, S, G_2, and M phases. The transition from one phase to another is regulated by cyclin-dependent protein kinases. The activity of the CDKs is regulated by cyclins and by protein phosphorylation.

During cytokinesis, the phragmoplast gives rise to the cell plate in a multistep process that involves vesicle fusion. After cytokinesis, primary cell walls are deposited. The cytosol of adjacent cells is continuous through the cell walls because of the presence of membrane-lined channels called plasmodesmata, which play a role in cell–cell communication.

Web Material

Web Topics

1.1 The Plant Kingdom
The major groups of the plant kingdom are surveyed and described.

1.2 Flower Structure and the Angiosperm Life Cycle
The steps in the reproductive style of angiosperms are discussed and illustrated.

1.3 Plant Tissue Systems: Dermal, Ground, and Vascular
A more detailed treatment of plant anatomy is given.

1.4 The Structures of Chloroplast Glycosylglycerides
The chemical structures of the chloroplast lipids are illustrated.

1.5 The Multiple Steps in Construction of the Cell Plate Following Mitosis
Details of the production of the cell plate during cytokinesis in plants are described.

Chapter References

Alberts, B., Johnson, A., Lewis, J., Raff, M., Roberts, K., and Walter, P. (2002) *Molecular Biology of the Cell*, 4th ed. Garland, New York.

Buchanan, B. B., Gruissem, W., and Jones, R. L. (eds.) (2000) *Biochemistry and Molecular Biology of Plants.* Amer. Soc. Plant Physiologists, Rockville, MD.

Ding, B., Turgeon, R., and Parthasarathy, M. V. (1992) Substructure of freeze substituted plasmodesmata. *Protoplasma* 169: 28–41.

Driouich, A., Levy, S., Staehelin, L. A., and Faye, L. (1994) Structural and functional organization of the Golgi apparatus in plant cells. *Plant Physiol. Biochem.* 32: 731–749.

Esau, K. (1960) *Anatomy of Seed Plants.* Wiley, New York.

Esau, K. (1977) *Anatomy of Seed Plants*, 2nd ed. Wiley, New York.

Faye, L., Fitchette-Lainé, A. C., Gomord, V., Chekkafi, A., Delaunay, A. M., and Driouich, A. (1992) Detection, biosynthesis and some functions of glycans N-linked to plant secreted proteins. In *Post-translational Modifications in Plants* (SEB Seminar Series, no. 53), N. H. Battey, H. G. Dickinson, and A. M. Heatherington, eds., Cambridge University Press, Cambridge, pp. 213–242.

Frederick, S. E., Mangan, M. E., Carey, J. B., and Gruber, P. J. (1992) Intermediate filament antigens of 60 and 65 kDa in the nuclear matrix of plants: Their detection and localization. *Exp. Cell Res.* 199: 213—222.

Gunning, B. E. S., and Steer, M. W. (1996) *Plant Cell Biology: Structure and Function of Plant Cells.* Jones and Bartlett, Boston.

Harwood, J. L. (1997) Plant lipid metabolism. In *Plant Biochemistry*, P. M. Dey and J. B. Harborne, eds., Academic Press, San Diego, CA, pp. 237–272.

Huang, A. H. C. (1987) Lipases in *The Biochemistry of Plants: A Comprehensive Treatise.* In Vol. 9, *Lipids: Structure and Function*, P. K. Stumpf, ed. Academic Press, New York, pp. 91–119.

Lucas, W. J., and Wolf, S. (1993) Plasmodesmata: The intercellular organelles of green plants. *Trends Cell Biol.* 3: 308–315.

O'Brien, T. P., and McCully, M. E. (1969) *Plant Structure and Development: A Pictorial and Physiological Approach.* Macmillan, New York.

Radford, J., and White, R. G. (1996) Preliminary localization of myosin to plasmodesmata. Third International Workshop on Basic and Applied Research in Plasmodesmal Biology, Zichron-Takov, Israel, March 10–16, 1996, pp. 37–38.

Renaudin, J.-P., Doonan, J. H., Freeman, D., Hashimoto, J., Hirt, H., Inze, D., Jacobs, T., Kouchi, H., Rouze, P., Sauter, M., et al. (1996) Plant cyclins: A unified nomenclature for plant A-, B- and D-type cyclins based on sequence organization. *Plant Mol. Biol.* 32: 1003–1018.

Robards, A. W., and Lucas, W. J. (1990) Plasmodesmata. *Annu. Rev. Plant Physiol. Plant Mol. Biol.* 41: 369–420.

Tilney, L. G., Cooke, T. J., Connelly, P. S., and Tilney, M. S. (1991) The structure of plasmodesmata as revealed by plasmolysis, detergent extraction, and protease digestion. *J. Cell Biol.* 112: 739–748.

White, R. G., Badelt, K., Overall, R. L., and Vesk, M. (1994) Actin associated with plasmodesmata. *Protoplasma* 180: 169–184.

Yang, C., Min, G. W., Tong, X. J., Luo, Z., Liu, Z. F., and Zhai, Z. H. (1995) The assembly of keratins from higher plant cells. *Protoplasma* 188: 128–132.

Chapter

3 Water and Plant Cells

WATER PLAYS A CRUCIAL ROLE in the life of the plant. For every gram of organic matter made by the plant, approximately 500 g of water is absorbed by the roots, transported through the plant body and lost to the atmosphere. Even slight imbalances in this flow of water can cause water deficits and severe malfunctioning of many cellular processes. Thus, every plant must delicately balance its uptake and loss of water. This balancing is a serious challenge for land plants. To carry on photosynthesis, they need to draw carbon dioxide from the atmosphere, but doing so exposes them to water loss and the threat of dehydration.

A major difference between plant and animal cells that affects virtually all aspects of their relation with water is the existence in plants of the cell wall. Cell walls allow plant cells to build up large internal hydrostatic pressures, called **turgor pressure**, which are a result of their normal water balance. Turgor pressure is essential for many physiological processes, including cell enlargement, gas exchange in the leaves, transport in the phloem, and various transport processes across membranes. Turgor pressure also contributes to the rigidity and mechanical stability of nonlignified plant tissues. In this chapter we will consider how water moves into and out of plant cells, emphasizing the molecular properties of water and the physical forces that influence water movement at the cell level. But first we will describe the major functions of water in plant life.

WATER IN PLANT LIFE

Water makes up most of the mass of plant cells, as we can readily appreciate if we look at microscopic sections of mature plant cells: Each cell contains a large water-filled vacuole. In such cells the cytoplasm makes up only 5 to 10% of the cell volume; the remainder is vacuole. Water typically constitutes 80 to 95% of the mass of growing plant tissues. Common vegetables such as carrots and lettuce may contain 85 to 95% water. Wood, which is composed mostly of dead cells, has a lower water content; sapwood, which functions in transport in the xylem, contains 35 to

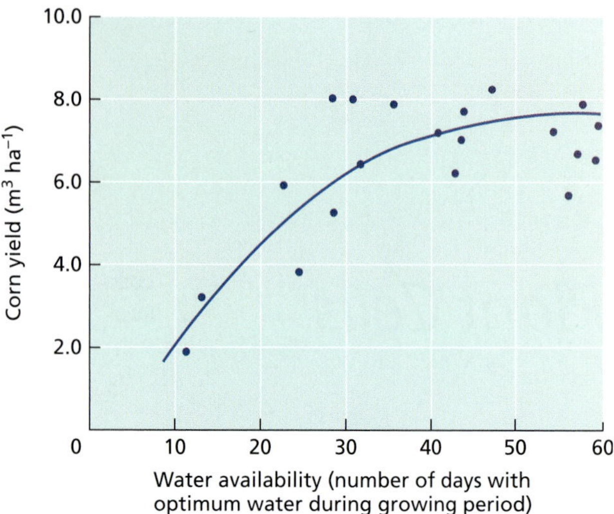

FIGURE 3.1 Corn yield as a function of water availability. The data plotted here were gathered at an Iowa farm over a 4-year period. Water availability was assessed as the number of days without water stress during a 9-week growing period. (Data from *Weather and Our Food Supply* 1964.)

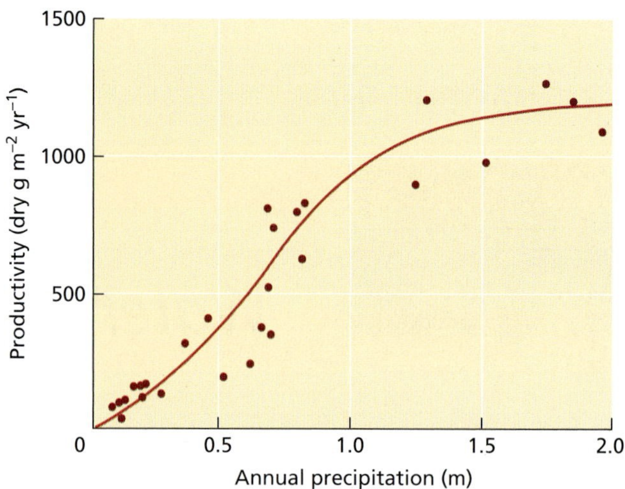

FIGURE 3.2 Productivity of various ecosystems as a function of annual precipitation. Productivity was estimated as net aboveground accumulation of organic matter through growth and reproduction. (After Whittaker 1970.)

75% water; and heartwood has a slightly lower water content. Seeds, with a water content of 5 to 15%, are among the driest of plant tissues, yet before germinating they must absorb a considerable amount of water.

Water is the most abundant and arguably the best solvent known. As a solvent, it makes up the medium for the movement of molecules within and between cells and greatly influences the structure of proteins, nucleic acids, polysaccharides, and other cell constituents. Water forms the environment in which most of the biochemical reactions of the cell occur, and it directly participates in many essential chemical reactions.

Plants continuously absorb and lose water. Most of the water lost by the plant evaporates from the leaf as the CO_2 needed for photosynthesis is absorbed from the atmosphere. On a warm, dry, sunny day a leaf will exchange up to 100% of its water in a single hour. During the plant's lifetime, water equivalent to 100 times the fresh weight of the plant may be lost through the leaf surfaces. Such water loss is called **transpiration**.

Transpiration is an important means of dissipating the heat input from sunlight. Heat dissipates because the water molecules that escape into the atmosphere have higher-than-average energy, which breaks the bonds holding them in the liquid. When these molecules escape, they leave behind a mass of molecules with lower-than-average energy and thus a cooler body of water. For a typical leaf, nearly half of the net heat input from sunlight is dissipated by transpiration. In addition, the stream of water taken up by the roots is an important means of bringing dissolved soil minerals to the root surface for absorption.

Of all the resources that plants need to grow and function, water is the most abundant and at the same time the most limiting for agricultural productivity (Figure 3.1). The fact that water is limiting is the reason for the practice of crop irrigation. Water availability likewise limits the productivity of natural ecosystems (Figure 3.2). Thus an understanding of the uptake and loss of water by plants is very important.

We will begin our study of water by considering how its structure gives rise to some of its unique physical properties. We will then examine the physical basis for water movement, the concept of water potential, and the application of this concept to cell–water relations.

THE STRUCTURE AND PROPERTIES OF WATER

Water has special properties that enable it to act as a solvent and to be readily transported through the body of the plant. These properties derive primarily from the polar structure of the water molecule. In this section we will examine how the formation of hydrogen bonds contributes to the properties of water that are necessary for life.

The Polarity of Water Molecules Gives Rise to Hydrogen Bonds

The water molecule consists of an oxygen atom covalently bonded to two hydrogen atoms. The two O—H bonds form an angle of 105° (Figure 3.3). Because the oxygen atom is more **electronegative** than hydrogen, it tends to attract the electrons of the covalent bond. This attraction results in a partial negative charge at the oxygen end of the molecule and a partial positive charge at each hydrogen.

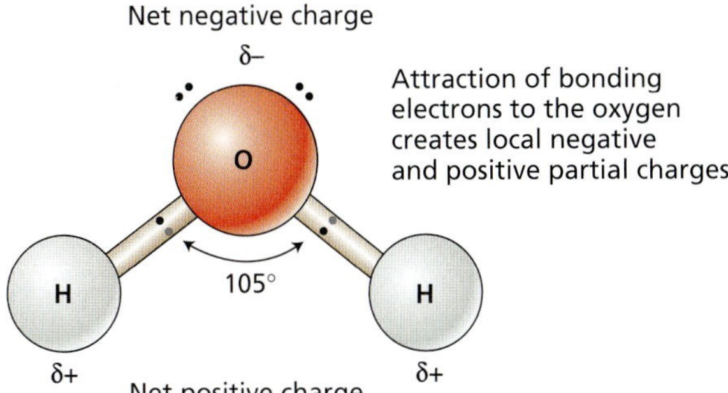

FIGURE 3.3 Diagram of the water molecule. The two intramolecular hydrogen–oxygen bonds form an angle of 105°. The opposite partial charges (δ– and δ+) on the water molecule lead to the formation of intermolecular hydrogen bonds with other water molecules. Oxygen has six electrons in the outer orbitals; each hydrogen has one.

These partial charges are equal, so the water molecule carries no *net* charge.

This separation of partial charges, together with the shape of the water molecule, makes water a *polar molecule*, and the opposite partial charges between neighboring water molecules tend to attract each other. The weak electrostatic attraction between water molecules, known as a **hydrogen bond**, is responsible for many of the unusual physical properties of water.

Hydrogen bonds can also form between water and other molecules that contain electronegative atoms (O or N). In aqueous solutions, hydrogen bonding between water molecules leads to local, ordered clusters of water that, because of the continuous thermal agitation of the water molecules, continually form, break up, and re-form (Figure 3.4).

(A) Correlated configuration

(B) Random configuration

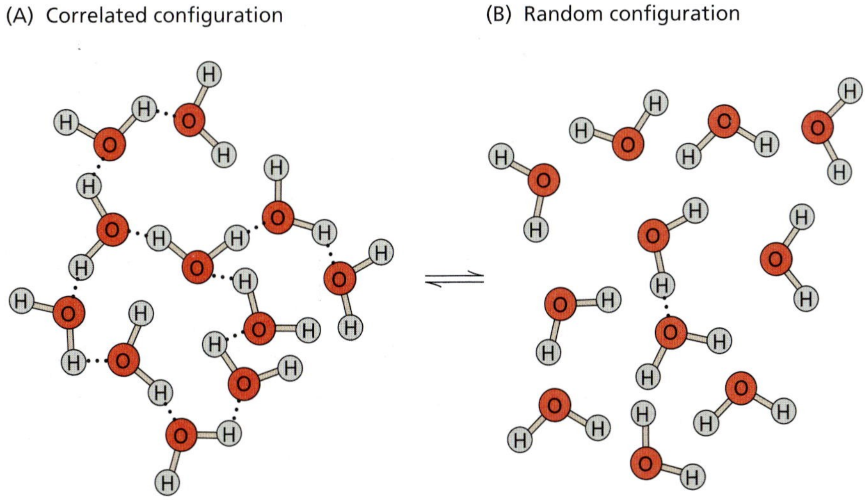

FIGURE 3.4 (A) Hydrogen bonding between water molecules results in local aggregations of water molecules. (B) Because of the continuous thermal agitation of the water molecules, these aggregations are very short-lived; they break up rapidly to form much more random configurations.

The Polarity of Water Makes It an Excellent Solvent

Water is an excellent solvent: It dissolves greater amounts of a wider variety of substances than do other related solvents. This versatility as a solvent is due in part to the small size of the water molecule and in part to its polar nature. The latter makes water a particularly good solvent for ionic substances and for molecules such as sugars and proteins that contain polar —OH or —NH_2 groups.

Hydrogen bonding between water molecules and ions, and between water and polar solutes, in solution effectively decreases the electrostatic interaction between the charged substances and thereby increases their solubility. Furthermore, the polar ends of water molecules can orient themselves next to charged or partially charged groups in macromolecules, forming **shells of hydration**. Hydrogen bonding between macromolecules and water reduces the interaction between the macromolecules and helps draw them into solution.

The Thermal Properties of Water Result from Hydrogen Bonding

The extensive hydrogen bonding between water molecules results in unusual thermal properties, such as high specific heat and high latent heat of vaporization. **Specific heat** is the heat energy required to raise the temperature of a substance by a specific amount.

When the temperature of water is raised, the molecules vibrate faster and with greater amplitude. To allow for this motion, energy must be added to the system to break the hydrogen bonds between water molecules. Thus, compared with other liquids, water requires a relatively large energy input to raise its temperature. This large energy input requirement is important for plants because it helps buffer temperature fluctuations.

Latent heat of vaporization is the energy needed to separate molecules from the liquid phase and move them into the gas phase at constant temperature—a process that occurs during transpiration. For water at 25°C, the heat of vaporization is 44 kJ mol^{-1}—the highest value known for any liquid. Most of this energy is used to break hydrogen bonds between water molecules.

The high latent heat of vaporization of water enables plants to cool themselves by evaporating water from leaf surfaces, which are prone to heat up because of the radiant input from the sun. Transpiration is an important component of temperature regulation in plants.

The Cohesive and Adhesive Properties of Water Are Due to Hydrogen Bonding

Water molecules at an air–water interface are more strongly attracted to neighboring water molecules than to the gas phase in contact with the water surface. As a consequence of this unequal attraction, an air–water interface minimizes its surface area. To increase the area of an air–water interface, hydrogen bonds must be broken, which requires an input of energy. The energy required to increase the surface area is known as **surface tension**. Surface tension not only influences the shape of the surface but also may create a pressure in the rest of the liquid. As we will see later, surface tension at the evaporative surfaces of leaves generates the physical forces that pull water through the plant's vascular system.

The extensive hydrogen bonding in water also gives rise to the property known as **cohesion**, the mutual attraction between molecules. A related property, called **adhesion**, is the attraction of water to a solid phase such as a cell wall or glass surface. Cohesion, adhesion, and surface tension give rise to a phenomenon known as **capillarity**, the movement of water along a capillary tube.

In a vertically oriented glass capillary tube, the upward movement of water is due to (1) the attraction of water to the polar surface of the glass tube (adhesion) and (2) the surface tension of water, which tends to minimize the area of the air–water interface. Together, adhesion and surface tension pull on the water molecules, causing them to move up the tube until the upward force is balanced by the weight of the water column. The smaller the tube, the higher the capillary rise. For calculations related to capillary rise, see **Web Topic 3.1**.

Water Has a High Tensile Strength

Cohesion gives water a high **tensile strength**, defined as the maximum force per unit area that a continuous column of water can withstand before breaking. We do not usually think of liquids as having tensile strength; however, such a property must exist for a water column to be pulled up a capillary tube.

We can demonstrate the tensile strength of water by placing it in a capped syringe (Figure 3.5). When we *push* on the plunger, the water is compressed and a positive **hydrostatic pressure** builds up. Pressure is measured in units called *pascals* (Pa) or, more conveniently, *megapascals* (MPa). One MPa equals approximately 9.9 atmospheres. Pressure is equivalent to a force per unit area (1 Pa = 1 N m^{-2}) and to an energy per unit volume (1 Pa = 1 J m^{-3}). A newton (N) = 1 kg m s^{-1}. Table 3.1 compares units of pressure.

If instead of pushing on the plunger we *pull* on it, a tension, or *negative hydrostatic pressure*, develops in the water to resist the pull. How hard must we pull on the plunger before the water molecules are torn away from each other and the water column breaks? Breaking the water column requires sufficient energy to break the hydrogen bonds that attract water molecules to one another.

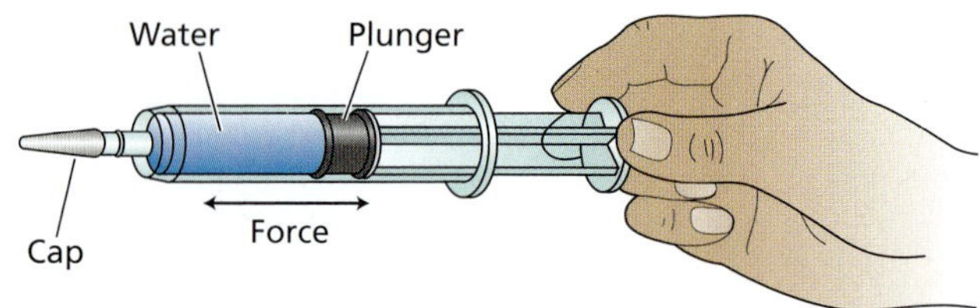

FIGURE 3.5 A sealed syringe can be used to create positive and negative pressures in a fluid like water. Pushing on the plunger compresses the fluid, and a positive pressure builds up. If a small air bubble is trapped within the syringe, it shrinks as the pressure increases. Pulling on the plunger causes the fluid to develop a tension, or negative pressure. Any air bubbles in the syringe will expand as the pressure is reduced.

Careful studies have demonstrated that water in small capillaries can resist tensions more negative than –30 MPa (the negative sign indicates tension, as opposed to compression). This value is only a fraction of the theoretical tensile strength of water computed on the basis of the strength of hydrogen bonds. Nevertheless, it is quite substantial.

The presence of gas bubbles reduces the tensile strength of a water column. For example, in the syringe shown in Figure 3.5, expansion of microscopic bubbles often interferes with the ability of the water to resist the pull exerted by the plunger. If a tiny gas bubble forms in a column of water under tension, the gas bubble may expand indefinitely, with the result that the tension in the liquid phase collapses, a phenomenon known as **cavitation**. As we will see in Chapter 4, cavitation can have a devastating effect on water transport through the xylem.

WATER TRANSPORT PROCESSES

When water moves from the soil through the plant to the atmosphere, it travels through a widely variable medium (cell wall, cytoplasm, membrane, air spaces), and the mechanisms of water transport also vary with the type of medium. For many years there has been much uncertainty

TABLE 3.1
Comparison of units of pressure

1 atmosphere	= 14.7 pounds per square inch
	= 760 mm Hg (at sea level, 45° latitude)
	= 1.013 bar
	= 0.1013 Mpa
	= 1.013×10^5 Pa

A car tire is typically inflated to about 0.2 MPa.
The water pressure in home plumbing is typically 0.2–0.3 MPa.
The water pressure under 15 feet (5 m) of water is about 0.05 MPa.

about how water moves across plant membranes. Specifically it was unclear whether water movement into plant cells was limited to the diffusion of water molecules across the plasma membrane's lipid bilayer or also involved diffusion through protein-lined pores (Figure 3.6).

Some studies indicated that diffusion directly across the lipid bilayer was not sufficient to account for observed rates of water movement across membranes, but the evidence in support of microscopic pores was not compelling. This uncertainty was put to rest with the recent discovery of **aquaporins** (see Figure 3.6). Aquaporins are integral membrane proteins that form water-selective channels across the membrane. Because water diffuses faster through such channels than through a lipid bilayer, aquaporins facilitate water movement into plant cells (Weig et al. 1997; Schäffner 1998; Tyerman et al. 1999). Note that although the presence of aquaporins may alter the *rate* of water movement across the membrane, they do not change the direction of transport or the driving force for water movement. The mode of action of aquaporins is being acitvely investigated (Tajkhorshid et al. 2002).

We will now consider the two major processes in water transport: molecular diffusion and bulk flow.

Diffusion Is the Movement of Molecules by Random Thermal Agitation

Water molecules in a solution are not static; they are in continuous motion, colliding with one another and exchanging kinetic energy. The molecules intermingle as a result of their random thermal agitation. This random motion is called **diffusion**. As long as other forces are not acting on the molecules, diffusion causes the net movement of molecules from regions of high concentration to regions of low concentration—that is, down a concentration gradient (Figure 3.7).

In the 1880s the German scientist Adolf Fick discovered that the rate of diffusion is directly proportional to the concentration gradient ($\Delta c_s/\Delta x$)—that is, to the difference in concentration of substance *s* (Δc_s) between two points separated by the distance Δx. In symbols, we write this relation as Fick's first law:

$$J_s = -D_s \frac{\Delta c_s}{\Delta x} \tag{3.1}$$

The rate of transport, or the **flux density** (J_s), is the amount of substance *s* crossing a unit area per unit time (e.g., J_s may have units of moles per square meter per second [mol m^{-2} s^{-1}]). The **diffusion coefficient** (D_s) is a proportionality constant that measures how easily substance *s* moves through a particular medium. The diffusion coefficient is a characteristic of the substance (larger molecules have smaller diffusion coefficients) and depends on the medium (diffusion in air is much faster than diffusion in a liquid, for example). The negative sign in the equation indicates that the flux moves down a concentration gradient.

Fick's first law says that a substance will diffuse faster when the concentration gradient becomes steeper (Δc_s is large) or when the diffusion coefficient is increased. This equation accounts only for movement in response to a concentration gradient, and not for movement in response to other forces (e.g., pressure, electric fields, and so on).

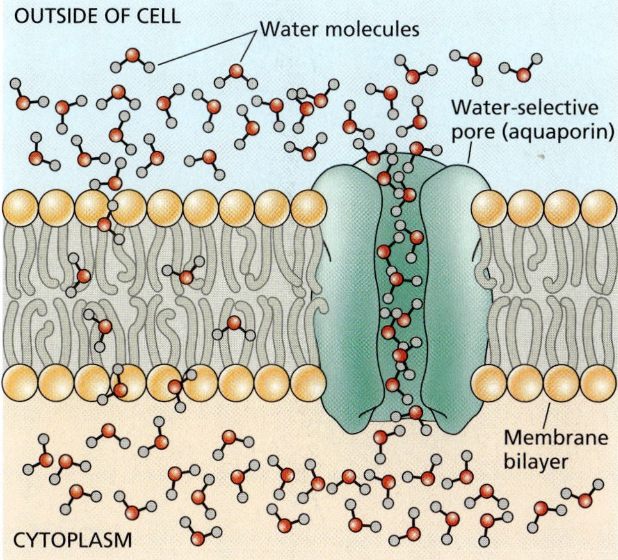

FIGURE 3.6 Water can cross plant membranes by diffusion of individual water molecules through the membrane bilayer, as shown on the left, and by microscopic bulk flow of water molecules through a water-selective pore formed by integral membrane proteins such as aquaporins.

Diffusion Is Rapid over Short Distances but Extremely Slow over Long Distances

From Fick's first law, one can derive an expression for the time it takes for a substance to diffuse a particular distance. If the initial conditions are such that all the solute molecules are concentrated at the starting position (Figure 3.8A), then the concentration front moves away from the starting position, as shown for a later time point in Figure 3.8B. As the substance diffuses away from the starting point, the concentration gradient becomes less steep (Δc_s decreases), and thus net movement becomes slower.

The average time needed for a particle to diffuse a distance L is equal to L^2/D_s, where D_s is the diffusion coefficient, which depends on both the identity of the particle and the medium in which it is diffusing. Thus the average time required for a substance to diffuse a given distance increases in proportion to the *square* of that distance. The diffusion coefficient for glucose in water is about 10^{-9} m^2 s^{-1}. Thus the average time required for a glucose molecule to diffuse across a cell with a diameter of 50 μm is 2.5 s. However, the average time needed for the same glucose molecule to diffuse a distance of 1 m in water is approxi-

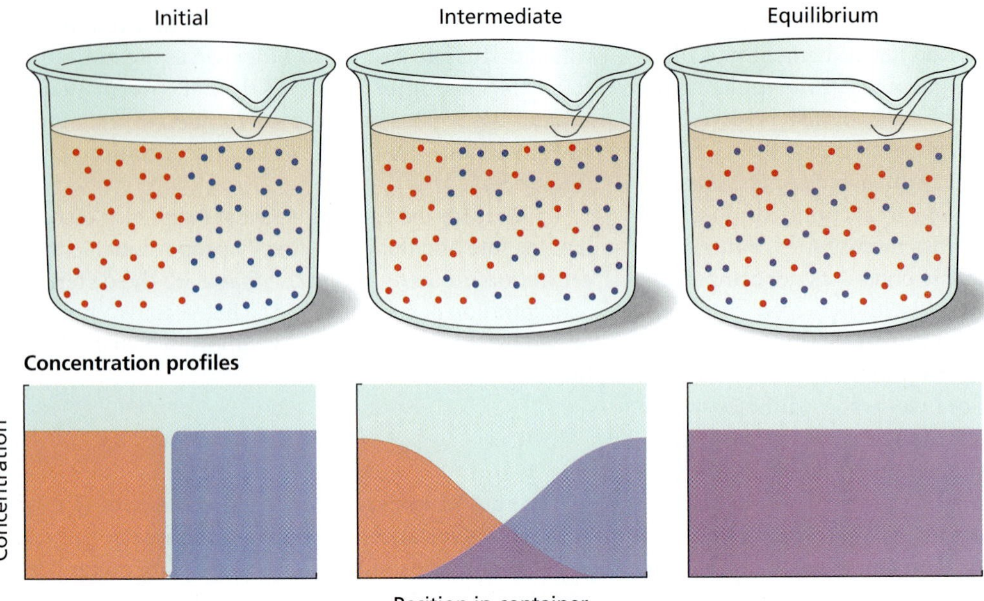

FIGURE 3.7 Thermal motion of molecules leads to diffusion—the gradual mixing of molecules and eventual dissipation of concentration differences. Initially, two materials containing different molecules are brought into contact. The materials may be gas, liquid, or solid. Diffusion is fastest in gases, slower in liquids, and slowest in solids. The initial separation of the molecules is depicted graphically in the upper panels, and the corresponding concentration profiles are shown in the lower panels as a function of position. With time, the mixing and randomization of the molecules diminishes net movement. At equilibrium the two types of molecules are randomly (evenly) distributed.

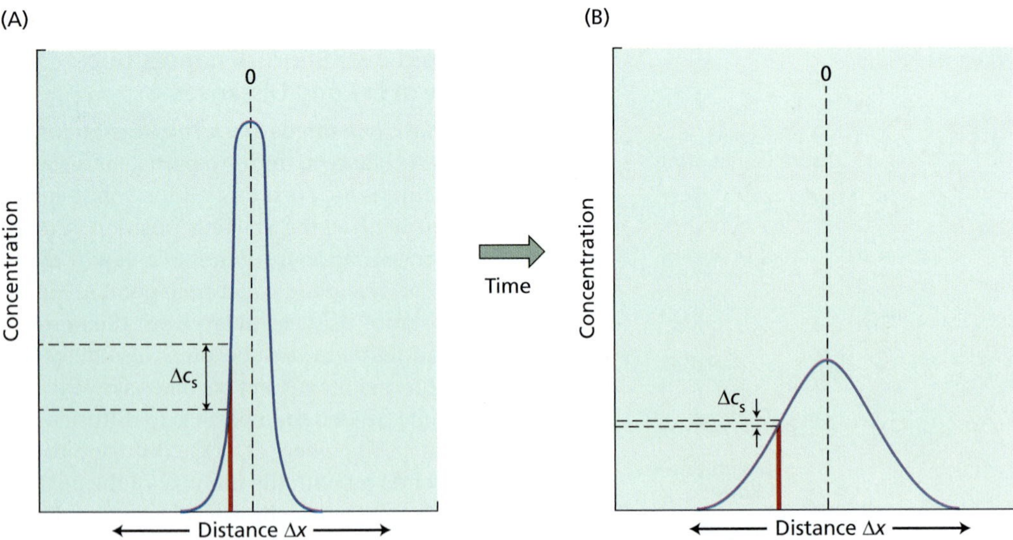

FIGURE 3.8 Graphical representation of the concentration gradient of a solute that is diffusing according to Fick's law. The solute molecules were initially located in the plane indicated on the x-axis. (A) The distribution of solute molecules shortly after placement at the plane of origin. Note how sharply the concentration drops off as the distance, x, from the origin increases. (B) The solute distribution at a later time point. The average distance of the diffusing molecules from the origin has increased, and the slope of the gradient has flattened out. (After Nobel 1999.)

mately 32 years. These values show that diffusion in solutions can be effective within cellular dimensions but is far too slow for mass transport over long distances. For additional calculations on diffusion times, see **Web Topic 3.2**.

Pressure-Driven Bulk Flow Drives Long-Distance Water Transport

A second process by which water moves is known as **bulk flow** or **mass flow**. Bulk flow is the concerted movement of groups of molecules en masse, most often in response to a pressure gradient. Among many common examples of bulk flow are water moving through a garden hose, a river flowing, and rain falling.

If we consider bulk flow through a tube, the rate of volume flow depends on the radius (r) of the tube, the viscosity (η) of the liquid, and the pressure gradient ($\Delta\Psi_p/\Delta x$) that drives the flow. Jean-Léonard-Marie Poiseuille (1797–1869) was a French physician and physiologist, and the relation just described is given by one form of Poiseuille's equation:

$$\text{Volume flow rate} = \left(\frac{\pi r^4}{8\eta}\right)\left(\frac{\Delta\Psi_p}{\Delta x}\right) \qquad (3.2)$$

expressed in cubic meters per second ($m^3\ s^{-1}$). This equation tells us that pressure-driven bulk flow is very sensitive to the radius of the tube. If the radius is doubled, the volume flow rate increases by a factor of 16 (2^4).

Pressure-driven bulk flow of water is the predominant mechanism responsible for long-distance transport of water in the xylem. It also accounts for much of the water flow through the soil and through the cell walls of plant tissues. In contrast to diffusion, pressure-driven bulk flow is independent of solute concentration gradients, as long as viscosity changes are negligible.

Osmosis Is Driven by a Water Potential Gradient

Membranes of plant cells are **selectively permeable**; that is, they allow the movement of water and other small uncharged substances across them more readily than the movement of larger solutes and charged substances (Stein 1986).

Like molecular diffusion and pressure-driven bulk flow, **osmosis** occurs spontaneously in response to a driving force. In simple diffusion, substances move down a concentration gradient; in pressure-driven bulk flow, substances move down a pressure gradient; in osmosis, both types of gradients influence transport (Finkelstein 1987). *The direction and rate of water flow across a membrane are determined not solely by the concentration gradient of water or by the pressure gradient, but by the sum of these two driving forces.*

We will soon see how osmosis drives the movement of water across membranes. First, however, let's discuss the concept of a composite or total driving force, representing the free-energy gradient of water.

The Chemical Potential of Water Represents the Free-Energy Status of Water

All living things, including plants, require a continuous input of free energy to maintain and repair their highly organized structures, as well as to grow and reproduce. Processes such as biochemical reactions, solute accumulation, and long-distance transport are all driven by an input of free energy into the plant. (For a detailed discussion of the thermodynamic concept of free energy, see Chapter 2 on the web site.)

The **chemical potential** of water is a quantitative expression of the free energy associated with water. In thermodynamics, free energy represents the potential for performing work. Note that chemical potential is a relative quantity: It is expressed as the difference between the potential of a substance in a given state and the potential of the same substance in a standard state. The unit of chemical potential is energy per mole of substance (J mol^{-1}).

For historical reasons, plant physiologists have most often used a related parameter called **water potential**, defined as the chemical potential of water divided by the partial molal volume of water (the volume of 1 mol of water): $18 \times 10^{-6}\ m^3\ mol^{-1}$. Water potential is a measure of the free energy of water per unit volume (J m^{-3}). These units are equivalent to pressure units such as the pascal, which is the common measurement unit for water potential. Let's look more closely at the important concept of water potential.

Three Major Factors Contribute to Cell Water Potential

The major factors influencing the water potential in plants are *concentration*, *pressure*, and *gravity*. Water potential is symbolized by Ψ_w (the Greek letter psi), and the water potential of solutions may be dissected into individual components, usually written as the following sum:

$$\Psi_w = \Psi_s + \Psi_p + \Psi_g \qquad (3.3)$$

The terms Ψ_s and Ψ_p and Ψ_g denote the effects of solutes, pressure, and gravity, respectively, on the free energy of water. (Alternative conventions for components of water potential are discussed in **Web Topic 3.3**.) The reference state used to define water potential is pure water at ambient pressure and temperature. Let's consider each of the terms on the right-hand side of Equation 3.3.

Solutes. The term Ψ_s, called the **solute potential** or the **osmotic potential**, represents the effect of dissolved solutes on water potential. Solutes reduce the free energy of water by diluting the water. This is primarily an entropy effect; that is, the mixing of solutes and water increases the disorder of the system and thereby lowers free energy. This means that the osmotic potential is independent of the specific nature of the solute. For dilute solutions of nondissociating substances, like sucrose, the osmotic potential may be estimated by the **van't Hoff equation**:

$$\Psi_s = -RTc_s \tag{3.4}$$

where R is the gas constant (8.32 J mol^{-1} K^{-1}), T is the absolute temperature (in degrees Kelvin, or K), and c_s is the solute concentration of the solution, expressed as **osmolality** (moles of total dissolved solutes per liter of water [mol L^{-1}]). The minus sign indicates that dissolved solutes reduce the water potential of a solution relative to the reference state of pure water.

Table 3.2 shows the values of RT at various temperatures and the Ψ_s values of solutions of different solute concentrations. For ionic solutes that dissociate into two or more particles, c_s must be multiplied by the number of dissociated particles to account for the increased number of dissolved particles.

Equation 3.4 is valid for "ideal" solutions at dilute concentration. Real solutions frequently deviate from the ideal, especially at high concentrations—for example, greater than 0.1 mol L^{-1}. In our treatment of water potential, we will assume that we are dealing with ideal solutions (Friedman 1986; Nobel 1999).

Pressure. The term Ψ_p is the **hydrostatic pressure** of the solution. Positive pressures raise the water potential; negative pressures reduce it. Sometimes Ψ_p is called *pressure potential*. The positive hydrostatic pressure within cells is the pressure referred to as *turgor pressure*. The value of Ψ_p can also be negative, as is the case in the xylem and in the walls between cells, where a *tension*, or *negative hydrostatic pressure*, can develop. As we will see, negative pressures outside cells are very important in moving water long distances through the plant.

Hydrostatic pressure is measured as the deviation from ambient pressure (for details, see **Web Topic 3.5**). Remember that water in the reference state is at ambient pressure, so by this definition Ψ_p = 0 MPa for water in the standard state. Thus the value of Ψ_p for pure water in an open beaker is 0 MPa, even though its absolute pressure is approximately 0.1 MPa (1 atmosphere).

Gravity. Gravity causes water to move downward unless the force of gravity is opposed by an equal and opposite force. The term Ψ_g depends on the height (h) of the water above the reference-state water, the density of water (ρ_w), and the acceleration due to gravity (g). In symbols, we write the following:

$$\Psi_g = \rho_w g h \tag{3.5}$$

where $\rho_w g$ has a value of 0.01 MPa m^{-1}. Thus a vertical distance of 10 m translates into a 0.1 MPa change in water potential.

When dealing with water transport at the cell level, the gravitational component (Ψ_g) is generally omitted because it is negligible compared to the osmotic potential and the hydrostatic pressure. Thus, in these cases Equation 3.3 can be simplified as follows:

$$\Psi_w = \Psi_s + \Psi_p \tag{3.6}$$

In discussions of dry soils, seeds, and cell walls, one often finds reference to another component of Ψ_w, the matric potential, which is discussed in **Web Topic 3.4**.

Water potential in the plant. Cell growth, photosynthesis, and crop productivity are all strongly influenced by water potential and its components. Like the body temperature of humans, water potential is a good overall indicator of plant health. Plant scientists have thus expended considerable effort in devising accurate and reliable methods for evaluating the water status of plants. Some of the instruments that have been used to measure Ψ_w, Ψ_s, and Ψ_p are described in **Web Topic 3.5**.

Water Enters the Cell along a Water Potential Gradient

In this section we will illustrate the osmotic behavior of plant cells with some numerical examples. First imagine an open beaker full of pure water at 20°C (Figure 3.9A). Because the water is open to the atmosphere, the hydrostatic pressure of the water is the same as atmospheric pressure (Ψ_p = 0 MPa). There are no solutes in the water, so Ψ_s = 0 MPa; therefore the water potential is 0 MPa ($\Psi_w = \Psi_s + \Psi_p$).

TABLE 3.2
Values of *RT* and osmotic potential of solutions at various temperatures

Temperature (°C)	RT[a] (L MPa mol^{-1})	Osmotic potential (MPa) of solution with solute concentration in mol L^{-1} water: 0.01	0.10	1.00	Osmotic potential of seawater (MPa)
0	2.271	−0.0227	−0.227	−2.27	−2.6
20	2.436	−0.0244	−0.244	−2.44	−2.8
25	2.478	−0.0248	−0.248	−2.48	−2.8
30	2.519	−0.0252	−0.252	−2.52	−2.9

[a]R = 0.0083143 L MPa mol^{-1} K^{-1}.

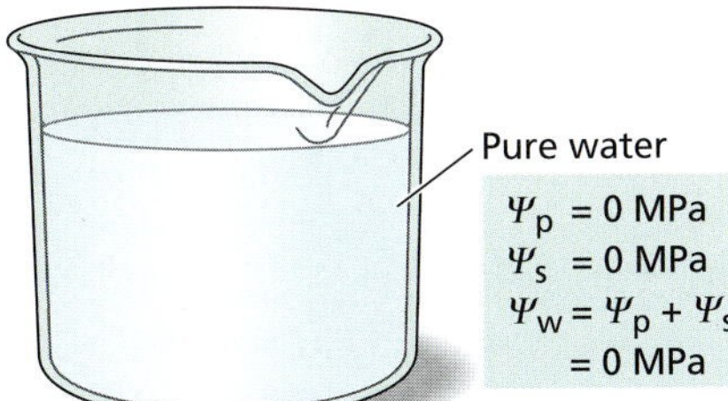

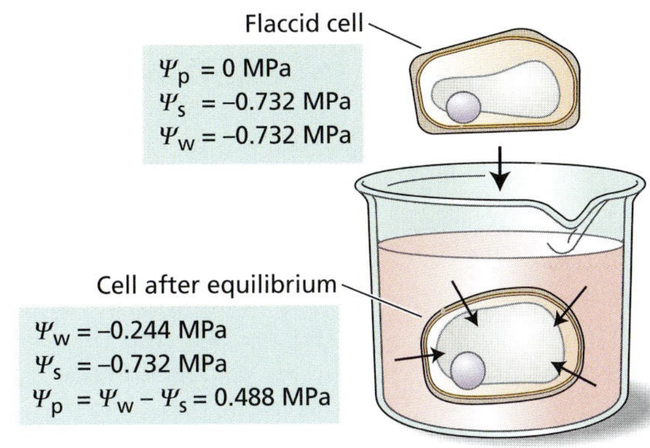

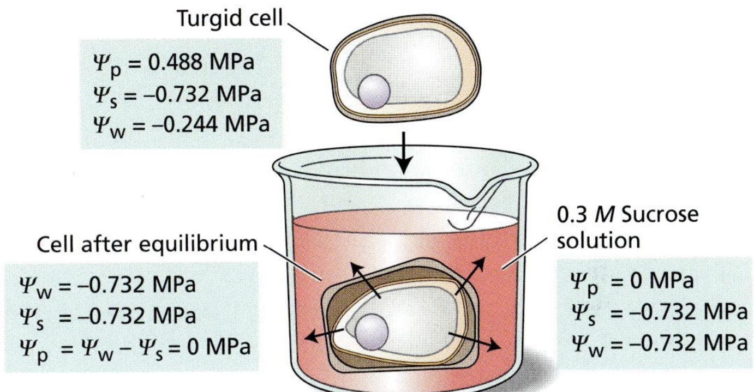

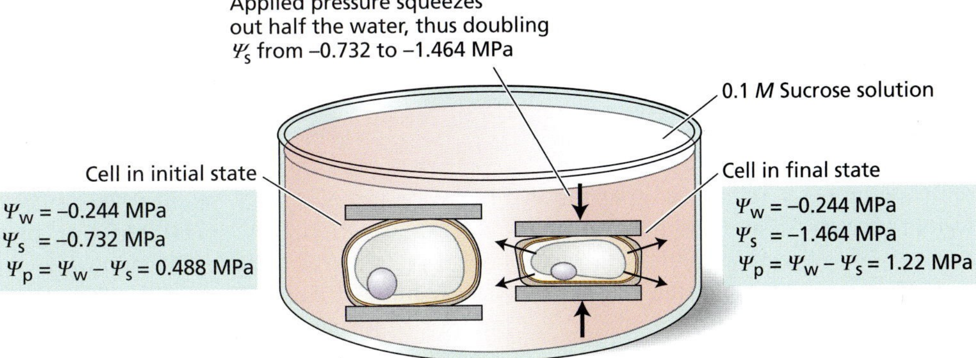

FIGURE 3.9 Five examples illustrating the concept of water potential and its components. (A) Pure water. (B) A solution containing 0.1 *M* sucrose. (C) A flaccid cell (in air) is dropped in the 0.1 *M* sucrose solution. Because the starting water potential of the cell is less than the water potential of the solution, the cell takes up water. After equilibration, the water potential of the cell rises to equal the water potential of the solution, and the result is a cell with a positive turgor pressure. (D) Increasing the concentration of sucrose in the solution makes the cell lose water. The increased sucrose concentration lowers the solution water potential, draws water out from the cell, and thereby reduces the cell's turgor pressure. In this case the protoplast is able to pull away from the cell wall (i.e, the cell plasmolyzes) because sucrose molecules are able to pass through the relatively large pores of the cell walls. In contrast, when a cell desiccates in air (e.g., the flaccid cell in panel C) plasmolysis does not occur because the water held by capillary forces in the cell walls prevents air from infiltrating into any void between the plasma membrane and the cell wall. (E) Another way to make the cell lose water is to press it slowly between two plates. In this case, half of the cell water is removed, so cell osmotic potential increases by a factor of 2.

Now imagine dissolving sucrose in the water to a concentration of 0.1 *M* (Figure 3.9B). This addition lowers the osmotic potential (Ψ_s) to –0.244 MPa (see Table 3.2) and decreases the water potential (Ψ_w) to –0.244 MPa.

Next consider a flaccid, or limp, plant cell (i.e., a cell with no turgor pressure) that has a total internal solute concentration of 0.3 *M* (Figure 3.9C). This solute concentration gives an osmotic potential (Ψ_s) of –0.732 MPa. Because the cell is flaccid, the internal pressure is the same as ambient pressure, so the hydrostatic pressure (Ψ_p) is 0 MPa and the water potential of the cell is –0.732 MPa.

What happens if this cell is placed in the beaker containing 0.1 *M* sucrose (see Figure 3.9C)? Because the water potential of the sucrose solution (Ψ_w = –0.244 MPa; see Figure 3.9B) is greater than the water potential of the cell (Ψ_w = –0.732 MPa), water will move from the sucrose solution to the cell (from high to low water potential).

Because plant cells are surrounded by relatively rigid cell walls, even a slight increase in cell volume causes a large increase in the hydrostatic pressure within the cell. As water enters the cell, the cell wall is stretched by the contents of the enlarging protoplast. The wall resists such stretching by pushing back on the cell. This phenomenon is analogous to inflating a basketball with air, except that air is compressible, whereas water is nearly incompressible.

As water moves into the cell, the hydrostatic pressure, or turgor pressure (Ψ_p), of the cell increases. Consequently, the cell water potential (Ψ_w) increases, and the difference between inside and outside water potentials ($\Delta\Psi_w$) is reduced. Eventually, cell Ψ_p increases enough to raise the cell Ψ_w to the same value as the Ψ_w of the sucrose solution. At this point, equilibrium is reached ($\Delta\Psi_w$ = 0 MPa), and net water transport ceases.

Because the volume of the beaker is much larger than that of the cell, the tiny amount of water taken up by the cell does not significantly affect the solute concentration of the sucrose solution. Hence Ψ_s, Ψ_p, and Ψ_w of the sucrose solution are not altered. Therefore, at equilibrium, $\Psi_{w(cell)} = \Psi_{w(solution)}$ = –0.244 MPa.

The exact calculation of cell Ψ_p and Ψ_s requires knowledge of the change in cell volume. However, if we assume that the cell has a very rigid cell wall, then the increase in cell volume will be small. Thus we can assume to a first approximation that $\Psi_{s(cell)}$ is unchanged during the equilibration process and that $\Psi_{s(solution)}$ remains at –0.732 MPa. We can obtain cell hydrostatic pressure by rearranging Equation 3.6 as follows: $\Psi_p = \Psi_w - \Psi_s$ = (–0.244) – (–0.732) = 0.488 MPa.

Water Can Also Leave the Cell in Response to a Water Potential Gradient

Water can also leave the cell by osmosis. If, in the previous example, we remove our plant cell from the 0.1 *M* sucrose solution and place it in a 0.3 *M* sucrose solution (Figure 3.9D), $\Psi_{w(solution)}$ (–0.732 MPa) is more negative than $\Psi_{w(cell)}$ (–0.244 MPa), and water will move from the turgid cell to the solution.

As water leaves the cell, the cell volume decreases. As the cell volume decreases, cell Ψ_p and Ψ_w decrease also until $\Psi_{w(cell)} = \Psi_{w(solution)}$ = –0.732 MPa. From the water potential equation (Equation 3.6) we can calculate that at equilibrium, Ψ_p = 0 MPa. As before, we assume that the change in cell volume is small, so we can ignore the change in Ψ_s.

If we then slowly squeeze the turgid cell by pressing it between two plates (Figure 3.9E), we effectively raise the cell Ψ_p, consequently raising the cell Ψ_w and creating a $\Delta\Psi_w$ such that water now flows *out* of the cell. If we continue squeezing until half the cell water is removed and then hold the cell in this condition, the cell will reach a new equilibrium. As in the previous example, at equilibrium, $\Delta\Psi_w$ = 0 MPa, and the amount of water added to the external solution is so small that it can be ignored. The cell will thus return to the Ψ_w value that it had before the squeezing procedure. However, the components of the cell Ψ_w will be quite different.

Because half of the water was squeezed out of the cell while the solutes remained inside the cell (the plasma membrane is selectively permeable), the cell solution is concentrated twofold, and thus Ψ_s is lower (–0.732 × 2 = –1.464 MPa). Knowing the final values for Ψ_w and Ψ_s, we can calculate the turgor pressure, using Equation 3.6, as $\Psi_p = \Psi_w - \Psi_s$ = (–0.244) – (–1.464) = 1.22 MPa. In our example we used an external force to change cell volume without a change in water potential. In nature, it is typically the water potential of the cell's environment that changes, and the cell gains or loses water until its Ψ_w matches that of its surroundings.

One point common to all these examples deserves emphasis: *Water flow is a passive process. That is, water moves in response to physical forces, toward regions of low water potential or low free energy.* There are no metabolic "pumps" (reactions driven by ATP hydrolysis) that push water from one place to another. This rule is valid as long as water is the only substance being transported. When solutes are transported, however, as occurs for short distances across membranes (see Chapter 6) and for long distances in the phloem (see Chapter 10), then water transport may be coupled to solute transport and this coupling may move water against a water potential gradient.

For example, the transport of sugars, amino acids, or other small molecules by various membrane proteins can "drag" up to 260 water molecules across the membrane per molecule of solute transported (Loo et al. 1996). Such transport of water can occur even when the movement is against the usual water potential gradient (i.e., toward a larger water potential) because the loss of free energy by the solute more than compensates for the gain of free energy by the water. The net change in free energy remains negative. In the phloem, the bulk flow of solutes and water within sieve tubes occurs along gradients in hydrostatic

(turgor) pressure rather than by osmosis. Thus, within the phloem, water can be transported from regions with lower water potentials (e.g., leaves) to regions with higher water potentials (e.g., roots). *These situations notwithstanding, in the vast majority of cases water in plants moves from higher to lower water potentials.*

Small Changes in Plant Cell Volume Cause Large Changes in Turgor Pressure

Cell walls provide plant cells with a substantial degree of volume homeostasis relative to the large changes in water potential that they experience as the everyday consequence of the transpirational water losses associated with photosynthesis (see Chapter 4). Because plant cells have fairly rigid walls, a change in cell Ψ_w is generally accompanied by a large change in Ψ_p, with relatively little change in cell (protoplast) volume.

This phenomenon is illustrated in plots of Ψ_w, Ψ_p, and Ψ_s as a function of relative cell volume. In the example of a hypothetical cell shown in Figure 3.10, as Ψ_w decreases from 0 to about –2 MPa, the cell volume is reduced by only 5%. Most of this decrease is due to a reduction in Ψ_p (by about 1.2 MPa); Ψ_s decreases by about 0.3 MPa as a result of water loss by the cell and consequent increased concentration of cell solutes. Contrast this with the volume changes of a cell lacking a wall.

Measurements of cell water potential and cell volume (see Figure 3.10) can be used to quantify how cell walls influence the water status of plant cells.

1. Turgor pressure ($\Psi_p > 0$) exists only when cells are relatively well hydrated. Turgor pressure in most cells approaches zero as the relative cell volume decreases by 10 to 15%. However, for cells with very rigid cell walls (e.g., mesophyll cells in the leaves of many palm trees), the volume change associated with turgor loss can be much smaller, whereas in cells with extremely elastic walls, such as the water-storing cells in the stems of many cacti, this volume change may be substantially larger.
2. The Ψ_p curve of Figure 3.10 provides a way to measure the relative rigidity of the cell wall, symbolized by ε (the Greek letter epsilon): $\varepsilon = \Delta\Psi_p/\Delta(\text{relative volume})$. ε is the slope of the Ψ_p curve. ε is not constant but decreases as turgor pressure is lowered because nonlignified plant cell walls usually are rigid only when turgor pressure puts them under tension. Such cells act like a basketball: The wall is stiff (has high ε) when the ball is inflated but becomes soft and collapsible ($\varepsilon = 0$) when the ball loses pressure.
3. When ε and Ψ_p are low, changes in water potential are dominated by changes in Ψ_s (note how Ψ_w and Ψ_s curves converge as the relative cell volume approaches 85%).

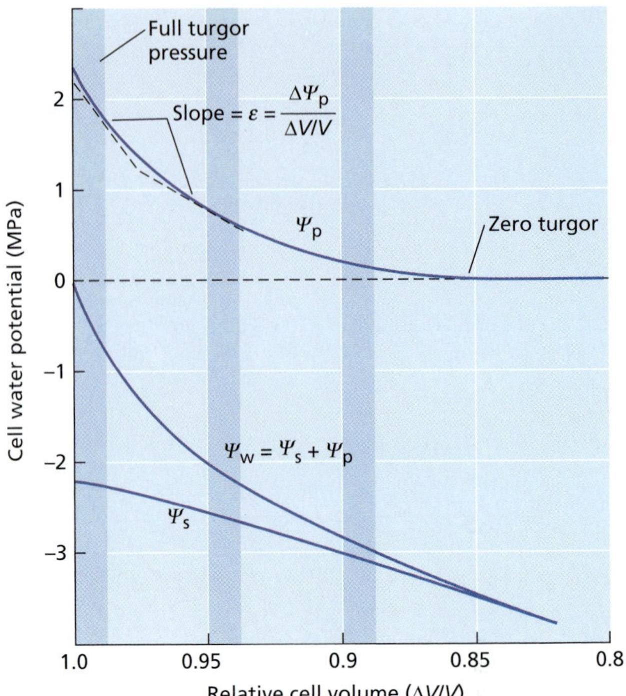

FIGURE 3.10 Relation between cell water potential (Ψ_w) and its components (Ψ_p and Ψ_s), and relative cell volume ($\Delta V/V$). The plots show that turgor pressure (Ψ_p) decreases steeply with the initial 5% decrease in cell volume. In comparison, osmotic potential (Ψ_s) changes very little. As cell volume decreases below 0.9 in this example, the situation reverses: Most of the change in water potential is due to a drop in cell Ψ_s accompanied by relatively little change in turgor pressure. The slope of the curve that illustrates Ψ_p versus volume relationship is a measure of the cell's elastic modulus (ε) (a measurement of wall rigidity). Note that ε is not constant but decreases as the cell loses turgor. (After Tyree and Jarvis 1982, based on a shoot of Sitka spruce.)

Water Transport Rates Depend on Driving Force and Hydraulic Conductivity

So far, we have seen that water moves across a membrane in response to a water potential gradient. The direction of flow is determined by the direction of the Ψ_w gradient, and the rate of water movement is proportional to the magnitude of the driving gradient. However, for a cell that experiences a change in the water potential of its surroundings (e.g., see Figure 3.9), the movement of water across the cell membrane will decrease with time as the internal and external water potentials converge (Figure 3.11). The rate approaches zero in an exponential manner (see Dainty 1976), with a half-time (half-times conveniently characterize processes that change exponentially with time) given by the following equation:

$$t_{1/2} = \left(\frac{0.693}{(A)(Lp)}\right)\left(\frac{V}{\varepsilon - \Psi_s}\right) \quad (3.7)$$

where V and A are, respectively, the volume and surface of

(A)

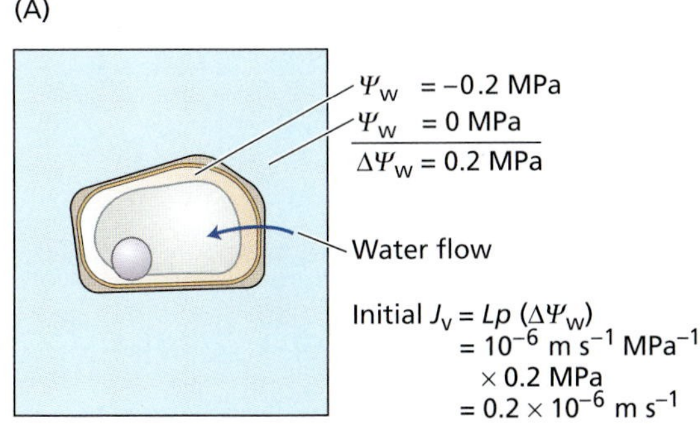

(B)

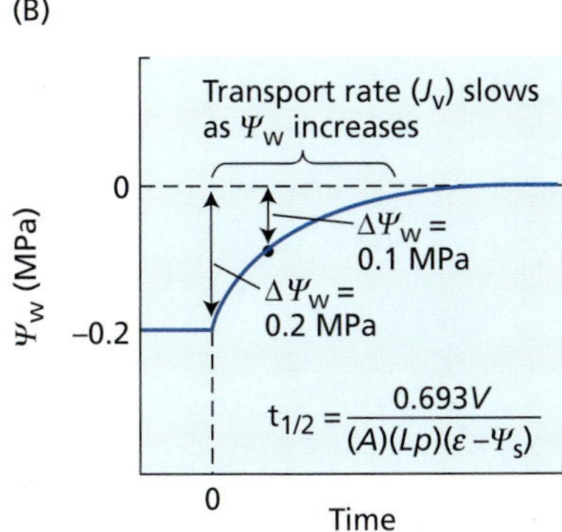

FIGURE 3.11 The rate of water transport into a cell depends on the water potential difference ($\Delta\Psi_w$) and the hydraulic conductivity of the cell membranes (Lp). In this example, (A) the initial water potential difference is 0.2 MPa and Lp is 10^{-6} m s^{-1} MPa^{-1}. These values give an initial transport rate (J_v) of 0.2×10^{-6} m s^{-1}. (B) As water is taken up by the cell, the water potential difference decreases with time, leading to a slowing in the rate of water uptake. This effect follows an exponentially decaying time course with a half-time ($t_{1/2}$) that depends on the following cell parameters: volume (V), surface area (A), Lp, volumetric elastic modulus (ε), and cell osmotic potential (Ψ_s).

the cell, and Lp is the **hydraulic conductivity** of the cell membrane. Hydraulic conductivity describes how readily water can move across a membrane and has units of volume of water per unit area of membrane per unit time per unit driving force (i.e., m^3 m^{-2} s^{-1} MPa^{-1}). For additional discussion on hydraulic conductivity, see 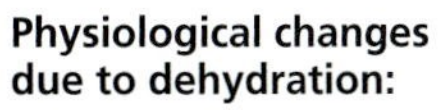**Web Topic 3.6**.

A short half-time means fast equilibration. Thus, cells with large surface-to-volume ratios, high membrane hydraulic conductivity, and stiff cell walls (large ε) will come rapidly into equilibrium with their surroundings. Cell half-times typically range from 1 to 10 s, although some are much shorter (Steudle 1989). These low half-times mean that single cells come to water potential equilibrium with their surroundings in less than 1 minute. For multicellular tissues, the half-times may be much larger.

The Water Potential Concept Helps Us Evaluate the Water Status of a Plant

The concept of water potential has two principal uses: First, water potential governs transport across cell membranes, as we have described. Second, water potential is often used as a measure of the *water status* of a plant. Because of transpirational water loss to the atmosphere, plants are seldom fully hydrated. They suffer from water deficits that lead to inhibition of plant growth and photosynthesis, as well as to other detrimental effects. Figure 3.12 lists some of the physiological changes that plants experience as they become dry.

The process that is most affected by water deficit is cell growth. More severe water stress leads to inhibition of cell division, inhibition of wall and protein synthesis, accumu-

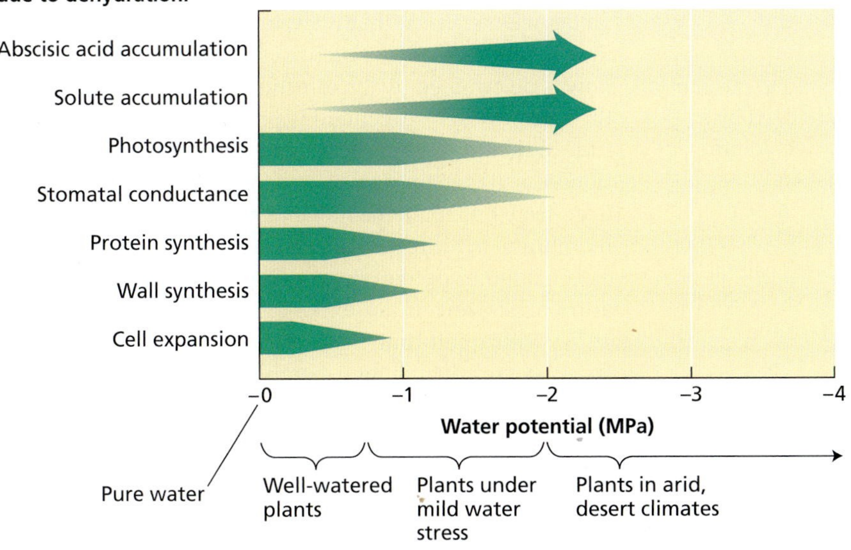

FIGURE 3.12 Water potential of plants under various growing conditions, and sensitivity of various physiological processes to water potential. The intensity of the bar color corresponds to the magnitude of the process. For example, cell expansion decreases as water potential falls (becomes more negative). Abscisic acid is a hormone that induces stomatal closure during water stress (see Chapter 23). (After Hsiao 1979.)

lation of solutes, closing of stomata, and inhibition of photosynthesis. Water potential is one measure of how hydrated a plant is and thus provides a relative index of the *water stress* the plant is experiencing (see Chapter 25).

Figure 3.12 also shows representative values for Ψ_w at various stages of water stress. In leaves of well-watered plants, Ψ_w ranges from –0.2 to about –1.0 MPa, but the leaves of plants in arid climates can have much lower values, perhaps –2 to –5 MPa under extreme conditions. Because water transport is a passive process, plants can take up water only when the plant Ψ_w is less than the soil Ψ_w. As the soil becomes drier, the plant similarly becomes less hydrated (attains a lower Ψ_w). If this were not the case, the soil would begin to extract water from the plant.

The Components of Water Potential Vary with Growth Conditions and Location within the Plant

Just as Ψ_w values depend on the growing conditions and the type of plant, so too, the values of Ψ_s can vary considerably. Within cells of well-watered garden plants (examples include lettuce, cucumber seedlings, and bean leaves), Ψ_s may be as high as –0.5 MPa, although values of –0.8 to –1.2 MPa are more typical. The upper limit for cell Ψ_s is set probably by the minimum concentration of dissolved ions, metabolites, and proteins in the cytoplasm of living cells.

At the other extreme, plants under drought conditions sometimes attain a much lower Ψ_s. For instance, water stress typically leads to an accumulation of solutes in the cytoplasm and vacuole, thus allowing the plant to maintain turgor pressure despite low water potentials.

Plant tissues that store high concentrations of sucrose or other sugars, such as sugar beet roots, sugarcane stems, or grape berries, also attain low values of Ψ_s. Values as low as –2.5 MPa are not unusual. Plants that grow in saline environments, called **halophytes**, typically have very low values of Ψ_s. A low Ψ_s lowers cell Ψ_w enough to extract water from salt water, without allowing excessive levels of salts to enter at the same time. Most crop plants cannot survive in seawater, which, because of the dissolved salts, has a lower water potential than the plant tissues can attain while maintaining their functional competence.

Although Ψ_s *within* cells may be quite negative, the apoplastic solution surrounding the cells—that is, in the cell walls and in the xylem—may contain only low concentrations of solutes. Thus, Ψ_s of this phase of the plant is typically much higher—for example, –0.1 to 0 MPa. Negative water potentials in the xylem and cell walls are usually due to negative Ψ_p. Values for Ψ_p within cells of well-watered garden plants may range from 0.1 to perhaps 1 MPa, depending on the value of Ψ_s inside the cell.

A positive turgor pressure (Ψ_p) is important for two principal reasons. First, growth of plant cells requires turgor pressure to stretch the cell walls. The loss of Ψ_p under water deficits can explain in part why cell growth is so sensitive to water stress (see Chapter 25). The second reason positive turgor is important is that turgor pressure increases the mechanical rigidity of cells and tissues. This function of cell turgor pressure is particularly important for young, nonlignified tissues, which cannot support themselves mechanically without a high internal pressure. A plant **wilts** (becomes flaccid) when the turgor pressure inside the cells of such tissues falls toward zero. **Web Topic 3.7** discusses plasmolysis, the shrinking of the protoplast away from the cell wall, which occurs when cells in solution lose water.

Whereas the solution inside cells may have a positive and large Ψ_p, the water outside the cell may have negative values for Ψ_p. In the xylem of rapidly transpiring plants, Ψ_p is negative and may attain values of –1 MPa or lower. The magnitude of Ψ_p in the cell walls and xylem varies considerably, depending on the rate of transpiration and the height of the plant. During the middle of the day, when transpiration is maximal, xylem Ψ_p reaches its lowest, most negative values. At night, when transpiration is low and the plant rehydrates, it tends to increase.

SUMMARY

Water is important in the life of plants because it makes up the matrix and medium in which most biochemical processes essential for life take place. The structure and properties of water strongly influence the structure and properties of proteins, membranes, nucleic acids, and other cell constituents.

In most land plants, water is continually lost to the atmosphere and taken up from the soil. The movement of water is driven by a reduction in free energy, and water may move by diffusion, by bulk flow, or by a combination of these fundamental transport mechanisms. Water diffuses because molecules are in constant thermal agitation, which tends to even out concentration differences. Water moves by bulk flow in response to a pressure difference, whenever there is a suitable pathway for bulk movement of water. Osmosis, the movement of water across membranes, depends on a gradient in free energy of water across the membrane—a gradient commonly measured as a difference in water potential.

Solute concentration and hydrostatic pressure are the two major factors that affect water potential, although when large vertical distances are involved, gravity is also important. These components of the water potential may be summed as follows: $\Psi_w = \Psi_s + \Psi_p + \Psi_g$. Plant cells come into water potential equilibrium with their local environment by absorbing or losing water. Usually this change in cell volume results in a change in cell Ψ_p, accompanied by minor changes in cell Ψ_s. The rate of water transport across a membrane depends on the water potential difference across the membrane and the hydraulic conductivity of the membrane.

In addition to its importance in transport, water potential is a useful measure of the water status of plants. As we will see in Chapter 4, diffusion, bulk flow, and osmosis all help move water from the soil through the plant to the atmosphere.

Web Material

Web Topics

3.1 Calculating Capillary Rise

Quantification of capillary rise allows us to assess the functional role of capillary rise in water movement of plants.

3.2 Calculating Half-Times of Diffusion

The assessment of the time needed for a molecule like glucose to diffuse across cells, tissues, and organs shows that diffusion has physiological significance only over short distances.

3.3 Alternative Conventions for Components of Water Potential

Plant physiologists have developed several conventions to define water potential of plants. A comparison of key definitions in some of these convention systems provides us with a better understanding of the water relations literature.

3.4 The Matric Potential

A brief discussion of the concept of matric potential, used to quantify the chemical potential of water in soils, seeds, and cell walls.

3.5 Measuring Water Potential

A detailed description of available methods to measure water potential in plant cells and tissues.

3.6 Understanding Hydraulic Conductivity

Hydraulic conductivity, a measurement of the membrane permeability to water, is one of the factors determining the velocity of water movements in plants.

3.7 Wilting and Plasmolysis

Plasmolysis is a major structural change resulting from major water loss by osmosis.

Chapter References

Dainty, J. (1976) Water relations of plant cells. In *Transport in Plants*, Vol. 2, Part A: *Cells* (Encyclopedia of Plant Physiology, New Series, Vol. 2.), U. Lüttge and M. G. Pitman, eds., Springer, Berlin, pp. 12–35.

Finkelstein, A. (1987) *Water Movement through Lipid Bilayers, Pores, and Plasma Membranes: Theory and Reality*. Wiley, New York.

Friedman, M. H. (1986) *Principles and Models of Biological Transport*. Springer Verlag, Berlin.

Hsiao, T. C. (1979) Plant responses to water deficits, efficiency, and drought resistance. *Agricult. Meteorol.* 14: 59–84.

Loo, D. D. F., Zeuthen, T., Chandy, G., and Wright, E. M. (1996) Cotransport of water by the Na^+/glucose cotransporter. *Proc. Natl. Acad. Sci. USA* 93: 13367–13370.

Nobel, P. S. (1999) *Physicochemical and Environmental Plant Physiology*, 2nd ed. Academic Press, San Diego, CA.

Schäffner, A. R. (1998) Aquaporin function, structure, and expression: Are there more surprises to surface in water relations? *Planta* 204: 131–139.

Stein, W. D. (1986) *Transport and Diffusion across Cell Membranes*. Academic Press, Orlando, FL.

Steudle, E. (1989) Water flow in plants and its coupling to other processes: An overview. *Methods Enzymol.* 174: 183–225.

Tajkhorshid, E., Nollert, P., Jensen, M. Ø., Miercke, L. H. W., O'Connell, J., Stroud, R. M., and Schulten, K. (2002) Control of the selectivity of the aquaporin water channel family by global orientation tuning. *Science* 296: 525–530.

Tyerman, S. D., Bohnert, H. J., Maurel, C., Steudle, E., and Smith, J. A. C. (1999) Plant aquaporins: Their molecular biology, biophysics and significance for plant–water relations. *J. Exp. Bot.* 50: 1055–1071.

Tyree, M. T., and Jarvis, P. G. (1982) Water in tissues and cells. In *Physiological Plant Ecology*, Vol. 2: *Water Relations and Carbon Assimilation* (Encyclopedia of Plant Physiology, New Series, Vol. 12B), O. L. Lange, P. S. Nobel, C. B. Osmond, and H. Ziegler, eds., Springer, Berlin, pp. 35–77.

Weather and Our Food Supply (CAED Report 20). (1964) Center for Agricultural and Economic Development, Iowa State University of Science and Technology, Ames, IA.

Weig, A., Deswarte, C., and Chrispeels, M. J. (1997) The major intrinsic protein family of *Arabidopsis* has 23 members that form three distinct groups with functional aquaporins in each group. *Plant Physiol.* 114: 1347–1357.

Whittaker R. H. (1970) *Communities and Ecosystems*. Macmillan, New York.

Chapter

4 Water Balance of Plants

LIFE IN EARTH'S ATMOSPHERE presents a formidable challenge to land plants. On the one hand, the atmosphere is the source of carbon dioxide, which is needed for photosynthesis. Plants therefore need ready access to the atmosphere. On the other hand, the atmosphere is relatively dry and can dehydrate the plant. To meet the contradictory demands of maximizing carbon dioxide uptake while limiting water loss, plants have evolved adaptations to control water loss from leaves, and to replace the water lost to the atmosphere.

In this chapter we will examine the mechanisms and driving forces operating on water transport within the plant and between the plant and its environment. Transpirational water loss from the leaf is driven by a gradient in water vapor concentration. Long-distance transport in the xylem is driven by pressure gradients, as is water movement in the soil. Water transport through cell layers such as the root cortex is complex, but it responds to water potential gradients across the tissue.

Throughout this journey water transport is passive in the sense that the free energy of water decreases as it moves. Despite its passive nature, water transport is finely regulated by the plant to minimize dehydration, largely by regulating transpiration to the atmosphere. We will begin our examination of water transport by focusing on water in the soil.

WATER IN THE SOIL

The water content and the rate of water movement in soils depend to a large extent on soil type and soil structure. Table 4.1 shows that the physical characteristics of different soils can vary greatly. At one extreme is sand, in which the soil particles may be 1 mm or more in diameter. Sandy soils have a relatively low surface area per gram of soil and have large spaces or channels between particles.

At the other extreme is clay, in which particles are smaller than 2 μm in diameter. Clay soils have much greater surface areas and smaller

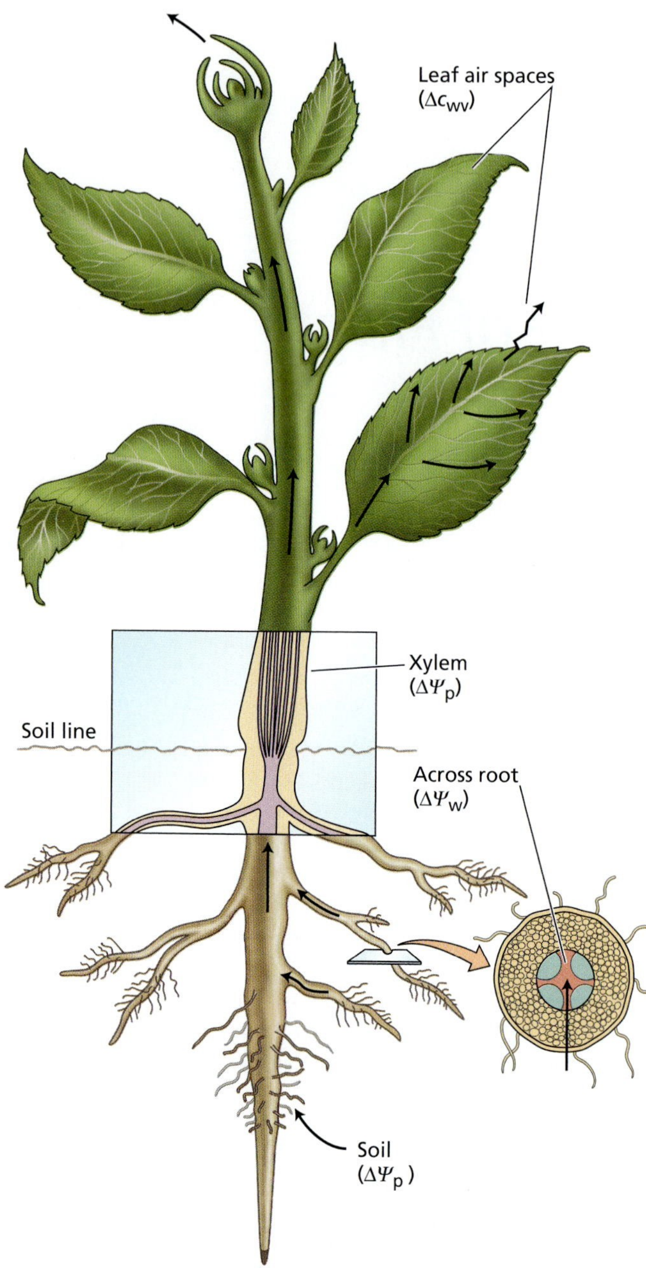

FIGURE 4.1 Main driving forces for water flow from the soil through the plant to the atmosphere: differences in water vapor concentration (Δc_{wv}), hydrostatic pressure ($\Delta \Psi_p$), and water potential ($\Delta \Psi_w$).

channels between particles. With the aid of organic substances such as humus (decomposing organic matter), clay particles may aggregate into "crumbs" that help improve soil aeration and infiltration of water.

TABLE 4.1
Physical characteristics of different soils

Soil	Particle diameter (µm)	Surface area per gram (m^2)
Coarse sand	2000–200	<1–10
Fine sand	200–20	
Silt	20–2	10–100
Clay	<2	100–1000

When a soil is heavily watered by rain or by irrigation, the water percolates downward by gravity through the spaces between soil particles, partly displacing, and in some cases trapping, air in these channels. Water in the soil may exist as a film adhering to the surface of soil particles, or it may fill the entire channel between particles.

In sandy soils, the spaces between particles are so large that water tends to drain from them and remain only on the particle surfaces and at interstices between particles. In clay soils, the channels are small enough that water does not freely drain from them; it is held more tightly (see **Web Topic 4.1**). The moisture-holding capacity of soils is called the **field capacity**. Field capacity is the water content of a soil after it has been saturated with water and excess water has been allowed to drain away. Clay soils or soils with a high humus content have a large field capacity. A few days after being saturated, they might retain 40% water by volume. In contrast, sandy soils typically retain 3% water by volume after saturation.

In the following sections we will examine how the negative pressure in soil water alters soil water potential, how water moves in the soil, and how roots absorb the water needed by the plant.

A Negative Hydrostatic Pressure in Soil Water Lowers Soil Water Potential

Like the water potential of plant cells, the water potential of soils may be dissected into two components, the osmotic potential and the hydrostatic pressure. The osmotic potential (Ψ_s; see Chapter 3) of soil water is generally negligible because solute concentrations are low; a typical value might be –0.02 MPa. For soils that contain a substantial concentration of salts, however, Ψ_s is significant, perhaps –0.2 MPa or lower.

The second component of soil water potential is hydrostatic pressure (Ψ_p) (Figure 4.1). For wet soils, Ψ_p is very close to zero. As a soil dries out, Ψ_p decreases and can become quite negative. Where does the negative pressure in soil water come from?

Recall from our discussion of capillarity in Chapter 3 that water has a high surface tension that tends to minimize air–water interfaces. As a soil dries out, water is first removed from the center of the largest spaces between particles. Because of adhesive forces, water tends to cling to the surfaces of soil particles, so a large surface area between soil water and soil air develops (Figure 4.2).

As the water content of the soil decreases, the water recedes into the interstices between soil particles, and the air–water surface develops curved air–water interfaces.

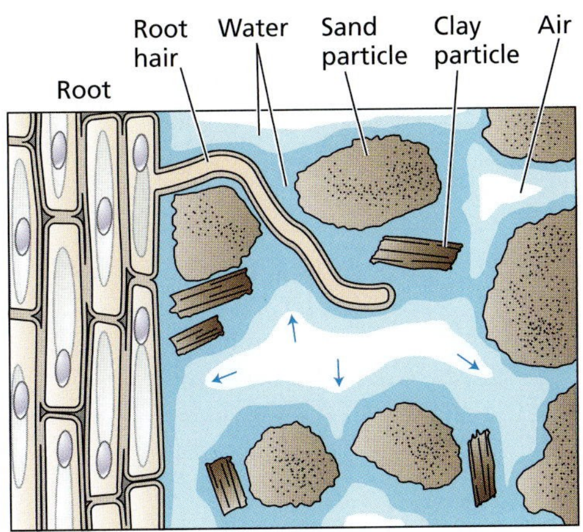

FIGURE 4.2 Root hairs make intimate contact with soil particles and greatly amplify the surface area that can be used for water absorption by the plant. The soil is a mixture of particles (sand, clay, silt, and organic material), water, dissolved solutes, and air. Water is adsorbed to the surface of the soil particles. As water is absorbed by the plant, the soil solution recedes into smaller pockets, channels, and crevices between the soil particles. At the air–water interfaces, this recession causes the surface of the soil solution to develop concave menisci (curved interfaces between air and water marked in the figure by arrows), and brings the solution into tension (negative pressure) by surface tension. As more water is removed from the soil, more acute menisci are formed, resulting in greater tensions (more negative pressures).

Water under these curved surfaces develops a negative pressure that may be estimated by the following formula:

$$\Psi_p = \frac{-2T}{r} \quad (4.1)$$

where T is the surface tension of water (7.28×10^{-8} MPa m) and r is the radius of curvature of the air–water interface.

The value of Ψ_p in soil water can become quite negative because the radius of curvature of air–water surfaces may become very small in drying soils. For instance, a curvature $r = 1$ μm (about the size of the largest clay particles) corresponds to a Ψ_p value of –0.15 MPa. The value of Ψ_p may easily reach –1 to –2 MPa as the air–water interface recedes into the smaller cracks between clay particles.

Soil scientists often describe soil water potential in terms of a matric potential (Jensen et al. 1998). For a discussion of the relation between matric potential and water potential see **Web Topic 3.3**.

Water Moves through the Soil by Bulk Flow

Water moves through soils predominantly by bulk flow driven by a pressure gradient. In addition, diffusion of water vapor accounts for some water movement. As plants absorb water from the soil, they deplete the soil of water near the surface of the roots. This depletion reduces Ψ_p in the water near the root surface and establishes a pressure gradient with respect to neighboring regions of soil that have higher Ψ_p values. *Because the water-filled pore spaces in the soil are interconnected, water moves to the root surface by bulk flow through these channels down the pressure gradient.*

The rate of water flow in soils depends on two factors: the size of the pressure gradient through the soil, and the hydraulic conductivity of the soil. **Soil hydraulic conductivity** is a measure of the ease with which water moves through the soil, and it varies with the type of soil and water content. Sandy soils, with their large spaces between particles, have a large hydraulic conductivity, whereas clay soils, with the minute spaces between their particles, have an appreciably smaller hydraulic conductivity.

As the water content (and hence the water potential) of a soil decreases, the hydraulic conductivity decreases drastically (see **Web Topic 4.2**). This decrease in soil hydraulic conductivity is due primarily to the replacement of water in the soil spaces by air. When air moves into a soil channel previously filled with water, water movement through that channel is restricted to the periphery of the channel. As more of the soil spaces become filled with air, water can flow through fewer and narrower channels, and the hydraulic conductivity falls.

In very dry soils, the water potential (Ψ_w) may fall below what is called the **permanent wilting point**. At this point the water potential of the soil is so low that plants cannot regain turgor pressure even if all water loss through transpiration ceases. This means that the water potential of the soil (Ψ_w) is less than or equal to the osmotic potential (Ψ_s) of the plant. Because cell Ψ_s varies with plant species, the permanent wilting point is clearly not a unique property of the soil; it depends on the plant species as well.

WATER ABSORPTION BY ROOTS

Intimate contact between the surface of the root and the soil is essential for effective water absorption by the root. This contact provides the surface area needed for water uptake and is maximized by the growth of the root and of root hairs into the soil. **Root hairs** are microscopic extensions of root epidermal cells that greatly increase the surface area of the root, thus providing greater capacity for absorption of ions and water from the soil. When 4-month-old rye (*Secale*) plants were examined, their root hairs were found to constitute more than 60% of the surface area of the roots (see Figure 5.6).

Water enters the root most readily in the apical part of the root that includes the root hair zone. More mature regions of the root often have an outer layer of protective tissue, called an *exodermis* or *hypodermis*, that contains hydrophobic materials in its walls and is relatively impermeable to water.

The intimate contact between the soil and the root surface is easily ruptured when the soil is disturbed. It is for this reason that newly transplanted seedlings and plants

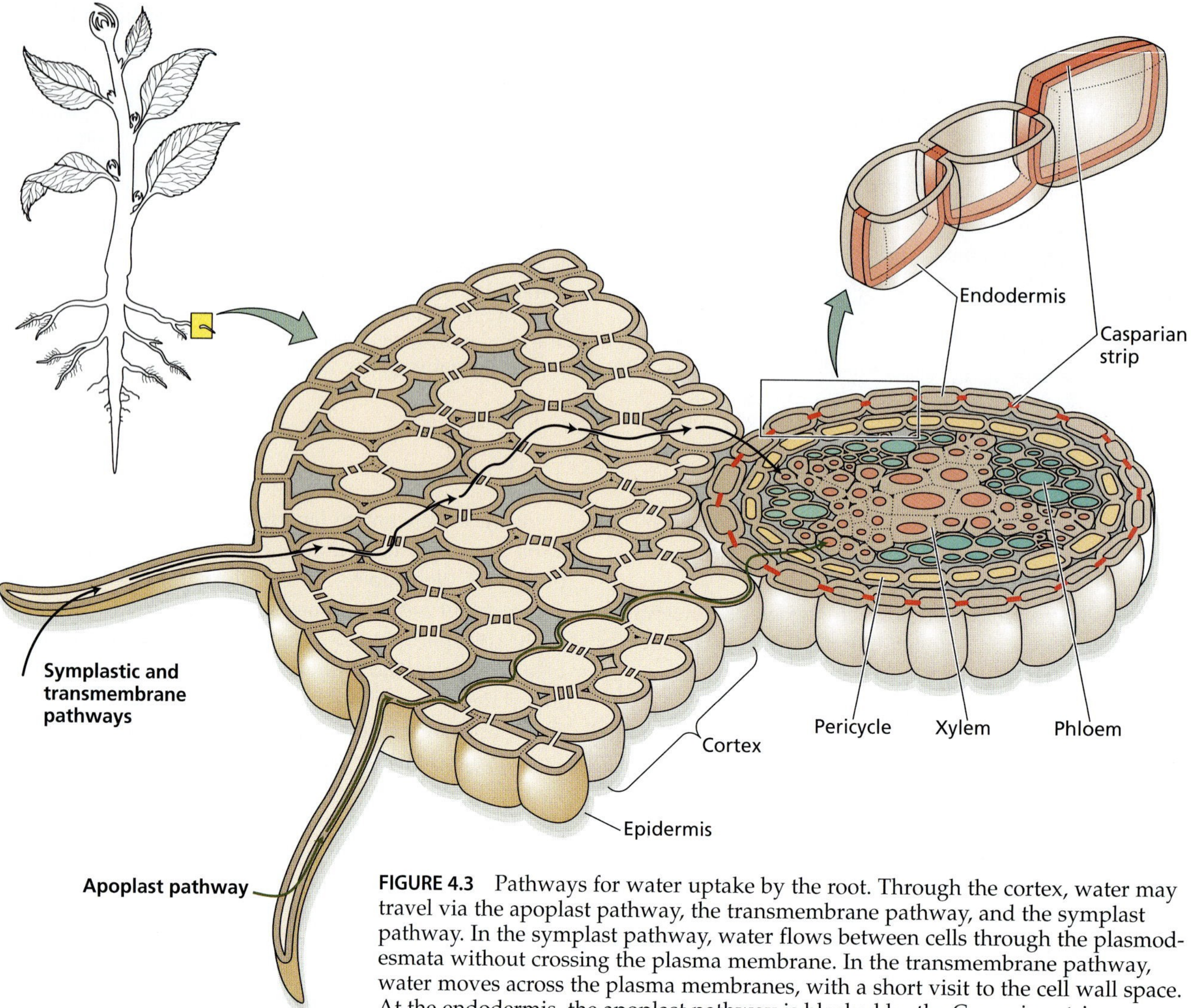

FIGURE 4.3 Pathways for water uptake by the root. Through the cortex, water may travel via the apoplast pathway, the transmembrane pathway, and the symplast pathway. In the symplast pathway, water flows between cells through the plasmodesmata without crossing the plasma membrane. In the transmembrane pathway, water moves across the plasma membranes, with a short visit to the cell wall space. At the endodermis, the apoplast pathway is blocked by the Casparian strip.

need to be protected from water loss for the first few days after transplantation. Thereafter, new root growth into the soil reestablishes soil–root contact, and the plant can better withstand water stress.

Let's consider how water moves within the root, and the factors that determine the rate of water uptake into the root.

Water Moves in the Root via the Apoplast, Transmembrane, and Symplast Pathways

In the soil, water is transported predominantly by bulk flow. However, when water comes in contact with the root surface, the nature of water transport becomes more complex. From the epidermis to the endodermis of the root, there are three pathways through which water can flow (Figure 4.3): the apoplast, transmembrane, and symplast pathways.

1. In the apoplast pathway, water moves exclusively through the cell wall without crossing any membranes. The apoplast is the continuous system of cell walls and intercellular air spaces in plant tissues.
2. The transmembrane pathway is the route followed by water that sequentially enters a cell on one side, exits the cell on the other side, enters the next in the series, and so on. In this pathway, water crosses at least two membranes for each cell in its path (the plasma membrane on entering and on exiting). Transport across the tonoplast may also be involved.
3. In the symplast pathway, water travels from one cell to the next via the plasmodesmata (see Chapter 1). The symplast consists of the entire network of cell cytoplasm interconnected by plasmodesmata.

Although the relative importance of the apoplast, transmembrane, and symplast pathways has not yet been clearly established, experiments with the pressure probe technique (see **Web Topic 3.6**) indicate that the apoplast pathway is particularly important for water uptake by young corn roots (Frensch et al. 1996; Steudle and Frensch 1996).

At the endodermis, water movement through the apoplast pathway is obstructed by the Casparian strip (see Figure 4.3). The **Casparian strip** is a band of radial cell walls in the endodermis that is impregnated with the wax-like, hydrophobic substance **suberin**. Suberin acts as a barrier to water and solute movement. The endodermis becomes suberized in the nongrowing part of the root, several millimeters behind the root tip, at about the same time that the first protoxylem elements mature (Esau 1953). The Casparian strip breaks the continuity of the apoplast pathway, and forces water and solutes to cross the endodermis by passing through the plasma membrane. Thus, despite the importance of the apoplast pathway in the root cortex and the stele, water movement across the endodermis occurs through the symplast.

Another way to understand water movement through the root is to consider the root as a single pathway having a single hydraulic conductance. Such an approach has led to the development of the concept of **root hydraulic conductance** (see **Web Topic 4.3** for details).

The apical region of the root is most permeable to water. Beyond this point, the exodermis becomes suberized, limiting water uptake (Figure 4.4). However, some water absorption may take place through older roots, perhaps through breaks in the cortex associated with the outgrowth of secondary roots.

Water uptake decreases when roots are subjected to low temperature or anaerobic conditions, or treated with respiratory inhibitors (such as cyanide). These treatments inhibit root respiration, and the roots transport less water. The exact explanation for this effect is not yet clear. On the other hand, the decrease in water transport in the roots provides an explanation for the wilting of plants in waterlogged soils: Submerged roots soon run out of oxygen, which is normally provided by diffusion through the air spaces in the soil (diffusion through gas is 10^4 times faster than diffusion through water). The anaerobic roots transport less water to the shoots, which consequently suffer net water loss and begin to wilt.

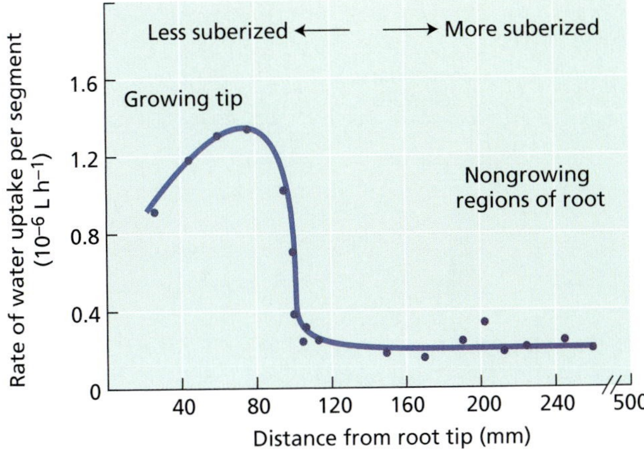

FIGURE 4.4 Rate of water uptake at various positions along a pumpkin root. (After Kramer and Boyer 1995.)

Solute Accumulation in the Xylem Can Generate "Root Pressure"

Plants sometimes exhibit a phenomenon referred to as **root pressure**. For example, if the stem of a young seedling is cut off just above the soil, the stump will often exude sap from the cut xylem for many hours. If a manometer is sealed over the stump, positive pressures can be measured. These pressures can be as high as 0.05 to 0.5 MPa.

Roots generate positive hydrostatic pressure by absorbing ions from the dilute soil solution and transporting them into the xylem. The buildup of solutes in the xylem sap leads to a decrease in the xylem osmotic potential (Ψ_s) and thus a decrease in the xylem water potential (Ψ_w). This lowering of the xylem Ψ_w provides a driving force for water absorption, which in turn leads to a positive hydrostatic pressure in the xylem. In effect, the whole root acts like an osmotic cell; the multicellular root tissue behaves as an osmotic membrane does, building up a positive hydrostatic pressure in the xylem in response to the accumulation of solutes.

Root pressure is most likely to occur when soil water potentials are high and transpiration rates are low. When transpiration rates are high, water is taken up so rapidly into the leaves and lost to the atmosphere that a positive pressure never develops in the xylem.

Plants that develop root pressure frequently produce liquid droplets on the edges of their leaves, a phenomenon known as **guttation** (Figure 4.5). Positive xylem pressure

FIGURE 4.5 Guttation in leaves from strawberry (*Fragaria grandiflora*). In the early morning, leaves secrete water droplets through the hydathodes, located at the margins of the leaves. Young flowers may also show guttation. (Photograph courtesy of R. Aloni.)

causes exudation of xylem sap through specialized pores called *hydathodes* that are associated with vein endings at the leaf margin. The "dewdrops" that can be seen on the tips of grass leaves in the morning are actually guttation droplets exuded from such specialized pores. Guttation is most noticeable when transpiration is suppressed and the relative humidity is high, such as during the night.

WATER TRANSPORT THROUGH THE XYLEM

In most plants, the xylem constitutes the longest part of the pathway of water transport. In a plant 1 m tall, more than 99.5% of the water transport pathway through the plant is within the xylem, and in tall trees the xylem represents an even greater fraction of the pathway. Compared with the complex pathway across the root tissue, the xylem is a simple pathway of low resistance. In the following sections we will examine how water movement through the xylem is optimally suited to carry water from the roots to the leaves, and how negative hydrostatic pressure generated by leaf transpiration pulls water through the xylem.

The Xylem Consists of Two Types of Tracheary Elements

The conducting cells in the xylem have a specialized anatomy that enables them to transport large quantities of water with great efficiency. There are two important types of **tracheary elements** in the xylem: tracheids and vessel elements (Figure 4.6). Vessel elements are found only in angiosperms, a small group of gymnosperms called the Gnetales, and perhaps some ferns. Tracheids are present in both angiosperms and gymnosperms, as well as in ferns and other groups of vascular plants.

The maturation of both tracheids and vessel elements involves the "death" of the cell. Thus, functional water-conducting cells have no membranes and no organelles. What re-

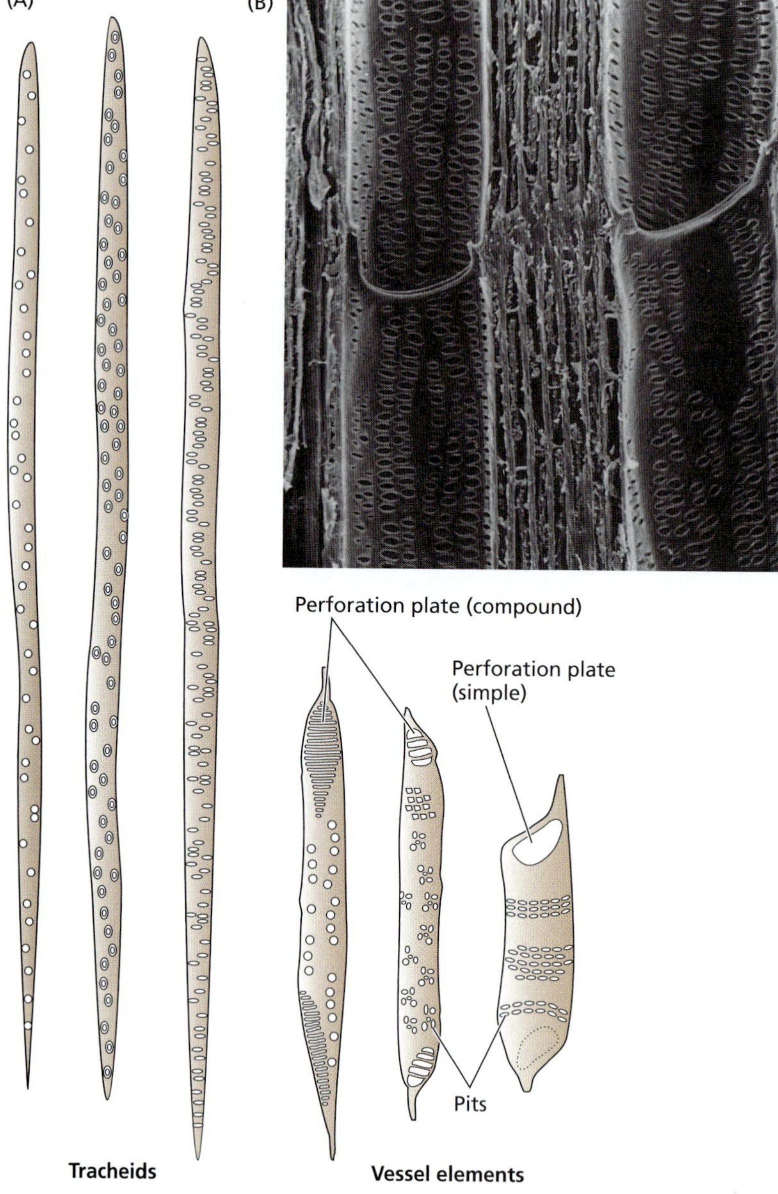

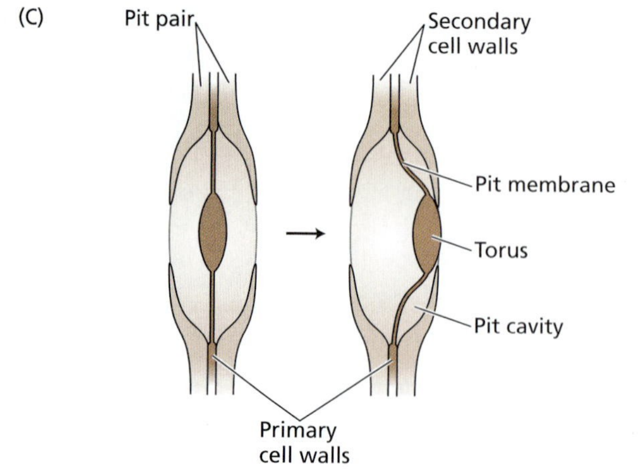

FIGURE 4.6 Tracheary elements and their interconnections. (A) Structural comparison of tracheids and vessel elements, two classes of tracheary elements involved in xylem water transport. Tracheids are elongate, hollow, dead cells with highly lignified walls. The walls contain numerous pits—regions where secondary wall is absent but primary wall remains. The shape and pattern of wall pitting vary with species and organ type. Tracheids are present in all vascular plants. Vessels consist of a stack of two or more vessel elements. Like tracheids, vessel elements are dead cells and are connected to one another through perforation plates—regions of the wall where pores or holes have developed. Vessels are connected to other vessels and to tracheids through pits. Vessels are found in most angiosperms and are lacking in most gymnosperms. (B) Scanning electron micrograph of oak wood showing two vessel elements that make up a portion of a vessel. Large pits are visible on the side walls, and the end walls are open at the perforation plate. (420×) (C) Diagram of a bordered pit with a torus either centered in the pit cavity or lodged to one side of the cavity, thereby blocking flow. (B © G. Shih-R. Kessel/Visuals Unlimited; C after Zimmermann 1983.)

mains are the thick, lignified cell walls, which form hollow tubes through which water can flow with relatively little resistance.

Tracheids are elongated, spindle-shaped cells (Figure 4.6A) that are arranged in overlapping vertical files. Water flows between tracheids by means of the numerous **pits** in their lateral walls (Figure 4.6B). Pits are microscopic regions where the secondary wall is absent and the primary wall is thin and porous (Figure 4.6C). Pits of one tracheid are typically located opposite pits of an adjoining tracheid, forming **pit pairs**. Pit pairs constitute a low-resistance path for water movement between tracheids. The porous layer between pit pairs, consisting of two primary walls and a middle lamella, is called the **pit membrane**.

Pit membranes in tracheids of some species of conifers have a central thickening, called a **torus** (pl. *tori*) (see Figure 4.6C). The torus acts like a valve to close the pit by lodging itself in the circular or oval wall thickenings bordering these pits. Such lodging of the torus is an effective way of preventing dangerous gas bubbles from invading neighboring tracheids (we will discuss this formation of bubbles, a process called cavitation, shortly).

Vessel elements tend to be shorter and wider than tracheids and have perforated end walls that form a **perforation plate** at each end of the cell. Like tracheids, vessel elements have pits on their lateral walls (see Figure 4.6B). Unlike tracheids, the perforated end walls allow vessel members to be stacked end to end to form a larger conduit called a **vessel** (again, see Figure 4.6B). Vessels vary in length both within and between species. Maximum vessel lengths range from 10 cm to many meters. Because of their open end walls, vessels provide a very efficient low-resistance pathway for water movement. The vessel members found at the extreme ends of a vessel lack perforations at the end walls and communicate with neighboring vessels via pit pairs.

Water Movement through the Xylem Requires Less Pressure Than Movement through Living Cells

The xylem provides a low-resistance pathway for water movement, thus reducing the pressure gradients needed to transport water from the soil to the leaves. Some numerical values will help us appreciate the extraordinary efficiency of the xylem. We will calculate the driving force required to move water through the xylem at a typical velocity and compare it with the driving force that would be needed to move water through a cell-to-cell pathway.

For the purposes of this comparison, we will use a figure of 4 mm s^{-1} for the xylem transport velocity and 40 μm as the vessel radius. This is a high velocity for such a narrow vessel, so it will tend to exaggerate the pressure gradient required to support water flow in the xylem. Using a version of Poiseuille's equation (see Equation 3.2), we can calculate the pressure gradient needed to move water at a velocity of 4 mm s^{-1} through an *ideal* tube with a uniform inner radius of 40 μm. The calculation gives a value of 0.02 MPa m^{-1}. Elaboration of the assumptions, equations, and calculations can be found in **Web Topic 4.4**.

Of course, *real* xylem conduits have irregular inner wall surfaces, and water flow through perforation plates and pits adds additional resistance. Such deviations from an ideal tube will increase the frictional drag above that calculated from Poiseuille's equation. However, measurements show that the actual resistance is greater by approximately a factor of 2 (Nobel 1999). Thus our estimate of 0.02 MPa m^{-1} is in the correct range for pressure gradients found in real trees.

Let's now compare this value (0.02 MPa m^{-1}) with the driving force that would be necessary to move water at the same velocity from cell to cell, crossing the plasma membrane each time. Using Poiseuille's equation, as described in **Web Topic 4.4**, the driving force needed to move water through a layer of cells at 4 mm s^{-1} is calculated to be 2×10^8 MPa m^{-1}. This is ten orders of magnitude greater than the driving force needed to move water through our 40-μm-radius xylem vessel. Our calculation clearly shows that water flow through the xylem is vastly more efficient than water flow across the membranes of living cells.

What Pressure Difference Is Needed to Lift Water 100 Meters to a Treetop?

With the foregoing example in mind, let's see how large of a pressure gradient is needed to move water up to the top of a very tall tree. The tallest trees in the world are the coast redwoods (*Sequoia sempervirens*) of North America and *Eucalyptus regnans* of Australia. Individuals of both species can exceed 100 m. If we think of the stem of a tree as a long pipe, we can estimate the pressure difference that is needed

The Cell Walls of Guard Cells Have Specialized Features

Guard cells can be found in leaves of all vascular plants, and they are also present in organs from more primitive plants, such as the liverworts and the mosses (Ziegler 1987). Guard cells show considerable morphological diversity, but we can distinguish two main types: One is typical of grasses and a few other monocots, such as palms; the other is found in all dicots, in many monocots, and in mosses, ferns, and gymnosperms.

In grasses (see Figure 4.13A), guard cells have a characteristic dumbbell shape, with bulbous ends. The pore proper is a long slit located between the two "handles" of the dumbbells. These guard cells are always flanked by a

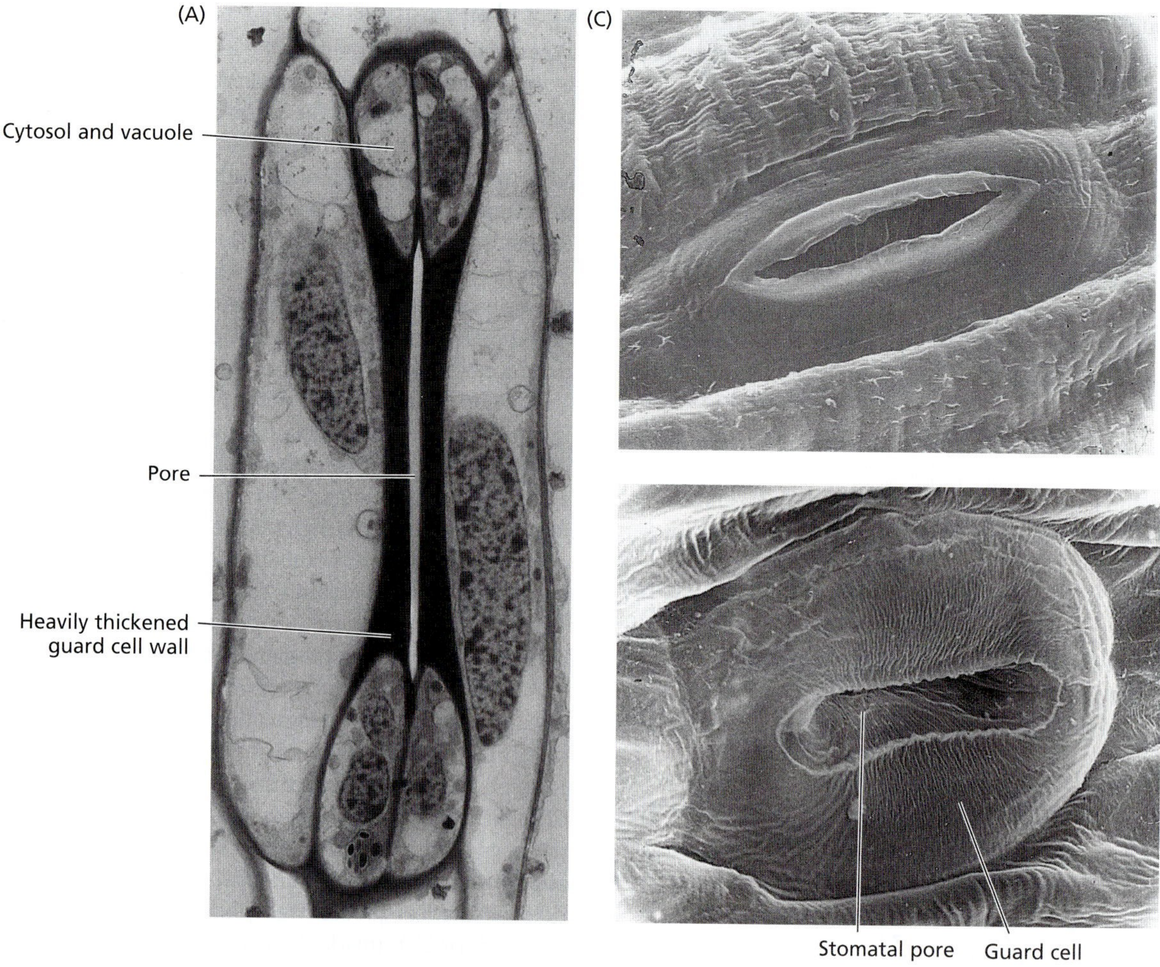

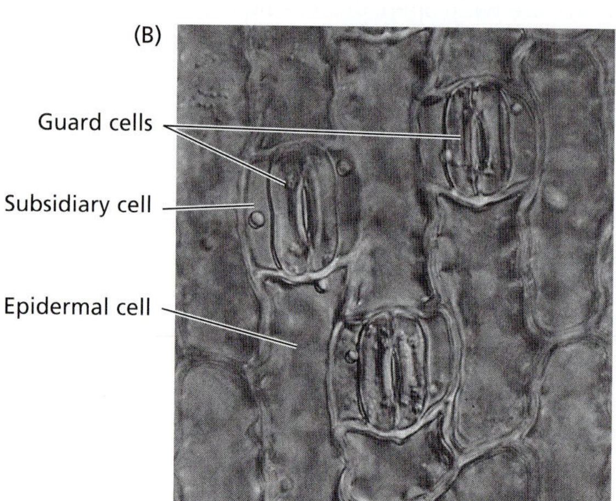

FIGURE 4.13 Electron micrographs of stomata. (A) A stoma from a grass. The bulbous ends of each guard cell show their cytosolic content and are joined by the heavily thickened walls. The stomatal pore separates the two midportions of the guard cells. (2560×) (B) Stomatal complexes of the sedge, *Carex,* viewed with differential interference contrast light microscopy. Each complex consists of two guard cells surrounding a pore and two flanking subsidiary cells. (550×) (C) Scanning electron micrographs of onion epidermis. The top panel shows the outside surface of the leaf, with a stomatal pore inserted in the cuticle. The bottom panel shows a pair of guard cells facing the stomatal cavity, toward the inside of the leaf. (1640×) (A from Palevitz 1981, B from Jarvis and Mansfield 1981, A and B courtesy of B. Palevitz; micrographs in C from Zeiger and Hepler 1976 [top] and E. Zeiger and N. Burnstein [bottom].)

pair of differentiated epidermal cells called **subsidiary cells**, which help the guard cells control the stomatal pores (see Figure 4.13B). The guard cells, subsidiary cells, and pore are collectively called the **stomatal complex**.

In dicot plants and nongrass monocots, kidney-shaped guard cells have an elliptical contour with the pore at its center (see Figure 4.13C). Although subsidiary cells are not uncommon in species with kidney-shaped stomata, they are often absent, in which case the guard cells are surrounded by ordinary epidermal cells.

A distinctive feature of the guard cells is the specialized structure of their walls. Portions of these walls are substantially thickened (Figure 4.14) and may be up to 5 μm across, in contrast to the 1 to 2 μm typical of epidermal cells. In kidney-shaped guard cells, a differential thickening pattern results in very thick inner and outer (lateral) walls, a thin dorsal wall (the wall in contact with epidermal cells), and a somewhat thickened ventral (pore) wall (see Figure 4.14). The portions of the wall that face the atmosphere extend into well-developed ledges, which form the pore proper.

The alignment of **cellulose microfibrils**, which reinforce all plant cell walls and are an important determinant of cell shape (see Chapter 15), plays an essential role in the opening and closing of the stomatal pore. In ordinary cells having a cylindrical shape, cellulose microfibrils are oriented transversely to the long axis of the cell. As a result, the cell expands in the direction of its long axis because the cellulose reinforcement offers the least resistance at right angles to its orientation.

In guard cells the microfibril organization is different. Kidney-shaped guard cells have cellulose microfibrils fanning out radially from the pore (Figure 4.15A). Thus the cell girth is reinforced like a steel-belted radial tire, and the guard cells curve outward during stomatal opening (Sharpe et al. 1987). In grasses, the dumbbell-shaped guard cells function like beams with inflatable ends. As the bulbous ends of the cells increase in volume and swell, the beams are separated from each other and the slit between them widens (Figure 4.15B).

An Increase in Guard Cell Turgor Pressure Opens the Stomata

Guard cells function as multisensory hydraulic valves. Environmental factors such as light intensity and quality, temperature, relative humidity, and intracellular CO_2 concentra-

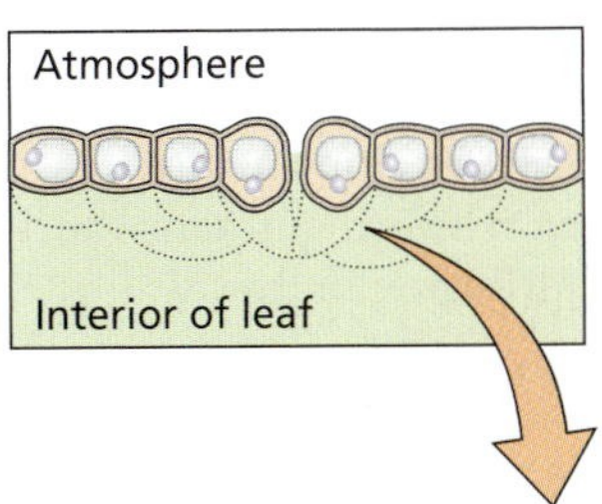

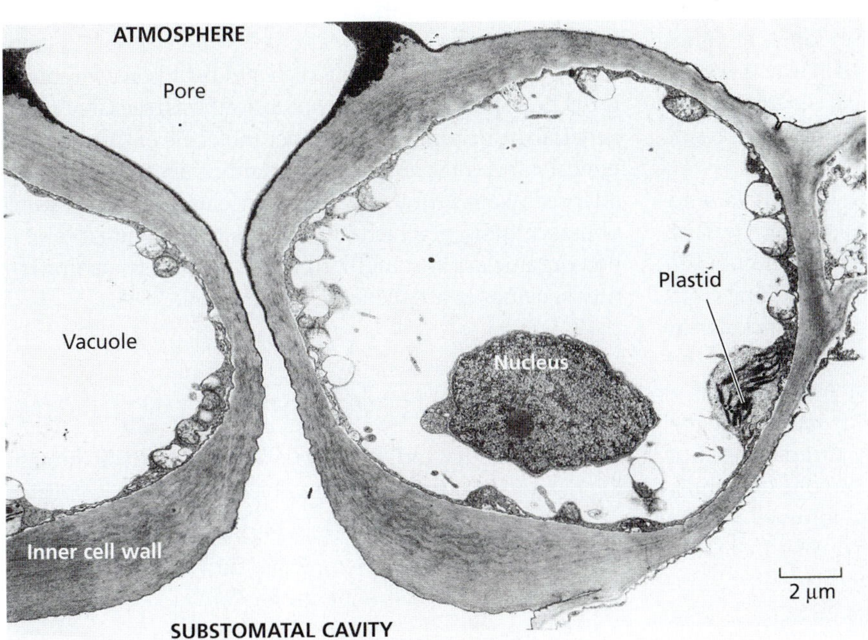

FIGURE 4.14 Electron micrograph showing a pair of guard cells from the dicot *Nicotiana tabacum* (tobacco). The section was made perpendicular to the main surface of the leaf. The pore faces the atmosphere; the bottom faces the substomatal cavity inside the leaf. Note the uneven thickening pattern of the walls, which determines the asymmetric deformation of the guard cells when their volume increases during stomatal opening. (From Sack 1987, courtesy of F. Sack.)

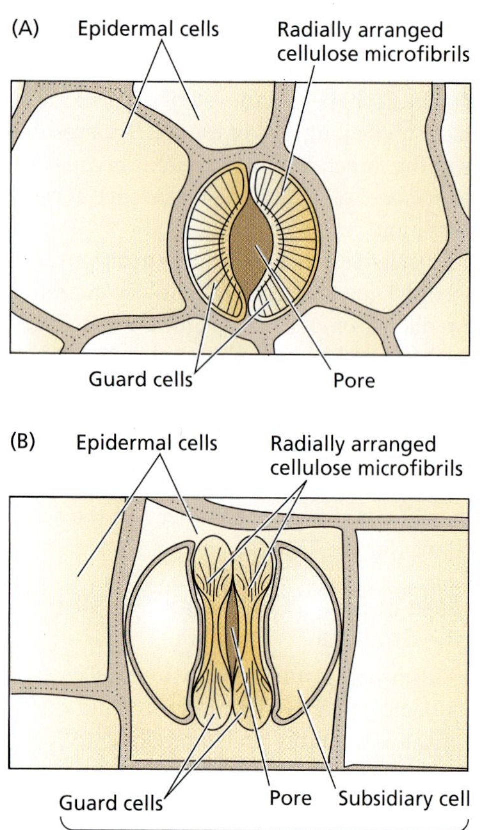

FIGURE 4.15 The radial alignment of the cellulose microfibrils in guard cells and epidermal cells of (A) a kidney-shaped stoma and (B) a grasslike stoma. (From Meidner and Mansfield 1968.)

tions are sensed by guard cells, and these signals are integrated into well-defined stomatal responses. If leaves kept in the dark are illuminated, the light stimulus is perceived by the guard cells as an opening signal, triggering a series of responses that result in opening of the stomatal pore.

The early aspects of this process are ion uptake and other metabolic changes in the guard cells, which will be discussed in detail in Chapter 18. Here we will note the effect of decreases in osmotic potential (Ψ_s) resulting from ion uptake and from biosynthesis of organic molecules in the guard cells. Water relations in guard cells follow the same rules as in other cells. As Ψ_s decreases, the water potential decreases and water consequently moves into the guard cells. As water enters the cell, turgor pressure increases. Because of the elastic properties of their walls, guard cells can reversibly increase their volume by 40 to 100%, depending on the species. Because of the differential thickening of guard cell walls, such changes in cell volume lead to opening or closing of the stomatal pore.

The Transpiration Ratio Measures the Relationship between Water Loss and Carbon Gain

The effectiveness of plants in moderating water loss while allowing sufficient CO_2 uptake for photosynthesis can be assessed by a parameter called the **transpiration ratio**. This value is defined as the amount of water transpired by the plant, divided by the amount of carbon dioxide assimilated by photosynthesis.

For typical plants in which the first stable product of carbon fixation is a three-carbon compound (such plants are called C_3 plants; see Chapter 8), about 500 molecules of water are lost for every molecule of CO_2 fixed by photosynthesis, giving a transpiration ratio of 500. (Sometimes the reciprocal of the transpiration ratio, called the *water use efficiency*, is cited. Plants with a transpiration ratio of 500 have a water use efficiency of 1/500, or 0.002.)

The large ratio of H_2O efflux to CO_2 influx results from three factors:

1. The concentration gradient driving water loss is about 50 times larger than that driving the influx of CO_2. In large part, this difference is due to the low concentration of CO_2 in air (about 0.03%) and the relatively high concentration of water vapor within the leaf.
2. CO_2 diffuses about 1.6 times more slowly through air than water does (the CO_2 molecule is larger than H_2O and has a smaller diffusion coefficient).
3. CO_2 uptake must cross the plasma membrane, the cytoplasm, and the chloroplast envelope before it is assimilated in the chloroplast. These membranes add to the resistance of the CO_2 diffusion pathway.

Some plants are adapted for life in particularly dry environments or seasons of the year. These plants, designated the C_4 and CAM plants, utilize variations in the usual photosynthetic pathway for fixation of carbon dioxide. Plants with C_4 photosynthesis (in which a four-carbon compound is the first stable product of photosynthesis; see Chapter 8) generally transpire less water per molecule of CO_2 fixed; a typical transpiration ratio for C_4 plants is about 250. Desert-adapted plants with CAM (crassulacean acid metabolism) photosynthesis, in which CO_2 is initially fixed into four-carbon organic acids at night, have even lower transpiration ratios; values of about 50 are not unusual.

OVERVIEW: THE SOIL–PLANT–ATMOSPHERE CONTINUUM

We have seen that movement of water from the soil through the plant to the atmosphere involves different mechanisms of transport:

- In the soil and the xylem, water moves by bulk flow in response to a pressure gradient ($\Delta\Psi_p$).

- In the vapor phase, water moves primarily by diffusion, at least until it reaches the outside air, where convection (a form of bulk flow) becomes dominant.
- When water is transported across membranes, the driving force is the water potential difference across the membrane. Such osmotic flow occurs when cells absorb water and when roots transport water from the soil to the xylem.

In all of these situations, *water moves toward regions of low water potential or free energy.* This phenomenon is illustrated in Figure 4.16, which shows representative values for water potential and its components at various points along the water transport pathway.

Water potential decreases continuously from the soil to the leaves. However, the components of water potential can be quite different at different parts of the pathway. For example, inside the leaf cells, such as in the mesophyll, the water potential is approximately the same as in the neighboring xylem, yet the components of Ψ_w are quite different. The dominant component of Ψ_w in the xylem is the negative pressure (Ψ_p), whereas in the leaf cell Ψ_p is generally positive. This large difference in Ψ_p occurs across the plasma membrane of the leaf cells. Within the leaf cells, water potential is reduced by a high concentration of dissolved solutes (low Ψ_s).

SUMMARY

Water is the essential medium of life. Land plants are faced with potentially lethal desiccation by water loss to the atmosphere. This problem is aggravated by the large surface area of leaves, their high radiant-energy gain, and their need to have an open pathway for CO_2 uptake. Thus there is a conflict between the need for water conservation and the need for CO_2 assimilation.

The need to resolve this vital conflict determines much of the structure of land plants: (1) an extensive root system

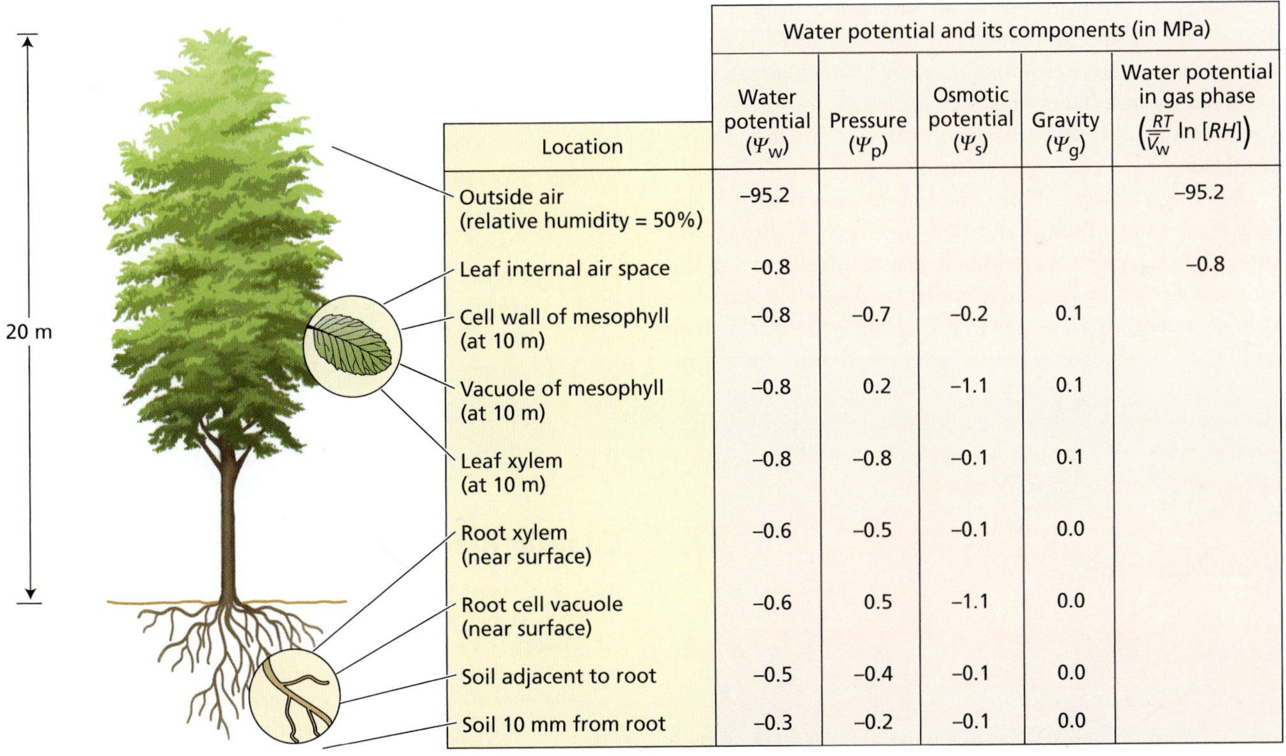

Location	Water potential and its components (in MPa): Water potential (Ψ_w)	Pressure (Ψ_p)	Osmotic potential (Ψ_s)	Gravity (Ψ_g)	Water potential in gas phase ($\frac{RT}{\bar{V}_w}\ln[RH]$)
Outside air (relative humidity = 50%)	–95.2				–95.2
Leaf internal air space	–0.8				–0.8
Cell wall of mesophyll (at 10 m)	–0.8	–0.7	–0.2	0.1	
Vacuole of mesophyll (at 10 m)	–0.8	0.2	–1.1	0.1	
Leaf xylem (at 10 m)	–0.8	–0.8	–0.1	0.1	
Root xylem (near surface)	–0.6	–0.5	–0.1	0.0	
Root cell vacuole (near surface)	–0.6	0.5	–1.1	0.0	
Soil adjacent to root	–0.5	–0.4	–0.1	0.0	
Soil 10 mm from root	–0.3	–0.2	–0.1	0.0	

FIGURE 4.16 Representative overview of water potential and its components at various points in the transport pathway from the soil through the plant to the atmosphere. Water potential (Ψ_w) can be measured through this continuum, but the components vary. In the liquid part of the pathway, pressure (Ψ_p), osmotic potential (Ψ_s), and gravity (Ψ_g), determine Ψ_w. In the air, only the relative humidity ($RT/\bar{V}_w \times \ln[RH]$) is important. Note that although the water potential is the same in the vacuole of the mesophyll cell and in the surrounding cell wall, the components of Ψ_w can differ greatly (e.g., in this case Ψ_p is 0.2 MPa inside the mesophyll cell and –0.7 MPa outside). (After Nobel 1999.)

to extract water from the soil; (2) a low-resistance pathway through the xylem vessel elements and tracheids to bring water to the leaves; (3) a hydrophobic cuticle covering the surfaces of the plant to reduce evaporation; (4) microscopic stomata on the leaf surface to allow gas exchange; and (5) guard cells to regulate the diameter (and diffusional resistance) of the stomatal aperture.

The result is an organism that transports water from the soil to the atmosphere purely in response to physical forces. No energy is expended directly by the plant to translocate water, although development and maintenance of the structures needed for efficient and controlled water transport require considerable energy input.

The mechanisms of water transport from the soil through the plant body to the atmosphere include diffusion, bulk flow, and osmosis. Each of these processes is coupled to different driving forces.

Water in the plant can be considered a continuous hydraulic system, connecting the water in the soil with the water vapor in the atmosphere. Transpiration is regulated principally by the guard cells, which regulate the stomatal pore size to meet the photosynthetic demand for CO_2 uptake while minimizing water loss to the atmosphere. Water evaporation from the cell walls of the leaf mesophyll cells generates large negative pressures (or tensions) in the apoplastic water. These negative pressures are transmitted to the xylem, and they pull water through the long xylem conduits.

Although aspects of the cohesion–tension theory of sap ascent are intermittently debated, an overwhelming body of evidence supports the idea that water transport in the xylem is driven by pressure gradients. When transpiration is high, negative pressures in the xylem water may cause cavitation (embolisms) in the xylem. Such embolisms can block water transport and lead to severe water deficits in the leaf. Water deficits are commonplace in plants, necessitating a host of adaptive responses that modify the physiology and development of plants.

Web Material

Web Topics

4.1 Irrigation

A discussion of some widely used irrigation methods and their impact on crop yield and soil salinity.

4.2 Soil Hydraulic Conductivity and Water Potential

Soil hydraulic conductivity determines the ease with which water moves through the soil, and it is closely related to soil water potential.

4.3 Root Hydraulic Conductance

A discussion of root hydraulic conductance and an example of its quantification.

4.4 Calculating Velocities of Water Movement in the Xylem and in Living Cells

Calculations of velocities of water movement through the xylem, up a tree trunk, and across cell membranes in a tissue, and their implications for water transport mechanism.

4.5 Leaf Transpiration and Water Vapor Gradients

An analysis of leaf transpiration and stomatal conductance, and their relationship with leaf and air water vapor concentrations.

Web Essays

4.1 A Brief History of the Study of Water Movement in the Xylem

The history of our understanding of sap ascent in plants, especially in trees, is a beautiful example of how knowledge about plants is acquired.

4.2 The Cohesion–Tension Theory at Work

A detailed discussion of the Cohesion–Tension theory on sap ascent in plants, and some alternative explanations.

4.3 How Water Climbs to the Top of a 112-Meter-Tall Tree

Measurements of photosynthesis and transpiration in 112-meter tall trees show that some of the conditions experienced by the top foliage compares to that of extreme deserts.

4.4 Cavitation and Refilling

A possible mechanism for cavitation repair is under active investigation.

Chapter References

Balling, A., and Zimmermann, U. (1990) Comparative measurements of the xylem pressure of *Nicotiana* plants by means of the pressure bomb and pressure probe. *Planta* 182: 325–338.

Bange, G. G. J. (1953) On the quantitative explanation of stomatal transpiration. *Acta Botanica Neerlandica* 2: 255–296.

Canny, M. J. (1998) Transporting water in plants. *Am. Sci.* 86: 152–159.

Davis, S. D., Sperry, J. S., and Hacke, U. G. (1999) The relationship between xylem conduit diameter and cavitation caused by freezing. *Am. J. Bot.* 86: 1367–1372.

Esau, K. (1953) *Plant Anatomy*. John Wiley & Sons, Inc. New York.

Frensch, J., Hsiao, T. C., and Steudle, E. (1996) Water and solute transport along developing maize roots. *Planta* 198: 348–355.

Hacke, U. G., Stiller, V., Sperry, J. S., Pittermann, J., and McCulloh, K. A. (2001) Cavitation fatigue: Embolism and refilling cycles can weaken the cavitation resistance of xylem. *Plant Physiol.* 125: 779–786.

Holbrook, N. M., Ahrens, E. T., Burns, M. J., and Zwieniecki, M. A. (2001) In vivo observation of cavitation and embolism repair using magnetic resonance imaging. *Plant Physiol.* 126: 27–31.

Holbrook, N. M., Burns, M. J., and Field, C. B. (1995) Negative xylem pressures in plants: A test of the balancing pressure technique. *Science* 270: 1193–1194.

Jackson, G. E., Irvine, J., and Grace, J. (1999) Xylem acoustic emissions and water relations of *Calluna vulgaris* L. at two climatological regions of Britain. *Plant Ecol.* 140: 3–14.

Jarvis, P. G., and Mansfield, T. A. (1981) *Stomatal Physiology.* Cambridge University Press, Cambridge.

Jensen, C. R., Mogensen, V. O., Poulsen, H.-H., Henson, I. E., Aagot, S., Hansen, E., Ali, M., and Wollenweber, B. (1998) Soil water matric potential rather than water content determines drought responses in field-grown lupin (*Lupinus angustifolius*). *Aust. J. Plant Physiol.* 25: 353–363.

Kramer, P. J., and Boyer, J. S. (1995) *Water Relations of Plants and Soils.* Academic Press, San Diego, CA.

Meidner, H., and Mansfield, D. (1968) *Stomatal Physiology.* McGraw-Hill, London.

Melcher, P. J., Meinzer, F. C., Yount, D. E., Goldstein, G., and Zimmermann, U. (1998) Comparative measurements of xylem pressure in transpiring and non-transpiring leaves by means of the pressure chamber and the xylem pressure probe. *J. Exp. Bot.* 49: 1757–1760.

Nobel, P. S. (1999) *Physicochemical and Environmental Plant Physiology,* 2nd ed. Academic Press, San Diego, CA.

Palevitz, B. A. (1981) The structure and development of guard cells. In *Stomatal Physiology,* P. G. Jarvis and T. A. Mansfield, eds., Cambridge University Press, Cambridge, pp. 1–23.

Pockman, W. T., Sperry, J. S., and O'Leary, J. W. (1995) Sustained and significant negative water pressure in xylem. *Nature* 378: 715–716.

Sack, F. D. (1987) The development and structure of stomata. In *Stomatal Function,* E. Zeiger, G. Farquhar, and I. Cowan, eds., Stanford University Press, Stanford, CA, pp. 59–90.

Sharpe, P. J. H., Wu, H.-I., and Spence, R. D. (1987) Stomatal mechanics. In *Stomatal Function,* E. Zeiger, G. Farquhar, and I. Cowan, eds., Stanford University Press, Stanford, CA, pp. 91–114.

Steudle, E. (2001) The cohesion-tension mechanism and the acquisition of water by plant roots. *Annu. Rev. Plant Physiol. Plant Mol. Biol.* 52: 847–875.

Steudle, E., and Frensch, J. (1996) Water transport in plants: Role of the apoplast. *Plant and Soil* 187: 67–79.

Tyree, M. T. (1997) The cohesion-tension theory of sap ascent: Current controversies. *J. Exp. Bot.* 48: 1753–1765.

Tyree, M. T., and Sperry, J. S. (1989) Vulnerability of xylem to cavitation and embolism. *Annu. Rev. Plant Physiol. Plant Mol. Biol.* 40: 19–38.

Wei, C., Tyree, M. T., and Steudle, E. (1999) Direct measurement of xylem pressure in leaves of intact maize plants: A test of the cohesion-tension theory taking hydraulic architecture into consideration. *Plant Physiol. Plant Mol. Biol.* 121: 1191–1205.

Zeiger, E., and Hepler, P. K. (1976) Production of guard cell protoplasts from onion and tobacco. *Plant Physiol.* 58: 492–498.

Ziegler, H. (1987) The evolution of stomata. In *Stomatal Function,* E. Zeiger, G. Farquhar, and I. Cowan, eds., Stanford University Press, Stanford, CA, pp. 29–58.

Zimmermann, M. H. (1983) *Xylem Structure and the Ascent of Sap.* Springer, Berlin.

Chapter

5 *Mineral Nutrition*

MINERAL NUTRIENTS ARE ELEMENTS acquired primarily in the form of inorganic ions from the soil. Although mineral nutrients continually cycle through all organisms, they enter the biosphere predominantly through the root systems of plants, so in a sense plants act as the "miners" of Earth's crust (Epstein 1999). The large surface area of roots and their ability to absorb inorganic ions at low concentrations from the soil solution make mineral absorption by plants a very effective process. After being absorbed by the roots, the mineral elements are translocated to the various parts of the plant, where they are utilized in numerous biological functions. Other organisms, such as mycorrhizal fungi and nitrogen-fixing bacteria, often participate with roots in the acquisition of nutrients.

The study of how plants obtain and use mineral nutrients is called **mineral nutrition**. This area of research is central to modern agriculture and environmental protection. High agricultural yields depend strongly on fertilization with mineral nutrients. In fact, yields of most crop plants increase linearly with the amount of fertilizer that they absorb (Loomis and Conner 1992). To meet increased demand for food, world consumption of the primary fertilizer mineral elements—nitrogen, phosphorus, and potassium—rose steadily from 112 million metric tons in 1980 to 143 million metric tons in 1990 and has remained constant through the last decade.

Crop plants, however, typically use less than half of the fertilizer applied (Loomis and Conner 1992). The remaining minerals may leach into surface waters or groundwater, become attached to soil particles, or contribute to air pollution. As a consequence of fertilizer leaching, many water wells in the United States no longer meet federal standards for nitrate concentrations in drinking water (Nolan and Stoner 2000). On a brighter note, plants are the traditional means for recycling animal wastes and are proving useful for removing deleterious minerals from toxic-waste dumps (Macek et al. 2000). Because of the complex nature of plant–soil–atmosphere relationships, studies in the area of mineral nutrition involve atmospheric chemists, soil scientists, hydrologists, microbiologists, and ecologists, as well as plant physiologists.

In this chapter we will discuss first the nutritional needs of plants, the symptoms of specific nutritional deficiencies, and the use of fertilizers to ensure proper plant nutrition. Then we will examine how soil and root structure influence the transfer of inorganic nutrients from the environment into a plant. Finally, we will introduce the topic of mycorrhizal associations. Chapters 6 and 12 address additional aspects of solute transport and nutrient assimilation, respectively.

ESSENTIAL NUTRIENTS, DEFICIENCIES, AND PLANT DISORDERS

Only certain elements have been determined to be essential for plant growth. An **essential element** is defined as one whose absence prevents a plant from completing its life cycle (Arnon and Stout 1939) or one that has a clear physiological role (Epstein 1999). If plants are given these essential elements, as well as energy from sunlight, they can synthesize all the compounds they need for normal growth. Table 5.1 lists the elements that are considered to be essential for most, if not all, higher plants. The first three elements—hydrogen, carbon, and oxygen—are not considered mineral nutrients because they are obtained primarily from water or carbon dioxide.

Essential mineral elements are usually classified as macronutrients or micronutrients, according to their relative concentration in plant tissue. In some cases, the differences in tissue content of macronutrients and micronutrients are not as great as those indicated in Table 5.1. For example, some plant tissues, such as the leaf mesophyll, have almost as much iron or manganese as they do sulfur or magnesium. Many elements often are present in concentrations greater than the plant's minimum requirements.

Some researchers have argued that a classification into macronutrients and micronutrients is difficult to justify physiologically. Mengel and Kirkby (1987) have proposed that the essential elements be classified instead according to their biochemical role and physiological function. Table 5.2 shows such a classification, in which plant nutrients have been divided into four basic groups:

1. The first group of essential elements forms the organic (carbon) compounds of the plant. Plants assimilate these nutrients via biochemical reactions involving oxidation and reduction.
2. The second group is important in energy storage reactions or in maintaining structural integrity. Elements in this group are often present in plant tissues as phosphate, borate, and silicate esters in which the elemental group is bound to the hydroxyl group of an organic molecule (i.e., sugar–phosphate).
3. The third group is present in plant tissue as either free ions or ions bound to substances such as the pectic acids present in the plant cell wall. Of particular importance are their roles as enzyme cofactors and in the regulation of osmotic potentials.
4. The fourth group has important roles in reactions involving electron transfer.

Naturally occurring elements, other than those listed in Table 5.1, can also accumulate in plant tissues. For example, aluminum is not considered to be an essential element, but plants commonly contain from 0.1 to 500 ppm aluminum, and addition of low levels of aluminum to a nutrient solution may stimulate plant growth (Marschner 1995).

TABLE 5.1
Adequate tissue levels of elements that may be required by plants

Element	Chemical symbol	Concentration in dry matter (% or ppm)[a]	Relative number of atoms with respect to molybdenum
Obtained from water or carbon dioxide			
Hydrogen	H	6	60,000,000
Carbon	C	45	40,000,000
Oxygen	O	45	30,000,000
Obtained from the soil			
Macronutrients			
Nitrogen	N	1.5	1,000,000
Potassium	K	1.0	250,000
Calcium	Ca	0.5	125,000
Magnesium	Mg	0.2	80,000
Phosphorus	P	0.2	60,000
Sulfur	S	0.1	30,000
Silicon	Si	0.1	30,000
Micronutrients			
Chlorine	Cl	100	3,000
Iron	Fe	100	2,000
Boron	B	20	2,000
Manganese	Mn	50	1,000
Sodium	Na	10	400
Zinc	Zn	20	300
Copper	Cu	6	100
Nickel	Ni	0.1	2
Molybdenum	Mo	0.1	1

Source: Epstein 1972, 1999.
[a] The values for the nonmineral elements (H, C, O) and the macronutrients are percentages. The values for micronutrients are expressed in parts per million.

TABLE 5.2
Classification of plant mineral nutrients according to biochemical function

Mineral nutrient	Functions
Group 1	**Nutrients that are part of carbon compounds**
N	Constituent of amino acids, amides, proteins, nucleic acids, nucleotides, coenzymes, hexoamines, etc.
S	Component of cysteine, cystine, methionine, and proteins. Constituent of lipoic acid, coenzyme A, thiamine pyrophosphate, glutathione, biotin, adenosine-5′-phosphosulfate, and 3-phosphoadenosine.
Group 2	**Nutrients that are important in energy storage or structural integrity**
P	Component of sugar phosphates, nucleic acids, nucleotides, coenzymes, phospholipids, phytic acid, etc. Has a key role in reactions that involve ATP.
Si	Deposited as amorphous silica in cell walls. Contributes to cell wall mechanical properties, including rigidity and elasticity.
B	Complexes with mannitol, mannan, polymannuronic acid, and other constituents of cell walls. Involved in cell elongation and nucleic acid metabolism.
Group 3	**Nutrients that remain in ionic form**
K	Required as a cofactor for more than 40 enzymes. Principal cation in establishing cell turgor and maintaining cell electroneutrality.
Ca	Constituent of the middle lamella of cell walls. Required as a cofactor by some enzymes involved in the hydrolysis of ATP and phospholipids. Acts as a second messenger in metabolic regulation.
Mg	Required by many enzymes involved in phosphate transfer. Constituent of the chlorophyll molecule.
Cl	Required for the photosynthetic reactions involved in O_2 evolution.
Mn	Required for activity of some dehydrogenases, decarboxylases, kinases, oxidases, and peroxidases. Involved with other cation-activated enzymes and photosynthetic O_2 evolution.
Na	Involved with the regeneration of phosphoenolpyruvate in C_4 and CAM plants. Substitutes for potassium in some functions.
Group 4	**Nutrients that are involved in redox reactions**
Fe	Constituent of cytochromes and nonheme iron proteins involved in photosynthesis, N_2 fixation, and respiration.
Zn	Constituent of alcohol dehydrogenase, glutamic dehydrogenase, carbonic anhydrase, etc.
Cu	Component of ascorbic acid oxidase, tyrosinase, monoamine oxidase, uricase, cytochrome oxidase, phenolase, laccase, and plastocyanin.
Ni	Constituent of urease. In N_2-fixing bacteria, constituent of hydrogenases.
Mo	Constituent of nitrogenase, nitrate reductase, and xanthine dehydrogenase.

Source: After Evans and Sorger 1966 and Mengel and Kirkby 1987.

Many species in the genera *Astragalus*, *Xylorhiza*, and *Stanleya* accumulate selenium, although plants have not been shown to have a specific requirement for this element.

Cobalt is part of cobalamin (vitamin B_{12} and its derivatives), a component of several enzymes in nitrogen-fixing microorganisms. Thus cobalt deficiency blocks the development and function of nitrogen-fixing nodules. Nonetheless, plants that do not fix nitrogen, as well as nitrogen-fixing plants that are supplied with ammonium or nitrate, do not require cobalt. Crop plants normally contain only relatively small amounts of nonessential elements.

Special Techniques Are Used in Nutritional Studies

To demonstrate that an element is essential requires that plants be grown under experimental conditions in which only the element under investigation is absent. Such conditions are extremely difficult to achieve with plants grown in a complex medium such as soil. In the nineteenth century, several researchers, including Nicolas-Théodore de Saussure, Julius von Sachs, Jean-Baptiste-Joseph-Dieudonné Boussingault, and Wilhelm Knop, approached this problem by growing plants with their roots immersed in a **nutrient solution** containing only inorganic salts. Their demonstration that plants could grow normally with no soil or organic matter proved unequivocally that plants can fulfill all their needs from only inorganic elements and sunlight.

The technique of growing plants with their roots immersed in nutrient solution without soil is called solution culture or **hydroponics** (Gericke 1937). Successful hydroponic culture (Figure 5.1A) requires a large volume of nutrient solution or frequent adjustment of the nutrient solution to prevent nutrient uptake by roots from producing radical changes in nutrient concentrations and pH of the medium. A sufficient supply of oxygen to the root sys-

(A) **Hydroponic growth system**

(B) **Nutrient film growth system**

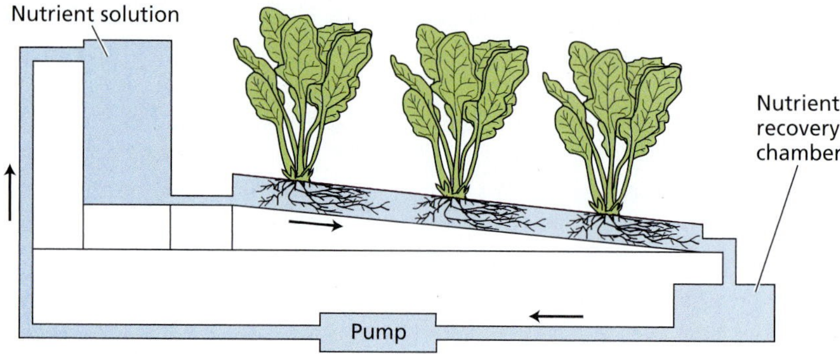

(C) **Aeroponic growth system**

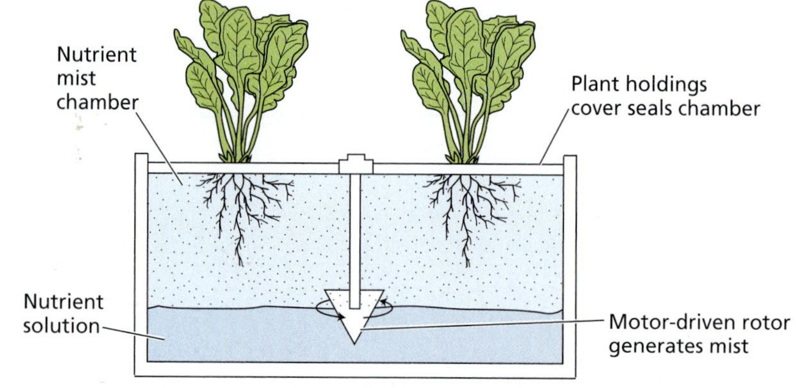

FIGURE 5.1 Hydroponic and aeroponic systems for growing plants in nutrient solutions in which composition and pH can be automatically controlled. (A) In a hydroponic system, the roots are immersed in the nutrient solution, and air is bubbled through the solution. (B) An alternative hydroponic system, often used in commercial production, is the nutrient film growth system, in which the nutrient solution is pumped as a thin film down a shallow trough surrounding the plant roots. In this system the composition and pH of the nutrient solution can be controlled automatically. (C) In the aeroponic system, the roots are suspended over the nutrient solution, which is whipped into a mist by a motor-driven rotor. (C after Weathers and Zobel 1992.)

tem—also critical—may be achieved by vigorous bubbling of air through the medium.

Hydroponics is used in the commercial production of many greenhouse crops. In one form of commercial hydroponic culture, plants are grown in a supporting material such as sand, gravel, vermiculite, or expanded clay (i.e., kitty litter). Nutrient solutions are then flushed through the supporting material, and old solutions are removed by leaching. In another form of hydroponic culture, plant roots lie on the surface of a trough, and nutrient solutions flow in a thin layer along the trough over the roots (Cooper 1979, Asher and Edwards 1983). This **nutrient film growth system** ensures that the roots receive an ample supply of oxygen (Figure 5.1B).

Another alternative, which has sometimes been heralded as the medium of the future, is to grow the plants **aeroponically** (Weathers and Zobel 1992). In this technique, plants are grown with their roots suspended in air while being sprayed continuously with a nutrient solution (Figure 5.1C). This approach provides easy manipulation of the gaseous environment around the root, but it requires higher levels of nutrients than hydroponic culture does to sustain rapid plant growth. For this reason and other technical difficulties, the use of aeroponics is not widespread.

Nutrient Solutions Can Sustain Rapid Plant Growth

Over the years, many formulations have been used for nutrient solutions. Early formulations developed by Knop in Germany included only KNO_3, $Ca(NO_3)_2$, KH_2PO_4, $MgSO_4$, and an iron salt. At the time this nutrient solution was believed to contain all the minerals required by the plant, but these experiments were carried out with chemicals that were contaminated with other elements that are now known to be essential (such as boron or molybdenum). Table 5.3 shows a more modern formulation for a nutrient solution. This formulation is called a modified **Hoagland solution**, named after Dennis R. Hoagland, a researcher who was prominent in the development of modern mineral nutrition research in the United States.

TABLE 5.3
Composition of a modified Hoagland nutrient solution for growing plants

Compound	Molecular weight	Concentration of stock solution	Concentration of stock solution	Volume of stock solution per liter of final solution	Element	Final concentration of element	
	g mol^{-1}	mM	g L^{-1}	mL		μM	ppm
Macronutrients							
KNO_3	101.10	1,000	101.10	6.0	N	16,000	224
$Ca(NO_3)_2 \cdot 4H_2O$	236.16	1,000	236.16	4.0	K	6,000	235
$NH_4H_2PO_4$	115.08	1,000	115.08	2.0	Ca	4,000	160
$MgSO_4 \cdot 7H_2O$	246.48	1,000	246.49	1.0	P	2,000	62
					S	1,000	32
					Mg	1,000	24
Micronutrients							
KCl	74.55	25	1.864	2.0 (combined for KCl through H_2MoO_4)	Cl	50	1.77
H_3BO_3	61.83	12.5	0.773		B	25	0.27
$MnSO_4 \cdot H_2O$	169.01	1.0	0.169		Mn	2.0	0.11
$ZnSO_4 \cdot 7H_2O$	287.54	1.0	0.288		Zn	2.0	0.13
$CuSO_4 \cdot 5H_2O$	249.68	0.25	0.062		Cu	0.5	0.03
H_2MoO_4 (85% MoO_3)	161.97	0.25	0.040		Mo	0.5	0.05
NaFeDTPA (10% Fe)	468.20	64	30.0	0.3–1.0	Fe	16.1–53.7	1.00–3.00
Optional[a]							
$NiSO_4 \cdot 6H_2O$	262.86	0.25	0.066	2.0	Ni	0.5	0.03
$Na_2SiO_3 \cdot 9H_2O$	284.20	1,000	284.20	1.0	Si	1,000	28

Source: After Epstein 1972.

Note: The macronutrients are added separately from stock solutions to prevent precipitation during preparation of the nutrient solution. A combined stock solution is made up containing all micronutrients except iron. Iron is added as sodium ferric diethylenetriaminepentaacetate (NaFeDTPA, trade name Ciba-Geigy Sequestrene 330 Fe; see Figure 5.2); some plants, such as maize, require the higher level of iron shown in the table.

[a] Nickel is usually present as a contaminant of the other chemicals, so it may not need to be added explicitly. Silicon, if included, should be added first and the pH adjusted with HCl to prevent precipitation of the other nutrients.

A modified Hoagland solution contains all of the known mineral elements needed for rapid plant growth. The concentrations of these elements are set at the highest possible levels without producing toxicity symptoms or salinity stress and thus may be several orders of magnitude higher than those found in the soil around plant roots. For example, whereas phosphorus is present in the soil solution at concentrations normally less than 0.06 ppm, here it is offered at 62 ppm (Epstein 1972). Such high initial levels permit plants to be grown in a medium for extended periods without replenishment of the nutrients. Many researchers, however, dilute their nutrient solutions severalfold and replenish them frequently to minimize fluctuations of nutrient concentration in the medium and in plant tissue.

Another important property of the modified Hoagland formulation is that nitrogen is supplied as both ammonium (NH_4^+) and nitrate (NO_3^-). Supplying nitrogen in a balanced mixture of cations and anions tends to reduce the rapid rise in the pH of the medium that is commonly observed when the nitrogen is supplied solely as nitrate anion (Asher and Edwards 1983). Even when the pH of the medium is kept neutral, most plants grow better if they have access to both NH_4^+ and NO_3^- because absorption and assimilation of the two nitrogen forms promotes cation–anion balance within the plant (Raven and Smith 1976; Bloom 1994).

A significant problem with nutrient solutions is maintaining the availability of iron. When supplied as an inorganic salt such as $FeSO_4$ or $Fe(NO_3)_2$, iron can precipitate out of solution as iron hydroxide. If phosphate salts are present, insoluble iron phosphate will also form. Precipitation of the iron out of solution makes it physically unavailable to the plant, unless iron salts are added at frequent intervals. Earlier researchers approached this problem by adding iron together with citric acid or tartaric acid. Compounds such as these are called **chelators** because they form soluble complexes with cations such as iron and calcium in which the cation is held by ionic forces, rather than by covalent bonds. Chelated cations thus are physically more available to a plant.

More modern nutrient solutions use the chemicals ethylenediaminetetraacetic acid (EDTA) or diethylenetriaminepentaacetic acid (DTPA, or pentetic acid) as chelating agents (Sievers and Bailar 1962). Figure 5.2 shows the structure of DTPA. The fate of the chelation complex during iron uptake by the root cells is not clear; iron may be released from the chelator when it is reduced from Fe^{3+} to Fe^{2+} at the root surface. The chelator may then diffuse back into the nutrient (or soil) solution and react with another Fe^{3+} ion or other metal ions. After uptake, iron is kept soluble by chelation with organic compounds present in plant cells. Citric acid may play a major role in iron chelation and its long-distance transport in the xylem.

Mineral Deficiencies Disrupt Plant Metabolism and Function

Inadequate supply of an essential element results in a nutritional disorder manifested by characteristic deficiency symptoms. In hydroponic culture, withholding of an essential element can be readily correlated with a given set of symptoms for acute deficiencies. Diagnosis of soil-grown plants can be more complex, for the following reasons:

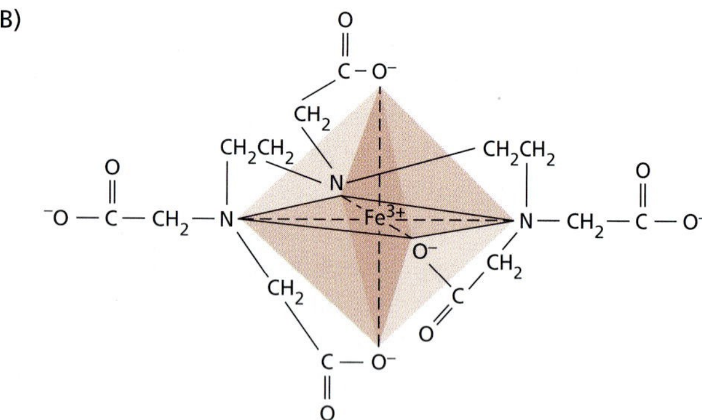

FIGURE 5.2 Chemical structure of the chelator DTPA by itself (A) and chelated to an Fe^{3+} ion (B). Iron binds to DTPA through interaction with three nitrogen atoms and the three ionized oxygen atoms of the carboxylate groups (Sievers and Bailar 1962). The resulting ring structure clamps the metallic ion and effectively neutralizes its reactivity in solution. During the uptake of iron at the root surface, Fe^{3+} appears to be reduced to Fe^{2+}, which is released from the DTPA–iron complex. The chelator can then bind to other available Fe^{3+} ions.

- Both chronic and acute deficiencies of several elements may occur simultaneously.
- Deficiencies or excessive amounts of one element may induce deficiencies or excessive accumulations of another.
- Some virus-induced plant diseases may produce symptoms similar to those of nutrient deficiencies.

Nutrient deficiency symptoms in a plant are the expression of metabolic disorders resulting from the insufficient supply of an essential element. These disorders are related to the roles played by essential elements in normal plant metabolism and function. Table 5.2 lists some of the roles of essential elements.

Even though each essential element participates in many different metabolic reactions, some general statements about the functions of essential elements in plant metabolism are possible. In general, the essential elements function in plant structure, metabolic function, and osmoregulation of plant cells. More specific roles may be related to the ability of divalent cations such as calcium or magnesium to modify the permeability of plant membranes. In addition, research continues to reveal specific roles of these elements in plant metabolism; for example, calcium acts as a signal to regulate key enzymes in the cytosol (Hepler and Wayne 1985; Sanders et al. 1999). Thus, most essential elements have multiple roles in plant metabolism.

When relating acute deficiency symptoms to a particular essential element, an important clue is the extent to which an element can be recycled from older to younger leaves. Some elements, such as nitrogen, phosphorus, and potassium, can readily move from leaf to leaf; others, such as boron, iron, and calcium, are relatively immobile in most plant species (Table 5.4). If an essential element is mobile, deficiency symptoms tend to appear first in older leaves. Deficiency of an immobile essential element will become evident first in younger leaves. Although the precise mechanisms of nutrient mobilization are not well understood,

TABLE 5.4
Mineral elements classified on the basis of their mobility within a plant and their tendency to retranslocate during deficiencies

Mobile	Immobile
Nitrogen	Calcium
Potassium	Sulfur
Magnesium	Iron
Phosphorus	Boron
Chlorine	Copper
Sodium	
Zinc	
Molybdenum	

Note: Elements are listed in the order of their abundance in the plant.

plant hormones such as cytokinins appear to be involved (see Chapter 21). In the discussion that follows, we will describe the specific deficiency symptoms and functional roles for the mineral essential elements as they are grouped in Table 5.2.

Group 1: Deficiencies in mineral nutrients that are part of carbon compounds. This first group consists of nitrogen and sulfur. Nitrogen availability in soils limits plant productivity in most natural and agricultural ecosystems. By contrast, soils generally contain sulfur in excess. Nonetheless, nitrogen and sulfur share the property that their oxidation–reduction states range widely (see Chapter 12). Some of the most energy-intensive reactions in life convert the highly oxidized, inorganic forms absorbed from the soil into the highly reduced forms found in organic compounds such as amino acids.

NITROGEN. Nitrogen is the mineral element that plants require in greatest amounts. It serves as a constituent of many plant cell components, including amino acids and nucleic acids. Therefore, nitrogen deficiency rapidly inhibits plant growth. If such a deficiency persists, most species show **chlorosis** (yellowing of the leaves), especially in the older leaves near the base of the plant (for pictures of nitrogen deficiency and the other mineral deficiencies described in this chapter, see **Web Topic 5.1**). Under severe nitrogen deficiency, these leaves become completely yellow (or tan) and fall off the plant. Younger leaves may not show these symptoms initially because nitrogen can be mobilized from older leaves. Thus a nitrogen-deficient plant may have light green upper leaves and yellow or tan lower leaves.

When nitrogen deficiency develops slowly, plants may have markedly slender and often woody stems. This woodiness may be due to a buildup of excess carbohydrates that cannot be used in the synthesis of amino acids or other nitrogen compounds. Carbohydrates not used in nitrogen metabolism may also be used in anthocyanin synthesis, leading to accumulation of that pigment. This condition is revealed as a purple coloration in leaves, petioles, and stems of some nitrogen-deficient plants, such as tomato and certain varieties of corn.

SULFUR. Sulfur is found in two amino acids and is a constituent of several coenzymes and vitamins essential for metabolism. Many of the symptoms of sulfur deficiency are similar to those of nitrogen deficiency, including chlorosis, stunting of growth, and anthocyanin accumulation. This similarity is not surprising, since sulfur and nitrogen are both constituents of proteins. However, the chlorosis caused by sulfur deficiency generally arises initially in mature and young leaves, rather than in the old leaves as in nitrogen deficiency, because unlike nitrogen, sulfur is not easily remobilized to the younger leaves in most species. Nonetheless, in many plant species sulfur chlorosis may occur simultaneously in all leaves or even initially in the older leaves.

Group 2: Deficiencies in mineral nutrients that are important in energy storage or structural integrity. This group consists of phosphorus, silicon, and boron. Phosphorus and silicon are found at concentrations within plant tissue that warrant their classification as macronutrients, whereas boron is much less abundant and considered a micronutrient. These elements are usually present in plants as ester linkages to a carbon molecule.

PHOSPHORUS. Phosphorus (as phosphate, PO_4^{3-}) is an integral component of important compounds of plant cells, including the sugar–phosphate intermediates of respiration and photosynthesis, and the phospholipids that make up plant membranes. It is also a component of nucleotides used in plant energy metabolism (such as ATP) and in DNA and RNA. Characteristic symptoms of phosphorus deficiency include stunted growth in young plants and a dark green coloration of the leaves, which may be malformed and contain small spots of dead tissue called **necrotic spots** (for a picture, see **Web Topic 5.1**).

As in nitrogen deficiency, some species may produce excess anthocyanins, giving the leaves a slight purple coloration. In contrast to nitrogen deficiency, the purple coloration of phosphorus deficiency is not associated with chlorosis. In fact, the leaves may be a dark greenish purple. Additional symptoms of phosphorus deficiency include the production of slender (but not woody) stems and the death of older leaves. Maturation of the plant may also be delayed.

SILICON. Only members of the family Equisetaceae—called *scouring rushes* because at one time their ash, rich in gritty silica, was used to scour pots—require silicon to complete their life cycle. Nonetheless, many other species accumulate substantial amounts of silicon within their tissues and show enhanced growth and fertility when supplied with adequate amounts of silicon (Epstein 1999).

Plants deficient in silicon are more susceptible to lodging (falling over) and fungal infection. Silicon is deposited primarily in the endoplasmic reticulum, cell walls, and intercellular spaces as hydrated, amorphous silica ($SiO_2 \cdot nH_2O$). It also forms complexes with polyphenols and thus serves as an alternative to lignin in the reinforcement of cell walls. In addition, silicon can ameliorate the toxicity of many heavy metals.

BORON. Although the precise function of boron in plant metabolism is unclear, evidence suggests that it plays roles in cell elongation, nucleic acid synthesis, hormone responses, and membrane function (Shelp 1993). Boron-deficient plants may exhibit a wide variety of symptoms, depending on the species and the age of the plant.

A characteristic symptom is black necrosis of the young leaves and terminal buds. The necrosis of the young leaves occurs primarily at the base of the leaf blade. Stems may be unusually stiff and brittle. Apical dominance may also be lost, causing the plant to become highly branched; however, the terminal apices of the branches soon become necrotic because of inhibition of cell division. Structures such as the fruit, fleshy roots, and tubers may exhibit necrosis or abnormalities related to the breakdown of internal tissues.

Group 3: Deficiencies in mineral nutrients that remain in ionic form. This group includes some of the most familiar mineral elements: The macronutrients potassium, calcium, and magnesium, and the micronutrients chlorine, manganese, and sodium. They may be found in solution in the cytosol or vacuoles, or they may be bound electrostatically or as ligands to larger carbon-containing compounds.

POTASSIUM. Potassium, present within plants as the cation K^+, plays an important role in regulation of the osmotic potential of plant cells (see Chapters 3 and 6). It also activates many enzymes involved in respiration and photosynthesis. The first observable symptom of potassium deficiency is mottled or marginal chlorosis, which then develops into necrosis primarily at the leaf tips, at the margins, and between veins. In many monocots, these necrotic lesions may initially form at the leaf tips and margins and then extend toward the leaf base.

Because potassium can be mobilized to the younger leaves, these symptoms appear initially on the more mature leaves toward the base of the plant. The leaves may also curl and crinkle. The stems of potassium-deficient plants may be slender and weak, with abnormally short internodal regions. In potassium-deficient corn, the roots may have an increased susceptibility to root-rotting fungi present in the soil, and this susceptibility, together with effects on the stem, results in an increased tendency for the plant to be easily bent to the ground (lodging).

CALCIUM. Calcium ions (Ca^{2+}) are used in the synthesis of new cell walls, particularly the middle lamellae that separate newly divided cells. Calcium is also used in the mitotic spindle during cell division. It is required for the normal functioning of plant membranes and has been implicated as a second messenger for various plant responses to both environmental and hormonal signals (Sanders et al. 1999). In its function as a second messenger, calcium may bind to **calmodulin**, a protein found in the cytosol of plant cells. The calmodulin–calcium complex regulates many metabolic processes.

Characteristic symptoms of calcium deficiency include necrosis of young meristematic regions, such as the tips of roots or young leaves, where cell division and wall formation are most rapid. Necrosis in slowly growing plants may be preceded by a general chlorosis and downward hooking of the young leaves. Young leaves may also appear deformed. The root system of a calcium-deficient plant may appear brownish, short, and highly branched. Severe stunting may result if the meristematic regions of the plant die prematurely.

MAGNESIUM. In plant cells, magnesium ions (Mg^{2+}) have a specific role in the activation of enzymes involved in respiration, photosynthesis, and the synthesis of DNA and RNA. Magnesium is also a part of the ring structure of the chlorophyll molecule (see Figure 7.6A). A characteristic symptom of magnesium deficiency is chlorosis between the leaf veins, occurring first in the older leaves because of the mobility of this element. This pattern of chlorosis results because the chlorophyll in the vascular bundles remains unaffected for longer periods than the chlorophyll in the cells between the bundles does. If the deficiency is extensive, the leaves may become yellow or white. An additional symptom of magnesium deficiency may be premature leaf abscission.

CHLORINE. The element chlorine is found in plants as the chloride ion (Cl^-). It is required for the water-splitting reaction of photosynthesis through which oxygen is produced (see Chapter 7) (Clarke and Eaton-Rye 2000). In addition, chlorine may be required for cell division in both leaves and roots (Harling et al. 1997). Plants deficient in chlorine develop wilting of the leaf tips followed by general leaf chlorosis and necrosis. The leaves may also exhibit reduced growth. Eventually, the leaves may take on a bronzelike color ("bronzing"). Roots of chlorine-deficient plants may appear stunted and thickened near the root tips.

Chloride ions are very soluble and generally available in soils because seawater is swept into the air by wind and is delivered to soil when it rains. Therefore, chlorine deficiency is unknown in plants grown in native or agricultural habitats. Most plants generally absorb chlorine at levels much higher than those required for normal functioning.

MANGANESE. Manganese ions (Mn^{2+}) activate several enzymes in plant cells. In particular, decarboxylases and dehydrogenases involved in the tricarboxylic acid (Krebs) cycle are specifically activated by manganese. The best-defined function of manganese is in the photosynthetic reaction through which oxygen is produced from water (Marschner 1995). The major symptom of manganese deficiency is intervenous chlorosis associated with the development of small necrotic spots. This chlorosis may occur on younger or older leaves, depending on plant species and growth rate.

SODIUM. Most species utilizing the C_4 and CAM pathways of carbon fixation (see Chapter 8) require sodium ions (Na^+). In these plants, sodium appears vital for regenerating phosphoenolpyruvate, the substrate for the first carboxylation in the C_4 and CAM pathways (Johnstone et al. 1988). Under sodium deficiency, these plants exhibit chlorosis and necrosis, or even fail to form flowers. Many C_3 species also benefit from exposure to low levels of sodium ions. Sodium stimulates growth through enhanced cell expansion, and it can partly substitute for potassium as an osmotically active solute.

Group 4: Deficiencies in mineral nutrients that are involved in redox reactions. This group of five micronutrients includes the metals iron, zinc, copper, nickel, and molybdenum. All of these can undergo reversible oxidations and reductions (e.g., $Fe^{2+} \rightleftharpoons Fe^{3+}$) and have important roles in electron transfer and energy transformation. They are usually found in association with larger molecules such as cytochromes, chlorophyll, and proteins (usually enzymes).

IRON. Iron has an important role as a component of enzymes involved in the transfer of electrons (redox reactions), such as cytochromes. In this role, it is reversibly oxidized from Fe^{2+} to Fe^{3+} during electron transfer. As in magnesium deficiency, a characteristic symptom of iron deficiency is intervenous chlorosis. In contrast to magnesium deficiency symptoms, these symptoms appear initially on the younger leaves because iron cannot be readily mobilized from older leaves. Under conditions of extreme or prolonged deficiency, the veins may also become chlorotic, causing the whole leaf to turn white.

The leaves become chlorotic because iron is required for the synthesis of some of the chlorophyll–protein complexes in the chloroplast. The low mobility of iron is probably due to its precipitation in the older leaves as insoluble oxides or phosphates or to the formation of complexes with phytoferritin, an iron-binding protein found in the leaf and other plant parts (Oh et al. 1996). The precipitation of iron diminishes subsequent mobilization of the metal into the phloem for long-distance translocation.

ZINC. Many enzymes require zinc ions (Zn^{2+}) for their activity, and zinc may be required for chlorophyll biosynthesis in some plants. Zinc deficiency is characterized by a reduction in internodal growth, and as a result plants display a rosette habit of growth in which the leaves form a circular cluster radiating at or close to the ground. The leaves may also be small and distorted, with leaf margins having a puckered appearance. These symptoms may result from loss of the capacity to produce sufficient amounts of the auxin indoleacetic acid. In some species (corn, sorghum, beans), the older leaves may become intervenously chlorotic and then develop white necrotic spots. This chlorosis may be an expression of a zinc requirement for chlorophyll biosynthesis.

COPPER. Like iron, copper is associated with enzymes involved in redox reactions being reversibly oxidized from Cu^+ to Cu^{2+}. An example of such an enzyme is plastocyanin, which is involved in electron transfer during the light reactions of photosynthesis (Haehnel 1984). The initial symptom of copper deficiency is the production of dark green leaves, which may contain necrotic spots. The necrotic spots appear first at the tips of the young leaves and then extend toward the leaf base along the margins. The leaves may also be twisted or malformed. Under extreme copper deficiency, leaves may abscise prematurely.

NICKEL. Urease is the only known nickel-containing enzyme in higher plants, although nitrogen-fixing microorganisms require nickel for the enzyme that reprocesses some of the hydrogen gas generated during fixation (hydrogen uptake hydrogenase) (see Chapter 12). Nickel-deficient plants accumulate urea in their leaves and, consequently, show leaf tip necrosis. Plants grown in soil seldom, if ever, show signs of nickel deficiency because the amounts of nickel required are minuscule.

MOLYBDENUM. Molybdenum ions (Mo^{4+} through Mo^{6+}) are components of several enzymes, including nitrate reductase and nitrogenase. Nitrate reductase catalyzes the reduction of nitrate to nitrite during its assimilation by the plant cell; nitrogenase converts nitrogen gas to ammonia in nitrogen-fixing microorganisms (see Chapter 12). The first indication of a molybdenum deficiency is general chlorosis between veins and necrosis of the older leaves. In some plants, such as cauliflower or broccoli, the leaves may not become necrotic but instead may appear twisted and subsequently die (whiptail disease). Flower formation may be prevented, or the flowers may abscise prematurely.

Because molybdenum is involved with both nitrate assimilation and nitrogen fixation, a molybdenum deficiency may bring about a nitrogen deficiency if the nitrogen source is primarily nitrate or if the plant depends on symbiotic nitrogen fixation. Although plants require only small amounts of molybdenum, some soils supply inadequate levels. Small additions of molybdenum to such soils can greatly enhance crop or forage growth at negligible cost.

Analysis of Plant Tissues Reveals Mineral Deficiencies

Requirements for mineral elements change during the growth and development of a plant. In crop plants, nutrient levels at certain stages of growth influence the yield of the economically important tissues (tuber, grain, and so on). To optimize yields, farmers use analyses of nutrient levels in soil and in plant tissue to determine fertilizer schedules.

Soil analysis is the chemical determination of the nutrient content in a soil sample from the root zone. As discussed later in the chapter, both the chemistry and the biology of soils are complex, and the results of soil analyses vary with sampling methods, storage conditions for the samples, and nutrient extraction techniques. Perhaps more important is that a particular soil analysis reflects the levels of nutrients *potentially* available to the plant roots from the soil, but soil analysis does not tell us how much of a particular mineral nutrient the plant actually needs or is able to absorb. This additional information is best determined by plant tissue analysis.

Proper use of **plant tissue analysis** requires an understanding of the relationship between plant growth (or yield) and the mineral concentration of plant tissue samples (Bouma 1983). As the data plot in Figure 5.3 shows, when the nutrient concentration in a tissue sample is low, growth is reduced. In this **deficiency zone** of the curve, an increase in nutrient availability is directly related to an increase in growth or yield. As the nutrient availability continues to increase, a point is reached at which further addition of nutrients is no longer related to increases in growth or yield but is reflected in increased tissue concentrations. This region of the curve is often called the **adequate zone**.

The transition between the deficiency and adequate zones of the curve reveals the **critical concentration** of the nutrient (see Figure 5.3), which may be defined as the minimum tissue content of the nutrient that is correlated with maximal growth or yield. As the nutrient concentration of the tissue increases beyond the adequate zone, growth or yield declines because of toxicity (this is the **toxic zone**).

To evaluate the relationship between growth and tissue nutrient concentration, researchers grow plants in soil or nutrient solution in which all the nutrients are present in adequate amounts except the nutrient under consideration. At the start of the experiment, the limiting nutrient is added in increasing concentrations to different sets of plants, and the concentrations of the nutrient in specific tissues are correlated with a particular measure of growth or yield. Several curves are established for each element, one for each tissue and tissue age.

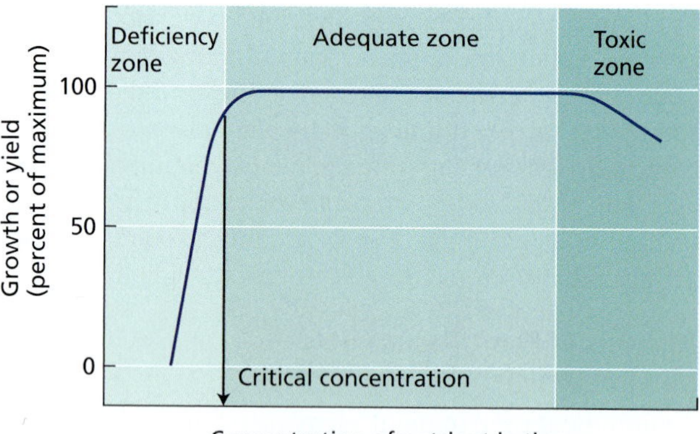

FIGURE 5.3 Relationship between yield (or growth) and the nutrient content of the plant tissue. The yield parameter may be expressed in terms of shoot dry weight or height. Three zones—deficiency, adequate, and toxic—are indicated on the graph. To yield data of this type, plants are grown under conditions in which the concentration of one essential nutrient is varied while all others are in adequate supply. The effect of varying the concentration of this nutrient during plant growth is reflected in the growth or yield. The critical concentration for that nutrient is the concentration below which yield or growth is reduced.

Because agricultural soils are often limited in the elements nitrogen, phosphorus, and potassium, many farmers routinely use, at a minimum, curves for these elements. If a nutrient deficiency is suspected, steps are taken to correct the deficiency before it reduces growth or yield. Plant analysis has proven useful in establishing fertilizer schedules that sustain yields and ensure the food quality of many crops.

TREATING NUTRITIONAL DEFICIENCIES

Many traditional and subsistence farming practices promote the recycling of mineral elements. Crop plants absorb the nutrients from the soil, humans and animals consume locally grown crops, and crop residues and manure from humans and animals return the nutrients to the soil. The main losses of nutrients from such agricultural systems ensue from leaching that carries dissolved ions away with drainage water. In acid soils, leaching may be decreased by the addition of lime—a mix of CaO, $CaCO_3$, and $Ca(OH)_2$—to make the soil more alkaline because many mineral elements form less soluble compounds when the pH is higher than 6 (Figure 5.4).

In the high-production agricultural systems of industrial countries, the unidirectional removal of nutrients from the soil to the crop can become significant because a large portion of crop biomass leaves the area of cultivation. Plants synthesize all their components from basic inorganic substances and sunlight, so it is important to restore these lost nutrients to the soil through the addition of fertilizers.

Crop Yields Can Be Improved by Addition of Fertilizers

Most chemical fertilizers contain inorganic salts of the macronutrients nitrogen, phosphorus, and potassium (see Table 5.1). Fertilizers that contain only one of these three nutrients are termed **straight fertilizers**. Some examples of straight fertilizers are superphosphate, ammonium nitrate, and muriate of potash (a source of potassium). Fertilizers that contain two or more mineral nutrients are called **compound fertilizers** or **mixed fertilizers**, and the numbers on the package label, such as 10-14-10, refer to the effective percentages of N, P_2O_5, and K_2O, respectively, in the fertilizer.

With long-term agricultural production, consumption of micronutrients can reach a point at which they, too, must be added to the soil as fertilizers. Adding micronutrients to the soil may also be necessary to correct a preexisting deficiency. For example, some soils in the United States are

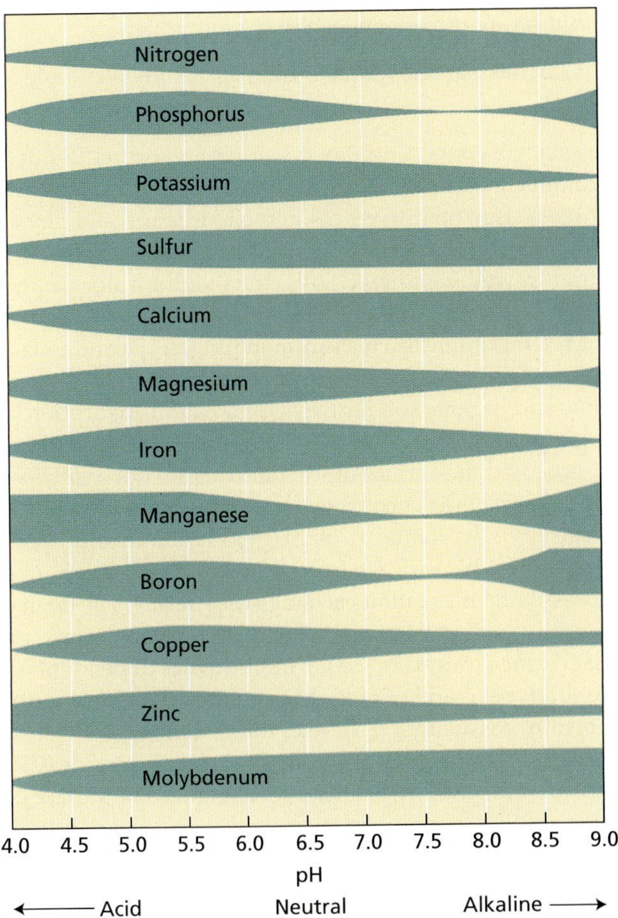

FIGURE 5.4 Influence of soil pH on the availability of nutrient elements in organic soils. The width of the shaded areas indicates the degree of nutrient availability to the plant root. All of these nutrients are available in the pH range of 5.5 to 6.5. (From Lucas and Davis 1961.)

deficient in boron, copper, zinc, manganese, molybdenum, or iron (Mengel and Kirkby 1987) and can benefit from nutrient supplementation.

Chemicals may also be applied to the soil to modify soil pH. As Figure 5.4 shows, soil pH affects the availability of all mineral nutrients. Addition of lime, as mentioned previously, can raise the pH of acidic soils; addition of elemental sulfur can lower the pH of alkaline soils. In the latter case, microorganisms absorb the sulfur and subsequently release sulfate and hydrogen ions that acidify the soil.

Organic fertilizers, in contrast to chemical fertilizers, originate from the residues of plant or animal life or from natural rock deposits. Plant and animal residues contain many of the nutrient elements in the form of organic compounds. Before crop plants can acquire the nutrient elements from these residues, the organic compounds must be broken down, usually by the action of soil microorganisms through a process called **mineralization**. Mineralization depends on many factors, including temperature, water and oxygen availability, and the type and number of microorganisms present in the soil.

As a consequence, the rate of mineralization is highly variable, and nutrients from organic residues become available to plants over periods that range from days to months to years. The slow rate of mineralization hinders efficient fertilizer use, so farms that rely solely on organic fertilizers may require the addition of substantially more nitrogen or phosphorus and suffer even higher nutrient losses than farms that use chemical fertilizers. Residues from organic fertilizers do improve the physical structure of most soils, enhancing water retention during drought and increasing drainage in wet weather.

Some Mineral Nutrients Can Be Absorbed by Leaves

In addition to nutrients being added to the soil as fertilizers, some mineral nutrients can be applied to the leaves as sprays, in a process known as **foliar application**, and the leaves can absorb the applied nutrients. In some cases, this method can have agronomic advantages over the application of nutrients to the soil. Foliar application can reduce the lag time between application and uptake by the plant, which could be important during a phase of rapid growth. It can also circumvent the problem of restricted uptake of a nutrient from the soil. For example, foliar application of mineral nutrients such as iron, manganese, and copper may be more efficient than application through the soil, where they are adsorbed on soil particles and hence are less available to the root system.

Nutrient uptake by plant leaves is most effective when the nutrient solution remains on the leaf as a thin film (Mengel and Kirkby 1987). Production of a thin film often requires that the nutrient solutions be supplemented with surfactant chemicals, such as the detergent Tween 80, that reduce surface tension. Nutrient movement into the plant seems to involve diffusion through the cuticle and uptake by leaf cells. Although uptake through the stomatal pore could provide a pathway into the leaf, the architecture of the pore (see Figures 4.13 and 4.14) largely prevents liquid penetration (Ziegler 1987).

For foliar nutrient application to be successful, damage to the leaves must be minimized. If foliar sprays are applied on a hot day, when evaporation is high, salts may accumulate on the leaf surface and cause burning or scorching. Spraying on cool days or in the evening helps to alleviate this problem. Addition of lime to the spray diminishes the solubility of many nutrients and limits toxicity. Foliar application has proved economically successful mainly with tree crops and vines such as grapes, but it is also used with cereals. Nutrients applied to the leaves could save an orchard or vineyard when soil-applied nutrients would be too slow to correct a deficiency. In wheat, nitrogen applied to the leaves during the later stages of growth enhances the protein content of seeds.

SOIL, ROOTS, AND MICROBES

The soil is a complex physical, chemical, and biological substrate. It is a heterogeneous material containing solid, liquid, and gaseous phases (see Chapter 4). All of these phases interact with mineral elements. The inorganic particles of the solid phase provide a reservoir of potassium, calcium, magnesium, and iron. Also associated with this solid phase are organic compounds containing nitrogen, phosphorus, and sulfur, among other elements. The liquid phase of the soil constitutes the soil solution, which contains dissolved mineral ions and serves as the medium for ion movement to the root surface. Gases such as oxygen, carbon dioxide, and nitrogen are dissolved in the soil solution, but in roots gases are exchanged predominantly through the air gaps between soil particles.

From a biological perspective, soil constitutes a diverse ecosystem in which plant roots and microorganisms compete strongly for mineral nutrients. In spite of this competition, roots and microorganisms can form alliances for their mutual benefit (**symbioses**, singular *symbiosis*). In this section we will discuss the importance of soil properties, root structure, and mycorrhizal symbiotic relationships to plant mineral nutrition. Chapter 12 addresses symbiotic relationships with nitrogen-fixing bacteria.

Negatively Charged Soil Particles Affect the Adsorption of Mineral Nutrients

Soil particles, both inorganic and organic, have predominantly negative charges on their surfaces. Many inorganic soil particles are crystal lattices that are tetrahedral arrangements of the cationic forms of aluminum and silicon (Al^{3+} and Si^{4+}) bound to oxygen atoms, thus forming aluminates and silicates. When cations of lesser charge replace Al^{3+} and Si^{4+}, inorganic soil particles become negatively charged.

Organic soil particles originate from the products of the microbial decomposition of dead plants, animals, and microorganisms. The negative surface charges of organic particles result from the dissociation of hydrogen ions from the carboxylic acid and phenolic groups present in this component of the soil. Most of the world's soil particles, however, are inorganic.

Inorganic soils are categorized by particle size:

- Gravel has particles larger than 2 mm.
- Coarse sand has particles between 0.2 and 2 mm.
- Fine sand has particles between 0.02 and 0.2 mm.
- Silt has particles between 0.002 and 0.02 mm.
- Clay has particles smaller than 0.002 mm (see Table 4.1).

The silicate-containing clay materials are further divided into three major groups—kaolinite, illite, and montmorillonite—based on differences in their structure and physical properties (Table 5.5). The kaolinite group is generally found in well-weathered soils; the montmorillonite and illite groups are found in less weathered soils.

Mineral cations such as ammonium (NH_4^+) and potassium (K^+) adsorb to the negative surface charges of inorganic and organic soil particles. This cation adsorption is an important factor in soil fertility. Mineral cations adsorbed on the surface of soil particles are not easily lost when the soil is leached by water, and they provide a nutrient reserve available to plant roots. Mineral nutrients adsorbed in this way can be replaced by other cations in a process known as **cation exchange** (Figure 5.5). The degree to which a soil can adsorb and exchange ions is termed its *cation exchange capacity* (*CEC*) and is highly dependent on the soil type. A soil with higher cation exchange capacity generally has a larger reserve of mineral nutrients.

Mineral anions such as nitrate (NO_3^-) and chloride (Cl^-) tend to be repelled by the negative charge on the surface of soil particles and remain dissolved in the soil solution. Thus the anion exchange capacity of most agricultural soils is small compared to the cation exchange capacity. Among anions, nitrate remains mobile in the soil solution, where it is susceptible to leaching by water moving through the soil.

Phosphate ions ($H_2PO_2^-$) may bind to soil particles containing aluminum or iron because the positively charged iron and aluminum ions (Fe^{2+}, Fe^{3+}, and Al^{3+}) have hydroxyl (OH^-) groups that exchange with phosphate. As a result, phosphate can be tightly bound, and its mobility and availability in soil can limit plant growth.

Sulfate (SO_4^{2-}) in the presence of calcium (Ca^{2+}) forms gypsum ($CaSO_4$). Gypsum is only slightly soluble, but it releases sufficient sulfate to support plant growth. Most

TABLE 5.5
Comparison of properties of three major types of silicate clays found in the soil

Property	Type of clay: Montmorillonite	Illite	Kaolinite
Size (μm)	0.01–1.0	0.1–2.0	0.1–5.0
Shape	Irregular flakes	Irregular flakes	Hexagonal crystals
Cohesion	High	Medium	Low
Water-swelling capacity	High	Medium	Low
Cation exchange capacity (milliequivalents 100 g^{-1})	80–100	15–40	3–15

Source: After Brady 1974.

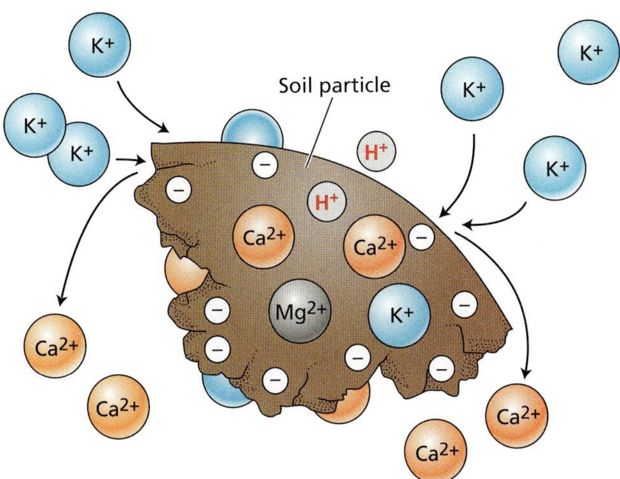

FIGURE 5.5 The principle of cation exchange on the surface of a soil particle. Cations are bound to the surface of soil particles because the surface is negatively charged. Addition of a cation such as potassium (K^+) can displace another cation such as calcium (Ca^{2+}) from its binding on the surface of the soil particle and make it available for uptake by the root.

nonacid soils contain substantial amounts of calcium; consequently, sulfate mobility in these soils is low, so sulfate is not highly susceptible to leaching.

Soil pH Affects Nutrient Availability, Soil Microbes, and Root Growth

Hydrogen ion concentration (pH) is an important property of soils because it affects the growth of plant roots and soil microorganisms. Root growth is generally favored in slightly acidic soils, at pH values between 5.5 and 6.5. Fungi generally predominate in acidic soils; bacteria become more prevalent in alkaline soils. Soil pH determines the availability of soil nutrients (see Figure 5.4). Acidity promotes the weathering of rocks that releases K^+, Mg^{2+}, Ca^{2+}, and Mn^{2+} and increases the solubility of carbonates, sulfates, and phosphates. Increasing the solubility of nutrients facilitates their availability to roots.

Major factors that lower the soil pH are the decomposition of organic matter and the amount of rainfall. Carbon dioxide is produced as a result of the decomposition of organic material and equilibrates with soil water in the following reaction:

$$CO_2 + H_2O \rightleftharpoons H^+ + HCO_3^-$$

This reaction releases hydrogen ions (H^+), lowering the pH of the soil. Microbial decomposition of organic material also produces ammonia and hydrogen sulfide that can be oxidized in the soil to form the strong acids nitric acid (HNO_3) and sulfuric acid (H_2SO_4), respectively. Hydrogen ions also displace K^+, Mg^{2+}, Ca^{2+}, and Mn^{2+} from the cation exchange complex in a soil. Leaching then may remove these ions from the upper soil layers, leaving a more acid soil. By contrast, the weathering of rock in arid regions releases K^+, Mg^{2+}, Ca^{2+}, and Mn^{2+} to the soil, but because of the low rainfall, these ions do not leach from the upper soil layers, and the soil remains alkaline.

Excess Minerals in the Soil Limit Plant Growth

When excess minerals are present in the soil, the soil is said to be saline, and plant growth may be restricted if these mineral ions reach levels that limit water availability or exceed the adequate zone for a particular nutrient (see Chapter 25). Sodium chloride and sodium sulfate are the most common salts in saline soils. Excess minerals in soils can be a major problem in arid and semiarid regions because rainfall is insufficient to leach the mineral ions from the soil layers near the surface. Irrigated agriculture fosters soil salinization if insufficient water is applied to leach the salt below the rooting zone. Irrigation water can contain 100 to 1000 g of minerals per cubic meter. An average crop requires about 4000 m^3 of water per acre. Consequently, 400 to 4000 kg of minerals may be added to the soil per crop (Marschner 1995).

In saline soil, plants encounter **salt stress**. Whereas many plants are affected adversely by the presence of relatively low levels of salt, other plants can survive high levels (**salt-tolerant plants**) or even thrive (**halophytes**) under such conditions. The mechanisms by which plants tolerate salinity are complex (see Chapter 25), involving molecular synthesis, enzyme induction, and membrane transport. In some species, excess minerals are not taken up; in others, minerals are taken up but excreted from the plant by salt glands associated with the leaves. To prevent toxic buildup of mineral ions in the cytosol, many plants may sequester them in the vacuole (Stewart and Ahmad 1983). Efforts are under way to bestow salt tolerance on salt-sensitive crop species using both classic plant breeding and molecular biology (Hasegawa et al. 2000).

Another important problem with excess minerals is the accumulation of heavy metals in the soil, which can cause severe toxicity in plants as well as humans (see **Web Essay 5.1**). Heavy metals include zinc, copper, cobalt, nickel, mercury, lead, cadmium, silver, and chromium (Berry and Wallace 1981).

Plants Develop Extensive Root Systems

The ability of plants to obtain both water and mineral nutrients from the soil is related to their capacity to develop an extensive root system. In the late 1930s, H. J. Dittmer examined the root system of a single winter rye plant after 16 weeks of growth and estimated that the plant had 13×10^6 primary and lateral root axes, extending more than 500 km in length and providing 200 m^2 of surface area (Dittmer 1937). This plant also had more than 10^{10} root hairs, providing another 300 m^2 of surface area.

In the desert, the roots of mesquite (genus *Prosopis*) may extend down more than 50 m to reach groundwater. Annual crop plants have roots that usually grow between 0.1 and 2.0 m in depth and extend laterally to distances of 0.3 to 1.0 m. In orchards, the major root systems of trees planted 1 m apart reach a total length of 12 to 18 km per tree. The annual production of roots in natural ecosystems may easily surpass that of shoots, so in many respects, the aboveground portions of a plant represent only "the tip of an iceberg."

Plant roots may grow continuously throughout the year. Their proliferation, however, depends on the availability of water and minerals in the immediate microenvironment surrounding the root, the so-called **rhizosphere**. If the rhizosphere is poor in nutrients or too dry, root growth is slow. As rhizosphere conditions improve, root growth increases. If fertilization and irrigation provide abundant nutrients and water, root growth may not keep pace with shoot growth. Plant growth under such conditions becomes carbohydrate limited, and a relatively small root system meets the nutrient needs of the whole plant (Bloom et al. 1993). Roots growing below the soil surface are studied by special techniques (see **Web Topic 5.2**).

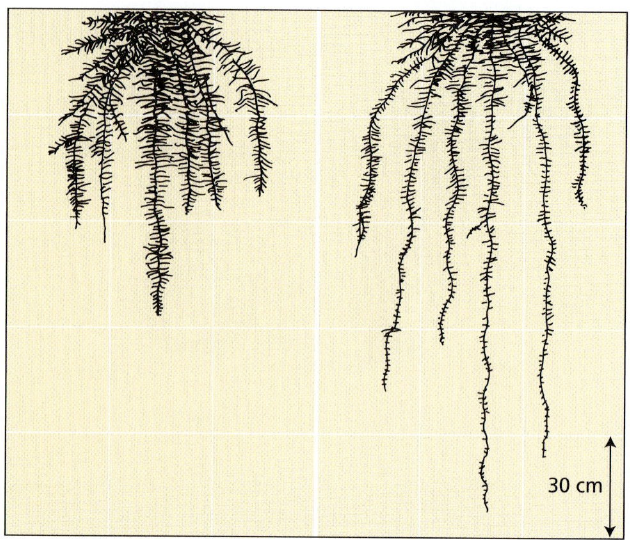

FIGURE 5.6 Fibrous root systems of wheat (a monocot). (A) The root system of a mature (3-month-old) wheat plant growing in dry soil. (B) The root system of a wheat plant growing in irrigated soil. It is apparent that the morphology of the root system is affected by the amount of water present in the soil. In a fibrous root system, the primary root axes are no longer distinguishable. (After Weaver 1926.)

Root Systems Differ in Form but Are Based on Common Structures

The *form* of the root system differs greatly among plant species. In monocots, root development starts with the emergence of three to six **primary** (or seminal) root axes from the germinating seed. With further growth, the plant extends new adventitious roots, called **nodal roots** or *brace roots*. Over time, the primary and nodal root axes grow and branch extensively to form a complex fibrous root system (Figure 5.6). In fibrous root systems, all the roots generally have the same diameter (except where environmental conditions or pathogenic interactions modify the root structure), so it is difficult to distinguish a main root axis.

In contrast to monocots, dicots develop root systems with a main single root axis, called a **taproot**, which may thicken as a result of secondary cambial activity. From this main root axis, lateral roots develop to form an extensively branched root system (Figure 5.7).

The development of the root system in both monocots and dicots depends on the activity of the root apical meristem and the production of lateral root meristems. Figure 5.8 shows a generalized diagram of the apical region of a plant root and identifies the three zones of activity: meristematic, elongation, and maturation.

In the **meristematic zone**, cells divide both in the direction of the root base to form cells that will differentiate into the tissues of the functional root and in the direction of the root apex to form the **root cap**. The root cap protects the delicate meristematic cells as the root moves through the soil. It also secretes a gelatinous material called *mucigel*, which commonly surrounds the root tip. The precise function of the mucigel is uncertain, but it has been suggested that it lubricates the penetration of the root through the soil, protects the root apex from desiccation, promotes the transfer of nutrients to the root, or affects the interaction between roots and soil microorganisms (Russell 1977). The root cap is central to the perception of gravity, the signal that directs the growth of roots downward. This process is termed the **gravitropic response** (see Chapter 19).

Cell division at the root apex proper is relatively slow; thus this region is called the **quiescent center**. After a few generations of slow cell divisions, root cells displaced from the apex by about 0.1 mm begin to divide more rapidly. Cell division again tapers off at about 0.4 mm from the apex, and the cells expand equally in all directions.

The **elongation zone** begins 0.7 to 1.5 mm from the apex (see Figure 5.8). In this zone, cells elongate rapidly and undergo a final round of divisions to produce a central ring of cells called the **endodermis**. The walls of this endodermal cell layer become thickened, and suberin (see Chapter 13) deposited on the radial walls forms the **Casparian strip**, a hydrophobic structure that prevents the apoplastic movement of water or solutes across the root (see Figure 4.3). The endodermis divides the root into two regions: the **cortex** toward the outside and the **stele** toward the inside. The stele contains the vascular elements of the root: the **phloem**, which transports metabolites from the shoot to the root, and the **xylem**, which transports water and solutes to the shoot.

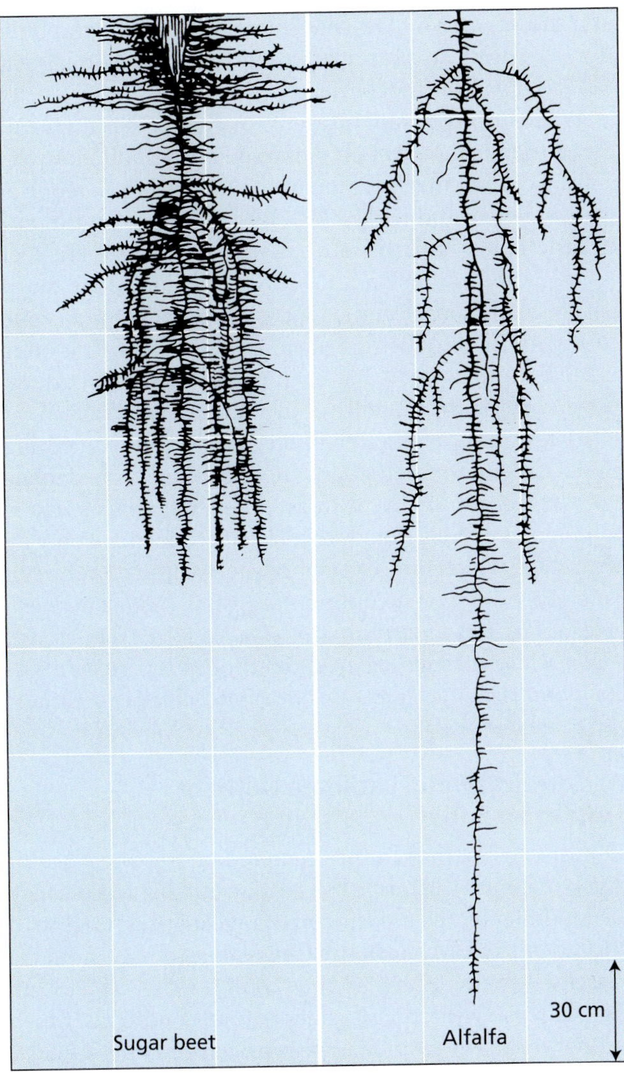

FIGURE 5.7 Taproot system of two adequately watered dicots: sugar beet and alfalfa. The sugar beet root system is typical of 5 months of growth; the alfalfa root system is typical of 2 years of growth. In both dicots, the root system shows a major vertical root axis. In the case of sugar beet, the upper portion of the taproot system is thickened because of its function as storage tissue. (After Weaver 1926.)

Phloem develops more rapidly than xylem, attesting to the fact that phloem function is critical near the root apex. Large quantities of carbohydrates must flow through the phloem to the growing apical zones in order to support cell division and elongation. Carbohydrates provide rapidly growing cells with an energy source and with the carbon skeletons required to synthesize organic compounds. Six-carbon sugars (hexoses) also function as osmotically active solutes in the root tissue. At the root apex, where the phloem is not yet developed, carbohydrate movement depends on symplastic diffusion and is relatively slow (Bret-Harte and Silk 1994). The low rates of cell division in the quiescent center may result from the fact that insufficient carbohydrates reach this centrally located region or that this area is kept in an oxidized state (see **Web Essay 5.2**).

Root hairs, with their large surface area for absorption of water and solutes, first appear in the **maturation zone** (see Figure 5.8), and it is here that the xylem develops the capacity to translocate substantial quantities of water and solutes to the shoot.

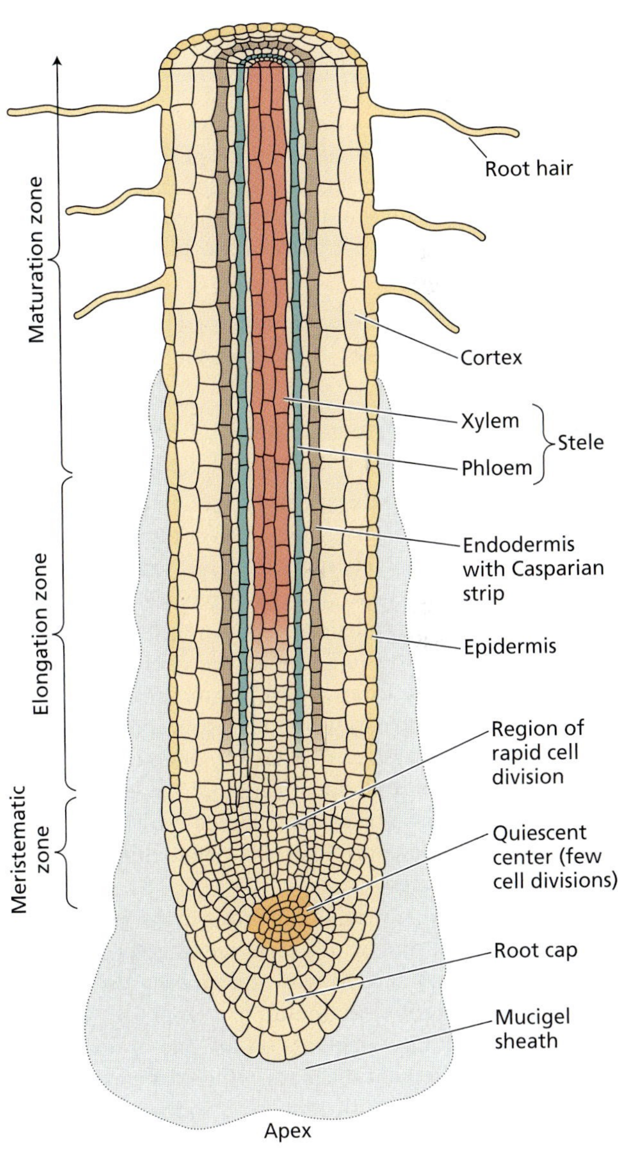

FIGURE 5.8 Diagrammatic longitudinal section of the apical region of the root. The meristematic cells are located near the tip of the root. These cells generate the root cap and the upper tissues of the root. In the elongation zone, cells differentiate to produce xylem, phloem, and cortex. Root hairs, formed in epidermal cells, first appear in the maturation zone.

Different Areas of the Root Absorb Different Mineral Ions

The precise point of entry of minerals into the root system has been a topic of considerable interest. Some researchers have claimed that nutrients are absorbed only at the apical regions of the root axes or branches (Bar-Yosef et al. 1972); others claim that nutrients are absorbed over the entire root surface (Nye and Tinker 1977). Experimental evidence supports both possibilities, depending on the plant species and the nutrient being investigated:

- Root absorption of calcium in barley appears to be restricted to the apical region.
- Iron may be taken up either at the apical region, as in barley (Clarkson 1985), or over the entire root surface, as in corn (Kashirad et al. 1973).
- Potassium, nitrate, ammonium, and phosphate can be absorbed freely at all locations of the root surface (Clarkson 1985), but in corn the elongation zone has the maximum rates of potassium accumulation (Sharp et al. 1990) and nitrate absorption (Taylor and Bloom 1998).
- In corn and rice, the root apex absorbs ammonium more rapidly than the elongation zone does (Colmer and Bloom 1998).
- In several species, root hairs are the most active in phosphate absorption (Fohse et al. 1991).

The high rates of nutrient absorption in the apical root zones result from the strong demand for nutrients in these tissues and the relatively high nutrient availability in the soil surrounding them. For example, cell elongation depends on the accumulation of solutes such as potassium, chloride, and nitrate to increase the osmotic pressure within the cell (see Chapter 15). Ammonium is the preferred nitrogen source to support cell division in the meristem because meristematic tissues are often carbohydrate limited, and the assimilation of ammonium consumes less energy than that of nitrate (see Chapter 12). The root apex and root hairs grow into fresh soil, where nutrients have not yet been depleted.

Within the soil, nutrients can move to the root surface both by bulk flow and by diffusion (see Chapter 3). In bulk flow, nutrients are carried by water moving through the soil toward the root. The amount of nutrient provided to the root by bulk flow depends on the rate of water flow through the soil toward the plant, which depends on transpiration rates and on nutrient levels in the soil solution. When both the rate of water flow and the concentrations of nutrients in the soil solution are high, bulk flow can play an important role in nutrient supply.

In diffusion, mineral nutrients move from a region of higher concentration to a region of lower concentration. Nutrient uptake by the roots lowers the concentration of nutrients at the root surface, generating concentration gradients in the soil solution surrounding the root. Diffusion of nutrients down their concentration gradient and bulk flow resulting from transpiration can increase nutrient availability at the root surface.

When absorption of nutrients by the roots is high and the nutrient concentration in the soil is low, bulk flow can supply only a small fraction of the total nutrient requirement (Mengel and Kirkby 1987). Under these conditions, diffusion rates limit the movement of nutrients to the root surface. When diffusion is too slow to maintain high nutrient concentrations near the root, a **nutrient depletion zone** forms adjacent to the root surface (Figure 5.9). This zone extends from about 0.2 to 2.0 mm from the root surface, depending on the mobility of the nutrient in the soil.

The formation of a depletion zone tells us something important about mineral nutrition: Because roots deplete the mineral supply in the rhizosphere, their effectiveness in mining minerals from the soil is determined not only by the rate at which they can remove nutrients from the soil solution, but by their continuous growth. *Without growth, roots would rapidly deplete the soil adjacent to their surface. Optimal nutrient acquisition therefore depends both on the capacity for nutrient uptake and on the ability of the root system to grow into fresh soil.*

Mycorrhizal Fungi Facilitate Nutrient Uptake by Roots

Our discussion thus far has centered on the direct acquisition of mineral elements by the root, but this process may be modified by the association of mycorrhizal fungi with the root system. **Mycorrhizae** (singular *mycorrhiza*, from the Greek words for "fungus" and "root") are not unusual; in fact, they are widespread under natural conditions. Much of the world's vegetation appears to have roots associated

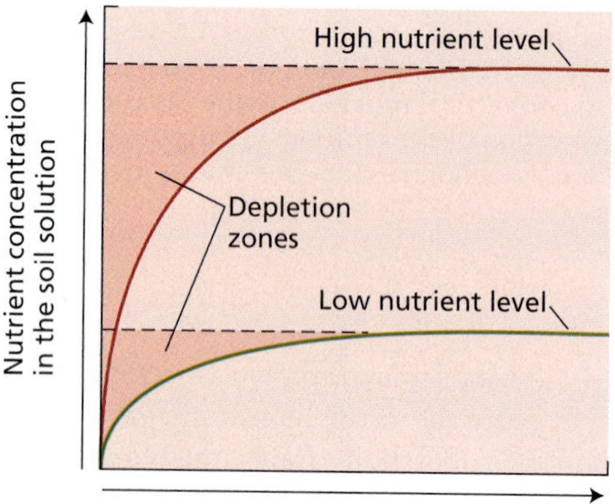

FIGURE 5.9 Formation of a nutrient depletion zone in the region of the soil adjacent to the plant root. A nutrient depletion zone forms when the rate of nutrient uptake by the cells of the root exceeds the rate of replacement of the nutrient by diffusion in the soil solution. This depletion causes a localized decrease in the nutrient concentration in the area adjacent to the root surface. (After Mengel and Kirkby 1987.)

with mycorrhizal fungi: 83% of dicots, 79% of monocots, and all gymnosperms regularly form mycorrhizal associations (Wilcox 1991).

On the other hand, plants from the families Cruciferae (cabbage), Chenopodiaceae (spinach), and Proteaceae (macadamia nuts), as well as aquatic plants, rarely if ever have mycorrhizae. Mycorrhizae are absent from roots in very dry, saline, or flooded soils, or where soil fertility is extreme, either high or low. In particular, plants grown under hydroponics and young, rapidly growing crop plants seldom have mycorrhizae.

Mycorrhizal fungi are composed of fine, tubular filaments called *hyphae* (singular *hypha*). The mass of hyphae that forms the body of the fungus is called the *mycelium* (plural *mycelia*). There are two major classes of mycorrhizal fungi: ectotrophic mycorrhizae and vesicular-arbuscular mycorrhizae (Smith et al. 1997). Minor classes of mycorrhizal fungi include the ericaceous and orchidaceous mycorrhizae, which may have limited importance in terms of mineral nutrient uptake.

Ectotrophic mycorrhizal fungi typically show a thick sheath, or "mantle," of fungal mycelium around the roots, and some of the mycelium penetrates between the cortical cells (Figure 5.10). The cortical cells themselves are not penetrated by the fungal hyphae but instead are surrounded by a network of hyphae called the **Hartig net**. Often the amount of fungal mycelium is so extensive that its total mass is comparable to that of the roots themselves. The fungal mycelium also extends into the soil, away from this compact mantle, where it forms individual hyphae or strands containing fruiting bodies.

The capacity of the root system to absorb nutrients is improved by the presence of external fungal hyphae that are much finer than plant roots and can reach beyond the areas of nutrient-depleted soil near the roots (Clarkson 1985). Ectotrophic mycorrhizal fungi infect exclusively tree species, including gymnosperms and woody angiosperms.

Unlike the ectotrophic mycorrhizal fungi, **vesicular-arbuscular mycorrhizal fungi** do not produce a compact mantle of fungal mycelium around the root. Instead, the hyphae grow in a less dense arrangement, both within the root itself and extending outward from the root into the surrounding soil (Figure 5.11). After entering the root through either the epidermis or a root hair, the hyphae not only extend through the regions between cells but also penetrate individual cells of the cortex. Within the cells, the hyphae can form oval structures called **vesicles** and branched structures called **arbuscules**. The arbuscules appear to be sites of nutrient transfer between the fungus and the host plant.

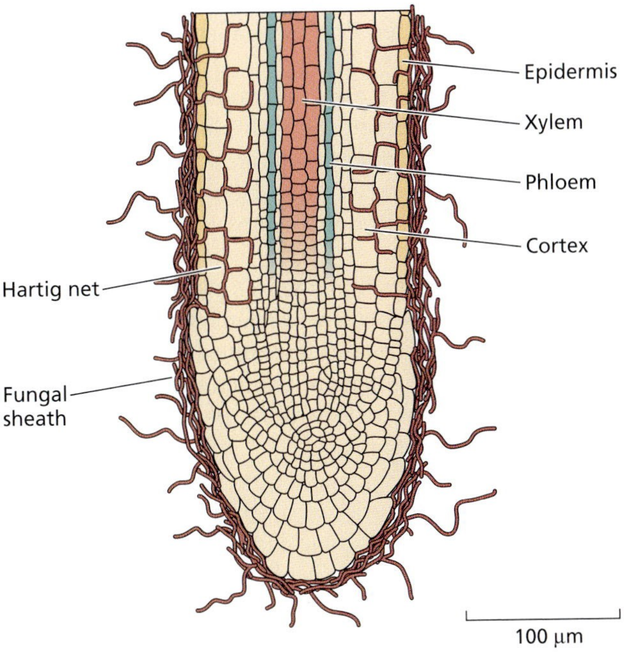

Epidermis
Xylem
Phloem
Cortex
Hartig net
Fungal sheath
100 μm

FIGURE 5.10 Root infected with ectotrophic mycorrhizal fungi. In the infected root, the fungal hyphae surround the root to produce a dense fungal sheath and penetrate the intercellular spaces of the cortex to form the Hartig net. The total mass of fungal hyphae may be comparable to the root mass itself. (After Rovira et al. 1983.)

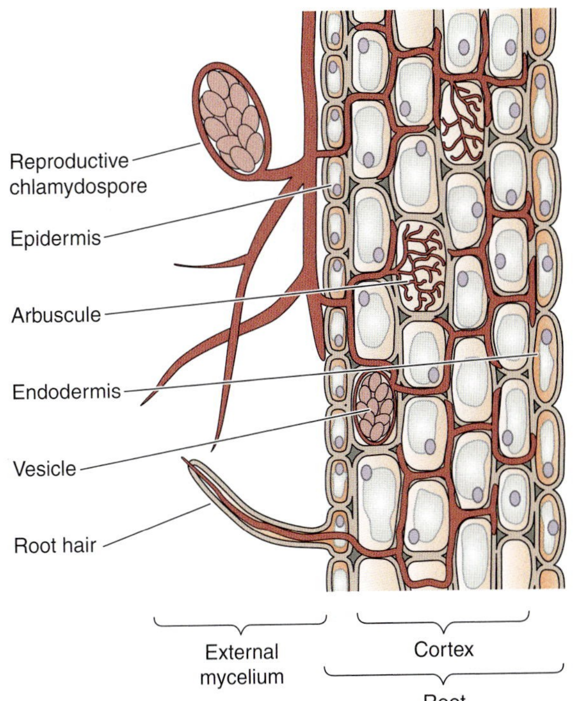

FIGURE 5.11 Association of vesicular-arbuscular mycorrhizal fungi with a section of a plant root. The fungal hyphae grow into the intercellular wall spaces of the cortex and penetrate individual cortical cells. As they extend into the cell, they do not break the plasma membrane or the tonoplast of the host cell. Instead, the hypha is surrounded by these membranes and forms structures known as arbuscules, which participate in nutrient ion exchange between the host plant and the fungus. (After Mauseth 1988.)

Outside the root, the external mycelium can extend several centimeters away from the root and may contain spore-bearing structures. Unlike the ectotrophic mycorrhizae, vesicular-arbuscular mycorrhizae make up only a small mass of fungal material, which is unlikely to exceed 10% of the root weight. Vesicular-arbuscular mycorrhizae are found in association with the roots of most species of herbaceous angiosperms (Smith et al. 1997).

The association of vesicular-arbuscular mycorrhizae with plant roots facilitates the uptake of phosphorus and trace metals such as zinc and copper. By extending beyond the depletion zone for phosphorus around the root, the external mycelium improves phosphorus absorption. Calculations show that a root associated with mycorrhizal fungi can transport phosphate at a rate more than four times higher than that of a root not associated with mycorrhizae (Nye and Tinker 1977). The external mycelium of the ectotrophic mycorrhizae can also absorb phosphate and make it available to the plant. In addition, it has been suggested that ectotrophic mycorrhizae proliferate in the organic litter of the soil and hydrolyze organic phosphorus for transfer to the root (Smith et al. 1997).

Nutrients Move from the Mycorrhizal Fungi to the Root Cells

Little is known about the mechanism by which the mineral nutrients absorbed by mycorrhizal fungi are transferred to the cells of plant roots. With ectotrophic mycorrhizae, inorganic phosphate may simply diffuse from the hyphae in the Hartig net and be absorbed by the root cortical cells. With vesicular-arbuscular mycorrhizae, the situation may be more complex. Nutrients may diffuse from intact arbuscules to root cortical cells. Alternatively, because some root arbuscules are continually degenerating while new ones are forming, degenerating arbuscules may release their internal contents to the host root cells.

A key factor in the extent of mycorrhizal association with the plant root is the nutritional status of the host plant. Moderate deficiency of a nutrient such as phosphorus tends to promote infection, whereas plants with abundant nutrients tend to suppress mycorrhizal infection.

Mycorrhizal association in well-fertilized soils may shift from a symbiotic relationship to a parasitic one in that the fungus still obtains carbohydrates from the host plant, but the host plant no longer benefits from improved nutrient uptake efficiency. Under such conditions, the host plant may treat mycorrhizal fungi as it does other pathogens (Brundrett 1991; Marschner 1995).

SUMMARY

Plants are autotrophic organisms capable of using the energy from sunlight to synthesize all their components from carbon dioxide, water, and mineral elements. Studies of plant nutrition have shown that specific mineral elements are essential for plant life. These elements are classified as macronutrients or micronutrients, depending on the relative amounts found in plant tissue.

Certain visual symptoms are diagnostic for deficiencies in specific nutrients in higher plants. Nutritional disorders occur because nutrients have key roles in plant metabolism. They serve as components of organic compounds, in energy storage, in plant structures, as enzyme cofactors, and in electron transfer reactions. Mineral nutrition can be studied through the use of hydroponics or aeroponics, which allow the characterization of specific nutrient requirements. Soil and plant tissue analysis can provide information on the nutritional status of the plant–soil system and can suggest corrective actions to avoid deficiencies or toxicities.

When crop plants are grown under modern high-production conditions, substantial amounts of nutrients are removed from the soil. To prevent the development of deficiencies, nutrients can be added back to the soil in the form of fertilizers. Fertilizers that provide nutrients in inorganic forms are called chemical fertilizers; those that derive from plant or animal residues are considered organic fertilizers. In both cases, plants absorb the nutrients primarily as inorganic ions. Most fertilizers are applied to the soil, but some are sprayed on leaves.

The soil is a complex substrate—physically, chemically, and biologically. The size of soil particles and the cation exchange capacity of the soil determine the extent to which a soil provides a reservoir for water and nutrients. Soil pH also has a large influence on the availability of mineral elements to plants.

If mineral elements, especially sodium or heavy metals, are present in excess in the soil, plant growth may be adversely affected. Certain plants are able to tolerate excess mineral elements, and a few species—for example, halophytes in the case of sodium—grow under these extreme conditions.

To obtain nutrients from the soil, plants develop extensive root systems. Roots have a relatively simple structure with radial symmetry and few differentiated cell types. Roots continually deplete the nutrients from the immediate soil around them, and such a simple structure may permit rapid growth into fresh soil.

Plant roots often form associations with mycorrhizal fungi. The fine hyphae of mycorrhizae extend the reach of roots into the surrounding soil and facilitate the acquisition of mineral elements, particularly those like phosphorus that are relatively immobile in the soil. In return, plants provide carbohydrates to the mycorrhizae. Plants tend to suppress mycorrhizal associations under conditions of high nutrient availability.

Web Material

Web Topics

5.1 Symptoms of Deficiency in Essential Minerals

Deficiency symptoms are characteristic of each essential element and can be used as diagnostic for the deficiency. These color pictures illustrate deficiency symptoms for each essential element in a tomato.

5.2 Observing Roots below Ground

The study of roots growing under natural conditions requires means to observe roots below ground. State-of-the-art techniques are described in this essay.

Web Essays

5.1 From Meals to Metals and Back

Heavy metal accumulation by plants is toxic. Understanding of the involved molecular process is helping to develop better phytoremediation crops.

5.2 Redox Control of the Root Quiescent Center

The redox status of the quiescent center seems to control the cell cycle of these cells.

Chapter References

Arnon, D. I., and Stout, P. R. (1939) The essentiality of certain elements in minute quantity for plants with special reference to copper. *Plant Physiol.* 14: 371–375.

Asher, C. J., and Edwards, D. G. (1983) Modern solution culture techniques. In *Inorganic Plant Nutrition* (Encyclopedia of Plant Physiology, New Series, Vol. 15B), A. Läuchli and R. L. Bieleski, eds., Springer, Berlin, pp. 94–119.

Bar-Yosef, B., Kafkafi, U., and Bresler, E. (1972) Uptake of phosphorus by plants growing under field conditions. I. Theoretical model and experimental determination of its parameters. *Soil Sci.* 36: 783–800.

Berry, W. L., and Wallace, A. (1981) Toxicity: The concept and relationship to the dose response curve. *J. Plant Nutr.* 3: 13–19.

Bloom, A. J. (1994) Crop acquisition of ammonium and nitrate. In *Physiology and Determination of Crop Yield*, K. J. Boote, J. M. Bennett, T. R. Sinclair, and G. M. Paulsen, eds., Soil Science Society of America, Inc., Crop Science Society of America, Inc., Madison, WI, pp. 303–309.

Bloom, A. J., Jackson, L. E., and Smart, D. R. (1993) Root growth as a function of ammonium and nitrate in the root zone. *Plant Cell Environ.* 16: 199–206.

Bouma, D. (1983) Diagnosis of mineral deficiencies using plant tests. In *Inorganic Plant Nutrition* (Encyclopedia of Plant Physiology, New Series, Vol. 15B), A. Läuchli and R. L. Bieleski, eds., Springer, Berlin, pp. 120–146.

Brady, N. C. (1974) *The Nature and Properties of Soils*, 8th ed. Macmillan, New York.

Bret-Harte, M. S., and Silk, W. K. (1994) Nonvascular, symplasmic diffusion of sucrose cannot satisfy the carbon demands of growth in the primary root tip of *Zea mays* L. *Plant Physiol.* 105: 19–33.

Brundrett, M. C. (1991) Mycorrhizas in natural ecosystems. *Adv. Ecol. Res.* 21: 171–313.

Clarke, S. M., and Eaton-Rye, J. J. (2000) Amino acid deletions in loop C of the chlorophyll a-binding protein CP47 alter the chloride requirement and/or prevent the assembly of photosystem II. *Plant Mol. Biol.* 44: 591–601.

Clarkson, D. T. (1985) Factors affecting mineral nutrient acquisition by plants. *Annu. Rev. Plant Physiol.* 36: 77–116.

Colmer, T. D., and Bloom, A. J. (1998) A comparison of net NH_4^+ and NO_3^- fluxes along roots of rice and maize. *Plant Cell Environ.* 21: 240–246.

Cooper, A. (1979) *The ABC of NFT: Nutrient Film Technique: The World's First Method of Crop Production without a Solid Rooting Medium.* Grower Books, London.

Dittmer, H. J. (1937) A quantitative study of the roots and root hairs of a winter rye plant (*Secale cereale*). *Am. J. Bot.* 24: 417–420.

Epstein, E. (1972) *Mineral Nutrition of Plants: Principles and Perspectives.* Wiley, New York.

Epstein, E. (1999) Silicon. *Annu. Rev. Plant Physiol. Plant Mol. Biol.* 50: 641–664.

Evans, H. J., and Sorger, G. J. (1966) Role of mineral elements with emphasis on the univalent cations. *Annu. Rev. Plant Physiol.* 17: 47–76.

Foehse, D., Claassen, N., and Jungk, A. (1991) Phosphorus efficiency of plants. II. Significance of root radius, root hairs and cation-anion balance for phosphorus influx in seven plant species. *Plant Soil* 132: 261–272.

Gericke, W. F. (1937) Hydroponics—Crop production in liquid culture media. *Science* 85: 177–178.

Haehnel, W. (1984) Photosynthetic electron transport in higher plants. *Annu. Rev. Plant Physiol.* 35: 659–693.

Harling, H., Czaja, I., Schell, J., and Walden, R. (1997) A plant cation-chloride co-transporter promoting auxin-independent tobacco protoplast division. *EMBO J.* 16: 5855–5866.

Hasegawa, P. M., Bressan, R. A., Zhu, J.-K., and Bohnert, H. J. (2000) Plant cellular and molecular responses to high salinity. *Annu. Rev. Plant Physiol. Plant Mol. Biol.* 51: 463–499.

Hepler, P. K., and Wayne, R. O. (1985) Calcium and plant development. *Annu. Rev. Plant Physiol.* 36: 397–440.

Johnstone, M., Grof, C. P. L., and Brownell, P. F. (1988) The effect of sodium nutrition on the pool sizes of intermediates of the C_4 photosynthetic pathway. *Aust. J. Plant Physiol.* 15: 749–760.

Kashirad, A., Marschner, H., and Richter, C. H. (1973) Absorption and translocation of ^{59}Fe from various parts of the corn plant. *Z. Pflanzenernähr. Bodenk.* 134: 136–147.

Loomis, R. S., and Connor, D. J. (1992) *Crop Ecology: Productivity and Management in Agricultural Systems.* Cambridge University Press, Cambridge.

Lucas, R. E., and Davis, J. F. (1961) Relationships between pH values of organic soils and availabilities of 12 plant nutrients. *Soil Sci.* 92: 177–182.

Macek, T., Mackova, M., and Kas, J. (2000) Exploitation of plants for the removal of organics in environmental remediation. *Biotech. Adv.* 18: 23–34.

Marschner, H. (1995) *Mineral Nutrition of Higher Plants*, 2nd ed. Academic Press, London.

Mauseth, J. D. (1988) *Plant Anatomy.* Benjamin/Cummings Pub. Co., Menlo Park, CA.

Mengel, K., and Kirkby, E. A. (1987) *Principles of Plant Nutrition.* International Potash Institute, Worblaufen-Bern, Switzerland.

Nolan, B. T. and Stoner, J. D. (2000) Nutrients in groundwater of the center conterminous United States 1992-1995. *Environ. Sci. Tech.* 34: 1156–1165.

Nye, P. H., and Tinker, P. B. (1977) *Solute Movement in the Soil-Root System*. University of California Press, Berkeley.

Oh, S.-H., Cho, S.-W., Kwon, T.-H., and Yang, M.-S. (1996) Purification and characterization of phytoferritin. *J. Biochem. Mol. Biol.* 29: 540–544.

Raven, J. A., and Smith, F. A. (1976) Nitrogen assimilation and transport in vascular land plants in relation to intracellular pH regulation. *New Phytol.* 76: 415–431.

Rovira, A. D., Bowen, C. D., and Foster, R. C. (1983) The significance of rhizosphere microflora and mycorrhizas in plant nutrition. In *Inorganic Plant Nutrition* (Encyclopedia of Plant Physiology, New Series, Vol. 15B) A. Läuchli and R. L. Bieleskis, eds., Springer, Berlin, pp. 61–93.

Russell, R. S. (1977) *Plant Root Systems: Their Function and Interaction with the Soil.* McGraw-Hill, London.

Sanders, D., Brownlee, C., and Harper J. F. (1999) Communicating with calcium. *Plant Cell* 11: 691–706.

Sharp, R. E., Hsiao, T. C., and Silk, W. K. (1990) Growth of the maize primary root at low water potentials. 2. Role of growth and deposition of hexose and potassium in osmotic adjustment. *Plant Physiol.* 93: 1337–1346.

Shelp, B. J. (1993) Physiology and biochemistry of boron in plants. In *Boron and Its Role in Crop Production*, U. C. Gupta, ed., CRC Press, Boca Raton, FL, pp. 53–85.

Sievers, R. E., and Bailar, J. C., Jr. (1962) Some metal chelates of ethylenediaminetetraacetic acid, diethylenetriaminepentaacetic acid, and triethylenetriaminehexaacetic acid. *Inorganic Chem.* 1: 174–182.

Smith, S. E., Read, D. J., and Harley, J. L. (1997) *Mycorrhizal Symbiosis.* Academic Press, San Diego, CA.

Stewart, G. R., and Ahmad, I. (1983) Adaptation to salinity in angiosperm halophytes. In *Metals and Micronutrients: Uptake and Utilization by Plants*, D. A. Robb and W. S. Pierpoint, eds., Academic Press, New York, pp. 33–50.

Taylor, A. R., and Bloom, A. J. (1998) Ammonium, nitrate and proton fluxes along the maize root. *Plant Cell Environ.* 21: 1255–1263.

Weathers, P. J., and Zobel, R. W. (1992) Aeroponics for the culture of organisms, tissues, and cells. *Biotech. Adv.* 10: 93–115.

Weaver, J. E. (1926) *Root Development of Field Crops*. McGraw-Hill, New York.

Wilcox, H. E. (1991) Mycorrhizae. In *Plant Roots: The Hidden Half*, Y. Waisel, A. Eshel, and U. Kafkafi, eds., Marcel Dekker, New York, pp. 731–765.

Ziegler, H. (1987) The evolution of stomata. In *Stomatal Function*, E. Zeiger, G. Farquhar, and I. Cowan, eds., Stanford University Press, Stanford, CA, pp. 29–57.

Chapter

6 Solute Transport

PLANT CELLS ARE SEPARATED from their environment by a plasma membrane that is only two lipid molecules thick. This thin layer separates a relatively constant internal environment from highly variable external surroundings. In addition to forming a hydrophobic barrier to diffusion, the membrane must facilitate and continuously regulate the inward and outward traffic of selected molecules and ions as the cell takes up nutrients, exports wastes, and regulates its turgor pressure. The same is true of the internal membranes that separate the various compartments within each cell.

As the cell's only contact with its surroundings, the plasma membrane must also relay information about its physical environment, about molecular signals from other cells, and about the presence of invading pathogens. Often these signal transduction processes are mediated by changes in ion fluxes across the membrane.

Molecular and ionic movement from one location to another is known as **transport**. Local transport of solutes into or within cells is regulated mainly by membranes. Larger-scale transport between plant and environment, or between leaves and roots, is also controlled by membrane transport at the cellular level. For example, the transport of sucrose from leaf to root through the phloem, referred to as *translocation,* is driven and regulated by membrane transport into the phloem cells of the leaf, and from the phloem to the storage cells of the root (see Chapter 10).

In this chapter we will consider first the physical and chemical principles that govern the movements of molecules in solution. Then we will show how these principles apply to membranes and to biological systems. We will also discuss the molecular mechanisms of transport in living cells and the great variety of membrane transport proteins that are responsible for the particular transport properties of plant cells. Finally, we will examine the pathway that ions take when they enter the root, as well as the mechanism of xylem loading, the process whereby ions are released into the vessel elements and tracheids of the stele.

PASSIVE AND ACTIVE TRANSPORT

According to Fick's first law (see Equation 3.1), the movement of molecules by diffusion always proceeds spontaneously, down a gradient of concentration or chemical potential (see Chapter 2 on the web site), until equilibrium is reached. The spontaneous "downhill" movement of molecules is termed **passive transport**. At equilibrium, no further net movements of solute can occur without the application of a driving force.

The movement of substances against or up a gradient of chemical potential (e.g., to a higher concentration) is termed **active transport**. It is not spontaneous, and it requires that work be done on the system by the application of cellular energy. One way (but not the only way) of accomplishing this task is to couple transport to the hydrolysis of ATP.

Recall from Chapter 3 that we can calculate the driving force for diffusion, or, conversely, the energy input necessary to move substances against a gradient, by measuring the potential-energy gradient, which is often a simple function of the difference in concentration. Biological transport can be driven by four major forces: concentration, hydrostatic pressure, gravity, and electric fields. (However, recall from Chapter 3 that in biological systems, gravity seldom contributes substantially to the force that drives transport.)

The **chemical potential** for any solute is defined as the sum of the concentration, electric, and hydrostatic potentials (and the chemical potential under standard conditions):

$$\tilde{\mu}_j = \mu_j^* + RT \ln C_j + z_j FE + \bar{V}_j P \tag{6.1}$$

$\tilde{\mu}_j$	=	μ_j^*	+	$RT \ln C_j$	+	$z_j FE$	+	$\bar{V}_j P$
Chemical potential for a given solute, *j*		Chemical potential of *j* under standard conditions		Concentration (activity) component		Electric-potential component		Hydrostatic-pressure component

Here $\tilde{\mu}_j$ is the chemical potential of the solute species *j* in joules per mole (J mol^{-1}), μ_j^* is its chemical potential under standard conditions (a correction factor that will cancel out in future equations and so can be ignored), *R* is the universal gas constant, *T* is the absolute temperature, and C_j is the concentration (more accurately the activity) of *j*.

The electrical term, $z_j FE$, applies only to ions; *z* is the electrostatic charge of the ion (+1 for monovalent cations, –1 for monovalent anions, +2 for divalent cations, and so on), *F* is Faraday's constant (equivalent to the electric charge on 1 mol of protons), and *E* is the overall electric potential of the solution (with respect to ground). The final term, $\bar{V}_j P$, expresses the contribution of the partial molal volume of *j* ($\bar{V}_j$) and pressure (*P*) to the chemical potential of *j*. (The partial molal volume of *j* is the change in volume per mole of substance *j* added to the system, for an infinitesimal addition.)

This final term, $\bar{V}_j P$, makes a much smaller contribution to $\tilde{\mu}_j$ than do the concentration and electrical terms, except in the very important case of osmotic water movements. As discussed in Chapter 3, the chemical potential of water (i.e., the water potential) depends on the concentration of dissolved solutes and the hydrostatic pressure on the system.

The importance of the concept of chemical potential is that it sums all the forces that may act on a molecule to drive net transport (Nobel 1991).

In general, diffusion (or passive transport) always moves molecules from areas of higher chemical potential downhill to areas of lower chemical potential. Movement against a chemical-potential gradient is indicative of active transport (Figure 6.1).

If we take the diffusion of sucrose across a permeable membrane as an example, we can accurately approximate the chemical potential of sucrose in any compartment by the concentration term alone (unless a solution is very concentrated, causing hydrostatic pressure to build up). From Equation 6.1, the chemical potential of sucrose inside a cell can be described as follows (in the next three equations, the subscript *s* stands for sucrose, and the superscripts *i* and *o* stand for inside and outside, respectively):

$$\tilde{\mu}_s^i = \mu_s^* + RT \ln C_s^i \tag{6.2}$$

$\tilde{\mu}_s^i$	=	μ_s^*	+	$RT \ln C_s^i$
Chemical potential of sucrose solution inside the cell		Chemical potential of sucrose solution under standard conditions		Concentration component

The chemical potential of sucrose outside the cell is calculated as follows:

$$\tilde{\mu}_s^o = \mu_s^* + RT \ln C_s^o \tag{6.3}$$

We can calculate the difference in the chemical potential of sucrose between the solutions inside and outside the cell, $\Delta\tilde{\mu}_s$, regardless of the mechanism of transport. To get the signs right, remember that for inward transport, sucrose is being removed (–) from outside the cell and added (+) to the inside, so the change in free energy in joules per mole of sucrose transported will be as follows:

$$\Delta\tilde{\mu}_s = \tilde{\mu}_s^i - \tilde{\mu}_s^o \tag{6.4}$$

Substituting the terms from Equations 6.2 and 6.3 into Equation 6.4, we get the following:

$$\begin{aligned}\Delta\tilde{\mu}_s &= \left(\mu_s^* + RT \ln C_s^i\right) - \left(\mu_s^* + RT \ln C_s^o\right)\\ &= RT\left(\ln C_s^i - \ln C_s^o\right)\\ &= RT \ln \frac{C_s^i}{C_s^o}\end{aligned} \tag{6.5}$$

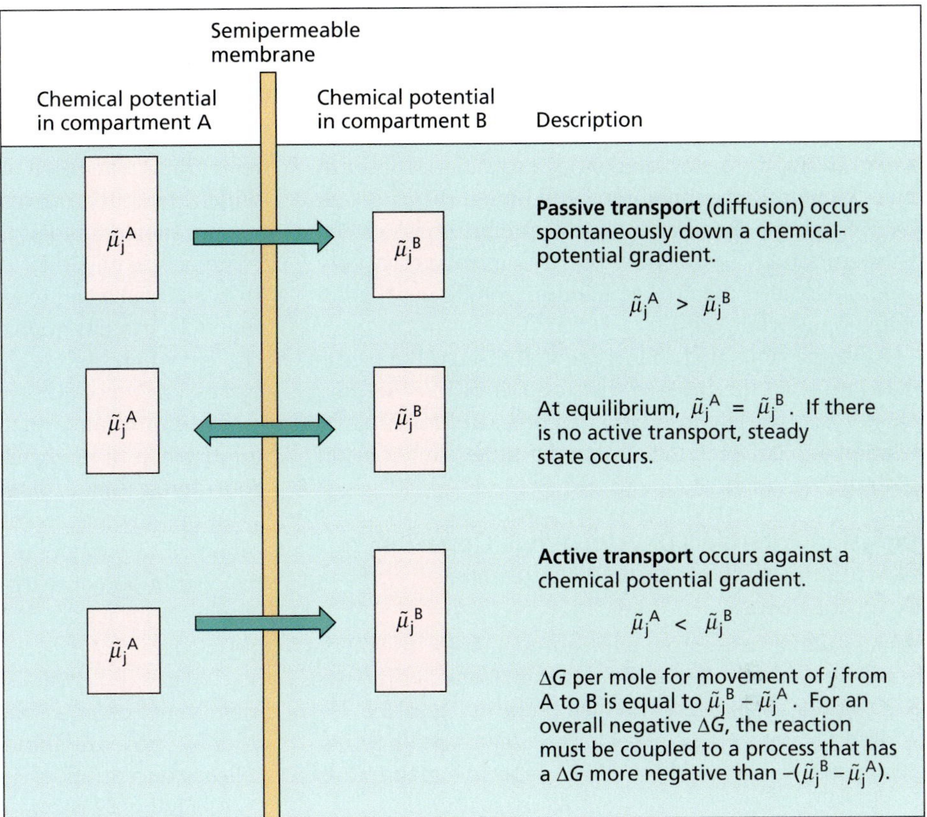

FIGURE 6.1 Relationship between the chemical potential, $\tilde{\mu}$, and the transport of molecules across a permeability barrier. The net movement of molecular species j between compartments A and B depends on the relative magnitude of the chemical potential of j in each compartment, represented here by the size of the boxes. Movement down a chemical gradient occurs spontaneously and is called passive transport; movement against or up a gradient requires energy and is called active transport.

If this difference in chemical potential is negative, sucrose could diffuse inward spontaneously (provided the membrane had a finite permeability to sucrose; see the next section). In other words, the driving force ($\Delta\tilde{\mu}_s$) for solute diffusion is related to the magnitude of the concentration gradient ($C_s^{\,i}/C_s^{\,o}$).

If the solute carries an electric charge (as does the potassium ion), the electrical component of the chemical potential must also be considered. Suppose the membrane is permeable to K^+ and Cl^- rather than to sucrose. Because the ionic species (K^+ and Cl^-) diffuse independently, each has its own chemical potential. Thus for inward K^+ diffusion,

$$\Delta\tilde{\mu}_K = \tilde{\mu}_K^{\,i} - \tilde{\mu}_K^{\,o} \tag{6.6}$$

Substituting the appropriate terms from Equation 6.1 into Equation 6.6, we get

$$\Delta\tilde{\mu}_s = (RT \ln [K^+]^i + zFE^i) - (RT \ln [K^+]^o + zFE^o) \tag{6.7}$$

and because the electrostatic charge of K^+ is +1, $z = +1$ and

$$\Delta\tilde{\mu}_K = RT \ln\frac{[K^+]^i}{[K^+]^o} + F(E^i - E^o) \tag{6.8}$$

The magnitude and sign of this expression will indicate the driving force for K^+ diffusion across the membrane, and its direction. A similar expression can be written for Cl^- (but remember that for Cl^-, $z = -1$).

Equation 6.8 shows that ions, such as K^+, diffuse in response to both their concentration gradients ($[K^+]^i/[K^+]^o$) and any electric-potential difference between the two compartments ($E^i - E^o$). One very important implication of this equation is that ions can be driven passively against their concentration gradients if an appropriate voltage (electric field) is applied between the two compartments. Because of the importance of electric fields in biological transport, $\tilde{\mu}$ is often called the **electrochemical potential**, and $\Delta\tilde{\mu}$ is the difference in electrochemical potential between two compartments.

TRANSPORT OF IONS ACROSS A MEMBRANE BARRIER

If the two KCl solutions in the previous example are separated by a biological membrane, diffusion is complicated by the fact that the ions must move through the membrane as well as across the open solutions. The extent to which a membrane permits the movement of a substance is called **membrane permeability**. As will be discussed later, permeability depends on the composition of the membrane, as well as on the chemical nature of the solute. In a loose sense, permeability can be expressed in terms of a diffusion coefficient for the solute in the membrane. However, permeability is influenced by several additional factors, such as the ability of a substance to enter the membrane, that are difficult to measure.

Despite its theoretical complexity, we can readily measure permeability by determining the rate at which a solute passes through a membrane under a specific set of conditions. Generally the membrane will hinder diffusion and thus reduce the speed with which equilibrium is reached. The permeability or resistance of the membrane itself, however, cannot alter the final equilibrium conditions. Equilibrium occurs when $\Delta\tilde{\mu}_j = 0$.

In the sections that follow we will discuss the factors that influence the passive distribution of ions across a membrane. These parameters can be used to predict the relationship between the electrical gradient and the concentration gradient of an ion.

Diffusion Potentials Develop When Oppositely Charged Ions Move across a Membrane at Different Rates

When salts diffuse across a membrane, an electric membrane potential (voltage) can develop. Consider the two KCl solutions separated by a membrane in Figure 6.2. The K^+ and Cl^- ions will permeate the membrane independently as they diffuse down their respective gradients of electrochemical potential. And unless the membrane is very porous, its permeability for the two ions will differ.

As a consequence of these different permeabilities, K^+ and Cl^- initially will diffuse across the membrane at different rates. The result will be a slight separation of charge, which instantly creates an electric potential across the membrane. In biological systems, membranes are usually more permeable to K^+ than to Cl^-. Therefore, K^+ will diffuse out of the cell (compartment A in Figure 6.2) faster than Cl^-, causing the cell to develop a negative electric charge with respect to the medium. A potential that develops as a result of diffusion is called a **diffusion potential**.

An important principle that must always be kept in mind when the movement of ions across membranes is considered is the principle of electrical neutrality. Bulk solutions always contain equal numbers of anions and cations. The existence of a membrane potential implies that the distribution of charges across the membrane is uneven; however, the actual number of unbalanced ions is negligible in chemical terms. For example, a membrane potential of –100 mV (millivolts), like that found across the plasma membranes of many plant cells, results from the presence of only one extra anion out of every 100,000 within the cell—a concentration difference of only 0.001%!

As Figure 6.2 shows, all of these extra anions are found immediately adjacent to the surface of the membrane; there is no charge imbalance throughout the bulk of the cell. In our example of KCl diffusion across a membrane, electrical neutrality is preserved because as K^+ moves ahead of Cl^- in the membrane, the resulting diffusion potential retards the movement of K^+ and speeds that of Cl^-. Ultimately, both ions diffuse at the same rate, but the diffusion potential persists and can be measured. As the system moves toward equilibrium and the concentration gradient collapses, the diffusion potential also collapses.

Initial conditions: $[KCl]_A > [KCl]_B$

Diffusion potential exists until chemical equilibrium is reached.

Equilibrium conditions: $[KCl]_A = [KCl]_B$

At chemical equilibrium, diffusion potential equals zero.

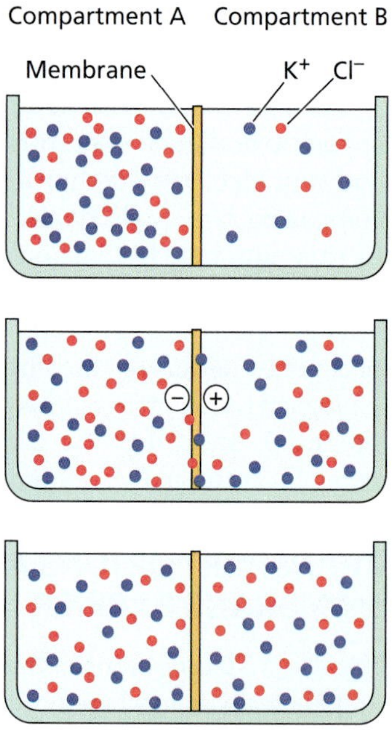

FIGURE 6.2 Development of a diffusion potential and a charge separation between two compartments separated by a membrane that is preferentially permeable to potassium. If the concentration of potassium chloride is higher in compartment A ($[KCl]_A > [KCl]_B$), potassium and chloride ions will diffuse at a higher rate into compartment B, and a diffusion potential will be established. When membranes are more permeable to potassium than to chloride, potassium ions will diffuse faster than chloride ions, and charge separation (+ and –) will develop.

The Nernst Equation Relates the Membrane Potential to the Distribution of an Ion at Equilibrium

Because the membrane is permeable to both K^+ and Cl^- ions, equilibrium in the preceding example will not be reached for either ion until the concentration gradients decrease to zero. However, if the membrane were permeable to only K^+, diffusion of K^+ would carry charges across the membrane until the membrane potential balanced the concentration gradient. Because a change in potential requires very few ions, this balance would be reached instantly. Transport would then be at equilibrium, even though the concentration gradients were unchanged.

When the distribution of any solute across a membrane reaches equilibrium, the passive flux, J (i.e., the amount of solute crossing a unit area of membrane per unit time), is the same in the two directions—outside to inside and inside to outside:

$$J_{o \to i} = J_{i \to o}$$

Fluxes are related to $\Delta\tilde{\mu}$ (for a discussion on fluxes and $\Delta\tilde{\mu}$, see Chapter 2 on the web site); thus at equilibrium, the electrochemical potentials will be the same:

$$\tilde{\mu}_j^o = \tilde{\mu}_j^i$$

and for any given ion (the ion is symbolized here by the subscript j):

$$\mu_j^* + RT \ln C_j^o + z_j FE^o = \mu_j^* + RT \ln C_j^i + z_j FE^i \quad (6.9)$$

By rearranging Equation 6.9, we can obtain the difference in electric potential between the two compartments at equilibrium ($E^i - E^o$):

$$E^i - E^o = \frac{RT}{z_j F}\left(\ln \frac{C_j^o}{C_j^i}\right)$$

This electric-potential difference is known as the **Nernst potential** (ΔE_j) for that ion:

$$\Delta E_j = E^i - E^o$$

and

$$\Delta E_j = \frac{RT}{z_j F}\left(\ln \frac{C_j^o}{C_j^i}\right) \quad (6.10)$$

or

$$\Delta E_j = \frac{2.3RT}{z_j F}\left(\log \frac{C_j^o}{C_j^i}\right)$$

This relationship, known as the **Nernst equation**, states that at equilibrium the difference in concentration of an ion between two compartments is balanced by the voltage difference between the compartments. The Nernst equation can be further simplified for a univalent cation at 25°C:

$$\Delta E_j = 59\text{mV} \log \frac{C_j^o}{C_j^i} \quad (6.11)$$

Note that a tenfold difference in concentration corresponds to a Nernst potential of 59 mV ($C_o/C_i = 10/1$; log 10 = 1). That is, a membrane potential of 59 mV would maintain a tenfold concentration gradient of an ion that is transported by passive diffusion. Similarly, if a tenfold concentration gradient of an ion existed across the membrane, passive diffusion of that ion down its concentration gradient (if it were allowed to come to equilibrium) would result in a difference of 59 mV across the membrane.

All living cells exhibit a membrane potential that is due to the asymmetric ion distribution between the inside and outside of the cell. We can readily determine these membrane potentials by inserting a microelectrode into the cell and measuring the voltage difference between the inside of the cell and the external bathing medium (Figure 6.3).

The Nernst equation can be used at any time to determine whether a given ion is at equilibrium across a membrane. However, a distinction must be made between equilibrium and steady state. **Steady state** is the condition in which influx and efflux of a given solute are equal and therefore the ion concentrations are constant with respect to time. Steady state is not the same as equilibrium (see Figure 6.1); in steady state, the existence of active transport across the membrane prevents many diffusive fluxes from ever reaching equilibrium.

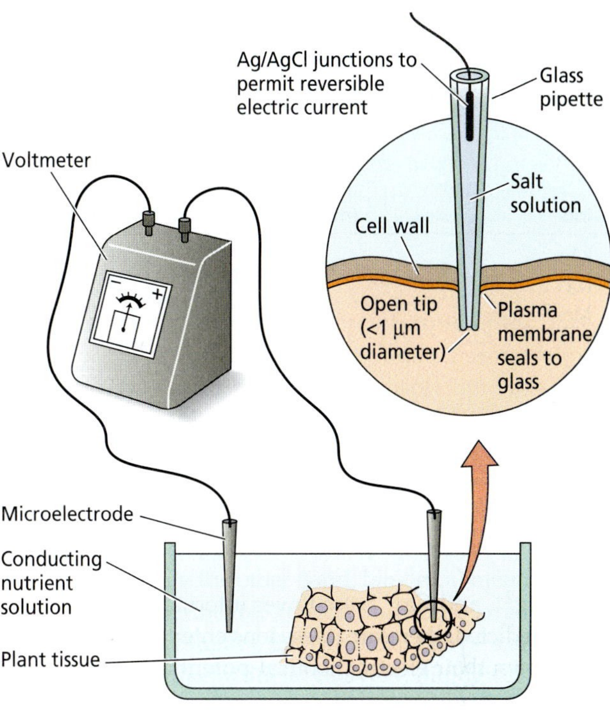

FIGURE 6.3 Diagram of a pair of microelectrodes used to measure membrane potentials across cell membranes. One of the glass micropipette electrodes is inserted into the cell compartment under study (usually the vacuole or the cytoplasm), while the other is kept in an electrolytic solution that serves as a reference. The microelectrodes are connected to a voltmeter, which records the electric-potential difference between the cell compartment and the solution. Typical membrane potentials across plant cell membranes range from –60 to –240 mV. The insert shows how electrical contact with the interior of the cell is made through the open tip of the glass micropipette, which contains an electrically conducting salt solution.

The Nernst Equation Can Be Used to Distinguish between Active and Passive Transport

Table 6.1 shows how the experimentally measured ion concentrations at steady state for pea root cells compare with predicted values calculated from the Nernst equation (Higinbotham et al. 1967). In this example, the external concentration of each ion in the solution bathing the tissue, and the measured membrane potential, were substituted into the Nernst equation, and a predicted internal concentration was calculated for that ion.

Notice that, of all the ions shown in Table 6.1, only K^+ is at or near equilibrium. The anions NO_3^-, Cl^-, $H_2PO_4^-$, and SO_4^{2-} all have higher internal concentrations than predicted, indicating that their uptake is active. The cations

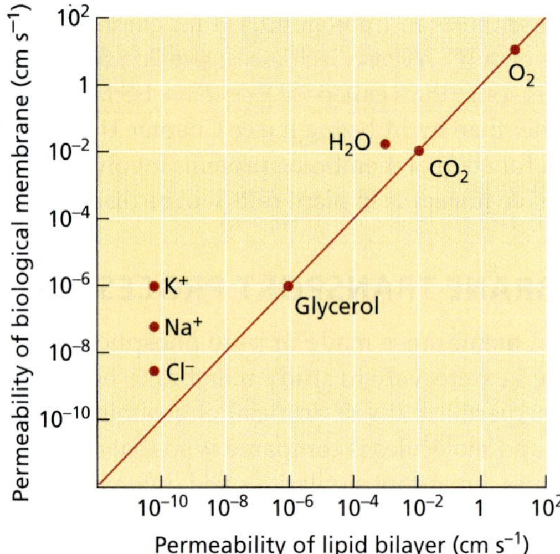

FIGURE 6.6 Typical values for the permeability, P, of a biological membrane to various substances, compared with those for an artificial phospholipid bilayer. For nonpolar molecules such as O_2 and CO_2, and for some small uncharged molecules such as glycerol, P values are similar in both systems. For ions and selected polar molecules, including water, the permeability of biological membranes is increased by one or more orders of magnitude, because of the presence of transport proteins. Note the logarithmic scale.

brane transport. In *Arabidopsis*, 849 genes, or 4.8% of all genes, code for proteins involved in membrane transport.

Although a particular transport protein is usually highly specific for the kinds of substances it will transport, its specificity is not absolute: It generally also transports a small family of related substances. For example, in plants a K^+ transporter on the plasma membrane may transport Rb^+ and Na^+ in addition to K^+, but K^+ is usually preferred. On the other hand, the K^+ transporter is completely ineffective in transporting anions such as Cl^- or uncharged solutes such as sucrose. Similarly, a protein involved in the transport of neutral amino acids may move glycine, alanine, and valine with equal ease but not accept aspartic acid or lysine.

In the next several pages we will consider the structures, functions, and physiological roles of the various membrane transporters found in plant cells, especially on the plasma membrane and tonoplast. We begin with a discussion of the role of certain transporters (channels and carriers) in promoting the diffusion of solutes across membranes. We then distinguish between primary and secondary active transport, and we discuss the roles of the electrogenic H^+-ATPase and various symporters (proteins that transport two substances in the same direction simultaneously) in driving proton-coupled secondary active transport.

Channel Transporters Enhance Ion and Water Diffusion across Membranes

Three types of membrane transporters enhance the movement of solutes across membranes: *channels*, *carriers*, and *pumps* (Figure 6.7). **Channels** are transmembrane proteins

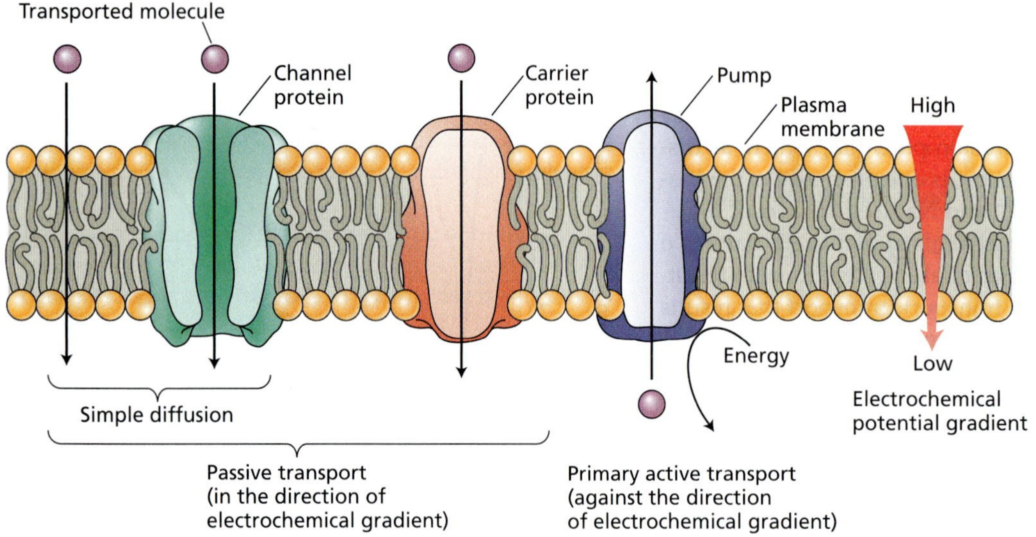

FIGURE 6.7 Three classes of membrane transport proteins: channels, carriers, and pumps. Channels and carriers can mediate the passive transport of solutes across membranes (by simple diffusion or facilitated diffusion), down the solute's gradient of electrochemical potential. Channel proteins act as membrane pores, and their specificity is determined primarily by the biophysical properties of the channel. Carrier proteins bind the transported molecule on one side of the membrane and release it on the other side. Primary active transport is carried out by pumps and uses energy directly, usually from ATP hydrolysis, to pump solutes against their gradient of electrochemical potential.

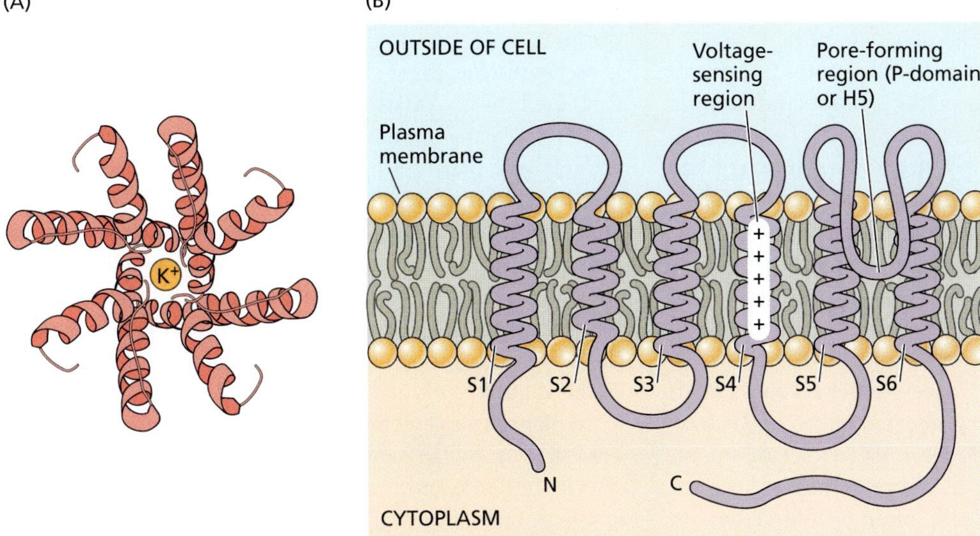

FIGURE 6.8 Models of K^+ channels in plants. (A) Top view of channel, looking through the pore of the protein. Membrane-spanning helices of four subunits come together in an inverted teepee with the pore at the center. The pore-forming regions of the four subunits dip into the membrane, with a K^+ selectivity finger region formed at the outer (near) part of the pore (more details on the structure of this channel can be found in Web Essay 6.1). (B) Side view of the inward rectifying K^+ channel, showing a polypeptide chain of one subunit, with six membrane-spanning helices. The fourth helix contains positively-charged amino acids and acts as a voltage-sensor. The pore-forming region is a loop between helices 5 and 6. (A after Leng et al. 2002; B after Buchanan et al. 2000.)

that function as selective pores, through which molecules or ions can diffuse across the membrane. The size of a pore and the density of surface charges on its interior lining determine its transport specificity. Transport through channels is always passive, and because the specificity of transport depends on pore size and electric charge more than on selective binding, channel transport is limited mainly to ions or water (Figure 6.8).

Transport through a channel may or may not involve transient binding of the solute to the channel protein. In any case, as long as the channel pore is open, solutes that can penetrate the pore diffuse through it extremely rapidly: about 10^8 ions per second through each channel protein. Channels are not open all the time: Channel proteins have structures called **gates** that open and close the pore in response to external signals (see Figure 6.8B). Signals that can open or close gates include voltage changes, hormone binding, or light. For example, voltage-gated channels open or close in response to changes in the membrane potential.

Individual ion channels can be studied in detail by the technique of patch clamp electrophysiology (see Web Topic 6.2), which can detect the electric current carried by ions diffusing through a single channel. Patch clamp studies reveal that, for a given ion, such as potassium, a given membrane has a variety of different channels. These channels may open in different voltage ranges, or in response to different signals, which may include K^+ or Ca^{2+} concentrations, pH, protein kinases and phosphatases, and so on. This specificity enables the transport of each ion to be fine-tuned to the prevailing conditions. Thus the ion permeability of a membrane is a variable that depends on the mix of ion channels that are open at a particular time.

As we saw in the experiment of Table 6.1, the distribution of most ions is not close to equilibrium across the membrane. Anion channels will always function to allow anions to diffuse out of the cell, and other mechanisms are needed for anion uptake. Similarly, calcium channels can function only in the direction of calcium release into the cytosol, and calcium must be expelled by active transport. The exception is potassium, which can diffuse either inward or outward, depending on whether the membrane potential is more negative or more positive than E_K, the potassium equilibrium potential.

K^+ channels that open only at more negative potentials are specialized for inward diffusion of K^+ and are known as **inward-rectifying,** or simply **inward,** K^+ channels. Conversely, K^+ channels that open only at more positive potentials are **outward-rectifying,** or **outward,** K^+ channels (see Web Essay 6.1). Whereas inward K^+ channels function in the accumulation of K^+ from the environment, or in the opening of stomata, various outward K^+ channels function in the closing of stomata, in the release of K^+ into the xylem or in regulation of the membrane potential.

Carriers Bind and Transport Specific Substances

Unlike channels, **carrier** proteins do not have pores that extend completely across the membrane. In transport mediated by a carrier, the substance being transported is initially bound to a specific site on the carrier protein. This requirement for binding allows carriers to be highly selective for a particular substrate to be transported. Carriers therefore specialize in the transport of specific organic metabolites. Binding causes a conformational change in the protein, which exposes the substance to the solution on the other side of the membrane. Transport is complete when the substance dissociates from the carrier's binding site.

Because a conformational change in the protein is required to transport individual molecules or ions, the rate of transport by a carrier is many orders of magnitude slower than through a channel. Typically, carriers may transport 100 to 1000 ions or molecules per second, which is about 10^6 times slower than transport through a channel. The binding and release of a molecule at a specific site on a protein that occur in carrier-mediated transport are similar to the binding and release of molecules from an enzyme in an enzyme-catalyzed reaction. As will be discussed later in the chapter, enzyme kinetics has been used to characterize transport carrier proteins (for a detailed discussion on kinetics, see Chapter 2 on the web site).

Carrier-mediated transport (unlike transport through channels) can be either passive or active, and it can transport a much wider range of possible substrates. Passive transport on a carrier is sometimes called **facilitated diffusion**, although it resembles diffusion only in that it transports substances down their gradient of electrochemical potential, without an additional input of energy. (This term might seem more appropriately applied to transport through channels, but historically it has not been used in this way.)

Primary Active Transport Is Directly Coupled to Metabolic or Light Energy

To carry out active transport, a carrier must couple the uphill transport of the solute with another, energy-releasing, event so that the overall free-energy change is negative. **Primary active transport** is coupled directly to a source of energy other than $\Delta\tilde{\mu}_j$, such as ATP hydrolysis, an oxidation–reduction reaction (the electron transport chain of mitochondria and chloroplasts), or the absorption of light by the carrier protein (in halobacteria, bacteriorhodopsin).

The membrane proteins that carry out primary active transport are called **pumps** (see Figure 6.7). Most pumps transport ions, such as H^+ or Ca^{2+}. However, as we will see later in the chapter, pumps belonging to the "ATP-binding cassette" family of transporters can carry large organic molecules.

Ion pumps can be further characterized as either electrogenic or electroneutral. In general, **electrogenic transport** refers to ion transport involving the net movement of charge across the membrane. In contrast, **electroneutral transport**, as the name implies, involves no net movement of charge. For example, the Na^+/K^+-ATPase of animal cells pumps three Na^+ ions out for every two K^+ ions in, resulting in a net outward movement of one positive charge. The Na^+/K^+-ATPase is therefore an electrogenic ion pump. In contrast, the H^+/K^+-ATPase of the animal gastric mucosa pumps one H^+ out of the cell for every one K^+ in, so there is no net movement of charge across the membrane. Therefore, the H^+/K^+-ATPase is an electroneutral pump.

In the plasma membranes of plants, fungi, and bacteria, as well as in plant tonoplasts and other plant and animal endomembranes, H^+ is the principal ion that is electrogenically pumped across the membrane. The **plasma membrane H^+-ATPase** generates the gradient of electrochemical potentials of H^+ across the plasma membranes, while the **vacuolar H^+-ATPase** and the **H^+-pyrophosphatase (H^+-PPase)** electrogenically pump protons into the lumen of the vacuole and the Golgi cisternae.

In plant plasma membranes, the most prominent pumps are for H^+ and Ca^{2+}, and the direction of pumping is outward. Therefore another mechanism is needed to drive the active uptake of most mineral nutrients. The other important way that solutes can be actively transported across a membrane against their gradient of electrochemical potential is by coupling of the uphill transport of one solute to the downhill transport of another. This type of carrier-mediated cotransport is termed **secondary active transport**, and it is driven indirectly by pumps.

Secondary Active Transport Uses the Energy Stored in Electrochemical-Potential Gradients

Protons are extruded from the cytosol by electrogenic H^+-ATPases operating in the plasma membrane and at the vacuole membrane. Consequently, a membrane potential and a pH gradient are created at the expense of ATP hydrolysis. This gradient of electrochemical potential for H^+, $\Delta\tilde{\mu}_{H^+}$, or (when expressed in other units) the **proton motive force (PMF)**, or Δp, represents stored free energy in the form of the H^+ gradient (see **Web Topic 6.3**).

The proton motive force generated by electrogenic H^+ transport is used in secondary active transport to drive the transport of many other substances against their gradient of electrochemical potentials. Figure 6.9 shows how secondary transport may involve the binding of a substrate (S) and an ion (usually H^+) to a carrier protein, and a conformational change in that protein.

There are two types of secondary transport: symport and antiport. The example shown in Figure 6.9 is called **symport** (and the protein involved is called a *symporter*) because the two substances are moving in the same direction through the membrane (see also Figure 6.10A). **Antiport** (facilitated by a protein called an *antiporter*) refers to coupled transport in which the downhill movement of protons drives the active (uphill) transport of a solute in the opposite direction (Figure 6.10B).

In both types of secondary transport, the ion or solute being transported simultaneously with the protons is moving against its gradient of electrochemical potential, so its transport is active. However, the energy driving this transport is provided by the proton motive force rather than directly by ATP hydrolysis.

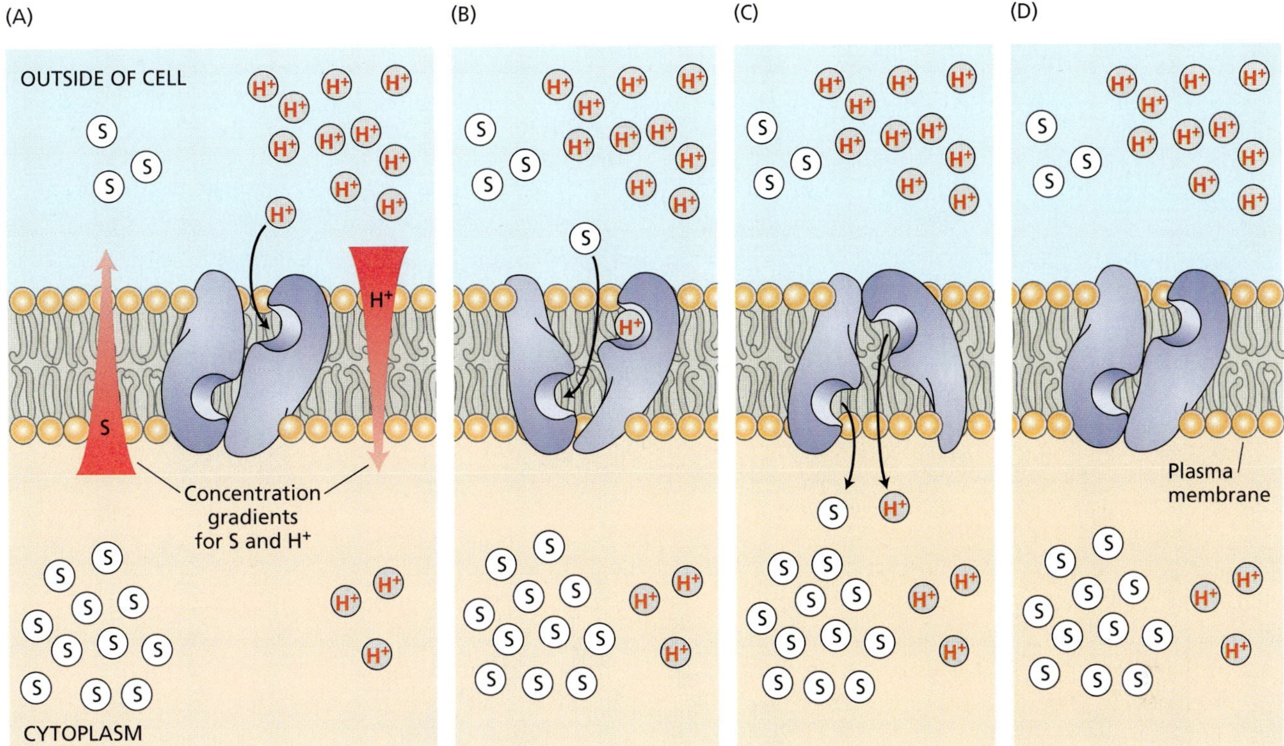

FIGURE 6.9 Hypothetical model for secondary active transport. The energy that drives the process has been stored in a $\Delta\tilde{\mu}_{H^+}$ (symbolized by the red arrow on the right in A) and is being used to take up a substrate (S) against its concentration gradient (left-hand red arrow). (A) In the initial conformation, the binding sites on the protein are exposed to the outside environment and can bind a proton. (B) This binding results in a conformational change that permits a molecule of S to be bound. (C) The binding of S causes another conformational change that exposes the binding sites and their substrates to the inside of the cell. (D) Release of a proton and a molecule of S to the cell's interior restores the original conformation of the carrier and allows a new pumping cycle to begin.

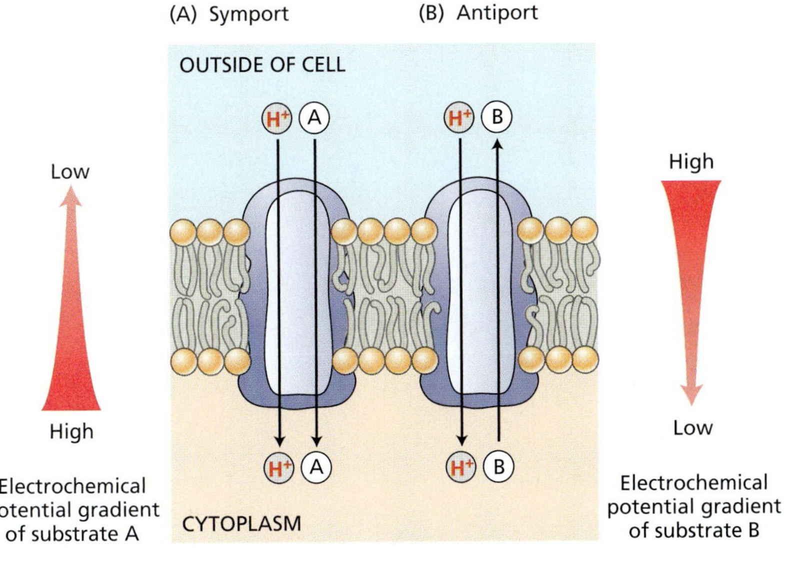

FIGURE 6.10 Two examples of secondary active transport coupled to a primary proton gradient. (A) In a symport, the energy dissipated by a proton moving back into the cell is coupled to the uptake of one molecule of a substrate (e.g., a sugar) into the cell. (B) In an antiport, the energy dissipated by a proton moving back into the cell is coupled to the active transport of a substrate (for example, a sodium ion) out of the cell. In both cases, the substrate under consideration is moving against its gradient of electrochemical potential. Both neutral and charged substrates can be transported by such secondary active transport processes.

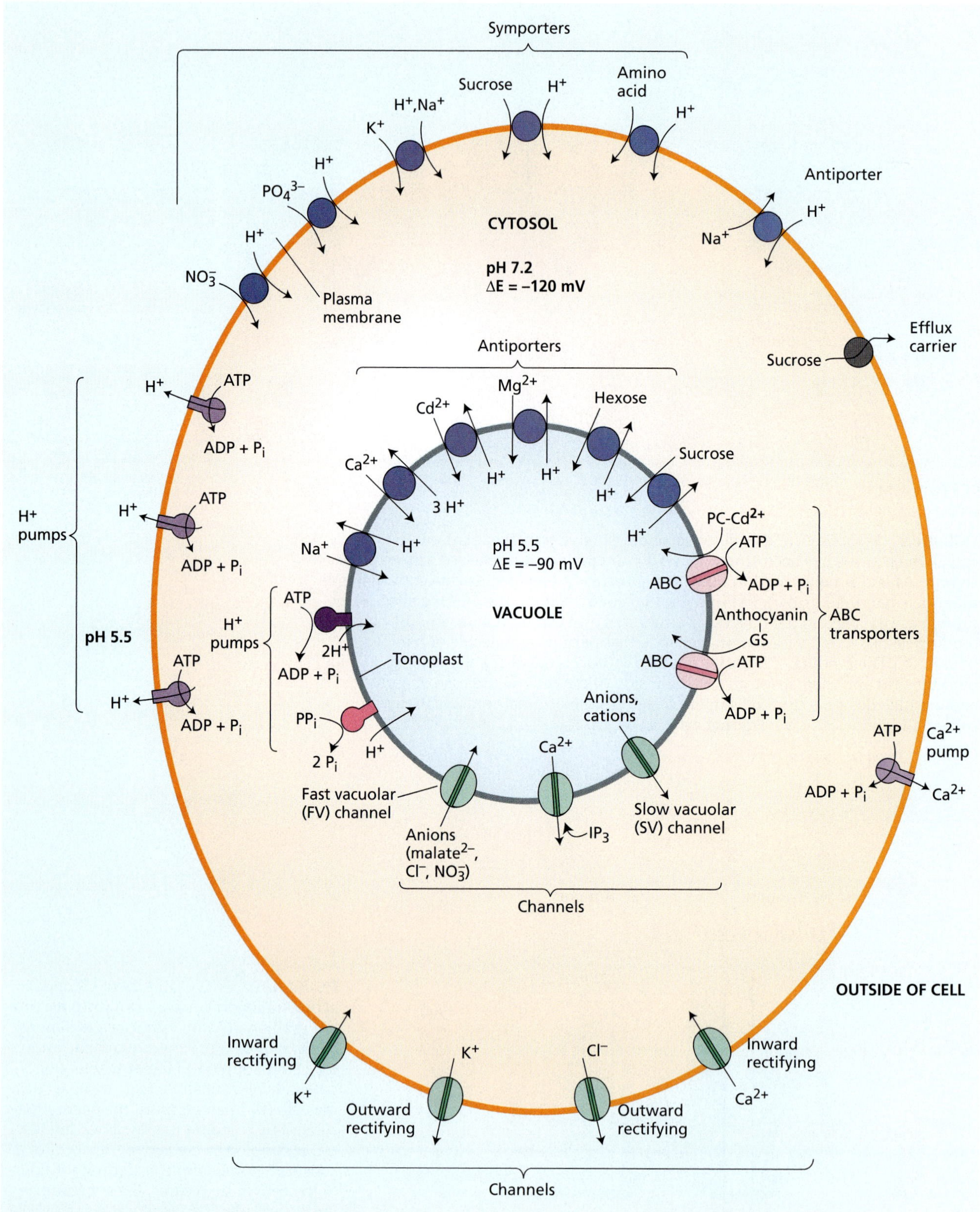

FIGURE 6.11 Overview of the various transport processes on the plasma membrane and tonoplast of plant cells.

Typically, transport across a biological membrane is energized by one primary active transport system coupled to ATP hydrolysis. The transport of that ion—for example, H^+—generates an ion gradient and an electrochemical potential. Many other ions or organic substrates can then be transported by a variety of secondary active-transport proteins, which energize the transport of their respective substrates by simultaneously carrying one or two H^+ ions down their energy gradient. Thus H^+ ions circulate across the membrane, outward through the primary active transport proteins, and back into the cell through the secondary transport proteins. In plants and fungi, sugars and amino acids are taken up by symport with protons.

Most of the ionic gradients across membranes of higher plants are generated and maintained by electrochemical-potential gradients of H^+ (Tazawa et al. 1987). In turn, these H^+ gradients are generated by the electrogenic proton pumps. Evidence suggests that in plants, Na^+ is transported out of the cell by a Na^+–H^+ antiporter and that Cl^-, NO_3^-, $H_2PO_4^-$, sucrose, amino acids, and other substances enter the cell via specific proton symporters.

What about K^+? At very low external concentrations, K^+ can be taken up by active symport proteins, but at higher concentrations it can enter the cell by diffusion through specific K^+ channels. However, even influx through channels is driven by the H^+-ATPase, in the sense that K^+ diffusion is driven by the membrane potential, which is maintained at a value more negative than the K^+ equilibrium potential by the action of the electrogenic H^+ pump. Conversely, K^+ efflux requires the membrane potential to be maintained at a value more positive than E_K, which can be achieved if efflux of Cl^- through Cl^- channels is allowed. Several representative transport processes located on the plasma membrane and the tonoplast are illustrated in Figure 6.11.

MEMBRANE TRANSPORT PROTEINS

We have seen in preceding sections that some transmembrane proteins operate as channels for the controlled diffusion of ions. Other membrane proteins act as carriers for other substances (mostly molecules and ions). Active transport utilizes carrier-type proteins that are energized directly by ATP hydrolysis or indirectly as symporters and antiporters. The latter systems use the energy of ion gradients (often a H^+ gradient) to drive the uphill transport of another ion or molecule. In the pages that follow we will examine in more detail the molecular properties, cellular locations, and genetic manipulations of some of these transport proteins.

Kinetic Analyses Can Elucidate Transport Mechanisms

Thus far, we have described cellular transport in terms of its energetics. However, cellular transport can also be studied by use of enzyme kinetics because transport involves the binding and dissociation of molecules at active sites on transport proteins. One advantage of the kinetic approach is that it gives new insights into the regulation of transport.

In kinetic experiments the effects of external ion (or other solute) concentrations on transport rates are measured. The kinetic characteristics of the transport rates can then be used to distinguish between different transporters. The maximum rate (V_{max}) of carrier-mediated transport, and often channel transport as well, cannot be exceeded, regardless of the concentration of substrate (Figure 6.12). V_{max} is approached when the substrate-binding site on the carrier is always occupied. The concentration of carrier, not the concentration of solute, becomes rate limiting. Thus V_{max} is a measure of the number of molecules of the specific carrier protein that are functioning in the membrane.

The constant K_m (which is numerically equal to the solute concentration that yields half the maximal rate of transport) tends to reflect the properties of the particular binding site (for a detailed discussion on K_m and V_{max} see Chapter 2 on the web site). Low K_m values indicate high affinity of the transport site for the transported substance. Such values usually imply the operation of a carrier system. Higher values of K_m indicate a lower affinity of the transport site for the solute. The affinity is often so low that in practice V_{max} is never reached. In such cases, kinetics alone cannot distinguish between carriers and channels.

Usually transport displays both high-affinity and low-affinity components when a wide range of solute concentrations are studied. Figure 6.13 shows sucrose uptake by soybean cotyledon protoplasts as a function of the external

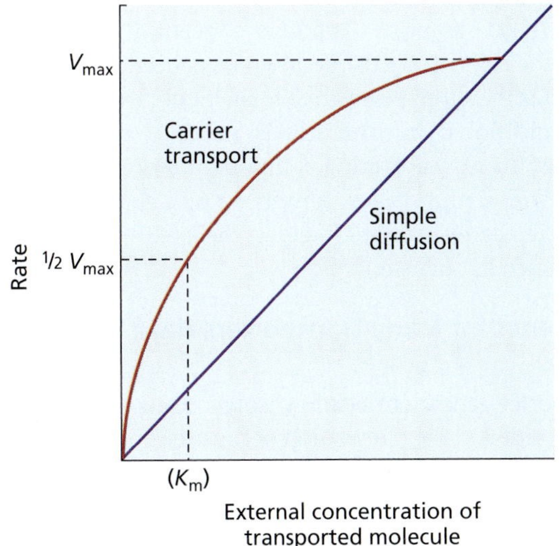

FIGURE 6.12 Carrier transport often shows saturation kinetics (V_{max}) (see Chapter 2 on the web site), because of saturation of a binding site. Ideally, diffusion through channels is directly proportional to the concentration of the transported solute, or for an ion, to the difference in electrochemical potential across the membrane.

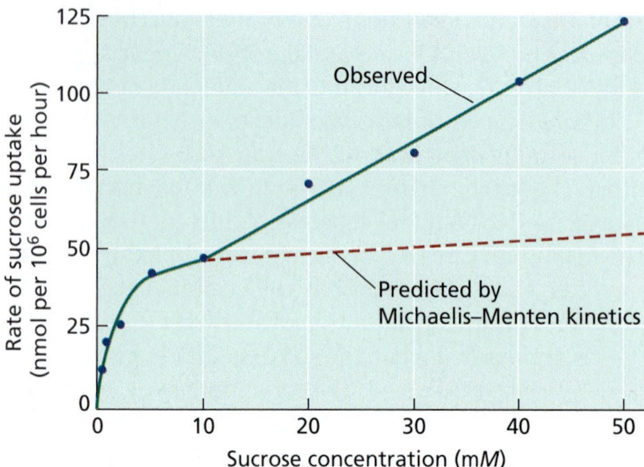

FIGURE 6.13 The transport properties of a solute can change at different solute concentrations. For example, at low concentrations (1 to 10 m*M*), the rate of uptake of sucrose by soybean cells shows saturation kinetics typical of carriers. A curve fit-ted to these data is predicted to approach a maximal rate (V_{max}) of 57 nmol per 10^6 cells per hour. Instead, at higher sucrose concentrations the uptake rate continues to increase linearly over a broad range of concentrations, suggesting the existence of other sucrose transporters, which might be carriers with very low affinity for the substrate. (From Lin et al. 1984.)

sucrose concentration (Lin et al. 1984). Uptake increases sharply with concentration and begins to saturate at about 10 m*M*. At concentrations above 10 m*M*, uptake becomes linear and nonsaturable. Inhibition of ATP synthesis with metabolic poisons blocks the saturable component but not the linear one. The interpretation is that sucrose uptake at low concentrations is an active carrier-mediated process (sucrose–H^+ symport). At higher concentrations, sucrose enters the cells by diffusion down its concentration gradient and is therefore insensitive to metabolic poisons. However, additional information is needed to investigate whether the nonsaturating component represents uptake by a carrier with very low affinity, or by a channel. (Transport by a carrier is more likely in the case of a molecular solute such as sucrose.)

The Genes for Many Transporters Have Been Cloned

Transporter gene identification, isolation, and cloning have greatly aided in the elucidation of the molecular properties of transporter proteins. Nitrate transport is an example that is of interest not only because of its nutritional importance, but also because of its complexity. Kinetic analysis shows that nitrate transport, like the sucrose transport shown in Figure 6.13, has both high-affinity (low K_m) and low-affinity (high K_m) components. In contrast with sucrose, nitrate is negatively charged, and such an electric charge imposes an energy requirement for the transport of the nitrate ion at all concentrations. The energy is provided by symport with H^+.

Nitrate transport is also strongly regulated according to nitrate availability: The enzymes required for nitrate transport, as well as nitrate assimilation (see Chapter 12), are induced in the presence of nitrate in the environment, and uptake can also be repressed if nitrate accumulates in the cells.

Mutants in nitrate transport or nitrate reduction can be selected by growth in the presence of chlorate (ClO_3^-). Chlorate is a nitrate analog that is taken up and reduced in wild-type plants to the toxic product chlorite. If plants resistant to chlorate are selected, they are likely to show mutations that block nitrate transport or reduction.

Several such mutations have been identified in *Arabidopsis*, a small crucifer that is ideal for genetic studies. The first transport gene identified in this way encodes a low-affinity inducible nitrate–proton symporter. As more genes for nitrate transport have been identified and characterized, the picture has become more complex. Each component of transport may involve more than one gene product, and at least one gene encodes a dual-affinity carrier that contributes to both high-affinity and low-affinity transport (Chrispeels et al. 1999).

The emerging picture of plant transporter genes shows that a family of genes, rather than an individual gene, exists in the plant genome for each transport function. Within a gene family, variations in transport characteristics such as K_m, in mode of regulation, and in differential tissue expression give plants a remarkable plasticity to acclimate to a broad range of environmental conditions.

The identification of regions of sequence similarity between plant transport genes and the transport genes of other organisms, such as yeast, has enabled the cloning of plant transport genes (Kochian 2000). In some cases, it has been possible to identify the gene after purifying the transport protein, but often sequence similarity is limited, and individual transport proteins represent too small a fraction of total protein. Another way to identify transport genes is to screen plant cDNA (complementary DNA) libraries for genes that complement (i.e., compensate for) transport deficiencies in yeast. Many yeast transport mutants are known and have been used to identify corresponding plant genes by complementation.

In the case of genes for ion channels, researchers have studied the behavior of the channel proteins by expressing the genes in oocytes of the toad *Xenopus*, which, because of their large size, are convenient for electrophysiological studies. Genes for both inward- and outward-rectifying K^+ channels have been cloned and studied in this way. Of the inward K^+ channel genes identified so far, one is expressed strongly in stomatal guard cells, another in roots, and a third in leaves. These channels are considered to be responsible for low-affinity K^+ uptake into plant cells.

An outward K^+ channel responsible for K^+ flux from root stelar cells into the dead xylem vessels has been cloned, and several genes for high-affinity K^+ carriers have been identified. Further research is needed to determine to what extent they each contribute to K^+ uptake, and how they obtain their energy (see **Web Topic 6.4**). Genes for plant vacuolar H^+–Ca^{2+} antiporters and genes for the proton symport of several amino acids and sugars have also been identified through various genetic techniques (Hirshi et al. 1996; Tanner and Caspari 1996; Kuehn et al. 1999).

Genes for Specific Water Channels Have Been Identified

Aquaporins are a class of proteins that is relatively abundant in plant membranes (see Chapter 3). Aquaporins reveal no ion currents when expressed in oocytes, but when the osmolarity of the external medium is reduced, expression of these proteins results in swelling and bursting of the oocytes. The bursting results from rapid influx of water across the oocyte plasma membrane, which normally has a very low water permeability. These results show that aquaporins form water channels in membranes (see Figure 3.6).

The existence of aquaporins was a surprise at first because it was thought that the lipid bilayer is itself sufficiently permeable to water. Nevertheless, aquaporins are common in plant and animal membranes, and their expression and activity appear to be regulated, possibly by protein phosphorylation, in response to water availability (Tyerman et al. 2002).

The Plasma Membrane H^+-ATPase Has Several Functional Domains

The outward, active transport of H^+ across the plasma membrane creates gradients of pH and electric potential that drive the transport of many other substances (ions and molecules) through the various secondary active-transport proteins. Figure 6.14 illustrates how a membrane H^+-ATPase might work.

Plant and fungal plasma membrane H^+-ATPases and Ca^{2+}-ATPases are members of a class known as P-type ATPases, which are phosphorylated as part of the catalytic cycle that hydrolyzes ATP. Because of this phosphorylation step, the plasma membrane ATPases are strongly inhibited by orthovanadate (HVO_4^{2-}), a phosphate (HPO_4^{2-}) analog that competes with phosphate from ATP for the aspartic acid phosphorylation site on the enzyme. The high affinity of the enzyme for vanadate is attributed to the fact that vanadate can mimic the transitional structure of phosphate during hydrolysis.

Plasma membrane H^+-ATPases are encoded by a family of about ten genes. Each gene encodes an isoform of the enzyme (Sussman 1994). The isoforms are tissue specific, and they are preferentially expressed in the root, the seed, the phloem, and so on. The functional specificity of each isoform is not yet understood; it may alter the pH optimum of some isoforms and allow transport to be regulated in different ways for each tissue.

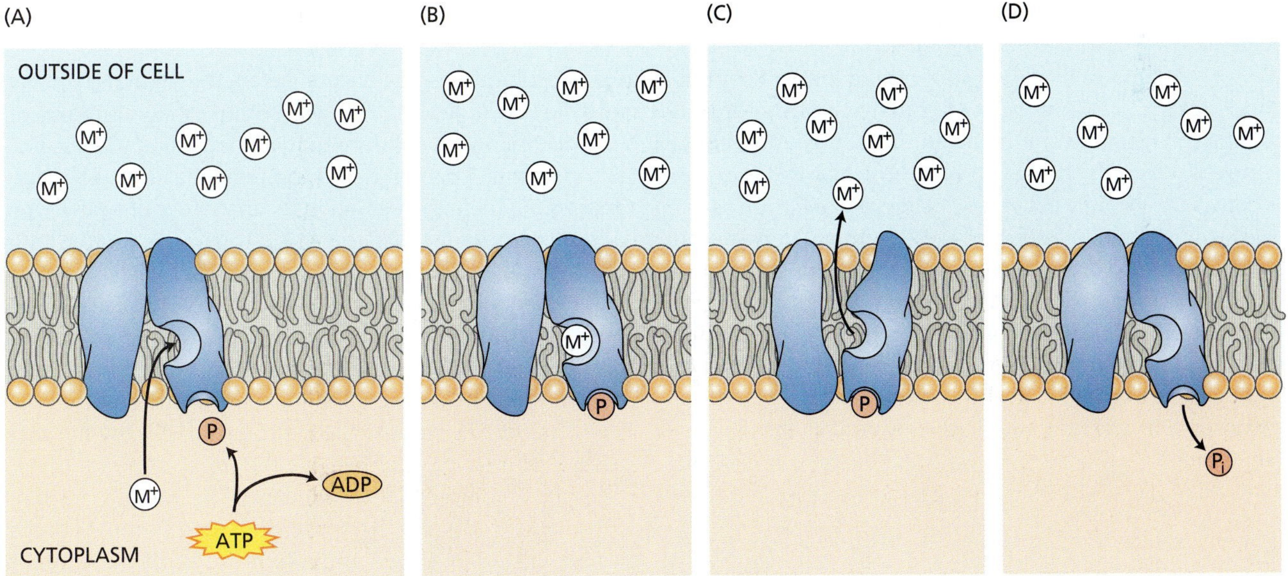

FIGURE 6.14 Hypothetical steps in the transport of a cation (the hypothetical M^+) against its chemical gradient by an electrogenic pump. The protein, embedded in the membrane, binds the cation on the inside of the cell (A) and is phosphorylated by ATP (B). This phosphorylation leads to a conformational change that exposes the cation to the outside of the cell and makes it possible for the cation to diffuse away (C). Release of the phosphate ion (P) from the protein into the cytosol (D) restores the initial configuration of the membrane protein and allows a new pumping cycle to begin.

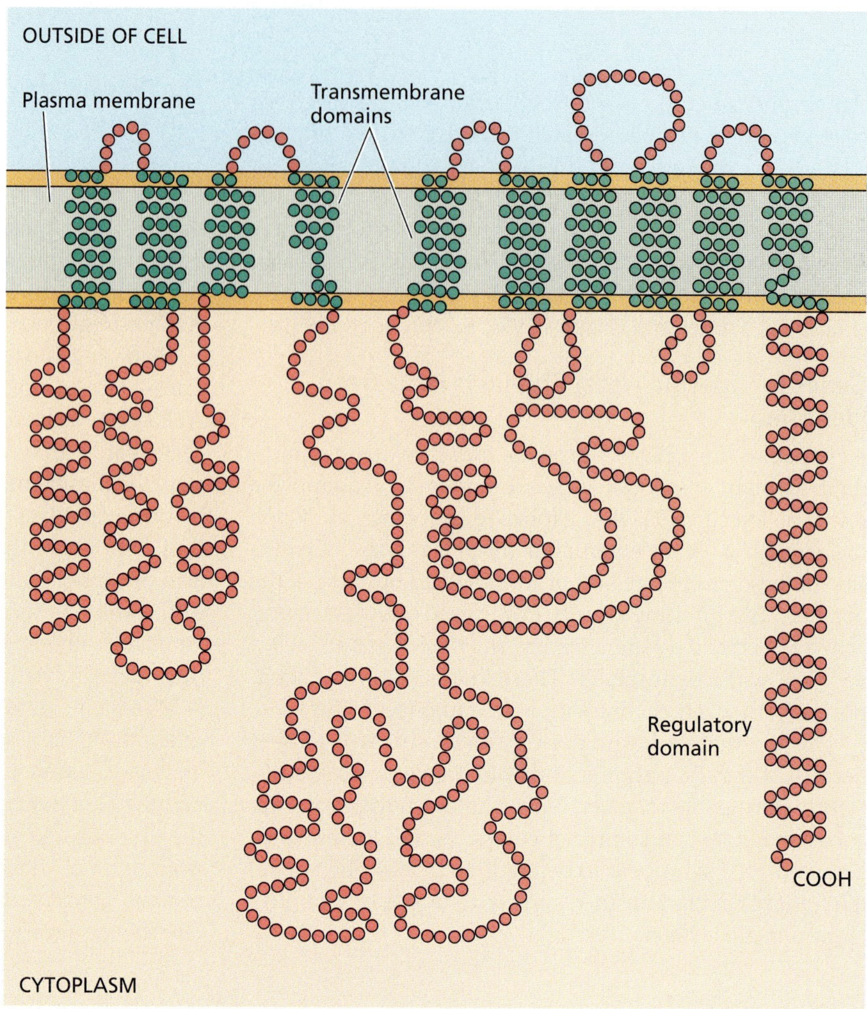

FIGURE 6.15 Two-dimensional representation of the plasma membrane H^+-ATPase. The H^+-ATPase has 10 transmembrane segments. The regulatory domain is the autoinhibitory domain. (From Palmgren 2001.)

Figure 6.15 shows a model of the functional domains of the plasma membrane H^+-ATPase of yeast, which is similar to that of plants. The protein has ten membrane-spanning domains that cause it to loop back and forth across the membrane. Some of the membrane-spanning domains make up the pathway through which protons are pumped. The catalytic domain, including the aspartic acid residue that becomes phosphorylated during the catalytic cycle, is on the cytosolic face of the membrane.

Like other enzymes, the plasma membrane ATPase is regulated by the concentration of substrate (ATP), pH, temperature, and other factors. In addition, H^+-ATPase molecules can be reversibly activated or deactivated by specific signals, such as light, hormones, pathogen attack, and the like. This type of regulation is mediated by a specialized autoinhibitory domain at the C-terminal end of the polypeptide chain, which acts to regulate the activity of the proton pump (see Figure 6.15). If the autoinhibitory domain is removed through the action of a protease, the enzyme becomes irreversibly activated (Palmgren 2001).

The autoinhibitory effect of the C-terminal domain can also be regulated through the action of protein kinases and phosphatases that add or remove phosphate groups to serine or threonine residues on the autoinhibitory domain of the enzyme. For example, one mechanism of response to pathogens in tomato involves the activation of protein phosphatases that dephosphorylate residues on the plasma membrane H^+-ATPase, thereby activating it (Vera-Estrella et al. 1994). This is one step in a cascade of responses that activate plant defenses.

The Vacuolar H^+-ATPase Drives Solute Accumulation into Vacuoles

Because plant cells increase their size primarily by taking up water into large, central vacuoles, the osmotic pressure of the vacuole must be maintained sufficiently high for water to enter from the cytoplasm. The tonoplast regulates the traffic of ions and metabolites between the cytosol and the vacuole, just as the plasma membrane regulates uptake into the cell. Tonoplast transport became a vigorous area of research following the development of methods for the isolation of intact vacuoles and tonoplast vesicles (see **Web Topic 6.5**). These studies led to the discovery of a new type of proton-pumping ATPase, which transports protons into the vacuole (see Figure 6.11).

The vacuolar H^+-ATPase (also called **V-ATPase**) differs both structurally and functionally from the plasma membrane H^+-ATPase. The vacuolar ATPase is more closely related to the F-ATPases of mitochondria and chloroplasts (see Chapter 11). Because the hydrolysis of ATP by the vacuolar ATPase does not involve the formation of a phosphorylated intermediate, vacuolar ATPases are insensitive to vanadate, the inhibitor of plasma membrane ATPases discussed earlier. Vacuolar ATPases are specifically inhibited by the antibiotic bafilomycin, as well as by high concentrations of nitrate, neither of which inhibit plasma membrane ATPases. Use of these selective inhibitors makes it possible to identify different types of ATPases, and to assay their activity.

Vacuolar ATPases belong to a general class of ATPases that are present on the endomembrane systems of all

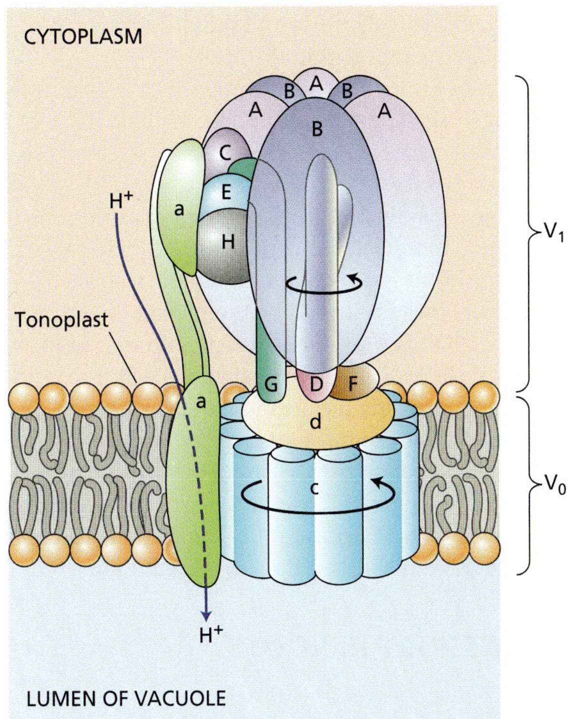

FIGURE 6.16 Model of the V-ATPase rotary motor. Many polypeptide subunits come together to make this complex enzyme. The V_1 catalytic complex is easily dissociated from the membrane, and contains the nucleotide-binding and catalytic sites. Components of V_1 are designated by uppercase letters. The intrinsic membrane complex mediating H^+ transport is designated V_0, and its subunits are given lowercase letters. It is proposed that ATPase reactions catalyzed by each of the A subunits, acting in sequence, drive the rotation of the shaft D and the six c subunits. The rotation of the c subunits relative to subunit a is thought to drive the transport of H^+ across the membrane. (Based on an illustration courtesy of M. F. Manolson.)

eukaryotes. They are large enzyme complexes, about 750 kDa, composed of at least ten different subunits (Lüttge and Ratajczak 1997). These subunits are organized into a peripheral catalytic complex, V_1, and an integral membrane channel complex, V_0 (Figure 6.16). Because of their similarities to F-ATPases, vacuolar ATPases are assumed to operate like tiny rotary motors (see Chapter 11).

Vacuolar ATPases are electrogenic proton pumps that transport protons from the cytoplasm to the vacuole and generate a proton motive force across the tonoplast. The electrogenic proton pumping accounts for the fact that the vacuole is typically 20 to 30 mV more positive than the cytoplasm, although it is still negative relative to the external medium. To maintain bulk electrical neutrality, anions such as Cl^- or malate^{2-} are transported from the cytoplasm into the vacuole through channels in the membrane (Barkla and Pantoja 1996). Without the simultaneous movement of anions along with the pumped protons, the charge buildup across the tonoplast would make the pumping of additional protons energetically impossible.

The conservation of bulk electrical neutrality by anion transport makes it possible for the vacuolar H^+-ATPase to generate a large concentration (pH) gradient of protons across the tonoplast. This gradient accounts for the fact that the pH of the vacuolar sap is typically about 5.5, while the cytoplasmic pH is 7.0 to 7.5. Whereas the electrical component of the proton motive force drives the uptake of anions into the vacuole, the electrochemical-potential gradient for H^+ ($\Delta\tilde{\mu}_{H^+}$) is harnessed to drive the uptake of cations and sugars into the vacuole via secondary transport (antiporter) systems (see Figure 6.11).

Although the pH of most plant vacuoles is mildly acidic (about 5.5), the pH of the vacuoles of some species is much lower—a phenomenon termed *hyperacidification*. Vacuolar hyperacidification is the cause of the sour taste of certain fruits (lemons) and vegetables (rhubarb). Some extreme examples are listed in Table 6.2. Biochemical studies with lemon fruits have suggested that the low pH of the lemon fruit vacuoles (specifically, those of the juice sac cells) is due to a combination of factors:

- The low permeability of the vacuolar membrane to protons permits a steeper pH gradient to build up.
- A specialized vacuolar ATPase is able to pump protons more efficiently (with less wasted energy) than normal vacuolar ATPases can (Müller et al. 1997).

TABLE 6.2
The vacuolar pH of some hyperacidifying plant species

Tissue	Species	pH[a]
Fruits		
	Lime (*Citrus aurantifolia*)	1.7
	Lemon (*Citrus limonia*)	2.5
	Cherry (*Prunus cerasus*)	2.5
	Grapefruit (*Citrus paradisi*)	3.0
Leaves		
	Rosette oxalis (*Oxalis deppei*)	1.3
	Wax begonia (*Begonia semperflorens*)	1.5
	Begonia 'Lucerna'	0.9 – 1.4
	Oxalis sp.	1.9 – 2.6
	Sorrel (*Rumex* sp.)	2.6
	Prickly Pear (*Opuntia phaeacantha*)[b]	1.4 (6:45 A.M.) 5.5 (4:00 P.M.)

Source: Data from Small 1946.

[a] The values represent the pH of the juice or expressed sap of each tissue, usually a good indicator of vacuolar pH.

[b] The vacuolar pH of the cactus *Opuntia phaeacantha* varies with the time of day. As will be discussed in Chapter 8, many desert succulents have a specialized type of photosynthesis, called crassulacean acid metabolism (CAM), that causes the pH of the vacuoles to decrease during the night.

- The accumulation of organic acids such as citric, malic, and oxalic acids helps maintain the low pH of the vacuole by acting as buffers.

Plant Vacuoles Are Energized by a Second Proton Pump, the H^+-Pyrophosphatase

Another type of proton pump, an H^+-pyrophosphatase (H^+-PPase) (Rea et al. 1998), appears to work in parallel with the vacuolar ATPase to create a proton gradient across the tonoplast (see Figure 6.11). This enzyme consists of a single polypeptide that has a molecular mass of 80 kDa. The H^+-PPase harnesses its energy from the hydrolysis of inorganic pyrophosphate (PP_i).

The free energy released by PP_i hydrolysis is less than that from ATP hydrolysis. However, the vacuolar H^+-PPase transports only one H^+ ion per PP_i molecule hydrolyzed, whereas the vacuolar ATPase appears to transport two H^+ ions per ATP hydrolyzed. Thus the energy available per H^+ ion transported appears to be the same, and the two enzymes appear to be able to generate comparable H^+ gradients.

In some plants the synthesis of the vacuolar H^+-PPase is induced by low O_2 levels (hypoxia), Pi starvation, or by chilling. This indicates that the vacuolar H^+-PPase might function as a backup system to maintain essential cell metabolism under conditions in which ATP supply is depleted because of the inhibition (see **Web Essay 11.1**). It is of interest that the plant vacuolar H^+-PPase is not found in animals or yeast, although a similar enzyme is present in some bacteria and protists.

Large metabolites such as flavonoids, anthocyanins and secondary products of metabolism are sequestered in the vacuole. These large molecules are transported into vacuoles by **ATP-binding cassette (ABC) transporters**. Transport processes by the ABC transporters consume ATP and do not depend on a primary electrochemical gradient (see **Web Topic 6.6**). Recent studies have shown that ABC transporters can also be found at the plasma membrane and in mitochondria (Theodoulou 2000).

Calcium Pumps, Antiports, and Channels Regulate Intracellular Calcium

Calcium is another important ion whose concentration is strongly regulated. Calcium concentrations in the cell wall and the apoplastic (extracellular) spaces are usually in the millimolar range; free cytosolic Ca^{2+} concentrations are maintained at the micromolar (10^{-6} *M*) range, against the large electrochemical-potential gradient that drives Ca^{2+} diffusion into the cell.

Small fluctuations in cytosolic Ca^{2+} concentration drastically alter the activities of many enzymes, making calcium an important second messenger in signal transduction. Most of the calcium in the cell is stored in the central vacuole, where it is taken up via Ca^{2+}–H^+ antiporters, which use the electrochemical potential of the proton gradient to energize the accumulation of calcium into the vacuole (Bush 1995). Mitochondria and the endoplasmic reticulum also store calcium within the cells.

Calcium efflux from the vacuole into the cytosol may in some cells be triggered by inositol trisphosphate (IP_3). IP_3, which appears to act as a "second messenger" in certain signal transduction pathways, induces the opening of IP_3-gated calcium channels on the tonoplast and endoplasmic reticulum (ER). (For a more detailed description of these sensory transduction pathways see Chapter 14 on the web site.)

Calcium ATPases are found at the plasma membrane (Chung et al. 2000) and in some endomembranes of plant cells (see Figure 6.11). Plant cells regulate cytosolic Ca^{2+} concentrations by controlling the opening of Ca^{2+} channels that allow calcium to diffuse in, as well as by modulating the activity of pumps that drive Ca^{2+} out of the cytoplasm back into the extracellular spaces. Whereas the plasma membrane calcium pumps move calcium out of the cell, the calcium pumps on the ER transport calcium into the ER lumen.

ION TRANSPORT IN ROOTS

Mineral nutrients absorbed by the root are carried to the shoot by the transpiration stream moving through the xylem (see Chapter 4). Both the initial uptake of nutrients and the subsequent movement of mineral ions from the root surface across the cortex and into the xylem are highly specific, well-regulated processes.

Ion transport across the root obeys the same biophysical laws that govern cellular transport. However, as we have seen in the case of water movement (see Chapter 4), the anatomy of roots imposes some special constraints on the pathway of ion movement. In this section we will discuss the pathways and mechanisms involved in the radial movement of ions from the root surface to the tracheary elements of the xylem.

Solutes Move through Both Apoplast and Symplast

Thus far, our discussion of cellular ion transport has not included the cell wall. In terms of the transport of small molecules, the cell wall is an open lattice of polysaccharides through which mineral nutrients diffuse readily. Because all plant cells are separated by cell walls, ions can diffuse across a tissue (or be carried passively by water flow) entirely through the cell wall space without ever entering a living cell. This continuum of cell walls is called the *extracellular space*, or *apoplast* (see Figure 4.3).

We can determine the apoplastic volume of a slice of plant tissue by comparing the uptake of ^{3}H-labeled water and ^{14}C-labeled mannitol. Mannitol is a nonpermeating sugar alcohol that diffuses within the extracellular space but cannot enter the cells. Water, on the other hand, freely penetrates both the cells and the cell walls. Measurements of this type usually show that 5 to 20% of the plant tissue volume is occupied by cell walls.

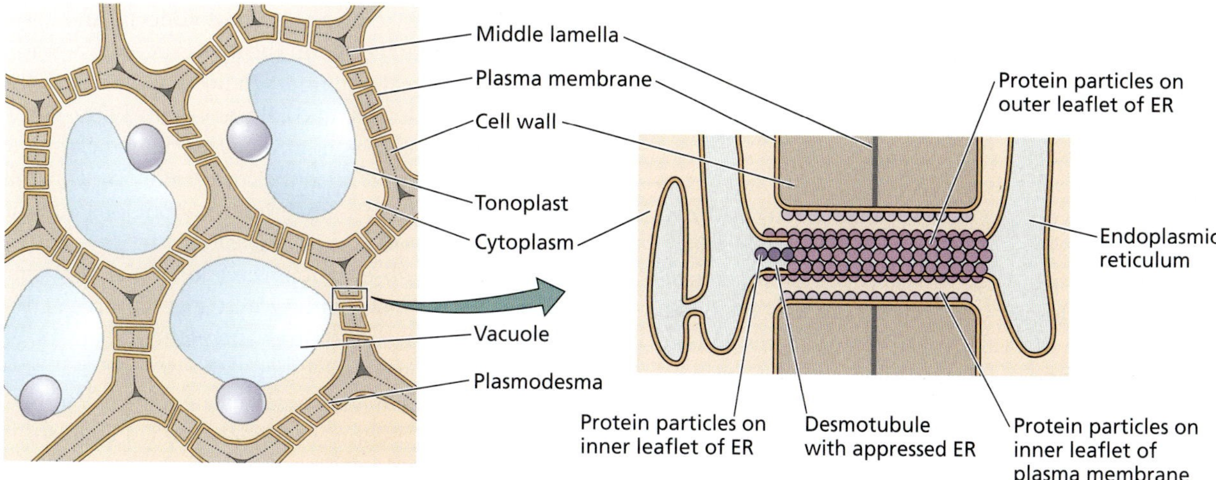

FIGURE 6.17 Diagram illustrating how plasmodesmata connect the cytoplasms of neighboring cells. Plasmodesmata are about 40 nm in diameter and allow diffusion of water and small molecules from one cell to the next. In addition, the size of the opening can be regulated by rearrangements of the internal proteins to allow the passage of larger molecules.

Just as the cell walls form a continuous phase, so do the cytoplasms of neighboring cells, collectively referred to as the *symplast*. Plant cells are interconnected by cytoplasmic bridges called plasmodesmata (see Chapter 1), cylindrical pores 20 to 60 nm in diameter (see Figure 1.27). Each plasmodesma is lined with a plasma membrane and contains a narrow tubule, the desmotubule, that is a continuation of the endoplasmic reticulum.

In tissues where significant amounts of intercellular transport occur, neighboring cells contain large numbers of plasmodesmata, up to 15 per square micrometer of cell surface (Figure 6.17). Specialized secretory cells, such as floral nectaries and leaf salt glands, appear to have high densities of plasmodesmata; so do the cells near root tips, where most nutrient absorption occurs.

By injecting dyes or by making electrical-resistance measurements on cells containing large numbers of plasmodesmata, investigators have shown that ions, water, and small solutes can move from cell to cell through these pores. Because each plasmodesma is partly occluded by the desmotubule and associated proteins (see Chapter 1), the movement of large molecules such as proteins through the plasmodesmata requires special mechanisms (Ghoshroy et al. 1997). Ions, on the other hand, appear to move from cell to cell through the entire plant by simple diffusion through the symplast (see Chapter 4).

Ions Moving through the Root Cross Both Symplastic and Apoplastic Spaces

Ion absorption by the roots (see Chapter 5) is more pronounced in the root hair zone than in the meristem and elongation zones. Cells in the root hair zone have completed their elongation but have not yet begun secondary growth. The root hairs are simply extensions of specific epidermal cells that greatly increase the surface area available for ion absorption.

An ion that enters a root may immediately enter the symplast by crossing the plasma membrane of an epidermal cell, or it may enter the apoplast and diffuse between the epidermal cells through the cell walls. From the apoplast of the cortex, an ion may either cross the plasma membrane of a cortical cell, thus entering the symplast, or diffuse radially all the way to the endodermis via the apoplast. In all cases, ions must enter the symplast before they can enter the stele, because of the presence of the Casparian strip.

The apoplast forms a continuous phase from the root surface through the cortex. At the boundary between the vascular cylinder (the stele) and the cortex is a layer of specialized cells, the endodermis. As discussed in Chapters 4 and 5, a suberized cell layer in the endodermis, known as the Casparian strip, effectively blocks the entry of water and mineral ions into the stele via the apoplast.

Once an ion has entered the stele through the symplastic connections across the endodermis, it continues to diffuse from cell to cell into the xylem. Finally, the ion reenters the apoplast as it diffuses into a xylem tracheid or vessel element. Again, the Casparian strip prevents the ion from diffusing back out of the root through the apoplast. The presence of the Casparian strip allows the plant to maintain a higher ionic concentration in the xylem than exists in the soil water surrounding the roots.

Xylem Parenchyma Cells Participate in Xylem Loading

Once ions have been taken up into the symplast of the root at the epidermis or cortex, they must be loaded into the tracheids or vessel elements of the stele to be translocated to the shoot. The stele consists of dead tracheary elements and the living xylem parenchyma. Because the xylem tracheary elements are dead cells, they lack cytoplasmic continuity with surrounding xylem parenchyma. To enter the tracheary elements, the ions must exit the symplast by crossing a plasma membrane a second time.

The process whereby ions exit the symplast and enter the conducting cells of the xylem is called **xylem loading**. The mechanism of xylem loading has long baffled scientists. Ions could enter the tracheids and vessel elements of the xylem by simple passive diffusion. In this case, the movement of ions from the root surface to the xylem would take only a single step requiring metabolic energy. The site of this single-step, energy-dependent uptake would be the plasma membrane surfaces of the root epidermal, cortical, or endodermal cells. According to the passive-diffusion model, ions move passively into the stele via the symplast down a gradient of electrochemical potential, and then leak out of the living cells of the stele (possibly because of lower oxygen availability in the interior of the root) into the nonliving conducting cells of the xylem.

Support for the passive-diffusion model was provided by use of ion-specific microelectrodes to measure the electrochemical potentials of various ions across maize roots (Figure 6.18) (Dunlop and Bowling 1971). Data from this and other studies indicate that K^+, Cl^-, Na^+, SO_4^{2-}, and NO_3^- are all taken up actively by the epidermal and cortical cells and are maintained in the xylem against a gradient of electrochemical potential when compared with the external medium (Lüttge and Higinbotham 1979). However, none of these ions is at a higher electrochemical potential in the xylem than in the cortex or living portions of the stele. Therefore, the final movement of ions into the xylem could be due to passive diffusion.

However, other observations have led to the view that this final step of xylem loading may also involve active processes within the stele (Lüttge and Higinbotham 1979). With the type of apparatus shown in Figure 6.19, it is possible to make simultaneous measurements of ion uptake into the epidermal or cortical cytoplasm and of ion loading into the xylem.

By using treatments with inhibitors and plant hormones, investigators have shown that ion uptake by the cortex and ion loading into the xylem operate independently. For example, treatment with the protein synthesis inhibitor cycloheximide or with the cytokinin benzyladenine inhibits xylem loading without affecting uptake by the cortex. This result indicates that efflux from the stelar cells is regulated independently from uptake by the cortical cells.

Recent biochemical studies have supported a role for the xylem parenchyma cells in xylem loading. The plasma

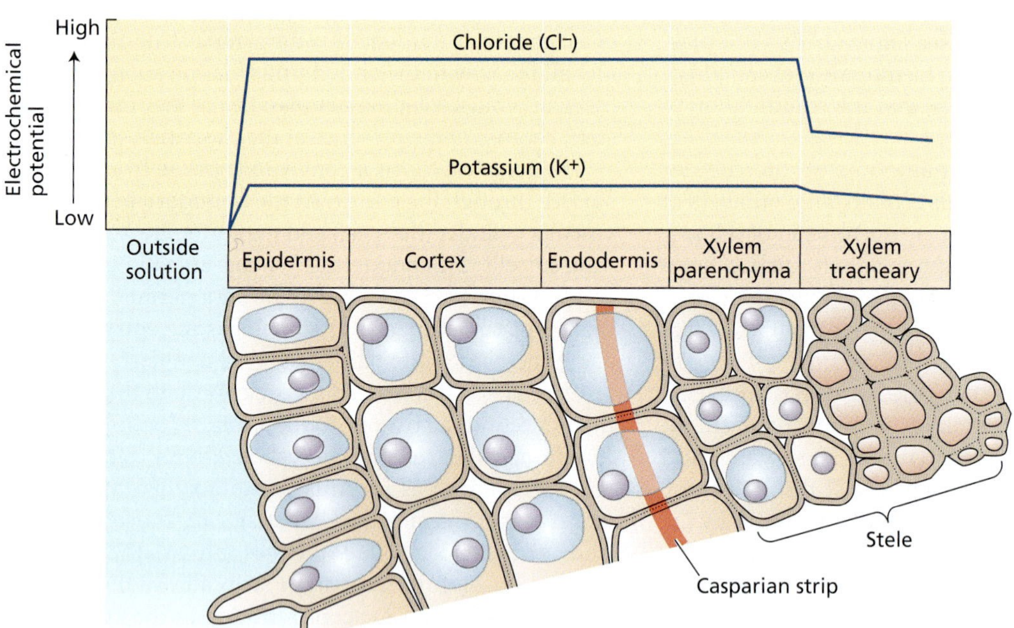

FIGURE 6.18 Diagram showing electrochemical potentials of K^+ and Cl^- across a maize root. To determine the electrochemical potentials, the root was bathed in a solution containing 1 m*M* KCl and 0.1 m*M* $CaCl_2$. A reference electrode was positioned in the bathing solution, and an ion-sensitive measuring electrode was inserted in different cells of the root. The horizontal axis shows the different tissues found in a root cross section. The substantial increase in electrochemical potential for both K^+ and Cl^- between the bathing medium and the epidermis indicates that ions are taken up into the root by an active transport process. In contrast, the potentials decrease at the xylem vessels, suggesting that ions are transported into the xylem by passive diffusion down the gradient of electrochemical potential. (After Dunlop and Bowling 1971.)

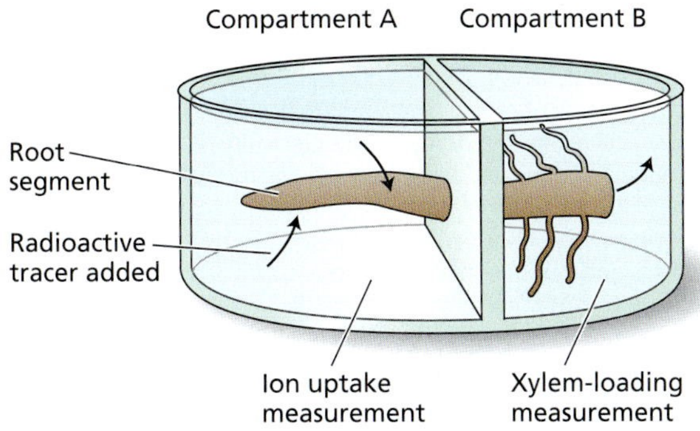

FIGURE 6.19 We can measure the relationship between ion uptake into the root and xylem loading by placing a root segment across two compartments and adding a radioactive tracer to one of them (in this case compartment A). The rate of disappearance of the tracer from compartment A gives a measure of ion uptake, and the rate of appearance in compartment B provides a measurement of xylem loading. (After Lüttge and Higinbotham 1979.)

membranes of xylem parenchyma cells contain proton pumps, water channels, and a variety of ion channels specialized for influx or efflux (Maathuis et al. 1997). In barley xylem parenchyma, two types of cation efflux channels have been identified: K^+-specific efflux channels and nonselective cation efflux channels. These channels are regulated by both the membrane potential and the cytosolic calcium concentration (De Boer and Wegner 1997). This finding suggests that the flux of ions from the xylem parenchyma cells into the xylem tracheary elements, rather than being due to simple leakage, is under tight metabolic control through regulation of the plasma membrane H^+-ATPase and ion efflux channels.

SUMMARY

The movement of molecules and ions from one location to another is known as transport. Plants exchange solutes and water with their environment and among their tissues and organs. Both local and long-distance transport processes in plants are controlled largely by cellular membranes.

Forces that drive biological transport, which include concentration gradients, electric-potential gradients, and hydrostatic pressures, are integrated by an expression called the electrochemical potential. Transport of solutes down a chemical gradient (e.g., by diffusion) is known as passive transport. Movement of solutes against a chemical-potential gradient is known as active transport and requires energy input.

The extent to which a membrane permits or restricts the movement of a substance is called membrane permeability. The permeability depends on the chemical properties of the particular solute and on the lipid composition of the membrane, as well as on the membrane proteins that facilitate the transport of specific substances.

When cations and anions move passively across a membrane at different rates, the electric potential that develops is called the diffusion potential. For each ion, the relationship between the voltage difference across the membrane and the distribution of the ion at equilibrium is described by the Nernst equation. The Nernst equation shows that at equilibrium the difference in concentration of an ion between two compartments is balanced by the voltage difference between the compartments. That voltage difference, or membrane potential, is seen in all living cells because of the asymmetric ion distributions between the inside and outside of the cells.

The electrical effects of different ions diffusing simultaneously across a cell membrane are summed by the Goldman equation. Electrogenic pumps, which carry out active transport and carry a net charge, change the membrane potential from the value created by diffusion.

Membranes contain specialized proteins—channels, carriers, and pumps—that facilitate solute transport. Channels are transport proteins that span the membrane, forming pores through which solutes diffuse down their gradient of electrochemical potentials. Carriers bind a solute on one side of the membrane and release it on the other side. Transport specificity is determined largely by the properties of channels and carriers.

A family of H^+-pumping ATPases provides the primary driving force for transport across the plasma membrane of plant cells. Two other kinds of electrogenic proton pumps serve this purpose at the tonoplast. Plant cells also have calcium-pumping ATPases that participate in the regulation of intracellular calcium concentrations, as well as ATP-binding cassette transporters that use the energy of ATP to transport large anionic molecules. The gradient of electrochemical potential generated by H^+ pumping is used to drive the transport of other substances in a process called secondary transport.

Genetic studies have revealed many genes, and their corresponding transport proteins, that account for the versatility of plant transport. Patch clamp electrophysiology provides unique information on ion channels, and it enables measurement of the permeability and gating of individual channel proteins.

Solutes move between cells either through the extracellular spaces (the apoplast) or from cytoplasm to cytoplasm (via the symplast). Cytoplasms of neighboring cells are connected by plasmodesmata, which facilitate symplastic transport. When an ion enters the root, it may be taken up into the cytoplasm of an epidermal cell, or it may diffuse through the apoplast into the root cortex and enter the symplast through a cortical cell. From the symplast, the ion is loaded into the xylem and transported to the shoot.

Web Material

Web Topics

6.1 Relating the Membrane Potential to the Distribution of Several Ions across the Membrane: The Goldman Equation

A brief explanation of the use of the Goldman equation to calculate the membrane permeability of more than one ion.

6.2 Patch Clamp Studies in Plant Cells

The electrophysiological method of patch clamping as applied to plant cells is described, with some specific examples.

6.3 Chemiosmosis in Action

The chemiosmotic theory explains how electrical and concentration gradients are used to perform cellular work.

6.4 Kinetic Analysis of Multiple Transporter Systems

Application of principles on enzyme kinetics to transport systems provides an effective way to characterize different carriers.

6.5 Transport Studies with Isolated Vacuoles and Membrane Vesicles

Certain experimental techniques enable the isolation of tonoplasts and plasma membranes for study.

6.6 ABC Transporters in Plants

ATP-binding cassette (ABC) transporters are a large family of active transport proteins energized directly by ATP.

Web Essay

6.1 Potassium Channels

Several plant K^+ channels have been characterized.

Chapter References

Barkla, B. J., and Pantoja, O. (1996) Physiology of ion transport across the tonoplast of higher plants. *Annu. Rev. Plant Physiol. Plant Mol. Biol.* 47: 159–184.

Buchanan, B. B., Gruissem, W., and Jones, R. L., eds. (2000) *Biochemistry and Molecular Biology of Plants*. Amer. Soc. Plant Physiologists, Rockville, MD.

Bush, D. S. (1995) Calcium regulation in plant cells and its role in signaling. *Annu. Rev. Plant Physiol. Plant Mol. Biol.* 46: 95–122.

Chrispeels, M. J., Crawford, N. M., and Schroeder, J. I. (1999) Proteins for transport of water and mineral nutrients across the membranes of plant cells. *Plant Cell* 11: 661–675.

Chung, W. S., Lee, S. H., Kim, J. C., Heo, W. D., Kim, M. C., Park, C. Y., Park, H. C., Lim, C. O., Kim, W. B., Harper, J. F., and Cho, M. J. (2000) Identification of a calmodulin-regulated soybean Ca^{2+}-ATPase (SCA1) that is located in the plasma membrane. *Plant Cell* 12: 1393–1407.

De Boer, A. H., and Wegner, L. H. (1997) Regulatory mechanisms of ion channels in xylem parenchyma cells. *J. Exp. Bot.* 48: 441–449.

Dunlop, J., and Bowling, D. J. F. (1971) The movement of ions to the xylem exudate of maize roots. *J. Exp. Bot.* 22: 453–464.

Ghoshroy, S., Lartey, R., Sheng, J., and Citovsky, V. (1997) Transport of proteins and nucleic acids through plasmodesmata. *Annu. Rev. Plant Physiol. Plant Mol. Biol.* 48: 27–50.

Higinbotham, N., Etherton, B., and Foster, R. J. (1967) Mineral ion contents and cell transmembrane electropotentials of pea and oat seedling tissue. *Plant Physiol.* 42: 37–46.

Higinbotham, N., Graves, J. S., and Davis, R. F. (1970) Evidence for an electrogenic ion transport pump in cells of higher plants. *J. Membr. Biol.* 3: 210–222.

Hirshi, K. D., Zhen, R.-G., Rea, P. A., and Fink, G. R. (1996) CAX1, an H^+/Ca^{2+} antiporter from *Arabidopsis*. *Proc. Natl Acad. Sci. USA* 93: 8782–8786.

Kochian, L. V. (2000) Molecular physiology of mineral nutrient acquisition, transport and utilization. In *Biochemistry and Molecular Biology of Plants*, B. Buchanan, W. Gruissem, and R. Jones, eds., American Society of Plant Physiologists, Rockville, MD, pp. 1204–1249.

Kuehn, C., Barker, L., Buerkle, L., and Frommer, W. B. (1999) Update on sucrose transport in higher plants. *J. Exp. Bot.* 50: 935–953.

Leng, Q., Mercier, R. W., Hua, B-G., Fromm, H., and Berkowitz, G. A. (2002) Electrophysical analysis of cloned cyclic nucleotide-gated ion channels. *Plant Physiol.* 128: 400–410.

Lin, W., Schmitt, M. R., Hitz, W. D., and Giaquinta, R. T. (1984) Sugar transport into protoplasts isolated from developing soybean cotyledons. *Plant Physiol.* 75: 936–940.

Lüttge, U., and Higinbotham, N. (1979) *Transport in Plants.* Springer-Verlag, New York.

Lüttge, U., and Ratajczak, R. (1997) The physiology, biochemistry and molecular biology of the plant vacuolar ATPase. *Adv. Bot. Res.* 25: 253–296.

Maathuis, F. J. M., Ichida, A. M., Sanders, D., and Schroeder, J. I. (1997) Roles of higher plant K^+ channels. *Plant Physiol.* 114: 1141–1149.

Müller, M., Irkens-Kiesecker, U., Kramer, D., and Taiz, L. (1997) Purification and reconstitution of the vacuolar H^+-ATPases from lemon fruits and epicotyls. *J. Biol. Chem.* 272: 12762–12770.

Nobel, P. (1991) *Physicochemical and Environmental Plant Physiology*. Academic Press, San Diego, CA.

Palmgren, M. G. (2001) Plant plasma membrane H^+-ATPases: Powerhouses for nutrient uptake. *Annu. Rev. Plant Physiol. Plant Mol. Biol.* 52: 817–845.

Rea, P. A., Li, Z-S., Lu, Y-P., and Drozdowicz, Y. M.(1998) From vacuolar Gs-X pumps to multispecific ABC transporters. *Annu. Rev. Plant Physiol. Plant Mol. Biol.* 49: 727–760.

Small, J. (1946) *pH and Plants, an Introduction to Beginners*. D. Van Nostrand, New York.

Spanswick, R. M. (1981) Electrogenic ion pumps. *Annu. Rev. Plant Physiol.* 32: 267–289.

Sussman, M. R. (1994) Molecular analysis of proteins in the plant plasma membrane. *Annu. Rev. Plant Physiol. Plant Mol. Biol.* 45: 211–234.

Tanner, W., and Caspari, T. (1996) Membrane transport carriers. *Annu. Rev. Plant Physiol. Plant Mol. Biol.* 47: 595–626.

Tazawa, M., Shimmen, T., and Mimura, T. (1987) Membrane control in the *Characeae*. *Annu. Rev. Plant Phsyiol.* 38: 95–117.

Theodoulou, F. L. (2000) Plant ABC transporters. *Biochim. Biophys. Acta* 1465: 79–103.

Tyerman, S. D., Niemietz, C. M., and Bramley, H. (2002) Plant aquaporins: Multifunctional water and solute channels with expanding roles. *Plant Cell Envir.* 25: 173–194.

Vera-Estrella, R., Barkla, B. J., Higgins, V. J., and Blumwald, E. (1994) Plant defense response to fungal pathogens. Activation of host-plasma-membrane H^+-ATPase by elicitor-induced enzyme dephosphorylation. *Plant Physiol.* 104: 209–215.

UNIT

III

Biochemistry and Metabolism

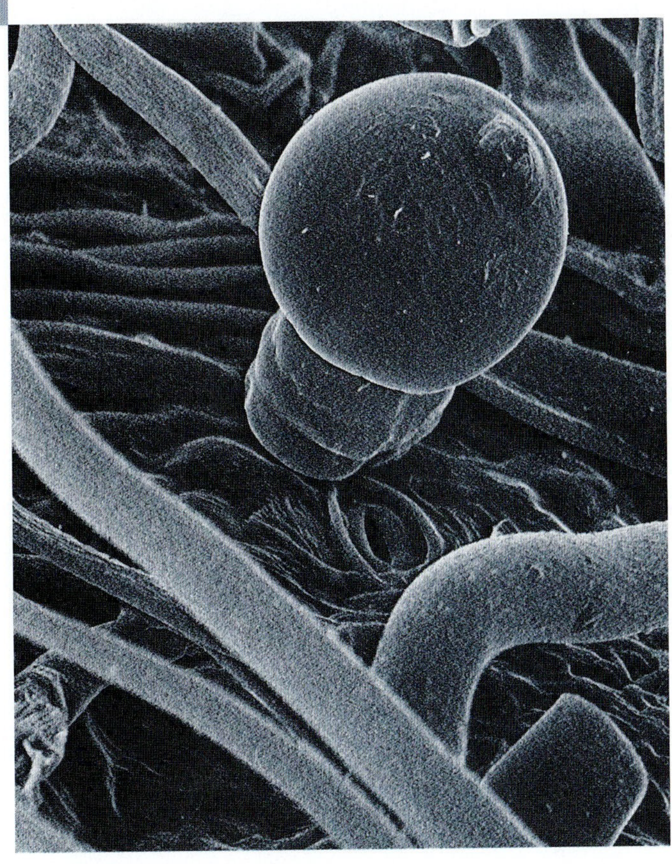

UNIT

II

Biochemistry and Metabolism

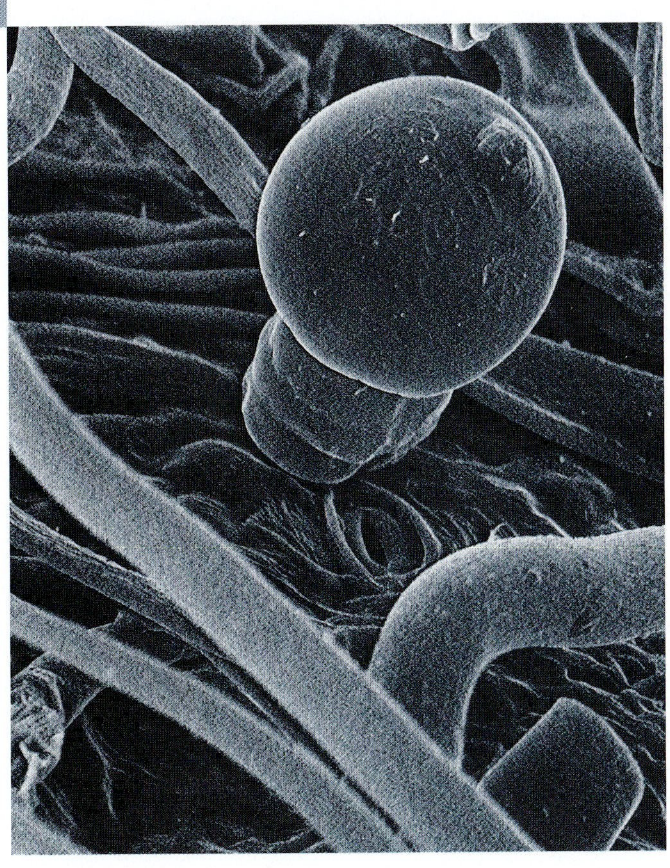

Chapter 7

Photosynthesis: The Light Reactions

LIFE ON EARTH ULTIMATELY DEPENDS ON ENERGY derived from the sun. Photosynthesis is the only process of biological importance that can harvest this energy. In addition, a large fraction of the planet's energy resources results from photosynthetic activity in either recent or ancient times (fossil fuels). This chapter introduces the basic physical principles that underlie photosynthetic energy storage and the current understanding of the structure and function of the photosynthetic apparatus (Blankenship 2002).

The term *photosynthesis* means literally "synthesis using light." As we will see in this chapter, photosynthetic organisms use solar energy to synthesize carbon compounds that cannot be formed without the input of energy. More specifically, light energy drives the synthesis of carbohydrates from carbon dioxide and water with the generation of oxygen:

$$\underset{\text{Carbon dioxide}}{6\,CO_2} + \underset{\text{Water}}{6\,H_2O} \rightarrow \underset{\text{Carbohydrate}}{C_6H_{12}O_6} + \underset{\text{Oxygen}}{6\,O_2}$$

Energy stored in these molecules can be used later to power cellular processes in the plant and can serve as the energy source for all forms of life.

This chapter deals with the role of light in photosynthesis, the structure of the photosynthetic apparatus, and the processes that begin with the excitation of chlorophyll by light and culminate in the synthesis of ATP and NADPH.

PHOTOSYNTHESIS IN HIGHER PLANTS

The most active photosynthetic tissue in higher plants is the mesophyll of leaves. Mesophyll cells have many chloroplasts, which contain the specialized light-absorbing green pigments, the **chlorophylls**. In photosynthesis, the plant uses solar energy to oxidize water, thereby releasing oxygen, and to reduce carbon dioxide, thereby forming large carbon compounds, primarily sugars. The complex series of reactions that culminate in the reduction of CO_2 include the thylakoid reactions and the carbon fixation reactions.

The **thylakoid reactions** of photosynthesis take place in the specialized internal membranes of the chloroplast called thylakoids (see Chapter 1). The end products of these thylakoid reactions are the high-energy compounds ATP and NADPH, which are used for the synthesis of sugars in the **carbon fixation reactions**. These synthetic processes take place in the stroma of the chloroplasts, the aqueous region that surrounds the thylakoids. The thylakoid reactions of photosynthesis are the subject of this chapter; the carbon fixation reactions are discussed in Chapter 8.

In the chloroplast, light energy is converted into chemical energy by two different functional units called *photosystems*. The absorbed light energy is used to power the transfer of electrons through a series of compounds that act as electron donors and electron acceptors. The majority of electrons ultimately reduce $NADP^+$ to NADPH and oxidize H_2O to O_2. Light energy is also used to generate a proton motive force (see Chapter 6) across the thylakoid membrane, which is used to synthesize ATP.

GENERAL CONCEPTS

In this section we will explore the essential concepts that provide a foundation for an understanding of photosynthesis. These concepts include the nature of light, the properties of pigments, and the various roles of pigments.

Light Has Characteristics of Both a Particle and a Wave

A triumph of physics in the early twentieth century was the realization that light has properties of both particles and waves. A wave (Figure 7.1) is characterized by a **wavelength**, denoted by the Greek letter lambda (λ), which is the distance between successive wave crests. The **frequency**, represented by the Greek letter nu (ν), is the number of wave crests that pass an observer in a given time. A simple equation relates the wavelength, the frequency, and the speed of any wave:

$$c = \lambda\nu \tag{7.1}$$

where c is the speed of the wave—in the present case, the speed of light (3.0×10^8 m s^{-1}). The light wave is a transverse (side-to-side) electromagnetic wave, in which both electric and magnetic fields oscillate perpendicularly to the direction of propagation of the wave and at 90° with respect to each other.

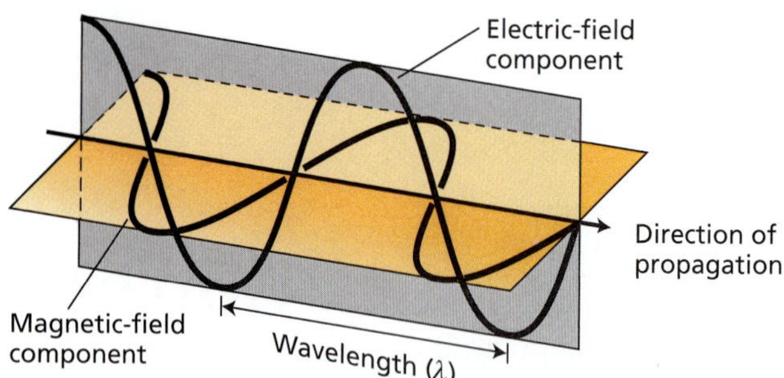

FIGURE 7.1 Light is a transverse electromagnetic wave, consisting of oscillating electric and magnetic fields that are perpendicular to each other and to the direction of propagation of the light. Light moves at a speed of 3×10^8 m s^{-1}. The wavelength (λ) is the distance between successive crests of the wave.

Light is also a particle, which we call a **photon**. Each photon contains an amount of energy that is called a **quantum** (plural *quanta*). The energy content of light is not continuous but rather is delivered in these discrete packets, the quanta. The energy (E) of a photon depends on the frequency of the light according to a relation known as Planck's law:

$$E = h\nu \tag{7.2}$$

where h is Planck's constant (6.626×10^{-34} J s).

Sunlight is like a rain of photons of different frequencies. Our eyes are sensitive to only a small range of frequencies—the visible-light region of the electromagnetic spectrum (Figure 7.2). Light of slightly higher frequencies (or

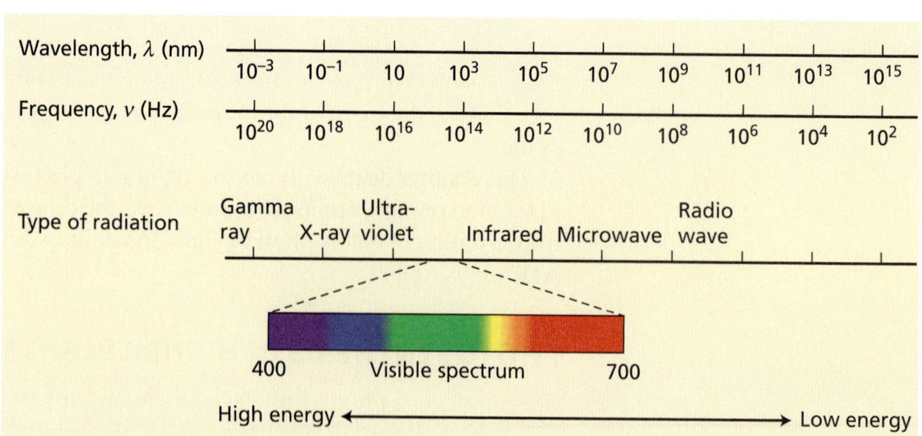

FIGURE 7.2 Electromagnetic spectrum. Wavelength (λ) and frequency (ν) are inversely related. Our eyes are sensitive to only a narrow range of wavelengths of radiation, the visible region, which extends from about 400 nm (violet) to about 700 nm (red). Short-wavelength (high-frequency) light has a high energy content; long-wavelength (low-frequency) light has a low energy content.

shorter wavelengths) is in the ultraviolet region of the spectrum, and light of slightly lower frequencies (or longer wavelengths) is in the infrared region. The output of the sun is shown in Figure 7.3, along with the energy density that strikes the surface of Earth. The absorption spectrum of chlorophyll *a* (curve C in Figure 7.3) indicates approximately the portion of the solar output that is utilized by plants.

An **absorption spectrum** (plural *spectra*) provides information about the amount of light energy taken up or absorbed by a molecule or substance as a function of the wavelength of the light. The absorption spectrum for a particular substance in a nonabsorbing solvent can be determined by a spectrophotometer as illustrated in Figure 7.4. Spectrophotometry, the technique used to measure the absorption of light by a sample, is more completely discussed in **Web Topic 7.1**.

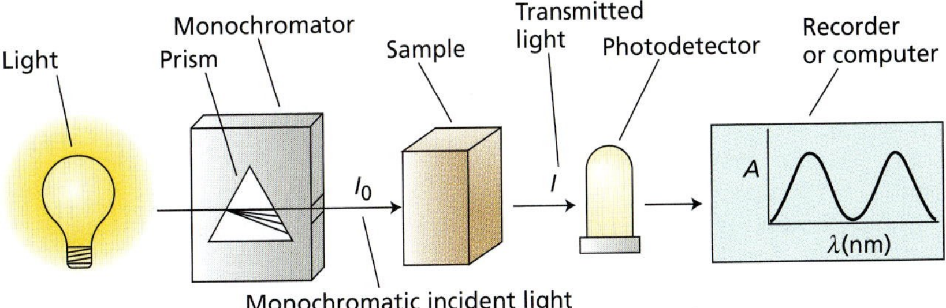

FIGURE 7.4 Schematic diagram of a spectrophotometer. The instrument consists of a light source, a monochromator that contains a wavelength selection device such as a prism, a sample holder, a photodetector, and a recorder or computer. The output wavelength of the monochromator can be changed by rotation of the prism; the graph of absorbance (*A*) versus wavelength (λ) is called a spectrum.

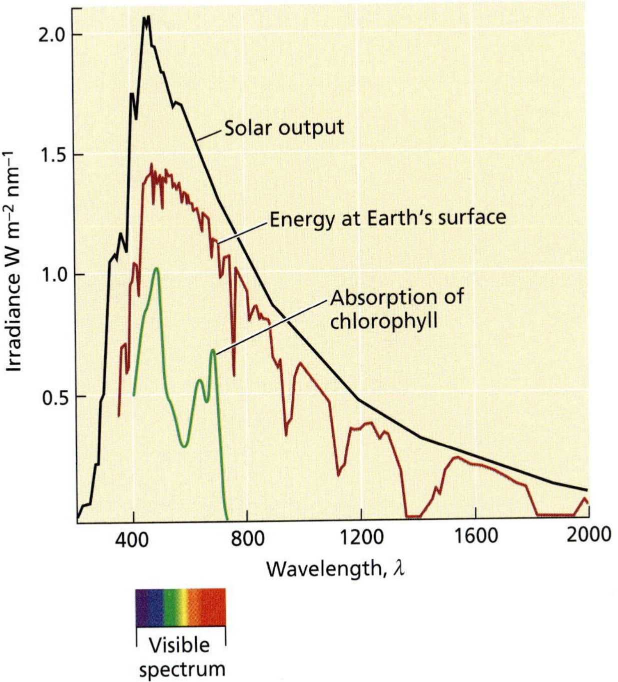

FIGURE 7.3 The solar spectrum and its relation to the absorption spectrum of chlorophyll. Curve A is the energy output of the sun as a function of wavelength. Curve B is the energy that strikes the surface of Earth. The sharp valleys in the infrared region beyond 700 nm represent the absorption of solar energy by molecules in the atmosphere, chiefly water vapor. Curve C is the absorption spectrum of chlorophyll, which absorbs strongly in the blue (about 430 nm) and the red (about 660 nm) portions of the spectrum. Because the green light in the middle of the visible region is not efficiently absorbed, most of it is reflected into our eyes and gives plants their characteristic green color.

When Molecules Absorb or Emit Light, They Change Their Electronic State

Chlorophyll appears green to our eyes because it absorbs light mainly in the red and blue parts of the spectrum, so only some of the light enriched in green wavelengths (about 550 nm) is reflected into our eyes (see Figure 7.3).

The absorption of light is represented by Equation 7.3, in which chlorophyll (Chl) in its lowest-energy, or ground, state absorbs a photon (represented by $h\nu$) and makes a transition to a higher-energy, or excited, state (Chl*):

$$\text{Chl} + h\nu \rightarrow \text{Chl}^* \tag{7.3}$$

The distribution of electrons in the excited molecule is somewhat different from the distribution in the ground-state molecule (Figure 7.5) Absorption of blue light excites the chlorophyll to a higher energy state than absorption of red light because the energy of photons is higher when their wavelength is shorter. In the higher excited state, chlorophyll is extremely unstable, very rapidly gives up some of its energy to the surroundings as heat, and enters the lowest excited state, where it can be stable for a maximum of several nanoseconds (10^{-9} s). Because of this inherent instability of the excited state, any process that captures its energy must be extremely rapid.

In the lowest excited state, the excited chlorophyll has four alternative pathways for disposing of its available energy.

1. Excited chlorophyll can re-emit a photon and thereby return to its ground state—a process known as **fluorescence**. When it does so, the wavelength of fluorescence is slightly longer (and of lower energy) than the wavelength of absorption because a portion of the excitation energy is converted into heat before the fluorescent photon is emitted. Chlorophylls fluoresce in the red region of the spectrum.

2. The excited chlorophyll can return to its ground state by directly converting its excitation energy into heat, with no emission of a photon.

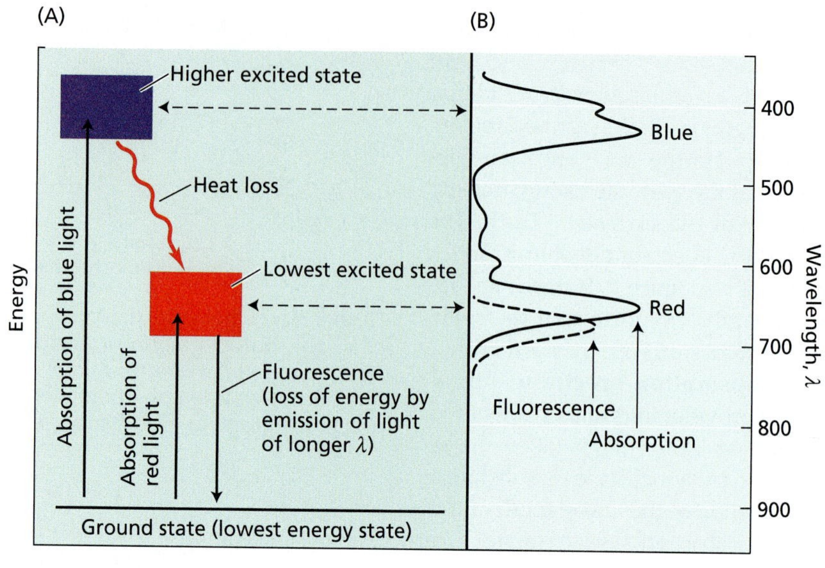

FIGURE 7.5 Light absorption and emission by chlorophyll. (A) Energy level diagram. Absorption or emission of light is indicated by vertical lines that connect the ground state with excited electron states. The blue and red absorption bands of chlorophyll (which absorb blue and red photons, respectively) correspond to the upward vertical arrows, signifying that energy absorbed from light causes the molecule to change from the ground state to an excited state. The downward-pointing arrow indicates fluorescence, in which the molecule goes from the lowest excited state to the ground state while re-emitting energy as a photon. (B) Spectra of absorption and fluorescence. The long-wavelength (red) absorption band of chlorophyll corresponds to light that has the energy required to cause the transition from the ground state to the first excited state. The short-wavelength (blue) absorption band corresponds to a transition to a higher excited state.

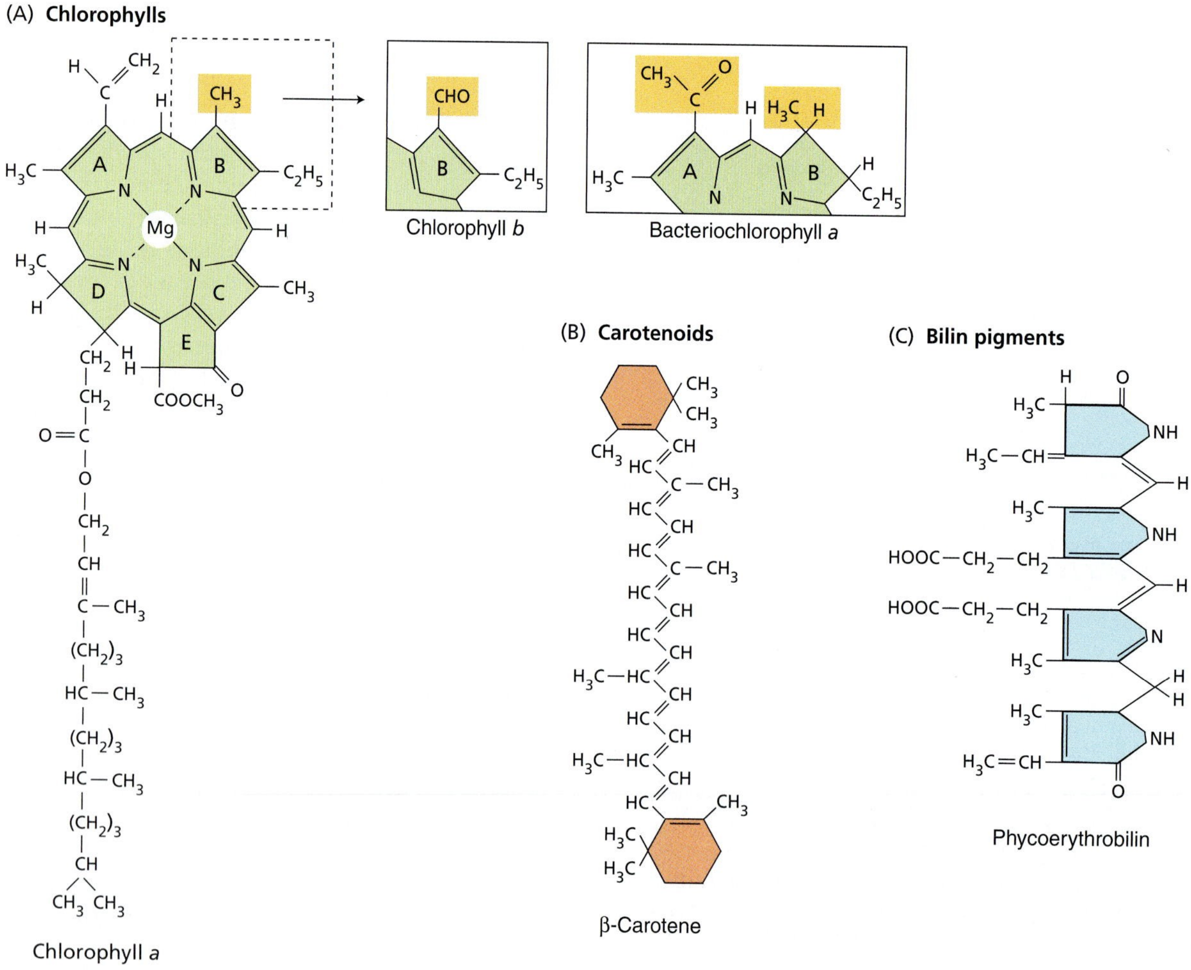

(A) **Chlorophylls**

Chlorophyll *a*

Chlorophyll *b*

Bacteriochlorophyll *a*

(B) **Carotenoids**

β-Carotene

(C) **Bilin pigments**

Phycoerythrobilin

3. Chlorophyll may participate in **energy transfer**, during which an excited chlorophyll transfers its energy to another molecule.

4. A fourth process is **photochemistry**, in which the energy of the excited state causes chemical reactions to occur. The photochemical reactions of photosynthesis are among the fastest known chemical reactions. This extreme speed is necessary for photochemistry to compete with the three other possible reactions of the excited state just described.

Photosynthetic Pigments Absorb the Light That Powers Photosynthesis

The energy of sunlight is first absorbed by the pigments of the plant. All pigments active in photosynthesis are found in the chloroplast. Structures and absorption spectra of several photosynthetic pigments are shown in Figures 7.6 and 7.7, respectively. The chlorophylls and **bacteriochlorophylls** (pigments found in certain bacteria) are the typical pigments of photosynthetic organisms, but all organisms contain a mixture of more than one kind of pigment, each serving a specific function.

Chlorophylls *a* and *b* are abundant in green plants, and *c* and *d* are found in some protists and cyanobacteria. A number of different types of bacteriochlorophyll have been found; type *a* is the most widely distributed. **Web Topic 7.2** shows the distribution of pigments in different types of photosynthetic organisms.

All chlorophylls have a complex ring structure that is chemically related to the porphyrin-like groups found in hemoglobin and cytochromes (see Figure 7.6A). In addition, a long hydrocarbon tail is almost always attached to the ring structure. The tail anchors the chlorophyll to the hydrophobic portion of its environment. The ring structure contains some loosely bound electrons and is the part of the molecule involved in electron transitions and redox reactions.

The different types of **carotenoids** found in photosynthetic organisms are all linear molecules with multiple conjugated double bonds (see Figure 7.6B). Absorption bands in the 400 to 500 nm region give carotenoids their characteristic orange color. The color of carrots, for example, is due to the carotenoid β-carotene, whose structure and absorption spectrum are shown in Figures 7.6 and 7.7, respectively.

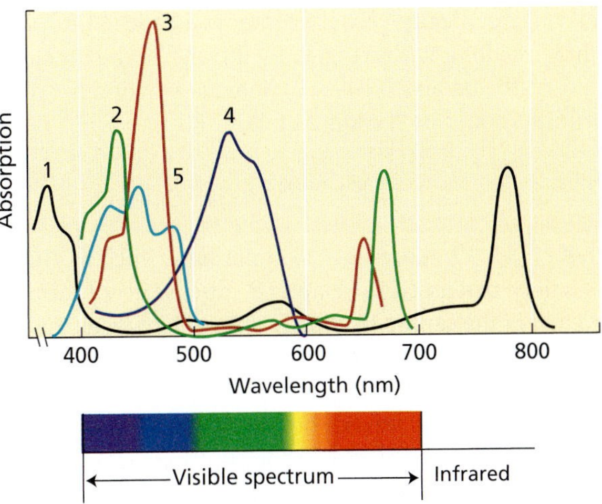

FIGURE 7.7 Absorption spectra of some photosynthetic pigments. Curve 1, bacteriochlorophyll *a*; curve 2, chlorophyll *a*; curve 3, chlorophyll *b*; curve 4, phycoerythrobilin; curve 5, β-carotene. The absorption spectra shown are for pure pigments dissolved in nonpolar solvents, except for curve 4, which represents an aqueous buffer of phycoerythrin, a protein from cyanobacteria that contains a phycoerythrobilin chromophore covalently attached to the peptide chain. In many cases the spectra of photosynthetic pigments in vivo are substantially affected by the environment of the pigments in the photosynthetic membrane. (After Avers 1985.)

Carotenoids are found in all photosynthetic organisms, except for mutants incapable of living outside the laboratory. Carotenoids are integral constituents of the thylakoid membrane and are usually associated intimately with both antenna and reaction center pigment proteins. The light absorbed by the carotenoids is transferred to chlorophyll for photosynthesis; because of this role they are called **accessory pigments**.

FIGURE 7.6 Molecular structure of some photosynthetic pigments. (A) The chlorophylls have a porphyrin-like ring structure with a magnesium atom (Mg) coordinated in the center and a long hydrophobic hydrocarbon tail that anchors them in the photosynthetic membrane. The porphyrin-like ring is the site of the electron rearrangements that occur when the chlorophyll is excited and of the unpaired electrons when it is either oxidized or reduced. Various chlorophylls differ chiefly in the substituents around the rings and the pattern of double bonds. (B) Carotenoids are linear polyenes that serve as both antenna pigments and photoprotective agents. (C) Bilin pigments are open-chain tetrapyrroles found in antenna structures known as phycobilisomes that occur in cyanobacteria and red algae.

KEY EXPERIMENTS IN UNDERSTANDING PHOTOSYNTHESIS

Establishing the overall chemical equation of photosynthesis required several hundred years and contributions by many scientists (literature references for historical developments can be found on the web site). In 1771, Joseph Priestley observed that a sprig of mint growing in air in which a candle had burned out improved the air so that another candle could burn. He had discovered oxygen evolution by plants. A Dutchman, Jan Ingenhousz, documented the essential role of light in photosynthesis in 1779.

Other scientists established the roles of CO_2 and H_2O and showed that organic matter, specifically carbohydrate, is a product of photosynthesis along with oxygen. By the end of the nineteenth century, the balanced overall chemical reaction for photosynthesis could be written as follows:

$$6\,CO_2 + 6\,H_2O \xrightarrow{\text{Light, plant}} C_6H_{12}O_6 + 6\,O_2 \qquad (7.4)$$

where $C_6H_{12}O_6$ represents a simple sugar such as glucose. As will be discussed in Chapter 8, glucose is not the actual product of the carbon fixation reactions. However, the energetics for the actual products is approximately the same, so the representation of glucose in Equation 7.4 should be regarded as a convenience but not taken literally.

The chemical reactions of photosynthesis are complex. In fact, at least 50 intermediate reaction steps have now been identified, and undoubtedly additional steps will be discovered. An early clue to the chemical nature of the essential chemical process of photosynthesis came in the 1920s from investigations of photosynthetic bacteria that did not produce oxygen as an end product. From his studies on these bacteria, C. B. van Niel concluded that photosynthesis is a redox (reduction–oxidation) process. This conclusion has been confirmed, and it has served as a fundamental concept on which all subsequent research on photosynthesis has been based.

We now turn to the relationship between photosynthetic activity and the spectrum of absorbed light. We will discuss some of the critical experiments that have contributed to our present understanding of photosynthesis, and we will consider equations for essential chemical reactions of photosynthesis.

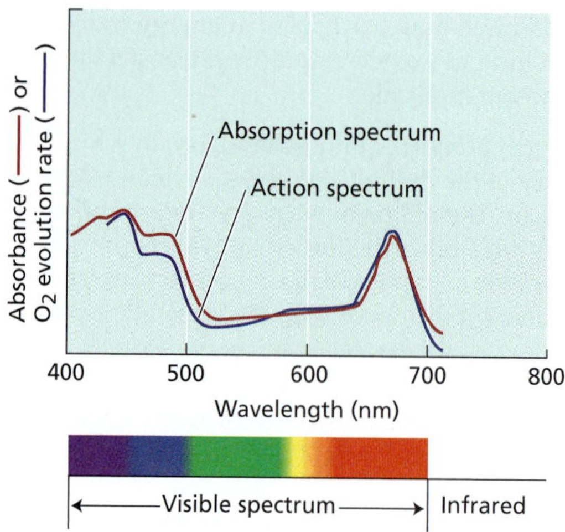

FIGURE 7.8 Action spectrum compared with an absorption spectrum. The absorption spectrum is measured as shown in Figure 7.4. An action spectrum is measured by plotting a response to light such as oxygen evolution, as a function of wavelength. If the pigment used to obtain the absorption spectrum is the same as those that cause the response, the absorption and action spectra will match. In the example shown here, the action spectrum for oxygen evolution matches the absorption spectrum of intact chloroplasts quite well, indicating that light absorption by the chlorophylls mediates oxygen evolution. Discrepancies are found in the region of carotenoid absorption, from 450 to 550 nm, indicating that energy transfer from carotenoids to chlorophylls is not as effective as energy transfer between chlorophylls.

Action Spectra Relate Light Absorption to Photosynthetic Activity

The use of action spectra has been central to the development of our current understanding of photosynthesis. An **action spectrum** depicts the magnitude of a response of a biological system to light, as a function of wavelength. For example, an action spectrum for photosynthesis can be constructed from measurements of oxygen evolution at different wavelengths (Figure 7.8). Often an action spectrum can identify the chromophore (pigment) responsible for a particular light-induced phenomenon.

Some of the first action spectra were measured by T. W. Engelmann in the late 1800s (Figure 7.9). Engelmann used a prism to disperse sunlight into a rainbow that was allowed to fall on an aquatic algal filament. A population of O_2-seeking bacteria was introduced into the system. The

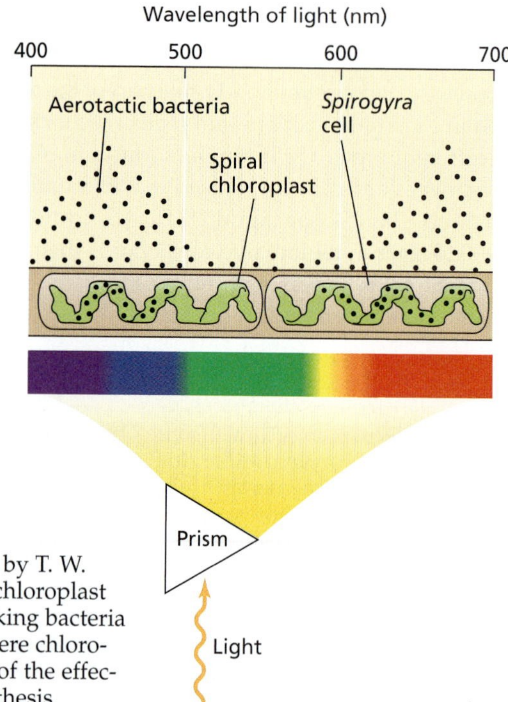

FIGURE 7.9 Schematic diagram of the action spectrum measurements by T. W. Engelmann. Engelmann projected a spectrum of light onto the spiral chloroplast of the filamentous green alga *Spirogyra* and observed that oxygen-seeking bacteria introduced into the system collected in the region of the spectrum where chlorophyll pigments absorb. This action spectrum gave the first indication of the effectiveness of light absorbed by accessory pigments in driving photosynthesis.

bacteria congregated in the regions of the filaments that evolved the most O_2. These were the regions illuminated by blue light and red light, which are strongly absorbed by chlorophyll. Today, action spectra can be measured in room-sized spectrographs in which a huge monochromator bathes the experimental samples in monochromatic light. But the principle of the experiment is the same as that of Engelmann's experiments.

Action spectra were very important for the discovery of two distinct photosystems operating in O_2-evolving photosynthetic organisms. Before we introduce the two photosystems, however, we need to describe the light-gathering antennas and the energy needs of photosynthesis.

Photosynthesis Takes Place in Complexes Containing Light-Harvesting Antennas and Photochemical Reaction Centers

A portion of the light energy absorbed by chlorophylls and carotenoids is eventually stored as chemical energy via the formation of chemical bonds. This conversion of energy from one form to another is a complex process that depends on cooperation between many pigment molecules and a group of electron transfer proteins.

The majority of the pigments serve as an **antenna complex**, collecting light and transferring the energy to the **reaction center complex**, where the chemical oxidation and reduction reactions leading to long-term energy storage take place (Figure 7.10). Molecular structures of some of the antenna and reaction center complexes are discussed later in the chapter.

How does the plant benefit from this division of labor between antenna and reaction center pigments? Even in bright sunlight, a chlorophyll molecule absorbs only a few photons each second. If every chlorophyll had a complete reaction center associated with it, the enzymes that make up this system would be idle most of the time, only occasionally being activated by photon absorption. However, if many pigments can send energy into a common reaction center, the system is kept active a large fraction of the time.

In 1932, Robert Emerson and William Arnold performed a key experiment that provided the first evidence for the cooperation of many chlorophyll molecules in energy conversion during photosynthesis. They delivered very brief (10^{-5} s) flashes of light to a suspension of the green alga *Chlorella pyrenoidosa* and measured the amount of oxygen produced. The flashes were spaced about 0.1 s apart, a time that Emerson and Arnold had determined in earlier work was long enough for the enzymatic steps of the process to be completed before the arrival of the next flash. The investigators varied the energy of the flashes and found that at high energies the oxygen production did not increase when a more intense flash was given: The photosynthetic system was saturated with light (Figure 7.11).

In their measurement of the relationship of oxygen production to flash energy, Emerson and Arnold were surprised to find that under saturating conditions, only one molecule of oxygen was produced for each 2500 chlorophyll molecules in the sample. We know now that several hundred pigments are associated with each reaction center and that each reaction center must operate four times

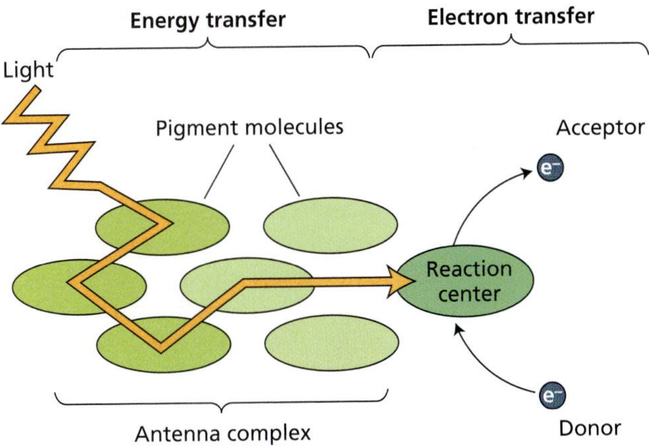

FIGURE 7.10 Basic concept of energy transfer during photosynthesis. Many pigments together serve as an antenna, collecting light and transferring its energy to the reaction center, where chemical reactions store some of the energy by transferring electrons from a chlorophyll pigment to an electron acceptor molecule. An electron donor then reduces the chlorophyll again. The transfer of energy in the antenna is a purely physical phenomenon and involves no chemical changes.

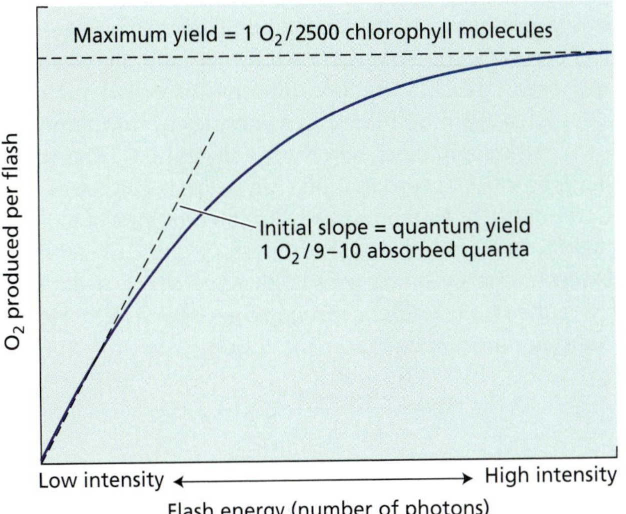

FIGURE 7.11 Relationship of oxygen production to flash energy, the first evidence for the interaction between the antenna pigments and the reaction center. At saturating energies, the maximum amount of O_2 produced is 1 molecule per 2500 chlorophyll molecules.

to produce one molecule of oxygen—hence the value of 2500 chlorophylls per O_2.

The reaction centers and most of the antenna complexes are integral components of the photosynthetic membrane. In eukaryotic photosynthetic organisms, these membranes are found within the chloroplast; in photosynthetic prokaryotes, the site of photosynthesis is the plasma membrane or membranes derived from it.

The graph shown in Figure 7.11 permits us to calculate another important parameter of the light reactions of photosynthesis, the quantum yield. The **quantum yield** of photosynthesis (Φ) is defined as follows:

$$\Phi = \frac{\text{Number of photochemical products}}{\text{Total number of quanta absorbed}} \tag{7.5}$$

In the linear portion (low light intensity) of the curve, an increase in the number of photons stimulates a proportional increase in oxygen evolution. Thus the slope of the curve measures the quantum yield for oxygen production. The quantum yield for a particular process can range from 0 (if that process does not respond to light) to 1.0 (if every photon absorbed contributes to the process). A more detailed discussion of quantum yields can be found in **Web Topic 7.3**.

In functional chloroplasts kept in dim light, the quantum yield of photochemistry is approximately 0.95, the quantum yield of fluorescence is 0.05 or lower, and the quantum yields of other processes are negligible. The vast majority of excited chlorophyll molecules therefore lead to photochemistry.

The Chemical Reaction of Photosynthesis Is Driven by Light

It is important to realize that equilibrium for the chemical reaction shown in Equation 7.4 lies very far in the direction of the reactants. The equilibrium constant for Equation 7.4, calculated from tabulated free energies of formation for each of the compounds involved, is about 10^{-500}. This number is so close to zero that one can be quite confident that in the entire history of the universe no molecule of glucose has formed spontaneously from H_2O and CO_2 without external energy being provided. The energy needed to drive the photosynthetic reaction comes from light. Here's a simpler form of Equation 7.4:

$$CO_2 + H_2O \xrightarrow{\text{Light, plant}} (CH_2O) + O_2 \tag{7.6}$$

where (CH_2O) is one-sixth of a glucose molecule. About nine or ten photons of light are required to drive the reaction of Equation 7.6.

Although the photochemical quantum yield under optimum conditions is nearly 100%, the *efficiency* of the conversion of light into chemical energy is much less. If red light of wavelength 680 nm is absorbed, the total energy input (see Equation 7.2) is 1760 kJ per mole of oxygen formed. This amount of energy is more than enough to drive the reaction in Equation 7.6, which has a standard-state free-energy change of +467 kJ mol^{-1}. The efficiency of conversion of light energy at the optimal wavelength into chemical energy is therefore about 27%, which is remarkably high for an energy conversion system. Most of this stored energy is used for cellular maintenance processes; the amount diverted to the formation of biomass is much less (see Figure 9.2).

There is no conflict between the fact that the photochemical quantum efficiency (quantum yield) is nearly 1 (100%) and the energy conversion efficiency is only 27%. The *quantum efficiency* is a measure of the fraction of absorbed photons that engage in photochemistry; the *energy efficiency* is a measure of how much energy in the absorbed photons is stored as chemical products. The numbers indicate that almost all the absorbed photons engage in photochemistry, but only about a fourth of the energy in each photon is stored, the remainder being converted to heat.

Light Drives the Reduction of NADP and the Formation of ATP

The overall process of photosynthesis is a redox chemical reaction, in which electrons are removed from one chemical species, thereby oxidizing it, and added to another species, thereby reducing it. In 1937, Robert Hill found that in the light, isolated chloroplast thylakoids reduce a variety of compounds, such as iron salts. These compounds serve as oxidants in place of CO_2, as the following equation shows:

$$4\ Fe^{3+} + 2\ H_2O \rightarrow 4\ Fe^{2+} + O_2 + 4\ H^+ \tag{7.7}$$

Many compounds have since been shown to act as artificial electron acceptors in what has come to be known as the Hill reaction. Their use has been invaluable in elucidating the reactions that precede carbon reduction.

We now know that during the normal functioning of the photosynthetic system, light reduces nicotinamide adenine dinucleotide phosphate (NADP), which in turn serves as the reducing agent for carbon fixation in the Calvin cycle (see Chapter 8). ATP is also formed during the electron flow from water to NADP, and it, too, is used in carbon reduction.

The chemical reactions in which water is oxidized to oxygen, NADP is reduced, and ATP is formed are known as the *thylakoid reactions* because almost all the reactions up to NADP reduction take place within the thylakoids. The carbon fixation and reduction reactions are called the *stroma reactions* because the carbon reduction reactions take place in the aqueous region of the chloroplast, the stroma.

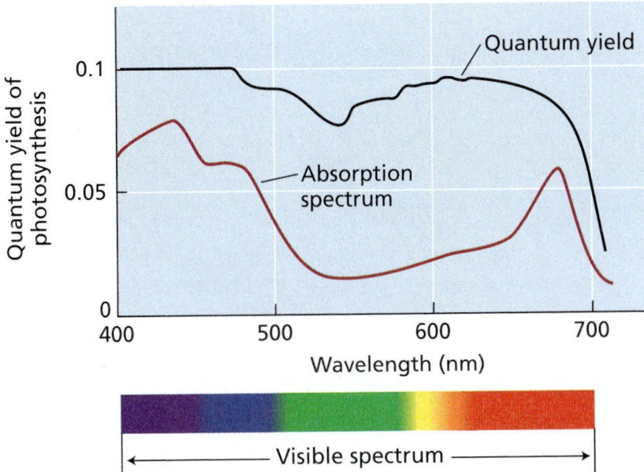

FIGURE 7.12 Red drop effect. The quantum yield of photosynthesis (black curve) falls off drastically for far-red light of wavelengths greater than 680 nm, indicating that far-red light alone is inefficient in driving photosynthesis. The slight dip near 500 nm reflects the somewhat lower efficiency of photosynthesis using light absorbed by accessory pigments, carotenoids.

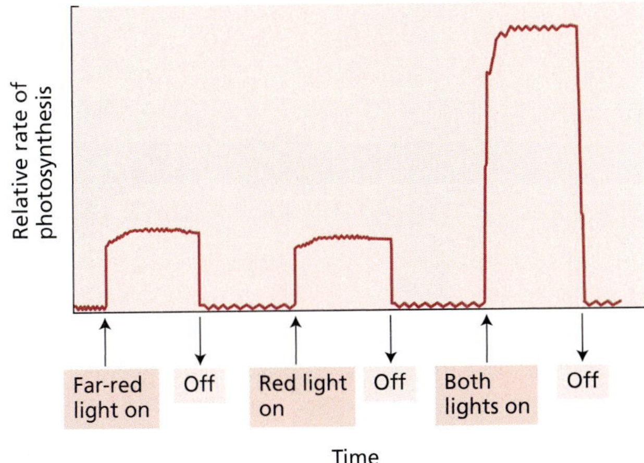

FIGURE 7.13 Enhancement effect. The rate of photosynthesis when red and far-red light are given together is greater than the sum of the rates when they are given apart. The enhancement effect provided essential evidence in favor of the concept that photosynthesis is carried out by two photochemical systems working in tandem but with slightly different wavelength optima.

Although this division is somewhat arbitrary, it is conceptually useful.

Oxygen-Evolving Organisms Have Two Photosystems That Operate in Series

By the late 1950s, several experiments were puzzling the scientists who studied photosynthesis. One of these experiments carried out by Emerson, measured the quantum yield of photosynthesis as a function of wavelength and revealed an effect known as the red drop (Figure 7.12).

If the quantum yield is measured for the wavelengths at which chlorophyll absorbs light, the values found throughout most of the range are fairly constant, indicating that any photon absorbed by chlorophyll or other pigments is as effective as any other photon in driving photosynthesis. However, the yield drops dramatically in the far-red region of chlorophyll absorption (greater than 680 nm).

This drop cannot be caused by a decrease in chlorophyll absorption because the quantum yield measures only light that has actually been absorbed. Thus, light with a wavelength greater than 680 nm is much less efficient than light of shorter wavelengths.

Another puzzling experimental result was the **enhancement effect**, also discovered by Emerson. He measured the rate of photosynthesis separately with light of two different wavelengths and then used the two beams simultaneously (Figure 7.13). When red and far-red light were given together, the rate of photosynthesis was greater than the sum of the individual rates. This was a startling and surprising observation.

These observations were eventually explained by experiments performed in the 1960s (see **Web Topic 7.4**) that led to the discovery that two photochemical complexes, now known as **photosystems I** and **II** (**PSI and PSII**), operate in series to carry out the early energy storage reactions of photosynthesis.

Photosystem I preferentially absorbs far-red light of wavelengths greater than 680 nm; photosystem II preferentially absorbs red light of 680 nm and is driven very poorly by far-red light. This wavelength dependence explains the enhancement effect and the red drop effect. Another difference between the photosystems is that

- Photosystem I produces a strong reductant, capable of reducing $NADP^+$, and a weak oxidant.
- Photosystem II produces a very strong oxidant, capable of oxidizing water, and a weaker reductant than the one produced by photosystem I.

The reductant produced by photosystem II re-reduces the oxidant produced by photosystem I. These properties of the two photosystems are shown schematically in Figure 7.14.

The scheme of photosynthesis depicted in Figure 7.14, called the Z (for *zigzag*) *scheme*, has become the basis for understanding O_2-evolving (oxygenic) photosynthetic organisms. It accounts for the operation of two physically and chemically distinct photosystems (I and II), each with its own antenna pigments and photochemical reaction center. The two photosystems are linked by an electron transport chain.

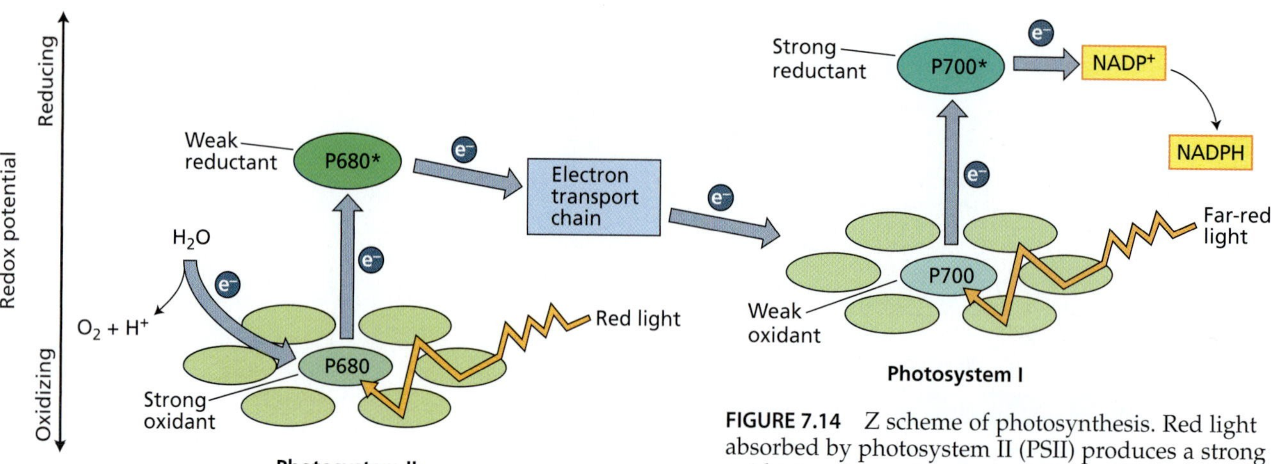

FIGURE 7.14 Z scheme of photosynthesis. Red light absorbed by photosystem II (PSII) produces a strong oxidant and a weak reductant. Far-red light absorbed by photosystem I (PSI) produces a weak oxidant and a strong reductant. The strong oxidant generated by PSII oxidizes water, while the strong reductant produced by PSI reduces $NADP^+$. This scheme is basic to an understanding of photosynthetic electron transport. P680 and P700 refer to the wavelengths of maximum absorption of the reaction center chlorophylls in PSII and PSI, respectively.

ORGANIZATION OF THE PHOTOSYNTHETIC APPARATUS

The previous section explained some of the physical principles underlying photosynthesis, some aspects of the functional roles of various pigments, and some of the chemical reactions carried out by photosynthetic organisms. We now turn to the architecture of the photosynthetic apparatus and the structure of its components.

The Chloroplast Is the Site of Photosynthesis

In photosynthetic eukaryotes, photosynthesis takes place in the subcellular organelle known as the chloroplast. Figure 7.15 shows a transmission electron micrograph of a thin section from a pea chloroplast. The most striking aspect of the structure of the chloroplast is the extensive system of internal membranes known as **thylakoids**. All the chlorophyll is contained within this membrane system, which is the site of the light reactions of photosynthesis.

The carbon reduction reactions, which are catalyzed by water-soluble enzymes, take place in the **stroma** (plural *stromata*), the region of the chloroplast outside the thylakoids. Most of the thylakoids appear to be very closely associated with each other. These stacked membranes are known as **grana lamellae** (singular *lamella*; each stack is called a *granum*), and the exposed membranes in which stacking is absent are known as **stroma lamellae**.

Two separate membranes, each composed of a lipid bilayer and together known as the **envelope**, surround most types of chloroplasts (Figure 7.16). This double-membrane system contains a variety of metabolite transport systems.

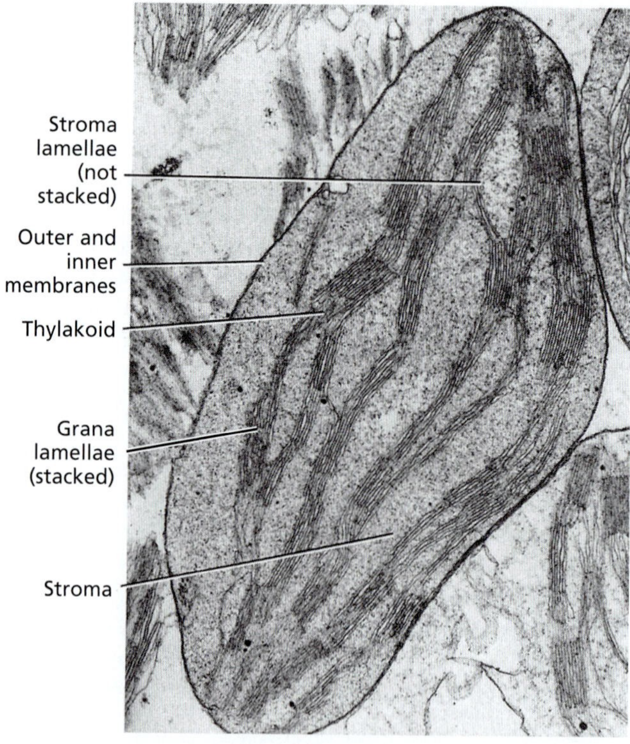

FIGURE 7.15 Transmission electron micrograph of a chloroplast from pea (*Pisum sativum*), fixed in glutaraldehyde and OsO_4, embedded in plastic resin, and thin-sectioned with an ultramicrotome. (14,500×) (Courtesy of J. Swafford.)

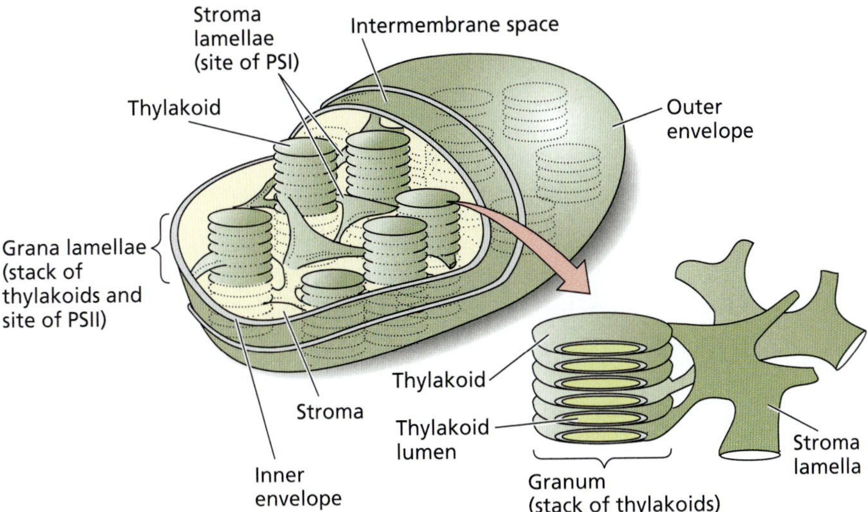

FIGURE 7.16 Schematic picture of the overall organization of the membranes in the chloroplast. The chloroplast of higher plants is surrounded by the inner and outer membranes (envelope). The region of the chloroplast that is inside the inner membrane and surrounds the thylakoid membranes is known as the stroma. It contains the enzymes that catalyze carbon fixation and other biosynthetic pathways. The thylakoid membranes are highly folded and appear in many pictures to be stacked like coins, although in reality they form one or a few large interconnected membrane systems, with a well-defined interior and exterior with respect to the stroma. The inner space within a thylakoid is known as the lumen. (After Becker 1986.)

The chloroplast also contains its own DNA, RNA, and ribosomes. Many of the chloroplast proteins are products of transcription and translation within the chloroplast itself, whereas others are encoded by nuclear DNA, synthesized on cytoplasmic ribosomes, and then imported into the chloroplast. This remarkable division of labor, extending in many cases to different subunits of the same enzyme complex, will be discussed in more detail later in this chapter. For some dynamic structures of chloroplasts see **Web Essay 7.1**.

Thylakoids Contain Integral Membrane Proteins

A wide variety of proteins essential to photosynthesis are embedded in the thylakoid membranes. In many cases, portions of these proteins extend into the aqueous regions on both sides of the thylakoids. These **integral membrane proteins** contain a large proportion of hydrophobic amino acids and are therefore much more stable in a nonaqueous medium such as the hydrocarbon portion of the membrane (see Figure 1.5A).

The reaction centers, the antenna pigment–protein complexes, and most of the electron transport enzymes are all integral membrane proteins. In all known cases, integral membrane proteins of the chloroplast have a unique orientation within the membrane. Thylakoid membrane proteins have one region pointing toward the stromal side of the membrane and the other oriented toward the interior portion of the thylakoid, known as the *lumen* (see Figures 7.16 and 7.17).

The chlorophylls and accessory light-gathering pigments in the thylakoid membrane are always associated in a noncovalent but highly specific way with proteins. Both antenna and reaction center chlorophylls are associated with proteins that are organized within the membrane so as to optimize energy transfer in antenna complexes and electron transfer in reaction centers, while at the same time minimizing wasteful processes.

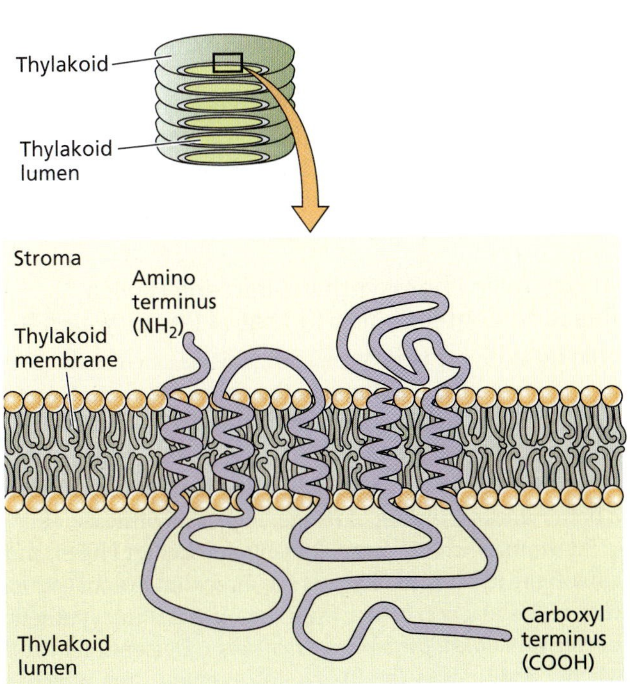

FIGURE 7.17 Predicted folding pattern of the D1 protein of the PSII reaction center. The hydrophobic portion of the membrane is traversed five times by the peptide chain rich in hydrophobic amino acid residues. The protein is asymmetrically arranged in the thylakoid membrane, with the amino (NH_2) terminus on the stromal side of the membrane and the carboxyl (COOH) terminus on the lumen side. (After Trebst 1986.)

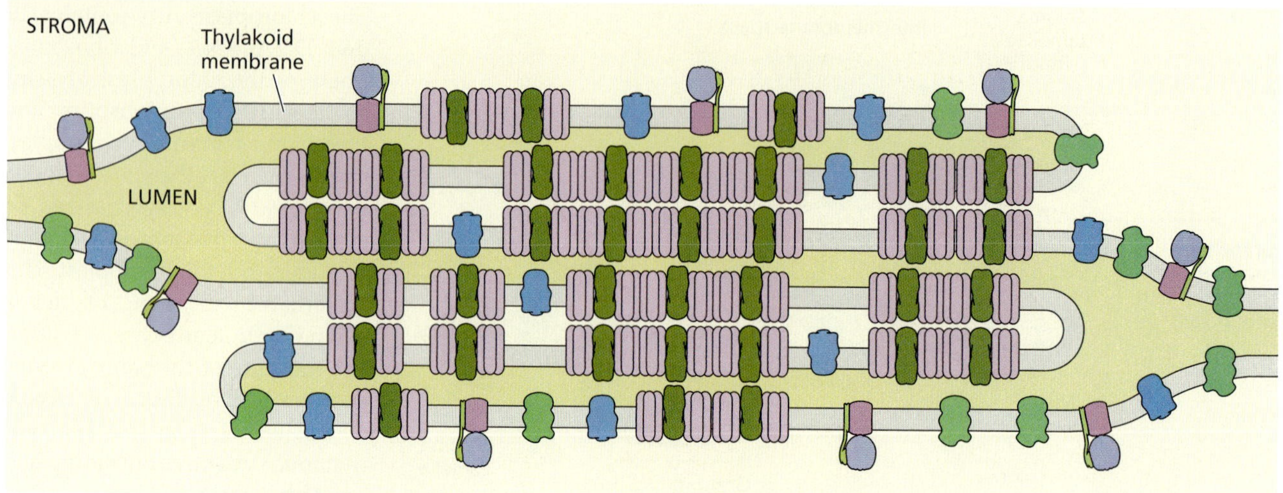

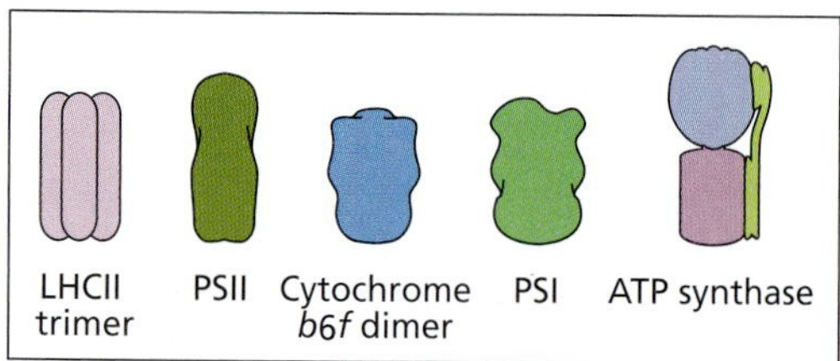

FIGURE 7.18 Organization of the protein complexes of the thylakoid membrane. Photosystem II is located predominantly in the stacked regions of the thylakoid membrane; photosystem I and ATP synthase are found in the unstacked regions protruding into the stroma. Cytochrome b_6f complexes are evenly distributed. This lateral separation of the two photosystems requires that electrons and protons produced by photosystem II be transported a considerable distance before they can be acted on by photosystem I and the ATP-coupling enzyme. (After Allen and Forsberg 2001.)

Photosystems I and II Are Spatially Separated in the Thylakoid Membrane

The PSII reaction center, along with its antenna chlorophylls and associated electron transport proteins, is located predominantly in the grana lamellae (Figure 7.18) (Allen and Forsberg 2001).

The PSI reaction center and its associated antenna pigments and electron transfer proteins, as well as the coupling-factor enzyme that catalyzes the formation of ATP, are found almost exclusively in the stroma lamellae and at the edges of the grana lamellae. The cytochrome b_6f complex of the electron transport chain that connects the two photosystems (see Figure 7.21) is evenly distributed between stroma and grana.

Thus the two photochemical events that take place in O_2-evolving photosynthesis are spatially separated. This separation implies that one or more of the electron carriers that function between the photosystems diffuses from the grana region of the membrane to the stroma region, where electrons are delivered to photosystem I.

In PSII, the oxidation of two water molecules produces four electrons, four protons, and a single O_2 (see Equation 7.8). The protons produced by this oxidation of water must also be able to diffuse to the stroma region, where ATP is synthesized. The functional role of this large separation (many tens of nanometers) between photosystems I and II is not entirely clear but is thought to improve the efficiency of energy distribution between the two photosystems (Trissl and Wilhelm 1993; Allen and Forsberg 2001).

The spatial separation between photosystems I and II indicates that a strict one-to-one stoichiometry between the two photosystems is not required. Instead, PSII reaction centers feed reducing equivalents into a common intermediate pool of soluble electron carriers (plastoquinone), which will be described in detail later in the chapter. The PSI reaction centers remove the reducing equivalents from the common pool, rather than from any specific PSII reaction center complex.

Most measurements of the relative quantities of photosystems I and II have shown that there is an excess of photosystem II in chloroplasts. Most commonly, the ratio of PSII to PSI is about 1.5:1, but it can change when plants are grown in different light conditions.

Anoxygenic Photosynthetic Bacteria Have a Reaction Center Similar to That of Photosystem II

Non-O_2-evolving (anoxygenic) organisms, such as the purple photosynthetic bacteria of the genera *Rhodobacter* and *Rhodopseudomonas*, contain only a single photosystem. These simpler organisms have been very useful for detailed structural and functional studies that have contributed to a better understanding of oxygenic photosynthesis.

Hartmut Michel, Johann Deisenhofer, Robert Huber, and coworkers in Munich resolved the three-dimensional structure of the reaction center from the purple photosynthetic bacterium *Rhodopseudomonas viridis* (Deisenhofer and Michel 1989). This landmark achievement, for which a Nobel Prize was awarded in 1988, was the first high-resolution, X-ray structural determination for an integral membrane protein, and the first structural determination for a reaction center complex (see Figures 7.5.A and 7.5.B in **Web Topic 7.5**). Detailed analysis of these structures, along with the characterization of numerous mutants, has revealed many of the principles involved in the energy storage processes carried out by all reaction centers.

The structure of the bacterial reaction center is thought to be similar in many ways to that found in photosystem II from oxygen-evolving organisms, especially in the electron acceptor portion of the chain. The proteins that make up the core of the bacterial reaction center are relatively similar in sequence to their photosystem II counterparts, implying an evolutionary relatedness.

ORGANIZATION OF LIGHT-ABSORBING ANTENNA SYSTEMS

The antenna systems of different classes of photosynthetic organisms are remarkably varied, in contrast to the reaction centers, which appear to be similar in even distantly related organisms. The variety of antenna complexes reflects evolutionary adaptation to the diverse environments in which different organisms live, as well as the need in some organisms to balance energy input to the two photosystems (Grossman et al. 1995; Green and Durnford 1996).

Antenna systems function to deliver energy efficiently to the reaction centers with which they are associated (van Grondelle et al. 1994; Pullerits and Sundström 1996). The size of the antenna system varies considerably in different organisms, ranging from a low of 20 to 30 bacteriochlorophylls per reaction center in some photosynthetic bacteria, to generally 200 to 300 chlorophylls per reaction center in higher plants, to a few thousand pigments per reaction center in some types of algae and bacteria. The molecular structures of antenna pigments are also quite diverse, although all of them are associated in some way with the photosynthetic membrane.

The physical mechanism by which excitation energy is conveyed from the chlorophyll that absorbs the light to the reaction center is thought to be **resonance transfer**. By this mechanism the excitation energy is transferred from one molecule to another by a nonradiative process.

A useful analogy for resonance transfer is the transfer of energy between two tuning forks. If one tuning fork is struck and properly placed near another, the second tuning fork receives some energy from the first and begins to vibrate. As in resonance energy transfer in antenna complexes, the efficiency of energy transfer between the two tuning forks depends on their distance from each other and their relative orientation, as well as their pitches or vibrational frequencies.

Energy transfer in antenna complexes is very efficient: Approximately 95 to 99% of the photons absorbed by the antenna pigments have their energy transferred to the reaction center, where it can be used for photochemistry. There is an important difference between energy transfer among pigments in the antenna and the electron transfer that occurs in the reaction center: Whereas energy transfer is a purely physical phenomenon, electron transfer involves chemical changes in molecules.

The Antenna Funnels Energy to the Reaction Center

The sequence of pigments within the antenna that funnel absorbed energy toward the reaction center has absorption maxima that are progressively shifted toward longer red wavelengths (Figure 7.19). This red shift in absorption maximum means that the energy of the excited state is somewhat lower nearer the reaction center than in the more peripheral portions of the antenna system.

As a result of this arrangement, when excitation is transferred, for example, from a chlorophyll *b* molecule absorbing maximally at 650 nm to a chlorophyll *a* molecule absorbing maximally at 670 nm, the difference in energy between these two excited chlorophylls is lost to the environment as heat.

For the excitation to be transferred back to the chlorophyll *b*, the energy lost as heat would have to be resupplied. The probability of reverse transfer is therefore smaller simply because thermal energy is not sufficient to make up the deficit between the lower-energy and higher-energy pigments. This effect gives the energy-trapping process a degree of directionality or irreversibility and makes the delivery of excitation to the reaction center more efficient. In essence, the system sacrifices some energy from each quantum so that nearly all of the quanta can be trapped by the reaction center.

Many Antenna Complexes Have a Common Structural Motif

In all eukaryotic photosynthetic organisms that contain both chlorophyll *a* and chlorophyll *b*, the most abundant antenna proteins are members of a large family of structurally related proteins. Some of these proteins are associated primarily with photosystem II and are called **light-harvesting complex II** (**LHCII**) proteins; others are associated with photosystem I and are called *LHCI* proteins. These antenna complexes are also known as **chlorophyll *a*/*b* antenna proteins** (Paulsen 1995; Green and Durnford 1996).

The structure of one of the LHCII proteins has been determined by a combination of electron microscopy and electron crystallography (Figure 7.20) (Kühlbrandt et al. 1994). The protein contains three α-helical regions and binds about 15 chlorophyll *a* and *b* molecules, as well as a few carotenoids. Only some of these pigments are visible in the resolved structure. The structure of the LHCI proteins has not yet been determined but is probably similar to that of the LHCII proteins. All of these proteins have significant sequence similarity and are almost certainly descendants of a common ancestral protein (Grossman et al. 1995; Green and Durnford 1996).

Light absorbed by carotenoids or chlorophyll *b* in the LHC proteins is rapidly transferred to chlorophyll *a* and

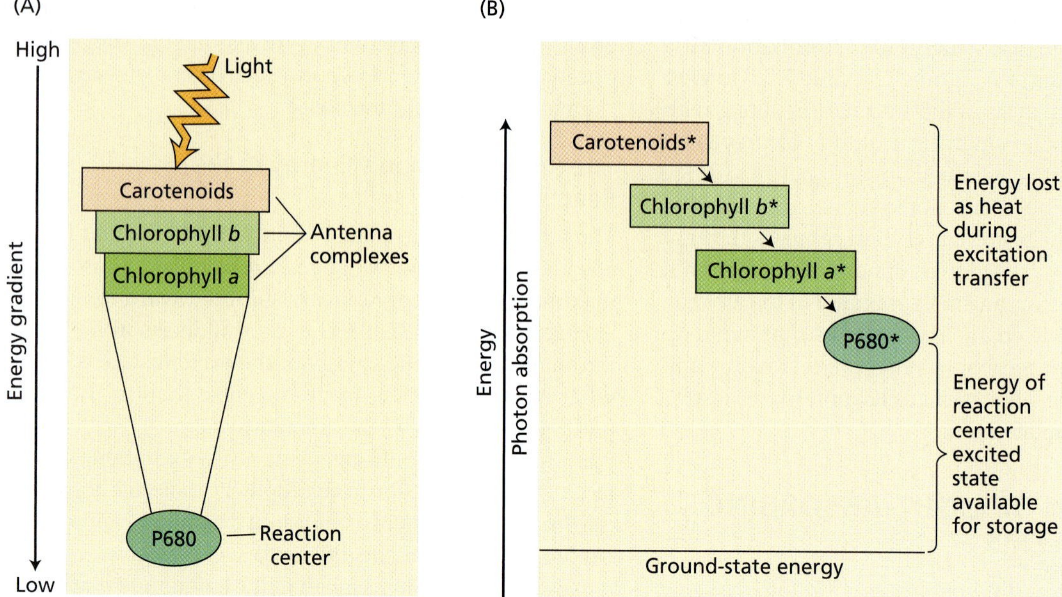

FIGURE 7.19 Funneling of excitation from the antenna system toward the reaction center. (A) The excited-state energy of pigments increases with distance from the reaction center; that is, pigments closer to the reaction center are lower in energy than those farther from the reaction center. This energy gradient ensures that excitation transfer toward the reaction center is energetically favorable and that excitation transfer back out to the peripheral portions of the antenna is energetically unfavorable. (B) Some energy is lost as heat to the environment by this process, but under optimal conditions almost all the excitations absorbed in the antenna complexes can be delivered to the reaction center. The asterisks denote an excited state.

then to other antenna pigments that are intimately associated with the reaction center. The LHCII complex is also involved in regulatory processes, which are discussed later in the chapter.

MECHANISMS OF ELECTRON TRANSPORT

Some of the evidence that led to the idea of two photochemical reactions operating in series was discussed earlier in this chapter. Here we will consider in detail the chemical reactions involved in electron transfer during photosynthesis. We will discuss the excitation of chlorophyll by light and the reduction of the first electron acceptor, the flow of electrons through photosystems II and I, the oxidation of water as the primary source of electrons, and the reduction of the final electron acceptor ($NADP^+$). The chemiosmotic mechanism that mediates ATP synthesis will be discussed in detail later in the chapter (see "Proton Transport and ATP Synthesis in the Chloroplast").

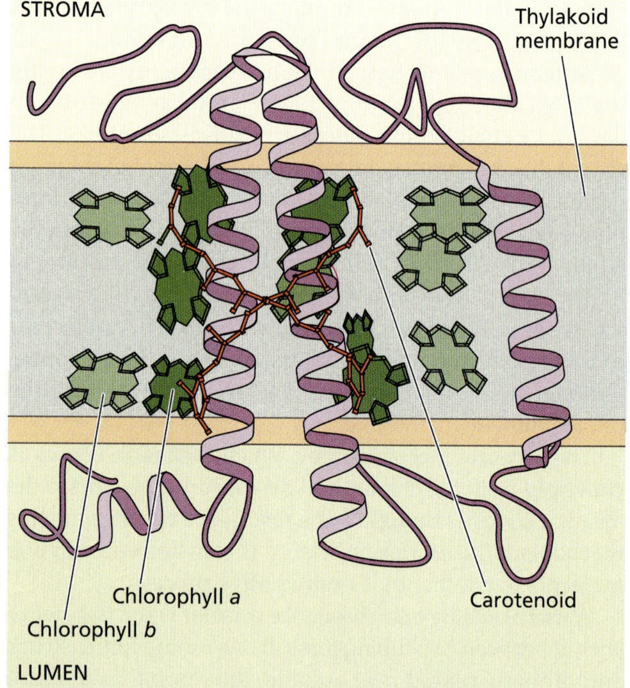

FIGURE 7.20 Two-dimensional view of the structure of the LHCII antenna complex from higher plants, determined by a combination of electron microscopy and electron crystallography. Like X-ray crystallography, electron crystallography uses the diffraction patterns of soft-energy electrons to resolve macromolecule structures. The antenna complex is a transmembrane pigment protein, with three helical regions that cross the nonpolar part of the membrane. Approximately 15 chlorophyll *a* and *b* molecules are associated with the complex, as well as several carotenoids. The positions of several of the chlorophylls are shown, and two of the carotenoids form an X in the middle of the complex. In the membrane, the complex is trimeric and aggregates around the periphery of the PSII reaction center complex. (After Kühlbrandt et al. 1994.)

Electrons Ejected from Chlorophyll Travel Through a Series of Electron Carriers Organized in the "Z Scheme"

Figure 7.21 shows a current version of the Z scheme, in which all the electron carriers known to function in electron flow from H_2O to $NADP^+$ are arranged vertically at their midpoint redox potentials (see **Web Topic 7.6** for further detail). Components known to react with each other are connected by arrows, so the Z scheme is really a synthesis of both kinetic and thermodynamic information. The large vertical arrows represent the input of light energy into the system.

Photons excite the specialized chlorophyll of the reaction centers (P680 for PSII, and P700 for PSI), and an electron is ejected. The electron then passes through a series of electron carriers and eventually reduces P700 (for electrons from PSII) or $NADP^+$ (for electrons from PSI). Much of the following discussion describes the journeys of these electrons and the nature of their carriers.

Almost all the chemical processes that make up the light reactions of photosynthesis are carried out by four major protein complexes: photosystem II, the cytochrome b_6f complex, photosystem I, and the ATP synthase. These four integral membrane complexes are vectorially oriented in the thylakoid membrane to function as follows (Figure 7.22):

- Photosystem II oxidizes water to O_2 in the thylakoid lumen and in the process releases protons into the lumen.
- Cytochrome b_6f receives electrons from PSII and delivers them to PSI. It also transports additional protons into the lumen from the stroma.
- Photosystem I reduces $NADP^+$ to NADPH in the stroma by the action of ferredoxin (Fd) and the flavoprotein ferredoxin–NADP reductase (FNR).
- ATP synthase produces ATP as protons diffuse back through it from the lumen into the stroma.

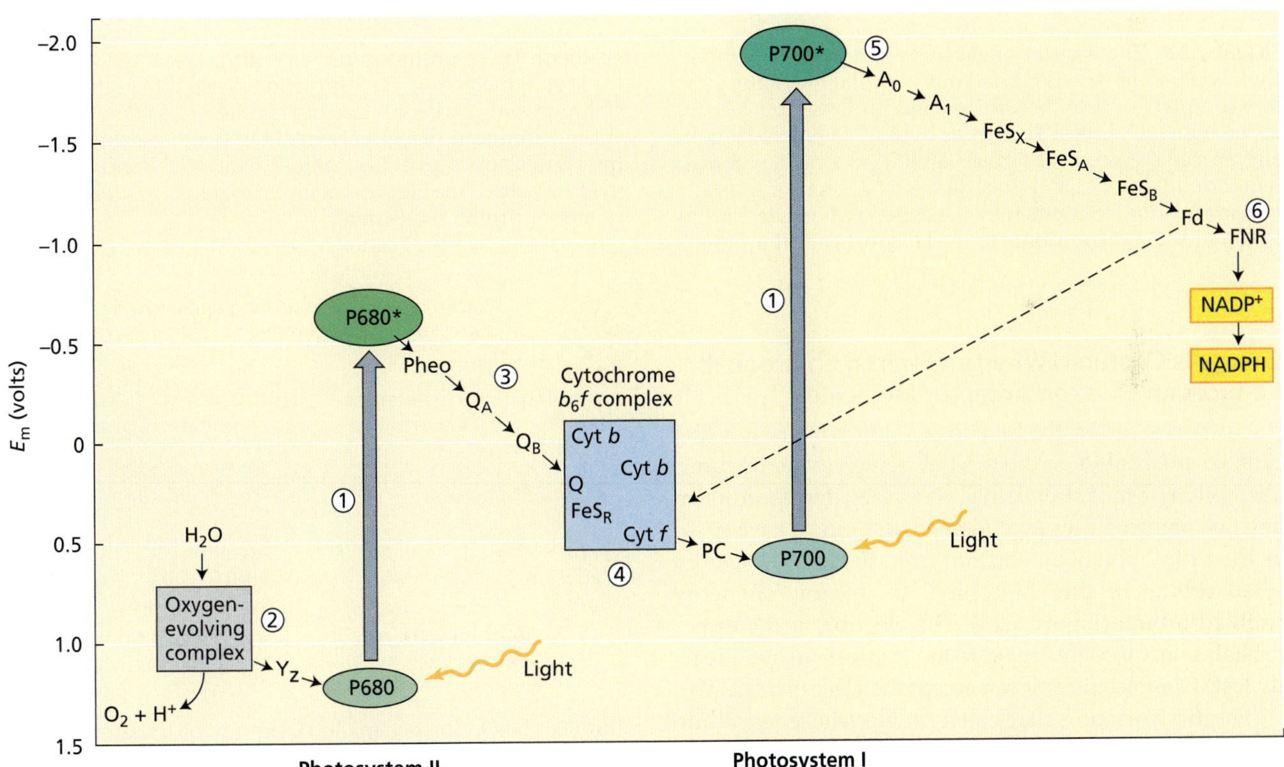

FIGURE 7.21 Detailed Z scheme for O_2-evolving photosynthetic organisms. The redox carriers are placed at their midpoint redox potentials (at pH 7). (1) The vertical arrows represent photon absorption by the reaction center chlorophylls: P680 for photosystem II (PSII) and P700 for photosystem I (PSI). The excited PSII reaction center chlorophyll, P680*, transfers an electron to pheophytin (Pheo). (2) On the oxidizing side of PSII (to the left of the arrow joining P680 with P680*), P680 oxidized by light is re-reduced by Y_Z, that has received electrons from oxidation of water. (3) On the reducing side of PSII (to the right of the arrow joining P680 with P680*), pheophytin transfers electrons to the acceptors Q_A and Q_B, which are plastoquinones. (4) The cytochrome b_6f complex transfers electrons to plastocyanin (PC), a soluble protein, which in turn reduces $P700^+$ (oxidized P700). (5) The acceptor of electrons from P700* (A_0) is thought to be a chlorophyll, and the next acceptor (A_1) is a quinone. A series of membrane-bound iron–sulfur proteins (FeS_X, FeS_A, and FeS_B) transfers electrons to soluble ferredoxin (Fd). (6) The soluble flavoprotein ferredoxin–NADP reductase (FNR) reduces $NADP^+$ to NADPH, which is used in the Calvin cycle to reduce CO_2 (see Chapter 8). The dashed line indicates cyclic electron flow around PSI. (After Blankenship and Prince 1985.)

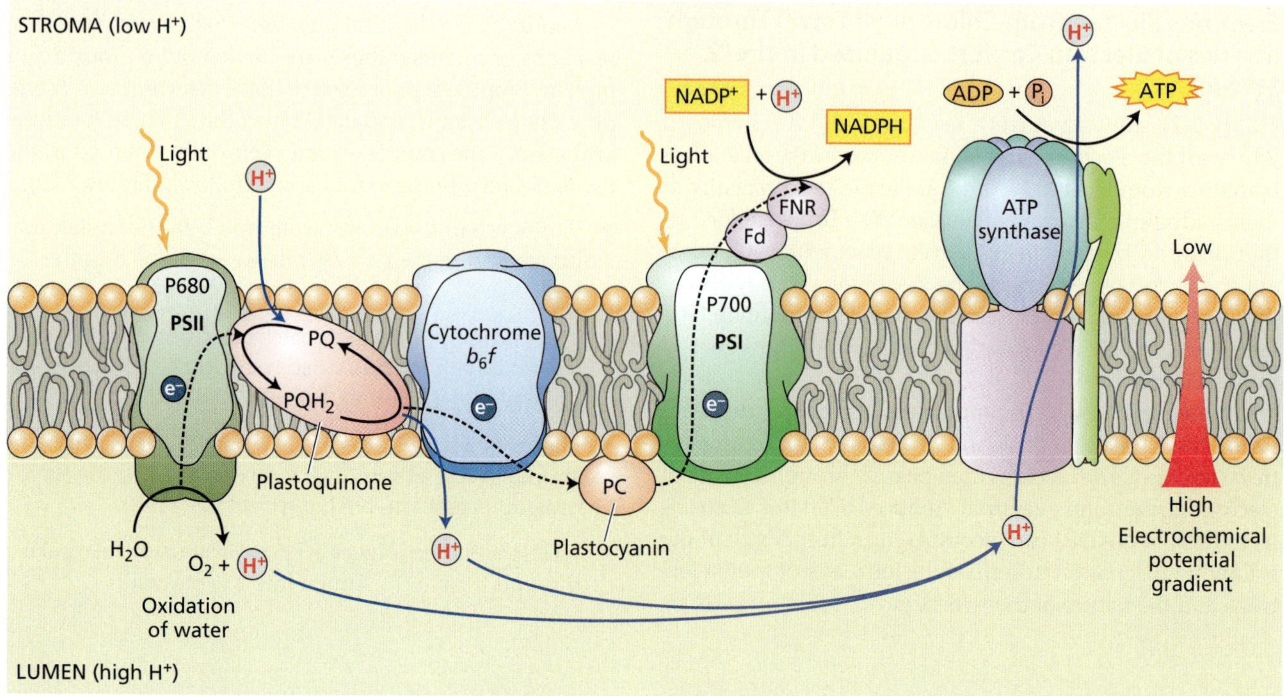

FIGURE 7.22 The transfer of electrons and protons in the thylakoid membrane is carried out vectorially by four protein complexes. Water is oxidized and protons are released in the lumen by PSII. PSI reduces $NADP^+$ to NADPH in the stroma, via the action of ferredoxin (Fd) and the flavoprotein ferredoxin–NADP reductase (FNR). Protons are also transported into the lumen by the action of the cytochrome b_6f complex and contribute to the electrochemical proton gradient. These protons must then diffuse to the ATP synthase enzyme, where their diffusion down the electrochemical potential gradient is used to synthesize ATP in the stroma. Reduced plastoquinone (PQH_2) and plastocyanin transfer electrons to cytochrome b_6f and to PSI, respectively. Dashed lines represent electron transfer; solid lines represent proton movement.

Energy Is Captured When an Excited Chlorophyll Reduces an Electron Acceptor Molecule

As discussed earlier, the function of light is to excite a specialized chlorophyll in the reaction center, either by direct absorption or, more frequently, via energy transfer from an antenna pigment. This excitation process can be envisioned as the promotion of an electron from the highest-energy filled orbital of the chlorophyll to the lowest-energy unfilled orbital (Figure 7.23). The electron in the upper orbital is only loosely bound to the chlorophyll and is easily lost if a molecule that can accept the electron is nearby.

The first reaction that converts electron energy into chemical energy—that is, the primary photochemical event—is the transfer of an electron from the excited state of a chlorophyll in the reaction center to an acceptor molecule. An equivalent way to view this process is that the absorbed photon causes an electron rearrangement in the reaction center chlorophyll, followed by an electron transfer process in which part of the energy in the photon is captured in the form of redox energy.

Immediately after the photochemical event, the reaction center chlorophyll is in an oxidized state (electron deficient, or positively charged) and the nearby electron acceptor mol-

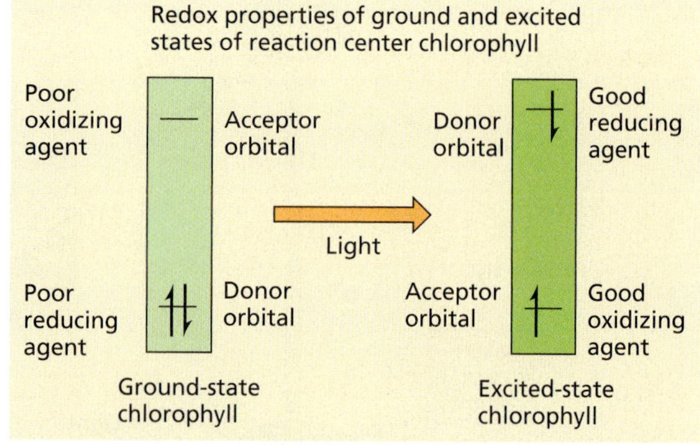

FIGURE 7.23 Orbital occupation diagram for the ground and excited states of reaction center chlorophyll. In the ground state the molecule is a poor reducing agent (loses electrons from a low-energy orbital) and a poor oxidizing agent (accepts electrons only into a high-energy orbital). In the excited state the situation is reversed, and an electron can be lost from the high-energy orbital, making the molecule an extremely powerful reducing agent. This is the reason for the extremely negative excited-state redox potential shown by P680* and P700* in Figure 7.21. The excited state can also act as a strong oxidant by accepting an electron into the lower-energy orbital, although this pathway is not significant in reaction centers. (After Blankenship and Prince 1985.)

ecule is reduced (electron rich, or negatively charged). The system is now at a critical juncture. The lower-energy orbital of the positively charged oxidized reaction center chlorophyll shown in Figure 7.23 has a vacancy and can accept an electron. If the acceptor molecule donates its electron back to the reaction center chlorophyll, the system will be returned to the state that existed before the light excitation, and all the absorbed energy will be converted into heat.

This wasteful *recombination* process, however, does not appear to occur to any substantial degree in functioning reaction centers. Instead, the acceptor transfers its extra electron to a secondary acceptor and so on down the electron transport chain. The oxidized reaction center of the chlorophyll that had donated an electron is re-reduced by a secondary donor, which in turn is reduced by a tertiary donor. In plants, the ultimate electron donor is H_2O, and the ultimate electron acceptor is $NADP^+$ (see Figure 7.21).

The essence of photosynthetic energy storage is thus the initial transfer of an electron from an excited chlorophyll to an acceptor molecule, followed by a very rapid series of secondary chemical reactions that separate the positive and negative charges. These secondary reactions separate the charges to opposite sides of the thylakoid membrane in approximately 200 picoseconds (1 picosecond = 10^{-12} s).

With the charges thus separated, the reversal reaction is many orders of magnitude slower, and the energy has been captured. Each of the secondary electron transfers is accompanied by a loss of some energy, thus making the process effectively irreversible. The quantum yield for the production of stable products in purified reaction centers from photosynthetic bacteria has been measured as 1.0; that is, every photon produces stable products, and no reversal reactions occur.

Although these types of measurements have not been made on purified reaction centers from higher plants, the measured quantum requirements for O_2 production under optimal conditions (low-intensity light) indicate that the values for the primary photochemical events are very close to 1.0. The structure of the reaction center appears to be extremely fine-tuned for maximal rates of productive reactions and minimal rates of energy-wasting reactions.

The Reaction Center Chlorophylls of the Two Photosystems Absorb at Different Wavelengths

As discussed earlier in the chapter, PSI and PSII have distinct absorption characteristics. Precise measurements of absorption maxima were made possible by optical changes in the reaction center chlorophylls in the reduced and oxidized states. The reaction center chlorophyll is transiently in an oxidized state after losing an electron and before being re-reduced by its electron donor.

In the oxidized state, the strong light absorbance in the red region of the spectrum that is characteristic of chlorophylls is lost, or **bleached**. It is therefore possible to monitor the redox state of these chlorophylls by time-resolved optical absorbance measurements in which this bleaching is monitored directly (see **Web Topic 7.1**).

Using such techniques, Bessel Kok found that the reaction center chlorophyll of photosystem I absorbs maximally at 700 nm in its reduced state. Accordingly, this chlorophyll is named **P700** (the P stands for *pigment*). H. T. Witt and coworkers found the analogous optical transient of photosystem II at 680 nm, so its reaction center chlorophyll is known as **P680**. Earlier, Louis Duysens had identified the reaction center bacteriochlorophyll from purple photosynthetic bacteria as **P870**.

The X-ray structure of the bacterial reaction center (see Figures 7.5.A and 7.5.B in **Web Topic 7.5**) clearly indicates that P870 is a closely coupled pair or dimer of bacteriochlorophylls, rather than a single molecule. The primary donor of photosystem I, P700, is a dimer of chlorophyll *a* molecules. Photosystem II also contains a dimer of chlorophylls, although the primary donor, P680, may not reside entirely on these pigments. In the oxidized state, reaction center chlorophylls contain an unpaired electron. Molecules with unpaired electrons often can be detected by a magnetic-resonance technique known as **electron spin resonance** (**ESR**). ESR studies, along with the spectroscopic measurements already described, have led to the discovery of many intermediate electron carriers in the photosynthetic electron transport system.

The Photosystem II Reaction Center Is a Multisubunit Pigment–Protein Complex

Photosystem II is contained in a multisubunit protein supercomplex (Figure 7.24) (Barber et al. 1999). In higher plants, the multisubunit protein supercomplex has two complete reaction centers and some antenna complexes. The core of the reaction center consists of two membrane proteins known as D1 and D2, as well as other proteins, as shown in Figure 7.25 (Zouni et al. 2001).

The primary donor chlorophyll (P680), additional chlorophylls, carotenoids, pheophytins, and plastoquinones (two electron acceptors described in the following section) are bound to the membrane proteins D1 and D2. These proteins have some sequence similarity to the L and M peptides of purple bacteria. Other proteins serve as antenna complexes or are involved in oxygen evolution. Some, such as cytochrome b_{559}, have no known function but may be involved in a protective cycle around photosystem II.

Water Is Oxidized to Oxygen by Photosystem II

Water is oxidized according to the following chemical reaction (Hoganson and Babcock 1997):

$$2\,H_2O \rightarrow O_2 + 4\,H^+ + 4\,e^- \qquad (7.8)$$

This equation indicates that four electrons are removed from two water molecules, generating an oxygen molecule and four hydrogen ions. (For more on oxidation–reduction reactions, see Chapter 2 on the web site and **Web Topic 7.6**.)

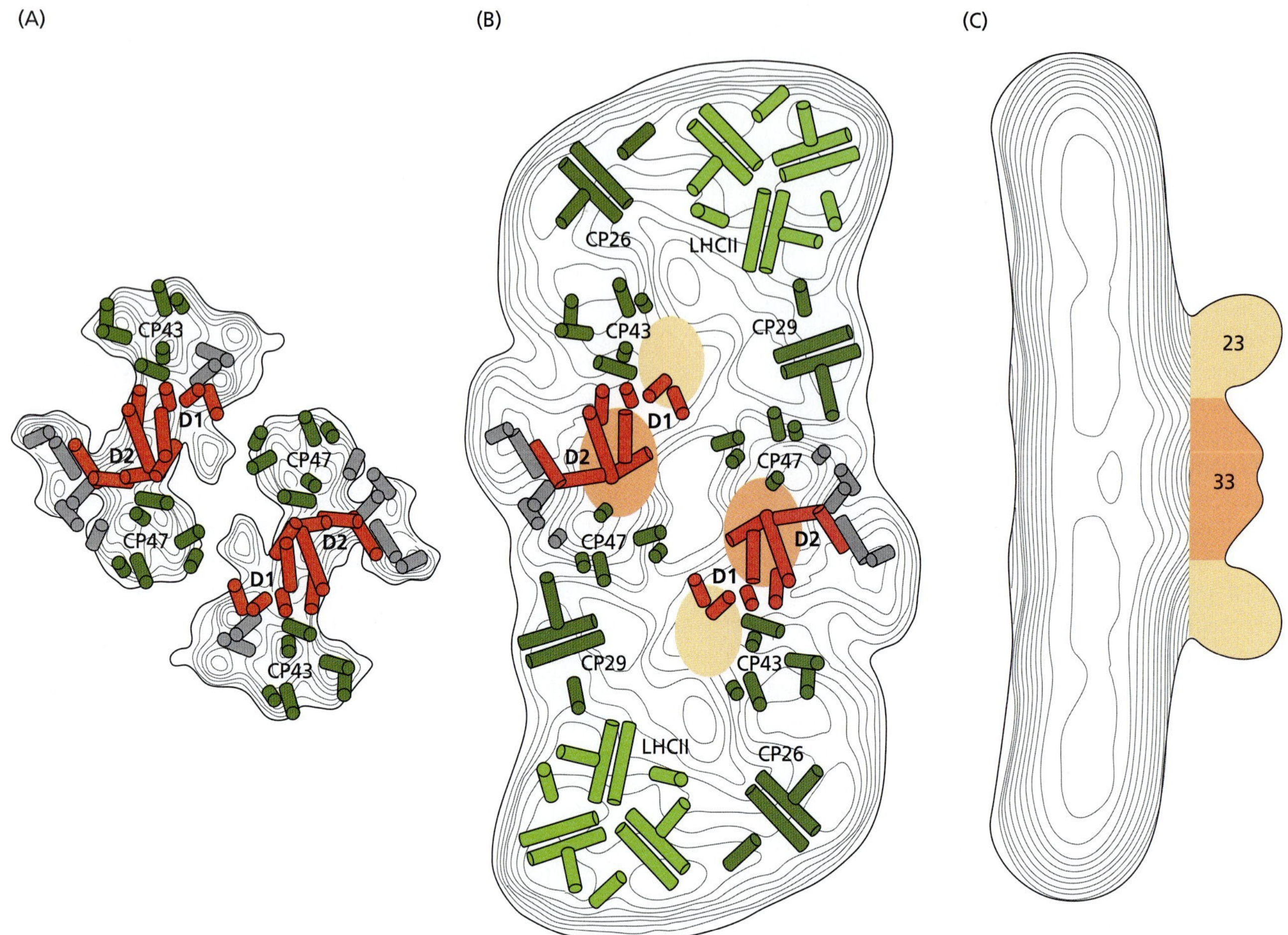

FIGURE 7.24 Structure of dimeric multisubunit protein supercomplex of photosystem II from higher plants, as determined by electron microscopy. The figure shows two complete reaction centers, each of which is a dimeric complex. (A) Helical arrangement of the D1 and D2 (red) and CP43 and CP47 (green) core subunits. (B) View from the lumenal side of the supercomplex, including additional antenna complexes, LHCII, CP26 and CP29, and extrinsic oxygen-evolving complex, shown as orange and yellow circles. Unassigned helices are shown in gray. (C) Side view of the complex illustrating the arrangement of the extrinsic proteins of the oxygen-evolving complex. (After Barber et al. 1999.)

Water is a very stable molecule. Oxidation of water to form molecular oxygen is very difficult, and the photosynthetic oxygen-evolving complex is the only known biochemical system that carries out this reaction. Photosynthetic oxygen evolution is also the source of almost all the oxygen in Earth's atmosphere.

The chemical mechanism of photosynthetic water oxidation is not yet known, although many studies have provided a substantial amount of information about the process (see **Web Topic 7.7** and Figure 7.26). The protons produced by water oxidation are released into the lumen of the thylakoid, not directly into the stromal compartment (see Figure 7.22). They are released into the lumen because of the vectorial nature of the membrane and the fact that the oxygen-evolving complex is localized on the interior surface of the thylakoid. These protons are eventually transferred from the lumen to the stroma by translocation through ATP synthase. In this way the protons released during water oxidation contribute to the electrochemical potential driving ATP formation.

It has been known for many years that manganese (Mn) is an essential cofactor in the water-oxidizing process (see Chapter 5), and a classic hypothesis in photosynthesis research postulates that Mn ions undergo a series of oxidations—which are known as *S states,* and are labeled S_0, S_1, S_2, S_3, and S_4 (see **Web Topic 7.7**)—that are perhaps linked to H_2O oxidation and the generation of O_2 (see Figure 7.26). This hypothesis has received strong support from a variety of experiments, most notably X-ray absorption and ESR studies, both of which detect the manganese directly (Yachandra

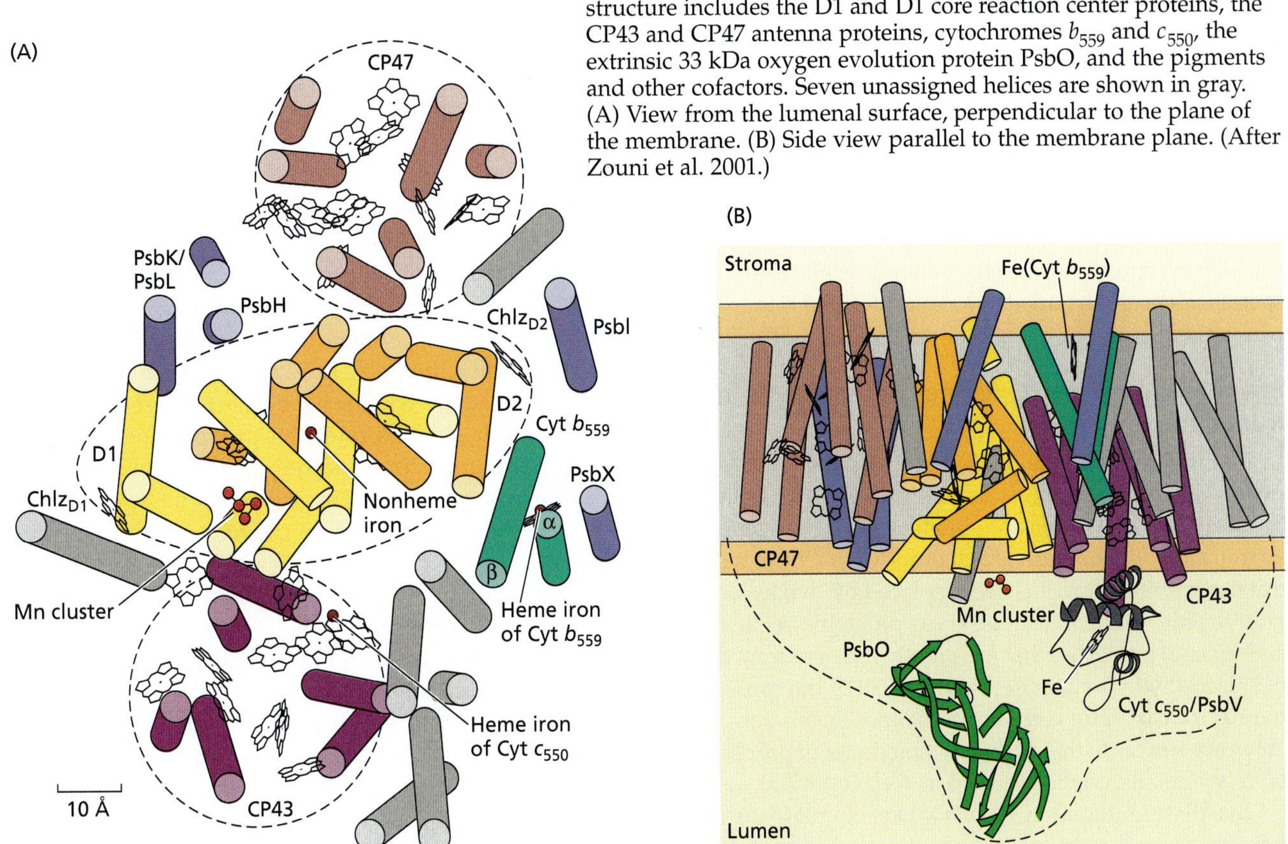

FIGURE 7.25 Structure of the photosystem II reaction center from the cyanobacterium *Synechococcus elongatus*, resolved at 3.8 Å. The structure includes the D1 and D1 core reaction center proteins, the CP43 and CP47 antenna proteins, cytochromes b_{559} and c_{550}, the extrinsic 33 kDa oxygen evolution protein PsbO, and the pigments and other cofactors. Seven unassigned helices are shown in gray. (A) View from the lumenal surface, perpendicular to the plane of the membrane. (B) Side view parallel to the membrane plane. (After Zouni et al. 2001.)

FIGURE 7.26 Model of the S state cycle of oxygen evolution in PSII. Successive stages in the oxidation of water via the Mn oxygen-evolving complex are shown. Y_z is a tyrosine radical that is an intermediate electron carrier between P680 and the Mn cluster. (After Tommos and Babcock 1998.)

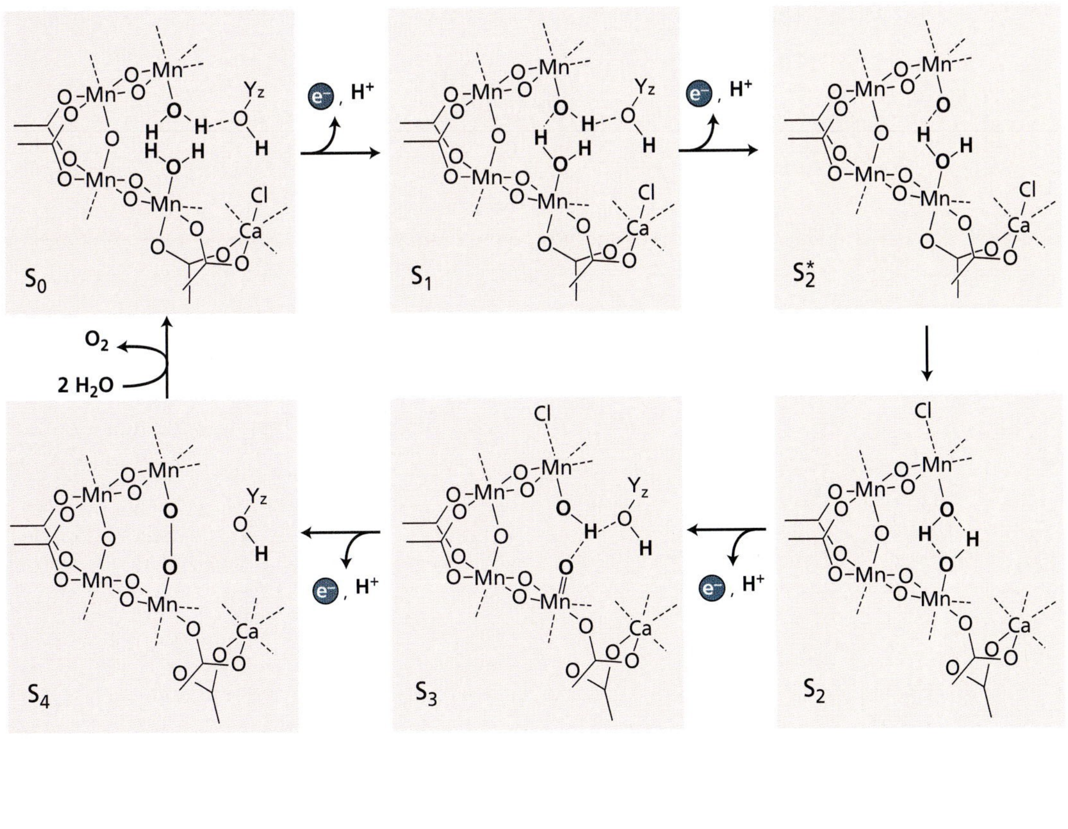

et al. 1996). Analytical experiments indicate that four Mn ions are associated with each oxygen-evolving complex. Other experiments have shown that Cl^- and Ca^{2+} ions are essential for O_2 evolution (see Figure 7.26 and **Web Topic 7.7**).

One electron carrier, generally identified as Y_Z, functions between the oxygen-evolving complex and P680 (see Figures 7.21 and 7.26). To function in this region, Y_Z needs to have a very strong tendency to retain its electrons. This species has been identified as a radical formed from a tyrosine residue in the D1 protein of the PSII reaction center.

Pheophytin and Two Quinones Accept Electrons from Photosystem II

Evidence from spectral and ESR studies indicates that pheophytin acts as an early acceptor in photosystem II, followed by a complex of two plastoquinones in close proximity to an iron atom. **Pheophytin** is a chlorophyll in which the central magnesium atom has been replaced by two hydrogen atoms. This chemical change gives pheophytin chemical and spectral properties that are slightly different from those of chlorophyll. The precise arrangement of the carriers in the electron acceptor complex is not known, but it is probably very similar to that of the reaction center of purple bacteria (for details, see Figure 7.5.B in **Web Topic 7.5**).

Two plastoquinones (Q_A and Q_B) are bound to the reaction center and receive electrons from pheophytin in a sequential fashion (Okamura et al. 2000). Transfer of the two electrons to Q_B reduces it to Q_B^{2-}, and the reduced Q_B^{2-} takes two protons from the stroma side of the medium, yielding a fully reduced **plastohydroquinone** (QH_2) (Figure 7.27). The plastohydroquinone then dissociates from the reaction center complex and enters the hydrocarbon portion of the membrane, where it in turn transfers its electrons to the cytochrome b_6f complex. Unlike the large protein complexes of the thylakoid membrane, hydroquinone is a small, nonpolar molecule that diffuses readily in the nonpolar core of the membrane bilayer.

Electron Flow through the Cytochrome b_6f Complex Also Transports Protons

The **cytochrome b_6f complex** is a large multisubunit protein with several prosthetic groups (Cramer et al. 1996; Berry et al. 2000). It contains two *b*-type hemes and one *c*-type heme (**cytochrome *f***). In *c*-type cytochromes the heme is covalently attached to the peptide; in *b*-type cytochromes the chemically similar protoheme group is not covalently attached (Figure 7.28). In addition, the complex contains a **Rieske iron–sulfur protein** (named for the scientist who discovered it), in which two iron atoms are bridged by two sulfur atoms.

The structures of cytochrome *f* and the related cytochrome bc_1 complex have been determined and suggest a mechanism for electron and proton flow. The precise way by which electrons and protons flow through the cytochrome b_6f complex is not yet fully understood, but a mechanism known as the **Q cycle** accounts for most of the observations. In this mechanism, plastohydroquinone (QH_2) is oxidized, and one of the two electrons is passed along a linear electron transport chain toward photosystem I, while the other electron goes through a cyclic process that increases the number of protons pumped across the membrane (Figure 7.29).

In the linear electron transport chain, the oxidized Rieske protein (**FeS_R**) accepts an electron from plastohydroquinone (QH_2) and transfers it to cytochrome *f* (see Figure 7.29A). Cytochrome *f* then transfers an electron to the blue-colored copper protein plastocyanin (PC), which in turn reduces oxidized P700 of PSI. In the cyclic part of the process (see Figure 7.29B), the plastosemiquinone (see Figure 7.27) transfers its other electron to one of the *b*-type hemes, releasing both of its protons to the lumenal side of the membrane.

The *b*-type heme transfers its electron through the second *b*-type heme to an oxidized quinone molecule, reducing it to the semiquinone form near the stromal surface of

(A)

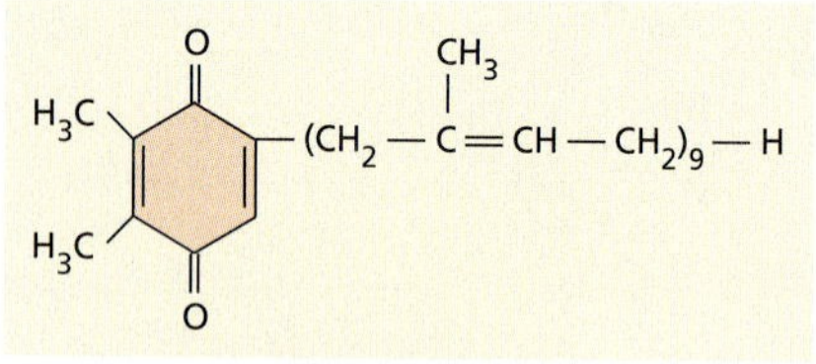

H_3C, O, CH_3, $(CH_2-C=CH-CH_2)_9-H$, H_3C, O

Plastoquinone

(B)

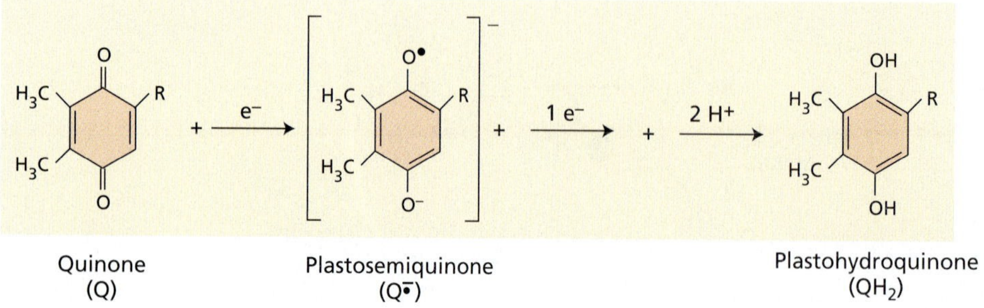

Quinone (Q) + e⁻ → [Plastosemiquinone ($Q^{\bar{\cdot}}$)]⁻ + 1 e⁻ + 2 H+ → Plastohydroquinone (QH_2)

FIGURE 7.27 Structure and reactions of plastoquinone that operate in photosystem II. (A) The plastoquinone consists of a quinoid head and a long nonpolar tail that anchors it in the membrane. (B) Redox reactions of plastoquinone. The fully oxidized quinone (Q), anionic semiquinone ($Q^{\bar{\cdot}}$), and reduced hydroquinone (QH_2) forms are shown; R represents the side chain.

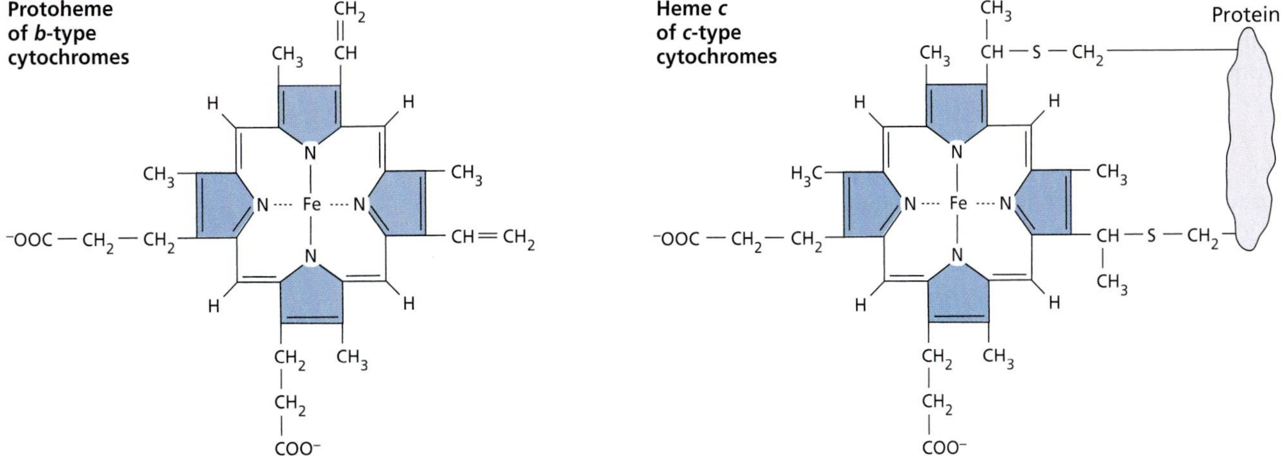

FIGURE 7.28 Structure of prosthetic groups of *b*- and *c*-type cytochromes. The protoheme group (also called protoporphyrin IX) is found in *b*-type cytochromes, the heme *c* group in *c*-type cytochromes. The heme *c* group is covalently attached to the protein by thioether linkages with two cysteine residues in the protein; the protoheme group is not covalently attached to the protein. The Fe ion is in the 2+ oxidation state in reduced cytochromes and in the 3+ oxidation state in oxidized cytochromes.

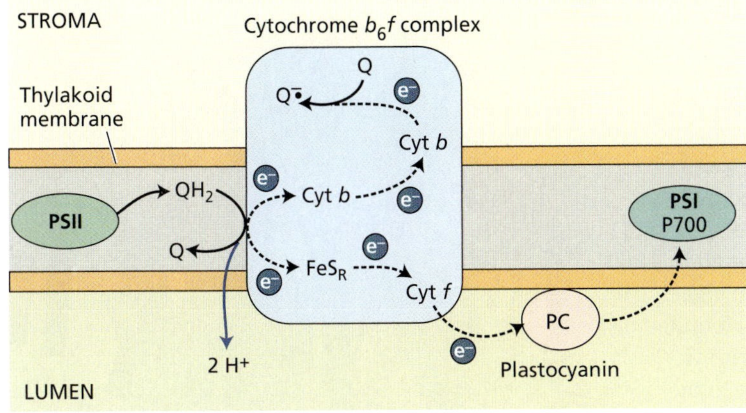

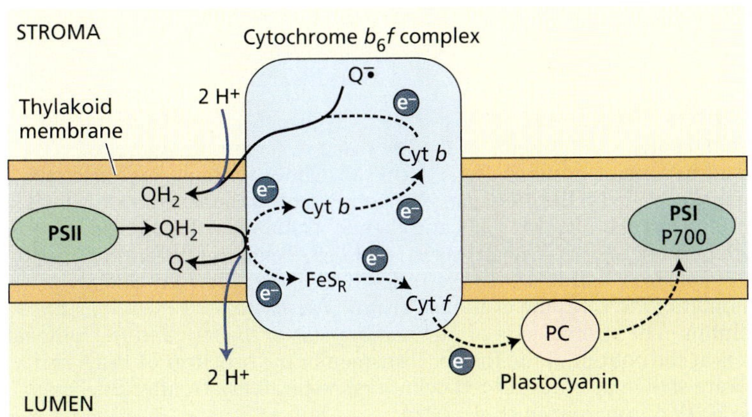

FIGURE 7.29 Mechanism of electron and proton transfer in the cytochrome b_6f complex. This complex contains two *b*-type cytochromes (Cyt *b*), a *c*-type cytochrome (Cyt *c*, historically called cytochrome *f*), a Rieske Fe–S protein (FeS_R), and two quinone oxidation–reduction sites. (A) The noncyclic or linear processes: A plastohydroquinone (QH_2) molecule produced by the action of PSII (see Figure 7.27) is oxidized near the lumenal side of the complex, transferring its two electrons to the Rieske Fe–S protein and one of the *b*-type cytochromes and simultaneously expelling two protons to the lumen. The electron transferred to FeS_R is passed to cytochrome *f* (Cyt *f*) and then to plastocyanin (PC), which reduces P700 of PSI. The reduced *b*-type cytochrome transfers an electron to the other *b*-type cytochrome, which reduces a quinone (Q) to the semiquinone (Q•) state (see Figure 7.27). (B) The cyclic processes: A second QH_2 is oxidized, with one electron going from FeS_R to PC and finally to P700. The second electron goes through the two *b*-type cytochromes and reduces the semiquinone to the plastohydroquinone, at the same time picking up two protons from the stroma. Overall, four protons are transported across the membrane for every two electrons delivered to P700.

the complex. Another similar sequence of electron flow fully reduces the plastoquinone, which picks up protons from the stromal side of the membrane and is released from the b_6f complex as plastohydroquinone.

The net result of two turnovers of the complex is that two electrons are transferred to P700, two plastohydroquinones are oxidized to the quinone form, and one oxidized plastoquinone is reduced to the hydroquinone form. In addition, four protons are transferred from the stromal to the lumenal side of the membrane.

By this mechanism, electron flow connecting the acceptor side of the PSII reaction center to the donor side of the PSI reaction center also gives rise to an electrochemical potential across the membrane, due in part to H^+ concentration differences on the two sides of the membrane. This electrochemical potential is used to power the synthesis of ATP. The cyclic electron flow through the cytochrome *b* and plastoquinone increases the number of protons pumped per electron beyond what could be achieved in a strictly linear sequence.

Plastoquinone and Plastocyanin Carry Electrons between Photosystems II and I

The location of the two photosystems at different sites on the thylakoid membranes (see Figure 7.18) requires that at least one component be capable of moving along or within the membrane in order to deliver electrons produced by photosystem II to photosystem I. The cytochrome b_6f complex is distributed equally between the grana and the stroma regions of the membranes, but its large size makes it unlikely that it is the mobile carrier. Instead, plastoquinone or plastocyanin or possibly both are thought to serve as mobile carriers to connect the two photosystems.

Plastocyanin is a small (10.5 kDa), water-soluble, copper-containing protein that transfers electrons between the cytochrome b_6f complex and P700. This protein is found in the lumenal space (see Figure 7.29). In certain green algae and cyanobacteria, a *c*-type cytochrome is sometimes found instead of plastocyanin; which of these two proteins is synthesized depends on the amount of copper available to the organism.

The Photosystem I Reaction Center Reduces NADP+

The PSI reaction center complex is a large multisubunit complex (Figure 7.30) (Jordan et al. 2001). In contrast to PSII, a core antenna consisting of about 100 chlorophylls is a part of the PSI reaction center, P700. The core antenna and P700 are bound to two proteins, PsaA and PsaB, with molecular masses in the range of 66 to 70 kDa (Brettel 1997; Chitnis 2001; see also **Web Topic 7.8**).

The antenna pigments form a bowl surrounding the electron transfer cofactors, which are in the center of the complex. In

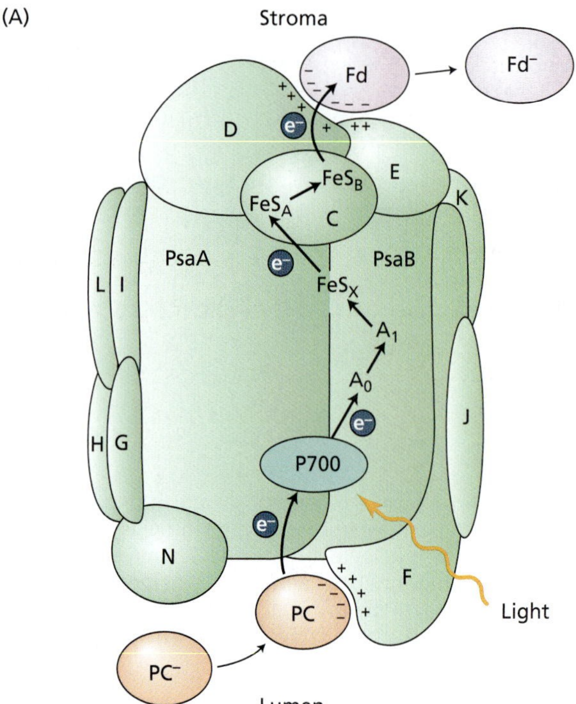

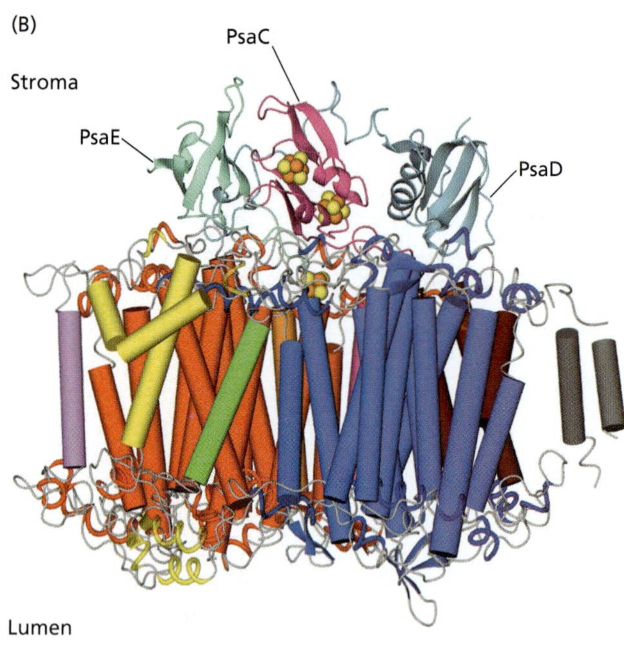

FIGURE 7.30 Structure of photosystem I. (A) Structural model of the PSI reaction center. Components of the PSI reaction center are organized around two major proteins, PsaA and PsaB. Minor proteins PsaC to PsaN are labelled C to N. Electrons are transferred from plastocyanin (PC) to P700 (see Figures 7.21 and 7.22) and then to a chlorophyll molecule, A_0, to phylloquinone, A_1, to the FeS_X, FeS_A, and FeS_B Fe–S centers, and finally to the soluble iron–sulfur protein, ferrodoxin (Fd). (B) Side view of one monomer of PSI from the cyanobacterium *Synechococcus elongatus*, at 2.5 Å resolution. The stromal side of the membrane is at the top, and the lumenal side is at the bottom of the figure. Transmembrane α-helices of PsaA and PsaB are shown as blue and red cylinders, respectively. (A after Buchanan et al. 2000; B from Jordan et al. 2001.)

their reduced form, the electron carriers that function in the acceptor region of photosystem I are all extremely strong reducing agents. These reduced species are very unstable and thus difficult to identify. Evidence indicates that one of these early acceptors is a chlorophyll molecule, and another is a quinone species, phylloquinone, also known as vitamin K_1.

Additional electron acceptors include a series of three membrane-associated iron–sulfur proteins, or bound ferredoxins, also known as **Fe–S centers FeS_X, FeS_A,** and **FeS_B** (see Figure 7.30). Fe–S center X is part of the P700-binding protein; centers A and B reside on an 8 kDa protein that is part of the PSI reaction center complex. Electrons are transferred through centers A and B to **ferredoxin (Fd)**, a small, water-soluble iron–sulfur protein (see Figures 7.21 and 7.30). The membrane-associated flavoprotein **ferredoxin–NADP reductase (FNR)** reduces $NADP^+$ to NADPH, thus completing the sequence of noncyclic electron transport that begins with the oxidation of water (Karplus et al. 1991).

In addition to the reduction of $NADP^+$, reduced ferredoxin produced by photosystem I has several other functions in the chloroplast, such as the supply of reductants to reduce nitrate and the regulation of some of the carbon fixation enzymes (see Chapter 8).

Cyclic Electron Flow Generates ATP but no NADPH

Some of the cytochrome b_6f complexes are found in the stroma region of the membrane, where photosystem I is located. Under certain conditions **cyclic electron flow** from the reducing side of photosystem I, through the b_6f complex and back to P700, is known to occur. This cyclic electron flow is coupled to proton pumping into the lumen, which can be utilized for ATP synthesis but does not oxidize water or reduce $NADP^+$. Cyclic electron flow is especially important as an ATP source in the bundle sheath chloroplasts of some plants that carry out C_4 carbon fixation (see Chapter 8).

Some Herbicides Block Electron Flow

The use of herbicides to kill unwanted plants is widespread in modern agriculture. Many different classes of herbicides have been developed, and they act by blocking amino acid, carotenoid, or lipid biosynthesis or by disrupting cell division. Other herbicides, such as DCMU (dichlorophenyl-dimethylurea) and paraquat, block photosynthetic electron flow (Figure 7.31). DCMU is also known as diuron. Paraquat has acquired public notoriety because of its use on marijuana crops.

Many herbicides, DCMU among them, act by blocking electron flow at the quinone acceptors of photosystem II, by competing for the binding site of plastoquinone that is normally occupied by Q_B. Other herbicides, such as paraquat, act by accepting electrons from the early acceptors of photosystem I and then reacting with oxygen to form superoxide, O_2^-, a species that is very damaging to chloroplast components, especially lipids.

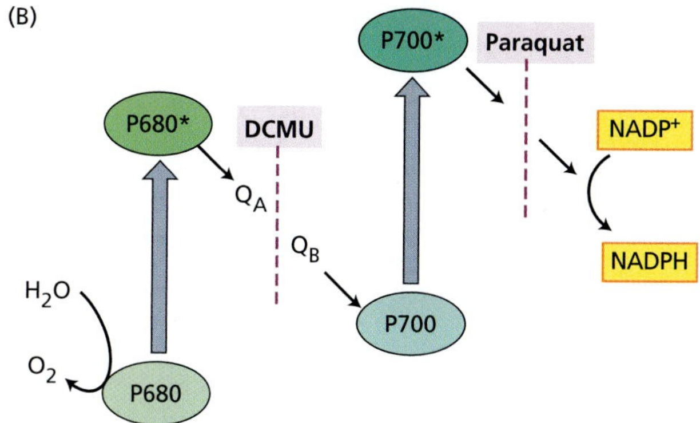

FIGURE 7.31 Chemical structure and mechanism of action of two important herbicides. (A) Chemical structure of dichlorophenyl-dimethylurea (DCMU) and methyl viologen (paraquat), two herbicides that block photosynthetic electron flow. DCMU is also known as diuron. (B) Sites of action of the two herbicides. DCMU blocks electron flow at the quinone acceptors of photosystem II, by competing for the binding site of plastoquinone. Paraquat acts by accepting electrons from the early acceptors of photosystem I.

PROTON TRANSPORT AND ATP SYNTHESIS IN THE CHLOROPLAST

In the preceding sections we learned how captured light energy is used to reduce $NADP^+$ to NADPH. Another fraction of the captured light energy is used for light-dependent ATP synthesis, which is known as **photophosphorylation**. This process was discovered by Daniel Arnon and his coworkers in the 1950s. In normal cellular conditions, photophosphorylation requires electron flow, although under some conditions electron flow and photophosphorylation can take place independently of each other. Electron flow without accompanying phosphorylation is said to be **uncoupled**.

It is now widely accepted that photophosphorylation works via the **chemiosmotic mechanism**, first proposed in the 1960s by Peter Mitchell. The same general mechanism drives phosphorylation during aerobic respiration in bacteria and mitochondria (see Chapter 11), as well as the transfer of many ions and metabolites across membranes (see Chapter 6). Chemiosmosis appears to be a unifying aspect of membrane processes in all forms of life.

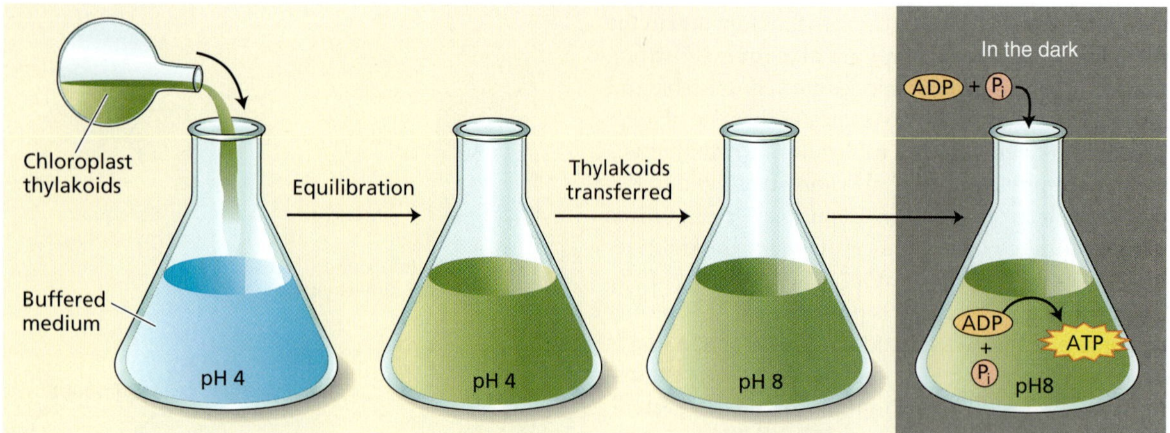

FIGURE 7.32 Summary of the experiment carried out by Jagendorf and coworkers. Isolated chloroplast thylakoids kept previously at pH 8 were equilibrated in an acid medium at pH 4. The thylakoids were then transferred to a buffer at pH 8 that contained ADP and P_i. The proton gradient generated by this manipulation provided a driving force for ATP synthesis in the absence of light. This experiment verified a prediction of the chemiosmotic theory stating that a chemical potential across a membrane can provide energy for ATP synthesis.

In Chapter 6 we discussed the role of ATPases in chemiosmosis and ion transport at the cell's plasma membrane. The ATP used by the plasma membrane ATPase is synthesized by photophosphorylation in the chloroplast and oxidative phosphorylation in the mitochondrion. Here we are concerned with chemiosmosis and transmembrane proton concentration differences used to make ATP in the chloroplast.

The basic principle of chemiosmosis is that ion concentration differences and electric-potential differences across membranes are a source of free energy that can be utilized by the cell. As described by the second law of thermodynamics (see Chapter 2 on the web site for a detailed discussion), any nonuniform distribution of matter or energy represents a source of energy. Differences in **chemical potential** of any molecular species whose concentrations are not the same on opposite sides of a membrane provide such a source of energy.

The asymmetric nature of the photosynthetic membrane and the fact that proton flow from one side of the membrane to the other accompanies electron flow were discussed earlier. The direction of proton translocation is such that the stroma becomes more alkaline (fewer H^+ ions) and the lumen becomes more acidic (more H^+ ions) as a result of electron transport (see Figures 7.22 and 7.29).

Some of the early evidence supporting a chemiosmotic mechanism of photosynthetic ATP formation was provided by an elegant experiment carried out by André Jagendorf and coworkers (Figure 7.32). They suspended chloroplast thylakoids in a pH 4 buffer, and the buffer diffused across the membrane, causing the interior, as well as the exterior, of the thylakoid to equilibrate at this acidic pH. They then rapidly transferred the thylakoids to a pH 8 buffer, thereby creating a pH difference of 4 units across the thylakoid membrane, with the inside acidic relative to the outside.

They found that large amounts of ATP were formed from ADP and P_i by this process, with no light input or electron transport. This result supports the predictions of the chemiosmotic hypothesis, described in the paragraphs that follow.

Mitchell proposed that the total energy available for ATP synthesis, which he called the **proton motive force** (Δp), is the sum of a proton chemical potential and a transmembrane electric potential. These two components of the proton motive force from the outside of the membrane to the inside are given by the following equation:

$$\Delta p = \Delta E - 59(\text{pH}_i - \text{pH}_o) \qquad (7.9)$$

where ΔE is the transmembrane electric potential, and $\text{pH}_i - \text{pH}_o$ (or ΔpH) is the pH difference across the membrane. The constant of proportionality (at 25°C) is 59 mV per pH unit, so a transmembrane pH difference of 1 pH unit is equivalent to a membrane potential of 59 mV.

Under conditions of steady-state electron transport in chloroplasts, the membrane electric potential is quite small because of ion movement across the membrane, so Δp is built almost entirely by ΔpH. The stoichiometry of protons translocated per ATP synthesized has recently been found to be four H^+ ions per ATP (Haraux and De Kouchkovsky 1998).

In addition to the need for mobile electron carriers discussed earlier, the uneven distribution of photosystems II and I, and of ATP synthase at the thylakoid membrane (see Figure 7.18), poses some challenges for the formation of ATP. ATP synthase is found only in the stroma lamellae and at the edges of the grana stacks. Protons pumped across the membrane by the cytochrome b_6f complex or protons produced by water oxidation in the middle of the grana must move laterally up to several tens of nanometers to reach ATP synthase.

The ATP is synthesized by a large (400 kDa) enzyme complex known by several names: **ATP synthase**, **ATPase** (after the reverse reaction of ATP hydrolysis), and **CF_o–CF_1** (Boyer 1997). This enzyme consists of two parts: a hydrophobic membrane-bound portion called CF_o and a portion that sticks out into the stroma called CF_1 (Figure 7.33).

CF_o appears to form a channel across the membrane through which protons can pass. CF_1 is made up of several peptides, including three copies of each of the α and β peptides arranged alternately much like the sections of an orange. Whereas the catalytic sites are located largely on the β polypeptide, many of the other peptides are thought to have primarily regulatory functions. CF_1 is the portion of the complex that synthesizes ATP.

The molecular structure of the mitochondrial ATP synthase has been determined by X-ray crystallography (Stock et al. 1999). Although there are significant differences between the chloroplast and mitochondrial enzymes, they have the same overall architecture and probably nearly identical catalytic sites. In fact, there are remarkable similarities in the way electron flow is coupled to proton translocation in chloroplasts, mitochondria, and purple bacteria (Figure 7.34). Another remarkable aspect of the mechanism of the ATP synthase is that the internal stalk and probably much of the CF_o portion of the enzyme rotate during catalysis (Yasuda et al. 2001). The enzyme is actually a tiny molecular motor (see **Web Topics 7.9 and 11.4**).

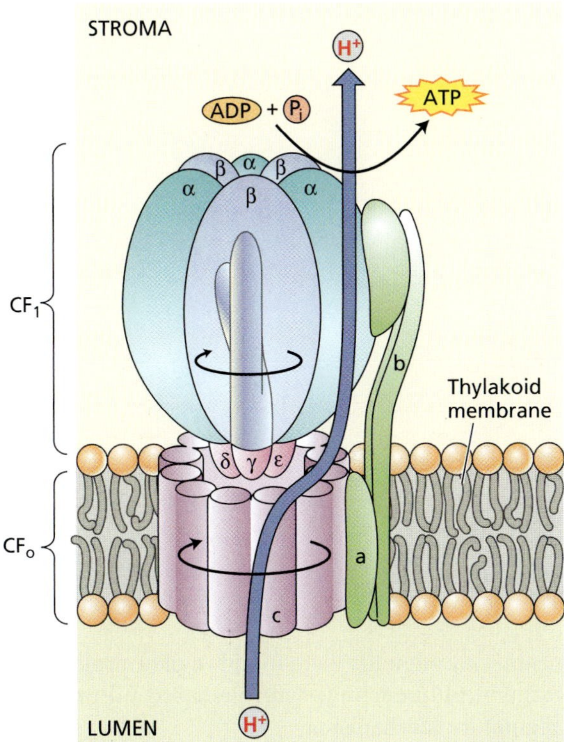

FIGURE 7.33 Structure of ATP synthase. This enzyme consists of a large multisubunit complex, CF_1, attached on the stromal side of the membrane to an integral membrane portion, known as CF_o. CF_1 consists of five different polypeptides, with a stoichiometry of α_3, β_3, γ, δ, ε. CF_o contains probably four different polypeptides, with a stoichiometry of a, b, b′, c_{12}.

REPAIR AND REGULATION OF THE PHOTOSYNTHETIC MACHINERY

Photosynthetic systems face a special challenge. They are designed to absorb large amounts of light energy and process it into chemical energy. At the molecular level, the energy in a photon can be damaging, particularly under unfavorable conditions. In excess, light energy can lead to the production of toxic species, such as superoxide, singlet oxygen, and peroxide, and damage can occur if the light energy is not dissipated safely (Horton et al. 1996; Asada 1999; Müller et al. 2001). Photosynthetic organisms therefore contain complex regulatory and repair mechanisms. Some of these mechanisms regulate energy flow in the antenna system, to avoid excess excitation of the reaction centers and ensure that the two photosystems are equally driven. Although very effective, these processes are not entirely fail-safe, and sometimes toxic compounds are produced. Additional mechanisms are needed to dissipate these compounds—in particular, toxic oxygen species.

Despite these protective and scavenging mechanisms, damage can occur, and additional mechanisms are required to repair the system. Figure 7.35 provides an overview of the several levels of the regulation and repair systems.

Carotenoids Serve as Photoprotective Agents

In addition to their role as accessory pigments, carotenoids play an essential role in **photoprotection**. The photosynthetic membrane can easily be damaged by the large amounts of energy absorbed by the pigments if this energy cannot be stored by photochemistry; this is why a protection mechanism is needed. The photoprotection mechanism can be thought of as a safety valve, venting excess energy before it can damage the organism. When the energy stored in chlorophylls in the excited state is rapidly dissipated by excitation transfer or photochemistry, the excited state is said to be **quenched**.

If the excited state of chlorophyll is not rapidly quenched by excitation transfer or photochemistry, it can react with molecular oxygen to form an excited state of oxygen known as **singlet oxygen** ($^1O_2^*$). The extremely reactive singlet oxygen goes on to react with and damage many cellular components, especially lipids. Carotenoids exert their photoprotective action by rapidly quenching the excited state of chlorophyll. The excited state of carotenoids does not have

(A) **Purple bacteria**

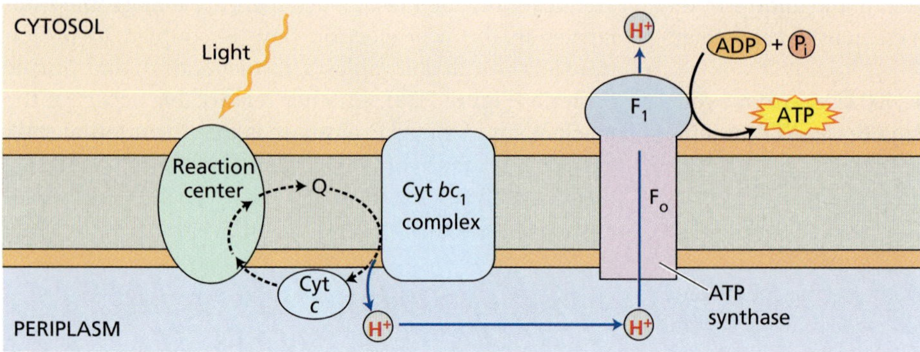

(B) **Chloroplasts**

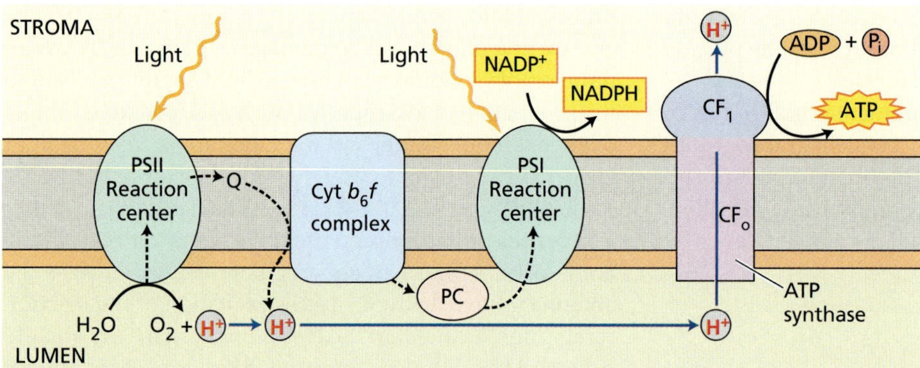

(C) **Mitochondria**

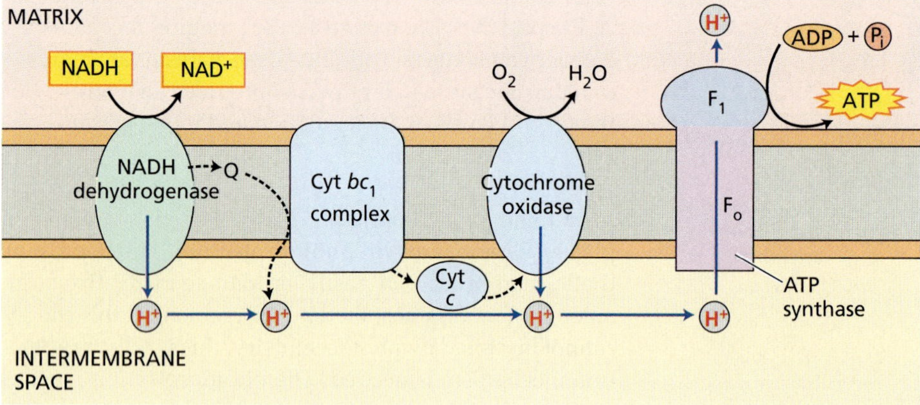

FIGURE 7.34 Similarities of photosynthetic and respiratory electron flow in bacteria, chloroplasts, and mitochondria. In all three, electron flow is coupled to proton translocation, creating a transmembrane proton motive force (Δp). The energy in the proton motive force is then used for the synthesis of ATP by ATP synthase. (A) A reaction center (RC) in purple photosynthetic bacteria carries out cyclic electron flow, generating a proton potential by the action of the cytochrome bc_1 complex. (B) Chloroplasts carry out noncyclic electron flow, oxidizing water and reducing $NADP^+$. Protons are produced by the oxidation of water and by the oxidation of PQH_2 (Q) by the cytochrome b_6f complex. (C) Mitochondria oxidize NADH to NAD^+ and reduce oxygen to water. Protons are pumped by the enzyme NADH dehydrogenase, the cytochrome bc_1 complex, and cytochrome oxidase. The ATP synthases in the three systems are very similar in structure.

sufficient energy to form singlet oxygen, so it decays back to its ground state while losing its energy as heat.

Mutant organisms that lack carotenoids cannot live in the presence of both light and molecular oxygen—a rather difficult situation for an O_2-evolving photosynthetic organism. For non-O_2-evolving photosynthetic bacteria, mutants that lack carotenoids can be maintained under laboratory conditions if oxygen is excluded from the growth medium.

Recently carotenoids were found to play a role in nonphotochemical quenching, which is a second protective and regulatory mechanism.

Some Xanthophylls Also Participate in Energy Dissipation

Nonphotochemical quenching, a major process regulating the delivery of excitation energy to the reaction center, can be thought of as a "volume knob" that adjusts the flow of

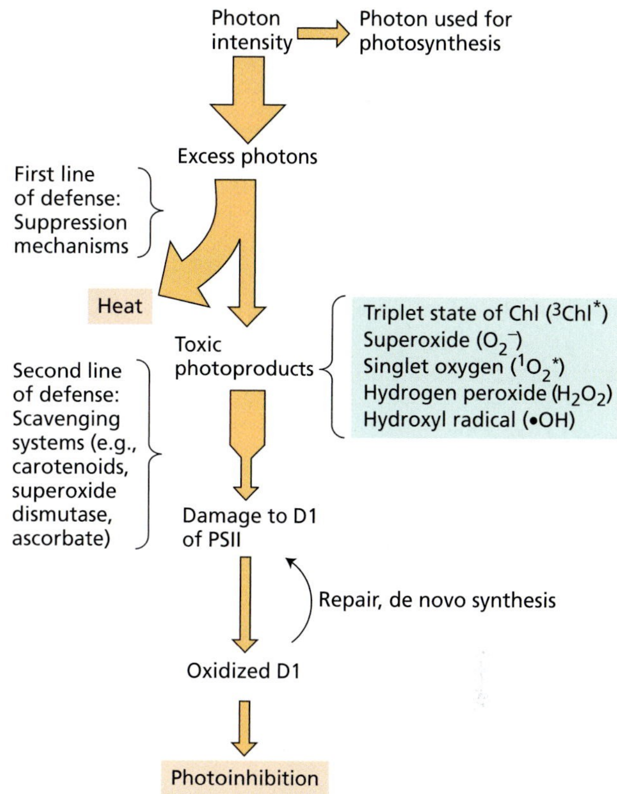

FIGURE 7.35 Overall picture of the regulation of photon capture and the protection and repair of photodamage. Protection against photodamage is a multilevel process. The first line of defense is suppression of damage by quenching of excess excitation as heat. If this defense is not sufficient and toxic photoproducts form, a variety of scavenging systems eliminate the reactive photoproducts. If this second line of defense also fails, the photoproducts can damage the D1 protein of photosystem II. This damage leads to photoinhibition. The D1 protein is then excised from the PSII reaction center and degraded. A newly synthesized D1 is reinserted into the PSII reaction center to form a functional unit. (After Asada 1999.)

excitations to the PSII reaction center to a manageable level, depending on the light intensity and other conditions. The process appears to be an essential part of the regulation of antenna systems in most algae and plants.

Nonphotochemical quenching is the quenching of chlorophyll fluorescence (see Figure 7.5) by processes other than photochemistry. As a result of nonphotochemical quenching, a large fraction of the excitations in the antenna system caused by intense illumination are quenched by conversion into heat (Krause and Weis 1991). Nonphotochemical quenching is thought to be involved in protecting the photosynthetic machinery against overexcitation and subsequent damage.

The molecular mechanism of nonphotochemical quenching is not well understood, although it is clear that the pH of the thylakoid lumen and the state of aggregation of the antenna complexes are important factors. Three carotenoids, called **xanthophylls**, are involved in nonphotochemical quenching: violaxanthin, antheraxanthin, and zeaxanthin (Figure 7.36).

In high light, violaxanthin is converted into zeaxanthin, via the intermediate antheraxanthin, by the enzyme violaxanthin de-epoxidase. When light intensity decreases, the process is reversed. Binding of protons and zeaxanthin to light-harvesting antenna proteins is thought to cause conformational changes that lead to quenching and heat dissipation (Demmig-

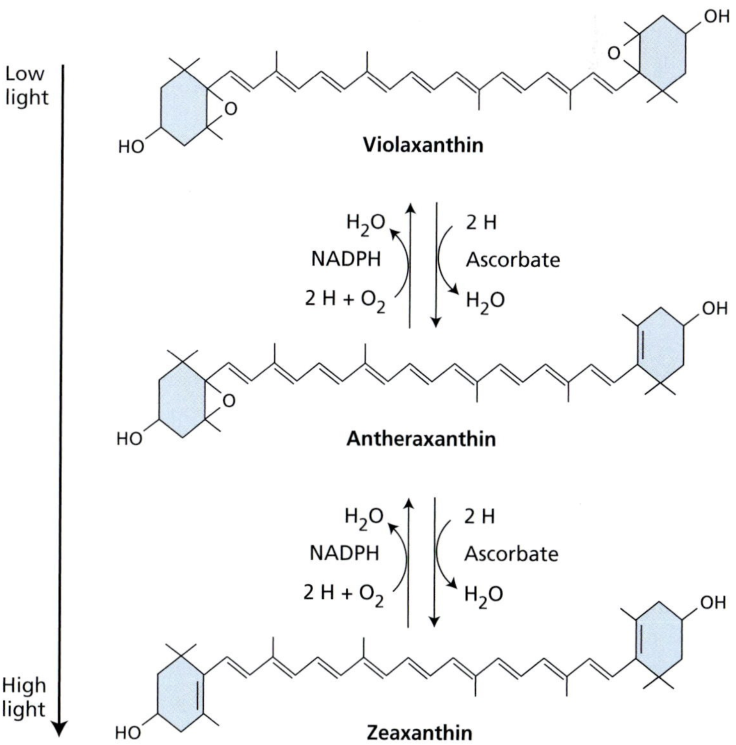

FIGURE 7.36 Chemical structure of violaxanthin, antheraxanthin, and zeaxanthin. The highly quenched state of photosystem II is associated with zeaxanthin, the unquenched state with violaxanthin. Enzymes interconvert these two carotenoids, with antheraxanthin as the intermediate, in response to changing conditions, especially changes in light intensity. Zeaxanthin formation uses ascorbate as a cofactor, and violaxanthin formation requires NADPH. (After Pfündel and Bilger 1994.)

Low light
Violaxanthin
H_2O NADPH $2\,H + O_2$ | 2 H Ascorbate H_2O
Antheraxanthin
H_2O NADPH $2\,H + O_2$ | 2 H Ascorbate H_2O
Zeaxanthin
High light

Adams and Adams 1992; Horton et al. 1996). Nonphotochemical quenching appears to be preferentially associated with a peripheral antenna complex of photosystem II, the PsbS protein (Li et al. 2000).

The Photosystem II Reaction Center Is Easily Damaged

Another effect that appears to be a major factor in the stability of the photosynthetic apparatus is photoinhibition, which occurs when excess excitation arriving at the PSII reaction center leads to its inactivation and damage (Long et al. 1994). **Photoinhibition** is a complex set of molecular processes, defined as the inhibition of photosynthesis by excess light.

As will be discussed in detail in Chapter 9, photoinhibition is reversible in early stages. Prolongued inhibition, however, results in damage to the system such that the PSII reaction center must be disassembled and repaired (Melis 1999). The main target of this damage is the D1 protein that makes up part of the PSII reaction center complex (see Figure 7.24). When D1 is damaged by excess light, it must be removed from the membrane and replaced with a newly synthesized molecule. The other components of the PSII reaction center are not damaged by excess excitation and are thought to be recycled, so the D1 protein is the only component that needs to be synthesized.

Photosystem I Is Protected from Active Oxygen Species

Photosystem I is particularly vulnerable to damage from active oxygen species. The ferredoxin acceptor of PSI is a very strong reductant that can easily reduce molecular oxygen to form superoxide (O_2^-). This reduction competes with the normal channeling of electrons to the reduction of $NADP^+$ and other processes. Superoxide is one of a series of active oxygen species that can be very damaging to biological membranes. Superoxide formed in this way can be eliminated by the action of a series of enzymes, including superoxide dismutase and ascorbate peroxidase (Asada 1999).

Thylakoid Stacking Permits Energy Partitioning between the Photosystems

The fact that photosynthesis in higher plants is driven by two photosystems with different light-absorbing properties poses a special problem. If the rate of delivery of energy to PSI and PSII is not precisely matched and conditions are such that the rate of photosynthesis is limited by the available light (low light intensity), the rate of electron flow will be limited by the photosystem that is receiving less energy. In the most efficient situation, the input of energy would be the same to both photosystems. However, no single arrangement of pigments would satisfy this requirement because at different times of day the light intensity and spectral distribution tend to favor one photosystem or the other (Trissl and Wilhelm 1993; Allen and Forsberg 2001).

This problem can be solved by a mechanism that shifts energy from one photosystem to the other in response to different conditions. Such a regulating mechanism has been shown to operate in different experimental conditions. The observation that the overall quantum yield of photosynthesis is nearly independent of wavelength (see Figure 7.12) strongly suggests that such a mechanism exists.

Thylakoid membranes contain a protein kinase that can phosphorylate a specific threonine residue on the surface of LHCII, one of the membrane-bound antenna pigment proteins described earlier in the chapter (see Figure 7.20). When LHCII is not phosphorylated, it delivers more energy to photosystem II, and when it is phosphorylated, it delivers more energy to photosystem I (Haldrup et al. 2001).

The kinase is activated when plastoquinone, one of the electron carriers between PSI and PSII, accumulates in the reduced state. Reduced plastoquinone accumulates when PSII is being activated more frequently than PSI. The phosphorylated LHCII then migrates out of the stacked regions of the membrane into the unstacked regions (see Figure 7.18), probably because of repulsive interactions with negative charges on adjacent membranes.

The lateral migration of LHCII shifts the energy balance toward photosystem I, which is located in the stroma lamellae, and away from photosystem II, which is located in the stacked membranes of the grana. This situation is called *state 2*. If plastoquinone becomes more oxidized because of excess excitation of photosystem I, the kinase is deactivated and the level of phosphorylation of LHCII is decreased by the action of a membrane-bound phosphatase. LHCII then moves back to the grana, and the system is in *state 1*. The net result is a very precise control of the energy distribution between the photosystems, allowing the most efficient use of the available energy.

GENETICS, ASSEMBLY, AND EVOLUTION OF PHOTOSYNTHETIC SYSTEMS

Chloroplasts have their own DNA, mRNA, and protein synthesis machinery, but some chloroplast proteins are encoded by nuclear genes and imported into the chloroplast. In this section we will consider the genetics, assembly, and evolution of the main chloroplast components.

Chloroplast, Cyanobacterial, and Nuclear Genomes Have Been Sequenced

The complete chloroplast genomes of several organisms have been sequenced. Chloroplast DNA is circular and ranges in size from 120 to 160 kilobases. The chloroplast genome contains coding sequences for approximately 120 proteins. Some of these DNA sequences code for proteins that are yet to be characterized. It is uncertain whether all these genes are transcribed into mRNA and translated into protein, but it seems likely that some chloroplast proteins remain to be identified.

The complete genome of the cyanobacterium *Synechocystis* (strain PCC 6803) and the higher plant *Arabidopsis* have been sequenced, and genomes of important crop plants such as rice and maize have been completed (Kotani and Tabata 1998; Arabidopsis Genome Initiative 2000). Genomic data for both chloroplast and nuclear DNA will provide new insights into the mechanism of photosynthesis, as well as many other plant processes.

Chloroplast Genes Exhibit Non-Mendelian Patterns of Inheritance

Chloroplasts and mitochondria reproduce by division rather than by **de novo synthesis**. This mode of reproduction is not surprising, since these organelles contain genetic information that is not present in the nucleus. During cell division, chloroplasts are divided between the two daughter cells. In most sexual plants, however, only the maternal plant contributes chloroplasts to the zygote. In these plants the normal Mendelian pattern of inheritance does not apply to chloroplast-encoded genes because the offspring receive chloroplasts from only one parent. The result is **non-Mendelian**, or **maternal, inheritance**. Numerous traits are inherited in this way; one example is the herbicide resistance trait discussed in **Web Topic 7.10**.

Many Chloroplast Proteins Are Imported from the Cytoplasm

Chloroplast proteins can be encoded by either chloroplastic or nuclear DNA. The chloroplast-encoded proteins are synthesized on chloroplast ribosomes; the nucleus-encoded proteins are synthesized on cytoplasmic ribosomes and then transported into the chloroplast. Many nuclear genes contain introns—that is, base sequences that do not code for protein. The mRNA is processed to remove the introns, and the proteins are then synthesized in the cytoplasm.

The genes needed for chloroplast function are distributed in the nucleus and in the chloroplast genome with no evident pattern, but both sets are essential for the viability of the chloroplast. Some chloroplast genes are necessary for other cellular functions, such as heme and lipid synthesis. Control of the expression of the nuclear genes that code for chloroplast proteins is complex, involving light-dependent regulation mediated by both phytochrome (see Chapter 17) and blue light (see Chapter 18), as well as other factors (Bruick and Mayfield 1999; Wollman et al. 1999).

The transport of chloroplast proteins that are synthesized in the cytoplasm is a tightly regulated process (Chen and Schnell 1999). For example, the enzyme rubisco (see Chapter 8), which functions in carbon fixation, has two types of subunits, a chloroplast-encoded large subunit and a nucleus-encoded small subunit. Small subunits of rubisco are synthesized in the cytoplasm and transported into the chloroplast, where the enzyme is assembled.

In this and other known cases, the nucleus-encoded chloroplast proteins are synthesized as precursor proteins containing an N-terminal amino acid sequence known as a **transit peptide**. This terminal sequence directs the precursor protein to the chloroplast, facilitates its passage through both the outer and the inner envelope membranes, and is then clipped off. The electron carrier plastocyanin is a water-soluble protein that is encoded in the nucleus but functions in the lumen of the chloroplast. It therefore must cross three membranes to reach its destination in the lumen. The transit peptide of plastocyanin is very large and is processed in more than one step.

The Biosynthesis and Breakdown of Chlorophyll Are Complex Pathways

Chlorophylls are complex molecules exquisitely suited to the light absorption, energy transfer, and electron transfer functions that they carry out in photosynthesis (see Figure 7.6). Like all other biomolecules, chlorophylls are made by a biosynthetic pathway in which simple molecules are used as building blocks to assemble more complex molecules (Porra 1997; Beale 1999). Each step in the biosynthetic pathway is enzymatically catalyzed.

The chlorophyll biosynthetic pathway consists of more than a dozen steps (see **Web Topic 7.11**). The process can be divided into several phases (Figure 7.37), each of which can be considered separately, but which in the cell are highly coordinated and regulated. This regulation is essential because free chlorophyll and many of the biosynthetic intermediates are damaging to cellular components. The damage results largely because chlorophylls absorb light efficiently, but in the absence of accompanying proteins, they lack a pathway for disposing of the energy, with the result that toxic singlet oxygen is formed.

The breakdown pathway of chlorophyll in senescent leaves is quite different from the biosynthetic pathway (Matile et al. 1996). The first step is removal of the phytol tail by an enzyme known as chlorophyllase, followed by removal of the magnesium by magnesium de-chelatase. Next the porphyrin structure is opened by an oxygen-dependent oxygenase enzyme to form an open-chain tetrapyrrole.

The tetrapyrrole is further modified to form water-soluble, colorless products. These colorless metabolites are then exported from the senescent chloroplast and transported to the vacuole, where they are permanently stored. The chlorophyll metabolites are not further processed or recycled, although the proteins associated with them in the chloroplast are subsequently recycled into new proteins. The recycling of proteins is thought to be important for the nitrogen economy of the plant.

Complex Photosynthetic Organisms Have Evolved from Simpler Forms

The complicated photosynthetic apparatus found in plants and algae is the end product of a long evolutionary sequence. Much can be learned about this evolutionary

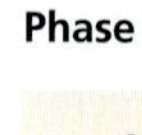

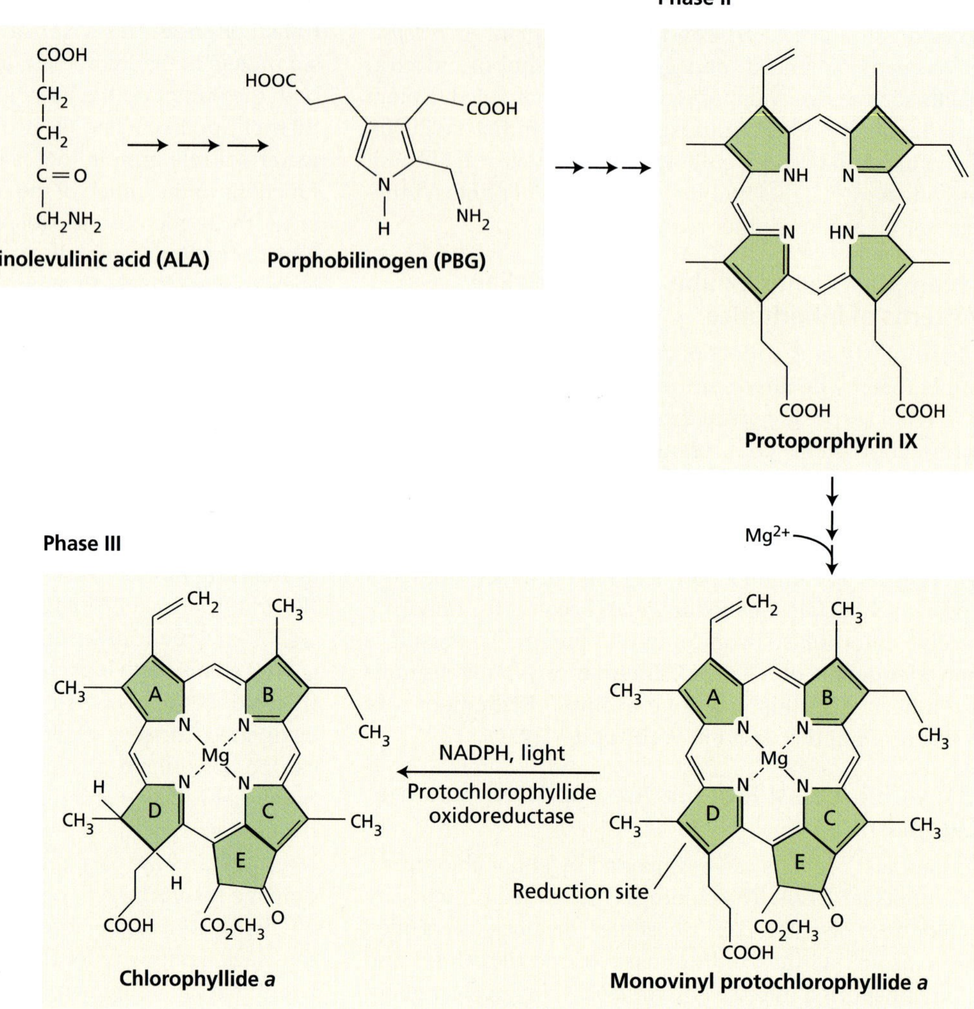

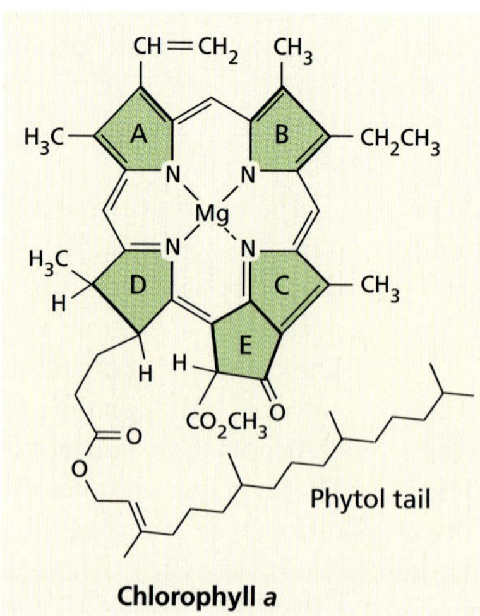

FIGURE 7.37 The biosynthetic pathway of chlorophyll. The pathway begins with glutamic acid, which is converted to 5-aminolevulinic acid (ALA). Two molecules of ALA are condensed to form porphobilinogen (PBG). Four PBG molecules are linked to form protoporphyrin IX. The magnesium (Mg) is then inserted, and the light-dependent cyclization of ring E, the reduction of ring D, and the attachment of the phytol tail complete the process. Many steps in the process are omitted in this figure.

process from analysis of simpler prokaryotic photosynthetic organisms, including the anoxygenic photosynthetic bacteria and the cyanobacteria.

The chloroplast is a semiautonomous cell organelle, with its own DNA and a complete protein synthesis apparatus. Many of the proteins that make up the photosynthetic apparatus, as well as all the chlorophylls and lipids, are synthesized in the chloroplast. Other proteins are imported from the cytoplasm and are encoded by nuclear genes. How did this curious division of labor come about? Most experts now agree that the chloroplast is the descendant of a symbiotic relationship between a cyanobacterium and a simple nonphotosynthetic eukaryotic cell. This type of relationship is called **endosymbiosis** (Cavalier-Smith 2000).

Originally the cyanobacterium was capable of independent life, but over time much of its genetic information needed for normal cellular functions was lost, and a substantial amount of information needed to synthesize the photosynthetic apparatus was transferred to the nucleus. So the chloroplast was no longer capable of life outside its host and eventually became an integral part of the cell.

In some types of algae, chloroplasts are thought to have arisen by endosymbiosis of eukaryotic photosynthetic organisms (Palmer and Delwiche 1996). In these organisms the chloroplast is surrounded by three and in some cases four membranes, which are thought to be remnants of the plasma membranes of the earlier organisms. Mitochondria are also thought to have originated by endosymbiosis in a separate event much earlier than chloroplast formation.

The answers to other questions related to the evolution of photosynthesis are less clear. These include the nature of the earliest photosynthetic systems, how the two photosystems became linked, and the evolutionary origin of the oxygen evolution complex (Blankenship and Hartman 1998; Xiong et al. 2000).

SUMMARY

Photosynthesis is the storage of solar energy carried out by plants, algae, and photosynthetic bacteria. Absorbed photons excite chlorophyll molecules, and these excited chlorophylls can dispose of this energy as heat, fluorescence, energy transfer, or photochemistry. Light is absorbed mainly in the antenna complexes, which comprise chlorophylls, accessory pigments, and proteins and are located at the thylakoid membranes of the chloroplast.

Photosynthetic antenna pigments transfer the energy to a specialized chlorophyll–protein complex known as a reaction center. The reaction center contains multisubunit protein complexes and hundreds or, in some organisms, thousands of chlorophylls. The antenna complexes and the reaction centers are integral components of the thylakoid membrane. The reaction center initiates a complex series of chemical reactions that capture energy in the form of chemical bonds.

The relationship between the amount of absorbed quanta and the yield of a photochemical product made in a light-dependent reaction is given by the quantum yield. The quantum yield of the early steps of photosynthesis is approximately 0.95, indicating that nearly every photon that is absorbed yields a charge separation at the reaction center.

Plants and some photosynthetic prokaryotes have two reaction centers, photosystem I and photosystem II, that function in series. The two photosystems are spatially separated: PSI is found exclusively in the nonstacked stroma membranes, PSII largely in the stacked grana membranes. The reaction center chlorophylls of PSI absorb maximally at 700 nm, those of PSII at 680 nm. Photosystems II and I carry out noncyclic electron transport, oxidize water to molecular oxygen, and reduce $NADP^+$ to NADPH. It is energetically very difficult to oxidize water to form molecular oxygen, and the photosynthetic oxygen-evolving system is the only known biochemical system that can oxidize water, thus providing almost all the oxygen in Earth's atmosphere. The photooxidation of water is modeled by the five-step S state mechanism. Manganese is an essential cofactor in the water-oxidizing process, and the five S states appear to represent successive oxidized states of a manganese-containing enzyme.

A tyrosine residue of the D1 protein of the PSII reaction center functions as an electron carrier between the oxygen-evolving complex and P680. Pheophytin and two plastoquinones are electron carriers between P680 and the large cytochrome b_6f complex. Plastocyanin is the electron carrier between cytochrome b_6f and P700. The electron carriers that accept electrons from P700 are very strong reducing agents, and they include a quinone and three membrane-bound iron–sulfur proteins known as bound ferredoxins. The electron flow ends with the reduction of $NADP^+$ to NADPH by a membrane-bound, ferrodoxin–NADP reductase.

A portion of the energy of photons is also initially stored as chemical-potential energy, largely in the form of a pH difference across the thylakoid membrane. This energy is quickly converted into chemical energy during ATP formation by action of an enzyme complex known as the ATP synthase. The photophosphorylation of ADP by the ATP synthase is driven by a chemiosmotic mechanism. Photosynthetic electron flow is coupled to proton translocation across the thylakoid membrane, and the stroma becomes more alkaline and the lumen more acidic. This proton gradient drives ATP synthesis with a stoichiometry of four H^+ ions per ATP. NADPH and ATP formed by the light reactions provide the energy for carbon reduction.

Excess light energy can damage photosynthetic systems, and several mechanisms minimize such damage. Carotenoids work as photoprotective agents by rapidly quenching the excited state of chlorophyll. Changes in the phosphorylated state of antenna pigment proteins can change the energy distribution between photosystems I and II when there is an imbalance between the energy absorbed by each photosystem. The xanthophyll cycle also contributes to the dissipation of excess energy by nonphotochemical quenching.

Chloroplasts contain DNA and encode and synthesize some of the proteins that are essential for photosynthesis. Additional proteins are encoded by nuclear DNA, synthesized in the cytosol, and imported into the chloroplast. Chlorophylls are synthesized in a biosynthetic pathway involving more than a dozen steps, each of which is very carefully regulated. Once synthesized, proteins and pigments are assembled into the thylakoid membrane.

Web Material

Web Topics

7.1 Principles of Spectrophotometry
Spectroscopy is a key technique to study light reactions.

7.2 The Distribution of Chlorophylls and Other Photosynthetic Pigments
The content of chlorophylls and other photosynthetic pigments varies among plant kingdoms.

7.3 Quantum Yield
Quantum yields measure how effectively light drives a photobiological process.

7.4 Antagonistic Effects of Light on Cytochrome Oxidation
Photosystems I and II were discovered in some ingenious experiments.

7.5 Structures of Two Bacterial Reaction Centers
X-ray diffraction studies resolved the atomic structure of the reaction center of photosystem II.

7.6 Midpoint Potentials and Redox Reactions
The measurement of midpoint potentials is useful for analyzing electron flow through photosystem II.

7.7 Oxygen Evolution
The S state mechanism is a valuable model for water splitting in PSII.

7.8 Photosystem I
The PSI reaction is a multiprotein complex.

7.9 ATP Synthase
The ATP synthase functions as a molecular motor.

7.10 Mode of Action of Some Herbicides
Some herbicides kill plants by blocking photosynthetic electron flow.

7.11 Chlorophyll Biosynthesis
Chlorophyll and heme share early steps of their biosynthetic pathways.

Web Essay

7.1 A novel view of chloroplast structure
Stromules extend the reach of the chloroplasts.

Chapter References

Allen, J. F., and Forsberg, J. (2001) Molecular recognition in thylakoid structure and function. *Trends Plant Sci.* 6: 317–326.

Arabidopsis Genome Initiative. (2000) Analysis of the genome sequence of the flowering plant *Arabidopsis thaliana*. *Nature* 408: 796–815.

Asada, K. (1999) The water–water cycle in chloroplasts: Scavenging of active oxygens and dissipation of excess photons. *Annu. Rev. Plant Physiol. Plant Mol. Biol.* 50: 601–639.

Avers, C. J. (1985) *Molecular Cell Biology.* Addison-Wesley, Reading, MA.

Barber, J., Nield, N., Morris, E. P., and Hankamer, B. (1999) Subunit positioning in photosystem II revisited. *Trends Biochem. Sci.* 24: 43–45.

Beale, S. I. (1999) Enzymes of chlorophyll biosynthesis. *Photosynth. Res.* 60: 43–73.

Becker, W. M. (1986) *The World of the Cell.* Benjamin/Cummings, Menlo Park, CA.

Berry, E. A., Guergova-Kuras, M., Huang, L.-S., and Crofts, A. R. (2000) Structure and function of cytochrome *bc* complexes. *Annu. Rev. Biochem.* 69: 1005–1075.

Blankenship, R. E. (2002) *Molecular Mechanisms of Photosynthesis.* Blackwell Science, Oxford.

Blankenship, R. E., and Hartman, H. (1998) The origin and evolution of oxygenic photosynthesis. *Trends Biochem. Sci.* 23: 94–97.

Blankenship, R. E., and Prince, R. C. (1985) Excited-state redox potentials and the Z scheme of photosynthesis. *Trends Biochem. Sci.* 10: 382–383.

Boyer, P. D. (1997) The ATP synthase: A splendid molecular machine. *Annu. Rev. Biochem.* 66: 717–749.

Brettel, K. (1997) Electron transfer and arrangement of the redox cofactors in photosystem I. *Biochim. Biophys. Acta* 1318: 322–373.

Bruick, R. K., and Mayfield, S. P. (1999) Light-activated translation of chloroplast mRNAs. *Trends Plant Sci.* 4: 190–195.

Buchanan, B. B., Gruissem., W., and Jones, R. L., eds. (2000) *Biochemistry and Molecular Biology of Plants.* Amer. Soc. Plant Physiologists, Rockville, MD.

Cavalier-Smith, T. (2000) Membrane heredity and early chloroplast evolution. *Trends Plant Sci.* 5: 174–182.

Chen, X., and Schnell, D. J. (1999) Protein import into chloroplasts. *Trends Cell Biol.* 9: 222–227.

Chitnis, P. R. (2001) Photosystem I: Function and physiology. *Annu. Rev. Plant Physiol. Plant Mol. Biol.* 52: 593–626.

Cramer, W. A., Soriano, G. M., Ponomarev, M., Huang, D., Zhang, H., Martinez, S. E., and Smith, J. L. (1996) Some new structural aspects and old controversies concerning the cytochrome b_6f complex of oxygenic photosynthesis. *Annu. Rev. Plant Physiol. Plant Mol. Biol.* 47: 477–508.

Deisenhofer, J., and Michel, H. (1989) The photosynthetic reaction center from the purple bacterium *Rhodopseudomonas viridis*. *Science* 245: 1463–1473.

Demmig-Adams, B., and Adams, W. W., III. (1992) Photoprotection and other responses of plants to high light stress. *Annu. Rev. Plant Physiol. Plant Mol. Biol.* 43: 599–626.

Green, B. R., and Durnford, D. G. (1996) The chlorophyll-carotenoid proteins of oxygenic photosynthesis. *Annu. Rev. Plant Physiol. Plant Mol. Biol.* 47: 685–714.

Grossman, A. R., Bhaya, D., Apt, K. E., and Kehoe, D. M. (1995) Light-harvesting complexes in oxygenic photosynthesis: Diversity, control, and evolution. *Annu. Rev. Genet.* 29: 231–288.

Haldrup, A., Jensen, P. E., Lunde, C., and Scheller, H. V. (2001) Balance of power: A view of the mechanism of photosynthetic state transitions. *Trends Plant Sci.* 6: 301–305.

Haraux, F., and De Kouchkovsky, Y. (1998) Energy coupling and ATP synthase. *Photosynth. Res.* 57: 231–251.

Hoganson, C. W., and Babcock, G. T. (1997) A metalloradical mechanism for the generation of oxygen from water in photosynthesis. *Science* 277: 1953–1956.

Horton, P., Ruban, A. V., and Walters, R. G. (1996) Regulation of light harvesting in green plants. *Annu. Rev. Plant Physiol. Plant Mol. Biol.* 47: 655–684.

Jordan, P., Fromme, P., Witt, H. T., Klukas, O., Saenger, W., and Krauss, N. (2001) Three-dimensional structure of cyanobacterial photosystem I at 2.5 Å resolution. *Nature* 411: 909–917.

Karplus, P. A., Daniels, M. J., and Herriott, J. R. (1991) Atomic structure of ferredoxin-$NADP^+$ reductase: Prototype for a structurally novel flavoenzyme family. *Science* 251: 60–66.

Kotani, H., and Tabata, S. (1998) Lessons from sequencing of the genome of a unicellular cyanobacterium, *Synechocystis* sp. PCC6803. *Annu. Rev. Plant Physiol. Plant Mol. Biol.* 49: 151–171.

Krause, G. H., and Weis, E. (1991) Chlorophyll fluorescence and photosynthesis: The basics. *Annu. Rev. Plant Physiol. Plant Mol. Biol.* 42: 313–350.

Kühlbrandt, W., Wang, D. N., and Fujiyoshi, Y. (1994) Atomic model of plant light-harvesting complex by electron crystallography. *Nature* 367: 614–621.

Li, X. P., Bjorkman, O., Shih, C., Grossman, A. R., Rosenquist, M., Jansson, S., and Niyogi, K. K. (2000) A pigment-binding protein essential for regulation of photosynthetic light harvesting. *Nature* 403: 391–395.

Long, S. P., Humphries, S., and Falkowski, P. G. (1994) Photoinhibition of photosynthesis in nature. *Annu. Rev. Plant Physiol. Plant Mol. Biol.* 45: 633–662.

Matile, P., Hörtensteiner, S., Thomas, H., and Kräutler, B. (1996) Chlorophyll breakdown in senescent leaves. *Plant Physiol.* 112: 1403–1409.

Melis, A. (1999) Photosystem-II damage and repair cycle in chloroplasts: What modulates the rate of photodamage in vivo? *Trends Plant Sci.* 4: 130–135.

Müller, P., Li, X.-P., and Niyogi, K. K. (2001) Non-photochemical quenching: A response to excess light energy. *Plant Physiol.* 125: 1558–1566.

Okamura, M. Y., Paddock, M. L., Graige, M. S., and Feher, G. (2000) Proton and electron transfer in bacterial reaction centers. *Biochim. Biophys. Acta* 1458: 148–163.

Palmer, J. D., and Delwiche, C. F. (1996) Second-hand chloroplasts and the case of the disappearing nucleus. *Proc. Natl. Acad. Sci. USA* 93: 7432–7435.

Paulsen, H. (1995) Chlorophyll *a/b*-binding proteins. *Photochem. Photobiol.* 62: 367–382.

Pfündel, E., and Bilger, W. (1994) Regulation and the possible function of the violaxanthin cycle. *Photosynth. Res.* 42: 89–109.

Porra, R. J. (1997) Recent progress in porphyrin and chlorophyll biosynthesis. *Photochem. Photobiol.* 65: 492–516.

Pullerits, T., and Sundström, V. (1996) Photosynthetic light-harvesting pigment-protein complexes: Toward understanding how and why. *Acc. Chem. Res.* 29: 381–389.

Stock, D., Leslie, A. G. W., and Walker, J. E. (1999) Molecular architecture of the rotary motor in ATP synthase. *Science* 286: 1700–1705.

Tommos, C., and Babcock, G. T. (1999) Oxygen production in nature: A light-driven metalloradical enzyme process. *Acc. Chem. Res.* 37: 18–25.

Trebst, A. (1986) The topology of the plastoquinone and herbicide binding peptides of photosystem II in the thylakoid membrane. *Z. Naturforsch. Teil C.* 240–245.

Trissl, H.-W., and Wilhelm, C. (1993) Why do thylakoid membranes from higher plants form grana stacks? *Trends Biochem. Sci.* 18: 415–419.

van Grondelle, R., Dekker, J. P., Gillbro, T., and Sundström, V. (1994) Energy transfer and trapping in photosynthesis. *Biochim. Biophys. Acta* 1187: 1–65.

Wollman, F.-A., Minai, L., and Nechushtai, R. (1999) The biogenesis and assembly of photosynthetic proteins in thylakoid membranes. *Biochim. Biophys. Acta* 1411: 21–85.

Xiong, J., Fisher, W., Inoue, K., Nakahara, M., and Bauer, C. E. (2000) Molecular evidence for the early evolution of photosynthesis. *Science* 289: 1724–1730.

Yachandra, V. K., Sauer, K., and Klein, M. P. (1996) Manganese cluster in photosynthesis: Where plants oxidize water to dioxygen. *Chem. Rev.* 96: 2927–2950.

Yasuda, R., Noji, H., Yoshida, M., Kinosita, K., and Itoh, H. (2001) Resolution of distinct rotational substeps by submillisecond kinetic analysis of F 1-ATPase. *Nature* 410: 898–904.

Zouni, A., Witt, H.-T., Kern, J., Fromme, P., Krauss, N., Saenger, W., and Orth, P. (2001) Crystal structure of photosystem II from *Synechococcus elongatus* at 3.8 Å resolution. *Nature* 409: 739–743.

Chapter

Photosynthesis: Carbon Reactions

IN CHAPTER 5 WE DISCUSSED plants' requirements for mineral nutrients and light in order to grow and complete their life cycle. Because living organisms interact with one another and their environment, mineral nutrients cycle through the biosphere. These cycles involve complex interactions, and each cycle is critical in its own right. Because the amount of matter in the biosphere remains constant, energy must be supplied to keep the cycles operational. Otherwise increasing entropy dictates that the flow of matter would ultimately stop.

Autotrophic organisms have the ability to convert physical and chemical sources of energy into carbohydrates in the absence of organic substrates. Most of the external energy is consumed in transforming CO_2 to a reduced state that is compatible with the needs of the cell (—CHOH—).

Recent estimates indicate that about 200 billion tons of CO_2 are converted to biomass each year. About 40% of this mass originates from the activities of marine phytoplankton. The bulk of the carbon is incorporated into organic compounds by the carbon reduction reactions associated with photosynthesis.

In Chapter 7 we saw how the photochemical oxidation of water to molecular oxygen is coupled to the generation of ATP and reduced pyridine nucleotide (NADPH) by reactions taking place in the chloroplast thylakoid membrane. The reactions catalyzing the reduction of CO_2 to carbohydrate are coupled to the consumption of NADPH and ATP by enzymes found in the stroma, the soluble phase of chloroplasts.

These stroma reactions were long thought to be independent of light and, as a consequence, were referred to as the *dark reactions*. However, because these stroma-localized reactions depend on the products of the photochemical processes, and are also directly regulated by light, they are more properly referred to as the *carbon reactions of photosynthesis*.

In this chapter we will examine the cyclic reactions that accomplish fixation and reduction of CO_2, then consider how the phenomenon of photorespiration catalyzed by the carboxylating enzyme alters the effi-

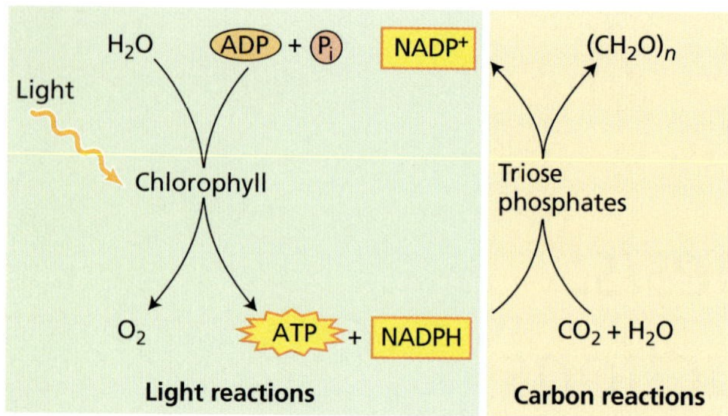

FIGURE 8.1 The light and carbon reactions of photosynthesis. Light is required for the generation of ATP and NADPH. The ATP and NADPH are consumed by the carbon reactions, which reduce CO_2 to carbohydrate (triose phosphates).

ciency of photosynthesis. This chapter will also describe biochemical mechanisms for concentrating carbon dioxide that allow plants to mitigate the impact of photorespiration: CO_2 pumps, C_4 metabolism, and crassulacean acid metabolism (CAM). We will close the chapter with a consideration of the synthesis of sucrose and starch.

THE CALVIN CYCLE

All photosynthetic eukaryotes, from the most primitive alga to the most advanced angiosperm, reduce CO_2 to carbohydrate via the same basic mechanism: the photosynthetic carbon reduction cycle originally described for C_3 species (the **Calvin cycle**, or **reductive pentose phosphate [RPP] cycle**). Other metabolic pathways associated with the photosynthetic fixation of CO_2, such as the C_4 photosynthetic carbon assimilation cycle and the photorespiratory carbon oxidation cycle, are either auxiliary to or dependent on the basic Calvin cycle.

In this section we will examine how CO_2 is fixed by the Calvin cycle through the use of ATP and NADPH generated by the light reactions (Figure 8.1), and how the Calvin cycle is regulated.

The Calvin Cycle Has Three Stages: Carboxylation, Reduction, and Regeneration

The Calvin cycle was elucidated as a result of a series of elegant experiments by Melvin Calvin and his colleagues in the 1950s, for which a Nobel Prize was awarded in 1961 (see **Web Topic 8.1**). In the Calvin cycle, CO_2 and water from the environment are enzymatically combined with a five-carbon acceptor molecule to generate two molecules of a three-carbon intermediate. This intermediate (3-phosphoglycerate) is reduced to carbohydrate by use of the ATP and NADPH generated photochemically. The cycle is completed by regeneration of the five-carbon acceptor (ribulose-1,5-bisphosphate, abbreviated RuBP).

The Calvin cycle proceeds in three stages (Figure 8.2):

1. *Carboxylation* of the CO_2 acceptor ribulose-1,5-bisphosphate, forming two molecules of 3-phosphoglycerate, the first stable intermediate of the Calvin cycle
2. *Reduction* of 3-phosphoglycerate, forming gyceraldehyde-3-phosphate, a carbohydrate
3. *Regeneration* of the CO_2 acceptor ribulose-1,5-bisphosphate from glyceraldehyde-3-phosphate

The carbon in CO_2 is the most oxidized form found in nature (+4). The carbon of the first stable intermediate, 3-phosphoglycerate, is more reduced (+3), and it is further reduced in the glyceraldehyde-3-phosphate product (+1). Overall, the early reactions of the Calvin cycle complete the reduction of atmospheric carbon and, in so doing, facilitate its incorporation into organic compounds.

The Carboxylation of Ribulose Bisphosphate Is Catalyzed by the Enzyme Rubisco

CO_2 enters the Calvin cycle by reacting with ribulose-1,5-bisphosphate to yield two molecules of 3-phosphoglycerate (Figure 8.3 and Table 8.1), a reaction catalyzed by the chloroplast enzyme ribulose bisphosphate carboxylase/oxygenase, referred to as **rubisco** (see **Web Topic 8.2**). As indi-

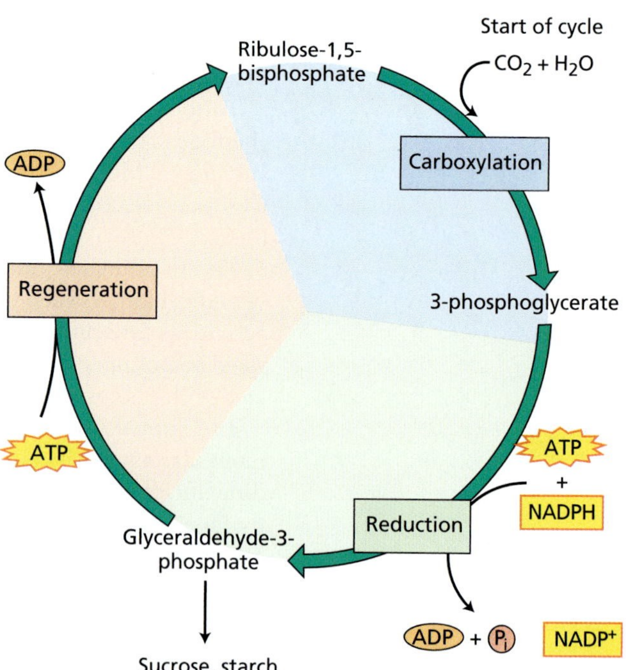

FIGURE 8.2 The Calvin cycle proceeds in three stages: (1) carboxylation, during which CO_2 is covalently linked to a carbon skeleton; (2) reduction, during which carbohydrate is formed at the expense of the photochemically derived ATP and reducing equivalents in the form of NADPH; and (3) regeneration, during which the CO_2 acceptor ribulose-1,5-bisphosphate re-forms.

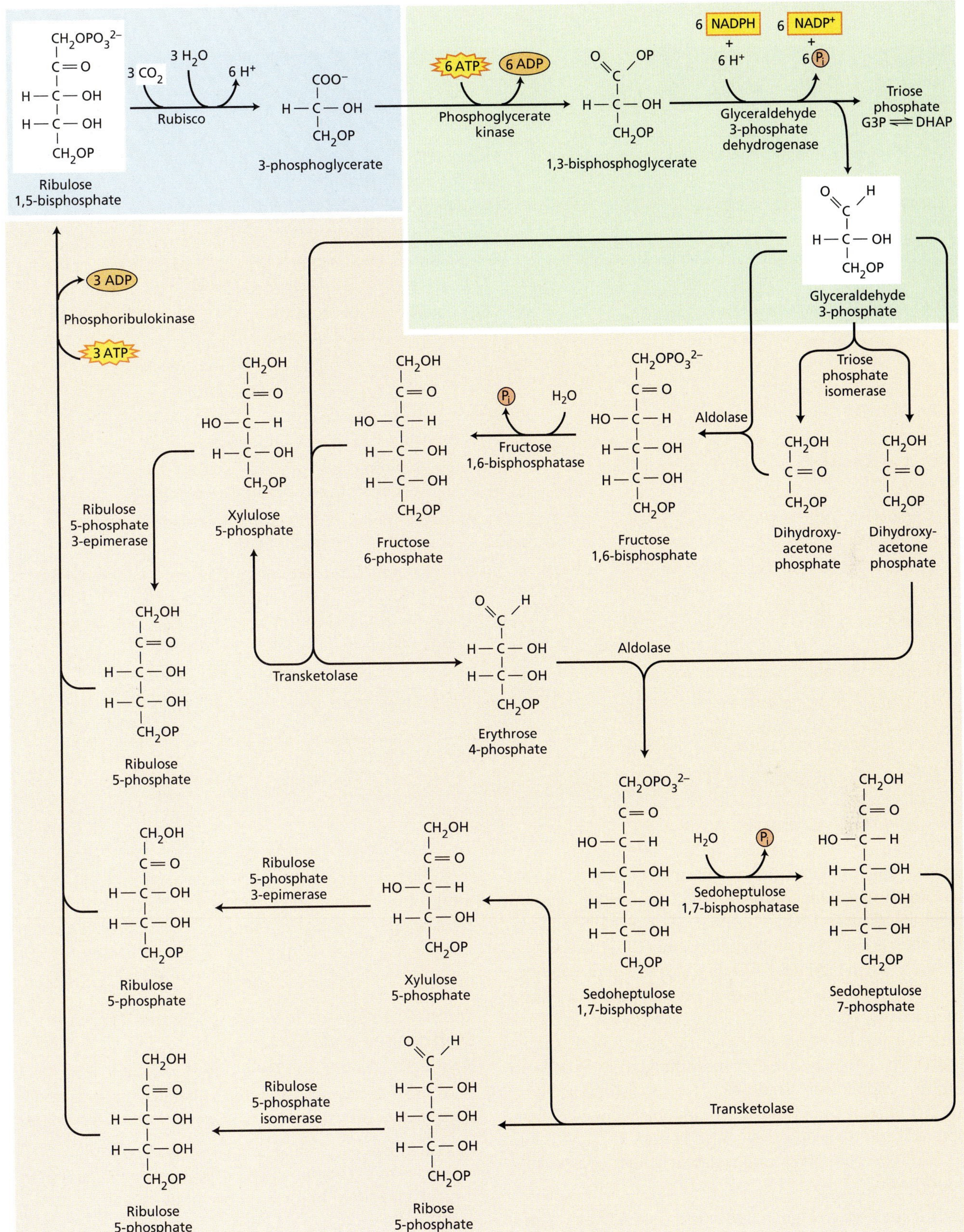

FIGURE 8.3 The Calvin cycle. The carboxylation of three molecules of ribulose-1,5-bisphosphate leads to the *net* synthesis of one molecule of glyceraldehyde-3-phosphate and the regeneration of the three molecules of starting material. This process starts and ends with three molecules of ribulose-1,5-bisphosphate, reflecting the cyclic nature of the pathway.

TABLE 8.1
Reactions of the Calvin cycle

Enzyme	Reaction
1. Ribulose-1,5-bisphosphate carboxylase/oxygenase	6 Ribulose-1,5-bisphosphate + 6 CO_2 + 6 H_2O → 12 (3-phosphoglycerate) + 12 H^+
2. 3-Phosphoglycerate kinase	12 (3-Phosphoglycerate) + 12 ATP → 12 (1,3-bisphosphoglycerate) + 12 ADP
3. NADP:glyceraldehyde-3-phosphate dehydrogenase	12 (1,3-Bisphosphoglycerate) + 12 NADPH + 12 H^+ → 12 glyceraldehye-3-phosphate + 12 $NADP^+$ + 12 P_i
4. Triose phosphate isomerase	5 Glyceraldehyde-3-phosphate → 5 dihydroxyacetone-3-phosphate
5. Aldolase	3 Glyceraldehyde-3-phosphate + 3 dihydroxyacetone-3-phosphate → 3 fructose-1,6-bisphosphate
6. Fructose-1,6-bisphosphatase	3 Fructose-1,6-bisphosphate + 3 H_2O → 3 fructose-6-phosphate + 3 P_i
7. Transketolase	2 Fructose-6-phosphate + 2 glyceraldehyde-3-phosphate → 2 erythrose-4-phosphate + 2 xylulose-5-phosphate
8. Aldolase	2 Erythrose-4-phosphate + 2 dihydroxyacetone-3-phosphate → 2 sedoheptulose-1,7-bisphosphate
9. Sedoheptulose-1,7,bisphosphatase	2 Sedoheptulose-1,7-bisphosphate + 2 H_2O → 2 sedoheptulose-7-phosphate + 2 P_i
10. Transketolase	2 Sedoheptulose-7-phosphate + 2 glyceraldehyde-3-phosphate → 2 ribose-5-phosphate + 2 xylulose-5-phosphate
11a. Ribulose-5-phosphate epimerase	4 Xylulose-5-phosphate → 4 ribulose-5-phosphate
11b. Ribose-5-phosphate isomerase	2 Ribose-5-phosphate → 2 ribulose-5-phosphate
12. Ribulose-5-phosphate kinase	6 Ribulose-5-phosphate + 6 ATP → 6 ribulose-1,5-bisphosphate + 6 ADP + 6 H^+
Net: 6 CO_2 + 11 H_2O + 12 NADPH + 18 ATP → Fructose-6-phosphate + 12 $NADP^+$ + 6 H^+ + 18 ADP + 17 P_i	

Note: P_i stands for inorganic phosphate.

cated by the full name, the enzyme also has an oxygenase activity in which O_2 competes with CO_2 for the common substrate ribulose-1,5-bisphosphate (Lorimer 1983). As we will discuss later, this property limits net CO_2 fixation.

As shown in Figure 8.4, CO_2 is added to carbon 2 of ribulose-1,5-bisphosphate, yielding an unstable, enzyme-bound intermediate, which is hydrolyzed to yield two molecules of the stable product 3-phosphoglycerate (see Table 8.1, reaction 1). The two molecules of 3-phosphoglycerate—labeled "upper" and "lower" on the figure—are distinguished by the fact that the upper molecule contains the newly incorporated carbon dioxide, designated here as $^*CO_2$.

Two properties of the carboxylase reaction are especially important:

1. The negative change in free energy (see Chapter 2 on the web site for a discussion of free energy) associated with the carboxylation of ribulose-1,5-bisphosphate is large; thus the forward reaction is strongly favored.

2. The affinity of rubisco for CO_2 is sufficiently high to ensure rapid carboxylation at the low concentrations of CO_2 found in photosynthetic cells.

Rubisco is very abundant, representing up to 40% of the total soluble protein of most leaves. The concentration of rubisco active sites within the chloroplast stroma is calculated to be about 4 m*M*, or about 500 times greater than the concentration of its CO_2 substrate (see **Web Topic 8.3**).

Triose Phosphates Are Formed in the Reduction Step of the Calvin Cycle

Next in the Calvin cycle (Figure 8.3 and Table 8.1), the 3-phosphoglycerate formed in the carboxylation stage undergoes two modifications:

1. It is first phosphorylated via 3-phosphoglycerate kinase to 1,3-bisphosphoglycerate through use of the ATP generated in the light reactions (Table 8.1, reaction 2).

2. Then it is reduced to glyceraldehyde-3-phosphate through use of the NADPH generated by the light reactions (Table 8.1, reaction 3). The chloroplast enzyme NADP:glyceraldehyde-3-phosphate dehydrogenase catalyzes this step. Note that the enzyme is similar to that of glycolysis (which will be dis-

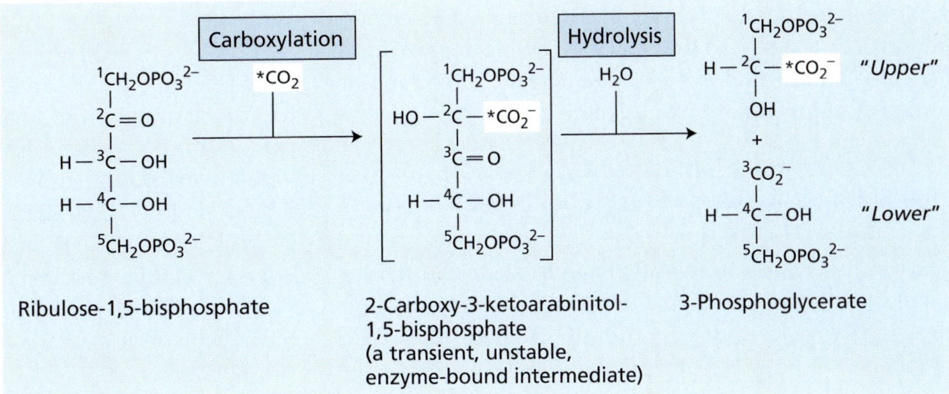

FIGURE 8.4 The carboxylation of ribulose-1,5-bisphosphate by rubisco.

cussed in Chapter 11), except that NADP rather than NAD is the coenzyme. An NADP-linked form of the enzyme is synthesized during chloroplast development (greening), and this form is preferentially used in biosynthetic reactions.

Operation of the Calvin Cycle Requires the Regeneration of Ribulose-1,5-Bisphosphate

The continued uptake of CO_2 requires that the CO_2 acceptor, ribulose-1,5-bisphosphate, be constantly regenerated. To prevent depletion of Calvin cycle intermediates, three molecules of ribulose-1,5-bisphosphate (15 carbons total) are formed by reactions that reshuffle the carbons from the five molecules of triose phosphate ($5 \times 3 = 15$ carbons). This reshuffling consists of reactions 4 through 12 in Table 8.1 (see also Figure 8.3):

1. One molecule of glyceraldehyde-3-phosphate is converted via triose phosphate isomerase to dihydroxyacetone-3-phosphate in an isomerization reaction (reaction 4).
2. Dihydroxyacetone-3-phosphate then undergoes aldol condensation with a second molecule of glyceraldehyde-3-phosphate, a reaction catalyzed by aldolase to give fructose-1,6-bisphosphate (reaction 5).
3. Fructose-1,6-bisphosphate occupies a key position in the cycle and is hydrolyzed to fructose-6-phosphate (reaction 6), which then reacts with the enzyme transketolase.
4. A two-carbon unit (C-1 and C-2 of fructose-6-phosphate) is transferred via transketolase to a third molecule of glyceraldehyde-3-phosphate to give erythrose-4-phosphate (from C-3 to C-6 of the fructose) and xylulose-5-phosphate (from C-2 of the fructose and the glyceraldehyde-3-phosphate) (reaction 7).
5. Erythrose-4-phosphate then combines via aldolase with a fourth molecule of triose phosphate (dihydroxyacetone-3-phosphate) to yield the seven-carbon sugar sedoheptulose-1,7-bisphosphate (reaction 8).
6. This seven-carbon bisphosphate is then hydrolyzed by way of a specific phosphatase to give sedoheptulose-7-phosphate (reaction 9).
7. Sedoheptulose-7-phosphate donates a two-carbon unit to the fifth (and last) molecule of glyceraldehyde-3-phosphate via transketolase and produces ribose-5-phosphate (from C-3 to C-7 of sedoheptulose) and xylulose-5-phosphate (from C-2 of the sedoheptulose and the glyceraldehyde-3-phosphate) (reaction 10).
8. The two molecules of xylulose-5-phosphate are converted to two molecules of ribulose-5-phosphate sugars by a ribulose-5-phosphate epimerase (reaction 11a). The third molecule of ribulose-5-phosphate is formed from ribose-5-phosphate by ribose-5-phosphate isomerase (reaction 11b).
9. Finally, ribulose-5-phosphate kinase catalyzes the phosphorylation of ribulose-5-phosphate with ATP, thus regenerating the three needed molecules of the initial CO_2 acceptor, ribulose-1,5-bisphosphate (reaction 12).

The Calvin Cycle Regenerates Its Own Biochemical Components

The Calvin cycle reactions regenerate the biochemical intermediates that are necessary to maintain the operation of the cycle. But more importantly, the rate of operation of the Calvin cycle can be enhanced by increases in the concentration of its intermediates; that is, the cycle is **autocatalytic**. As a consequence, the Calvin cycle has the metabolically desirable feature of producing more substrate than is consumed, as long as triose phosphate is not being diverted elsewhere:

$$5\ RuBP^{4-} + 5\ CO_2 + 9\ H_2O + 16\ ATP^{4-} + 10\ NADPH \rightarrow$$
$$6\ RuBP^{4-} + 14\ P_i + 6\ H^+ + 16\ ADP^{3-} + 10\ NADP^+$$

The importance of this autocatalytic property is shown by experiments in which previously darkened leaves or isolated chloroplasts are illuminated. In such experiments, CO_2 fixation starts only after a lag, called the *induction period*, and the rate of photosynthesis increases with time in the first few minutes after the onset of illumination. The increase in the rate of photosynthesis during the induction period is due in part to the activation of enzymes by light (discussed later), and in part to an increase in the concentration of intermediates of the Calvin cycle.

Calvin Cycle Stoichiometry Shows That Only One-Sixth of the Triose Phosphate Is Used for Sucrose or Starch

The synthesis of carbohydrates (starch, sucrose) provides a sink ensuring an adequate flow of carbon atoms through the Calvin cycle under conditions of continuous CO_2 uptake. An important feature of the cycle is its overall stoichiometry. At the onset of illumination, most of the triose phosphates are drawn back into the cycle to facilitate the buildup of an adequate concentration of metabolites. When photosynthesis reaches a steady state, however, five-sixths of the triose phosphate contributes to regeneration of the ribulose-1,5-bisphosphate, and one-sixth is exported to the cytosol for the synthesis of sucrose or other metabolites that are converted to starch in the chloroplast.

An input of energy, provided by ATP and NADPH, is required in order to keep the cycle functioning in the fixation of CO_2. The calculation at the end of Table 8.1 shows that in order to synthesize the equivalent of 1 molecule of hexose, 6 molecules of CO_2 are fixed at the expense of 18 ATP and 12 NADPH. In other words, the Calvin cycle consumes two molecules of NADPH and three molecules of ATP for every molecule of CO_2 fixed into carbohydrate.

We can compute the maximal overall thermodynamic efficiency of photosynthesis if we know the energy content of the light, the minimum quantum requirement (moles of quanta absorbed per mole of CO_2 fixed; see Chapter 7), and the energy stored in a mole of carbohydrate (hexose).

Red light at 680 nm contains 175 kJ (42 kcal) per quantum mole of photons. The minimum quantum requirement is usually calculated to be 8 photons per molecule of CO_2 fixed, although the number obtained experimentally is 9 to 10 (see Chapter 7). Therefore, the minimum light energy needed to reduce 6 moles of CO_2 to a mole of hexose is approximately $6 \times 8 \times 175$ kJ = 8400 kJ (2016 kcal). However, a mole of a hexose such as fructose yields only 2804 kJ (673 kcal) when totally oxidized.

Comparing 8400 and 2804 kJ, we see that the maximum overall thermodynamic efficiency of photosynthesis is about 33%. However, most of the unused light energy is lost in the generation of ATP and NADPH by the light reactions (see Chapter 7) rather than during operation of the Calvin cycle.

We can calculate the efficiency of the Calvin cycle more directly by computing the changes in free energy associated with the hydrolysis of ATP and the oxidation of NADPH, which are 29 and 217 kJ (7 and 52 kcal) per mole, respectively. We saw in the list summarizing the Calvin cycle reactions that the synthesis of 1 molecule of fructose-6-phosphate from 6 molecules of CO_2 uses 12 NADPH and 18 ATP molecules. Therefore the Calvin cycle consumes $(12 \times 217) + (18 \times 29) = 3126$ kJ (750 kcal) in the form of NADPH and ATP, resulting in a thermodynamic efficiency close to 90%.

An examination of these calculations shows that the bulk of the energy required for the conversion of CO_2 to carbohydrate comes from NADPH. That is, 2 mol NADPH $\times$ 52 kcal mol^{-1} = 104 kcal, but 3 mol ATP $\times$ 7 kcal mol^{-1} = 21 kcal. Thus, 83% (104 of 125 kcal) of the energy stored comes from the reductant NADPH.

The Calvin cycle does not occur in all autotrophic cells. Some anaerobic bacteria use other pathways for autotrophic growth:

- The ferredoxin-mediated synthesis of organic acids from acetyl– and succinyl– CoA derivatives via a reversal of the citric acid cycle (the reductive carboxylic acid cycle of green sulfur bacteria)
- The glyoxylate-producing cycle (the hydroxypropionate pathway of green nonsulfur bacteria)
- The linear route (acetyl-CoA pathway) of acetogenic, methanogenic bacteria

Thus although the Calvin cycle is quantitatively the most important pathway of autotrophic CO_2 fixation, others have been described.

REGULATION OF THE CALVIN CYCLE

The high energy efficiency of the Calvin cycle indicates that some form of regulation ensures that all intermediates in the cycle are present at adequate concentrations and that the cycle is turned off when it is not needed in the dark. In general, variation in the concentration or in the specific activity of enzymes modulates catalytic rates, thereby adjusting the level of metabolites in the cycle.

Changes in gene expression and protein biosynthesis regulate enzyme concentration. Posttranslational modification of proteins contributes to the regulation of enzyme activity. At the genetic level the amount of each enzyme present in the chloroplast stroma is regulated by mechanisms that control expression of the nuclear and chloroplast genomes (Maier et al. 1995; Purton 1995).

Short-term regulation of the Calvin cycle is achieved by several mechanisms that optimize the concentration of intermediates. These mechanisms minimize reactions operating in opposing directions, which would waste resources (Wolosiuk et al. 1993). Two general mechanisms can change the kinetic properties of enzymes:

1. The transformation of covalent bonds such as the reduction of disulfides and the carbamylation of amino groups, which generate a chemically modified enzyme.
2. The modification of noncovalent interactions, such as the binding of metabolites or changes in the composi-

tion of the cellular milieu (e.g., pH). In addition, the binding of the enzymes to the thylakoid membranes enhances the efficiency of the Calvin cycle, thereby achieving a higher level of organization that favors the channeling and protection of substrates.

Light-Dependent Enzyme Activation Regulates the Calvin Cycle

Five light-regulated enzymes operate in the Calvin cycle:

1. Rubisco
2. NADP:glyceraldehyde-3-phosphate dehydrogenase
3. Fructose-1,6-bisphosphatase
4. Sedoheptulose-1,7-bisphosphatase
5. Ribulose-5-phosphate kinase

The last four enzymes contain one or more disulfide (—S—S—) groups. Light controls the activity of these four enzymes via the **ferredoxin–thioredoxin system**, a covalent thiol-based oxidation–reduction mechanism identified by Bob Buchanan and colleagues (Buchanan 1980; Wolosiuk et al. 1993; Besse and Buchanan 1997; Schürmann and Jacquot 2000). In the dark these residues exist in the oxidized state (—S—S—), which renders the enzyme inactive or subactive. In the light the —S—S— group is reduced to the sulfhydryl state (—SH HS—). This redox change leads to activation of the enzyme (Figure 8.5). The resolution of the crystal structure of each member of the ferredoxin–thioredoxin system and of the target enzymes fructose-1,6-bisphosphatase and NADP:malate dehydrogenase (Dai et al. 2000) have provided valuable information about the mechanisms involved.

This sulfhydryl (also called dithiol) signal of the regulatory protein thioredoxin is transmitted to specific target enzymes, resulting in their activation (see **Web Topic 8.4**). In some cases (such as fructose-1,6-bisphosphatase), the thioredoxin-linked activation is enhanced by an effector (e.g., fructose-1,6-bisphosphate substrate).

Inactivation of the target enzymes observed upon darkening appears to take place by a reversal of the reduction (activation) pathway. That is, oxygen converts the thioredoxin and target enzyme from the reduced state (—SH HS—) to the oxidized state (—S—S—) and, in so doing, leads to inactivation of the enzyme (see Figure 8.5; see also **Web Topic 8.4**). The last four of the enzymes listed here are regulated directly by thioredoxin; the first, rubisco, is regulated indirectly by a thioredoxin accessory enzyme, rubisco activase (see the next section).

Rubisco Activity Increases in the Light

The activity of rubisco is also regulated by light, but the enzyme itself does not respond to thioredoxin. George Lorimer and colleagues found that rubisco is activated when activator CO_2 (a different molecule from the substrate CO_2 that becomes fixed) reacts slowly with an uncharged ε-NH_2 group of lysine within the active site of the enzyme. The resulting carbamate derivative (a new anionic site) then rapidly binds Mg^{2+} to yield the activated complex (Figure 8.6).

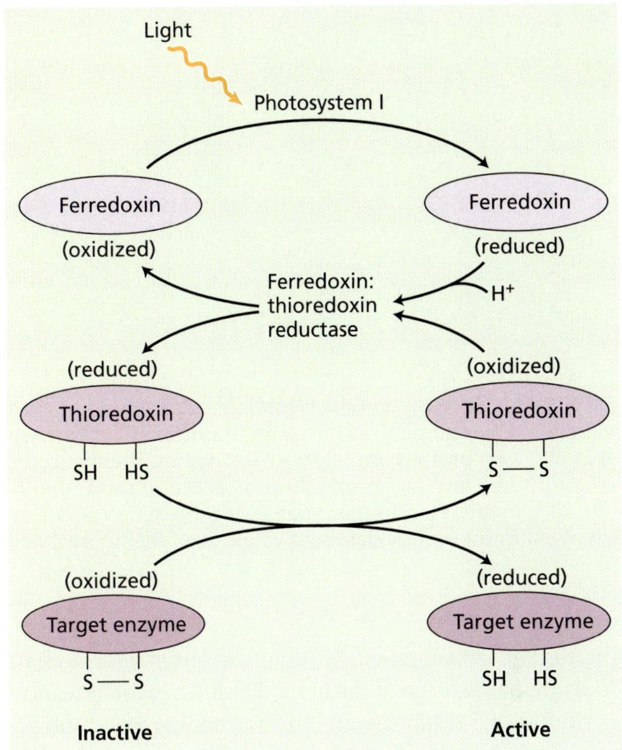

FIGURE 8.5 The ferredoxin–thioredoxin system reduces specific enzymes in the light. Upon reduction, biosynthetic enzymes are converted from an inactive to an active state. The activation process starts in the light by a reduction of ferredoxin by photosystem I (see Chapter 7). The reduced ferredoxin plus two protons are used to reduce a catalytically active disulfide (—S—S—) group of the iron–sulfur enzyme ferredoxin:thioredoxin reductase, which in turn reduces the highly specific disulfide (—S—S—) bond of the small regulatory protein thioredoxin (see Web Topic 8.4 for details). The reduced form (—SH HS—) of thioredoxin then reduces the critical disulfide bond (converts —S—S— to —SH HS—) of a target enzyme and thereby leads to activation of that enzyme. The light signal is thus converted to a sulfhydryl, or —SH, signal via ferredoxin and the enzyme ferredoxin:thioredoxin reductase.

Two protons are released during the formation of the ternary complex rubisco–CO_2–Mg^{2+}, so activation is promoted by an increase in both pH and Mg^{2+} concentration. Thus, light-dependent stromal changes in pH and Mg^{2+} (see the next section) appear to facilitate the observed activation of rubisco by light.

In the active state, rubisco binds another molecule of CO_2, which reacts with the 2,3-enediol form of ribulose-1,5-bisphosphate (P—O—CH_2—COH=COH—CHOH—CH_2O—P) yielding 2-carboxy-3-ketoribitol 1,5-bisphos-

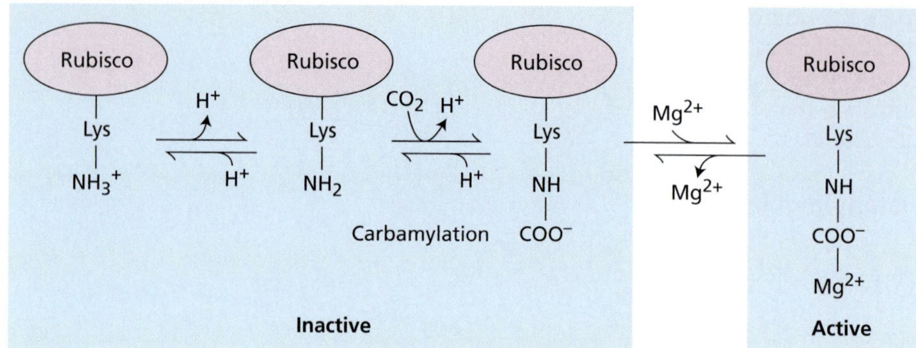

FIGURE 8.6 One way in which rubisco is activated involves the formation of a carbamate–Mg^{2+} complex on the ε-amino group of a lysine within the active site of the enzyme. Two protons are released. Activation is enhanced by the increase in Mg^{2+} concentration and higher pH (low H^+ concentration) that result from illumination. The CO_2 involved in the carbamate–Mg^{2+} reaction is not the same as the CO_2 involved in the carboxylation of ribulose-1,5-bisphosphate.

phate. The extreme instability of the latter intermediate leads to the cleavage of the bond that links carbons 2 and 3 of ribulose-1,5-bisphosphate, and as a consequence, rubisco releases two molecules of 3-phosphoglycerate.

The binding of sugar phosphates, such as ribulose-1,5-bisphosphate, to rubisco prevents carbamylation. The sugar phosphates can be removed by the enzyme rubisco activase, in a reaction that requires ATP. The primary role of rubisco activase is to accelerate the release of bound sugar phosphates, thus preparing rubisco for carbamylation (Salvucci and Ogren 1996, see also **Web Topic 8.5**).

Rubisco is also regulated by a natural sugar phosphate, carboxyarabinitol-1-phosphate, that closely resembles the six-carbon transition intermediate of the carboxylation reaction. This inhibitor is present at low concentrations in leaves of many species and at high concentrations in leaves of legumes such as soybean and bean. Carboxyarabinitol-1-phosphate binds to rubisco at night, and it is removed by the action of rubisco activase in the morning, when photon flux density increases.

Recent work has shown that in some plants rubisco activase is regulated by the ferredoxin–thioredoxin system (Zhang and Portis 1999). In addition to connecting thioredoxin to all five regulatory enzymes of the Calvin cycle, this finding provides a new mechanism for linking light to the regulation of enzyme activity.

Light-Dependent Ion Movements Regulate Calvin Cycle Enzymes

Light causes reversible ion changes in the stroma that influence the activity of rubisco and other chloroplast enzymes. Upon illumination, protons are pumped from the stroma into the lumen of the thylakoids. The proton efflux is coupled to Mg^{2+} uptake into the stroma. These ion fluxes decrease the stromal concentration of H^+ (pH 7 → 8) and increase that of Mg^{2+}. These changes in the ionic composition of the chloroplast stroma are reversed upon darkening.

Several Calvin cycle enzymes (rubisco, fructose-1,6-bisphosphatase, sedoheptulose-1,7-bisphosphatase, and ribulose-5-phosphate kinase) are more active at pH 8 than at pH 7 and require Mg^{2+} as a cofactor for catalysis. Hence these light-dependent ion fluxes enhance the activity of key enzymes of the Calvin cycle (Heldt 1979).

Light-Dependent Membrane Transport Regulates the Calvin Cycle

The rate at which carbon is exported from the chloroplast plays a role in regulation of the Calvin cycle. Carbon is exported as triose phosphates in exchange for orthophosphate via the phosphate translocator in the inner membrane of the chloroplast envelope (Flügge and Heldt 1991). To ensure continued operation of the Calvin cycle, at least five-sixths of the triose phosphate must be recycled (see Table 8.1 and Figure 8.3). Thus, at most one-sixth can be exported for sucrose synthesis in the cytosol or diverted to starch synthesis within the chloroplast. The regulation of this aspect of photosynthetic carbon metabolism will be discussed further when the syntheses of sucrose and starch are considered in detail later in this chapter.

THE C_2 OXIDATIVE PHOTOSYNTHETIC CARBON CYCLE

An important property of rubisco is its ability to catalyze both the carboxylation and the oxygenation of RuBP. Oxygenation is the primary reaction in a process known as **photorespiration**. Because photosynthesis and photorespiration work in diametrically opposite directions, photorespiration results in loss of CO_2 from cells that are simultaneously fixing CO_2 by the Calvin cycle (Ogren 1984; Leegood et al. 1995).

In this section we will describe the C_2 oxidative photosynthetic carbon cycle—the reactions that result in the partial recovery of carbon lost through oxidation.

Photosynthetic CO_2 Fixation and Photorespiratory Oxygenation Are Competing Reactions

The incorporation of one molecule of O_2 into the 2,3-enediol isomer of ribulose-1,5-bisphosphate generates an unstable intermediate that rapidly splits into 2-phosphoglycolate and 3-phosphoglycerate (Figure 8.7 and Table 8.2, reaction 1). The ability to catalyze the oxygenation of ribulose-1,5-bisphosphate is a property of all rubiscos, regard-

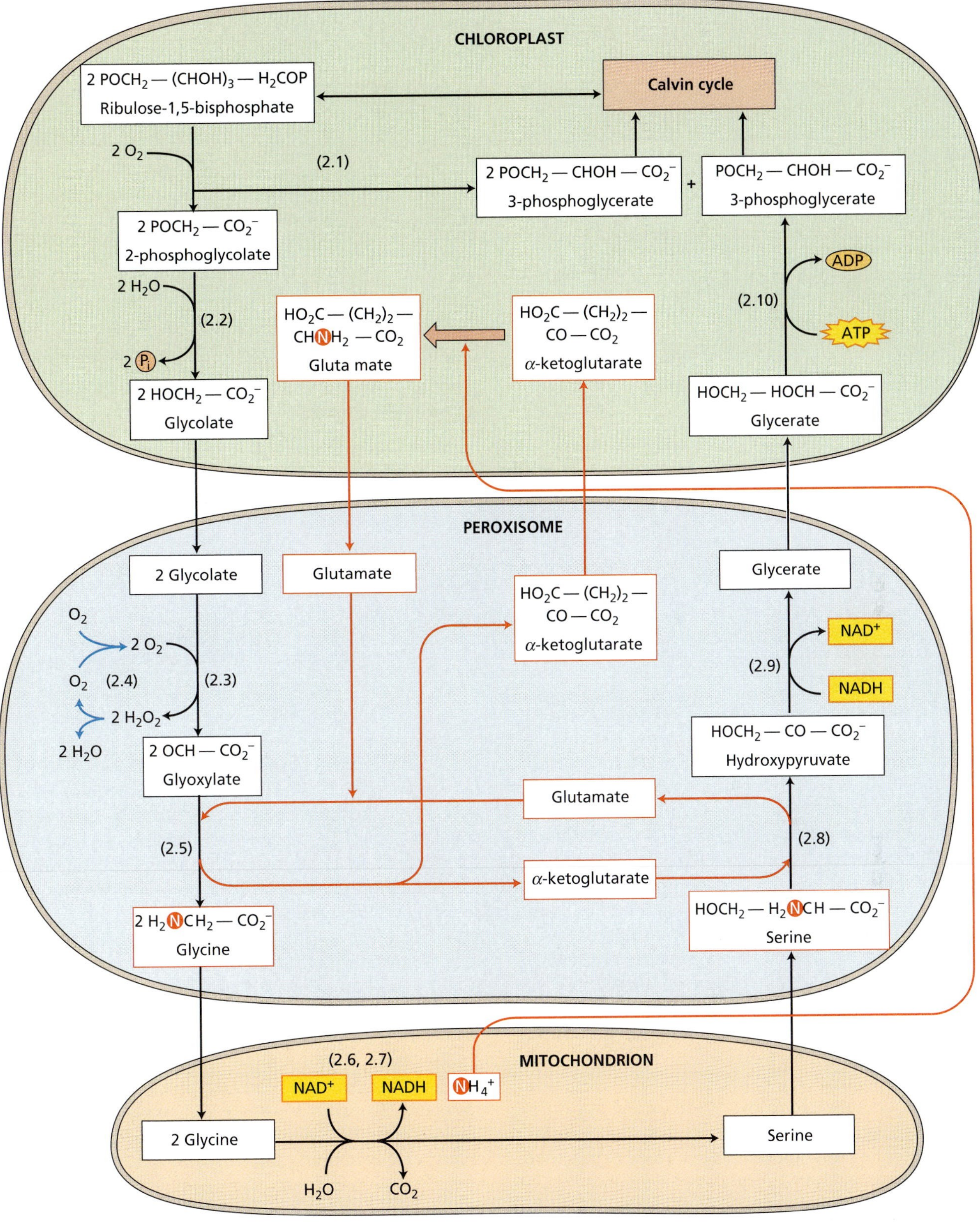

FIGURE 8.7 The main reactions of the photorespiratory cycle. Operation of the C_2 oxidative photosynthetic cycle involves the cooperative interaction among three organelles: chloroplasts, mitochondria, and peroxisomes. Two molecules of glycolate (four carbons) transported from the chloroplast into the peroxisome are converted to glycine, which in turn is exported to the mitochondrion and transformed to serine (three carbons) with the concurrent release of carbon dioxide (one carbon). Serine is transported to the peroxisome and transformed to glycerate. The latter flows to the chloroplast where it is phosphorylated to 3-phosphoglycerate and incorporated into the Calvin cycle. Inorganic nitrogen (ammonia) released by the mitochondrion is captured by the chloroplast for the incorporation into amino acids by using appropiate skeletons (α-ketoglutarate). The heavy arrow in red marks the assimilation of ammonia into glutamate catalyzed by glutamine synthetase. In addition, the uptake of oxygen in the peroxisome supports a short oxygen cycle coupled to oxidative reactions. The flow of carbon, nitrogen and oxygen are indicated in black, red and blue, respectively. See Table 8.2 for a description of each numbered reaction.

TABLE 8.2
Reactions of the C_2 oxidative photosynthetic carbon cycle

Enzyme	Reaction
1. Ribulose-1,5-bisphosphate carboxylase/oxygenase (chloroplast)	2 Ribulose-1,5-bisphosphate + 2 $O_2 \rightarrow$ 2 phosphoglycolate + 2 3-phosphoglycerate + 4 H^+
2. Phosphoglycolate phosphatase (chloroplast)	2 Phosphoglycolate + 2 $H_2O \rightarrow$ 2 glycolate + 2 P_i
3. Glycolate oxidase (peroxisome)	2 Glycolate + 2 $O_2 \rightarrow$ 2 glyoxylate + 2 H_2O_2
4. Catalase (peroxisome)	2 $H_2O_2 \rightarrow$ 2 H_2O + O_2
5. Glyoxylate:glutamate aminotransferase (peroxisome)	2 Glyoxylate + 2 glutamate $\rightarrow$ 2 glycine + 2 α-ketoglutarate
6. Glycine decarboxylase (mitochondrion)	Glycine + NAD^+ + H^+ + H_4-folate $\rightarrow$ NADH + CO_2 + NH_4^+ + methylene-H_4-folate
7. Serine hydroxymethyltransferase (mitochondrion)	Methylene-H_4-folate + H_2O + glycine $\rightarrow$ serine + H_4-folate
8. Serine aminotransferase (peroxisome)	Serine + α-ketoglutarate $\rightarrow$ hydroxypyruvate + glutamate
9. Hydroxypyruvate reductase (peroxisome)	Hydroxypyruvate + NADH + $H^+ \rightarrow$ glycerate + NAD^+
10. Glycerate kinase (chloroplast)	Glycerate + ATP $\rightarrow$ 3-phosphoglycerate + ADP + H^+

Note: Upon the release of glycolate from the chloroplast (reactions 2 $\rightarrow$ 3), the interplay of this organelle with the peroxisome and the mitochondrion drives the following overall reaction:

$$2 \text{ Glycolate} + \text{glutamate} + O_2 \rightarrow \text{glycerate} + \alpha\text{-ketoglutarate} + NH_4^+ + CO_2 + H_2O$$

The 3-phosphoglycerate formed in the chloroplast (reaction 10) is converted to ribulose-1,5-bisphosphate via the reductive and regenerative reactions of the Calvin cycle. The ammonia and α-ketoglutarate are converted to glutamate in the chloroplast by ferrodoxin-linked glutamate synthase (GOGAT).

P_i stands for inorganic phosphate.

less of taxonomic origin. Even the rubisco from anaerobic, autotrophic bacteria catalyzes the oxygenase reaction when exposed to oxygen.

As alternative substrates for rubisco, CO_2 and O_2 compete for reaction with ribulose-1,5-bisphosphate because carboxylation and oxygenation occur within the same active site of the enzyme. Offered equal concentrations of CO_2 and O_2 in a test tube, angiosperm rubiscos fix CO_2 about 80 times faster than they oxygenate. However, an aqueous solution in equilibrium with air at 25°C has a CO_2:O_2 ratio of 0.0416 (see **Web Topics 8.2 and 8.3**). At these concentrations, carboxylation in air outruns oxygenation by a scant three to one.

The C_2 oxidative photosynthetic carbon cycle acts as a scavenger operation to recover fixed carbon lost during photorespiration by the oxygenase reaction of rubisco (**Web Topic 8.6**). The 2-phosphoglycolate formed in the chloroplast by oxygenation of ribulose-1,5-bisphosphate is rapidly hydrolyzed to glycolate by a specific chloroplast phosphatase (Figure 8.7 and Table 8.2, reaction 2). Subsequent metabolism of the glycolate involves the cooperation of two other organelles: peroxisomes and mitochondria (see Chapter 1) (Tolbert 1981).

Glycolate leaves the chloroplast via a specific transporter protein in the envelope membrane and diffuses to the peroxisome. There it is oxidized to glyoxylate and hydrogen peroxide (H_2O_2) by a flavin mononucleotide-dependent oxidase: glycolate oxidase (Figure 8.7 and Table 8.2, reaction 3). The vast amounts of hydrogen peroxide released in the peroxisome are destroyed by the action of catalase (Table 8.2, reaction 4) while the glyoxylate undergoes transamination (reaction 5). The amino donor for this transamination is probably glutamate, and the product is the amino acid glycine.

Glycine leaves the peroxisome and enters the mitochondrion (see Figure 8.7). There the glycine decarboxylase multienzyme complex catalyzes the conversion of two molecules of glycine and one of NAD^+ to one molecule each of serine, NADH, NH_4^+ and CO_2 (Table 8.2, reactions 6 and 7). This multienzyme complex, present in large concentrations in the matrix of plant mitochondria, comprises four proteins, named H-protein (a lipoamide-containing polypeptide), P-protein (a 200 kDa, homodimer, pyridoxal phosphate-containing protein), T-protein (a folate-dependent protein), and L-protein (a flavin adenine nucleotide–containing protein).

The ammonia formed in the oxidation of glycine diffuses rapidly from the matrix of mitochondria to chloroplasts, where glutamine synthetase combines it with carbon skeletons to form amino acids. The newly formed serine leaves the mitochondria and enters the peroxisome, where it is converted first by transamination to hydroxypyruvate (Table 8.2, reaction 8) and then by an NADH-dependent reduction to glycerate (reaction 9).

A malate-oxaloacetate shuttle transfers NADH from the cytoplasm into the peroxisome, thus maintaining an adequate concentration of NADH for this reaction. Finally, glycerate reenters the chloroplast, where it is phosphorylated to yield 3-phosphoglycerate (Table 8.2, reaction 10).

In photorespiration, various compounds are circulated in concert through two cycles. In one of the cycles, carbon exits the chloroplast in two molecules of glycolate and returns in one molecule of glycerate. In the other cycle, nitrogen exits the chloroplast in one molecule of glutamate and returns in one molecule of ammonia (together with one molecule of α-ketoglutarate) (see Figure 8.7).

Thus overall, two molecules of phosphoglycolate (four carbon atoms), lost from the Calvin cycle by the oxygenation of RuBP, are converted into one molecule of 3-phosphoglycerate (three carbon atoms) and one CO_2. In other words, 75% of the carbon lost by the oxygenation of ribulose-1,5-bisphosphate is recovered by the C_2 oxidative photosynthetic carbon cycle and returned to the Calvin cycle (Lorimer 1981).

On the other hand, the total organic nitrogen remains unchanged because the formation of inorganic nitrogen (NH_4^+) in the mitochondrion is balanced by the synthesis of glutamine in the chloroplast. Similarly, the use of NADH in the peroxisome (by hydroxypyruvate reductase) is balanced by the reduction of NAD^+ in the mitochondrion (by glycine decarboxylase).

Competition between Carboxylation and Oxygenation Decreases the Efficiency of Photosynthesis

Because photorespiration is concurrent with photosynthesis, it is difficult to measure the rate of photorespiration in intact cells. Two molecules of 2-phosphoglycolate (four carbon atoms) are needed to make one molecule of 3-phosphoglycerate, with the release of one molecule of CO_2; so theoretically one-fourth of the carbon entering the C_2 oxidative photosynthetic carbon cycle is released as CO_2.

Measurements of CO_2 release by sunflower leaves support this calculated value. This result indicates that the actual rate of photosynthesis is approximately 120 to 125% of the measured rate. The ratio of carboxylation to oxygenation in air at 25°C is computed to be between 2.5 and 3. Further calculations indicate that photorespiration lowers the efficiency of photosynthetic carbon fixation from 90% to approximately 50%.

This decreased efficiency can be measured as an increase in the quantum requirement for CO_2 fixation under photorespiratory conditions (air with high O_2 and low CO_2) as opposed to nonphotorespiratory conditions (low O_2 and high CO_2).

Carboxylation and Oxygenation Are Closely Interlocked in the Intact Leaf

Photosynthetic carbon metabolism in the intact leaf reflects the integrated balance between two mutually opposing and interlocking cycles (Figure 8.8). The Calvin cycle can operate independently, but the C_2 oxidative photosynthetic carbon cycle depends on the Calvin cycle for a supply of ribulose-1,5-bisphosphate. The balance between the two cycles is determined by three factors: the kinetic properties of rubisco, the concentrations of the substrates CO_2 and O_2, and temperature.

As the temperature increases, the concentration of CO_2 in a solution in equilibrium with air decreases more than the concentration of O_2 does (see **Web Topic 8.3**). Consequently, the concentration ratio of CO_2 to O_2 decreases as the temperature rises. As a result of this property, photorespiration (oxygenation) increases relative to photosynthesis (carboxylation) as the temperature rises. This effect is enhanced by the kinetic properties of rubisco, which also result in a relative increase in oxygenation at higher temperatures (Ku and Edwards 1978). Overall, then, increasing temperatures progressively tilt the balance away from the Calvin cycle and toward the oxidative photosynthetic carbon cycle (see Chapter 9).

The Biological Function of Photorespiration Is Unknown

Although the C_2 oxidative photosynthetic carbon cycle recovers 75% of the carbon originally lost from the Calvin cycle as 2-phosphoglycolate, why does 2-phosphoglycolate form at all? One possible explanation is that the formation

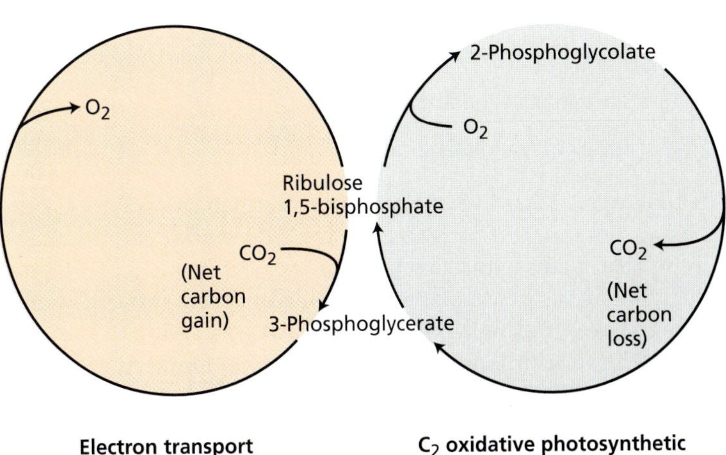

FIGURE 8.8 The flow of carbon in the leaf is determined by the balance between two mutually opposing cycles. Whereas the Calvin cycle is capable of independent operation in the presence of adequate substrates generated by photosynthetic electron transport, the C_2 oxidative photosynthetic carbon cycle requires continued operation of the Calvin cycle to regenerate its starting material, ribulose-1,5-bisphosphate.

of 2-phosphoglycolate is a consequence of the chemistry of the carboxylation reaction, which requires an intermediate that can react with both CO_2 and O_2.

Such a reaction would have had little consequence in early evolutionary times if the ratio of CO_2 to O_2 in air were higher than it is today. However, the low CO_2:O_2 ratios prevalent in modern times are conducive to photorespiration, with no other function than the recovery of some of the carbon present in 2-phosphoglycolate.

Another possible explanation is that photorespiration is important, especially under conditions of high light intensity and low intercellular CO_2 concentration (e.g., when stomata are closed because of water stress), to dissipate excess ATP and reducing power from the light reactions, thus preventing damage to the photosynthetic apparatus. *Arabidopsis* mutants that are unable to photorespire grow normally under 2% CO_2, but they die rapidly if transferred to normal air. There is evidence from work with transgenic plants that photorespiration protects C_3 plants from photooxidation and photoinhibition (Kozaki and Takeba 1996). Further work is needed to improve our understanding of the function of photorespiration.

CO_2-CONCENTRATING MECHANISMS I: ALGAL AND CYANOBACTERIAL PUMPS

Many plants either do not photorespire at all, or they do so to only a limited extent. These plants have normal rubiscos, and their lack of photorespiration is a consequence of mechanisms that concentrate CO_2 in the rubisco environment and thereby suppress the oxygenation reaction.

In this and the two following sections we will discuss three mechanisms for concentrating CO_2 at the site of carboxylation:

1. C_4 photosynthetic carbon fixation (C_4)
2. Crassulacean acid metabolism (CAM)
3. CO_2 pumps at the plasma membrane

The first two of these CO_2-concentrating mechanisms are found in some angiosperms and involve "add-ons" to the Calvin cycle. Plants with C_4 metabolism are often found in hot environments; CAM plants are typical of desert environments. We will examine each of these two systems after we consider the third mechanism: a CO_2 pump found in aquatic plants that has been studied extensively in unicellular cyanobacteria and algae.

When algal and cyanobacterial cells are grown in air enriched with 5% CO_2 and then transferred to a low-CO_2 medium, they display symptoms typical of photorespiration (O_2 inhibition of photosynthesis at low concentration of CO_2). But if the cells are grown in air containing 0.03% CO_2, they rapidly develop the ability to concentrate inorganic carbon (CO_2 plus HCO_3^-) internally. Under these low-CO_2 conditions, the cells no longer photorespire.

At the concentrations of CO_2 found in aquatic environments, rubisco operates far below its maximal specific activity. Marine and freshwater organisms overcome this drawback by accumulating inorganic carbon by the use of CO_2 and HCO_3^- pumps at the plasma membrane. ATP derived from the light reactions provides the energy necessary for the active uptake of CO_2 and HCO_3^-. Total inorganic carbon inside some cyanobacterial cells can reach concentrations of 50 m*M* (Ogawa and Kaplan 1987). Recent work indicates that a single gene encoding a transcription factor can regulate the expression of genes that encode the components of the CO_2-concentrating mechanism in algae (Xiang et al. 2001).

The proteins that function as CO_2–HCO_3^- pumps are not present in cells grown in high concentrations of CO_2 but are induced upon exposure to low concentrations of CO_2. The accumulated HCO_3^- is converted to CO_2 by the enzyme carbonic anhydrase, and the CO_2 enters the Calvin cycle.

The metabolic consequence of this CO_2 enrichment is suppression of the oxygenation of ribulose bisphosphate and hence also suppression of photorespiration. The energetic cost of this adaptation is the additional ATP needed for concentrating the CO_2.

CO_2-CONCENTRATING MECHANISMS II: THE C_4 CARBON CYCLE

There are differences in leaf anatomy between plants that have a C_4 carbon cycle (called *C_4 plants*) and those that photosynthesize solely via the Calvin photosynthetic cycle (*C_3 plants*). A cross section of a typical C_3 leaf reveals one major cell type that has chloroplasts, the **mesophyll**. In contrast, a typical C_4 leaf has two distinct chloroplast-containing cell types: mesophyll and **bundle sheath** (or *Kranz*, German for "wreath") cells (Figure 8.9).

There is considerable anatomic variation in the arrangement of the bundle sheath cells with respect to the mesophyll and vascular tissue. In all cases, however, operation of the C_4 cycle requires the cooperative effort of both cell types. No mesophyll cell of a C_4 plant is more than two or three cells away from the nearest bundle sheath cell (see Figure 8.9A). In addition, an extensive network of plasmodesmata (see Figure 1.27) connects mesophyll and bundle sheath cells, thus providing a pathway for the flow of metabolites between the cell types.

Malate and Aspartate Are Carboxylation Products of the C_4 Cycle

Early labeling of C_4 acids was first observed in $^{14}CO_2$ labeling studies of sugarcane by H. P. Kortschack and colleagues and of maize by Y. Karpilov and coworkers. When leaves were exposed for a few seconds to $^{14}CO_2$ in the light, 70 to 80% of the label was found in the C_4 acids malate and aspartate—a pattern very different from the one observed in leaves that photosynthesize solely via the Calvin cycle.

(A)

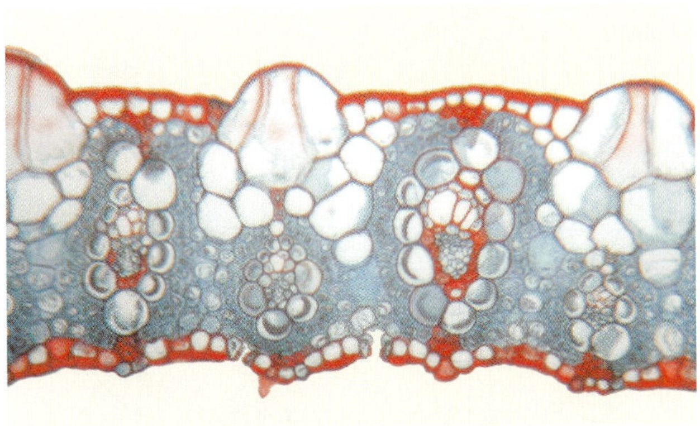

(B)

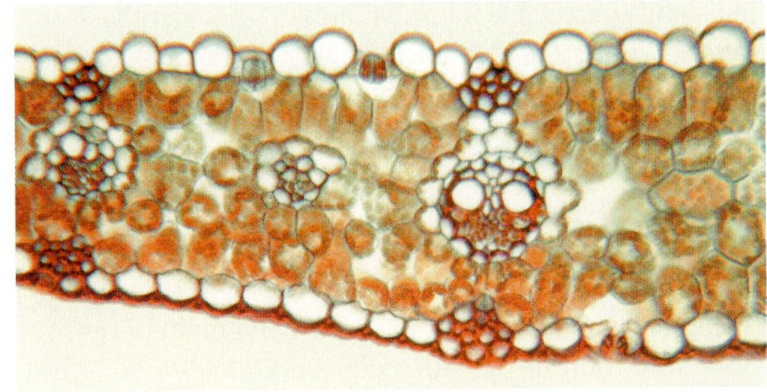

(C)

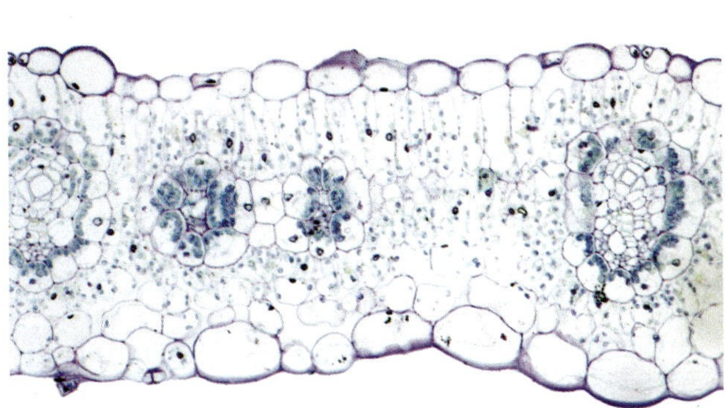

(D)

(E)

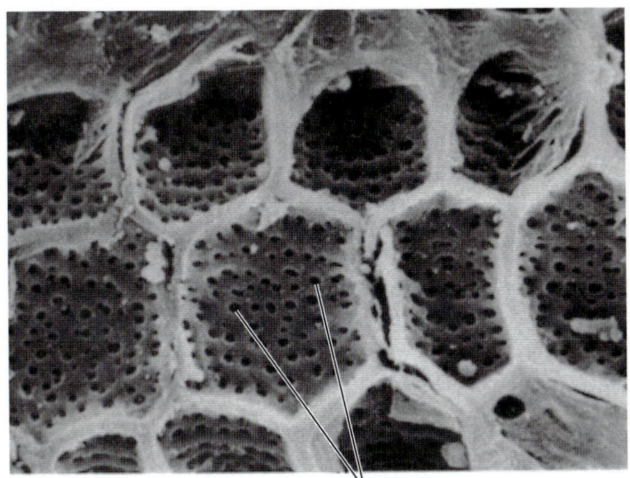

FIGURE 8.9 Cross-sections of leaves, showing the anatomic differences between C_3 and C_4 plants. (A) A C_4 monocot, *Saccharum officinarum* (sugarcane). (135×) (B) A C_3 monocot, Poa sp. (a grass). (240×) (C) A C_4 dicot, *Flaveria australasica* (Asteraceae). (740×) The bundle sheath cells are large in C_4 leaves (A and C), and no mesophyll cell is more than two or three cells away from the nearest bundle sheath cell. These anatomic features are absent in the C_3 leaf (B). (D) Three-dimensional model of a C_4 leaf. (E) Scanning electron micrograph of a C_4 leaf from *Triodia irritans*, showing the plasmodesmata pits in the bundle sheath cell walls through which metabolites of the C_4 carbon cycle are thought to be transported. (A and B © David Webb; C courtesy of Athena McKown; D after Lüttge and Higinbotham; E from Craig and Goodchild 1977.)

In pursuing these initial observations, M. D. Hatch and C. R. Slack elucidated what is now known as the C_4 photosynthetic carbon cycle (C_4 cycle) (Figure 8.10). They established that the C_4 acids malate and aspartate are the first stable, detectable intermediates of photosynthesis in leaves of sugarcane and that carbon atom 4 of malate subsequently becomes carbon atom 1 of 3-phosphoglycerate (Hatch and Slack 1966). The primary carboxylation in these leaves is catalyzed not by rubisco, but by PEP (phosphoenylpyruvate) carboxylase (Chollet et al. 1996).

The manner in which carbon is transferred from carbon atom 4 of malate to carbon atom 1 of 3-phosphoglycerate became clear when the involvement of mesophyll and bundle sheath cells was elucidated. The participating enzymes occur in one of the two cell types: PEP carboxylase and pyruvate–orthophosphate dikinase are restricted to mesophyll cells; the decarboxylases and the enzymes of the complete Calvin cycle are confined to the bundle sheath cells. With this knowledge, Hatch and Slack were able to formulate the basic model of the cycle (Figure 8.11 and Table 8.3).

The C_4 Cycle Concentrates CO_2 in Bundle Sheath Cells

The basic C_4 cycle consists of four stages:

1. Fixation of CO_2 by the carboxylation of phosphoenolpyruvate in the mesophyll cells to form a C_4 acid (malate and/or aspartate)
2. Transport of the C_4 acids to the bundle sheath cells
3. Decarboxylation of the C_4 acids within the bundle sheath cells and generation of CO_2, which is then reduced to carbohydrate via the Calvin cycle

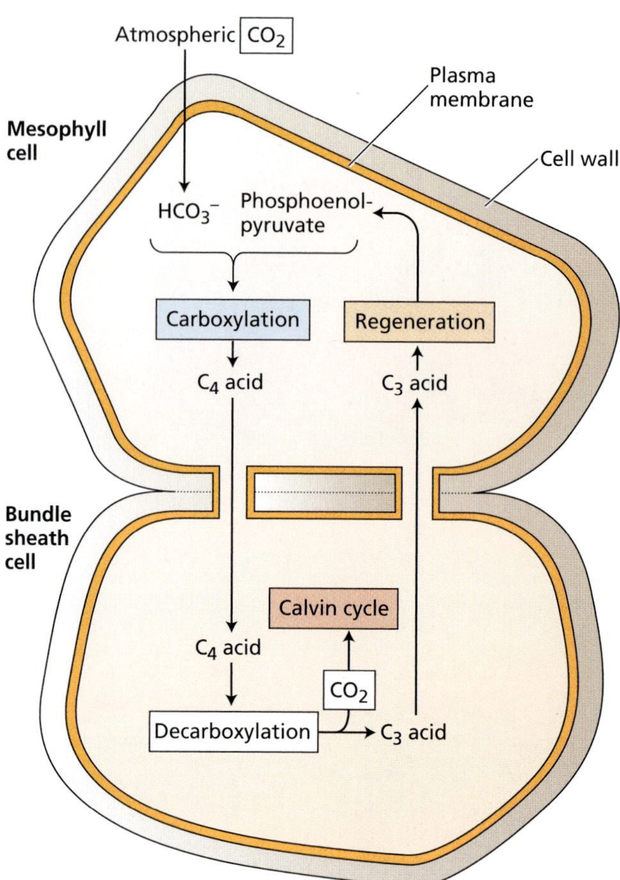

FIGURE 8.10 The basic C_4 photosynthetic carbon cycle involves four stages in two different cell types: (1) Fixation of CO_2 into a four-carbon acid in a mesophyll cell; (2) Transport of the four-carbon acid from the mesophyll cell to a bundle sheath cell; (3) Decarboxylation of the four-carbon acid, and the generation of a high CO_2 concentration in the bundle sheath cell. The CO_2 released is fixed by rubisco and converted to carbohydrate by the Calvin cycle.(4) Transport of the residual three-carbon acid back to the mesophyll cell, where the original CO_2 acceptor, phosphoenolpyruvate, is regenerated.

TABLE 8.3
Reactions of the C_4 photosynthetic carbon cycle

Enzyme	Reaction
1. Phosphoenolpyruvate (PEP) carboxylase	Phosphoenolpyruvate + HCO_3^- → oxaloacetate + P_i
2. NADP:malate dehydrogenase	Oxaloacetate + NADPH + H^+ → malate + $NADP^+$
3. Aspartate aminotransferase	Oxaloacetate + glutamate → aspartate + α-ketoglutarate
4. NAD(P) malic enzyme	Malate + $NAD(P)^+$ → pyruvate + CO_2 + NAD(P)H + H^+
5. Phosphoenolpyruvate carboxykinase	Oxaloacetate + ATP → phosphoenolpyruvate + CO_2 + ADP
6. Alanine aminotransferase	Pyruvate + glutamate ↔ alanine + α-ketoglutarate
7. Adenylate kinase	AMP + ATP → 2 ADP
8. Pyruvate–orthophosphate dikinase	Pyruvate + P_i + ATP → phosphoenolpyruvate + AMP + PP_i
9. Pyrophosphatase	PP_i + H_2O → 2 P_i

Note: P_i and PP_i stand for inorganic phosphate and pyrophosphate, respectively.

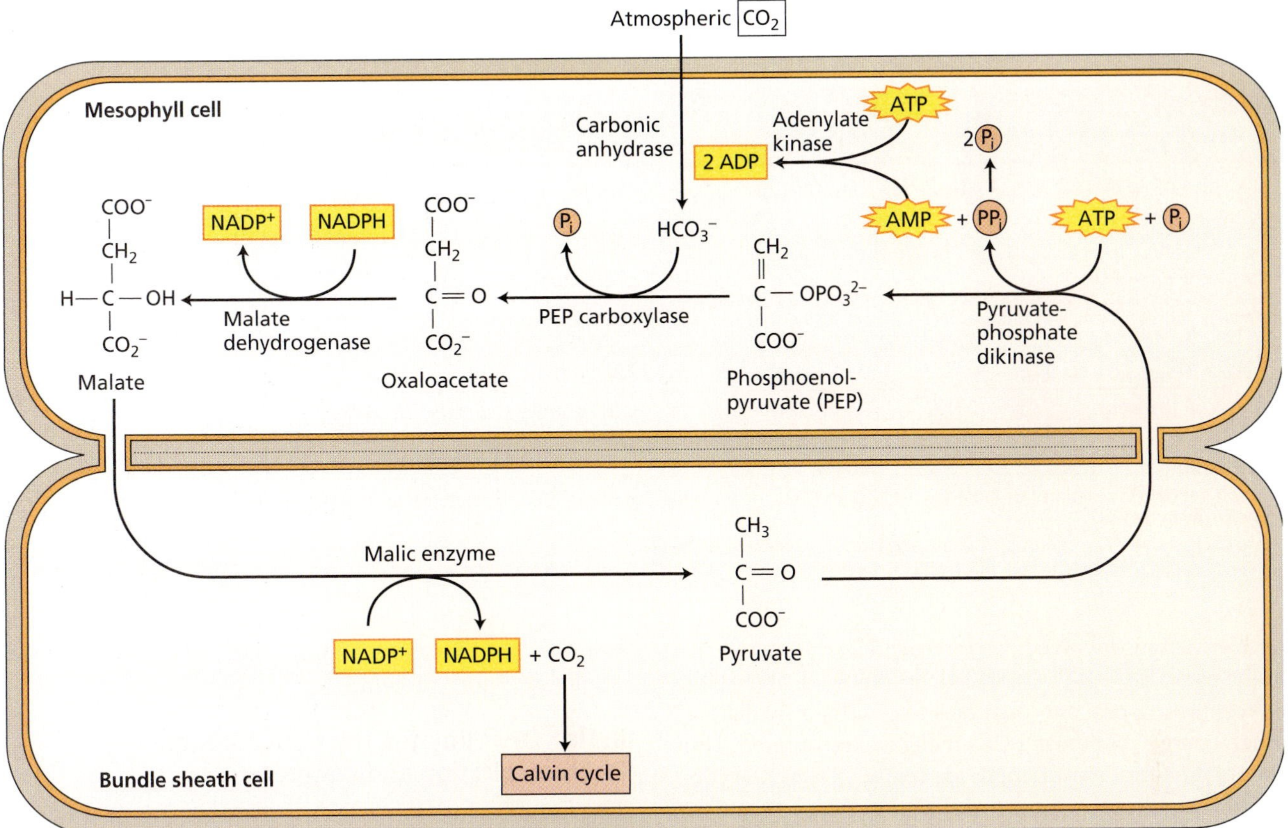

FIGURE 8.11 The C_4 photosynthetic pathway. The hydrolysis of two ATP drives the cycle in the direction of the arrows, thus pumping CO_2 from the atmosphere to the Calvin cycle of the chloroplasts from bundle sheath cells.

4. Transport of the C_3 acid (pyruvate or alanine) that is formed by the decarboxylation step back to the mesophyll cell and regeneration of the CO_2 acceptor phosphoenolpyruvate

One interesting feature of the cycle is that regeneration of the primary acceptor—phosphoenolpyruvate—consumes two "high-energy" phosphate bonds: one in the reaction catalyzed by pyruvate–orthophosphate dikinase (Table 8.3, reaction 8) and another in the conversion of PP_i to $2P_i$ catalyzed by pyrophosphatase (reaction 9; see also Figure 8.11).

Shuttling of metabolites between mesophyll and bundle sheath cells is driven by diffusion gradients along numerous plasmodesmata, and transport within the cells is regulated by concentration gradients and the operation of specialized translocators at the chloroplast envelope. The cycle thus effectively shuttles CO_2 from the atmosphere into the bundle sheath cells. This transport process generates a much higher concentration of CO_2 in the bundle sheath cells than would occur in equilibrium with the external atmosphere. This elevated concentration of CO_2 at the carboxylation site of rubisco results in suppression of the oxygenation of ribulose-1,5-bisphosphate and hence of photorespiration.

Discovered in the tropical grasses, sugarcane, and maize, the C_4 cycle is now known to occur in 16 families of both monocotyledons and dicotyledons, and it is particularly prominent in Gramineae (corn, millet, sorghum, sugarcane), Chenopodiaceae (*Atriplex*), and Cyperaceae (sedges). About 1% of all known species have C_4 metabolism (Edwards and Walker 1983).

There are three variations of the basic C_4 pathway that occur in different species (see **Web Topic 8.7**). The variations differ principally in the C_4 acid (malate or aspartate) transported into the bundle sheath cells and in the manner of decarboxylation.

The Concentration of CO_2 in Bundle Sheath Cells Has an Energy Cost

The net effect of the C_4 cycle is to convert a dilute solution of CO_2 in the mesophyll cells into a concentrated CO_2 solution in cells of the bundle sheath. Studies of a PEP carboxylase–deficient mutant of *Amaranthus edulis* clearly showed that the lack of an effective mechanism for concentrating CO_2 in the bundle sheath markedly enhances photorespiration in a C_4 plant (Dever et al. 1996).

Thermodynamics tells us that work must be done to establish and maintain the CO_2 concentration gradient in the bundle sheath (for a detailed discussion of theomodynamics, see Chapter 2 on the web site). This principle also applies to the operation of the C_4 cycle. From a summation

TABLE 8.4
Energetics of the C_4 photosynthetic carbon cycle

Phosphoenolpyruvate + H_2O + NADPH + CO_2 (mesophyll)	→	malate + $NADP^+$ + P_i (mesophyll)
Malate + $NADP^+$	→	pyruvate + NADPH + CO_2 (bundle sheath)
Pyruvate + P_i+ ATP	→	phosphoenolpyruvate + AMP + PP_i (mesophyll)
PP_i + H_2O	→	2 P_i (mesophyll)
AMP + ATP	→	2ADP
Net: CO_2 (mesophyll) + ATP + 2 H_2O	→	CO_2 (bundle sheath) + 2ADP + 2 P_i
Cost of concentrating CO_2 within the bundle sheath cell = 2 ATP per CO_2		

Note: As shown in reaction 1 of Table 8.3, the H_2O and CO_2 shown in the first line of this table actually react with phosphoenolpyruvate as HCO_3^-.
P_i and PP_i stand for inorganic phosphate and pyrophosphate, respectively.

of the reactions involved, we can calculate the energy cost to the plant (Table 8.4). The calculation shows that the CO_2-concentrating process consumes two ATP equivalents (2 "high-energy" bonds) per CO_2 molecule transported. Thus the total energy requirement for fixing CO_2 by the combined C_4 and Calvin cycles (calculated in Tables 8.4 and 8.1, respectively) is five ATP plus two NADPH per CO_2 fixed.

Because of this higher energy demand, C_4 plants photosynthesizing under nonphotorespiratory conditions (high CO_2 and low O_2) require more quanta of light per CO_2 than C_3 leaves do. In normal air, the quantum requirement of C_3 plants changes with factors that affect the balance between photosynthesis and photorespiration, such as temperature. By contrast, owing to the mechanisms built in to avoid photorespiration, the quantum requirement of C_4 plants remains relatively constant under different environmental conditions (see Figure 9.23).

Light Regulates the Activity of Key C_4 Enzymes

Light is essential for the operation of the C_4 cycle because it regulates several specific enzymes. For example, the activities of PEP carboxylase, NADP:malate dehydrogenase, and pyruvate–orthophosphate dikinase (see Table 8.3) are regulated in response to variations in photon flux density by two different processes: reduction–oxidation of thiol groups and phosphorylation–dephosphorylation.

NADP:malate dehydrogenase is regulated via the thioredoxin system of the chloroplast (see Figure 8.5). The enzyme is reduced (activated) upon illumination of leaves and is oxidized (inactivated) upon darkening. PEP carboxylase is activated by a light-dependent phosphorylation–dephosphorylation mechanism yet to be characterized.

The third regulatory member of the C_4 pathway, pyruvate–orthophosphate dikinase, is rapidly inactivated by an unusual ADP-dependent phosphorylation of the enzyme when the photon flux density drops (Burnell and Hatch 1985). Activation is accomplished by phosphorolytic cleavage of this phosphate group. Both reactions, phosphorylation and dephosphorylation, appear to be catalyzed by a single regulatory protein.

In Hot, Dry Climates, the C_4 Cycle Reduces Photorespiration and Water Loss

Two features of the C_4 cycle in C_4 plants overcome the deleterious effects of higher temperature on photosynthesis that were noted earlier. First, the affinity of PEP carboxylase for its substrate, HCO_3^-, is sufficiently high that the enzyme is saturated by HCO_3^- in equlibrium with air levels of CO_2. Furthermore, because the substrate is HCO_3^-, oxygen is not a competitor in the reaction. This high activity of PEP carboxylase enables C_4 plants to reduce the stomatal aperture and thereby conserve water while fixing CO_2 at rates equal to or greater than those of C_3 plants. The second beneficial feature is the suppression of photorespiration resulting from the concentration of CO_2 in bundle sheath cells (Marocco et al. 1998).

These features enable C_4 plants to photosynthesize more efficiently at high temperatures than C_3 plants, and they are probably the reason for the relative abundance of C_4 plants in drier, hotter climates. Depending on their natural environment, some plants show properties intermediate between strictly C_3 and C_4 species.

CO_2-CONCENTRATING MECHANISMS III: CRASSULACEAN ACID METABOLISM

A third mechanism for concentrating CO_2 at the site of rubisco is found in crassulacean acid metabolism (CAM). Despite its name, CAM is not restricted to the family Crassulaceae (*Crassula, Kalanchoe, Sedum*); it is found in numerous angiosperm families. Cacti and euphorbias are CAM plants, as well as pineapple, vanilla, and agave.

The CAM mechanism enables plants to improve water use efficiency. Typically, a CAM plant loses 50 to 100 g of water for every gram of CO_2 gained, compared with values of 250 to 300 g and 400 to 500 g for C_4 and C_3 plants, respectively (see Chapter 4). Thus, CAM plants have a competitive advantage in dry environments.

The CAM mechanism is similar in many respects to the C_4 cycle. In C_4 plants, formation of the C_4 acids in the mesophyll is spatially separated from decarboxylation of the C_4 acids and from refixation of the resulting CO_2 by the Calvin cycle in the bundle sheath. In CAM plants, formation of the C_4 acids is both temporally and spatially separated. At night, CO_2 is captured by PEP carboxylase in the cytosol, and the malate that forms from the oxaloacetate product is stored in the vacuole (Figure 8.12). During the day, the stored malate is transported to the chloroplast and decarboxylated by NADP-malic enzyme, the released CO_2 is fixed by the Calvin cycle, and the NADPH is used for converting the decarboxylated triose phosphate product to starch.

The Stomata of CAM Plants Open at Night and Close during the Day

CAM plants such as cacti achieve their high water use efficiency by opening their stomata during the cool, desert nights and closing them during the hot, dry days. Closing the stomata during the day minimizes water loss, but because H_2O and CO_2 share the same diffusion pathway, CO_2 must then be taken up at night.

CO_2 is incorporated via carboxylation of phosphoenolpyruvate to oxaloacetate, which is then reduced to malate. The malate accumulates and is stored in the large vacuoles that are a typical, but not obligatory, anatomic feature of the leaf cells of CAM plants (see Figure 8.12). The accumulation of substantial amounts of malic acid, equivalent to the amount of CO_2 assimilated at night, has long been recognized as a nocturnal acidification of the leaf (Bonner and Bonner 1948).

With the onset of day, the stomata close, preventing loss of water and further uptake of CO_2. The leaf cells deacidify as the reserves of vacuolar malic acid are consumed. Decarboxylation is usually achieved by the action of NADP-malic enzyme on malate (Drincovich et al. 2001). Because the stomata are closed, the internally released CO_2 cannot escape from the leaf and instead is fixed and converted to carbohydrate by the Calvin cycle.

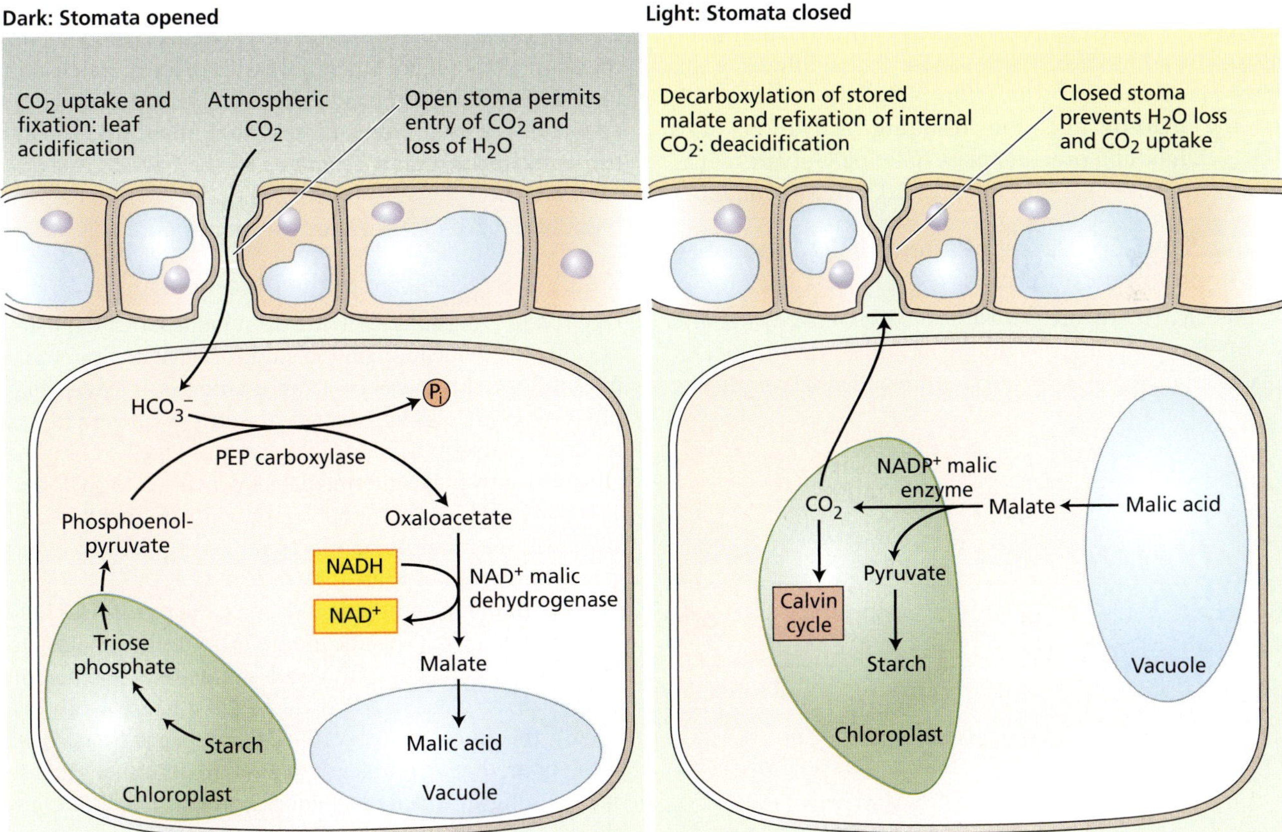

FIGURE 8.12 Crassulacean acid metabolism (CAM). Temporal separation of CO_2 uptake from photosynthetic reactions: CO_2 uptake and fixation take place at night, and decarboxylation and refixation of the internally released CO_2 occur during the day. The adaptive advantage of CAM is the reduction of water loss by transpiration, achieved by the stomatal opening during the night.

The elevated internal concentration of CO_2 effectively suppresses the photorespiratory oxygenation of ribulose bisphosphate and favors carboxylation. The C_3 acid resulting from the decarboxylation is thought to be converted first to triose phosphate and then to starch or sucrose, thus regenerating the source of the original carbon acceptor.

Phosphorylation Regulates the Activity of PEP Carboxylase in C_4 and CAM Plants

The CAM mechanism that we have outlined in this discussion requires separation of the initial carboxylation from the subsequent decarboxylation, to avoid a futile cycle. In addition to the spatial and temporal separation exhibited by C_4 and CAM plants, respectively, a futile cycle is avoided by the regulation of PEP carboxylase (Figure 8.13). In C_4 plants the carboxylase is "switched on," or active, during the day and in CAM plants during the night. In both C_4 and CAM plants, PEP carboxylase is inhibited by malate and activated by glucose-6-phosphate (see **Web Essay 8.1** for a detailed discussion of the regulation of PEP carboxylase).

Phosphorylation of a single serine residue of the CAM enzyme diminishes the malate inhibition and enhances the action of glucose-6-phosphate so that the enzyme becomes catalytically more active (Chollet et al. 1996; Vidal and Chollet 1997) (see Figure 8.13). The phosphorylation is catalyzed by a PEP carboxylase-kinase. The synthesis of this kinase is stimulated by the efflux of Ca^{2+} from the vacuole to the cytosol and the resulting activation of a Ca^{2+}/calmodulin protein kinase (Giglioli-Guivarc'h et al. 1996; Coursol et al. 2000; Nimmo 2000; Bakrim et al. 2001).

Some Plants Adjust Their Pattern of CO_2 Uptake to Environmental Conditions

Plants have many mechanisms that maximize water and CO_2 supply during development and reproduction. C_3 plants regulate the stomatal aperture of their leaves during the day, and stomata close during the night. C_4 and CAM plants utilize PEP carboxylase to fix CO_2, and they separate that enzyme from rubisco either spatially (C_4 plants) or temporally (CAM plants).

Some CAM plants show longer-term regulation and are able to adjust their pattern of CO_2 uptake to environmental conditions. Facultative CAM plants such as the ice plant (*Mesembryanthemum crystallinum*) carry on C_3 metabolism under unstressed conditions, and they shift to CAM in response to heat, water, or salt stress. This form of regulation requires the expression of numerous CAM genes in response to stress signals (Adams et al. 1998; Cushman 2001).

In aquatic environments, cyanobacteria and green algae have abundant water but find low CO_2 concentrations in their surroundings and actively concentrate inorganic CO_2 intracellularly. In diatoms, which abound in the phytoplankton, a CO_2-concentrating mechanism operates simultaneously with a C_4 pathway (Reinfelder et al. 2000). Diatoms are a fine example of photosynthetic organisms that have the capacity to use different CO_2-concentrating mechanisms in response to environmental fluctuations.

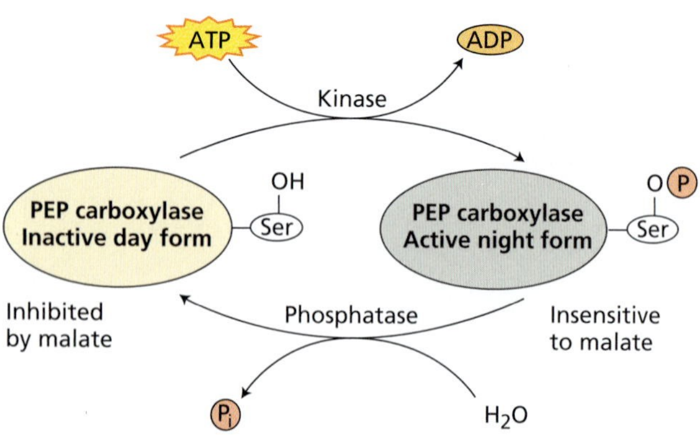

FIGURE 8.13 Diurnal regulation of CAM phosphoenolpyruvate (PEP) carboxylase. Phosphorylation of the serine residue (Ser-OP) yields a form of the enzyme which is active during the night and relatively insensitive to malate. During the day, dephosphorylation of the serine (Ser-OH) gives a form of the enzyme which is inhibited by malate.

SYNTHESIS OF STARCH AND SUCROSE

In most species, sucrose is the principal form of carbohydrate translocated throughout the plant by the phloem. Starch is an insoluble stable carbohydrate reserve that is present in almost all plants. Both starch and sucrose are synthesized from the triose phosphate that is generated by the Calvin cycle (see Table 8.1) (Beck and Ziegler 1989). The pathways for the synthesis of starch and sucrose are shown in Figure 8.14.

Starch Is Synthesized in the Chloroplast

Electron micrographs showing prominent starch deposits, as well as enzyme localization studies, leave no doubt that the chloroplast is the site of starch synthesis in leaves (Figure 8.15). Starch is synthesized from triose phosphate via fructose-1,6-bisphosphate (Table 8.5 and Figure 8.14). The glucose-1-phosphate intermediate is converted to ADP-glucose via ADP-glucose pyrophosphorylase (Figure 8.14 and Table 8.5, reaction 5) in a reaction that requires ATP and generates pyrophosphate (PP_i, or $H_2P_2O_7^{2-}$).

As in many biosynthetic reactions, the pyrophosphate is hydrolyzed via a specific inorganic pyrophosphatase to two orthophosphate (P_i) molecules (Table 8.5, reaction 6), thereby driving reaction 5 toward ADP-glucose synthesis. Finally, the glucose moiety of ADP-glucose is transferred to the nonreducing end (carbon 4) of the terminal glucose of a growing starch chain (Table 8.5, reaction 7), thus completing the reaction sequence.

Sucrose Is Synthesized in the Cytosol

The site of sucrose synthesis has been studied by cell fractionation, in which the organelles are isolated and separated from one another. Enzyme analyses have shown that sucrose is synthesized in the cytosol from triose phosphates

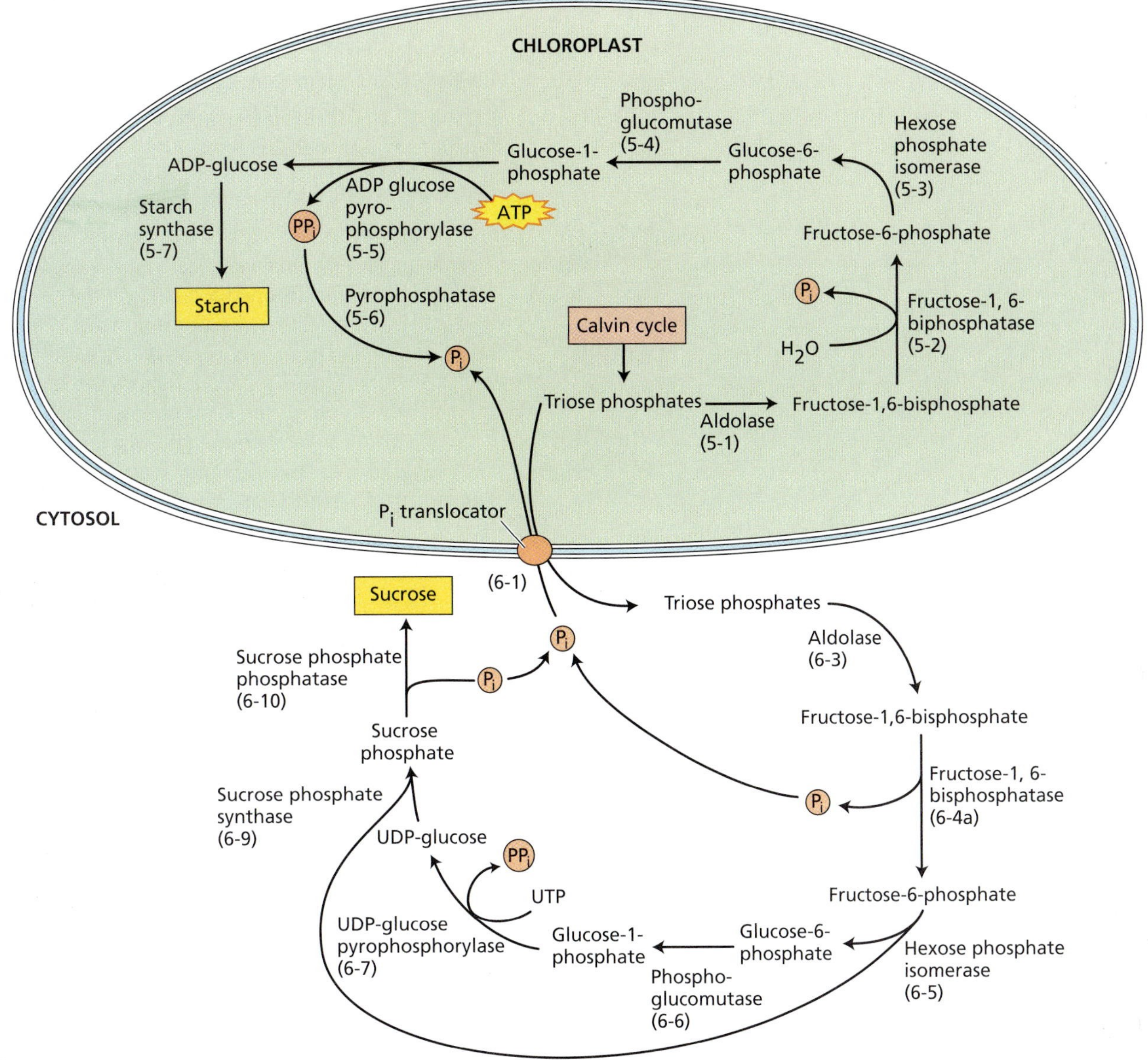

FIGURE 8.14 The syntheses of starch and sucrose are competing processes that occur in the chloroplast and the cytosol, respectively. When the cytosolic P_i concentration is high, chloroplast triose phosphate is exported to the cytosol via the P_i in exchange for P_i, and sucrose is synthesized. When the cytosolic P_i concentration is low, triose phosphate is retained within the chloroplast, and starch is synthesized. The numbers facing the arrows are keyed to Tables 8.5 and 8.6.

by a pathway similar to that of starch—that is, by way of fructose-1,6-bisphosphate and glucose-1-phosphate (Figure 8.14 and Table 8.6, reactions 2–6).

In sucrose synthesis, the glucose-1-phosphate is converted to UDP-glucose via a specific UDP-glucose pyrophosphorylase (Table 8.6, reaction 7) that is analogous to the ADP-glucose pyrophosphorylase of chloroplasts. At this stage, two consecutive reactions complete the synthesis of sucrose (Huber and Huber 1996). First, sucrose-6-phosphate synthase catalyzes the reaction of UDP-glucose with fructose-6-phosphate to yield sucrose-6-phosphate and UDP (Table 8.6, reaction 9). Second, the sucrose-6-phosphate phosphatase (phosphohydrolase) cleaves the phosphate from sucrose-6-phosphate, yielding sucrose (Table 8.6, reaction 10). The latter reaction, which is essentially irreversible, pulls the former in the direction of sucrose synthesis.

As in starch synthesis, the pyrophosphate formed in the reaction catalyzed by UDP-glucose pyrophosphorylase (Table 8.6, reaction 7) is hydrolyzed, but not immediately as in the chloroplasts. Because of the absence of an inorganic pyrophosphatase, the pyrophosphate can be used by other enzymes, in transphosphorylation reactions. One example is fructose-6-phosphate phosphotransferase, an enzyme that catalyzes a reaction like the one catalyzed by phosphofructokinase (Table 8.6, reaction 4b) except that pyrophosphate replaces ATP as the phosphoryl donor.

A comparison of the reactions in Tables 8.5 and 8.6 (as illustrated in Figure 8.14) reveals that the conversion of triose phosphates to glucose-1-phosphate in the pathways

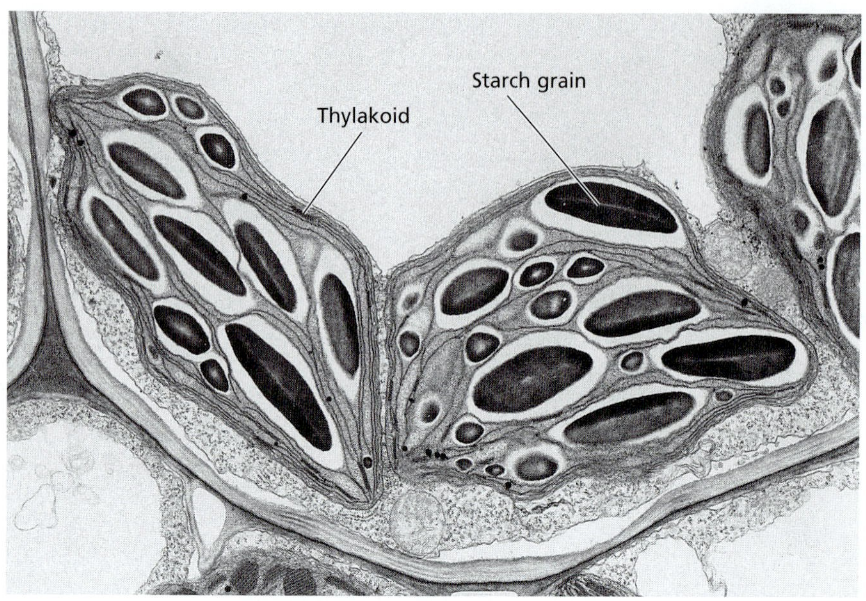

FIGURE 8.15 Electron micrograph of a bundle sheath cell from maize, showing the starch grains in the chloroplasts. (15,800×) (Photo by S. E. Frederick, courtesy of E. H. Newcomb.)

leading to the synthesis of starch and sucrose have several steps in common. However, these pathways utilize isozymes (different forms of enzymes catalyzing the same reaction) that are unique to the chloroplast or cytosol.

The isozymes show markedly different properties. For example, the chloroplastic fructose-1,6-bisphosphatase is regulated by the thioredoxin system but not by fructose-2,6-bisphosphate and AMP. Conversely, the cytosolic form of the enzyme is regulated by fructose-2,6-bisphosphate (see the next section), is sensitive to AMP especially in the presence of fructose-2,6-bisphosphate, and is unaffected by thioredoxin.

Aside from the cytosolic fructose-1,6-bisphosphatase, sucrose synthesis is regulated at the level of sucrose phosphate synthase, an allosteric enzyme that is activated by glucose-6-phosphate and inhibited by orthophosphate. The enzyme is inactivated in the dark by phosphorylation of a specific serine residue via a protein kinase and activated in the light by dephosphorylation via a protein phosphatase. Glucose-6-phosphate inhibits the kinase, and P_i inhibits the phosphatase.

The recent purification and cloning of sucrose-6-phosphate phosphatase from rice leaves (Lund et al. 2000) is providing new information on the molecular and functional properties of this enzyme. These studies indicate that sucrose-6-phosphate synthase and sucrose-6-phosphatase exist as a supramolecular complex showing an enzymatic activity that is higher than that of the isolated constituent enzymes (Salerno et al. 1996). This noncovalent interaction of the two enzymes involved in the last two steps of sucrose synthesis points to a novel regulatory feature of carbohydrate metabolism in plants.

The Syntheses of Sucrose and Starch Are Competing Reactions

The relative concentrations of orthophosphate and triose phosphate are major factors that control whether photosynthetically fixed carbon is partitioned as starch in the chloroplast or as sucrose in the cytosol. The two compartments communicate with one another via the phosphate/triose phosphate translocator, also called the phosphate translocator (see Table 8.6, reaction 1), a strict stoichiometric antiporter.

The phosphate translocator catalyzes the movement of orthophosphate and triose phosphate in opposite directions between chloroplast and cytosol. A low concentration of orthophosphate in the cytosol limits the export of triose phosphate from the chloroplast through the translocator, thereby promoting the synthesis of starch. Conversely, an abundance of orthophosphate in the cytosol inhibits starch synthesis within the chloroplast and promotes the export of triose phosphate into the cytosol, where it is converted to sucrose.

Orthophosphate and triose phosphate control the activity of several regulatory enzymes in the sucrose and starch biosynthetic pathways. The chloroplast enzyme ADP-glucose pyrophosphorylase (see Table 8.5, reaction 5) is the key enzyme that regulates the synthesis of starch from glucose-1-phosphate. This enzyme is stimulated by 3-phosphoglycerate and inhibited by orthophosphate. A high concentration ratio of 3-phosphoglycerate to orthophosphate is typically found in illuminated chloroplasts that are actively synthesizing starch. Reciprocal conditions prevail in the dark.

Fructose-2,6-bisphosphate is a key control molecule that allows increased synthesis of sucrose in the light and decreased synthesis in the dark. It is found in the cytosol in minute concentrations, and it exerts a regulatory effect on the cytosolic interconversion of fructose-1,6-bisphosphate and fructose-6-phosphate (Huber 1986; Stitt 1990):

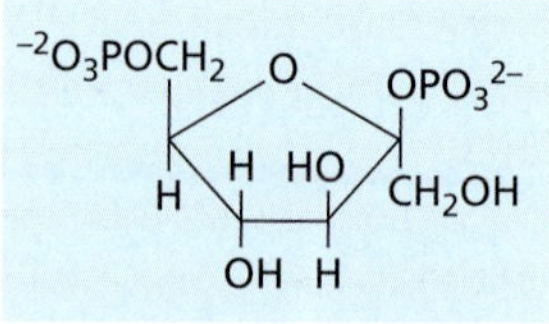

Fructose-2,6-bisphosphate (a regulatory metabolite)

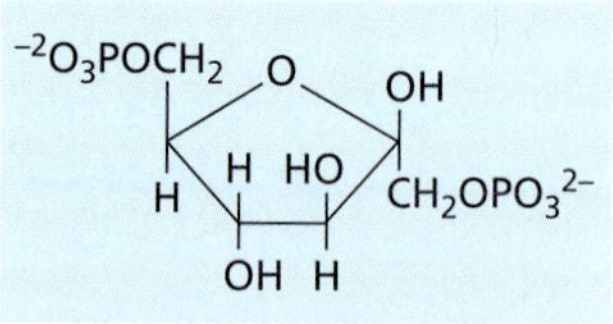

Fructose-1,6-bisphosphate (an intermediary metabolite)

TABLE 8.5
Reactions of starch synthesis from triose phosphate in chloroplasts

1. *Fructose-1,6,bisphosphate aldolase*
   Dihydroxyacetone-3-phosphate + glyceraldehyde-3-phosphate→ fructose-1,6-bisphosphate

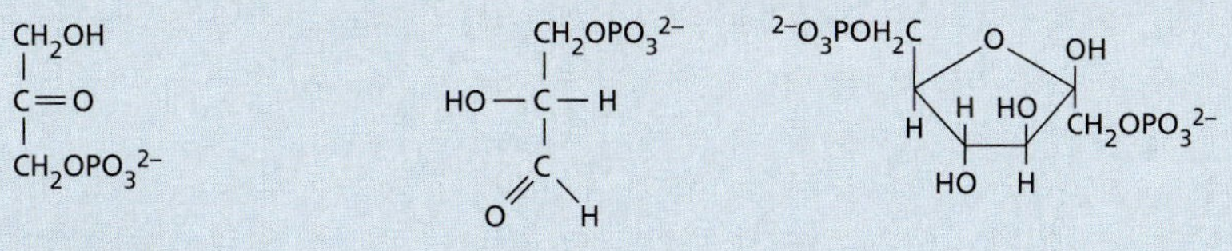

2. *Fructose-1,6-bisphosphatase*
   Fructose-1,6-bisphosphate + H_2O → fructose-6-phosphate + P_i

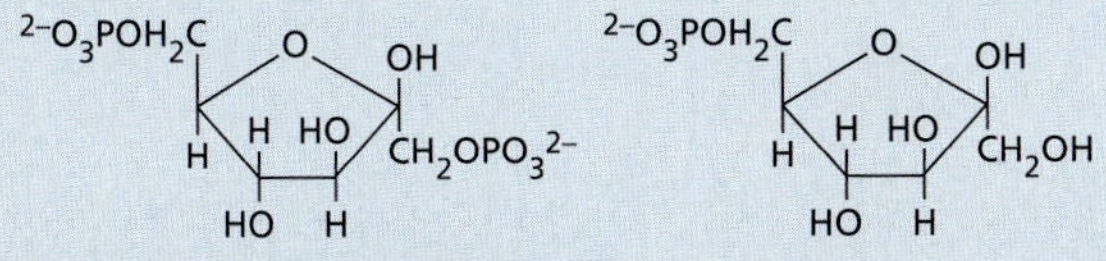

3. *Hexose phosphate isomerase*
   Fructose-6-phosphate → glucose-6-phosphate

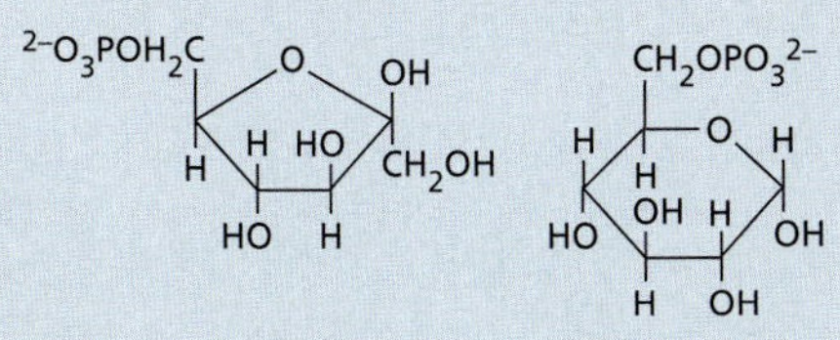

4. *Phosphoglucomutase*
   Glucose-6-phosphate → glucose-1-phosphate

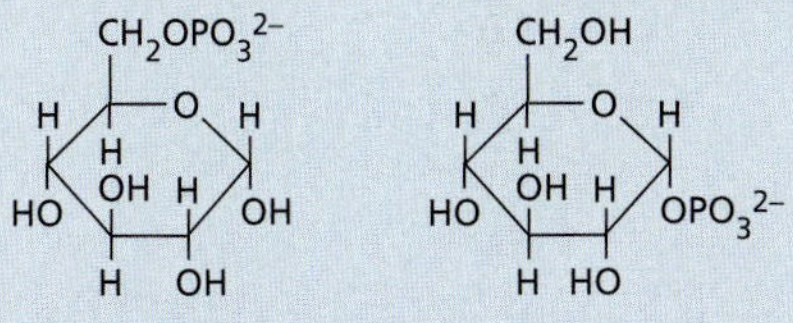

5. *ADP-glucose pyrophosphorylase*
   Glucose-1-phosphate + ATP → ADP-glucose + PP_i

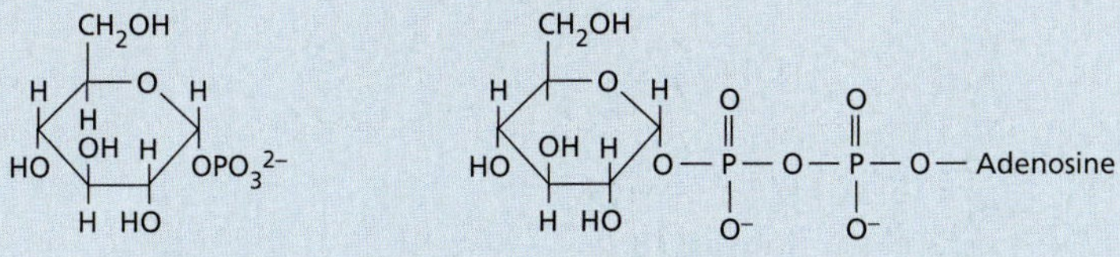

6. *Pyrophosphatase*

   $PP_i + H_2O \rightarrow 2\ P_i + 2H^+$

7. *Starch synthase*
   ADP-glucose + $(1,4\text{-}\alpha\text{-D-glucosyl})_n$ → ADP + $(1,4\text{-}\alpha\text{-D-glucosyl})_{n+1}$

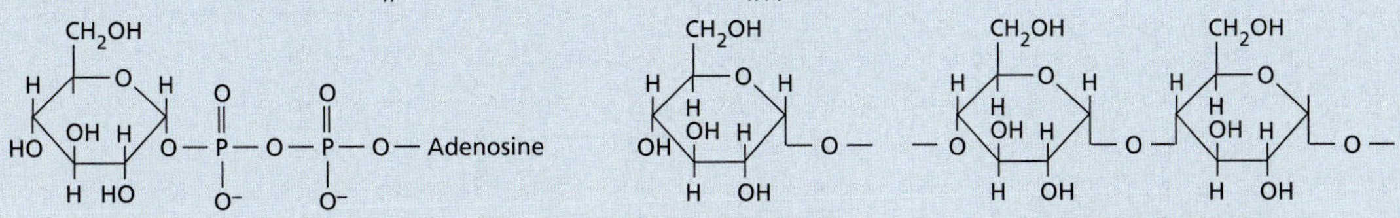

   Nonreducing end of a starch chain with n residues

   Elongated starch with $n + 1$ residues

Note: Reaction 6 is irreversible and "pulls" the preceding reaction to the right.
P_i and PP_i stand for inorganic phosphate and pyrophosphate, respectively.

TABLE 8.6
Reactions of sucrose synthesis from triose phosphate in the cytosol

1. *Phosphate/triose phosphate translocator*
Triose phosphate (chloroplast) + P_i (cytosol) → triose phosphate (cytosol) + P_i (chloroplast)

2. *Triose phosphate isomerase*
Dihydroxyacetone-3-phosphate → glyceraldehyde-3-phosphate

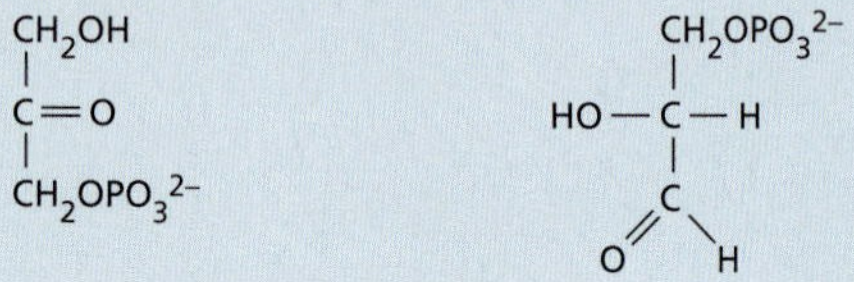

3. *Fructose-1,6-bisphosphate aldolase*
Dihydroxyacetone-3-phosphate + glyceraldehyde-3-phosphate → fructose-1,6-bisphosphate

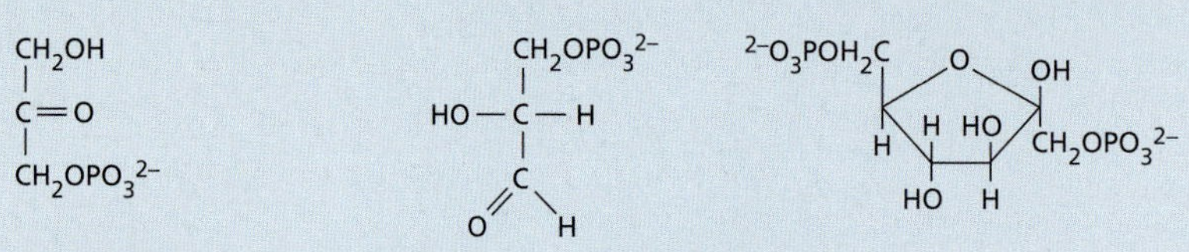

4a. *Fructose-1,6-phosphatase*
Fructose-1,6-bisphosphate + H_2O → fructose-6-phosphate + P_i

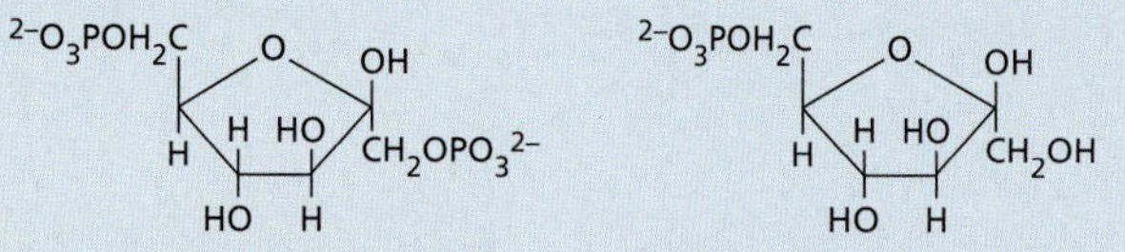

4b. *PP_i-linked phosphofructokinase*
Fructose-6-phosphate + PP_i → fructose-1,6-bisphosphate + P_i

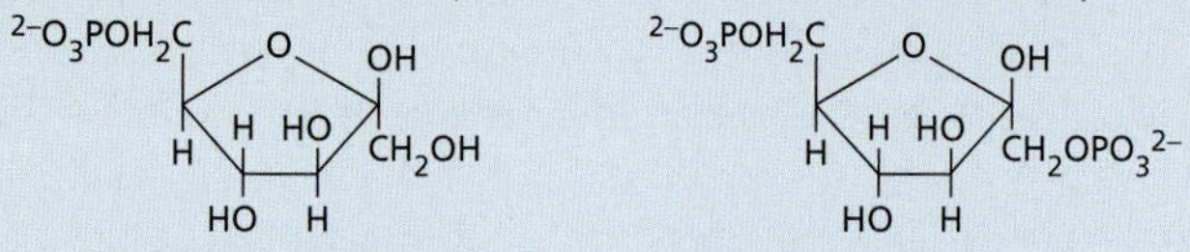

5. *Hexose phosphate isomerase*
Fructose-6-phosphate → glucose-6-phosphate

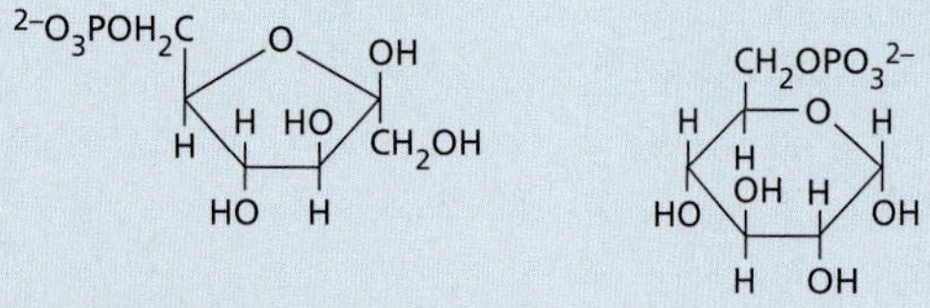

6. *Phosphoglucomutase*
Glucose-6-phosphate → glucose-1-phosphate

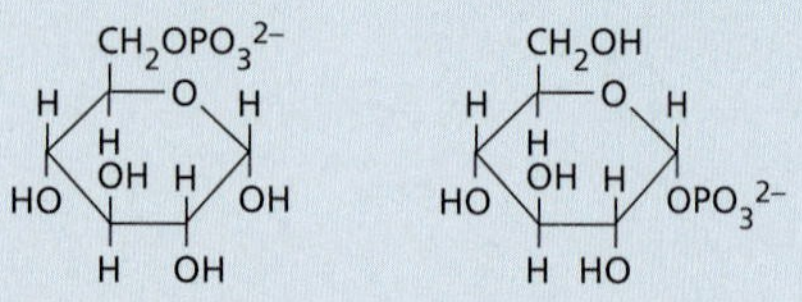

7. *UDP-glucose pyrophosphorylase*
Glucose-1-phosphate + UTP → UDP-glucose + PP_i

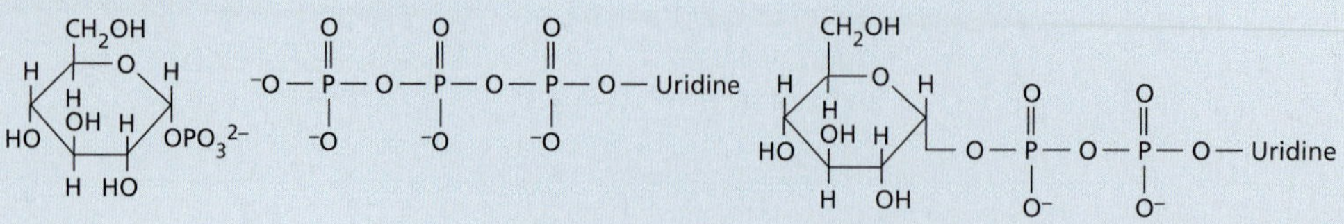

TABLE 8.6 (continued)
Reactions of sucrose synthesis from triose phosphate in the cytosol

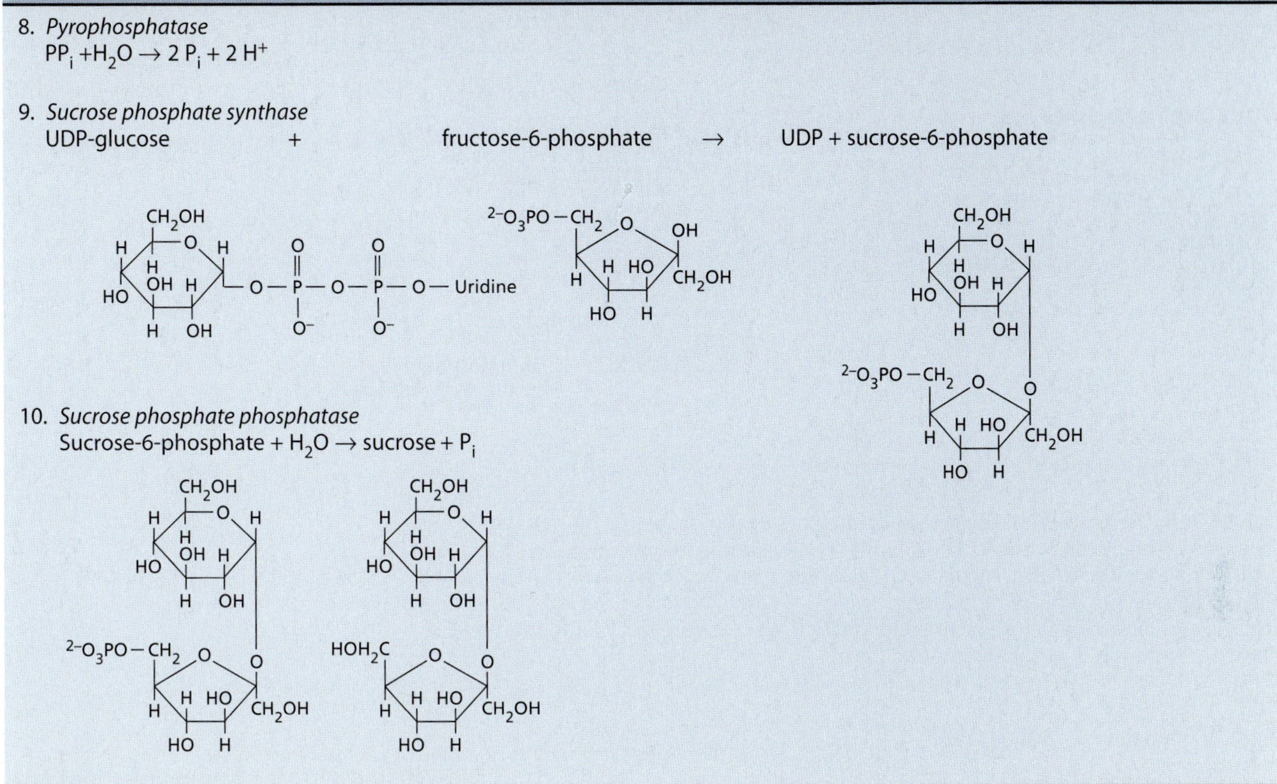

8. *Pyrophosphatase*
$PP_i + H_2O \rightarrow 2\ P_i + 2\ H^+$

9. *Sucrose phosphate synthase*
UDP-glucose + fructose-6-phosphate → UDP + sucrose-6-phosphate

10. *Sucrose phosphate phosphatase*
Sucrose-6-phosphate + $H_2O \rightarrow$ sucrose + P_i

Note: Reaction 1 takes place on the chloroplast inner envelope membrane. Reactions 2 through 10 take place in the cytosol. Reaction 8 is irreversible and "pulls" the preceding reaction to the right.
P_i and PP_i stand for inorganic phosphate and pyrophosphate, respectively.

Increased cytosolic fructose-2,6-bisphosphate is associated with decreased rates of sucrose synthesis because fructose-2,6-bisphosphate is a powerful inhibitor of cytosolic fructose-1,6-bisphosphatase (see Table 8.6, reaction 4a) and an activator of the pyrophosphate-dependent (PP_i-linked) phosphofructokinase (reaction 4b). But what, in turn, controls the cytosolic concentration of fructose-2,6-bisphosphate?

Fructose-2,6-bisphosphate is synthesized from fructose-6-phosphate by a special fructose-6-phosphate 2-kinase (not to be confused with the fructose-6-phosphate 1-kinase of glycolysis) and is degraded specifically by fructose-2,6-bisphosphatase (not to be confused with fructose-1,6-bisphosphatase of the Calvin cycle). Recent evidence suggests that, as in animal cells, both plant activities reside on a single polypeptide chain.

The kinase and phosphatase activities are controlled by orthophosphate and triose phosphate. Orthophosphate stimulates fructose-6-phosphate 2-kinase and inhibits fructose-2,6-bisphosphatase; triose phosphate inhibits the 2-kinase (Figure 8.16). Consequently, a low cytosolic ratio of triose phosphate to orthophosphate promotes the formation of fructose-2,6-bisphosphate, which in turn inhibits the hydrolysis of cytosolic fructose-1,6-bisphosphate and slows the rate of sucrose synthesis. A high cytosolic ratio of triose phosphate to orthophosphate has the opposite effect.

Light regulates the concentration of these activators and inhibitors through the reactions associated with photosynthesis and thereby controls the concentration of fructose-2,6-bisphosphate in the cytosol. The glycolytic enzyme phosphofructokinase also functions in the conversion of fructose-6-phosphate to fructose-1,6-bisphosphate, but in plants it is not appreciably affected by fructose-2,6-bisphosphate.

The activity of phosphofructokinase in plants appears to be regulated by the relative concentrations of ATP, ADP, and AMP. The remarkable plasticity of plants was once again illustrated by recent gene deletion experiments with transformed tobacco plants. This experiment shows that the transformed plants can grow without a functional pyrophosphate-dependent fructose-6-phosphate kinase enzyme. In this case the conversion of fructose-6-phosphate to fructose-1,6-bisphosphate is apparently catalyzed exclusively by phosphofructokinase (Paul et al. 1995).

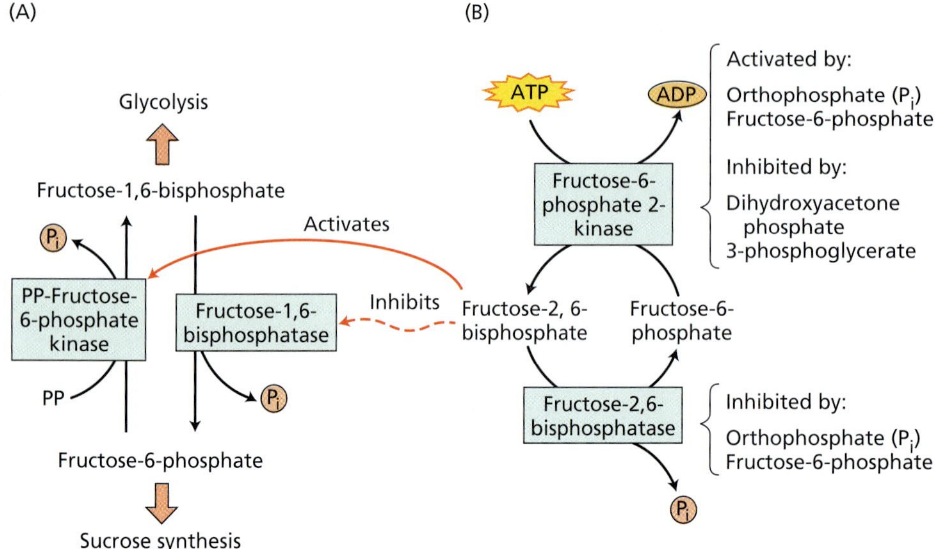

FIGURE 8.16 Regulation of the cytosolic interconversion of fructose-6-phosphate and fructose-1,6-bisphosphate. (A) The key metabolites in the allocation between glycolysis and sucrose synthesis. The regulatory metabolite fructose 2,6-bisphosphate regulates the interconversion by inhibiting the phosphatase and activating the kinase, as shown. (B) The synthesis of fructose-2,6-bisphosphate itself is under strict regulation by the activators and inhibitors shown in the figure.

SUMMARY

The reduction of CO_2 to carbohydrate via the carbon-linked reactions of photosynthesis is coupled to the consumption of NADPH and ATP synthesized by the light reactions of thylakoid membranes. Photosynthetic eukaryotes reduce CO_2 via the Calvin cycle that takes place in the stroma, or soluble phase, of chloroplasts. Here, CO_2 and water are combined with ribulose-1,5-bisphosphate to form two molecules of 3-phosphoglycerate, which are reduced and converted to carbohydrate. The continued operation of the cycle is ensured by the regeneration of ribulose-1,5-bisphosphate. The Calvin cycle consumes two molecules of NADPH and three molecules of ATP for every CO_2 fixed and, provided these substrates, has a thermodynamic efficiency close to 90%.

Several light-dependent systems act jointly to regulate the Calvin cycle: changes in ions (Mg^{2+} and H^+), effector metabolites (enzyme substrates), and protein-mediated systems (rubisco activase, ferredoxin–thioredoxin system).

The ferredoxin–thioredoxin control system plays a versatile role by linking light to the regulation of other chloroplast processes, such as carbohydrate breakdown, photophosphorylation, fatty acid biosynthesis, and mRNA translation. Control of these reactions by light separates opposing biosynthetic from degradative processes and thereby minimizes the waste of resources that would occur if the processes operated concurrently.

Rubisco, the enzyme that catalyzes the carboxylation of ribulose-1,5-bisphosphate, also acts as an oxygenase. In both cases the enzyme must be carbamylated to be fully active. The carboxylation and oxygenation reactions take place at the active site of rubisco. When reacting with oxygen, rubisco produces 2-phosphoglycolate and 3-phosphoglycerate from ribulose-1,5-bisphosphate rather than two 3-phosphoglycerates as with CO_2, thereby decreasing the efficiency of photosynthesis.

The C_2 oxidative photosynthetic carbon cycle rescues the carbon lost as 2-phosphoglycolate by rubisco oxygenase activity. The dissipative effects of photorespiration are avoided in some plants by mechanisms that concentrate CO_2 at the carboxylation sites in the chloroplast. These mechanisms include a C_4 photosynthetic carbon cycle, CAM metabolism, and "CO_2 pumps" of algae and cyanobacteria.

The carbohydrates synthesized by the Calvin cycle are converted into storage forms of energy and carbon: sucrose and starch. Sucrose, the transportable form of carbon and energy in most plants, is synthesized in the cytosol, and its synthesis is regulated by phosphorylation of sucrose phosphate synthase. Starch is synthesized in the chloroplast. The balance between the biosynthetic pathways for sucrose and starch is determined by the relative concentrations of metabolite effectors (orthophosphate, fructose-6-phosphate, 3-phosphoglycerate, and dihydroxyacetone phosphate).

These metabolite effectors function in the cytosol by way of the enzymes synthesizing and degrading fructose-2,6-bisphosphate, the regulatory metabolite that plays a primary role in controlling the partitioning of photosynthetically fixed carbon between sucrose and starch. Two of these effectors, 3-phosphoglycerate and orthophosphate, also act on starch synthesis in the chloroplast by allosterically regulating the activity of ADP-glucose pyrophosphorylase. In this way the synthesis of starch from triose phosphates during the day can be separated from its breakdown, which is required to provide energy to the plant at night.

Web Material

Web Topics

8.1 How the Calvin Cycle Was Elucidated

Experiments carried out in the 1950s led to the discovery of the path of CO_2 fixation.

8.2 Rubisco: A Model Enzyme for Studying Structure and Function

As the most abundant enzyme on Earth, rubisco was obtained in quantities sufficient for elucidating its structure and catalytic properties.

8.3 Carbon Dioxide: Some Important Physicochemical Properties

Plants have adapted to the properties of CO_2 by altering the reactions catalyzing its fixation.

8.4 Thioredoxins

First found to regulate chloroplast enzymes, thioredoxins are now known to play a regulatory role in all types of cells.

8.5 Rubisco Activase

Rubisco is unique among Calvin cycle enzymes in its regulation by a specific protein, rubisco activase.

8.6 Operation of the C_2 Oxidative Photosynthetic Carbon Cycle

The enzymes of the C_2 oxidative photosynthetic carbon cycle are localized in three different organelles.

8.7 Three Variations of C_4 Metabolism

Certain reactions of the C_4 photosynthetic pathway differ among plant species.

Web Essay

8.1 Modulation of Phosphoenolpyruvate Carboxylase in C_4 and CAM Plants

The CO_2-fixing enzyme, phosphoenolpyruvate carboxylase is regulated differently in C_4 and CAM species.

Chapter References

Adams, P., Nelson, D. E., Yamada, S., Chmara, W., Jensen, R. G., Bohnert, H. J., and Griffiths, H. (1998) Tansley Review No. 97; Growth and development of *Mesembryanthemum crystallinum*. *New Phytol.* 138:171–190.

Bakrim, N., Brulfert, J., Vidal, J., and Chollet, R. (2001) Phosphoenolpyruvate carboxylase kinase is controlled by a similar signaling cascade in CAM and C_4 plants. *Biochem. Biophys. Res. Commun.* 286: 1158–1162.

Beck, E., and Ziegler, P. (1989) Biosynthesis and degradation of starch in higher plants. *Annu. Rev. Plant Physiol. Plant Mol. Biol.* 40: 95–118.

Besse, I., and Buchanan, B. B. (1997) Thioredoxin-linked plant and animal processes: The new generation. *Bot. Bull. Acad. Sinica* 38: 1–11.

Bonner, W., and Bonner, J. (1948) The role of carbon dioxide in acid formation by succulent plants. *Am. J. Bot.* 35: 113–117.

Buchanan, B. B. (1980) Role of light in the regulation of chloroplast enzymes. *Annu. Rev. Plant Phsyiol.* 31: 341–394.

Burnell, J. N., and Hatch, M. D. (1985) Light–dark modulation of leaf pyruvate, P_i dikinase. *Trends Biochem. Sci.* 10: 288–291.

Chollet, R., Vidal, J., and O'Leary, M. H. (1996) Phosphoenolpyruvate carboxylase: A ubiquitous, highly regulated enzyme in plants. *Annu. Rev. Plant Physiol. Plant Mol. Biol.* 47: 273–298.

Coursol, S., Giglioli-Guivarc'h, N., Vidal, J., and Pierre J.-N. (2000) An increase in the phosphoinositide-specific phospholipase C activity precedes induction of C_4 phosphoenolpyruvate carboxylase phosphorylation in illuminated and NH_4Cl-treated protoplasts from *Digitaria sanguinalis*. *Plant J.* 23: 497–506.

Craig, S., and Goodchild, D. J. (1977) Leaf ultrastructure of *Triodia irritans*: A C_4 grass possessing an unusual arrangement of photosynthetic tissues. *Aust. J. Bot.* 25: 277–290.

Cushman, J. C. (2001) Crassulacean acid metabolism: A plastic photosynthetic adaptation to arid environments. *Plant Physiol.* 127: 1439–1448.

Dai, S., Schwendtmayer, C., Schürmann, P., Ramaswamy, S., and Eklund, H. (2000) Redox signaling in chloroplasts: Cleavage of disulfides by an iron-sulfur cluster. *Science* 287: 655–658.

Dever, L. V., Bailey, K. J., Lacuesta, M., Leegood, R. C., and Lea P. J. (1996) The isolation and characterization of mutants of the C_4 plant *Amaranthus edulis*. *Comp. Rend. Acad. Sci., III.* 919–959.

Drincovich, M. F., Casati, P., and Andreo, C. S. (2001) NADP-malic enzyme from plants: A ubiquitous enzyme involved in different metabolic pathways. *FEBS Lett.* 490: 1–6.

Edwards, G. E., and Walker, D. (1983) *C_3, C_4: Mechanisms and Cellular and Environmental Regulation of Photosynthesis.* University of California Press, Berkeley.

Flügge, U. I., and Heldt,. H. W. (1991) Metabolite translocators of the chloroplast envelope. *Annu. Rev. Plant Physiol. Plant Mol. Biol.* 42: 129–144.

Frederick, S. E., and Newcomb, E. H. (1969) Cytochemical localization of catalase in leaf microbodies (peroxisomes). *J. Cell Biol.* 43: 343–353.

Giglioli-Guivarc'h, N., Pierre, J.-N., Brown, S., Chollet, R., Vidal, J., and Gadal, P. (1996) The light-dependent transduction pathway controlling the regulatory phosphorylation of C_4 phosphoenolpyruvate carboxylase in protoplasts from *Digitaria sanguinalis*. *Plant Cell* 8: 573–586.

Hatch, M. D., and Slack, C. R. (1966) Photosynthesis by sugarcane leaves. A new carboxylation reaction and the pathway of sugar formation. *Biochem. J.* 101: 103–111.

Heldt, H. W. (1979) Light-dependent changes of stromal H^+ and Mg^{2+} concentrations controlling CO_2 fixation. In *Photosynthesis II (Encyclopedia of Plant Physiology,* New Series, vol. 6) M. Gibbs and E. Latzko, eds. Springer, Berlin, pp. 202–207.

Huber, S. C. (1986) Fructose-2,6-bisphosphate as a regulatory metabolite in plants. *Annu. Rev. Plant Physiol.* 37: 233–246.

Huber, S. C., and Huber, J. L. (1996) Role and regulation of sucrose-phosphate synthase in higher plants. *Annu. Rev. Plant Physiol. Plant Mol. Biol.* 47: 431–444.

Kozaki, A., and Takeba, G. (1996) Photorespiration protects C_3 plants from photooxidation. *Nature* 384: 557–560.

Ku, S. B., and Edwards, G. E. (1978) Oxygen inhibition of photosynthesis. III. Temperature dependence of quantum yield and its relation to O_2/CO_2 solubility ratio. *Planta* 140: 1–6.

Leegood, R. C. Lea, P. J., Adcock, M. D., and Haeusler, R. D. (1995) The regulation and control of photorespiration. *J. Exp. Bot.* 46: 1397–1414.

Lorimer, G. H. (1981) The carboxylation and oxygenation of ribulose 1,5-bisphosphate: The primary events in photosynthesis and photorespiration. *Annu. Rev. Plant Physiol.* 32 349–383.

Lorimer G. H. (1983) Ribulose-1,5-bisphosphate oxygenase. *Annu. Rev. Biochem.* 52: 507–535.

Lund, J. E., Ashton, A. R., Hatch, M. D., and Heldt, H. W. (2000) Purification, molecular cloning, and sequence analysis of sucrose-6^F-phosphate phosphohydrolase from plants. *Proc. Natl. Acad. Sci. USA* 97: 12914–12919.

Lüttge, U., and Higinbotham, N. (1979) *Transport in Plants.* Springer-Verlag, New York.

Maier, R. M., Neckermann, K., Igloi, G. L., and Koessel, H. (1995) Complete sequence of the maize chloroplast genome: Gene content, hotspots of divergence and fine tuning of genetic information by transcript editing. *J. Mol. Biol.* 251: 614–628.

Maroco, J. P., Ku, M. S. B., Lea P. J., Dever, L. V., Leegood, R. C., Furbank, R. T., and Edwards, G. E. (1998) Oxygen requirement and inhibition of C_4 photosynthesis: An analysis of C_4 plants deficient in the C_3 and C_4 cycles. *Plant Physiol.* 116: 823–832.

Nimmo, H. G. (2000) The regulation of phosphoenolpyruvate carboxylase in CAM plants. *Trends Plant Sci.* 5: 75–80.

Ogawa, T., and Kaplan, A. (1987) The stoichiometry between CO_2 and H^+ fluxes involved in the transport of inorganic carbon in cyanobacteria. *Plant Physiol.* 83: 888–891.

Ogren, W. L. (1984) Photorespiration: Pathways, regulation and modification. *Annu. Rev. Plant Physiol.* 35: 415–422.

Paul, M., Sonnewald, U., Hajirezaei, M., Dennis, D., and Stitt, M. (1995) Transgenic tobacco plants with strongly decreased expression of pyrophosphate: Fructose-6-phosphate 1-phosphotransferase do not differ significantly from wild type in photosynthate partitioning, plant growth or their ability to cope with limiting phosphate, limiting nitrogen and suboptimal temperatures. *Planta* 196: 277–283.

Purton, S. (1995) The chloroplast genome of *Chlamydomonas. Sci. Prog.* 78: 205–216.

Reinfelder, J. R., Kraepiel, A. M. L., and Morel, F. M. M. (2000) Unicellular C_4 photosynthesis in a marine diatom. *Nature* 407: 996–999.

Salerno, G. L., Echeverria, E., and Pontis, H. G. (1996) Activation of sucrose-phosphate synthase by a protein factor/sucrose-phosphate phosphatase. *Cell. Mol. Biol.* 42: 665–672.

Salvucci, M. E., and Ogren, W. L. (1996) The mechanism of Rubisco activase: Insights from studies of the properties and structure of the enzyme. *Photosynth. Res.* 47: 1–11.

Schürmann, P., and Jacquot, J.-P. (2000) Plant thioredoxin systems revisited. *Annu. Rev. Plant Physiol. Plant Mol. Biol.* 51: 371–400.

Stitt, M. (1990) Fructose-2,6-bisphosphate as a regulatory molecule in plants. *Annu. Rev. Plant Physiol. Plant Mol. Biol.* 41: 153–185.

Tolbert, N. E. (1981) Metabolic pathways in peroxisomes and glyoxysomes. *Annu. Rev. Biochem.* 50: 133–157.

Vidal, J., and Chollet, R. (1997) Regulatory phosphorylation of C_4 PEP carboxylase. *Trends Plant Sci.* 2: 230–237.

Wolosiuk, R. A., Ballicora, M. A., and Hagelin, K. (1993) The reductive pentose phosphate cycle for photosynthetic carbon dioxide assimilation: Enzyme modulation. *FASEB J.* 7: 622–637.

Xiang, Y., Zhang, J., and Weeks, D. P. (2001) The Cia5 gene controls formation of the carbon concentrating mechanism in *Chlamydomonas reinhardtii. Proc. Natl. Acad. Sci. USA* 98: 5341–5346.

Zhang, N., and Portis, A. R. (1999) Mechanism of light regulation of Rubisco: A specific role for the larger Rubisco activase isoform involving reductive activation by thioredoxin-f. *Proc. Natl. Acad. Sci. USA* 96: 9438–9443.

Chapter

Photosynthesis: Physiological and Ecological Considerations

THE CONVERSION OF SOLAR ENERGY to the chemical energy of organic compounds is a complex process that includes electron transport and photosynthetic carbon metabolism (see Chapters 7 and 8). Earlier discussions of the photochemical and biochemical reactions of photosynthesis should not overshadow the fact that, under natural conditions, the photosynthetic process takes place in intact organisms that are continuously responding to internal and external changes. This chapter addresses some of the photosynthetic responses of the intact leaf to its environment. Additional photosynthetic responses to different types of stress are covered in Chapter 25.

The impact of the environment on photosynthesis is of interest to both plant physiologists and agronomists. From a physiological standpoint, we wish to understand how photosynthesis responds to environmental factors such as light, ambient CO_2 concentrations, and temperature. The dependence of photosynthetic processes on environment is also important to agronomists because plant productivity, and hence crop yield, depends strongly on prevailing photosynthetic rates in a dynamic environment.

In studying the environmental dependence of photosynthesis, a central question arises: How many environmental factors can limit photosynthesis at one time? The British plant physiologist F. F. Blackman hypothesized in 1905 that, under any particular conditions, the rate of photosynthesis is limited by the slowest step, the so-called *limiting factor*.

The implication of this hypothesis is that at any given time, photosynthesis can be limited either by light or by CO_2 concentration, but not by both factors. This hypothesis has had a marked influence on the approach used by plant physiologists to study photosynthesis—that is, varying one factor and keeping all other environmental conditions constant.

TABLE 9.1
Some characteristics of limitations to the rate of photosynthesis

	Conditions that lead to this limitation		Response of photosynthesis under this limitation to		
Limiting factor	CO_2	Light	CO_2	O_2	Light
Rubisco activity	Low	High	Strong	Strong	Absent
RuBP regeneration	High	Low	Moderate	Moderate	Strong

In the intact leaf, three major metabolic steps have been identified as important for optimal photosynthetic performance:

1. Rubisco activity
2. Regeneration of ribulose bisphosphate (RuBP)
3. Metabolism of the triose phosphates

The first two steps are the most prevalent under natural conditions. Table 9.1 provides some examples of how light and CO_2 can affect these key metabolic steps. In the following sections, biophysical, biochemical, and environmental aspects of photosynthesis in leaves are discussed in detail.

LIGHT, LEAVES, AND PHOTOSYNTHESIS

Scaling up from the chloroplast (the focus of Chapters 7 and 8) to the leaf adds new levels of complexity to photosynthesis. At the same time, the structural and functional properties of the leaf make possible other levels of regulation.

We will start by examining how leaf anatomy, and movements by chloroplasts and leaves, control the absorption of light for photosynthesis. Then we will describe how chloroplasts and leaves adapt to their light environment and how the photosynthetic response of leaves grown under low light reflects their adaptation to a low-light environment. Leaves also adapt to high light conditions, illustrating that plants are physiologically flexible and that they adapt to their immediate environment.

Both the amount of light and the amount of CO_2 determine the photosynthetic response of leaves. In some situations, photosynthesis is limited by an inadequate supply of light or CO_2. In other situations, absorption of too much light can cause severe problems, and special mechanisms protect the photosynthetic system from excessive light. Multiple levels of control over photosynthesis allow plants to grow successfully in a constantly changing environment and different habitats.

CONCEPTS AND UNITS IN THE MEASUREMENT OF LIGHT

Three light parameters are especially important in the measurement of light: (1) spectral quality, (2) amount, and (3) direction. Spectral quality was discussed in Chapter 7 (see Figures 7.2 and 7.3, and **Web Topic 7.1**). A discussion of the amount and direction of light reaching the plant requires consideration of the geometry of the part of the plant that receives the light: Is the plant organ flat or cylindrical?

Flat, or planar, light sensors are best suited for flat leaves. The light reaching the plant can be measured as energy, and the amount of energy that falls on a flat sensor of known area per unit time is quantified as **irradiance** (see Table 9.2). Units can be expressed in terms of energy, such as watts per square meter (W m^{-2}). Time (seconds) is contained within the term watt: 1 W = 1 joule (J) s^{-1}.

Light can also be measured as the number of incident **quanta** (singular *quantum*). In this case, units can be expressed in moles per square meter per second (mol m^{-2} s^{-1}), where *moles* refers to the number of photons (1 mol of light = 6.02×10^{23} photons, Avogadro's number). This measure is called **photon irradiance**. Quanta and energy units can be interconverted relatively easily, provided that the wavelength of the light, λ, is known. The energy of a photon is related to its wavelength as follows:

$$E = \frac{hc}{\lambda}$$

where c is the speed of light (3×10^8 m s^{-1}), h is Planck's constant (6.63×10^{-34} J s), and λ is the wavelength

TABLE 9.2
Concepts and units for the quantification of light

	Energy measurements (W m^{-2})	Photon measurements (mol $m^{-2}s^{-1}$)
Flat light sensor	Irradiance	Photon irradiance
	Photosynthetically active radiation (PAR, 400-700 nm, energy units)	PAR (quantum units)
	—	Photosynthetic photon flux density (PPFD)
Spherical light sensor	Fluence rate (energy units) Scalar irradiance	Fluence rate (quantum units) Quantum scalar irradiance

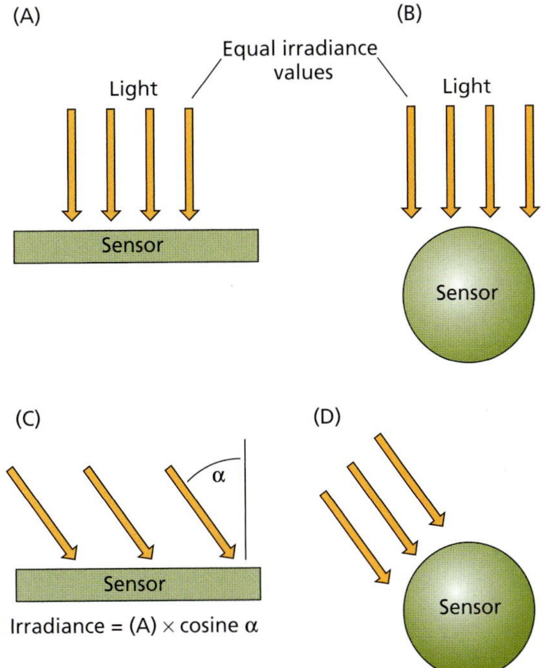

FIGURE 9.1 Flat and spherical light sensors. Equivalent amounts of collimated light strike a flat irradiance-type sensor (A) and a spherical sensor (B) that measure fluence rate. With collimated light, A and B will give the same light readings. When the light direction is changed 45°, the spherical sensor (D) will measure the same quantity as in B. In contrast, the flat irradiance sensor (C) will measure an amount equivalent to the irradiance in A multiplied by the cosine of the angle α in C. (After Björn and Vogelmann 1994.)

of light, usually expressed in nm (1 nm = 10^{-9} m). From this equation it can be shown that a photon at 400 nm has twice the energy of a photon at 800 nm (see **Web Topic 9.1**).

Now let's turn our attention to the direction of light. Light can strike a flat surface directly from above or obliquely. When light deviates from perpendicular, irradiance is proportional to the cosine of the angle at which the light rays hit the sensor (Figure 9.1).

There are many examples in nature in which the light-intercepting object is not flat (e.g., complex shoots, whole plants, chloroplasts). In addition, in some situations light can come from many directions simultaneously (e.g., direct light from the sun plus the light that is reflected upward from sand, soil, or snow). In these situations it makes more sense to measure light with a spherical sensor that takes measurements omnidirectionally (from all directions).

The term for this omnidirectional measurement is **fluence rate** (see Table 9.2) (Rupert and Letarjet 1978), and this quantity can be expressed in watts per square meter (W m^{-2}) or moles per square meter per second (mol m^{-2} s^{-1}). The units clearly indicate whether light is being measured as energy (W) or as photons (mol).

In contrast to a flat sensor's sensitivity, the sensitivity to light of a spherical sensor is independent of direction (see Figure 9.1). Depending on whether the light is collimated (rays are parallel) or diffuse (rays travel in random directions), values for fluence rate versus irradiance measured with a flat or a spherical sensor can provide different values (see Figure 9.1) (for a detailed discussion, see Björn and Vogelmann 1994).

Photosynthetically active radiation (**PAR**, 400–700 nm) may also be expressed in terms of energy (W m^{-2}) or quanta (mol m^{-2} s^{-1}) (McCree 1981). Note that PAR is an irradiance-type measurement. In research on photosynthesis, when PAR is expressed on a quantum basis, it is given the special term **photosynthetic photon flux density** (**PPFD**). However, it has been suggested that the term *density* be discontinued because within the International System of Units (SI units, where *SI* stands for *Système International*), *density* can mean area or volume.

In summary, when choosing how to quantify light, it is important to match sensor geometry and spectral response with that of the plant. Flat, cosine-corrected sensors are ideally suited to measure the amount of light that strikes the surface of a leaf; spherical sensors are more appropriate in other situations, such as in studies of a chloroplast suspension or a branch from a tree (see Table 9.2).

How much light is there on a sunny day, and what is the relationship between PAR irradiance and PAR fluence rate? Under direct sunlight, PAR irradiance and fluence rate are both about 2000 μmol m^{-2} s^{-1}, though higher values can be measured at high altitudes. The corresponding value in energy units is about 400 W m^{-2}.

Leaf Anatomy Maximizes Light Absorption

Roughly 1.3 kW m^{-2} of radiant energy from the sun reaches Earth, but only about 5% of this energy can be converted into carbohydrates by a photosynthesizing leaf (Figure 9.2). The reason this percentage is so low is that a major fraction of the incident light is of a wavelength either too short or too long to be absorbed by the photosynthetic pigments (see Figure 7.3). Of the absorbed light energy, a significant fraction is lost as heat, and a smaller amount is lost as fluorescence (see Chapter 7).

Recall from Chapter 7 that radiant energy from the sun consists of many different wavelengths of light. Only photons of wavelengths from 400 to 700 nm are utilized in photosynthesis, and about 85 to 90% of this PAR is absorbed by the leaf; the remainder is either reflected at the leaf surface or transmitted through the leaf (Figure 9.3). Because chlorophyll absorbs very strongly in the blue and the red regions of the spectrum (see Figure 7.3), the transmitted and reflected light are vastly enriched in green—hence the green color of vegetation.

The anatomy of the leaf is highly specialized for light absorption (Terashima and Hikosaka 1995). The outermost cell layer, the epidermis, is typically transparent to visible light, and the individual cells are often convex. Convex epidermal cells can act as lenses and can focus light so that the amount reaching some of the chloroplasts can be many times greater than the amount of ambient light (Vogel-

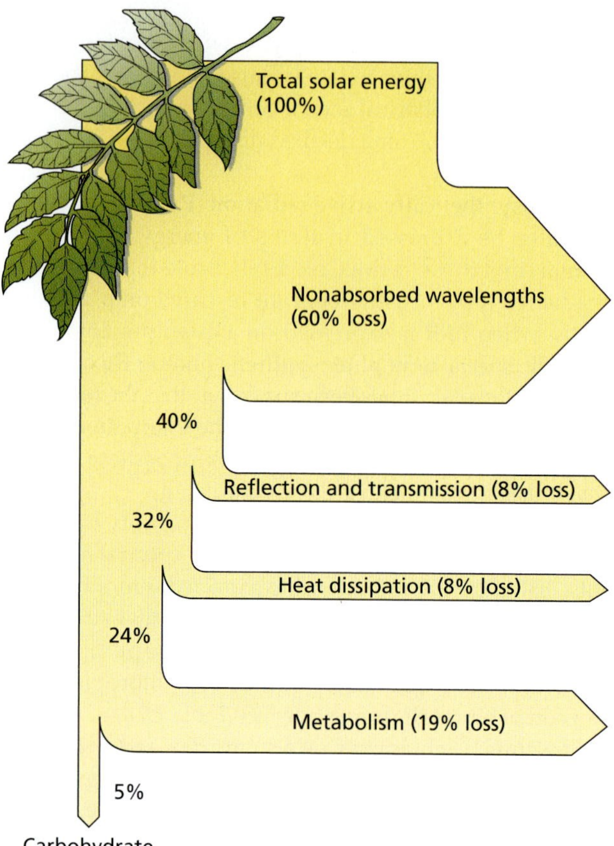

FIGURE 9.2 Conversion of solar energy into carbohydrates by a leaf. Of the total incident energy, only 5% is converted into carbohydrates.

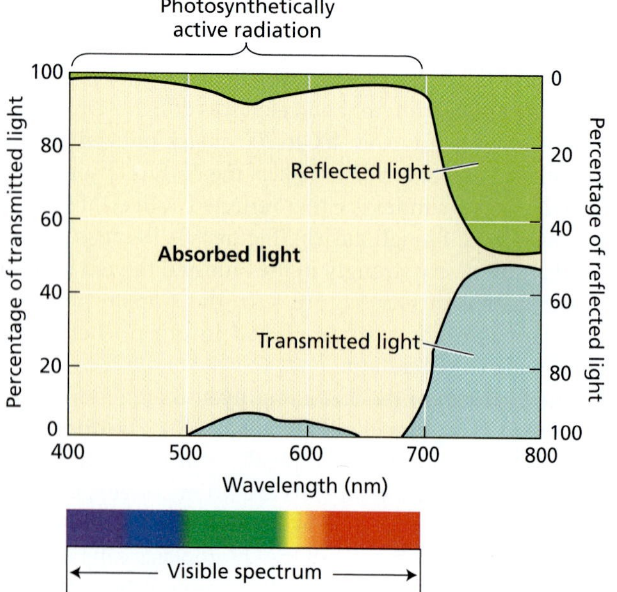

mann et al. 1996). Epidermal focusing is common among herbaceous plants and is especially prominent among tropical plants that grow in the forest understory, where light levels are very low.

Below the epidermis, the top layers of photosynthetic cells are called **palisade cells**; they are shaped like pillars that stand in parallel columns one to three layers deep (Figure 9.4). Some leaves have several layers of columnar palisade cells, and we may wonder how efficient it is for a plant to invest energy in the development of multiple cell layers when the high chlorophyll content of the first layer would appear to allow little transmission of the incident light to the leaf interior. In fact, more light than might be expected penetrates the first layer of palisade cells because of the sieve effect and light channeling.

The **sieve effect** is due to the fact that chlorophyll is not uniformly distributed throughout cells but instead is confined to the chloroplasts. This packaging of chlorophyll results in shading between the chlorophyll molecules and creates gaps between the chloroplasts, where light is not absorbed—hence the reference to a sieve. Because of the sieve effect, the total absorption of light by a given amount of chlorophyll in a palisade cell is less than the light absorbed by the same amount of chlorophyll in a solution.

Light channeling occurs when some of the incident light is propagated through the central vacuole of the palisade cells and through the air spaces between the cells, an arrangement that facilitates the transmission of light into the leaf interior (Vogelmann 1993).

Below the palisade layers is the **spongy mesophyll**, where the cells are very irregular in shape and are surrounded by large air spaces (see Figure 9.4). The large air spaces generate many interfaces between air and water that reflect and refract the light, thereby randomizing its direction of travel. This phenomenon is called *light scattering*.

Light scattering is especially important in leaves because the multiple reflections between cell–air interfaces greatly increase the length of the path over which photons travel, thereby increasing the probability for absorption. In fact, photon path lengths within leaves are commonly four times or more longer than the thickness of the leaf (Richter and Fukshansky 1996). Thus the palisade cell properties that allow light to pass through, and the spongy mesophyll cell properties that are conducive to light scattering, result in more uniform light absorption throughout the leaf.

Some environments, such as deserts, have so much light that it is potentially harmful to leaves. In these environments leaves often have special anatomic features, such as

FIGURE 9.3 Optical properties of a bean leaf. Shown here are the percentages of light absorbed, reflected, and transmitted, as a function of wavelength. The transmitted and reflected green light in the wave band at 500 to 600 nm gives leaves their green color. Note that most of the light above 700 nm is not absorbed by the leaf. (From Smith 1986.)

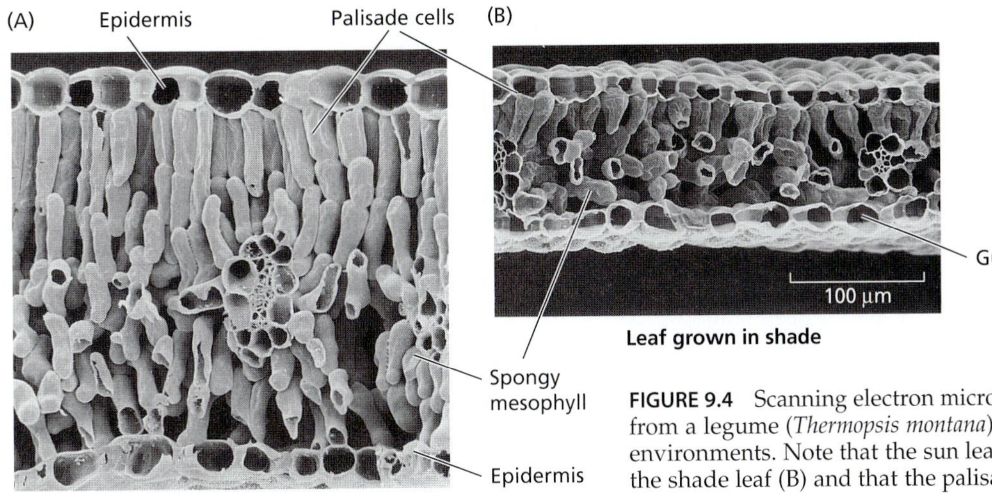

FIGURE 9.4 Scanning electron micrographs of the leaf anatomy from a legume (*Thermopsis montana*) grown in different light environments. Note that the sun leaf (A) is much thicker than the shade leaf (B) and that the palisade (columnlike) cells are much longer in the leaves grown in sunlight. Layers of spongy mesophyll cells can be seen below the palisade cells. (Micrographs courtesy of T. Vogelmann.)

hairs, salt glands, and epicuticular wax that increase the reflection of light from the leaf surface, thereby reducing light absorption (Ehleringer et al. 1976). Such adaptations can decrease light absorption by as much as 40%, minimizing heating and other problems associated with the absorption of too much light.

Chloroplast Movement and Leaf Movement Can Control Light Absorption

Chloroplast movement is widespread among algae, mosses, and leaves of higher plants (Haupt and Scheuerlein 1990). If chloroplast orientation and location are controlled, leaves can regulate how much of the incident light is absorbed. Under low light (Figure 9.5B), chloroplasts gather at the cell surfaces parallel to the plane of the leaf so that they are aligned perpendicularly to the incident light—a position that maximizes absorption of light.

Under high light (Figure 9.5C), the chloroplasts move to the cell surfaces that are parallel to the incident light, thus avoiding excess absorption of light. Such chloroplast rearrangement can decrease the amount of light absorbed by the leaf by about 15% (Gorton et al. 1999). Chloroplast movement in leaves is a typical blue-light response (see Chapter 18). Blue light also controls chloroplast orientation

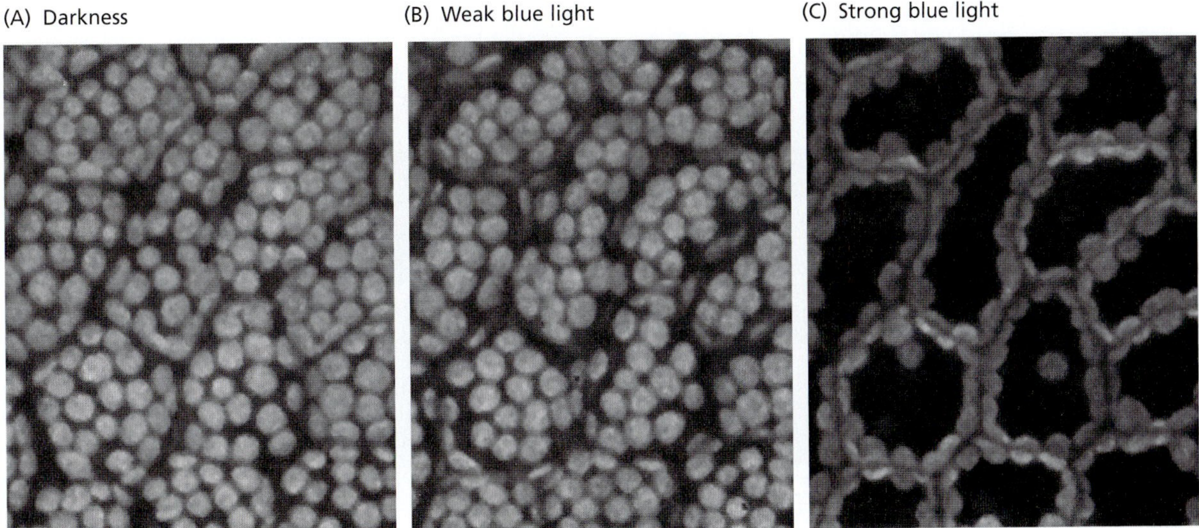

FIGURE 9.5 Chloroplast distribution in photosynthesizing cells of the duckweed *Lemna*. These surface views show the same cells under three conditions: (A) darkness, (B) weak blue light, and (C) strong blue light. In A and B, chloroplasts are positioned near the upper surface of the cells, where they can absorb maximum amounts of light. When the cells were irradiated with strong blue light (C), the chloroplasts move to the side walls, where they shade each other, thus minimizing the absorption of excess light. (Micrographs courtesy of M. Tlalka and M. D. Fricker.)

in many of the lower plants, but in some algae, chloroplast movement is controlled by phytochrome (Haupt and Scheuerlein 1990). In leaves, chloroplasts move along actin microfilaments in the cytoplasm, and calcium regulates their movement (Tlalka and Fricker 1999).

Leaves have the highest light absorption when the leaf blade, or lamina, is perpendicular to the incident light. Some plants control light absorption by **solar tracking** (Koller 2000); that is, their leaves continuously adjust the orientation of their laminae such that they remain perpendicular to the sun's rays (Figure 9.6). Alfalfa, cotton, soybean, bean, lupine, and some wild species of the mallow family (Malvaceae) are examples of the numerous plant species that are capable of solar tracking.

Solar-tracking leaves keep a nearly vertical position at sunrise, facing the eastern horizon, where the sun will rise. The leaf blades then lock on to the rising sun and follow its movement across the sky with an accuracy of ±15° until sunset, when the laminae are nearly vertical, facing the west, where the sun will set. During the night the leaf takes a horizontal position and reorients just before dawn so that it faces the eastern horizon in anticipation of another sunrise. Leaves track the sun only on clear days, and they stop when a cloud obscures the sun. In the case of intermittent cloud cover, some leaves can reorient as rapidly as 90° per hour and thus can catch up to the new solar position when the sun emerges from behind a cloud (Koller 1990).

Solar tracking is another blue-light response, and the sensing of blue light in solar-tracking leaves occurs in specialized regions. In species of *Lavatera* (Malvaceae), the photosensitive region is located in or near the major leaf veins (Koller 1990). In lupines, (*Lupinus*, Fabaceae), leaves consist of five or more leaflets, and the photosensitive region is located in the basal part of each leaflet lamina.

In many cases, leaf orientation is controlled by a specialized organ called the **pulvinus** (plural *pulvini*), found at the junction between the blade and petiole. The pulvinus contains motor cells that change their osmotic potential and generate mechanical forces that determine laminar orientation. In other plants, leaf orientation is controlled by small mechanical changes along the length of the petiole and by movements of the younger parts of the stem.

Some solar-tracking plants can also move their leaves such that they avoid full exposure to sunlight, thus minimizing heating and water loss. Building on the term **heliotropism** (bending toward the sun), which is often used to describe sun-induced leaf movements, these sun-avoiding leaves are called *paraheliotropic*, and leaves that maximize light interception by solar tracking are called *diaheliotropic*. Some plant species can display diaheliotropic movements when they are well watered and paraheliotropic movements when they experience water stress.

Since full sunlight usually exceeds the amount of light that can be utilized for photosynthesis, what advantage is gained by solar tracking? By keeping leaves perpendicular to the sun, solar-tracking plants maintain maximum photosynthetic rates throughout the day, including early morning and late afternoon. Moreover, air temperature is lower during the early morning and late afternoon, so water stress is lower. Solar tracking therefore gives an advantage to plants that grow in arid regions.

Plants Adapt to Sun and Shade

Some plants have enough developmental plasticity to adapt to a range of light regimes, growing as sun plants in sunny areas and as shade plants in shady habitats. Some shady habitats receive less than 1% of the PAR available in an exposed habitat. Leaves that are adapted to very sunny

(A)

(B)

FIGURE 9.6 Leaf movement in sun-tracking plants. (A) Initial leaf orientation in the lupine *Lupinus succulentus*. (B) Leaf orientation 4 hours after exposure to oblique light. The direction of the light beam is indicated by the arrows. Movement is generated by asymmetric swelling of a pulvinus, found at the junction between the lamina and the petiole. In natural conditions, the leaves track the sun's trajectory in the sky. (From Vogelmann and Björn 1983, courtesy of T. Vogelmann.)

or very shady environments are often unable to survive in the other type of habitat (see Figure 9.10). Sun and shade leaves have some contrasting characteristics:

- *Shade leaves* have more total chlorophyll per reaction center, have a higher ratio of chlorophyll *b* to chlorophyll *a*, and are usually thinner than sun leaves.
- *Sun leaves* have more rubisco, and a larger pool of xanthophyll cycle components than shade leaves (see Chapter 7).

Contrasting anatomic characteristics can also be found in leaves of the same plant that are exposed to different light regimes. Figure 9.4 shows some anatomic differences between a leaf grown in the sun and a leaf grown in the shade. Sun-grown leaves are thicker and have longer palisade cells than leaves growing in the shade. Even different parts of a single leaf show adaptations to their light microenvironment. Cells in the upper surface of the leaf, which are exposed to the highest prevailing photon flux, have characteristics of cells from leaves grown in full sunlight; cells in the lower surface of the leaf have characteristics of cells found in shade-grown leaves (Terashima 1992).

These morphological and biochemical modifications are associated with specific functions. Far-red light is absorbed primarily by PSI, and altering the ratio of PSI to PSII or changing the light-harvesting antennae associated with the photosystems makes it possible to maintain a better balance of energy flow through the two photosystems (Melis 1996). These adaptations are found in nature; some shade plants show a 3:1 ratio of photosystem II to photosystem I reaction centers, compared with the 2:1 ratio found in sun plants (Anderson 1986). Other shade plants, rather than altering the ratio of PSI to PSII, add more antennae chlorophyll to PSII. These adaptations appear to enhance light absorption and energy transfer in shady environments, where far-red light is more abundant.

Sun and shade plants also differ in their respiration rates, and these differences alter the relationship between respiration and photosynthesis, as we'll see a little later in this chapter.

Plants Compete for Sunlight

Plants normally compete for sunlight. Held upright by stems and trunks, leaves configure a canopy that absorbs light and influences photosynthetic rates and growth beneath them.

Leaves that are shaded by other leaves have much lower photosynthetic rates. Some plants have very thick leaves that transmit little, if any, light. Other plants, such as those of the dandelion (*Taraxacum* sp.), have a rosette growth habit, in which leaves grow radially very close to each other and to the stem, thus preventing the growth of any leaves below them.

Trees represent an outstanding adaptation for light interception. The elaborate branching structure of trees vastly increases the interception of sunlight. Very little PAR penetrates the canopy of many forests; almost all of it is absorbed by leaves (Figure 9.7).

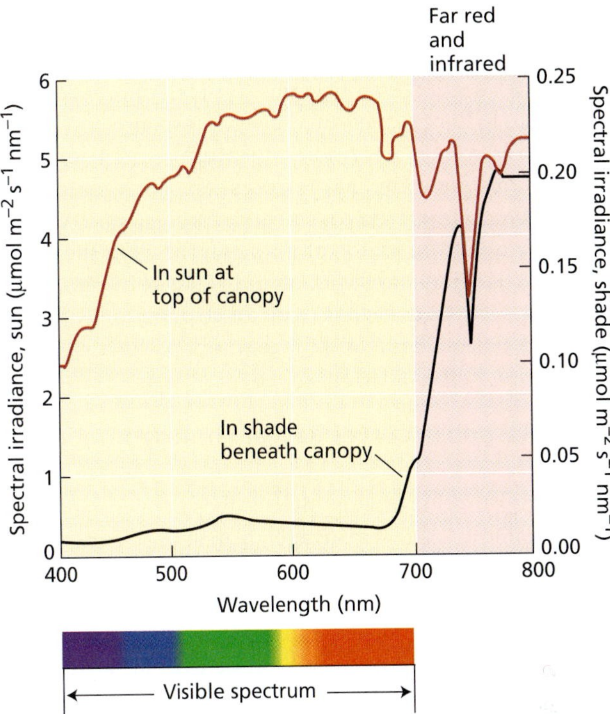

FIGURE 9.7 The spectral distribution of sunlight at the top of a canopy and under the canopy. For unfiltered sunlight, the total irradiance was 1900 µmol m^{-2} s^{-1}; for shade, 17.7 µmol m^{-2} s^{-1}. Most of the photosynthetically active radiation was absorbed by leaves in the canopy. (From Smith 1994.)

Another feature of the shady habitat is **sunflecks**, patches of sunlight that pass through small gaps in the leaf canopy and move across shaded leaves as the sun moves. In a dense forest, sunflecks can change the photon flux impinging on a leaf in the forest floor more than tenfold within seconds. For some of these leaves, a sunfleck contains nearly 50% of the total light energy available during the day, but this critical energy is available for only a few minutes in a very high dose.

Sunflecks also play a role in the carbon metabolism of lower leaves in dense crops that are shaded by the upper leaves of the plant. Rapid responses of both the photosynthetic apparatus and the stomata to sunflecks have been of substantial interest to plant physiologists and ecologists (Pearcy et al. 1997) because they represent unique physiological responses specialized for capturing a short burst of sunlight.

PHOTOSYNTHETIC RESPONSES TO LIGHT BY THE INTACT LEAF

Light is a critical resource for plants that can often limit growth and reproduction. The photosynthetic properties of the leaf provide valuable information about plant adaptations to their light environment.

In this section we describe typical photosynthetic responses to light as measured in light-response curves. We also consider how an important feature of light-response curves, the light compensation point, explains contrasting physiological properties of sun and shade plants. We then describe quantum yields of photosynthesis in the intact leaf, and the differences in quantum yields between C_3 and C_4 plants. The section closes with descriptions of leaf adaptations to excess light, and the different pathways of heat dissipation in the leaf.

Light-Response Curves Reveal Photosynthetic Properties

Measuring CO_2 fixation in intact leaves at increasing photon flux allows us to construct light-response curves (Figure 9.8) that provide useful information about the photosynthetic properties of leaves. In the dark there is no photosynthetic carbon assimilation, and CO_2 is given off by the plant because of respiration (see Chapter 11). By convention, CO_2 assimilation is negative in this part of the light-response curve. As the photon flux increases, photosynthetic CO_2 assimilation increases until it equals CO_2 release by mitochondrial respiration. The point at which CO_2 uptake exactly balances CO_2 release is called the **light compensation point**.

The photon flux at which different leaves reach the light compensation point varies with species and developmental conditions. One of the more interesting differences is found between plants grown in full sunlight and those grown in the shade (Figure 9.9). Light compensation points of sun plants range from 10 to 20 μmol m^{-2} s^{-1}; corresponding values for shade plants are 1 to 5 μmol m^{-2} s^{-1}.

The values for shade plants are lower because respiration rates in shade plants are very low, so little net photosynthesis suffices to bring the net rates of CO_2 exchange to zero. Low respiratory rates seem to represent a basic adaptation that allows shade plants to survive in light-limited environments.

Increasing photon flux above the light compensation point results in a proportional increase in photosynthetic rate (see Figure 9.8), yielding a linear relationship between photon flux and photosynthetic rate. Such a linear rela-

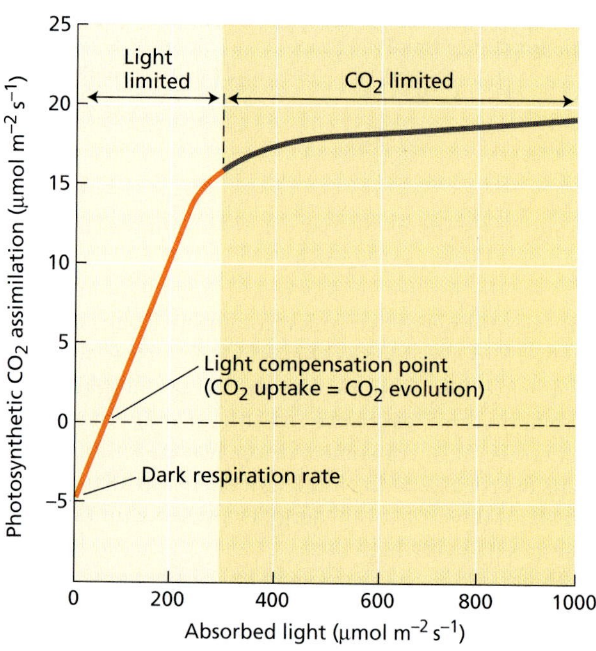

FIGURE 9.8 Response of photosynthesis to light in a C_3 plant. In darkness, respiration causes a net efflux of CO_2 from the plant. The light compensation point is reached when photosynthetic CO_2 assimilation equals the amount of CO_2 evolved by respiration. Increasing light above the light compensation point proportionally increases photosynthesis indicating that photosynthesis is limited by the rate of electron transport, which in turn is limited by the amount of available light. This portion of the curve is referred to as light-limited. Further increases in photosynthesis are eventually limited by the carboxylation capacity of rubisco or the metabolism of triose phosphates. This part of the curve is referred to as CO_2 limited.

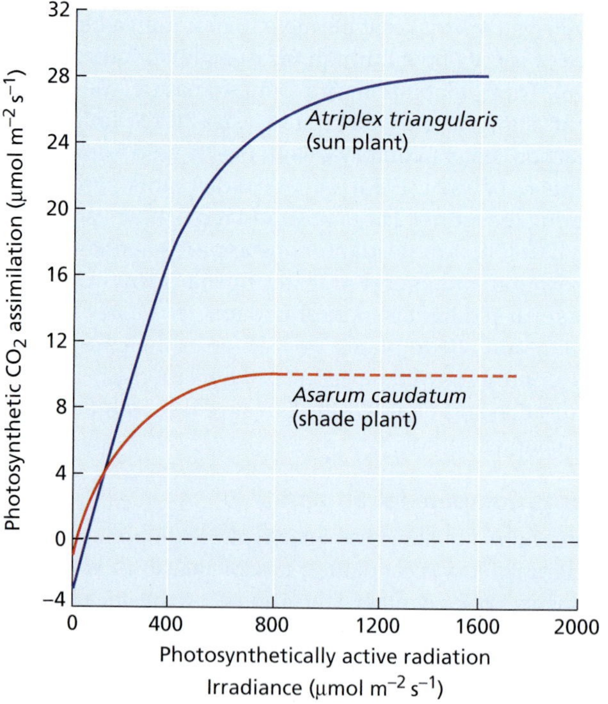

FIGURE 9.9 Light–response curves of photosynthetic carbon fixation in sun and shade plants. *Atriplex triangularis* (triangle orache) is a sun plant, and *Asarum caudatum* (a wild ginger) is a shade plant. Typically, shade plants have a low light compensation point and have lower maximal photosynthetic rates than sun plants. The dashed line has been extrapolated from the measured part of the curve. (From Harvey 1979.)

tionship comes about because photosynthesis is *light limited* at those levels of incident light, so more light stimulates more photosynthesis.

In this linear portion of the curve, the slope of the line reveals the **maximum quantum yield** of photosynthesis for the leaf. Recall that quantum yield is the relation between a given light-dependent product (in this case CO_2 assimilation) and the number of absorbed photons (see Equation 7.5).

Quantum yields vary from 0, where none of the light energy is used in photosynthesis, to 1, where all the absorbed light is used. Recall from Chapter 7 that the quantum yield of photochemistry is about 0.95, and the quantum yield of oxygen evolution by isolated chloroplasts is about 0.1 (10 photons per molecule of O_2).

In the intact leaf, measured quantum yields for CO_2 fixation vary between 0.04 and 0.06. Healthy leaves from many species of C_3 plants, kept under low O_2 concentrations that inhibit photorespiration, usually show a quantum yield of 0.1. In normal air, the quantum yield of C_3 plants is lower, typically 0.05.

Quantum yield varies with temperature and CO_2 concentration because of their effect on the ratio of the carboxylase and oxygenase reactions of rubisco (see Chapter 8). Below 30°C, quantum yields of C_3 plants are generally higher than those of C_4 plants; above 30°C, the situation is usually reversed (see Figure 9.23). Despite their different growth habitats, sun and shade plants show similar quantum yields.

At higher photon fluxes, the photosynthetic response to light starts to level off (see Figure 9.8) and reaches *saturation*. Once the saturation point is reached, further increases in photon flux no longer affect photosynthetic rates, indicating that factors other than incident light, such as electron transport rate, rubisco activity, or the metabolism of triose phosphates, have become limiting to photosynthesis.

After the saturation point, photosynthesis is commonly referred to as *CO_2 limited*, reflecting the inability of the Calvin cycle enzymes to keep pace with the absorbed light energy. Light saturation levels for shade plants are substantially lower than those for sun plants (see Figure 9.9). These levels usually reflect the maximal photon flux to which the leaf was exposed during growth (Figure 9.10).

The light-response curve of most leaves saturates between 500 and 1000 μmol m^{-2} s^{-1}, photon fluxes well below full sunlight (which is about 2000 μmol m^{-2} s^{-1}). Although individual leaves are rarely able to utilize full sunlight, whole plants usually consist of many leaves that shade each other. For example, only a small fraction of a tree's leaves are exposed to full sun at any given time of the day. The rest of the leaves receive subsaturating photon fluxes in the form of small patches of light that pass through gaps in the leaf canopy or in the form of light transmitted through other leaves. Because the photosynthetic response of the intact plant is the sum of the photosynthetic activity of all the leaves, only rarely is photosynthesis saturated at the level of the whole plant.

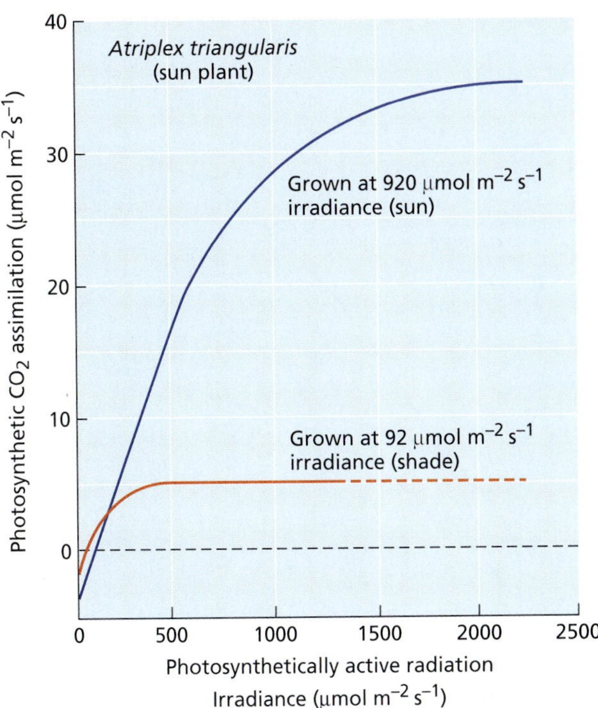

FIGURE 9.10 Light–response of photosynthesis of a sun plant gown under sun or shade conditions. The upper curve represents an *Atriplex triangularis* leaf grown at an irradiance ten times higher than that of the lower curve. In the leaf grown at the lower light levels, photosynthesis saturates at a substantially lower irradiance, indicating that the photosynthetic properties of a leaf depend on its growing conditions. The dashed line has been extrapolated from the measured part of the curve. (From Björkman 1981.)

Light-response curves of individual trees and of the forest canopy show that photosynthetic rate increases with photon flux and photosynthesis usually does not saturate, even in full sunlight (Figure 9.11). Along these lines, crop productivity is related to the total amount of light received during the growing season, and given enough water and nutrients, the more light a crop receives, the higher the biomass (Ort and Baker 1988).

Leaves Must Dissipate Excess Light Energy

When exposed to excess light, leaves must dissipate the surplus absorbed light energy so that it does not harm the photosynthetic apparatus (Figure 9.12). There are several routes for energy dissipation involving *nonphotochemical quenching* (see Chapter 7), which is the quenching of chlorophyll fluorescence by mechanisms other than photochemistry. The most important example involves the transfer of absorbed light energy away from electron transport toward heat production. Although the molecular mechanisms are not yet fully understood, the xanthophyll cycle appears to be an important avenue for dissipation of excess light energy (see **Web Essay 9.1**).

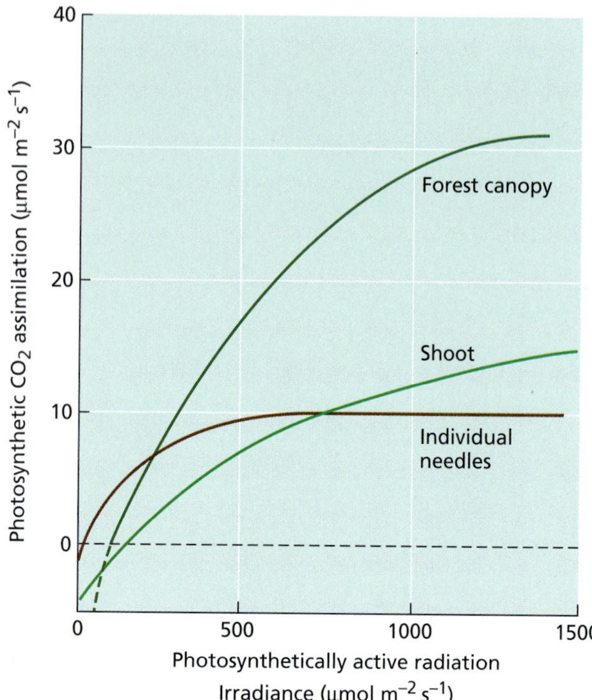

FIGURE 9.11 Changes in photosynthesis (expressed on a per-square-meter basis) in individual needles, a complex shoot, and a forest canopy of Sitka spruce (*Picea sitchensis*) as a function of irradiance. Complex shoots consist of groupings of needles that often shade each other, similar to the situation in a canopy where branches often shade other branches. As a result of shading, much higher irradiance levels are needed to saturate photosynthesis. The dashed line has been extrapolated from the measured part of the curve. (From Jarvis and Leverenz 1983.)

The xanthophyll cycle. Recall from Chapter 7 that the xanthophyll cycle, which comprises the three carotenoids violaxanthin, antheraxanthin, and zeaxanthin, is involved in the dissipation of excess light energy in the leaf (see Figure 7.36). Under high light, violaxanthin is converted to antheraxanthin and then to zeaxanthin. Note that the two aromatic rings of violaxanthin have a bound oxygen atom in them, antheraxanthin has one, and zeaxanthin has none (again, see Figure 7.36). Experiments have shown that zeaxanthin is the most effective of the three xanthophylls in heat dissipation, and antheraxanthin is only half as effective. Whereas the levels of antheraxanthin remain relatively constant throughout the day, the zeaxanthin content increases at high irradiances and decreases at low irradiances.

In leaves growing under full sunlight, zeaxanthin and antheraxanthin can make up 60% of the total xanthophyll cycle pool at maximal irradiance levels attained at midday (Figure 9.13). In these conditions a substantial amount of excess light energy absorbed by the thylakoid membranes can be dissipated as heat, thus preventing damage to the photosynthetic machinery of the chloroplast (see Chapter 7). The fraction of light energy that is dissipated depends on irradiance, species, growth conditions, nutrient status, and ambient temperature (Demmig-Adams and Adams 1996).

The xanthophyll cycle in sun and shade leaves. Leaves that grow in full sunlight contain a substantially larger xanthophyll pool than shade leaves, so they can dissipate higher amounts of excess light energy. Nevertheless, the xanthophyll cycle also operates in plants that grow in the low light of the forest understory, where they are only occasionally exposed to high light when sunlight passes through gaps in the overlying leaf canopy, forming sunflecks (which were described earlier in the chapter). Exposure to one sunfleck results in the conversion of much of the violaxanthin in the leaf to zeaxanthin. In contrast to typical leaves, in which violaxanthin levels increase again when irradiances drop, the zeaxanthin formed in shade leaves of the forest understory is retained and protects the leaf against exposure to subsequent sunflecks.

The xanthophyll cycle is also found in species such as conifers, the leaves of which remain green during winter, when photosynthetic rates are very low yet light absorption remains high. Contrary to the diurnal cycling of the xanthophyll pool observed in the summer, zeaxanthin lev-

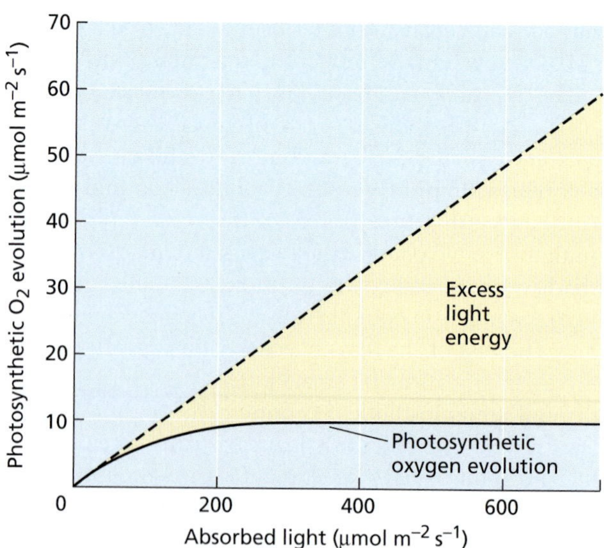

FIGURE 9.12 Excess light energy in relation to a light–response curve of photosynthetic evolution. The broken line shows theoretical oxygen evolution in the absence of any rate limitation to photosynthesis. At levels of photon flux up to 150 μmol m^{-2} s^{-1}, a shade plant is able to utilize the absorbed light. Above 150 μmol m^{-2} s^{-1}, however, photosynthesis saturates, and an increasingly larger amount of the absorbed light energy must be dissipated. At higher irradiances there is a large difference between the fraction of light used by photosynthesis versus that which must be dissipated (excess light energy). The differences are much higher in a shade plant than in a sun plant. (After Osmond 1994.)

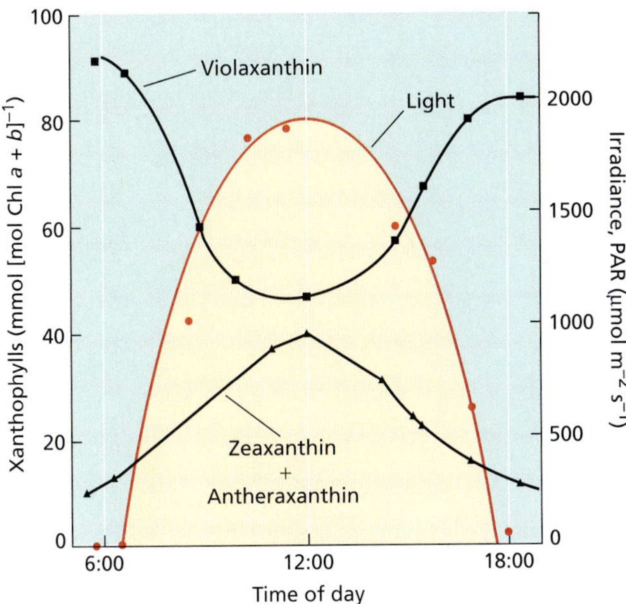

FIGURE 9.13 Diurnal changes in xanthophyll content as a function of irradiance in sunflower (*Helianthus annuus*). As the amount of light incident to a leaf increases, a greater proportion of violaxanthin is converted to antheraxanthin and zeaxanthin, thereby dissipating excess excitation energy and protecting the photosynthetic apparatus. (After Demmig-Adams and Adams 1996.)

els remain high all day during the winter. Presumably this mechanism maximizes dissipation of light energy, thereby protecting the leaves against photooxidation during winter (Adams et al. 2001).

In addition to protecting the photosynthetic system against high light, the xanthophyll cycle may help protect against high temperatures. Chloroplasts are more tolerant of heat when they accumulate zeaxanthin (Havaux et al. 1996). Thus, plants may employ more than one biochemical mechanism to guard against the deleterious effect of excess heat.

Leaves Must Dissipate Vast Quantities of Heat

The heat load on a leaf exposed to full sunlight is very high. In fact, a leaf with an effective thickness of water of 300 µm would warm up by 100°C every minute if all available solar energy were absorbed and no heat were lost. However, this enormous heat load is dissipated by the emission of long-wave radiation, by sensible (i.e., perceptible) heat loss, and by evaporative (or latent) heat loss (Figure 9.14):

- Air circulation around the leaf removes heat from the leaf surfaces if the temperature of the leaf is higher than that of the air; this phenomenon is called **sensible heat loss**.
- **Evaporative heat loss** occurs because the evaporation of water requires energy. Thus as water evaporates from a leaf, it withdraws heat from the leaf and cools it. The human body is cooled by the same principle, through perspiration.

Sensible heat loss and evaporative heat loss are the most important processes in the regulation of leaf temperature, and the ratio of the two is called the **Bowen ratio** (Campbell 1977):

$$\text{Bowen ratio} = \frac{\text{Sensible heat loss}}{\text{Evaporative heat loss}}$$

In well-watered crops, transpiration (see Chapter 4), and hence water evaporation from the leaf, is high, so the Bowen ratio is low (see **Web Topic 9.2**). On the other hand, when evaporative cooling is limited, the Bowen ratio is large. For example, in some cacti, stomata closure prevents evaporative cooling; all the heat is dissipated by sensible heat loss, and the Bowen ratio is infinite.

Plants with very high Bowen ratios conserve water but have to endure very high leaf temperatures in order to maintain a sufficient temperature gradient between the leaf and the air. Slow growth is usually correlated with these adaptations.

Isoprene Synthesis Helps Leaves Cope with Heat

We have seen how the xanthophyll cycle can protect against high light, but how do chloroplasts cope with the

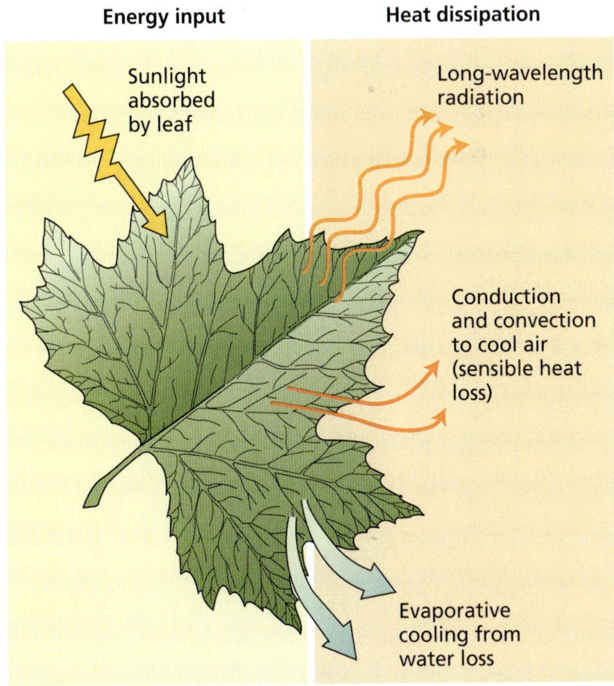

FIGURE 9.14 The absorption and dissipation of energy from sunlight by the leaf. The imposed heat load must be dissipated in order to avoid damage to the leaf. The heat load is dissipated by emission of long-wavelength radiation, by sensible heat loss to the air surrounding the leaf, and by the evaporative cooling caused by transpiration.

high leaf temperatures that usually accompany high light? Isoprene synthesis appears to confer stability to photosynthetic membranes at high light and temperatures. Many plants, including American oak (*Quercus* sp.), aspen (*Populus* sp.), and kudzu (*Pueraria lobata*) emit gaseous five-carbon molecules such as isoprene (2-methyl-1,3-butadiene; see Chapter 13).

On a global scale, these emissions amount to 5×10^{14} g released to the atmosphere each year. These gaseous hydrocarbons are responsible for the pine scent (α- and β-pinene) in coniferous forests and can form a blue haze above forests on hot days. Because isoprene and related hydrocarbons play an important role in atmospheric chemistry, they have attracted much attention from atmospheric scientists.

Isoprene emission from leaves can constitute a significant fraction of the carbon assimilated in photosynthesis. For example, up to 2% of the carbon fixed by photosynthesis in aspen and oak leaves at 30°C is released as isoprene (Sharkey 1996). Sun leaves synthesize more isoprene than shade leaves, and synthesis is proportional to leaf temperature and water stress.

Evidence that isoprene confers stability to photosynthetic membranes under high temperatures comes from three types of experimental results:

1. Whereas preventing isoprene emission with an inhibitor increases susceptibility to damage by heat, adding isoprene to plants that do not produce isoprene confers heat stability (Sharkey et al. 2001).
2. Mutant plants unable to emit isoprene are more easily damaged by high temperatures than are wild-type plants (Sharkey and Singsaas 1995).
3. Isoprene is rapidly synthesized enzymatically in response to elevated leaf temperatures.

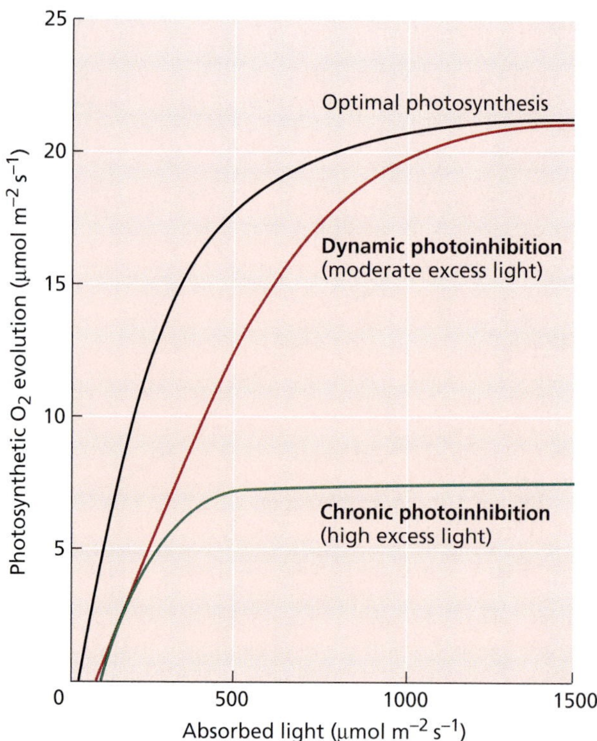

FIGURE 9.15 Changes in the light–response curves of photosynthesis caused by photoinhibition. Exposure to moderate levels of excess light can decrease quantum efficiency (reduced slope of curve) without reducing maximum photosynthetic rate, a condition called dynamic photoinhibition. Exposure to high levels of excess light leads to chronic photoinhibition, where damage to the chloroplast decreases both quantum efficiency and maximum photosynthetic rate. (After Osmond 1994.)

Absorption of Too Much Light Can Lead to Photoinhibition

Recall from Chapter 7 that when leaves are exposed to more light than they can utilize (see Figure 9.12), the reaction center of PSII is inactivated and damaged, in a phenomenon called **photoinhibition**. The characteristics of photoinhibition in the intact leaf depend on the amount of light to which the plant is exposed (Figure 9.15), and two types of photoinhibition are identified: dynamic photoinhibition and chronic photoinhibition (Osmond 1994).

Under moderate excess light, **dynamic photoinhibition** is observed. Quantum efficiency decreases (contrast the slopes of the curves in Figure 9.15), but the maximum photosynthetic rate remains unchanged. Dynamic photoinhibition is caused by the diversion of absorbed light energy toward heat dissipation—hence the decrease in quantum efficiency. This decrease is often temporary, and quantum efficiency can return to its initial higher value when photon flux decreases below saturation levels.

Chronic photoinhibition results from exposure to high levels of excess light that damage the photosynthetic system and decrease both quantum efficiency and maximum photosynthetic rate (see Figure 9.15). Chronic photoinhibition is associated with damage and replacement of the D1 protein from the reaction center of PSII (see Chapter 7). In contrast to dynamic photoinhibition, these effects are relatively long-lasting, persisting for weeks or months.

Early researchers of photoinhibition interpreted all decreases in quantum efficiency as damage to the photosynthetic apparatus. It is now recognized that short-term decreases in quantum efficiency seem to reflect protective mechanisms (see Chapter 7), whereas chronic photoinhibition represents actual damage to the chloroplast resulting from excess light, or a failure of the protective mechanisms.

How significant is photoinhibition in nature? Dynamic photoinhibition appears to occur normally at midday, when leaves are exposed to maximum amounts of light and there is a corresponding reduction in carbon fixation. Photoinhibition is more pronounced at low temperatures, and it becomes chronic under more extreme climatic conditions.

Studies of natural willow populations, and crops of *Brassica napus* (oilseed rape) and *Zea mays* (maize), have shown that the cumulative effects of a daily depression in photosynthetic rates caused by photoinhibition decrease biomass by 10% at the end of the growing season (Long et al. 1994). This may not seem a particularly large effect, but it could be significant in natural plant populations competing for limited resources—conditions under which any reduction in carbon allocated to reproduction can adversely affect reproductive success and survival.

PHOTOSYNTHETIC RESPONSES TO CARBON DIOXIDE

We have discussed how plant growth and leaf anatomy are influenced by light. Now we turn our attention to how CO_2 concentration affects photosynthesis. CO_2 diffuses from the atmosphere into leaves—first through stomata, then through the intercellular air spaces, and ultimately into cells and chloroplasts. In the presence of adequate amounts of light, higher CO_2 concentrations support higher photosynthetic rates. The reverse is also true; that is, low CO_2 concentration can limit the amount of photosynthesis.

In this section we will discuss the concentration of atmospheric CO_2 in recent history, and its availability for carbon-fixing processes. Then we'll consider the limitations that CO_2 places on photosynthesis and the impact of the CO_2-concentrating mechanisms of C_4 plants.

Atmospheric CO_2 Concentration Keeps Rising

Carbon dioxide is a trace gas in the atmosphere, presently accounting for about 0.037%, or 370 parts per million (ppm), of air. The partial pressure of ambient CO_2 (C_a) varies with atmospheric pressure and is approximately 36 pascals (Pa) at sea level (see **Web Topic 9.3**). Water vapor usually accounts for up to 2% of the atmosphere and O_2 for about 20%. The bulk of the atmosphere, nearly 80%, is nitrogen.

The current atmospheric concentration of CO_2 is almost twice the concentration that has prevailed during most of the last 160,00 years, as measured from air bubbles trapped in glacial ice in Antarctica (Figure 9.16A). Except for the last 200 years, CO_2 concentrations during the recent geological past have been low, fluctuating between 180 and 260 ppm. These low concentrations were typical of times extending back to the Cretaceous, when Earth was much warmer and the CO_2 concentration may have been as high as 1200 to 2800 ppm (Ehleringer et al. 1991).

The current CO_2 concentration of the atmosphere is increasing by about 1 ppm each year, primarily because of the burning of fossil fuels (see Figure 9.16C). Since 1958, when systematic measurements of CO_2 began at Mauna Loa, Hawaii, atmospheric CO_2 concentrations have increased by more than 17% (Keeling et al. 1995), and by 2020 the atmospheric CO_2 concentration could reach 600 ppm.

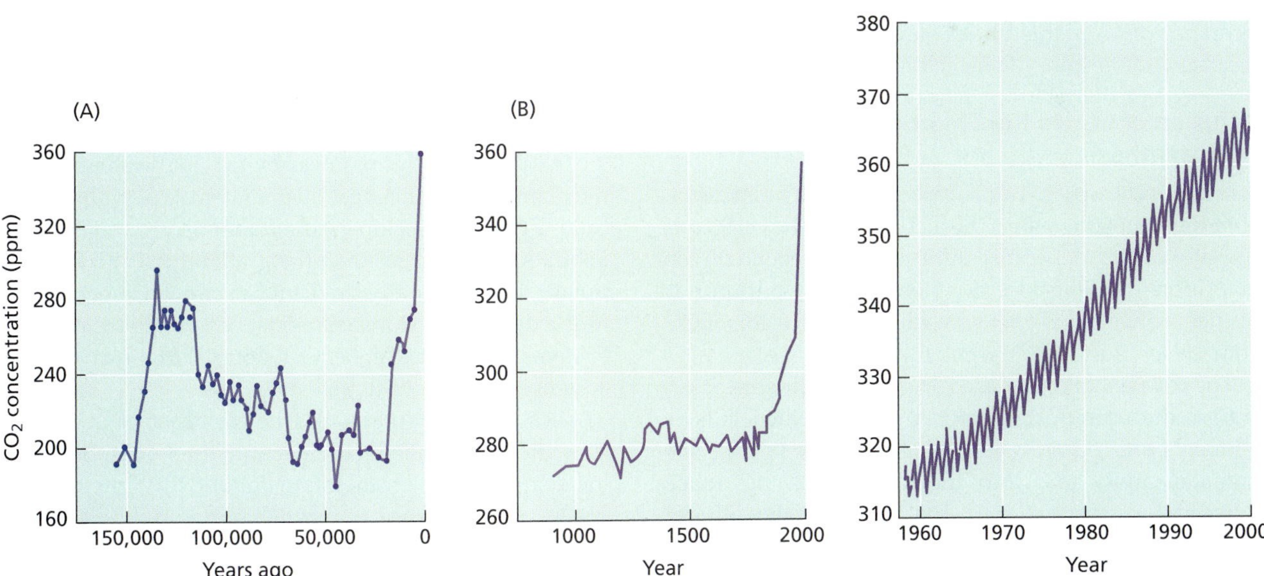

FIGURE 9.16 Concentration of atmospheric CO_2 from the present to 160,000 years ago. (A) Past atmospheric CO_2 concentrations, determined from bubbles trapped in glacial ice in Antarctica, were much lower than current levels. (B) In the last 1000 years, the rise in CO_2 concentration coincides with the Industrial Revolution and the increased burning of fossil fuels. (C) Current atmospheric concentrations of CO_2 measured at Mauna Loa, Hawaii, continue to rise. The wavy nature of the trace is caused by change in atmospheric CO_2 concentrations associated with the growth of agricultural crops. Each year the highest CO_2 concentration is observed in May, just before the Northern Hemisphere growing season, and the lowest concentration is observed in October. (After Barnola et al. 1994, Keeling and Whorf 1994, Neftel et al. 1994, and Keeling et al. 1995.)

The greenhouse effect. The consequences of this increase in atmospheric CO_2 are under intense scrutiny by scientists and government agencies, particularly because of predictions that the **greenhouse effect** is altering the world's climate. The term *greenhouse effect* refers to the resulting warming of Earth's climate, which is caused by the trapping of long-wavelength radiation by the atmosphere.

A greenhouse roof transmits visible light, which is absorbed by plants and other surfaces inside the greenhouse. The absorbed light energy is converted to heat, and part of it is re-emitted as long-wavelength radiation. Because glass transmits long-wavelength radiation very poorly, this radiation cannot leave the greenhouse through the glass roof, and the greenhouse heats up.

Certain gases in the atmosphere, particularly CO_2 and methane, play the same role as the glass roof in a greenhouse. The increased CO_2 concentration and temperature associated with the greenhouse effect can influence photosynthesis. At current atmospheric CO_2 concentrations, photosynthesis in C_3 plants is CO_2 limited (as we will discuss later in the chapter), but this situation could change as atmospheric CO_2 concentrations continue to rise. Under laboratory conditions, most C_3 plants grow 30 to 60% faster when CO_2 concentration is doubled (to 600–700 ppm), and the growth rate changes depend on nutrient status (Bowes 1993). In some plants the enhanced growth is only temporary.

For many crops, such as tomatoes, lettuce, cucumbers, and roses growing in greenhouses under optimal nutrition, carbon dioxide enrichment in the greenhouse environment results in increased productivity. The photosynthetic performance of C_3 plants under elevated CO_2 is enhanced because photorespiration decreases (see Chapter 8).

Diffusion of CO_2 to the Chloroplast Is Essential to Photosynthesis

For photosynthesis to occur, carbon dioxide must diffuse from the atmosphere into the leaf and into the carboxylation site of rubisco. Because diffusion rates depend on concentration gradients (see Chapters 3 and 6), appropriate gradients are needed to ensure adequate diffusion of CO_2 from the leaf surface to the chloroplast.

The cuticle that covers the leaf is nearly impermeable to CO_2, so the main port of entry of CO_2 into the leaf is the stomatal pore. CO_2 diffuses through the pore into the substomatal cavity and into the intercellular air spaces between the mesophyll cells. This portion of the diffusion path of CO_2 into the chloroplast is a gaseous phase. The remainder of the diffusion path to the chloroplast is a liquid phase, which begins at the water layer that wets the walls of the mesophyll cells and continues through the plasma membrane, the cytosol, and the chloroplast. (For the properties of CO_2 in solution, see **Web Topic 8.3**.)

Each portion of this diffusion pathway imposes a resistance to CO_2 diffusion, so the supply of CO_2 for photosynthesis meets a series of different points of resistance (Figure 9.17). An evaluation of the magnitude of each point of resistance is helpful for understanding CO_2 limitations to photosynthesis.

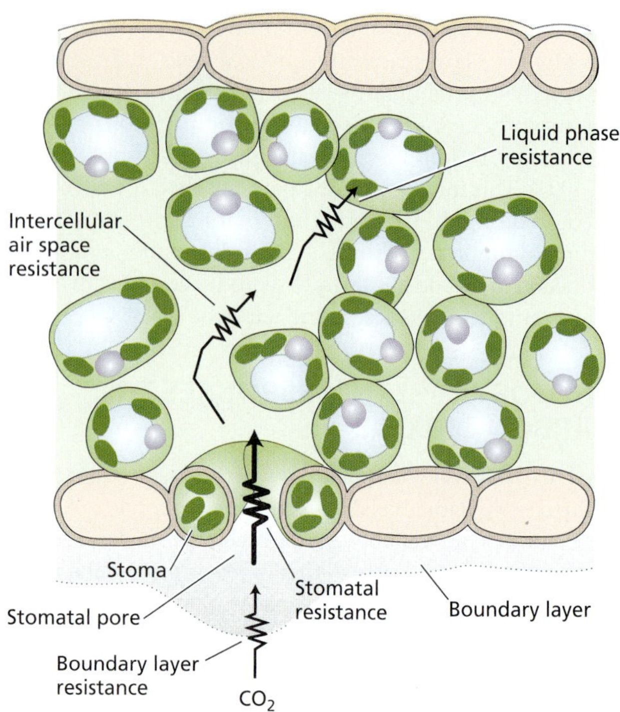

FIGURE 9.17 Points of resistance to the diffusion of CO_2 from outside the leaf to the chloroplasts. The stomatal pore is the major point of resistance to CO_2 diffusion.

Carbon dioxide enters the intercellular air spaces of the leaf through the stomatal pores. From the air spaces it dissolves in the water of wet cell walls and diffuses into the cell and chloroplast. The same path is traveled in the reverse direction by H_2O.

The sharing of this pathway by CO_2 and water presents the plant with a functional dilemma. In air of high relative humidity, the diffusion gradient that drives water loss is about 50 times larger than the gradient that drives CO_2 uptake. In drier air, this gradient can be even larger. Therefore, a decrease in stomatal resistance through the opening of stomata facilitates higher CO_2 uptake but is unavoidably accompanied by substantial water loss.

Recall from Chapter 4 that the gas phase of CO_2 diffusion into the leaf can be divided into three components—the boundary layer, the stomata, and the intercellular spaces of the leaf—each of which imposes a resistance to CO_2 diffusion (see Figure 9.17).

The boundary layer consists of relatively unstirred air at the leaf surface, and its resistance to diffusion is called the **boundary layer resistance**. The magnitude of the boundary layer resistance decreases with leaf size and wind speed. The boundary layer resistance to water and CO_2 diffusion is physically related to the boundary layer resistance to sensible heat loss discussed earlier.

Smaller leaves have a lower boundary layer resistance to CO_2 and water diffusion, and to sensible heat loss. Leaves of desert plants are usually small, facilitating sensible heat loss. The large leaves often found in the humid Tropics can have large boundary layer resistances, but these leaves can dissipate the radiation heat load by evaporative cooling because of the high transpiration rates made possible by the abundant water supply in these habitats.

After diffusing through the boundary layer, CO_2 enters the leaf through the stomatal pores, which impose the next type of resistance in the diffusion pathway, the **stomatal resistance**. Under most conditions in nature, in which the air around a leaf is seldom completely still, the boundary layer resistance is much smaller than the stomatal resistance, and the main limitation to CO_2 diffusion is imposed by the stomatal resistance.

There is also a resistance to CO_2 diffusion in the air spaces that separate the substomatal cavity from the walls of the mesophyll cells, called the **intercellular air space resistance**. This resistance is also usually small—causing a drop of 0.5 Pa or less in partial pressure of CO_2, compared with the 36 Pa outside the leaf.

The resistance to CO_2 diffusion of the liquid phase—the **liquid phase resistance**, also called **mesophyll resistance**—encompasses diffusion from the intercellular leaf spaces to the carboxylation sites in the chloroplast. This point of resistance to CO_2 diffusion has been calculated as approximately one-tenth of the combined boundary layer resistance and stomatal resistance when the stomata are fully open. This low resistance value can be attributed in part to the large surface area of mesophyll cells exposed to the intercellular air spaces, which can be as much as 10 to 30 times the projected leaf area (Syvertsen et al. 1995). In addition, the localization of chloroplasts near the cell periphery minimizes the distance that CO_2 diffuses to carboxylation sites within the chloroplast.

The positioning of chloroplasts and the relatively large percentage of intercellular air space (about 20–40%) are special anatomic features that facilitate the internal diffusion and uptake of CO_2 by leaves (Evans 1999). Because the stomatal pores usually impose the largest resistance to CO_2 uptake and water loss in the diffusion pathway, this regulation provides the plant with an effective way to control gas exchange between the leaf and the atmosphere. In experimental measurements of gas exchange from leaves, the boundary layer resistance and the intercellular air space resistance are usually ignored, and the stomatal resistance is used as the single parameter describing the gas phase resistance to CO_2 (see **Web Topic 9.4**).

Patterns of Light Absorption Generate Gradients of CO_2 Fixation within the Leaf

We have discussed how leaf anatomy is specialized for capturing light and how it also facilitates the internal diffusion of CO_2, but where in the leaf do maximum rates of photosynthesis occur? In most leaves, light is preferentially absorbed at the upper surface, whereas CO_2 enters through the lower surface. Given that light and CO_2 enter from opposing sides of the leaf, does photosynthesis occur uniformly within the leaf tissues, or is there a gradient in photosynthesis across the leaf? The photosynthetic properties of a leaf are determined by the following:

- Profiles of light absorption across the mesophyll
- Photosynthetic capacity of those tissues
- Internal CO_2 supply

For most leaves, internal CO_2 diffusion is rapid, so limitations on photosynthetic performance within the leaf are imposed by factors other than CO_2 supply. When white light enters the upper surface of a leaf, blue and red photons are preferentially absorbed by chloroplasts near the irradiated surface (Figure 9.18), owing to the strong absorption bands of chlorophyll in the blue and red regions of the spectrum (see Figure 7.5). Green light, on the other hand, penetrates deeper into the leaf. Compared to blue and red,

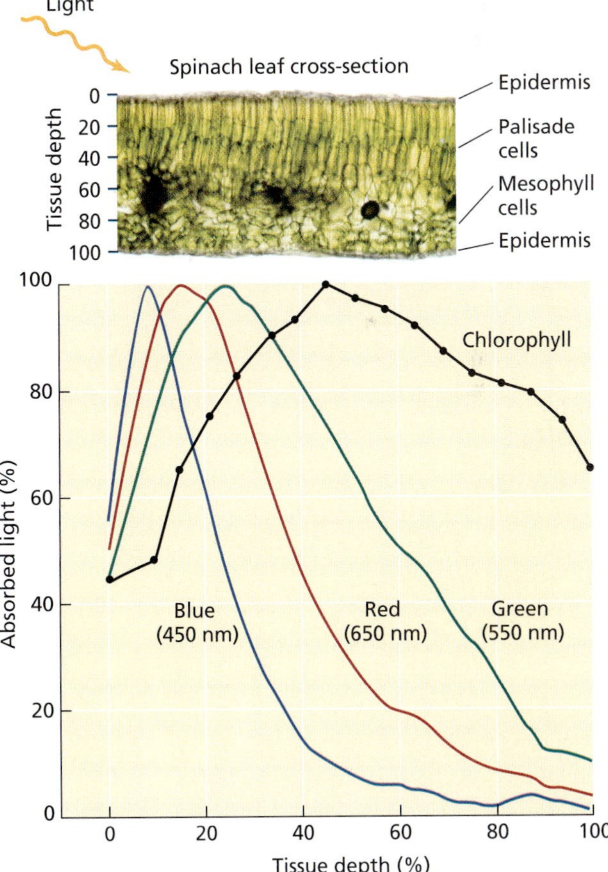

FIGURE 9.18 Distribution of absorbed light in spinach sun leaves. Irradiation with blue, green or red light results in different profiles of absorbed light in the leaf. The micrograph above the graph shows a cross-section of a spinach leaf, with rows of palisade cells occupying nearly half of the leaf thickness. The shapes of the curves are in part a result of the unequal distribution of chlorophyll within the leaf tissues. (From Nishio et al. 1993 and Vogelmann and Han 2000; micrograph courtesy of T. Vogelmann.)

chlorophyll absorbs poorly in the green (again, see Figure 7.5), yet green light is very effective in supplying energy for photosynthesis in the tissues within the leaf depleted from blue and red photons.

The capacity of the leaf tissue for photosynthetic CO_2 assimilation depends to a large extent on its rubisco content. In spinach and the faba bean (*Vicia faba*), rubisco content starts out low at the top of the leaf, increases toward the middle, and then decreases again toward the bottom. As a result, the distribution of carbon fixation within the leaf is bell shaped (Figure 9.19). The spongy mesophyll (see Figure 9.4) fixes about 40% of the total carbon in spinach. The functional significance of the rubisco distribution and the profiles of carbon assimilation within leaves is not yet known, although it is likely that photosynthesis profiles vary in leaves with different anatomy and in leaves adapted to different environments.

CO_2 Imposes Limitations on Photosynthesis

Expressing photosynthetic rate as a function of the partial pressure of CO_2 in the intercellular air space (C_i) within the leaf (see **Web Topic 9.4**) makes it possible to evaluate limitations to photosynthesis imposed by CO_2 supply. At very low intercellular CO_2 concentrations, photosynthesis is strongly limited by the low CO_2, while respiratory rates are unaffected. As a result, there is a negative balance between CO_2 fixed by photosynthesis and CO_2 produced by respiration, and a net efflux of CO_2 from the plant.

Increasing intercellular CO_2 to the concentration at which these two processes balance each other defines the **CO_2 compensation point**, at which the net efflux of CO_2 from the plant is zero (Figure 9.20A). This concept is analogous to that of the light compensation point discussed earlier in the chapter: *The CO_2 compensation point reflects the balance between photosynthesis and respiration as a function of CO_2 concentration, and the light compensation point reflects that balance as a function of photon flux.*

In C_3 plants, increasing CO_2 above the compensation point stimulates photosynthesis over a wide concentration

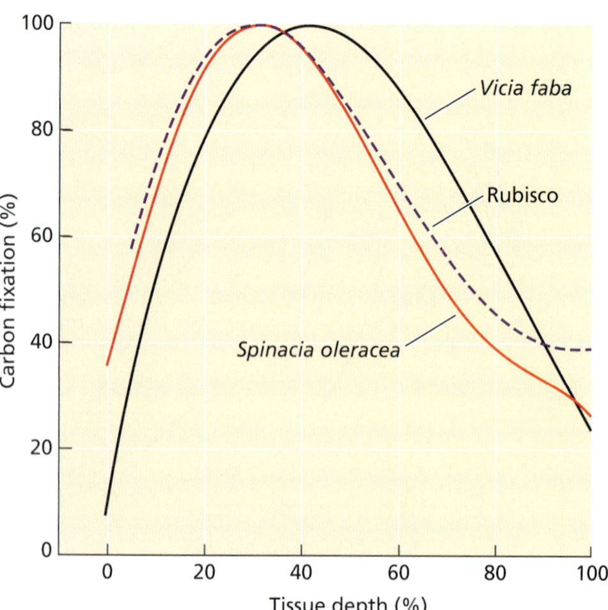

FIGURE 9.19 Distribution of rubisco and carbon fixation within leaves. Carbon fixation (solid line) within spinach leaves closely follows the internal distribution of rubisco (dashed line). Carbon fixation profiles are similar between *Vicia* and spinach. (From Nishio et al. 1993 and Jeje and Zimmermann 1983.)

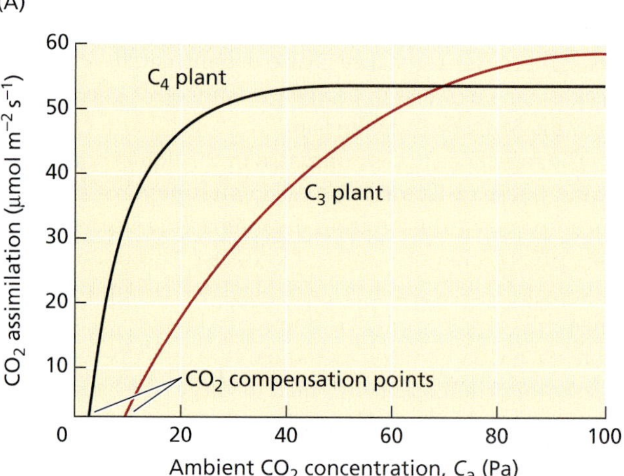

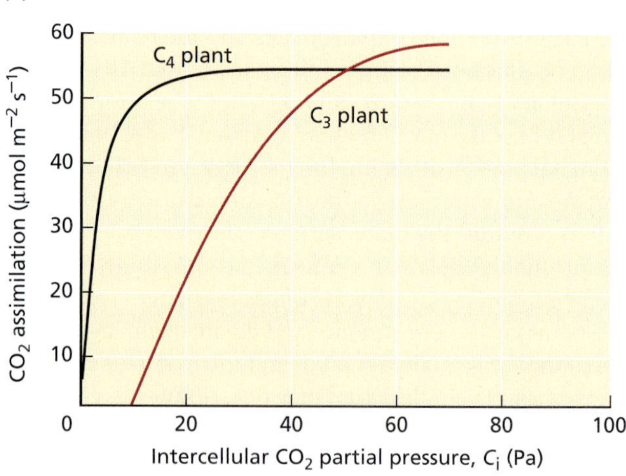

FIGURE 9.20 Changes in photosynthesis as a function of ambient intercellular CO_2 concentrations in *Tidestromia oblongifolia* (Arizona honeysweet), a C_4 plant, and *Larrea divaricata* (creosote bush), a C_3 plant. Photosynthetic rate is plotted against (A) partial pressure of CO_2 in ambient air and (B) calculated intercellular partial pressure of CO_2 inside the leaf (see Equation 5 in **Web Topic 9.4**). The partial pressure at which CO_2 assimilation is zero defines the CO_2 compensation point. (From Berry and Downton 1982.)

range (see Figure 9.20A). At low to intermediate CO_2 concentrations, photosynthesis is limited by the carboxylation capacity of rubisco. At high CO_2 concentrations, photosynthesis is limited by the capacity of Calvin cycle to regenerate the acceptor molecule ribulose-1,5-bisphosphate, which depends on electron transport rates. By regulating stomatal conductance, most leaves appear to regulate their C_i (internal partial pressure for CO_2) such that it is intermediate between limitations imposed by carboxylation capacity and the capacity to regenerate ribulose-1,5-bisphosphate.

A plot of CO_2 assimilation as a function intercellular partial pressures of CO_2 tells us how photosynthesis is regulated by CO_2, independent of the functioning of stomata (Figure 9.20B). Inspection of such a plot for C_3 and C_4 plants reveals interesting differences between the two types of carbon metabolism:

- In *C_4 plants*, photosynthetic rates saturate at C_i values of about 15 Pa, reflecting the effective CO_2-concentrating mechanisms operating in these plants (see Chapter 8).
- In *C_3 plants*, increasing C_i levels continue to stimulate photosynthesis over a much broader range.

These results indicate that C_3 plants may benefit more from ongoing increases in atmospheric CO_2 concentrations (see Figure 9.16). In contrast, photosynthesis in C_4 plants is CO_2 saturated at low concentrations, and as a result C_4 plants do not benefit from increases in atmospheric CO_2 concentrations. Figure 9.20 also shows that plants with C_4 metabolism have a CO_2 compensation point of zero or nearly zero, reflecting their very low levels of photorespiration (see Chapter 8). This difference between C_3 and C_4 plants is not seen when the experiments are conducted at low oxygen concentrations because oxygenation is also suppressed in C_3 plants.

CO_2-Concentrating Mechanisms Affect Photosynthetic Responses of Leaves

Because of the operating CO_2-concentrating mechanisms in C_4 plants, CO_2 concentration at the carboxylation sites within C_4 chloroplasts is often saturating for rubisco activity. As a result, plants with C_4 metabolism need less rubisco than C_3 plants need to achieve a given rate of photosynthesis, and require less nitrogen to grow (von Caemmerer 2000).

In addition, the CO_2-concentrating mechanism allows the leaf to maintain high photosynthetic rates at lower C_i values, which require lower rates of stomatal conductance for a given rate of photosynthesis. Thus, C_4 plants can use water and nitrogen more efficiently than C_3 plants can. On the other hand, the additional energy cost of the concentrating mechanism (see Chapter 8) makes C_4 plants less efficient in their utilization of light. This is probably one of the reasons that most shade-adapted plants are C_3 plants.

Many cacti and other succulent plants with CAM metabolism open their stomata at night and close them during the day (Figure 9.21). The CO_2 taken up during the night is fixed into malate (see Chapter 8). Because air temperatures are much lower at night than during the day, water loss is low and a significant amount of water is saved relative to the amount of CO_2 fixed.

The main constraint on CAM metabolism is that the capacity to store malic acid is limited, and this limitation restricts the amount of CO_2 uptake. However, many CAM plants can fix CO_2 via the Calvin cycle at the end of the day, when temperature gradients are less extreme.

Cladodes (flattened stems) of cacti can survive after detachment from the plant for several months without

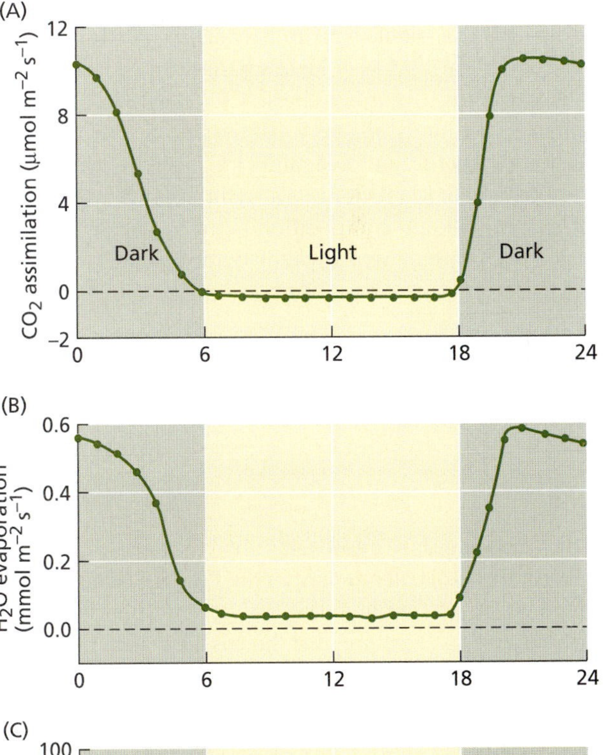

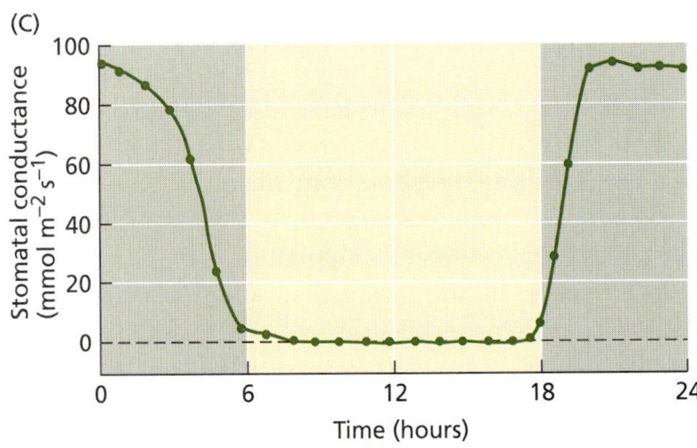

FIGURE 9.21 Photosynthetic carbon assimilation, evaporation, and stomatal conductance of a CAM plant, the cactus *Opuntia ficus-indica*, during a 24-hour period. The whole plant was kept in a gas exchange chamber in the laboratory. The dark period is indicated by shaded areas. In contrast to plants with C_3 or C_4 metabolism, CAM plants open their stomata and fix CO_2 at night. (From Gibson and Nobel 1986.)

water. Their stomata are closed all the time, and the CO_2 released by respiration is refixed into malate. This process, which has been called *CAM idling*, allows the plant to survive for prolonged periods of time while losing remarkably little water.

Discrimination of Carbon Isotopes Reveals Different Photosynthetic Pathways

Atmospheric CO_2 contains the naturally occurring carbon isotopes ^{12}C, ^{13}C, and ^{14}C in the proportions 98.9%, 1.1%, and 10^{-10}%, respectively. $^{14}CO_2$ is present in such small quantities that it has no physiological relevance, but $^{13}CO_2$ is different. The chemical properties of $^{13}CO_2$ are identical to those of $^{12}CO_2$, but because of the slight difference in mass (2.3%), most plants assimilate less $^{13}CO_2$ than $^{12}CO_2$. In other words, plants discriminate against the heavier isotope of carbon, and they have smaller ratios of ^{13}C to ^{12}C than are found in atmospheric CO_2. How effective are plants at distinguishing between the two carbon isotopes? Although discrimination against ^{13}C is subtle, the isotope composition of plants reveals a wealth of information.

Carbon isotope composition is measured by use of a mass spectrometer, which yields the following ratio:

$$R = \frac{^{13}CO_2}{^{12}CO_2} \tag{9.1}$$

The **isotope composition** of plants, $\delta^{13}C$, is quantified on a per mil (–‰) basis:

$$\delta^{13}C\,‰ = \left(\frac{R_{\text{sample}}}{R_{\text{standard}}} - 1\right) \times 1000 \tag{9.2}$$

where the standard represents the carbon isotopes contained in a fossil belemnite from the Pee Dee limestone formation of South Carolina. The $\delta^{13}C$ of atmospheric CO_2 has a value of –8 ‰, meaning that there is less ^{13}C in the atmospheric CO_2 than is found in the carbonate of the belemnite standard. What are some typical values for carbon isotope ratios of plants? C_3 plants have a $\delta^{13}C$ of about –28 ‰; C_4 plants have an average value of –14 ‰ (Farquhar et al. 1989). Both C_3 and C_4 plants have less ^{13}C than the isotope standard, which means that there has been a discrimination against ^{13}C during the photosynthetic process.

Because the per mil calculation involves multiplying by 1000, the actual isotope discrimination is small. Nonetheless, differences in carbon isotope discrimination are easily detectable with mass spectrometers. For example, measuring the $\delta^{13}C$ of table sugar (sucrose) makes it possible to determine if the sucrose came from sugar beet (a C_3 plant) or sugarcane (a C_4 plant).

What is the physiological basis for ^{13}C depletion in plants? One reason in both C_3 and C_4 plants is diffusion. CO_2 diffuses from air outside of the leaf to the carboxylation sites within leaves. Because $^{12}CO_2$ is lighter than $^{13}CO_2$, it diffuses slightly faster toward the carboxylation site, creating an effective diffusion discrimination of –4.4 ‰. However, the largest isotope discrimination step is the carboxylation reaction catalyzed by rubisco (Farquhar et al. 1989).

Rubisco has an intrinsic discrimination value against ^{13}C of –30 ‰. By contrast, PEP carboxylase, the primary CO_2 fixation enzyme of C_4 plants, has a much smaller isotope discrimination effect (about –2 to –6 ‰). Thus the inherent difference between the discrimination effects of the two carboxylating enzymes causes the different isotope compositions observed in C_3 and C_4 plants (Farquhar et al. 1989).

Other physiological characteristics of plants affect isotope composition. One factor is the partial pressure of CO_2 in the intercellular air spaces of leaves (C_i). In C_3 plants the potential discrimination by rubisco of –30 ‰ is not fully expressed because the availability of CO_2 at the carboxylation site becomes a limiting factor restricting the discrimination by rubisco. More discrimination occurs when C_i is high, as when stomata are open. Open stomata also facilitate water loss. Thus, lower water use efficiency is correlated with greater discrimination against ^{13}C (Farquhar et al. 1989).

Fossil fuels have a $\delta^{13}C$ of about –26 ‰ because the carbon in these deposits came from organisms that had a C_3 carbon fixation pathway. Furthermore, measuring $\delta^{13}C$ in fossil, carbonate-containing soils and fossil teeth makes it possible to determine that C_4 photosynthesis developed and became prevalent relatively recently (see **Web Topic 9.5**).

CAM plants can have $\delta^{13}C$ values that are intermediate between those of C_3 and C_4 plants. In CAM plants that fix CO_2 at night via PEP carboxylase, $\delta^{13}C$ is similar to that of C_4 plants. However, when some CAM plants are well watered, they switch to C_3 mode by opening their stomata and fixing CO_2 during the day via rubisco. Under these conditions the isotope composition shifts more toward that of C_3 plants. Thus the $^{13}C/^{12}C$ values of CAM plants reflect how much carbon is fixed via the C_3 pathway versus the C_4 pathway (see **Web Topic 9.5**).

Plants also fractionate other isotopes, such as $^{18}O/^{16}O$ and $^{15}N/^{14}N$, and the various patterns of isotope enrichment or depletion can be used as indicators of particular metabolic pathways or features.

PHOTOSYNTHETIC RESPONSES TO TEMPERATURE

When photosynthetic rate is plotted as a function of temperature, the curve has a characteristic bell shape (Figure 9.22). The ascending arm of the curve represents a temperature-dependent stimulation of photosynthesis up to an optimum; the descending arm is associated with deleterious effects, some of which are reversible while others are not.

Temperature affects all biochemical reactions of photosynthesis, so it is not surprising that the responses to temperature are complex. We can gain insight into the underlying mechanisms by comparing photosynthetic rates in air at normal and at high CO_2 concentrations. At high CO_2 (see Figure 9.22A), there is an ample supply of CO_2 at the car-

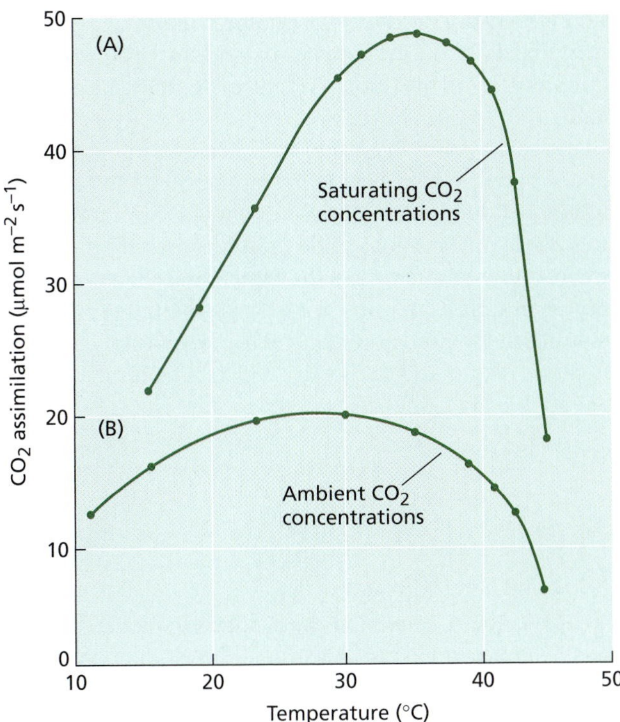

FIGURE 9.22 Changes in photosynthesis as a function of temperature at CO_2 concentrations that saturate photosynthetic CO_2 assimilation (A) and at normal atmospheric CO_2 concentrations (B). Photosynthesis depends strongly on temperature at saturating CO_2 concentrations. Note the significantly higher photosynthetic rates at saturating CO_2 concentrations. (Redrawn from Berry and Björkman 1980.)

boxylation sites, and the rate of photosynthesis is limited primarily by biochemical reactions connected with electron transport (see Chapter 7). In these conditions, temperature changes have large effects on fixation rates.

At ambient CO_2 concentrations (see Figure 9.22B), photosynthesis is limited by the activity of rubisco, and the response reflects two conflicting processes: an increase in carboxylation rate with temperature and a decrease in the affinity of rubisco for CO_2 as the temperature rises (see Chapter 8). These opposing effects dampen the temperature response of photosynthesis at ambient CO_2 concentrations.

Respiration rates also increase as a function of temperature, and the interaction between photorespiration and photosynthesis becomes apparent in temperature responses. Figure 9.23 shows changes in quantum yield as a function of temperature in a C_3 plant and in a C_4 plant. In the C_4 plant the quantum yield remains constant with temperature, reflecting typical low rates of photorespiration. In the C_3 plant the quantum yield decreases with temperature, reflecting a stimulation of photorespiration by temperature and an ensuing higher energy demand per net CO_2 fixed.

At low temperatures, photosynthesis is often limited by phosphate availability at the chloroplast (Sage and Sharkey 1987). When triose phosphates are exported from the chloroplast to the cytosol, an equimolar amount of inorganic phosphate is taken up via translocators in the chloroplast membrane.

If the rate of triose phosphate utilization in the cytosol decreases, phosphate uptake into the chloroplast is inhibited and photosynthesis becomes phosphate limited (Geiger and Servaites 1994). Starch synthesis and sucrose synthesis decrease rapidly with temperature, reducing the demand for triose phosphates and causing the phosphate limitation observed at low temperatures.

The highest photosynthetic rates seen in temperature responses represent the so-called *optimal temperature response*. When these temperatures are exceeded, photosynthetic rates decrease again. It has been argued that this optimal temperature is the point at which the capacities of the various steps of photosynthesis are optimally balanced, with some of the steps becoming limiting as the temperature decreases or increases.

Optimal temperatures have strong genetic and physiological components. Plants of different species growing in habitats with different temperatures have different optimal temperatures for photosynthesis, and plants of the same

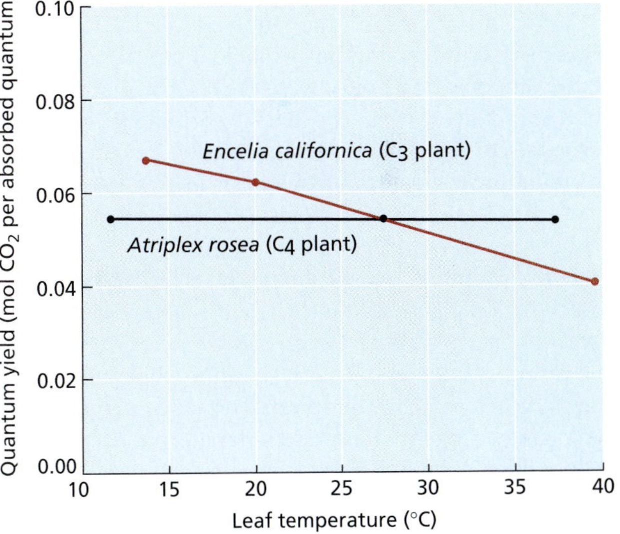

FIGURE 9.23 The quantum yield of photosynthetic carbon fixation in a C_3 plant and in a C_4 plant as a function of leaf temperature. In normal air, photorespiration increases with temperature in C_3 plants, and the energy cost of net CO_2 fixation increases accordingly. This higher energy cost is expressed in lower quantum yields at higher temperatures. Because of the CO_2 concentrating mechanisms of C_4 plants, photorespiration is low in these plants, and the quantum yield does not show a temperature dependence. Note that at lower temperatures the quantum yield of C_3 plants is higher than that of C_4 plants, indicating that photosynthesis in C_3 plants is more efficient at lower temperatures. (From Ehleringer and Björkman 1977.)

species, grown at different temperatures and then tested for their photosynthetic responses, show temperature optima that correlate with the temperature at which they were grown. Plants growing at low temperatures maintain higher photosynthetic rates at low temperatures than plants grown at high temperatures.

These changes in photosynthetic properties in response to temperature play an important role in plant adaptations to different environments. Plants are remarkably plastic in their adaptations to temperature. In the lower temperature range, plants growing in alpine areas are capable of net CO_2 uptake at temperatures close to 0°C; at the other extreme, plants living in Death Valley, California, have optimal rates of photosynthesis at temperatures approaching 50°C.

SUMMARY

Photosynthetic activity in the intact leaf is an integral process that depends on many biochemical reactions. Different environmental factors can limit photosynthetic rates.

Leaf anatomy is highly specialized for light absorption, and the properties of palisade and mesophyll cells ensure uniform light absorption throughout the leaf. In addition to the anatomic features of the leaf, chloroplast movements within cells and solar tracking by the leaf blade help maximize light absorption. Light transmitted through upper leaves is absorbed by leaves growing beneath them.

Many properties of the photosynthetic apparatus change as a function of the available light, including the light compensation point, which is higher in sun leaves than in shade leaves. The linear portion of the light-response curve for photosynthesis provides a measure of the quantum yield of photosynthesis in the intact leaf. In temperate areas, quantum yields of C_3 plants are generally higher than those of C_4 plants.

Sunlight imposes a substantial heat load on the leaf, which is dissipated back into the air by long-wavelength radiation, by sensible heat loss, or by evaporative heat loss. Increasing CO_2 concentrations in the atmosphere are increasing the heat load on the biosphere. This process could cause damaging changes in the world's climate, but it could also reduce the CO_2 limitations on photosynthesis. At high photon flux, photosynthesis in most plants is CO_2 limited, but the limitation is substantially lower in C_4 and CAM plants because of their CO_2-concentrating mechanisms.

Diffusion of CO_2 into the leaf is constrained by a series of different points of resistance. The largest resistance is usually that imposed by the stomata, so modulation of stomatal apertures provides the plant with an effective means of controlling water loss and CO_2 uptake. Both stomatal and nonstomatal factors affect CO_2 limitations on photosynthesis.

Temperature responses of photosynthesis reflect the temperature sensitivity of the biochemical reactions of photosynthesis and are most pronounced at high CO_2 concentrations. Because of the role of photorespiration, the quantum yield is strongly dependent on temperature in C_3 plants but is nearly independent of temperature in C_4 plants.

Leaves growing in cold climates can maintain higher photosynthetic rates at low temperatures than leaves growing in warmer climates. Leaves grown at high temperatures perform better at high temperatures than leaves grown at low temperatures do. Functional changes in the photosynthetic apparatus in response to prevailing temperatures in their environment have an important effect on the capacity of plants to live in diverse habitats.

Web Material

Web Topics

9.1 Working with Light

Amount, direction, and spectral quality are important parameters for the measurement of light.

9.2 Heat Dissipation from Leaves: The Bowen Ratio

Sensible heat loss and evaporative heat loss are the most important processes in the regulation of leaf temperature.

9.3 Working with Gases

This web topic explains how to work with mole fractions and other physical parameters of gases.

9.4 Calculating Important Parameters in Leaf Gas Exchange

Gas exchange methods allow us to measure photosynthesis and stomatal conductance in the intact leaf.

9.5 Isotope Discrimination

The carbon isotope composition of plants reveals a wealth of information.

Web Essay

9.1 The Xanthophyll Cycle

Molecular and biophysical studies are revealing the role of the xanthophyll cycle on the photoprotection of leaves.

Chapter References

Adams, W. W., Demmig-Adams, B., Rosenstiel, T. N., and Ebbert, V. (2001) Dependence of photosynthesis and energy dissipation activity upon growth form and light environment during the winter. *Photosynth. Res.* 67: 51–62.

Anderson, J. M. (1986) Photoregulation of the composition, function, and structure of thylakoid membranes. *Annu. Rev. Plant Physiol.* 37: 93–136.

Barnola, J. M., Raynaud, D., Lorius, C., and Korothevich, Y. S. (1994) Historical CO_2 record from the Vostok ice core. In *Trends '93: A Compendium of Data on Global Change* (ORNL/CDIAC-65), T. A. Boden, D. P. Kaiser, R. J. Sepanski, and F. W. Stoss, eds., Carbon Dioxide Information Center, Oak Ridge National Laboratory, Oak Ridge, TN, pp. 7–10.

Berry, J., and Björkman, O. (1980) Photosynthetic response and adaptation to temperature in higher plants. *Annu. Rev. Plant Physiol.* 31: 491–543.

Berry, J. A., and Downton, J. S. (1982) Environmental regulation of photosynthesis. In *Photosynthesis: Development, Carbon Metabolism and Plant Productivity*, Vol. II, Govindjee, ed., Academic Press, New York, pp. 263–343.

Björkman, O. (1981) Responses to different quantum flux densities. In *Encyclopedia of Plant Physiology*, New Series, Vol. 12A, O. L. Lange, P. S. Nobel, C. B. Osmond, and H. Zeigler, eds., Springer, Berlin, pp. 57–107.

Björn, L. O., and Vogelmann, T. C. (1994) Quantification of light. In *Photomorphogenesis in Plants*, 2nd ed., R. E. Kendrick and G. H. M. Kronenberg, eds., Kluwer, Dordrecht, Netherlands, pp. 17–25.

Bowes, G. (1993) Facing the inevitable: Plants and increasing atmospheric CO_2. *Annu. Rev. Plant Physiol. Plant Mol. Biol.* 44: 309–332.

Campbell, G. S. (1977) *An Introduction to Environmental Biophysics*. Springer-Verlag, New York.

Demmig-Adams, B., and Adams, W. (1996) The role of xanthophyll cycle carotenoids in the protection of photosynthesis. *Trends Plant Sci.* 1: 21–26.

Ehleringer, J. R., Björkman, O., and Mooney, H. A. (1976) Leaf pubescence: Effects on absorptance and photosynthesis in a desert shrub. *Science* 192: 376–377.

Ehleringer, J. R., and Björkman, O. (1977) Quantum yields for CO_2 uptake in C_3 and C_4 plants. *Plant Physiol.* 59: 86–90.

Ehleringer, J. R., Sage, R. F., Flanagan, L. B., and Pearcy, R. W. (1991) Climate change and the evolution of C_4 photosynthesis. *Trends Ecol. Evol.* 6: 95–99.

Evans, J. R. (1999) Leaf anatomy enables more equal access to light and CO_2 between chloroplasts. *New Phytol.* 143: 93–104.

Farquhar, G. D., Ehleringer, J. R., and Hubick, K. T. (1989) Carbon isotope discrimination and photosynthesis. *Annu. Rev. Plant Physiol. Plant Mol. Biol.* 40: 503–538.

Geiger, D. R., and Servaites, J. C. (1994) Diurnal regulation of photosynthetic carbon metabolism in C_3 plants. *Annu. Rev. Plant Physiol. Plant Mol. Biol.* 45: 235–256.

Gibson, A. C., and Nobel, P. S. (1986) *The Cactus Primer*. Harvard University Press, Cambridge, MA.

Gorton, H. L., Williams, W. E., and Vogelmann, T. C. (1999) Chloroplast movement in *Alocasia macrorrhiza*. *Physiol. Plant.* 106: 421–428.

Harvey, G. W. (1979) Photosynthetic performance of isolated leaf cells from sun and shade plants. *Carnegie Inst. Washington Yearbook* 79: 161–164.

Haupt, W., and Scheuerlein, R. (1990) Chloroplast movement. *Plant Cell Environ.* 13: 595–614.

Havaux, M., Tardy, F., Ravenel, J., Chanu, D., and Parot, P. (1996) Thylakoid membrane stability to heat stress studied by flash spectroscopic measurements of the electrochromic shift in intact potato leaves: Influence of the xanthophyll content. *Plant Cell Environ.* 19: 1359–1368.

Jarvis, P. G., and Leverenz, J. W. (1983) Productivity of temperate, deciduous and evergreen forests. In *Encyclopedia of Plant Physiology*, New Series, Vol. 12D, O. L. Lange, P. S. Nobel, C. B. Osmond, and H. Zeigler, eds., Springer, Berlin, pp. 233–280.

Jeje, A., and Zimmermann, M. (1983) The anisotropy of the mesophyll and CO_2 capture sites in *Vicia faba* L. leaves at low light intensities. *J. Exp. Bot.* 34: 1676–1694.

Keeling, C. D., and Whorf, T. P. (1994) Atmospheric CO_2 records from sites in the SIO air sampling network. In *Trends '93: A Compendium of Data on Global Change* (ORNL/CDIAC-65), T. A. Boden, D. P. Kaiser, R. J. Sepanski, and F. W. Stoss, eds., Carbon Dioxide Information Center, Oak Ridge National Laboratory, Oak Ridge, TN, pp. 16–26.

Keeling, C. D., Whorf, T. P., Wahlen, M., and Van der Plicht, J. (1995) Interannual extremes in the rate of rise of atmospheric carbon dioxide since 1980. *Nature* 375: 666–670.

Koller, D. (1990) Light-driven leaf movements. *Plant Cell Environ.* 13: 615–632.

Koller, D. (2000) Plants in search of sunlight. *Adv. Bot. Res.* 33: 35–131.

Long, S. P., Humphries, S., and Falkowski, P. G. (1994) Photoinhibition of photosynthesis in nature. *Annu. Rev. Plant Physiol. Plant Mol. Biol.* 45: 633–662.

McCree, K. J. (1981) Photosynthetically active radiation. In *Encyclopedia of Plant Physiology*, New Series, Vol. 12A, O. L. Lange, P. S. Nobel, C. B. Osmond, and H. Zeigler, eds., Springer, Berlin, pp. 41–55.

Melis, A. (1996) Excitation energy transfer: Functional and dynamic aspects of Lhc (cab) proteins. In *Oxygenic Photosynthesis: The Light Reactions*, D. R. Ort and C. F. Yocum, eds., Kluwer, Dordrecht, Netherlands, pp. 523–538.

Neftel, A., Friedle, H., Moor, E., Lötscher, H., Oeschger, H., Siegenthaler, U., and Stauffer, B. (1994) Historical CO_2 record from the Siple Station ice core. In *Trends '93: A Compendium of Data on Global Change* (ORNL/CDIAC-65), T. A. Boden, D. P. Kaiser, R. J. Sepanski, and F. W. Stoss, eds., Carbon Dioxide Information Center, Oak Ridge National Laboratory, Oak Ridge, TN, pp. 11–15.

Nishio, J. N., Sun, J., and Vogelmann, T. C. (1993) Carbon fixation gradients across spinach leaves do not follow internal light gradient. *Plant Cell* 5: 953–961.

O'Leary, M. H. (1988) Carbon isotopes in photosynthesis. *BioScience* 38: 328–333.

Ort, D. R., and Baker, N. R. (1988) Consideration of photosynthetic efficiency at low light as a major determinant of crop photosynthetic performance. *Plant Physiol. Biochem.* 26: 555–565.

Osmond, C. B. (1994) What is photoinhibition? Some insights from comparisons of shade and sun plants. In *Photoinhibition of Photosynthesis: From Molecular Mechanisms to the Field*. N. R. Baker and J. R. Bowyer, eds., BIOS Scientific, Oxford, pp. 1–24.

Pearcy, R. W., Gross, L. J., and He, D. (1997) An improved dynamic model of photosynthesis for estimation of carbon gain in sunfleck light regimes. *Plant Cell Environ.* 20: 411–424.

Richter, T., and Fukshansky, L. (1996) Optics of a bifacial leaf: 2. Light regime as affected by leaf structure and the light source. *Photochem. Photobiol.* 63: 517–527.

Rupert, C. S., and Letarjet, R. (1978) Toward a nomenclature and dosimetric scheme applicable to all radiations. *Photochem. Photobiol.* 28: 3–5.

Sage, R. F., and Sharkey, T. D. (1987) The effect of temperature on the occurrence of O_2 and CO_2 insensitive photosynthesis in field grown plants. *Plant Physiol.* 84: 658–664.

Sharkey, T. D. (1996) Emission of low molecular mass hydrocarbons from plants. *Trends Plant Sci.* 1: 78–82.

Sharkey, T. D., and Singsaas, E. L. (1995) Why plants emit isoprene. *Nature* 374: 769.

Sharkey, T. D., Chen, X., and Yeh, S. (2001) Isoprene increases thermotolerance of fosmidomycin-fed leaves. *Plant Physiol.* 125: 2001–2006.

Smith, H. (1986). The perception of light quality. In *Photomorphogenesis in Plants*, R. E. Kendrick and G. H. M. Kronenberg, eds., Nijhoff, Dordrecht, Netherlands, pp. 187–217.

Smith, H. (1994). Sensing the light environment: The functions of the phytochrome family. In *Photomorphogenesis in Plants*, 2nd ed., R. E. Kendrick and G. H. M. Kronenberg, eds., Nijhoff, Dordrecht, Netherlands, pp. 377–416.

Syvertsen, J. P., Lloyd, J., McConchie, C., Kriedemann, P. E., and Farquhar, G. D. (1995) On the relationship between leaf anatomy and CO_2 diffusion through the mesophyll of hypostomatous leaves. *Plant Cell Environ.* 18: 149–157.

Terashima, I. (1992) Anatomy of non-uniform leaf photosynthesis. *Photosynth. Res.* 31: 195–212.

Terashima, I., and Hikosaka, K. (1995) Comparative ecophysiology of leaf and canopy photosynthesis. *Plant Cell Environ.* 18: 1111–1128.

Tlalka, M., and Fricker, M. (1999) The role of calcium in blue-light-dependent chloroplast movement in *Lemna trisulca* L. *Plant J.* 20: 461–473.

Vogelmann, T. C. (1993) Plant tissue optics. *Annu. Rev. Plant Physiol. Plant Mol. Biol.* 44: 231–251.

Vogelmann, T. C., and Björn, L. O. (1983) Response to directional light by leaves of a sun-tracking lupine (*Lupinus succulentus*). *Physiol. Plant.* 59: 533–538.

Vogelmann, T. C., and Han, T. (2000) Measurement of gradients of absorbed light in spinach leaves from chlorophyll fluorescence profiles. *Plant Cell Environ.* 23: 1303–1311.

Vogelmann, T. C., Bornman, J. F., and Yates, D. J. (1996) Focusing of light by leaf epidermal cells. *Physiol. Plant.* 98: 43–56.

von Caemmerer, S. (2000) *Biochemical Models of Leaf Photosynthesis*. CSIRO, Melbourne, Australia.

Chapter

10

Translocation in the Phloem

SURVIVAL ON LAND POSES SOME SERIOUS CHALLENGES to terrestrial plants, foremost of which is the need to acquire and retain water. In response to these environmental pressures, plants evolved roots and leaves. Roots anchor the plant and absorb water and nutrients; leaves absorb light and exchange gases. As plants increased in size, the roots and leaves became increasingly separated from each other in space. Thus, systems evolved for long-distance transport that allowed the shoot and the root to efficiently exchange products of absorption and assimilation.

You will recall from Chapters 4 and 6 that the xylem is the tissue that transports water and minerals from the root system to the aerial portions of the plant. The **phloem** is the tissue that translocates the products of photosynthesis from mature leaves to areas of growth and storage, including the roots. As we will see, the phloem also redistributes water and various compounds throughout the plant body. These compounds, some of which initially arrive in the mature leaves via the xylem, can be either transferred out of the leaves without modification or metabolized before redistribution.

The discussion that follows emphasizes translocation in the phloem of angiosperms because most of the research has been conducted on that group of plants. Gymnosperms will be compared briefly to angiosperms in terms of the anatomy of their conducting cells and possible differences in their mechanism of translocation. First we will examine some aspects of translocation in the phloem that have been researched extensively and are thought to be well understood. These include the pathway and patterns of translocation, materials translocated in the phloem, and rates of movement.

In the second part of the chapter we will explore aspects of translocation in the phloem that need further investigation. Some of these areas, such as phloem loading and unloading and the allocation and partitioning of photosynthetic products, are being studied intensively at present.

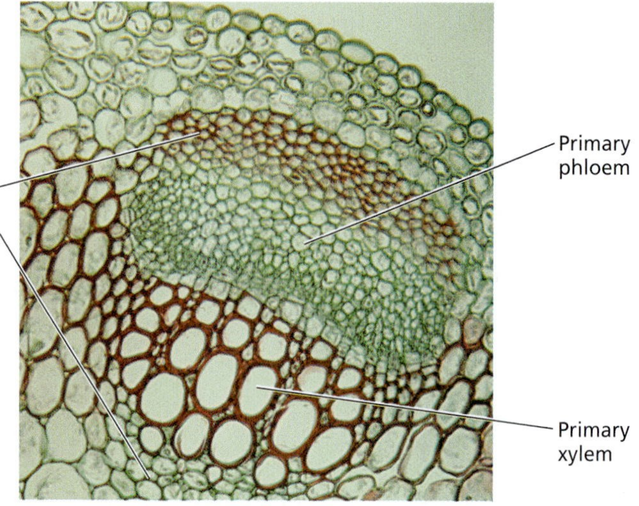

FIGURE 10.1 Transverse section of a vascular bundle of trefoil, a clover (*Trifolium*). (130×) The primary phloem is toward the outside of the stem. Both the primary phloem and the primary xylem are surrounded by a bundle sheath of thick-walled sclerenchyma cells, which isolate the vascular tissue from the ground tissue. (© J. N. A. Lott/Biological Photo Service.)

PATHWAYS OF TRANSLOCATION

The two long-distance transport pathways—the phloem and the xylem—extend throughout the plant body. The phloem is generally found on the outer side of both primary and secondary vascular tissues (Figures 10.1 and 10.2). In plants with secondary growth the phloem constitutes the inner bark.

The cells of the phloem that conduct sugars and other organic materials throughout the plant are called **sieve elements**. *Sieve element* is a comprehensive term that includes both the highly differentiated **sieve tube elements** typical of the angiosperms and the relatively unspecialized **sieve cells** of gymnosperms. In addition to sieve elements, the phloem tissue contains companion cells (discussed below) and parenchyma cells (which store and release food molecules). In some cases the phloem tissue also includes fibers and sclereids (for protection and strengthening of the tissue) and laticifers (latex-containing cells). However, only the sieve elements are directly involved in translocation.

The small veins of leaves and the primary vascular bundles of stems are often surrounded by a **bundle sheath** (see Figure 10.1), which consists of one or more layers of compactly arranged cells. (You will recall the bundle sheath cells involved in C_4 metabolism discussed in Chapter 8.) In the vascular tissue of leaves, the bundle sheath surrounds the small veins all the way to their ends, isolating the veins from the intercellular spaces of the leaf.

We will begin our discussion of translocation pathways with the experimental evidence demonstrating that the sieve elements are the conducting cells in the phloem. Then we will examine the structure and physiology of these unusual plant cells.

Sugar Is Translocated in Phloem Sieve Elements

Early experiments on phloem transport date back to the nineteenth century, indicating the importance of long-distance transport in plants (see **Web Topic 10.1**). These classical experiments demonstrated that removal of a ring of bark around the trunk of a tree, which removes the phloem, effectively stops sugar transport from the leaves to the roots without altering water transport through the xylem. When radioactive compounds became available, radiolabeled $^{14}CO_2$ was used to show that sugars made in the photosynthetic process are translocated through the phloem sieve elements (see **Web Topic 10.1**).

Mature Sieve Elements Are Living Cells Highly Specialized for Translocation

Detailed knowledge of the ultrastructure of sieve elements is critical to any discussion of the mechanism of translocation in the phloem. Mature sieve elements are unique among living plant cells (Figures 10.3 and 10.4). They lack

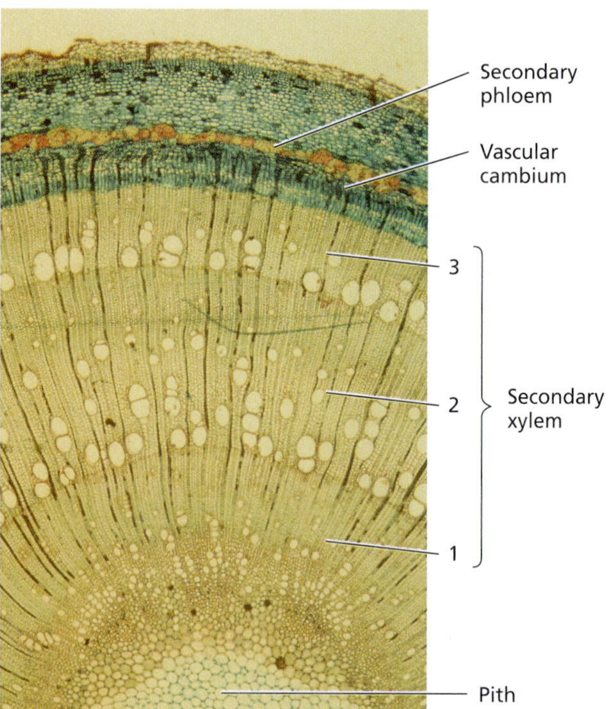

FIGURE 10.2 Transverse section of a 3-year-old stem of an ash (*Fraxinus excelsior*) tree. (27×) The numbers 1, 2, and 3 indicate growth rings in the secondary xylem. The old secondary phloem has been crushed by expansion of the xylem. Only the most recent (innermost) layer of secondary phloem is functional. (© P. Gates/Biological Photo Service.)

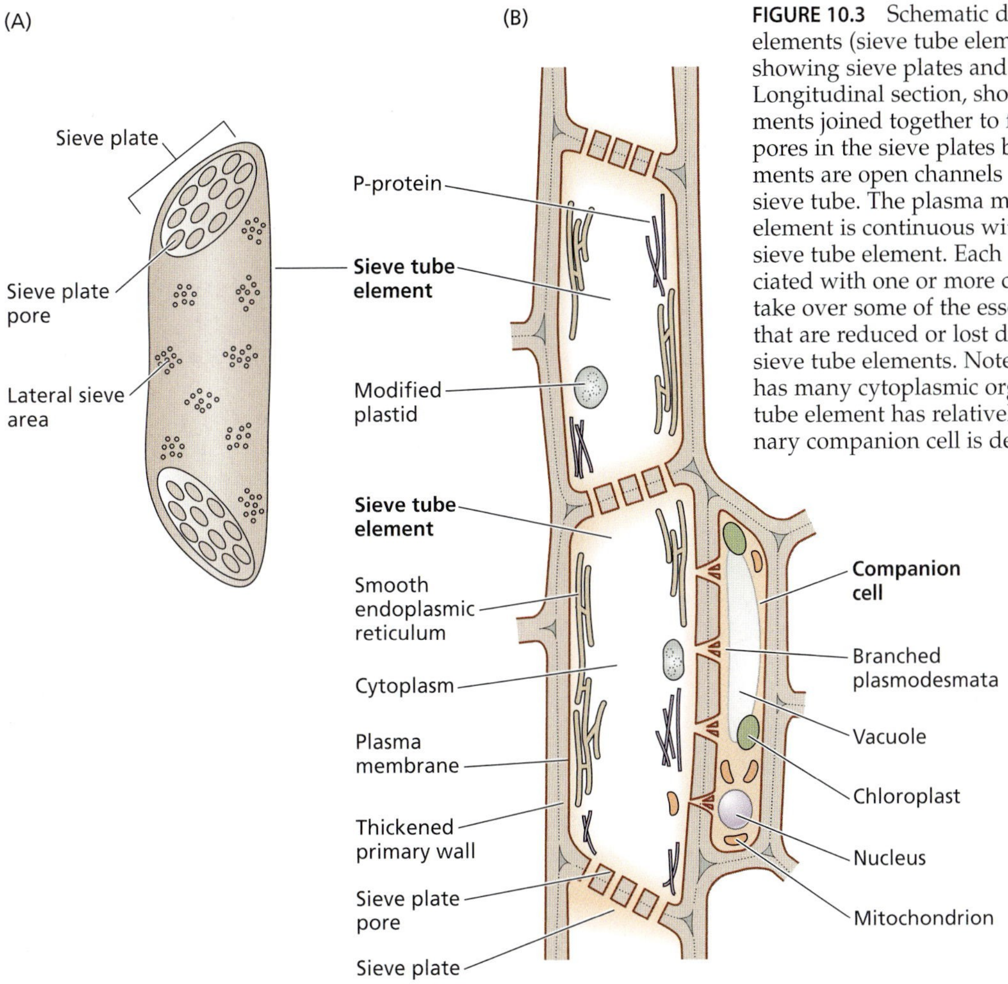

(A)

Sieve plate
Sieve plate pore
Lateral sieve area

(B)

P-protein
Sieve tube element
Modified plastid
Sieve tube element
Smooth endoplasmic reticulum
Cytoplasm
Plasma membrane
Thickened primary wall
Sieve plate pore
Sieve plate
Companion cell
Branched plasmodesmata
Vacuole
Chloroplast
Nucleus
Mitochondrion

FIGURE 10.3 Schematic drawings of mature sieve elements (sieve tube elements). (A) External view, showing sieve plates and lateral sieve areas. (B) Longitudinal section, showing two sieve tube elements joined together to form a sieve tube. The pores in the sieve plates between the sieve tube elements are open channels for transport through the sieve tube. The plasma membrane of a sieve tube element is continuous with that of its neighboring sieve tube element. Each sieve tube element is associated with one or more companion cells, which take over some of the essential metabolic functions that are reduced or lost during differentiation of the sieve tube elements. Note that the companion cell has many cytoplasmic organelles, whereas the sieve tube element has relatively few organelles. An ordinary companion cell is depicted here.

many structures normally found in living cells, even the undifferentiated cells from which mature sieve elements are formed. For example, sieve elements lose their nuclei and tonoplasts (vacuolar membrane) during development. Microfilaments, microtubules, Golgi bodies, and ribosomes are also absent from the mature cells. In addition to the plasma membrane, organelles that are retained include somewhat modified mitochondria, plastids, and smooth endoplasmic reticulum. The walls are nonlignified, though they are secondarily thickened in some cases.

Thus the sieve elements have a cellular structure different from that of tracheary elements of the xylem (which are dead at maturity), lack a plasma membrane, and have lignified secondary walls. As we will see, living cells are critical to the mechanism of translocation in the phloem.

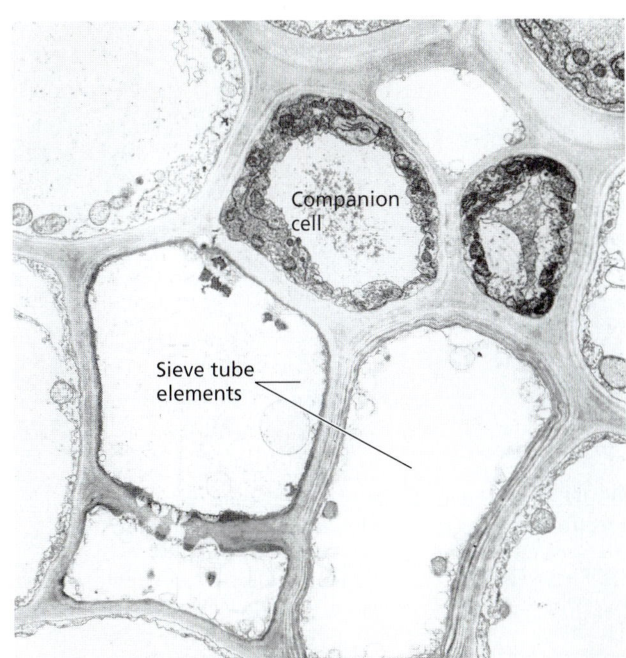

FIGURE 10.4 Electron micrograph of a transverse section of ordinary companion cells and mature sieve tube elements. (3600×) The cellular components are distributed along the walls of the sieve tube elements. (From Warmbrodt 1985.)

Sieve Areas Are the Prominent Feature of Sieve Elements

Sieve elements (sieve cells and sieve tube elements) have characteristic sieve areas in their cell walls, where pores interconnect the conducting cells (see Figure 10.5). The sieve area pores range in diameter from less than 1 µm to approximately 15 µm. Unlike sieve areas of gymnosperms, the sieve areas of angiosperms can differentiate into **sieve plates** (see Figure 10.5 and Table 10.1).

Sieve plates have larger pores than the other sieve areas in the cell and are generally found on the end walls of sieve tube elements, where the individual cells are joined together to form a longitudinal series called a **sieve tube** (see Figure 10.3). Furthermore, the sieve plate pores of sieve tube elements are open channels that allow transport between cells (see Figure 10.5).

In contrast, all of the sieve areas are more or less the same in gymnosperms such as conifers. The pores of gymnosperm sieve areas meet in large median cavities in the middle of the wall. Smooth endoplasmic reticulum (SER) covers the sieve areas (Figure 10.6) and is continuous through the sieve pores and median cavity, as indicated by ER-specific staining. Observation of living material with confocal laser scanning microscopy confirms that the observed distribution of SER is not an artifact of fixation (Schulz 1992).

TABLE 10.1
Characteristics of the two types of sieve elements in seed plants

Sieve tube elements found in angiosperms
1. Some sieve areas are differentiated into sieve plates; individual sieve tube elements are joined together into a sieve tube.
2. Sieve plate pores are open channels.
3. P-protein is present in all dicots and many monocots.
4. Companion cells are sources of ATP and perhaps other compounds and, in some species, are transfer cells or intermediary cells.
Sieve cells found in gymnosperms
1. There are no sieve plates; all sieve areas are similar.
2. Pores in sieve areas appear blocked with membranes
3. There is no P-protein.
4. Albuminous cells sometimes function as companion cells.

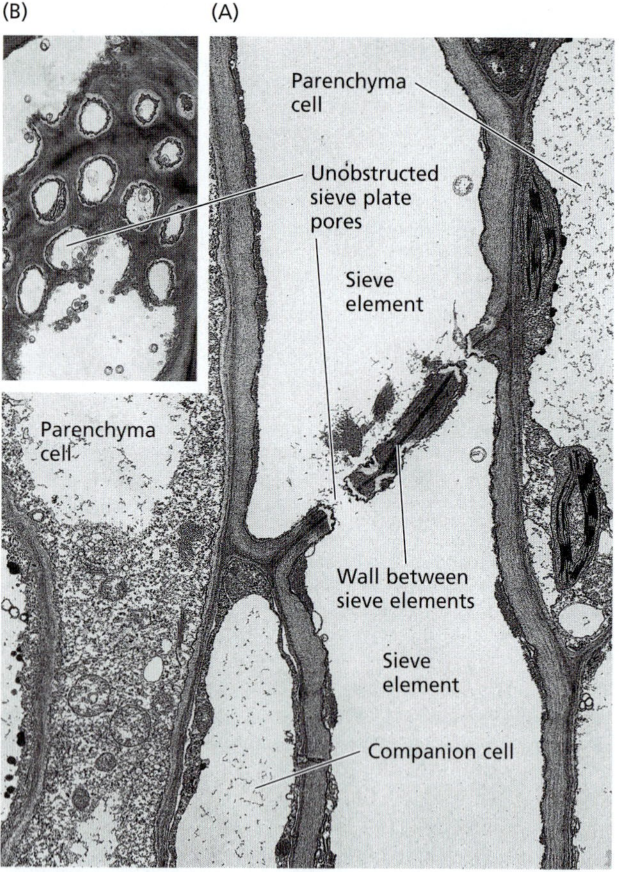

FIGURE 10.5 Sieve elements and open sieve plate pores. (A) Electron micrograph of a longitudinal section of two mature sieve elements (sieve tube elements), showing the wall between the sieve elements (called a sieve plate) in the hypocotyl of winter squash (*Cucurbita maxima*). (3685×) (B) The inset shows sieve plate pores in face view (4280×). In both images A and B, the sieve plate pores are open—that is, unobstructed by P-protein. (From Evert 1982.)

Deposition of P-Protein and Callose Seals Off Damaged Sieve Elements

The sieve tube elements of most angiosperms are rich in a phloem protein called **P-protein** (see Figure 10.3B) (Clark et al. 1997). (In classical literature, P-protein was called *slime*.) P-protein is found in all dicots and in many monocots, and it is absent in gymnosperms. It occurs in several different forms (tubular, fibrillar, granular, and crystalline) depending on the species and maturity of the cell.

In immature cells, P-protein is most evident as discrete bodies in the cytosol known as **P-protein bodies**. P-protein bodies may be spheroidal, spindle-shaped, or twisted and coiled. They generally disperse into tubular or fibrillar forms during cell maturation.

P-proteins have been characterized at the molecular level. For example, P-proteins from the genus *Cucurbita* consist of two major proteins: PP1, the phloem filament protein, and PP2, the phloem lectin. The gene that encodes PP1 in pumpkin (*Cucurbita maxima*) has sequence similarity to genes encoding cysteine proteinase inhibitors, suggesting a possible role in defense against phloem-feeding insects. Both PP1 and PP2 are thought to be synthesized in companion cells (discussed in the next section) and transported via the plasmodesmata to the sieve elements, where they associate to form P-protein filaments and P-protein bodies (Clark et al. 1997).

P-protein appears to function in sealing off damaged sieve elements by plugging up the sieve plate pores. Sieve tubes are under very high internal turgor pressure, and the sieve elements in a sieve tube are connected through open sieve plate pores. When a sieve tube is cut or punctured,

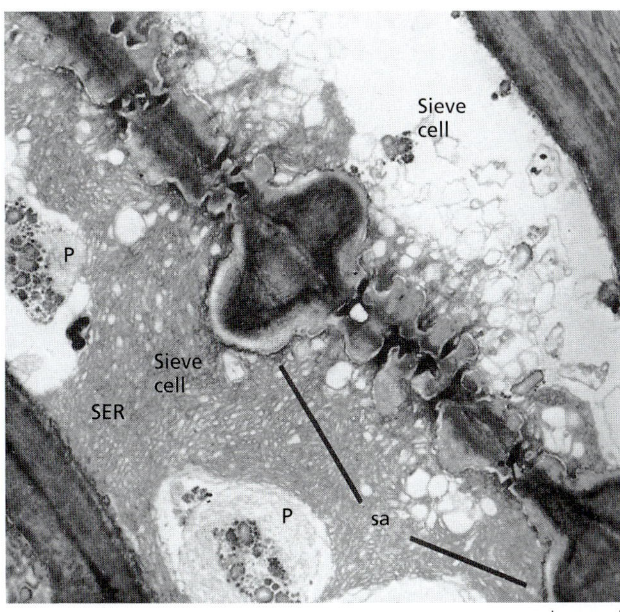

FIGURE 10.6 Electron micrograph showing a sieve area (sa) linking two sieve cells of a conifer (*Pinus resinosa*). Smooth endoplasmic reticulum (SER) covers the sieve area on both sides and is also found within the pores and the extended median cavity. Plastids (P) are enclosed by the SER. (From Schulz 1990.)

the release of pressure causes the contents of the sieve elements to surge toward the cut end, from which the plant could lose much sugar-rich phloem sap if there were no sealing mechanism. (*Sap* is a general term used to refer to the fluid contents of plant cells.) When surging occurs, however, P-protein and other cellular inclusions are trapped on the sieve plate pores, helping to seal the sieve element and to prevent further loss of sap.

A longer-term solution to sieve tube damage is the production of **callose** in the sieve pores. Callose, a β-1,3-glucan, is synthesized by an enzyme in the plasma membrane and is deposited between the plasma membrane and the cell wall. Callose is synthesized in functioning sieve elements in response to damage and other stresses, such as mechanical stimulation and high temperatures, or in preparation for normal developmental events, such as dormancy. The deposition of **wound callose** in the sieve pores efficiently seals off damaged sieve elements from surrounding intact tissue. As the sieve elements recover from damage, the callose disappears from these pores.

Companion Cells Aid the Highly Specialized Sieve Elements

Each sieve tube element is associated with one or more **companion cells** (see Figures 10.3B, 10.4, and 10.5). The division of a single mother cell forms the sieve tube element and the companion cell. Numerous plasmodesmata (see Chapter 1) penetrate the walls between sieve tube elements and their companion cells, suggesting a close functional relationship and a ready exchange of solutes between the two cells. The plasmodesmata are often complex and branched on the companion cell side.

Companion cells play a role in the transport of photosynthetic products from producing cells in mature leaves to the sieve elements in the minor (small) veins of the leaf. They are also thought to take over some of the critical metabolic functions, such as protein synthesis, that are reduced or lost during differentiation of the sieve elements (Bostwick et al. 1992). In addition, the numerous mitochondria in companion cells may supply energy as ATP to the sieve elements.

There are at least three different types of companion cells in the minor veins of mature, exporting leaves: "ordinary" companion cells, transfer cells, and intermediary cells. All three cell types have dense cytoplasm and abundant mitochondria.

Ordinary companion cells (Figure 10.7A) have chloroplasts with well-developed thylakoids and a cell wall with a smooth inner surface. Of most significance, relatively few plasmodesmata connect this type of companion cell to any of the surrounding cells except its own sieve element. As a result, the symplast of the sieve element and its companion cell is relatively, if not entirely, symplastically isolated from that of surrounding cells.

Transfer cells are similar to ordinary companion cells, except for the development of fingerlike wall ingrowths, particularly on the cell walls that face away from the sieve element (Figure 10.7B). These wall ingrowths greatly increase the surface area of the plasma membrane, thus increasing the potential for solute transfer across the membrane.

Because of the scarcity of cytoplasmic connections to surrounding cells and the wall ingrowths in transfer cells, the ordinary companion cell and the transfer cell are thought to be specialized for taking up solutes from the apoplast or cell wall space. Xylem parenchyma cells can also be modified as transfer cells, probably serving to retrieve and reroute solutes moving in the xylem, which is also part of the apoplast.

Though ordinary companion cells and transfer cells are relatively isolated symplastically from surrounding cells, there are some plasmodesmata in the walls of these cells. The function of these plasmodesmata is not known. The fact that they are present indicates that they must have a function, and an important one, since the cost of having them is high: They are the avenues by which viruses become systemic in the plant. They are, however, difficult to study because they are so inaccessible.

Intermediary cells appear well suited for taking up solutes via cytoplasmic connections (Figure 10.7C). Intermediary cells have numerous plasmodesmata connecting them to surrounding cells, particularly to the bundle sheath cells. Although the presence of many plasmodesmatal connections to surrounding cells is their most characteristic feature, intermediary cells are also distinctive in having

(A)

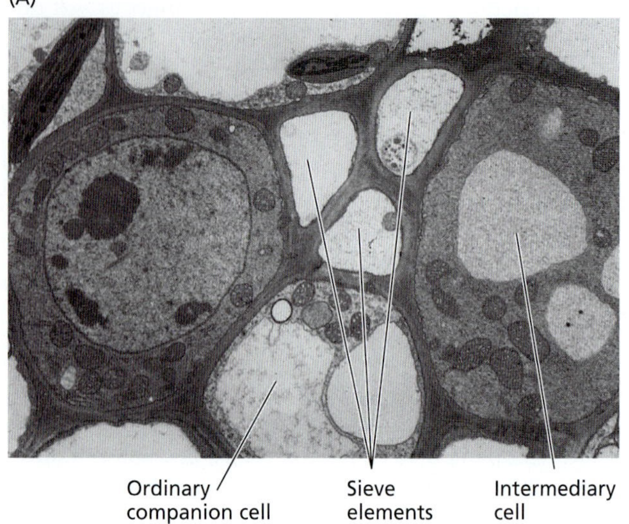

(B)

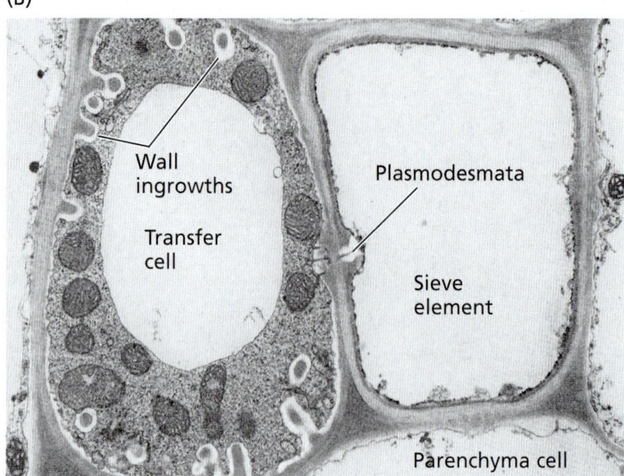

(C)

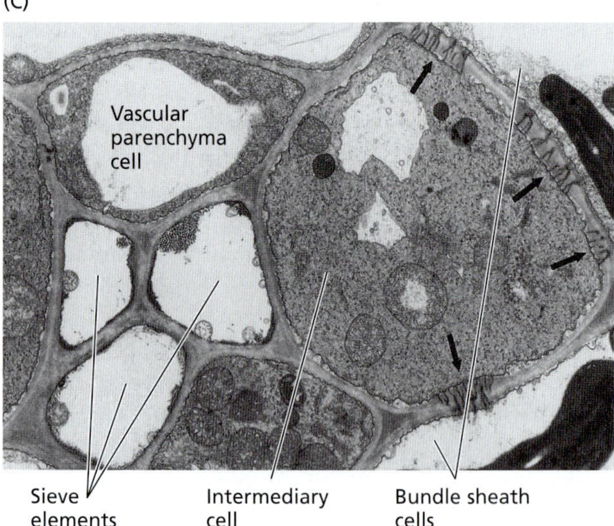

FIGURE 10.7 Electron micrographs of companion cells in minor veins of mature leaves. (A) Three sieve elements abut two intermediary cells and a more lightly stained ordinary companion cell in a minor vein from *Mimulus cardinalis*. (6585×) (B) A sieve element adjacent to a transfer cell with numerous wall ingrowths in pea (*Pisum sativum*). (8020×) Such ingrowths greatly increase the surface area of the transfer cell's plasma membrane, thus increasing the transfer of materials from the mesophyll to the sieve elements. (C) A typical intermediary cell with numerous fields of plasmodesmata (arrows) connecting it to neighboring bundle sheath cells. These plasmodesmata are branched on both sides, but the branches are longer and narrower on the intermediary cell side. Minor-vein phloem was taken from heartleaf maskflower (*Alonsoa warscewiczii*). (4700×) (A and C from Turgeon et al. 1993, courtesy of R. Turgeon; B from Brentwood 1978.)

numerous small vacuoles, as well as poorly developed thylakoids and a lack of starch grains in the chloroplasts.

In general, ordinary companion cells and transfer cells are found in plants that feature an apoplastic step in the transfer of sugars from mesophyll cells to sieve elements. Companion cells and transfer cells transfer sugars from the apoplast to the symplast of the sieve elements and companion cells in the source. Intermediary cells, on the other hand, function in symplastic transport of sugars from mesophyll cells to sieve elements in plants where no apoplastic step appears to occur in the source leaf.

PATTERNS OF TRANSLOCATION: SOURCE TO SINK

Sap in the phloem is not translocated exclusively in either an upward or a downward direction, and translocation in the phloem is not defined with respect to gravity. Rather, sap is translocated from areas of supply, called *sources*, to areas of metabolism or storage, called *sinks*.

Sources include any exporting organs, typically mature leaves, that are capable of producing photosynthate in excess of their own needs. The term *photosynthate* refers to products of photosynthesis. Another type of source is a storage organ during the exporting phase of its development. For example, the storage root of the biennial wild beet (*Beta maritima*) is a sink during the growing season of the first year, when it accumulates sugars received from the source leaves. During the second growing season the same root becomes a source; the sugars are remobilized and utilized to produce a new shoot, which ultimately becomes reproductive.

It is noteworthy that cultivated varieties of beets have been selected for the capacity of their roots to act as sinks during all phases of development. Thus, roots of the cultivated sugar beet (*Beta vulgaris*) can increase in dry mass during both the first and the second growing seasons, so the leaves serve as sources during both flowering and fruiting stages.

Sinks include any nonphotosynthetic organs of the plant and organs that do not produce enough photosynthetic products to support their own growth or storage needs. Roots, tubers, developing fruits, and immature leaves, which must import carbohydrate for normal development, are all examples of sink tissues. Both girdling and labeling studies support the source-to-sink pattern of translocation in the phloem.

Source-to-Sink Pathways Follow Anatomic and Developmental Patterns

Although the overall pattern of transport in the phloem can be stated simply as source-to-sink movement, the specific pathways involved are often more complex. Not all sources supply all sinks on a plant; rather, certain sources preferentially supply specific sinks. In the case of herbaceous plants, such as sugar beet and soybean, the following generalizations can be made.

Proximity. The proximity of the source to the sink is a significant factor. The upper mature leaves on a plant usually provide photosynthates to the growing shoot tip and young, immature leaves; the lower leaves supply predominantly the root system. Intermediate leaves export in both directions, bypassing the intervening mature leaves.

Development. The importance of various sinks may shift during plant development. Whereas the root and shoot apices are usually the major sinks during vegetative growth, fruits generally become the dominant sinks during reproductive development, particularly for adjacent and other nearby leaves.

Vascular connections. Source leaves preferentially supply sinks with which they have direct vascular connections. In the shoot system, for example, a given leaf is generally connected via the vascular system to other leaves directly above or below it on the stem. Such a vertical row of leaves is called an **orthostichy**. The number of internodes between leaves on the same orthostichy varies with the species. Figure 10.8A shows the three-dimensional structure of the phloem in an internode of dahlia (*Dahlia pinnata*).

Modification of translocation pathways. Interference with a translocation pathway by wounding or pruning can alter the patterns established by proximity and vascular

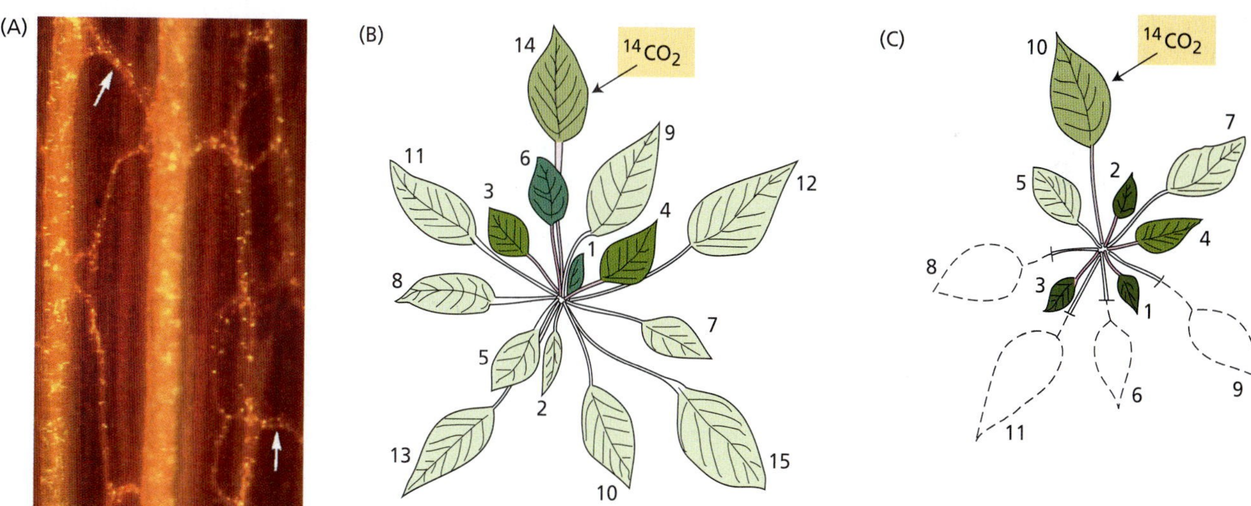

FIGURE 10.8 (A) Longitudinal view of a typical three-dimensional structure of the phloem in a thick section (from an internode of dahlia [*Dahlia pinnata*]). View here after clearing, staining with aniline blue, and observing under an epifluorescent microscope; the sieve plates are seen as numerous small dots because of the yellow staining of callosa in the sieve areas. Two large longitudinal vascular bundles are prominent. This staining reveals the delicate sieve tubes forming the phloem network; two phloem anastomoses are marked by arrows. (B) Distribution of radioactivity from a single labeled source leaf in an intact plant. The distribution of radioactivity in leaves of a sugar beet plant (*Beta vulgaris*) was determined 1 week after $^{14}CO_2$ was supplied for 4 hours to a single source leaf (arrow). The degree of radioactive labeling is indicated by the intensity of shading of the leaves. Leaves are numbered according to their age; the youngest, newly emerged leaf is designated 1. The ^{14}C label was translocated mainly to the sink leaves directly above the source leaf (that is, sink leaves on the same orthostichy as the source; for example, leaves 1 and 6 are sink leaves directly above source leaf 14). (C) Same as B, except all source leaves on the side of the plant opposite the labeled leaf were removed 24 hours before labeling. Sink leaves on both sides of the plant now receive ^{14}C-labeled assimilates from the source. (A courtesy of R. Aloni; B and C based on data from Joy 1964.)

connections that have been outlined here. In the absence of direct connections between source and sink, vascular interconnections, called **anastomoses** (singular *anastomosis*) (see Figure 10.8A), can provide an alternative pathway. In sugar beet, for example, removing source leaves from one side of the plant can bring about cross-transfer of photosynthates to young leaves (sink leaves) on the pruned side (Figure 10.8C). Removal of the lower source leaves on a plant can force the upper source leaves to translocate materials to the roots, and removal of the upper source leaves can force lower source leaves to translocate materials to the upper parts of the plant.

The plasticity of the translocation pathway depends on the extent of the interconnections between vascular bundles and thus on the species and organs studied. In some species the leaves on a branch with no fruits cannot transport photosynthate to the fruits on an adjacent defoliated branch. But in other plants, such as soybean (*Glycine max*), photosynthate is transferred readily from a partly defruited side to a partly defoliated side.

MATERIALS TRANSLOCATED IN THE PHLOEM: SUCROSE, AMINO ACIDS, HORMONES, AND SOME INORGANIC IONS

Water is the most abundant substance transported in the phloem. Dissolved in the water are the translocated solutes, mainly carbohydrates (Table 10.2). Sucrose is the sugar most commonly transported in sieve elements. There is always some sucrose in sieve element sap, and it can reach concentrations of 0.3 to 0.9 *M*.

Nitrogen is found in the phloem largely in amino acids and amides, especially glutamate and aspartate and their respective amides, glutamine and asparagine. Reported levels of amino acids and organic acids vary widely, even for the same species, but they are usually low compared with carbohydrates.

Almost all the endogenous plant hormones, including auxin, gibberellins, cytokinins, and abscisic acid (see Chapters 19, 20, 21, and 23), have been found in sieve elements. The long-distance transport of hormones is thought to occur at least partly in the sieve elements. Nucleotide phosphates and proteins have also been found in phloem sap.

Proteins found in the phloem include filamentous P-proteins (which are involved in the sealing of wounded sieve elements), protein kinases (protein phosphorylation), thioredoxin (disulfide reduction), ubiquitin (protein turnover), chaperones (protein folding), and protease inhibitors (protection of phloem proteins from degradation and defense against phloem-feeding insects) (Schobert et al. 1995; Yoo et al. 2000).

Inorganic solutes that move in the phloem include potassium, magnesium, phosphate, and chloride (see Table 10.2). In contrast, nitrate, calcium, sulfur, and iron are relatively immobile in the phloem.

TABLE 10.2
The composition of phloem sap from castor bean (*Ricinus communis*), collected as an exudate from cuts in the phloem

Component	Concentration (mg mL^{-1})
Sugars	80.0–106.0
Amino acids	5.2
Organic acids	2.0–3.2
Protein	1.45–2.20
Potassium	2.3–4.4
Chloride	0.355–0.675
Phosphate	0.350–0.550
Magnesium	0.109–0.122

Source: Hall and Baker 1972.

We will begin the discussion of phloem content with a look at the methods used to identify materials translocated in the phloem. We will then examine the translocated sugars and the complexities of nitrogen transport in the plant.

Phloem Sap Can Be Collected and Analyzed

The collection of phloem sap has been experimentally challenging (see **Web Topic 10.2**). A few species exude phloem sap from wounds that sever sieve elements, making it possible to collect relatively pure samples of phloem sap. Another approach is to use the stylet of an aphid as a "natural syringe."

Aphids are small insects that feed by inserting their mouthparts, consisting of four tubular stylets, into a single sieve element of a leaf or stem. Sap can be collected from aphid stylets cut from the body of the insect, usually with a laser, after the aphid has been anesthetized with CO_2. The high turgor pressure in the sieve element forces the cell contents through the stylet to the cut end, where they can be collected. Exudate from severed stylets provides a fairly accurate picture of the composition of phloem sap (see **Web Topic 10.2**). Exudation from severed stylets can continue for hours, suggesting that the aphid prevents the plant's normal sealing mechanisms from operating.

Sugars Are Translocated in Nonreducing Form

Results from analyses of collected sap indicate that the translocated carbohydrates are all nonreducing sugars. Reducing sugars, such as glucose and fructose, contain an exposed aldehyde or ketone group (Figure 10.9A). In a nonreducing sugar, such as sucrose, the ketone or aldehyde group is reduced to an alcohol or combined with a similar group on another sugar (Figure 10.9B). Most researchers believe that the nonreducing sugars are the major compounds translocated in the phloem because they are less reactive than their reducing counterparts.

Sucrose is the most commonly translocated sugar; many of the other mobile carbohydrates contain sucrose bound to varying numbers of galactose molecules. Raffinose consists

FIGURE 10.9 Structures of compounds not normally translocated in the phloem (A) and of compounds commonly translocated in the phloem (B).

(A) **Reducing sugars, which are not generally translocated in the phloem**

The reducing groups are aldehyde (glucose and mannose) and ketone (fructose) groups.

Aldehyde

H—C=O
H—C—OH
HO—C—H
H—C—OH
H—C—OH
CH_2OH
D-Glucose

Aldehyde

H—C=O
HO—C—H
HO—C—H
H—C—OH
H—C—OH
CH_2OH
D-Mannose

Ketone

CH_2OH
C=O
HO—C—H
H—C—OH
H—C—OH
CH_2OH
D-Fructose

(B) **Compounds commonly translocated in the phloem**

Sucrose is a disaccharide made up of one glucose and one fructose molecule. Raffinose, stachyose, and verbascose contain sucrose bound to one, two, or three galactose molecules, respectively.

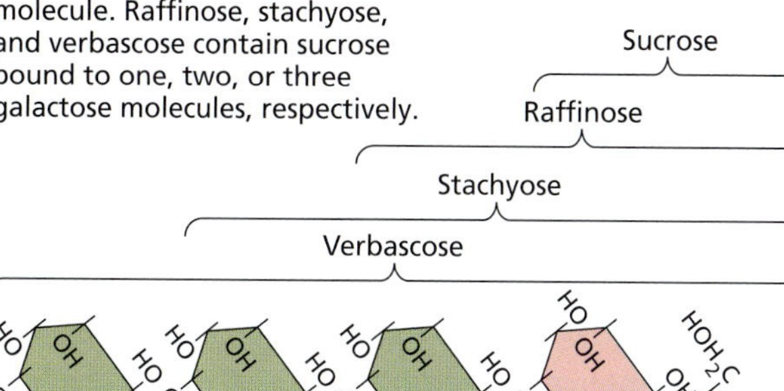

Nonreducing sugar

Mannitol is a sugar alcohol formed by the reduction of the aldehyde group of mannose.

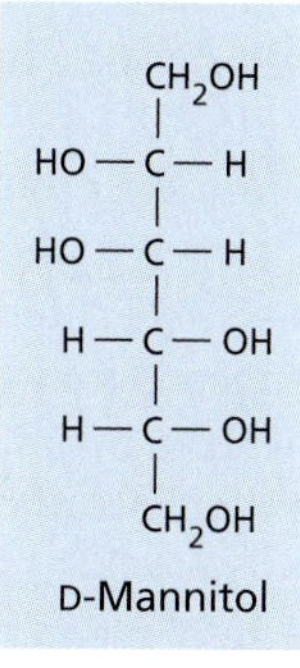

Sugar alcohol

Glutamic acid, an amino acid, and glutamine, its amide, are important nitrogenous compounds in the phloem, in addition to aspartate and asparagine.

HO—C(=O)—CH₂—CH₂—CH(NH₂)—C(=O)—OH

Glutamic acid

Amino acid

H_2N—C(=O)—CH₂—CH₂—CH(NH₂)—C(=O)—OH

Glutamine

Amide

Species with nitrogen-fixing nodules also utilize ureides as transport forms of nitrogen.

Allantoic acid

Allantoin

H_2N—C(=O)—NH—$CH_2CH_2CH_2$CH(NH_2)—COOH

Citrulline

Ureides

of sucrose and one galactose molecule, stachyose consists of sucrose and two galactose molecules, and verbascose consists of sucrose and three galactose molecules (see Figure 10.9B). Translocated sugar alcohols include mannitol and sorbitol.

Phloem and Xylem Interact to Transport Nitrogenous Compounds

Nitrogen is transported throughout the plant in either inorganic or organic form, with the predominant form depending on several factors, including the transport pathway. Whereas nitrogen is transported in the phloem almost entirely in organic form, in the xylem it can be transported either as nitrate or as part of an organic molecule. (see Chapter 12). Usually the same group of organic molecules carries nitrogen in both the xylem and the phloem.

The form in which nitrogen is transported in the xylem depends on the species studied. Species that do not form a symbiotic association with nitrogen-fixing microorganisms usually depend on soil nitrate as their major nitrogen source (see Chapter 12). In the xylem of these species, nitrogen is usually present in the form of both nitrate and nitrogen-rich organic molecules, particularly the amides asparagine and glutamine (see Figure 10.9B).

Species with nitrogen-fixing nodules on their roots (see Chapter 12) depend on atmospheric nitrogen, rather than on soil nitrate, as their major nitrogen source. After being converted to an organic form, this nitrogen is transported in the xylem to the shoot, usually in the form of amides or ureides such as allantoin, allantoic acid, or citrulline (see Figure 10.9B).

Whenever nitrogen is assimilated into organic compounds in the roots, both the energy and the carbon skeletons required for assimilation are derived from photosynthates transported to the roots via the phloem. Nitrogen levels in mature leaves are quite stable, indicating that at least some of the excess nitrogen continuously arriving via the xylem is redistributed via the phloem to fruits or younger leaves. (See **Web Topic 10.3** for information on nitrogen transport in the soybean.)

Finally, levels of nitrogenous compounds in the phloem are quite high during leaf senescence. In woody species, senescing leaves mobilize and export nitrogenous compounds to the woody tissues for storage; in herbaceous plants nitrogen is exported generally to the seeds. Other solutes, such as mineral ions, are redistributed from senescing leaves in the same manner.

RATES OF MOVEMENT

The rate of movement of materials in the sieve elements can be expressed in two ways: as **velocity**, the linear distance traveled per unit time, or as **mass transfer rate**, the quantity of material passing through a given cross section of phloem or sieve elements per unit time. Mass transfer rates based on the cross-sectional area of the sieve elements are preferred because the sieve elements are the conducting cells of the phloem. Values for mass transfer rate range from 1 to 15 g h^{-1} cm^{-2} of sieve elements (see **Web Topic 10.4**).

In early publications reporting on rates of transport in the phloem, the units of velocity were centimeters per hour (cm h^{-1}), and the units of mass transfer were grams per hour per square centimeter (g h^{-1} cm^{-2}) of phloem or sieve elements. The currently preferred units (SI units) are meters (m) or millimeters (mm) for length, seconds (s) for time, and kilograms (kg) for mass.

Velocities of Phloem Transport Far Exceed the Rate of Diffusion

Both velocities and mass transfer rates can be measured with radioactive tracers. (Methods of measuring mass transfer rates are described in **Web Topic 10.4**.) In the simplest type of experiment for measuring velocity, ^{11}C- or ^{14}C-labeled CO_2 is applied for a brief period of time to a source leaf (pulse labeling), and the arrival of label at a sink tissue or at a particular point along the pathway is monitored with an appropriate detector.

The length of the translocation pathway divided by the time interval required for label to be first detected at the sink yields a measure of velocity. A more accurate measurement of velocity is obtained from monitoring the arrival of label at two points along the pathway. This method excludes from the measurement the time required for fixation of labeled carbon by photosynthesis, for its incorporation into transport sugar, and for accumulation of sugar in the sieve elements of the source leaf.

In general, velocities measured by a variety of techniques average about 1 m h^{-1} and range from 0.3 to 1.5 m h^{-1} (30–150 cm h^{-1}). Transport velocities in the phloem are clearly quite high, well in excess of the rate of diffusion over long distances. Any proposed mechanism of phloem translocation must account for these high velocities.

THE MECHANISM OF TRANSLOCATION IN THE PHLOEM: THE PRESSURE-FLOW MODEL

The mechanism of phloem translocation in angiosperms is best explained by the pressure-flow model, which accounts for most of the experimental and structural data currently available. We will see in this discussion that the pressure-flow model explains phloem translocation as a flow of solution (bulk flow) driven by an osmotically generated pressure gradient between source and sink.

In early research on phloem translocation, both active and passive mechanisms were considered. All theories, both active and passive, assume an energy requirement in both sources and sinks. In sources, energy is necessary to move photosynthate from producing cells into the sieve elements. This movement of photosynthate is called *phloem loading*, and it is discussed in detail later in the chapter. In sinks, energy is essential for some aspects of movement from sieve elements to sink cells, which store or metabolize the sugar. This movement of photosynthate from sieve elements to sink cells is called *phloem unloading* and will also be discussed later.

The passive mechanisms of phloem transport further assume that energy is required in the sieve elements of the path between sources and sinks simply to maintain structures such as the cell plasma membrane and to recover sugars lost from the phloem by leakage. The pressure-flow model is an example of a passive mechanism. The active theories, on the other hand, postulate an additional expenditure of energy by path sieve elements in order to drive translocation itself (Zimmermann and Milburn 1975).

A Pressure Gradient Drives Translocation

Diffusion is far too slow to account for the velocities of solute movement observed in the phloem. Translocation velocities average 1 m h^{-1}; the rate of diffusion is 1 m per 32 years! (See Chapter 3 for a discussion of diffusion velocities and the distances over which diffusion is an effective transport mechanism.)

The **pressure-flow model**, first proposed by Ernst Münch in 1930, states that a flow of solution in the sieve elements is driven by an osmotically generated *pressure gradient* between source and sink ($\Delta\Psi_p$). The pressure gradient is established as a consequence of phloem loading at the source and phloem unloading at the sink.

Recall from Chapter 3 (Equation 3.6) that $\Psi_w = \Psi_s + \Psi_p$; that is, $\Psi_p = \Psi_w - \Psi_s$. In source tissues, energy-driven phloem loading leads to an accumulation of sugars in the sieve elements, generating a low (negative) solute potential ($\Delta\Psi_s$) and causing a steep drop in the water potential ($\Delta\Psi_w$). In response to the water potential gradient, water enters the sieve elements and causes the turgor pressure (Ψ_p) to increase.

At the receiving end of the translocation pathway, phloem unloading leads to a lower sugar concentration in the sieve elements, generating a higher (more positive) solute potential in the sieve elements of sink tissues. As the water potential of the phloem rises above that of the xylem, water tends to leave the phloem in response to the water potential gradient, causing a decrease in turgor pressure in the sieve elements of the sink. Figure 10.10 illustrates the pressure-flow hypothesis.

If no cross-walls were present in the translocation pathway—that is, if the entire pathway were a single membrane-enclosed compartment—the different pressures at the source and sink would rapidly equilibrate. The presence of sieve plates greatly increases the resistance along the pathway and results in the generation and maintenance of a substantial pressure gradient in the sieve elements between source and sink. The sieve element contents are physically pushed along the translocation pathway as a bulk flow, much like water flowing through a garden hose.

Close inspection of the water potential values shown in Figure 10.10 shows *that water in the phloem is moving against a water potential gradient from source to sink*. Such water movement does not violate the laws of thermodynamics because the water is moving by bulk flow rather than by osmosis. That is, no membranes are crossed during transport from one sieve tube to another, and solutes are moving at the same rate as the water molecules.

Under these conditions, the solute potential, Ψ_s, cannot contribute to the driving force for water movement, although it still influences the water potential. Water movement in the translocation pathway is therefore driven by the pressure gradient rather than by the water potential gradient. Of course, the passive, pressure-driven, long-distance translocation in the sieve tubes ultimately depends on the active, short-distance transport mechanisms involved in phloem loading and unloading. These active mechanisms are responsible for setting up the pressure gradient.

The Predictions of the Pressure-Flow Model Have Been Confirmed

Some important predictions emerge from the pressure-flow model:

- The sieve plate pores must be unobstructed. If P-protein or other materials blocked the pores, the resistance to flow of the sieve element sap would be too great.
- True *bidirectional transport* (i.e., simultaneous transport in both directions) in a single sieve element cannot occur. A mass flow of solution precludes such bidirectional movement because a solution can flow in only one direction in a pipe at any one time. Solutes within the phloem can move bidirectionally, but in different sieve elements or at different times.
- Great expenditures of energy are not required in order to drive translocation in the tissues along the path, although energy is required to maintain the structure of the sieve elements and to reload any sugars lost to the apoplast by leakage. Therefore, treatments that restrict the supply of ATP in the path, such as low temperature, anoxia, and metabolic inhibitors, should not stop translocation.
- The pressure-flow hypothesis demands the presence of a positive pressure gradient. Turgor pressure must be higher in sieve elements of sources than in sieve elements of sinks, and the pressure difference must be large enough to overcome the resistance of the pathway and to maintain flow at the observed velocities.

The available evidence testing these predictions supports the pressure-flow hypothesis.

Sieve Plate Pores Are Open Channels

Ultrastructural studies of sieve elements are challenging because of the high internal pressure in these cells. When

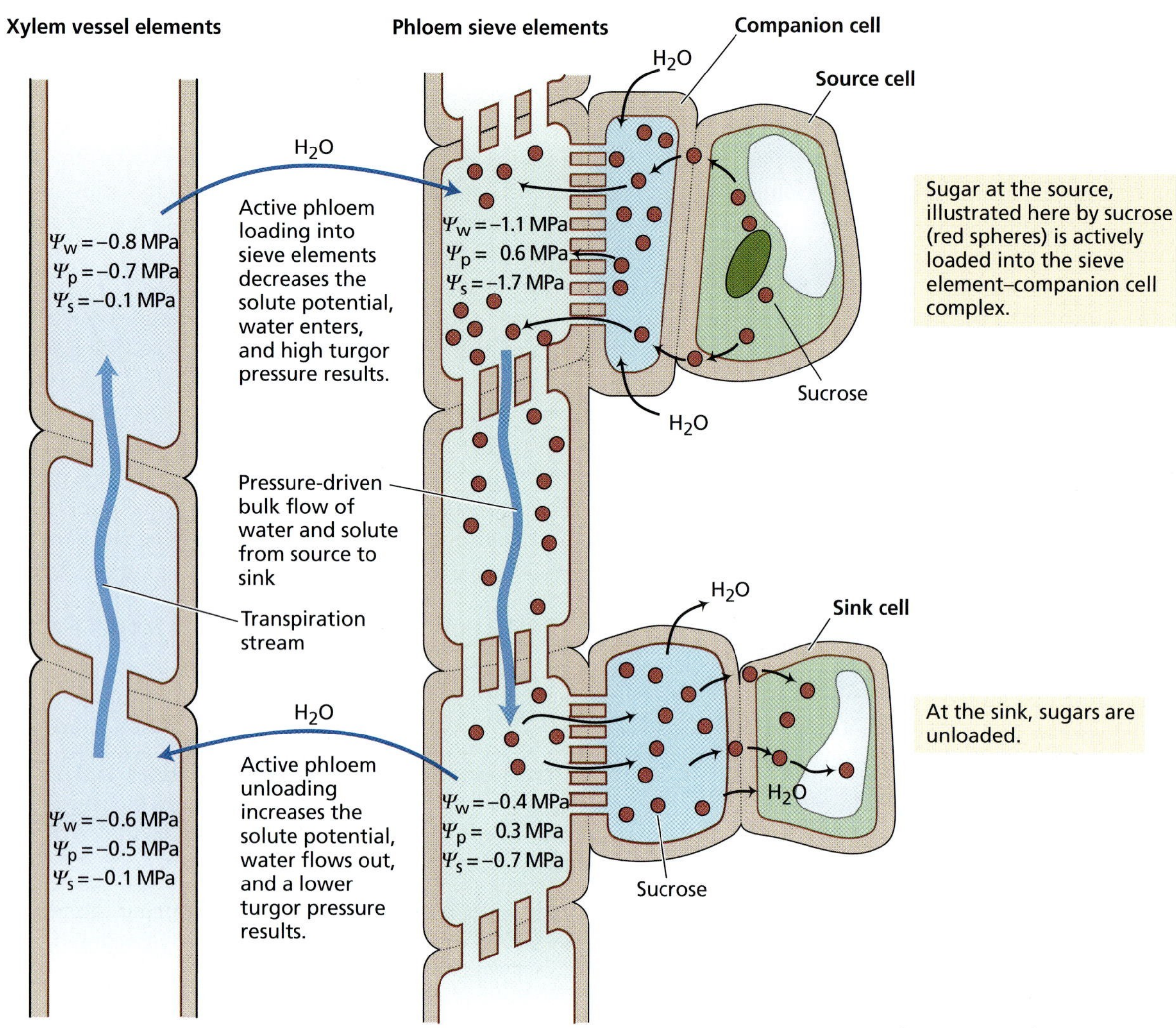

FIGURE 10.10 Pressure-flow model of translocation in the phloem. Possible values for Ψ_w, Ψ_p, and Ψ_s in the xylem and phloem are illustrated. (After Nobel 1991.)

the phloem is excised or killed slowly with chemical fixatives, the turgor pressure in the sieve elements is released. The contents of the cell, including P-protein, surge toward the point of pressure release and, in the case of sieve tube elements, accumulate on the sieve plates. This accumulation is probably the reason that many earlier electron micrographs show sieve plates that are obstructed.

Newer, rapid freezing and fixation techniques provide reliable pictures of undisturbed sieve elements. Electron micrographs of sieve tube elements prepared by such techniques show that P-protein is usually found along the periphery of the sieve tube elements (see Figures 10.3, 10.4, and 10.5), or it is evenly distributed throughout the lumen of the cell. Furthermore, the pores contain P-protein in similar positions, lining the pore or in a loose network. The open condition of the pores, seen in many species, such as cucurbits, sugar beet, and bean (e.g., see Figure 10.5), supports the pressure-flow model.

In addition to obtaining the structural evidence provided by electron microscopy, it is important to determine whether the sieve plate pores are open in the intact tissue. The use of confocal laser scanning microscopy, which allows for the direct observation of translocation through living sieve elements, addresses this question (Knoblauch and van Bel 1998). Such experiments show that the sieve plate pores of living, translocating sieve elements are open (Figure 10.11).

Bidirectional Transport Cannot Be Seen in Single Sieve Elements

Researchers have investigated bidirectional transport by applying two different radiotracers to two source leaves, one above the other (Eschrich 1975). Each leaf receives one

(A)

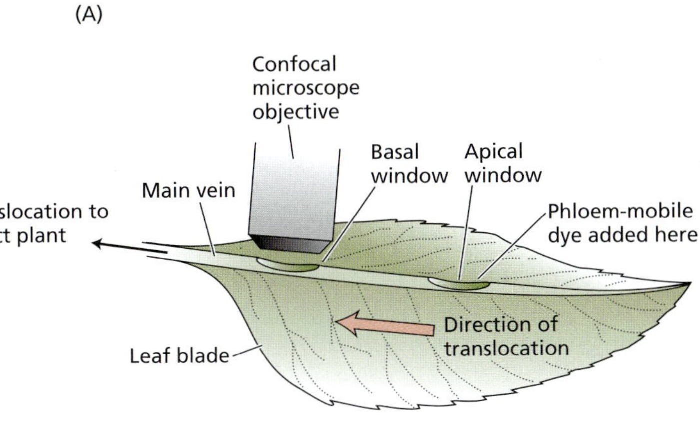

(B)

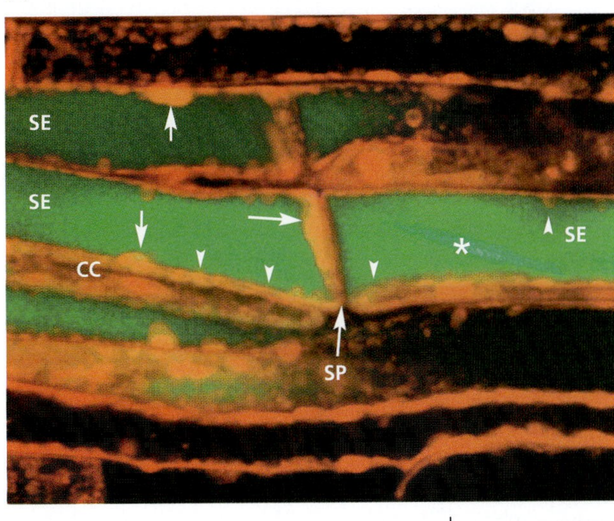

(C)

15 µm

FIGURE 10.11 Translocation in living, functional sieve elements of a leaf attached to an intact broad bean (*Vicia faba*) plant. (A) Two windows were sliced parallel to the epidermis on the lower side of the main vein of a mature leaf, exposing the phloem tissue. The objective of the laser confocal microscope was positioned over the basal window. A phloem-mobile fluorescent dye was added at the apical window. If translocation occurred, the dye would become visible in the microscope at the basal window of the leaf. In this way it could be demonstrated that the sieve elements being observed were alive and functional. (B, C) Phloem tissue of bean doubly stained with a locally applied fluorescent dye (red) that primarily stains membranes, and a translocated fluorescent dye (green). Protein (arrows) deposited against the plasma membrane and the sieve plate does not impede translocation. A crystalline P-protein body (asterisk) is stained by the green dye. Plastids (arrowheads) are evenly distributed around the periphery of the sieve element. CC = companion cell, SP = sieve plate. See also **Web Topic 10.8**. (From Knoblauch and van Bel 1998; courtesy of A. van Bel.)

of the tracers, and a point between the two sources is monitored for the presence of both tracers.

Transport in two directions has often been detected in sieve elements of different vascular bundles in stems. Transport in two directions has also been seen in adjacent sieve elements of the same bundle in petioles. Bidirectional transport in adjacent sieve elements can occur in the petiole of a leaf that is undergoing the transition from sink to source and simultaneously importing and exporting photosynthates through its petiole. However, simultaneous bidirectional transport in a single sieve element has never been demonstrated.

Translocation Rate Is Typically Insensitive to the Energy Supply of the Path Tissues

In plants that can survive periods of low temperature, such as sugar beet, rapidly chilling a short segment of the petiole of a source leaf to approximately 1°C does not cause sustained inhibition of mass transport out of the leaf (Figure 10.12). Rather, there is a brief period of inhibition, after which transport slowly returns to the control rate. Chilling reduces respiration rate and both the synthesis and the consumption of ATP in the petiole by about 90%, at a time when translocation has recovered and is proceeding normally. These experiments show that the energy requirement for transport through the pathway of these plants is small, consistent with the pressure-flow hypothesis.

Extreme treatments that inhibit all energy metabolism do inhibit translocation. For example, in bean (*Phaseolus vulgaris*), treating the petiole of a source leaf with a metabolic inhibitor (cyanide) inhibited translocation out of the

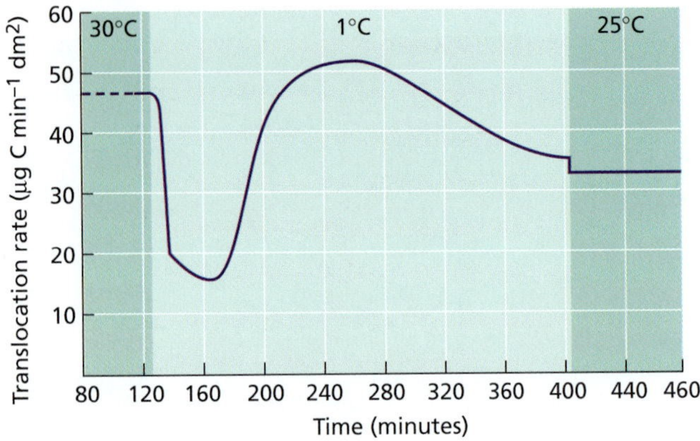

FIGURE 10.12 Loss of metabolic energy resulting from the chilling of the leaf petiole partially reduces the rate of translocation in sugar beet (*Beta vulgaris*), although translocation rates recover with time. The fact that translocation recovers when ATP production and utilization are largely inhibited by chilling indicates that the energy requirement for translocation is small. $^{14}CO_2$ was supplied to a source leaf, and a 2 cm portion of its petiole was chilled to 1°C. Translocation was monitored by the arrival of ^{14}C at a sink leaf. (dm [decimeter] = 0.1 meter) (Data from Geiger and Sovonick 1975.)

leaf. However, examination of the treated tissue by electron microscopy revealed blockage of the sieve plate pores by cellular debris (Giaquinta and Geiger 1977). Clearly, these results do not bear on the question of whether energy is required for translocation along the pathway.

Pressure Gradients Are Sufficient to Drive a Mass Flow of Solution

Turgor pressure in sieve elements can be either calculated from the water potential and solute potential ($\Psi_p = \Psi_w - \Psi_s$) or measured directly. The most effective technique uses micromanometers or pressure transducers sealed over exuding aphid stylets (see Figure 10.2.A in **Web Topic 10.2**) (Wright and Fisher 1980). The data obtained are accurate because aphids pierce only a single sieve element, and the plasma membrane apparently seals well around the aphid stylet. When the turgor pressure of sieve elements is measured by this technique, the pressure at the source is higher than that at the sink.

In soybean, the observed pressure difference between source and sink has been shown to be sufficient to drive a mass flow of solution through the pathway, taking into account the path resistance (caused mainly by the sieve plate pores), the path length, and the velocity of translocation (Fisher 1978). The actual pressure difference between source and sink was calculated from the water potential and solute potential to be 0.41 MPa, and the pressure difference required for translocation by pressure flow was calculated to be 0.12 to 0.46 MPa. Thus the observed pressure difference appears to be sufficient to drive mass flow through the phloem.

We can therefore conclude that all the experiments and data described here support the operation of pressure flow in angiosperm phloem. The lack of an energy requirement in the pathway and the presence of open sieve plate pores provide definitive evidence for a mechanism in which the path phloem is relatively passive. The failure to detect bidirectional transport or motility proteins, as well as the positive data on pressure gradients, is in accord with the pressure-flow hypothesis.

The Mechanism of Phloem Transport in Gymnosperms May Be Different

Although pressure flow explains translocation in angiosperms, it may not be sufficient for gymnosperms. Very little physiological information on gymnosperm phloem is available, and speculation about translocation in these species is based almost entirely on interpretations of electron micrographs. As discussed previously, the sieve cells of gymnosperms are similar in many respects to sieve tube elements of angiosperms, but the sieve areas of sieve cells are relatively unspecialized and do not appear to consist of open pores (see Figure 10.6).

The pores in gymnosperms are filled with numerous membranes that are continuous with the smooth endoplasmic reticulum adjacent to the sieve areas. Such pores are clearly inconsistent with the requirements of the pressure-flow hypothesis. Although these electron micrographs might be artifactual and fail to show conditions in the intact tissue, translocation in gymnosperms might involve a different mechanism—a possibility that requires further investigation.

PHLOEM LOADING: FROM CHLOROPLASTS TO SIEVE ELEMENTS

Several transport steps are involved in the movement of photosynthate from the mesophyll chloroplasts to the sieve elements of mature leaves, which is called **phloem loading** (Oparka and van Bel 1992):

1. Triose phosphate formed by photosynthesis during the day (see Chapter 8) is transported from the chloroplast to the cytosol, where it is converted to sucrose. During the night, carbon from stored starch exits the chloroplast probably in the form of glucose and is converted to sucrose. (Other transport sugars are later synthesized from sucrose in some species.)

2. Sucrose moves from the mesophyll cell to the vicinity of the sieve elements in the smallest veins of the leaf (Figure 10.13). This **short-distance transport** pathway usually covers a distance of only two or three cell diameters.

3. In a process called **sieve element loading**, sugars are transported into the sieve elements and companion cells. In most of the species studied so far, sugars become more concentrated in the sieve elements and

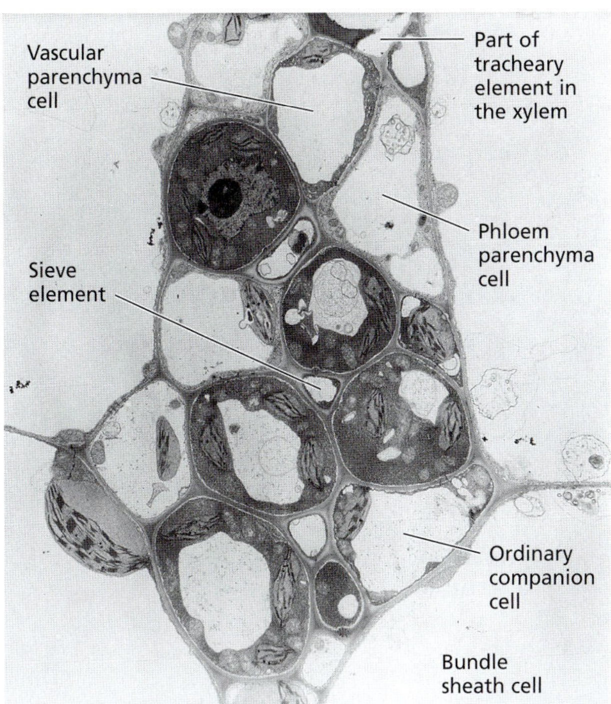

FIGURE 10.13 Electron micrograph showing the relationship between the various cell types of a small vein in a source leaf of sugar beet (*Beta vulgaris).* Photosynthetic cells (mesophyll cells) surround the compactly arranged cells of the bundle sheath layer. Photosynthate from the mesophyll must move a distance equivalent to several cell diameters before being loaded into the sieve elements. (From Evert and Mierzwa 1985, courtesy of R. Evert.)

companion cells than in the mesophyll. Note that with respect to loading, the sieve elements and companion cells are often considered a functional unit, called the *sieve element–companion cell complex.* Once inside the sieve elements, sucrose and other solutes are translocated away from the source, a process known as **export**. Translocation through the vascular system to the sink is referred to as **long-distance transport**.

As discussed earlier, the processes of loading at the source and unloading at the sink provide the driving force that generates the pressure gradient pushing phloem sap in long-distance transport and are thus of considerable basic, as well as agricultural, importance. A thorough understanding of these mechanisms should provide the basis of technology aimed at enhancing crop productivity by increasing the accumulation of photosynthate by edible sink tissues, such as cereal grains.

Photosynthate Can Move from Mesophyll Cells to the Sieve Elements via the Apoplast or the Symplast

We have seen that solutes (mainly sugars) in source leaves must move from the photosynthesizing cells in the mesophyll to the veins. Sugars might move entirely through the symplast (cytoplasm) via the plasmodesmata, or they might enter the apoplast at some point en route to the phloem (Figure 10.14). (See Figure 4.3 for a general description of the symplast and apoplast.) In the latter case, the sugars are actively loaded from the apoplast into the sieve elements and companion cells by an energy-driven, selective transporter located in the plasma membranes of these cells. In fact, the apoplastic and symplastic routes are used in different species.

Early research on phloem loading focused on the apoplastic pathway. Apoplastic phloem loading leads to three basic predictions (Grusak et al. 1996): (1) Transported sugars should be found in the apoplast; (2) in experiments in which sugars are supplied to the apoplast, the exogenously supplied sugars should accumulate in sieve elements and companion cells; and (3) inhibition of sugar uptake from the apoplast should result in inhibition of export from the leaf. Many studies devoted to testing these predictions have provided solid evidence for apoplastic loading in several species (see **Web Topic 10.5**).

Sucrose Uptake in the Apoplastic Pathway Requires Metabolic Energy

In source leaves, sugars become more concentrated in the sieve elements and companion cells than in the mesophyll. This difference in solute concentration, found in most of the species studied, can be demonstrated through measurement of the osmotic potential (Ψ_s) of the various cell types in the leaf (see Chapter 3).

In sugar beet, the osmotic potential of the mesophyll is approximately –1.3 MPa, and the osmotic potential of the sieve elements and companion cells is about –3.0 MPa (Geiger et al. 1973). Most of this difference in osmotic potential is thought to result from accumulated sugar, specifically sucrose because sucrose is the major transport sugar in this species. Experimental studies have also demonstrated that both externally supplied sucrose and sucrose made from photosynthetic products accumulate in the sieve elements and companion cells of the minor veins of sugar beet source leaves (Figure 10.15).

The fact that sucrose is at a higher concentration in the sieve element–companion cell complex than in surrounding cells indicates that sucrose is actively transported against its chemical-potential gradient. The dependence of sucrose accumulation on active transport is supported by the fact that treating source tissue with respiratory inhibitors both decreases ATP concentration and inhibits loading of exogenous sugar. On the other hand, other metabolites, such as organic acids and hormones, may enter sieve elements passively (see **Web Topic 10.6**).

In the Apoplastic Pathway, Sieve Element Loading Involves a Sucrose–H^+ Symporter

A sucrose–H^+ symporter is thought to mediate the transport of sucrose from the apoplast into the sieve element–companion cell complex. Recall from Chapter 6

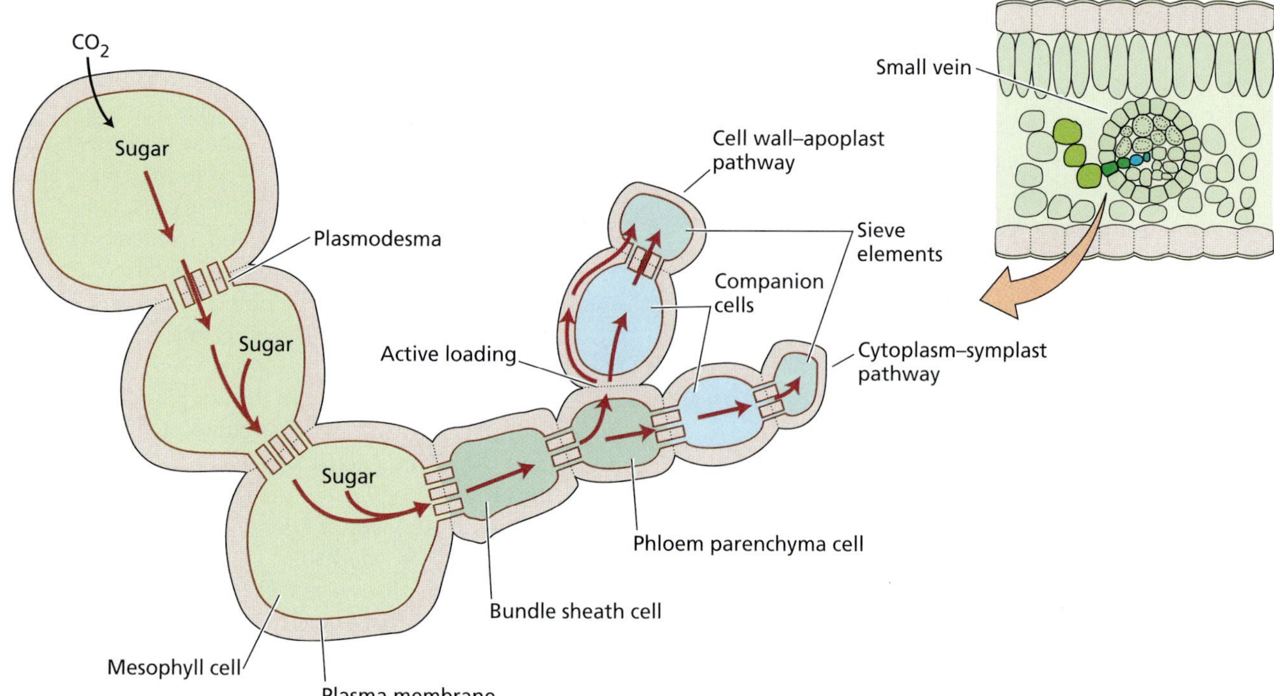

FIGURE 10.14 Schematic diagram of pathways of phloem loading in source leaves. In the totally symplastic pathway, sugars move from one cell to another in the plasmodesmata, all the way from the mesophyll to the sieve elements. In the partly apoplastic pathway, sugars enter the apoplast at some point. For simplicity, sugars are shown here entering the apoplast near the sieve element–companion cell complex, but they could also enter the apoplast earlier in the path and then move to the small veins. In any case, the sugars are actively loaded into the companion cells and sieve elements from the apoplast. Sugars loaded into the companion cells are thought to move through plasmodesmata into the sieve elements.

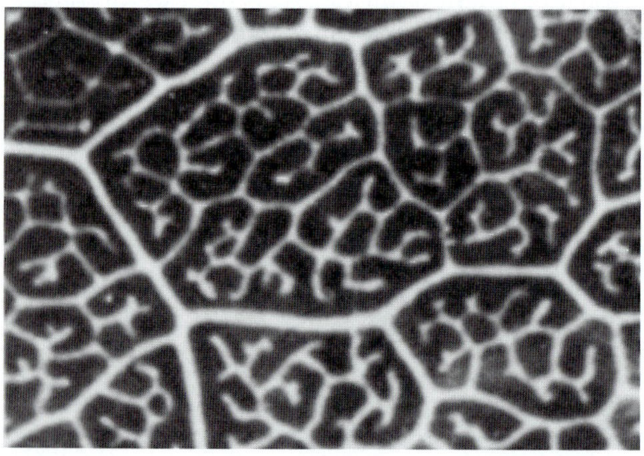

FIGURE 10.15 This autoradiograph shows that labeled sugar moves from the apoplast into sieve elements and companion cells against its concentration gradient. A solution of ^{14}C-labeled sucrose was applied for 30 minutes to the upper surface of a sugar beet (*Beta vulgaris*) leaf that had previously been kept in darkness for 3 hours. The leaf cuticle was removed to allow penetration of the solution to the interior of the leaf. Label accumulates in the small veins, sieve elements, and companion cells of the source leaf, indicating the ability of these cells to transport sucrose against its concentration gradient. (From Fondy 1975, courtesy of D. Geiger.)

that symport is a secondary transport process that uses the energy generated by the proton pump (see Figure 6.10A). The energy dissipated by protons moving back into the cell is coupled to the uptake of a substrate, in this case sucrose (Figure 10.16).

High pH (low H^+ concentration) in the apoplast reduces the uptake of exogenous sucrose into the sieve elements and companion cells of broad bean. This effect occurs because a low proton concentration in the apoplast reduces the driving force for proton diffusion into the symplast and for the sucrose–H^+ symporter.

Data from molecular studies support the operation of a sucrose–H^+ symporter in sieve element loading. Proton-pumping ATPases, localized by immunological techniques, have been found in the plasma membranes of companion cells of *Arabidopsis* and in transfer cells of broad bean. In transfer cells, the H^+-ATPase molecules are most concentrated in the plasma membrane infoldings that face the bundle sheath and phloem parenchyma cells (for details, see **Web Topic 10.7**).

Such localization suggests that the function of these H^+-ATPases is to energize the transport of photosynthate from

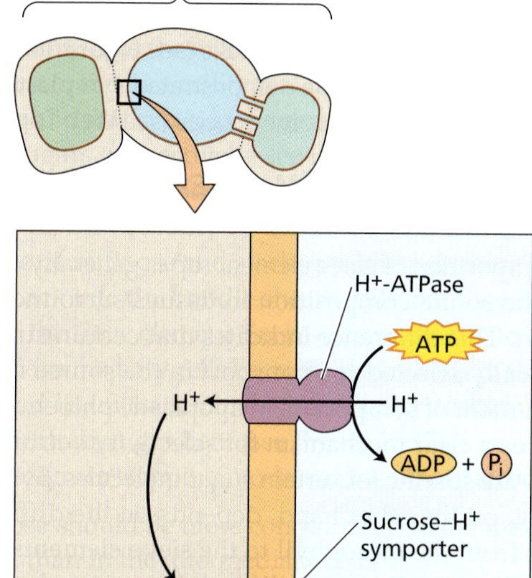

FIGURE 10.16 ATP-dependent sucrose transport in sieve element loading. In the cotransport model of sucrose loading into the symplast of the sieve element–companion cell complex, the plasma membrane ATPase pumps protons out of the cell into the apoplast, establishing a high proton concentration there. The energy in this proton gradient is then used to drive the transport of sucrose into the symplast of the sieve element–companion cell complex through a sucrose–H^+ symporter.

the apoplast to the sieve elements (Bouche-Pillon et al. 1994). Furthermore, the distribution of the H^+-ATPases in companion cells of *Arabidopsis* appears to be correlated with the distribution of a sucrose–H^+ symporter called *SUC2* (DeWitt and Sussman 1995; Truernit and Sauer 1995). The SUC2 transporter has also been localized in companion cells of broad-leaved plantain, *Plantago major* (see **Web Topic 10.7**). H^+-ATPases and sucrose–H^+ symporters are sometimes co-localized in the plasma membranes of sieve elements (Langhans et al. 2001) rather than companion cells.

SUC2 is one of several sucrose–H^+ symporters that have been cloned and localized in the phloem (Table 10.3). The carriers are found in plasma membranes of either sieve elements (SUT1, SUT2, and SUT4) or companion cells (SUC2). Work with SUT1 has shown that the messenger RNAs for symporters found in the sieve element membrane are synthesized in the companion cells (Kuhn et al. 1997). This finding agrees with the fact that sieve elements lack nuclei. The symporter protein is probably also synthesized in the companion cells, since ribosomes do not appear to persist in mature sieve elements.

The roles played by the various carriers listed in Table 10.3 are still being elucidated. Most of the transporters are found in source, path, and sink tissues. SUT1, characterized as a high-affinity/low-capacity transporter found in the minor veins of source tissues, appears to be important in phloem loading. Potato plants transformed with antisense DNA to SUT1 showed reduced transporter activity, a reduction in root and tuber growth, and accumulation of starch and lipids in source leaves (Schulz et al. 1998).

SUT1 is also thought to play a role in the retrieval of sucrose lost in transit. The important role of SUT1 in phloem loading appears to be complemented by SUT4, a low-affinity/high-capacity carrier (Weise et al. 2000). SUT2, on the other hand, appears to function as a sucrose sensor. This is indicated by findings showing that SUT2 is more highly expressed in sink and path tissues than in source leaves, and by the similarity between many structural features of SUT2 and yeast sugar sensors (Lalonde et al. 1999; Barker et al. 2000). Finally, uptake into companion cells appears to be the function of SUC2.

Regulating sucrose loading. The mechanisms that regulate the loading of sucrose from the apoplast to the sieve elements by the sucrose–H^+ symporter await characterization. Possible regulatory factors include the following:

- *The solute potential or, more likely, the turgor pressure of the sieve elements.* A decrease in sieve element turgor below a certain threshold would lead to a compensatory increase in loading.

TABLE 10.3
Sucrose–H^+ symporters in the phloem

Carrier	Location	Species	Affinity	Source
SUT1	Sieve elements	Tobacco, tomato, potato	High	Kuhn et al. 1997
SUT2	Sieve elements	Tomato	Sensor	Barker et al. 2000
SUT4	Sieve elements	*Arabidopsis*, tomato, potato	Low	Weise et al. 2000
SUC2	Companion cells	*Arabidopsis*, plantain	—	Truernit and Sauer, 1995; Stadler et al. 1995

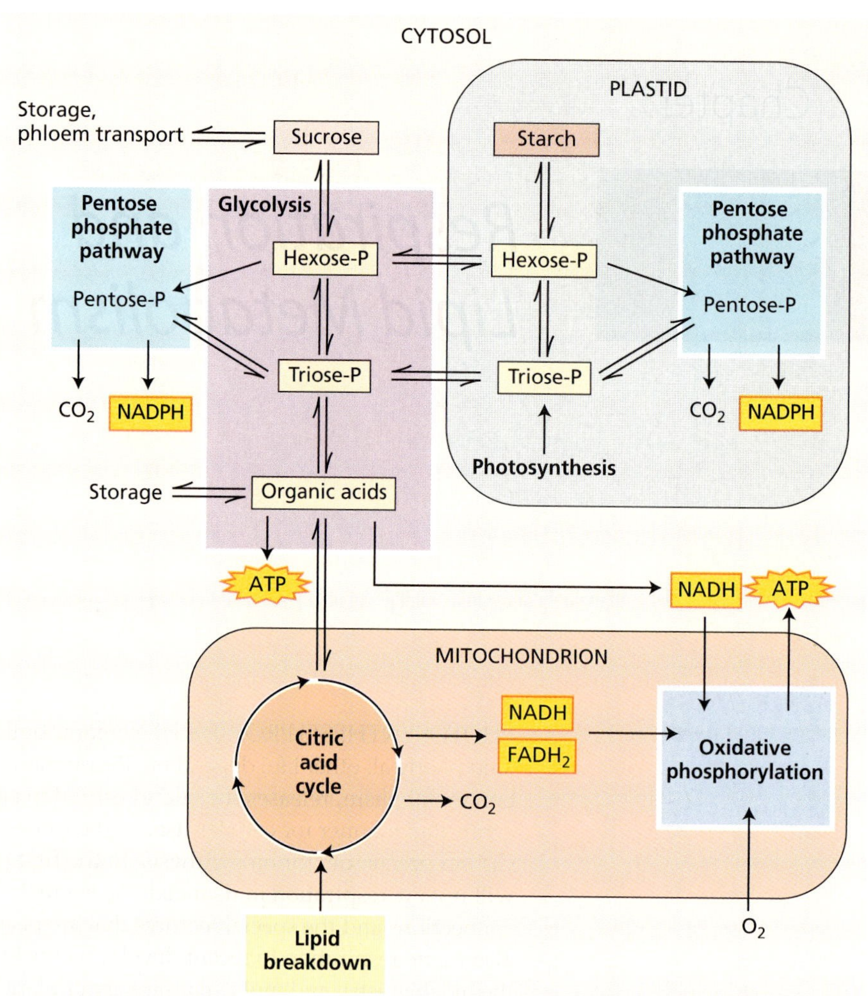

FIGURE 11.1 Overview of respiration. Substrates for respiration are generated by other cellular processes and enter the respiratory pathways. Glycolysis and the pentose phosphate pathways in the cytosol and plastid convert sugars to organic acids, via hexose phosphates and triose phosphates, generating NADH or NADPH and ATP. The organic acids are oxidized in the mitochondrial citric acid cycle, and the NADH and $FADH_2$ produced provide the energy for ATP synthesis by the electron transport chain and ATP synthase in oxidative phosphorylation. In gluconeogenesis, carbon from lipid breakdown is broken down in the glyoxysomes, metabolized in the citric acid cycle, and then used to synthesize sugars in the cytosol by reverse glycolysis.

From a chemical standpoint, plant respiration can be expressed as the oxidation of the 12-carbon molecule sucrose and the reduction of 12 molecules of O_2:

$$C_{12}H_{22}O_{11} + 13\ H_2O \rightarrow 12\ CO_2 + 48\ H^+ + 48\ e^-$$

$$12\ O_2 + 48\ H^+ + 48\ e^- \rightarrow 24\ H_2O$$

giving the following net reaction:

$$C_{12}H_{22}O_{11} + 12\ O_2 \rightarrow 12\ CO_2 + 11\ H_2O$$

This reaction is the reversal of the photosynthetic process; it represents a coupled redox reaction in which sucrose is completely oxidized to CO_2 while oxygen serves as the ultimate electron acceptor, being reduced to water. The standard free-energy decrease for the reaction as written is 5760 kJ (1380 kcal) per mole (342 g) of sucrose oxidized. The controlled release of this free energy, along with its coupling to the synthesis of ATP, is the primary, though by no means only, role of respiratory metabolism.

To prevent damage (incineration) of cellular structures, the cell mobilizes the large amount of free energy released in the oxidation of sucrose in a series of step-by-step reactions. These reactions can be grouped into four major processes: glycolysis, the citric acid cycle, the reactions of the pentose phosphate pathway, and oxidative phosphorylation. The substrates of respiration enter the respiratory process at different points in the pathways, as summarized in Figure 11.1:

- **Glycolysis** involves a series of reactions carried out by a group of soluble enzymes located in both the cytosol and the plastid. A sugar—for example, sucrose—is partly oxidized via six-carbon sugar phosphates (hexose phosphates) and three-carbon sugar phosphates (triose phosphates) to produce an organic acid—for example, pyruvate. The process yields a small amount of energy as ATP, and reducing power in the form of a reduced pyridine nucleotide, NADH.
- In the **pentose phosphate pathway**, also located both in the cytosol and the plastid, the six-carbon glucose-6-phosphate is initially oxidized to the five-carbon ribulose-5-phosphate. The carbon is lost as CO_2, and reducing power is conserved in the form of two molecules of another reduced pyridine nucleotide, NADPH. In the following near-equilibrium reactions, ribulose-5-phosphate is converted into three- to seven-carbon sugars.

- In the **citric acid cycle**, pyruvate is oxidized completely to CO_2, and a considerable amount of reducing power (16 NADH + 4 $FADH_2$ equivalents per sucrose) is generated in the process. With one exception (succinate dehydrogenase), these reactions involve a series of enzymes located in the internal aqueous compartment, or matrix, of the mitochondrion (see Figure 11.5). As we will discuss later, succinate dehydrogenase is localized in the inner of the two mitochondrial membranes.
- In oxidative phosphorylation, electrons are transferred along an **electron transport chain**, consisting of a collection of electron transport proteins bound to the inner of the two mitochondrial membranes. This system transfers electrons from NADH (and related species)—produced during glycolysis, the pentose phosphate pathway, and the citric acid cycle—to oxygen. This electron transfer releases a large amount of free energy, much of which is conserved through the synthesis of ATP from ADP and P_i (inorganic phosphate) catalyzed by the enzyme **ATP synthase**. Collectively the redox reactions of the electron transport chain and the synthesis of ATP are called **oxidative phosphorylation**. This final stage completes the oxidation of sucrose.

Nicotinamide adenine dinucleotide (NAD^+/NADH) is an organic cofactor (coenzyme) associated with many enzymes that catalyze cellular redox reactions. NAD^+ is the oxidized form of the cofactor, and it undergoes a reversible two-electron reaction that yields NADH (Figure 11.2):

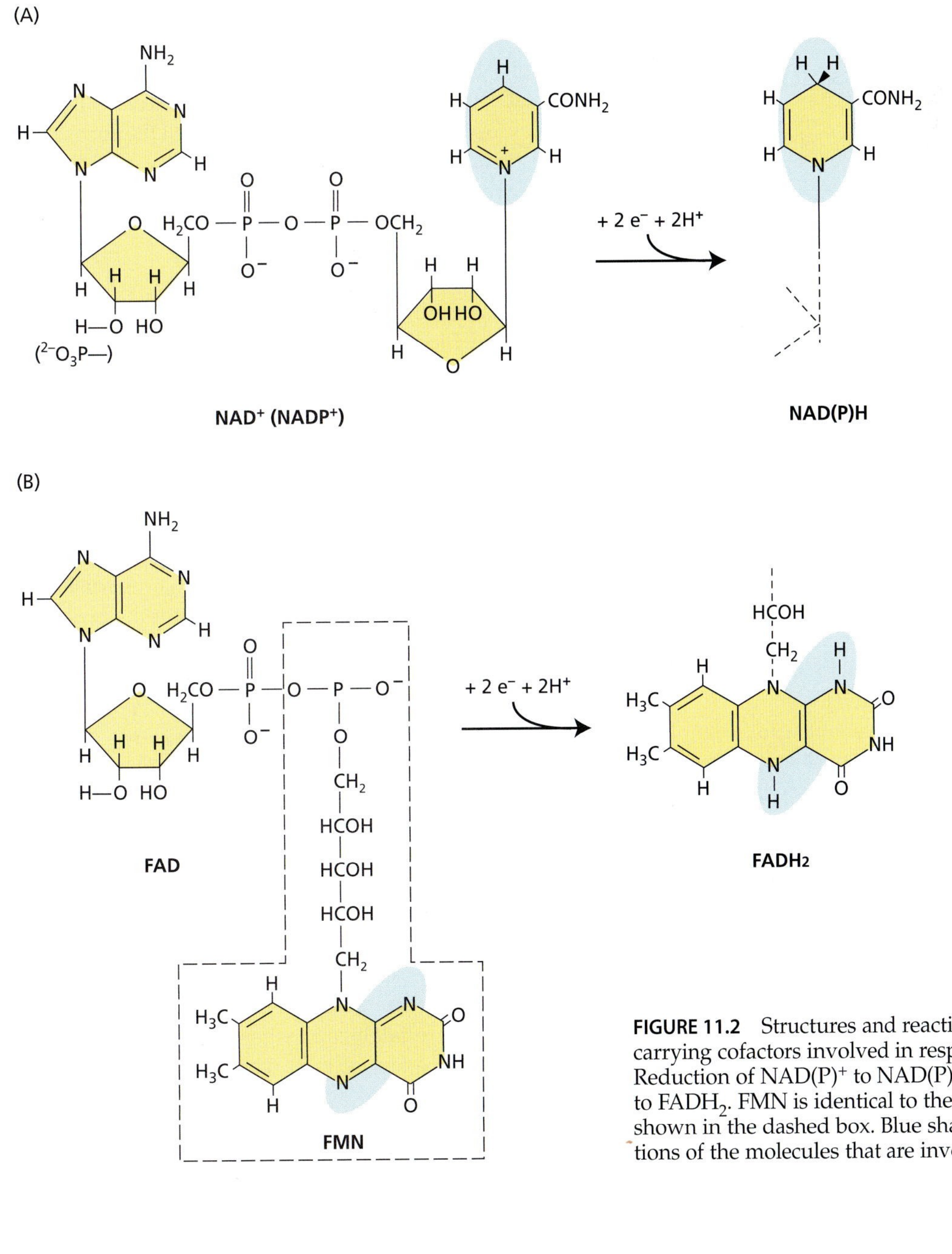

(A)

NAD^+ ($NADP^+$) $\xrightarrow{+ 2e^- + 2H^+}$ NAD(P)H

(B)

FAD $\xrightarrow{+ 2e^- + 2H^+}$ $FADH_2$

FMN

FIGURE 11.2 Structures and reactions of the major electron-carrying cofactors involved in respiratory bioenergetics. (A) Reduction of $NAD(P)^+$ to NAD(P)H; (B) Reduction of FAD to $FADH_2$. FMN is identical to the flavin part of FAD and is shown in the dashed box. Blue shaded areas show the portions of the molecules that are involved in the redox reaction.

$$NAD^+ + 2\,e^- + H^+ \rightarrow NADH$$

The standard reduction potential for this redox couple is about –320 mV, which makes it a relatively strong reductant (i.e., electron donor). NADH is thus a good molecule in which to conserve the free energy carried by electrons released during the stepwise oxidations of glycolysis and the citric acid cycle. A related compound, nicotinamide adenine dinucleotide phosphate ($NADP^+$/NADPH), functions in redox reactions of photosynthesis (see Chapter 8) and of the oxidative pentose phosphate pathway; it also takes part in mitochondrial metabolism (Møller and Rasmusson 1998). This will be discussed later in the chapter.

The oxidation of NADH by oxygen via the electron transport chain releases free energy (220 kJ mol^{-1}, or 52 kcal mol^{-1}) that drives the synthesis of ATP. We can now formulate a more complete picture of respiration as related to its role in cellular energy metabolism by coupling the following two reactions:

$$C_{12}H_{22}O_{11} + 12\,O_2 \rightarrow 12\,CO_2 + 11\,H_2O$$

$$60\,ADP + 60\,P_i \rightarrow 60\,ATP + 60\,H_2O$$

Keep in mind that not all the carbon that enters the respiratory pathway ends up as CO_2. Many respiratory intermediates are the starting points for pathways that assimilate nitrogen into organic form, pathways that synthesize nucleotides and lipids, and many others (see Figure 11.13).

GLYCOLYSIS: A CYTOSOLIC AND PLASTIDIC PROCESS

In the early steps of glycolysis (from the Greek words *glykos*, "sugar," and *lysis*, "splitting"), carbohydrates are converted to hexose phosphates, which are then split into two triose phosphates. In a subsequent energy-conserving phase, the triose phosphates are oxidized and rearranged to yield two molecules of pyruvate, an organic acid. Besides preparing the substrate for oxidation in the citric acid cycle, glycolysis yields a small amount of chemical energy in the form of ATP and NADH.

When molecular oxygen is unavailable—for example, in plant roots in flooded soils—glycolysis can be the main source of energy for cells. For this to work, the **fermentation pathways**, which are localized in the cytosol, reduce pyruvate to recycle the NADH produced by glycolysis. In this section we will describe the basic glycolytic and fermentative pathways, emphasizing features that are specific for plant cells. We will end by discussing the pentose phosphate pathway.

Glycolysis Converts Carbohydrates into Pyruvate, Producing NADH and ATP

Glycolysis occurs in all living organisms (prokaryotes and eukaryotes). The principal reactions associated with the classic glycolytic and fermentative pathways in plants are almost identical with those of animal cells (Figure 11.3). However, plant glycolysis has unique regulatory features, as well as a parallel partial glycolytic pathway in plastids and alternative enzymatic routes for several cytosolic steps. In animals the substrate of glycolysis is glucose and the end product pyruvate. Because sucrose is the major translocated sugar in most plants and is therefore the form of carbon that most nonphotosynthetic tissues import, sucrose (not glucose) can be argued to be the true sugar substrate for plant respiration. The end products of plant glycolysis include another organic acid, malate.

In the early steps of glycolysis, sucrose is broken down into the two monosaccharides—glucose and fructose—which can readily enter the glycolytic pathway. Two pathways for the degradation of sucrose are known in plants, both of which also take part in the unloading of sucrose from the phloem (see Chapter 10).

In most plant tissues sucrose synthase, localized in the cytosol, is used to degrade sucrose by combining sucrose with UDP to produce fructose and UDP-glucose. UDP-glucose pyrophosphorylase then converts UDP-glucose and pyrophosphate (PP_i) into UTP and glucose-6-phosphate (see Figure 11.3). In some tissues, invertases present in the cell wall, vacuole, or cytosol hydrolyze sucrose to its two component hexoses (glucose and fructose). The hexoses are then phosphorylated in a reaction that uses ATP. Whereas the sucrose synthase reaction is close to equilibrium, the invertase reaction releases sufficient energy to be essentially irreversible.

Plastids such as chloroplasts or amyloplasts (see Chapter 1) can also supply substrates for glycolysis. Starch is synthesized and catabolized only in plastids (see Chapter 8), and carbon obtained from starch degradation enters the glycolytic pathway in the cytosol primarily as hexose phosphate (which is translocated out of amyloplasts) or triose phosphate (which is translocated out of chloroplasts). Photosynthetic products can also directly enter the glycolytic pathway as triose phosphate (Hoefnagel et al. 1998).

Plastids convert starch into triose phosphates using a separate set of glycolytic isozymes that convert hexose phosphates to triose phosphates. All the enzymes shown in Figure 11.3 have been measured at levels sufficient to support the respiration rates observed in intact plant tissues.

In the initial phase of glycolysis, each hexose unit is phosphorylated twice and then split, eventually producing two molecules of triose phosphate. This series of reactions consumes two to four molecules of ATP per sucrose unit, depending on whether the sucrose is split by sucrose synthase or invertase. These reactions also include two of the three essentially irreversible reactions of the glycolytic pathway that are catalyzed by hexokinase and phosphofructokinase (see Figure 11.3). The phosphofructokinase reaction is one of the control points of glycolysis in both plants and animals.

The energy-conserving phase of glycolysis. The reactions discussed thus far transfer carbon from the various substrate pools into triose phosphates. Once glyceraldehyde-3-phosphate is formed, the glycolytic pathway can begin to extract usable energy in the energy-conserving phase. The enzyme glyceraldehyde-3-phosphate dehydrogenase catalyzes the oxidation of the aldehyde to a carboxylic acid, reducing NAD^+ to NADH. This reaction releases sufficient free energy to allow the phosphorylation (using inorganic phosphate) of glyceraldehyde-3-phosphate to produce 1,3-bisphosphoglycerate. The phosphorylated carboxylic acid on carbon 1 of 1,3-bisphosphoglycerate (see Figure 11.3) has a large standard free energy of hydrolysis (–49.3 kJ mol^{-1}, or –11.8 kcal mol^{-1}). Thus, 1,3-bisphosphoglycerate is a strong donor of phosphate groups.

In the next step of glycolysis, catalyzed by phosphoglycerate kinase, the phosphate on carbon 1 is transferred to a molecule of ADP, yielding ATP and 3-phosphoglycerate. For each sucrose entering the pathway, four ATPs are generated by this reaction—one for each molecule of 1,3-bisphosphoglycerate.

This type of ATP synthesis, traditionally referred to as **substrate-level phosphorylation**, involves the direct transfer of a phosphate group from a substrate molecule to ADP, to form ATP. As we will see, ATP synthesis by substrate-level phosphorylation is mechanistically distinct from ATP synthesis by ATP synthases involved in the oxidative phosphorylation in mitochondria (which will be described later in this chapter) or photophosphorylation in chloroplasts (see Chapter 7).

In the following two reactions, the phosphate on 3-phosphoglycerate is transferred to carbon 2 and then a molecule of water is removed, yielding the compound phosphoenylpyruvate (PEP). The phosphate group on PEP has a high standard free energy of hydrolysis (–61.9 kJ mol^{-1}, or –14.8 kcal mol^{-1}), which makes PEP an extremely good phosphate donor for ATP formation. Using PEP as substrate, the enzyme pyruvate kinase catalyzes a second substrate-level phosphorylation to yield ATP and pyruvate. This final step, which is the third essentially irreversible step in glycolysis, yields four additional molecules of ATP for each sucrose that enters the pathway.

Plants Have Alternative Glycolytic Reactions

The sequence of reactions leading to the formation of pyruvate from glucose occurs in all organisms that carry out glycolysis. In addition, organisms can operate this pathway in the opposite direction to synthesize sugar from organic acids. This process is known as **gluconeogenesis**.

Gluconeogenesis is not common in plants, but it does operate in the seeds of some plants, such as castor bean and sunflower, that store a significant quantity of their carbon reserves in the form of oils (triacylglycerols). After the seed germinates, much of the oil is converted by gluconeogenesis to sucrose, which is then used to support the growing seedling. In the initial phase of glycolysis, gluconeogenesis overlaps with the pathway for synthesis of sucrose from photosynthetic triose phosphate described in Chapter 8, which is typical for plants.

Because the glycolytic reaction catalyzed by ATP-dependent phosphofructokinase is essentially irreversible (see Figure 11.3), an additional enzyme, fructose-1,6-bisphosphatase, converts fructose-1,6-bisphosphate to fructose-6-phosphate and P_i during gluconeogenesis. ATP-dependent phosphofructokinase and fructose-1,6-bisphosphatase represent a major control point of carbon flux through the glycolytic/gluconeogenic pathways in both plants and animals, as well as in sucrose synthesis in plants (see Chapter 8).

In plants, the interconversion of fructose-6-phosphate and fructose-1,6-bisphosphate is made more complex by the presence of an additional (cytosolic) enzyme, a PP_i-dependent phosphofructokinase (pyrophosphate:fructose-6-phosphate 1-phosphotransferase), which catalyzes the following reversible reaction (see Figure 11.3):

$$\text{Fructose-6-P} + PP_i \leftrightarrow \text{fructose-1,6-}P_2 + P_i$$

where P represents phosphate and P_2 bisphosphate. PP_i-dependent phosphofructokinase is found in the cytosol of most plant tissues at levels that are considerably higher than those of the ATP-dependent phosphofructokinase (Kruger 1997). Suppression of the PP_i-dependent phosphofructokinase in transgenic potato has indicated that it contributes to glycolytic flux, but that it is not essential for plant survival, indicating that other enzymes can take over its function.

The reaction catalyzed by the PP_i-dependent phosphofructokinase is readily reversible, but it is unlikely to operate in sucrose synthesis (Dennis and Blakely 2000). Like ATP-dependent phosphofructokinase and fructose bisphosphatase, this enzyme appears to be regulated by fluctuations in cell metabolism (discussed later in the chapter), suggesting that under some circumstances operation of the glycolytic pathway in plants differs from that in many other organisms (see **Web Essay 11.1**).

At the end of the glycolytic sequence, plants have alternative pathways for metabolizing PEP. In one pathway PEP is carboxylated by the ubiquitous cytosolic enzyme PEP carboxylase to form the organic acid oxaloacetate (OAA). The OAA is then reduced to malate by the action of malate dehydrogenase, which uses NADH as the source of electrons, and this performs a role similar to that of the dehydrogenases during fermentative metabolism (see Figure 11.3). The resulting malate can be stored by export to the vacuole or transported to the mitochondrion, where it can enter the citric acid cycle. Thus the operation of pyruvate kinase and PEP carboxylase can produce alternative organic acids—pyruvate or malate—for mitochondrial respiration, though pyruvate dominates in most tissues.

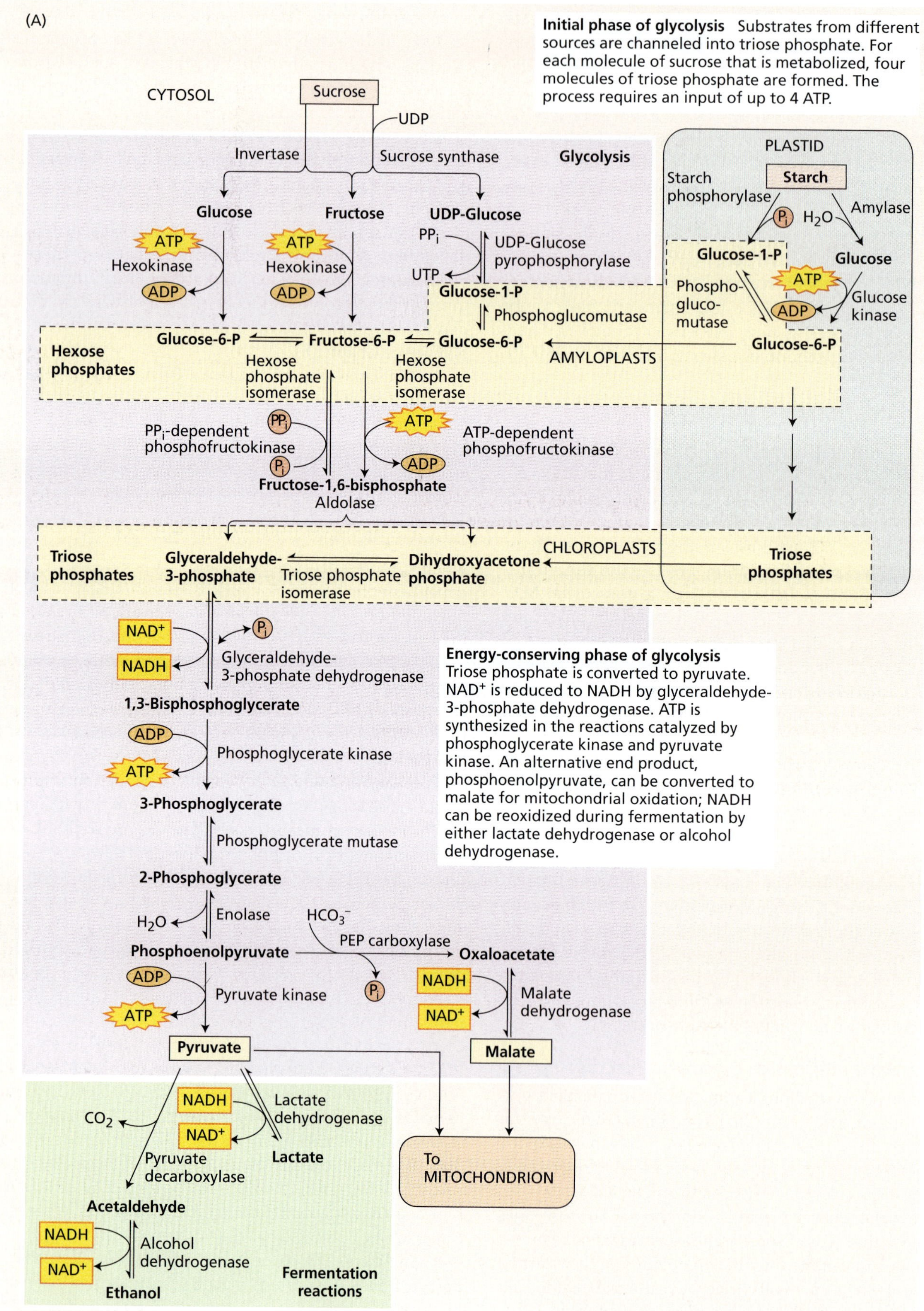
(A)
Initial phase of glycolysis Substrates from different sources are channeled into triose phosphate. For each molecule of sucrose that is metabolized, four molecules of triose phosphate are formed. The process requires an input of up to 4 ATP.
CYTOSOL
Sucrose
UDP
Invertase
Sucrose synthase
Glycolysis
PLASTID
Starch phosphorylase
Starch
Amylase
P_i
H_2O
Glucose
Fructose
UDP-Glucose
ATP
Hexokinase
ADP
PP_i
UTP
UDP-Glucose pyrophosphorylase
Glucose-1-P
Phosphoglucomutase
Phospho-gluco-mutase
Glucose kinase
Hexose phosphates
Glucose-6-P
Fructose-6-P
Hexose phosphate isomerase
AMYLOPLASTS
PP_i-dependent phosphofructokinase
ATP-dependent phosphofructokinase
Fructose-1,6-bisphosphate
Aldolase
CHLOROPLASTS
Triose phosphates
Glyceraldehyde-3-phosphate
Triose phosphate isomerase
Dihydroxyacetone phosphate
NAD^+
NADH
Glyceraldehyde-3-phosphate dehydrogenase
Energy-conserving phase of glycolysis Triose phosphate is converted to pyruvate. NAD^+ is reduced to NADH by glyceraldehyde-3-phosphate dehydrogenase. ATP is synthesized in the reactions catalyzed by phosphoglycerate kinase and pyruvate kinase. An alternative end product, phosphoenolpyruvate, can be converted to malate for mitochondrial oxidation; NADH can be reoxidized during fermentation by either lactate dehydrogenase or alcohol dehydrogenase.
1,3-Bisphosphoglycerate
Phosphoglycerate kinase
3-Phosphoglycerate
Phosphoglycerate mutase
2-Phosphoglycerate
Enolase
H_2O
HCO_3^-
PEP carboxylase
Phosphoenolpyruvate
Oxaloacetate
Pyruvate kinase
Malate dehydrogenase
Pyruvate
Malate
Lactate dehydrogenase
CO_2
Pyruvate decarboxylase
Lactate
To MITOCHONDRION
Acetaldehyde
Alcohol dehydrogenase
Ethanol
Fermentation reactions

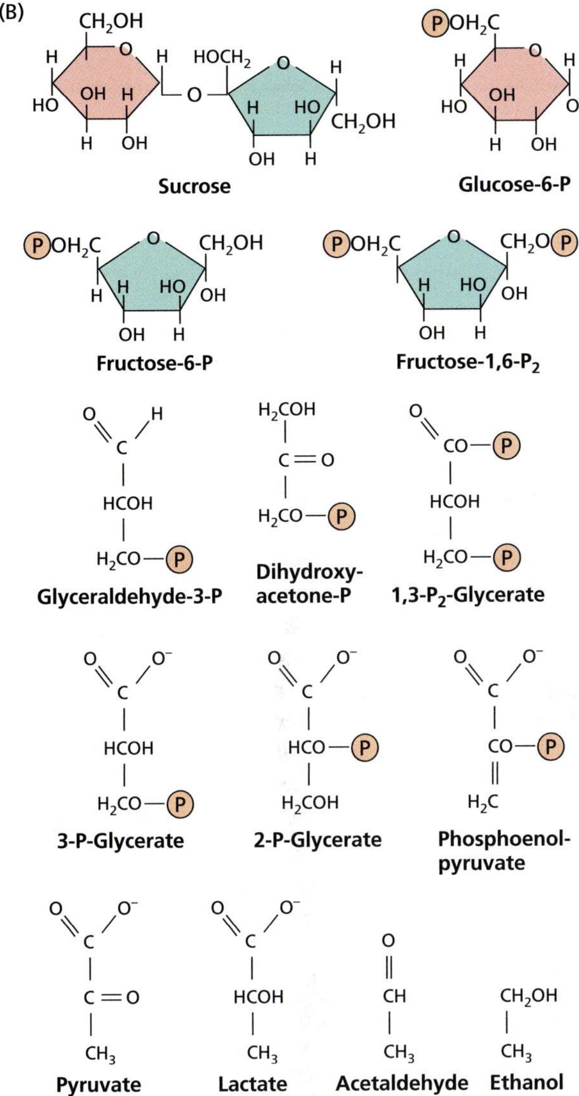

FIGURE 11.3 Reactions of plant glycolysis and fermentation. (A) In the main pathway, sucrose is oxidized to the organic acid pyruvate. The double arrows denote reversible reactions; the single arrows, essentially irreversible reactions. (B) The structures of the intermediates. P, phosphate; P_2, bisphosphate.

In the Absence of O_2, Fermentation Regenerates the NAD+ Needed for Glycolysis

In the absence of oxygen, the citric acid cycle and oxidative phosphorylation cannot function. Glycolysis thus cannot continue to operate because the cell's supply of NAD+ is limited, and once all the NAD+ becomes tied up in the reduced state (NADH), the reaction catalyzed by glyceraldehyde-3-phosphate dehydrogenase cannot take place. To overcome this problem, plants and other organisms can further metabolize pyruvate by carrying out one or more forms of **fermentative metabolism** (see Figure 11.3).

In alcoholic fermentation (common in plants, but more widely known from brewer's yeast), the two enzymes pyruvate decarboxylase and alcohol dehydrogenase act on pyruvate, ultimately producing ethanol and CO_2 and oxidizing NADH in the process. In lactic acid fermentation (common to mammalian muscle but also found in plants), the enzyme lactate dehydrogenase uses NADH to reduce pyruvate to lactate, thus regenerating NAD+.

Under some circumstances, plant tissues may be subjected to low (hypoxic) or zero (anoxic) concentrations of ambient oxygen, forcing them to carry out fermentative metabolism. The best-studied example involves flooded or waterlogged soils in which the diffusion of oxygen is sufficiently reduced to cause root tissues to become hypoxic.

In corn the initial response to low oxygen is lactic acid fermentation, but the subsequent response is alcoholic fermentation. Ethanol is thought to be a less toxic end product of fermentation because it can diffuse out of the cell, whereas lactate accumulates and promotes acidification of the cytosol. In numerous other cases plants function under near-anaerobic conditions by carrying out some form of fermentation.

Fermentation Does Not Liberate All the Energy Available in Each Sugar Molecule

Before we leave the topic of glycolysis, we need to consider the efficiency of fermentation. *Efficiency* is defined here as the energy conserved as ATP relative to the energy potentially available in a molecule of sucrose. The standard free-energy change ($\Delta G^{0\prime}$) for the complete oxidation of sucrose is –5760 kJ mol^{-1} (1380 kcal mol^{-1}). The value of $\Delta G^{0\prime}$ for the synthesis of ATP is 32 kJ mol^{-1} (7.7 kcal mol^{-1}). However, under the nonstandard conditions that normally exist in both mammalian and plant cells, the synthesis of ATP requires an input of free energy of approximately 50 kJ mol^{-1} (12 kcal mol^{-1}). (For a discussion of free energy, see Chapter 2 on the web site.)

Given the net synthesis of four molecules of ATP for each sucrose molecule that is converted to ethanol (or lactate), the efficiency of anaerobic fermentation is only about 4%. Most of the energy available in sucrose remains in the reduced by-product of fermentation: lactate or ethanol. During aerobic respiration, the pyruvate produced by glycolysis is transported into mitochondria, where it is further oxidized, resulting in a much more efficient conversion of the free energy originally available in the sucrose.

Because of the low efficiency of energy conservation under fermentation, an increased rate of glycolysis is needed to sustain the ATP production necessary for cell survival. This is called the *Pasteur effect* after the French microbiologist Louis Pasteur, who first noted it when yeast switched from aerobic respiration to anaerobic alcoholic fermentation. The higher rates of glycolysis result from changes in glycolytic metabolite levels, as well as from increased expression of genes encoding enzymes of glycolysis and fermentation (Sachs et al. 1996).

Plant Glycolysis Is Controlled by Its Products

In vivo, glycolysis appears to be regulated at the level of fructose-6-phosphate phosphorylation and PEP turnover. In contrast to animals, AMP and ATP are not major effectors of plant phosphofructokinase and pyruvate kinase. The cytosolic concentration of PEP, which is a potent inhibitor of the plant ATP-dependent phosphofructokinase, is a more important regulator of plant glycolysis.

This inhibitory effect of PEP on phosphofructokinase is strongly decreased by inorganic phosphate, making the cytosolic ratio of PEP to P_i a critical factor in the control of plant glycolytic activity. Pyruvate kinase and PEP carboxylase, the enzymes that metabolize PEP in the last steps of glycolysis (see Figure 11.3), are in turn sensitive to feedback inhibition by citric acid cycle intermediates and their derivatives, including malate, citrate, 2-oxoglutarate, and glutamate.

In plants, therefore, the control of glycolysis comes from the "bottom up" (see Figure 11.12), with primary regulation at the level of PEP metabolism by pyruvate kinase and PEP carboxylase and secondary regulation exerted by PEP at the conversion of fructose-6-phosphate to fructose-1,6-bisphosphate (see Figure 11.3). In animals, the primary control operates at the phosphofructokinase, and secondary control at the pyruvate kinase.

One conceivable benefit of bottom-up control of glycolysis is that it permits plants to control net glycolytic flux to pyruvate independently of related metabolic processes such as the Calvin cycle and sucrose–triose phosphate–starch interconversion (Plaxton 1996). Another benefit of this control mechanism is that glycolysis may adjust to the demand for biosynthetic precursors.

The presence of two enzymes metabolizing PEP in plant cells—pyruvate kinase and PEP carboxylase—has consequences for the control of glycolysis that are not quite clear. Though the two enzymes are inhibited by similar metabolites, the PEP carboxylase can under some conditions perform a bypass reaction around the pyruvate kinase. The resulting malate can then enter the mitochondrial citric acid cycle. Hence, the bottom-up regulation enables a high flexibility in the control of plant glycolysis.

Experimental support for multiple pathways of PEP metabolism comes from the study of transgenic tobacco plants with less than 5% of the normal level of cytosolic pyruvate kinase in their leaves (Plaxton 1996). In these plants, rates of both leaf respiration and photosynthesis were unaffected relative to controls having wild-type levels of pyruvate kinase. However, reduced root growth in the transgenic plants indicated that the pyruvate kinase reaction could not be circumvented without some detrimental effects.

The regulation of the conversion of fructose-6-phosphate to fructose-1,6-bisphosphate is also complex. Fructose-2,6-bisphosphate, another hexose bisphosphate, is present at varying levels in the cytosol (see Chapter 8). It markedly inhibits the activity of cytosolic fructose-1,6-bisphosphatase but stimulates the activity of PP_i-dependent phosphofructokinase. These observations suggest that fructose-2,6-bisphosphate plays a central role in partitioning flux between ATP-dependent and PP_i-dependent pathways of fructose phosphate metabolism at the crossing point between sucrose synthesis and glycolysis.

Understanding of the fine levels of glycolysis regulation requires the study of temporal changes in metabolite levels (Givan 1999). Methods are now available by rapid extraction and simultaneous analyses of many metabolites—for example, by mass spectrometry—an approach called *metabolic profiling* (see **Web Essay 11.2**).

The Pentose Phosphate Pathway Produces NADPH and Biosynthetic Intermediates

The glycolytic pathway is not the only route available for the oxidation of sugars in plant cells. Sharing common metabolites, the **oxidative pentose phosphate pathway** (also known as the *hexose monophosphate shunt*) can also accomplish this task (Figure 11.4). The reactions are carried out by soluble enzymes present in the cytosol and in plastids. Generally, the pathway in plastids predominates over the cytosolic pathway (Dennis et al. 1997).

The first two reactions of this pathway involve the oxidative events that convert the six-carbon glucose-6-phosphate to a five-carbon sugar, ribulose-5-phosphate, with loss of a CO_2 molecule and generation of two molecules of NADPH (not NADH). The remaining reactions of the pathway convert ribulose-5-phosphate to the glycolytic intermediates glyceraldehyde-3-phosphate and fructose-6-phosphate. Because glucose-6-phosphate can be regenerated from glyceraldehyde-3-phosphate and fructose-6-phosphate by glycolytic enzymes, for six turns of the cycle we can write the reaction as follows:

$$\text{6 glucose-6-P} + 12\ NADP^+ + 7\ H_2O \rightarrow$$
$$\text{5 glucose-6-P} + 6\ CO_2 + P_i + 12\ NADPH + 12\ H^+$$

The net result is the complete oxidation of one glucose-6-phosphate molecule to CO_2 with the concomitant synthesis of 12 NADPH molecules.

Studies of the release of $^{14}CO_2$ from isotopically labeled glucose indicate that glycolysis is the more dominant breakdown pathway, accounting for 80 to 95% of the total carbon flux in most plant tissues. However, the pentose phosphate pathway does contribute to the flux, and developmental studies indicate that its contribution increases as plant cells develop from a meristematic to a more differentiated state (Ap Rees 1980). The oxidative pentose phosphate pathway plays several roles in plant metabolism:

- The product of the two oxidative steps is NADPH, and this NADPH is thought to drive reductive steps associated with various biosynthetic reactions that occur in the cytosol. In nongreen plastids, such as amyloplasts, and in chloroplasts functioning in the

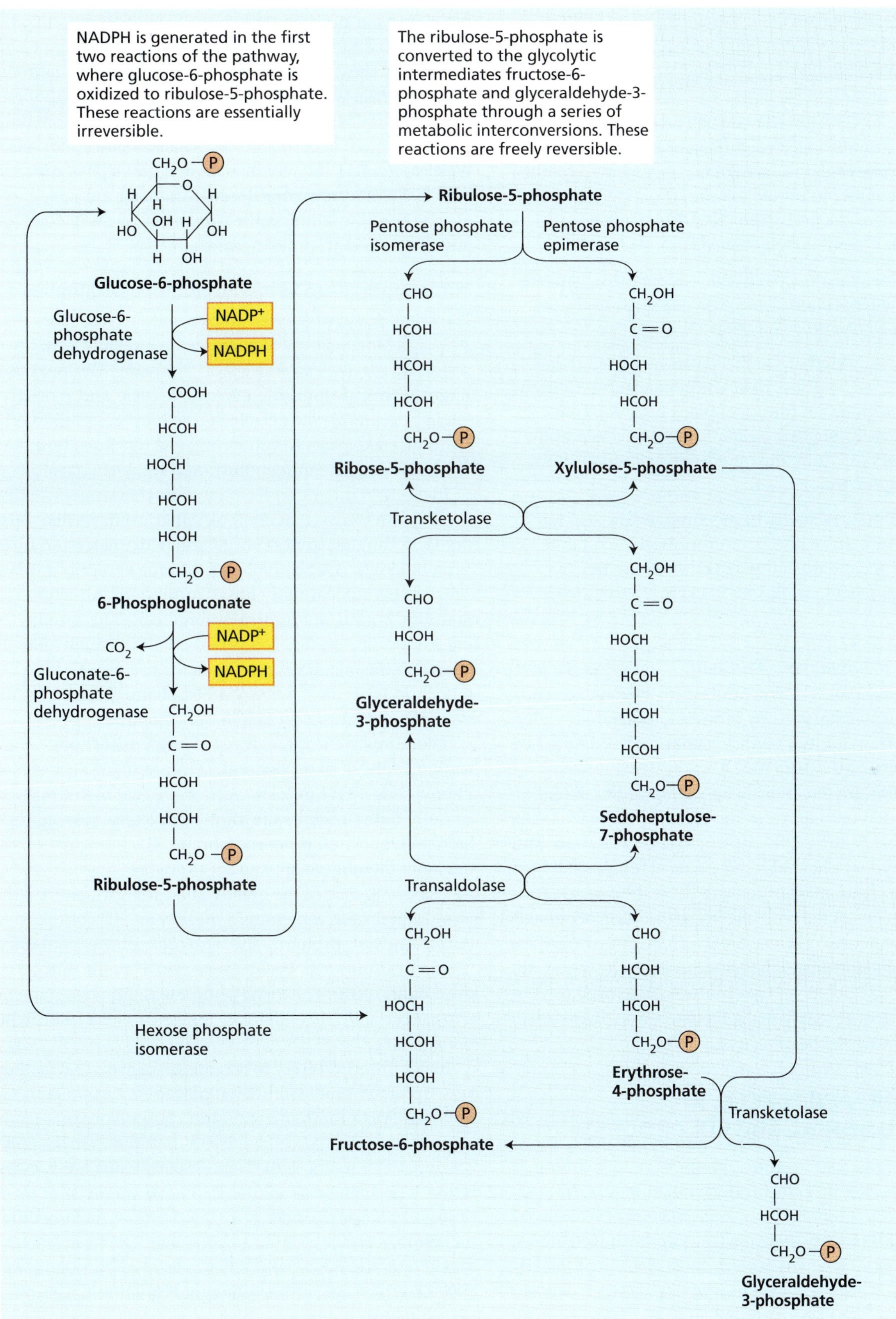

FIGURE 11.4 Reactions of the oxidative pentose phosphate pathway in higher plants. P, phosphate.

dark, the pathway may also supply NADPH for biosynthetic reactions such as lipid biosynthesis and nitrogen assimilation.

- Because plant mitochondria are able to oxidize cytosolic NADPH via an NADPH dehydrogenase localized on the external surface of the inner membrane, some of the reducing power generated by this pathway may contribute to cellular energy metabolism; that is, electrons from NADPH may end up reducing O_2 and generating ATP.
- The pathway produces ribose-5-phosphate, a precursor of the ribose and deoxyribose needed in the synthesis of RNA and DNA, respectively.
- Another intermediate in this pathway, the four-carbon erythrose-4-phosphate, combines with PEP in the initial reaction that produces plant phenolic compounds, including the aromatic amino acids and the precursors of lignin, flavonoids, and phytoalexins (see Chapter 13).
- During the early stages of greening, before leaf tissues become fully photoautotrophic, the oxidative pentose phosphate pathway is thought to be involved in generating Calvin cycle intermediates.

Control of the oxidative pathway. The oxidative pentose phosphate pathway is controlled by the initial reaction of the pathway catalyzed by glucose-6-phosphate dehydrogenase, the activity of which is markedly inhibited by a high ratio of NADPH to $NADP^+$.

In the light, however, little operation of the oxidative pathway is likely to occur in the chloroplast because the end products of the pathway, fructose-6-phosphate and glyceraldehyde-3-phosphate, are being synthesized by the Calvin cycle. Thus, mass action will drive the nonoxidative interconversions of the pathway in the direction of pentose synthesis. Moreover, glucose-6-phosphate dehydrogenase will be inhibited during photosynthesis by the high ratio of NADPH to $NADP^+$ in the chloroplast, as well as by a reductive inactivation involving the ferredoxin–thioredoxin system (see Chapter 8).

THE CITRIC ACID CYCLE: A MITOCHONDRIAL MATRIX PROCESS

During the nineteenth century, biologists discovered that in the absence of air, cells produce ethanol or lactic acid, whereas in the presence of air, cells consume O_2 and produce CO_2 and H_2O. In 1937 the German-born British biochemist Hans A. Krebs reported the discovery of the **citric acid cycle**—also called the *tricarboxylic acid cycle* or *Krebs cycle*. The elucidation of the citric acid cycle not only explained how pyruvate is broken down to CO_2 and H_2O; it also highlighted the key concept of cycles in metabolic pathways. For his discovery, Hans Krebs was awarded the Nobel Prize in physiology and medicine in 1953.

Because the citric acid cycle is localized in the matrix of mitochondria, we will begin with a general description of mitochondrial structure and function, knowledge obtained mainly through experiments on isolated mitochondria (see **Web Topic 11.1**). We will then review the steps of the citric acid cycle, emphasizing the features that are specific to plants. For all plant-specific properties, we will consider how they affect respiratory function.

Mitochondria Are Semiautonomous Organelles

The breakdown of sucrose to pyruvate releases less than 25% of the total energy in sucrose; the remaining energy is stored in the two molecules of pyruvate. The next two stages of respiration (the citric acid cycle and oxidative phosphorylation—i.e., electron transport coupled to ATP synthesis) take place within an organelle enclosed by a double membrane, the **mitochondrion** (plural *mitochondria*).

In electron micrographs, plant mitochondria—whether in situ or in vitro—usually look spherical or rodlike (Figure 11.5), ranging from 0.5 to 1.0 μm in diameter and up to 3 μm in length (Douce 1985). With some exceptions, plant cells have a substantially lower number of mitochondria than that found in a typical animal cell. The number of mitochondria per plant cell varies, and it is usually directly related to the metabolic activity of the tissue, reflecting the mitochondrial role in energy metabolism. Guard cells, for example, are unusually rich in mitochondria.

The ultrastructural features of plant mitochondria are similar to those of mitochondria in nonplant tissues (see Figure 11.5). Plant mitochondria have two membranes: a smooth **outer membrane** that completely surrounds a highly invaginated **inner membrane**. The invaginations of the inner membrane are known as **cristae** (singular *crista*). As a consequence of the greatly enlarged surface area, the inner membrane can contain more than 50% of the total mitochondrial protein. The aqueous phase contained within the inner membrane is referred to as the mitochondrial **matrix** (plural *matrices*), and the region between the two mitochondrial membranes is known as the **intermembrane space**.

Intact mitochondria are osmotically active; that is, they take up water and swell when placed in a hypo-osmotic medium. Most inorganic ions and charged organic molecules are not able to diffuse freely into the matrix space. The inner membrane is the osmotic barrier; the outer membrane is permeable to solutes that have a molecular mass of less than approximately 10,000 Da (i.e., most cellular metabolites and ions, but not proteins). The lipid fraction of both membranes is primarily made up of phospholipids, 80% of which are either phosphatidylcholine or phosphatidylethanolamine.

Like chloroplasts, mitochondria are semiautonomous organelles because they contain ribosomes, RNA, and

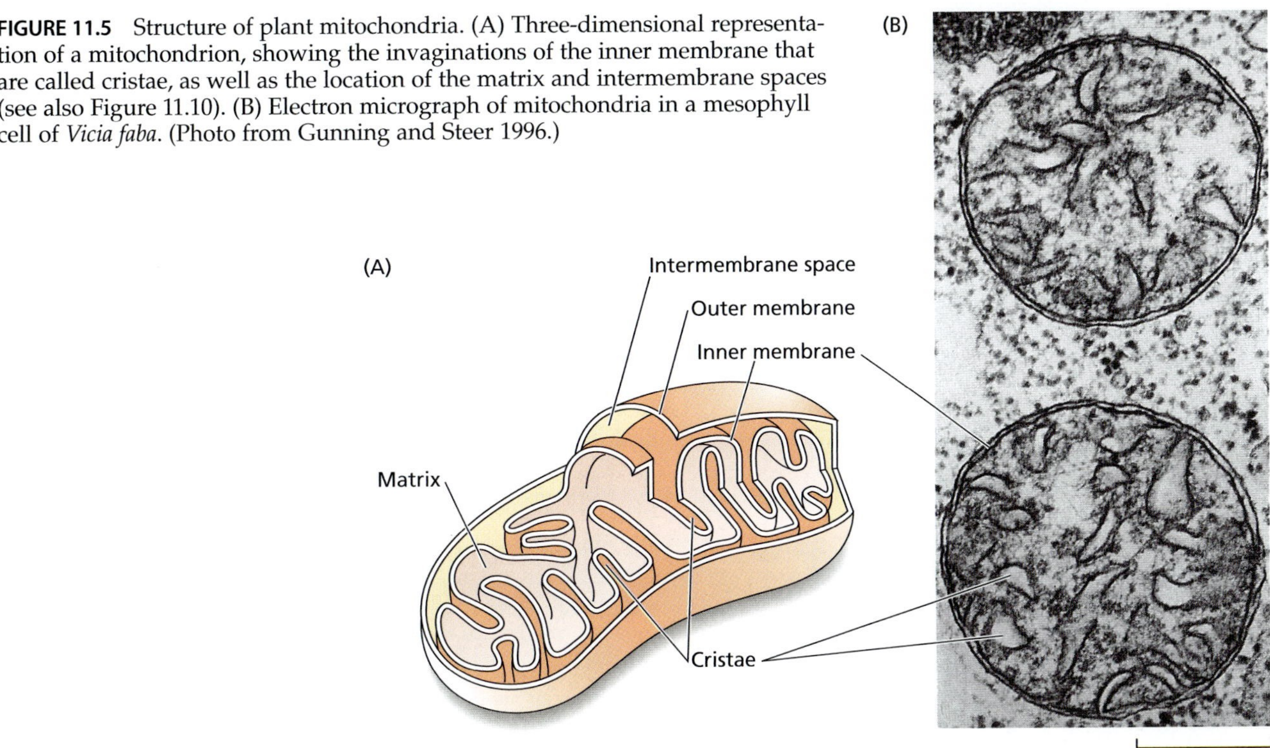

FIGURE 11.5 Structure of plant mitochondria. (A) Three-dimensional representation of a mitochondrion, showing the invaginations of the inner membrane that are called cristae, as well as the location of the matrix and intermembrane spaces (see also Figure 11.10). (B) Electron micrograph of mitochondria in a mesophyll cell of *Vicia faba*. (Photo from Gunning and Steer 1996.)

DNA, which encodes a limited number of mitochondrial proteins. Plant mitochondria are thus able to carry out the various steps of protein synthesis and to transmit their genetic information. Mitochondria proliferate through the division by fission of preexisting mitochondria and not through de novo biogenesis of the organelle.

Pyruvate Enters the Mitochondrion and Is Oxidized via the Citric Acid Cycle

As already noted, the citric acid cycle is also known as the tricarboxylic acid cycle, because of the importance of the tricarboxylic acids citric acid (citrate) and isocitric acid (isocitrate) as early intermediates (Figure 11.6). This cycle constitutes the second stage in respiration and takes place in the mitochondrial matrix. Its operation requires that the pyruvate generated in the cytosol during glycolysis be transported through the impermeable inner mitochondrial membrane via a specific transport protein (as will be described shortly).

Once inside the mitochondrial matrix, pyruvate is decarboxylated in an oxidation reaction by the enzyme pyruvate dehydrogenase. The products are NADH (from NAD^+), CO_2, and acetic acid in the form of acetyl-CoA, in which a thioester bond links the acetic acid to a sulfur-containing cofactor, coenzyme A (CoA) (see Figure 11.6). Pyruvate dehydrogenase exists as a large complex of several enzymes that catalyze the overall reaction in a three-step process: decarboxylation, oxidation, and conjugation to CoA.

In the next reaction the enzyme citrate synthase combines the acetyl group of acetyl-CoA with a four-carbon dicarboxylic acid (oxaloacetate, OAA) to give a six-carbon tricarboxylic acid (citrate). Citrate is then isomerized to isocitrate by the enzyme aconitase.

The following two reactions are successive oxidative decarboxylations, each of which produces one NADH and releases one molecule of CO_2, yielding a four-carbon molecule, succinyl-CoA. At this point, three molecules of CO_2 have been produced for each pyruvate that entered the mitochondrion, or 12 CO_2 for each molecule of sucrose oxidized.

During the remainder of the citric acid cycle, succinyl-CoA is oxidized to OAA, allowing the continued operation of the cycle. Initially the large amount of free energy available in the thioester bond of succinyl-CoA is conserved through the synthesis of ATP from ADP and P_i via a substrate-level phosphorylation catalyzed by succinyl-CoA synthetase. (Recall that the free energy available in the thioester bond of acetyl-CoA was used to form a carbon–carbon bond in the step catalyzed by citrate synthase.) The resulting succinate is oxidized to fumarate by succinate dehydrogenase, which is the only membrane-associated enzyme of the citric acid cycle and also part of the electron transport chain (which is the next major topic to be discussed in this chapter).

The electrons and protons removed from succinate end up not on NAD^+ but on another cofactor involved in redox reactions: FAD (flavin adenine dinucleotide). FAD is covalently bound to the active site of succinate dehydrogenase and undergoes a reversible two-electron reduction to produce $FADH_2$ (see Figure 11.2).

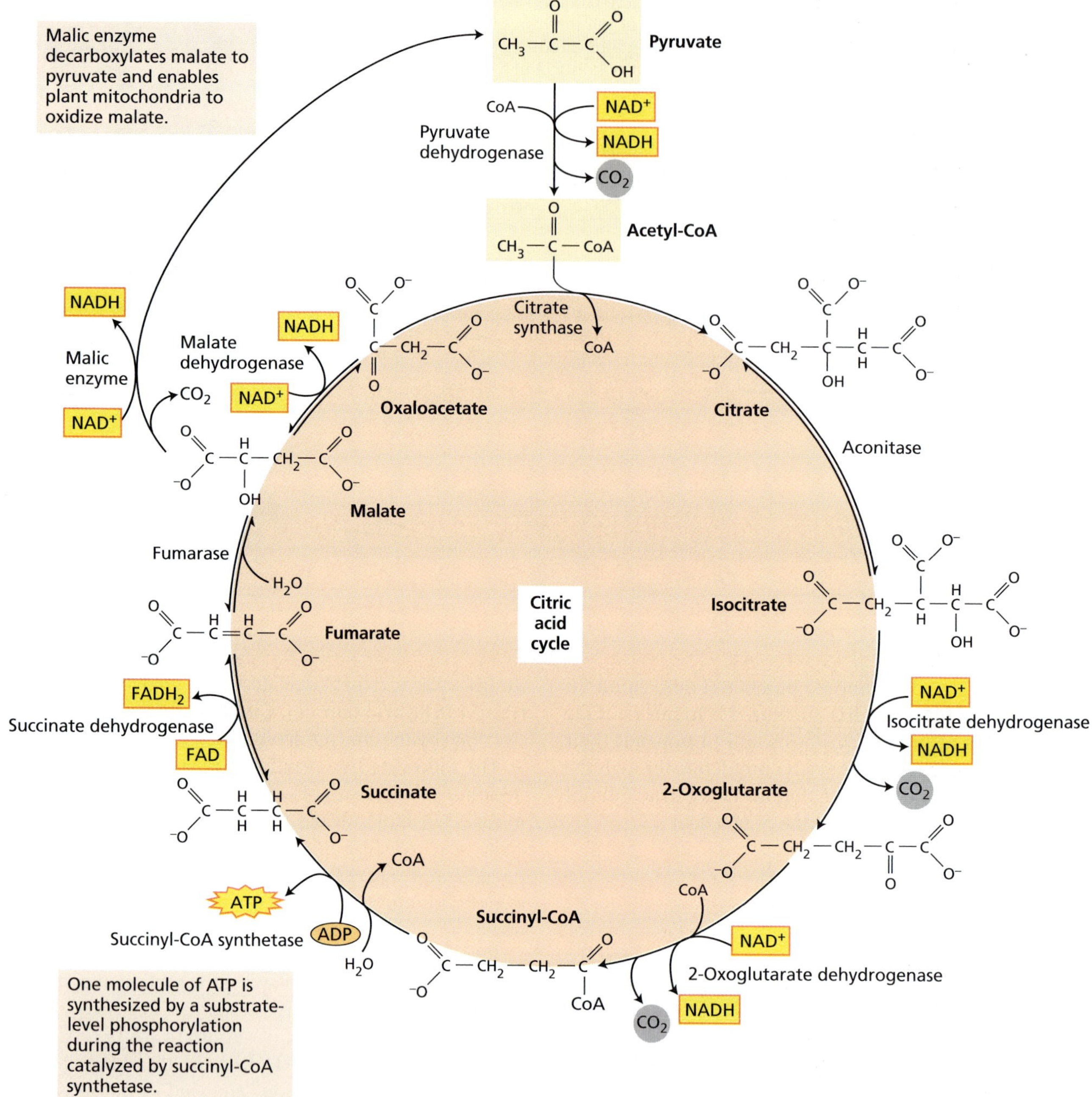

FIGURE 11.6 Reactions and enzymes of the plant citric acid cycle. Pyruvate is completely oxidized to three molecules of CO_2. The electrons released during these oxidations are used to reduce four molecules of NAD^+ to NADH and one molecule of FAD to $FADH_2$.

In the final two reactions of the citric acid cycle, fumarate is hydrated to produce malate, which is subsequently oxidized by malate dehydrogenase to regenerate OAA and produce another molecule of NADH. The OAA produced is now able to react with another acetyl-CoA and continue the cycling.

The stepwise oxidation of one molecule of pyruvate in the mitochondrion gives rise to three molecules of CO_2, and much of the free energy released during these oxidations is conserved in the form of four NADH and one $FADH_2$. In addition, one molecule of ATP is produced by a substrate-level phosphorylation during the citric acid cycle.

All the enzymes associated with the citric acid cycle are found in plant mitochondria. Some of them may be associated in multienzyme complexes, which would facilitate movement of metabolites between the enzymes.

The Citric Acid Cycle of Plants Has Unique Features

The citric acid cycle reactions outlined in Figure 11.6 are not all identical with those carried out by animal mitochondria. For example, the step catalyzed by succinyl-CoA synthetase produces ATP in plants and GTP in animals.

A feature of the plant citric acid cycle that is absent in many other organisms is the significant activity of NAD^+ malic enzyme, which has been found in the matrix of all plant mitochondria analyzed to date. This enzyme catalyzes the oxidative decarboxylation of malate:

$$\text{Malate} + NAD^+ \rightarrow \text{pyruvate} + CO_2 + NADH$$

The presence of NAD^+ malic enzyme enables plant mitochondria to operate alternative pathways for the metabolism of PEP derived from glycolysis (see **Web Essay 11.1**). As already described, malate can be synthesized from PEP in the cytosol via the enzymes PEP carboxylase and malate dehydrogenase (see Figure 11.3). Malate is then transported into the mitochondrial matrix, where NAD^+ malic enzyme can oxidize it to pyruvate. This reaction makes possible the complete net oxidation of citric acid cycle intermediates such as malate (Figure 11.7A) or citrate (Figure 11.7B) (Oliver and McIntosh 1995).

Alternatively, the malate produced via the PEP carboxylase can replace citric acid cycle intermediates used in biosynthesis. Reactions that can replenish intermediates in a metabolic cycle are known as *anaplerotic*. For example, export of 2-oxoglutarate for nitrogen assimilation in the chloroplast will cause a shortage of malate needed in the citrate synthase reaction. This malate can be replaced through the PEP carboxylase pathway (Figure 11.7C).

The presence of an alternative pathway for the oxidation of malate is consistent with the observation that many plants, in addition to those that carry out crassulacean acid metabolism (see Chapter 8), store significant levels of malate in their central vacuole.

ELECTRON TRANSPORT AND ATP SYNTHESIS AT THE INNER MITOCHONDRIAL MEMBRANE

ATP is the energy carrier used by cells to drive living processes, and chemical energy conserved during the citric acid cycle in the form of NADH and $FADH_2$ (redox equivalents with high-energy electrons) must be converted to ATP to perform useful work in the cell. This O_2-dependent process, called **oxidative phosphorylation**, occurs in the inner mitochondrial membrane.

In this section we will describe the process by which the energy level of the electrons is lowered in a stepwise fashion and conserved in the form of an electrochemical proton gradient across the inner mitochondrial membrane. Although fundamentally similar in all aerobic cells, the electron transport chain of plants (and fungi) contains mul-

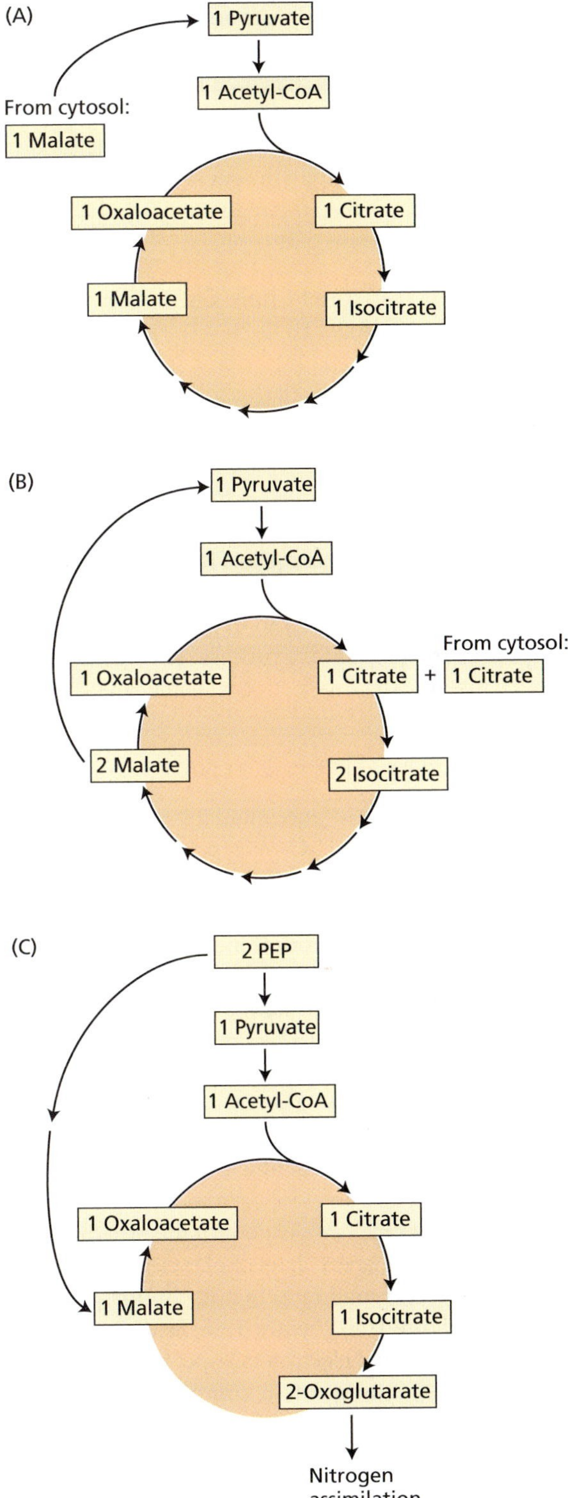

FIGURE 11.7 Malic enzyme and PEP carboxylase provide plants with metabolic flexibility for the metabolism of phosphoenolpyruvate. Malic enzyme makes it possible for plant mitochondria to oxidize both malate (A) and citrate (B) to CO_2 without involving pyruvate delivered by glycolysis. The joint action of PEP carboxylase and pyruvate kinase can convert glycolytic PEP to 2-oxoglutarate, which is used for nitrogen assimilation (C).

tiple NAD(P)H dehydrogenases and an alternative oxidase not found in mammalian mitochondria.

We will also examine the enzyme that uses the energy of the proton gradient to synthesize ATP: the F_oF_1-ATP synthase. After examining the various stages in the production of ATP, we will summarize the energy conservation steps at each stage, as well as the regulatory mechanisms that coordinate the different pathways.

The Electron Transport Chain Catalyzes a Flow of Electrons from NADH to O_2

For each molecule of sucrose oxidized through glycolysis and the citric acid cycle pathways, 4 molecules of NADH are generated in the cytosol and 16 molecules of NADH plus 4 molecules of $FADH_2$ (associated with succinate dehydrogenase) are generated in the mitochondrial matrix. These reduced compounds must be reoxidized or the entire respiratory process will come to a halt.

The electron transport chain catalyzes an electron flow from NADH (and $FADH_2$) to oxygen, the final electron acceptor of the respiratory process. For the oxidation of NADH, the overall two-electron transfer can be written as follows:

$$NADH + H^+ + \frac{1}{2} O_2 \rightarrow NAD^+ + H_2O$$

From the reduction potentials for the NADH–NAD^+ pair (–320 mV) and the H_2O–½O_2 pair (+810 mV), it can be calculated that the standard free energy released during this overall reaction ($-nF\Delta E^{0\prime}$) is about 220 kJ mol^{-1} (52 kcal mol^{-1}) per two electrons (for a detailed discussion on standard free energy see Chapter 2 on the web site). Because the succinate–fumarate reduction potential is higher (+30 mV), only 152 kJ mol^{-1} (36 kcal mol^{-1}) of energy is released for each two electrons generated during the oxidation of succinate. The role of the electron transport chain is to bring about the oxidation of NADH (and $FADH_2$) and, in the process, utilize some of the free energy released to generate an electrochemical proton gradient, $\Delta\tilde{\mu}_{H^+}$, across the inner mitochondrial membrane.

The electron transport chain of plants contains the same set of electron carriers found in mitochondria from other organisms (Figure 11.8) (Siedow 1995; Siedow and Umbach 1995). The individual electron transport proteins are organized into four multiprotein complexes (identified by Roman numerals I through IV), all of which are localized in the inner mitochondrial membrane:

Complex I (NADH dehydrogenase). Electrons from NADH generated in the mitochondrial matrix during the citric acid cycle are oxidized by complex I (an NADH dehydrogenase). The electron carriers in complex I include a tightly bound cofactor (flavin mononucleotide [FMN], which is chemically similar to FAD; see Figure 11.2B) and several iron–sulfur centers. Complex I then transfers these electrons to ubiquinone. Four protons are pumped from the matrix to the intermembrane space for every electron pair passing through the complex.

Ubiquinone, a small lipid-soluble electron and proton carrier, is located within the inner membrane. It is not tightly associated with any protein, and it can diffuse within the hydrophobic core of the membrane bilayer.

Complex II (succinate dehydrogenase). Oxidation of succinate in the citric acid cycle is catalyzed by this complex, and the reducing equivalents are transferred via the $FADH_2$ and a group of iron–sulfur proteins into the ubiquinone pool. This complex does not pump protons.

Complex III (cytochrome bc_1 complex). This complex oxidizes reduced ubiquinone (ubiquinol) and transfers the electrons via an iron–sulfur center, two *b*-type cytochromes (b_{565} and b_{560}), and a membrane-bound cytochrome c_1 to cytochrome *c*. Four protons per electron pair are pumped by complex III.

Cytochrome *c* is a small protein loosely attached to the outer surface of the inner membrane and serves as a mobile carrier to transfer electrons between complexes III and IV.

Complex IV (cytochrome c oxidase). This complex contains two copper centers (Cu_A and Cu_B) and cytochromes *a* and a_3. Complex IV is the terminal oxidase and brings about the four-electron reduction of O_2 to two molecules of H_2O. Two protons are pumped per electon pair (see Figure 11.8).

Both structurally and functionally, ubiquinone and the cytochrome bc_1 complex are very similar to plastoquinone and the cytochrome b_6f complex, respectively, in the photosynthetic electron transport chain (see Chapter 7).

Some Electron Transport Enzymes Are Unique to Plant Mitochondria

In addition to the set of electron carriers described in the previous section, plant mitochondria contain some components not found in mammalian mitochondria (see Figure 11.8). Note that none of these additional enzymes pump protons and that energy conservation is therefore lower whenever they are used:

- Two NAD(P)H dehydrogenases, both Ca^{2+}-dependent, attached to the outer surface of the inner membrane facing the intermembrane space can oxidize cytosolic NADH and NADPH. Electrons from these external NAD(P)H dehydrogenases—ND_{ex}(NADH) and ND_{ex}(NADPH)—enter the main electron transport chain at the level of the ubiquinone pool (see **Web Topic 11.2**) (Møller 2001).
- Plant mitochondria have two pathways for oxidizing matrix NADH. Electron flow through complex I, described in the previous section, is sensitive to inhibition by several compounds, including rotenone and piericidin. In addition, plant mitochondria have a rotenone-resistant dehydrogenase, ND_{in}(NADH), for

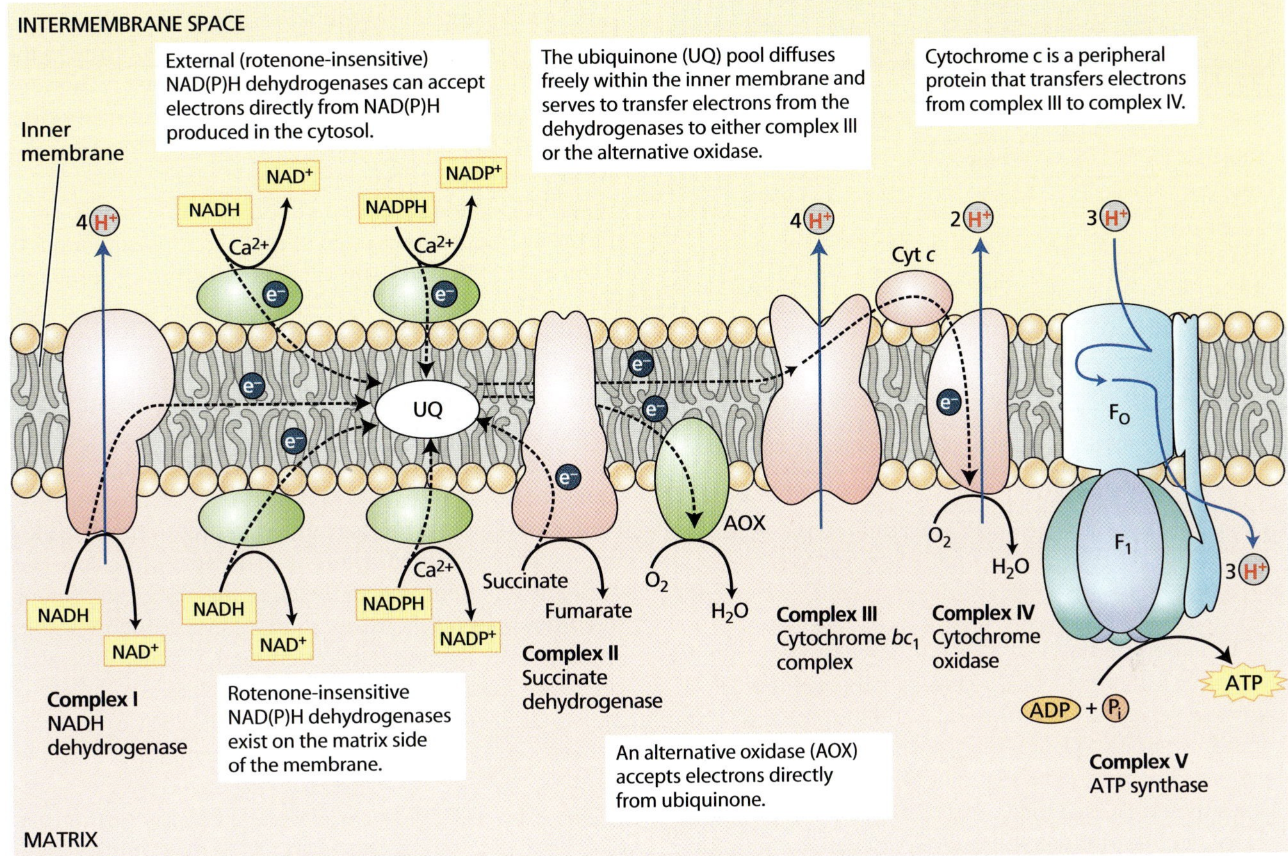

FIGURE 11.8 Organization of the electron transport chain and ATP synthesis in the inner membrane of plant mitochondria. In addition to the five standard protein complexes found in nearly all other mitochondria, the electron transport chain of plant mitochondria contains five additional enzymes marked in green. None of these additional enzymes pumps protons. Specific inhibitors, rotenone for complex I, antimycin for complex III, cyanide for complex IV, and salicylhydroxamic acid (SHAM) for the alternative oxidase, are important tools to investigate the electron transport chain of plant mitochondria.

the oxidation of NADH derived from citric acid cycle substrates. The role of this pathway may well be as a bypass being engaged when complex I is overloaded (Møller and Rasmusson 1998; Møller 2001), such as under photorespiratory conditions, as we will see shortly (see also **Web Topic 11.2**).

- An NADPH dehydrogenase, ND_{in}(NADPH), is present on the matrix surface. Very little is known about this enzyme.
- Most, if not all, plants have an "alternative" respiratory pathway for the reduction of oxygen. This pathway involves the so-called alternative oxidase that, unlike cytochrome *c* oxidase, is insensitive to inhibition by cyanide, azide, or carbon monoxide (see **Web Topic 11.3**).

The nature and physiological significance of these plant-specific enzymes will be considered more fully later in the chapter.

ATP Synthesis in the Mitochondrion Is Coupled to Electron Transport

In oxidative phosphorylation, the transfer of electrons to oxygen via complexes I to IV is coupled to the synthesis of ATP from ADP and P_i via the ATP synthase (complex V). The number of ATPs synthesized depends on the nature of the electron donor.

In experiments conducted with the use of isolated mitochondria, electrons derived from internal (matrix) NADH give ADP:O ratios (the number of ATPs synthesized per two electrons transferred to oxygen) of 2.4 to 2.7 (Table 11.1). Succinate and externally added NADH each give values in the range of 1.6 to 1.8, while ascorbate, which serves as an artificial electron donor to cytochrome *c*, gives values of 0.8 to 0.9. Results such as these (for both plant and animal mitochondria) have led to the general concept that there are three sites of energy conservation along the electron transport chain, at complexes I, III, and IV.

TABLE 11.1
Theoretical and experimental ADP:O ratios in isolated plant mitochondria

Substrate	ADP:O ratio	
	Theoretical[a]	Experimental
Malate	2.5	2.4–2.7
Succinate	1.5	1.6–1.8
NADH (external)	1.5	1.6–1.8
Ascorbate	1.0[b]	0.8–0.9

[a]It is assumed that complexes I, III, and IV pump 4, 4, and 2 H^+ per 2 electrons, respectively; that the cost of synthesizing one ATP and exporting it to the cytosol is 4 H^+ (Brand 1994); and that the non-phosphorylating pathways are not active.

[b]Cytochrome *c* oxidase pumps only two protons when it is measured with ascorbate as electron donor. However, two electrons move from the outer surface of the inner membrane (where the electrons are donated) across the inner membrane to the inner, matrix side. As a result, 2 H^+ are consumed on the matrix side. This means that the net movement of H^+ and charges is equivalent to the movement of a total of 4 H^+, giving an ADP:O ratio of 1.0.

The experimental ADP:O ratios agree quite well with the values calculated on the basis of the number of H^+ pumped by complexes I, III, and IV and the cost of 4 H^+ for synthesizing one ATP (see next section and Table 11.1). For instance, electrons from external NADH pass only complexes III and IV, so a total of 6 H^+ are pumped, giving 1.5 ATP (when the alternative oxidase pathway is not used).

The mechanism of mitochondrial ATP synthesis is based on the chemiosmotic hypothesis, described in **Web Topic 6.3** and Chapter 7, which was first proposed in 1961 by Nobel laureate Peter Mitchell as a general mechanism of energy conservation across biological membranes (Nicholls and Ferguson 2002). According to the chemiosmotic theory, the orientation of electron carriers within the mitochondrial inner membrane allows for the transfer of protons (H^+) across the inner membrane during electron flow. Numerous studies have confirmed that mitochondrial electron transport is associated with a net transfer of protons from the mitochondrial matrix to the intermembrane space (see Figure 11.8) (Whitehouse and Moore 1995).

Because the inner mitochondrial membrane is impermeable to H^+, an electrochemical proton gradient can build up. As discussed in Chapters 6 and 7, the free energy associated with the formation of an electrochemical proton gradient ($\Delta\tilde{\mu}_{H^+}$, also referred to as a proton motive force, Δp, when expressed in units of volts) is made up of an electric transmembrane potential component (ΔE) and a chemical-potential component (ΔpH) according to the following equation:

$$\Delta p = \Delta E - 59\Delta pH$$

where

$$\Delta E = E_{inside} - E_{outside}$$

and

$$\Delta pH = pH_{inside} - pH_{outside}$$

ΔE results from the asymmetric distribution of a charged species (H^+) across the membrane, and ΔpH is due to the proton concentration difference across the membrane. Because protons are translocated from the mitochondrial matrix to the intermembrane space, the resulting ΔE across the inner mitochondrial membrane is negative.

As this equation shows, both ΔE and ΔpH contribute to the proton motive force in plant mitochondria, although ΔE is consistently found to be of greater magnitude, probably because of the large buffering capacity of both cytosol and matrix, which prevent large pH changes. This situation contrasts to that in the chloroplast, where almost all of the proton motive force across the thylakoid membrane is made up by a proton gradient (see Chapter 7).

The free-energy input required to generate $\Delta\tilde{\mu}_{H^+}$ comes from the free energy released during electron transport. How electron transport is coupled to proton translocation is not well understood in all cases. Because of the low permeability (conductance) of the inner membrane to protons, the proton electrochemical gradient is reasonably stable, once generated, and the free energy $\Delta\tilde{\mu}_{H^+}$ can be utilized to carry out chemical work (ATP synthesis). The $\Delta\tilde{\mu}_{H^+}$ is coupled to the synthesis of ATP by an additional protein complex associated with the inner membrane, the F_oF_1-ATP synthase.

The **F_oF_1-ATP synthase** (also called *complex V*) consists of two major components, F_1 and F_o (see Figure 11.8). **F_1** is a peripheral membrane protein complex that is composed of at least five different subunits and contains the catalytic site for converting ADP and P_i to ATP. This complex is attached to the matrix side of the inner membrane. **F_o** is an integral membrane protein complex that consists of at least three different polypeptides that form the channel through which protons cross the inner membrane.

The passage of protons through the channel is coupled to the catalytic cycle of the F_1 component of the ATP synthase, allowing the ongoing synthesis of ATP and the simultaneous utilization of the $\Delta\tilde{\mu}_{H^+}$. For each ATP synthesized, 3 H^+ pass through the F_o from the intermembrane space to the matrix down the electrochemical proton gradient.

A high-resolution X-ray structure of most of the F_1 complex of the mammalian mitochondrial ATP synthase supports a "rotational model" for the catalytic mechanism of ATP synthesis (see **Web Topic 11.4**) (Abrahams et al. 1994). The structure and function of the mitochondrial ATP synthase is similar to that of the CF_o–CF_1 ATP synthase in photophosphorylation (see Chapter 7).

The operation of a chemiosmotic mechanism of ATP synthesis has several implications. First, the true site of ATP formation on the mitochondrial inner membrane is the ATP synthase, not complex I, III, or IV. These complexes serve as sites of energy conservation whereby electron transport is coupled to the generation of a $\Delta\tilde{\mu}_{H^+}$.

Second, the chemiosmotic theory explains the action mechanism of uncouplers, a wide range of chemically unrelated compounds (including 2,4-dinitrophenol and FCCP [*p*-trifluoromethoxycarbonylcyanide phenylhydrazone]) that decreases mitochondrial ATP synthesis but often stimulates the rate of electron transport (see **Web Topic 11.5**). All of these compounds make the inner membrane leaky to protons, which prevents the buildup of a sufficiently large $\Delta\tilde{\mu}_{H^+}$ to drive ATP synthesis.

In experiments on isolated mitochondria, higher rates of electron flow (measured as the rate of oxygen uptake in the presence of a substrate such as succinate) are observed upon addition of ADP (referred to as *state 3*) than in its absence (Figure 11.9). ADP provides a substrate that stimulates dissipation of the $\Delta\tilde{\mu}_{H^+}$ through the F_oF_1-ATP synthase during ATP synthesis. Once all the ADP has been converted to ATP, the $\Delta\tilde{\mu}_{H^+}$ builds up again and reduces the rate of electron flow (*state 4*). The ratio of the rates with and without ADP (state 3:state 4) is referred to as the *respiratory control ratio*.

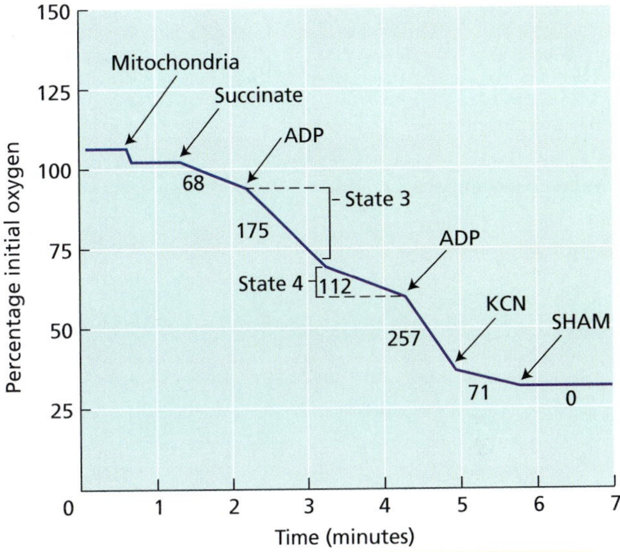

FIGURE 11.9 Regulation of respiratory rate by ADP during succinate oxidation in isolated mitochondria from mung bean (*Vigna radiata*). The numbers below the traces are the rates of oxygen uptake expressed as O_2 consumed (nmol min^{-1} mg $protein^{-1}$). (Data courtesy of Steven J. Stegink.)

Transporters Exchange Substrates and Products

The electrochemical proton gradient also plays a role in the movement of the organic acids of the citric acid cycle and of substrates and products of ATP synthesis in and out of mitochondria. Although ATP is synthesized in the mitochondrial matrix, most of it is used outside the mitochondrion, so an efficient mechanism is needed for moving ADP in and ATP out of the organelle.

Adenylate transport involves another inner-membrane protein, the ADP/ATP (adenine nucleotide) transporter, which catalyzes an exchange of ADP and ATP across the inner membrane (Figure 11.10). The movement of the more negatively charged ATP^{4-} out of the mitochondria in exchange for ADP^{3-}—that is, one net negative charge out—is driven by the electric-potential gradient (ΔE, positive outside) generated by proton pumping.

The uptake of inorganic phosphate (P_i) involves an active phosphate transporter protein that uses the proton gradient component (ΔpH) of the proton motive force to drive the electroneutral exchange of P_i^- (in) for OH^- (out). As long as a ΔpH is maintained across the inner membrane, the P_i content within the matrix will remain high. Similar reasoning applies to the uptake of pyruvate, which is driven by the electroneutral exchange of pyruvate for OH^-, leading to continued uptake of pyruvate from the cytosol (see Figure 11.10).

The total cost of taking up a phosphate (1 OH^- out, which is the same as 1 H^+ in) and exchanging ADP for ATP (one negative charge out, which is the same as one positive charge in) is 1 H^+. This proton should also be included in calculation of the cost of synthesizing one ATP. Thus the total cost is 3 H^+ used by the ATP synthase plus 1 H^+ for the exchange across the membrane, or a total of 4 H^+.

The inner membrane also contains transporters for dicarboxylic acids (malate or succinate) exchanged for P_i^{2-} and for the tricarboxylic acid citrate exchanged for malate (see Figure 11.10 and **Web Topic 11.5**).

Aerobic Respiration Yields about 60 Molecules of ATP per Molecule of Sucrose

The complete oxidation of a sucrose molecule leads to the net formation of

- 8 molecules of ATP by substrate-level phosphorylation (4 during glycolysis and 4 in the citric acid cycle)
- 4 molecules of NADH in the cytosol
- 16 molecules of NADH plus 4 molecules of $FADH_2$ (via succinate dehydrogenase) in the mitochondrial matrix

On the basis of theoretical ADP:O values (see Table 11.1), a total of approximately 52 molecules of ATP will be generated

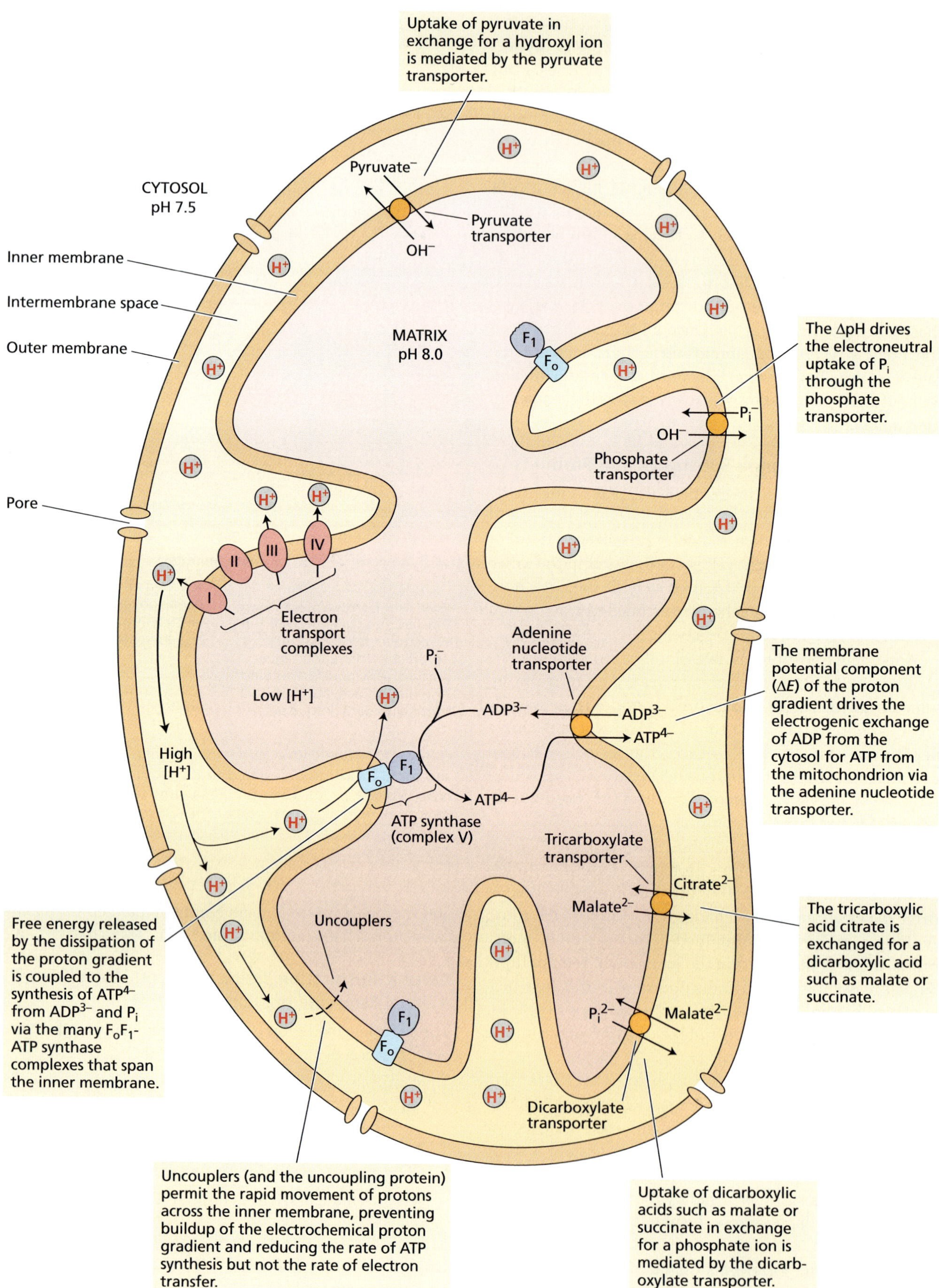
Uptake of pyruvate in exchange for a hydroxyl ion is mediated by the pyruvate transporter.
CYTOSOL
pH 7.5
Inner membrane
Intermembrane space
Outer membrane
Pyruvate$^-$
Pyruvate transporter
OH^-
MATRIX
pH 8.0
F_1
F_o
H^+
The ΔpH drives the electroneutral uptake of P_i through the phosphate transporter.
P_i^-
OH^-
Phosphate transporter
Pore
I
II
III
IV
Electron transport complexes
Adenine nucleotide transporter
P_i^-
Low [H^+]
ADP^{3-}
ADP^{3-}
ATP^{4-}
High [H^+]
ATP^{4-}
ATP synthase (complex V)
The membrane potential component (ΔE) of the proton gradient drives the electrogenic exchange of ADP from the cytosol for ATP from the mitochondrion via the adenine nucleotide transporter.
Tricarboxylate transporter
Citrate^{2-}
Malate^{2-}
Uncouplers
The tricarboxylic acid citrate is exchanged for a dicarboxylic acid such as malate or succinate.
Free energy released by the dissipation of the proton gradient is coupled to the synthesis of ATP^{4-} from ADP^{3-} and P_i via the many F_oF_1-ATP synthase complexes that span the inner membrane.
P_i^{2-}
Malate^{2-}
Dicarboxylate transporter
Uncouplers (and the uncoupling protein) permit the rapid movement of protons across the inner membrane, preventing buildup of the electrochemical proton gradient and reducing the rate of ATP synthesis but not the rate of electron transfer.
Uptake of dicarboxylic acids such as malate or succinate in exchange for a phosphate ion is mediated by the dicarboxylate transporter.

◀ **FIGURE 11.10** Transmembrane transport in plant mitochondria. An electrochemical proton gradient ($\Delta\tilde{\mu}_{H^+}$) consisting of a membrane potential (ΔE, –200mV, negative inside) and a ΔpH (alkaline inside) is established across the inner mitochondrial membrane during electron transport as outlined in the text. Specific metabolites are moved across the inner membrane by specialized proteins, called transporters or carriers (After Douce 1985).

per sucrose by oxidative phosphorylation. The result is a total of about 60 ATPs synthesized per sucrose (Table 11.2).

Using 50 kJ mol^{-1} (12 kcal mol^{-1}) as the actual free energy of formation of ATP in vivo, we find that about 3010 kJ mol^{-1} (720 kcal mol^{-1}) of free energy is conserved in the form of ATP per mole of sucrose oxidized during aerobic respiration. This amount represents about 52% of the standard free energy available from the complete oxidation of sucrose; the rest is lost as heat. This is a vast improvement over the conversion of only 4% of the energy available in sucrose to ATP that is associated with fermentative metabolism.

Several Subunits of Respiratory Complexes Are Encoded by the Mitochondrial Genome

With the first complete sequencing of plant mitochondrial DNA (mtDNA) in *Arabidopsis thaliana* (Marienfeld et al. 1999), our knowledge about the mitochondrial genome has taken a great leap forward.

Some characteristics of the plant mitochondrial genetic system are not generally found in the mitochondria of animals, protozoans, or even fungi. Most notably, RNA processing differs between plant mitochondria and mitochondria from most other organisms. Several plant mitochondrial genes contain introns, and some genes are even split between separate transcript molecules, which must be joined by splicing. Plant mtDNA also lacks strict complementarity to translated mRNA (see **Web Topic 11.6**). Another characteristic feature of the plant mitochondrial genetic system is that it strictly observes the universal genetic code, showing none of the deviations found in mtDNA in all other kingdoms.

Plant mitochondrial genomes are generally much larger than those of animals. The plant mtDNA is in the range of 200 to 2400 kilobase pairs (kb) in size, with large variations even between closely related plant species. This size compares with the compact and uniform 16 kb genome found in mammalian mitochondria. The size differences are due mainly to the presence of much noncoding sequence, including numerous introns, in plant mtDNA. Mammalian mtDNA encodes only 13 proteins, in contrast to the 35 known proteins encoded by the *Arabidopsis* mtDNA. Both plant and mammalian mitochondria encode rRNAs and tRNAs.

The genes of the mtDNA can be divided into two main groups: those needed for expression of mitochondrial genes (tRNA, rRNA, and ribosome proteins) and those for oxidative phosphorylation complexes. Plant mtDNA encodes nine subunits for complex I, one for complex III, three for complex IV, three for ATP synthase, and five proteins for biogenesis of cytochromes (Marienfeld et al. 1999). The mitochondrially encoded subunits are essential for the activity of the respiratory complexes, a feature also evident in the sequence conservation to their bacterial homologs. The nuclear genome encodes all proteins not encoded in mtDNA, and the nuclear-encoded proteins are the large majority—for example, all proteins in the citric acid cycle. The nuclear-encoded mitochondrial proteins are synthesized by cytosolic ribosomes and imported via translocators in the outer and inner mitochondrial membrane. Therefore, oxidative phosphorylation is dependent on expression of genes located in two separate genomes. Any change in expression in response to a stimulus or for developmental reasons must be coordinated.

Whereas the expression of nuclear genes for mitochondrial proteins appears to be regulated as other nuclear genes, much less is known about the expression of mitochondrial genes. The master circle of plant mtDNA is normally split into several smaller subgenomic segments, and genes can be down-regulated by decreased copy number for a segment of the mtDNA (Leon et al. 1998). The gene promoters in mtDNA are of several kinds and show different transcriptional activity. However, a main control of mitochondrial gene expression appears to take place at the posttranslational level, by degradation of excess polypeptides (McCabe et al. 2000).

TABLE 11.2
The maximum yield of cytosolic ATP from the complete oxidation of sucrose to CO_2 via aerobic glycolysis and the citric acid cycle

Part reaction	ATP per sucrose[a]	
Glycolysis		
4 substrate-level phosphorylations		4
4 NADH	4 ×1.5	6
Citric acid cycle		
4 substrate level phosphorylations		4
4 $FADH_2$	4 × 1.5	6
16 NADH	16 × 2.5	40
Total		60

Source: Adapted from Brand 1994.
Note: Cytosolic NADH is assumed oxidized by the external NADH dehydrogenase. The nonphosphorylating pathways are assumed not to be engaged.
[a]Calculated using the theoretical values from Table 11.1

Plants Have Several Mechanisms That Lower the ATP Yield

As we have seen, a complex machinery is required for a high efficiency of energy conservation in oxidative phosphorylation. So it is perhaps surprising that plant mitochondria have several functional proteins that reduce this efficiency. Probably plants are less limited by the energy supply (sunlight) than by other factors in the environment (e.g., access to nitrogen or phosphate). As a consequence, adaptational flexibility may be more important than energetic efficiency.

In the following subsections we will discuss the role of the nonphosphorylating mechanisms and their possible usefulness in the life of the plant.

The alternative oxidase. If cyanide (1 m*M*) is added to actively respiring animal tissues, cytochrome *c* oxidase is inhibited and the respiration rate quickly drops to less than 1% of its initial level. However, most plant tissues display a level of cyanide-resistant respiration that can represent 10 to 25%, and in some tissues up to 100%, of the uninhibited control rate. The enzyme responsible for this oxygen uptake has been identified as a cyanide-resistant oxidase component of the plant mitochondrial electron transport chain called the **alternative oxidase** (see Figure 11.8 and **Web Topic 11.3**) (Vanlerberghe and McIntosh 1997).

Electrons feed off the main electron transport chain into the alternative pathway at the level of the ubiquinone pool (see Figure 11.8). The alternative oxidase, the only component of the alternative pathway, catalyzes a four-electron reduction of oxygen to water and is specifically inhibited by several compounds, most notably salicylhydroxamic acid (SHAM).

When electrons pass to the alternative pathway from the ubiquinone pool, two sites of proton pumping (at complexes III and IV) are bypassed. Because there is no energy conservation site in the alternative pathway between ubiquinone and oxygen, the free energy that would normally be conserved as ATP is lost as heat when electrons are shunted through the alternative pathway.

How can a process as seemingly energetically wasteful as the alternative pathway contribute to plant metabolism? One example of the functional usefulness of the alternative oxidase is its activity during floral development in certain members of the Araceae (the arum family)—for example, the voodoo lily (*Sauromatum guttatum*). Just before pollination, tissues of the clublike inflorescence, called the *appendix*, which bears male and female flowers, exhibit a dramatic increase in the rate of respiration via the alternative pathway. As a result, the temperature of the upper appendix increases by as much as 25°C over the ambient temperature for a period of about 7 hours.

During this extraordinary burst of heat production, certain amines, indoles, and terpenes are volatilized, and the plant therefore gives off a putrid odor that attracts insect pollinators. Salicylic acid, a phenolic compound related to aspirin (see Chapter 13), has been identified as the chemical signal responsible for initiating this thermogenic event in the voodoo lily (Raskin et al. 1989) (see **Web Essay 11.3**). In most plants, however, both the respiratory rates and the rate of cyanide-resistant respiration are too low to generate sufficient heat to raise the temperature significantly, so what other role(s) does the alternative pathway play?

It has been suggested that the alternative pathway can function as an "energy overflow" pathway, oxidizing respiratory substrates that accumulate in excess of those needed for growth, storage, or ATP synthesis (Lambers 1985). This view suggests that electrons flow through the alternative pathway only when the activity of the main pathway is saturated. Such saturation is reached in the test tube in state 4 (see Figure 11.9); in vivo, saturation may occur if the respiration rate exceeds the cell's demand for ATP (i.e., if ADP levels are very low). However, it is now clear that the alternative oxidase can be active before the cytochrome pathway is saturated. Thus the alternative oxidase makes it possible for the mitochondrion to adjust the relative rates of ATP production and synthesis of carbon skeletons for use in biosynthetic reactions.

Another possible function of the alternative pathway is in the response of plants to a variety of stresses (phosphate deficiency, chilling, drought, osmotic stress, and so on), many of which can inhibit mitochondrial respiration (see Chapter 25 and **Web Essay 11.1**) (Wagner and Krab 1995).

By draining off electrons from the electron transport chain, the alternative pathway prevents a potential overreduction of the ubiquinone pool (see Figure 11.8), which, if left unchecked, can lead to the generation of destructive reactive oxygen species such as superoxide anions and hydroxyl radicals. In this way the alternative pathway may lessen the detrimental effects of stress on respiration (see **Web Essay 11.4**) (Wagner and Krab 1995; Møller 2001).

The uncoupling protein. A protein found in the inner membrane of mammalian mitochondria, the **uncoupling protein**, can dramatically increase the proton permeability of the membrane and thus act as an uncoupler. As a result, less ATP and more heat is generated. Heat production appears to be one of the uncoupling protein's main functions in mammalian cells.

It has long been thought that the alternative oxidase in plants and the uncoupling protein in mammals were simply two different means of achieving the same end. It was therefore surprising when a protein similar to the uncoupling protein was discovered in plant mitochondria (Vercesi et al. 1995; Laloi et al. 1997). This protein is stress induced and, like the alternative oxidase, may function to prevent overreduction of the electron transport chain (see **Web Topic 11.3** and **Web Essay 11.4**). It remains unclear, however, why plant mitochondria require both mechanisms.

The internal, rotenone-insensitive NADH dehydrogenase, ND_{in}(NADH). This is one of the multiple NAD(P)H dehydrogenases found in plant mitochondria (see Figure 11.8). It has been suggested to work as a non-proton-pumping bypass when complex I is overloaded. Complex I has a higher affinity for NADH (ten times lower K_m), than ND_{in}(NADH). At lower NADH levels in the matrix, typically when ADP is available (state 3), complex I will dominate, whereas when ADP is rate limiting (state 4), NADH levels will increase and ND_{in}(NADH) will be more active. The physiological importance of this enzyme is, however, still unclear.

Mitochondrial Respiration Is Controlled by Key Metabolites

The substrates of ATP synthesis—ADP and P_i—appear to be key regulators of the rates of glycolysis in the cytosol, as well as the citric acid cycle and oxidative phosphorylation in the mitochondria. Control points exist at all three stages of respiration; here we will give just a brief overview of some major features.

The best-characterized site of regulation of the citric acid cycle is at the pyruvate dehydrogenase complex, which is reversibly phosphorylated by a regulatory kinase and a phosphatase. Pyruvate dehydrogenase is inactive in the phosphorylated state, and the regulatory kinase is inhibited by pyruvate, allowing the enzyme to be active when substrate is available (Figure 11.11). In addition, several citric acid cycle enzymes, including pyruvate dehydrogenase and 2-oxoglutarate dehydrogenase, are directly inhibited by NADH.

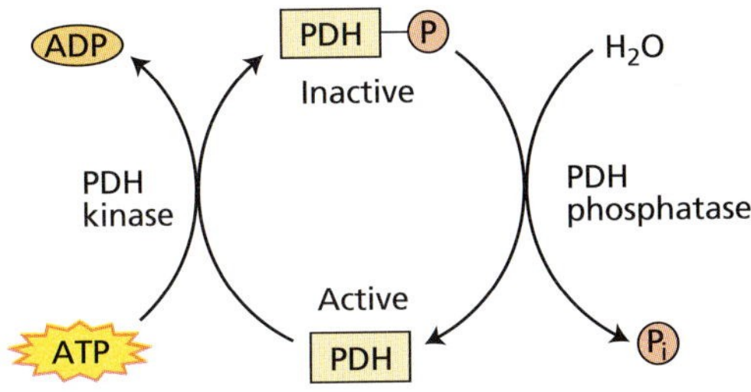

Pyruvate + CoA + NAD^+ ⟶ Acetyl-CoA + CO_2 + NADH

Effect on PDH activity	Mechanism
Activating	
Pyruvate	Inhibits kinase
ADP	Inhibits kinase
Mg^{2+} (or Mn^{2+})	Stimulates phosphatase
Inactivating	
NADH	Inhibits PDH Stimulates kinase
Acetyl CoA	Inhibits PDH Stimulates kinase
NH_4^+	Inhibits PDH Stimulates kinase

FIGURE 11.11 Regulation of pyruvate dehydrogenase (PDH) activity by reversible phosphorylation and by other metabolites.

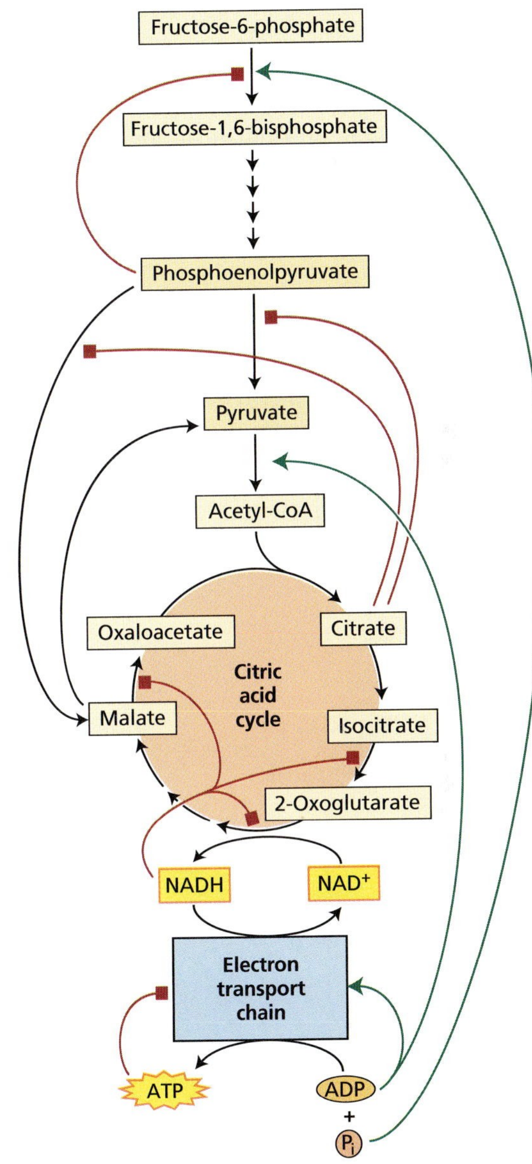

FIGURE 11.12 Concept of bottom-up regulation of plant respiration. Several substrates for respiration (e.g., ADP) stimulate enzymes in early steps of the pathways (green arrows . In contrast, accumulation of products (e.g., ATP) inhibits (red squares) earlier reactions in a stepwise fashion. For instance, ATP inhibits the electron transport chain leading to an accumulation of NADH. NADH inhibits citric acid enzymes such as isocitrate dehydrogenase and 2-oxoglutarate dehydrogenase. Then, citric acid cycle intermediates like citrate inhibit the PEP-metabolizing enzymes in the cytosol. Finally, PEP inhibits the conversion of fructose-6-phosphate to fructose-1,6-biphosphate and restricts carbon feeding into glycolysis.

The citric acid cycle oxidations, and subsequently respiration, are dynamically controlled by the cellular level of adenine nucleotides. As the cell's demand for ATP in the cytosol decreases relative to the rate of synthesis of ATP in the mitochondria, less ADP will be available, and the electron transport chain will operate at a reduced rate (see Figure 11.10). This slowdown could be signaled to citric acid cycle enzymes through an increase in matrix NADH, inhibiting the activity of several citric acid cycle dehydrogenases (Oliver and McIntosh 1995).

The buildup of citric acid cycle intermediates and their derivates, such as citrate and glutamate, inhibits the action of cytosolic pyruvate kinase, increasing the cytosolic PEP concentration, which in turn reduces the rate of conversion of fructose-6-phosphate to fructose-1,6-bisphosphate, thus inhibiting glycolysis.

In summary, plant respiratory rates are controlled from the "bottom up" by the cellular level of ADP (Figure 11.12). ADP initially regulates the rate of electron transfer and ATP synthesis, which in turn regulates citric acid cycle activity, which, finally, regulates the rate of the glycolytic reactions.

Respiration Is Tightly Coupled to Other Pathways

Glycolysis, the pentose phosphate pathway, and the citric acid cycle are linked to several other important metabolic pathways, some of which will be covered in greater detail in Chapter 13. The respiratory pathways are central to the production of a wide variety of plant metabolites, including amino acids, lipids and related compounds, isoprenoids, and porphyrins (Figure 11.13). Indeed, much of the reduced carbon that is metabolized by glycolysis and the citric acid cycle is diverted to biosynthetic purposes and not oxidized to CO_2.

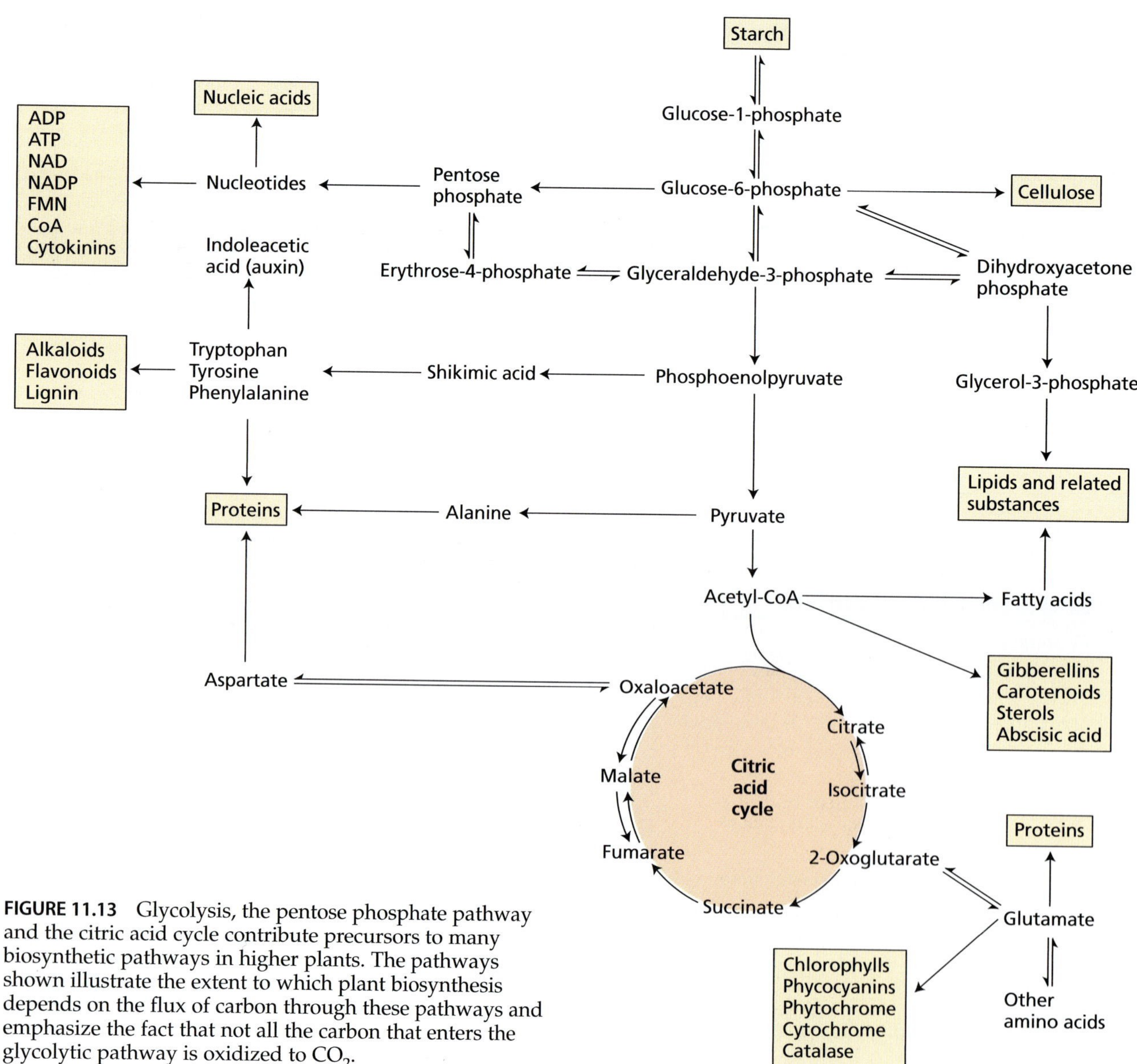

FIGURE 11.13 Glycolysis, the pentose phosphate pathway and the citric acid cycle contribute precursors to many biosynthetic pathways in higher plants. The pathways shown illustrate the extent to which plant biosynthesis depends on the flux of carbon through these pathways and emphasize the fact that not all the carbon that enters the glycolytic pathway is oxidized to CO_2.

RESPIRATION IN INTACT PLANTS AND TISSUES

Many rewarding studies of plant respiration and its regulation have been carried out on isolated organelles and on cell-free extracts of plant tissues. But how does this knowledge relate to the function of the whole plant in a natural or agricultural setting?

In this section we'll examine respiration and mitochondrial function in the context of the whole plant under a variety of conditions. First, when green tissues are exposed to light, respiration and photosynthesis operate simultaneously and interact in complex ways. Next we will discuss different rates of tissue respiration, which may be under developmental control, as well as the very interesting case of cytoplasmic male sterility. Finally, we will look at the influence of various environmental factors on respiration rates.

Plants Respire Roughly Half of the Daily Photosynthetic Yield

Many factors can affect the respiration rate of an intact plant or of its individual organs. Relevant factors include the species and growth habit of the plant, the type and age of the specific organ, and environmental variables such as the external oxygen concentration, temperature, and nutrient and water supply (see Chapter 25, **Web Topic 11.7**, and **Web Essay 11.5**).

Whole-plant respiration rates, particularly when considered on a fresh-weight basis, are generally lower than respiration rates reported for animal tissues. This difference is due in large part to the presence, in plant cells, of a large central vacuole and cell wall compartments, neither of which contains mitochondria. Nonetheless, respiration rates in some plant tissues are as high as those observed in actively respiring animal tissues, so the plant respiratory process is not inherently slower than respiration in animals. In fact, isolated plant mitochondria respire faster than mammalian mitochondria, when expressed on a per mg protein basis.

Even though plants generally have low respiration rates, the contribution of respiration to the overall carbon economy of the plant can be substantial (see **Web Topic 11.7**). Whereas only green tissues photosynthesize, all tissues respire, and they do so 24 hours a day. Even in photosynthetically active tissues, respiration, if integrated over the entire day, can represent a substantial fraction of gross photosynthesis. A survey of several herbaceous species indicated that 30 to 60% of the daily gain in photosynthetic carbon was lost to respiration, although these values tended to decrease in older plants (Lambers 1985).

Young trees lose roughly a third of their daily photosynthate as respiration, and this loss can double in older trees as the ratio of photosynthetic to nonphotosynthetic tissue decreases. In tropical areas, 70 to 80% of the daily photosynthetic gain can be lost to respiration because of the high dark respiration rates associated with elevated night temperatures.

Respiration Operates during Photosynthesis

Mitochondria are involved in the metabolism of photosynthesizing leaves. The glycine generated by photorespiration is oxidized to serine in the mitochondrion (see Chapter 8). At the same time, mitochondria in photosynthesizing tissue also carry out respiration via the citric acid cycle (often called *dark respiration* because it does not require light). Relative to the maximum rate of photosynthesis, dark respiration rates measured in green tissues are far slower, generally by a factor ranging from 6- to 20-fold. Given that rates of photorespiration can often reach 20 to 40% of the gross photosynthetic rate, citric acid cycle-mediated mitochondrial respiration operates at rates also well below the rate of photorespiration.

A question that has not been adequately answered is how much mitochondrial respiration (apart from the involvement of mitochondria in the photorespiratory carbon oxidation cycle) operates simultaneously with photosynthesis in illuminated green tissues. The activity of pyruvate dehydrogenase, one of the ports of entry into the citric acid cycle, decreases in the light to 25% of the dark activity (Budde and Randall 1990). The overall rate of respiration decreases in the light, but the extent of the decrease remains uncertain at present. It is clear, however, that the mitochondrion is a major supplier of ATP to the cytosol even in illuminated leaves (Krömer 1995).

Another role of mitochondrial respiration during photosynthesis is to supply carbon metabolites for biosynthetic reactions—for example, by formation of 2-oxoglutarate needed for nitrogen assimilation. Leaf mitochondria typically have high capacities of nonphosphorylating pathways in the electron transport chain. By oxidizing NADH with lower ATP yield, mitochondria can maintain a higher 2-oxoglutarate production by the respiratory pathways without being restricted by the cytosolic demand for ATP (see Figures 11.7C and 11.12) (Hoefnagel et al. 1998; Noctor and Foyer 1998).

Additional evidence for the involvement of mitochondrial respiration in photosynthesizing leaves has been obtained in studies with mitochondrial mutants defective in respiratory complexes, showing that leaf development and photosynthesis are negatively affected (Vedel et al. 1999).

Different Tissues and Organs Respire at Different Rates

A useful rule of thumb is that the greater the overall metabolic activity of a given tissue, the higher its respiration rate. Developing buds usually show very high rates of respiration (on a dry-weight basis), and respiration rates of vegetative tissues usually decrease from the point of growth (e.g., the leaf tip in dicotyledons and the leaf base in monocotyledons) to more differentiated regions. A well-studied example is the growing barley leaf (Thompson et al. 1998). In mature vegetative tissues, stems generally have the lowest respiration rates, and leaf and root respiration varies with the plant species and the conditions under which the plants are growing.

When a plant tissue has reached maturity, its respiration rate will either remain roughly constant or decrease slowly as the tissue ages and ultimately senesces. An exception to this pattern is the marked rise in respiration, known as the *climacteric*, that accompanies the onset of ripening in many fruits (avocado, apple, banana) and senescence in detached leaves and flowers. Both ripening and the climacteric respiratory rise are triggered by the endogenous production of ethylene, as well as by an exogenous application of ethylene (see Chapter 22). In general, ethylene-induced respiration is associated with an active cyanide-resistant alternative pathway, but the role of this pathway in ripening is not clear (Tucker 1993).

Mitochondrial Function Is Crucial during Pollen Development

A physiological feature directly linked to the plant mitochondrial genome is a phenomenon known as **cytoplasmic male sterility**, or ***cms***. Plant lines that display *cms* do not form viable pollen—hence the designation *male sterility*. The term *cytoplasmic* here refers to the fact that this trait is transmitted in a non-Mendelian fashion; the *cms* genotype is always maternally inherited with the mitochondrial genome. *cms* is a very important trait in plant breeding because a stable male sterile line can facilitate the production of hybrid seed stock. For this use, *cms* traits that produce no major effects throughout the plant's life cycle, except for male sterility, have been found for many species.

All plants carrying the *cms* trait that have been characterized at the molecular level show the presence of distinct rearrangements in their mtDNA, relative to wild-type plants. These rearrangements create novel open reading frames and have been strongly correlated with *cms* phenotypes in various systems. Nuclear restorer genes can overcome the effects of the mtDNA rearrangements and restore fertility to plants with the *cms* genotype. Such restorer genes are essential for the commercial utilization of *cms* if seeds are the harvested product.

An interesting consequence of the use of the *cms* gene occurred in the late 1960s, at which time 85% of the hybrid feed corn grown in the United States was derived from the use of a *cms* line of maize called *cms*-T (Texas). In *cms*-T maize, the mtDNA rearrangements give rise to a unique 13 kDa protein, URF13 (Levings and Siedow 1992). How the URF13 protein acts to bring about male sterility is not known, but in the late 1960s a disease appeared, caused by a race of the fungus *Bipolaris maydis* (also called *Cochliobolus heterostrophus*). This specific race synthesizes a compound (HmT-toxin) that specifically interacts with the URF13 protein to produce pores in the inner mitochondrial membrane, with the result that selective permeability is lost.

The interaction between HmT-toxin and URF13 made *Bipolaris maydis* race T a particularly virulent pathogen on *cms*-T maize and led to an epidemic in the corn-growing regions of the United States that was known as southern corn leaf blight. As a result of this epidemic, the use of *cms*-T in the production of hybrid maize was discontinued. No other *cms* maize has been found to be a suitable replacement, so current production of hybrid corn seed has reverted to manual detasseling that prevents self-pollination.

As compared to other organs, the amount of mitochondria per cell and the expression of respiratory proteins are very high in developing anthers, where pollen development is an energy-demanding process (Huang et al. 1994). Male sterility is a common phenotype of mutations in mitochondrial genes for subunits of the complexes of oxidative phosphorylation (Vedel et al. 1999). Such mutants can be viable because of the existence of the alternative nonphosphorylating respiratory pathways.

Programmed cell death (**PCD**) is part of normal anther development. There are now indications that mitochondria are involved in plant PCD and that PCD is premature in anthers of *cms* sunflower (see **Web Essay 11.6**).

Environmental Factors Alter Respiration Rates

Many environmental factors can alter the operation of metabolic pathways and respiratory rates. Here we will examine the roles of environmental oxygen (O_2), temperature, and carbon dioxide (CO_2).

Oxygen. Oxygen can affect plant respiration because of its role as a substrate in the overall process. At 25°C, the equilibrium concentration of O_2 in an air-saturated (21% O_2), aqueous solution is about 250 μ*M*. The K_m value for oxygen in the reaction catalyzed by cytochrome *c* oxidase is well below 1 μ*M*, so there should be no apparent dependence of the respiration rate on external O_2 concentrations (see Chapter 2 on the web site for a discussion of K_m). However, respiration rates decrease if the atmospheric oxygen concentration is below 5% for whole tissues or below 2 to 3% for tissue slices. These findings show that oxygen diffusion through the aqueous phase in the tissue imposes a limitation on plant respiration.

The diffusion limitation imposed by an aqueous phase emphasizes the importance of the intercellular air spaces found in plant tissues for oxygen availability in the mitochondria. If there were no gaseous diffusion pathway throughout the plant, the cellular respiration rates of many plants would be limited by an insufficient oxygen supply (see **Web Essay 11.3**).

Water saturation/low O_2. Diffusion limitation is even more significant when plant organs are growing in an aqueous medium. When plants are grown hydroponically, the solutions must be aerated vigorously to keep oxygen levels high in the vicinity of the roots. The problem of oxygen supply also arises with plants growing in very wet or flooded soils (see Chapter 25).

Some plants, particularly trees, have a restricted geographic distribution because of the need to maintain a supply of oxygen to their roots. For instance, dogwood and tulip tree poplar can survive only in well-drained, aerated soils because their roots cannot tolerate more than a limited exposure to a flooded condition. On the other hand, many plant species are adapted to grow in flooded soils. Herbaceous species such as rice and sunflower often rely on a network of intercellular air spaces (aerenchyma) running from the leaves to the roots to provide a continuous, gaseous pathway for the movement of oxygen to the flooded roots.

Limitation in oxygen supply can be more severe for trees having very deep roots that grow in wet soils. Such roots must survive on anaerobic (fermentative) metabolism or develop structures that facilitate the movement of oxygen to the roots. Examples of such structures are outgrowths of the roots, called *pneumatophores*, that protrude out of the water and provide a gaseous pathway for oxygen diffusion into the roots. Pneumatophores are found in *Avicennia* and *Rhizophora*, trees that grow in mangrove swamps under continuously flooded conditions.

Temperature. Respiration typically increases with temperature (see, however, **Web Essay 11.3**). Between 0 and 30°C, the increase in respiration rate for every 10°C increase in ambient temperature (commonly referred to as the dimensionless, temperature coefficient, Q_{10}) is about 2. Above 30°C the respiration rate often increases more slowly, reaches a plateau at 40 to 50°C and decreases at even higher temperatures. High night temperatures are thought to account for the high respiratory rates of tropical plants.

Low temperatures are utilized to retard postharvest respiration rates during the storage of fruits and vegetables. However, complications may arise from such storage. For instance, when potato tubers are stored at temperatures above 10°C, respiration and ancillary metabolic activities are sufficient to allow sprouting. Below 5°C, respiration rates and sprouting are reduced in most tissues, but the breakdown of stored starch and its conversion to sucrose impart an unwanted sweetness to the tubers. As a compromise, potatoes are stored at 7 to 9°C, which prevents the breakdown of starch while minimizing respiration and germination.

CO_2 concentration. It is common practice in the commercial storage of fruits to take advantage of the effects of atmospheric oxygen and temperature on respiration, and to store fruits at low temperatures under 2 to 3% oxygen and 3 to 5% CO_2. The reduced temperature lowers the respiration rate, as does the reduced oxygen. Low levels of oxygen are used instead of anoxic conditions to avoid lowering tissue oxygen tensions to the point that stimulates fermentative metabolism.

Carbon dioxide has a limited direct inhibitory effect on the respiration rate at a concentration of 3 to 5%, which is well in excess of the 0.036% (360 ppm) normally found in the atmosphere. The atmospheric CO_2 concentration is increasing rapidly as a result of human activities, and it is projected to double, to 700 ppm, before the end of the twenty-first century (see Chapter 9).

Compared to plants grown at 350 ppm CO_2, plants grown at 700 ppm CO_2 have been reported to have a 15 to 20% slower dark respiration rate (on a dry-weight basis) (Drake et al. 1999), but this has been questioned (Jahnke 2001; Bruhn et al. 2002). The number of mitochondria per unit cell area actually doubles in the high CO_2 environment. These data imply that the respiratory activity in the light instead may increase at higher ambient CO_2 concentrations (Griffin et al. 2001). Thus it is presently a matter of debate how plants growing at an increased CO_2 concentration will contribute to the global carbon cycle.

LIPID METABOLISM

Whereas animals use fats for energy storage, plants use them mainly for carbon storage. Fats and oils are important storage forms of reduced carbon in many seeds, including those of agriculturally important species such as soybean, sunflower, peanut, and cotton. Oils often serve a major storage function in nondomesticated plants that produce small seeds. Some fruits, such as olives and avocados, also store fats and oils.

In this final part of the chapter we describe the biosynthesis of two types of glycerolipids: the *triacylglycerols* (the fats and oils stored in seeds) and the *polar glycerolipids* (which form the lipid bilayers of cellular membranes) (Figure 11.14). We will see that the biosynthesis of triacylglycerols and polar glycerolipids requires the cooperation of two organelles: the plastids and the endoplasmic reticulum. Plants can also use fats and oils for energy production. We will thus examine the complex process by which germinating seeds obtain metabolic energy from the oxidation of fats and oils.

Fats and Oils Store Large Amounts of Energy

Fats and oils belong to the general class *lipids*, a structurally diverse group of hydrophobic compounds that are soluble in organic solvents and highly insoluble in water. Lipids represent a more reduced form of carbon than carbohydrates, so the complete oxidation of 1 g of fat or oil (which contains about 40 kJ, or 9.3 kcal, of energy) can produce considerably more ATP than the oxidation of 1 g of starch (about 15.9 kJ, or 3.8 kcal). Conversely, the biosynthesis of

FIGURE 11.14 Structural features of triacylglycerols and polar glycerolipids in higher plants. The carbon chain lengths of the fatty acids, which always have an even number of carbons, range from 12 to 20 but are typically 16 or 18. Thus, the value of *n* is usually 14 or 16.

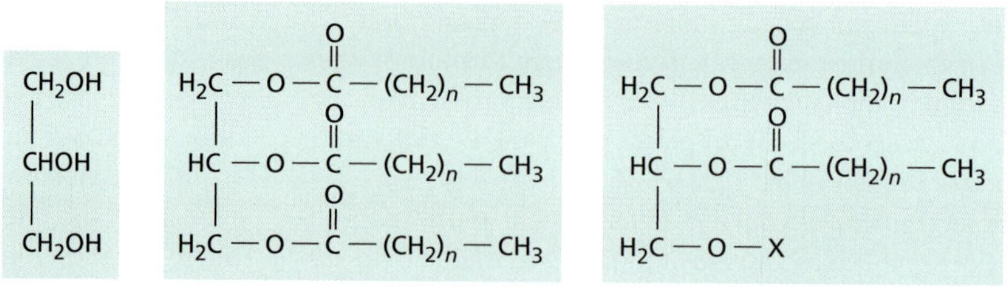

Glycerol **Triacylglycerol** (the major stored lipid) **Glycerolipid**

X = H	Diacylglycerol (DAG)
$X = HPO_3^{2-}$	Phosphatidic acid
$X = PO_3^{2-}—CH_2—CH_2—\overset{+}{N}(CH_3)_3$	Phosphatidylcholine
$X = PO_3^{2-}—CH_2—CH_2—NH_2$	Phosphatidylethanolamine
X = galactose	Galactolipids

fats, oils, and related molecules, such as the phospholipids of membranes, requires a correspondingly large investment of metabolic energy.

Other lipids are important for plant structure and function but are not used for energy storage. These include waxes, which make up the protective cuticle that reduces water loss from exposed plant tissues, and terpenoids (also known as isoprenoids), which include carotenoids involved in photosynthesis and sterols present in many plant membranes (see Chapter 13).

Triacylglycerols Are Stored in Oleosomes

Fats and oils exist mainly in the form of triacylglycerols (*acyl* refers to the fatty acid portion), or triglycerides, in which fatty acid molecules are linked by ester bonds to the three hydroxyl groups of glycerol (see Figure 11.14).

The fatty acids in plants are usually straight-chain carboxylic acids having an even number of carbon atoms. The carbon chains can be as short as 12 units and as long as 20, but more commonly they are 16 or 18 carbons long. *Oils* are liquid at room temperature, primarily because of the presence of unsaturated bonds in their component fatty acids; *fats*, which have a higher proportion of saturated fatty acids, are solid at room temperature. The major fatty acids in plant lipids are shown in Table 11.3.

The composition of fatty acids in plant lipids varies with the species. For example, peanut oil is about 9% palmitic acid, 59% oleic acid, and 21% linoleic acid, and cottonseed oil is 20% palmitic acid, 30% oleic acid, and 45% linoleic acid. The biosynthesis of these fatty acids will be discussed shortly.

Triacylglycerols in most seeds are stored in the cytoplasm of either cotyledon or endosperm cells in organelles known as **oleosomes** (also called *spherosomes* or *oil bodies*) (see Chapter 1). Oleosomes have an unusual membrane barrier that separates the triglycerides from the aqueous cytoplasm. A single layer of phospholipids (i.e., a half-bilayer) surrounds the oil body with the hydrophilic ends of the phospholipids exposed to the cytosol and the hydrophobic acyl hydrocarbon chains facing the triacylglycerol interior (see Chapter 1). The oleosome is stabilized

TABLE 11.3
Common fatty acids in higher plant tissues

Name[a]	Structure
Saturated Fatty Acids	
Lauric acid (12:0)	$CH_3(CH_2)_{10}CO_2H$
Myristic acid (14:0)	$CH_3(CH_2)_{12}CO_2H$
Palmitic acid (16:0)	$CH_3(CH_2)_{14}CO_2H$
Stearic acid (18:0)	$CH_3(CH_2)_{16}CO_2H$
Unsaturated Fatty Acids	
Oleic acid (18:1)	$CH_3(CH_2)_7CH{=}CH(CH_2)_7CO_2H$
Linoleic acid (18:2)	$CH_3(CH_2)_4CH{=}CH—CH_2—CH{=}CH(CH_2)_7CO_2H$
Linolenic acid (18:3)	$CH_3CH_2CH{=}CH—CH_2—CH{=}CH—CH_2—CH{=}CH—(CH_2)_7CO_2H$

[a]Each fatty acid has a numerical abbreviation. The number before the colon represents the total number of carbons; the number after the colon is the number of double bonds.

by the presence of specific proteins, called *oleosins*, that coat the surface and prevent the phospholipids of adjacent oil bodies from coming in contact and fusing.

This unique membrane structure for oleosomes results from the pattern of triacylglycerol biosynthesis. Triacylglycerol synthesis is completed by enzymes located in the membranes of the endoplasmic reticulum (ER), and the resulting fats accumulate between the two monolayers of the ER membrane bilayer. The bilayer swells apart as more fats are added to the growing structure, and ultimately a mature oil body buds off from the ER (Napier et al. 1996).

Polar Glycerolipids Are the Main Structural Lipids in Membranes

As outlined in Chapter 1, each membrane in the cell is a bilayer of *amphipathic* (i.e., having both hydrophilic and hydrophobic regions) lipid molecules in which a polar head group interacts with the aqueous phase while hydrophobic fatty acid chains form the center of the membrane. This hydrophobic core prevents random diffusion of solutes between cell compartments and thereby allows the biochemistry of the cell to be organized.

The main structural lipids in membranes are the polar glycerolipids (see Figure 11.14), in which the hydrophobic portion consists of two 16-carbon or 18-carbon fatty acid chains esterified to positions 1 and 2 of a glycerol backbone. The polar head group is attached to position 3 of the glycerol. There are two categories of polar glycerolipids:

1. **Glyceroglycolipids**, in which sugars form the head group (Figure 11.15A)
2. **Glycerophospholipids**, in which the head group contains phosphate (Figure 11.15B)

Plant membranes have additional structural lipids, including sphingolipids and sterols (see Chapter 13), but these are minor components. Other lipids perform specific roles in photosynthesis and other processes. Included among these lipids are chlorophylls, plastoquinone, carotenoids, and tocopherols, which together account for about one-third of the lipids in plant leaves.

Figure 11.15 shows the nine major glycerolipid classes in plants, each of which can be associated with many different fatty acid combinations. The structures shown in Figure 11.15 illustrate some of the more common molecular species.

Chloroplast membranes, which account for 70% of the membrane lipids in photosynthetic tissues, are dominated by glyceroglycolipids; other membranes of the cell contain glycerophospholipids (Table 11.4). In nonphotosynthetic tissues, phospholipids are the major membrane glycerolipids.

Fatty Acid Biosynthesis Consists of Cycles of Two-Carbon Addition

Fatty acid biosynthesis involves the cyclic condensation of two-carbon units in which acetyl-CoA is the precursor. In plants, fatty acids are synthesized exclusively in the plastids; in animals, fatty acids are synthesized primarily in the cytosol.

The enzymes of the pathway are thought to be held together in a complex that is collectively referred to as *fatty acid synthase*. The complex probably allows the series of reactions to occur more efficiently than it would if the enzymes were physically separated from each other. In addition, the growing acyl chains are covalently bound to a low-molecular-weight, acidic protein called **acyl carrier protein (ACP)**. When conjugated to the acyl carrier protein, the fatty acid chain is referred to as **acyl-ACP**.

The first committed step in the pathway (i.e., the first step unique to the synthesis of fatty acids) is the synthesis of malonyl-CoA from acetyl-CoA and CO_2 by the enzyme acetyl-CoA carboxylase (Figure 11.16) (Sasaki et al. 1995). The tight regulation of acetyl-CoA carboxylase appears to control the overall rate of fatty acid synthesis (Ohlrogge and Jaworski 1997). The malonyl-CoA then reacts with ACP to yield malonyl-ACP:

1. In the first cycle of fatty acid synthesis, the acetate group from acetyl-CoA is transferred to a specific cys-

TABLE 11.4
Glycerolipid components of cellular membranes

	Lipid composition (percentage of total)		
	Chloroplast	Endoplasmic reticulum	Mitochondrion
Phosphatidylcholine	4	47	43
Phosphatidylethanolamine	—	34	35
Phosphatidylinositol	1	17	6
Phosphatidylglycerol	7	2	3
Diphosphatidylglycerol	—	—	13
Monogalactosyldiacylglycerol	55	—	—
Digalactosyldiacylglycerol	24	—	—
Sulfolipid	8	—	—

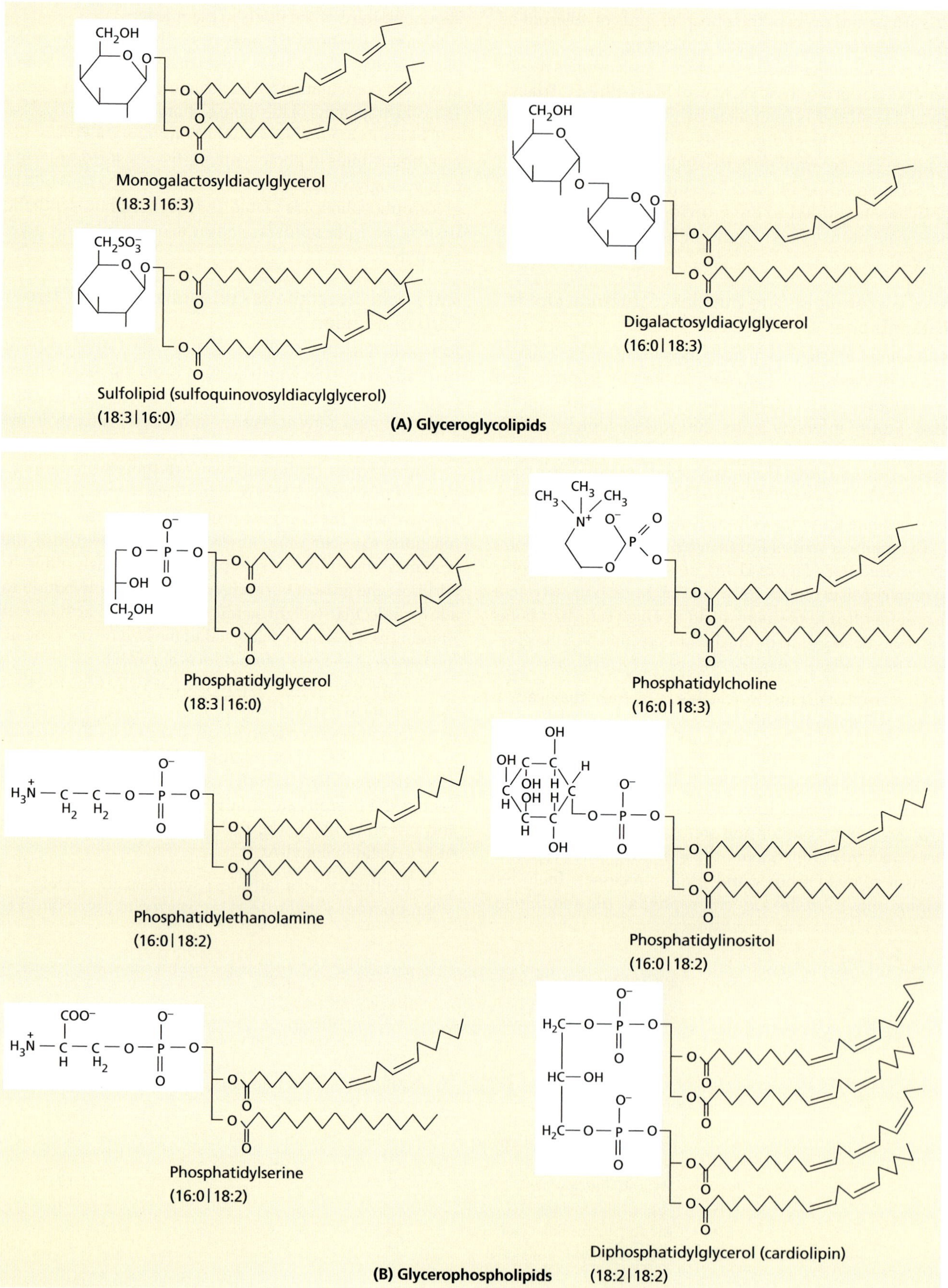

CH_2OH

Monogalactosyldiacylglycerol
(18:3 | 16:3)

CH_2OH

Digalactosyldiacylglycerol
(16:0 | 18:3)

$CH_2SO_3^-$

Sulfolipid (sulfoquinovosyldiacylglycerol)
(18:3 | 16:0)

(A) Glyceroglycolipids

Phosphatidylglycerol
(18:3 | 16:0)

Phosphatidylcholine
(16:0 | 18:3)

Phosphatidylethanolamine
(16:0 | 18:2)

Phosphatidylinositol
(16:0 | 18:2)

Phosphatidylserine
(16:0 | 18:2)

Diphosphatidylglycerol (cardiolipin)
(18:2 | 18:2)

(B) Glycerophospholipids

$CH_3-C(=O)-SCoA$

Acetyl-CoA

1. First committed step in fatty acid biosynthetic pathway.

ATP

CO_2

ADP + P_i

Acetyl-CoA carboxylase

$^{-}OOC-CH_2-C(=O)-SCoA$

Malonyl-CoA

2. Malonyl group is transferred to acyl carrier protein.

ACP

$^{-}OOC-CH_2-C(=O)-SACP$

Malony-ACP

3. The first cycle of fatty acid synthesis begins here.

Condensing enzyme

Decarboxylation step

CO_2

5. The second cycle of fatty acid synthesis begins here.

Decarboxylation step

ACP, CO_2

Condensing enzyme

$CH_3-C(=O)-CH_2-C(=O)-SACP$

Acetoacetyl-ACP

(Continues to 16- to 18-carbon chain length)

ACP

Completed fatty acid

7. ACP is removed from completed fatty acid in a transferase reaction.

6. The cycle continues multiple times adding acetate (2-carbon) units from malonyl-ACP.

$CH_3-CH_2-CH_2-C(=O)-SACP$

Butyryl-ACP
(acyl-ACP)

2 NADPH

2 $NADPH^+$

4. The keto group at carbon 3 is removed in three steps.

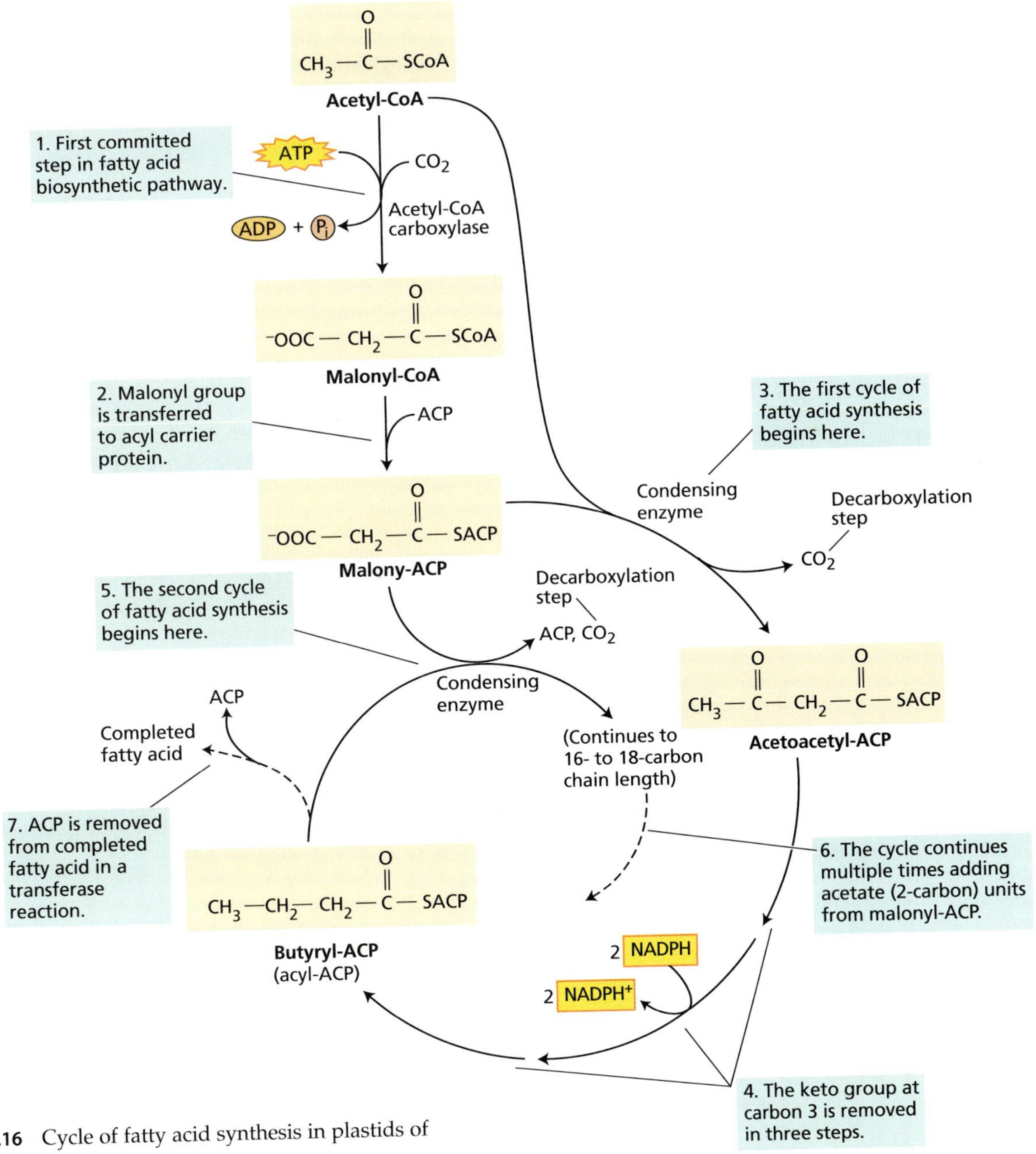

FIGURE 11.16 Cycle of fatty acid synthesis in plastids of plant cells.

teine of *condensing enzyme* (3-ketoacyl-ACP synthase) and then combined with malonyl-ACP to form acetoacetyl-ACP.

2. Next the keto group at carbon 3 is removed (reduced) by the action of three enzymes to form a new acyl chain (butyryl-ACP), which is now four carbons long (see Figure 11.16).

3. The four-carbon acid and another molecule of malonyl-ACP then become the new substrates for condensing enzyme, resulting in the addition of another two-carbon unit to the growing chain, and the cycle continues until 16 or 18 carbons have been added.

4. Some 16:0-ACP is released from the fatty acid synthase machinery, but most molecules that are elongated to 18:0-ACP are efficiently converted to 18:1-ACP by a desaturase enzyme. The repetition of this sequence of events makes 16:0-ACP and 18:1-ACP the major products of fatty acid synthesis in plastids (Figure 11.17).

Fatty acids may undergo further modification after they are linked with glycerol to form glycerolipids. Additional double bonds are placed in the 16:0 and 18:1 fatty acids by a series of desaturase isozymes. Desaturase isozymes are integral membrane proteins found in the chloroplast and the endoplasmic reticulum (ER). Each desaturase inserts a double bond at a specific position in the fatty acid chain, and the enzymes act sequentially to produce the final 18:3 and 16:3 products (Ohlrogge and Browse 1995).

Glycerolipids Are Synthesized in the Plastids and the ER

The fatty acids synthesized in the plastid are next used to make the glycerolipids of membranes and oleosomes. The first steps of glycerolipid synthesis are two acylation reactions that transfer fatty acids from acyl-ACP or acyl-CoA to glycerol-3-phosphate to form **phosphatidic acid**.

The action of a specific phosphatase produces **diacylglycerol (DAG)** from phosphatidic acid. Phosphatidic acid can also be converted directly to phosphatidylinositol or phosphatidylglycerol; DAG can give rise to phosphatidylethanolamine or phosphatidylcholine (see Figure 11.17).

The localization of the enzymes of glycerolipid synthesis reveals a complex and highly regulated interaction between the chloroplast, where fatty acids are synthesized, and other membrane systems of the cell. In simple terms, the biochemistry involves two pathways referred to as the *prokaryotic* (or chloroplast) pathway and the *eukaryotic* (or ER) pathway.

1. In chloroplasts, the **prokaryotic pathway** utilizes the 16:0- and 18:1-ACP products of chloroplast fatty acid synthesis to synthesize phosphatidic acid and its derivatives. Alternatively, the fatty acids may be exported to the cytoplasm as CoA esters.

2. In the cytoplasm, the **eukaryotic pathway** uses a separate set of acyltransferases in the ER to incorporate the fatty acids into phosphatidic acid and its derivatives.

A simplified version of this model is depicted in Figure 11.17.

In some higher plants, including *Arabidopsis* and spinach, the two pathways contribute almost equally to chloroplast lipid synthesis. In many other angiosperms, however, phosphatidylglycerol is the only product of the prokaryotic pathway, and the remaining chloroplast lipids are synthesized entirely by the eukaryotic pathway.

The biochemistry of triacylglycerol synthesis in oilseeds is generally the same as described for the glycerolipids.

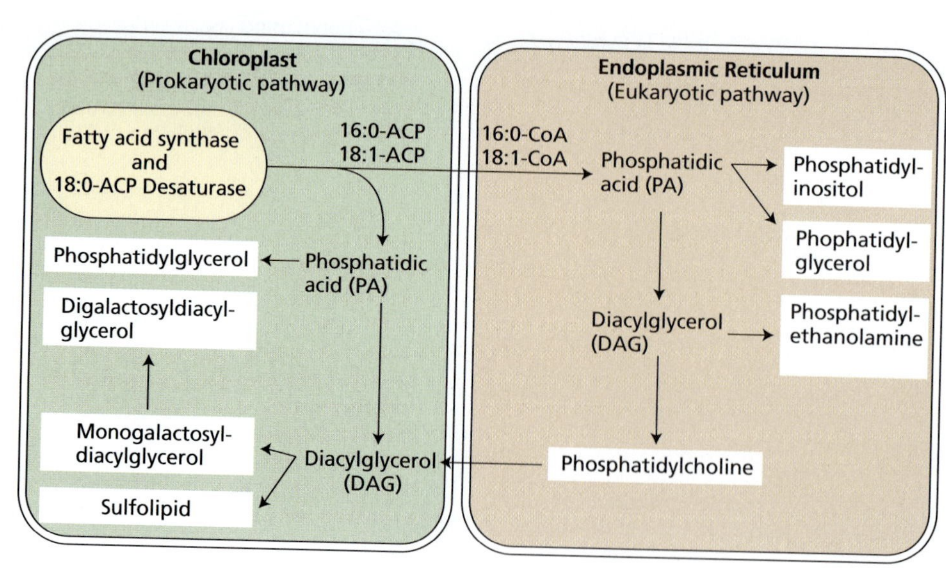

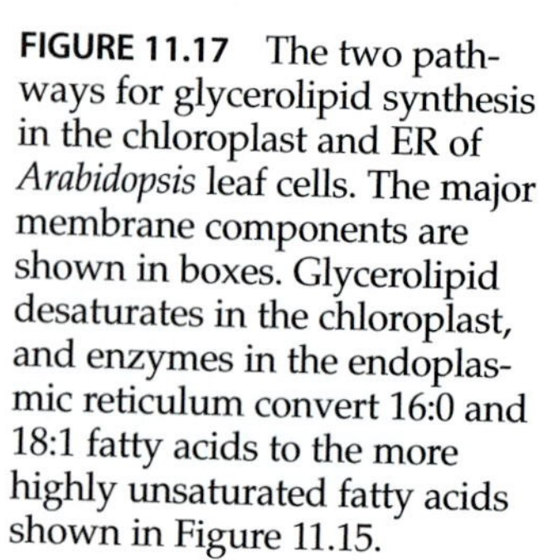
FIGURE 11.17 The two pathways for glycerolipid synthesis in the chloroplast and ER of *Arabidopsis* leaf cells. The major membrane components are shown in boxes. Glycerolipid desaturates in the chloroplast, and enzymes in the endoplasmic reticulum convert 16:0 and 18:1 fatty acids to the more highly unsaturated fatty acids shown in Figure 11.15.

16:0- and 18:1-ACP are synthesized in the plastids of the cell and exported as CoA thioesters for incorporation into DAG in the endoplasmic reticulum (see Figure 11.17).

The key enzymes in oilseed metabolism (not shown in Figure 11.17), are acyl-CoA:DAG acyltransferase and PC:DAG acyltransferase, which catalyze triacylglycerol synthesis (Dahlqvist et al. 2000). As noted earlier, triacylglycerol molecules accumulate in specialized subcellular structures—the oleosomes—from which they can be mobilized during germination and converted to sugar.

Lipid Composition Influences Membrane Function

A central question in membrane biology is the functional reason behind lipid diversity. Each membrane system of the cell has a characteristic and distinct complement of lipid types, and within a single membrane each class of lipids has a distinct fatty acid composition. Our understanding of a membrane is one in which lipids make up the fluid, semipermeable bilayer that is the matrix for the functional membrane proteins.

Since this bulk lipid role could be satisfied by a single unsaturated species of phosphatidylcholine, obviously such a simple model is unsatisfactory. Why is lipid diversity needed? One aspect of membrane biology that might offer answers to this central question is the relationship between lipid composition and the ability of organisms to adjust to temperature changes (Wolter et al. 1992). For example, chill-sensitive plants experience sharp reductions in growth rate and development at temperatures between 0 and 12°C (see Chapter 25). Many economically important crops, such as cotton, soybean, maize, rice, and many tropical and subtropical fruits, are classified as chill sensitive. In contrast, most plants that originate from temperate regions are able to grow and develop at chilling temperatures and are classified as chill-resistant plants.

It has been suggested that because of the decrease in lipid fluidity at lower temperatures, the primary event of chilling injury is a transition from a liquid-crystalline phase to a gel phase in the cellular membranes. According to this proposal, this transition would result in alterations in the metabolism of chilled cells and lead to injury and death of the chill-sensitive plants. The degree of unsaturation of the fatty acids would determine the temperature at which such damage occurred.

Recent research, however, suggests that the relationship between membrane unsaturation and plant responses to temperature is more subtle and complex (see **Web Topic 11.8**). The responses of *Arabidopsis* mutants with increased saturation of fatty acids to low temperature appear quite distinct from what is predicted by the chilling sensitivity hypothesis, suggesting that normal chilling injury may not be strictly related to the level of unsaturation of membrane lipids.

On the other hand, experiments with transgenic tobacco plants that are chill sensitive show opposite results. The transgenic expression of exogenous genes in tobacco has been used specifically to decrease the level of saturated phosphatidylglycerol or to bring about a general increase in membrane unsaturation. In each case, damage caused by chilling was alleviated to some extent.

These new findings make it clear that the extent of membrane unsaturation or the presence of particular lipids, such as disaturated phosphatidylglycerol, can affect the responses of plants to low temperature. As discussed in **Web Topic 11.8**, more work is required to fully understand the relationship between lipid composition and membrane function.

Membrane Lipids Are Precursors of Important Signaling Compounds

Plants, animals, and microbes all use membrane lipids as precursors for compounds that are used for intracellular or long-range signaling. For example, jasmonate derived from linolenic acid (18:3) activates plant defenses against insects and many fungal pathogens. In addition, jasmonate regulates other aspects of plant growth, including the development of anthers and pollen (Stintzi and Browse 2000). **Phosphatidylinositol-4,5-bisphosphate (PIP_2)** is the most important of several phosphorylated derivatives of phosphatidylinositol known as *phosphoinositides*. In animals, receptor-mediated activation of phospholipase C leads to the hydrolysis of PIP_2 to inositol trisphosphate (IP_3) and diacylglycerol, which both act as intracellular secondary messengers.

The action of IP_3 in releasing Ca^{2+} into the cytoplasm (through calcium-sensitive channels in the tonoplast and other membranes) and thereby regulating cellular processes has been demonstrated in several plant systems, including the stomatal guard cells (Schroeder et al. 2001). Information about other types of lipid signaling in plants is becoming available through biochemical and molecular genetic studies of phospholipases (Wang 2001) and other enzymes involved in the generation of these signals.

Storage Lipids Are Converted into Carbohydrates in Germinating Seeds

After germinating, oil-containing seeds metabolize stored triacylglycerols by converting lipids to sucrose. Plants are not able to transport fats from the endosperm to the root and shoot tissues of the germinating seedling, so they must convert stored lipids to a more mobile form of carbon, generally sucrose. This process involves several steps that are located in different cellular compartments: oleosomes, glyoxysomes, mitochondria, and cytosol.

Overview: Lipids to sucrose. The conversion of lipids to sucrose in oilseeds is triggered by germination and begins with the hydrolysis of triacylglycerols stored in the oil bodies to free fatty acids, followed by oxidation of the fatty acids to produce acetyl-CoA (Figure 11.18). The fatty

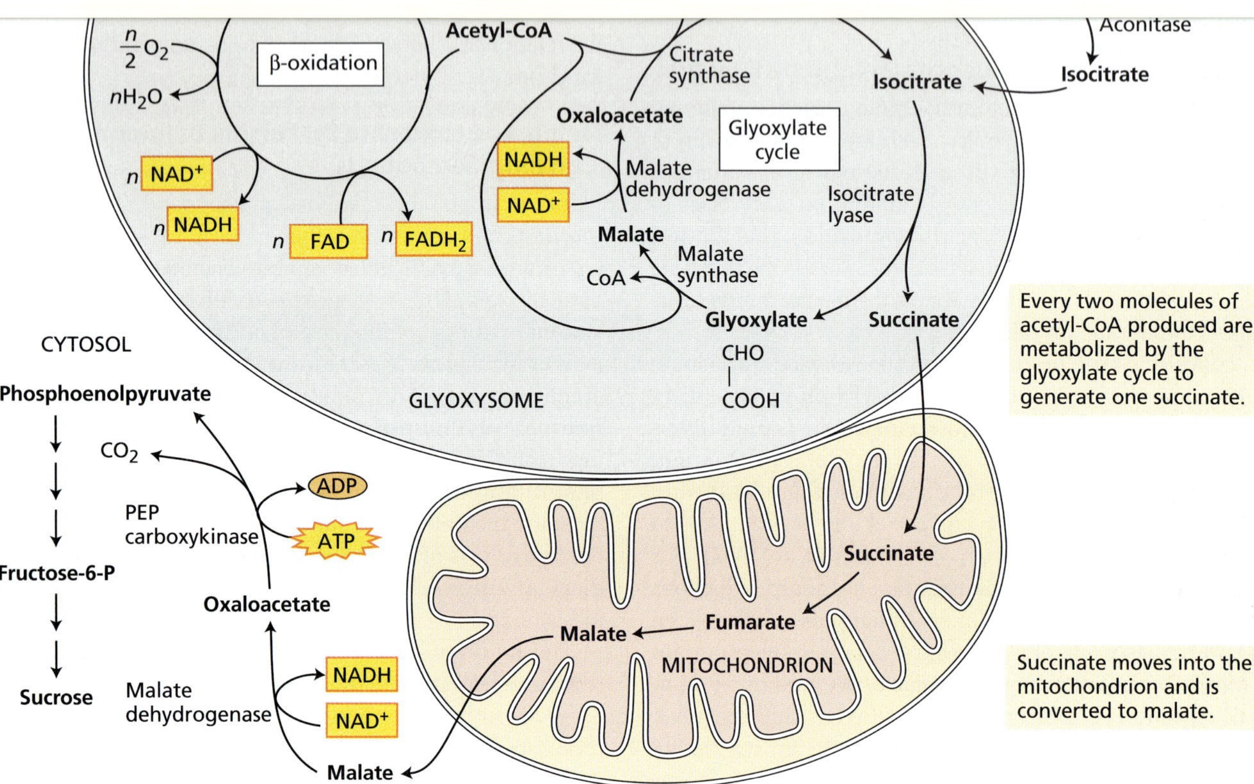

(B)

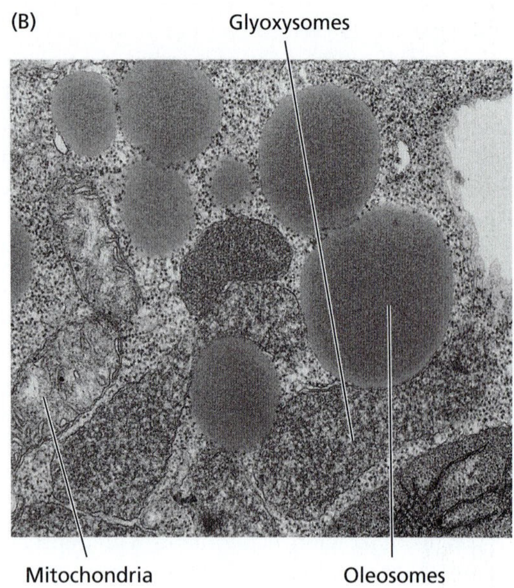

FIGURE 11.18 The conversion of fats to sugars during germination in oil-storing seeds. (A) Carbon flow during fatty acid breakdown and gluconeogenesis (refer to Figures 11.2, 11.3, and 11.6 for structures). (B) Electron micrograph of a cell from the oil-storing cotyledon of a cucumber seedling, showing glyoxysomes, mitochondria, and oleosomes. (Photo courtesy of R. N. Trelease.)

acids are oxidized in a type of peroxisome called a **glyoxysome,** an organelle enclosed by a single bilayer membrane that is found in the oil-rich storage tissues of seeds. Acetyl-CoA is metabolized in the glyoxysome (see Figure 11.18A) to produce succinate, which is transported from the glyoxysome to the mitochondrion, where it is converted first to oxaloacetate and then to malate. The process ends in the cytosol with the conversion of malate to glucose via gluconeogenesis, and then to sucrose.

Although some of this fatty acid–derived carbon is diverted to other metabolic reactions in certain oilseeds, in castor bean (*Ricinus communis*) the process is so efficient that each gram of lipid metabolized results in the formation of 1 g of carbohydrate, which is equivalent to a 40% recovery of free energy in the form of carbon bonds ([15.9 kJ/40 kJ] × 100 = 40%).

Lipase hydrolysis. The initial step in the conversion of lipids to carbohydrate is the breakdown of triglycerides stored in the oil bodies by the enzyme lipase, which, at least in castor bean endosperm, is located on the half-membrane that serves as the outer boundary of the oil body. The lipase hydrolyzes triacylglycerols to three molecules of fatty acid and glycerol. Corn and cotton also contain a lipase activity in the oil body, but peanut, soybean, and cucumber show lipase activity in the glyoxysome instead. During the breakdown of lipids, oil bodies and glyoxysomes are generally in close physical association (see Figure 11.18B).

β-Oxidation of fatty acids. After hydrolysis of the triacylglycerols, the resulting fatty acids enter the glyoxysome, where they are activated by conversion to fatty-acyl-CoA by the enzyme fatty-acyl-CoA synthase. Fatty-acyl-CoA is the initial substrate for the β-oxidation series of reactions, in which C_n fatty acids (fatty acids composed of *n* number of carbons) are sequentially broken down to *n*/2 molecules of acetyl-CoA (see Figure 11.18A). This reaction sequence involves the reduction of ½ O_2 to H_2O and the formation of 1 NADH and 1 $FADH_2$ for each acetyl-CoA produced.

In mammalian tissues, the four enzymes associated with β-oxidation are present in the mitochondrion; in plant seed storage tissues, they are localized exclusively in the glyoxysome. Interestingly, in plant vegetative tissues (e.g., mung bean hypocotyl and potato tuber), the β-oxidation reactions are localized in a related organelle, the peroxisome (see Chapter 1).

The glyoxylate cycle. The function of the glyoxylate cycle is to convert two molecules of acetyl-CoA to succinate. The acetyl-CoA produced by β-oxidation is further metabolized in the glyoxysome through a series of reactions that make up the glyoxylate cycle (see Figure 11.18A). Initially, the acetyl-CoA reacts with oxaloacetate to give citrate, which is then transferred to the cytoplasm for isomerization to isocitrate by aconitase. Isocitrate is reimported into the peroxisome and converted to malate by two reactions that are unique to the glyoxylate pathway.

1. First isocitrate (C_6) is cleaved by the enzyme isocitrate lyase to give succinate (C_4) and glyoxylate (C_2). This succinate is exported to the motochondria.
2. Next malate synthase combines a second molecule of acetyl-CoA with glyoxylate to produce malate.

Malate is then oxidized by malate dehydrogenase to oxaloacetate, which can combine with another acetyl-CoA to continue the cycle (see Figure 11.18A). The glyoxylate produced keeps the cycle operating in the glyoxysome, but the succinate is exported to the mitochondria for further processing.

The mitochondrial role. Moving from the glyoxysomes to the mitochondria, the succinate is converted to malate by the normal citric acid cycle reactions. The resulting malate can be exported from the mitochondria in exchange for succinate via the dicarboxylate transporter located in the inner mitochondrial membrane. Malate is then oxidized to oxaloacetate by malate dehydrogenase in the cytosol, and the resulting oxaloacetate is converted to carbohydrate.

This conversion requires circumventing the irreversibility of the pyruvate kinase reaction (see Figure 11.3) and is facilitated by the enzyme PEP carboxykinase, which utilizes the phosphorylating ability of ATP to convert oxaloacetate to PEP and CO_2 (see Figure 11.18A). From PEP, gluconeogenesis can proceed to the production of glucose, as described earlier. Sucrose is the final product of this process, and the primary form of reduced carbon translocated from the cotyledons to the growing seedling tissues. Not all seeds quantitatively convert fat to sugar (see **Web Topic 11.9**).

SUMMARY

In plant respiration, reduced cellular carbon generated during photosynthesis is oxidized to CO_2 and water, and this oxidation is coupled to the synthesis of ATP. Respiration takes place in three main stages: glycolysis, the citric acid cycle, and oxidative phosphorylation. The latter comprises the electron transport chain and ATP synthesis.

In glycolysis, carbohydrate is converted in the cytosol to pyruvate, and a small amount of ATP is synthesized via substrate-level phosphorylation. Pyruvate is subsequently oxidized within the mitochondrial matrix through the citric acid cycle, generating a large number of reducing equivalents in the form of NADH and $FADH_2$.

In the third stage, oxidative phosphorylation, electrons from NADH and $FADH_2$ pass through the electron transport chain in the inner mitochondrial membrane to reduce oxygen. The chemical energy is conserved in the form of an electrochemical proton gradient, which is created by the coupling of electron flow to proton pumping from the matrix to the intermembrane space. This energy is then converted into chemical energy in the form of ATP by the F_0F_1-ATP synthase, also located in the inner membrane, which couples ATP synthesis from ADP and P_i to the flow of protons back into the matrix down their electrochemical gradient.

Aerobic respiration in plants has several unique features, including the presence of a cyanide-resistant alternative oxidase and multiple NAD(P)H dehydrogenases, none of which pumps protons. Substrate oxidation during respiration is regulated at control points in glycolysis, the citric acid cycle, and the electron transport chain, but ultimately substrate oxidation is controlled by the level of cellular ADP. Carbohydrates can also be oxidized via the oxidative pentose phosphate pathway, in which the reducing power is produced in the form of NADPH mainly for biosynthetic purposes. Numerous glycolytic and citric acid cycle intermediates also provide the starting material for a multitude of biosynthetic pathways.

More than 50% of the daily photosynthetic yield can be respired by a plant, but many factors can affect the respiration rate observed at the whole-plant level. These factors include the nature and age of the plant tissue, as well as environmental factors such as light, oxygen concentration, temperature, and CO_2 concentration.

Lipids play a major role in plants: Amphipathic lipids serve as the primary nonprotein components of plant membranes; fats and oils are an efficient storage form of reduced carbon, particularly in seeds. Glycerolipids play important roles as structural components of membranes. Fatty acids are synthesized in plastids using acetyl-CoA. Fatty acids from the plastid can be transported to the ER, where they are further modified.

Membrane function may be influenced by the lipid composition. The degree of unsaturation of the fatty acids influences the sensitivity of plants to cold but does not seem to be involved in normal chilling injury. On the other hand, certain membrane lipid breakdown products, such as jasmonic acid, can act as signaling agents in plant cells.

Triacylglycerol is synthesized in the ER and accumulates within the phospholipid bilayer, forming oil bodies. During germination in oil-storing seeds, the stored lipids are metabolized to carbohydrate in a series of reactions that involve a metabolic sequence known as the glyoxylate cycle. This cycle takes place in glyoxysomes, and subsequent steps occur in the mitochondria. The reduced carbon generated during lipid breakdown in the glyoxysomes is ultimately converted to carbohydrate in the cytosol by gluconeogenesis.

Web Material

Web Topics

11.1 Isolation of Mitochondria

Methods for the isolation of intact, functional mitochondria have been developed.

11.2 The Electron Transport Chain of Plant Mitochondria Contains Multiple NAD(P)H Dehydrogenases

Mitochondrial NAD(P)H dehydrogenases oxidize NADH or NADPH and pass the electrons to ubiquinone.

11.3 The Alternative Oxidase

The alternative oxidase is an oxidoreductase localized at the inner membrane of plant mitochondria.

11.4 F_0F_1-ATP Synthases: The World's Smallest Rotary Motors

Rotation of the γ subunit brings about the conformational changes that allow the release of ATP from the enzyme.

11.5 Transport In and Out of Plant Mitochondria

Plant mitochondria operate different transport mechanisms.

11.6 The Genetic System of Plant Mitochondria Has Some Special Features

The mitochondrial genome encodes about 40 mitochondrial proteins.

11.7 Does Respiration Reduce Crop Yields?

Empirical relations between plant respiration rates and crop yield have been established.

11.8 The Lipid Composition of Membranes Affects the Cell Biology and Physiology of Plants

Lipid mutants are expanding our understanding of the ability of organisms to adapt to temperature changes.

11.9 Utilization of Oil Reserves in Cotyledons

In some species, only part of the stored lipid in the cotyledons is exported as carbohydrate.

Web Essays

11.1 Metabolic Flexibility Helps Plants Survive Stress

The ability of plants to carry out a metabolic step in different ways increases plant survival under stress.

11.2 Metabolic Profiling of Plant Cells
Metabolic profiling complements genomics and proteomics.

11.3 Temperature Regulation by Thermogenic Flowers
In thermogenic flowers, such as the Arum lilies, temperature can increase up to 20°C above their surroundings.

11.4 Reactive Oxygen Species (ROS) and Plant Mitochondria
The production of damaging reactive oxygen species is an unavoidable consequence of aerobic respiration.

11.5 The Role of Respiration in Desiccation Tolerance
Respiration has both positive and negative effects on the survival of plant cells under water stress.

11.6 Balancing Life and Death; The Role of Mitochondria in Programmed Cell Death
Programmed cell death is an integral part of the life cycle of plants, directly involving mitochondria.

Chapter References

Abrahams, J. P., Leslie, A. G. W., Lutter, R., and Walker, J. E. (1994) Structure at 2.8 Å resolution of F_1-ATPase from bovine heart mitochondria. *Nature* 370: 621–628.

Ap Rees, T. (1980) Assessment of the contributions of metabolic pathways to plant respiration. In *The Biochemistry of Plants*, Vol. 2, D. D. Davies, ed., Academic Press, New York, pp. 1–29.

Brand, M. D. (1994) The stoichiometry of proton pumping and ATP synthesis in mitochondria. *Biochemist* 16(4): 20–24.

Bruhn, D., Mikkelsen, T. N., and Atkin, O. K. (2002) Does the direct effect of atmospheric CO_2 concentration on leaf respiration vary with temperature? Responses in two species of *Plantago* that differ in relative growth rate. *Physiol. Plant.* 114: 57–64.

Budde, R. J. A., and Randall, D. D. (1990) Pea leaf mitochondrial pyruvate dehydrogenase complex is inactivated *in vivo* in a light-dependent manner. *Proc. Natl. Acad. Sci. USA* 87: 673–676.

Dahlqvist, A., Stahl, U., Lenman, M., Banas, A., Lee, M., Sandager, L., Ronne, H., and Stymne, S. (2000) Phospholipid:diacylglycerol acyltransferase: An enzyme that catalyzes the acyl-CoA-independent formation of triacylglycerol in yeast and plants. *Proc. Natl. Acad. Sci. USA* 97: 6487–6492.

Dennis, D. T., and Blakely, S. D. (2000) Carbohydrate metabolism. In *Biochemistry & Molecular Biology of Plants*, B. Buchanan, W. Gruissem, and R. Jones, eds., American Society of Plant Physiologists, Rockville, MD, pp. 630–674.

Dennis, D. T., Huang, Y., and Negm, F. B. (1997) Glycolysis, the pentose phosphate pathway and anaerobic respiration. In *Plant Metabolism*, 2nd ed., D. T. Dennis, D. H. Turpin, D. D. Lefebvre, and D. B. Layzell, eds., Longman, Singapore, pp. 105–123.

Douce, R. (1985) *Mitochondria in Higher Plants: Structure, Function, and Biogenesis*. Academic Press, Orlando, FL.

Drake, B. G., Azcon-Bieto, J., Berry, J., Bunce, J., Dijkstra, P., Farrar, J., Gifford, R. M., Gonzalez-Meler, M. A., Koch, G., Lambers, H., Siedow, J., and Wullschleger, S. (1999) Does elevated atmospheric CO_2 concentration inhibit mitochondrial respiration in green plants? *Plant Cell Environ.* 22: 649–657.

Givan, C. V. (1999) Evolving concepts in plant glycolysis: Two centuries of progress. *Biol. Rev.* 74: 277–309.

Griffin, K. L., Anderson, O. R., Gastrich, M. D., Lewis, J. D., Lin, G., Schuster, W., Seemann, J. R., Tissue, D. T., Turnbull, M. H., and Whitehead, D. (2001) Plant growth in elevated CO_2 alters mitochondrial number and chloroplast fine structure. *Proc. Natl. Acad. Sci. USA* 98: 2473–2478.

Gunning, B. E. S., and Steer, M. W. (1996) *Plant Cell Biology: Structure and Function of Plant Cells.* Jones and Bartlett, Boston.

Hoefnagel, M. H. N., Atkin, O. K., and Wiskich, J. T. (1998) Interdependence between chloroplasts and mitochondria in the light and the dark. *Biochim. Biophys. Acta* 1366: 235–255.

Huang, J., Struck, F., Matzinger, D. F., and Levings, C. S. (1994) Flower-enhanced expression of a nuclear-encoded mitochondrial respiratory protein is associated with changes in mitochondrion number. *Plant Cell* 6: 439–448.

Jahnke, S. (2001) Atmospheric CO_2 concentration does not directly affect leaf respiration in bean or poplar. *Plant Cell Environ.* 24: 1139–1151.

Krömer, S. (1995) Respiration during photosynthesis. *Annu. Rev. Plant Physiol. Plant Mol. Biol.* 46: 45–70.

Kruger, N. J. (1997) Carbohydrate synthesis and degradation. In *Plant Metabolism*, 2nd ed., D. T. Dennis, D. H. Turpin, D. D. Lefebvre, and D. B. Layzell, eds., Longman, Singapore, pp. 83–104.

Laloi, M., Klein, M., Riesmeier, J. W., Müller-Röber, B., Fleury, C., Bouillaud, F., and Ricquier, D. (1997) A plant cold-induced uncoupling protein. *Nature* 389: 135–136.

Lambers, H. (1985) Respiration in intact plants and tissues. Its regulation and dependence on environmental factors, metabolism and invaded organisms. In *Higher Plant Cell Respiration* (Encyclopedia of Plant Physiology, New Series, Vol. 18), R. Douce and D. A. Day, eds., Springer, Berlin, pp. 418–473.

Leon, P., Arroyo, A., and Mackenzie, S. (1998) Nuclear control of plastid and mitochondrial development in higher plants. *Annu. Rev. Plant Physiol. Plant Mol. Biol.* 49: 453–480.

Levings, C. S., III, and Siedow, J. N. (1992) Molecular basis of disease susceptibility in the Texas cytoplasm of maize. *Plant Mol. Biol.* 19: 135–147.

Marienfeld, J., Unseld, M., and Brennicke, A. (1999) The mitochondrial genome of *Arabidopsis* is composed of both native and immigrant information. *Trends Plant Sci.* 4: 495–502.

McCabe, T. C., Daley, D., and Whelan, J. (2000) Regulatory, developmental and tissue aspects of mitochondrial biogenesis in plants. *Plant Biol.* 2: 121–135.

Møller, I. M. (2001) Plant mitochondria and oxidative stress. Electron transport, NADPH turnover and metabolism of reactive oxygen species. *Annu. Rev. Plant Physiol. Plant Mol. Biol.* 52: 561–591.

Møller, I. M., and Rasmusson, A. G. (1998) The role of NADP in the mitochondrial matrix. *Trends Plant Sci.* 3: 21–27.

Napier, J. A., Stobart, A. K., and Shewry, P. R. (1996) The structure and biogenesis of plant oil bodies: The role of the ER membrane and the oleosin class of proteins. *Plant Mol. Biol.* 31: 945–956.

Nicholls, D. G., and Ferguson, S. J. (2002) *Bioenergetics 3*, 3rd ed. Academic Press, San Diego, CA.

Noctor, G., and Foyer, C. H. (1998) A re-evaluation of the ATP:NADPH budget during C3 photosynthesis: A contribution from nitrate assimilation and its associated respiratory activity? *J. Exp. Bot.* 49: 1895–1908.

Ohlrogge, J. B., and Browse, J. A. (1995) Lipid biosynthesis. *Plant Cell* 7: 957–970.

Ohlrogge, J. B., and Jaworski, J. G. (1997) Regulation of fatty acid synthesis. *Annu. Rev. Plant Physiol. Plant Mol. Biol.* 48: 109–136.

Oliver, D. J., and McIntosh, C. A. (1995) The biochemistry of the mitochondrial matrix. In *The Molecular Biology of Plant Mitochondria*, C. S. Levings III and I. Vasil, eds., Kluwer, Dordrecht, Netherlands, pp. 237–280.

Plaxton, W. C. (1996) The organization and regulation of plant glycolysis. *Annu. Rev. Plant Physiol. Plant Mol. Biol.* 47: 185–214.

Raskin, I., Turner, I. M., and Melander, W. R. (1989) Regulation of heat production in the inflorescences of an *Arum* lily by endogenous salicylic acid. *Proc. Natl. Acad. Sci. USA* 86: 2214–2218.

Sachs, M. M., Subbaiah, C. C., and Saab, I. N. (1996) Anaerobic gene expression and flooding tolerance in maize. *J. Exp. Bot.* 47: 1–15.

Sasaki, Y., Konishi, T., and Nagano, Y. (1995) The compartmentation of acetyl-coenzyme A carboxylase in plants. *Plant Physiol.* 108: 445–449.

Schroeder, J. I., Allen, G. J., Hugouvieux, V., Kwak, J. M., and Waner, D. (2001) Guard cell signal transduction. *Annu. Rev. Plant Physiol. Plant Mol. Biol.* 52: 627–658.

Siedow, J. N. (1995) Bioenergetics: The plant mitochondrial electron transfer chain. In *The Molecular Biology of Plant Mitochondria*, C. S. Levings III and I. Vasil, eds., Kluwer, Dordrecht, Netherlands, pp. 281–312.

Siedow, J. N., and Umbach, A. L. (1995) Plant mitochondrial electron transfer and molecular biology. *Plant Cell* 7: 821–831.

Stintzi, A., and Browse, J. (2000) The *Arabidopsis* male-sterile mutant, *opr3*, lacks the 12-oxophytodienoic acid reductase required for jasmonate synthesis. *Proc. Natl. Acad. Sci. USA* 97: 10625–10630.

Thompson, P., Bowsher, C. G., and Tobin, A. K. (1998) Heterogeneity of mitochondrial protein biogenesis during primary leaf development in barley. *Plant Physiol.* 118: 1089–1099.

Tucker, G. A. (1993) Introduction. In *Biochemistry of Fruit Ripening*, G. Seymour, J. Taylor, and G. Tucker, eds., Chapman & Hall, London, pp. 1–51.

Vanlerberghe, G. C., and McIntosh, L. (1997) Alternative oxidase: From gene to function. *Annu. Rev. Plant Physiol. Plant Mol. Biol.* 48: 703–734.

Vedel, F., Lalanne, É., Sabar, M., Chétrit, P., and De Paepe, R. (1999) The mitochondrial respiratory chain and ATP synthase complexes: Composition, structure and mutational studies. *Plant Physiol. Biochem.* 37: 629–643.

Vercesi, A. E., Martins I. S., Silva, M. P., and Leite, H. M. F. (1995) PUMPing plants. *Nature* 375: 24.

Wagner, A. M., and Krab, K. (1995) The alternative respiration pathway in plants: Role and regulation. *Physiol. Plant.* 95: 318–325.

Wang, X. (2001) Plant phospholipases. *Annu. Rev. Plant Physiol. Plant Mol. Biol.* 52: 211–231.

Whitehouse, D. G., and Moore, A. L. (1995) Regulation of oxidative phosphorylation in plant mitochondria. In *The Molecular Biology of Plant Mitochondria*, C. S. Levings III and I. K. Vasil, eds., Kluwer, Dordrecht, Netherlands, pp. 313–344.

Wolter, F. P., Schmidt, R., and Heinz, E. (1992) Chilling sensitivity of *Arabidopsis thaliana* with genetically engineered membrane lipids. *EMBO J.* 11: 4685–4692.

Chapter

12 Assimilation of Mineral Nutrients

HIGHER PLANTS ARE AUTOTROPHIC ORGANISMS that can synthesize their organic molecular components out of inorganic nutrients obtained from their surroundings. For many mineral nutrients, this process involves absorption from the soil by the roots (see Chapter 5) and incorporation into the organic compounds that are essential for growth and development. This incorporation of mineral nutrients into organic substances such as pigments, enzyme cofactors, lipids, nucleic acids, and amino acids is termed **nutrient assimilation**.

Assimilation of some nutrients—particularly nitrogen and sulfur—requires a complex series of biochemical reactions that are among the most energy-requiring reactions in living organisms:

- In nitrate (NO_3^-) assimilation, the nitrogen in NO_3^- is converted to a higher-energy form in nitrite (NO_2^-), then to a yet higher-energy form in ammonium (NH_4^+), and finally into the amide nitrogen of glutamine. This process consumes the equivalent of 12 ATPs per nitrogen (Bloom et al. 1992).
- Plants such as legumes form symbiotic relationships with nitrogen-fixing bacteria to convert molecular nitrogen (N_2) into ammonia (NH_3). Ammonia (NH_3) is the first stable product of natural fixation; at physiological pH, however, ammonia is protonated to form the ammonium ion (NH_4^+). The process of biological nitrogen fixation, together with the subsequent assimilation of NH_3 into an amino acid, consumes about 16 ATPs per nitrogen (Pate and Layzell 1990; Vande Broek and Vanderleyden 1995).
- The assimilation of sulfate (SO_4^{2-}) into the amino acid cysteine via the two pathways found in plants consumes about 14 ATPs (Hell 1997).

For some perspective on the enormous energies involved, consider that if these reactions run rapidly in reverse—say, from NH_4NO_3 (ammonium nitrate) to N_2—they become explosive, liberating vast amounts of energy as motion, heat, and light. Nearly all explosives are based on the rapid oxidation of nitrogen or sulfur compounds.

Assimilation of other nutrients, especially the macronutrient and micronutrient cations (see Chapter 5), involves the formation of complexes with organic compounds. For example, Mg^{2+} associates with chlorophyll pigments, Ca^{2+} associates with pectates within the cell wall, and Mo^{6+} associates with enzymes such as nitrate reductase and nitrogenase. These complexes are highly stable, and removal of the nutrient from the complex may result in total loss of function.

This chapter outlines the primary reactions through which the major nutrients (nitrogen, sulfur, phosphate, cations, and oxygen) are assimilated. We emphasize the physiological implications of the required energy expenditures and introduce the topic of symbiotic nitrogen fixation.

NITROGEN IN THE ENVIRONMENT

Many biochemical compounds present in plant cells contain nitrogen (see Chapter 5). For example, nitrogen is found in the nucleoside phosphates and amino acids that form the building blocks of nucleic acids and proteins, respectively. Only the elements oxygen, carbon, and hydrogen are more abundant in plants than nitrogen. Most natural and agricultural ecosystems show dramatic gains in productivity after fertilization with inorganic nitrogen, attesting to the importance of this element.

In this section we will discuss the biogeochemical cycle of nitrogen, the crucial role of nitrogen fixation in the conversion of molecular nitrogen into ammonium and nitrate, and the fate of nitrate and ammonium in plant tissues.

Nitrogen Passes through Several Forms in a Biogeochemical Cycle

Nitrogen is present in many forms in the biosphere. The atmosphere contains vast quantities (about 78% by volume) of molecular nitrogen (N_2) (see Chapter 9). For the most part, this large reservoir of nitrogen is not directly available to living organisms. Acquisition of nitrogen from the atmosphere requires the breaking of an exceptionally stable triple covalent bond between two nitrogen atoms ($N\equiv N$) to produce ammonia (NH_3) or nitrate (NO_3^-). These reactions, known as **nitrogen fixation,** can be accomplished by both industrial and natural processes.

Under elevated temperature (about 200°C) and high pressure (about 200 atmospheres), N_2 combines with hydrogen to form ammonia. The extreme conditions are required to overcome the high activation energy of the reaction. This nitrogen fixation reaction, called the *Haber–Bosch process*, is a starting point for the manufacture of many industrial and agricultural products. Worldwide industrial production of nitrogen fertilizers amounts to more than 80×10^{12} g yr^{-1} (FAOSTAT 2001).

Natural processes fix about 190×10^{12} g yr^{-1} of nitrogen (Table 12.1) through the following processes (Schlesinger 1997):

- *Lightning*. Lightning is responsible for about 8% of the nitrogen fixed. Lightning converts water vapor and

TABLE 12.1
The major processes of the biogeochemical nitrogen cycle

Process	Definition	Rate (10^{12} g yr^{-1})[a]
Industrial fixation	Industrial conversion of molecular nitrogen to ammonia	80
Atmospheric fixation	Lightning and photochemical conversion of molecular nitrogen to nitrate	19
Biological fixation	Prokaryotic conversion of molecular nitrogen to ammonia	170
Plant acquisition	Plant absorption and assimilation of ammonium or nitrate	1200
Immobilization	Microbial absorption and assimilation of ammonium or nitrate	N/C
Ammonification	Bacterial and fungal catabolism of soil organic matter to ammonium	N/C
Nitrification	Bacterial (*Nitrosomonas* sp.) oxidation of ammonium to nitrite and subsequent bacterial (*Nitrobacter* sp.) oxidation of nitrite to nitrate	N/C
Mineralization	Bacterial and fungal catabolism of soil organic matter to mineral nitrogen through ammonification or nitrification	N/C
Volatilization	Physical loss of gaseous ammonia to the atmosphere	100
Ammonium fixation	Physical embedding of ammonium into soil particles	10
Denitrification	Bacterial conversion of nitrate to nitrous oxide and molecular nitrogen	210
Nitrate leaching	Physical flow of nitrate dissolved in groundwater out of the topsoil and eventually into the oceans	36

Note: Terrestrial organisms, the soil, and the oceans contain about 5.2×10^{15} g, 95×10^{15} g, and 6.5×10^{15} g, respectively, of organic nitrogen that is active in the cycle. Assuming that the amount of atmospheric N_2 remains constant (inputs = outputs), the *mean residence time* (the average time that a nitrogen molecule remains in organic forms) is about 370 years [(pool size)/(fixation input) = $(5.2 \times 10^{15}$ g + 95×10^{15} g)/(80×10^{12} g yr^{-1} + 19×10^{12} g yr^{-1} + 170×10^{12} g yr^{-1})] (Schlesinger 1997).

[a]N/C, not calculated.

oxygen into highly reactive hydroxyl free radicals, free hydrogen atoms, and free oxygen atoms that attack molecular nitrogen (N_2) to form nitric acid (HNO_3). This nitric acid subsequently falls to Earth with rain.

- *Photochemical reactions*. Approximately 2% of the nitrogen fixed derives from photochemical reactions between gaseous nitric oxide (NO) and ozone (O_3) that produce nitric acid (HNO_3).
- *Biological nitrogen fixation*. The remaining 90% results from biological nitrogen fixation, in which bacteria or blue-green algae (cyanobacteria) fix N_2 into ammonium (NH_4^+).

From an agricultural standpoint, biological nitrogen fixation is critical because industrial production of nitrogen fertilizers seldom meets agricultural demand (FAOSTAT 2001).

Once fixed in ammonium or nitrate, nitrogen enters a biogeochemical cycle and passes through several organic or inorganic forms before it eventually returns to molecular nitrogen (Figure 12.1; see also Table 12.1). The ammonium (NH_4^+) and nitrate (NO_3^-) ions that are generated through fixation or released through decomposition of soil organic matter become the object of intense competition among plants and microorganisms. To remain competitive, plants have developed mechanisms for scavenging these ions from the soil solution as quickly as possible (see Chapter 5). Under the elevated soil concentrations that occur after fertilization, the absorption of ammonium and nitrate by the roots may exceed the capacity of a plant to assimilate these ions, leading to their accumulation within the plant's tissues.

Stored Ammonium or Nitrate Can Be Toxic

Plants can store high levels of nitrate, or they can translocate it from tissue to tissue without deleterious effect. However, if livestock and humans consume plant material that is high in nitrate, they may suffer methemoglobinemia, a disease in which the liver reduces nitrate to nitrite, which combines with hemoglobin and renders the hemoglobin unable to bind oxygen. Humans and other animals may also convert nitrate into nitrosamines, which are potent carcinogens. Some countries limit the nitrate content in plant materials sold for human consumption.

In contrast to nitrate, high levels of ammonium are toxic to both plants and animals. Ammonium dissipates transmembrane proton gradients (Figure 12.2) that are required for both photosynthetic and respiratory electron transport (see Chapters 7 and 11) and for sequestering metabolites in

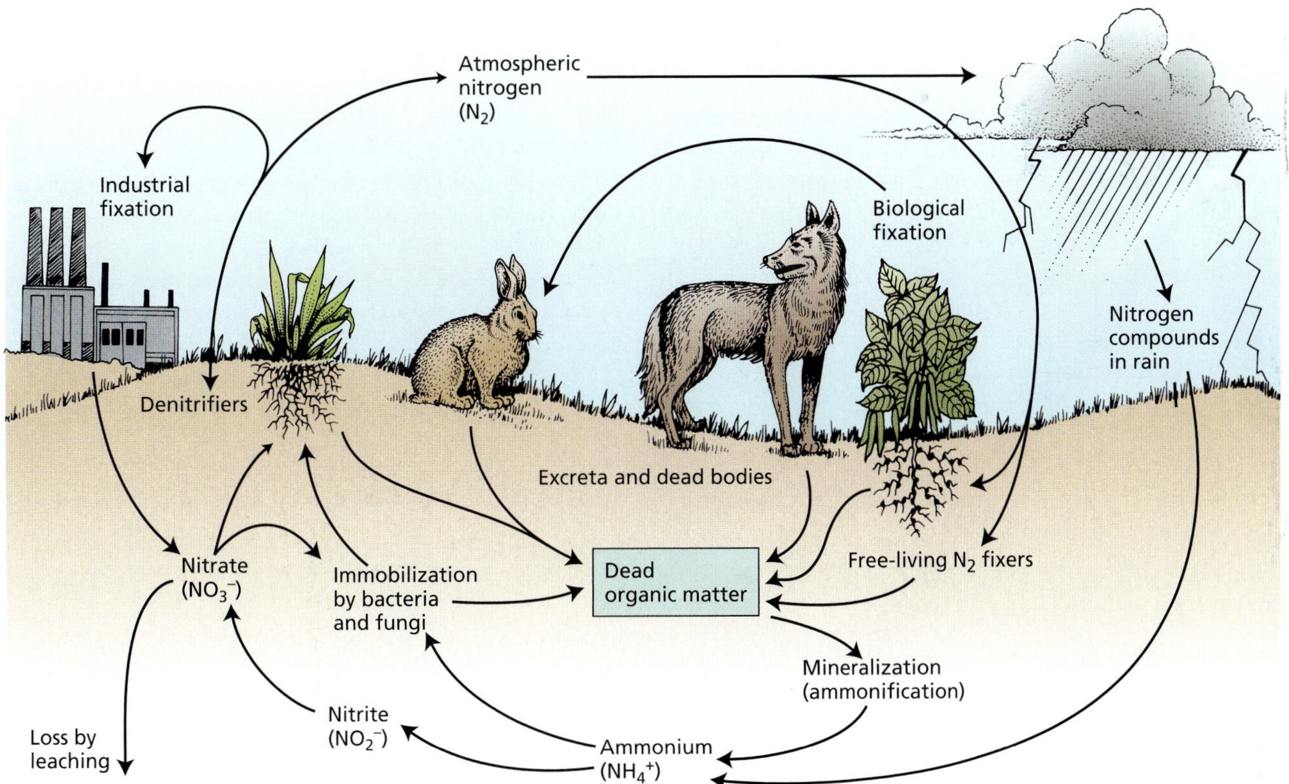

FIGURE 12.1 Nitrogen cycles through the atmosphere as it changes from a gaseous form to reduced ions before being incorporated into organic compounds in living organisms. Some of the steps involved in the nitrogen cycle are shown.

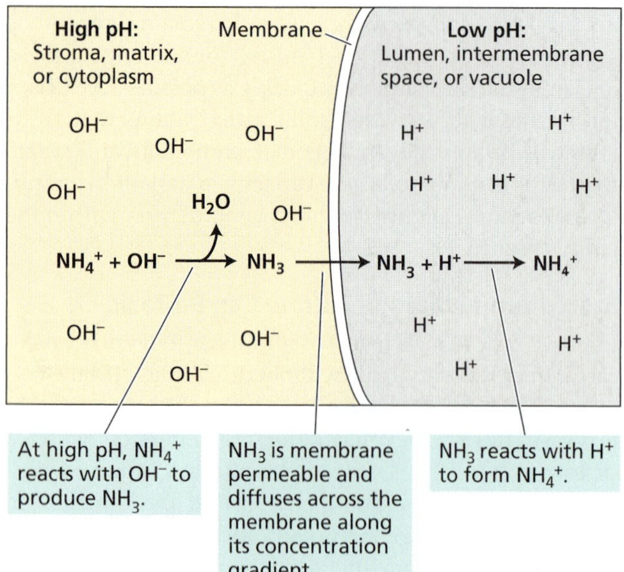

FIGURE 12.2 NH_4^+ toxicity can dissipate pH gradients. The left side represents the stroma, matrix, or cytoplasm, where the pH is high; the right side represents the lumen, intermembrane space, or vacuole, where the pH is low; and the membrane represents the thylakoid, inner mitochondrial, or tonoplast membrane for a chloroplast, mitochondrion, or root cell, respectively. The net result of the reaction shown is that both the OH^- concentration on the left side and the H^+ concentration on the right side have been diminished; that is, the pH gradient has been dissipated. (After Bloom 1997.)

the vacuole (see Chapter 6). Because high levels of ammonium are dangerous, animals have developed a strong aversion to its smell. The active ingredient in smelling salts, a medicinal vapor released under the nose to revive a person who has fainted, is ammonium carbonate. Plants assimilate ammonium near the site of absorption or generation and rapidly store any excess in their vacuoles, thus avoiding toxic effects on membranes and the cytosol.

In the next section we will discuss the process by which the nitrate absorbed by the roots via an H^+–NO_3^- symporter (see Chapter 6 for a discussion of symport) is assimilated into organic compounds, and the enzymatic processes mediating the reduction of nitrate first into nitrite and then into ammonium.

NITRATE ASSIMILATION

Plants assimilate most of the nitrate absorbed by their roots into organic nitrogen compounds. The first step of this process is the reduction of nitrate to nitrite in the cytosol (Oaks 1994). The enzyme **nitrate reductase** catalyzes this reaction:

$$NO_3^- + NAD(P)H + H^+ + 2\,e^- \rightarrow NO_2^- + NAD(P)^+ + H_2O \qquad (12.1)$$

where NAD(P)H indicates NADH or NADPH. The most common form of nitrate reductase uses only NADH as an electron donor; another form of the enzyme that is found predominantly in nongreen tissues such as roots can use either NADH or NADPH (Warner and Kleinhofs 1992).

The nitrate reductases of higher plants are composed of two identical subunits, each containing three prosthetic groups: FAD (flavin adenine dinucleotide), heme, and a molybdenum complexed to an organic molecule called a *pterin* (Mendel and Stallmeyer 1995; Campbell 1999).

A pterin (fully oxidized)

Nitrate reductase is the main molybdenum-containing protein in vegetative tissues, and one symptom of molybdenum deficiency is the accumulation of nitrate that results from diminished nitrate reductase activity.

Comparison of the amino acid sequences for nitrate reductase from several species with those of other well-characterized proteins that bind FAD, heme, or molybdenum has led to the three-domain model for nitrate reductase shown in Figure 12.3. The FAD-binding domain accepts two electrons from NADH or NADPH. The electrons then pass through the heme domain to the molybdenum complex, where they are transferred to nitrate.

Nitrate, Light, and Carbohydrates Regulate Nitrate Reductase

Nitrate, light, and carbohydrates influence nitrate reductase at the transcription and translation levels (Sivasankar and Oaks 1996). In barley seedlings, nitrate reductase mRNA was detected approximately 40 minutes after addition of nitrate, and maximum levels were attained within 3 hours (Figure 12.4). In contrast to the rapid mRNA accumulation,

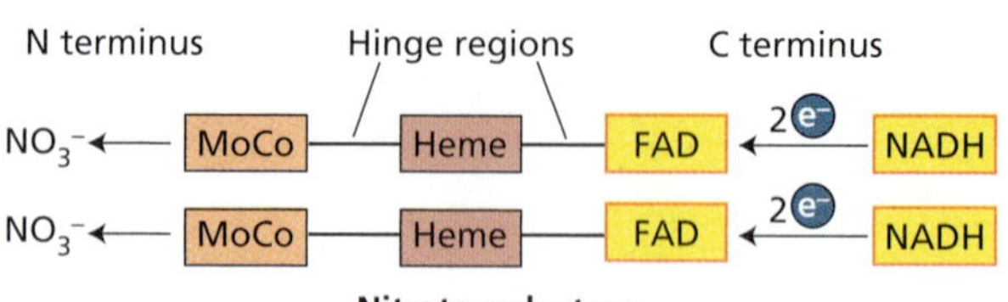

FIGURE 12.3 A model of the nitrate reductase dimer, illustrating the three binding domains whose polypeptide sequences are similar in eukaryotes: molybdenum complex (MoCo), heme, and FAD. The NADH binds at the FAD-binding region of each subunit and initiates a two-electron transfer from the carboxyl (C) terminus, through each of the electron transfer components, to the amino (N) terminus. Nitrate is reduced at the molybdenum complex near the amino terminus. The polypeptide sequences of the hinge regions are highly variable among species.

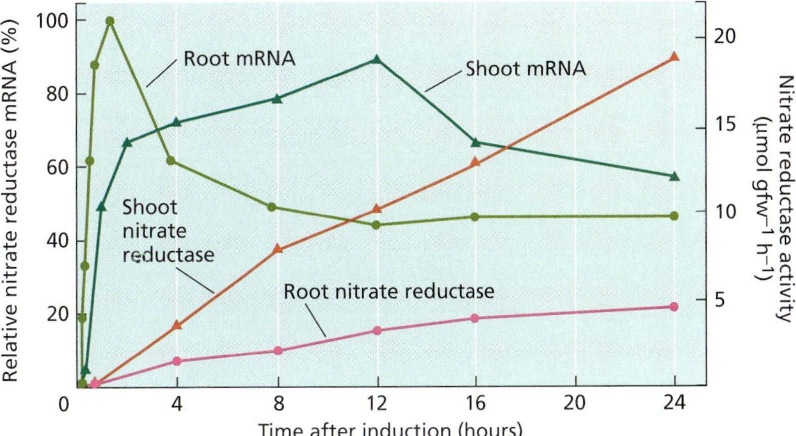

FIGURE 12.4 Stimulation of nitrate reductase activity follows the induction of nitrate reductase mRNA in shoots and roots of barley; gfw, grams fresh weight. (From Kleinhofs et al. 1989.)

there was a gradual linear increase in nitrate reductase activity, reflecting the slower synthesis of the protein.

In addition, the protein is subject to posttranslational modulation (involving a reversible phosphorylation) that is analogous to the regulation of sucrose phosphate synthase (see Chapters 8 and 10). Light, carbohydrate levels, and other environmental factors stimulate a protein phosphatase that dephosphorylates several serine residues on the nitrate reductase protein and thereby activates the enzyme.

Operating in the reverse direction, darkness and Mg^{2+} stimulate a protein kinase that phosphorylates the same serine residues, which then interact with a 14-3-3 inhibitor protein, and thereby inactivate nitrate reductase (Kaiser et al. 1999). *Regulation of nitrate reductase activity through phosphorylation and dephosphorylation provides more rapid control than can be achieved through synthesis or degradation of the enzyme (minutes versus hours).*

Nitrite Reductase Converts Nitrite to Ammonium

Nitrite (NO_2^-) is a highly reactive, potentially toxic ion. Plant cells immediately transport the nitrite generated by nitrate reduction (see Equation 12.1) from the cytosol into chloroplasts in leaves and plastids in roots. In these organelles, the enzyme nitrite reductase reduces nitrite to ammonium according to the following overall reaction:

$$NO_2^- + 6\,Fd_{red} + 8\,H^+ + 6\,e^- \rightarrow NH_4^+ + 6\,Fd_{ox} + 2\,H_2O \qquad (12.2)$$

where Fd is ferredoxin, and the subscripts *red* and *ox* stand for *reduced* and *oxidized*, respectively. Reduced ferredoxin derives from photosynthetic electron transport in the chloroplasts (see Chapter 7) and from NADPH generated by the oxidative pentose phosphate pathway in nongreen tissues (see Chapter 11).

Chloroplasts and root plastids contain different forms of the enzyme, but both forms consist of a single polypeptide containing two prosthetic groups: an iron–sulfur cluster (Fe_4S_4) and a specialized heme (Siegel and Wilkerson 1989). These groups acting together bind nitrite and reduce it directly to ammonium, without accumulation of nitrogen compounds of intermediate redox states. The electron flow through ferredoxin, Fe_4S_4, and heme can be represented as in Figure 12.5.

Nitrite reductase is encoded in the nucleus and synthesized in the cytoplasm with an N-terminal transit peptide that targets it to the plastids (Wray 1993). Whereas NO_3^- and light induce the transcription of nitrite reductase mRNA, the end products of the process—asparagine and glutamine—repress this induction.

Plants Can Assimilate Nitrate in Both Roots and Shoots

In many plants, when the roots receive small amounts of nitrate, nitrate is reduced primarily in the roots. As the supply of nitrate increases, a greater proportion of the

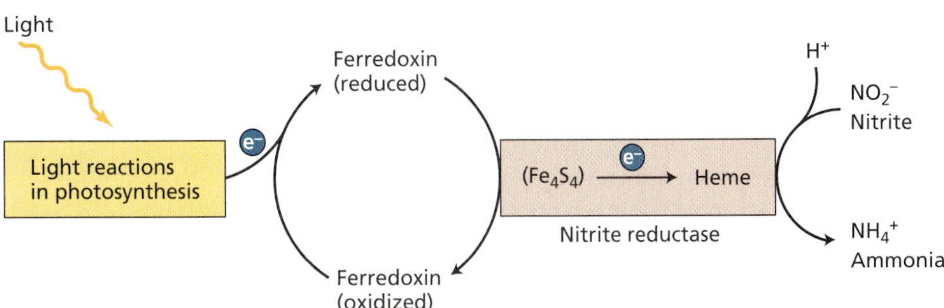

FIGURE 12.5 Model for coupling of photosynthetic electron flow, via ferredoxin, to the reduction of nitrite by nitrite reductase. The enzyme contains two prosthetic groups, Fe_4S_4 and heme, which participate in the reduction of nitrite to ammonium.

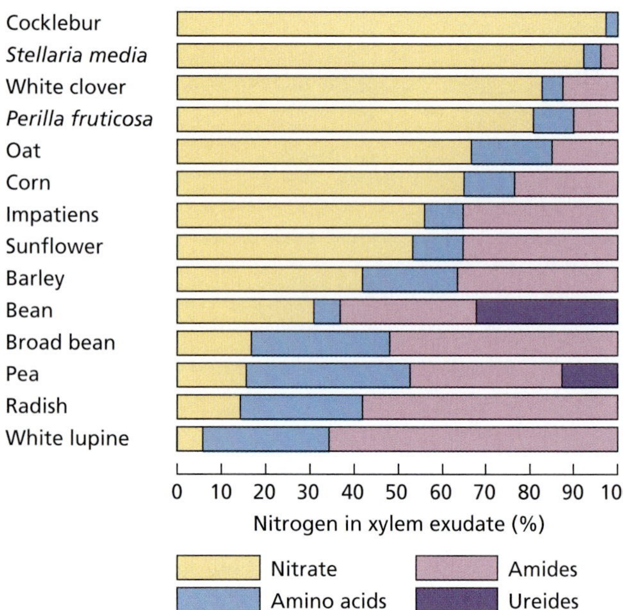

FIGURE 12.6 Relative amounts of nitrate and other nitrogen compounds in the xylem exudate of various plant species. The plants were grown with their roots exposed to nitrate solutions, and xylem sap was collected by severing of the stem. Note the presence of ureides, specialized nitrogen compounds, in bean and pea (which will be discussed later in the text). (After Pate 1983.)

absorbed nitrate is translocated to the shoot and assimilated there (Marschner 1995). Even under similar conditions of nitrate supply, the balance between root and shoot nitrate metabolism—as indicated by the proportion of nitrate reductase activity in each of the two organs or by the relative concentrations of nitrate and reduced nitrogen in the xylem sap—varies from species to species.

In plants such as the cocklebur (*Xanthium strumarium*), nitrate metabolism is restricted to the shoot; in other plants, such as white lupine (*Lupinus albus*), most nitrate is metabolized in the roots (Figure 12.6). Generally, species native to temperate regions rely more heavily on nitrate assimilation by the roots than do species of tropical or subtropical origins.

AMMONIUM ASSIMILATION

Plant cells avoid ammonium toxicity by rapidly converting the ammonium generated from nitrate assimilation or photorespiration (see Chapter 8) into amino acids. The primary pathway for this conversion involves the sequential actions of glutamine synthetase and glutamate synthase (Lea et al. 1992). In this section we will discuss the enzymatic processes that mediate the assimilation of ammonium into essential amino acids, and the role of amides in the regulation of nitrogen and carbon metabolism.

Conversion of Ammonium to Amino Acids Requires Two Enzymes

Glutamine synthetase (GS) combines ammonium with glutamate to form glutamine (Figure 12.7A):

$$\text{Glutamate} + NH_4^+ + \text{ATP} \rightarrow \text{glutamine} + \text{ADP} + P_i \quad (12.3)$$

This reaction requires the hydrolysis of one ATP and involves a divalent cation such as Mg^{2+}, Mn^{2+}, or Co^{2+} as a cofactor. Plants contain two classes of GS, one in the cytosol and the other in root plastids or shoot chloroplasts. The cytosolic forms are expressed in germinating seeds or in the vascular bundles of roots and shoots and produce glutamine for intracellular nitrogen transport. The GS in root plastids generates amide nitrogen for local consumption; the GS in shoot chloroplasts reassimilates photorespiratory NH_4^+ (Lam et al. 1996). Light and carbohydrate levels alter the expression of the plastid forms of the enzyme, but they have little effect on the cytosolic forms.

Elevated plastid levels of glutamine stimulate the activity of **glutamate synthase** (also known as *glutamine:2-oxoglutarate aminotransferase*, or **GOGAT**). This enzyme transfers the amide group of glutamine to 2-oxoglutarate, yielding two molecules of glutamate (see Figure 12.7A). Plants contain two types of GOGAT: One accepts electrons from NADH; the other accepts electrons from ferredoxin (Fd):

$$\text{Glutamine} + \text{2-oxoglutarate} + \text{NADH} + H^+ \rightarrow 2\ \text{glutamate} + \text{NAD}^+ \quad (12.4)$$

$$\text{Glutamine} + \text{2-oxoglutarate} + \text{Fd}_{red} \rightarrow 2\ \text{glutamate} + \text{Fd}_{ox} \quad (12.5)$$

The NADH type of the enzyme (NADH-GOGAT) is located in plastids of nonphotosynthetic tissues such as roots or vascular bundles of developing leaves. In roots, NADH-GOGAT is involved in the assimilation of NH_4^+ absorbed from the rhizosphere (the soil near the surface of the roots); in vascular bundles of developing leaves, NADH-GOGAT assimilates glutamine translocated from roots or senescing leaves.

The ferredoxin-dependent type of glutamate synthase (Fd-GOGAT) is found in chloroplasts and serves in photorespiratory nitrogen metabolism. Both the amount of protein and its activity increase with light levels. Roots, particularly those under nitrate nutrition, have Fd-GOGAT in plastids. Fd-GOGAT in the roots presumably functions to incorporate the glutamine generated during nitrate assimilation.

Ammonium Can Be Assimilated via an Alternative Pathway

Glutamate dehydrogenase (GDH) catalyzes a reversible reaction that synthesizes or deaminates glutamate (Figure 12.7B):

$$\text{2-Oxoglutarate} + NH_4^+ + \text{NAD(P)H} \leftrightarrow \text{glutamate} + H_2O + \text{NAD(P)}^+ \quad (12.6)$$

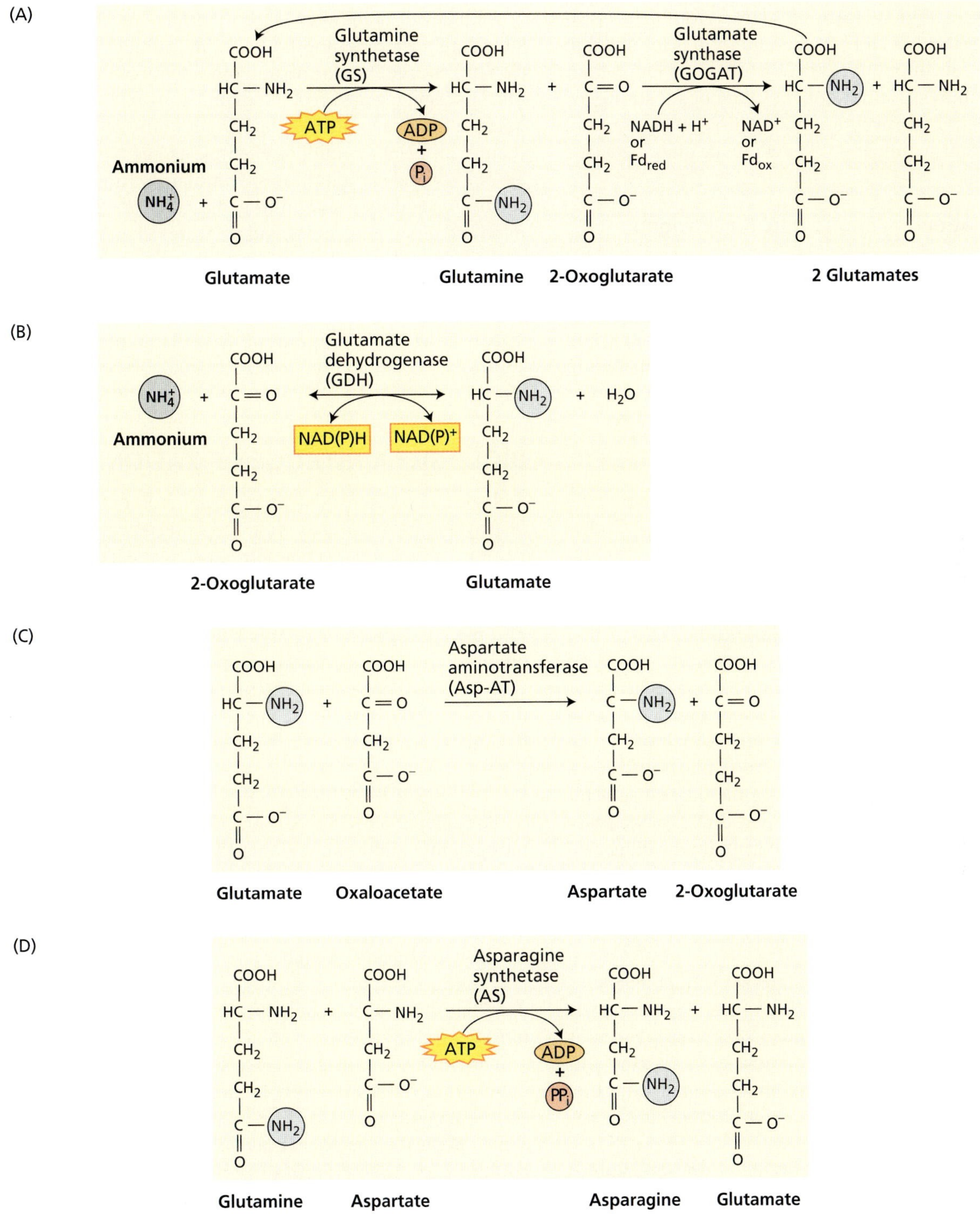

FIGURE 12.7 Structure and pathways of compounds involved in ammonium metabolism. Ammonium can be assimilated by one of several processes. (A) The GS-GOGAT pathway that forms glutamine and glutamate. A reduced cofactor is required for the reaction: ferredoxin in green leaves and NADH in nonphotosynthetic tissue. (B) The GDH pathway that forms glutamate using NADH or NADPH as a reductant. (C) Transfer of the amino group from glutamate to oxaloacetate to form aspartate (catalyzed by aspartate aminotransferase). (D) Synthesis of asparagine by transfer of an amino acid group from glutamine to aspartate (catalyzed by asparagine synthesis).

An NADH-dependent form of GDH is found in mitochondria, and an NADPH-dependent form is localized in the chloroplasts of photosynthetic organs. Although both forms are relatively abundant, they cannot substitute for the GS–GOGAT pathway for assimilation of ammonium, and their primary function is to deaminate glutamate (see Figure 12.7B).

Transamination Reactions Transfer Nitrogen

Once assimilated into glutamine and glutamate, nitrogen is incorporated into other amino acids via transamination reactions. The enzymes that catalyze these reactions are known as aminotransferases. An example is **aspartate aminotransferase** (**Asp-AT**), which catalyzes the following reaction (Figure 12.7C):

$$\text{Glutamate} + \text{oxaloacetate} \rightarrow \text{aspartate} + \text{2-oxoglutarate} \quad (12.7)$$

in which the amino group of glutamate is transferred to the carboxyl atom of aspartate. Aspartate is an amino acid that participates in the malate–aspartate shuttle to transfer reducing equivalents from the mitochondrion and chloroplast into the cytosol (see Chapter 11) and in the transport of carbon from mesophyll to bundle sheath for C_4 carbon fixation (see Chapter 8). All transamination reactions require pyridoxal phosphate (vitamin B_6) as a cofactor.

Aminotransferases are found in the cytoplasm, chloroplasts, mitochondria, glyoxysomes, and peroxisomes. The aminotransferases localized in the chloroplasts may have a significant role in amino acid biosynthesis because plant leaves or isolated chloroplasts exposed to radioactively labeled carbon dioxide rapidly incorporate the label into glutamate, aspartate, alanine, serine, and glycine.

Asparagine and Glutamine Link Carbon and Nitrogen Metabolism

Asparagine, isolated from asparagus as early as 1806, was the first amide to be identified (Lam et al. 1996). It serves not only as a protein precursor, but as a key compound for nitrogen transport and storage because of its stability and high nitrogen-to-carbon ratio (2 N to 4 C for asparagine, versus 2 N to 5 C for glutamine or 1 N to 5 C for glutamate).

The major pathway for asparagine synthesis involves the transfer of the amide nitrogen from glutamine to asparagine (Figure 12.7D):

$$\text{Glutamine} + \text{aspartate} + \text{ATP} \rightarrow \text{asparagine} + \text{glutamate} + \text{AMP} + \text{PP}_i \quad (12.8)$$

Asparagine synthetase (**AS**), the enzyme that catalyzes this reaction, is found in the cytosol of leaves and roots and in nitrogen-fixing nodules (see the next section). In maize roots, particularly those under potentially toxic levels of ammonia, ammonium may replace glutamine as the source of the amide group (Sivasankar and Oaks 1996).

High levels of light and carbohydrate—conditions that stimulate plastid GS and Fd-GOGAT—inhibit the expression of genes coding for AS and the activity of the enzyme. The opposing regulation of these competing pathways helps balance the metabolism of carbon and nitrogen in plants (Lam et al. 1996). Conditions of ample energy (i.e., high levels of light and carbohydrates) stimulate GS and GOGAT, inhibit AS, and thus favor nitrogen assimilation into glutamine and glutamate, compounds that are rich in carbon and participate in the synthesis of new plant materials.

By contrast, energy-limited conditions inhibit GS and GOGAT, stimulate AS, and thus favor nitrogen assimilation into asparagine, a compound that is rich in nitrogen and sufficiently stable for long-distance transport or long-term storage.

BIOLOGICAL NITROGEN FIXATION

Biological nitrogen fixation accounts for most of the fixation of atmospheric N_2 into ammonium, thus representing the key entry point of molecular nitrogen into the biogeochemical cycle of nitrogen (see Figure 12.1). In this section we will describe the properties of the nitrogenase enzymes that fix nitrogen, the symbiotic relations between nitrogen-fixing organisms and higher plants, the specialized structures that form in roots when infected by nitrogen-fixing bacteria, and the genetic and signaling interactions that regulate nitrogen fixation by symbiotic prokaryotes and their hosts.

Free-Living and Symbiotic Bacteria Fix Nitrogen

Some bacteria, as stated earlier, can convert atmospheric nitrogen into ammonium (Table 12.2). Most of these nitrogen-fixing prokaryotes are free-living in the soil. A few form symbiotic associations with higher plants in which the prokaryote directly provides the host plant with fixed nitrogen in exchange for other nutrients and carbohydrates (top portion of Table 12.2). Such symbioses occur in nodules that form on the roots of the plant and contain the nitrogen-fixing bacteria.

The most common type of symbiosis occurs between members of the plant family Leguminosae and soil bacteria of the genera *Azorhizobium*, *Bradyrhizobium*, *Photorhizobium*, *Rhizobium*, and *Sinorhizobium* (collectively called **rhizobia**; Table 12.3 and Figure 12.8). Another common type of symbiosis occurs between several woody plant species, such as alder trees, and soil bacteria of the genus *Frankia*. Still other types involve the South American herb *Gunnera* and the tiny water fern *Azolla*, which form associations with the cyanobacteria *Nostoc* and *Anabaena*, respectively (see Table 12.2 and Figure 12.9).

Nitrogen Fixation Requires Anaerobic Conditions

Because oxygen irreversibly inactivates the **nitrogenase** enzymes involved in nitrogen fixation, nitrogen must be fixed under anaerobic conditions. Thus each of the nitro-

TABLE 12.2
Examples of organisms that can carry out nitrogen fixation

Symbiotic nitrogen fixation	
Host plant	**N-fixing symbionts**
Leguminous: legumes, *Parasponia*	*Azorhizobium, Bradyrhizobium, Photorhizobium, Rhizobium, Sinorhizobium*
Actinorhizal: alder (tree), *Ceanothus* (shrub), *Casuarina* (tree), *Datisca* (shrub)	*Frankia*
Gunnera	*Nostoc*
Azolla (water fern)	*Anabaena*
Sugarcane	*Acetobacter*
Free-living nitrogen fixation	
Type	**N-fixing genera**
Cyanobacteria (blue-green algae)	*Anabaena, Calothrix, Nostoc*
Other bacteria	
Aerobic	*Azospirillum, Azotobacter, Beijerinckia, Derxia*
Facultative	*Bacillus, Klebsiella*
Anaerobic	
Nonphotosynthetic	*Clostridium, Methanococcus* (archaebacterium)
Photosynthetic	*Chromatium, Rhodospirillum*

gen-fixing organisms listed in Table 12.2 either functions under natural anaerobic conditions or can create an internal anaerobic environment in the presence of oxygen.

In cyanobacteria, anaerobic conditions are created in specialized cells called *heterocysts* (see Figure 12.9). Heterocysts are thick-walled cells that differentiate when filamentous cyanobacteria are deprived of NH_4^+. These cells lack photosystem II, the oxygen-producing photosystem of chloroplasts (see Chapter 7), so they do not generate oxygen (Burris 1976). Heterocysts appear to represent an adaptation for nitrogen fixation, in that they are widespread among aerobic cyanobacteria that fix nitrogen.

Cyanobacteria can fix nitrogen under anaerobic conditions such as those that occur in flooded fields. In Asian countries, nitrogen-fixing cyanobacteria of both the heterocyst and nonheterocyst types are a major means for maintaining an adequate nitrogen supply in the soil of rice fields. These microorganisms fix nitrogen when the fields are flooded and die as the fields dry, releasing the fixed nitrogen to the soil. Another important

TABLE 12.3
Associations between host plants and rhizobia

Plant host	**Rhizobial symbiont**
Parasponia (a nonlegume, formerly called *Trema*)	*Bradyrhizobium* spp.
Soybean (*Glycine max*)	*Bradyrhizobium japonicum* (slow-growing type); *Sinorhizobium fredii* (fast-growing type)
Alfalfa (*Medicago sativa*)	*Sinorhizobium meliloti*
Sesbania (aquatic)	*Azorhizobium* (forms both root and stem nodules; the stems have adventitious roots)
Bean (*Phaseolus*)	*Rhizobium leguminosarum* bv. *phaseoli*; *Rhizobium tropicii; Rhizobium etli*
Clover (*Trifolium*)	*Rhizobium leguminosarum* bv. *trifolii*
Pea (*Pisum sativum*)	*Rhizobium leguminosarum* bv. *viciae*
Aeschenomene (aquatic)	*Photorhizobium* (photosynthetically active rhizobia that form stem nodules, probably associated with adventitious roots)

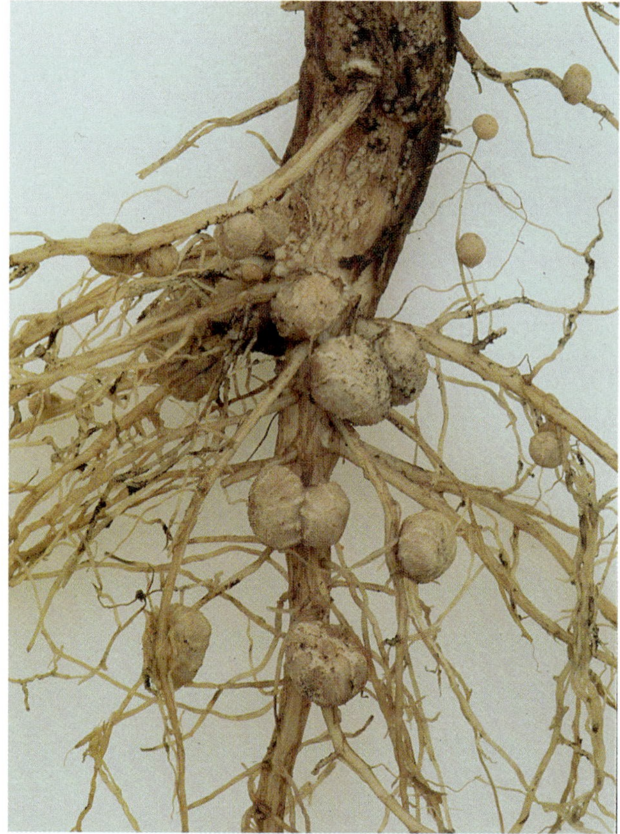

FIGURE 12.8 Root nodules on soybean. The nodules are a result of infection by *Rhizobium japonicum*. (© Wally Eberhart/Visuals Unlimited.)

source of available nitrogen in flooded rice fields is the water fern *Azolla*, which associates with the cyanobacterium *Anabaena*. The *Azolla–Anabaena* association can fix as much as 0.5 kg of atmospheric nitrogen per hectare per day, a rate of fertilization that is sufficient to attain moderate rice yields.

Free-living bacteria that are capable of fixing nitrogen are aerobic, facultative, or anaerobic (see Table 12.2, bottom):

- *Aerobic* nitrogen-fixing bacteria such as *Azotobacter* are thought to maintain reduced oxygen conditions (microaerobic conditions) through their high levels of respiration (Burris 1976). Others, such as *Gloeothece*, evolve O_2 photosynthetically during the day and fix nitrogen during the night.
- *Facultative* organisms, which are able to grow under both aerobic and anaerobic conditions, generally fix nitrogen only under anaerobic conditions.
- For *anaerobic* nitrogen-fixing bacteria, oxygen does not pose a problem, because it is absent in their habitat. These anaerobic organisms can be either photosynthetic (e.g., *Rhodospirillum*), or nonphotosynthetic (e.g., *Clostridium*).

Symbiotic Nitrogen Fixation Occurs in Specialized Structures

Symbiotic nitrogen-fixing prokaryotes dwell within **nodules**, the special organs of the plant host that enclose the nitrogen-fixing bacteria (see Figure 12.8). In the case of *Gunnera*, these organs are existing stem glands that develop independently of the symbiont. In the case of legumes and actinorhizal plants, the nitrogen-fixing bacteria induce the plant to form root nodules.

Grasses can also develop symbiotic relationships with nitrogen-fixing organisms, but in these associations root nodules are not produced. Instead, the nitrogen-fixing bacteria seem to colonize plant tissues or anchor to the root surfaces, mainly around the elongation zone and the root hairs (Reis et al. 2000). For example, the nitrogen-fixing

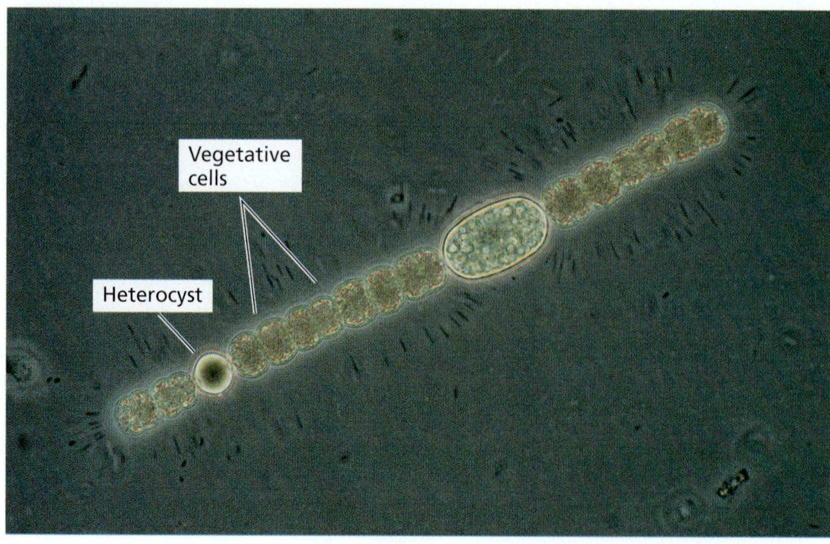

FIGURE 12.9 A heterocyst in a filament of the nitrogen-fixing cyanobacterium *Anabaena*. The thick-walled heterocysts, interspaced among vegetative cells, have an anaerobic inner environment that allows cyanobacteria to fix nitrogen in aerobic conditions. (© Paul W. Johnson/Biological Photo Service.)

bacterium *Acetobacter diazotrophicus* lives in the apoplast of stem tissues in sugarcane and may provide its host with sufficient nitrogen to grant independence from nitrogen fertilization (Dong et al. 1994). The potential for applying *Azospirillum* to corn and other grains has been explored, but *Azospirillum* seems to fix little nitrogen when associated with plants (Vande Broek and Vanderleyden 1995).

Legumes and actinorhizal plants regulate gas permeability in their nodules, maintaining a level of oxygen within the nodule that can support respiration but is sufficiently low to avoid inactivation of the nitrogenase (Kuzma et al. 1993). Gas permeability increases in the light and decreases under drought or upon exposure to nitrate. The mechanism for regulating gas permeability is not yet known.

Nodules contain an oxygen-binding heme protein called **leghemoglobin**. Leghemoglobin is present in the cytoplasm of infected nodule cells at high concentrations (700 μM in soybean nodules) and gives the nodules a pink color. The host plant produces the globin portion of leghemoglobin in response to infection by the bacteria (Marschner 1995); the bacterial symbiont produces the heme portion. Leghemoglobin has a high affinity for oxygen (a K_m of about 0.01 μM), about ten times higher than the β chain of human hemoglobin.

Although leghemoglobin was once thought to provide a buffer for nodule oxygen, recent studies indicate that it stores only enough oxygen to support nodule respiration for a few seconds (Denison and Harter 1995). Its function is to help transport oxygen to the respiring symbiotic bacterial cells in a manner analogous to hemoglobin transporting oxygen to respiring tissues in animals (Ludwig and de Vries 1986).

Establishing Symbiosis Requires an Exchange of Signals

The symbiosis between legumes and rhizobia is not obligatory. Legume seedlings germinate without any association with rhizobia, and they may remain unassociated throughout their life cycle. Rhizobia also occur as free-living organisms in the soil. Under nitrogen-limited conditions, however, the symbionts seek out one another through an elaborate exchange of signals. This signaling, the subsequent infection process, and the development of nitrogen-fixing nodules involve specific genes in both the host and the symbionts.

Plant genes specific to nodules are called **nodulin** (*Nod*) genes; rhizobial genes that participate in nodule formation are called **nodulation** (*nod*) genes (Heidstra and Bisseling 1996). The *nod* genes are classified as common *nod* genes or host-specific *nod* genes. The common *nod* genes—*nodA*, *nodB*, and *nodC*—are found in all rhizobial strains; the host-specific *nod* genes—such as *nodP*, *nodQ*, and *nodH*; or *nodF*, *nodE*, and *nodL*—differ among rhizobial species and determine the host range. Only one of the *nod* genes, the regulatory *nodD*, is constitutively expressed, and as we will explain in detail, its protein product (NodD) regulates the transcription of the other *nod* genes.

The first stage in the formation of the symbiotic relationship between the nitrogen-fixing bacteria and their host is migration of the bacteria toward the roots of the host plant. This migration is a chemotactic response mediated by chemical attractants, especially (iso)flavonoids and betaines, secreted by the roots. These attractants activate the rhizobial NodD protein, which then induces transcription of the other *nod* genes (Phillips and Kapulnik 1995). The promoter region of all *nod* operons, except that of *nodD*, contains a highly conserved sequence called the *nod* box. Binding of the activated NodD to the *nod* box induces transcription of the other *nod* genes.

Nod Factors Produced by Bacteria Act as Signals for Symbiosis

The *nod* genes activated by NodD code for nodulation proteins, most of which are involved in the biosynthesis of Nod factors. **Nod factors** are lipochitin oligosaccharide signal molecules, all of which have a chitin β-1→4-linked *N*-acetyl-D-glucosamine backbone (varying in length from three to six sugar units) and a fatty acyl chain on the C-2 position of the nonreducing sugar (Figure 12.10).

Three of the *nod* genes (*nodA*, *nodB*, and *nodC*) encode enzymes (NodA, NodB, and NodC, respectively) that are required for synthesizing this basic structure (Stokkermans et al. 1995):

1. NodA is an *N*-acyltransferase that catalyzes the addition of a fatty acyl chain.
2. NodB is a chitin-oligosaccharide deacetylase that removes the acetyl group from the terminal nonreducing sugar.

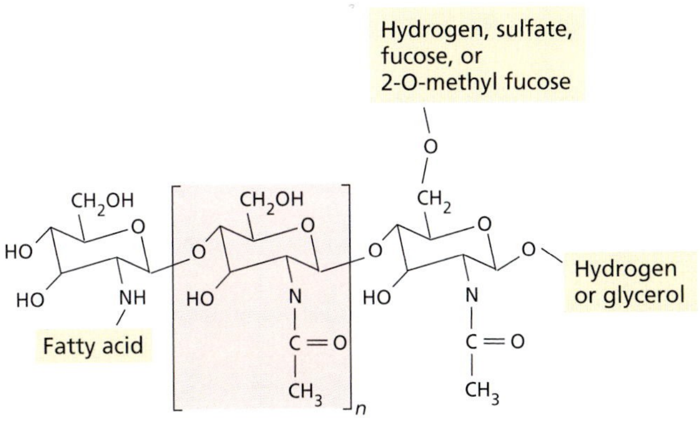

FIGURE 12.10 Nod factors are lipochitin oligosaccharides. The fatty acid chain typically has 16 to 18 carbons. The number of repeated middle sections (n) is usually 2 to 3. (After Stokkermans et al. 1995.)

3. NodC is a chitin-oligosaccharide synthase that links *N*-acetyl-D-glucosamine monomers.

Host-specific *nod* genes that vary among rhizobial species are involved in the modification of the fatty acyl chain or the addition of groups important in determining host specificity (Carlson et al. 1995):

- NodE and NodF determine the length and degree of saturation of the fatty acyl chain; those of *Rhizobium leguminosarum* bv. *viciae* and *R. meliloti* result in the synthesis of an 18:4 and a 16:2 fatty acyl group, respectively. (Recall from Chapter 11 that the number before the colon gives the total number of carbons in the fatty acyl chain, and the number after the colon gives the number of double bonds.)
- Other enzymes, such as NodL, influence the host specificity of Nod factors through the addition of specific substitutions at the reducing or nonreducing sugar moieties of the chitin backbone.

A particular legume host responds to a specific Nod factor. The legume receptors for Nod factors appear to be special lectins (sugar-binding proteins) produced in the root hairs (van Rhijn et al. 1998; Etzler et al. 1999). Nod factors activate these lectins, increasing their hydrolysis of phosphoanhydride bonds of nucleoside di- and triphosphates. This lectin activation directs particular rhizobia to appropriate hosts and facilitates attachment of the rhizobia to the cell walls of a root hair.

Nodule Formation Involves Several Phytohormones

Two processes—infection and nodule organogenesis—occur simultaneously during root nodule formation. During the infection process, rhizobia that are attached to the root hairs release Nod factors that induce a pronounced curling of the root hair cells (Figure 12.11A and B). The rhizobia become enclosed in the small compartment formed by the curling. The cell wall of the root hair degrades in these regions, also in response to Nod factors, allowing the bacterial cells direct access to the outer surface of the plant plasma membrane (Lazarowitz and Bisseling 1997).

The next step is formation of the **infection thread** (Figure 12.11C), an internal tubular extension of the plasma membrane that is produced by the fusion of Golgi-derived membrane vesicles at the site of infection. The thread grows at its tip by the fusion of secretory vesicles to the end of the tube. Deeper into the root cortex, near the xylem, cortical cells dedifferentiate and start dividing, forming a distinct area within the cortex, called a *nodule primordium*, from which the nodule will develop. The nodule primordia form opposite the protoxylem poles of the root vascular bundle (Timmers et al. 1999) (See **Web Topic 12.1**).

Different signaling compounds, acting either positively or negatively, control the position of nodule primordia. The nucleoside uridine diffuses from the stele into the cortex in the protoxylem zones of the root and stimulates cell division (Lazarowitz and Bisseling 1997). Ethylene is synthesized in the region of the pericycle, diffuses into the cortex, and blocks cell division opposite the phloem poles of the root.

The infection thread filled with proliferating rhizobia elongates through the root hair and cortical cell layers, in the direction of the nodule primordium. When the infection thread reaches specialized cells within the nodule, its tip fuses with the plasma membrane of the host cell, releasing bacterial cells that are packaged in a membrane derived from the host cell plasma membrane (see Figure 12.11D). Branching of the infection thread inside the nodule enables the bacteria to infect many cells (see Figure 12.11E and F) (Mylona et al. 1995).

At first the bacteria continue to divide, and the surrounding membrane increases in surface area to accommodate this growth by fusing with smaller vesicles. Soon thereafter, upon an undetermined signal from the plant, the bacteria stop dividing and begin to enlarge and to differentiate into nitrogen-fixing endosymbiotic organelles called **bacteroids**. The membrane surrounding the bacteroids is called the *peribacteroid membrane*.

The nodule as a whole develops such features as a vascular system (which facilitates the exchange of fixed nitrogen produced by the bacteroids for nutrients contributed by the plant) and a layer of cells to exclude O_2 from the root nodule interior. In some temperate legumes (e.g., peas), the nodules are elongated and cylindrical because of the presence of a *nodule meristem*. The nodules of tropical legumes, such as soybeans and peanuts, lack a persistent meristem and are spherical (Rolfe and Gresshoff 1988).

FIGURE 12.11 The infection process during nodule organogenesis. (A) Rhizobia bind to an emerging root hair in response to chemical attractants sent by the plant. (B) In response to factors produced by the bacteria, the root hair exhibits abnormal curling growth, and rhizobia cells proliferate within the coils. (C) Localized degradation of the root hair wall leads to infection and formation of the infection thread from Golgi secretory vesicles of root cells. (D) The infection thread reaches the end of the cell, and its membrane fuses with the plasma membrane of the root hair cell. (E) Rhizobia are released into the apoplast and penetrate the compound middle lamella to the subepidermal cell plasma membrane, leading to the initiation of a new infection thread, which forms an open channel with the first. (F) The infection thread extends and branches until it reaches target cells, where vesicles composed of plant membrane that enclose bacterial cells are released into the cytosol.

The Nitrogenase Enzyme Complex Fixes N_2

Biological nitrogen fixation, like industrial nitrogen fixation, produces ammonia from molecular nitrogen. The overall reaction is

$$N_2 + 8\,e^- + 8\,H^+ + 16\,ATP \rightarrow 2\,NH_3 + H_2 + 16\,ADP + 16\,P_i \qquad (12.9)$$

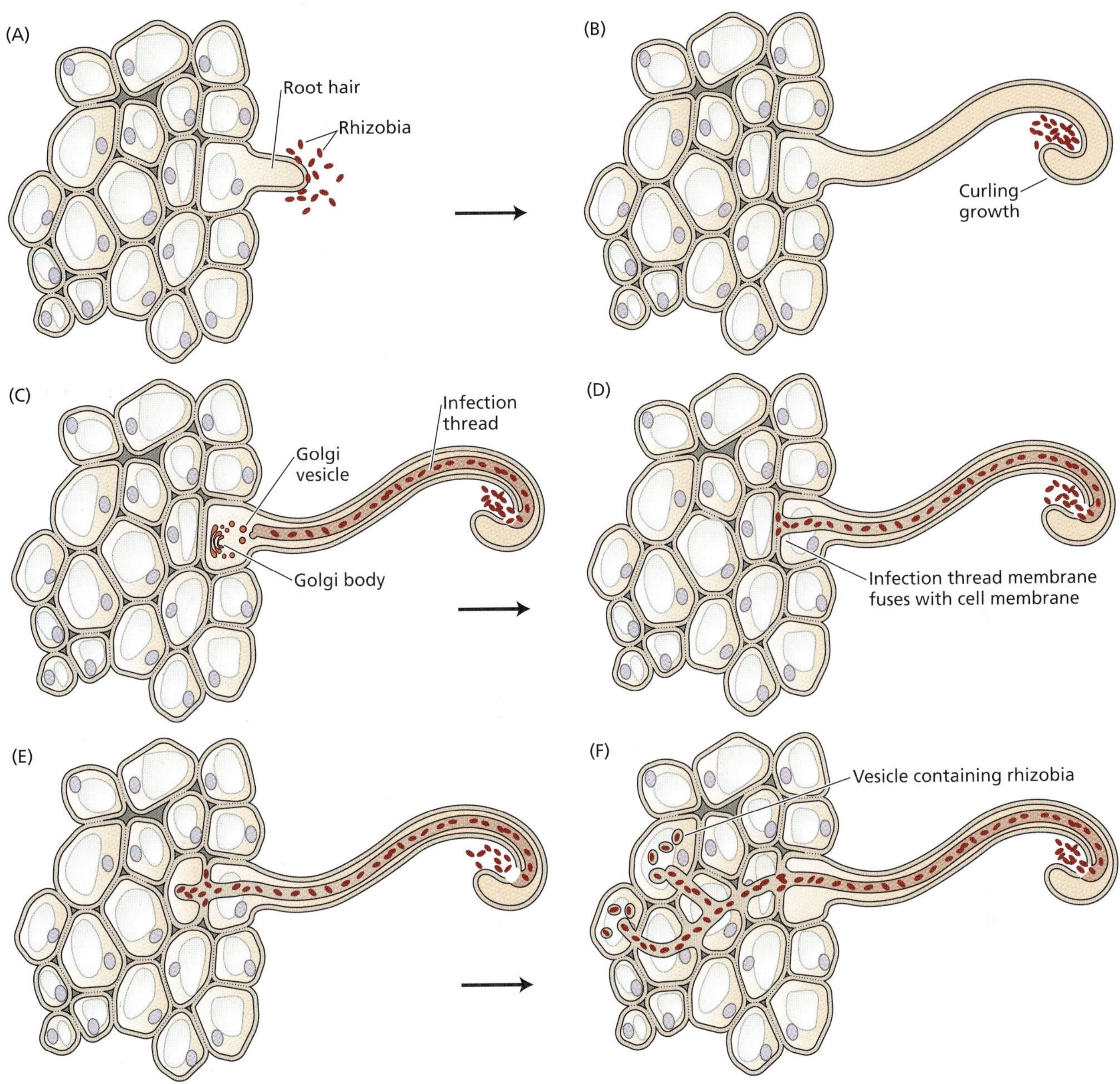

Note that the reduction of N_2 to 2 NH_3, a six-electron transfer, is coupled to the reduction of two protons to evolve H_2. The **nitrogenase enzyme complex** catalyzes this reaction.

The nitrogenase enzyme complex can be separated into two components—the Fe protein and the MoFe protein—neither of which has catalytic activity by itself (Figure 12.12):

- The Fe protein is the smaller of the two components and has two identical subunits of 30 to 72 kDa each, depending on the organism. Each subunit contains an iron–sulfur cluster (4 Fe and 4 S^{2-}) that participates in the redox reactions involved in the conversion of N_2 to NH_3. The Fe protein is irreversibly inactivated by O_2 with typical half-decay times of 30 to 45 seconds (Dixon and Wheeler 1986).
- The MoFe protein has four subunits, with a total molecular mass of 180 to 235 kDa, depending on the species. Each subunit has two Mo–Fe–S clusters. The MoFe protein is also inactivated by oxygen, with a half-decay time in air of 10 minutes.

In the overall nitrogen reduction reaction (see Figure 12.12), ferredoxin serves as an electron donor to the Fe protein, which in turn hydrolyzes ATP and reduces the MoFe protein. The MoFe protein then can reduce numerous substrates (Table 12.4), although under natural conditions it reacts only with N_2 and H^+. One of the reactions catalyzed by nitrogenase, the reduction of acetylene to ethylene, is used in estimating nitrogenase activity (see **Web Topic 12.2**).

The energetics of nitrogen fixation is complex. The production of NH_3 from N_2 and H_2 is an exergonic reaction

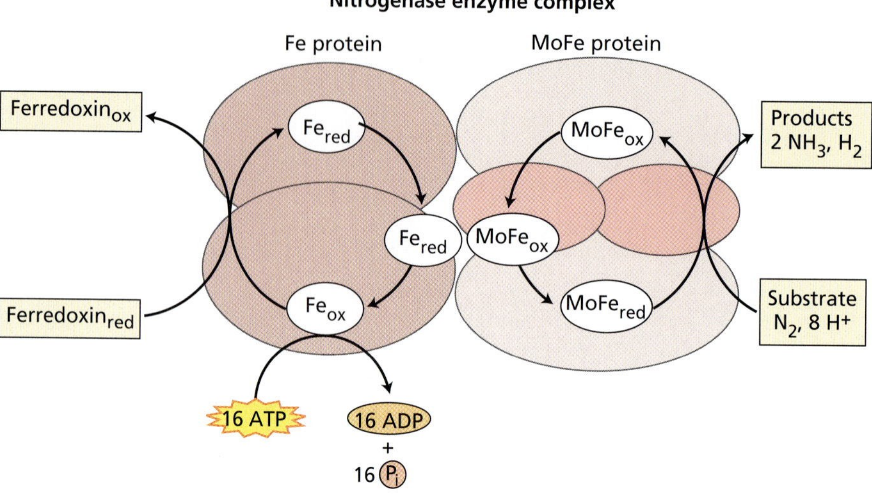

FIGURE 12.12 The reaction catalyzed by nitrogenase. Ferredoxin reduces the Fe protein. Binding and hydrolysis of ATP to the Fe protein is thought to cause a conformational change of the Fe protein that facilitates the redox reactions. The Fe protein reduces the MoFe protein, and the MoFe protein reduces the N_2. (After Dixon and Wheeler 1986, and Buchanan et al. 2000.)

(see Chapter 2 on the website for a discussion of exergonic reactions), with a $\Delta G^{0\prime}$ (change in free energy) of –27 kJ mol^{-1}. However, industrial production of NH_3 from N_2 and H_2 is *endergonic*, requiring a very large energy input because of the activation energy needed to break the triple bond in N_2. For the same reason, the enzymatic reduction of N_2 by nitrogenase also requires a large investment of energy (see Equation 12.9), although the exact changes in free energy are not yet known.

Calculations based on the carbohydrate metabolism of legumes show that a plant consumes 12 g of organic carbon per gram of N_2 fixed (Heytler et al. 1984). On the basis of Equation 12.9, the $\Delta G^{0\prime}$ for the overall reaction of biological nitrogen fixation is about –200 kJ mol^{-1}. Because the overall reaction is highly exergonic, ammonium production is limited by the slow operation (number of N_2 molecules reduced per unit time) of the nitrogenase complex (Ludwig and de Vries 1986).

Under natural conditions, substantial amounts of H^+ are reduced to H_2 gas, and this process can compete with N_2 reduction for electrons from nitrogenase. In rhizobia, 30 to 60% of the energy supplied to nitrogenase may be lost as H_2, diminishing the efficiency of nitrogen fixation. Some rhizobia, however, contain hydrogenase, an enzyme that can split the H_2 formed and generate electrons for N_2 reduction, thus improving the efficiency of nitrogen fixation (Marschner 1995).

TABLE 12.4
Reactions catalyzed by nitrogenase

$N_2 \rightarrow NH_3$	Molecular nitrogen fixation
$N_2O \rightarrow N_2 + H_2O$	Nitrous oxide reduction
$N_3^- \rightarrow N_2 + NH_3$	Azide reduction
$C_2H_2 \rightarrow C_2H_4$	Acetylene reduction
$2\ H^+ \rightarrow H_2$	H_2 production
$ATP \rightarrow ADP + P_i$	ATP hydrolytic activity

Source: After Burris 1976.

Amides and Ureides Are the Transported Forms of Nitrogen

The symbiotic nitrogen-fixing prokaryotes release ammonia that, to avoid toxicity, must be rapidly converted into organic forms in the root nodules before being transported to the shoot via the xylem. Nitrogen-fixing legumes can be divided into amide exporters or ureide exporters on the basis of the composition of the xylem sap. Amides (principally the amino acids asparagine or glutamine) are exported by temperate-region legumes, such as pea (*Pisum*), clover (*Trifolium*), broad bean (*Vicia*), and lentil (*Lens*).

Ureides are exported by legumes of tropical origin, such as soybean (*Glycine*), kidney bean (*Phaseolus*), peanut (*Arachis*), and southern pea (*Vigna*). The three major ureides are allantoin, allantoic acid, and citrulline (Figure 12.13). Allantoin is synthesized in peroxisomes from uric acid, and allantoic acid is synthesized from allantoin in the endoplasmic reticulum. The site of citrulline synthesis from the amino acid ornithine has not yet been determined. All three compounds are ultimately released into the xylem and transported to the shoot, where they are rapidly catabolized to ammonium. This ammonium enters the assimilation pathway described earlier.

SULFUR ASSIMILATION

Sulfur is among the most versatile elements in living organisms (Hell 1997). Disulfide bridges in proteins play structural and regulatory roles (see Chapter 8). Sulfur participates in electron transport through iron–sulfur clusters (see Chapters 7 and 11). The catalytic sites for several enzymes and coenzymes, such as urease and coenzyme A, contain sulfur. Secondary metabolites (compounds that are not involved in primary pathways of growth and develop-

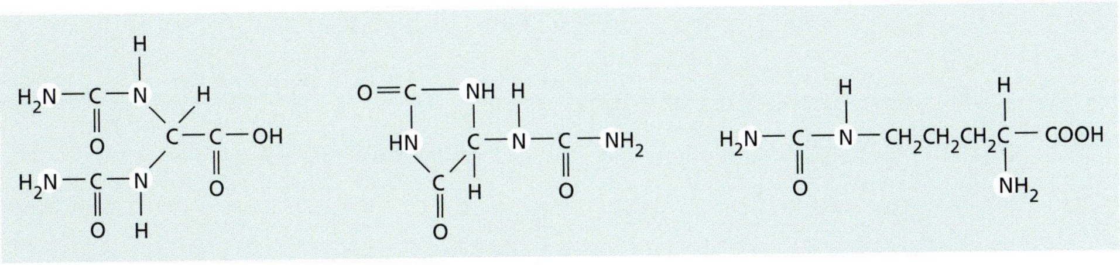

Allantoic acid Allantoin Citrulline

FIGURE 12.13 The major ureide compounds used to transport nitrogen from sites of fixation to sites where their deamination will provide nitrogen for amino acid and nucleoside synthesis.

ment) that contain sulfur range from the rhizobial Nod factors discussed in the previous section to antiseptic alliin in garlic and anticarcinogen sulforaphane in broccoli.

The versatility of sulfur derives in part from the property that it shares with nitrogen: *multiple stable oxidation states*. In this section we discuss the enzymatic steps that mediate sulfur assimilation, and the biochemical reactions that catalyze the reduction of sulfate into the two sulfur-containing amino acids, cysteine and methionine.

Sulfate Is the Absorbed Form of Sulfur in Plants

Most of the sulfur in higher-plant cells derives from sulfate (SO_4^{2-}) absorbed via an H^+–SO_4^{2-} symporter (see Chapter 6) from the soil solution. Sulfate in the soil comes predominantly from the weathering of parent rock material. Industrialization, however, adds an additional source of sulfate: atmospheric pollution. The burning of fossil fuels releases several gaseous forms of sulfur, including sulfur dioxide (SO_2) and hydrogen sulfide (H_2S), which find their way to the soil in rain.

When dissolved in water, SO_2 is hydrolyzed to become sulfuric acid (H_2SO_4), a strong acid, which is the major source of acid rain. Plants can also metabolize sulfur dioxide taken up in the gaseous form through their stomata. Nonetheless, prolonged exposure (more than 8 hours) to high atmospheric concentrations (greater than 0.3 ppm) of SO_2 causes extensive tissue damage because of the formation of sulfuric acid.

Sulfate Assimilation Requires the Reduction of Sulfate to Cysteine

The first step in the synthesis of sulfur-containing organic compounds is the reduction of sulfate to the amino acid cysteine (Figure 12.14). Sulfate is very stable and thus needs to be activated before any subsequent reactions may proceed. Activation begins with the reaction between sulfate and ATP to form 5′-adenylylsulfate (which is sometimes referred to as adenosine-5′-phosphosulfate and thus is abbreviated APS) and pyrophosphate (PP_i) (see Figure 12.14):

$$SO_4^{2-} + \text{Mg-ATP} \rightarrow \text{APS} + PP_i \qquad (12.10)$$

The enzyme that catalyzes this reaction, ATP sulfurylase, has two forms: The major one is found in plastids, and a minor one is found in the cytoplasm (Leustek et al. 2000). The activation reaction is energetically unfavorable. To drive this reaction forward, the products APS and PP_i must be converted immediately to other compounds. PP_i is hydrolyzed to inorganic phosphate (P_i) by inorganic pyrophosphatase according to the following reaction:

$$PP_i + H_2O \rightarrow 2\ P_i \qquad (12.11)$$

The other product, APS, is rapidly reduced or sulfated. Reduction is the dominant pathway (Leustek et al. 2000).

The reduction of APS is a multistep process that occurs exclusively in the plastids. First, APS reductase transfers two electrons apparently from reduced glutathione (GSH) to produce sulfite (SO_3^{2-}):

$$\text{APS} + 2\ \text{GSH} \rightarrow SO_3^{2-} + 2\ H^+ + \text{GSSG} + \text{AMP} \qquad (12.12)$$

where GSSG stands for oxidized glutathione. (The *SH* in GSH and the *SS* in GSSG stand for S—H and S—S bonds, respectively.)

Second, sulfite reductase transfers six electrons from ferredoxin (Fd_{red}) to produce sulfide (S^{2-}):

$$SO_3^{2-} + 6\ Fd_{red} \rightarrow S^{2-} + 6\ Fd_{ox} \qquad (12.13)$$

The resultant sulfide then reacts with *O*-acetylserine (OAS) to form cysteine and acetate. The *O*-acetylserine that reacts with S^{2-} is formed in a reaction catalyzed by serine acetyltransferase:

$$\text{Serine} + \text{acetyl-CoA} \rightarrow \text{OAS} + \text{CoA} \qquad (12.14)$$

The reaction that produces cysteine and acetate is catalyzed by OAS thiol-lyase:

$$\text{OAS} + S^{2-} \rightarrow \text{cysteine} + \text{acetate} \qquad (12.15)$$

The sulfation of APS, localized in the cytosol, is the alternative pathway. First, APS kinase catalyzes a reaction of APS with ATP to form 3′-phosphoadenosine-5′-phosphosulfate (PAPS).

$$\text{APS} + \text{ATP} \rightarrow \text{PAPS} + \text{ADP} \qquad (12.16)$$

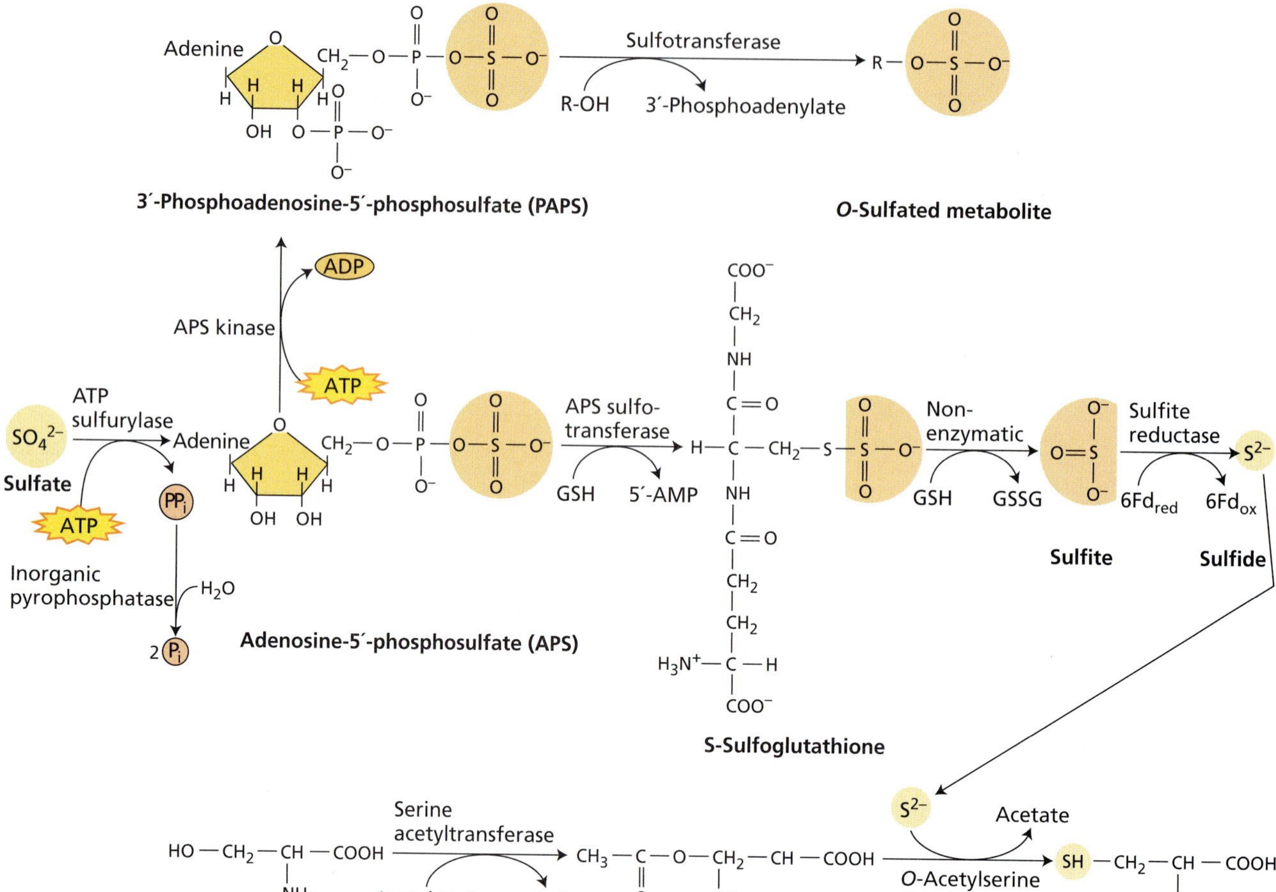

FIGURE 12.14 Structure and pathways of compounds involved in sulfur assimilation. The enzyme ATP sulfurylase cleaves pyrophosphate from ATP and replaces it with sulfate. Sulfide is produced from APS through reactions involving reduction by glutathione and ferredoxin. The sulfide or thiosulfide reacts with *O*-acetylserine to form cysteine. Fd, ferredoxin; GSH, glutathione, reduced; GSSG, glutathione, oxidized.

Sulfotransferases then may transfer the sulfate group from PAPS to various compounds, including choline, brassinosteroids, flavonol, gallic acid glucoside, glucosinolates, peptides, and polysaccharides (Leustek and Saito 1999).

Sulfate Assimilation Occurs Mostly in Leaves

The reduction of sulfate to cysteine changes the oxidation number of sulfur from +6 to –4, thus entailing the transfer of 10 electrons. Glutathione, ferredoxin, NAD(P)H, or *O*-acetylserine may serve as electron donors at various steps of the pathway (see Figure 12.14).

Leaves are generally much more active than roots in sulfur assimilation, presumably because photosynthesis provides reduced ferredoxin and photorespiration generates serine that may stimulate the production of *O*-acetylserine (see Chapter 8). Sulfur assimilated in leaves is exported via the phloem to sites of protein synthesis (shoot and root apices, and fruits) mainly as glutathione (Bergmann and Rennenberg 1993):

Glycine
Cysteine
Glutamate

Reduced glutathione

Glutathione also acts as a signal that coordinates the absorption of sulfate by the roots and the assimilation of sulfate by the shoot.

Methionine Is Synthesized from Cysteine

Methionine, the other sulfur-containing amino acid found in proteins, is synthesized in plastids from cysteine (see **Web Topic 12.3** for further detail). After cysteine and methionine are synthesized, sulfur can be incorporated into proteins and a variety of other compounds, such as acetyl-CoA and *S*-adenosylmethionine. The latter compound is important in the synthesis of ethylene (see Chapter 22) and in reactions involving the transfer of methyl groups, as in lignin synthesis (see Chapter 13).

PHOSPHATE ASSIMILATION

Phosphate (HPO_4^{2-}) in the soil solution is readily absorbed by plant roots via an H^+–HPO_4^{2-} symporter (see Chapter 6) and incorporated into a variety of organic compounds, including sugar phosphates, phospholipids, and nucleotides. The main entry point of phosphate into assimilatory pathways occurs during the formation of ATP, the energy "currency" of the cell. In the overall reaction for this process, inorganic phosphate is added to the second phosphate group in adenosine diphosphate to form a phosphate ester bond.

In mitochondria, the energy for ATP synthesis derives from the oxidation of NADH by oxidative phosphorylation (see Chapter 11). ATP synthesis is also driven by light-dependent photophosphorylation in the chloroplasts (see Chapter 7). In addition to these reactions in mitochondria and chloroplasts, reactions in the cytosol also assimilate phosphate.

Glycolysis incorporates inorganic phosphate into 1,3-bisphosphoglyceric acid, forming a high-energy acyl phosphate group. This phosphate can be donated to ADP to form ATP in a substrate-level phosphorylation reaction (see Chapter 11). Once incorporated into ATP, the phosphate group may be transferred via many different reactions to form the various phosphorylated compounds found in higher-plant cells.

CATION ASSIMILATION

Cations taken up by plant cells form complexes with organic compounds in which the cation becomes bound to the complex by noncovalent bonds (for a discussion of noncovalent bonds, see Chapter 2 on the web site). Plants assimilate macronutrient cations such as potassium, magnesium, and calcium, as well as micronutrient cations such as copper, iron, manganese, cobalt, sodium, and zinc, in this manner. In this section we will describe coordination bonds and electrostatic bonds, which mediate the assimilation of several cations that plants require as nutrients, and the special requirements for the absorption of iron by roots and subsequent assimilation of iron within plants.

Cations Form Noncovalent Bonds with Carbon Compounds

The noncovalent bonds formed between cations and carbon compounds are of two types: coordination bonds and electrostatic bonds. In the formation of a coordination complex, several oxygen or nitrogen atoms of a carbon compound donate unshared electrons to form a bond with the cation nutrient. As a result, the positive charge on the cation is neutralized.

Coordination bonds typically form between polyvalent cations and carbon molecules—for example, complexes between copper and tartaric acid (Figure 12.15A) or magnesium and chlorophyll *a* (Figure 12.15B). The nutrients that are assimilated as coordination complexes include copper, zinc, iron, and magnesium. Calcium can also form coordination complexes with the polygalacturonic acid of cell walls (Figure 12.15C).

Electrostatic bonds form because of the attraction of a positively charged cation for a negatively charged group such as carboxylate ($—COO^-$) on a carbon compound. Unlike the situation in coordination bonds, the cation in an electrostatic bond retains its positive charge. Monovalent cations such as potassium (K^+) can form electrostatic bonds with the carboxylic groups of many organic acids (Figure 12.16A). Nonetheless, much of the potassium that is accumulated by plant cells and functions in osmotic regulation and enzyme activation remains in the cytosol and the vacuole as the free ion. Divalent ions such as calcium form electrostatic bonds with pectates (Figure 12.16B) and the carboxylic groups of polygalacturonic acid (see Chapter 15).

In general, cations such as magnesium (Mg^{2+}) and calcium (Ca^{2+}) are assimilated by the formation of both coordination complexes and electrostatic bonds with amino acids, phospholipids, and other negatively charged molecules.

Roots Modify the Rhizosphere to Acquire Iron

Iron is important in iron–sulfur proteins (see Chapter 7) and as a catalyst in enzyme-mediated redox reactions (see Chapter 5), such as those of nitrogen metabolism discussed earlier. Plants obtain iron from the soil, where it is present primarily as ferric iron (Fe^{3+}) in oxides such as $Fe(OH)^{2+}$, $Fe(OH)_3$, and $Fe(OH)_4^-$. At neutral pH, ferric iron is highly insoluble. To absorb sufficient amounts of iron from the soil solution, roots have developed several mechanisms that increase iron solubility and thus its availability. These mechanisms include:

- Soil acidification that increases the solubility of ferric iron.
- Reduction of ferric iron to the more soluble ferrous form (Fe^{2+}).
- Release of compounds that form stable, soluble complexes with iron (Marschner 1995). Recall from Chapter 5 that such compounds are called iron chelators (see Figure 5.2).

Roots generally acidify the soil around them. They extrude protons during the absorption and assimilation of

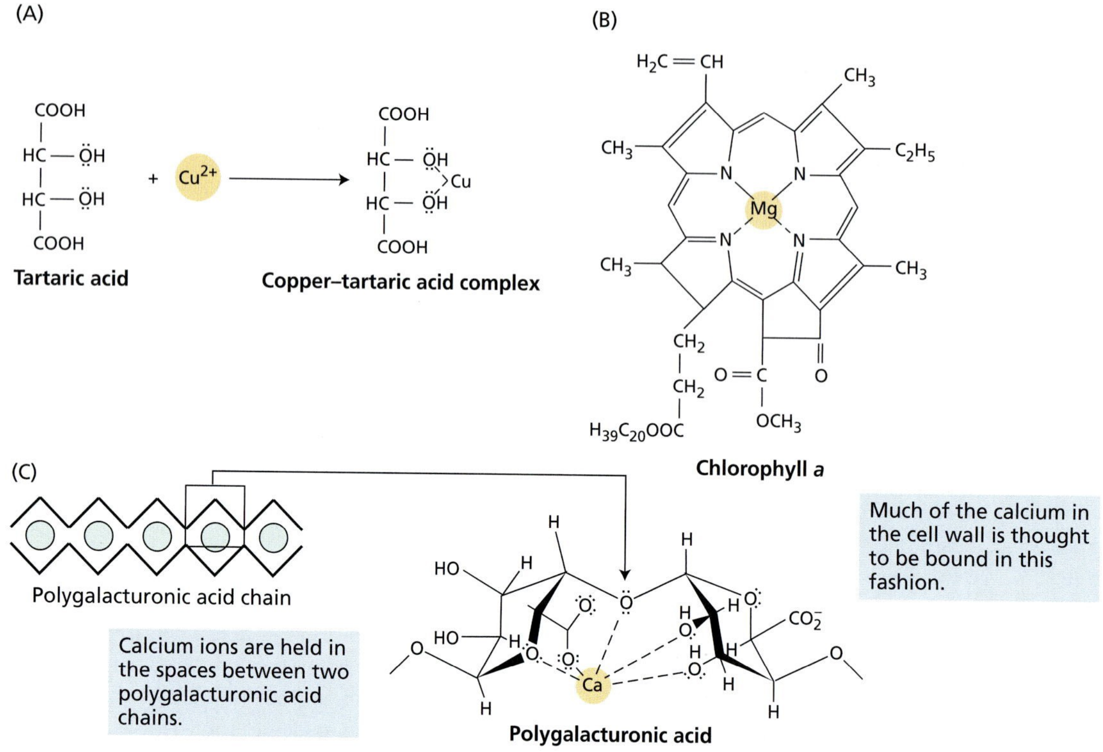

FIGURE 12.15 Examples of coordination complexes. Coordination complexes form when oxygen or nitrogen atoms of a carbon compound donate unshared electron pairs (represented by dots) to form a bond with a cation. (A) Copper ions share electrons with the hydroxyl oxygens of tartaric acid. (B) Magnesium ions share electrons with nitrogen atoms in chlorophyll *a*. Dashed lines represent a coordination bond between unshared electrons from the nitrogen atoms and the magnesium cation. (C) The "egg box" model of the interaction of polygalacturonic acid, a major constituent of pectins in cell walls, and calcium ions. At right is an enlargement of a single calcium ion forming a coordination complex with the hydroxyl oxygens of the galacturonic acid residues. (After Rees 1977.)

cations, particularly ammonium, and release organic acids such as malic acid and citric acid that enhance iron and phosphate availability (see Figure 5.4). Iron deficiencies stimulate the extrusion of protons by roots. In addition, plasma membranes in roots contain an enzyme, called *iron-chelating reductase*, that reduces ferric iron to the ferrous form, with NADH or NADPH serving as the electron donor. The activity of this enzyme increases under iron deprivation.

Several compounds secreted by roots form stable chelates with iron. Examples include malic acid, citric acid, phenolics, and piscidic acid. Grasses produce a special class

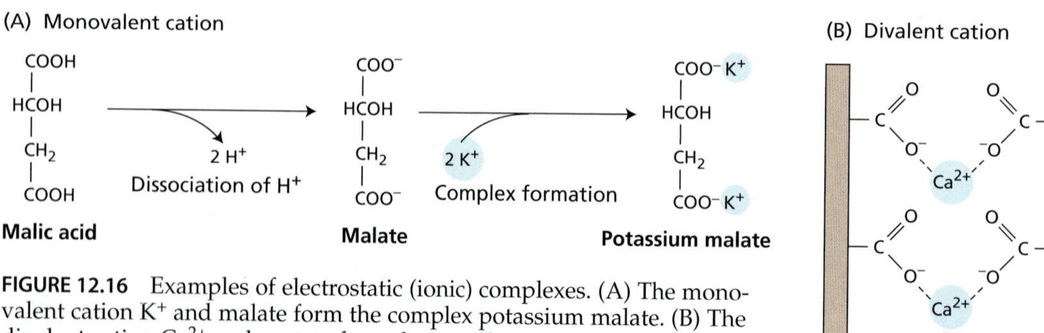

(B) Divalent cation

Calcium pectate

FIGURE 12.16 Examples of electrostatic (ionic) complexes. (A) The monovalent cation K^+ and malate form the complex potassium malate. (B) The divalent cation Ca^{2+} and pectate form the complex calcium pectate. Divalent cations can form cross-links between parallel strands that contain negatively charged carboxyl groups. Calcium cross-links play a structural role in the cell walls.

of iron chelators called *phytosiderophores*. Phytosiderophores are made of amino acids that are not found in proteins, such as mugineic acid, and form highly stable complexes with Fe^{3+}. Root cells of grasses have Fe^{3+}–phytosiderophore transport systems in their plasma membrane that bring the chelate into the cytoplasm. Under iron deficiency, grass roots release more phytosiderophores into the soil and increase the capacity of their Fe^{3+}–phytosiderophore transport system.

Iron Forms Complexes with Carbon and Phosphate

Once the roots absorb iron or an iron chelate, they oxidize it to a ferric form and translocate much of it to the leaves as an electrostatic complex with citrate.

Most of the iron in the plant is found in the heme molecule of cytochromes within the chloroplasts and mitochondria (see Chapter 7). An important assimilatory reaction for iron is its insertion into the porphyrin precursor of heme. This reaction is catalyzed by the enzyme ferrochelatase (Figure 12.17) (Jones 1983). In addition, iron–sulfur proteins of the electron transport chain (see Chapter 7) contain nonheme Fe covalently bound to the sulfur atoms of cysteine residues in the apoprotein. Iron is also found in Fe_2S_2 centers, which contain two irons (each complexed with the sulfur atoms of cysteine residues) and two inorganic sulfides.

Free iron (iron that is not complexed with carbon compounds) may interact with oxygen to form superoxide anions (O_2^-), which can damage membranes by degrading unsaturated lipid components. Plant cells may limit such damage by storing surplus iron in an iron–protein complex called **phytoferritin** (Bienfait and Van der Mark 1983). Phytoferritin consists of a protein shell with 24 identical subunits forming a hollow sphere that has a molecular mass of about 480 kDa. Within this sphere is a core of 5400 to 6200 iron atoms present as a ferric oxide–phosphate complex.

How iron is released from phytoferritin is uncertain, but breakdown of the protein shell appears to be involved. The level of free iron in plant cells regulates the de novo biosynthesis of phytoferritin (Lobreaux et al. 1992).

OXYGEN ASSIMILATION

Respiration accounts for the bulk (about 90%) of the oxygen (O_2) assimilated by plant cells (see Chapter 11). Another major pathway for the assimilation of O_2 into organic compounds involves the incorporation of O_2 from water (see reaction 1 in Table 8.1). A small proportion of oxygen can be directly assimilated into organic compounds in the process of *oxygen fixation*.

In oxygen fixation, molecular oxygen is added directly to an organic compound in reactions carried out by enzymes known as *oxygenases*. Recall from Chapter 8 that oxygen is directly incorporated into an organic compound during photorespiration in a reaction that involves the oxygenase activity of ribulose-1,5-bisphosphate carboxylase/oxygenase (rubisco), the enzyme of CO_2 fixation (Ogren 1984). The first stable product that contains oxygen originating from molecular oxygen is 2-phosphoglycolate.

In general, oxygenases are classified as dioxygenases or monooxygenases, according to the number of atoms of oxygen that are transferred to a carbon compound in the catalyzed reaction. In **dioxygenase** reactions, both oxygen atoms are incorporated into one or two carbon compounds (Figure 12.18A and B). Examples of dioxygenases in plant cells are lipoxygenase, which catalyzes the addition of two atoms of oxygen to unsaturated fatty acids (see Figure 12.18A), and prolyl hydroxylase, the enzyme that converts proline to the less common amino acid hydroxyproline (see Figure 12.18B).

Hydroxyproline is an important component of the cell wall protein extensin (see Chapter 15). The synthesis of hydroxyproline from proline differs from the synthesis of all other amino acids in that the reaction occurs after the proline has been incorporated into protein and is therefore a posttranslational modification reaction. Prolyl hydroxylase is localized in the endoplasmic reticulum, suggesting that most proteins containing hydroxyproline are found in the secretory pathway.

Monooxygenases add one of the atoms in molecular oxygen to a carbon compound; the other oxygen atom is converted into water. Monooxygenases are sometimes referred to as *mixed-function oxidases* because of their ability to catalyze simultaneously both the oxygenation reaction and the oxidase reaction (reduction of oxygen to water). The monooxygenase reaction also requires a reduced substrate (NADH or NADPH) as an electron donor, according to the following equation:

$$A + O_2 + BH_2 \rightarrow AO + H_2O + B$$

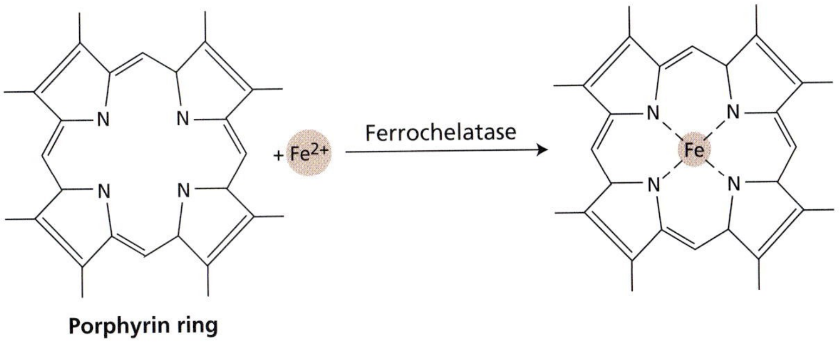

FIGURE 12.17 The ferrochelatase reaction. The enzyme ferrochelatase catalyzes the insertion of iron into the porphyrin ring to form a coordination complex. See Figure 7.37 for illustration of the biosynthesis of the porphyrin ring.

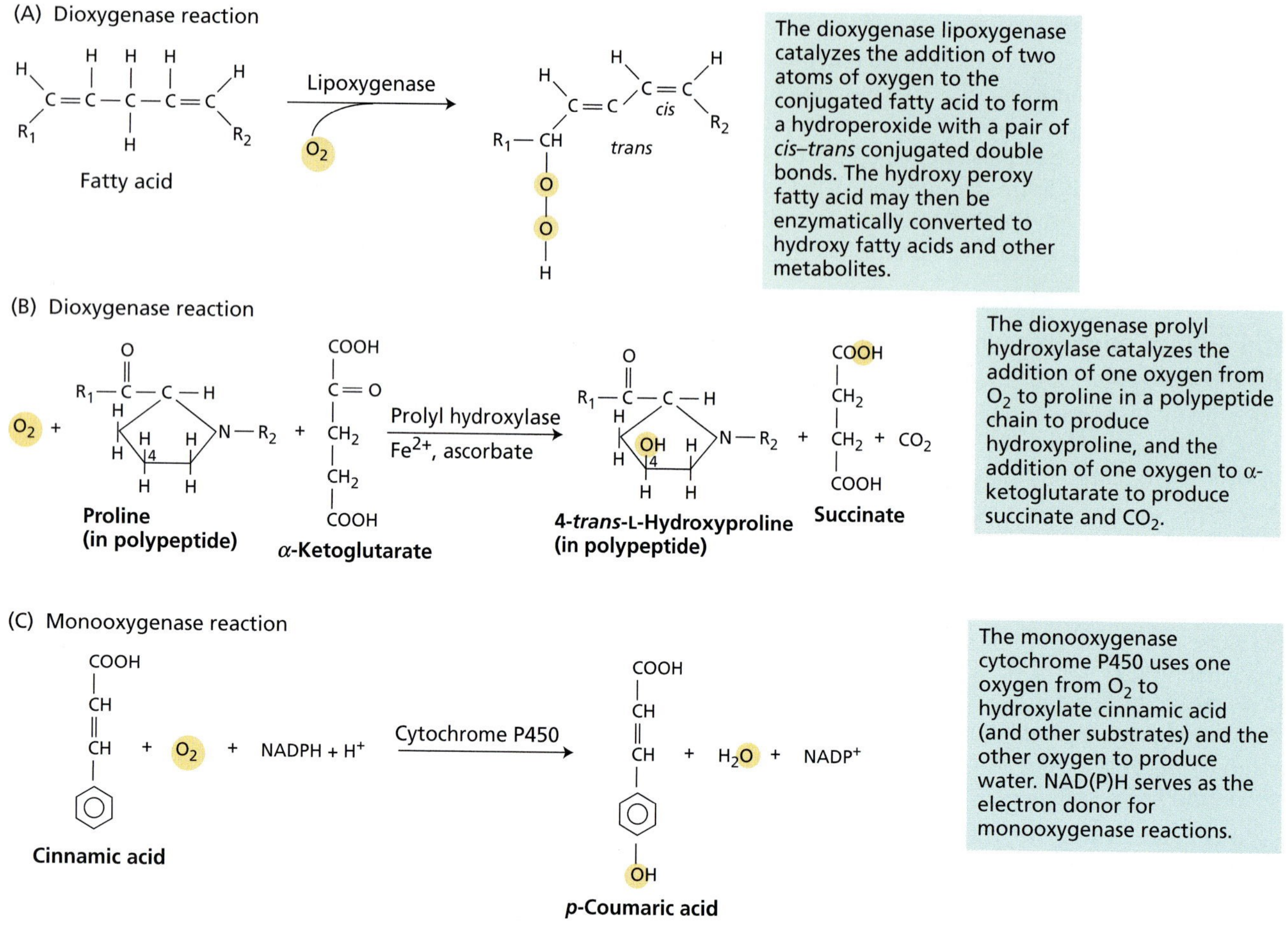

(A) Dioxygenase reaction

The dioxygenase lipoxygenase catalyzes the addition of two atoms of oxygen to the conjugated fatty acid to form a hydroperoxide with a pair of *cis–trans* conjugated double bonds. The hydroxy peroxy fatty acid may then be enzymatically converted to hydroxy fatty acids and other metabolites.

(B) Dioxygenase reaction

The dioxygenase prolyl hydroxylase catalyzes the addition of one oxygen from O_2 to proline in a polypeptide chain to produce hydroxyproline, and the addition of one oxygen to α-ketoglutarate to produce succinate and CO_2.

(C) Monooxygenase reaction

The monooxygenase cytochrome P450 uses one oxygen from O_2 to hydroxylate cinnamic acid (and other substrates) and the other oxygen to produce water. NAD(P)H serves as the electron donor for monooxygenase reactions.

FIGURE 12.18 Examples of the two types of oxygenase reactions in cells of higher plants.

where A represents an organic compound and B represents the electron donor.

An important monooxygenase in plants is the family of heme proteins collectively called cytochrome P450, which catalyzes the hydroxylation of cinnamic acid to *p*-coumaric acid (Figure 12.18C). In monooxygenases, the oxygen is first activated by being combined with the iron atom of the heme group; NADPH serves as the electron donor. The mixed-function oxidase system is localized on the endoplasmic reticulum and is capable of oxidizing a variety of substrates, including mono- and diterpenes and fatty acids.

THE ENERGETICS OF NUTRIENT ASSIMILATION

Nutrient assimilation generally requires large amounts of energy to convert stable, low-energy inorganic compounds into high-energy organic compounds. For example, the reduction of nitrate to nitrite and then to ammonium requires the transfer of about ten electrons and accounts for about 25% of the total energy expenditures in both roots and shoots (Bloom 1997). Consequently, a plant may use one-fourth of its energy to assimilate nitrogen, a constituent that accounts for less than 2% of the total dry weight of the plant.

Many of these assimilatory reactions occur in the stroma of the chloroplast, where they have ready access to powerful reducing agents such as NADPH, thioredoxin, and ferredoxin generated during photosynthetic electron transport. This process—coupling nutrient assimilation to photosynthetic electron transport—is called **photoassimilation** (Figure 12.19).

Photoassimilation and the Calvin cycle occur in the same compartment but only when photosynthetic electron transport generates reductant in excess of the needs of the Calvin cycle (e.g., under conditions of high light and low CO_2), does photoassimilation proceed (Robinson 1988). High levels of CO_2 inhibit photoassimilation (Figure 12.20 see **Web Essay 12.1**). As a result, C_4 plants (see Chapter 8) conduct the majority of their photoassimilation in mesophyll cells, where the CO_2 concentrations are lower (Becker et al. 1993).

The mechanisms that regulate the partitioning of reductant between the Calvin cycle and photoassimilation warrant investigation because atmospheric levels of CO_2 are

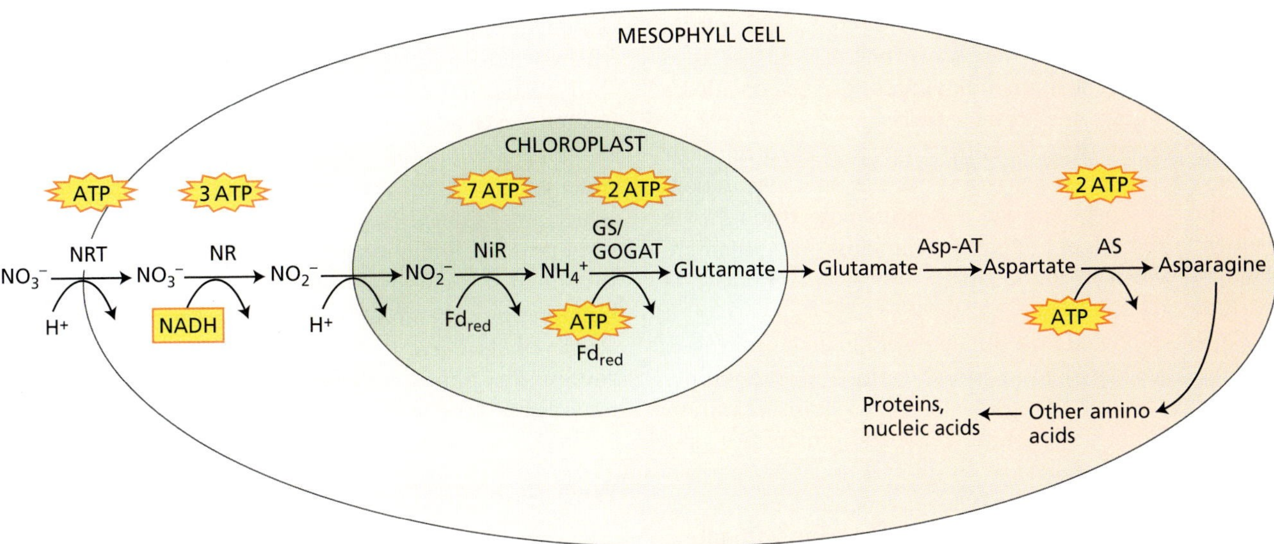

FIGURE 12.19 Summary of the processes involved in the assimilation of mineral nitrogen in the leaf. Nitrate translocated from the roots through the xylem is absorbed by a mesophyll cell via one of the nitrate–proton symporters (NRT) into the cytoplasm. There it is reduced to nitrite via nitrate reductase (NR). Nitrite is translocated into the stroma of the chloroplast along with a proton. In the stroma, nitrite is reduced to ammonium via nitrite reductase (NiR) and this ammonium is converted into glutamate via the sequential action of glutamine synthetase (GS) and glutamate synthase (GOGAT). Once again in the cytoplasm, the glutamate is transaminated to aspartate via aspartate aminotransferase (Asp-AT). Finally, asparagine synthetase (AS) converts aspartate into asparagine. The approximate amounts of ATP equivalents are given above each reaction.

expected to double during the next century (see Chapter 9), so this phenomenon may affect plant–nutrient relations.

SUMMARY

Nutrient assimilation is the process by which nutrients acquired by plants are incorporated into the carbon constituents necessary for growth and development. These processes often involve chemical reactions that are highly energy intensive and thus may depend directly on reductant generated through photosynthesis.

For nitrogen, assimilation is but one in a series of steps that constitute the nitrogen cycle. The nitrogen cycle encompasses the various states of nitrogen in the biosphere and their interconversions. The principal sources of nitrogen available to plants are nitrate (NO_3^-) and ammonium (NH_4^+).

The nitrate absorbed by roots is assimilated in either roots or shoots, depending on nitrate availability and plant

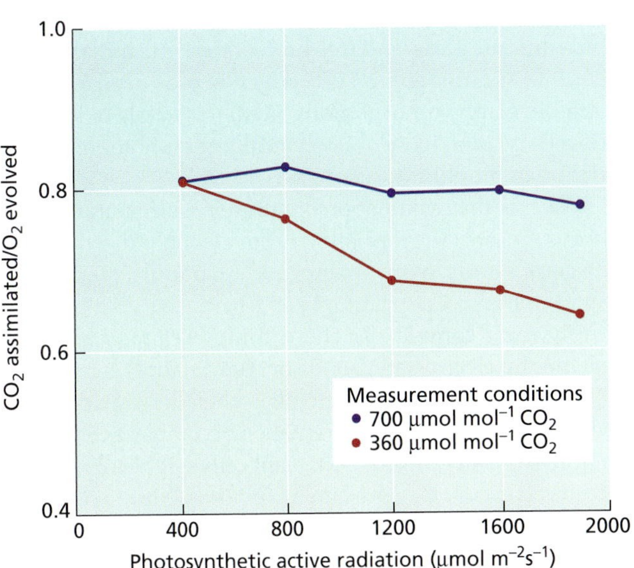

FIGURE 12.20 The assimilatory quotient (AQ = CO_2 assimilated/O_2 evolved) of wheat seedlings as a function of light level (photosynthetic active radiation). Nitrate photoassimilation is directly related to assimilatory quotient because transfer of electrons to nitrate and nitrite during photoassimilation increases O_2 evolution from the light-dependent reactions of photosynthesis, while CO_2 assimilation by the light-independent reactions continues at similar rates. Therefore, plants that are photoassimilating nitrate exhibit a lower AQ. In measurements carried out at ambient, 360 μmol mol^{-1} CO_2 concentrations (red trace), the AQ decreased as a function of incident radiation, indicating that photoassimilation rates increased. At elevated (700 μmol mol^{-1} CO_2, blue trace) the AQ remains constant at all light levels used, indicating that the CO_2-fixing reactions are competing for reductant, and inhibit photoassimilation. (After Bloom et al. 2002.)

species. In nitrate assimilation, nitrate is reduced to nitrite (NO_2^-) in the cytosol via the enzyme nitrate reductase; then nitrite is reduced to ammonium in root plastids or chloroplasts via the enzyme nitrite reductase.

Ammonium, derived either from root absorption or generated through nitrate assimilation or photorespiration, is converted to glutamine and glutamate through the sequential actions of glutamine synthetase and glutamate synthase, which are located in the cytosol and root plastids or chloroplasts.

Once assimilated into glutamine or glutamate, nitrogen may be transferred to many other organic compounds through various reactions, including the transamination reactions. Interconversion between glutamine and asparagine by asparagine synthetase balances carbon metabolism and nitrogen metabolism within a plant.

Many plants form a symbiotic relationship with nitrogen-fixing bacteria that contain an enzyme complex, nitrogenase, that can reduce atmospheric nitrogen to ammonia. Legumes and actinorhizal plants form associations with rhizobia and *Frankia*, respectively. These associations result from a finely tuned interaction between the symbiont and host plant that involves the recognition of specific signals, the induction of a specialized developmental program within the plant, the uptake of the bacteria by the plant, and the development of nodules, unique organs that house the bacteria within plant cells. Some nitrogen-fixing prokaryotic microorganisms do not form symbiotic relationships with higher plants but benefit plants by enriching the nitrogen content of the soil.

Like nitrate, sulfate (SO_4^{2-}) must be reduced by assimilation. In sulfate reduction, an activated form of sulfate called 5′-adenylylsulfate (APS) forms. Sulfide (S^{2-}), the end product of sulfate reduction, does not accumulate in plant cells, but is instead rapidly incorporated into the amino acids cysteine and methionine.

Phosphate (HPO_4^{2-}) is present in a variety of compounds found in plant cells, including sugar phosphates, lipids, nucleic acids, and free nucleotides. The initial product of its assimilation is ATP, which is produced by substrate-level phosphorylations in the cytosol, oxidative phosphorylation in the mitochondria, and photophosphorylation in the chloroplasts.

Whereas the assimilation of nitrogen, sulfur, and phosphorus requires the formation of covalent bonds with carbon compounds, many macro- and micronutrient cations (e.g., K^+, Mg^{2+}, Ca^{2+}, Cu^{2+}, Fe^{3+}, Mn^{2+}, Co^{2+}, Na^+, Zn^{2+}) simply form complexes. These complexes may be held together by electrostatic bonds or coordination bonds.

Iron assimilation may involve chelation, oxidation–reduction reactions, and the formation of complexes. In order to store large amounts of iron, plant cells synthesize phytoferritin, an iron storage protein. An important function of iron in plant cells is to act as a redox component in the active site of enzymes, often as an iron–porphyrin complex. Iron is inserted into a porphyrin group in the ferrochelatase reaction.

In addition to being utilized in respiration, molecular oxygen can be assimilated in the process of oxygen fixation, the direct addition of oxygen to organic compounds. This process is catalyzed by enzymes known as oxygenases, which are classified as monooxygenases or dioxygenases.

Nutrient assimilation requires large amounts of energy to convert stable, low-energy, inorganic compounds into high-energy organic compounds. A plant may use one-fourth of its energy to assimilate nitrogen. Plants use energy from photosynthesis to assimilate inorganic compounds in a process called photoassimilation.

Web Material

Web Topics

12.1 Development of a Root Nodule

Nodule primordia form opposite to the protoxylem poles of the root vascular bundles.

12.2 Measurement of Nitrogen Fixation

Acetylene reduction is used as an indirect measurement of nitrogen reduction.

12.3 The Synthesis of Methionine

Methionine is synthesized in plastids from cysteine.

Web Essay

12.1 Elevated CO_2 and Nitrogen Photoassimilation

In leaves grown under high CO_2 concentrations, CO_2 inhibits nitrogen photoassimilation because it competes for reductant.

Chapter References

Becker, T. W., Perrot-Rechenmann, C., Suzuki, A., and Hirel, B. (1993) Subcellular and immunocytochemical localization of the enzymes involved in ammonia assimilation in mesophyll and bundle-sheath cells of maize leaves. *Planta* 191: 129–136.

Bergmann, L., and Rennenberg, H. (1993) Glutathione metabolism in plants. In *Sulfur Nutrition and Assimilation in Higher Plants. Regulatory, Agricultural and Environmental Aspects*, L. J. De Kok, I. Stulen, H. Rennenberg, C. Brunold, and W. E. Rauser, eds., SPB Acad. Pub., The Hague, Netherlands, pp. 109–123.

Bienfait, H. F., and Van der Mark, F. (1983) Phytoferritin and its role in iron metabolism. In *Metals and Micronutrients: Uptake and Utilization by Plants*, D. A. Robb and W. S. Pierpoint, eds., Academic Press, New York, pp. 111–123.

Bloom, A. J. (1997) Nitrogen as a limiting factor: Crop acquisition of ammonium and nitrate. In *Ecology in Agriculture*, L. E. Jackson, ed., Academic Press, San Diego, CA, pp. 145–172.

Bloom, A. J., Smart, D. R., Nguyen, D. T., and Searles, P. S. (2002) Nitrogen assimilation and growth of wheat under elevated carbon dioxide. *Proc. Natl. Acad. Sci. USA* 99: 1730–1735.

Bloom, A. J., Sukrapanna, S. S., and Warner, R. L. (1992) Root respiration associated with ammonium and nitrate absorption and assimilation by barley. *Plant Physiol.* 99: 1294–1301.

Buchanan, B., Gruissem, W., and Jones, R., eds. (2000) *Biochemistry and Molecular Biology of Plants.* American Society of Plant Physiologists. Rockville, MD.

Burris, R. H. (1976) Nitrogen fixation. In *Plant Biochemistry*, 3rd ed., J. Bonner and J. Varner, eds., Academic Press, New York, pp. 887–908.

Campbell, W. H. (1999) Nitrate reductase structure, function and regulation: Bridging the gap between biochemistry and physiology. *Annu. Rev. Plant Physiol. Plant Mol. Biol.* 50: 277–303.

Carlson, R. W., Forsberg, L. S., Price, N. P. J., Bhat, U. R., Kelly, T. M., and Raetz, C. R. H. (1995) The structure and biosynthesis of *Rhizobium leguminosarum* lipid A. In *Progress in Clinical and Biological Research*, Vol. 392: *Bacterial Endotoxins: Lipopolysaccharides from Genes to Therapy: Proceedings of the Third Conference of the International Endotoxin Society, held in Helsinki, Finland, on August 15-18, 1994*, J. Levin et al., eds., John Wiley and Sons, New York, pp. 25–31.

Denison, R. F., and Harter, B. L. (1995) Nitrate effects on nodule oxygen permeability and leghemoglobin. *Plant Physiol.* 107: 1355–1364.

Dixon, R. O. D., and Wheeler, C. T. (1986) *Nitrogen Fixation in Plants.* Chapman and Hall, New York.

Dong, Z., Canny, M. J., McCully, M. E., Roboredo, M. R., Cabadilla, C. F., Ortega, E., and Rodes, R. (1994) A nitrogen-fixing endophyte of sugarcane stems: A new role for the apoplast. *Plant Physiol.* 105: 1139–1147.

Etzler, M. E., Kalsi, G., Ewing, N. N., Roberts, N. J., Day, R. B., and Murphy, J. B. (1999) A nod factor binding lectin with apyrase activity from legume roots. *Proc. Natl. Acad. Sci. USA* 96: 5856–5861.

FAOSTAT. (2001) *Agricultural Data.* Food and Agricultural Organization of the United Nations, Rome.

Heidstra, R., and Bisseling, T. (1996) Nod factor-induced host responses and mechanisms of Nod factor perception. *New Phytol.* 133: 25–43.

Hell, R. (1997) Molecular physiology of plant sulfur metabolism. *Planta* 202: 138–148.

Heytler, P. G., Reddy, G. S., and Hardy, R. W. F. (1984) *In vivo* energetics of symbiotic nitrogen fixation in soybeans. In *Nitrogen Fixation and CO_2 Metabolism*, P. W. Ludden and I. E. Burris, eds., Elsevier, New York, pp. 283–292.

Jones, O. T. G. (1983) Ferrochelatase. In *Metals and Micronutrients: Uptake and Utilization by Plants*, D. A. Robb and W. S. Pierpoint, eds., Academic Press, New York, pp. 125–144.

Kaiser, W. M., Weiner, H., and Huber, S. C. (1999) Nitrate reductase in higher plants: A case study for transduction of environmental stimuli into control of catalytic activity. *Physiol. Plant.* 105: 385–390.

Kleinhofs, A., Warner, R. L., Lawrence, J. M., Melzer, J. M., Jeter, J. M., and Kudrna, D. A. (1989) Molecular genetics of nitrate reductase in barley. In *Molecular and Genetic Aspects of Nitrate Assimilation*, J. L. Wray and J. R. Kinghorn, eds., Oxford Science, New York, pp. 197–211.

Kuzma, M. M., Hunt, S., and Layzell, D. B. (1993) Role of oxygen in the limitation and inhibition of nitrogenase activity and respiration rate in individual soybean nodules. *Plant Physiol.* 101: 161–169.

Lam, H.-M., Coschigano, K. T., Oliveira, I. C., Melo-Oliveira, R., and Coruzzi, G. M. (1996) The molecular-genetics of nitrogen assimilation into amino acids in higher plants. *Annu. Rev. Plant Physiology Plant Mol. Biol.* 47: 569–593.

Lazarowitz, S. G., and Bisseling, T. (1997) Plant development from the cellular perspective: Integrating the signals (Cellular Integration of Signaling Pathways in Plant Development, Acquafredda de Maratea, Italy, May 20–30, 1997). *Plant Cell* 9: 1884–1900.

Lea, P. J., Blackwell, R. D., and Joy, K. W. (1992) Ammonia assimilation in higher plants. In *Nitrogen Metabolism of Plants* (Proceedings of the Phytochemical Society of Europe 33), K. Mengel and D. J. Pilbeam, eds., Clarendon, Oxford, pp. 153–186.

Leustek, T., and Saito, K. (1999) Sulfate transport and assimilation in plants. *Plant Physiol.* 120: 637–643.

Leustek, T., Martin, M. N., Bick, J.-A., and Davies, J. P. (2000) Pathways and regulation of sulfur metabolism revealed through molecular and genetic studies. *Annu. Rev. Plant Physiol. Plant Mol. Biol.* 51: 141–165.

Lobreaux, S., Massenet, O., and Briat, J. -F. (1992) Iron induces ferritin synthesis in maize plantlets. *Plant Mol. Biol.* 19: 563–575.

Ludwig, R. A., and de Vries, G. E. (1986) Biochemical physiology of *Rhizobium* dinitrogen fixation. In *Nitrogen Fixation*, Vol. 4: *Molecular Biology*, W. I. Broughton and S. Puhler, eds., Clarendon, Oxford, pp. 50–69.

Marschner, H. (1995) *Mineral Nutrition of Higher Plants*, 2nd ed. Academic Press, London.

Mendel, R. R., and Stallmeyer, B. (1995) Molybdenum cofactor (nitrate reductase) biosynthesis in plants: First molecular analysis. In *Current Plant Science and Biotechnology in Agriculture*, Vol. 22: *Current Issues in Plant Molecular and Cellular Biology: Proceedings of the VIIIth International Congress on Plant Tissue and Cell Culture, Florence, Italy, 12–17 June, 1994*, M. Terzi, R. Cella and A. Falavigna, eds., Kluwer, Dordrecht, Netherlands, pp. 577–582.

Mylona, P., Pawlowski, K., and Bisseling, T. (1995) Symbiotic nitrogen fixation. *Plant Cell* 7: 869–885.

Oaks, A. (1994) Primary nitrogen assimilation in higher plants and its regulation. *Can. J. Bot.* 72: 739–750.

Ogren, W. L. (1984) Photorespiration: Pathways, regulation, and modification. *Annu. Rev. Plant Physiol.* 35: 415–442.

Pate, J. S. (1983) Patterns of nitrogen metabolism in higher plants and their ecological significance. In *Nitrogen as an Ecological Factor: The 22nd Symposium of the British Ecological Society, Oxford 1981*, J. A. Lee, S. McNeill, and I. H. Rorison, eds., Blackwell, Boston, pp. 225–255.

Pate, J. S., and Layzell, D. B. (1990) Energetics and biological costs of nitrogen assimilation. In *The Biochemistry of Plants*, Vol. 16: *Intermediary Nitrogen Metabolism*, B. J. Miflin and P. J. Lea, eds., Academic Press, San Diego, CA, pp. 1–42.

Phillips, D. A., and Kapulnik, Y. (1995) Plant isoflavonoids, pathogens and symbionts. *Trends Microbiol.* 3: 58–64.

Rees, D. A. (1977) *Polysaccharide Shapes.* Chapman and Hall, London.

Reis, V. M., Baldani, J. I., Baldani, V. L. D., and Dobereiner, J. (2000) Biological dinitrogen fixation in Gramineae and palm trees. *Crit. Rev. Plant Sci.* 19: 227–247.

Robinson, J. M. (1988) Spinach leaf chloroplast carbon dioxide and nitrite photoassimilations do not compete for photogenerated reductant: Manipulation of reductant levels by quantum flux density titrations. *Plant Physiol.* 88: 1373–1380.

Rolfe, B. G., and Gresshoff, P. M. (1988) Genetic analysis of legume nodule initiation. *Annu. Rev. Plant Physiol. Plant Mol. Biol.* 39: 297–320.

Schlesinger, W. H. (1997) *Biogeochemistry: An Analysis of Global Change*, 2nd ed. Academic Press, San Diego, CA.

Siegel, L. M., and Wilkerson, J. Q. (1989) Structure and function of spinach ferredoxin-nitrite reductase. In *Molecular and Genetic Aspects of Nitrate Assimilation*, J. L. Wray and J. R. Kinghorn, eds., Oxford Science, Oxford, pp. 263–283.

Sivasankar, S., and Oaks, A. (1996) Nitrate assimilation in higher plants—The effect of metabolites and light. *Plant Physiol. Biochem.* 34: 609–620.

Stokkermans, T. J. W., Ikeshita, S., Cohn, J., Carlson, R. W., Stacey, G., Ogawa, T., and Peters, N. K. (1995) Structural requirements of synthetic and natural product lipo-chitin oligosaccharides for induction of nodule primordia on *Glycine soja*. *Plant Physiol.* 108: 1587–1595.

Timmers, A. C. J., Auriac, M. –C., and Truchet, G. (1999) Refined analysis of early symbiotic steps of the Rhizobium-Medicago: Interaction in relation with microtubular cytoskeleton rearrangements. *Development* 126: 3617-3628

Vande Broek, A., and Vanderleyden, J. (1995) Review: Genetics of the *Azospirillum*-plant root association. *Crit. Rev. Plant Sci.* 14: 445–466.

van Rhijn, P., Goldberg, R. B., and Hirsch, A. M. (1998) *Lotus corniculatus* nodulation specificity is changed by the presence of a soybean lectin gene. *Plant Cell* 10: 1233–1249.

Warner, R. L., and Kleinhofs, A. (1992) Genetics and molecular biology of nitrate metabolism in higher plants. *Physiol. Plant.* 85: 245–252.

Wray, J. L. (1993) Molecular biology, genetics and regulation of nitrite reduction in higher plants. *Physiol. Plant.* 89: 607–612.

Chapter

13 Secondary Metabolites and Plant Defense

IN NATURAL HABITATS, plants are surrounded by an enormous number of potential enemies. Nearly all ecosystems contain a wide variety of bacteria, viruses, fungi, nematodes, mites, insects, mammals, and other herbivorous animals. By their nature, plants cannot avoid these herbivores and pathogens simply by moving away; they must protect themselves in other ways.

The cuticle (a waxy outer layer) and the periderm (secondary protective tissue), besides retarding water loss, provide barriers to bacterial and fungal entry. In addition, a group of plant compounds known as secondary metabolites defend plants against a variety of herbivores and pathogenic microbes. Secondary compounds may serve other important functions as well, such as structural support, as in the case of lignin, or pigments, as in the case of the anthocyanins.

In this chapter we will discuss some of the mechanisms by which plants protect themselves against both herbivory and pathogenic organisms. We will begin with a discussion of the three classes of compounds that provide surface protection to the plant: cutin, suberin, and waxes. Next we will describe the structures and biosynthetic pathways for the three major classes of secondary metabolites: terpenes, phenolics, and nitrogen-containing compounds. Finally, we will examine specific plant responses to pathogen attack, the genetic control of host–pathogen interactions, and cell signaling processes associated with infection.

CUTIN, WAXES, AND SUBERIN

All plant parts exposed to the atmosphere are coated with layers of lipid material that reduce water loss and help block the entry of pathogenic fungi and bacteria. The principal types of coatings are cutin, suberin, and waxes. Cutin is found on most aboveground parts; suberin is present on underground parts, woody stems, and healed wounds. Waxes are associated with both cutin and suberin.

Cutin, Waxes, and Suberin Are Made Up of Hydrophobic Compounds

Cutin is a macromolecule, a polymer consisting of many long-chain fatty acids that are attached to each other by ester linkages, creating a rigid three-dimensional network. Cutin is formed from 16:0 and 18:1 fatty acids[1] with hydroxyl or epoxide groups situated either in the middle of the chain or at the end opposite the carboxylic acid function (Figure 13.1A).

Cutin is a principal constituent of the **cuticle**, a multilayered secreted structure that coats the outer cell walls of the epidermis on the aerial parts of all herbaceous plants (Figure 13.2). The cuticle is composed of a top coating of wax, a thick middle layer containing cutin embedded in wax (the cuticle proper), and a lower layer formed of cutin and wax blended with the cell wall substances pectin, cellulose, and other carbohydrates (the cuticular layer). Recent research suggests that, in addition to cutin, the cuticle may contain a second lipid polymer, made up of long-chain hydrocarbons, that has been named *cutan* (Jeffree 1996).

Waxes are not macromolecules, but complex mixtures of long-chain acyl lipids that are extremely hydrophobic. The most common components of wax are straight-chain alkanes and alcohols of 25 to 35 carbon atoms (see Figure 13.1B). Long-chain aldehydes, ketones, esters, and free fatty acids are also found. The waxes of the cuticle are synthesized by epidermal cells. They leave the epidermal cells as droplets that pass through pores in the cell wall by an unknown mechanism. The top coating of cuticle wax often crystallizes in an intricate pattern of rods, tubes, or plates (Figure 13.3).

Suberin is a polymer whose structure is very poorly understood. Like cutin, suberin is formed from hydroxy or epoxy fatty acids joined by ester linkages. However, suberin differs from cutin in that it has dicarboxylic acids (see Figure 13.1C), more long-chain components, and a significant proportion of phenolic compounds as part of its structure.

(A) Hydroxy fatty acids that polymerize to make **cutin**:

$HOCH_2(CH_2)_{14}COOH$

$CH_3(CH_2)_8CH(OH)(CH_2)_5COOH$

(B) Common **wax** components:

Straight-chain alkanes	$CH_3(CH_2)_{27}CH_3$
	$CH_3(CH_2)_{29}CH_3$
Fatty acid ester	$CH_3(CH_2)_{22}C(=O)—O(CH_2)_{25}CH_3$
Long-chain fatty acid	$CH_3(CH_2)_{22}COOH$
Long-chain alcohol	$CH_3(CH_2)_{24}CH_2OH$

(C) Hydroxy fatty acids that polymerize along with other constituents to make **suberin**:

$HOCH_2(CH_2)_{14}COOH$

$HOOC(CH_2)_{14}COOH$ (a dicarboxylic acid)

FIGURE 13.1 Constituents of (A) cutin, (B) waxes, and (C) suberin.

(A)

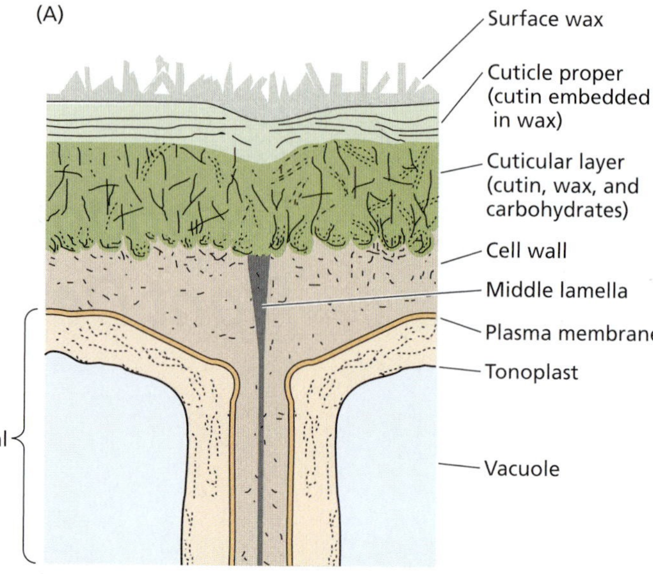

(B)

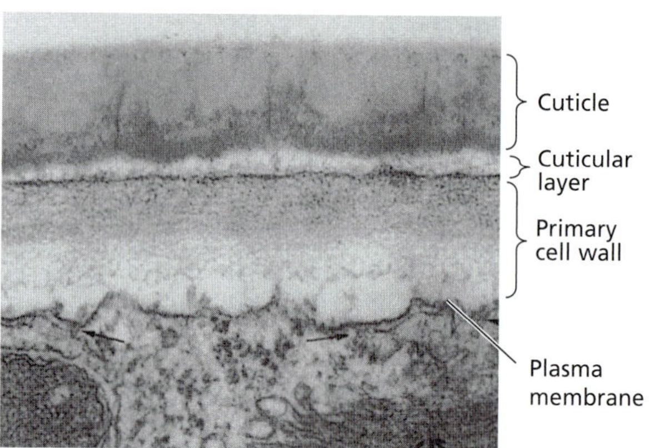

FIGURE 13.2 (A) Schematic drawing of the structure of the plant cuticle, the protective covering on the epidermis of leaves and young stems at the stage of full leaf expansion. (B) Electron micrograph of the cuticle of a glandular cell from a young leaf (*Lamium* sp.), showing the presence of the cuticle layers indicated in A, except for surface waxes, which are not visible. (51,000×) (A, after Jeffree 1996; B, from Gunning and Steer 1996.)

[1] Recall from Chapter 11 that the nomenclature for fatty acids is X:Y, where X is the number of carbon atoms and Y is the number of *cis* double bonds.

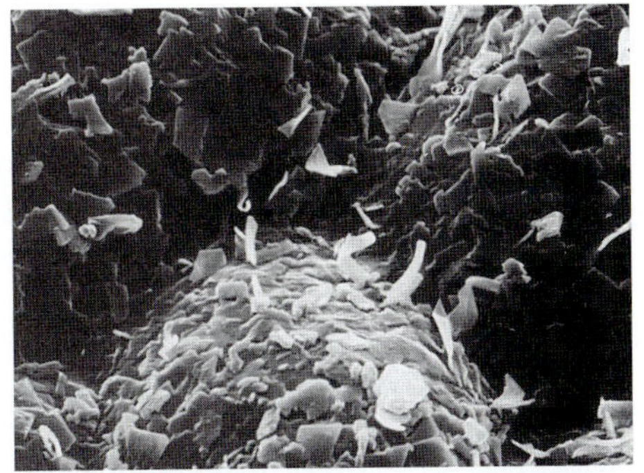

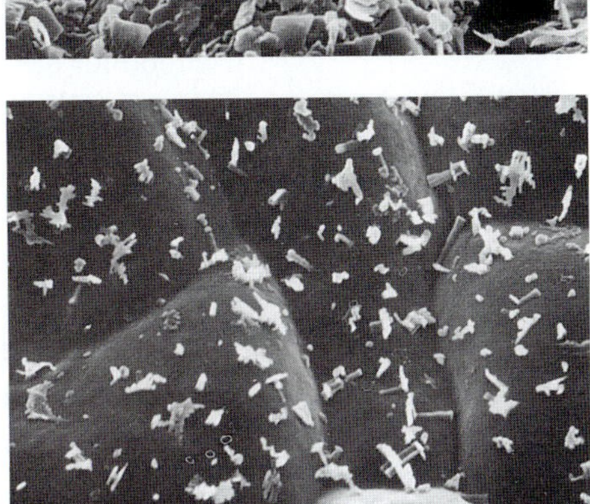

FIGURE 13.3 Surface wax deposits, which form the top layer of the cuticle, adopt different forms. These scanning electron micrographs show the leaf surfaces of two different lines of *Brassica oleracea*, which differ in wax crystal structure. (From Eigenbrode et al. 1991, courtesy of S. D. Eigenbrode, with permission from the Entomological Society of America.)

Suberin is a cell wall constituent found in many locations throughout the plant. We have already noted its presence in the Casparian strip of the root endodermis, which forms a barrier between the apoplast of the cortex and the stele (see Chapter 4). Suberin is a principal component of the outer cell walls of all underground organs and is associated with the cork cells of the **periderm**, the tissue that forms the outer bark of stems and roots during secondary growth of woody plants. Suberin also forms at sites of leaf abscission and in areas damaged by disease or wounding.

Cutin, Waxes, and Suberin Help Reduce Transpiration and Pathogen Invasion

Cutin, suberin, and their associated waxes form barriers between the plant and its environment that function to keep water in and pathogens out. The cuticle is very effective at limiting water loss from aerial parts of the plant but does not block transpiration completely because even with the stomata closed, some water is lost. The thickness of the cuticle varies with environmental conditions. Plant species native to arid areas typically have thicker cuticles than plants from moist habitats have, but plants from moist habitats often develop thick cuticles when grown under dry conditions.

The cuticle and suberized tissue are both important in excluding fungi and bacteria, although they do not appear to be as important in pathogen resistance as some of the other defenses we will discuss in this chapter. Many fungi penetrate directly through the plant surface by mechanical means. Others produce cutinase, an enzyme that hydrolyzes cutin and thus facilitates entry into the plant.

SECONDARY METABOLITES

Plants produce a large, diverse array of organic compounds that appear to have no direct function in growth and development. These substances are known as **secondary metabolites**, *secondary products*, or *natural products*. Secondary metabolites have no generally recognized, direct roles in the processes of photosynthesis, respiration, solute transport, translocation, protein synthesis, nutrient assimilation, differentiation, or the formation of carbohydrates, proteins, and lipids discussed elsewhere in this book.

Secondary metabolites also differ from primary metabolites (amino acids, nucleotides, sugars, acyl lipids) in having a restricted distribution in the plant kingdom. That is, particular secondary metabolites are often found in only one plant species or related group of species, whereas primary metabolites are found throughout the plant kingdom.

Secondary Metabolites Defend Plants against Herbivores and Pathogens

For many years the adaptive significance of most plant secondary metabolites was unknown. These compounds were thought to be simply functionless end products of metabolism, or metabolic wastes. Study of these substances was pioneered by organic chemists of the nineteenth and early twentieth centuries who were interested in these substances because of their importance as medicinal drugs, poisons, flavors, and industrial materials.

More recently, many secondary metabolites have been suggested to have important ecological functions in plants:

- They protect plants against being eaten by herbivores (herbivory) and against being infected by microbial pathogens.
- They serve as attractants for pollinators and seed-dispersing animals and as agents of plant–plant competition.

In the remainder of this chapter we will discuss the major types of plant secondary metabolites, their biosynthesis, and what is known about their functions in the plant, particularly their roles in defense.

Plant Defenses Are a Product of Evolution

We can begin by asking how plants came to have defenses. According to evolutionary biologists, plant defenses must have arisen through heritable mutations, natural selection, and evolutionary change. Random mutations in basic metabolic pathways led to the appearance of new compounds that happened to be toxic or deterrent to herbivores and pathogenic microbes.

As long as these compounds were not unduly toxic to the plants themselves and the metabolic cost of producing them was not excessive, they gave the plants that possessed them greater reproductive fitness than undefended plants had. Thus the defended plants left more descendants than undefended plants, and they passed their defensive traits on to the next generation.

Interestingly, the very defense compounds that increase the reproductive fitness of plants by warding off fungi, bacteria, and herbivores may also make them undesirable as food for humans. Many important crop plants have been artificially selected for producing relatively low levels of these compounds, which of course can make them more susceptible to insects and disease.

Secondary Metabolites Are Divided into Three Major Groups

Plant secondary metabolites can be divided into three chemically distinct groups: terpenes, phenolics, and nitrogen-containing compounds. Figure 13.4 shows in simpli-

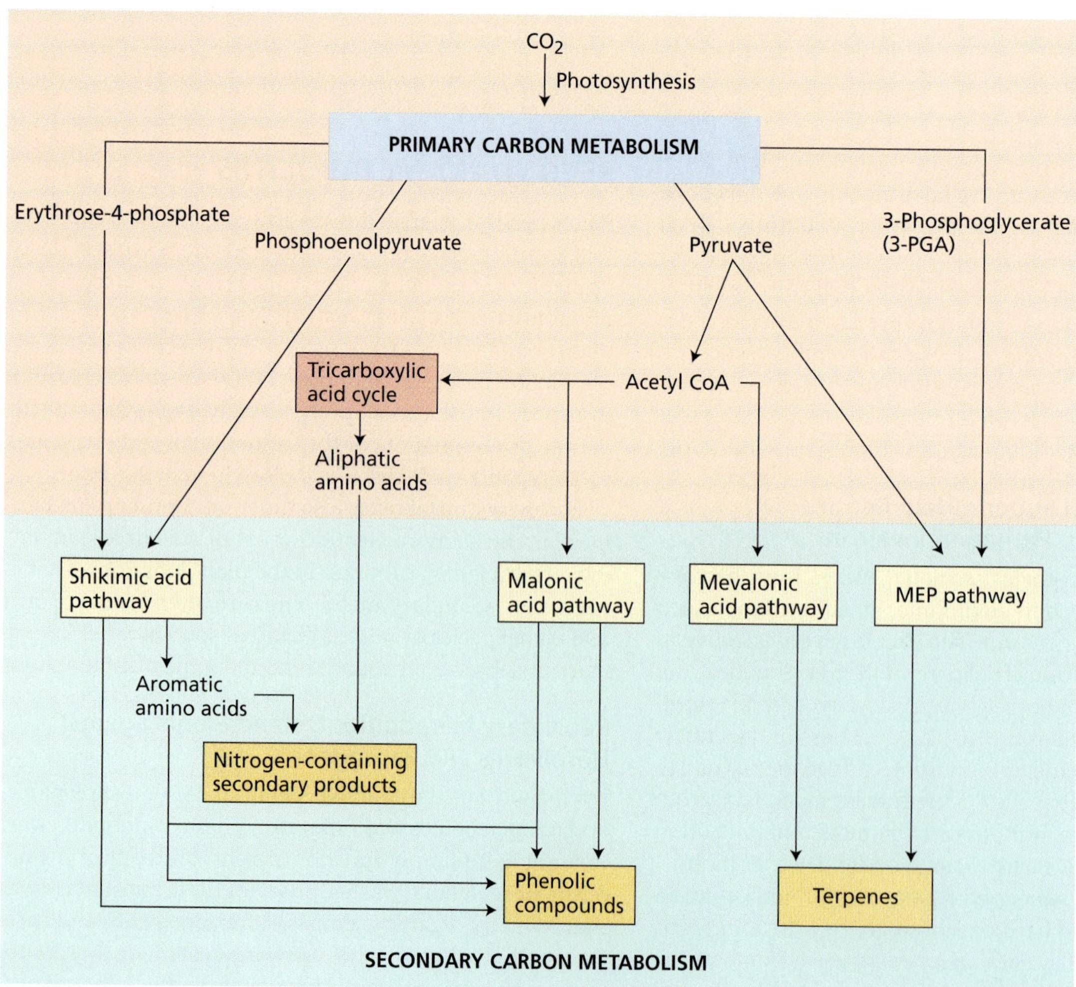

FIGURE 13.4 A simplified view of the major pathways of secondary-metabolite biosynthesis and their interrelationships with primary metabolism.

fied form the pathways involved in the biosynthesis of secondary metabolites and their interconnections with primary metabolism.

TERPENES

The **terpenes**, or *terpenoids*, constitute the largest class of secondary products. The diverse substances of this class are generally insoluble in water. They are biosynthesized from acetyl-CoA or glycolytic intermediates. After discussing the biosynthesis of terpenes, we'll examine how they act to repel herbivores and how some herbivores circumvent the toxic effects of terpenes.

Terpenes Are Formed by the Fusion of Five-Carbon Isoprene Units

All terpenes are derived from the union of five-carbon elements that have the branched carbon skeleton of isopentane:

$$(H_3C)_2CH-CH_2-CH_3$$

The basic structural elements of terpenes are sometimes called **isoprene units** because terpenes can decompose at high temperatures to give isoprene:

$$H_2C{=}C(CH_3)-CH{=}CH_2$$

Thus all terpenes are occasionally referred to as *isoprenoids*.

Terpenes are classified by the number of five-carbon units they contain, although extensive metabolic modifications can sometimes make it difficult to pick out the original five-carbon residues. Ten-carbon terpenes, which contain two C_5 units, are called *monoterpenes*; 15-carbon terpenes (three C_5 units) are *sesquiterpenes*; and 20-carbon terpenes (four C_5 units) are *diterpenes*. Larger terpenes include *triterpenes* (30 carbons), *tetraterpenes* (40 carbons), and *polyterpenoids* ($[C_5]_n$ carbons, where $n > 8$).

There Are Two Pathways for Terpene Biosynthesis

Terpenes are biosynthesized from primary metabolites in at least two different ways. In the well-studied **mevalonic acid pathway**, three molecules of acetyl-CoA are joined together stepwise to form mevalonic acid (Figure 13.5). This key six-carbon intermediate is then pyrophosphorylated, decarboxylated, and dehydrated to yield **isopentenyl diphosphate (IPP[2])**.

IPP is the activated five-carbon building block of terpenes. Recently, it was discovered that IPP also can be formed from intermediates of glycolysis or the photosynthetic carbon reduction cycle via a separate set of reactions called the **methylerythritol phosphate (MEP) pathway** that operates in chloroplasts and other plastids (Lichtenthaler 1999). Although all the details have not yet been elucidated, *glyceraldehyde-3-phosphate* and two carbon atoms derived from *pyruvate* appear to combine to generate an intermediate that is eventually converted to IPP.

Isopentenyl Diphosphate and Its Isomer Combine to Form Larger Terpenes

Isopentenyl diphosphate and its isomer, dimethylallyl diphosphate (DPP), are the activated five-carbon building blocks of terpene biosynthesis that join together to form larger molecules. First IPP and DPP react to give geranyl diphosphate (GPP), the 10-carbon precursor of nearly all the monoterpenes (see Figure 13.5). GPP can then link to another molecule of IPP to give the 15-carbon compound farnesyl diphosphate (FPP), the precursor of nearly all the sesquiterpenes. Addition of yet another molecule of IPP gives the 20-carbon compound geranylgeranyl diphosphate (GGPP), the precursor of the diterpenes. Finally, FPP and GGPP can dimerize to give the triterpenes (C_{30}) and the tetraterpenes (C_{40}), respectively.

Some Terpenes Have Roles in Growth and Development

Certain terpenes have a well-characterized function in plant growth or development and so can be considered primary rather than secondary metabolites. For example, the gibberellins, an important group of plant hormones, are diterpenes. Sterols are triterpene derivatives that are essential components of cell membranes, which they stabilize by interacting with phospholipids (see Chapter 11). The red, orange, and yellow carotenoids are tetraterpenes that function as accessory pigments in photosynthesis and protect photosynthetic tissues from photooxidation (see Chapter 7). The hormone abscisic acid (see Chapter 23) is a C_{15} terpene produced by degradation of a carotenoid precursor.

Long-chain polyterpene alcohols known as *dolichols* function as carriers of sugars in cell wall and glycoprotein synthesis (see Chapter 15). Terpene-derived side chains, such as the phytol side chain of chlorophyll (see Chapter 7), help anchor certain molecules in membranes. Thus various terpenes have important primary roles in plants. However, the vast majority of the different terpene structures produced by plants are secondary metabolites that are presumed to be involved in defense.

Terpenes Defend against Herbivores in Many Plants

Terpenes are toxins and feeding deterrents to many plant-feeding insects and mammals; thus they appear to play important defensive roles in the plant kingdom (Gershenzon and Croteau 1992). For example, the monoterpene esters called **pyrethroids** that occur in the leaves and flow-

[2] IPP is the abbreviation for isopentenyl *pyro*phosphate, an earlier name for this compound. The other pyrophosphorylated intermediates in the pathway are also now referred to as *di*phosphates.

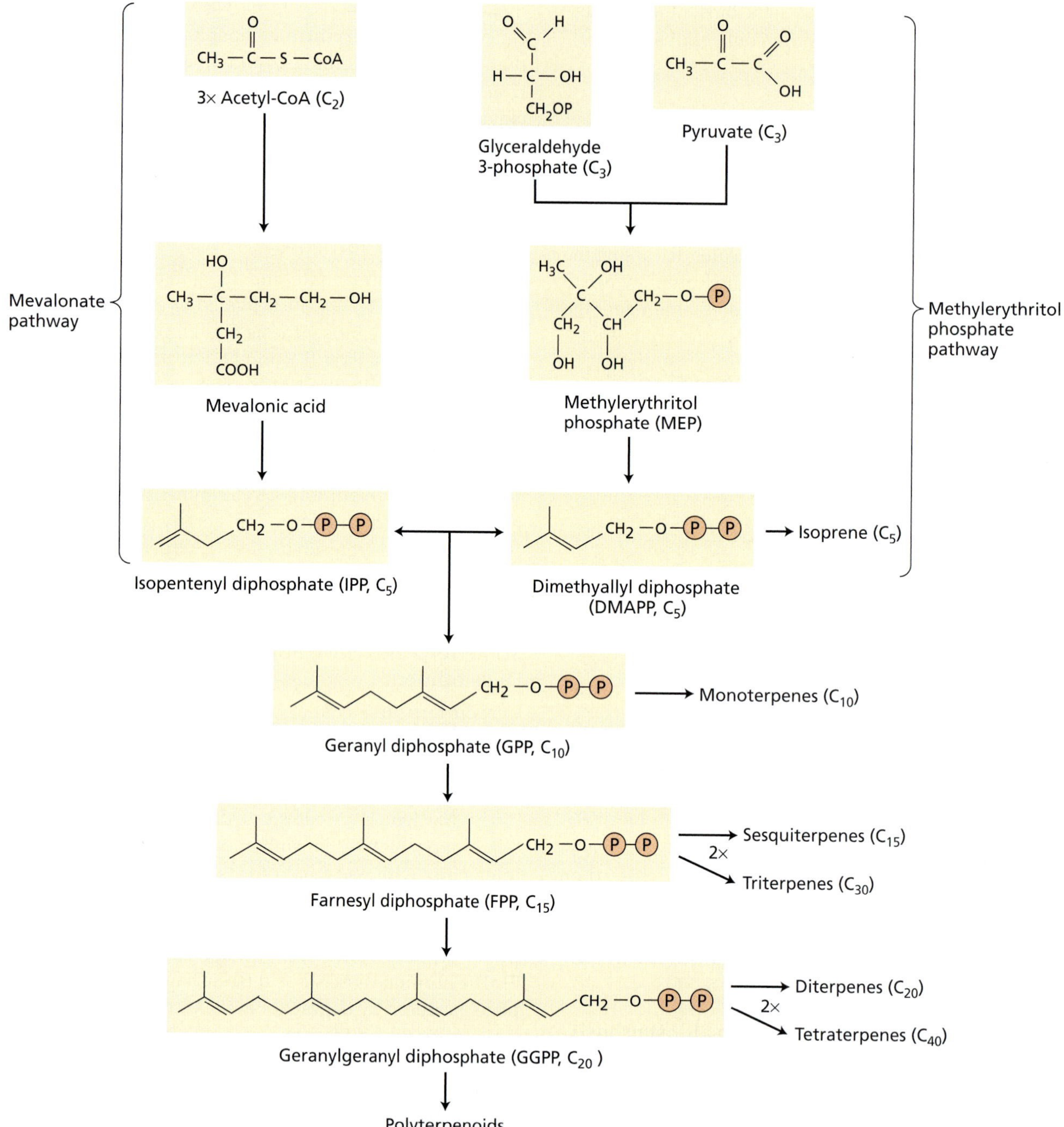

FIGURE 13.5 Outline of terpene biosynthesis. The basic 5-carbon units of terpenes are synthesized by two different pathways. The phosphorylated intermediates, IPP and DMAPP, are combined to make 10-carbon, 15-carbon and larger terpenes.

ers of *Chrysanthemum* species show very striking insecticidal activity. Both natural and synthetic pyrethroids are popular ingredients in commercial insecticides because of their low persistence in the environment and their negligible toxicity to mammals.

In conifers such as pine and fir, monoterpenes accumulate in resin ducts found in the needles, twigs, and trunk. These compounds are toxic to numerous insects, including bark beetles, which are serious pests of conifer species throughout the world. Many conifers respond to bark beetle infestation by producing additional quantities of monoterpenes (Trapp and Croteau 2001).

Many plants contain mixtures of volatile monoterpenes and sesquiterpenes, called **essential oils**, that lend a char-

FIGURE 13.6 Structures of limonene (A) and menthol (B). These two well-known monoterpenes serve as defenses against insects and other organisms that feed on these plants. (A, photo © Calvin Larsen/Photo Researchers, Inc.; B, photo © David Sieren/Visuals Unlimited.)

acteristic odor to their foliage. Peppermint, lemon, basil, and sage are examples of plants that contain essential oils. The chief monoterpene constituent of peppermint oil is menthol; that of lemon oil is limonene (Figure 13.6).

Essential oils have well-known insect repellent properties. They are frequently found in glandular hairs that project outward from the epidermis and serve to "advertise" the toxicity of the plant, repelling potential herbivores even before they take a trial bite. In the glandular hairs, the terpenes are stored in a modified extracellular space in the cell wall (Figure 13.7). Essential oils can be extracted from plants by steam distillation and are important commercially in flavoring foods and making perfumes.

Recent research has revealed an interesting twist on the role of volatile terpenes in plant protection. In corn, cotton, wild tobacco, and other species, certain monoterpenes and sesquiterpenes are produced and emitted only after insect feeding has already begun. These substances repel ovipositing herbivores and attract natural enemies, including predatory and parasitic insects, that kill plant-feeding insects and so help minimize further damage (Turlings et al. 1995; Kessler and Baldwin 2001). Thus, volatile terpenes are not only defenses in their own right, but also provide a way for plants to call for defensive help from other organisms. The ability of plants to attract natural enemies of plant-feeding insects shows promise as a new, ecologically sound means of pest control (see **Web Essay 13.1**).

Among the nonvolatile terpene antiherbivore compounds are the **limonoids**, a group of triterpenes (C_{30}) well known as bitter substances in citrus fruit. Perhaps the most powerful deterrent to insect feeding known is *azadirachtin* (Figure 13.8A), a complex limonoid from the neem tree (*Azadirachta indica*) of Africa and Asia. Azadirachtin is a feeding deterrent to some insects at doses as low as 50 parts per billion, and it exerts a variety of toxic effects (Aerts and Mordue 1997). It has considerable potential as a commercial insect control agent because of its low toxicity to mammals, and several preparations containing azadirachtin are now being marketed in North America and India.

The **phytoecdysones**, first isolated from the common fern, *Polypodium vulgare*, are a group of plant steroids that have the same basic structure as insect molting hormones (Figure 13.8B). Ingestion of phytoecdysones by insects disrupts molting and other developmental processes, often with lethal consequences.

Triterpenes that are active against vertebrate herbivores include cardenolides and saponins. **Cardenolides** are glycosides (compounds containing an attached sugar or sugars) that taste bitter and are extremely toxic to higher animals. In humans, they have dramatic effects on the heart muscle through their influence on Na^+/K^+-activated ATPases. In carefully regulated doses, they slow and strengthen the heartbeat. Cardenolides extracted from species of foxglove

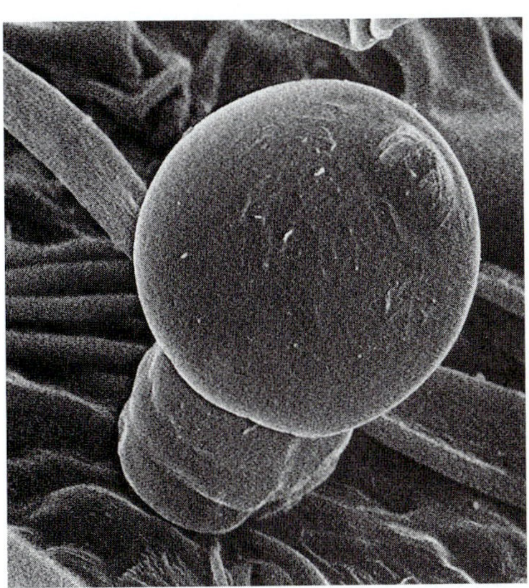

FIGURE 13.7 Monoterpenes and sesquiterpenes are commonly found in glandular hairs on the plant surface. This scanning electron micrograph shows a glandular hair on a young leaf of spring sunflower (*Balsamorhiza sagittata*). Terpenes are thought to be synthesized in the cells of the hair and are stored in the rounded cap at the top. This "cap" is an extracellular space that forms when the cuticle and a portion of the cell wall pull away from the remainder of the cell. (1105×) (© J. N. A. Lott/Biological Photo Service.)

FIGURE 13.8 Structure of two triterpenes, azadirachtin (A), and α-ecdysone (B), which serve as powerful feeding deterrents to insects. (A, photo © Inga Spence/Visuals Unlimited; B, photo ©Wally Eberhart/Visuals Unlimited.)

(A) Azadirachtin, a limonoid

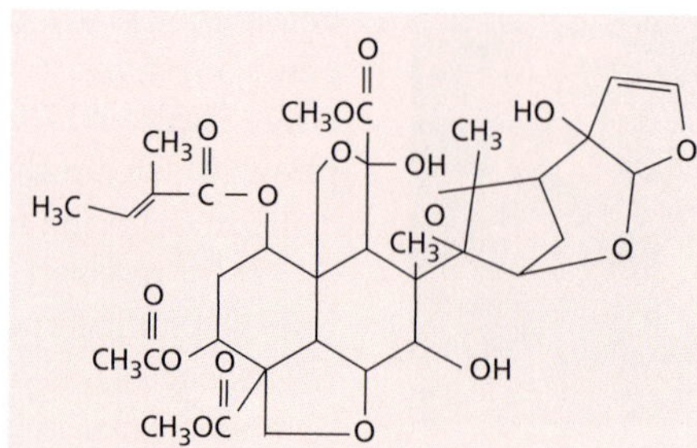

(B) α-Ecdysone, an insect molting hormone

(*Digitalis*) are prescribed to millions of patients for the treatment of heart disease (see **Web Topic 13.1**).

Saponins are steroid and triterpene glycosides, so named because of their soaplike properties. The presence of both lipid-soluble (the steroid or triterpene) and water-soluble (the sugar) elements in one molecule gives saponins detergent properties, and they form a soapy lather when shaken with water. The toxicity of saponins is thought to be a result of their ability to form complexes with sterols. Saponins may interfere with sterol uptake from the digestive system or disrupt cell membranes after being absorbed into the bloodstream.

PHENOLIC COMPOUNDS

Plants produce a large variety of secondary products that contain a phenol group—a hydroxyl functional group on an aromatic ring:

These substances are classified as phenolic compounds. Plant **phenolics** are a chemically heterogeneous group of nearly 10,000 individual compounds: Some are soluble only in organic solvents, some are water-soluble carboxylic acids and glycosides, and others are large, insoluble polymers.

In keeping with their chemical diversity, phenolics play a variety of roles in the plant. After giving a brief account of phenolic biosynthesis, we will discuss several principal groups of phenolic compounds and what is known about their roles in the plant. Many serve as defense compounds against herbivores and pathogens. Others function in mechanical support, in attracting pollinators and fruit dispersers, in absorbing harmful ultraviolet radiation, or in reducing the growth of nearby competing plants.

Phenylalanine Is an Intermediate in the Biosynthesis of Most Plant Phenolics

Plant phenolics are biosynthesized by several different routes and thus constitute a heterogeneous group from a metabolic point of view. Two basic pathways are involved: the shikimic acid pathway and the malonic acid pathway (Figure 13.9). The shikimic acid pathway participates in the biosynthesis of most plant phenolics. The malonic acid pathway, although an important source of phenolic secondary products in fungi and bacteria, is of less significance in higher plants.

The **shikimic acid pathway** converts simple carbohydrate precursors derived from glycolysis and the pentose phosphate pathway to the aromatic amino acids (see **Web Topic 13.2**) (Herrmann and Weaver 1999). One of the pathway intermediates is shikimic acid, which has given its name to this whole sequence of reactions. The well-known, broad-spectrum herbicide glyphosate (available commercially as Roundup) kills plants by blocking a step in this pathway (see Chapter 2 on the web site). The shikimic acid pathway is present in plants, fungi, and bacteria but is not found in animals. Animals have no way to synthesize the three aromatic amino acids—phenylalanine, tyrosine, and tryptophan—which are therefore essential nutrients in animal diets.

The most abundant classes of secondary phenolic compounds in plants are derived from phenylalanine via the

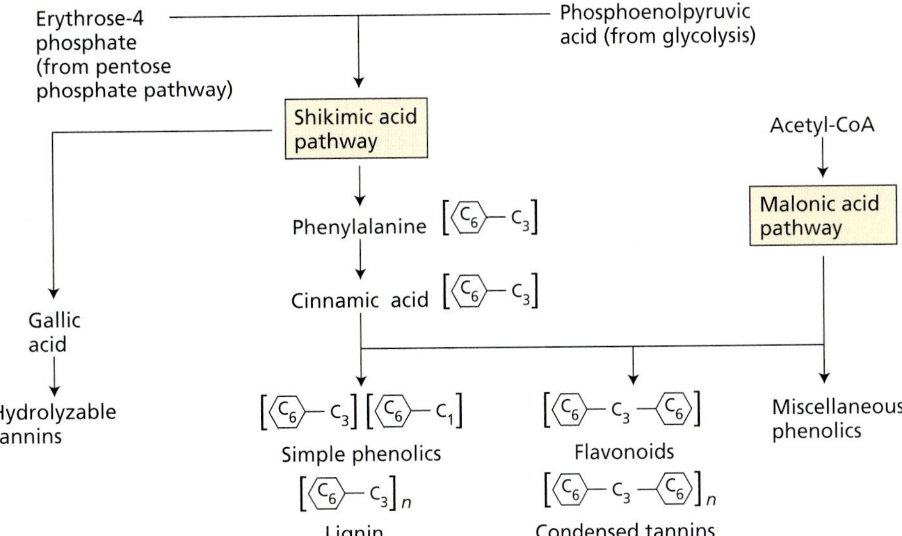

FIGURE 13.9 Plant phenolics are biosynthesized in several different ways. In higher plants, most phenolics are derived at least in part from phenylalanine, a product of the shikimic acid pathway. Formulas in brackets indicate the basic arrangement of carbon skeletons:

indicates a benzene ring, and C3 is a three-carbon chain. More detail on the pathway from phenylalanine onward is given in Figure 13.10.

elimination of an ammonia molecule to form cinnamic acid (Figure 13.10). This reaction is catalyzed by **phenylalanine ammonia lyase** (**PAL**), perhaps the most studied enzyme in plant secondary metabolism. PAL is situated at a branch point between primary and secondary metabolism, so the reaction that it catalyzes is an important regulatory step in the formation of many phenolic compounds.

The activity of PAL is increased by environmental factors, such as low nutrient levels, light (through its effect on phytochrome), and fungal infection. The point of control appears to be the initiation of transcription. Fungal invasion, for example, triggers the transcription of messenger RNA that codes for PAL, thus increasing the amount of PAL in the plant, which then stimulates the synthesis of phenolic compounds.

The regulation of PAL activity in plants is made more complex by the existence in many species of multiple PAL-encoding genes, some of which are expressed only in specific tissues or only under certain environmental conditions (Logemann et al. 1995).

Reactions subsequent to that catalyzed by PAL lead to the addition of more hydroxyl groups and other substituents. *Trans*-cinnamic acid, *p*-coumaric acid, and their derivatives are simple phenolic compounds called **phenylpropanoids** because they contain a benzene ring:

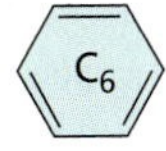

and a three-carbon side chain. Phenylpropanoids are important building blocks of the more complex phenolic compounds discussed later in this chapter.

Now that the biosynthetic pathways leading to most widespread phenolic compounds have been determined, researchers have turned their attention to studying how these pathways are regulated. In some cases, specific enzymes, such as PAL, are important in controlling flux through the pathway. Several transcription factors have been shown to regulate phenolic metabolism by binding to the promoter regions of certain biosynthetic genes and activating transcription. Some of these factors activate the transcription of large groups of genes (Jin and Martin 1999).

Some Simple Phenolics Are Activated by Ultraviolet Light

Simple phenolic compounds are widespread in vascular plants and appear to function in different capacities. Their structures include the following:

- Simple phenylpropanoids, such as *trans*-cinnamic acid, *p*-coumaric acid, and their derivatives, such as caffeic acid, which have a basic phenylpropanoid carbon skeleton (Figure 13.11A):

  C_6—C_3

- Phenylpropanoid lactones (cyclic esters) called *coumarins*, also with a phenylpropanoid skeleton (see Figure 13.11B)
- Benzoic acid derivatives, which have a C_6—C_1 skeleton: which is formed from phenylpropanoids by cleavage of a two-carbon fragment from the side chain (see Figure 13.11C) (see also Figure 13.10)

As with many other secondary products, plants can elaborate on the basic carbon skeleton of simple phenolic compounds to make more complex products.

Many simple phenolic compounds have important roles in plants as defenses against insect herbivores and fungi. Of special interest is the phototoxicity of certain coumarins called **furanocoumarins**, which have an attached furan ring (see Figure 13.11B).

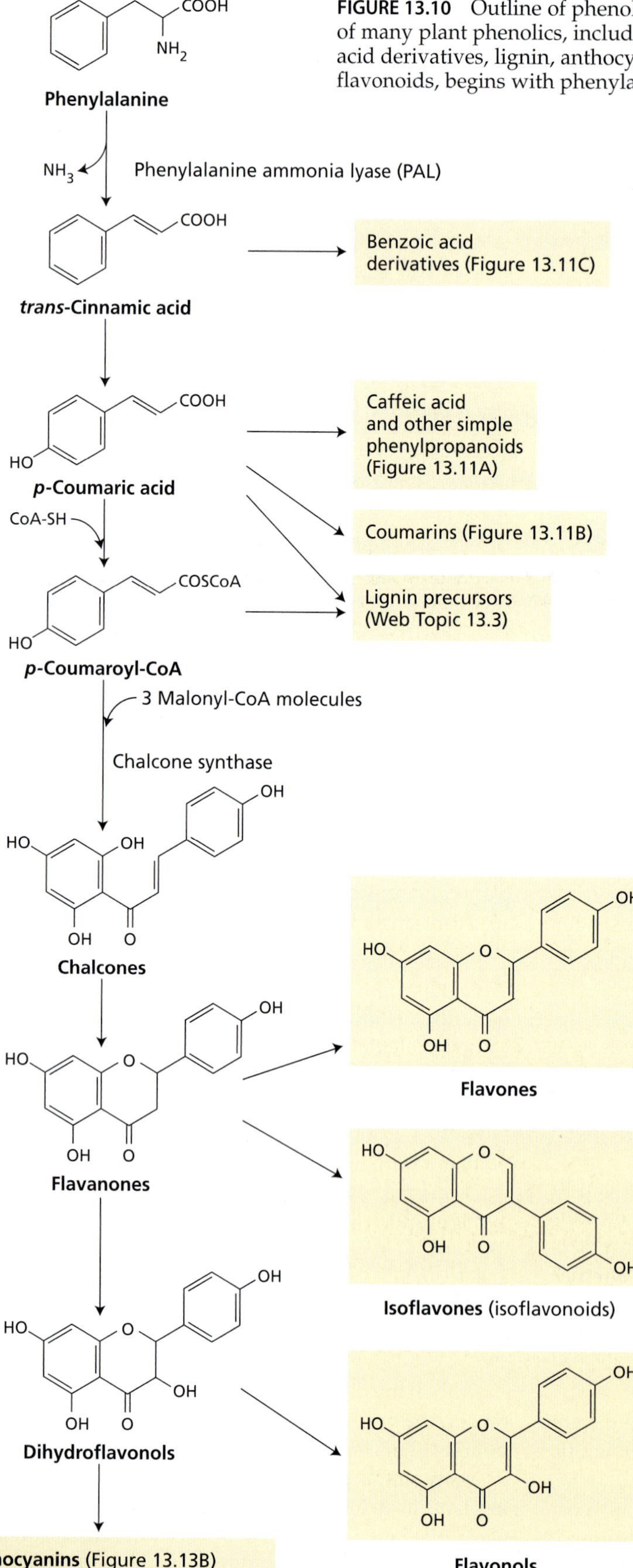

FIGURE 13.10 Outline of phenolic biosynthesis from phenylalanine. The formation of many plant phenolics, including simple phenylpropanoids, coumarins, benzoic acid derivatives, lignin, anthocyanins, isoflavones, condensed tannins, and other flavonoids, begins with phenylalanine.

These compounds are not toxic until they are activated by light. Sunlight in the ultraviolet A (UV-A) region (320–400 nm) causes some furanocoumarins to become activated to a high-energy electron state. Activated furanocoumarins can insert themselves into the double helix of DNA and bind to the pyrimidine bases cytosine and thymine, thus blocking transcription and repair and leading eventually to cell death.

Phototoxic furanocoumarins are especially abundant in members of the Umbelliferae family, including celery, parsnip, and parsley. In celery, the level of these compounds can increase about 100-fold if the plant is stressed or diseased. Celery pickers, and even some grocery shoppers, have been known to develop skin rashes from handling stressed or diseased celery. Some insects have adapted to survive on plants that contain furanocoumarins and other phototoxic compounds by living in silken webs or rolled-up leaves, which screen out the activating wavelengths (Sandberg and Berenbaum 1989).

The Release of Phenolics into the Soil May Limit the Growth of Other Plants

From leaves, roots, and decaying litter, plants release a variety of primary and secondary metabolites into the environment. Investigation of the effects of these compounds on neighboring plants is the study of **allelopathy**. If a plant can reduce the growth of nearby plants by releasing chemicals into the soil, it may increase its access to light, water, and nutrients and thus its evolutionary fitness. Generally speaking, the term *allelopathy* has come to be applied to the harmful effects of plants on their neighbors, although a precise definition also includes beneficial effects.

Simple phenylpropanoids and benzoic acid derivatives are frequently cited as having allelopathic activity. Compounds such as caffeic acid and ferulic acid (see Figure 13.11A) occur in soil in appreciable amounts and have been shown in laboratory experiments to inhibit the germination and growth of many plants (Inderjit et al. 1995).

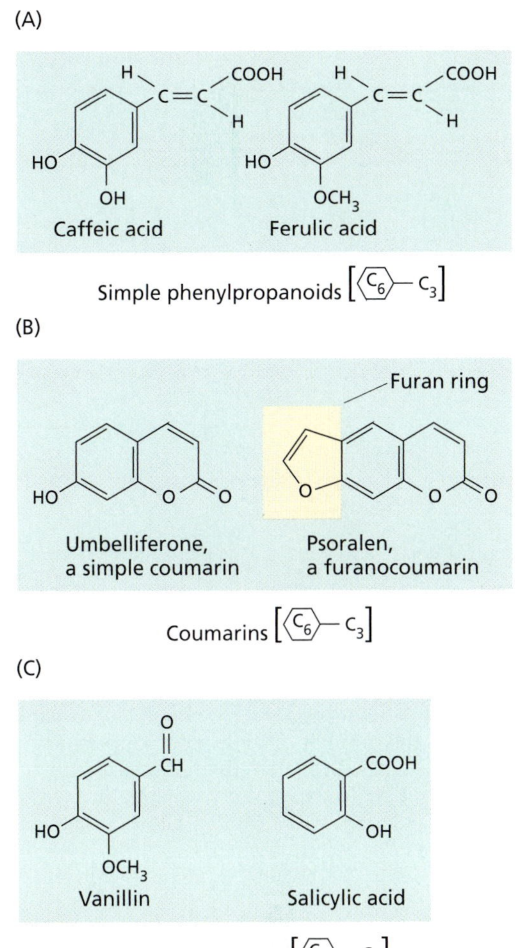

FIGURE 13.11 Simple phenolic compounds play a great diversity of roles in plants. (A) Caffeic acid and ferulic acid may be released into the soil and inhibit the growth of neighboring plants. (B) Psoralen is a furanocoumarin that exhibits phototoxicity to insect herbivores. (C) Salicylic acid is a plant growth regulator that is involved in systemic resistance to plant pathogens.

In spite of results such as these, the importance of allelopathy in natural ecosystems is still controversial. Many scientists doubt that allelopathy is a significant factor in plant–plant interactions because good evidence for this phenomenon has been hard to obtain. It is easy to show that extracts or purified compounds from one plant can inhibit the growth of other plants in laboratory experiments, but it has been very difficult to demonstrate that these compounds are present in the soil in sufficient concentration to inhibit growth. Furthermore, organic substances in the soil are often bound to soil particles and may be rapidly degraded by microbes.

In spite of the lack of supporting evidence, allelopathy is currently of great interest because of its potential agricultural applications. Reductions in crop yields caused by weeds or residues from the previous crop may in some cases be a result of allelopathy. An exciting future prospect is the development of crop plants genetically engineered to be allelopathic to weeds.

Lignin Is a Highly Complex Phenolic Macromolecule

After cellulose, the most abundant organic substance in plants is **lignin**, a highly branched polymer of phenylpropanoid groups

[C6—C3]

that plays both primary and secondary roles. The precise structure of lignin is not known because it is difficult to extract lignin from plants, where it is covalently bound to cellulose and other polysaccharides of the cell wall.

Lignin is generally formed from three different phenylpropanoid alcohols: coniferyl, coumaryl, and sinapyl, alcohols which are synthesized from phenylalanine via various cinnamic acid derivatives. The phenylpropanoid alcohols are joined into a polymer through the action of enzymes that generate free-radical intermediates. The proportions of the three monomeric units in lignin vary among species, plant organs, and even layers of a single cell wall. In the polymer, there are often multiple C—C and C—O—C bonds in each phenylpropanoid alcohol unit, resulting in a complex structure that branches in three dimensions. Unlike polymers such as starch, rubber, or cellulose, the units of lignin do not appear to be linked in a simple, repeating way. However, recent research suggests that a guiding protein may bind the individual phenylpropanoid units during lignin biosynthesis, giving rise to a scaffold that then directs the formation of a large, repeating unit (Davin and Lewis 2000; Hatfield and Vermerris 2001). (See **Web Topic 13.3** for the partial structure of a hypothetical lignin molecule.)

Lignin is found in the cell walls of various types of supporting and conducting tissue, notably the tracheids and vessel elements of the xylem. It is deposited chiefly in the thickened secondary wall but can also occur in the primary wall and middle lamella in close contact with the celluloses and hemicelluloses already present. The mechanical rigidity of lignin strengthens stems and vascular tissue, allowing upward growth and permitting water and minerals to be conducted through the xylem under negative pressure without collapse of the tissue. Because lignin is such a key component of water transport tissue, the ability to make lignin must have been one of the most important adaptations permitting primitive plants to colonize dry land.

Besides providing mechanical support, lignin has significant protective functions in plants. Its physical toughness deters feeding by animals, and its chemical durability makes it relatively indigestible to herbivores. By bonding to cellulose and protein, lignin also reduces the digestibility of these substances. Lignification blocks the growth of pathogens and is a frequent response to infection or wounding.

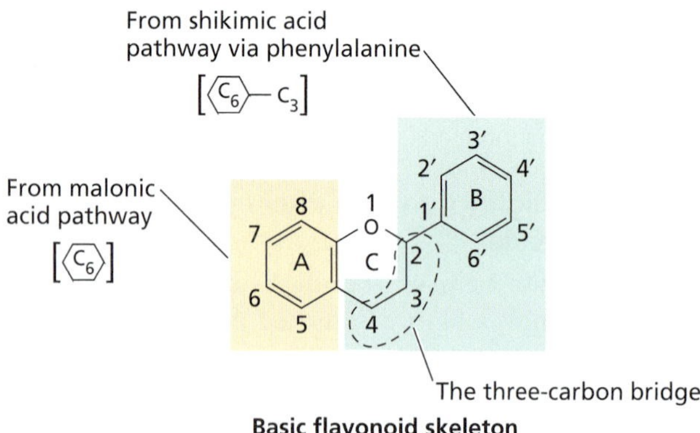

FIGURE 13.12 Basic flavonoid carbon skeleton. Flavonoids are biosynthesized from products of the shikimic acid and malonic acid pathways. Positions on the flavonoid ring system are numbered as shown.

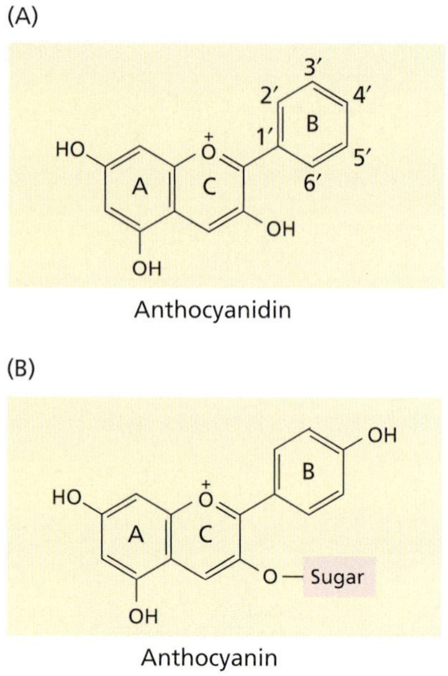

(A) Anthocyanidin

(B) Anthocyanin

FIGURE 13.13 The structures of anthocyanidins (A) and anthocyanin (B). The colors of anthocyanidins depend in part on the substituents attached to ring B (see Table 13.1). An increase in the number of hydroxyl groups shifts absorption to a longer wavelength and gives a bluer color. Replacement of a hydroxyl group with a methoxyl group (OCH_3) shifts absorption to a slightly shorter wavelength, resulting in a redder color.

There Are Four Major Groups of Flavonoids

The **flavonoids** are one of the largest classes of plant phenolics. The basic carbon skeleton of a flavonoid contains 15 carbons arranged in two aromatic rings connected by a three-carbon bridge:

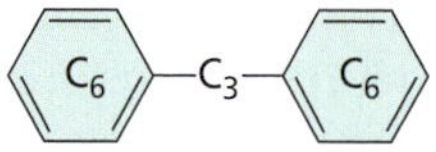

$$C_6—C_3—C_6$$

This structure results from two separate biosynthetic pathways: the shikimic acid pathway and the malonic acid pathway (Figure 13.12).

Flavonoids are classified into different groups, primarily on the basis of the degree of oxidation of the three-carbon bridge. We will discuss four of the groups shown in Figure 13.10: the anthocyanins, the flavones, the flavonols, and the isoflavones.

The basic flavonoid carbon skeleton may have numerous substituents. Hydroxyl groups are usually present at positions 4, 5, and 7, but they may also be found at other positions. Sugars are very common as well; in fact, the majority of flavonoids exist naturally as glycosides.

Whereas both hydroxyl groups and sugars increase the water solubility of flavonoids, other substituents, such as methyl ethers or modified isopentyl units, make flavonoids lipophilic (hydrophobic). Different types of flavonoids perform very different functions in the plant, including pigmentation and defense.

Anthocyanins Are Colored Flavonoids That Attract Animals

In addition to predator–prey interactions, there are mutualistic associations among plants and animals. In return for the reward of ingesting nectar or fruit pulp, animals perform extremely important services for plants as carriers of pollen and seeds. Secondary metabolites are involved in these plant–animal interactions, helping to attract animals to flowers and fruit by providing visual and olfactory signals.

The colored pigments of plants are of two principal types: carotenoids and flavonoids. *Carotenoids*, as we have already seen, are yellow, orange, and red terpenoid compounds that also serve as accessory pigments in photosynthesis (see Chapter 7). *Flavonoids* are phenolic compounds that include a wide range of colored substances.

The most widespread group of pigmented flavonoids is the **anthocyanins**, which are responsible for most of the red, pink, purple, and blue colors observed in plant parts. By coloring flowers and fruits, the anthocyanins are vitally important in attracting animals for pollination and seed dispersal.

Anthocyanins are glycosides that have sugars at position 3 (Figure 13.13B) and sometimes elsewhere. Without their sugars, anthocyanins are known as **anthocyanidins** (Figure 13.13A). Anthocyanin color is influenced by many factors, including the number of hydroxyl and methoxyl groups in ring B of the anthocyanidin (see Figure 13.13A), the presence of aromatic acids esterified to the main skeleton, and the pH of the cell vacuole in which these compounds are stored. Anthocyanins may also exist in supramolecular complexes along with chelated metal ions and flavone copigments. The blue pigment of dayflower (*Commelina communis*) was found

TABLE 13.1
Effects of ring substituents on anthocyanidin color

Anthocyanidin	Substituents	Color
Pelargonidin	4′— OH	Orange red
Cyanidin	3′— OH, 4′— OH	Purplish red
Delphinidin	3′— OH,4′— OH,5′— OH	Bluish purple
Peonidin	3′— OCH_3, 4′— OH	Rosy red
Petunidin	3′— OCH_3, 4′— OH, 5′— OCH_3	Purple

to consist of a large complex of six anthocyanin molecules, six flavones, and two associated magnesium ions (Kondo et al. 1992). The most common anthocyanidins and their colors are shown in Figure 13.13 and Table 13.1.

Considering the variety of factors affecting anthocyanin coloration and the possible presence of carotenoids as well, it is not surprising that so many different shades of flower and fruit color are found in nature. The evolution of flower color may have been governed by selection pressures for different sorts of pollinators, which often have different color preferences.

Color, of course, is just one type of signal used to attract pollinators to flowers. Volatile chemicals, particularly monoterpenes, frequently provide attractive scents.

Flavonoids May Protect against Damage by Ultraviolet Light

Two other major groups of flavonoids found in flowers are **flavones** and **flavonols** (see Figure 13.10). These flavonoids generally absorb light at shorter wavelengths than anthocyanins do, so they are not visible to the human eye. However, insects such as bees, which see farther into the ultraviolet range of the spectrum than humans do, may respond to flavones and flavonols as attractant cues (Figure 13.14). Flavonols in a flower often form symmetric patterns of stripes, spots, or concentric circles called *nectar guides* (Lunau 1992). These patterns may be conspicuous to insects and are thought to help indicate the location of pollen and nectar.

Flavones and flavonols are not restricted to flowers; they are also present in the leaves of all green plants. These two classes of flavonoids function to protect cells from excessive UV-B radiation (280–320 nm) because they accumulate in the epidermal layers of leaves and stems and absorb light strongly in the UV-B region while allowing the visible (photosynthetically active) wavelengths to pass through uninterrupted. In addition, exposure of plants to increased UV-B light has been demonstrated to increase the synthesis of flavones and flavonols.

Arabidopsis thaliana mutants that lack the enzyme chalcone synthase produce no flavonoids. Lacking flavonoids, these plants are much more sensitive to UV-B radiation than wild-type individuals are, and they grow very poorly under normal conditions. When shielded from UV light, however, they grow normally (Li et al. 1993). A group of simple phenylpropanoid esters are also important in UV protection in *Arabidopsis*.

(A)

(B)

FIGURE 13.14 Black-eyed Susan (*Rudbeckia* sp.) as seen by humans (A) and as it might appear to honeybees (B). (A) To humans, the golden-eye has yellow rays and a brown central disc. (B) To bees, the tips of the rays appear "light yellow," the inner portion of the rays "dark yellow," and the central disc "black." Ultraviolet-absorbing flavonols are found in the inner parts of the rays but not in the tips. The distribution of flavonols in the rays and the sensitivity of insects to part of the UV spectrum contribute to the "bull's-eye" pattern seen by honeybees, which presumably helps them locate pollen and nectar. Special lighting was used to simulate the spectral sensitivity of the honeybee visual system. (Courtesy of Thomas Eisner.)

Other functions of flavonoids have recently been discovered. For example, flavones and flavonols secreted into the soil by legume roots mediate the interaction of legumes and nitrogen-fixing symbionts, a phenomenon described in Chapter 12. As will be discussed in Chapter 19, recent work suggests that flavonoids also play a regulatory role in plant development as modulators of polar auxin transport.

Isoflavonoids Have Antimicrobial Activity

The **isoflavonoids** (isoflavones) are a group of flavonoids in which the position of one aromatic ring (ring B) is shifted (see Figure 13.10). Isoflavonoids are found mostly in legumes and have several different biological activities. Some, such as the rotenoids, have strong insecticidal actions; others have anti-estrogenic effects. For example, sheep grazing on clover rich in isoflavonoids often suffer from infertility. The isoflavonoid ring system has a three-dimensional structure similar to that of steroids (see Figure 13.8B), allowing these substances to bind to estrogen receptors. Isoflavonoids may also be responsible for the anticancer benefits of food prepared from soybeans.

In the past few years, isoflavonoids have become best known for their role as *phytoalexins*, antimicrobial compounds synthesized in response to bacterial or fungal infection that help limit the spread of the invading pathogen. Phytoalexins are discussed in more detail later in this chapter.

Tannins Deter Feeding by Herbivores

A second category of plant phenolic polymers with defensive properties, besides lignins, is the **tannins**. The term *tannin* was first used to describe compounds that could convert raw animal hides into leather in the process known as tanning. Tannins bind the collagen proteins of animal hides, increasing their resistance to heat, water, and microbes.

There are two categories of tannins: condensed and hydrolyzable. **Condensed tannins** are compounds formed by the polymerization of flavonoid units (Figure 13.15A). They are frequent constituents of woody plants. Because condensed tannins can often be hydrolyzed to anthocyanidins by treatment with strong acids, they are sometimes called *pro-anthocyanidins*.

Hydrolyzable tannins are heterogeneous polymers containing phenolic acids, especially gallic acid, and simple sugars (see Figure 13.15B). They are smaller than condensed tannins and may be hydrolyzed more easily; only dilute acid is needed. Most tannins have molecular masses between 600 and 3000.

Tannins are general toxins that significantly reduce the growth and survivorship of many herbivores when added to their diets. In addition, tannins act as feeding repellents to a great diversity of animals. Mammals such as cattle, deer, and apes characteristically avoid plants or parts of plants with high tannin contents. Unripe fruits, for

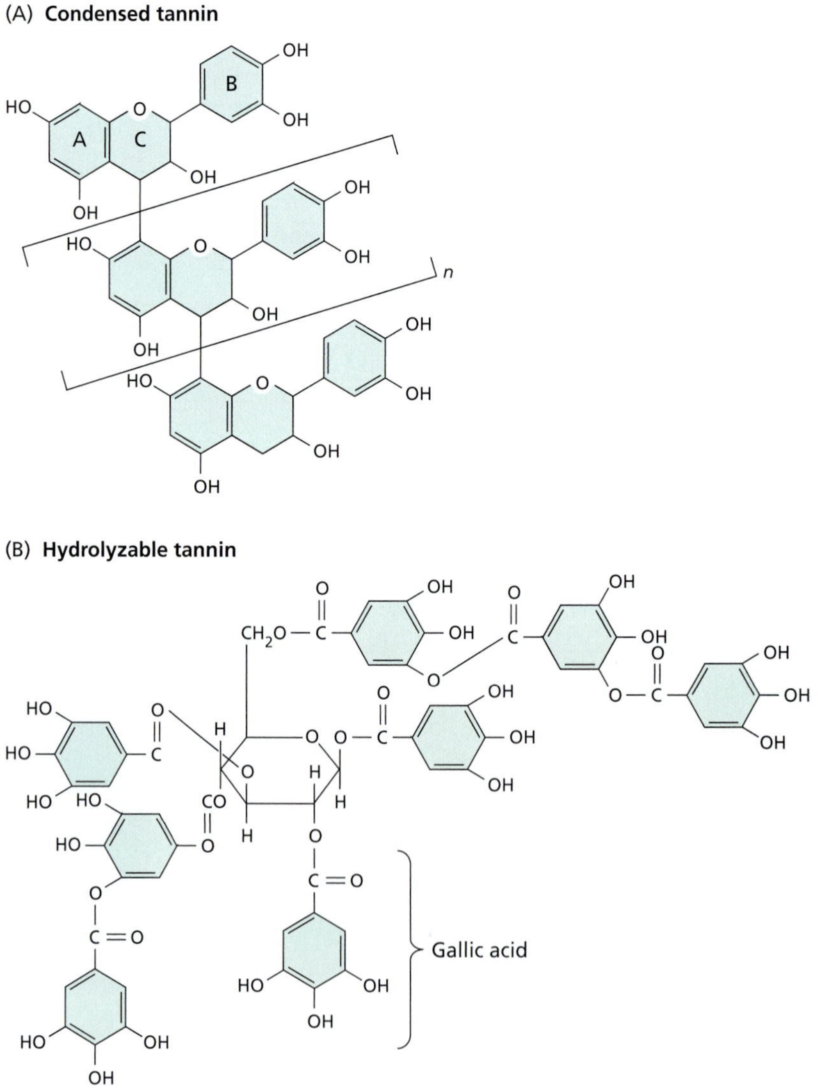

(A) Condensed tannin

(B) Hydrolyzable tannin

FIGURE 13.15 Structure of some tannins formed from phenolic acids or flavonoid units. (A) The general structure of a condensed tannin, where *n* is usually 1 to 10. There may also be a third —OH group on ring B. (B) The hydrolyzable tannin from sumac (*Rhus semialata*) consists of glucose and eight molecules of gallic acid.

instance, frequently have very high tannin levels, which may be concentrated in the outer cell layers.

Interestingly, humans often prefer a certain level of astringency in tannin-containing foods, such as apples, blackberries, tea, and red wine. Recently, polyphenols (tannins) in red wine were shown to block the formation of endothelin-1, a signaling molecule that makes blood vessels constrict (Corder et al. 2001). This effect of wine tannins may account for the often-touted health benefits of red wine, especially the reduction in the risk of heart disease associated with moderate red wine consumption.

Although moderate amounts of specific polyphenolics may have health benefits for humans, the defensive properties of most tannins are due to their toxicity, which is generally attributed to their ability to bind proteins nonspecifically. It has long been thought that plant tannins complex proteins in the guts of herbivores by forming hydrogen bonds between their hydroxyl groups and electronegative sites on the protein (Figure 13.16A).

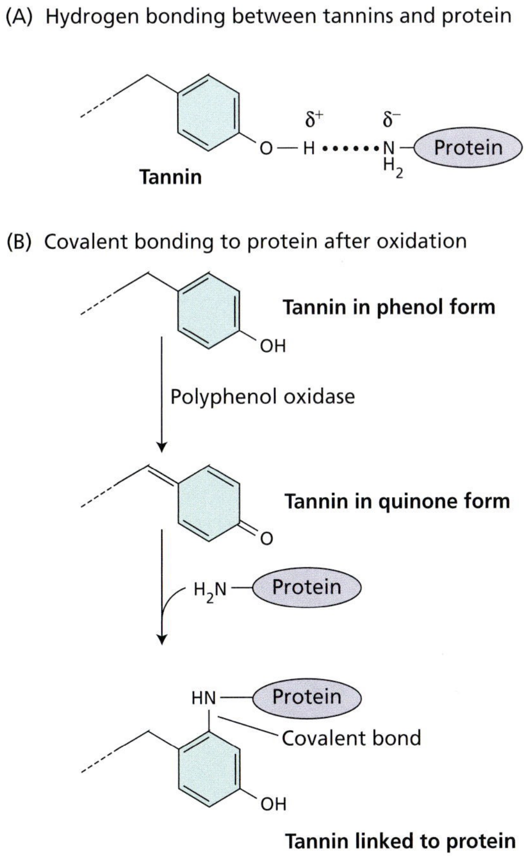

FIGURE 13.16 Proposed mechanisms for the interaction of tannins with proteins. (A) Hydrogen bonds may form between the phenolic hydroxyl groups of tannins and electronegative sites on the protein. (B) Phenolic hydroxyl groups may bind covalently to proteins following activation by oxidative enzymes, such as polyphenol oxidase.

More recent evidence indicates that tannins and other phenolics can also bind to dietary protein in a covalent fashion (see Figure 13.16B). The foliage of many plants contains enzymes that oxidize phenolics to their corresponding quinone forms in the guts of herbivores (Felton et al. 1989). Quinones are highly reactive electrophilic molecules that readily react with the nucleophilic —NH_2 and —SH groups of proteins (see Figure 13.16B). By whatever mechanism protein–tannin binding occurs, this process has a negative impact on herbivore nutrition. Tannins can inactivate herbivore digestive enzymes and create complex aggregates of tannins and plant proteins that are difficult to digest.

Herbivores that habitually feed on tannin-rich plant material appear to possess some interesting adaptations to remove tannins from their digestive systems. For example, some mammals, such as rodents and rabbits, produce salivary proteins with a very high proline content (25–45%) that have a high affinity for tannins. Secretion of these proteins is induced by ingestion of food with a high tannin content and greatly diminishes the toxic effects of tannins (Butler 1989). The large number of proline residues gives these proteins a very flexible, open conformation and a high degree of hydrophobicity that facilitates binding to tannins.

Plant tannins also serve as defenses against microorganisms. For example, the nonliving heartwood of many trees contains high concentrations of tannins that help prevent fungal and bacterial decay.

NITROGEN-CONTAINING COMPOUNDS

A large variety of plant secondary metabolites have nitrogen in their structure. Included in this category are such well-known antiherbivore defenses as alkaloids and cyanogenic glycosides, which are of considerable interest because of their toxicity to humans and their medicinal properties. Most nitrogenous secondary metabolites are biosynthesized from common amino acids.

In this section we will examine the structure and biological properties of various nitrogen-containing secondary metabolites, including alkaloids, cyanogenic glycosides, glucosinolates, and nonprotein amino acids. In addition, we will discuss the ability of *systemin*, a protein released from damaged cells, to serve as a wound signal to the rest of the plant.

Alkaloids Have Dramatic Physiological Effects on Animals

The **alkaloids** are a large family of more than 15,000 nitrogen-containing secondary metabolites found in approximately 20% of the species of vascular plants. The nitrogen atom in these substances is usually part of a **heterocyclic ring**, a ring that contains both nitrogen and carbon atoms. As a group, alkaloids are best known for their striking pharmacological effects on vertebrate animals.

As their name would suggest, most alkaloids are alkaline. At pH values commonly found in the cytosol (pH 7.2)

TABLE 13.2
Major types of alkaloids, their amino acid precursors, and well-known examples of each type

Alkaloid class	Structure	Biosynthetic precursor	Examples	Human uses
Pyrrolidine		Ornithine (aspartate)	Nicotine	Stimulant, depressant, tranquilizer
Tropane		Ornithine	Atropine	Prevention of intestinal spasms, antidote to other poisons, dilation of pupils for examination
			Cocaine	Stimulant of the central nervous system, local anesthetic
Piperidine		Lysine (or acetate)	Coniine	Poison (paralyzes motor neurons)
Pyrrolizidine		Ornithine	Retrorsine	None
Quinolizidine		Lysine	Lupinine	Restoration of heart rhythm
Isoquinoline		Tyrosine	Codeine	Analgesic (pain relief), treatment of coughs
			Morphine	Analgesic
Indole		Tryptophan	Psilocybin	Halucinogen
			Reserpine	Treatment of hypertension, treatment of psychoses
			Strychnine	Rat poison, treatment of eye disorders

or the vacuole (pH 5 to 6), the nitrogen atom is protonated; hence, alkaloids are positively charged and are generally water soluble.

Alkaloids are usually synthesized from one of a few common amino acids—in particular, lysine, tyrosine, and tryptophan. However, the carbon skeleton of some alkaloids contains a component derived from the terpene pathway. Table 13.2 lists the major alkaloid types and their amino acid precursors. Several different types, including nicotine and its relatives (Figure 13.17), are derived from ornithine, an intermediate in arginine biosynthesis. The B vitamin nicotinic acid (niacin) is a precursor of the pyridine (six-membered) ring of this alkaloid; the pyrrolidine (five-membered) ring of nicotine arises from ornithine (Figure 13.18). Nicotinic acid is also a constituent of NAD^+ and $NADP^+$, which serve as electron carriers in metabolism.

The role of alkaloids in plants has been a subject of speculation for at least 100 years. Alkaloids were once thought to be nitrogenous wastes (analogous to urea and uric acid in animals), nitrogen storage compounds, or growth regulators, but there is little evidence to support any of these functions. Most alkaloids are now believed to function as defenses against predators, especially mammals, because

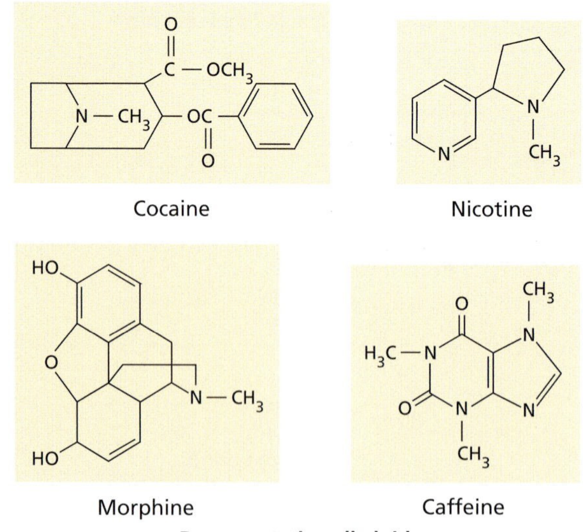

FIGURE 13.17 Examples of alkaloids, a diverse group of secondary metabolites that contain nitrogen, usually as part of a heterocyclic ring. Caffeine is a purine-type alkaloid similar to the nucleic acid bases adenine and guanine. The pyrrolidine (five-membered) ring of nicotine arises from ornithine; the pyridine (six-membered) ring is derived from nicotinic acid.

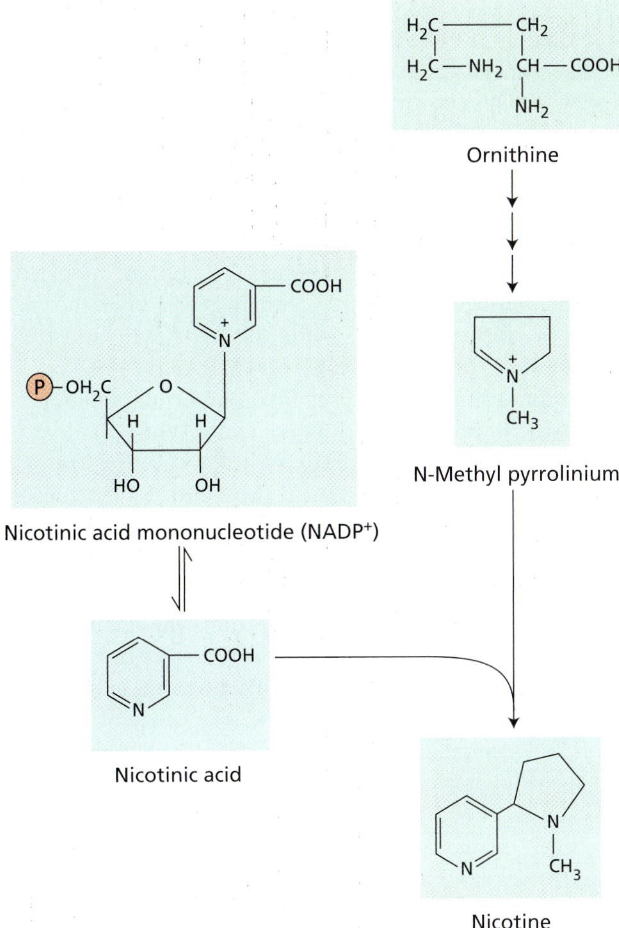

FIGURE 13.18 Nicotine biosynthesis begins with the biosynthesis of the nicotinic acid (niacin) from aspartate and glyceraldehyde-3-phosphate. Nicotinic acid is also a component of NAD^+ and $NADP^+$, important participants in biological oxidation–reduction reactions. The five-membered ring of nicotine is derived from ornithine, an intermediate in arginine biosynthesis.

Indeed, some livestock actually seem to prefer alkaloid-containing plants to less harmful forage.

Nearly all alkaloids are also toxic to humans when taken in sufficient quantity. For example, strychnine, atropine, and coniine (from poison hemlock) are classic alkaloid poisoning agents. At lower doses, however, many are useful pharmacologically. Morphine, codeine, and scopolamine are just a few of the plant alkaloids currently used in medicine. Other alkaloids, including cocaine, nicotine, and caffeine (see Figure 13.17), enjoy widespread nonmedical use as stimulants or sedatives.

On a cellular level, the mode of action of alkaloids in animals is quite variable. Many alkaloids interfere with components of the nervous system, especially the chemical transmitters; others affect membrane transport, protein synthesis, or miscellaneous enzyme activities.

One group of alkaloids, the pyrrolizidine alkaloids, illustrates how herbivores can become adapted to tolerate plant defensive substances and even use them in their own defense (Hartmann 1999). Within plants, pyrrolizidine alkaloids occur naturally as nontoxic N-oxides. In herbivore digestive tracts, however, they are quickly reduced to uncharged, hydrophobic tertiary alkaloids (Figure 13.19), which easily pass through membranes and are toxic. Nevertheless, some herbivores, such as cinnabar moth (*Tyria jacobeae*), have developed the ability to reconvert tertiary pyrrolizidine alkaloids to the nontoxic N-oxide form immediately after its absorption from the digestive tract. These herbivores may then store the N-oxides in their bodies as defenses against their own predators.

Not all of the alkaloids that appear in plants are produced by the plant itself. Many grasses harbor endogenous fungal symbionts that grow in the apoplast and synthesize a variety of different types of alkaloids. Grasses with fungal symbionts often grow faster and are better defended of their general toxicity and deterrence capability (Hartmann 1992).

Large numbers of livestock deaths are caused by the ingestion of alkaloid-containing plants. In the United States, a significant percentage of all grazing livestock animals are poisoned each year by consumption of large quantities of alkaloid-containing plants such as lupines (*Lupinus*), larkspur (*Delphinium*), and groundsel (*Senecio*). This phenomenon may be due to the fact that domestic animals, unlike wild animals, have not been subjected to natural selection for the avoidance of toxic plants.

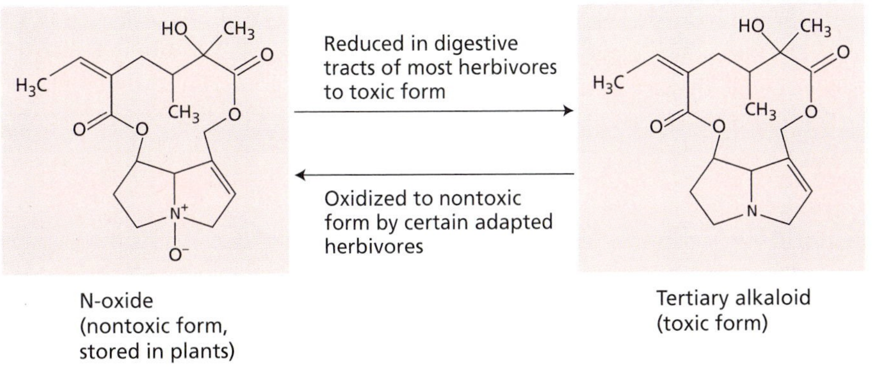

Reduced in digestive tracts of most herbivores to toxic form

Oxidized to nontoxic form by certain adapted herbivores

N-oxide (nontoxic form, stored in plants)

Tertiary alkaloid (toxic form)

FIGURE 13.19 Two forms of pyrrolizidine alkaloids occur in nature: the N-oxide form and the tertiary alkaloid. The nontoxic N-oxide found in plants is reduced to the toxic tertiary form in the digestive tracts of most herbivores. However, some adapted herbivores can convert the toxic tertiary alkaloid back to the nontoxic N-oxide. These forms are illustrated here for the alkaloid senecionine, found in species of ragwort (*Senecio*).

against insect and mammalian herbivores than those without symbionts. Unfortunately, certain grasses with symbionts, such as tall fescue, are important pasture grasses that may become toxic to livestock when their alkaloid content is too high. Efforts are under way to breed tall fescue with alkaloid levels that are not poisonous to livestock but still provide protection against insects (see **Web Essay 13.2**).

Like monoterpenes in conifer resin and many other antiherbivore defense compounds, alkaloids increase in response to initial herbivore damage, fortifying the plant against subsequent attack (Karban and Baldwin 1997). For example, *Nicotiana attenuata,* a wild tobacco that grows in the deserts of the Great Basin, produces higher levels of nicotine following herbivory. When it is attacked by nicotine-tolerant caterpillars, however, there is no increase in nicotine. Instead, volatile terpenes are released that attract enemies of the caterpillars. Clearly, wild tobacco and other plants must have ways of determining what type of herbivore is damaging their foliage. Herbivores might signal their presence by the type of damage they inflict or the distinctive chemical compounds they release. Recently, the oral secretions of caterpillars feeding on corn leaves were shown to contain a fatty acid–amino acid conjugate that induced the plant to produce defensive terpenes when applied to cut leaves.

Cyanogenic Glycosides Release the Poison Hydrogen Cyanide

Various nitrogenous protective compounds other than alkaloids are found in plants. Two groups of these substances—cyanogenic glycosides and glucosinolates—are not in themselves toxic but are readily broken down to give off volatile poisons when the plant is crushed. Cyanogenic glycosides release the well-known poisonous gas hydrogen cyanide (HCN).

The breakdown of cyanogenic glycosides in plants is a two-step enzymatic process. Species that make cyanogenic glycosides also make the enzymes necessary to hydrolyze the sugar and liberate HCN:

1. In the first step the sugar is cleaved by a glycosidase, an enzyme that separates sugars from other molecules to which they are linked (Figure 13.20).
2. In the second step the resulting hydrolysis product, called an α-hydroxynitrile or cyanohydrin, can decompose spontaneously at a low rate to liberate HCN. This second step can be accelerated by the enzyme hydroxynitrile lyase.

Cyanogenic glycosides are not normally broken down in the intact plant because the glycoside and the degradative enzymes are spatially separated, in different cellular compartments or in different tissues. In sorghum, for example, the cyanogenic glycoside dhurrin is present in the vacuoles of epidermal cells, while the hydrolytic and lytic enzymes are found in the mesophyll (Poulton 1990).

Under ordinary conditions this compartmentation prevents decomposition of the glycoside. When the leaf is damaged, however, as during herbivore feeding, the cell contents of different tissues mix and HCN forms. Cyanogenic glycosides are widely distributed in the plant kingdom and are frequently encountered in legumes, grasses, and species of the rose family.

Considerable evidence indicates that cyanogenic glycosides have a protective function in certain plants. HCN is a fast-acting toxin that inhibits metalloproteins, such as the iron-containing cytochrome oxidase, a key enzyme of mitochondrial respiration. The presence of cyanogenic glycosides deters feeding by insects and other herbivores, such as snails and slugs. As with other classes of secondary metabolites, however, some herbivores have adapted to feed on cyanogenic plants and can tolerate large doses of HCN.

The tubers of cassava (*Manihot esculenta*), a high-carbohydrate, staple food in many tropical countries, contain high levels of cyanogenic glycosides. Traditional processing methods, such as grating, grinding, soaking, and drying, lead to the removal or degradation of a large fraction of the cyanogenic glycosides present in cassava tubers. However, chronic cyanide poisoning leading to partial paralysis of the limbs is still widespread in regions where cassava is a major food source because the traditional detoxification methods employed to remove cyanogenic glycosides from cassava are not completely effective. In addition, many populations that consume cassava have poor nutrition, which aggravates the effects of the cyanogenic glycosides.

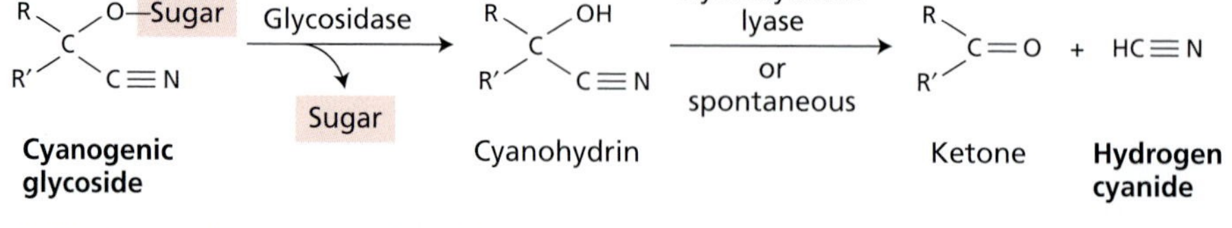

FIGURE 13.20 Enzyme-catalyzed hydrolysis of cyanogenic glycosides to release hydrogen cyanide. R and R′ represent various alkyl or aryl substituents. For example, if R is phenyl, R′ is hydrogen, and the sugar is the disaccharide β-gentiobiose, the compound is amygdalin (the common cyanogenic glycoside found in the seeds of almonds, apricots, cherries, and peaches).

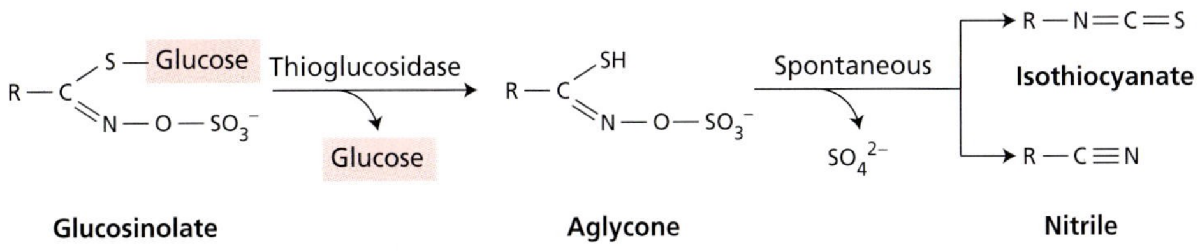

Glucosinolate — Thioglucosidase (releases Glucose) → Aglycone — Spontaneous (releases SO_4^{2-}) → Isothiocyanate ($R—N=C=S$) or Nitrile ($R—C\equiv N$)

FIGURE 13.21 Hydrolysis of glucosinolates to mustard-smelling volatiles. R represents various alkyl or aryl substituents. For example, if R is $CH_2=CH—CH_2^-$, the compound is sinigrin, a major glucosinolate of black mustard seeds and horseradish roots.

Efforts are currently under way to reduce the cyanogenic glycoside content of cassava through both conventional breeding and genetic engineering approaches. However, the complete elimination of cyanogenic glycosides may not be desirable because these substances are probably responsible for the fact that cassava can be stored for very long periods of time without being attacked by pests.

Glucosinolates Release Volatile Toxins

A second class of plant glycosides, called the **glucosinolates**, or mustard oil glycosides, break down to release volatile defensive substances. Found principally in the Brassicaceae and related plant families, glucosinolates give off the compounds responsible for the smell and taste of vegetables such as cabbage, broccoli, and radishes.

The release of these mustard-smelling volatiles from glucosinolates is catalyzed by a hydrolytic enzyme, called a thioglucosidase or myrosinase, that cleaves glucose from its bond with the sulfur atom (Figure 13.21). The resulting aglycone, the nonsugar portion of the molecule, rearranges with loss of the sulfate to give pungent and chemically reactive products, including isothiocyanates and nitriles, depending on the conditions of hydrolysis. These products function in defense as herbivore toxins and feeding repellents. Like cyanogenic glycosides, glucosinolates are stored in the intact plant separately from the enzymes that hydrolyze them, and they are brought into contact with these enzymes only when the plant is crushed.

As with other secondary metabolites, certain animals are adapted to feed on glucosinolate-containing plants without ill effects. For adapted herbivores, such as the cabbage butterfly, glucosinolates often serve as stimulants for feeding and egg laying, and the isothiocyanates produced after glucosinolate hydrolysis act as volatile attractants (Renwick et al. 1992).

Most of the recent research on glucosinolates in plant defense has concentrated on rape, or canola (*Brassica napus*), a major oil crop in both North America and Europe. Plant breeders have tried to lower the glucosinolate levels of rapeseed so that the high-protein seed meal remaining after oil extraction can be used as animal food. The first low-glucosinolate varieties tested in the field were unable to survive because of severe pest problems. However, more recently developed varieties with low glucosinolate levels in seeds but high glucosinolate levels in leaves are able to hold their own against pests and still provide a protein-rich seed residue for animal feeding.

Nonprotein Amino Acids Defend against Herbivores

Plants and animals incorporate the same 20 amino acids into their proteins. However, many plants also contain unusual amino acids, called **nonprotein amino acids**, that are not incorporated into proteins but are present instead in the free form and act as protective substances. Nonprotein amino acids are often very similar to common protein amino acids. Canavanine, for example, is a close analog of arginine, and azetidine-2-carboxylic acid has a structure very much like that of proline (Figure 13.22).

Nonprotein amino acids exert their toxicity in various ways. Some block the synthesis or uptake of protein amino

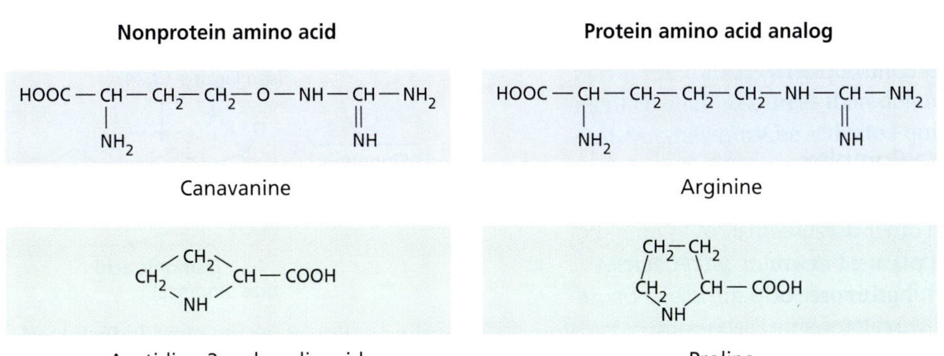

FIGURE 13.22 Nonprotein amino acids and their protein amino acid analogs. The nonprotein amino acids are not incorporated into proteins but are defensive compounds found in free form in plant cells.

set in motion that lead eventually to defense responses (see Figure 13.25). A common early element of these cascades is a transient change in the ion permeability of the plasma membrane. *R* gene activation stimulates an influx of Ca^{2+} and H^+ ions into the cell and an efflux of K^+ and Cl^- ions (Nürnberger and Scheel 2001). The influx of Ca^{2+} activates the oxidative burst that may act directly in defense (as already described), as well as signaling other defense reactions. Other components of pathogen-stimulated signal transduction pathways include nitric oxide, mitogen-activated protein (MAP) kinases, calcium-dependent protein kinases, jasmonic acid, and salicylic acid (see the next section).

A Single Encounter with a Pathogen May Increase Resistance to Future Attacks

When a plant survives the infection of a pathogen at one site, it often develops increased resistance to subsequent attacks at sites throughout the plant and enjoys protection against a wide range of pathogen species. This phenomenon, called **systemic acquired resistance** (**SAR**), develops over a period of several days following initial infection (Ryals et al. 1996). Systemic acquired resistance appears to result from increased levels of certain defense compounds that we have already mentioned, including chitinases and other hydrolytic enzymes.

Although the mechanism of SAR induction is still unknown, one of the endogenous signals is likely to be **salicylic acid**. The level of this benzoic acid derivative, a C_6–C_1 compound rises dramatically in the zone of infection after initial attack, and it is thought to establish SAR in other parts of the plant, although salicylic acid itself is not the mobile signal (Figure 13.27).

In addition to salicylic acid, recent studies suggest that its methyl ester, methyl salicylate, acts as a volatile SAR-inducing signal transmitted to distant parts of the plant and even to neighboring plants (Shulaev et al. 1997). Thus, even though plants lack immune systems like those present in many animals, they have developed elaborate mechanisms to protect themselves from disease-causing microbes.

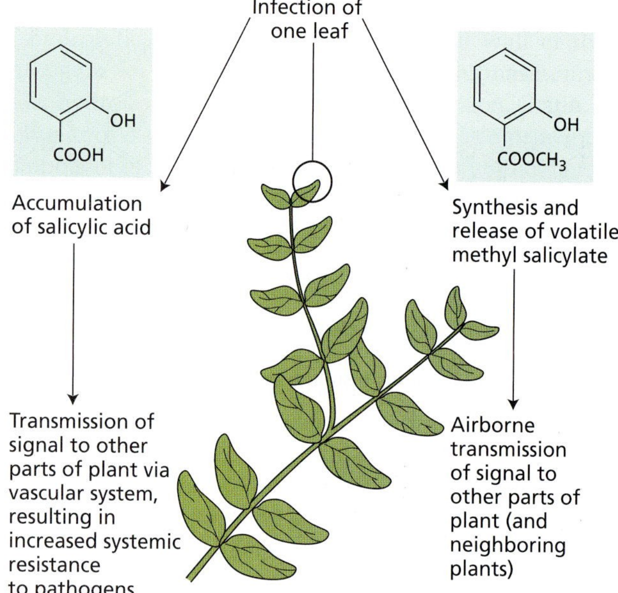

FIGURE 13.27 Initial pathogen infection may increase resistance to future pathogen attack through development of systemic acquired resistance.

SUMMARY

Plants produce an enormous diversity of substances that have no apparent roles in growth and development processes and so are classified under the heading of secondary metabolites. Scientists have long speculated that these compounds protect plants from predators and pathogens on the basis of their toxicity and repellency to herbivores and microbes when tested in vitro. Recent experiments on plants whose secondary-metabolite expression has been altered by modern molecular methods have begun to confirm these defensive roles.

There are three major groups of secondary metabolites: terpenes, phenolics, and nitrogen-containing compounds. Terpenes, composed of five-carbon isoprene units, are toxins and feeding deterrents to many herbivores.

Phenolics, which are synthesized primarily from products of the shikimic acid pathway, have several important roles in plants. Lignin mechanically strengthens cell walls. Flavonoid pigments function as shields against harmful ultraviolet radiation and as attractants for pollinators and fruit dispersers. Finally, lignin, flavonoids, and other phenolic compounds serve as defenses against herbivores and pathogens.

Members of the third major group, nitrogen-containing secondary metabolites, are synthesized principally from common amino acids. Compounds such as alkaloids, cyanogenic glycosides, glucosinolates, nonprotein amino acids, and proteinase inhibitors protect plants from a variety of herbivorous animals.

Plants have evolved multiple defense mechanisms against microbial pathogens. Besides antimicrobial secondary metabolites, some of which are preformed and some of which are induced by infection, other modes of defense include the construction of polymeric barriers to pathogen penetration and the synthesis of enzymes that degrade pathogen cell walls. In addition, plants employ specific recognition and signaling systems enabling the rapid detection of pathogen invasion and initiation of a vigorous defensive response. Once infected, some plants also develop an immunity to subsequent microbial attacks.

For millions of years, plants have produced defenses against herbivory and microbial attack. Well-defended plants have tended to leave more survivors than poorly defended plants, so the capacity to produce effective defensive products has become widely established in the plant kingdom. In response, many species of herbivores and microbes have evolved the ability to feed on or infect plants containing secondary products without being adversely affected, and this herbivore and pathogen pressure has in turn selected for new defensive products in plants.

The study of plant secondary metabolites has many practical applications. By virtue of their biological activities against herbivorous animals and microbes, many of these substances are employed commercially as insecticides, fungicides, and pharmaceuticals, while others find uses as fragrances, flavorings, medicinal drugs, and industrial materials. The breeding of increased levels of secondary metabolites into crop plants has made it possible to reduce the need for certain costly and potentially harmful pesticides. In some cases, however, it has been necessary to reduce the levels of naturally occurring secondary metabolites to minimize toxicity to humans and domestic animals.

Web Material

Web Topics

13.1 Structure of Various Triterpenes

The structures of several triterpenes are given.

13.2 The Shikimic Acid Pathway

The biochemical pathway for the synthesis of aromatic amino acids, the precursors of phenolic compounds, is presented.

13.3 Detailed Chemical Structure of a Portion of a Lignin Molecule

The partial structure of a hypothetical lignin molecule from European beech (*Fagus sylvatica*) is described.

Web Essays

13.1 Unraveling the Function of Secondary Metabolites

Wild tobacco plants use alkaloids and terpenes to defend themselves against herbivores.

13.2 Alkaloid-Making Fungal Symbionts

Fungal endophytes can enhance plant growth, increase resistance to various stresses, and act as "defensive mutualists" against herbivores.

Chapter References

Aerts, R. J., and Mordue, A. J. (1997) Feeding deterrence and toxocity of neem triterpenoids. *J. Chem. Ecol.* 23: 2117–2132.

Boller, T. (1995) Chemoperception of microbial signals in plant cells. *Annu. Rev. Plant Physiol. Plant Mol. Biol.* 46: 189–214.

Bradley, D. J., Kjellbom, P., and Lamb, C. J. (1992) Elicitor- and wound-induced oxidative cross-linking of a proline-rich plant cell wall protein: A novel, rapid defense response. *Cell* 70: 21–30.

Butler, L. G. (1989) Effects of condensed tannin on animal nutrition. In *Chemistry and Significance of Condensed Tannins*, R. W. Hemingway and J. J. Karchesy, eds., Plenum, New York, pp. 391–402.

Corder, R., Douthwaite, J. A., Lees, D. M., Khan, N. Q., Viseu dos Santos, A. C., Wood, E. G., and Carrier, M. J. (2001) Endothelin-1 synthesis reduced by red wine. *Nature* 414: 863–864.

Creelman, R. A., and Mullet, J. E. (1997) Biosynthesis and action of jasmonates in plants. *Annu. Rev. Plant Physiol. Plant Mol. Biol.* 48: 355–381.

Davin, L. B., and Lewis, N. G. (2000) Dirigent proteins and dirigent sites explain the mystery of specificity of radical precursor coupling in lignan and lignin biosynthesis. *Plant Physiol.* 123: 453–461.

Eigenbrode, S. D., Stoner, K. A., Shelton, A. M., and Kain, W. C. (1991) Characteristics of glossy leaf waxes associated with resistance to diamondback moth (Lepidoptera: Plutellidae) in *Brassica oleracea*. *J. Econ. Entomol.* 83: 1609–1618.

Felton, G. W., Donato, K., Del Vecchio, R. J., and Duffey, S. S. (1989) Activation of plant foliar oxidases by insect feeding reduces nutritive quality of foliage for noctuid herbivores. *J. Chem. Ecol.* 15: 2667–2694.

Gershenzon, J., and Croteau, R. (1992) Terpenoids. In *Herbivores: Their Interactions with Secondary Plant Metabolites*, Vol. 1: *The Chemical Participants*, 2nd ed., G. A. Rosenthal and M. R. Berenbaum, eds., Academic Press, San Diego, CA, pp. 165–219.

Gunning, B. E. S., and Steer, M. W. (1996) *Plant Cell Biology: Structure and Function of Plant Cells.* Jones and Bartlett, Boston.

Hain, R., Reif, H.-J., Krause, E., Langebartels, R., Kindl, H., Vornam, B., Wiese, W., Schmelzer, E., Schreier, P. H., Stoecker, R. H., and Stenzel, K. (1993) Disease resistance results from foreign phytoalexin expression in a novel plant. *Nature* 361: 153–156.

Hartmann, T. (1992) Alkaloids. In *Herbivores: Their Interactions with Secondary Plant Metabolites*, Vol. 1: *The Chemical Participants*, 2nd ed., G. A. Rosenthal and M. R. Berenbaum, eds., Academic Press, San Diego, CA, pp. 79–121.

Hartmann, T. (1999) Chemical ecology of pyrrolizidine alkaloids. *Planta* 207: 483–495.

Hatfield, R., and Vermerris, W. (2001) Lignin formation in plants. The dilemma of linkage specificity. *Plant Physiol.* 126: 1351–1357.

Herrmann, K. M., and Weaver, L. M. (1999) The shikimate pathway. *Annu. Rev. Plant Physiol. Plant Mol. Biol.* 50: 473–503.

Inderjit, Dakshini, K. M. M., and Einhellig, F. A., eds. (1995) *Allelopathy: Organisms, Processes, and Applications.* ACS Symposium series American Chemical Society, Washington, DC.

Jeffree, C. E. (1996) Structure and ontogeny of plant cuticles. In *Plant Cuticles: An Integrated Functional Approach*, G. Kerstiens, ed., BIOS Scientific, Oxford, pp. 33–85.

Jin, H., and Martin, C. (1999) Multifunctionality and diversity within the plant *MYB*-gene family. *Plant Mol. Biol.* 41: 577–585.

Johnson, R., Narvaez, J., An, G., and Ryan, C. (1989) Expression of proteinase inhibitors I and II in transgenic tobacco plants: Effects on natural defense against *Manduca sexta* larvae. *Proc. Natl. Acad. Sci. USA* 86: 9871–9875.

Karban, R., and Baldwin, I. T. (1997) *Induced Responses to Herbivory.* University of Chicago Press, Chicago.

Web Chapter

14 Gene Expression and Signal Transduction

Content available at www.plantphys.net

Each living cell contains a set of instructions for building the entire organism, consisting of genes linearly arranged in the form of chromosomes. This fundamental concept in biology began with Mendel's genetic studies with garden peas in 1865, and culminated with Watson and Crick's discovery of the structure of DNA in 1953. But the story did not end there. A new field of molecular biology arose focused on the structure, replication, and expression of genes. Genes encode proteins, and elucidation of the elaborate machinery involved in transcription and translation was one of the early triumphs of the new field of molecular biology. More recently, molecular biologists have sought to understand how gene expression is regulated, for it turns out that the genetic "instructions" found on the chromosomes are incomplete without a full complement of regulatory proteins from the cytoplasm to direct their activity. In this chapter we will review basic concepts in gene expression in prokaryotes and eukaryotes.

While molecular biologists were studying cell function from the gene outward, developmental biologists were tracking the signals that regulate development, both external and internal, from the "skin" inward. They discovered that developmental signals, such as light or hormones, involve specific receptors and typically require amplification in the form of "second messengers." Ultimately these second messengers regulate the activities of crucial processes, such as membrane transport or gene expression, which bring about the physiological or developmental response. Thus developmental and molecular biologists approach the same problem from opposite directions. The second part of this chapter provides an overview of various signaling mecha-

(Continued)

nisms found in living cells. The models presented are derived mainly from animal and microbial systems, in which they were first discovered. Related mechanisms in plants will be discussed in the various chapters of the text devoted to development, light, and hormones.

Chapter

15 *Cell Walls: Structure, Biogenesis, and Expansion*

PLANT CELLS, UNLIKE ANIMAL CELLS, are surrounded by a relatively thin but mechanically strong cell wall. This wall consists of a complex mixture of polysaccharides and other polymers that are secreted by the cell and are assembled into an organized network linked together by both covalent and noncovalent bonds. Plant cell walls also contain structural proteins, enzymes, phenolic polymers, and other materials that modify the wall's physical and chemical characteristics.

The cell walls of prokaryotes, fungi, algae, and plants are distinctive from each other in chemical composition and microscopic structure, yet they all serve two common primary functions: regulating cell volume and determining cell shape. As we will see, however, plant cell walls have acquired additional functions that are not apparent in the walls of other organisms. Because of these diverse functions, the structure and composition of plant cell walls are complex and variable.

In addition to these biological functions, the plant cell wall is important in human economics. As a natural product, the plant cell wall is used commercially in the form of paper, textiles, fibers (cotton, flax, hemp, and others), charcoal, lumber, and other wood products. Another major use of plant cell walls is in the form of extracted polysaccharides that have been modified to make plastics, films, coatings, adhesives, gels, and thickeners in a huge variety of products.

As the most abundant reservoir of organic carbon in nature, the plant cell wall also takes part in the processes of carbon flow through ecosystems. The organic substances that make up humus in the soil and that enhance soil structure and fertility are derived from cell walls. Finally, as an important source of roughage in our diet, the plant cell wall is a significant factor in human health and nutrition.

We begin this chapter with a description of the general structure and composition of cell walls and the mechanisms of the biosynthesis and secretion of cell wall materials. We then turn to the role of the primary cell wall in cell expansion. The mechanisms of tip growth will be contrasted with those of diffuse growth, particularly with respect to the establishment of cell polarity and the control of the rate of cell expansion. Finally, we will describe the dynamic changes in the cell wall that often accompany cell differentiation, along with the role of cell wall fragments as signaling molecules.

THE STRUCTURE AND SYNTHESIS OF PLANT CELL WALLS

Without a cell wall, plants would be very different organisms from what we know. Indeed, the plant cell wall is essential for many processes in plant growth, development, maintenance, and reproduction:

- Plant cell walls determine the mechanical strength of plant structures, allowing those structures to grow to great heights.
- Cell walls glue cells together, preventing them from sliding past one another. This constraint on cellular movement contrasts markedly to the situation in animal cells, and it dictates the way in which plants develop (see Chapter 16).
- A tough outer coating enclosing the cell, the cell wall acts as a cellular "exoskeleton" that controls cell shape and allows high turgor pressures to develop.
- Plant morphogenesis depends largely on the control of cell wall properties because the expansive growth of plant cells is limited principally by the ability of the cell wall to expand.
- The cell wall is required for normal water relations of plants because the wall determines the relationship between the cell turgor pressure and cell volume (see Chapter 3).
- The bulk flow of water in the xylem requires a mechanically tough wall that resists collapse by the negative pressure in the xylem.
- The wall acts as a diffusion barrier that limits the size of macromolecules that can reach the plasma membrane from outside, and it is a major structural barrier to pathogen invasion.

Much of the carbon that is assimilated in photosynthesis is channeled into polysaccharides in the wall. During specific phases of development, these polymers may be hydrolyzed into their constituent sugars, which may be scavenged by the cell and used to make new polymers. This phenomenon is most notable in many seeds, in which wall polysaccharides of the endosperm or cotyledons function primarily as food reserves. Furthermore, oligosaccharide components of the cell wall may act as important signaling molecules during cell differentiation and during recognition of pathogens and symbionts.

The diversity of functions of the plant cell wall requires a diverse and complex plant cell wall structure. In this section we will begin with a brief description of the morphology and basic architecture of plant cell walls. Then we will discuss the organization, composition, and synthesis of primary and secondary cell walls.

Plant Cell Walls Have Varied Architecture

Stained sections of plant tissues reveal that the cell wall is not uniform, but varies greatly in appearance and composition in different cell types (Figure 15.1). Cell walls of the cortical parenchyma are generally thin and have few distinguishing features. In contrast, the walls of some specialized cells, such as epidermal cells, collenchyma, phloem fibers, xylem tracheary elements, and other forms of sclerenchyma have thicker, multilayered walls. Often these walls are intricately sculpted and are impregnated with specific substances, such as lignin, cutin, suberin, waxes, silica, or structural proteins.

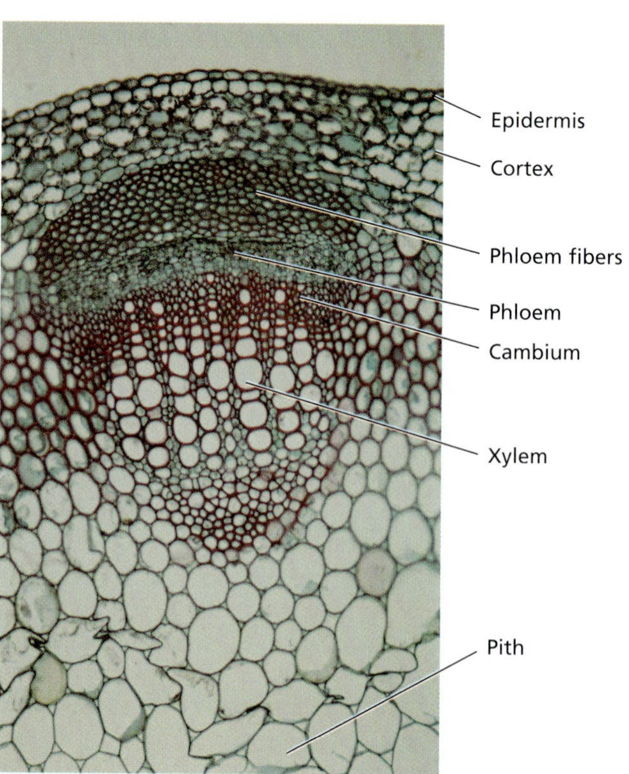

FIGURE 15.1 Cross section of a stem of *Trifolium* (clover), showing cells with varying wall morphology. Note the highly thickened walls of the phloem fibers. (Photo © James Solliday/Biological Photo Service.)

The individual sides of a wall surrounding a cell may also vary in thickness, embedded substances, sculpting, and frequency of pitting and plasmodesmata. For example, the outer wall of the epidermis is usually much thicker than the other walls of the cell; moreover, this wall lacks plasmodesmata and is impregnated with cutin and waxes. In guard cells, the side of the wall adjacent to the stomatal pore is much thicker than the walls on the other sides of the cell. Such variations in wall architecture for a single cell reflect the polarity and differentiated functions of the cell and arise from targeted secretion of wall components to the cell surface.

Despite this diversity in cell wall morphology, cell walls commonly are classified into two major types: primary walls and secondary walls. **Primary walls** are formed by growing cells and are usually considered to be relatively unspecialized and similar in molecular architecture in all cell types. Nevertheless, the ultrastructure of primary walls also shows wide variation. Some primary walls, such as those of the onion bulb parenchyma, are very thin (100 nm) and architecturally simple (Figure 15.2). Other primary walls, such as those found in collenchyma or in the epidermis (Figure 15.3), may be much thicker and consist of multiple layers.

Secondary walls are the cell walls that form after cell growth (enlargement) has ceased. Secondary walls may become highly specialized in structure and composition, reflecting the differentiated state of the cell. Xylem cells, such as those found in wood, are notable for possessing highly thickened secondary walls that are strengthened by **lignin** (see Chapter 13).

A thin layer of material, the **middle lamella** (plural *lamellae*), can usually be seen at the junction where the walls of neighboring cells come into contact. The composition of the middle lamella differs from the rest of the wall in that it is high in pectin and contains different proteins compared with the bulk of the wall. Its origin can be traced to the cell plate that formed during cell division.

As we saw in Chapter 1, the cell wall is usually penetrated by tiny membrane-lined channels, called **plasmodesmata** (singular *plasmodesma*), which connect neighboring cells. Plasmodesmata function in communication between cells, by allowing passive transport of small molecules and active transport of proteins and nucleic acids between the cytoplasms of adjacent cells.

The Primary Cell Wall Is Composed of Cellulose Microfibrils Embedded in a Polysaccharide Matrix

In primary cell walls, cellulose microfibrils are embedded in a highly hydrated matrix (Figure 15.4). This structure provides both strength and flexibility. In the case of cell walls, the **matrix** (plural *matrices*) consists of two major groups of polysaccharides, usually called hemicelluloses and pectins, plus a small amount of structural protein. The matrix polysaccharides consist of a variety of polymers that may vary according to cell type and plant species (Table 15.1).

(A)

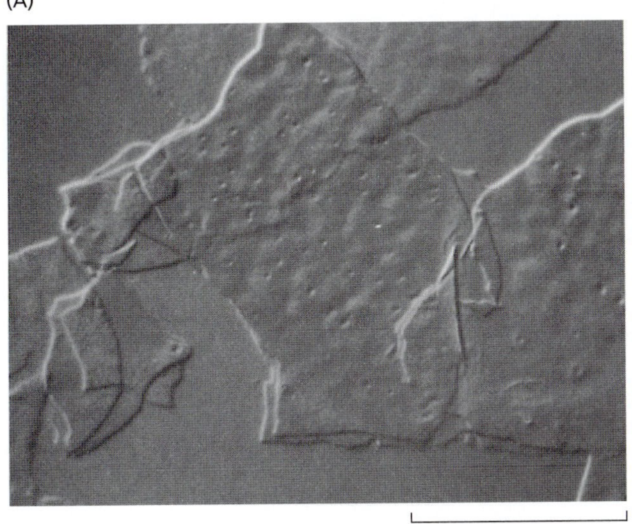

(B)

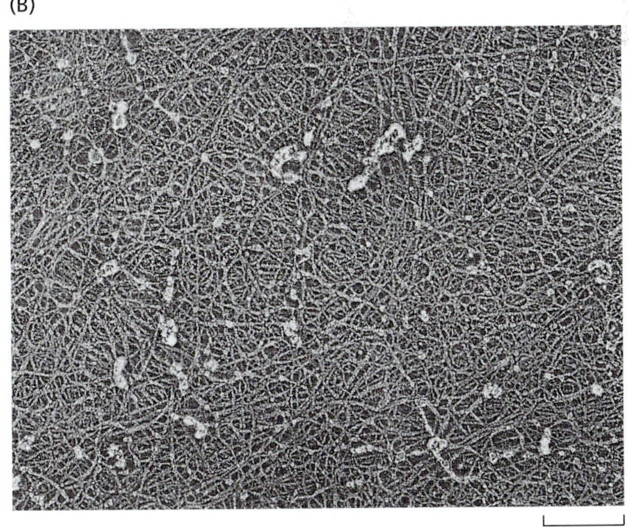

FIGURE 15.2 Primary cell walls from onion parenchyma. (A) This surface view of cell wall fragments was taken through the use of Nomarski optics. Note that the wall looks like a very thin sheet with small surface depressions; these depressions may be pit fields, places where plasmodesmatal connections between cells are concentrated. (B) This surface view of a cell wall was prepared by a freeze-etch replica technique. It shows the fibrillar nature of the cell wall. (From McCann et al. 1990, courtesy of M. McCann.)

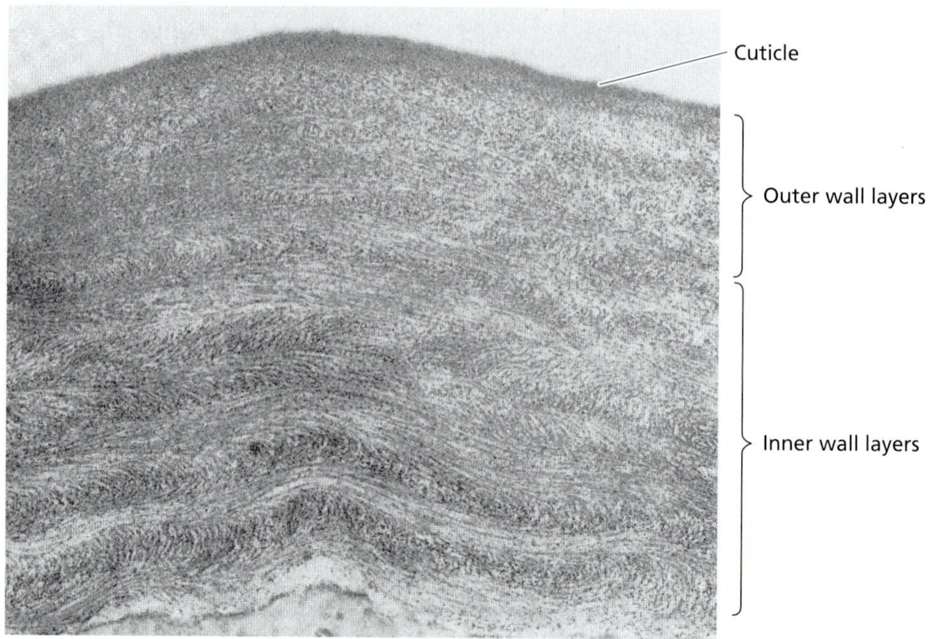

FIGURE 15.3 Electron micrograph of the outer epidermal cell wall from the growing region of a bean hypocotyl. Multiple layers are visible within the wall. The inner layers are thicker and more defined than the outer layers because the outer layers are the older regions of the wall and have been stretched and thinned by cell expansion. (From Roland et al. 1982.)

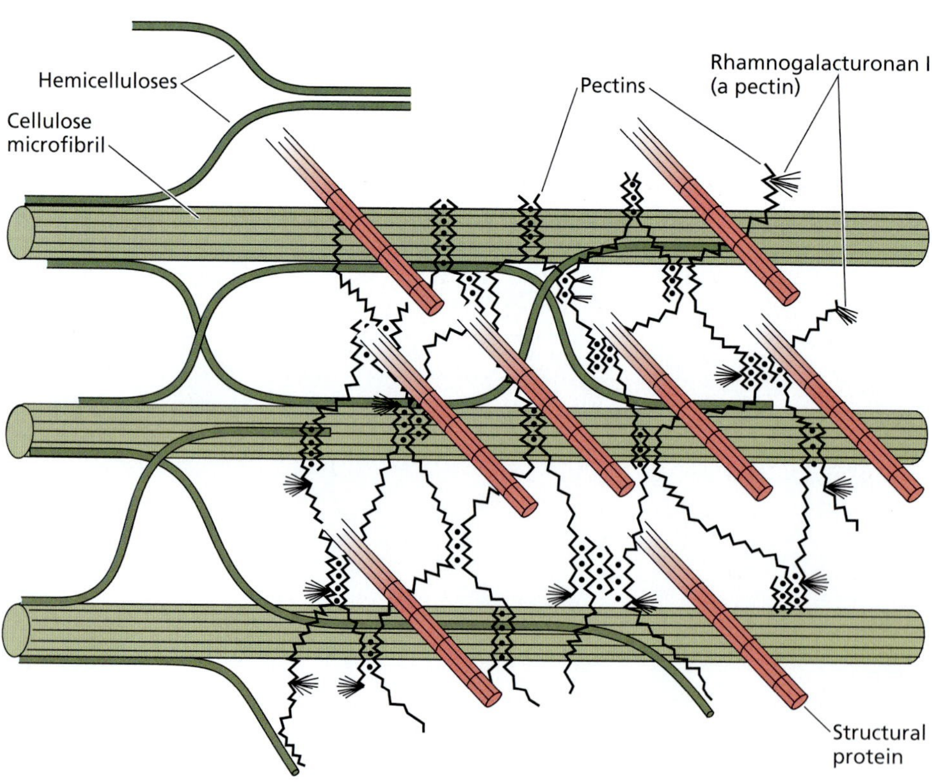

FIGURE 15.4 Schematic diagram of the major structural components of the primary cell wall and their likely arrangement. Cellulose microfibrils are coated with hemicelluloses (such as xyloglucan), which may also cross-link the microfibrils to one another. Pectins form an interlocking matrix gel, perhaps interacting with structural proteins. (From Brett and Waldron 1996.)

TABLE 15.1
Structural components of plant cell walls

Class	Examples
Cellulose	Microfibrils of (1→4)β-D-glucan
Matrix Polysaccharides	
Pectins	Homogalacturonan
	Rhamnogalacturonan
	Arabinan
	Galactan
Hemicelluloses	Xyloglucan
	Xylan
	Glucomannan
	Arabinoxylan
	Callose (1→3)β-D-glucan
	(1→3,1→4)β-D-glucan [grasses only]
Lignin	(see Chapter 13)
Structural proteins	(see Table 15.2)

These polysaccharides are named after the principal sugars they contain. For example, a *glucan* is a polymer made up of glucose, a *xylan* is a polymer made up of xylose, a *galactan* is made from galactose, and so on. *Glycan* is the general term for a polymer made up of sugars. For branched polysaccharides, the backbone of the polysaccharide is usually indicated by the last part of the name.

For example, *xyloglucan* has a glucan backbone (a linear chain of glucose residues) with xylose sugars attached to it in the side chains; *glucuronoarabinoxylan* has a xylan backbone (made up of xylose subunits) with glucuronic acid and arabinose side chains. However, a compound name does not necessarily imply a branched structure. For example, *glucomannan* is the name given to a polymer containing both glucose and mannose in its backbone.

Cellulose microfibrils are relatively stiff structures that contribute to the strength and structural bias of the cell wall. The individual glucans that make up the microfibril are closely aligned and bonded to each other to make a highly ordered (**crystalline**) ribbon that excludes water and is relatively inaccessible to enzymatic attack. As a result, cellulose is very strong and very stable and resists degradation.

Hemicelluloses are flexible polysaccharides that characteristically bind to the surface of cellulose. They may form tethers that bind cellulose microfibrils together into a cohesive network (see Figure 15.4), or they may act as a slippery coating to prevent direct microfibril–microfibril contact. Another term for these molecules is *cross-linking glucans*, but in this chapter we'll use the more traditional term, *hemicelluloses*. As described later, the term *hemicellulose* includes several different kinds of polysaccharides.

Pectins form a hydrated gel phase in which the cellulose–hemicellulose network is embedded. They act as hydrophilic filler, to prevent aggregation and collapse of the cellulose network. They also determine the porosity of the cell wall to macromolecules. Like hemicelluloses, pectins include several different kinds of polysaccharides.

The precise role of wall **structural proteins** is uncertain, but they may add mechanical strength to the wall and assist in the proper assembly of other wall components.

The primary wall is composed of approximately 25% cellulose, 25% hemicelluloses, and 35% pectins, with perhaps 1 to 8% structural protein, on a dry-weight basis. However, large deviations from these values may be found. For example, the walls of grass coleoptiles consist of 60 to 70% hemicelluloses, 20 to 25% cellulose, and only about 10% pectins. Cereal endosperm walls are mostly (about 85%) hemicelluloses. Secondary walls typically contain much higher cellulose contents.

In this chapter we will present a basic model of the primary wall, but be aware that plant cell walls are more diverse than this model suggests. The composition of matrix polysaccharides and structural proteins in walls varies significantly among different species and cell types (Carpita and McCann 2000). Most notably, in grasses and related species the major matrix polysaccharides differ from those that make up the matrix of most other land plants (Carpita 1996).

The primary wall also contains much water. This water is located mostly in the matrix, which is perhaps 75 to 80% water. The hydration state of the matrix is an important determinant of the physical properties of the wall; for example, removal of water makes the wall stiffer and less extensible. This stiffening effect of dehydration may play a role in growth inhibition by water deficits. We will examine the structure of each of the major polymers of the cell wall in more detail in the sections that follow.

Cellulose Microfibrils Are Synthesized at the Plasma Membrane

Cellulose is a tightly packed microfibril of linear chains of (1→4)-linked β-D-glucose (Figure 15.5 and **Web Topic 15.1**). Because of the alternating spatial configuration of the glucosidic bonds linking adjacent glucose residues, the repeating unit in cellulose is considered to be cellobiose, a (1→4)-linked β-D-glucose disaccharide.

Cellulose microfibrils are of indeterminate length and vary considerably in width and in degree of order, depending on the source. For instance, cellulose microfibrils in land plants appear under the electron microscope to be 5 to 12 nm wide, whereas those formed by algae may be up to 30 nm wide and more crystalline. This variety in width corresponds to a variation in the number of parallel chains that make up the cross section of a microfibril—estimated to consist of about 20 to 40 individual chains in the thinner microfibrils.

The precise molecular structure of the cellulose microfibril is uncertain. Current models of microfibril organization suggest that it has a substructure consisting of highly crystalline domains linked together by less organized "amor-

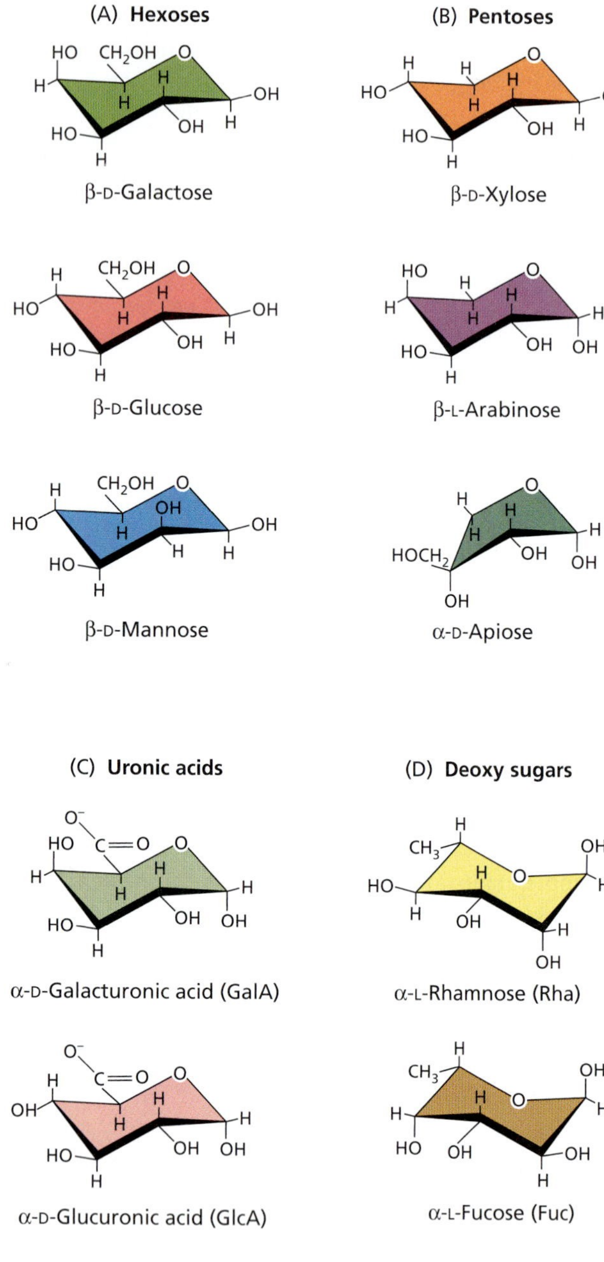

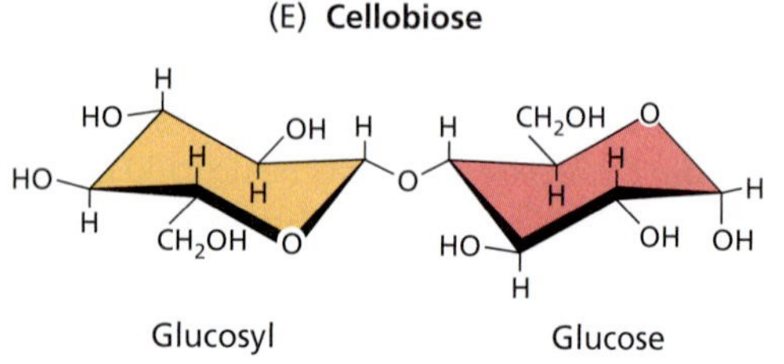

FIGURE 15.5 Conformational structures of sugars commonly found in plant cell walls. (A) Hexoses (six-carbon sugars). (B) Pentoses (five-carbon sugars). (C) Uronic acids (acidic sugars). (D) Deoxy sugars. (E) Cellobiose (showing the (1→4)β-D-linkage between two glucose residues in inverted orientation).

phous" regions (Figure 15.6). Within the crystalline domains, adjacent glucans are highly ordered and bonded to each other by noncovalent bonding, such as hydrogen bonds and hydrophobic interactions.

The individual glucan chains of cellulose are composed of 2000 to more than 25,000 glucose residues (Brown et al. 1996). These chains are long enough (about 1 to 5 μm long) to extend through multiple crystalline and amorphous regions within a microfibril. When cellulose is degraded—for example, by fungal cellulases—the amorphous regions are degraded first, releasing small crystallites that are thought to correspond to the crystalline domains of the microfibril.

The extensive noncovalent bonding between adjacent glucans within a cellulose microfibril gives this structure remarkable properties. Cellulose has a high tensile strength, equivalent to that of steel. Cellulose is also insoluble, chemically stable, and relatively immune to chemical and enzymatic attack. These properties make cellulose an excellent structural material for building a strong cell wall.

Evidence from electron microscopy indicates that cellulose microfibrils are synthesized by large, ordered protein complexes, called **particle rosettes** or **terminal complexes**, that are embedded in the plasma membrane (Figure 15.7) (Kimura et al. 1999). These structures contain many units of **cellulose synthase**, the enzyme that synthesizes the individual (1→4)β-D-glucans that make up the microfibril (see **Web Topic 15.2**).

Cellulose synthase, which is located on the cytoplasmic side of the plasma membrane, transfers a glucose residue from a sugar nucleotide donor to the growing glucan chain. Sterol-glucosides (sterols linked to a chain of two or three glucose residues) serve as the primers, or initial acceptors, to start the growth of the glucan chain (Peng et al. 2002). The sterol is clipped from the glucan by an endoglucanase, and the growing glucan chain is then extruded through the membrane to the exterior of the cell, where, together with other glucan chains, it crystallizes into a microfibril and interacts with xyloglucans and other matrix polysaccharides.

The sugar nucleotide donor is probably uridine diphosphate D-glucose (UDP-glucose). Recent evidence suggests that the glucose used for the synthesis of cellulose may be obtained from sucrose (a disaccharide composed of fructose and glucose) (Amor et al. 1995; Salnikov et al. 2001). According to this hypothesis, the enzyme **sucrose synthase** acts as a metabolic channel to transfer glucose taken from sucrose, via UDP-glucose, to the growing cellulose chain (Figure 15.8).

After many years of fruitless searching, the genes for cellulose synthase in higher plants have now been isolated (Pear et al. 1996; Arioli et al. 1998; Holland et al. 2000; Richmond and Somerville 2000). In *Arabidopsis*, the cellulose synthases are part of a large family of proteins whose function may be to synthesize the backbones of many cell wall polysaccharides.

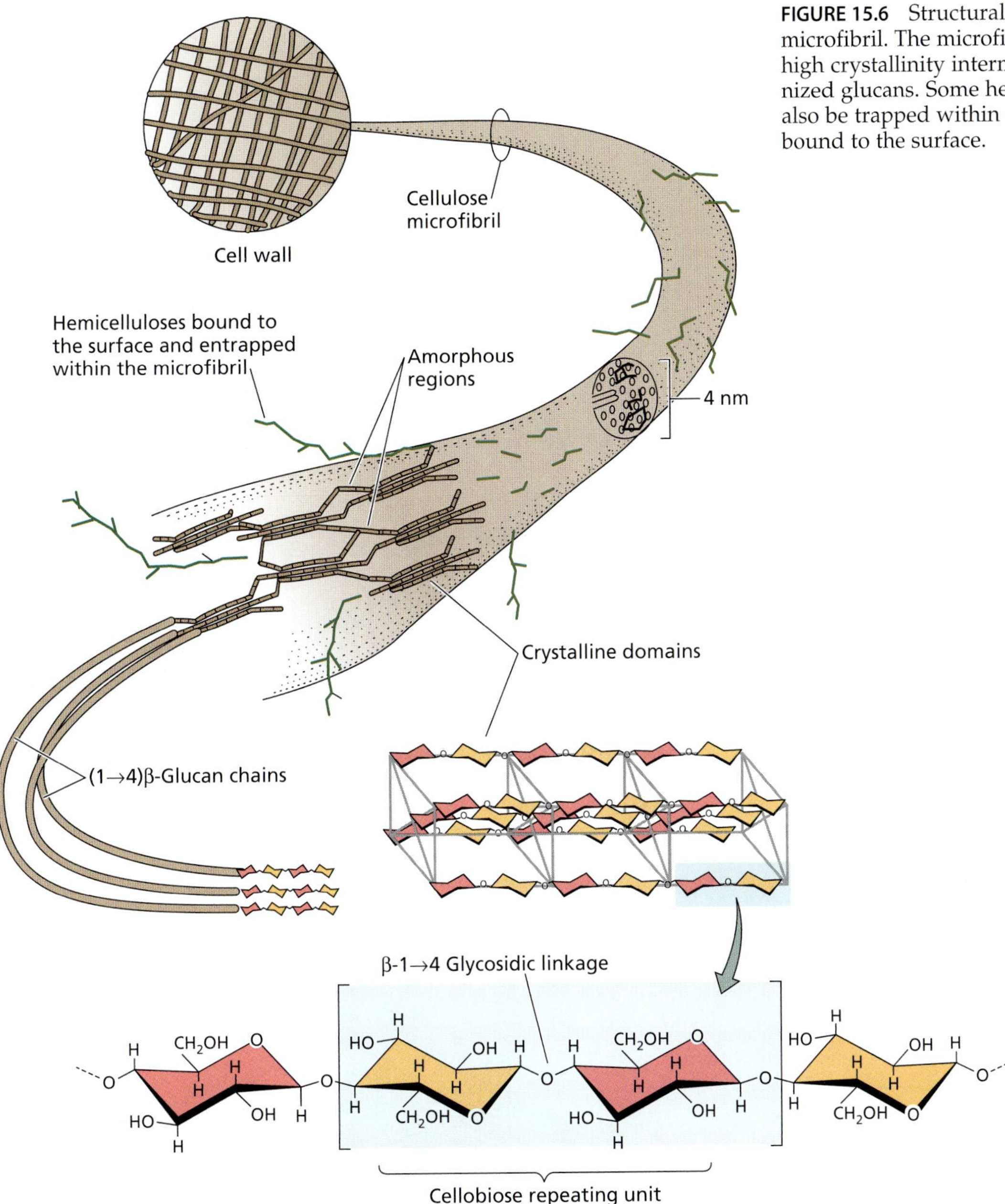

FIGURE 15.6 Structural model of a cellulose microfibril. The microfibril has regions of high crystallinity intermixed with less organized glucans. Some hemicelluloses may also be trapped within the microfibril and bound to the surface.

The formation of cellulose involves not only the synthesis of the glucan, but also the crystallization of multiple glucan chains into a microfibril. Little is known about the control of this process, except that the direction of microfibril deposition may be guided by microtubules adjacent to the membrane.

When the cellulose microfibril is synthesized, it is deposited into a milieu (the wall) that contains a high concentration of other polysaccharides that are able to interact with and perhaps modify the growing microfibril. In vitro binding studies have shown that hemicelluloses such as xyloglucan and xylan may bind to the surface of cellulose. Some hemicelluloses may also become physically entrapped within the microfibril during its formation, thereby reducing the crystallinity and order of the microfibril (Hayashi 1989).

Matrix Polymers Are Synthesized in the Golgi and Secreted in Vesicles

The matrix is a highly hydrated phase in which the cellulose microfibrils are embedded. The major polysaccharides of the matrix are synthesized by membrane-bound enzymes in the Golgi apparatus and are delivered to the cell wall via exocytosis of tiny vesicles (Figure 15.9 and **Web Topic 15.3**). The enzymes responsible for synthesis are *sugar-nucleotide polysaccharide glycosyltransferases*. These

FIGURE 15.7 Cellulose synthesis by the cell. (A) Electron micrograph showing newly synthesized cellulose microfibrils immediately exterior to the plasma membrane. (B) Freeze-fracture labeled replicas showing reactions with antibodies against cellulose synthase. A field of labeled rosettes (arrows) with seven clearly labeled rosettes and one unlabeled rosette. The inset shows an enlarged view of two selected rosettes (terminal complexes) with immunogold labels. (C) Schematic diagram showing cellulose being synthesized by membrane synthase complex ("rosette") and its presumed guidance by the underlying microtubules in the cytoplasm. (A and C from Gunning and Steer 1996 B from Kimura et al. 1999.)

(A)

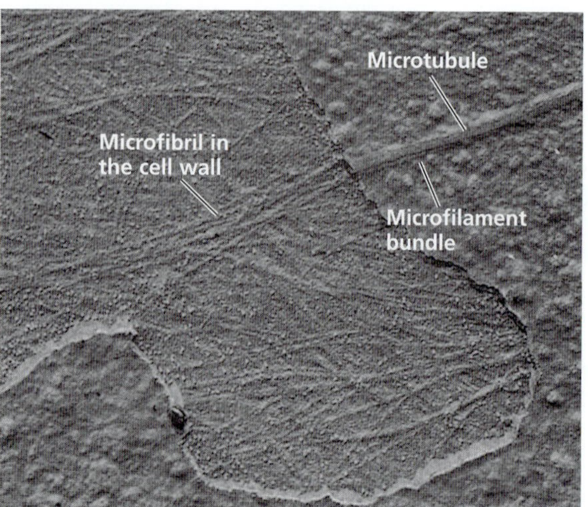

(B)

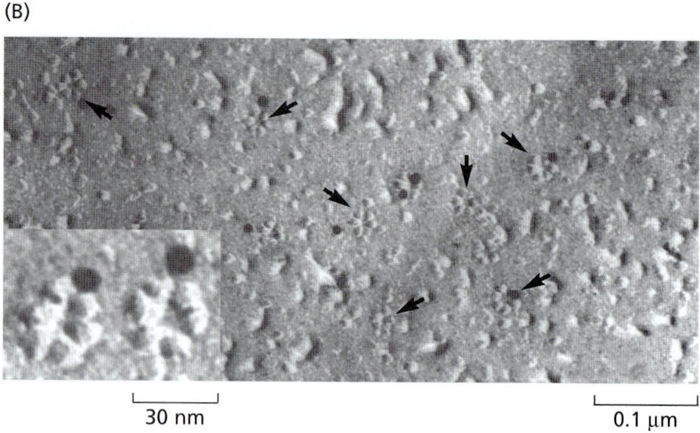

(C)

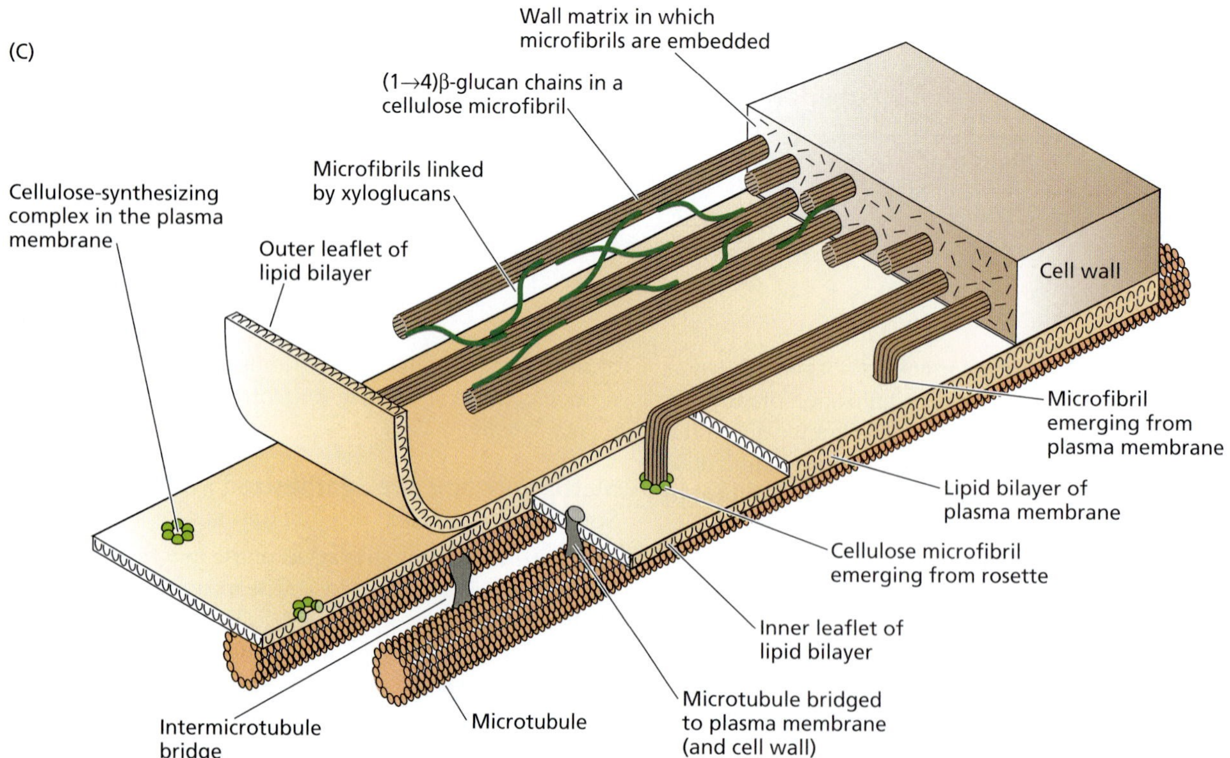

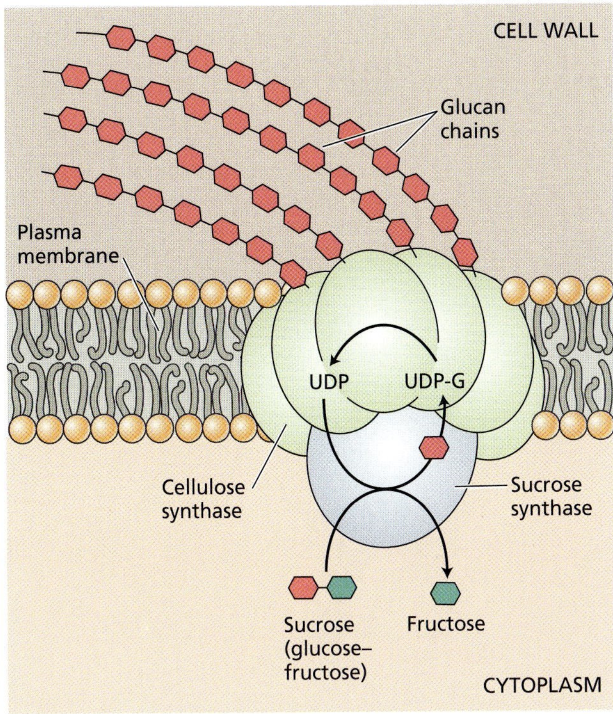

FIGURE 15.8 Model of cellulose synthesis by a multisubunit complex containing cellulose synthase. Glucose residues are donated to the growing glucan chains by UDP-glucose (UDP-G). Sucrose synthase may act as a metabolic channel to transfer glucose taken from sucrose to UDP-glucose, or UDP-glucose may be obtained directly from the cytoplasm. (After Delmer and Amor 1995.)

enzymes transfer monosaccharides from sugar nucleotides to the growing end of the polysaccharide chain.

Unlike cellulose, which forms a crystalline microfibril, the matrix polysaccharides are much less ordered and are often described as amorphous. This noncrystalline character is a consequence of the structure of these polysaccharides—their branching and their nonlinear conformation. Nevertheless, spectroscopy studies indicate that there is partial order in the orientation of hemicelluloses and pectins in the cell wall, probably as a result of a physical tendency for these polymers to become aligned along the long axis of cellulose (Séné et al. 1994; Wilson et al. 2000).

Hemicelluloses Are Matrix Polysaccharides That Bind to Cellulose

Hemicelluloses are a heterogeneous group of polysaccharides (Figure 15.10) that are bound tightly in the wall. Typically they are solubilized from depectinated walls by the use of a strong alkali (1–4 *M* NaOH). Several kinds of hemicelluloses are found in plant cell walls, and walls from different tissues and different species vary in their hemicellulose composition.

In the primary wall of dicotyledons (the best-studied example), the most abundant hemicellulose is **xyloglucan** (see Figure 15.10A). Like cellulose, this polysaccharide has a backbone of (1→4)-linked β-D-glucose residues. Unlike cellulose, however, xyloglucan has short side chains that contain xylose, galactose, and often, though not always, a terminal fucose.

By interfering with the linear alignment of the glucan backbones with one another, these side chains prevent the

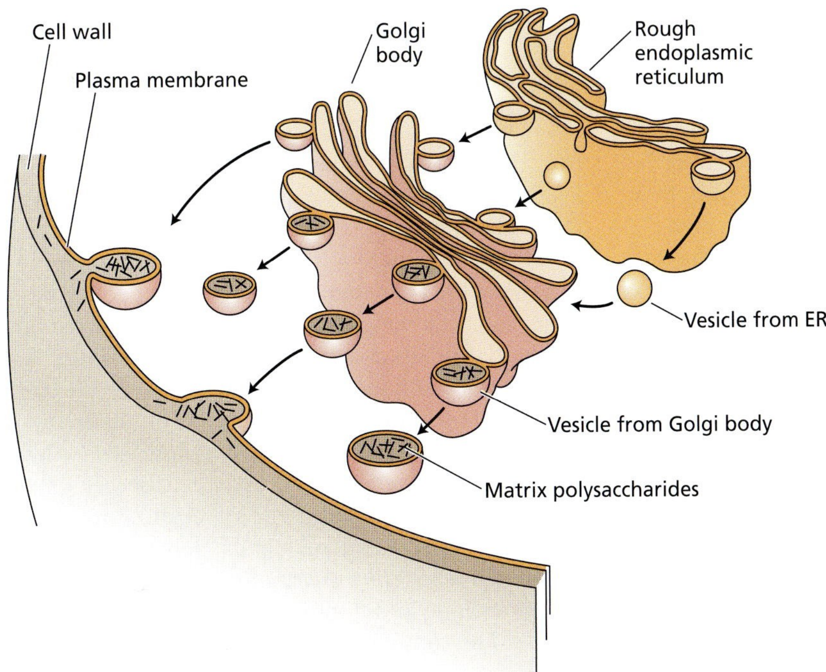

FIGURE 15.9 Scheme for the synthesis and delivery of matrix polysaccharides to the cell wall. Polysaccharides are synthesized by enzymes in the Golgi apparatus and then secreted to the wall by fusion of membrane vesicles to the plasma membrane.

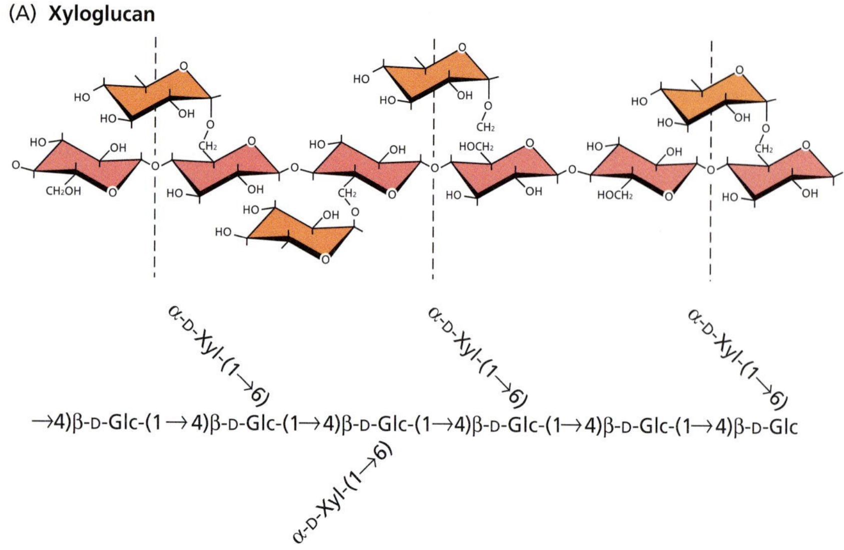

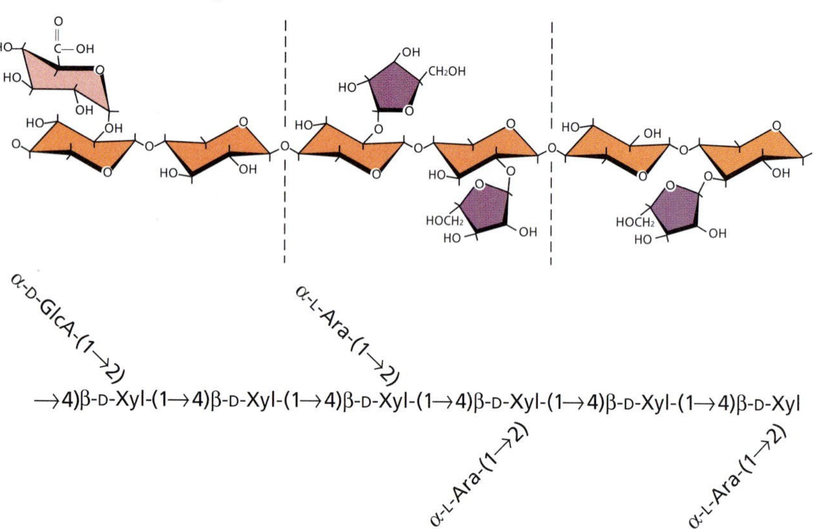

FIGURE 15.10 Partial structures of common hemicelluloses. (For details on carbohydrate nomenclature, see **Web Topic 15.1**.) (A) Xyloglucan has a backbone of (1→4)-linked β-D-glucose (Glc), with (1→6)-linked branches containing β-D-xylose (Xyl). In some cases galactose (Gal) and fucose (Fuc) are added to the xylose side chains. (B) Glucuronoarabinoxylans have a (1→4)-linked backbone of β-D-xylose (Xyl). They may also have side chains containing arabinose (Ara), 4-*O*-methyl-glucuronic acid (4-*O*-Me-α–D–GlcA), or other sugars. (From Carpita and McCann 2000.)

assembly of xyloglucan into a crystalline microfibril. Because xyloglucans are longer (about 50–500 nm) than the spacing between cellulose microfibrils (20–40 nm), they have the potential to link several microfibrils together.

Varying with the developmental state and plant species, the hemicellulose fraction of the wall may also contain large amounts of other important polysaccharides—for example, **glucuronoarabinoxylans** (see Figure 15.10B) and **glucomannans**. Secondary walls typically contain less xyloglucan and more xylans and glucomannans, which also bind tightly to cellulose. The cell walls of grasses contain only small amounts of xyloglucan and pectin, which are replaced by glucuronoarabinoxylan and (1→3,1→4)β-D-glucan.

Pectins Are Gel-Forming Components of the Matrix

Like the hemicelluloses, pectins constitute a heterogeneous group of polysaccharides (Figure 15.11), characteristically containing acidic sugars such as galacturonic acid and neutral sugars such as rhamnose, galactose, and arabinose. Pectins are the most soluble of the wall polysaccharides; they can be extracted with hot water or with calcium chelators. In the wall, pectins are very large and complex molecules composed of different kinds of pectic polysaccharides.

Some pectic polysaccharides have a relatively simple primary structure, such as *homogalacturonan* (see Figure 15.11A). This polysaccharide, also called *polygalacturonic acid*, is a (1→4)-linked polymer of α-D-glucuronic acid residues. Figure 15.12 shows a triple-fluorescence-labeled section of tobacco stem parenchyma cells showing the distribution of cellulose and pectic homogalacturonan.

One of the most abundant of the pectins is the complex polysaccharide *rhamnogalacturonan I* (*RG I*), which has a long backbone and a variety of side chains (see Figure 15.11B). This molecule is very large and is believed to contain highly branched ("hairy") regions (i.e., with arabinan,

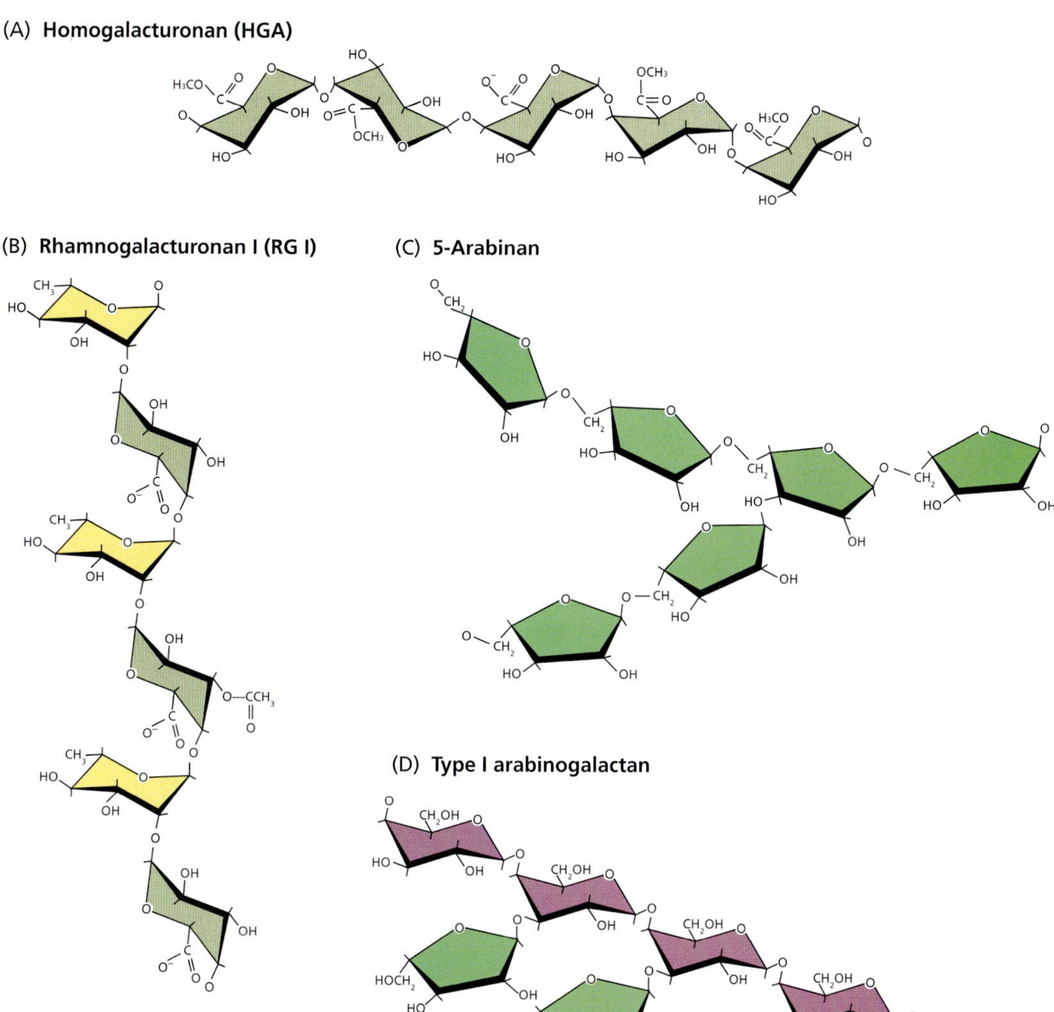

FIGURE 15.11 Partial structures of the most common pectins. (A) Homogalacturonan, also known as polygalacturonic acid or pectic acid, is made up of (1→4)-linked α-D-galacturonic acid (GalA) with occasional rhamnosyl residues that put a kink in the chain. The carboxyl residues are often methyl esterified. (B) Rhamnogalacturonan I (RG I) is a very large and heterogeneous pectin, with a backbone of alternating (1→4)α-D-galacturonic acid (GalA) and (1→2)α-D-rhamnose (Rha). Side chains are attached to rhamnose and are composed principally of arabinans (C), galactans, and arabinogalactans (D). These side chains may be short or quite long. The galacturonic acid residues are often methyl esterified. (From Carpita and McCann 2000.)

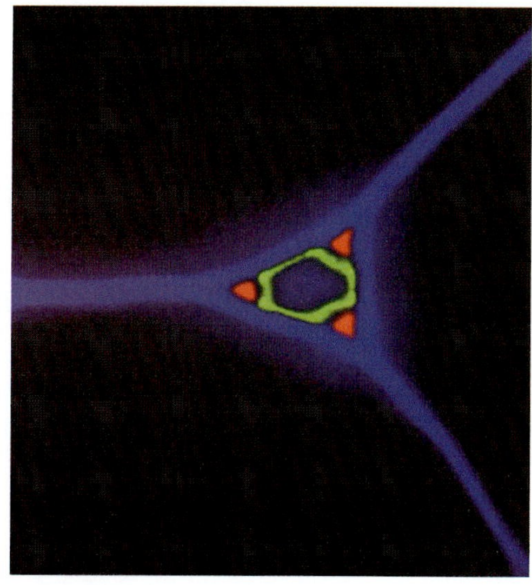

FIGURE 15.12 Triple-fluorescence-labeled section of tobacco stem showing the primary cell walls of three adjacent parenchyma cells bordering an intercellular space. The blue color is calcofluor (staining of cellulose), and the red and green colors indicate the binding of two monoclonal antibodies to different epitopes (immunologically distinct regions) of pectic homogalacturonan. (Courtesy of W. Willats.)

and galactan side shains) interspersed with unbranched ("smooth") regions of homogalacturonan (Figure 15.13A).

Pectic polysaccharides may be very complex. A striking example is a highly branched pectic polysaccharide called *rhamnogalacturonan II (RG II)* (see Figure 15.13C), which contains at least ten different sugars in a complicated pattern of linkages. Although RG I and RG II have similar names, *they have very different structures*. RG II units may be cross-linked by borate diesters (Ishi et al. 1999) and are important for wall structure. For example, *Arabidopsis* mutants that synthesize an altered RG II that is unable to be cross-linked by borate show substantial growth abnormalities (O'Neill et al. 2001).

Pectins typically form gels—loose networks formed by highly hydrated polymers. Pectins are what make fruit jams and jellies "gel," or solidify. In pectic gels, the charged carboxyl (COO^-) groups of neighboring pectin chains are linked

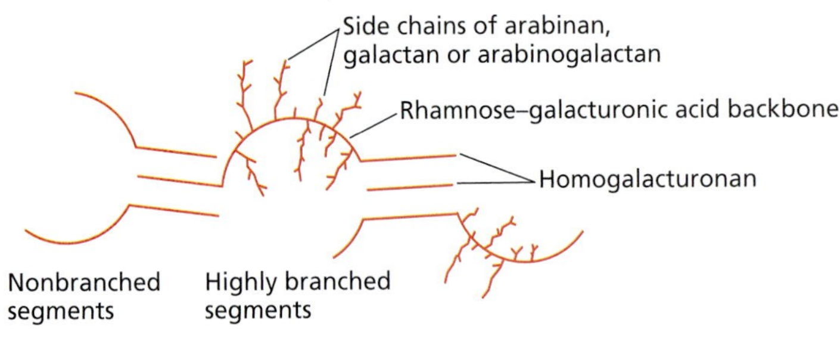

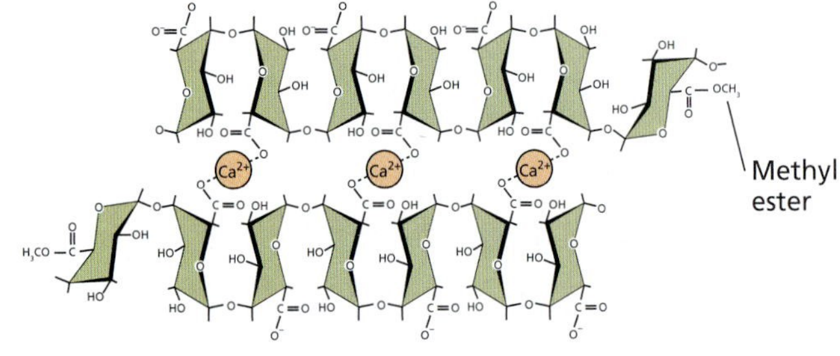

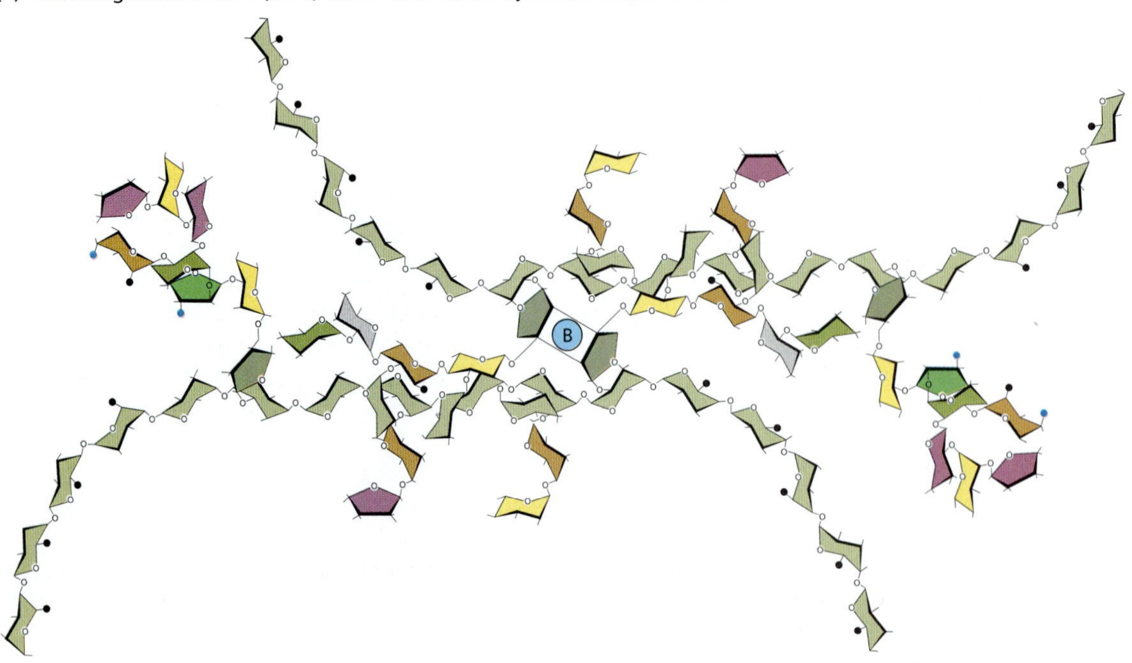

FIGURE 15.13 Pectin structure. (A) Proposed structure of rhamnogalacturonan I, containing highly branched segments interspersed with nonbranched segments, and a backbone of rhamnose and galacturonic acid. (B) Formation of a pectin network involves ionic bridging of the nonesterified carboxyl groups (COO^-) by calcium ions. When blocked by methyl-esterified groups, the carboxyl groups cannot participate in this type of interchain network formation. Likewise, the presence of side chains on the backbone interferes with network formation. (C) Structure of rhamnogalacturonan II (RG II). (B and C from Carpita and McCann 2000.)

TABLE 15.2
Structural proteins of the cell wall

Class of cell wall proteins	Percentage carbohydrate	Localization typically in:
HRGP (hydroxyproline-rich glycoprotein)	~55	Phloem, cambium, sclereids
PRP (proline-rich protein)	~0–20	Xylem, fibers, cortex
GRP (glycine-rich protein)	0	Xylem

together via Ca^{2+}, which forms a tight complex with pectin. A large calcium-bridged network may thus form, as illustrated in Figure 15.13B.

Pectins are subject to modifications that may alter their conformation and linkage in the wall. Many of the acidic residues are esterified with methyl, acetyl, and other unidentified groups during biosynthesis in the Golgi apparatus. Such esterification masks the charges of carboxyl groups and prevents calcium bridging between pectins, thereby reducing the gel-forming character of the pectin.

Once the pectin has been secreted into the wall, the ester groups may be removed by pectin esterases found in the wall, thus unmasking the charges of the carboxyl groups and increasing the ability of the pectin to form a rigid gel. By creating free carboxyl groups, de-esterification also increases the electric-charge density in the wall, which in turn may influence the concentration of ions in the wall and the activities of wall enzymes. In addition to being connected by calcium bridging, pectins may be linked to each other by various covalent bonds, including ester linkages between phenolic residues such as ferulic acid (see Chapter 13).

Structural Proteins Become Cross-Linked in the Wall

In addition to the major polysaccharides described in the previous section, the cell wall contains several classes of structural proteins. These proteins usually are classified according to their predominant amino acid composition—for example, hydroxyproline-rich glycoprotein (HRGP), glycine-rich protein (GRP), proline-rich protein (PRP), and so on (Table 15.2). Some wall proteins have sequences that are characteristic of more than one class. Many structural proteins of walls have highly repetitive primary structures and sometimes are highly glycosylated (Figure 15.14).

In vitro extraction studies have shown that newly secreted wall structural proteins are relatively soluble, but they become more and more insoluble during cell maturation or in response to wounding. The biochemical nature of the insolubilization process is uncertain, however.

Wall structural proteins vary greatly in their abundance, depending on cell type, maturation, and previous stimulation. Wounding, pathogen attack, and treatment with elicitors (molecules that activate plant defense responses; see Chapter 13) increase expression of the genes that code for many of these proteins. In histological studies, wall structural proteins are often localized to specific cell and tissue types. For example, HRGPs are associated mostly with cambium, phloem parenchyma, and various types of sclerenchyma. GRPs and PRPs are most often localized to xylem vessels and fibers and thus are more characteristic of a differentiated cell wall.

In addition to the structural proteins already listed, cell walls contain **arabinogalactan proteins** (**AGPs**) which usually amount to less than 1% of the dry mass of the wall. These water-soluble proteins are very heavily glycosylated: More than 90% of the mass of AGPs may be sugar residues—primarily galactose and arabinose (Figure 15.15) (Gaspar et al. 2001). Multiple AGP forms are found in plant tissues, either in the wall or associated with the plasma membrane, and they display tissue- and cell-specific expression patterns.

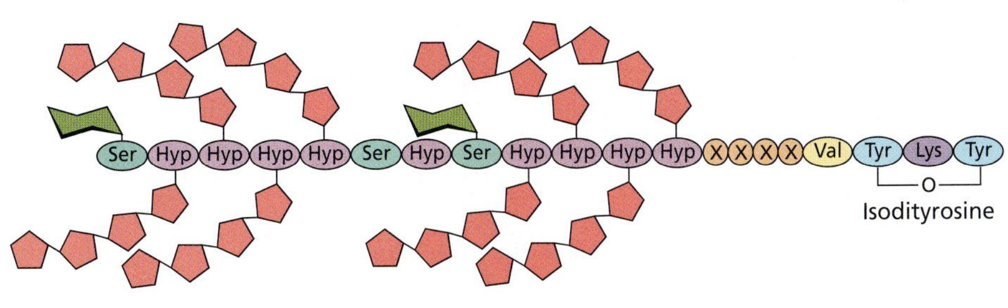

FIGURE 15.14 A repeated hydroxyproline-rich motif from a molecule of extensin from tomato, showing extensive glycosylation and the formation of intramolecular isodityrosine bonds. (From Carpita and McCann 2000.)

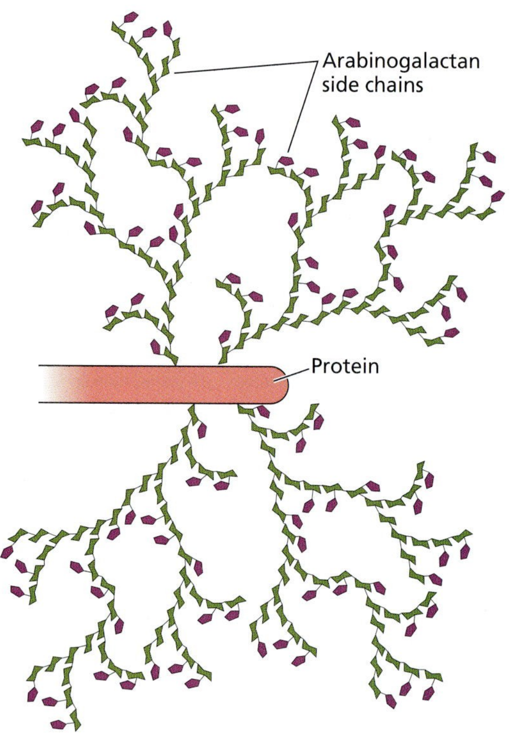

FIGURE 15.15 A highly branched arabinogalactan molecule. (From Carpita and McCann 2000.)

AGPs may function in cell adhesion and in cell signaling during cell differentiation. As evidence for the latter idea, treatment of suspension cultures with exogenous AGPs or with agents that specifically bind AGPs is reported to influence cell proliferation and embryogenesis. AGPs are also implicated in the growth, nutrition, and guidance of pollen tubes through stylar tissues, as well as in other developmental processes (Cheung et al. 1996; Gaspar et al. 2001). Finally, AGPs may also function as a kind of polysaccharide chaperone within secretory vesicles to reduce spontaneous association of newly synthesized polysaccharides until they are secreted to the cell wall.

New Primary Walls Are Assembled during Cytokinesis

Primary walls originate de novo during the final stages of cell division, when the newly formed **cell plate** separates the two daughter cells and solidifies into a stable wall that is capable of bearing a physical load from turgor pressure.

The cell plate forms when Golgi vesicles and ER cisternae aggregate in the spindle midzone area of a dividing cell. This aggregation is organized by the **phragmoplast**, a complex assembly of microtubules, membranes, and vesicles that forms during late anaphase or early telophase (see Chapter 1). The membranes of the vesicles fuse with each other, and with the lateral plasma membrane, to become the new plasma membrane separating the daughter cells. The contents of the vesicles are the precursors from which the new middle lamella and the primary wall are assembled.

After a wall forms, it can grow and mature through a process that may be outlined as follows:

Synthesis → secretion → assembly →
expansion (in growing cells) →
cross-linking and secondary wall formation

The synthesis and secretion of the major wall polymers were described earlier. Here we will consider the assembly and expansion of the wall.

After their secretion into the extracellular space, the wall polymers must be assembled into a cohesive structure; that is, the individual polymers must attain the physical arrangement and bonding relationships that are characteristic of the wall. Although the details of wall assembly are not understood, the prime candidates for this process are self-assembly and enzyme-mediated assembly.

Self-assembly. Self-assembly is attractive because it is mechanistically simple. Wall polysaccharides possess a marked tendency to aggregate spontaneously into organized structures. For example, isolated cellulose may be dissolved in strong solvents and then extruded to form stable fibers, called rayon.

Similarly, hemicelluloses may be dissolved in strong alkali; when the alkali is removed, these polysaccharides aggregate into concentric, ordered networks that resemble the native wall at the ultrastructural level. This tendency to aggregate can make the separation of hemicellulose into its component polymers technically difficult. In contrast, pectins are more soluble and tend to form dispersed, isotropic networks (gels). These observations indicate that the wall polymers have an inherent ability to aggregate into partly ordered structures.

Enzyme-mediated assembly. In addition to self-assembly, wall enzymes may take part in putting the wall together. A prime candidate for enzyme-mediated wall assembly is *xyloglucan endotransglycosylase* (*XET*). This enzyme has the ability to cut the backbone of a xyloglucan and to join one end of the cut xyloglucan with the free end of an acceptor xyloglucan (Figure 15.16). Such a transfer reaction integrates newly synthesized xyloglucans into the wall (Nishitani 1997; Thompson and Fry 2001).

Other wall enzymes that might aid in assembly of the wall include glycosidases, pectin methyl esterases, and various oxidases. Some glycosidases remove the side chains of hemicelluloses. This "debranching" activity increases the tendency of hemicelluloses to adhere to the surface of cellulose microfibrils. Pectin methyl esterases hydrolyze the methyl esters that block the carboxyl groups of pectins. By unblocking the carboxyl groups, these enzymes increase the concentration of acidic groups on the pectins and enhance the ability of pectins to form a Ca^{2+}-bridged gel network.

Oxidases such as peroxidase may catalyze cross-linking between phenolic groups (tyrosine, phenylalanine, ferulic

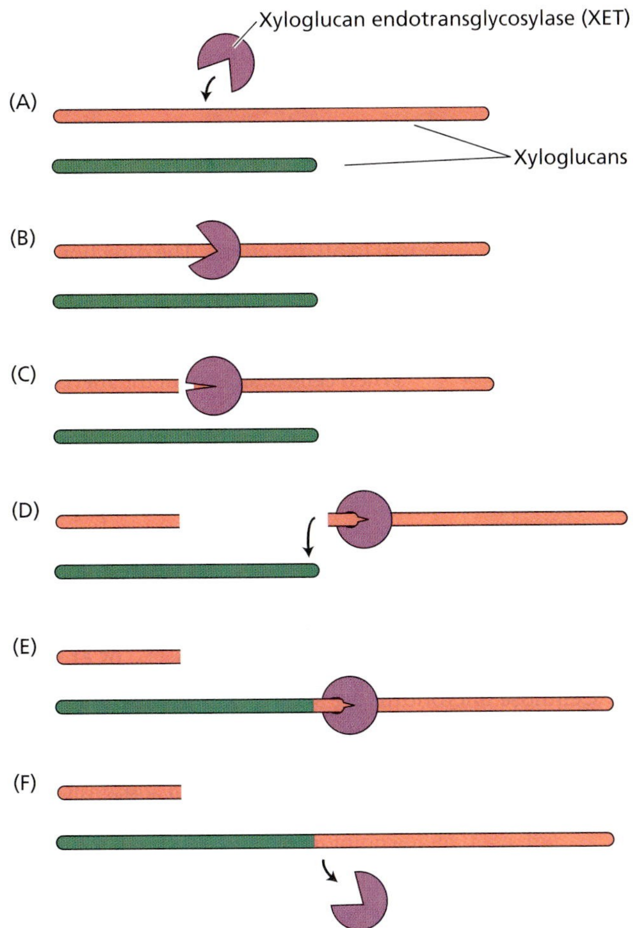

FIGURE 15.16 Action of xyloglucan endotransglycosylase (XET) to cut and stitch xyloglucan polymers into new configurations. Two xyloglucan chains are shown in (A) with two distinct patterns to emphasize their rearrangement. XET binds to the middle of one xyloglucan (B), cuts it (C), and transfers one end to the end of a second xyloglucan (D, E), resulting in one shorter and one longer xyloglucan (F). (After Smith and Fry 1991.)

acid) in wall proteins, pectins, and other wall polymers. Such phenolic coupling is clearly important for the formation of lignin cross-links, and it may likewise link diverse components of the wall together.

Secondary Walls Form in Some Cells after Expansion Ceases

After wall expansion ceases, cells sometimes continue to synthesize a wall, known as a secondary wall. Secondary walls are often quite thick, as in tracheids, fibers, and other cells that serve in mechanical support of the plant (Figure 15.17).

Often such secondary walls are multilayered and differ in structure and composition from the primary wall. For example, the secondary walls in wood contain xylans rather than xyloglucans, as well as a higher proportion of cellulose. The orientation of the cellulose microfibrils may be more neatly aligned parallel to each other in secondary walls than in primary walls. Secondary walls are often (but not always) impregnated with lignin.

Lignin is a phenolic polymer with a complex, irregular pattern of linkages that link the aromatic alcohol subunits together (see Chapter 13). These subunits are synthesized from phenylalanine and are secreted to the wall, where they are oxidized in place by the enzymes peroxidase and laccase. As lignin forms in the wall, it displaces water from the matrix and forms a hydrophobic network that bonds tightly to cellulose and prevents wall enlargement (Figure 15.18).

Lignin adds significant mechanical strength to cell walls and reduces the susceptibility of walls to attack by pathogens.

(A)

(B)

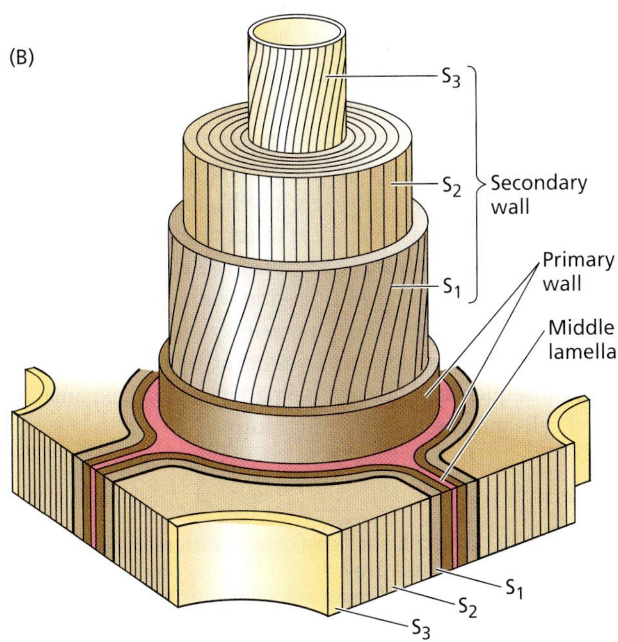

FIGURE 15.17 (A) Cross section of a *Podocarpus* sclereid, in which multiple layers in the secondary wall are visible. (B) Diagram of the cell wall organization often found in tracheids and other cells with thick secondary walls. Three distinct layers (S_1, S_2, S_3) are formed interior to the primary wall. (Photo ©David Webb.)

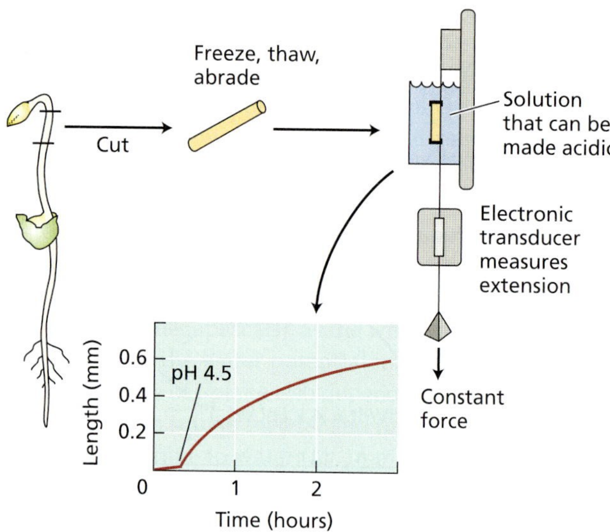

FIGURE 15.26 Acid-induced extension of isolated cell walls, measured in an extensometer. The wall sample from killed cells is clamped and put under tension in an extensometer that measures the length with an electronic transducer attached to a clamp. When the solution surrounding the wall is replaced with an acidic buffer (e.g., pH 4.5), the wall extends irreversibly in a time-dependent fashion (it creeps).

or other agents that denature proteins, they lose their acid growth ability. Such results indicate that acid growth is not due simply to the physical chemistry of the wall (e.g., a weakening of the pectin gel), but is catalyzed by one or more wall proteins.

The idea that proteins are required for acid growth was confirmed in reconstitution experiments, in which heat-inactivated walls were restored to nearly full acid growth responsiveness by addition of proteins extracted from growing walls (Figure 15.27). The active components proved to be a group of proteins that were named **expansins** (McQueen-Mason et al. 1992; Li et al. 1993). These proteins catalyze the pH-dependent extension and stress relaxation of cell walls. They are effective in catalytic amounts (about 1 part protein per 5000 parts wall, by dry weight).

The molecular basis for expansin action on wall rheology is still uncertain, but most evidence indicates that expansins cause wall creep by loosening noncovalent adhesion between wall polysaccharides (Cosgrove 2000; Li and Cosgrove 2001). Binding studies suggest that expansins may act at the interface between cellulose and one or more hemicelluloses.

With the completion of the *Arabidopsis* genome, we now know that *Arabidopsis* has a large collection of expansin genes, divided into two families: α-expansins and β-expansins. The two kinds of expansins act on different polymers of the cell wall (Cosgrove 2000). β-expansins have also been found in grass pollen, where they probably function to aid pollen tube penetration into the stigma and style (Li and Cosgrove 2001).

Glucanases and Other Hydrolytic Enzymes May Modify the Matrix

Several types of experiments implicate (1→4)β-D-glucanases in cell wall loosening, especially during auxin-induced cell elongation (see Chapter 19). For example, matrix glucans

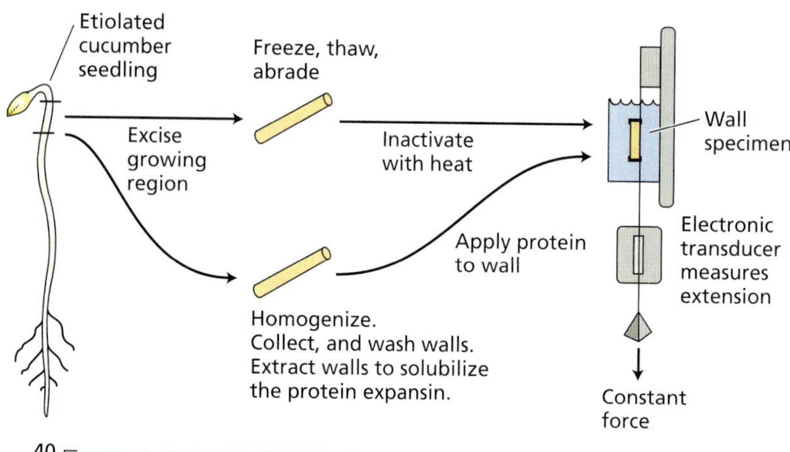

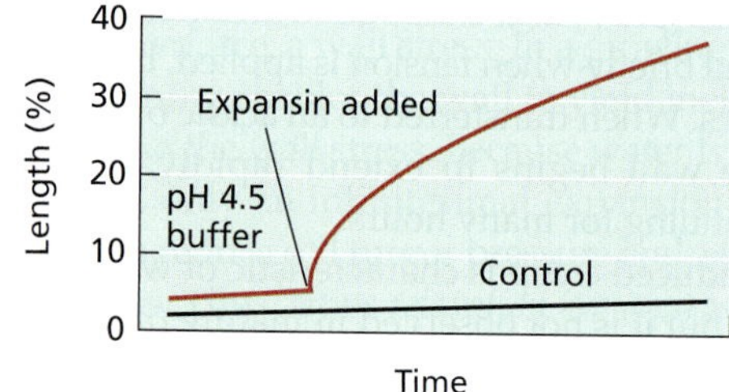

FIGURE 15.27 Scheme for the reconstitution of extensibility of isolated cell walls. (A) Cell walls are prepared as in Figure 15.21, and briefly heated to inactivate the endogenous acid extension response. To restore this response, proteins are extracted from growing walls and added to the solution surrounding the wall. (B) Addition of proteins containing expansins restores the acid extension properties of the wall. (After Cosgrove 1997.)

such as xyloglucan show enhanced hydrolysis and turnover in excised segments when growth is stimulated by auxin. Interference with this hydrolytic activity by antibodies or lectins reduces growth in excised segments.

Expression of (1→4)β-D-glucanases is associated with growing tissues, and application of glucanases to cells in vitro may stimulate growth. Such results support the idea that wall stress relaxation and expansion are the direct result of the activity of glucanases that digest xyloglucan in dicotyledons or (1→3,1→4)β-D-glucans in grass cell walls (Hoson 1993).

However, most glucanases and related wall hydrolases do not cause walls to extend in the same way that expansins do. Instead, treatment of walls with glucanases or pectinases may enhance the subsequent extension response to expansins (Cosgrove and Durachko 1994). These results suggest that wall hydrolytic enzymes such as (1→4)β-D-glucanases are not the principal catalysts of wall expansion, but they may act indirectly by modulating expansin-mediated polymer creep.

Xyloglucan endotransglycosylase has also been suggested as a potential wall-loosening enzyme. XET helps integrate newly secreted xyloglucan into the existing wall structure, but its function as a wall-loosening agent is still speculative.

Many Structural Changes Accompany the Cessation of Wall Expansion

The growth cessation that occurs during cell maturation is generally irreversible and is typically accompanied by a reduction in wall extensibility, as measured by various biophysical methods. These physical changes in the wall might come about by (1) a reduction in wall-loosening processes, (2) an increase in wall cross-linking, or (3) an alteration in the composition of the wall, making for a more rigid structure or one less susceptible to wall loosening. There is some evidence for each of these ideas (Cosgrove 1997).

Several modifications of the maturing wall may contribute to wall rigidification:

- Newly secreted matrix polysaccharides may be altered in structure so as to form tighter complexes with cellulose or other wall polymers, or they may be resistant to wall-loosening activities.
- Removal of mixed-link β-D-glucans is also coincident with growth cessation in these walls.
- De-esterification of pectins, leading to more rigid pectin gels, is similarly associated with growth cessation in both grasses and dicotyledons.
- Cross-linking of phenolic groups in the wall (such as tyrosine residues in HRGPs, ferulic acid residues attached to pectins, and lignin) generally coincides with wall maturation and is believed to be mediated by peroxidase, a putative wall rigidification enzyme.

Many structural changes occur in the wall during and after cessation of growth, and it has not yet been possible to identify the significance of individual processes for cessation of wall expansion.

WALL DEGRADATION AND PLANT DEFENSE

The plant cell wall is not simply an inert and static exoskeleton. In addition to acting as a mechanical restraint, the wall serves as an extracellular matrix that interacts with cell surface proteins, providing positional and developmental information. It contains numerous enzymes and smaller molecules that are biologically active and that can modify the physical properties of the wall, sometimes within seconds. In some cases, wall-derived molecules can also act as signals to inform the cell of environmental conditions, such as the presence of pathogens. This is an important aspect of the defense response of plants (see Chapter 13).

Walls may also be substantially modified long after growth has ceased. For instance, the cell wall may be massively degraded, such as occurs in ripening fruit or in the endosperm of germinating seeds. In cells that make up the abscission zones of leaves and fruits (see Chapter 22), the middle lamella may be selectively degraded, with the result that the cells become unglued and separate. Cells may also separate selectively during the formation of intercellular air spaces, during the emergence of the root from germinating seeds, and during other developmental processes. Plant cells may also modify their walls during pathogen attack as a form of defense.

In the sections that follow we will consider two types of dynamic changes that can occur in mature cell walls: hydrolysis and oxidative cross-linking. We will also discuss how fragments of the cell wall released during pathogen attack, or even during normal cell wall turnover, may act as cellular signals that influence metabolism and development.

Enzymes Mediate Wall Hydrolysis and Degradation

Hemicelluloses and pectins may be modified and broken down by a variety of enzymes that are found naturally in the cell wall. This process has been studied in greatest detail in ripening fruit, in which softening is thought to be the result of disassembly of the wall (Rose and Bennett 1999). Glucanases and related enzymes may hydrolyze the backbone of hemicelluloses. Xylosidases and related enzymes may remove the side branches from the backbone of xyloglucan. Transglycosylases may cut and join hemicelluloses together. Such enzymatic changes may alter the physical properties of the wall, for example, by changing the viscosity of the matrix or by altering the tendency of the hemicelluloses to stick to cellulose.

Messenger RNAs for expansin are expressed in ripening tomato fruit, suggesting that they play a role in wall disassembly (Rose et al. 1997). Similarly, softening fruits express high levels of pectin methyl esterase, which hydrolyzes the methyl esters from pectins. This hydrolysis makes the pectin more susceptible to subsequent hydrolysis by pectinases and related enzymes. The presence of these and related enzymes in the cell wall indicates that walls are capable of significant modification during development.

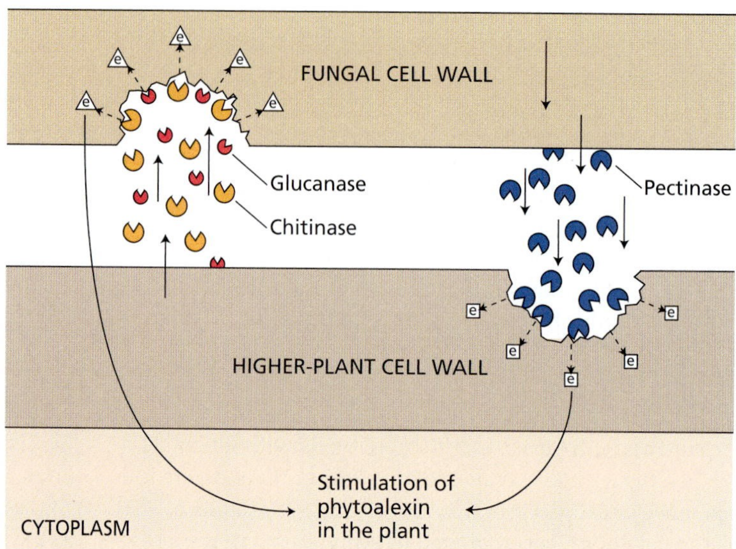

FIGURE 15.28 Scheme for the production of oligosaccharins during fungal invasion of plant cells. Enzymes secreted by the plant, such as chitinase and glucanase, attack the fungal wall, releasing oligosaccharins that elicit the production of defense compounds (phytoalexins) in the plant. Similarly, fungal pectinase releases biologically active oligosaccharins from the plant cell wall. (After Brett and Waldron 1996.)

Oxidative Bursts Accompany Pathogen Attack

When plant cells are wounded or treated with certain low-molecular-weight elicitors (see Chapter 13), they activate a defense response that results in the production of high concentrations of hydrogen peroxide, superoxide radicals, and other active oxygen species in the cell wall. This "oxidative burst" appears to be part of a defense response against invading pathogens (see Chapter 13) (Brisson et al. 1994; Otte and Barz 1996).

Active oxygen species may directly attack the pathogenic organisms, and they may indirectly deter subsequent invasion by the pathogenic organisms by causing a rapid cross-linking of phenolic components of the cell wall. In tobacco stems, for example, proline-rich structural proteins of the wall become rapidly insolubilized upon wounding or elicitor treatment, and this cross-linking is associated with an oxidative burst and with a mechanical stiffening of the cell walls.

Wall Fragments Can Act as Signaling Molecules

Degradation of cell walls can result in the production of biologically active fragments 10 to 15 residues long, called **oligosaccharins**, that may be involved in natural developmental responses and in defense responses (see **Web Topic 15.5**). Some of the reported physiological and developmental effects of oligosaccharins include stimulation of phytoalexin synthesis, oxidative bursts, ethylene synthesis, membrane depolarization, changes in cytoplasmic calcium, induced synthesis of pathogen-related proteins such as chitinase and glucanase, other systemic and local "wound" signals, and alterations in the growth and morphogenesis of isolated tissue samples (John et al. 1997).

The best-studied examples are oligosaccharide elicitors produced during pathogen invasion (see Chapter 13). For example, the fungus *Phytophthora* secretes an endopolygalacturonase (a type of pectinase) during its attack on plant tissues. As this enzyme degrades the pectin component of the plant cell wall, it produces pectin fragments—**oligogalacturonans**—that elicit multiple defense responses by the plant cell (Figure 15.28). The oligogalacturonans that are 10 to 13 residues long are most active in these responses.

Plant cell walls also contain a β-D-glucanase that attacks the β-D-glucan that is specific to the fungal cell wall. When this enzyme attacks the fungal wall, it releases glucan oligomers with potent elicitor activity. The wall components serve in this case as part of a sensitive system for the detection of pathogen invasion.

Oligosaccharins may also function during the normal control of cell growth and differentiation. For example, a specific nonasaccharide (an oligosaccharide containing nine sugar residues) derived from xyloglucan has been found to inhibit growth promotion by the auxin 2,4-dichlorophenoxyacetic acid (2,4-D). The nonasaccharide acts at an optimal concentration of 10^{-9} *M*. This xyloglucan oligosaccharin may act as a feedback inhibitor of growth; for example, when auxin-induced breakdown of xyloglucan is maximal, it may prevent excessive weakening of the cell wall. Related xyloglucan oligomers have also been reported to influence organogenesis in tissue cultures and may play a wider role in cell differentiation (Creelman and Mullet 1997).

SUMMARY

The architecture, mechanics, and function of plants depend crucially on the structure of the cell wall. The wall is secreted and assembled as a complex structure that varies in form and composition as the cell differentiates. Primary cell walls are synthesized in actively growing cells, and secondary cell walls are deposited in certain cells, such as xylem vessel elements and sclerenchyma, after cell expansion ceases.

The basic model of the primary wall is a network of cellulose microfibrils embedded in a matrix of hemicelluloses, pectins, and structural proteins. Cellulose microfibrils are highly ordered arrays of glucan chains synthesized on the membrane by protein complexes called particle rosettes. The genes for cellulose synthase in plants have recently been identified, bringing the realization that a large gene family encodes these and related proteins. The matrix is secreted into the wall via the Golgi apparatus. Hemicelluloses and proteins cross-link microfibrils, and pectins form hydrophilic gels that can become cross-linked by calcium ions. Wall assembly may be mediated by enzymes. For example, xyloglucan endotransglycosylase has the ability to carry out transglycosylation reactions that integrate newly synthesized xyloglucans into the wall.

Secondary walls differ from primary walls in that they contain a higher percentage of cellulose, they have different hemicelluloses, and lignin replaces pectins in the matrix. Secondary walls can also become highly thickened, sculpted, and embedded with specialized structural proteins.

In diffuse-growing cells, growth directionality is determined by wall structure, in particular the orientation of the cellulose microfibrils, which in turn is determined by the orientation of microtubules in the cytoplasm. Upon leaving the meristem, plant cells typically elongate greatly. Cell enlargement is limited by the ability of the cell wall to undergo polymer creep, which in turn is controlled in a complex way by the adhesion of wall polymers to one another and by the influence of pH on wall-loosening proteins such as expansins, glucanases, and other enzymes.

According to the acid growth hypothesis, proton extrusion by the plasma membrane H^+-ATPase acidifies the wall, activating the protein expansin. Expansins induce stress relaxation of the wall by loosening the bonds holding microfibrils together. The cessation of cell elongation appears to be due to cell wall rigidification caused by an increase in the number of cross-links.

Hydrolytic enzymes may degrade mature cell walls completely or selectively during fruit ripening, seed germination, and the formation of abscission layers. Cell walls can also undergo oxidative cross-linking in response to pathogen attack. In addition, pathogen attack may release cell wall fragments, and certain wall fragments have been shown to be capable of acting as cell signaling agents.

Web Material

Web Topics

15.1 Terminology for Polysaccharide Chemistry

A brief review of terms used to describe the structures, bonds, and polymers in polysaccharide chemistry is provided.

15.2 Molecular Model for the Synthesis of Cellulose and Other Wall Polysaccharides That Consist of a Disaccharide Repeat

A model is presented for the polymerization of cellubiose units into glucan chains by the enzyme cellulose synthase.

15.3 Matrix Components of the Cell Wall

The secretion of xyloglucan and glycosylated proteins by the Golgi can be demonstrated at the ultrastructural level.

15.4 The Mechanical Properties of Cell Walls: Studies with *Nitella*

Experiments demonstrating that the inner 25% of the cell wall determines the directionality of cell expansion are described.

15.5 Structure of Biologically Active Oligosaccharins

Some cell wall fragments have been demonstrated to have biological activity.

Web Essay

15.1 Calcium Gradients and Oscillations in Growing Pollen Tube

The role of calcium in regulating pollen tube tip growth is described.

Chapter References

Amor, Y., Haigler, C. H., Johnson, S., Wainscott, M., and Delmer, D. P. (1995) A membrane-associated form of sucrose synthase and its potential role in synthesis of cellulose and callose in plants. *Proc. Natl. Acad. Sci. USA* 92: 9353–9357.

Arioli, T., Peng, L., Betzner, A. S., Burn, J., Wittke, W., Herth, W., Camilleri, C., Hofte, H., Plazinski, J., Birch, R., Cork, A., Glover, J., Redmond, J., Williamson, R. E. (1998) Molecular analysis of cellulose biosynthesis in *Arapidopsis*. *Science* 279: 717–720.

Baskin, T. I., Wilson, J. E., Cork, A., and Williamson, R. E. (1994) Morphology and microtubule organization in *Arabidopsis* roots exposed to oryzalin or taxol. *Plant Cell Physiol.* 35: 935–942.

Bibikova, T. N., Jacob, T., Dahse, I., and Gilroy, S. (1998) Localized changes in apoplastic and cytoplasmic pH are associated with root hair development in *Arabidopsis thaliana*. *Development* 125: 2925–2934.

Brett, C. T., and Waldron, K. W. (1996) *Physiology and Biochemistry of Plant Cell Walls*, 2nd ed. Chapman and Hall, London.

Brisson, L. F., Tenhaken, R., and Lamb, C. (1994) Function of oxidative cross-linking of cell wall structural proteins in plant disease resistance. *Plant Cell* 6: 1703–1712.

Brown, R. M., Jr., Saxena, I. M., and Kudlicka, K. (1996) Cellulose biosynthesis in higher plants. *Trends Plant Sci.* 1: 149–155.

Buchanan, B. B., Gruissem, W., and Jones, R. L., eds. (2000) *Biochemistry, and Molecular Biology of Plants.* Amer. Soc. Plant Physiologists, Rockville, MD.

Carpita, N. C. (1996). Structure and biogenesis of the cell walls of grasses. *Annu. Rev. Plant Physiol. Plant Mol. Biol.* 47: 455–476.

Carpita, N. C., and McCann, M. (2000) The cell wall. In *Biochemistry and Molecular Biology of Plants*, B. B. Buchanan, W. Gruissem, and

R. L. Jones, eds., American Society of Plant Biologists, Rockville, MD, pp. 52–108.

Cheung, A. Y., Zhan, X. Y., Wang, H., and Wu, H.-M. (1996) Organ-specific and Agamous-regulated expression and glycosylation of a pollen tube growth-promoting protein. *Proc. Natl. Acad. Sci. USA* 93: 3853–3858.

Cosgrove, D. J. (1985) Cell wall yield properties of growing tissues. Evaluation by in vivo stress relaxation. *Plant Physiol.* 78: 347–356.

Cosgrove, D. J. (1997) Relaxation in a high-stress environment: The molecular bases of extensible cell walls and cell enlargement. *Plant Cell* 9: 1031–1041.

Cosgrove, D. J. (2000) Loosening of plant cell walls by expansins. *Nature* 407: 321–326.

Cosgrove, D. J., and Durachko, D. M. (1994) Autolysis and extension of isolated walls from growing cucumber hypocotyls. *J. Exp. Bot.* 45: 1711–1719.

Creelman, R. A., and Mullet, J. E. (1997) Oligosaccharins, brassinolides, and jasmonates: Nontraditional regulators of plant growth, development, and gene expression. *Plant Cell* 9: 1211–1223.

Darley, C. P., Forrester, A. M., and McQueen-Mason, S. J. (2001) The molecular basis of plant cell wall extension. *Plant Mol. Biol.* 47: 179–195.

Delmer, D. P., and Amor, Y. (1995) Cellulose biosynthesis. *Plant Cell* 7: 987–1000.

Gaspar, Y., Johnson, K. L., McKenna, J. A., Bacic, A., and Schultz, C. J. (2001) The complex structures of arabinogalactan-proteins and the journey towards understanding function. *Plant Mol. Biol.* 47: 161–176.

Gunning, B. S., and Steer, M. W. (1996) *Plant Cell Biology: Structure and Function*. Jones and Bartlett Publishers, Boston.

Hayashi, T. (1989) Xyloglucans in the primary cell wall. *Annu. Rev. Plant Physiol. Plant Mol. Biol.* 40: 139–168.

Holland, N., Holland, D., Helentjaris, T., Dhugga, K. S., Xoconostle-Cazares, B., and Delmer D. P. (2000) A comparative analysis of the plant cellulose synthase (CesA) gene family. *Plant Physiol.* 123: 1313–1324.

Hoson, T. (1993) Regulation of polysaccharide breakdown during auxin-induced cell wall loosening. *J. Plant Res.* 103: 369–381.

Ishii, T., Matsunaga, T., Pellerin, P., O'Neill, M. A., Darvill, A., and Albersheim, P. (1999) The plant cell wall polysaccharide rhamnogalacturonan II self-assembles into a covalently cross-linked dimer. *J. Biol. Chem.* 274: 13098–13104.

John, M., Röhrig, H., Schmidt, J., Walden, R., and Schell, J. (1997) Cell signalling by oligosaccharides. *Trends Plant Sci.* 2: 111–115.

Kimura, S., Laosinchai, W., Itoh, T., Cui, X. J., Linder, C. R., and Brown, R. M., Jr. (1999) Immunogold labeling of rosette terminal cellulose-synthesizing complexes in the vascular plant *Vigna angularis*. *Plant Cell* 11: 2075–2085.

Li, L.-C., and Cosgrove, D. J. (2001) Grass group I pollen allergens (beta-expansins) lack proteinase activity and do not cause wall loosening via proteolysis. *Eur. J. Biochem.* 268: 4217–4226.

Li, Z.-C., Durachko, D. M., and Cosgrove, D. J. (1993) An oat coleoptile wall protein that induces wall extension in vitro and that is antigenically related to a similar protein from cucumber hypocotyls. *Planta* 191: 349–356.

McCann, M. C., Wells, B., and Roberts, K. (1990) Direct visualization of cross-links in the primary plant cell wall. *J. Cell Sci.* 96: 323–334.

McQueen-Mason, S., Durachko, D. M., and Cosgrove, D. J. (1992) Two endogenous proteins that induce cell wall expansion in plants. *Plant Cell* 4: 1425–1433.

Nishitani, K. (1997) The role of endoxyloglucan transferase in the organization of plant cell walls. *Int. Rev. Cytol.* 173: 157–206.

O'Neill, M. A., Eberhard, S., Albersheim, P., and Darvill, A. G. (2001) Requirement of borate cross-linking of cell wall rhamnogalacturonan II for *Arabidopsis* growth. *Science* 294: 846–849.

Otte, O., and Barz, W. (1996) The elicitor-induced oxidative burst in cultured chickpea cells drives the rapid insolubilization of two cell wall structural proteins. *Planta* 200: 238–246.

Pear, J. R., Kawagoe, Y., Schreckengost, W. E., Delmer, D. P., and Stalker, D. M. (1996) Higher plants contain homologs of the bacterial celA genes encoding the catalytic subunit of cellulose synthase. *Proc. Natl. Acad. Sci. USA* 93: 12637–12642.

Peng, L., Kawagoe, Y., Hogan, P., and Delmer, D. (2002) Sitosterol-β-glucoside as primer for cellulose synthesis in plants. *Science* 295: 147–148.

Rayle, D. L., and Cleland, R. E. (1992) The acid growth theory of auxin-induced cell elongation is alive and well. *Plant Physiol.* 99: 1271–1274.

Richmond, T. A., and Somerville, C. R. (2000) The cellulose synthase superfamily. *Plant Physiol.* 124: 495–498.

Roland, J. C., Reis, D., Mosiniak, M., and Vian, B. (1982) Cell wall texture along the growth gradient of the mung bean hypocotyl: Ordered assembly and dissipative processes. *J. Cell Sci.* 56: 303–318.

Rose, J. K. C., and Bennett, A. B. (1999) Cooperative disassembly of the cellulose-xyloglucan network of plant cell walls: Parallels between cell expansion and fruit ripening. *Trends Plant Sci.* 4: 176–183.

Rose, J. K. C., Lee, H. H., and Bennett, A. B. (1997) Expression of a divergent expansin gene is fruit-specific and ripening-regulated. *Proc. Natl. Acad. Sci. USA* 94: 5955–5960.

Salnikov, V. V., Grimson, M. J., Delmer, D. P., and Haigler, C. H. (2001) Sucrose synthase localizes to cellulose synthesis sites in tracheary elements. *Phytochemistry* 57: 823–833.

Schopfer, P. (2001) Hydroxyl radical-induced cell-wall loosening in vitro and in vivo: Implications for the control of elongation growth. *Plant J.* 28: 679–688.

Séné, C. F. B., McCann, M. C., Wilson, R. H., and Grinter, R. (1994) Fourier-transform Raman and Fourier-transform infrared spectroscopy. An investigation of five higher plant cell walls and their components. *Plant Physiol.* 106: 1623–1631.

Smith, R. C., and Fry, S. C. (1991) Endotransglycosylation of xyloglucans in plant cell suspension cultures. *Biochem. J.* 279: 529–536.

Thompson, J. E., and Fry, S. C. (2001) Restructuring of wall-bound xyloglucan by transglycosylation in living plant cells. *Plant J.* 26: 23–34.

Wilson, R. H., Smith, A. C., Kacurakova, M., Saunders, P. K., Wellner, N., and Waldron, K. W. (2000) The mechanical properties and molecular dynamics of plant cell wall polysaccharides studied by Fourier-transform infrared spectroscopy. *Plant Physiol.* 124: 397–405.

Chapter

16

Growth and Development

THE VEGETATIVE PHASE OF DEVELOPMENT begins with embryogenesis, but development continues throughout the life of a plant. Plant developmental biologists are concerned with questions such as, How does a zygote give rise to an embryo, an embryo to a seedling? How do new plant structures arise from preexisting structures? Organs are generated by cell division and expansion, but they are also composed of tissues in which groups of cells have acquired specialized functions, and these tissues are arranged in specific patterns. How do these tissues form in a particular pattern, and how do cells differentiate? What are the basic principles that govern the size increase (growth) that occurs throughout plant development?

Understanding how growth, cell differentiation, and pattern formation are regulated at the cellular, biochemical, and molecular levels is the ultimate goal of developmental biologists. Such an understanding also must include the genetic basis of development. Ultimately, development is the unfolding of genetically encoded programs. Which genes are involved, what is their hierarchical order, and how do they bring about developmental change?

In this chapter we will explore what is known about these questions, beginning with embryogenesis. Embryogenesis initiates plant development, but unlike animal development, plant development is an ongoing process. Embryogenesis establishes the basic plant body plan and forms the meristems that generate additional organs in the adult.

After discussing the formation of the embryo, we will examine root and shoot meristems. Most plant development is postembryonic, and it occurs from meristems. Meristems can be considered to be cell factories in which the ongoing processes of cell division, expansion, and differentiation generate the plant body. Cells derived from meristems become the tissues and organs that determine the overall size, shape, and structure of the plant.

Vegetative meristems are highly repetitive—they produce the same or similar structures over and over again—and their activity can continue indefinitely, a phenomenon known as *indeterminate growth*. Some long-lived trees, such as bristlecone pines and the California redwoods, continue to grow for thousands of years. Others, particularly annual plants, may cease vegetative development with the initiation of flowering after only a few weeks or months of growth. Eventually the adult plant undergoes a transition from vegetative to reproductive development, culminating in the production of a zygote, and the process begins again. Reproductive development will be discussed in Chapter 24.

Cells derived from the apical meristems exhibit specific patterns of cell expansion, and these expansion patterns determine the overall shape and size of the plant. We will examine how plant growth is analyzed after discussing meristems, with an emphasis on growth patterns in space (relationship of plant structures) and time (when events occur).

Finally, despite their indeterminate growth habit, plants, like all other multicellular organisms, senesce and die. At the end of the chapter we will consider death as a developmental phenomenon, at both the cellular and organismal levels. For an historical overview of the study of plant development, see **Web Essay 16.1**.

EMBRYOGENESIS

The developmental process known as **embryogenesis** initiates plant development. Although embryogenesis usually begins with the union of a sperm with an egg, forming a single-celled *zygote*, somatic cells also may undergo embryogenesis under special circumstances. Fertilization also initiates three other developmental programs: endosperm, seed, and fruit development. Here we will focus on embryogenesis because it provides the key to understanding plant development.

Embryogenesis transforms a single-celled zygote into a multicellular, microscopic, embryonic plant. A completed **embryo** has the basic body plan of the mature plant and many of the tissue types of the adult, although these are present in a rudimentary form.

Double fertilization is unique to the flowering plants (see **Web Topics 1.1 and 1.2**). In plants, as in all other eukaryotes, the union of one sperm with the egg forms a single-celled zygote. In angiosperms, however, this event is accompanied by a second fertilization event, in which another sperm unites with two polar nuclei to form the triploid endosperm nucleus, from which the **endosperm** (the tissue that supplies food for the growing embryo) will develop.

Embryogenesis occurs within the **embryo sac** of the ovule while the ovule and associated structures develop into the **seed**. Embryogenesis and endosperm development typically occur in parallel with seed development, and the embryo is part of the seed. Endosperm may also be part of the mature seed, but in some species the endosperm disappears before seed development is completed. Embryogenesis and seed development are highly ordered, integrated processes, both of which are initiated by double fertilization. When completed, both the seed and the embryo within it become dormant and are able to survive long periods unfavorable for growth. The ability to form seeds is one of the keys to the evolutionary success of angiosperms as well as gymnosperms.

The fact that a zygote gives rise to an organized embryo with a predictable and species-specific structure tells us that the zygote is genetically programmed to develop in a particular way, and that cell division, cell expansion, and cell differentiation are tightly controlled during embryogenesis. If these processes were to occur at random in the embryo, the result would be a clump of disorganized cells with no definable form or function.

In this section we will examine these changes in greater detail. We will focus on molecular genetic studies that have been conducted with the model plant *Arabidopsis* that have provided insights into plant development. It is most likely that most angiosperms probably use similar developmental mechanisms that appeared early in the evolution of the flowering plants and that the diversity of plant form is brought about by relatively subtle changes in the time and place where the molecular regulators of development are expressed, rather than by different mechanisms altogether (Doebley and Lukens 1998).

Arabidopsis thaliana is a member of the Brassicaceae, or mustard family (Figure 16.1). It is a small plant, well suited for laboratory culture and experimentation. It has been called the *Drosophila* of plant biology because of its widespread use in the study of plant genetics and molecular genetic mechanisms, particularly in an effort to understand plant developmental change. It was the first higher plant to have its genome completely sequenced. Furthermore, there is a concerted international effort to understand the function of every gene in the *Arabidopsis* genome by the year 2010. As a result, we are much closer to an understanding of the molecular mechanisms governing *Arabidopsis* embryogenesis than of those for any other plant species.

Embryogenesis Establishes the Essential Features of the Mature Plant

Plants differ from most animals in that embryogenesis does not directly generate the tissues and organs of the adult. For example, angiosperm embryogenesis forms a rudimentary plant body, typically consisting of an embryonic axis and two cotyledons (if it is a dicot). Nevertheless, embryogenesis establishes the two basic developmental patterns that persist and can easily be seen in the adult plant:

1. The apical–basal axial developmental pattern.
2. The radial pattern of tissues found in stems and roots.

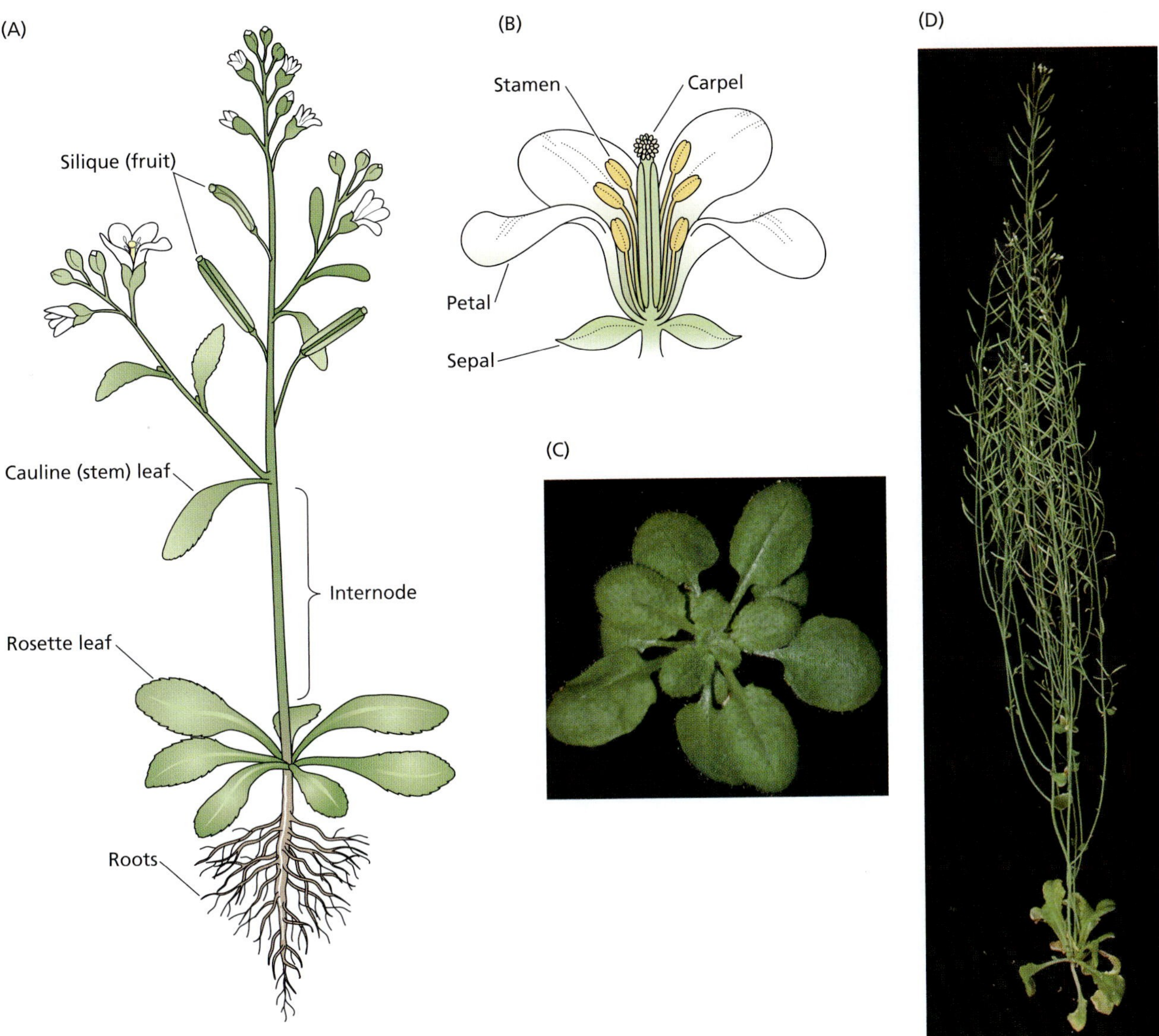

FIGURE 16.1 *Arabidopsis thaliana.* (A) Drawing of a mature *Arabidopsis* plant showing the various organs. (B) Drawing of a flower showing the floral organs. (C) An immature vegetative plant consisting of basal rosette leaves and a root system (not shown). (D) A mature plant after most of the flowers have matured and the siliques have developed. (A and B after Clark 2001; C and D courtesy of Caren Chang.)

Embryogenesis also establishes the **primary meristems**. Most of the structures that make up the adult plant are generated after embryogenesis through the activity of meristems. Although these primary meristems are established during embryogenesis, only upon germination will they become active and begin to generate the organs and tissues of the adult.

Axial patterning. Almost all plants exhibit an *axial polarity* in which the tissues and organs are arrayed in a precise order along a linear, or polarized, axis. The shoot apical meristem is at one end of the axis, the root apical meristem at the other. In the embryo and seedling, one or two cotyledons are attached just below the shoot apical meristem. Next in this linear array is the hypocotyl, followed by the root, the root apical meristem, and the root cap. This axial pattern is established during embryogenesis.

What may not be so obvious is the fact that any individual segment of either the root or the shoot also has apical and basal ends with different, distinct physiological and structural properties. For example, whereas adventitious roots develop from the basal ends of stem cuttings, buds develop from the apical ends, even if they are inverted (see Figure 19.12).

Radial patterning. Different tissues are organized in a precise pattern within plant organs. In stems and roots the tissues are arranged in a radial pattern extending from the outside of a stem or a root into its center. If we examine a root in cross section, for example, we see three concentric rings of tissues arrayed along a radial axis: An outermost

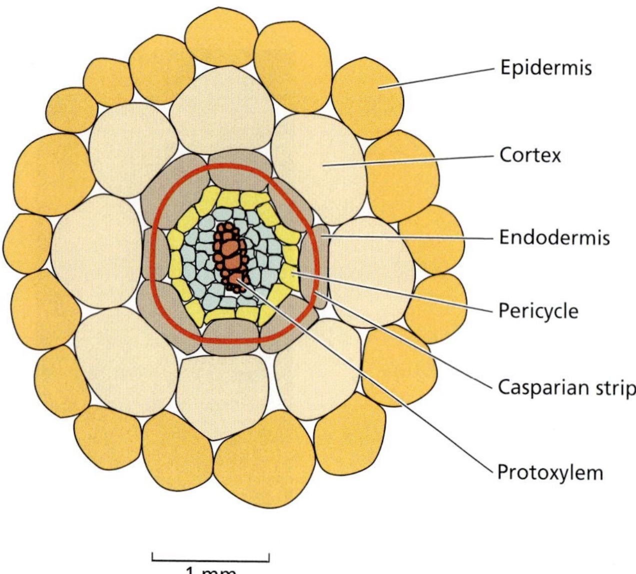

FIGURE 16.2 The radial pattern of tissues found in plant organs can be observed in a crosssection of the root. This crosssection of an *Arabidopsis* root was taken approximately 1 mm back from the root tip, a region in which the different tissues have formed.

layer of epidermal cells (the epidermis) covers a cylinder of cortical tissue (the cortex), which in turn overlies the vascular cylinder (the endodermis, pericycle, phloem, and xylem) (Figure 16.2) (see Chapter 1).

The **protoderm** is the meristem that gives rise to the epidermis, the **ground meristem** produces the future cortex and endodermis, and the **procambium** is the meristem that gives rise to the primary vascular tissue and vascular cambium.

Arabidopsis Embryos Pass through Four Distinct Stages of Development

The *Arabidopsis* pattern of embryogenesis has been studied extensively and is the one we will present here, but keep in mind that angiosperms exhibit many different patterns of embryonic development, and this is only one type.

The most important stages of embryogenesis in *Arabidopsis*, and many other angiosperms, are these:

1. *The* ***globular stage*** *embryo*. After the first zygotic division, the apical cell undergoes a series of highly ordered divisions, generating an eight-cell (**octant**) globular embryo by 30 hours after fertilization (Figure 16.3C). Additional precise cell divisions

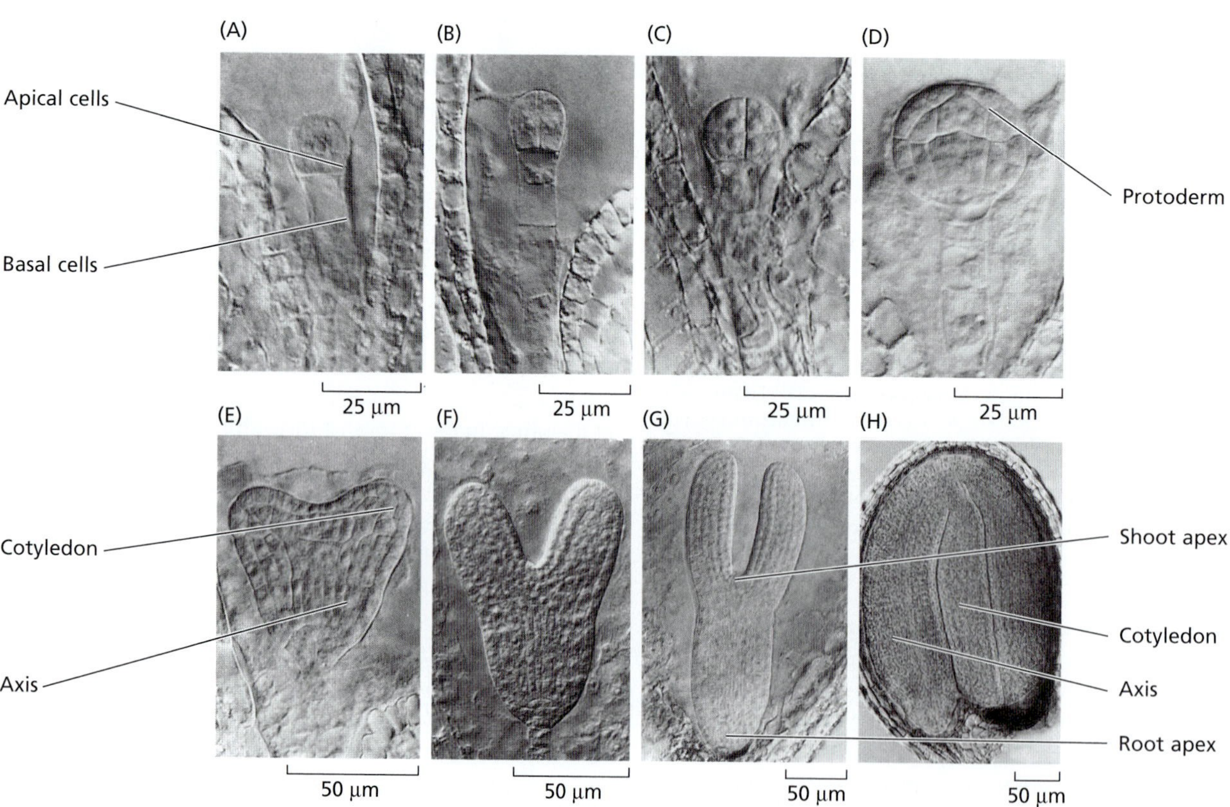

FIGURE 16.3 *Arabidopsis* embryogenesis is characterized by a precise pattern of cell division. Successive stages of embryogenesis are depicted here. (A) One-cell embryo after the first division of the zygote, which forms the apical and basal cells; (B) two-cell embryo; (C) eight-cell embryo; (D) early globular stage, which has developed a distinct protoderm (surface layer); (E) early heart stage; (F) late heart stage; (G) torpedo stage; (H) mature embryo. (From West and Harada 1993 photographs taken by K. Matsudaira Yee; courtesy of John Harada, © American Society of Plant Biologists, reprinted with permission.)

increase the number of cells in the sphere (Figure 16.3D).

2. *The **heart stage** embryo.* This stage forms through rapid cell divisions in two regions on either side of the future shoot apex. These two regions produce outgrowths that later will give rise to the cotyledons and give the embryo bilateral symmetry (Figure 16.3E and F).

3. *The **torpedo stage** embryo.* This stage forms as a result of cell elongation throughout the embryo axis and further development of the cotyledons (Figure 16.3G).

4. *The **maturation stage** embryo.* Toward the end of embryogenesis, the embryo and seed lose water and become metabolically quiescent as they enter dormancy (Figure 16.3H).

Cotyledons are food storage organs for many species, and during the cotyledon growth phase, proteins, starch, and lipids are synthesized and deposited in the cotyledons to be utilized by the seedling during the heterotrophic (nonphotosynthetic) growth that occurs after germination. Although food reserves are stored in the *Arabidopsis* cotyledons, the growth of the cotyledons is not as extensive in this species as it is in many other dicots. In monocots, the food reserves are stored mainly in the endosperm. In *Arabidopsis* and many other dicots, the endosperm develops rapidly early in embryogenesis but then is reabsorbed, and the mature seed lacks endosperm tissue.

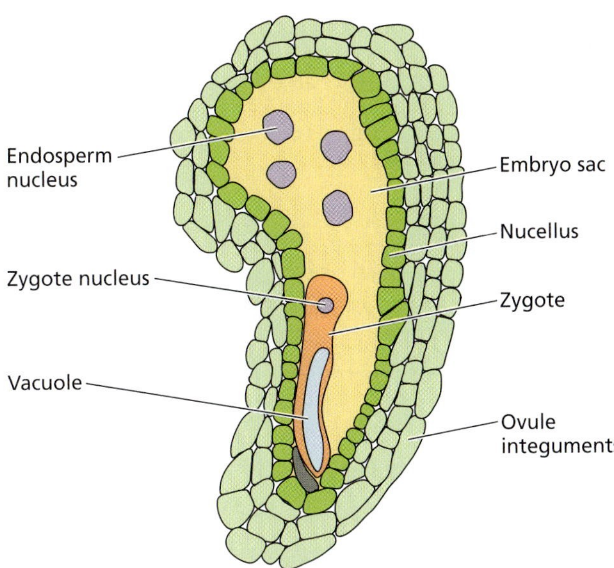

FIGURE 16.4 *Arabidopsis* ovule containing the embryo sac at about 4 hours after double fertilization. The zygote exhibits a marked polarization. The terminal half of the zygote has dense cytoplasm and a single large nucleus, while a large central vacuole occupies the basal half of the cell. At this stage, the embryo sac surrounding the zygote also contains 4 endosperm nuclei.

The Axial Pattern of the Embryo Is Established during the First Cell Division of the Zygote

Axial polarity is established very early in embryogenesis (see **Web Topic 16.1**). In fact, the zygote itself becomes polarized and elongates approximately threefold before its first division. The apical end of the zygote is densely cytoplasmic, but the basal half of the cell contains a large central vacuole (Figure 16.4).

The first division of the zygote is asymmetric and occurs at right angles to its long axis. This division creates two cells—an apical and a basal cell—that have very different fates (see Figure 16.3A). The smaller, apical daughter cell receives more cytoplasm than the larger, basal cell, which inherits the large zygotic vacuole. Almost all of the structures of the embryo, and ultimately the mature plant, are derived from the smaller apical cell. Two vertical divisions and one horizontal division of the apical cell generate the eight-celled (octant) globular embryo (see Figure 16.3C).

The basal cell also divides, but all of its divisions are horizontal, at right angles to the long axis. The result is a filament of six to nine cells known as the **suspensor** that attaches the embryo to the vascular system of the plant. Only one of the basal cell derivatives contributes to the embryo. The basal cell derivative nearest the embryo is known as the **hypophysis** (plural *hypophyses*), and it forms the **columella,** or central part of the root cap, and an essential part of the root apical meristem known as the *quiescent center*, which will be discussed later in the chapter (Figure 16.5).

Even though the embryo is spherical throughout the globular stage of embryogenesis (see Figure 16.3A–D), the cells within the apical and basal halves of the sphere have different identities and functions. As the embryo continues to grow and reaches the heart stage, its axial polarity becomes more distinct (see Figure 16.5), and three axial regions can readily be recognized:

1. The *apical region* gives rise to the cotyledons and shoot apical meristem.

2. The *middle region* gives rise to the hypocotyl, root, and most of the root meristem.

3. The *hypophysis* gives rise to the rest of the root meristem (see Figure 16.5).

The cells of the upper and lower tiers of the early globular stage embryo differ, and the embryo is divided into apical and basal halves, reflecting the axial pattern imposed on the embryo in the zygote.

The Radial Pattern of Tissue Differentiation Is First Visible at the Globular Stage

The radial pattern of tissue differentiation is first observed in the octant embryo (Figure 16.6). As cell division continues in the globular embryo, transverse divisions divide the

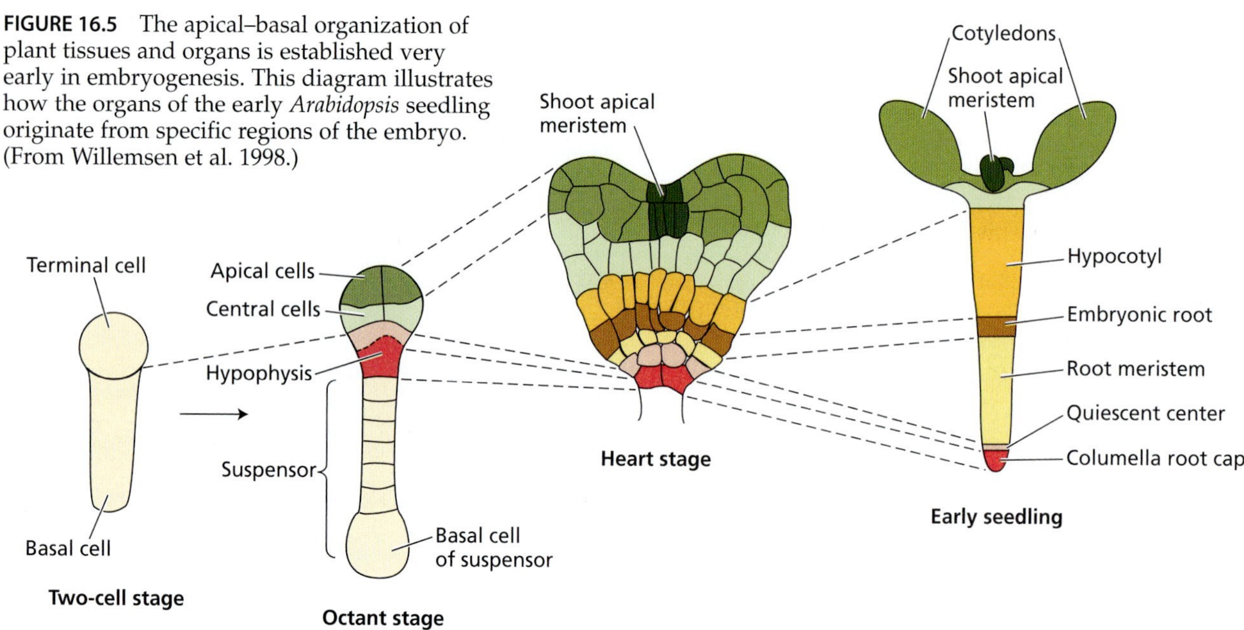

FIGURE 16.5 The apical–basal organization of plant tissues and organs is established very early in embryogenesis. This diagram illustrates how the organs of the early *Arabidopsis* seedling originate from specific regions of the embryo. (From Willemsen et al. 1998.)

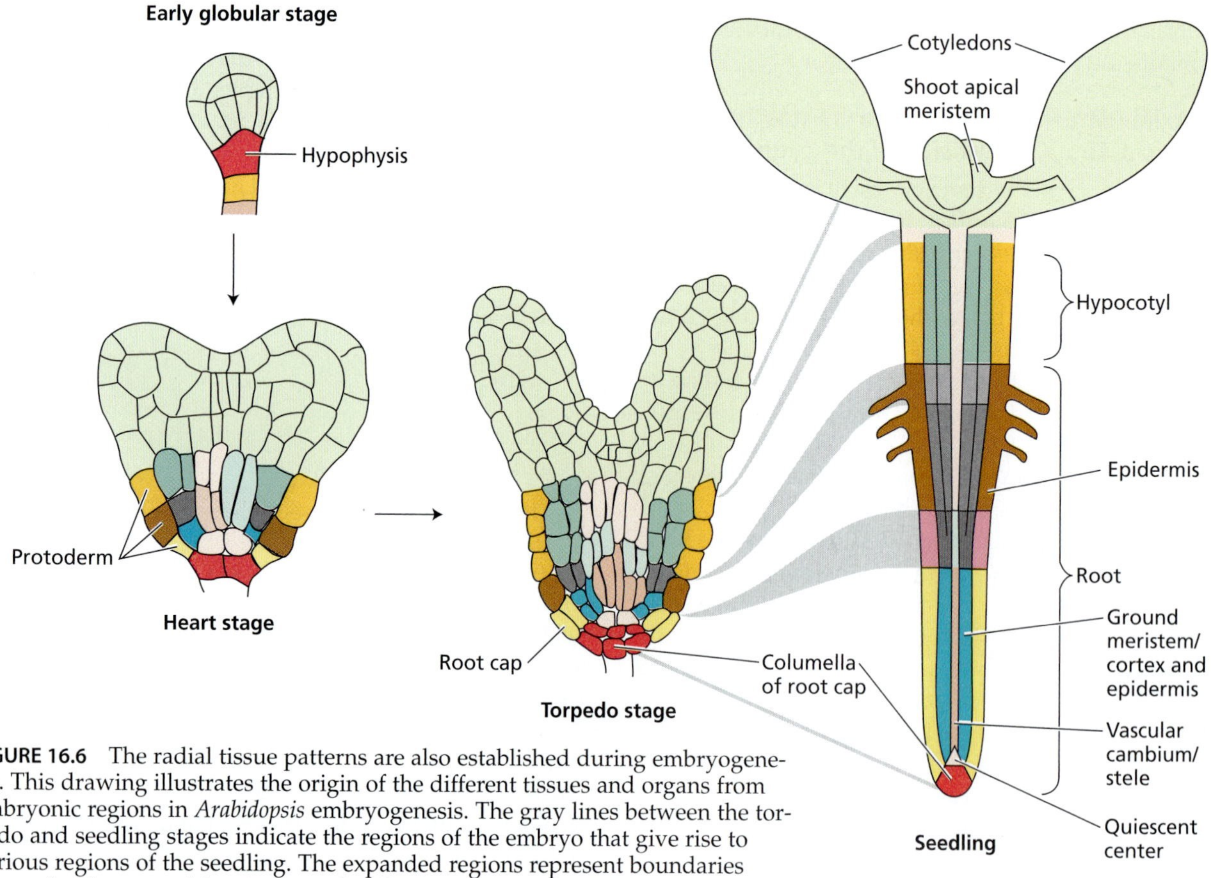

FIGURE 16.6 The radial tissue patterns are also established during embryogenesis. This drawing illustrates the origin of the different tissues and organs from embryonic regions in *Arabidopsis* embryogenesis. The gray lines between the torpedo and seedling stages indicate the regions of the embryo that give rise to various regions of the seedling. The expanded regions represent boundaries where developmental fate is somewhat flexible. (After Van Den Berg et al. 1995.)

lower tier of cells radially into three regions. These regions will become the radially arranged tissues of the root and stem axes. The outermost cells form a one-cell-thick surface layer, known as the **protoderm**. The protoderm covers both halves of the embryo and will generate the epidermis.

Cells that will become the ground meristem underlie the protoderm. The ground meristem gives rise to the **cortex** and, in the root and hypocotyl, it will also produce the **endodermis**. The procambium is the inner core of elongated cells that will generate the **vascular tissues** and, in the root, the **pericycle** (see Figure 16.2).

Embryogenesis Requires Specific Gene Expression

Analysis of *Arabidopsis* mutants that either fail to establish axial polarity or develop abnormally during embryogenesis has led to the identification of genes whose expression participates in tissue patterning during embryogenesis.

The **GNOM** ***gene: Axial patterning.*** Seedlings homozygous for mutations in the *GNOM* gene lack both roots and cotyledons (Figure 16.7A) (Mayer et al. 1993). Defects in *gnom* embryos first appear during the initial division of the zygote, and they persist throughout embryogenesis. In the most extreme mutants, *gnom* embryos are spherical and lack axial polarity entirely. We can conclude that *GNOM* gene expression is required for the establishment of axial polarity.[1]

(A) Wild type *gnom mutant*

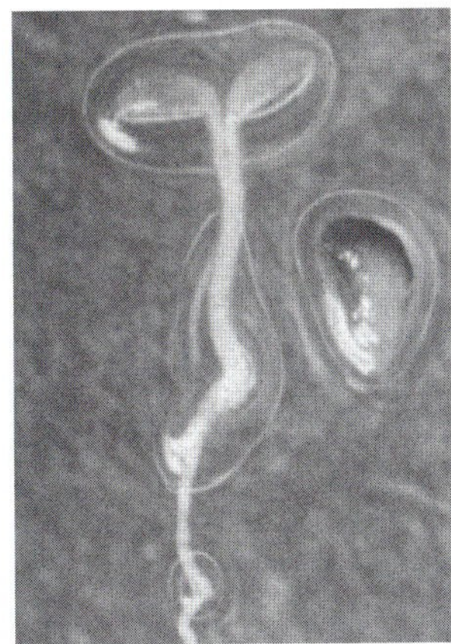

GNOM genes control apical–basal polarity

(B) Wild type *monopteros* mutant

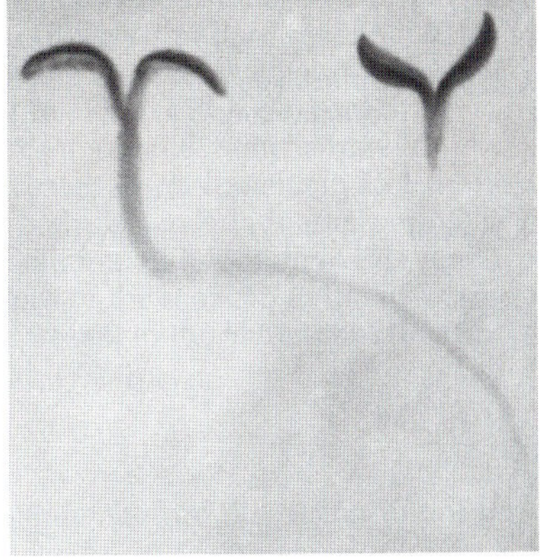

MONOPTEROS genes control formation of the primary root

FIGURE 16.7 Genes whose functions are essential for *Arabidopsis* embryogenesis have been identified by the selection of mutants in which a stage of embryogenesis is blocked, such as *gnom* and *monopteros*. The development of mutant seedlings is contrasted here with that of the wild type at the same stage of development. (A) The *GNOM* gene helps establish apical–basal polarity. A plant homozygous for *gnom* is shown on the right. (B) The *MONOPTEROS* gene is necessary for basal patterning and formation of the primary root. Plants homozygous for the *monopteros* mutation have a hypocotyl, a normal shoot apical meristem, and cotyledons, but they lack the primary root. (A from Willemsen et al. 1998; B from Berleth and Jürgens 1993.)

The **MONOPTEROS** ***gene: Primary root and vascular tissue.*** Mutations in the *MONOPTEROS* (*MP*) gene result in seedlings that lack both a hypocotyl and a root, although they do produce an apical region. The apical structures in the *mp* mutant embryos are not structurally normal, however, and the tissues of the cotyledons are disorganized (Figure 16.7B) (Berleth and Jürgens 1993). Embryos of *mp* mutants first show abnormalities at the octant stage, and they do not form a procambium in the lower part of the globular embryo, the part that should give rise to the hypocotyl and root. Later some vascular tissue does form in the cotyledons, but the strands are improperly connected.

Although the *mp* mutant embryos lack a primary root when they germinate, they will form adventitious roots as the seedlings grow into adult plants. The vascular tissues in all organs of these mutant plants are poorly developed, with frequent discontinuities. Thus the *MP* gene is required for the formation of the embryonic primary root, but not for root formation in the adult plant. The *MP* gene is important for the formation of vascular tissue in postembryonic development (Przemeck et al. 1996).

The **SHORT ROOT** *and* **SCARECROW** ***genes: Ground tissue development.*** Genes have been identified that function in the establishment of the radial tissue pattern in the root and hypocotyl during embryogenesis. These genes also are required for maintenance of the radial pattern during postembryonic development (Scheres et al. 1995; Di Laurenzio et al. 1996). To identify these genes, investigators isolated *Arabidopsis* mutants that caused roots to grow slowly (Figure 16.8B). Analysis of these mutants identified several that have defects in the radial tissue pattern. Two of the affected genes, *SHORT ROOT* (*SHR*) and *SCARECROW* (*SCR*), are necessary for tissue differentiation and cell differentiation not only in the embryo, but also in both primary and secondary roots and in the hypocotyl.

Mutants of *SHR* and *SCR* both produce roots with a single-celled layer of ground tissue (Figure 16.8D). Cells making up the single-celled layer of ground tissue have a mixed identity and show characteristics of both endodermal and cortical cells in plants with the *scr* mutation. These *scr* mutants also lack the cell layer called the **starch sheath**, a structure that is involved in the growth response to gravity (see Chapter 19). Roots of plants with the *shr* mutation also

[1] In discussions of plant and yeast genetics, wild-type (normal) genes are capitalized and italicized (in this case *GNOM*), and mutations are set in lowercase letters (here *gnom*).

- **Axillary meristems** are formed in the axils of leaves and are derived from the shoot apical meristem. The growth and development of axillary meristems produces branches from the main axis of the plant.
- **Intercalary meristems** are found within organs, often near their bases. The intercalary meristems of grass leaves and stems enables them to continue to grow despite mowing or grazing by cows.
- **Branch root meristems** have the structure of the primary root meristem, but they form from pericycle cells in mature regions of the root. Adventitious roots also can be produced from lateral root meristems that develop on stems, as when stem cuttings are rooted to propagate a plant.
- The **vascular cambium** (plural *cambia*) is a secondary meristem that differentiates along with the primary vascular tissue from the procambium within the vascular cylinder. It does not produce lateral organs, but only the woody tissues of stems and roots. The vascular cambium contains two types of meristematic cells: fusiform stem cells and ray stem cells. *Fusiform stem cells* are highly elongated, vacuolate cells that divide longitudinally to regenerate themselves, and whose derivatives differentiate into the conducting cells of the secondary xylem and phloem. *Ray stem cells* are small cells whose derivatives include the radially oriented files of parenchyma cells within wood known as rays.
- The **cork cambium** is a meristematic layer that develops within mature cells of the cortex and the secondary phloem. Derivatives of the cork cambium differentiate as cork cells that make up a protective layer called the *periderm*, or *bark*. The periderm forms the protective outer surface of the secondary plant body, replacing the epidermis in woody stems and roots.

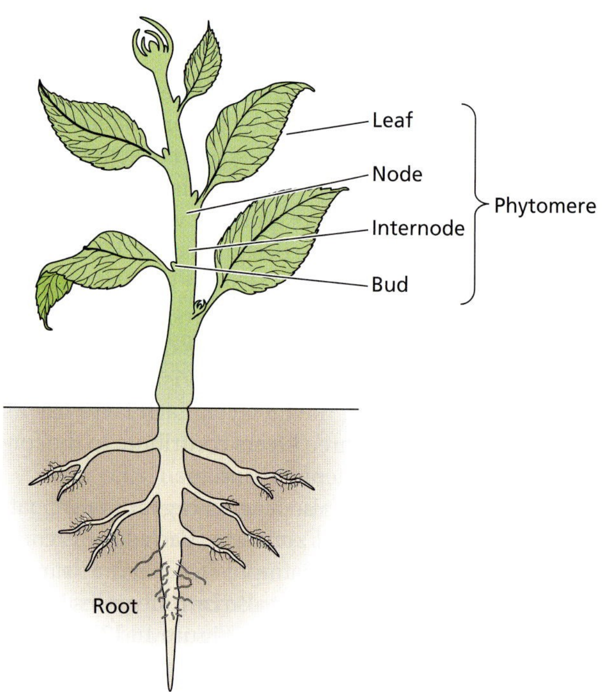

FIGURE 16.14 The shoot apical meristem repetitively forms units known as phytomeres. Each phytomere consists of one or more leaves, the node at which the leaves are attached, the internode immediately below the leaves, and one or more buds in the axils of the leaves.

Axillary, Floral, and Inflorescence Shoot Meristems Are Variants of the Vegetative Meristem

Several different types of shoot meristems can be distinguished on the basis of their developmental origin, the types of lateral organs they generate, and whether they are **determinate** (having a genetically programmed limit to their growth) or **indeterminate** (showing no predetermined limit to growth; growth continues so long as resources permit).

The vegetative shoot apical meristem usually is indeterminate in its development. It repetitively forms phytomeres as long as environmental conditions favor growth but do not generate a flowering stimulus. A **phytomere** is a developmental unit consisting of one or more leaves, the node to which the leaves are attached, the internode below the node, and one or more axillary buds (Figure 16.14). **Axillary buds** are secondary meristems; if they are also vegetative meristems, they will have a structure and developmental potential similar to that of the apical meristem.

Vegetative meristems may be converted directly into floral meristems when the plant is induced to flower (see Chapter 24). **Floral meristems** differ from vegetative meristems in that instead of leaves they produce floral organs: sepals, petals, stamens, and carpels. In addition, floral meristems are determinate: All meristematic activity stops after the last floral organs are produced.

In many cases, vegetative meristems are not directly converted to floral meristems. Instead, the vegetative meristem is first transformed into an **inflorescence meristem**. The types of lateral organs produced by an inflorescence meristem are different from the types produced by a floral meristem. The inflorescence meristem produces bracts and floral meristems in the axils of the bracts, instead of the sepals, petals, stamens, and ovules produced by floral meristems. Inflorescence meristems may be determinate or indeterminate, depending on the species.

LEAF DEVELOPMENT

The leaves of most plants are the organs of photosynthesis. This is where light energy is captured and used to drive the chemical reactions that are vital to the life of the plant. Although highly variable in size and shape from species to species, in general leaves are thin, flat structures with dorsiventral polarity. This pattern contrasts with that of the

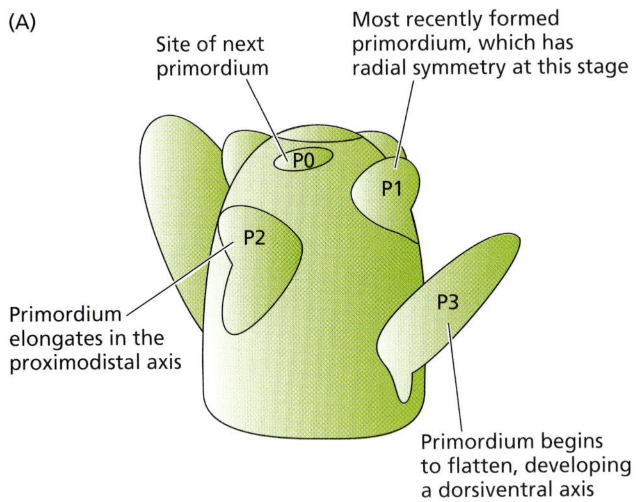

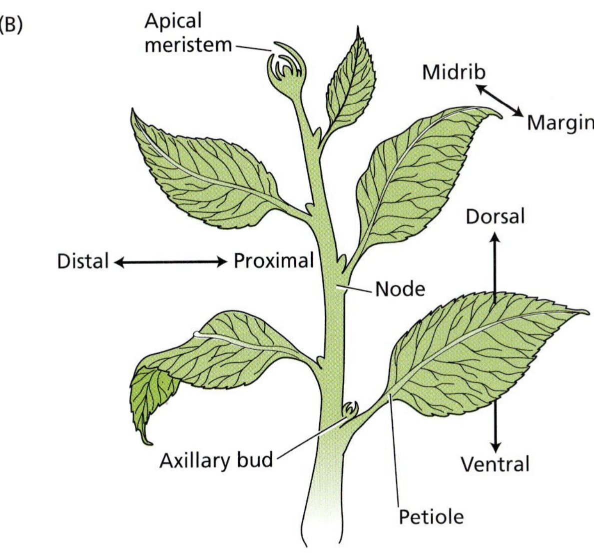

FIGURE 16.15 The origin of leaves at the shoot apex and their axes of symmetry on the stem (A) Leaf primordia in the flanks of the shoot apical meristem. (B) Diagram of a shoot showing the various axes along which development occurs. (After Christensen and Weigel 1998.)

shoot apical meristem and stem, both of which have radial symmetry. Another important difference is that leaf primordia exhibit determinate growth, while the vegetative shoot apical meristem is indeterminate. As described in the sections that follow, several distinct stages can be recognized in leaf development (Sinha 1999).

Stage 1: Organogenesis. A small number of cells in the L1 and L2 layers in the flanks of the apical dome of the shoot apical meristem acquire the **leaf founder cell** identity. These cells divide more rapidly than surrounding cells and produce the outgrowth that represents the **leaf primordium** (plural *primordia*) (Figure 16.15A). These primordia subsequently grow and develop into leaves.

Stage 2: Development of suborgan domains. Different regions of the primordium acquire identity as specific parts of the leaf. This differentiation occurs along three axes: **dorsiventral** (abaxial–adaxial), **proximodistal** (apical–basal), and **lateral** (margin–blade–midrib) (Figure 16.15B). The upper (adaxial) side of the leaf is specialized for light absorption; the lower (abaxial) surface is specialized for gas exchange. Leaf structure and maturation rates also vary along the proximodistal and lateral axes.

Stage 3: Cell and tissue differentiation. As the developing leaf grows, tissues and cells differentiate. Cells derived from the L1 layer differentiate as epidermis (epidermal cells, trichomes, and guard cells), derivatives of the L2 layer differentiate as the photosynthetic mesophyll cells, and vascular elements and bundle sheath cells are derived from the L3 layer. These cells differentiate in a genetically determined pattern that is characteristic of the species but to some degree modified in response to the environment.

The Arrangement of Leaf Primordia Is Genetically Programmed

The timing and pattern with which the primordia form is genetically determined and usually is a characteristic of the species. The number and order in which leaf primordia form is reflected in the subsequent arrangement of leaves around the stem, known as **phyllotaxy** (Figure 16.16). There are five main types of phyllotaxy:

1. ***Alternate phyllotaxy.*** A single leaf is initiated at each node (see Figure 16.16A).
2. ***Opposite phyllotaxy.*** Leaves are formed in pairs on opposite side of the stem (see Figure 16.16B).
3. ***Decussate phyllotaxy.*** Leaves are initiated in a pattern with two opposite leaves per node and with successive leaf pairs oriented at right angles to each other during vegetative development (see Figure 16.16C).
4. ***Whorled phyllotaxy.*** More than two leaves arise at each node (see Figure 16.16D).
5. ***Spiral phyllotaxy.*** A type of alternate phyllotaxy in which each leaf is initiated at a defined angle to the previous leaf, resulting in a spiral arrangement of leaves around the stem (see Figure 16.16E).

The positioning of leaf primordia must result from the precise spatial regulation of growth within the apex. We know little about how this positioning is regulated, or about the signals that initiate the formation of a primordium. One idea is that inhibitory fields generated by existing primordia influence the spacing of the next primordium.

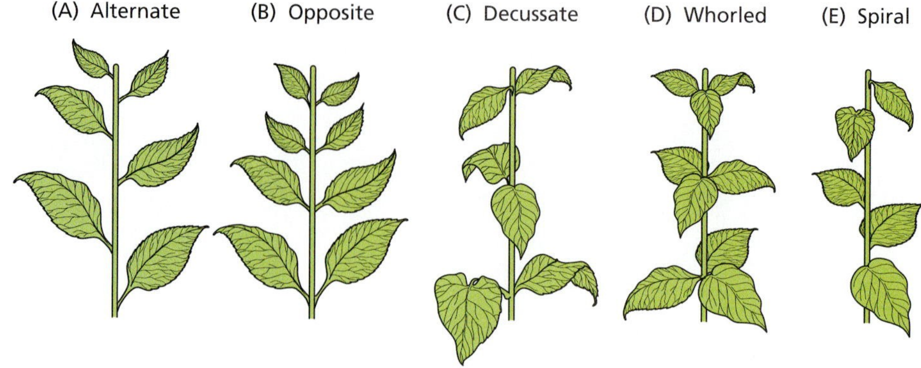

FIGURE 16.16 Five types of leaf arrangements (phyllotactic patterns) along the shoot axis. The same terms also are used for inflorescences and flowers.

ROOT DEVELOPMENT

Roots are adapted for growing through soil and absorbing the water and mineral nutrients in the capillary spaces between soil particles. These functions have placed constraints on the evolution of root structure. For example, lateral appendages would interfere with their penetration through the soil. As a result, roots have a streamlined axis, and no lateral organs are produced by the apical meristem. Branch roots arise internally and form only in mature, non-growing regions. Absorption of water and minerals is enhanced by fragile root hairs, which also form behind the growth zone. These long, threadlike cells greatly increase the root's absorptive surface area.

In this section we will discuss the origin of root form and structure (*root morphogenesis*), beginning with a description of the four developmental zones of the root tip. We will then turn to the apical meristem. The absence of leaves or buds makes cell lineages easier to follow in roots than in shoots, thus facilitating molecular genetic studies on the role of patterns of cell division in root development.

The Root Tip Has Four Developmental Zones

Roots grow and develop from their distal ends. Although the boundaries are not sharp, four developmental zones can be distinguished in a root tip: the root cap, the meristematic zone, the elongation zone, and the maturation zone (Figure 16.17). These four developmental zones occupy only a little more than a millimeter of the tip of the *Arabidopsis* root. The developing region is larger in other species, but growth is still confined to the tip. With the exception of the root cap, the boundaries of these zones overlap considerably:

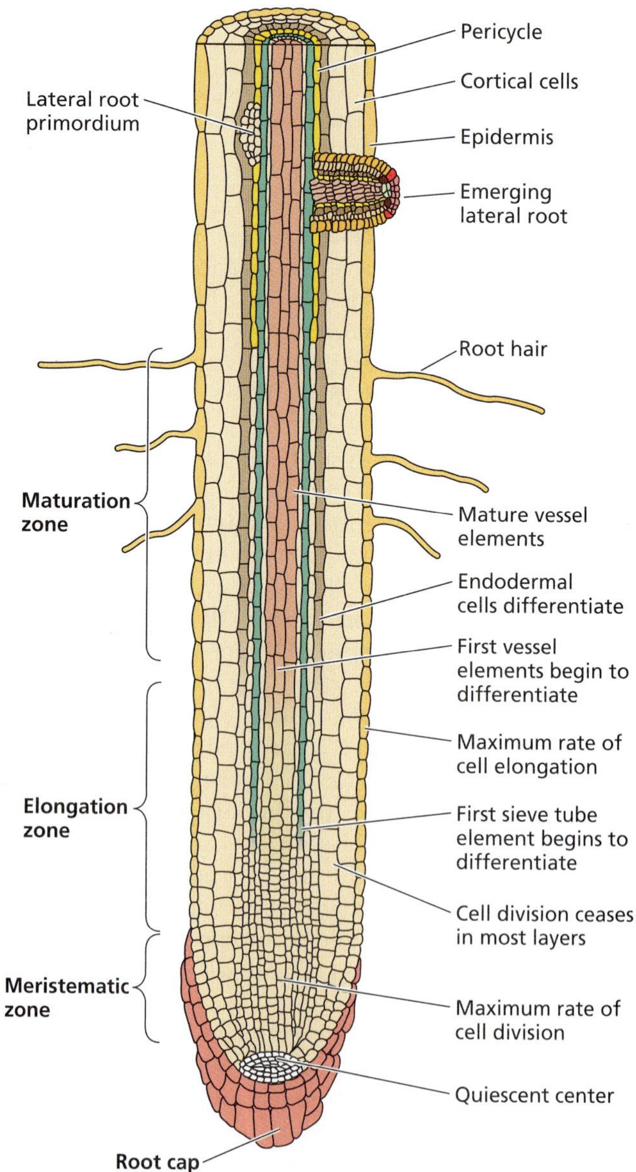

FIGURE 16.17 Simplified diagram of a primary root showing the root cap, the meristematic zone, the elongation zone, and the maturation zone. Cells in the meristematic zone have small vacuoles and expand and divide rapidly, generating many files of cells.

- The **root cap** protects the apical meristem from mechanical injury as the root pushes its way through the soil. Root cap cells form by specialized root cap stem cells. As the root cap stem cells produce new cells, older cells are progressively displaced toward the tip, where they are eventually sloughed off. As root cap cells differentiate, they acquire the ability to perceive gravitational stimuli and secrete mucopolysaccharides (slime) that help the root penetrate the soil.
- The **meristematic zone** lies just under the root cap, and in *Arabidopsis* it is about a quarter of a millimeter long. The root meristem generates only one organ, the primary root. It produces no lateral appendages.
- The **elongation zone**, as its name implies, is the site of rapid and extensive cell elongation. Although some cells may continue to divide while they elongate within this zone, the rate of division decreases progressively to zero with increasing distance from the meristem.
- The **maturation zone** is the region in which cells acquire their differentiated characteristics. Cells enter the maturation zone after division and elongation have ceased. Differentiation may begin much earlier, but cells do not achieve the mature state until they reach this zone. The radial pattern of differentiated tissues becomes obvious in the maturation zone. Later in the chapter we will examine the differentiation and maturation of one of these cell types, the tracheary element.

As discussed earlier, lateral or branch roots arise from the pericycle in mature regions of the root. Cell divisions in the pericycle establish secondary meristems that grow out through the cortex and epidermis, establishing a new growth axis (Figure 16.18). The primary and the secondary root meristems behave similarly in that divisions of the cells in the meristem give rise to progenitors of all the cells of the root.

Root Stem Cells Generate Longitudinal Files of Cells

Meristems are populations of dividing cells, but not all cells in the meristematic region divide at the same rate or with the same frequency. Typically, the central cells divide much more slowly than the surrounding cells. These rarely dividing cells are called the **quiescent center** of the root meristem (see Figure 16.17).

Cells are more sensitive to ionizing radiation when they are dividing. This is the basis of the use of radiation as a treatment for cancer in humans. As a result, the rapidly dividing cells of the meristem can be killed by doses of radiation that nondividing and slowly dividing cells, such as those of the quiescent center, can survive. If the rapidly dividing cells of the root are killed by ionizing radiation, in many cases the root can regenerate from the cells of the quiescent center. This ability suggests that quiescent-center cells are important for the patterning involved in forming a root.

The most striking structural feature of the root tip, when viewed in longitudinal section, is the presence of the long files of clonally related cells. Most cell divisions in the root tip are transverse, or **anticlinal**, with the plane of cytokinesis oriented at right angles to the axis of the root (such divisions tend to increase root length). There are relatively few **periclinal** divisions, in which the plane of division is parallel to the root axis (such divisions tend to increase root diameter).

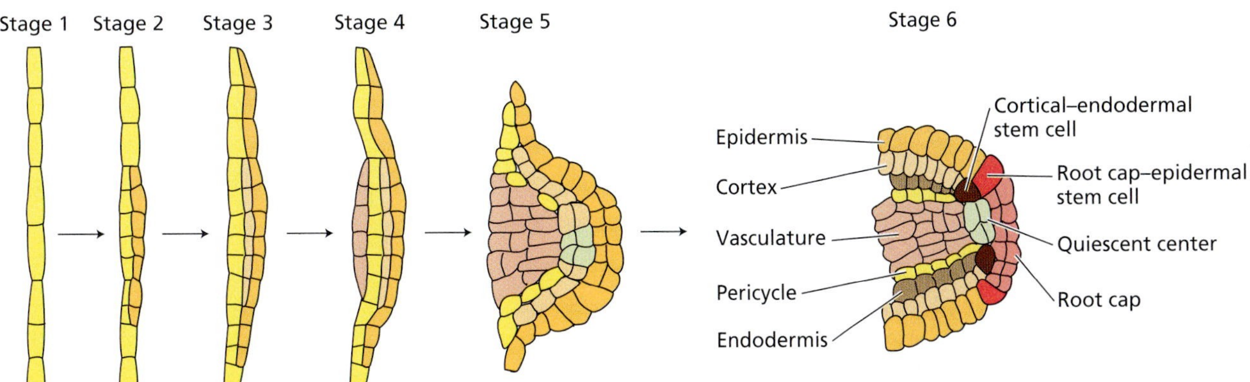

FIGURE 16.18 Model for lateral root formation in *Arabidopsis*. Six major stages are shown in the development of the primordium. The different tissue types are designated by colors. By stage 6, all tissues found in the primary root are present in the typical radial pattern of the branch root. (From Malamy and Benfey 1997.)

Periclinal divisions occur mostly near the root tip and establish new files of cells. As a result, the ultimate origin of any particular mature cell can be traced back to one or a few cells in the meristem. These are the stem cells of a particular file. In *Arabidopsis*, the stem cells surround the quiescent center, but they are not part of the quiescent center. The stem cells ultimately may be derived from quiescent-center cells, but this origin must occur during embryogenesis, since the quiescent-center cells do not divide after germination in normal development. Analysis of the cell division patterns in the roots of the water fern *Azolla* have contributed to our detailed understanding of meristem function. (For a discussion of this work, see **Web Topic 16.3**.)

Root Apical Meristems Contain Several Types of Stem Cells

The patterns of cellular organization found in the root meristems of seed plants are substantially different from those observed in more primitive vascular plants. All seed plants have several stem cells instead of the single stem cell found in plants such as the water fern *Azolla*. However, they are similar to *Azolla* in that it is possible to follow files of cells from the region of maturation into the meristem and, in some cases, to identify the stem cell from which the file was produced.

The *Arabidopsis* root apical meristem has the following structure (Figure 16.19):

- The **quiescent center** is composed of a group of four cells, also known as the center cells in the *Arabidopsis* root meristem. The quiescent-center cells in the *Arabidopsis* root usually do not divide after embryogenesis.
- The **cortical–endodermal stem cells** form a ring of cells that surround the quiescent center. These stem cells generate the cortical and endodermal layers. They undergo one anticlinal division (i.e., perpendicular to the longitudinal axis); then these daughters divide periclinally (i.e., parallel to the longitudinal axis) to establish the files that become the cortex and the endodermis, each of which constitutes only one cell layer in the *Arabidopsis* root (see also Figures 16.2 and 16.8C).
- The **columella stem cells** are the cells immediately above (apical to) the central cells. They divide anticlinally and periclinally to generate a sector of the root cap known as the columella.
- The **root cap–epidermal stem cells** are in the same tier as the columella stem cells but form a ring surrounding them. Anticlinal divisions of the root cap–epidermal stem cells generate the epidermal cell layer. Periclinal divisions of the same stem cells, followed by subsequent anticlinal divisions of the derivatives, produce the lateral root cap.

(A)

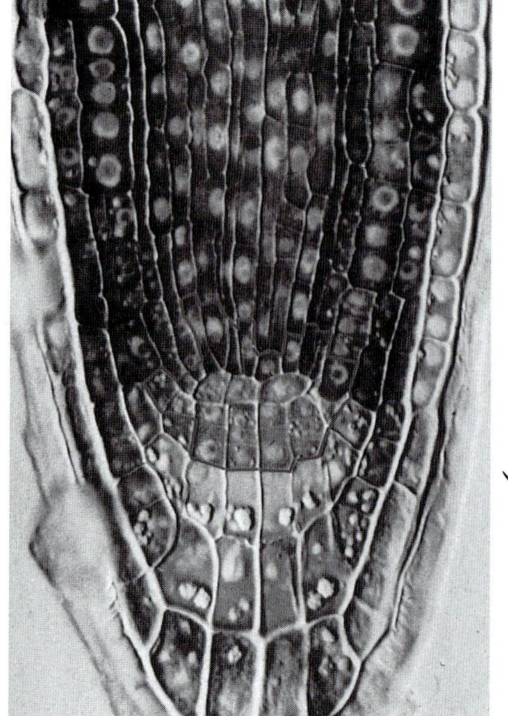

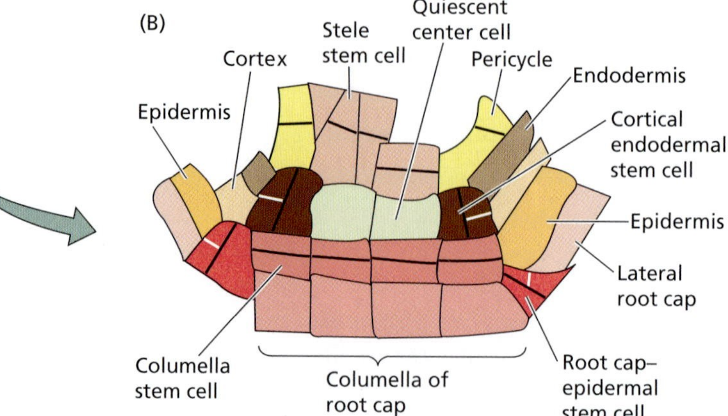

FIGURE 16.19 All the tissues in the *Arabidopsis* root are derived from a small number of stem cells in the root apical meristem. (A) Longitudinal section through the center of a root. The promeristem containing the stem cells that give rise to all the tissues of the root is outlined in green. (B) Diagram of the promeristem region outlined in A. Only two of the four quiescent-center cells are depicted in this section. The black lines indicate the cell division planes that occur in the stem cells. White lines indicate the secondary cell divisions that occur in the cortical–endodermal and lateral root cap–epidermal stem cells. (From Schiefelbein et al. 1997, courtesy of J. Schiefelbein, © the American Society of Plant Biologists, reprinted with permission.)

- The **stele stem cells** are a tier of cells just behind the quiescent-center cells. These cells generate the pericycle and vascular tissues.

The stem cells, together with their immediate derivatives in the apical meristem, are called the *promeristem*.

CELL DIFFERENTIATION

Differentiation is the process by which a cell acquires metabolic, structural, and functional properties that are distinct from those of its progenitor cell. In plants, unlike animals, cell differentiation is frequently reversible, particularly when differentiated cells are removed from the plant and placed in tissue culture. Under these conditions, cells dedifferentiate (i.e., lose their differentiated characteristics), reinitiate cell division, and in some cases, when provided with the appropriate nutrients and hormones, even regenerate whole plants.

This ability to dedifferentiate demonstrates that differentiated plant cells retain all the genetic information required for the development of a complete plant, a property termed **totipotency**. The only exceptions to this rule are cells that lose their nuclei, such as sieve tube elements of phloem, and cells that are dead at maturity, such as vessel elements and tracheids (collectively referred to as tracheary elements) in xylem.

As an example of the process of cell differentiation, we will discuss the formation of tracheary elements. The development of these cells from the meristematic to the fully differentiated state illustrates the types of control that plants exercise over cell specialization and provides an example of the cellular changes that are brought about by differentiation (Fukuda 1996).

A Secondary Cell Wall Forms during Tracheary Element Differentiation

As described in Chapter 4, tracheary elements are the conducting cells in which water and solutes move through the plant. They are dead at maturity, but before their death they are highly active and construct a secondary wall, often with an elaborate pattern, and they may grow extensively. Cell death (discussed later in this chapter) is the genetically programmed finale to tracheary element differentiation.

The formation of secondary walls during tracheary element differentiation involves the deposition of cellulose microfibrils and other noncellulosic polysaccharides at specific sites on the primary or secondary wall, resulting in characteristically patterned wall thickenings (see Chapter 15). The secondary walls of tracheary elements have a higher content of cellulose than primary walls, and they are impregnated with lignin, which is not usually present in primary walls.

In rapidly growing regions, the secondary-wall material is deposited as discrete annular rings, or in a spiral pattern, with the thickenings separated by bands of primary

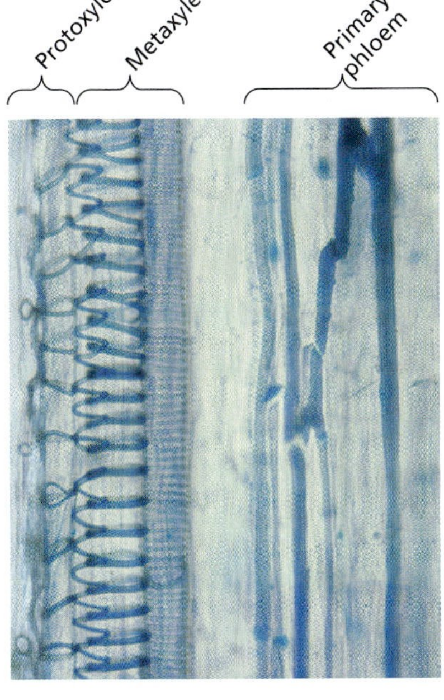

FIGURE 16.20 The formation of primary xylem and primary phloem in a developing strand in a young internode of cucumber (*Cucumis sativus*). The pattern of secondary-wall deposition during vessel element development varies according to the rate of cell elongation. The two first vessels to differentiate—the protoxylem—are observed on the left with secondary-wall thickening in the pattern of "annular rings." Because the first formed vessel was strongly stretched by internode growth, the narrow annular rings are pulled apart. The metaxylem vessels differentiate after the protoxylem and are characterized by spiral thickening. The early formed metaxylem vessel has a stretched helical thickening due to cell elongation, while the later formed vessel shows a dense helical thickening which has not been extended by elongation. The primary phloem sieve tubes are shown on the right, with typical delicate sieve elements. Their sieve plates are stained light blue, while the cytoplasm stains dark blue. (Courtesy of R. Aloni).

wall (Figure 16.20). As the cell grows, the primary wall extends and the rings or spirals are pulled apart. The tracheary elements that form after elongation stops usually have walls that are thickened. This thickening can be either uniformly or in a reticulate pattern. These cells cannot be stretched by growth.

Microtubules participate in determining the pattern of secondary-wall deposition. Before any alteration in the pattern of wall deposition is evident, cortical microtubules change from being more or less evenly distributed along the longitudinal walls of the cell to being clustered into bands (Figure 16.21A). Secondary wall is then deposited beneath the microtubule clusters (see Figure 16.21B).

The orientation of the cellulose microfibrils within the secondary-wall thickening is reflected in the alignment of microtubules in the cortical cytoplasm (Hepler 1981). If the microtubules are destroyed with an antimicrotubule agent such as colchicine, cell wall deposition can continue, but the cellulose microfibrils are no longer precisely ordered within the thickening, and the pattern of the secondary wall is disrupted (Figure 16.22).

FIGURE 16.21 Development of secondary-wall thickenings in vessel elements in roots of the water fern *Azolla*. (A) Electron micrograph of a grazing section through a differentiating cell. Groups of microtubules are seen in the cell cortex, forming bands at the site of wall thickening before the secondary wall begins to form. Many small vesicles lie along the microtubules. (B) Annular thickenings develop beneath the bands of microtubules and are hemispheric in profile. (Courtesy of A. Hardham.)

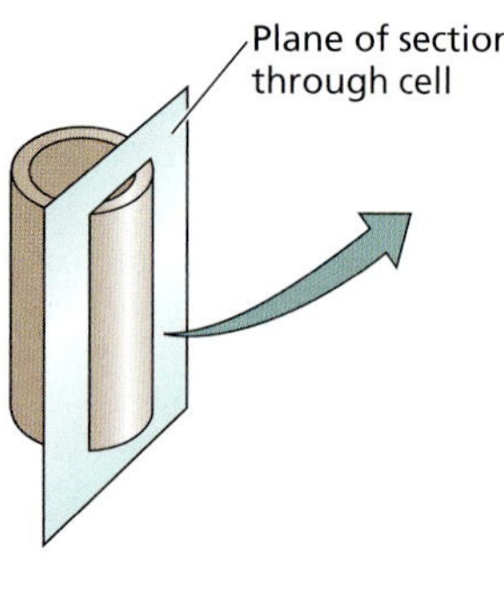

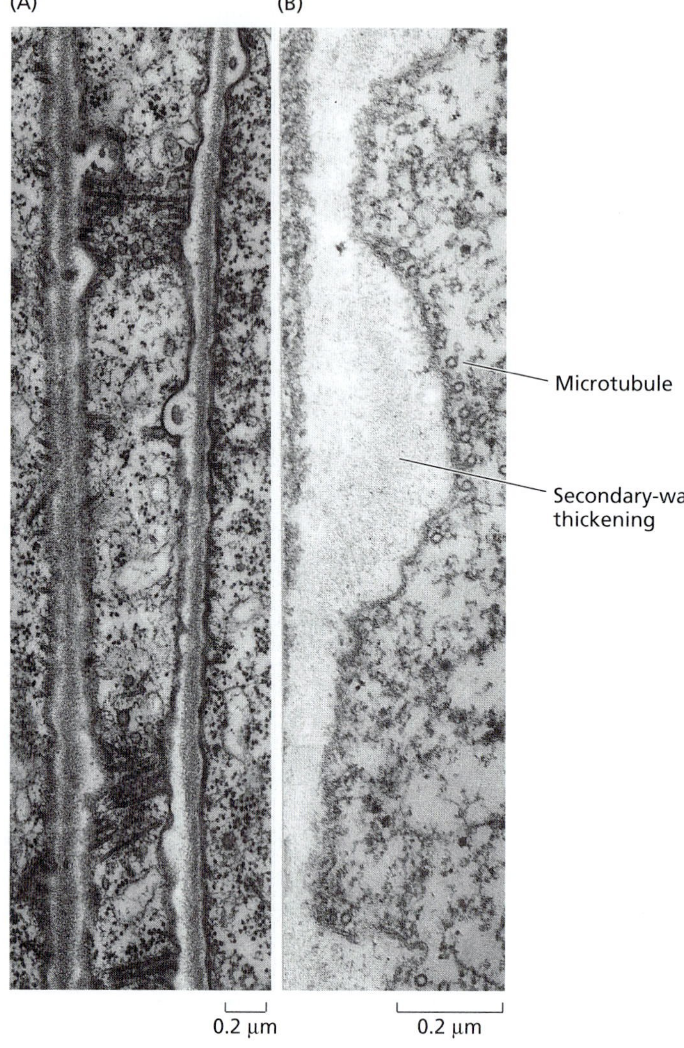

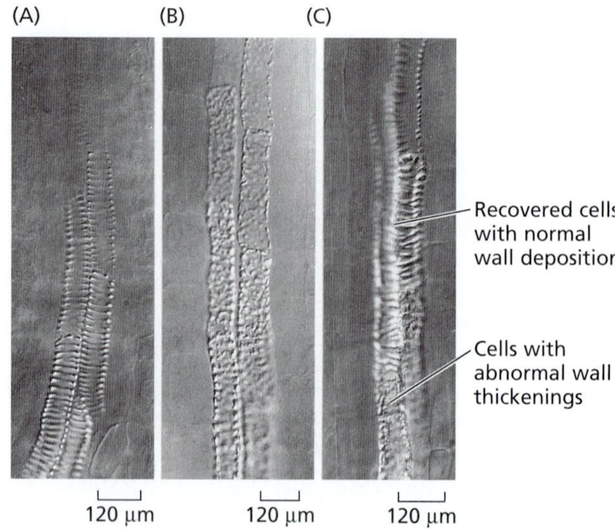

FIGURE 16.22 Colchicine treatments that destroy microtubules also disrupt the normal formation of secondary-wall thickenings in differentiating vessel elements. (A) During normal root growth in Azolla the wall thickenings are spaced evenly along the side walls. (B) In the presence of colchicine, secondary-wall materials are deposited in irregular patterns. (C) Normal growth resumes when the roots are transferred to fresh medium that lacks colchicine, and the newly differentiated vessel elements form with normal annular thickenings. (A from Hardham and Gunning 1979; B and C from Hardham and Gunning 1980.)

INITIATION AND REGULATION OF DEVELOPMENTAL PATHWAYS

Rapid progress has been made in identifying genes that play critical roles in regulating growth, cell differentiation, and pattern formation. This progress is largely a consequence of an intensive, international effort focused on *Arabidopsis*—first to sequence its genome, and subsequently to understand the function of all of its genes. However, many important discoveries have been made as a result of studies with other species, including *Antirrhinum*, maize, petunia, tomato, and tobacco.

In most cases, genes important for development were revealed by elaborate screens of the offspring of mutagenized plants to find mutant individuals with altered development (see the example in Figure 16.8B). These studies often involved heroic efforts to map, clone, and sequence the mutant gene, although now that its genome has been sequenced, the path to identifying any particular mutant gene and what it encodes is now much shorter in *Arabidopsis*.

At this point we have identified some of the players, but the rules of the game and the specific roles of most of the genes are still being worked out. However, many of these developmentally important genes have been found to encode either transcription factors (proteins with the ability to bind to specific DNA sequences and thus control the expression of other genes) or components of signaling pathways. The nature of these genes suggests some possible ways that development might be regulated.

Where these molecular genetic studies have been coupled with clonal analysis, cell biological, physiological, and/or biochemical studies, it has been possible to identify important principles of plant development. Although we are far from a complete understanding, these insights include the following:

- The expression of genes that encode transcription factors determines cell, tissue, and organ identity.
- The fate of a cell is determined by its position and not its clonal history.
- Developmental pathways are controlled by networks of interacting genes.
- Development is regulated by cell-to-cell signaling.

In the following discussion we will first examine the nature of some of the transcription factor and signal transduction component genes that have been shown to play key roles in development. Then we will outline in greater detail each of the developmental principles described here.

Transcription Factor Genes Control Development

With the completion of the sequencing of the *Arabidopsis* genome, it became apparent that approximately 1500 of its nearly 26,000 genes encode transcription factors (Riechmann et al. 2000). **Transcription factors** are proteins that have an affinity for DNA. They are able to turn the expression of genes on or off by binding to specific DNA sequences (see Chapter 14 on the web site).

These 1500 transcription factor genes belong to numerous families. Fewer than half of these families are found only in plants, but the majority are found in all eukaryotes. It is not known, or can even be estimated at this time, how many of these transcription factor genes regulate developmental pathways because only a small percentage of them have been studied. However, many members of two of these families—the MADS box and homeobox genes—have been found to be particularly important in plant development.

MADS box genes are key regulators of important biological functions in plants, animals, and fungi.[2] There are about 30 MADS box genes in the *Arabidopsis* genome, many of which control aspects of development. Specific MADS box genes are important for developmental events in the root, leaf, flower, ovule, and fruit (Riechmann and Meyerowitz 1997). They control the expression of specific sets of target genes, although at this point most of these downstream genes remain to be identified.

Any given MADS box gene is expressed in a specific temporally and spatially restricted manner, with its expression determined by other genes or signaling events. This has been established most clearly in the case of the development of the flower, where interacting sets of MADS box genes have been shown to determine floral organ identity (see Chapter 24).

Homeobox genes encode homeodomain proteins that act as transcription factors. **Homeodomain proteins** play a major role in regulating developmental pathways in all eukaryotes (see Chapter 14 on the web site). As with the MADS box genes, each homeobox gene participates in regulating a unique developmental event by controlling the expression of a unique set of target genes.

Homeodomain proteins belonging to the KNOTTED1 (KN1) class are involved in maintaining the indeterminacy of the shoot apical meristem. The original *knotted* (*kn1*) mutation was found in maize and is a gain-of-function mutation. In **gain-of-function**, or **dominant**, mutations, the phenotype results from the abnormal expression of a gene. In contrast, the phenotypes of **loss-of-function** mutations result from the loss of gene expression, and the mutations are therefore **recessive**.

Plants with the *kn1* mutation have small, irregular, tumorlike knots along the leaf veins. These knots result from abnormal cell divisions within the vascular tissues that distort the veins to form the knots, which protrude from the leaf surface (Figure 16.23) (Hake et al. 1989).

[2] The name *MADS* comes from the initials of the first four members of a family of transcription factors: *MCM1*, *AGAMOUS*, *DEFICIENS*, and *SRF*.

FIGURE 16.23 Inappropriate expression of the *KN1* gene during leaf development causes severe abnormalities around the leaf veins. The gain-of-function mutation *kn1* causes cell proliferation after normal cell division ceases; in addition, the division planes are abnormal, causing gross distortion of the blade surface. (From Sinha et al. 1993a, courtesy of S. Hake.)

Cell differentiation is relatively normal in the leaves of *kn1* mutant plants, except in the vicinity of the knots. The knots are similar to meristems in that they contain undifferentiated cells and continue to divide after cells around them have matured and ceased dividing. This behavior suggests that the *KN1* gene controls meristem function. The mutant phenotype results from the expression of the gene in the wrong tissues, rather than the loss of the normal developmental expression pattern. *KNOTTED1*-like homeobox, or ***KNOX***, genes have been found in several other plant species. *Arabidopsis* has three: *KNAT1*, *KNAT2*, and *SHOOTMERISTEMLESS* (*STM*) (Lincoln et al. 1994; Long et al. 1996).

Tobacco plants that have been transformed with the maize *KN1* gene, driven by a promoter that expresses the gene throughout the plant, develop numerous adventitious shoot meristems along leaf surfaces (Sinha et al. 1993b). These abnormalities are similar to the original gain-of-function *kn1* mutation. We can conclude from this that correct *KN1* gene expression is involved in defining meristem function.

Many Plant Signaling Pathways Utilize Protein Kinases

Protein kinases are ATP-dependent enzymes that add phosphate groups to proteins. Protein phosphorylation is a key regulatory mechanism that is utilized extensively to regulate the activity of enzymes and transcription factors. Although widely utilized by all eukaryotes, plant genomes are especially rich in genes that encode these enzymes. The *Arabidopsis* genome contains over 1200 genes that encode protein kinases. Of these, more than 600 encode *receptor protein kinases* (see Chapter 14 on the web site) (Shiu and Bleecker 2001).

The functions of most of these receptor protein kinases are unknown, but recently some have been shown to play important signaling roles in plant development. *Arabidopsis* has two such genes: *BRI1*, which encodes a receptor kinase that functions in brassinosteroid signaling (see **Web Topic 19.14**) and *CLAVATA1* (*CLV1*), which encodes a receptor kinase that participates in regulating the size of the uncommitted cell population in shoot apical meristem (we'll discuss *CLV1* a little later in the chapter).

Receptor kinases typically are integral membrane proteins. The receptor domain of these kinases resides outside the plasma membrane; the kinase catalytic domain is inside the cell, linked to the receptor domain by a transmembrane domain. The receptor domain has affinity for a signaling molecule, often a small protein or peptide, which is called the **receptor ligand**.

In the absence of the ligand, the kinase enzyme is inactive. The binding of the ligand to the receptor converts the protein to an active kinase (Figure 16.24). In the case of CLV1, ligand binding also triggers the formation of a complex consisting of a related protein, CLAVATA, a kinase-associated protein phosphatase (KAPP), and a rho GTPase-related protein. The ligand for CLV1 most likely is a small protein encoded by a third *CLAVATA* gene, *CLV3* (see Figure 16.24) (Clark et al. 1993; Clark 2001).

The *CLAVATA* genes were first identified as mutations that led to an increase in the size of the vegetative shoot apical meristem and floral meristems. One result was an increase in the number of lateral organs produced by the meristems of these mutants, which is particularly evident in the number of floral organs produced by the mutant meristems. Whereas *CLV1* encodes a typical receptor-like protein kinase, *CLV2* encodes a protein with a receptor domain similar to that of CLV1, but lacking a kinase domain. The protein encoded by the *CLV3* gene is unrelated to either CLV1 or CLV2.

A Cell's Fate Is Determined by Its Position

In both the root and shoot meristem, a small number of stem cells are the ultimate source of any particular tissue, and most of the cells in a given tissue are clonal, having arisen

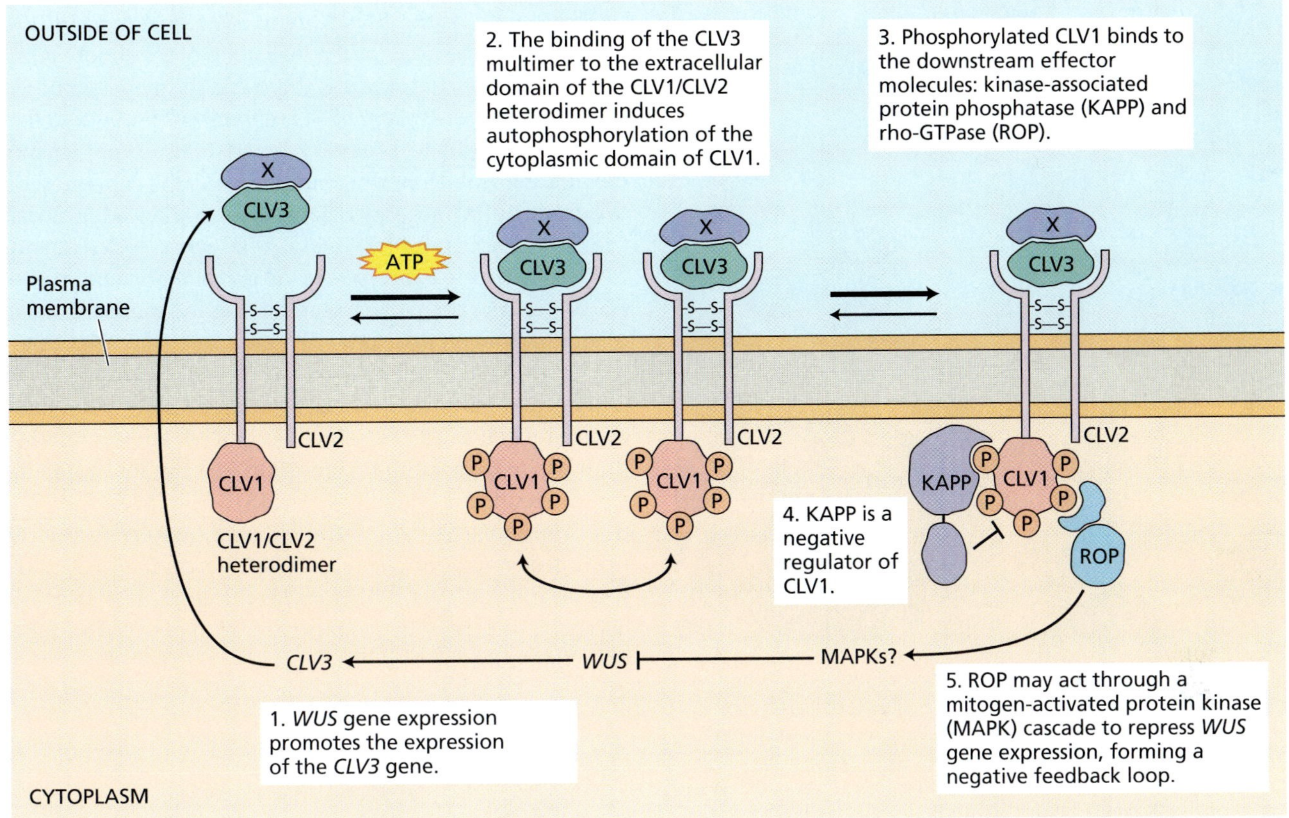

FIGURE 16.24 Model of the CLAVATA1/CLAVATA2 (CLV1/CLV2) receptor kinase signaling cascade, forming a negative feedback loop with the *WUS* gene. See Chapter 14 on the web site for further information about receptor kinase signaling pathways. (After Clark 2001.)

from the same stem cell. However, most evidence supports the view that *cell fate does not depend on cell lineage, but instead is determined by positional information* (Scheres 2001).

In the vast majority of cases, shoot epidermal cells are derived from a small number of stem cells in the L1 layer. However, the derivatives of the L1 layer are committed to become epidermal cells because they occupy the outermost layer and lie on top of the cortical cell layer, not because they were clonally derived from the stem cells in the L1 layer.

The plane in which a cell divides will determine the position of its daughter cells within a tissue, and this positioning in turn plays the most significant role in determining the fate of the daughter cells. The strongest evidence for the importance of position in determining a cell's ultimate fate comes from an examination of the fate of cells that are displaced from their usual position, such that they come to occupy a different layer.

The vast majority of the divisions in the L1 and L2 layers of the meristem are anticlinal, and anticlinal division is responsible for generating the layers in the first place. Nevertheless, occasional periclinal divisions occur, causing one derivative to occupy the adjacent layer. This periclinal division does not alter the composition of the tissue derived from this layer. Instead, the derivatives assume a function that is appropriate for a cell occupying that layer.

Further support for the importance of position in determining cell fate has been obtained through observations of cell differentiation in leaves of English ivy (*Hedera helix*), which have a mixture of mutant and wild-type cells. When a mutation occurs in a stem cell in the shoot apical meristem, all the cells in the plant derived from that stem cell will carry the mutation. Such a plant is said to be a **chimera**, a mixture of cells with a different genetic makeup. The analysis of chimeras is useful for studies on the clonal origin of different tissues.

When the mutation affects the ability of chloroplasts to differentiate, the presence of albino sectors shows that these sectors were derived from the stem cells carrying the mutation. In the ivy plant shown in Figure 16.25, the L2 layer carried a mutation causing albinism, and the L1 and L3 layers had a wild-type copy of the same gene. The L1 layer gives rise to the leaf and stem epidermis, but it is colorless because chloroplasts do not differentiate in most epidermal cells. Mesophyll tissue typically is derived from the L2 layer, so the leaves should be white because the L2 stem cells carried the mutant gene and passed it on to their derivatives.

FIGURE 16.25 Periclinal chimeras demonstrate that the mesophyll tissue has more than a single clonal origin in English ivy (*Hedera helix*). These variegated leaves provide clues on the clonal origins of different tissues. A mutation in a gene essential for chloroplast development occurred in some of the initial cells of the meristem, and cells derived from these mutated stem cells lack chloroplasts and are white, while cells derived from other stem cells have normal chloroplasts and appear green. (Courtesy of S. Poethig.)

Although a few of the leaves are white, or nearly so, most of the leaves show green patches. They are **variegated**. The green tissue in these leaves was derived from the cells originally in the L1 or L3 layer; the colorless regions were derived from the L2 layer. The variegation occurs because occasional periclinal divisions in the L1 or L3 layer early in leaf development establish clones of cells that can differentiate as green mesophyll cells. This is further evidence that cell differentiation is not dependent on cell lineage. The fate of a cell during development is determined by the position it occupies in the plant body.

Developmental Pathways Are Controlled by Networks of Interacting Genes

We have a great deal more to learn about the regulatory networks that control developmental pathways. However, several discoveries point to a model in which local and long-distance signaling events control the expression of genes that encode transcription factors. These transcription factors in turn determine the character or activities of a given tissue or cell. Often these mechanisms involve feedback loops in which two or more genes interact to regulate each other's expression. These interactions are seen most clearly in the case of the shoot apical meristem.

Expression of the *KNOX* gene *STM* (*SHOOTMERISTEMLESS*) is essential for the formation of the shoot apical meristem in the *Arabidopsis* embryo and for meristem function in the growing plant. *STM* is expressed throughout the apical dome of the vegetative meristem, except in the developing leaf primordia. Similarly, *STM* is expressed in the dome of the floral meristem, but it is silenced as floral organs appear. Two additional *KNOX* genes—*KNAT1* and *KNAT2*—also are expressed in the apical meristem of *Arabidopsis* and participate in maintaining the meristem cells in an undifferentiated state.

Because cells actively divide in the early stages of leaf and floral organ primordia development, *STM* is not necessary for cell division. Rather *KN1*, *STM*, and their functional homologs maintain meristem identity by suppressing differentiation. Another gene, *ASYMMETRIC LEAVES1* (*AS1*) promotes leaf development and is expressed in the primordia and young leaves of *Arabidopsis* (Figure 16.26) (Byrne et al. 2000). *STM* represses the expression of *AS1*, and *AS1* in turn represses the expression of *KNAT1* in the developing leaf primordia (Ori et al. 2000):

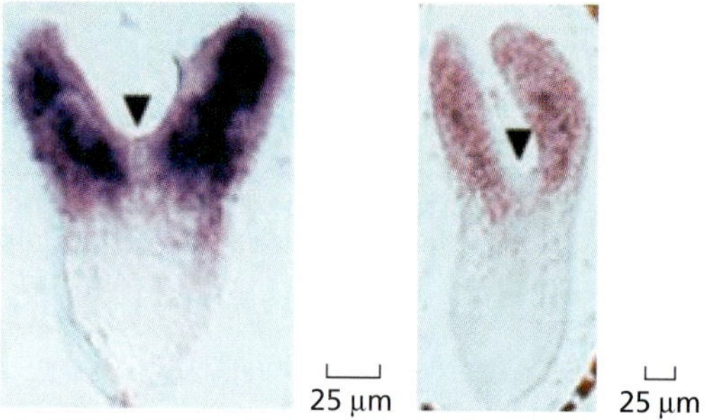

(B) *stm* mutant embryos

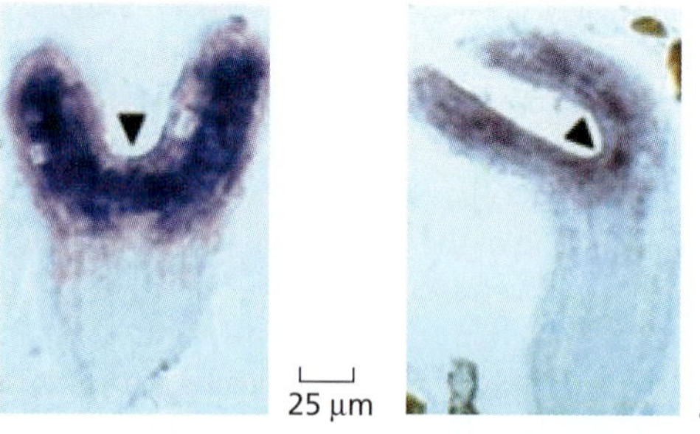

FIGURE 16.26 The meristem identity gene, *STM*, inhibits expression of the *ASYMMETRIC LEAVES1* (*AS1*) gene, which promotes leaf development in *Arabidopsis*. Arrows point to the shoot apical meristem–forming region. (A) Expression of the *STM* gene is normally confined to the shoot apical meristem in the wild type, and it confers meristem identity on the vegetative meristem. In contrast, the *AS1* gene is confined to leaf primordia and developing cotyledons in the wild type, as shown by in situ hybridization in embryos at two stages of development. (B) In *stm* mutants, expression of *AS1* expands into the region that would normally become the shoot apical meristem. As a result, the apical meristem does not form. (From Byrne et al. 2000.)

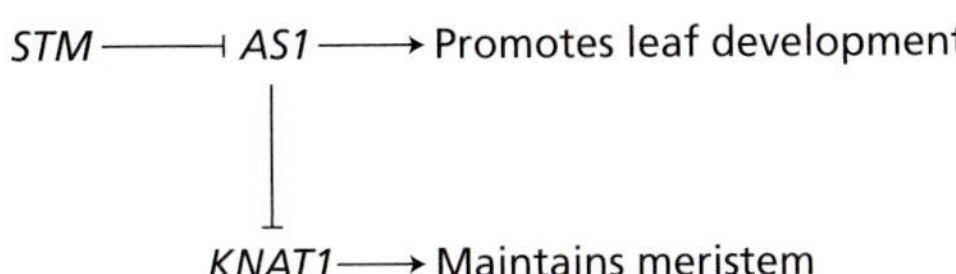

The *WUSCHEL* (*WUS*) gene, which encodes another homeodomain transcription factor, is a key regulator of stem cell indeterminacy (Laux et. al. 1996). In plants with loss-of-function *wus* mutations, either an apical meristem is lacking entirely, or their stem cells are used up after they have formed a few leaves. The *CLAVATA* genes negatively regulate *WUS* expression. *WUS* expression is expanded in both *clv1* and *clv3* mutants (Figure 16.27). Conversely, *WUS* expression positively regulates *CLV3* gene expression; (see Figure 16.24) (Brand et al. 2000).

Development Is Regulated by Cell-to-Cell Signaling

How do cells know where they are? If a cell's fate is determined by its position and not by clonal lineage, then cells must be able to sense their position relative to other cells, tissues, and organs. Neighboring cells and distant tissues and organs provide positional information. Cells in multicellular plants usually are in close contact with others around them, and the behavior of each cell is carefully coordinated with that of its neighbors throughout the life of the plant. Furthermore, each cell occupies a specific position within the tissue and organ to which it belongs.

Coordination of cellular activity requires cell–cell communication. That is, some developmentally important genes act *nonautonomously*. They do not have to be expressed in a given cell to affect the fate of that cell. A given gene or set of genes can exert an effect on development in neighboring cells or even cells in distant tissues through cell–cell communication, via at least three different mechanisms:

1. Ligand-induced signaling
2. Hormonal signaling
3. Signaling via trafficking of regulatory proteins and/or mRNAs

Ligand-induced signaling. There is evidence that cell wall components, particularly a class of glycoprotein macromolecules known as **arabinogalactan proteins**, or **AGPs**, may communicate positional information that will determine cell fate (see Chapter 15). AGPs would not be involved in signaling over a distance, but rather in telling a given cell who its neighbors were. That information then would program the cell to differentiate, or acquire a fate appropriate to its position.

Because plants have numerous, perhaps hundreds, of receptor kinases, we might expect many signaling events to be initiated by ligand-induced protein phosphorylation. At present, however, relatively few of the ligands activating protein kinases are known. But there is good evidence that the small protein encoded by the *CLV3* gene is the ligand that activates the CLV1 protein kinase.

The CLV3 protein contains fewer than 100 amino acids and contains a leader sequence suggesting that it would be excreted from the cells that produce it (Fletcher et al. 1999). Because of its small size and water solubility, it could freely diffuse through the extracellular space, or apoplast.

The **apoplast** consists mostly of the space occupied by the cell walls. Cell wall macromolecules are largely hydrophilic, and the wall contains passages between the macromolecules with an apparent pore size of 3.5 to 5 nm. This means that molecules with a mass of less than approximately 15 kDa can diffuse freely through the apoplast. With a molecular weight of approximately 11 kDa, the CLV3 protein easily could diffuse through the apoplast.

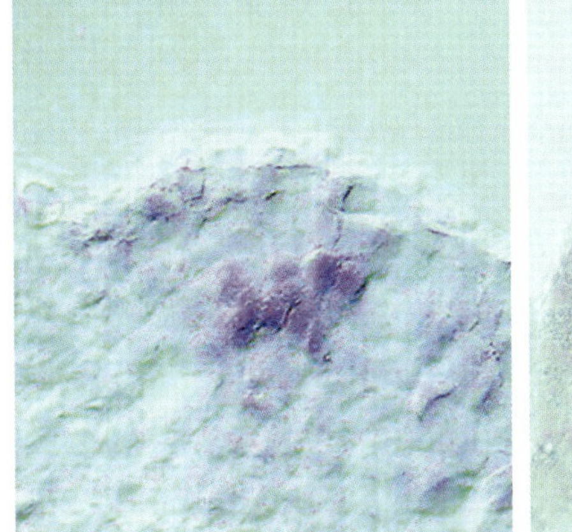

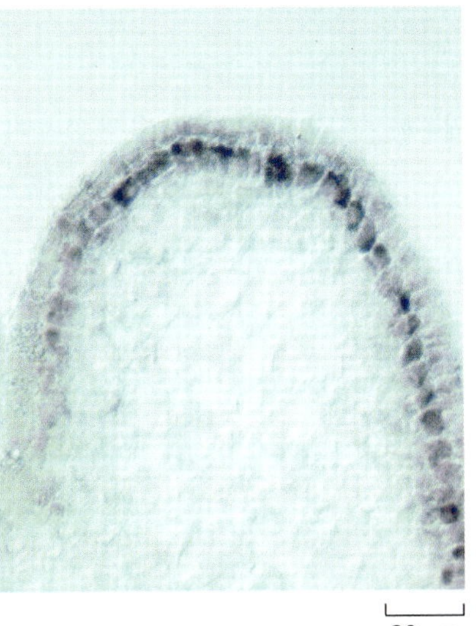

FIGURE 16.27 *WUS* gene expression in the shoot apical meristem of the wild type and the *clv3* mutant. The localization of *WUS* mRNA was detected by an in situ hybridization procedure. (A) In the wild type, *WUS* expression is confined to a small cluster of cells. (B) In the *clv3* mutant, *WUS* expression expands both apically and laterally, and the apical meristem itself is enlarged. (Brand et al. 2000.)

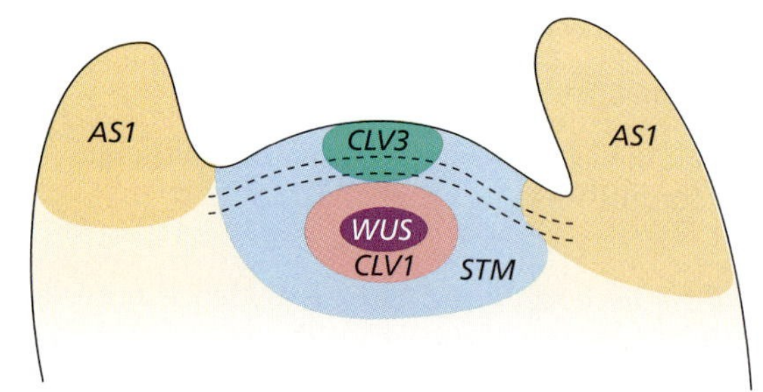

FIGURE 16.28 Patterns of expression of some developmentally important genes in the *Arabidopsis* shoot apical meristem. (From Clark 2001.)

The *CLV3* gene is expressed in cells of the L1 and L2 layers in the central zone of the shoot apical meristem, but not within the L3 layer or in the peripheral zone. In contrast, *CLV1* is expressed in deeper layers within the central zone in the L3 layer, as is the *WUS* gene. However, *CLV1* is expressed within a somewhat larger domain than *WUS* (Figure 16.28). Although *WUS* gene expression is required to maintain stem cell identity, *WUS* is expressed in only a small number of cells in the L3 layer of the meristem. It functions nonautonomously, acting on cells a short distance from the cells that express the gene.

The CLV3 protein controls the size of the stem cell population in the shoot apex by negatively regulating the expression of *WUS* in the L3 layer. The *CLV3* gene is expressed in cells in the central zone of the meristem, within the L1 and L2 layers. When *CLV1* or *CLV3* is knocked out by mutation, *WUS* gene expression spreads, and the number of undifferentiated stem cells expands (Brand et al. 2000). Because this expansion requires *CLV1*, it is likely that CLV3 protein diffuses from the L1 cells and binds to the receptor domain of CLV1 to activate its kinase domain to initiate a signal that represses *WUS* gene transcription.

WUS expression promotes *CLV3* expression, which in turn represses *WUS* expression. Thus the meristem has a sensitive feedback mechanism for controlling the size of the stem cell population.

Hormonal signaling. The plant hormones—auxin, ethylene, gibberellins, abscisic acid, cytokinins, and brassinosteroids—all play roles in regulating development. These roles will be presented in some detail in the chapters and sections devoted to these topics. In this discussion, however, we will focus on auxin signaling as an example of the types of mechanisms these roles might entail. This topic will be discussed in greater detail in Chapter 19.

Auxin signaling is essential for the development of axial polarity and the development of vascular tissue. Auxin has long been known to be the signal for the initiation of vascular tissue differentiation (see Chapter 19). This conclusion, however, is based largely on studies of the effects of applied auxins and auxin transport inhibitors. More recently, two *Arabidopsis* genes—*GNOM* and *MONOPTEROS*—known to be essential for the development of axial polarity and tissue differentiation during embryogenesis and adult plant development, have been found to be involved in auxin signaling. As presented earlier, the *Arabidopsis GNOM* gene was identified because embryos homozygous for mutations in this gene lack both roots and cotyledons and fail to develop axial polarity (see Figure 16.7A) (Mayer et al. 1993).

The *GNOM* gene product is required for correct localization of the auxin efflux carrier protein PIN1 (Figure 16.29).

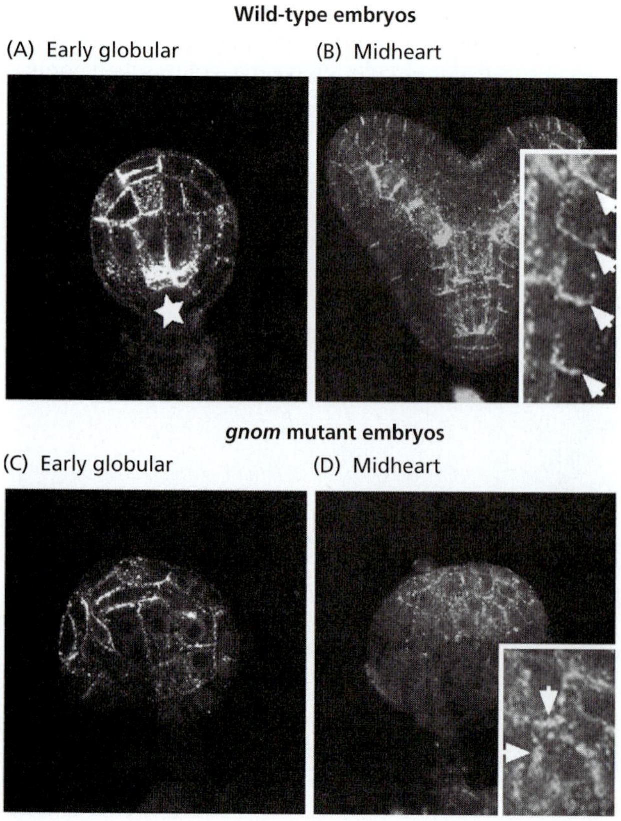

Wild-type embryos

(A) Early globular (B) Midheart

***gnom* mutant embryos**

(C) Early globular (D) Midheart

FIGURE 16.29 Comparison of the distribution patterns of the auxin efflux protein PIN1 in wild-type and *gnom* mutant *Arabidopsis* embryos. (A) Wild-type, early globular; PIN1 is localized in the provascular tissue early in the early globular stage, where the protein accumulates at the basal boundary of the four inner cells that will give rise to the provascular tissue. (B) Wild-type, midheart; in the heart stage, the provascular cells have accumulated PIN1 protein at their basal ends (see insert). (C) *gnom* mutant, early globular; PIN1 does not accumulate in the region where the provascular tissue will form in the early globular stage of the *gnom* mutant. (From Steinmann et al. 1999). (D) *gnom* mutant, midheart; formation of provascular tissue is blocked in the *gnom* mutant, and normal development is disrupted. PIN1 is still inserted in membranes in the mutant, but the localization is disorganized (see insert). (From Steinmann et al. 1999.)

GNOM encodes a guanine nucleotide exchange factor that is a component of the cellular machinery that establishes cell polarity. This machinery, and the GNOM protein in particular, are required for the correct localization of the auxin efflux carrier protein PIN1 at the basal end of the procambium cells during the globular stage of embryogenesis and subsequently in vascular cells throughout development (Steinmann et al. 1999; Grebe et al. 2000).

As we have seen, mutations in the *MONOPTEROS* (*MP*) gene result in seedlings that lack both a hypocotyl and root, although they do produce an apical region. The apical structures in the *mp* mutant embryos are not structurally normal, however, and the tissues of the cotyledons are disorganized (see Figure 16.7B) (Berleth and Jürgens 1993). Embryos of *mp* mutants first show abnormalities at the octant stage, and they do not form a procambium in the lower part of the globular embryo, the part that should give rise to the hypocotyl and root. Later some vascular tissue does form in the cotyledons, but the strands are improperly connected.

The *MP* gene encodes a protein related to the transcription factor known as **ARF** (***a**uxin **r**esponse **f**actor*) (Hardtke and Berleth 1998). Both ARF and MONOPTEROS bind to auxin response elements in the promoters of certain genes that are transcribed in the presence of auxin. Apparently the *MP* gene is required for expression of genes involved in vascular tissue differentiation.

Other evidence in support of auxin signaling during embryogenesis includes the finding that the putative auxin receptor protein, **ABP1**, is required for organized cell elongation and division in embryogenesis. *Arabidopsis* mutants homozygous for *abp1* do not form mature embryos, although they develop normally up to the early globular stage. These mutants cannot make the transition to bilateral symmetry, and cells fail to elongate (Chen et al. 2001).

Auxin signaling also participates in organogenesis from the shoot apical meristem and in the formation of lateral roots. *Arabidopsis* plants with mutations in the auxin efflux carrier gene *PIN1* develop a pinlike inflorescence that is devoid of lateral organs (Figure 16.30). In wild-type plants, *PIN1* gene expression is up-regulated in the early stages of primordium formation, before the primordia begin to bulge. The shoot apical meristem at the tip of the pinlike inflorescence in the *pin1* mutant plants has a normal structure, except that no organs are generated in the peripheral zone and the shoot produced lacks lateral appendages (Vernoux et al. 2000). Thus, auxin is likely to be required for signaling early events necessary for organogenesis from the shoot apical meristem.

This hypothesis is supported by work with tomato. When tomato apical meristems are cultured on medium containing the auxin transport inhibitor *N*-1-naphthylphthalamic acid (NPA), they continue to grow, but they develop into pinlike shoots lacking lateral appendages. When these NPA-induced pin meristems were treated with auxin at their tips, leaf initiation was restored (Reinhardt et al. 2000).

Other signaling mechanisms remain to be discovered. The mechanism by which cells communicate has not been established in other cases, although it is clear that positional information is exchanged between cells in different tissues. As presented earlier, the *SHR* and *SCR* genes are important for the establishment of the radial tissue patterns in roots. They encode rather similar transcription factors, but these two genes are expressed and function in different tissues.

SCR is required for the asymmetric cell division that forms the epidermis and cortex, and it also determines the endodermis cell fate. *SCR* is expressed in the stem cell that will give rise to the ground tissue before it divides asymmetrically to form the precursors of endodermis and cortex (Figure 16.31A). *SCR* continues to be expressed in the endodermis after the stem cell divides (Figure 16.31B).

SCR gene expression requires *SHR* expression, but the *SHR* gene is not expressed in either the cortex or the endodermis. Rather, *SHR* is expressed in the pericycle and the vascular cylinder (Figure 16.31C) (Helariutta et al. 2000). This implies that *SHR* gene expression generates a signal

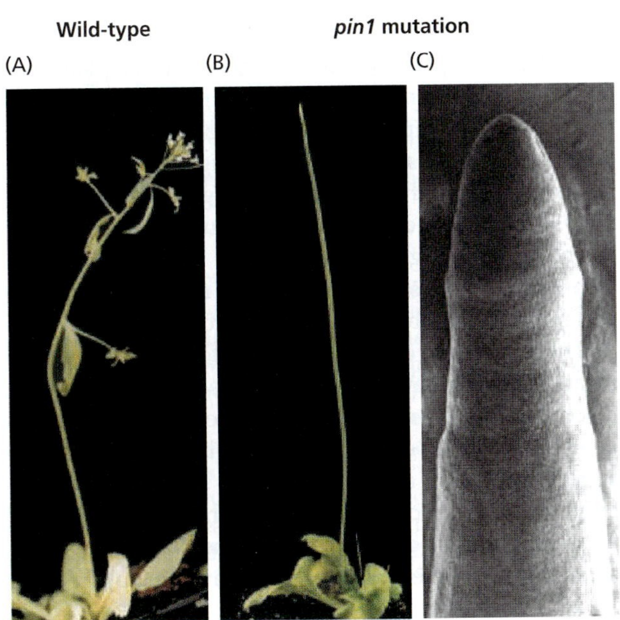

FIGURE 16.30 The *PIN1* gene is essential for the formation of lateral organs from the inflorescence meristem in *Arabidopsis*. (A) The inflorescence meristem generates a stem bearing cauline leaves and numerous floral buds in the wild type. (B) Plants with *pin1* mutations produce an inflorescence meristem, but it fails to generate lateral organs. (C) The inflorescence meristem produces only axial tissues, similar to the root apical meristem, as shown in this scanning electron micrograph. (From Vernoux et al. 2000.)

SUMMARY

The basic body plan of the mature plant is established during embryogenesis; in this process, tissues are arranged radially: an outer epidermal layer surrounding a cylinder of vascular tissue that is embedded within cortical or ground tissues. The apical–basal axial pattern of the mature plant, with root and shoot polar axes, also is established during embryogenesis, as are the primary meristems that will generate the adult plant.

One common type of angiosperm embryonic development, exemplified by *Arabidopsis thaliana*, is characterized by precise patterns of cell divisions, forming successive stages: the globular, heart, torpedo, and maturation stages. The axial body pattern is established during the first division of the zygote, and mutant genes eliminate part of the embryo. The radial tissue pattern is established during the globular stage, apparently as a result of the expression of genes that control cell identity. The *SHOOTMERISTEMLESS* (*STM*) gene is expressed in the region that gives rise to the shoot apical meristem during the heart stage of embryogenesis, and its continued expression suppresses differentiation of the cells of the shoot apical meristem. The *GNOM* gene is required for the establishment of axial polarity, and the *MONOPTEROS* gene is required for formation of the embryonic primary root as well as vascular development.

A complete explanation of the mechanisms responsible for establishing and maintaining these patterns is not possible at present, but there is evidence that an association of microtubules and microfilaments known as the preprophase band is important in determining the plane of cell division. Cell differentiation does not depend on cell lineage; however, the division of the stem cell is essential for this process. Expression of the *SCR* (*SCARECROW*) gene, which has been cloned and encodes a novel protein, is necessary for the division of the stem cell, and the *SHR* (*SHORTROOT*) gene must be expressed for the establishment of endodermal cell identity.

Meristems are populations of small, isodiametric cells that have "embryonic" characteristics. Vegetative meristems generate specific portions of the plant body, and they regenerate themselves. In many plants, the root and shoot apical meristems are capable of indefinite growth.

The vegetative shoot apical meristem repetitively generates lateral organs (leaves and lateral buds), as well as segments of the stem. Shoot apical meristems in angiosperms typically are organized into three distinct layers, designated L1, L2, and L3.

The root and shoot apical meristems are primary meristems formed during embryogenesis. Secondary meristems are initiated during postembryonic development and include the vascular cambium, cork cambium, axillary meristems, and secondary root meristems.

The repetitive activity of the vegetative shoot apical meristem generates a succession of developmental units, called phytomeres, each consisting of one or more leaves, the node, the internode, and one or more axillary buds. The vegetative shoot apical meristem is indeterminate in its activity in that it may function indefinitely, but it gives rise to leaf primordia that are determinate in their growth.

Leaves form in a characteristic pattern, with three stages: (1) organogenesis, (2) development of suborgan domains, (3) cell and tissue differentiation. The number and order in which leaf primordia form is reflected in the subsequent phyllotaxy (alternate, opposite, decussate, whorled, or spiral). The leaf primordia must be positioned as a result of the precise spatial regulation of cell division within the apex, but the factors controlling this activity are not known.

Roots grow from their distal ends. The root apical meristem is subterminal and covered by a root cap. Cell divisions in the root apex generate files of cells that subsequently elongate and differentiate to acquire specialized function. Four developmental zones are recognized in the root: root cap, meristematic zone, elongation zone, and maturation zone. In *Arabidopsis*, files of mature cells can be traced to stem cells within the meristem cell population. The *Arabidopsis* root apical meristem consists of a quiescent center, cortical–endodermal stem cells, columella stem cells, root cap–epidermal stem cells, and stele stem cells.

Differentiation is the process by which cells acquire metabolic, structural, and functional properties distinct from those of their progenitors. Tracheary element differentiation is an example of plant cell differentiation. Microtubules participate in determining the pattern in which the cellulose microfibrils are deposited in the secondary walls of tracheary elements.

MADS box genes are key regulators of important biological functions in plants, animals, and fungi. Homeobox genes encode homeodomain proteins that act as transcription factors. These transcription factors control the expression of other genes whose products transform and characterize the differentiated cell.

In the determination of a cell's fate, the cell's position is more important than its lineage. Plant cell fate is relatively plastic and can be changed when the positional signals necessary for its maintenance are altered.

The expression of homeobox genes similar to the maize genes *KNOTTED1* and *SHOOTMERISTEMLESS* is necessary for the continued indeterminate character of the shoot apical meristem, but the *WUSCHEL* gene determines stem cell identity. Loss of expression of *KNOX* genes in the leaf primordia appears to be important in the shift to determinate growth in these structures.

Cell position is communicated via cell–cell signaling, which may involve ligand-induced signaling, hormone signaling or trafficking of regulatory proteins and/or mRNAs through plasmodesmata. Molecules ranging in size up to about 1.6 nm (700–1000 Da) can pass from cell to cell through plasmodesmata connecting leaf epidermal cells. Plasmodesmata are, to some extent, gated so that passage through them can be regulated, and their size exclusion limit can be modified to permit the passage of much larger molecules, such as viruses.

Growth in plants is defined as an irreversible increase in volume. Plant growth can be quantitatively analyzed with kinematics, the study of particle movement and shape change.

Plant growth can be described in both spatial and material terms. Spatial descriptions focus on the patterns generated by all the cells located at different positions in the growth zones. Material analyses focus on the fate of the individual cells or tissue elements at various stages of development. A growth trajectory shows the distance of a tissue element from the apex over time, and is therefore a material description of growth. The growth velocity is the speed at which the tissue elements are being displaced from the apex. The relative elemental growth rate is a measure of the fractional increase in length of the axis per unit time and represents the magnitude of growth at a particular location.

Senescence and programmed cell death are essential aspects of plant development. Plants exhibit a variety of different senescence phenomena. Leaves are genetically programmed to senesce and die. Senescence is an active developmental process that is controlled by the plant's genetic program and initiated by specific environmental or developmental cues.

Senescence is an ordered series of cytological and biochemical events. The expression of most genes is reduced during senescence, but the expression of some genes (senescence-associated genes, or SAGs) is initiated. The newly active genes encode various hydrolytic enzymes, such as proteases, ribonucleases, lipases, and enzymes involved in the biosynthesis of ethylene, which carry out the degradative processes as the tissues die.

Programmed cell death (PCD) is a specialized type of senescence. One important function of PCD in plants is protection against pathogenic organisms in what is called the hypersensitive response, which has been demonstrated to be a genetically programmed process.

Web Material

Web Topics

16.1 Polarity of *Fucus* Zygotes

A wide variety of external gradients can polarize growth of cells that are initially apolar.

16.2 The Preprophase Band of Microtubules

Ultrastructural studies have elucidated the structure of the preprophase band of micro-tubules and its role in orienting the plane of cell division.

16.3 *Azolla* Root Development

Anatomical studies of the root of the aquatic fern, *Azolla*, have provided insights into cell fate during root development.

16.4 The Relative Elemental Growth Rate

The relative elemental growth rate at various points along a root can be evaluated by differentiation of the growth velocity with respect to position.

Web Essays

16.1 Plant Meristems: An Historical Overview

Scientists have used many approaches to unravel the secrets of plant meristems.

16.2 The Mermaids Wineglass

The giant marine green alga, *Acetabularia acetabulum*, holds a classic place in the history of biology.

16.3 Division Plane Determination in Plant Cells

Plant cells appear to utilize mechanisms different from those used by other eukaryotes to control their division planes.

Chapter References

Assaad, F., Mayer, U., Warner, G., and Jürgens, G. 1996. The *KEULE* gene is involved in cytokinesis in *Arabidopsis*. *Mol. Gen. Genet.* 253: 267–277.

Berleth, T., and Jürgens, G. (1993) The role of the *MONOPTEROS* gene in organising the basal body region of the *Arabidopsis* embryo. *Development* 118: 575–587.

Bowman, J. L., and Eshed, Y. (2000) Formation and maintenance of the shoot apical meristem. *Trends Plant Sci.* 5: 110–115.

Brand, U., Fletcher, J. C., Hobo, M., Meyerowitz, E. M., and Simon, R. (2000) Dependence of stem cell fate in *Arabidopsis* on a feedback loop regulated by *CLV3* activity. *Science* 289: 617–619.

Byrne, M. E., Barley, R., Curtis, M., Arroyo, J. M., Dunham, M., Hudson, A., and Martienssen, R. (2000) Asymmetric leaves1 mediates leaf patterning and stem cell function in *Arabidopsis*. *Nature* 408: 967–971.

Carpenter, R., and Coen, E. S. (1995) Transposon induced chimeras show that floricaula, a meristem identity gene, acts non-autonomously between cell layers. *Development* 121: 19–26.

Chen, J. -G., Ullah, H., Young, J. C., Sussman, M. R., and Jones, A. M. (2001) ABP1 is required for organized cell elongation and division in *Arabidopsis* embryogenesis. *Genes Dev.* 15: 902–911.

Christensen, D., and Weigel, D. (1998) Plant development: The making of a leaf. *Curr. Biol.* 8: R643–645.

Clark, S. E. (2001) Cell signaling at the shoot meristem. *Nature Rev. Mol. Cell. Biol.* 2: 276–284.

Clark, S. E., Running, M. P., and Meyerowitz, E. M. (1993) *CLAVATA1*, a regulator of meristem and flower development in *Arabidopsis*. *Development* 119: 397–418.

Di Laurenzio, L., Wysocka-Diller, J., Malamy, J. E., Pysh, L., Helariutta, Y., Freshour, G., Hahn, M. G., Fledman, K. A., and Benfey, P. N. (1996) The *SCARECROW* gene regulates an asymmetric cell division that is essential for generating the radial organization of the *Arabidopsis* root. *Cell* 86: 423–433.

Doebley, J., and Lukens, L. (1998) Transcriptional regulators and the evolution of plant form. *Plant Cell* 10: 1075–1082.

Doerner, P., Jorgensen, J.-E., You, R., Steppuhn, J., and Lamb, C. (1996) Control of root growth and development by cyclin expression. *Nature* 380: 520–523.

Fletcher, J. C., and Meyerowitz, E. M. (2000) Cell signaling within the shoot meristem. *Curr. Opin. Plant Biol.* 3: 23–30.

Fletcher, J. C., Brand, U., Running, M. P., Simon, R., and Meyerowitz, E. M. (1999) Signaling of cell fate decisions by *CLAVATA3* in *Arabidopsis* shoot meristems. *Science* 283: 1911–1914.

Fukuda, H. (1996) Xylogenesis: Initiation, progression and cell death. *Annu. Rev. Plant Physiol. Plant Mol. Biol.* 47: 299–325.

Grebe, M., Gadea, G., Steinmann, T., Kientz, M., Rahfeld, J.-U., Salchert, K., Koncz, C., and Jürgens, G. (2000) A conserved domain of the *Arabidopsis* GNOM protein mediates subunit interaction and cyclophilin 5 binding. *Plant Cell* 12: 343–356.

Hake, S., Vollbrecht, E., and Freeling, M. (1989) Cloning *KNOTTED*, the dominant morphological mutant in maize using Ds2 as a transposon tag. *EMBO J.* 8: 15–22.

Hardham, A. R., and Gunning, B. E. S. (1979) Interpolation of microtubules into cortical arrays during cell elongation and differentiation in roots of *Azolla pinnata*. *J. Cell Sci.* 37: 411–442.

Hardham, A. R., and Gunning, B. E. S. (1980) Some effects of colchicine on microtubules and cell division of *Azolla pinnata*. *Protoplasma* 102: 31–51.

Hardtke, C., and Berleth, T. (1998) The *Arabidopsis* gene *MONOPTEROS* encodes a transcription factor mediating embryo axis formation and vascular development. *EMBO J.* 17: 1405–1411.

Helariutta, Y., Fukaki, H., Wysocka-Diller, J., Nakajima, K., Sena, G., Hauser, M.-T., and Benfey, P. N. (2000) The *SHORT-ROOT* gene controls radial patterning of the *Arabidopsis* root through radial signaling. *Cell* 10: 555–567.

Hepler, P. K. (1981) Morphogenesis of tracheary elements and guard cells. In *Cytomorphogenesis in Plants*, O. Kiermayer, ed., Springer, Berlin, pp. 327–347.

Jackson, D., Veit, B., and Hake, S. (1994) Expression of maize KNOTTED1 related homeobox genes in the shoot apical meristem predicts patterns of morphogenesis in the vegetative shoot. *Development* 120: 405–413.

Laux, T., Mayer, Klaus, F. X., Berger, J., and Jürgens, G. (1996) The *WUSCHEL* gene is required for shoot and floral meristem integrity in *Arabidopsis. Development* 122: 87–96.

Lincoln, C., Long, J., Yamaguchi, J., Serikawa, K., and Hake, S. (1994) A *knotted1*-like homeobox gene in *Arabidopsis* is expressed in the vegetative meristem and dramatically alters leaf morphology when overexpressed in transgenic plants. *Plant Cell* 6: 1859–1876.

Long, J. A., Moan, E. I., Medford, J. I., and Barton, M. K. (1996) A member of the KNOTTED class of homeodomain proteins encoded by the *STM* gene of *Arabidopsis. Nature* 379: 66–69.

Lotan, T., Ohto, M.-A., Yee, K. M., West, M. A., Lo, R., Kwong, R. W., Yamagishi, K., Fisher, R. L., and Goldberg, R. B. (1998) *Arabidopsis LEAFY COTYLEDON1* is sufficient to induce embryo development in vegetative cells. *Cell* 93: 1195–1205.

Lucas, W. J., Bouche-Pillon, S., Jackson, D. P., Nguyen, L., Baker, L., Ding, B., and Hake, S. (1995) Selective trafficking of KNOTTED1 homeodomain protein and its mRNA through plasmodesmata. *Science* 270: 1980–1983.

Lukowitz, W., Mayer, U., and Jürgens, G. (1996) Cytokinesis in the *Arabidopsis* embryo involves the syntaxin-related *KNOLLE* gene product. *Cell* 84: 61–71.

Malamy, J. E. and Benfey, P. N. (1997) Organization and cell differentiation in lateral roots of *Arabidopsis thaliana. Development* 124: 33–44.

Mayer, U., Buettner, G., and Jürgens, G. (1993) Apical-basal pattern formation in the *Arabidopsis* embryo: Studies on the role of the gnom gene. *Development* 117: 149–162.

Nishimura, A., Tamaoki, M., Sato, Y., and Matsuoka, M. (1999) The expression of tobacco *knotted1*-type homeobox genes corresponds to regions predicted by the cytohistological zonation model. *Plant J.* 18: 337–347.

Ori, N., Eshed, Y., Chuck, G., Bowman, J. L., and Hake, S. (2000) Mechanisms that control *knox* gene expression in the *Arabidopsis* shoot. *Development* 127: 5523–5532.

Pennell, R. I., and Lamb, C. (1997) Programmed cell death in plants. *Plant Cell* 9: 1157–1168.

Przemeck, G. K. H., Mattsson, J., Hardtke, C. S., Sung, Z. R., and Berleth, T. (1996) Studies on the role of the *Arabidopsis* gene *MONOPTEROS* in vascular development and plant cell axialization. *Planta* 200: 229–237.

Reinhardt, D., Mandel, T., and Kuhlemeier, C. (2000) Auxin regulates the initiation and radial position of plant lateral organs. *Plant Cell* 12: 507–518.

Riechmann, J. L., and Meyerowitz, E. M. (1997) MADS domain proteins in plant development. *Biol. Chem.* 378: 1079–1101.

Riechmann, J. L., Herd, J., Martin, G, Reuber, L., Jiang, C. Z., Keddie, J., Adam, L., Pineda, O., Ratcliffe, O. J., Samaha, R. R., Creelman, R., Pilgrim, M., Broun, P., Zhang, J. Z., Ghandelhari, D., Sherman, B. K., and Yu, G.-L. (2000) *Arabidopsis* transcription factors: Genome-wide comparative analysis among eukaryotes. *Science* 290: 2105–2110.

Scheres, B. (2001) Plant cell identity. The role of position and lineage. *Plant Physiol.* 125: 112–114.

Scheres, B., Di Laurenzio, L., Willemsen, V., Hauser, M.-T., Janmaat, K., Weisbeek, P., and Benfey, P. N. (1995) Mutations affecting the radial organisation of the *Arabidopsis* root display specific defects throughout the embryonic axis. *Development* 121: 53–62.

Schiefelbein, J. W., Masucci, J. D., and Wang, H. (1997) Building a root: The control of patterning and morphogenesis during root development. *Plant Cell* 9: 1089–1098.

Shiu, S. H., and Bleecker, A. B. (2001) Receptor-like kinases from *Arabidopsis* form a monophyletic gene family related to animal receptor kinases. *Proc. Natl. Acad. Sci. USA* 98: 10763–10768.

Silk, W. K. (1994) Kinematics and dynamics of primary growth. *Biomimetics* 2: 199–213.

Sinha, N. (1999) Leaf development in angiosperms. *Annu. Rev. Plant Physiol. Plant Mol. Biol.* 50: 419–446.

Sinha, N., Hake, S., and Freeling, M. (1993a) Genetic and molecular analysis of leaf development. *Curr. Top. Dev. Biol.* 28: 47–80.

Sinha, N. R., Williams, R. E., and Hake, S. (1993b) Overexpression of the maize homeo box gene, *KNOTTED*-1, causes a switch from determinate to indeterminate cell fates. *Genes Dev.* 7: 787–795.

Steinmann, T., Geldner, N., Grebe, M., Mangold, S. A., Jackson, C. L., Paris, S., Galweiler, L., Palme, K., and Jürgens, G. (1999) Coordinated polar localization of auxin efflux carrier PIN1 by GNOM ARF GEF. *Science* 286: 316–318.

Torres-Ruiz, R. A., and Jürgens, G. (1994) Mutations in the *FASS* gene uncouple pattern formation and morphogenesis in *Arabidopsis* development. *Development* 120: 2967–2978.

Traas, J., Bellini, C., Nacry, P., Kronenberger, J., Bouchez, D., and Caboche, M. (1995) Normal differentiation patterns in plants lacking microtubular preprophase bands. *Nature* 375: 676–677.

Van Den Berg, C., Willemsen, V., Hage, W., Weisbeek, P., and Scheres, B. (1995) Cell fate in the *Arabidopsis* root meristem determined by directional signaling. *Nature* 378: 62–65.

Vernoux, T., Kronenberger, J., Grandjean, O., Laufs, P., and Traas, J. (2000) *PIN-FORMED1* regulates cell fate at the periphery of the shoot apical meristem. *Development* 127: 5157–5165.

Weigel, D., and Jürgens, G. (2002) Stem cells that make stems. *Nature* 415: 751–754.

West, M. A. L., and Harada, J. J. 1993. Embryogenesis in higher plants: An overview. *Plant Cell.* 5: 1361–1369.

Willemsen, V., Wolkenfelt, H., de Vrieze, G., Weisbeek, P., and Scheres, B. (1998) The *HOBBIT* gene is required for formation of the root meristem in the *Arabidopsis* embryo. *Development* 125: 521–531.

Zambryski, P., and Crawford, K. (2000) Plasmodesmata: Gatekeepers for cell-to-cell transport of developmental signals in plants. *Annu. Rev. Cell Dev. Biol.* 16: 393–421.

Chapter

17

Phytochrome and Light Control of Plant Development

HAVE YOU EVER LIFTED UP A BOARD that has been lying on a lawn for a few weeks and noticed that the grass growing underneath was much paler and spindlier than the surrounding grass? The reason this happens is that the board is opaque, keeping the underlying grass in darkness. Seedlings grown in the dark have a pale, unusually tall and spindly appearance. This form of growth, known as **etiolated growth**, is dramatically different from the stockier, green appearance of seedlings grown in the light (Figure 17.1).

Given the key role of photosynthesis in plant metabolism, one might be tempted to attribute much of this contrast to differences in the availability of light-derived metabolic energy. However, it takes very little light or time to initiate the transformation from the etiolated to the green state. So in the change from dark to light growth, light acts as a developmental trigger rather than a direct energy source.

If you were to remove the board and expose the pale patch of grass to light, it would appear almost the same shade of green as the surrounding grass within a week or so. Although not visible to the naked eye, these changes actually start almost immediately after exposure to light. For example, within hours of applying a single flash of relatively dim light to a dark-grown bean seedling in the laboratory, one can measure several developmental changes: a decrease in the rate of stem elongation, the beginning of apical-hook straightening, and the initiation of the synthesis of pigments that are characteristic of green plants.

Light has acted as a signal to induce a change in the form of the seedling, from one that facilitates growth beneath the soil, to one that is more adaptive to growth above ground. In the absence of light, the seedling uses primarily stored seed reserves for etiolated growth. However, seed plants, including grasses, don't store enough energy to sustain growth indefinitely. They require light energy not only to fuel photosynthesis, but to initiate the developmental switch from dark to light growth.

Photosynthesis cannot be the driving force of this transformation because chlorophyll is not present during this time. Full de-etiolation

FIGURE 17.1 Corn (*Zea mays*) (A and B) and bean (*Phaseolus vulgaris*) (C and D) seedlings grown either in the light (A and C) or the dark (B and D). Symptoms of etiolation in corn, a monocot, include the absence of greening, reduction in leaf size, failure of leaves to unroll, and elongation of the coleoptile and mesocotyl. In bean, a dicot, etiolation symptoms include absence of greening, reduced leaf size, hypocotyl elongation, and maintenance of the apical hook. (Photos © M. B. Wilkins.)

does require some photosynthesis, but the initial rapid changes are induced by a distinctly different light response, called **photomorphogenesis** (from Latin, meaning literally "light form begins").

Among the different pigments that can promote photomorphogenic responses in plants, the most important are those that absorb red and blue light. The blue-light photoreceptors will be discussed in relation to guard cells and phototropism in Chapter 18. The focus of this chapter is **phytochrome**, a protein pigment that absorbs red and far-red light most strongly, but that also absorbs blue light. As we will see in this chapter and in Chapter 24, phytochrome plays a key role in light-regulated vegetative and reproductive development.

We begin with the discovery of phytochrome and the phenomenon of red/far-red photoreversibility. Next we will discuss the biochemical and photochemical properties of phytochrome, and the conformational changes induced by light. Different types of phytochromes are encoded by different members of a multigene family, and different phytochromes regulate distinct processes in the plant. These different phytochrome responses can be classified according to the amount of light and light quality required to produce the effect. Finally, we will examine what is known about the mechanism of phytochrome action at the cellular and molecular levels, including signal transduction pathways and gene regulation.

THE PHOTOCHEMICAL AND BIOCHEMICAL PROPERTIES OF PHYTOCHROME

Phytochrome, a blue protein pigment with a molecular mass of about 125 kDa (kilodaltons), was not identified as a unique chemical species until 1959, mainly because of technical difficulties in isolating and purifying the protein. However, many of the biological properties of phytochrome had been established earlier in studies of whole plants.

The first clues regarding the role of phytochrome in plant development came from studies that began in the 1930s on red light–induced morphogenic responses, especially seed germination. The list of such responses is now enormous and includes one or more responses at almost every stage in the life history of a wide range of different green plants (Table 17.1).

A key breakthrough in the history of phytochrome was the discovery that the effects of *red light* (650–680 nm) on morphogenesis could be reversed by a subsequent irradiation with light of longer wavelengths (710–740 nm), called *far-red light*. This phenomenon was first demonstrated in germinating seeds, but was also observed in relation to stem and leaf growth, as well as floral induction (see Chapter 24).

The initial observation was that the germination of lettuce seeds is stimulated by red light and inhibited by far-red light. But the real breakthrough was made many years later when lettuce seeds were exposed to alternating treatments of red and far-red light. Nearly 100% of the seeds that received red light as the final treatment germinated; in seeds that received far-red light as the final treatment, however, germination was strongly inhibited (Figure 17.2) (Flint 1936).

Two interpretations of these results were possible. One is that there are two pigments, a red light–absorbing pigment and a far-red light–absorbing pigment, and the two pigments act antagonistically in the regulation of seed germination. Alternatively, there might be a single pigment that can exist in two interconvertible forms: a red

TABLE 17.1
Typical photoreversible responses induced by phytochrome in a variety of higher and lower plants

Group	Genus	Stage of development	Effect of red light
Angiosperms	*Lactuca* (lettuce)	Seed	Promotes germination
	Avena (oat)	Seedling (etiolated)	Promotes de-etiolation (e.g., leaf unrolling)
	Sinapis (mustard)	Seedling	Promotes formation of leaf primordia, development of primary leaves, and production of anthocyanin
	Pisum (pea)	Adult	Inhibits internode elongation
	Xanthium (cocklebur)	Adult	Inhibits flowering (photoperiodic response)
Gymnosperms	*Pinus* (pine)	Seedling	Enhances rate of chlorophyll accumulation
Pteridophytes	*Onoclea* (sensitive fern)	Young gametophyte	Promotes growth
Bryophytes	*Polytrichum* (moss)	Germling	Promotes replication of plastids
Chlorophytes	*Mougeotia* (alga)	Mature gametophyte	Promotes orientation of chloroplasts to directional dim light

light–absorbing form and a far-red light–absorbing form (Borthwick et al. 1952).

The model chosen—the one-pigment model—was the more radical of the two because there was no precedent for such a photoreversible pigment. Several years later phytochrome was demonstrated in plant extracts for the first time, and its unique photoreversible properties were exhibited in vitro, confirming the prediction (Butler et al. 1959).

In this section we will consider three broad topics:

1. Photoreversibility and its relationship to phytochrome responses
2. The structure of phytochrome, its synthesis and assembly, and the conformational changes associated with the interconversions of the two main forms of phytochrome: Pr and Pfr
3. The phytochrome gene family, the members of which have different functions in photomorphogenesis

Phytochrome Can Interconvert between Pr and Pfr Forms

In dark-grown or etiolated plants, phytochrome is present in a red light–absorbing form, referred to as **Pr** because it

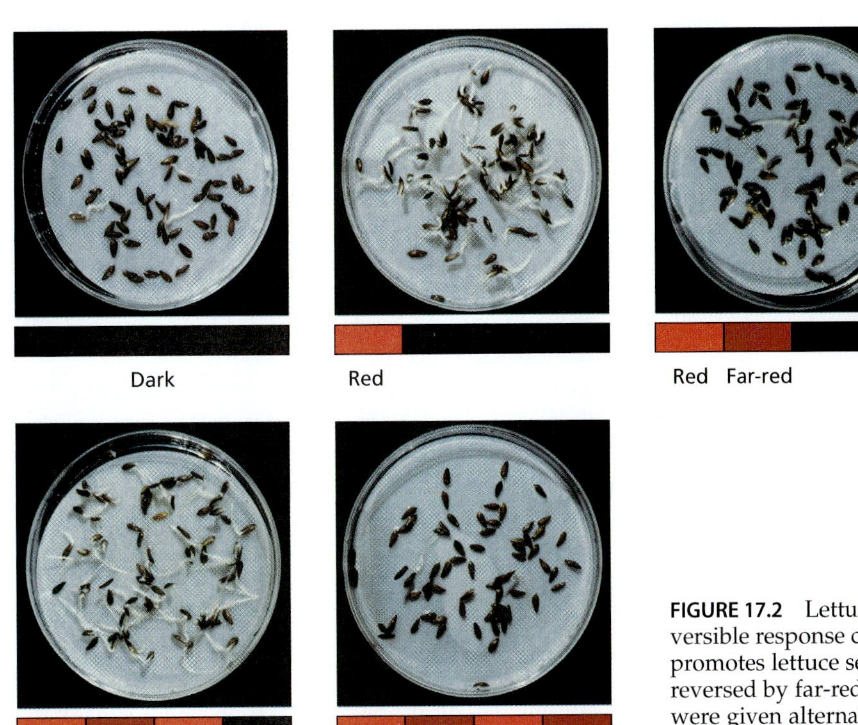

FIGURE 17.2 Lettuce seed germination is a typical photoreversible response controlled by phytochrome. Red light promotes lettuce seed germination, but this effect is reversed by far-red light. Imbibed (water-moistened) seeds were given alternating treatments of red followed by far-red light. The effect of the light treatment depended on the last treatment given. (Photos © M. B. Wilkins.)

is synthesized in this form. Pr, which to the human eye is blue, is converted by red light to a far-red light–absorbing form called **Pfr**, which is blue-green. Pfr, in turn, can be converted back to Pr by far-red light.

Known as **photoreversibility**, this conversion/reconversion property is the most distinctive property of phytochrome, and it may be expressed in abbreviated form as follows:

$$\text{Pr} \underset{\text{Far-red light}}{\overset{\text{Red light}}{\rightleftharpoons}} \text{Pfr}$$

The interconversion of the Pr and Pfr forms can be measured in vivo or in vitro. In fact, most of the spectral properties of carefully purified phytochrome measured in vitro are the same as those observed in vivo.

When Pr molecules are exposed to red light, most of them absorb it and are converted to Pfr, but some of the Pfr also absorbs the red light and is converted back to Pr because both Pr and Pfr absorb red light (Figure 17.3). Thus the proportion of phytochrome in the Pfr form after saturating irradiation by red light is only about 85%. Similarly, the very small amount of far-red light absorbed by Pr makes it impossible to convert Pfr entirely to Pr by broad-spectrum far-red light. Instead, an equilibrium of 97% Pr and 3% Pfr is achieved. This equilibrium is termed the **photostationary state**.

In addition to absorbing red light, both forms of phytochrome absorb light in the blue region of the spectrum (see Figure 17.3). Therefore, phytochrome effects can be elicited also by blue light, which can convert Pr to Pfr and vice versa. Blue-light responses can also result from the action of one or more specific blue-light photoreceptors (see Chapter 18). Whether phytochrome is involved in a response to blue light is often determined by a test of the ability of far-red light to reverse the response, since only phytochrome-induced responses are reversed by far-red light. Another way to discriminate between photoreceptors is to study mutants that are deficient in one of the photoreceptors.

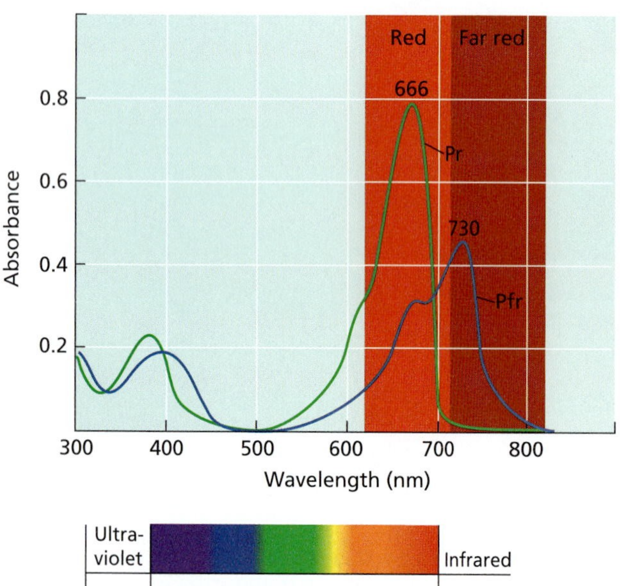

FIGURE 17.3 Absorption spectra of purified oat phytochrome in the Pr (green line) and Pfr (blue line) forms overlap. (After Vierstra and Quail 1983.)

Short-lived phytochrome intermediates. The photoconversions of Pr to Pfr, and of Pfr to Pr, are not one-step processes. By irradiating phytochrome with very brief flashes of light, we can observe absorption changes that occur in less than a millisecond.

Of course, sunlight includes a mixture of all visible wavelengths. Under such white-light conditions, both Pr and Pfr are excited, and phytochrome cycles continuously between the two. In this situation the intermediate forms of phytochrome accumulate and make up a significant fraction of the total phytochrome. Such intermediates could even play a role in initiating or amplifying phytochrome responses under natural sunlight, but this question has yet to be resolved.

Pfr Is the Physiologically Active Form of Phytochrome

Because phytochrome responses are induced by red light, they could in theory result either from the appearance of Pfr or from the disappearance of Pr. In most cases studied, a quantitative relationship holds between the magnitude of the physiological response and the amount of Pfr generated by light, but no such relationship holds between the physiological response and the loss of Pr.

Evidence such as this has led to the conclusion that Pfr is the physiologically active form of phytochrome. In cases in which it has been shown that a phytochrome response is not quantitatively related to the absolute amount of Pfr, it has been proposed that the ratio between Pfr and Pr, or between Pfr and the total amount of phytochrome, determines the magnitude of the response.

The conclusion that Pfr is the physiologically active form of phytochrome is supported by studies with mutants of *Arabidopsis* that are unable to synthesize phytochrome. In wild-type seedlings, hypocotyl elongation is strongly inhibited by white light, and phytochrome is one of the photoreceptors involved in this response. When grown under continuous white light, mutant seedlings with long hypocotyls were discovered and were termed *hy* mutants. Different *hy* mutants are designated by numbers: *hy1*, *hy2*, and so on. Because white light is a mixture of wavelengths (including red, far red, and blue), some, but not all, of the *hy* mutants have been shown to be deficient for one or more functional phytochrome(s).

The phenotypes of phytochrome-deficient mutants have been useful in identifying the physiologically active form of phytochrome. If the phytochrome-induced response to white light (hypocotyl growth inhibition) is caused by the absence of Pr, such phytochrome-deficient mutants (which have neither Pr nor Pfr) should have short hypocotyls in both darkness and white light. Instead, the opposite occurs; that is, they have long hypocotyls in both darkness and white light. It is the absence of Pfr that prevents the seedlings from responding to white light. In other words, Pfr brings about the physiological response.

Phytochrome Is a Dimer Composed of Two Polypeptides

Native phytochrome is a soluble protein with a molecular mass of about 250 kDa. It occurs as a dimer made up of two equivalent subunits. Each subunit consists of two components: a light-absorbing pigment molecule called the **chromophore**, and a polypeptide chain called the **apoprotein**. The apoprotein monomer has a molecular mass of about 125 kDa. Together, the apoprotein and its chromophore make up the **holoprotein**. In higher plants the chromophore of phytochrome is a linear tetrapyrrole termed **phytochromobilin**. There is only one chromophore per monomer of apoprotein, and it is attached to the protein through a thioether linkage to a cysteine residue (Figure 17.4).

Researchers have visualized the Pr form of phytochrome using electron microscopy and X-ray scattering, and the model shown in Figure 17.5 has been proposed (Nakasako et al. 1990). The polypeptide folds into two major domains separated by a "hinge" region. The larger N-terminal domain is approximately 70 kDa and bears the chromophore; the smaller C-terminal domain is approximately 55 kDa and contains the site where the two monomers associate with each other to form the dimer (see **Web Topic 17.1**).

Phytochromobilin Is Synthesized in Plastids

The phytochrome apoprotein alone cannot absorb red or far-red light. Light can be absorbed only when the polypeptide is covalently linked with phytochromobilin to form the holoprotein. Phytochromobilin is synthesized inside plastids and is derived from 5-aminolevulinic acid via a pathway that branches from the chlorophyll biosynthetic pathway (see **Web Topic 7.11**). It is thought to leak out of the plastid into the cytosol by a passive process.

Assembly of the phytochrome apoprotein with its chromophore is **autocatalytic**; that is, it occurs spontaneously when purified phytochrome polypeptide is mixed with purified chromophore in the test tube, with no additional proteins or cofactors (Li and Lagarias 1992). The resultant holoprotein has spectral properties similar to those observed for the holoprotein purified from plants, and it exhibits red/far-red reversibility (Li and Lagarias 1992).

Mutant plants that lack the ability to synthesize the chromophore are defective in processes that require the action of phytochrome, even though the apoprotein polypeptides are present. For example, several of the *hy* mutants noted earlier, in which white light fails to suppress hypocotyl elongation, have defects in chromophore biosynthesis. In *hy1* and *hy2* mutant plants, phytochrome apoprotein levels are normal, but there is little or no spectrally

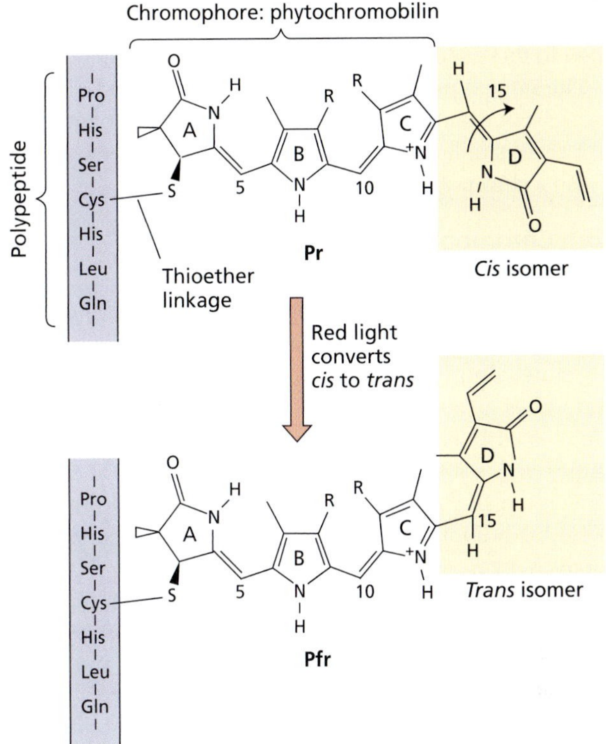

FIGURE 17.4 Structure of the Pr and Pfr forms of the chromophore (phytochromobilin) and the peptide region bound to the chromophore through a thioether linkage. The chromophore undergoes a *cis–trans* isomerization at carbon 15 in response to red and far-red light. (After Andel et al. 1997.)

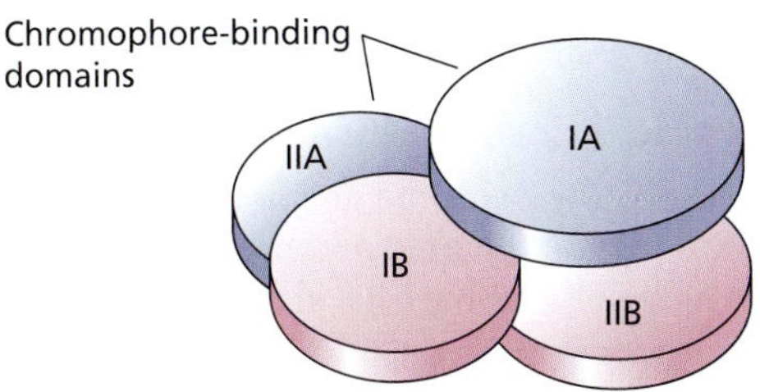

FIGURE 17.5 Structure of the phytochrome dimer. The monomers are labeled I and II. Each monomer consists of a chromophore-binding domain (A) and a smaller nonchromophore domain (B). The molecule as a whole has an ellipsoidal rather than globular shape. (After Tokutomi et al. 1989.)

active holoprotein. When a chromophore precursor is supplied to these seedlings, normal growth is restored.

The same type of mutation has been observed in other species. For example, the *yellow-green* mutant of tomato has properties similar to those of *hy* mutants, suggesting that it is also a chromophore mutant.

Both Chromophore and Protein Undergo Conformational Changes

Because the chromophore absorbs the light, conformational changes in the protein are initiated by changes in the chromophore. Upon absorption of light, the Pr chromophore undergoes a *cis–trans* isomerization of the double bond between carbons 15 and 16 and rotation of the C14–C15 single bond (see Figure 17.4) (Andel et al. 1997). During the conversion of Pr to Pfr, the protein moiety of the phytochrome holoprotein also undergoes a subtle conformational change.

Several lines of evidence suggest that the light-induced change in the conformation of the polypeptide occurs both in the N-terminal chromophore-binding domain and in the C-terminal region of the protein.

Two Types of Phytochromes Have Been Identified

Phytochrome is most abundant in etiolated seedlings; thus most biochemical studies have been carried out on phytochrome purified from nongreen tissues. Very little phytochrome is extractable from green tissues, and a portion of the phytochrome that can be extracted differs in molecular mass from the abundant form of phytochrome found in etiolated plants.

Research has shown that there are two different classes of phytochrome with distinct properties. These have been termed Type I and Type II phytochromes (Furuya 1993). Type I is about nine times more abundant than Type II in dark-grown pea seedlings; in light-grown pea seedlings the amounts of the two types are about equal. More recently, the two types have been shown to be distinct proteins.

The cloning of genes that encode different phytochrome polypeptides has clarified the distinct nature of the phytochromes present in etiolated and green seedlings. Even in etiolated seedlings, phytochrome is a mixture of related proteins encoded by different genes.

Phytochrome Is Encoded by a Multigene Family

The cloning of phytochrome genes made it possible to carry out a detailed comparison of the amino acid sequences of the related proteins. It also allowed the study of their expression patterns, at both the mRNA and the protein levels.

The first phytochrome sequences cloned were from monocots. These studies and subsequent research indicated that phytochromes are soluble proteins—a finding that is consistent with previous purification studies. A complementary-DNA clone encoding phytochrome from the dicot zucchini (*Cucurbita pepo*) was used to identify five structurally related phytochrome genes in *Arabidopsis* (Sharrock and Quail 1989). This phytochrome gene family is named *PHY*, and its five individual members are *PHYA*, *PHYB*, *PHYC*, *PHYD*, and *PHYE*.

The apoprotein by itself (without the chromophore) is designated PHY; the holoprotein (with the chromophore) is designated phy. By convention, phytochrome sequences from other higher plants are named according to their homology with the *Arabidopsis PHY* genes. Monocots appear to have representatives of only the *PHYA* through *PHYC* families, while dicots have others derived by gene duplication (Mathews and Sharrock 1997).

Some of the *hy* mutants have turned out to be selectively deficient in specific phytochromes. For example, *hy3* is deficient in phyB, and *hy1* and *hy2* are deficient in chromophore. These and other *phy* mutants have been useful in determining the physiological functions of the different phytochromes (as discussed later in this chapter).

PHY Genes Encode Two Types of Phytochrome

On the basis of their expression patterns, the products of members of the *PHY* gene family can be classified as either Type I or Type II phytochromes. *PHYA* is the only gene that encodes a Type I phytochrome. This conclusion is based on the expression pattern of the *PHYA* promoter, as well as on the accumulation of its mRNA and polypeptide in response to light. Additional studies of plants that contain mutated forms of the *PHYA* gene (termed *phyA* alleles) have confirmed this conclusion and have given some clues about the role of this phytochrome in whole plants.

The *PHYA* gene is transcriptionally active in dark-grown seedlings, but its expression is strongly inhibited in the light in monocots. In dark-grown oat, treatment with red light reduces phytochrome synthesis because the Pfr form of phytochrome inhibits the expression of its own gene. In addition, the *PHYA* mRNA is unstable, so once etiolated oat seedlings are transferred to the light, *PHYA* mRNA rapidly disappears. The inhibitory effect of light on *PHYA* transcription is less dramatic in dicots, and in *Arabidopsis* red light has no measurable effect on *PHYA*.

The amount of phyA in the cell is also regulated by protein destruction. The Pfr form of the protein encoded by the *PHYA* gene, called **PfrA**, is unstable. There is evidence that PfrA may become marked or tagged for destruction by the ubiquitin system (Vierstra 1994). As discussed in Chapter 14 on the web site, *ubiquitin* is a small polypeptide that binds covalently to proteins and serves as a recognition site for a large proteolytic complex, the *proteasome*.

Therefore, oats and other monocots rapidly lose most of their Type I phytochrome (phyA) in the light as a result of a combination of factors: inhibition of transcription, mRNA degradation, and proteolysis:

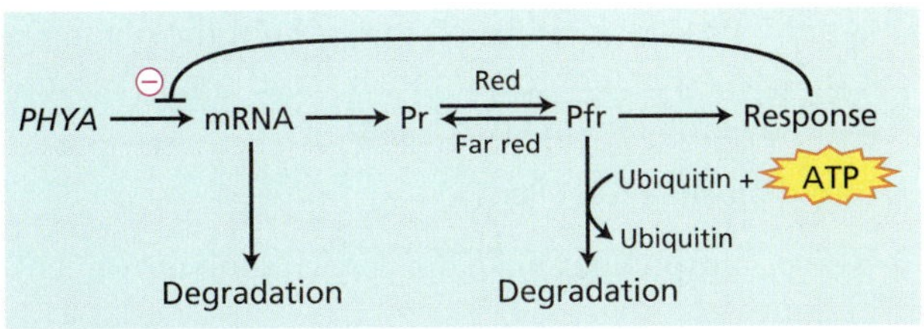

In dicots, phyA levels also decline in the light as a result of proteolysis, but not as dramatically.

The remaining *PHY* genes (*PHYB* through *PHYE*) encode the Type II phytochromes. Although detected in green plants, these phytochromes are also present in etiolated plants. The reason is that the expression of their mRNAs is not significantly changed by light, and the encoded phyB through phyE proteins are more stable in the Pfr form than is PfrA.

PHYB–E ⟶ mRNA ⟶ Pr ⇄ Pfr ⟶ Response (Red / Far red)

LOCALIZATION OF PHYTOCHROME IN TISSUES AND CELLS

Valuable insights into the function of a protein can be gained from a determination of where it is located. It is not surprising, therefore, that much effort has been devoted to the localization of phytochrome in organs and tissues, and within individual cells.

Phytochrome Can Be Detected in Tissues Spectrophotometrically

The unique photoreversible properties of phytochrome can be used to quantify the pigment in whole plants through the use of a spectrophotometer. Because its color is masked by chlorophyll, phytochrome is difficult to detect in green tissue. In dark-grown plants, where there is no chlorophyll, phytochrome has been detected in many angiosperm tissues—both monocot and dicot—as well as in gymnosperms, ferns, mosses, and algae.

In etiolated seedlings the highest phytochrome levels are usually found in meristematic regions or in regions that were recently meristematic, such as the bud and first node of pea (Figure 17.6), or the tip and node regions of the coleoptile in oat. However, differences in expression patterns between monocots and dicots and between Type I and Type II phytochromes are apparent when other, more sensitive methods are used.

Phytochrome Is Differentially Expressed In Different Tissues

The cloning of individual *PHY* genes has enabled researchers to determine the patterns of expression of individual phytochromes in specific tissues by several methods. The sequences can be used directly to probe mRNAs isolated from different tissues or to analyze transcriptional activity by means of a reporter gene, which visually reveals sites of gene expression. In the latter approach, the promoter of a *PHYA* or *PHYB* gene is joined to the coding portion of a reporter gene, such as the gene for the enzyme β-glucuronidase, which is

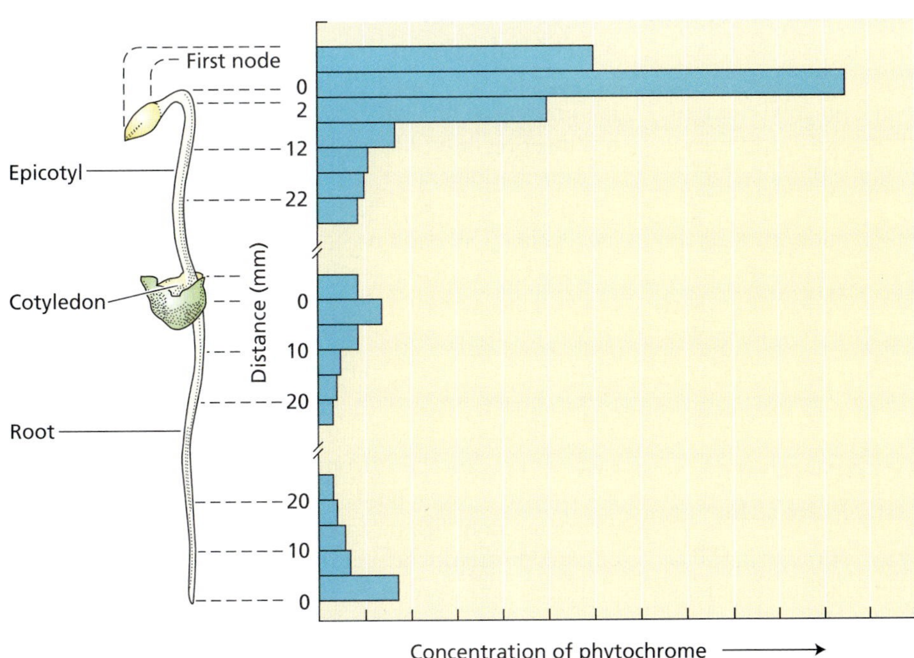

FIGURE 17.6 Phytochrome is most heavily concentrated in the regions where dramatic developmental changes are occurring: the apical meristems of the epicotyl and root. Shown here is the distribution of phytochrome in an etiolated pea seedling, as measured spectrophotometrically. (From Kendrick and Frankland 1983.)

called *GUS* (recall that the promoter is the sequence upstream of the gene that is required for transcription).

The advantage of using the *GUS* sequence is that it encodes an enzyme that, even in very small amounts, converts a colorless substrate to a colored precipitate when the substrate is supplied to the plant. Thus, cells in which the *PHYA* promoter is active will be stained blue, and other cells will be colorless. The hybrid, or fused, gene is then placed back into the plant through use of the Ti plasmid of *Agrobacterium tumefaciens* as a vector (see **Web Topic 21.5**).

When this method was used to examine the transcription of two different *PHYA* genes in tobacco, dark-grown seedlings were found to contain the highest amount of stain in the apical hook and the root tips, in keeping with earlier immunological studies (Adam et al. 1994). The pattern of staining in light-grown seedlings was similar but, as might be expected, was of much lower intensity. Similar studies with *Arabidopsis PHYA–GUS* and *PHYB–GUS* fusions placed back in *Arabidopsis* confirmed the *PHYA* results for tobacco and indicated that *PHYB–GUS* is expressed at much lower levels than *PHYA–GUS* in all tissues (Somers and Quail 1995).

A recent study comparing the expression patterns of *PHYB–GUS*, *PHYD–GUS*, and *PHYE–GUS* fusions in *Arabidopsis* has revealed that although these Type II promoters are less active than the Type I promoters, they do show distinct expression patterns (Goosey et al. 1997). Thus the general picture emerging from these studies is that the phytochromes are expressed in distinct but overlapping patterns.

In summary, phytochromes are most abundant in young, undifferentiated tissues, in the cells where the mRNAs are most abundant and the promoters are most active. The strong correlation between phytochrome abundance and cells that have the potential for dynamic developmental changes is consistent with the important role of phytochromes in controlling such developmental changes. However, note that the studies discussed here do not address whether the phytochromes are photoactive as apoproteins or holoproteins.

Because the expression patterns of individual phytochromes overlap, it is not surprising that they function cooperatively, although they probably also use distinct signal transduction pathways. Support for this idea also comes from the study of phytochrome mutants, which we will discuss later in this chapter.

CHARACTERISTICS OF PHYTOCHROME-INDUCED WHOLE-PLANT RESPONSES

The variety of different phytochrome responses in intact plants is extensive, in terms of both the kinds of responses (see Table 17.1) and the quantity of light needed to induce the responses. A survey of this variety will show how diversely the effects of a single photoevent—the absorption of light by Pr—are manifested throughout the plant. For ease of discussion, phytochrome-induced responses may be logically grouped into two types:

1. Rapid biochemical events
2. Slower morphological changes, including movements and growth

Some of the early biochemical reactions affect later developmental responses. The nature of these early biochemical events, which comprise signal transduction pathways, will be treated in detail later in the chapter. Here we will focus on the effects of phytochrome on whole-plant responses. As we will see, such responses can be classified into various types, depending on the amount and duration of light required, and on their action spectra.

Phytochrome Responses Vary in Lag Time and Escape Time

Morphological responses to the photoactivation of phytochrome may be observed visually after a *lag time*—the time between a stimulation and an observed response. The lag time may be as brief as a few minutes or as long as several weeks. The more rapid of these responses are usually reversible movements of organelles (see **Web Topic 17.2**) or reversible volume changes (swelling, shrinking) in cells, but even some growth responses are remarkably fast.

Red-light inhibition of the stem elongation rate of light-grown pigweed (*Chenopodium album*) is observed within 8 minutes after its relative level of Pfr is increased. Kinetic studies using *Arabidopsis* have confirmed this observation and further shown that phyA acts within minutes after exposure to red light (Parks and Spalding 1999). In these studies the primary contribution of phyA was found to be over by 3 hours, at which time phyA protein was no longer detectable through the use of antibodies, and the contribution of phyB increased (Morgan and Smith 1978). Longer lag times of several weeks are observed for the induction of flowering (see Chapter 24).

Information about the lag time for a phytochrome response helps researchers evaluate the kinds of biochemical events that could precede and cause the induction of that response. The shorter the lag time, the more limited the range of biochemical events that could have been involved.

Variety in phytochrome responses can also be seen in the phenomenon called **escape from photoreversibility**. Red light–induced events are reversible by far-red light for only a limited period of time, after which the response is said to have "escaped" from reversal control by light.

A model to explain this phenomenon assumes that phytochrome-controlled morphological responses are the result of a step-by-step sequence of linked biochemical reactions in the responding cells. Each of these sequences has a point of no return beyond which it proceeds irrevocably to the response. The escape time for different responses ranges from less than a minute to, remarkably, hours.

Phytochrome Responses Can Be Distinguished by the Amount of Light Required

In addition to being distinguished by lag times and escape times, phytochrome responses can be distinguished by the amount of light required to induce them. The amount of light is referred to as the **fluence**,[1] which is defined as the number of photons impinging on a unit surface area (see Chapter 9 and **Web Topic 9.1**). The most commonly used units for fluence are moles of quanta per square meter (mol m^{-2}). In addition to the fluence, some phytochrome responses are sensitive to the **irradiance**,[2] or *fluence rate*, of light. The units of irradiance in terms of photons are moles of quanta per square meter per second (mol m^{-2} s^{-1}).

Each phytochrome response has a characteristic range of light fluences over which the magnitude of the response is proportional to the fluence. As Figure 17.7 shows, these responses fall into three major categories based on the amount of light required: very-low-fluence responses (VLFRs), low-fluence responses (LFRs), and high-irradiance responses (HIRs).

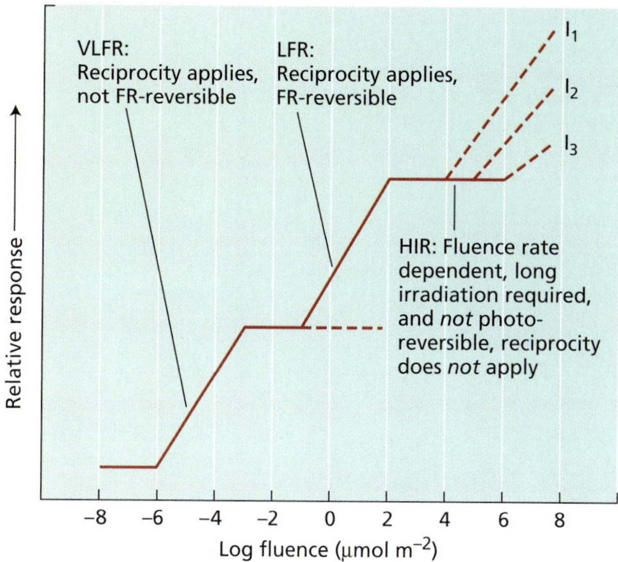

FIGURE 17.7 Three types of phytochrome responses, based on their sensitivities to fluence. The relative magnitudes of representative responses are plotted against increasing fluences of red light. Short light pulses activate VLFRs and LFRs. Because HIRs are also proportional to the irradiance, the effects of three different irradiances given continuously are illustrated ($I_1 > I_2 > I_3$). (From Briggs et al. 1984.)

Very-Low-Fluence Responses Are Nonphotoreversible

Some phytochrome responses can be initiated by fluences as low as 0.0001 μmol m^{-2} (one-tenth of the amount of light emitted from a firefly in a single flash), and they saturate (i.e., reach a maximum) at about 0.05 μmol m^{-2}. For example, in dark-grown oat seedlings, red light can stimulate the growth of the coleoptile and inhibit the growth of the mesocotyl (the elongated axis between the coleoptile and the root) at such low fluences. *Arabidopsis* seeds can be induced to germinate with red light in the range of 0.001 to 0.1 μmol m^{-2}. These remarkable effects of vanishingly low levels of illumination are called **very-low-fluence responses** (**VLFRs**).

The minute amount of light needed to induce VLFRs converts less than 0.02% of the total phytochrome to Pfr. Because the far-red light that would normally reverse a red-light effect converts 97% of the Pfr to Pr (as discussed earlier), about 3% of the phytochrome remains as Pfr—significantly more than is needed to induce VLFRs (Mandoli and Briggs 1984). Thus, far-red light cannot reverse VLFRs. The VLFR action spectrum matches the absorption spectrum of Pr, supporting the view that Pfr is the active form for these responses (Shinomura et al. 1996).

Ecological implications of the VLFR in seed germination are discussed in **Web Essay 17.1**

[1] For definitions of *fluence*, *irradiance*, and other terms involved in light measurement, see **Web Topic 9.1**.

[2] Irradiance is sometimes loosely equated with light intensity. The term *intensity*, however, refers to light emitted by the source, whereas *irradiance* refers to light that is incident on the object.

Low-Fluence Responses Are Photoreversible

Another set of phytochrome responses cannot be initiated until the fluence reaches 1.0 μmol m^{-2}, and they are saturated at 1000 μmol m^{-2}. These responses are referred to as **low-fluence responses** (**LFRs**), and they include most of the red/far-red photoreversible responses, such as the promotion of lettuce seed germination and the regulation of leaf movements, that are mentioned in Table 17.1. The LFR action spectrum for *Arabidopsis* seed germination is shown in Figure 17.8. LFR spectra include a main peak for stimulation in the red region (660 nm), and a major peak for inhibition in the far-red region (720 nm).

Both VLFRs and LFRs can be induced by brief pulses of light, provided that the total amount of light energy adds up to the required fluence. The total fluence is a function of two factors: the fluence rate (mol m^{-2} s^{-1}) and the irradiation time. Thus a brief pulse of red light will induce a response, provided that the light is sufficiently bright, and conversely, very dim light will work if the irradiation time is long enough. This reciprocal relationship between fluence rate and time is known as the **law of reciprocity**, which was first formulated by R. W. Bunsen and H. E. Roscoe in 1850. VLFRs and LFRs both obey the law of reciprocity.

High-Irradiance Responses Are Proportional to the Irradiance and the Duration

Phytochrome responses of the third type are termed **high-irradiance responses** (**HIRs**), several of which are listed in

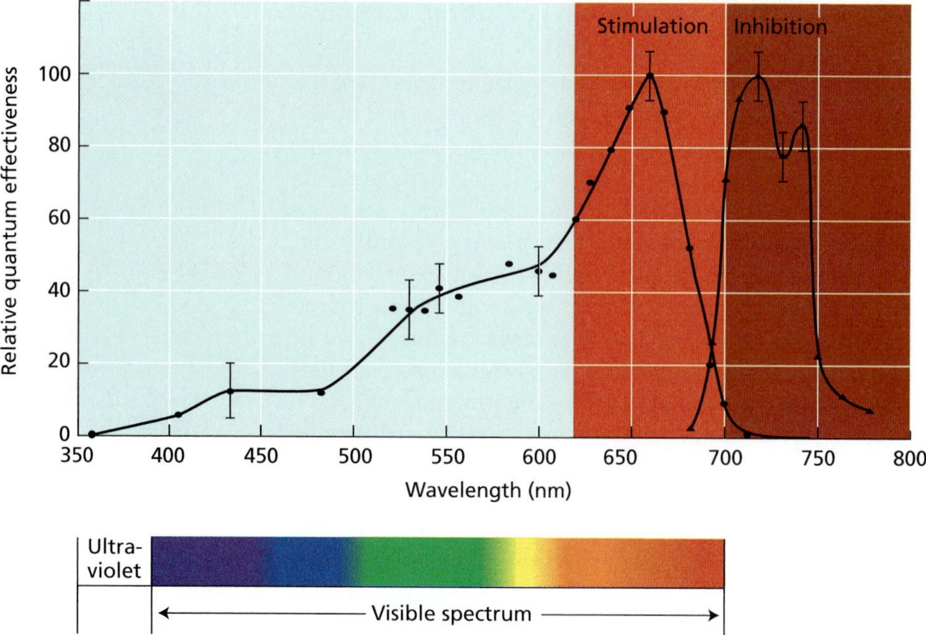

FIGURE 17.8 LFR action spectra for the photoreversible stimulation and inhibition of seed germination in *Arabidopsis*. (After Shropshire et al. 1961.)

Table 17.2. HIRs require prolonged or continuous exposure to light of relatively high irradiance, and the response is proportional to the irradiance within a certain range.

The reason that these responses are called high-irradiance responses rather than high-fluence responses is that they are proportional to irradiance (loosely speaking, the brightness of the light) rather than to fluence. HIRs saturate at much higher fluences than LFRs—at least 100 times higher—and are not photoreversible. Because neither continuous exposure to dim light nor transient exposure to bright light can induce HIRs, HIRs do not obey the law of reciprocity.

Many of the photoreversible LFRs listed in Table 17.1, particularly those involved in de-etiolation, also qualify as HIRs. For example, at low fluences the action spectrum for anthocyanin production in seedlings of white mustard (*Sinapis alba*) shows a single peak in the red region of the spectrum, the effect is reversible with far-red light, and the response obeys the law of reciprocity. However, if the dark-grown seedlings are instead exposed to high-irradiance light for several hours, the action spectrum now includes peaks in the far-red and blue regions (see the next section), the effect is no longer photoreversible, and the response becomes proportional to the irradiance. Thus the same effect can be either an LFR or an HIR, depending on its history of exposure to light.

TABLE 17.2
Some plant photomorphogenic responses induced by high irradiances

Synthesis of anthocyanin in various dicot seedings and in apple skin segments
Inhibition of hypocotyl elongation in mustard, lettuce, and petunia seedlings
Induction of flowering in henbane (*Hyoscyamus*)
Plumular hook opening in lettuce
Enlargement of cotyledons in mustard
Production of ethylene in sorghum

The HIR Action Spectrum of Etiolated Seedlings Has Peaks in the Far-Red, Blue, and UV-A Regions

HIRs, such as the inhibition of stem or hypocotyl growth, have usually been studied in dark-grown, etiolated seedlings. The HIR action spectrum for the inhibition of hypocotyl elongation in dark-grown lettuce seedlings is shown in Figure 17.9. For HIRs the main peak of activity is in the far-red region between the absorption maxima of Pr and Pfr, and there are peaks in the blue and UV-A regions as well. Because the absence of a peak in the red region is unusual for a phytochrome-mediated response, at first researchers believed that another pigment might be involved.

A large body of evidence now supports the view that phytochrome is one of the photoreceptors involved in HIRs (see **Web Topic 17.3**). However, it has long been suspected that the peaks in the UV-A and blue regions are due to a separate photoreceptor that absorbs UV-A and blue light.

As a test of this hypothesis, the HIR action spectrum for the inhibition of hypocotyl elongation was determined in dark-grown *hy2* mutants of *Arabidopsis*, which have little or no phytochrome holoprotein. As expected, the wild-type seedlings exhibited peaks in the UV-A, blue, and far-red regions of the spectrum. In contrast, the *hy2* mutant failed to respond to either far-red or red light. Although the phytochrome-deficient *hy2* mutant exhibited no peak in the far-red region, it showed a normal response to UV-A and blue light (Goto et al. 1993).

These results demonstrate that phytochrome is not involved in the HIR to either UV-A or blue light, and that a separate blue/UV-A photoreceptor is responsible for the response to these

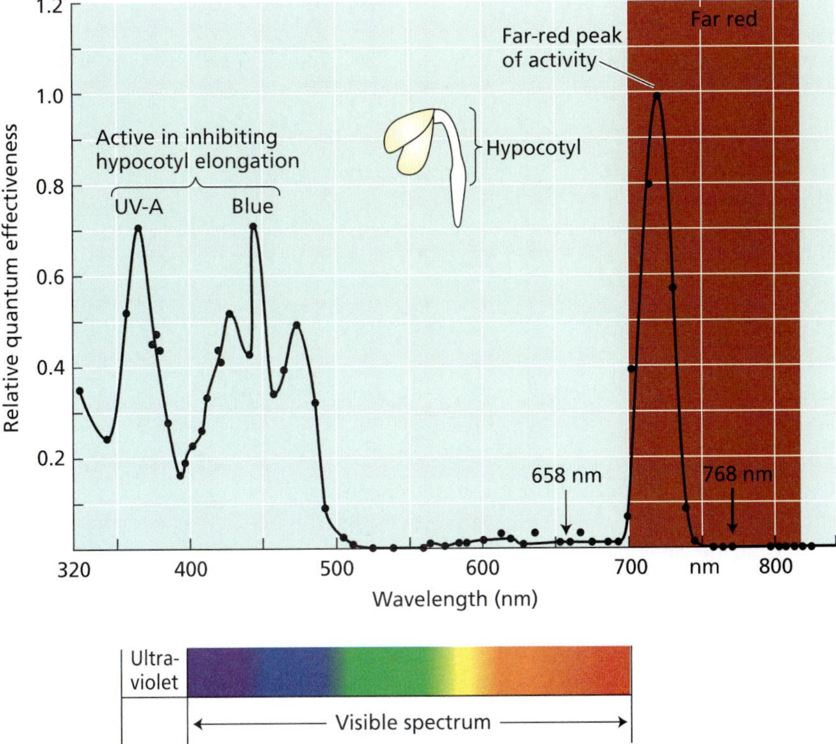

FIGURE 17.9 HIR action spectrum for the inhibition of hypocotyl elongation of dark-grown lettuce seedlings. The peaks of activity for the inhibition of hypocotyl elongation occur in the UV-A, blue, and far-red regions of the spectrum. (After Hartmann 1967.)

wavelengths. More recent studies indicate that the blue-light photoreceptors CRY1 and CRY2 are involved in blue-light inhibition of hypocotyl elongation.

The HIR Action Spectrum of Green Plants Has a Major Red Peak

During studies of the HIR of etiolated seedlings, it was observed that the response to continuous far-red light declines rapidly as the seedlings begin to green. For example, the action spectrum for the inhibition of hypocotyl growth of light-grown green *Sinapis alba* (white mustard) seedlings is shown in Figure 17.10. In general, HIR action spectra for light-grown plants exhibit a single major peak in the red, similar to the action spectra of LFRs (see Figure 17.8), except that the effect is nonphotoreversible.

The loss of responsiveness to continuous far-red light is strongly correlated with the depletion of the light-labile pool of Type I phytochrome, which consists mostly of phyA. This finding suggests that the HIR of etiolated seedlings to far-red light is mediated by phyA, whereas the HIR of green seedlings to red light is mediated by the Type II phytochrome phyB and possibly others.

ECOLOGICAL FUNCTIONS: SHADE AVOIDANCE

Thus far we have discussed phytochrome-regulated responses as studied in the laboratory. However, phytochrome plays important ecological roles for plants growing in the environment. In the discussion that follows we will learn how plants sense and respond to shading by other plants, and how phytochrome is involved in regulating various daily rhythms. We will also examine the specialized functions of the different phytochrome gene family members in these processes.

Phytochrome Enables Plants to Adapt to Changing Light Conditions

The presence of a red/far-red reversible pigment in all green plants, from algae to dicots, suggests that these wavelengths of light provide information that helps plants adjust to their environment. What environmental conditions change the relative levels of these two wavelengths of light in natural radiation?

The ratio of red light (R) to far-red light (FR) varies remarkably in different environments. This ratio can be defined as follows:

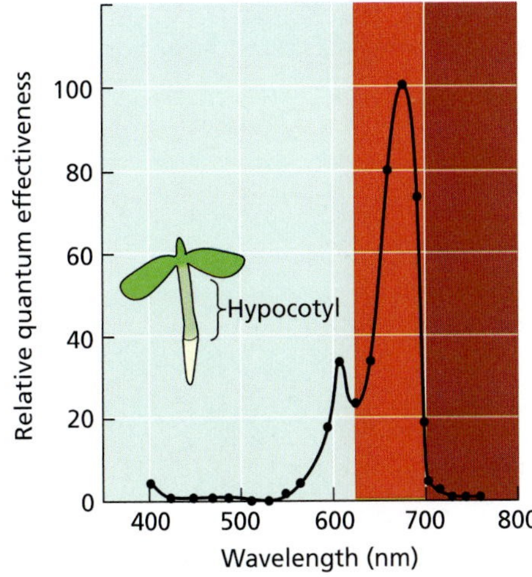

FIGURE 17.10 HIR action spectra for the inhibition of hypocotyl elongation of light-grown white mustard (*Sinapis alba*) seedlings. (After Beggs et al. 1980.)

$$R/FR = \frac{\text{Photon fluence rate in 10 nm band centered on 660 nm}}{\text{Photon fluence rate in 10 nm band centered on 730 nm}}$$

Table 17.3 compares both the total light intensity in photons (400–800 nm) and the R/FR values in eight natural environments. Both parameters vary greatly in different environments.

Compared with direct daylight, there is relatively more far-red light during sunset, under 5 mm of soil, or under the canopy of other plants (as on the floor of a forest). The canopy phenomenon results from the fact that green leaves absorb red light because of their high chlorophyll content but are relatively transparent to far-red light.

The R:FR ratio and shading. An important function of phytochrome is that it enables plants to sense shading by other plants. Plants that increase stem extension in response to shading are said to exhibit a **shade avoidance response**. As shading increases, the R:FR ratio decreases. The greater proportion of far-red light converts more Pfr to Pr, and the ratio of Pfr to total phytochrome (Pfr/Ptotal) decreases. When simulated natural radiation was used to vary the far-red content, it was found that for so-called sun plants (plants that normally grow in an open-field habitat), the higher the far-red content (i.e., the lower the Pfr:Ptotal ratio), the higher the rate of stem extension (Figure 17.11).

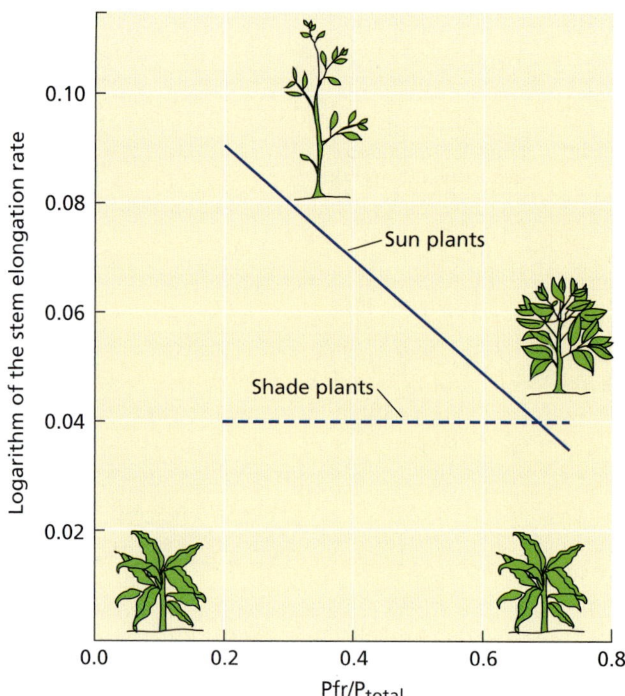

FIGURE 17.11 Role of phytochrome in shade perception in sun plants (solid line) versus shade plants (dashed line). (After Morgan and Smith 1979.)

In other words, simulated canopy shading (high levels of far-red light) induced these plants to allocate more of their resources to growing taller. This correlation did not hold for "shade plants," which normally grow in a shaded environment. Shade plants showed little or no reduction in their stem extension rate as they were exposed to higher R/FR values (see Figure 17.11). Thus there appears to be a systematic relationship between phytochrome-controlled growth and species habitat. Such results are taken as an indication of the involvement of phytochrome in shade perception.

For a "sun plant" or "shade-avoiding plant" there is a clear adaptive value in allocating its resources toward more rapid extension growth when it is shaded by another plant. In this way it can enhance its chances of growing above the canopy and acquiring a greater share of unfiltered, photosynthetically active light. The price for favoring internode elongation is usually reduced leaf area and reduced branching, but at least in the short run this adaptation to canopy shade seems to work.

TABLE 17.3
Ecologically important light parameters

	Photon flux density ($\mu mol\ m^{-2}\ s^{-1}$)	R/FR[a]
Daylight	1900	1.19
Sunset	26.5	0.96
Moonlight	0.005	0.94
Ivy canopy	17.7	0.13
Lakes, at a depth of 1 m		
Black Loch	680	17.2
Loch Leven	300	3.1
Loch Borralie	1200	1.2
Soil, at a depth of 5 mm	8.6	0.88

Source: Smith 1982, p. 493.

Note: The light intensity factor (400–800 nm) is given as the photon flux density, and phytochrome-active light is given as the R:FR ratio.

[a]Absolute values taken from spectroradiometer scans; the values should be taken to indicate the relationships between the various natural conditions and not as actual environmental means.

The R:FR ratio and seed germination. Light quality also plays a role in regulating the germination of some seeds. As discussed earlier, phytochrome was discovered in studies of light-dependent lettuce seed germination.

In general, large-seeded species, with ample food reserves to sustain prolonged seedling growth in darkness (e.g., underground), do not require light for germination. However, a light requirement is often observed in the small seeds of herbaceous and grassland species, many of which remain dormant, even while hydrated, if they are buried below the depth to which light penetrates. Even when such seeds are on or near the soil surface, their level of shading by the vegetation canopy (i.e., the R:FR ratio they receive) is likely to affect their germination. For example, it is well documented that far-red enrichment imparted by a leaf canopy inhibits germination in a range of small-seeded species.

For the small seeds of the tropical species trumpet tree (*Cecropia obtusifolia*) and Veracruz pepper (*Piper auritum*) planted on the floor of a deeply shaded forest, this inhibition can be reversed if a light filter is placed immediately above the seeds that permits the red component of the canopy-shaded light to pass through while blocking the far-red component. Although the canopy transmits very little red light, the level is enough to stimulate the seeds to germinate, probably because most of the inhibitory far-red light is excluded by the filter and the R:FR ratio is very high. These seeds would also be more likely to germinate in spaces receiving sunlight through gaps in the canopy than in densely shaded spaces. The sunlight would help ensure that the seedlings became photosynthetically self-sustaining before their seed food reserves were exhausted.

As will be discussed later in the chapter, recent studies on light-dependent lettuce seeds have shown that red light–induced germination is the result of an increase in the level of the biologically active form of the hormone gibberellin. Thus, phytochrome may promote seed germination through its effects on gibberellin biosynthesis (see Chapter 20).

ECOLOGICAL FUNCTIONS: CIRCADIAN RHYTHMS

Various metabolic processes in plants, such as oxygen evolution and respiration, cycle alternately through high-activity and low-activity phases with a regular periodicity of about 24 hours. These rhythmic changes are referred to as **circadian rhythms** (from the Latin *circa diem*, meaning "approximately a day"). The **period** of a rhythm is the time that elapses between successive peaks or troughs in the cycle, and because the rhythm persists in the absence of external controlling factors, it is considered to be **endogenous**.

The endogenous nature of circadian rhythms suggests that they are governed by an internal pacemaker, called an **oscillator**. The endogenous oscillator is coupled to a variety of physiological processes. An important feature of the oscillator is that it is unaffected by temperature, which enables the clock to function normally under a wide variety of seasonal and climatic conditions. The clock is said to exhibit **temperature compensation**.

Light is a strong modulator of rhythms in both plants and animals. Although circadian rhythms that persist under controlled laboratory conditions usually have periods one or more hours longer or shorter than 24 hours, in nature their periods tend to be uniformly closer to 24 hours because of the synchronizing effects of light at daybreak, referred to as **entrainment**. Both red and blue light are effective in entrainment. The red-light effect is photoreversible by far-red light, indicative of phytochrome; the blue-light effect is mediated by blue-light photoreceptor(s).

Phytochrome Regulates the Sleep Movements of Leaves

The sleep movements of leaves, referred to as **nyctinasty**, are a well-described example of a plant circadian rhythm that is regulated by light. In nyctinasty, leaves and/or leaflets extend horizontally (open) to face the light during the day and fold together vertically (close) at night (Figure 17.12). Nyctinastic leaf movements are exhibited by many legumes, such as *Mimosa*, *Albizia*, and *Samanea*, as well as members of the oxalis family. The change in leaf or leaflet angle is caused by rhythmic turgor changes in the cells of the **pulvinus** (plural *pulvini*), a specialized structure at the base of the petiole.

Once initiated, the rhythm of opening and closing persists even in constant darkness, both in whole plants and in isolated leaflets (Figure 17.13). The phase of the rhythm (see Chapter 24), however, can be shifted by various exogenous signals, including red or blue light.

(A)

(B)

FIGURE 17.12 Thigmonastic (touch-sensitive) leaf movements of *Mimosa pudica*. (A) Leaflets open. (B) Leaflets closed. Similar leaflet movements occur diurnally in nyctonasty. (Photos © David Sieren/Visuals Unlimited.)

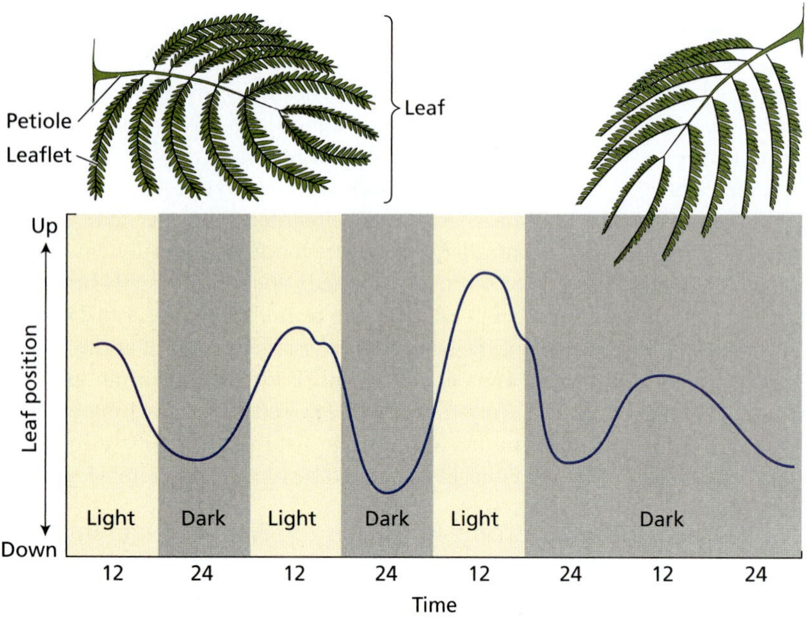

FIGURE 17.13 Circadian rhythm in the diurnal movements of *Albizia* leaves. The leaves are elevated in the morning and lowered in the evening. In parallel with the raising and lowering of the leaves, the leaflets open and close. The rhythm persists at a lower amplitude for a limited time in total darkness.

Light also directly affects movement: Blue light stimulates closed leaflets to open, and red light followed by darkness causes open leaflets to close. The leaflets begin to close within 5 minutes after being transferred to darkness, and closure is complete in 30 minutes. Because the effect of red light can be canceled by far-red light, phytochrome regulates leaflet closure.

The physiological mechanism of leaf movement is well understood. It results from turgor changes in cells located on opposite sides of the pulvinus, called **ventral motor cells** and **dorsal motor cells** (Figure 17.14). These changes in turgor pressure depend on K^+ and Cl^- fluxes across the plasma membranes of the dorsal and ventral motor cells. Leaflets open when the dorsal motor cells accumulate K^+ and Cl^-, causing them to swell, while the ventral motor cells release K^+ and Cl^-, causing them to shrink. Reversal of this process results in leaflet closure. Leaflet closure is therefore an example of a rapid response to phytochrome involving ion fluxes across membranes.

Gene expression and circadian rhythms. Phytochrome can also interact with circadian rhythms at the level of gene expression. The expression of genes in the *LHCB* family, encoding the light-harvesting chlorophyll *a/b*–binding proteins of photosystem II, is regulated at the transcriptional level by both circadian rhythms and phytochrome.

In leaves of pea and wheat, the level of *LHCB* mRNA has been found to oscillate during daily light–dark cycles, rising in the morning and falling in the evening. Since the rhythm persists even in continuous darkness, it appears to be a circadian rhythm. But phytochrome can perturb this cyclical pattern of expression.

When wheat plants are transferred from a cycle of 12 hours light and 12 hours dark to continuous darkness, the rhythm persists for a while, but it slowly *damps out* (i.e., reduces in amplitude until no peaks or troughs are discernible). If, however, the plants are given a pulse of red light before they are transferred to continuous darkness, no damping occurs (i.e., the levels of *LHCB* mRNA continue to oscillate as they do during the light–dark cycles).

In contrast, a far-red flash at the end of the day prevents the expression of *LHCB* in continuous darkness, and the effect of far red is reversed by red light. Note that it is not the oscillator that damps out under constant conditions, but the coupling of the oscillator to the physiological event being monitored. Red light restores the coupling between the oscillator and the physiological process.

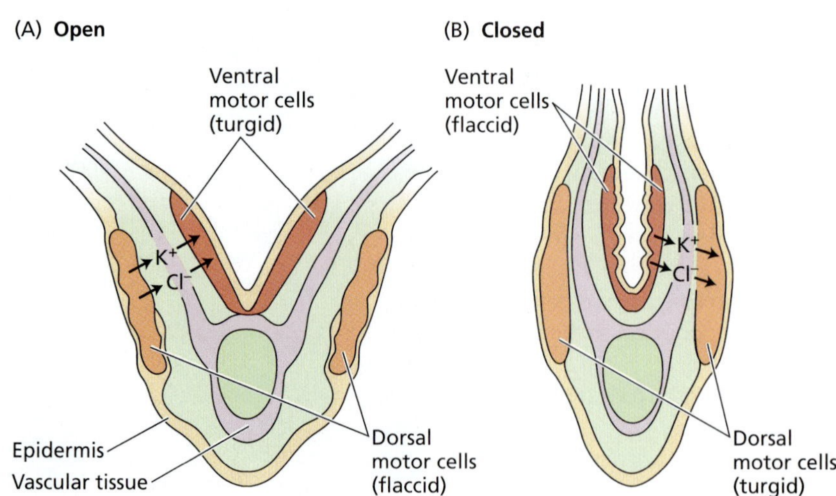

FIGURE 17.14 Ion fluxes between the dorsal and ventral motor cells of *Albizia pulvini* regulate leaflet opening and closing. (After Galston 1994.)

Circadian Clock Genes of *Arabidopsis* Have Been Identified

The isolation of clock mutants has been an important tool for the identification of clock genes in other organisms. Isolating clock mutants in plants requires a convenient assay that allows monitoring of the circadian rhythms of many thousands of individual plants to detect the rare abnormal phenotype.

To allow screening for clock mutants in *Arabidopsis*, the promoter region of the *LHCB* gene was fused to the gene that encodes luciferase, an enzyme that emits light in the presence of its substrate, luciferin. This reporter gene construct was then used to transform *Arabidopsis* with the Ti plasmid of *Agrobacterium* as a vector. Investigators were then able to monitor the temporal and spatial regulation of bioluminescence in individual seedlings in real time using a video camera (Millar et al. 1995).

A total of 21 independent ***toc*** (*timing of CAB* [*LHCB*] *expression*) mutants have been isolated, including both short-period and long-period lines. The *toc1* mutant in particular has been implicated in the core oscillator mechanism (Strayer et al. 2001). A model for the endogenous oscillator will be discussed later in the chapter.

ECOLOGICAL FUNCTIONS: PHYTOCHROME SPECIALIZATION

Phytochrome is encoded by a multigene family: *PHYA* through *PHYE*. Despite the great similarity in their structures, each of these phytochromes performs distinct roles in the life of the plant. In this section we will discuss the current state of our knowledge of the ecological functions of the different phytochromes, focusing primarily on phyA and phyB.

Phytochrome B Mediates Responses to Continuous Red or White Light

Phytochrome B was first suspected to play a role in responses to continuous light because the *hy3* mutant (now called *phyB*), which has long hypocotyls under continuous white light, was found to have an altered *PHYB* gene. In these mutants, *PHYB* mRNA was reduced in amount or was absent, and little or no phyB protein could be detected. In contrast, the levels of *PHYA* mRNA and phyA protein were normal.

Phytochrome B mediates shade avoidance by regulating hypocotyl length in response to red light given in low-fluence pulses or continuously, and as might be expected, the *phyB* mutant is unable to respond to shading by increasing hypocotyl extension. In addition, these plants do not extend their hypocotyls in response to far-red light given at the end of each photoperiod (called the *end-of-day far-red response*). Both of these responses are likely to involve perception of the Pfr:Ptotal ratio and occur in the low-fluence region of the spectrum. Although phyB is centrally involved in the shade avoidance response, evidence suggests that other phytochromes play important roles as well (Smith and Whitelam 1997).

The *phyB* mutant is deficient in chlorophyll and in some mRNAs that encode chloroplast proteins, and it is impaired in its ability to respond to plant hormones. Since a mutation in *PHYB* results in impaired perception of continuous red light, the presence of the other phytochromes must not be sufficient to confer responsiveness to continuous red or white light.

Phytochrome B also appears to regulate photoreversible seed germination, the phenomenon that originally led to the discovery of phytochrome. Wild-type *Arabidopsis* seeds require light for germination, and the response shows red/far-red reversibility in the low-fluence range. Mutants that lack phyA respond normally to red light; mutants deficient in phyB are unable to respond to low-fluence red light (Shinomura et al. 1996). This experimental evidence strongly suggests that phyB mediates photoreversible seed germination.

Phytochrome A Is Required for the Response to Continuous Far-Red Light

No phytochrome gene mutations other than *phyB* were found in the original *hy* collection, so the identification of *phyA* mutants required the development of more ingenious screens. As discussed previously, because the far-red HIRs were known to require light-labile (Type I) phytochrome, it was suspected that phyA must be the photoreceptor involved in the perception of continuous far-red light. If this is true, then the phyA mutants should fail to respond to continuous far-red light and grow tall and spindly under these light conditions. However, mutants lacking chromophore would also look like this because phyA can detect far-red light only when assembled with the chromophore into holophytochrome.

To select for just the phyA mutants, the seedlings that grew tall in continuous far-red light were then grown under continuous red light. The phyA-deficient mutants can grow normally under this regimen, but a chromophore-deficient mutant, which also lacks functional phyB, does not respond. The *phyA* mutant seedlings selected in this screen had no obvious phenotype when grown in normal white light, confirming that phyA has no discernible role in sensing white light.

This also explains why *phyA* mutants were not detected in the original long-hypocotyl screen. Thus, phyA appears to have a limited role in photomorphogenesis, restricted primarily to de-etiolation and far-red responses. For example, phyA would be important when seeds germinate under a canopy, which filters out much of the red light.

It is also clear from this constant far-red light phenotype that none of the other phytochromes is sufficient for the perception of constant far-red light, and despite the ability of all phytochromes to absorb red and far-red light, at least phyA and phyB have distinct roles in this regard.

TABLE 17.4
Comparison of the very-low-fluence (VLFR), low-fluence (LFR), and high-irradiance responses (HIR)

Type of Response	Photoreversibility	Reciprocity	Peaks of action spectra[a]	Photoreceptor
VLFR	No	Yes	Red, Blue	phyA, phyE[a]
LFR	Yes	Yes	Red, far red	phyB, phyD, phyE
HIR	No	No	Dark-grown: far red, blue, UV-A Light-grown: red	Dark-grown: phyA, cryptochrome Light-grown: phyB

[a] phyE is required for seed germination but not for other VLFR responses mediated by phyA

Phytochrome A also appears to be involved in the germination VLFR of *Arabidopsis* seeds. Thus, mutants lacking phyA cannot germinate in response to red light in the very-low-fluence range, but they show a normal response to red light in the low-fluence range (Shinomura et al. 1996). This result demonstrates that phyA functions as the primary photoreceptor for this VLFR, although recent evidence suggests that phyE is required for this component of seed germination (Hennig et al. 2002).

Table 17.4 summarizes the different roles of phyA, phyB, and other photoreceptors in the various phytochrome-mediated responses.

Developmental Roles for Phytochromes C, D, and E Are Also Emerging

Some of the roles of other phytochromes in plant growth and development have recently begun to be elucidated through experiments on mutant plants. Because these phytochromes have functions that overlap with those of phyA and phyB, it was necessary to screen for mutants in *phyAB* null mutant backgrounds to uncover mutations. For example, both phyD and phyE help mediate the shade avoidance response—a response mediated primarily by phyB.

The creation of double and triple mutants has made it possible to assess the relative role of each phytochrome in a given response. Thus it was found that, like phyB, phyD plays a role in regulating leaf petiole elongation, as well as in flowering time (see Chapter 24). Similar analyses support the idea that phyE acts redundantly with phyB and phyD in these processes, but also acts redundantly with phyA and phyB in inhibition of internode elongation.

Of the *Arabidopsis* phytochromes, phyC is the least well characterized. However, although *phyAphyBphyDphyE* quadruple mutants appear to have normal responses to the red:far red ratio, there are differences in phytochrome-regulated gene expression.

In summary, phyC, phyD, and phyE appear to play roles that are for the most part redundant with those of phyA and phyB. Whereas phyB appears to be involved in regulating all stages of development, the functions of the other phytochromes are restricted to specific developmental steps or responses.

Phytochrome Interactions Are Important Early in Germination

Figure 17.15A shows the action of constant red and far-red light absorbed separately by the phyA and phyB systems. Continuous red light absorbed by phyB stimulates de-eti-

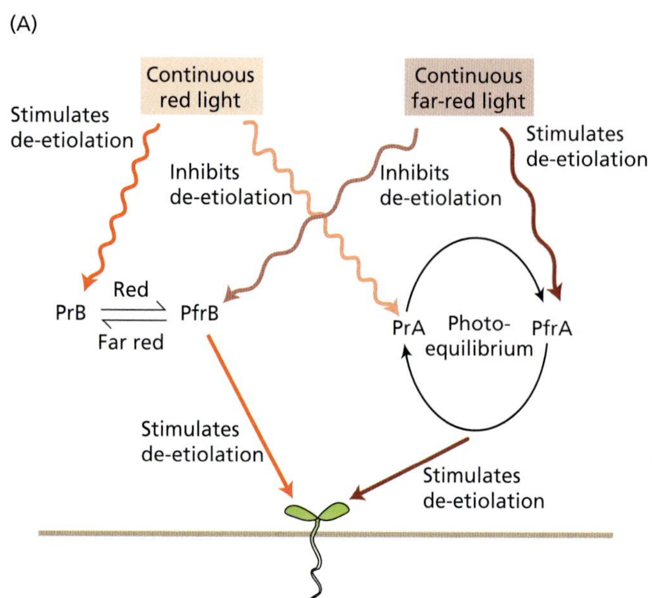

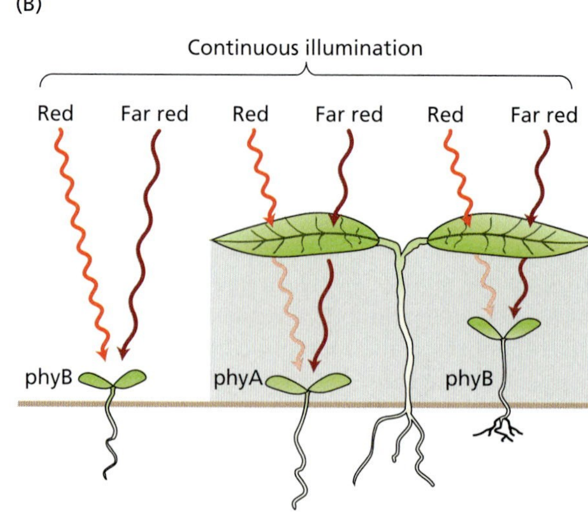

FIGURE 17.15 Mutually antagonistic roles of phyA and phyB. (After Quail et al. 1995.)

olation by maintaining high levels of PfrB. Continuous far-red light absorbed by PfrB prevents this stimulation by reducing the amount of PfrB. The stimulation of de-etiolation by phyA depends on the photostationary state of phytochrome (indicated in Figure 17.15A by the circular arrows). Continuous far-red light stimulates de-etiolation when absorbed by the phyA system; continuous red light inhibits the response.

The effects of phyA and phyB on seedling development in sunlight versus canopy shade (enriched in far-red light) are shown in Figure 17.15B. In open sunlight, which is enriched in red light compared with canopy shade, de-etiolation is mediated primarily by the phyB system (on the left in the figure). A seedling emerging under canopy shade, enriched in far-red light, initiates de-etiolation primarily through the phyA system (center). Because phyA is labile, however, the response is taken over by phyB (right). In switching over to phyB, the stem is released from growth inhibition (see Figure 17.15A), allowing for the accelerated rate of stem elongation that is part of the shade avoidance response (see **Web Topic 17.4**).

For a discussion of how plants sense their neighbors using reflected light, see **Web Essay 17.2**.

PHYTOCHROME FUNCTIONAL DOMAINS

Prior to the identification of the multiple forms of phytochrome, it was difficult to understand how a single photoreceptor could regulate such diverse processes in the cell. However, the discovery that phytochrome is encoded by members of a multigene family, each with its own pattern of expression, provided a more plausible alternative explanation: Each phytochrome-mediated response is regulated by a specific phytochrome, or by interactions between specific phytochromes. As discussed earlier, this hypothesis was supported by the phenotypes of mutants deficient in either phyA or phyB.

As a corollary to this hypothesis, it was further postulated that specific regions of the PHY proteins must be specialized to allow them to perform their distinct functions. Molecular biology provides the tools to answer such difficult questions. In this section we will describe what is known about the functional domains of the phytochrome holoprotein.

Just as mutations *reducing* the amount of a particular phytochrome have yielded information about its role, plants genetically engineered to *overexpress* a specific phytochrome are also useful. First, they allow an extension of the range of phytochrome levels testable in relation to function. Second, as we will see, a particular phytochrome sequence can be changed and reintroduced into a normal plant to test its phenotypic effects.

Usually plants overexpressing an introduced *PHYA* or *PHYB* gene have a dramatically altered phenotype. Such transgenic plants are often dwarfed, are dark green because of elevated chlorophyll levels, and show reduced apical dominance. This phenotype requires elevated levels of an intact, photoactive holoprotein because overexpression of a mutated form of phytochrome that is unable to combine with its chromophore has a normal phenotype. Similarly, plants expressing only the N-terminal domain of each phytochrome have a normal phenotype, even though elevated levels of the photoactive fragment accumulate.

Although protein overexpression greatly perturbs the normal metabolism of a cell and is therefore subject to certain artifacts, such studies of structure and function have helped build a picture of phytochrome as a molecule having two domains linked by a hinge: an N-terminal light-sensing domain in which the light specificity and stability reside, and a C-terminal domain that contains the signal-transmitting sequences (Figure 17.16).

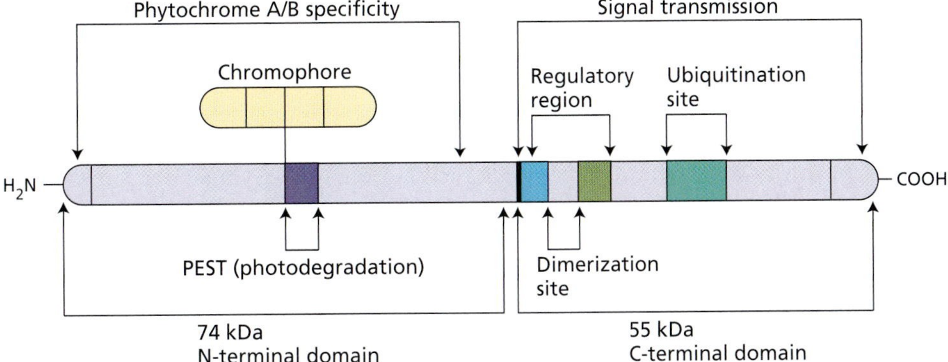

FIGURE 17.16 Schematic diagram of the phytochrome holoprotein, showing the various functional domains. The chromophore-binding site and PEST sequence are located in the N-terminal domain, which confers photosensory specificity to the molecule—that is, whether it responds to continuous red or far-red light. The C-terminal domain contains a dimerization site, a ubiquitination site, and a regulatory region. The C-terminal domain transmits signals to proteins that act downstream of phytochrome.

The C-terminal domain also contains the site for the formation of phytochrome dimers and the site for the addition of ubiquitin, a tag for degradation. (For a more detailed description of experiments that helped map the functional domains of phytochrome, see **Web Topic 17.5**.)

CELLULAR AND MOLECULAR MECHANISMS

All phytochrome-regulated changes in plants begin with absorption of light by the pigment. After light absorption, the molecular properties of phytochrome are altered, probably causing the signal-transmitting sequences in the C terminus to interact with one or more components of a signal transduction pathway that ultimately bring about changes in the growth, development, or position of an organ (see Table 17.1).

Some of the signal-transmitting motifs appear to interact with multiple signal transduction pathways; others appear to be unique to a specific pathway. Furthermore, it is reasonable to assume that the different phytochrome proteins utilize different sets of signal transduction pathways.

Molecular and biochemical techniques are helping to unravel the early steps in phytochrome action and the signal transduction pathways that lead to physiological or developmental responses. These responses fall into two general categories:

1. Relatively rapid turgor responses involving ion fluxes
2. Slower, long-term processes associated with photomorphogenesis, involving alterations in gene expression

In this section we will examine the effects of phytochrome on both membrane permeability and gene expression, as well as the possible chain of events constituting the signal transduction pathways that bring about these effects.

Phytochrome Regulates Membrane Potentials and Ion Fluxes

Phytochrome can rapidly alter the properties of membranes. We have already seen that low-fluence red light is required before the dark period to induce rapid leaflet closure during nyctinasty, and that fluxes of K^+ and Cl^- into and out of dorsal and ventral motor cells mediate the response. However, the rapidity of leaf closure in the dark (lag time about 5 minutes) would seem to rule out mechanisms based on gene expression. Instead, rapid phytochrome-induced changes in membrane permeability and transport appear to be involved.

During phytochrome-mediated leaflet closure, the apoplastic pH of the dorsal motor cells (the cells that swell during leaflet closure) decreases, while the apoplastic pH of the ventral motor cells (the cells that shrink during leaflet closure) increases. Thus the plasma membrane H^+ pump of the dorsal cells appears to be activated by darkness (provided that phytochrome is in the Pfr form), and the H^+ pump of the ventral cells appears to be deactivated under the same conditions (see Figure 17.14). The reverse pattern of apoplastic pH change is observed during leaflet opening.

Studies have also been carried out on phytochrome regulation of K^+ channels in isolated protoplasts (cells without their cell walls) of both dorsal and ventral motor cells from *Samanea* leaves (Kim et al. 1993). When the extracellular K^+ concentration was raised, K^+ entered the protoplasts and depolarized the membrane potential only if the K^+ channels were open. When the dorsal and ventral motor cell protoplasts were transferred to constant darkness, the state of the K^+ channels exhibited a circadian rhythmicity during a 21-hour incubation period, and the two cell types varied reciprocally, just as they do in vivo. That is, when the dorsal cell K^+ channels were open, the ventral cell K^+ channels were closed, and vice versa. Thus the circadian rhythm of leaf movements has its origins in the circadian rhythm of K^+ channel opening.

On the basis of the evidence thus far, we can conclude that phytochrome brings about leaflet closure by regulating the activities of the primary proton pumps and the K^+ channels of the dorsal and ventral motor cells. Although the effect is rapid, it is not instantaneous, and it is therefore unlikely to be due to a direct effect of phytochrome on the membrane. Instead, phytochrome acts indirectly via one or more signal transduction pathways, as in the case of the regulation of gene expression by phytochrome (see the next section).

However, some effects of red and far-red light on the membrane potential are so rapid that phytochrome may also interact directly with the membrane. Such rapid modulation has been measured in individual cells and has been inferred from the effects of red and far-red light on the surface potential of roots and oat (*Avena*) coleoptiles, where the lag between the production of Pfr and the onset of measurable potential changes is 4.5 s for hyperpolarization.

Changes in the bioelectric potential of cells imply changes in the flux of ions across the plasma membrane (see **Web Topic 17.6**). Membrane isolation studies provide evidence that a small portion of the total phytochrome is tightly bound to various organellar membranes.

These findings led some workers to suggest that membrane-bound phytochrome represents the physiologically active fraction, and that all the effects of phytochrome on gene expression are initiated by changes in membrane permeability. On the basis of sequence analysis, however, it is now clear that phytochrome is a hydrophilic protein without membrane-spanning domains. The current view is that it may be associated with microtubules located directly beneath the plasma membrane, at least in the case of the alga, *Mougeotia*, as described in **Web Topic 17.2**.

If phytochrome exerts its effects on membranes from some distance, no matter how small, involvement of a *second messenger* is implied, and calcium is a good candidate. Rapid changes in cytosolic free calcium have been implicated as second messengers in several signal transduction pathways, and there is evidence that calcium plays a role in chloroplast movement in *Mougeotia*.

Phytochrome Regulates Gene Expression

As the term *photomorphogenesis* implies, plant development is profoundly influenced by light. Etiolation symptoms include spindly stems, small leaves (in dicots), and the absence of chlorophyll. Complete reversal of these symptoms by light involves major long-term alterations in metabolism that can be brought about only by changes in gene expression.

The stimulation and repression of transcription by light can be very rapid, with lag times as short as 5 minutes. Such early-gene expression is likely to be regulated by the direct activation of transcription factors by one or more phytochrome-initiated signal transduction pathways. The activated transcription factors then enter the nucleus, where they stimulate the transcription of specific genes.

Some of these early gene products are transcription factors themselves, which activate the expression of other genes. Expression of the early genes, also called **primary response genes**, is independent of protein synthesis; expression of the late genes, or **secondary response genes**, requires the synthesis of new proteins.

The photoregulation of gene expression has focused on the nuclear genes that encode messages for chloroplast proteins: the small subunit of ribulose-1,6-bisphosphate carboxylase/oxygenase (rubisco) and the major light-harvesting chlorophyll *a/b*–binding proteins associated with the light-harvesting complex of photosystem II (LHCIIb proteins). These proteins play important roles in chloroplast development and greening; hence their regulation by phytochrome has been studied in detail. The genes for both of these proteins—*RBCS* and *LHCB* (also called *CAB* in some studies)—are present in multiple copies in the genome.

We can demonstrate phytochrome regulation of mRNA abundance (e.g., *RBCS* mRNAs) experimentally by giving etiolated plants a brief pulse of low-fluence red or far-red light, returning them to darkness to allow the signal transduction pathway to operate, and then measuring the abundance of specific mRNAs in total RNA prepared from each set of plants. If its abundance is regulated by phytochrome, the mRNA is absent or present at low levels in etiolated plants but is increased by red light. The red light–induced increase in expression can be reversed by immediate treatment with far-red light, but far-red light alone has little effect on mRNA abundance. The expression of some other genes is down-regulated under these conditions.

Recently red-light stimulation of lettuce seed germination has been correlated with an increase in the biologically active form of the hormone gibberellin. Red light causes a large increase in the expression of the gene coding for a key enzyme in the gibberellin biosynthetic pathway (Toyomasu et al. 1998). The effect of red light is reversed by a treatment with far-red light, indicative of phytochrome. Since gibberellin can substitute for red light in promoting lettuce seed germination, it appears that phytochrome promotes seed germination by increasing the biosynthesis of the hormone. Gibberellins are discussed in detail in Chapter 20.

For an expanded discussion see **Web Topic 17.7**.

Both Phytochrome and the Circadian Rhythm Regulate *LHCB*

A MYB-related transcription factor whose mRNA level increases rapidly when *Arabidopsis* is transferred from the dark to the light is involved in phytochrome-mediated expression of *LHCB* genes (Figure 17.17). (For information on MYB, see Chapter 14 on the web site.)

This transcription factor appears to bind to the promoter of certain *LHCB* genes and regulate their transcription, which, as Figure 17.17 shows, occurs later than the increase in the MYB-related protein (Wang et al. 1997). The gene that encodes the MYB-related protein is therefore probably a primary response gene, and the *LHCB* gene itself is probably a secondary response gene.

Recent work has indicated that this MYB-related protein, now known as *c*ircadian *c*lock *a*ssociated *1* (CCA1), also plays a role in the circadian regulation of *LHCB* gene expression. A second but distinct MYB-related protein, *l*ate elongated *hy*pocotyl (LHY), has also been identified as a potential clock gene. Expression of *CCA1* and *LHY* oscillates with a circadian rhythm. Constitutive expression of *CCA1* abolishes several circadian rhythms and suppresses both *CCA1* and *LHY* expression. When the *CCA1* gene is mutated so that no functional protein is produced, circadian and phytochrome regulation of four genes, including *LHCB*, is affected. These observations suggest that *CCA1* and *LHY* are associated with the circadian clock.

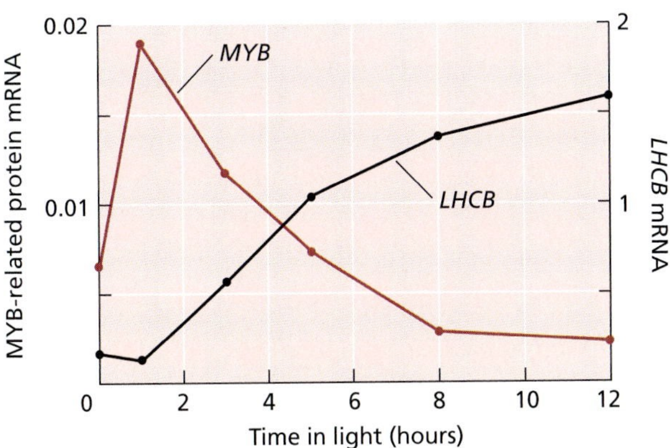

FIGURE 17.17 Time course for inducing transcription. Kinetics of the induction of transcripts for a MYB-related transcription factor (MYB) and the light-harvesting chlorophyll *a/b*–binding protein (LHCB) in *Arabidopsis* after transfer of the seedlings from darkness to continuous white light. (After Wang et al. 1997.)

A protein kinase (CK2) can interact with and phosphorylate CCA1. The CK2 kinase is a multisubunit protein with serine/threonine kinase activity. The regulatory subunit of CK2 (CKB3) has been shown to interact with, and phosphorylate, CCA1 in vitro. Mutations in CKB3 have also been found to perturb CK2 activity and, in turn, change the period of rhythmic expression of CCA1. These mutations affect many clock outputs, from gene expression to flowering time, suggesting that CK2 is involved in the regulation of the circadian clock via its interactions with CCA1 (Sugano et al. 1999).

The Circadian Oscillator Involves a Transcriptional Negative Feedback Loop

The circadian oscillators of cyanobacteria (*Synechococcus*), fungi (*Neurospora crassa*), insects (*Drosophila melanogaster*), and mouse (*Mus musculus*) have now been elucidated. In these four organisms, the oscillator is composed of several "clock genes" involved in a transcriptional–translational negative feedback loop.

So far, three major clock genes have been identified in *Arabidopsis*: *TOC1*, *LHY*, and *CCA1*. The protein products of these genes are all regulatory proteins. *TOC1* is not related to the clock genes of other organisms, suggesting that the plant oscillator is unique.

According to a recent model (Alabadi et al. 2001), light and the TOC1 regulatory protein activate *LHY* and *CCA1* expression at dawn (Figure 17.18). The increase in LHY and CCA1 represses the expression of the *TOC1* gene. Because TOC1 is a positive regulator of the *LHY* and *CCA1* genes, the repression of *TOC1* expression causes a progressive reduction in the levels of LHY and CCA1, which reach their minimum levels at the end of the day. As LHY and CCA1 levels decline, *TOC1* gene expression is released from inhibition. TOC1 reaches its maximum at the end of the day, when LHY and CCA1 are at their minimum. TOC1 then either directly or indirectly stimulates the expression of *LHY* and *CCA1*, and the cycle begins again.

The two MYB regulator proteins—LHY and CCA1—have dual functions. In addition to serving as components of the oscillator, they regulate the expression of other genes, such as *LHCB* and other "morning genes," and they repress genes expressed at night. Light acts to reinforce the effect of the *TOC1* gene in promoting *LHY* and *CCA1* expression. This reinforcement represents the underlying mechanism of *entrainment*. Other proteins, such as the CK2 kinase, affect the activity of CCA1, and thus regulate the clock. Phytochrome and the blue-light photoreceptor CRY2 (see Chapter 18) mediate the effects of red and blue light, respectively.

Regulatory Sequences Control Light-Regulated Transcription

The *cis*-acting regulatory sequences required to confer light regulation of gene expression have been studied extensively. Most eukaryotic promoters for genes that encode proteins comprise two functionally distinct regions: a short sequence that determines the transcription start site (the **TATA box**, named for its most abundant nucleotides) and upstream sequences, called ***cis*-acting regulatory elements**, that regulate the amount and pattern of transcription (see Chapter 14 on the web site). These regulatory sequences bind specific proteins, called ***trans*-acting factors**, that modulate the activity of the general

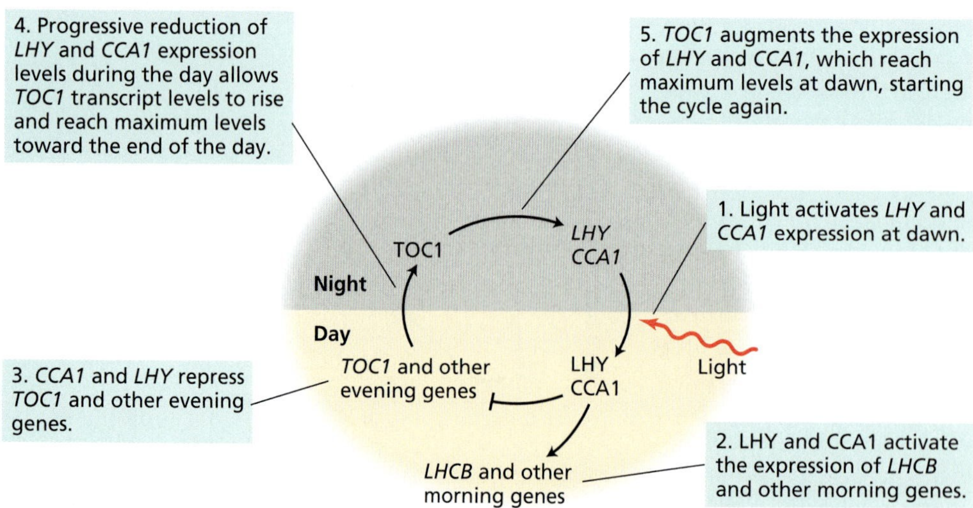

FIGURE 17.18 Circadian oscillator model showing the hypothetical interactions between the *TOC1* and *MYB* genes *LHY* and *CCA1*. Light acts at dawn to increase *LHY* and *CCA1* expression. *LHY* and *CCA1* act to regulate other daytime and evening genes.

transcription factors that assemble around the transcription start site with RNA polymerase II.

Overall, the picture emerging for light-regulated plant promoters is similar to that for other eukaryotic genes: a collection of modular elements, the number, position, flanking sequences, and binding activities of which can lead to a wide range of transcriptional patterns. No single DNA sequence or binding protein is common to all phytochrome-regulated genes.

At first it may appear paradoxical that light-regulated genes have such a range of elements, any combination of which can confer light-regulated expression. However, this array of sequences allows for the differential light- and tissue-specific regulation of many genes through the action of multiple photoreceptors. (For an expanded discussion, see **Web Topic 17.8**.)

Regulatory factors. As might be expected, the diverse range of phytochrome regulatory sequences can bind a wide variety of transcription factors. At least 50 of these regulatory factors have been identified recently by the use of genetic and molecular screens (Tepperman et al. 2001).

Although some of the early-acting signaling pathways are specific to phyA or phyB, it is clear that late-acting signaling pathways common to multiple photoreceptors must be used because different light qualities can trigger the same response (Chory and Wu 2001).

For example, SPA1 is a phyA-specific signaling intermediate that acts as a light-dependent repressor of photomorphogenesis in *Arabidopsis* seedlings (Hoecker and Quail 2001). The SPA1 protein has a coiled-coil protein domain that enables it to interact with another factor, COP1 (*constitutive photomorphogenesis 1*), that acts downstream of both phyA and phyB. The COP1 protein was identified in the screen for constitutive photomorphogenesis mutants that has yielded several other factors that act downstream of photoreceptors (see **Web Topic 17.9**). COP1 is an E3 ubiquitin ligase that targets other proteins for destruction by the 26S proteasome (see Chapter 14 on the web site).

The functions of many of these factors are probably modulated through the action of HY5, a protein first identified through the long-hypocotyl screen, discussed earlier in the chapter. HY5 is a basic leucine zipper–type transcription factor that is always located in the nucleus (see Chapter 14 on the web site). HY5 binds to the G-box motif of multiple light-inducible promoters and is necessary for optimal expression of the corresponding genes. In the dark, HY5 is ubiquitinated by COP1 and degraded by the 26S proteasome complex.

Phytochrome Moves to the Nucleus

It has long been a mystery as to how phytochrome could act in the nucleus when it is apparently localized in the cytosol. Recent exciting work has finally opened up the black box between phytochrome and gene expression. The most surprising finding is that in some cases phytochrome itself moves to the nucleus in a light-dependent manner.

Detection of this movement relied on the ability to fuse phytochrome to a visible marker, **green fluorescent protein (GFP)**, that can be activated by light of an appropriate wavelength being shone on plant cells. A big advantage of GFP fusions is that they can be visualized in living cells, making it possible to follow dynamic processes within the cell under the microscope.

Both phyA–GFP and phyB–GFP show light-activated import into the nucleus (Figure 17.19) (Sakamoto and Nagatani 1996; Sharma 2001). The phyB fusion moves to the nucleus in the Pfr form only, and transport is slow, taking several hours for full mobilization. In contrast, phyA–GFP can move in the Pfr or the Pr form, provided that it has cycled through Pfr first. Movement of phyA–GFP is much more rapid than that of phyB–GFP, taking only about 15 minutes.

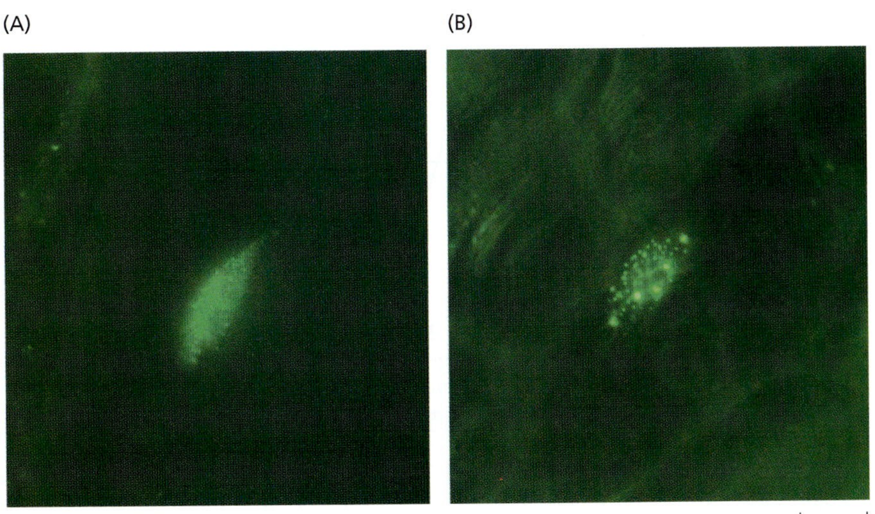

FIGURE 17.19 Nuclear localization of phy–GFP fusion proteins in epidermal cells of *Arabidopsis* hypocotyls. Transgenic *Arabidopsis* expressing phyA–GFP (left) or phyB–GFP (right) was observed under a fluorescence microscope. Only nuclei are visible. The plants were placed either under continuous far-red light (left) or white light (right) to induce the nuclear accumulation. The smaller bright green dots inside the nucleus are called "speckles." The significance of speckles is unknown. (From Yamaguchi et al. 1999, courtesy of A. Nagatani).

Sakamoto, K., and Nagatani, A. (1996) Nuclear localization activity of phytochrome B. *Plant J.* 10: 859–868.

Sharma, R. (2001) Phytochrome: A serine kinase illuminates the nucleus! *Current Science* 80: 178–188.

Sharrock, R. A., and Quail, P. H. (1989) Novel phytochrome sequences in *Arabidopsis thaliana*: Structure, evolution, and differential expression of a plant regulatory photoreceptor family. *Genes Dev.* 3: 1745–1757.

Shinomura, T., Nagatani, A., Hanzawa, H., Kubota, M., Watanabe, M., and Furuya, M. (1996) Action spectra for phytochrome A- and B-specific photoinduction of seed germination in *Arabidopsis thaliana*. *Proc. Natl. Acad. Sci. USA* 93: 8129–8133.

Shropshire, W., Jr., Klein, W. H., and Elstad, V. B. (1961) Action spectra of photomorphogenic induction and photoinactivation of germination in *Arabidopsis thaliana*. *Plant Cell Physiol.* 2: 63–69.

Smith, H. (1974) *Phytochrome and Photomorphogenesis: An Introduction to the Photocontrol of Plant Development*. McGraw-Hill, London.

Smith, H. (1982) Light quality photoperception and plant strategy. *Annu. Rev. Plant Physiol.* 33: 481–518.

Smith, H., and Whitelam, G. C. (1997) The shade avoidance syndrome: Multiple responses mediated by multiple phytochromes. *Plant Cell Environ.* 20: 840–844.

Somers, D. E., and Quail, P. H. (1995) Temporal and spatial expression patterns of *PHYA* and *PHYB* genes in *Arabidopsis*. *Plant J.* 7: 413–427.

Sugano, S., Andronis, C., Ong, M. S., Green, R. M., and Tobin, E. M. (1999) The protein kinase CK2 is involved in regulation of circadian rhythms in *Arabidopsis*. *Proc. Natl. Acad. Sci. USA* 96: 12362–12366.

Strayer, C., Oyama, T., Schultz, T. F., Raman, R., Somer, D. E., Mas, P., Panda, S., Kreps, J. A., and Kay, S. A. (2001) Cloning of the *Arabidopsis* clock gene *TOC1*, an autoregulatory response regulator homolog. *Science* 289: 768–771.

Tepperman, J. M., Zhu, T., Chang, H. S., Wang, X., and Quail, P. H. (2001) Multiple transcription factor genes are early targets of phytochrome A signaling. *Proc. Natl. Acad. Sci. USA* 98: 9437–9442.

Thümmler, F., Dufner, M., Kreisl, P., and Dittrich, P. (1992) Molecular cloning of a novel phytochrome gene of the moss *Ceratodon purpureus* which encodes a putative light-regulated protein kinase. *Plant Mol. Biol.* 20: 1003–1017.

Tokutomi, S., Nakasako, M., Sakai, J., Kataoka, M., Yamamoto, K. T., Wada, M., Tokunaga, F., and Furuya, M. (1989) A model for the dimeric molecular structure of phytochrome based on small angle x-ray scattering. *FEBS Lett.* 247: 139–142.

Toyomasu, T., Kawaide, H., Mitsuhashi, W., Inoue, Y. and Kamiya, Y. (1998) Phytochrome regulates gibberellin biosynthesis during germination of photoblastic lettuce seeds. *Plant Physiol.* 118: 1517–1523.

Vierstra, R. D. (1994) Phytochrome degradation. In *Photomorphogenesis in Plants*, 2nd ed., R. E. Kendrick and G. H. M. Kronenberg, eds., Martinus Nijhoff, Dordrecht, Netherlands, pp. 141–162.

Vierstra, R. D., and Quail, P. H. (1983) Purification and initial characterization of 124-kilodalton phytochrome from *Avena*. *Biochemistry* 22: 2498–2505.

Wang, Z.-Y., Kenigsbuch, D., Sun, L., Harel, E., Ong., M. S., and Tobin, E. M. (1997) A MYB-related transcription factor is involved in the phytochrome regulation of an *Arabidopsis Lhcb* gene. *Plant Cell* 9: 491–507.

Yamaguchi, R., Nakamura, M., Mochizuki, N., Kay, S. A., and Nagatani, A. (1999) Light-dependent translocation of a phytochrome B-GFP fusion protein to the nucleus in transgenic *Arabidopsis*. *J. Cell Biol.* 145: 437–445.

Chapter

18

Blue-Light Responses: Stomatal Movements and Morphogenesis

MOST OF US are familiar with the observation that house plants placed near a window have branches that grow toward the incoming light. This response, called *phototropism*, is an example of how plants alter their growth patterns in response to the direction of incident radiation. This response to light is intrinsically different from light trapping by photosynthesis. In photosynthesis, plants harness light and convert it into chemical energy (see Chapters 7 and 8). In contrast, phototropism is an example of the use of light as an *environmental signal*. There are two major families of plant responses to light signals: the phytochrome responses, which were covered in Chapter 17, and the **blue-light responses**.

Some blue-light responses were introduced in Chapter 9—for example, chloroplast movement within cells in response to incident photon fluxes, and sun tracking by leaves. As with the family of the phytochrome responses, there are numerous plant responses to blue light. Besides phototropism, they include inhibition of hypocotyl elongation, stimulation of chlorophyll and carotenoid synthesis, activation of gene expression, stomatal movements, phototaxis (the movement of motile unicellular organisms such as algae and bacteria toward or away from light), enhancement of respiration, and anion uptake in algae (Senger 1984). Blue-light responses have been reported in higher plants, algae, ferns, fungi, and prokaryotes.

Some responses, such as electrical events at the plasma membrane, can be detected within seconds of irradiation by blue light. More complex metabolic or morphogenetic responses, such as blue light–stimulated pigment biosynthesis in the fungus *Neurospora* or branching in the alga *Vaucheria*, might require minutes, hours, or even days (Horwitz 1994).

Readers may be puzzled by the different approaches to naming phytochrome and blue-light responses. The former are identified by a specific photoreceptor (phytochrome), the latter by the blue-light region of the visible spectrum. In the case of phytochrome, several of its spectroscopic and biochemical properties, particularly its red/far-red reversibility, made possible its early identification, and hundreds of photobiological responses of plants can be clearly attributed to the phytochrome photoreceptor (see Chapter 17).

In contrast, the spectroscopy of blue-light responses is complex. Both chlorophylls and phytochrome absorb blue light (400–500 nm) from the visible spectrum, and other chromophores and some amino acids, such as tryptophan, absorb light in the ultraviolet (250–400 nm) region. How, then, can we then distinguish specific responses to blue light? One important identification criterion is that in specific blue-light responses, blue light cannot be replaced by a red-light treatment, and there is no red/far-red reversibility. Red or far-red light would be effective if photosynthesis or phytochrome were involved.

Another key distinction is that *many blue-light responses of higher plants share a characteristic action spectrum*. You will recall from Chapter 7 that an action spectrum is a graph of the magnitude of the observed light response as a function of wavelength (see **Web Topic 7.1** for a detailed discussion of spectroscopy and action spectra). The action spectrum of the response can be compared with the *absorption spectra* of candidate photoreceptors. A close correspondence between action and absorption spectra provides a strong indication that the pigment under consideration is the photoreceptor mediating the light response under study (see Figure 7.8).

Action spectra for blue light–stimulated phototropism, stomatal movements, inhibition of hypocotyl elongation, and other key blue-light responses share a characteristic "three-finger" fine structure in the 400 to 500 nm region (Figure 18.1) that is not observed in spectra for responses

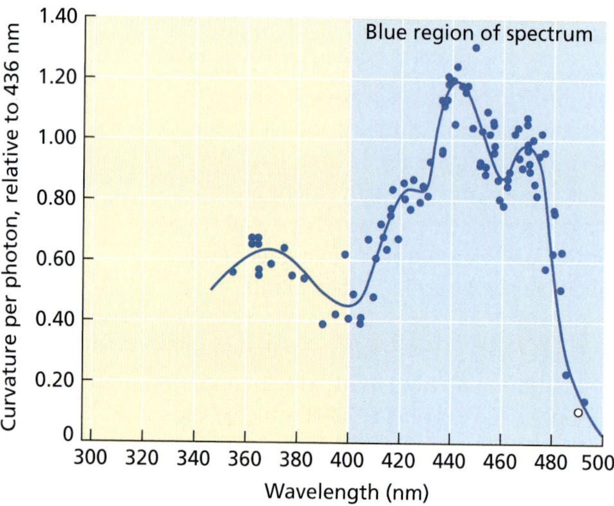

FIGURE 18.1 Action spectrum for blue light–stimulated phototropism in oat coleoptiles. An action spectrum shows the relationship between a biological response and the wavelengths of light absorbed. The "three-finger" pattern in the 400 to 500 nm region is characteristic of specific blue-light responses. (After Thimann and Curry 1960.)

to light that are mediated by photosynthesis, phytochrome, or other photoreceptors (Cosgrove 1994).

In this chapter we will describe representative blue-light responses in plants: phototropism, inhibition of stem elongation, and stomatal movements. The stomatal responses to blue light are discussed in detail because of the importance of stomata in leaf gas exchange (see Chapter 9) and in plant acclimations and adaptations to their environment. We will also discuss blue-light photoreceptors and the signal transduction cascade that links light perception with the final expression of blue-light sensing in the organism.

THE PHOTOPHYSIOLOGY OF BLUE-LIGHT RESPONSES

Blue-light signals are utilized by the plant in many responses, allowing the plant to sense the presence of light and its direction. This section describes the major morphological, physiological, and biochemical changes associated with typical blue-light responses.

Blue Light Stimulates Asymmetric Growth and Bending

Directional growth toward (or in special circumstances away from) the light, is called **phototropism**. It can be observed in fungi, ferns, and higher plants. Phototropism is a **photomorphogenetic** response that is particularly dramatic in dark-grown seedlings of both monocots and dicots. Unilateral light is commonly used in experimental studies, but phototropism can also be observed when a seedling is exposed to two unequally bright light sources (Figure 18.2), a condition that can occur in nature.

As it grows through the soil, the shoot of a grass is protected by a modified leaf that covers it, called a **coleoptile** (Figure 18.3; see also Figure 19.1). As discussed in detail in Chapter 19, unequal light perception in the coleoptile results in unequal concentrations of auxin in the lighted and shaded sides of the coleoptile, unequal growth, and bending.

Keep in mind that phototropic bending occurs only in *growing* organs, and that coleoptiles and shoots that have stopped elongating will not bend when exposed to unilateral light. In grass seedlings growing in soil under sunlight, coleoptiles stop growing as soon as the shoot has emerged from the soil and the first true leaf has pierced the tip of the coleoptile.

On the other hand, dark-grown, *etiolated* coleoptiles continue to elongate at high rates for several days and, depending on the species, can attain several centimeters in length. The large phototropic response of these etiolated coleoptiles (see Figure 18.3) has made them a classic model for studies of phototropism (Firn 1994).

The action spectrum shown in Figure 18.1 was obtained through measurement of the angles of curvature from oat coleoptiles that were irradiated with light of different

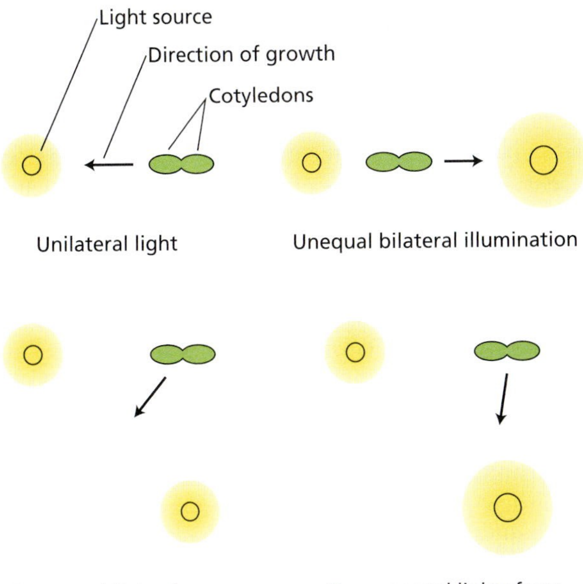

FIGURE 18.2 Relationship between direction of growth and unequal incident light. Cotyledons from a young seedling are shown as viewed from the top. The arrows indicate the direction of phototropic curvature. The diagrams illustrate how the direction of growth varies with the location and the intensity of the light source, but growth is always toward light. (After Firn 1994.)

wavelengths. The spectrum shows a peak at about 370 nm and the "three-finger" pattern in the 400 to 500 nm region discussed earlier. An action spectrum for phototropism in the dicot alfalfa (*Medicago sativa*) was found to be very similar to that of oat coleoptiles, suggesting that a common photoreceptor mediates phototropism in the two species.

Phototropism in sporangiophores of the mold *Phycomyces* has been studied to identify genes involved in phototropic responses. The sporangiophore consists of a sporangium (spore-bearing spherical structure) that develops on a stalk consisting of a long, single cell. Growth in the sporangiophore is restricted to a growing zone just below the sporangium.

When irradiated with unilateral blue light, the sporangiophore bends toward the light with an action spectrum similar to that of coleoptile phototropism (Cerda-Olmedo and Lipson 1987). These studies of *Phycomyces* have led to the isolation of many mutants with altered phototropic responses and the identification of several genes that are required for normal phototropism.

In recent years, phototropism of the stem of the small dicot *Arabidopsis* (Figure 18.4) has attracted much attention because of the ease with which advanced molecular techniques can be applied to *Arabidopsis* mutants. The genetics and the molecular biology of phototropism in *Arabidopsis* are discussed later in this chapter.

FIGURE 18.3 Time-lapse photograph of a corn coleoptile growing toward unilateral blue light given from the right. The consecutive exposures were made 30 minutes apart. Note the increasing angle of curvature as the coleoptile bends. (Courtesy of M. A. Quiñones.)

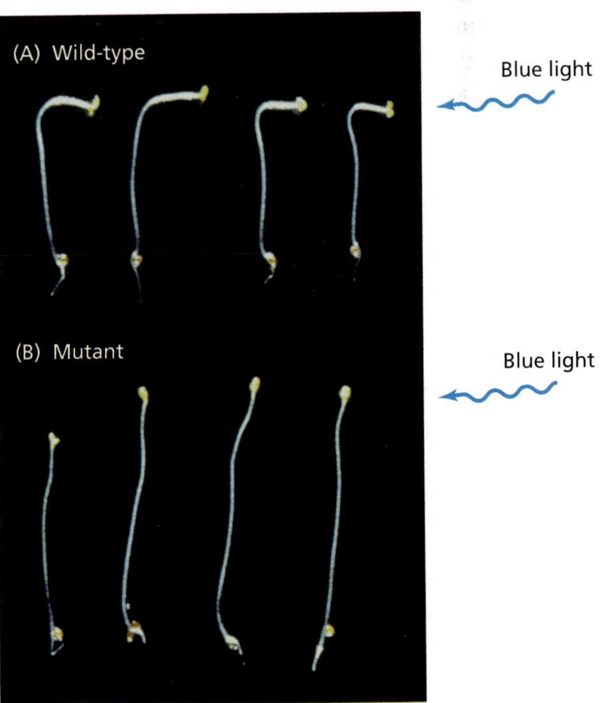

FIGURE 18.4 Phototropism in wild-type (A) and mutant (B) *Arabidopsis* seedlings. Unilateral light was applied from the right. (Courtesy of Dr. Eva Huala.)

How Do Plants Sense the Direction of the Light Signal?

Light gradients between lighted and shaded sides have been measured in coleoptiles and in hypocotyls from dicot seedlings irradiated with unilateral blue light. When a coleoptile is illuminated with 450 nm blue light, the ratio between the light that is incident to the surface of the illuminated side and the light that reaches the shaded side is 4:1 at the tip and the midregion of the coleoptile, and 8:1 at the base (Figure 18.5).

On the other hand, there is a *lens effect* in the sporangiophore of the mold *Phycomyces* irradiated with unilateral blue light, and as a result, the light measured at the distal cell surface of the sporangiophore is about twice the amount of light that is incident at the surface of the illuminated side. Light gradients and lens effects could play a role in how the bending organ senses the direction of the unilateral light (Vogelmann 1994).

Blue Light Rapidly Inhibits Stem Elongation

The stems of seedlings growing in the dark elongate very rapidly, and the inhibition of stem elongation by light is a key morphogenetic response of the seedling emerging from the soil surface (see Chapter 17). The conversion of Pr to Pfr (the red- and far red–absorbing forms of phytochrome, respectively) in etiolated seedlings causes a phytochrome-dependent, sharp decrease in elongation rates (see Figure 17.1).

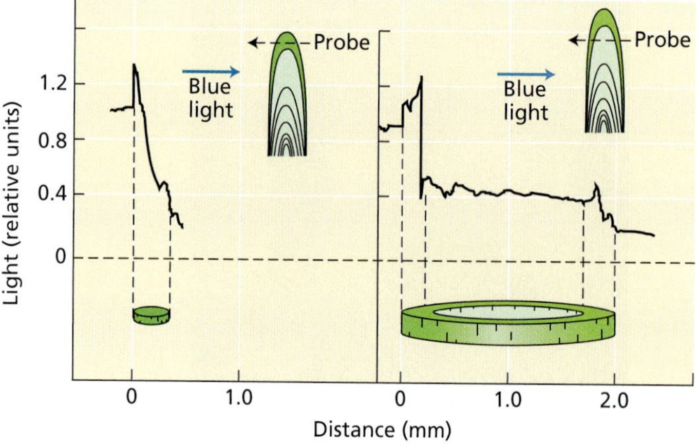

FIGURE 18.5 Distribution of transmitted, 450 nm blue light in an etiolated corn coleoptile. The diagram in the upper right of each frame shows the area of the coleoptile being measured by a fiber-optic probe. A cross section of the tissue appears at the bottom of each frame. The trace above it shows the amount of light sensed by the probe at each point. A sensing mechanism that depended on light gradients would sense the difference in the amount of light between the lighted and shaded sides of the coleoptile, and this information would be transduced into an unequal auxin concentration and bending. (After Vogelmann and Haupt 1985.)

However, action spectra for the decrease in elongation rate show strong activity in the blue region, which cannot be explained by the absorption properties of phytochrome (see Figure 17.9). In fact, the 400 to 500 nm blue region of the action spectrum for the inhibition of stem elongation closely resembles that of phototropism (compare the action spectra in Figures 17.10 and 18.1).

There are several ways to experimentally separate a reduction in elongation rates mediated by phytochrome from a reduction mediated by a specific blue-light response. If lettuce seedlings are given low fluence rates of blue light under a strong background of yellow light, their hypocotyl elongation rate is reduced by more than 50%. The background yellow light establishes a well-defined Pr:Pfr ratio (see Chapter 17). In such conditions, the low fluence rates of blue light added are too small to significantly change this ratio, ruling out a phytochrome effect on the reduction in elongation rate observed upon the addition of blue light.

Blue light– and phytochrome-mediated hypocotyl responses can also be distinguished by the swiftness of the response. Whereas phytochrome-mediated changes in elongation rates can be detected within 8 to 90 minutes, depending on the species, blue-light responses are rapid, and can be measured within 15 to 30 s (Figure 18.6). Interactions between phytochrome and the blue light–dependent sensory transduction cascade in the regulation of elongation rates will be described later in the chapter.

Another fast response elicited by blue light is a depolarization of the membrane of hypocotyl cells that precedes the inhibition of growth rate (see Figure 18.6). The membrane depolarization is caused by the activation of anion channels (see Chapter 6), which facilitates the efflux of anions such as chloride. Use of an anion channel blocker prevents the blue light–dependent membrane depolarization and decreases the inhibitory effect of blue light on hypocotyl elongation (Parks et al. 1998).

Blue Light Regulates Gene Expression

Blue light also regulates the expression of genes involved in several important morphogenetic processes. Some of these light-activated genes have been studied in detail—for example, the genes that code for the enzyme chalcone synthase, which catalyzes the first committed step in flavonoid biosynthesis, for the small subunit of rubisco, and for the proteins that bind chlorophylls *a* and *b* (see Chapters 13, 8, and 7, respectively). Most of the studies on light-activated genes show sensitivity to both blue and red light, as well as red/far-red reversibility, implicating both phytochrome and specific blue-light responses.

A recent study reported that *SIG5*, one of six *SIG* nuclear genes in *Arabidopsis* that play a regulatory role in the transcription of the chloroplast gene

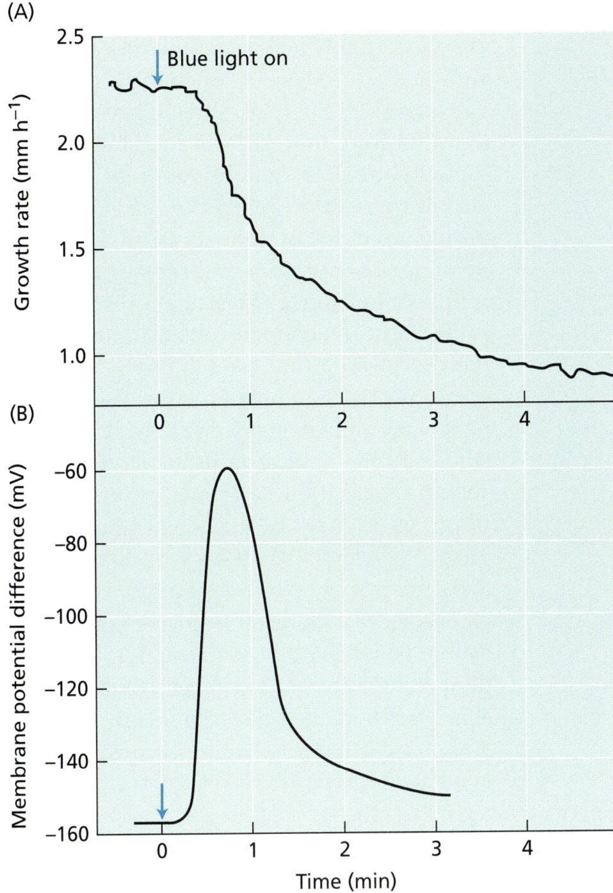

FIGURE 18.6 Blue light–induced (A) changes in elongation rates of etiolated cucumber seedlings and (B) transient membrane depolarization of hypocotyl cells. As the membrane depolarization (measured with intracellular electrodes) reaches its maximum, growth rate (measured with position transducers) declines sharply. Comparison of the two curves shows that the membrane starts to depolarize before the growth rate begins to decline, suggesting a cause–effect relation between the two phenomena. (After Spalding and Cosgrove 1989.)

psbD, which encodes the D2 subunit of the PSII reaction center (see Chapter 7), is specifically activated by blue light (Tsunoyama et al. 2002). In contrast, the other five *SIG* genes are activated by both blue and red light.

Another well-documented instance of gene expression that is mediated solely by a blue light–sensing system involves the *GSA* gene in the photosynthetic unicellular alga *Chlamydomonas reinhardtii* (Matters and Beale 1995). This gene encodes the enzyme glutamate-1-semialdehyde aminotransferase (GSA), a key enzyme in the chlorophyll biosynthesis pathway (see Chapter 7). The absence of phytochrome in *C. reinhardtii* simplifies the analysis of blue-light responses in this experimental system.

In synchronized cultures of *C. reinhardtii*, levels of *GSA* mRNA are strictly regulated by blue light, and 2 hours after the onset of illumination, *GSA* mRNA levels are 26-fold higher than they are in the dark (Figure 18.7). These blue light–mediated mRNA increases precede increases in chlorophyll content, indicating that chlorophyll biosynthesis is being regulated by activation of the *GSA* gene.

Blue Light Stimulates Stomatal Opening

We now turn our attention to the stomatal response to blue light. Stomata have a major regulatory role in gas exchange in leaves (see Chapter 9), and they can often affect yields of agricultural crops (see Chapter 25). Several characteristics of blue light–dependent stomatal movements make guard cells a valuable experimental system for the study of blue-light responses:

- The stomatal response to blue light is rapid and reversible, and it is localized in a single cell type, the guard cell.
- The stomatal response to blue light regulates stomatal movements throughout the life of the plant. This is unlike phototropism or hypocotyl elongation, which are functionally important at early stages of development.
- The signal transduction cascade that links the perception of blue light with the opening of stomata is understood in considerable detail.

In the following sections we will discuss two central aspects of the stomatal response to light, the osmoregulatory mechanisms that drive stomatal movements, and the role of a blue light–activated H$^+$-ATPase in ion uptake by guard cells.

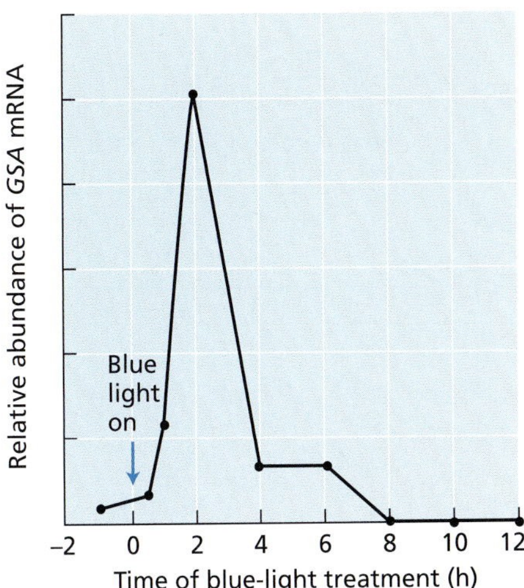

FIGURE 18.7 Time course of blue light–dependent gene expression in *Chlamydomonas reinhardtii*. The *GSA* gene encodes the enzyme glutamate-1-semialdehyde aminotransferase, which regulates an early step in chlorophyll biosynthesis. (After Matters and Beale 1995.)

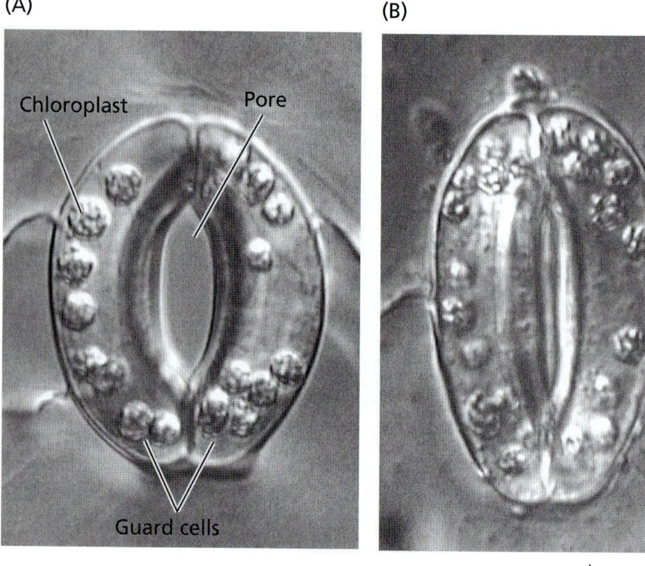

FIGURE 18.8 Light-stimulated stomatal opening in detached epidermis of *Vicia faba*. Open, light-treated stoma (A), is shown in the dark-treated, closed state in (B). Stomatal opening is quantified by microscopic measurement of the width of the stomatal pore. (Courtesy of E. Raveh.)

Light is the dominant environmental signal controlling stomatal movements in leaves of well-watered plants growing in natural environments. Stomata open as light levels reaching the leaf surface increase, and close as they decrease (Figure 18.8). In greenhouse-grown leaves of broad bean (*Vicia faba*), stomatal movements closely track incident solar radiation at the leaf surface (Figure 18.9).

Early studies of the stomatal response to light showed that DCMU (dichlorophenyldimethylurea), an inhibitor of photosynthetic electron transport (see Figure 7.31), causes a partial inhibition of light-stimulated stomatal opening. These results indicated that photosynthesis in the guard cell chloroplast plays a role in light-dependent stomatal opening, but the observation that the inhibition was only partial pointed to a nonphotosynthetic component of the stomatal response to light. Detailed studies of the light response of stomata have shown that light activates two distinct responses of guard cells: photosynthesis in the guard cell chloroplast (see **Web Essay 18.1**), and a specific blue-light response.

The specific stomatal response to blue light cannot be resolved properly under blue-light illumination because blue light simultaneously stimulates both the specific blue-light response and guard cell photosynthesis (for the photosynthetic response to blue light, see the action spectrum for photosynthesis in Figure 7.8). A clear-cut separation of the responses of the two light responses can be obtained in dual-beam experiments. High fluence rates of red light are used to *saturate* the photosynthetic response, and low photon fluxes of blue light are added after the response to the saturating red light has been completed (Figure 18.10). The addition of blue light causes substantial further stomatal opening that cannot be explained as a further stimulation of guard cell photosynthesis because photosynthesis is saturated by the background red light.

An action spectrum for the stomatal response to blue light under background red illumination shows the three-finger pattern discussed earlier (Figure 18.11). This action spectrum, typical of blue-light responses and distinctly different from the action spectrum for photosynthesis, further indicates that, in addition to photosynthesis, guard cells respond specifically to blue light.

When guard cells are treated with cellulolytic enzymes that digest the cell walls, *guard cell protoplasts* are released. Guard cell protoplasts *swell* when illuminated with blue light (Figure 18.12), indicating that blue light is sensed within the guard cells proper. The swelling of guard cell

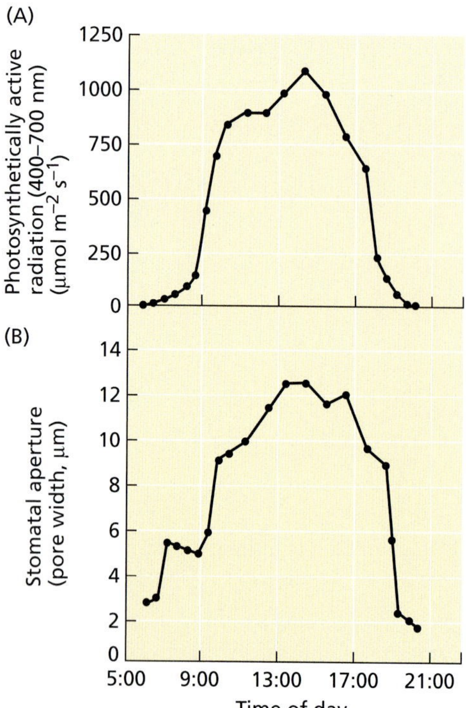

FIGURE 18.9 Stomatal opening tracks photosynthetic active radiation at the leaf surface. Stomatal opening in the lower surface of leaves of *Vicia faba* grown in a greenhouse, measured as the width of the stomatal pore (A), closely follows the levels of photosynthetically active radiation (400–700 nm) incident to the leaf (B), indicating that the response to light was the dominant response regulating stomatal opening. (After Srivastava and Zeiger 1995a.)

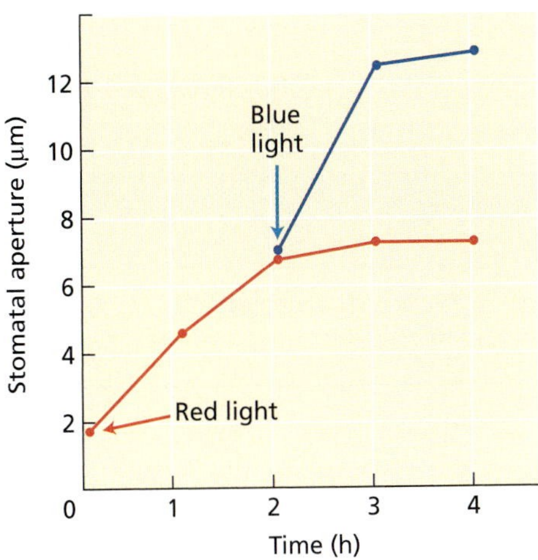

FIGURE 18.10 The response of stomata to blue light under a red-light background. Stomata from detached epidermis of *Commelina communis* (common dayflower) were treated with saturating photon fluxes of red light (red trace). In a parallel treatment, stomata illuminated with red light were also illuminated with blue light, as indicated by the arrow (blue trace). The increase in stomatal opening above the level reached in the presence of saturating red light indicates that a different photoreceptor system, stimulated by blue light, is mediating the additional increases in opening. (From Schwartz and Zeiger 1984.)

protoplasts also illustrates how intact guard cells function. The light-stimulated uptake of ions and the accumulation of organic solutes decrease the cell's osmotic potential (increase the osmotic pressure). Water flows in as a result, leading to an increase in turgor that in guard cells with intact walls is mechanically transduced into an increase in stomatal apertures (see Chapter 4). In the absence of a cell wall, the blue light–mediated increase in osmotic pressure causes the guard cell protoplast to swell.

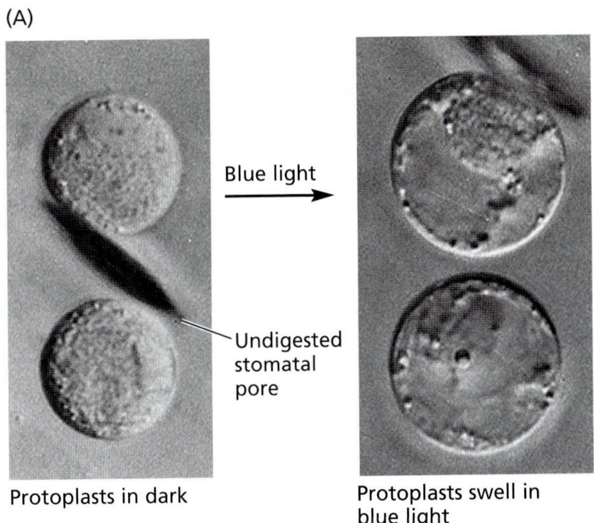

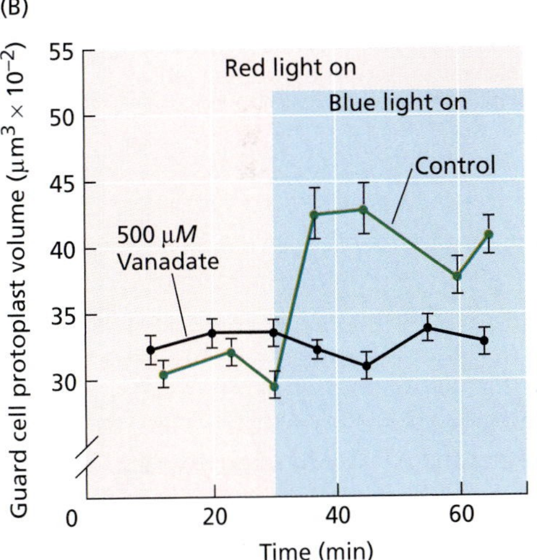

FIGURE 18.12 Blue light–stimulated swelling of guard cell protoplasts. (A) In the absence of a rigid cell wall, guard cell protoplasts of onion (*Allium* cepa) swell. (B) Blue light stimulates the swelling of guard cell protoplasts of broad bean (*Vicia faba*), and vanadate, an inhibitor of the H^+-ATPase, inhibits this swelling. Blue light stimulates ion and water uptake in the guard cell protoplasts, which in the intact guard cells provides a mechanical force that drives increases in stomatal apertures. (A from Zeiger and Hepler 1977; B after Amodeo et al. 1992.)

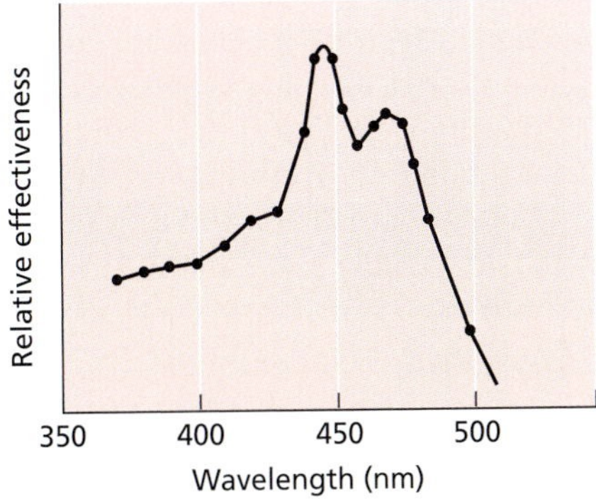

FIGURE 18.11 The action spectrum for blue light–stimulated stomatal opening (under a red-light background). (After Karlsson 1986.)

Blue Light Activates a Proton Pump at the Guard Cell Plasma Membrane

When guard cell protoplasts from broad bean (*Vicia faba*) are irradiated with blue light under background red-light illumination, the pH of the suspension medium becomes more acidic (Figure 18.13). This blue light–induced acidification is blocked by inhibitors that dissipate pH gradients, such as CCCP (discussed shortly), and by inhibitors of the proton-pumping H^+-ATPase, such as vanadate (see Figure 18.12C; see also Chapter 6).

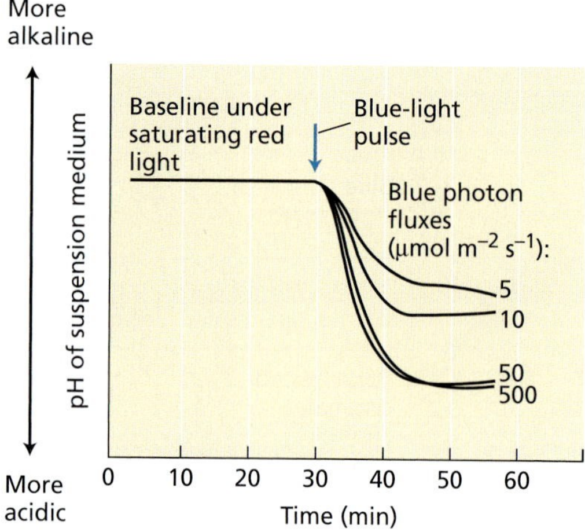

FIGURE 18.13 Acidification of a suspension medium of guard cell protoplasts of *Vicia faba* stimulated by a 30 s pulse of blue light. The acidification results from the stimulation of an H^+-ATPase at the plasma membrane by blue light, and it is associated with protoplast swelling (see Figure 18.12). (After Shimazaki et al. 1986.)

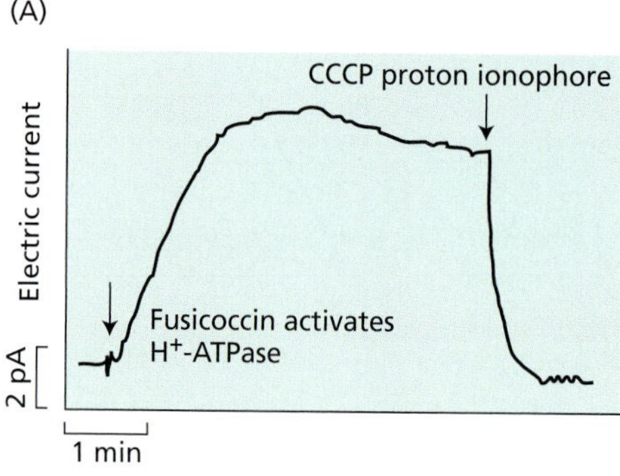

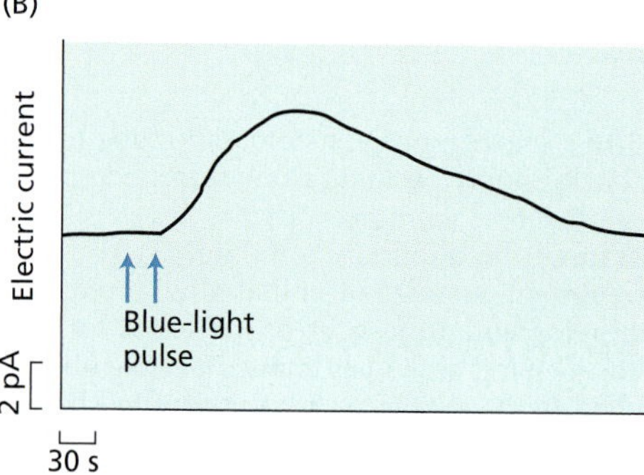

FIGURE 18.14 Activation of the H^+-ATPase at the plasma membrane of guard cell protoplasts by fusiccocin and blue light can be measured as electric current in patch clamp experiments. (A) Outward electric current (measured in picoamps, pA) at the plasma membrane of a guard cell protoplast stimulated by the fungal toxin fusicoccin, an activator of the H^+-ATPase. The current is abolished by the proton ionophore CCCP (carbonyl cyanide *m*-chlorophenylhydrazone). (B) Outward electric current at the plasma membrane of a guard cell protoplast stimulated by a blue-light pulse. These results indicate that blue light stimulates the H^+-ATPase. (A after Serrano et al. 1988; B after Assmann et al. 1985.)

This indicates that *the acidification results from the activation by blue light of a proton-pumping ATPase in the guard cell plasma membrane* that extrudes protons into the protoplast suspension medium and lowers its pH. In the intact leaf, this blue-light stimulation of proton pumping lowers the pH of the apoplastic space surrounding the guard cells. The plasma membrane ATPase from guard cells has been isolated and extensively characterized (Kinoshita et al. 2001).

The activation of electrogenic pumps such as the proton-pumping ATPase can be measured in patch-clamping experiments as an outward electric current at the plasma membrane (see **Web Topic 6.2** for a description of patch clamping). A patch clamp recording of a guard cell protoplast treated with the fungal toxin fusicoccin, a well-characterized activator of plasma membrane ATPases, is shown in Figure 18.14A. Exposure to fusicoccin stimulates an outward electric current, which is abolished by the proton ionophore carbonyl cyanide *m*-chlorophenylhydrazone (CCCP). This proton ionophore makes the plasma membrane highly permeable to protons, thus precluding the formation of a proton gradient across the membrane and abolishing net proton efflux.

The relationship between proton pumping at the guard cell plasma membrane and stomatal opening is evident from the observation that fusicoccin stimulates both proton extrusion from guard cell protoplasts and stomatal opening, and that CCCP inhibits the fusiccocin-stimulated opening. The increase in proton-pumping rates as a function of fluence rates of blue light (see Figure 18.13) indicates that the increasing rates of blue photons in the solar radiation reaching the leaf cause a larger stomatal opening.

The close relationship among the number of incident blue-light photons, proton pumping at the guard cell plasma membrane, and stomatal opening further suggests that the blue-light response of stomata might function as a sensor of photon fluxes reaching the guard cell.

Pulses of blue light given under a saturating red-light background also stimulate an outward electric current from guard cell protoplasts (see Figure 18.14B). The acidification measurements shown in Figure 18.13 indicate that the outward electric current measured in patch clamp experiments is carried by protons.

Blue-Light Responses Have Characteristic Kinetics and Lag Times

Some of the characteristics of the responses to blue-light pulses underscore some important properties of blue-light responses: the persistence of the response after the light signal has been switched off, and a significant lag time separating the onset of the light signal and the beginning of the response.

In contrast to typical photosynthetic responses, which are activated very quickly after a "light on" signal, and cease when the light goes off (see, for instance, Figure 7.13), blue-light responses proceed at maximal rates for several minutes after application of the pulse (see Figure 18.14B). This property can be explained by a physiologically inactive form of the blue-light photoreceptor that is converted to an active form by blue light, with the active form reverting slowly to the physiologically inactive form in the absence of blue light (Iino et al. 1985). The rate of the response to a blue-light pulse would thus depend on the time course of the reversion of the active form to the inactive one.

Another property of the response to blue-light pulses is a lag time, which lasts about 25 s in both the acidification response and the outward electric currents stimulated by blue light (see Figures 18.13 and 18.14). This amount of time is probably required for the signal transduction cascade to proceed from the photoreceptor site to the proton-pumping ATPase and for the proton gradient to form. Similar lag times have been measured for blue light–dependent inhibition of hypocotyl elongation, which was discussed earlier.

Blue Light Regulates Osmotic Relations of Guard Cells

Blue light modulates guard cell osmoregulation via its activation of proton pumping (described earlier) and via the stimulation of the synthesis of organic solutes. Before discussing these blue-light responses, let us briefly describe the major osmotically active solutes in guard cells.

The botanist Hugo von Mohl proposed in 1856 that turgor changes in guard cells provide the mechanical force for changes in stomatal apertures. The plant physiologist F. E. Lloyd hypothesized in 1908 that guard cell turgor is regulated by osmotic changes resulting from starch–sugar interconversions, a concept that led to a starch–sugar hypothesis of stomatal movements. The discovery of the changes in potassium concentrations in guard cells in the 1960s led to the modern theory of guard cell osmoregulation by potassium and its counterions.

Potassium concentration in guard cells increases severalfold when stomata open, from 100 m*M* in the closed state to 400 to 800 m*M* in the open state, depending on the plant species and the experimental conditions. These large concentration changes in the positively charged potassium ions are electrically balanced by the anions Cl^- and malate^{2-} (Figure 18.15A). In species of the genus *Allium*, such as onion (*Allium cepa*), K^+ ions are balanced solely by Cl^-. In most species, however, potassium fluxes are balanced by varying amounts of Cl^- and the organic anion malate^{2-} (Talbott et al. 1996).

The Cl^- ion is taken up into the guard cells during stomatal opening and extruded during stomatal closing. Malate, on the other hand, is synthesized in the guard cell cytosol, in a metabolic pathway that uses carbon skeletons generated by starch hydrolysis (see Figure 18.15B). The malate content of guard cells decreases during stomatal closing, but it remains to be established whether malate is catabolized in mitochondrial respiration or is extruded into the apoplast.

Potassium and chloride are taken up into guard cells via secondary transport mechanisms driven by the gradient of electrochemical potential for H^+, $\Delta\mu_{H^+}$, generated by the proton pump (see Chapter 6) discussed earlier in the chapter. Proton extrusion makes the electric-potential difference across the guard cell plasma membrane more negative; light-dependent hyperpolarizations as high as 50 mV have been measured. In addition, proton pumping generates a pH gradient of about 0.5 to 1 pH unit.

The electrical component of the proton gradient provides a driving force for the passive uptake of potassium ions via voltage-regulated potassium channels (see Chapter 6) (Schroeder et al. 2001). Chloride is thought to be taken up through anion channels. Thus, blue light–dependent stimulation of proton pumping plays a key role in guard cell osmoregulation during light-dependent stomatal movements

Guard cell chloroplasts (see Figure 18.8) contain large starch grains, and their starch content decreases during stomatal opening and increases during closing. Starch, an insoluble, high-molecular-weight polymer of glucose, does not contribute to the cell's osmotic potential, but the hydrolysis of starch into soluble sugars causes a decrease in the osmotic potential (or increase in osmotic pressure) of guard cells. In the reverse process, starch synthesis decreases the sugar concentration, resulting in an increase of the cell's osmotic potential, which the starch–sugar hypothesis predicted to be associated with stomatal closing.

With the discovery of the major role of potassium and its counterion in guard cell osmoregulation, the sugar–starch hypothesis was no longer considered important (Outlaw 1983). Recent studies, however, described in the next section, have characterized a major osmoregulatory phase of guard cells in which sucrose is the dominant osmotically active solute.

Sucrose Is an Osmotically Active Solute in Guard Cells

Studies of daily courses of stomatal movements in intact leaves have shown that the potassium content in guard cells increases in parallel with early-morning opening, but it decreases in the early afternoon under conditions in which apertures continue to increase. The sucrose content of guard cells increases slowly in the morning, but upon potassium efflux, sucrose becomes the dominant osmoti-

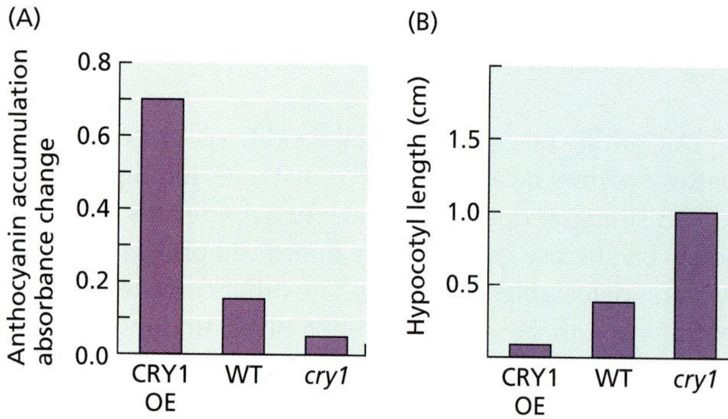

FIGURE 18.17 Blue light stimulates the accumulation of anthocyanin (A) and the inhibition of stem elongation (B) in transgenic and mutant seedlings of *Arabidopsis*. These bar graphs show a transgenic phenotype overexpressing the gene that encodes CRY1 (CRY1 OE), the wild type (WT), and *cry1* mutants. The enhanced blue-light response of the transgenic plant overexpressing the gene that encodes CRY1 demonstrates the important role of this gene product in stimulating anthocyanin biosynthesis and inhibiting stem elongation. (After Ahmad et al. 1998.)

Pterins are light-absorbing, pteridine derivatives that often function as pigments in insects, fishes, and birds (see Chapter 12 for pterin structure). When expressed in *Escherichia coli*, the cry1 protein binds FAD and a pterin, but it lacks detectable photolyase activity. No information is available on the chromophore(s) bound to cry1 in vivo, or on the nature of the photochemical reactions involving cry1, that would start the postulated sensory transduction cascade mediating the several blue-light responses mediated by cry1.

The most important evidence for a role of cry1 in blue light–mediated inhibition of stem elongation comes from overexpression studies. Overexpression of the CRY1 protein in transgenic tobacco or *Arabidopsis* plants results in a stronger blue light–stimulated inhibition of hypocotyl elongation than in the wild type, as well as increased production of anthocyanin, another blue-light response (Figure 18.17). Thus, overexpression of CRY1 caused an enhanced sensitivity to blue light in transgenic plants. Other blue-light responses, such as phototropism and blue light–dependent stomatal movements, appear to be normal in the *cry1* mutant phenotype.

A second gene product homologous to CRY1, named CRY2, has been isolated from *Arabidopsis* (Lin 2000). Both CRY1 and CRY2 appear ubiquitous throughout the plant kingdom. A major difference between them is that CRY2 is rapidly degraded in the light, whereas CRY1 is stable in light-grown seedlings.

Transgenic plants overexpressing the gene that encodes CRY2 show a small enhancement of the inhibition of hypocotyl elongation, indicating that unlike CRY1, CRY2 does not play a primary role in inhibiting stem elongation. On the other hand, the transgenic plants overexpressing the gene that encodes CRY2 show a large increase in blue light–stimulated cotyledon expansion, yet another blue-light response. In addition, CRY1 has been shown to be involved in the setting of the circadian clock in *Arabidopsis* (see Chapter 17), and both CRY1 and CRY2 have been shown to play a role in the induction of flowering (see Chapter 24). Cryptochrome homologs have been found to regulate the circadian clock in *Drosophila*, mouse, and humans.

Phototropins Are Involved in Phototropism and Chloroplast Movements

Some recently isolated *Arabidopsis* mutants impaired in blue light–dependent phototropism of the hypocotyl have provided valuable information about cellular events preceding bending. One of these mutants, the *nph1* (*n*on*p*hototropic *h*ypocotyl) mutant has been found to be genetically independent of the *hy4* (*cry1*) mutant discussed earlier: The *nph1* mutant lacks a phototropic response in the hypocotyl but has normal blue light–stimulated inhibition of hypocotyl elongation, while *hy4* has the converse phenotype. Recently the *nph1* gene was renamed *phot1*, and the protein it encodes was named **phototropin** (Briggs and Christie 2002).

The C-terminal half of phototropin is a serine/threonine kinase. The N-terminal half contains two repeated domains, of about 100 amino acids each, that have sequence similarities to other proteins involved in signaling in bacteria and mammals. Proteins with sequence similarity to the N terminus of phototropin bind flavin cofactors. These proteins are oxygen sensors in *Escherichia coli* and *Azotobacter*, and voltage sensors in potassium channels of *Drosophila* and vertebrates.

When expressed in insect cells, the N-terminal half of phototropin binds flavin mononucleotide (FMN) (see Figure 11.2B and **Web Essay 18.2**) and shows a blue light–dependent autophosphorylation reaction. This reaction resembles the blue light–dependent phosphorylation of a 120 kDa membrane protein found in growing regions of etiolated seedlings.

The *Arabidopsis* genome contains a second gene, *phot2*, which is related to *phot1*. The *phot1* mutant lacks hypocotyl phototropism in response to low-intensity blue light (0.01–1 μmol mol^{-2} s^{-1}) but retains a phototropic response at higher intensities (1–10 μmol m^{-2} s^{-1}). The *phot2* mutant has a normal phototropic response, but the *phot1/phot2* double mutant is severely impaired at both low and high intensities. These data indicate that both *phot1* and *phot2* are involved in the phototropic response, with *phot2* functioning at high light fluence rates.

Blue light–activated chloroplast movement. Leaves show an adaptive feature that can alter the intracellular distribution of their chloroplasts in order to control light absorption and prevent photodamage (see Figure 9.5). The action spectrum for chloroplast movement shows the "three finger" fine structure typical of blue-light responses. When incident radiation is weak, chloroplasts gather at the upper and lower surfaces of the mesophyll cells (the "accumulation" response; see Figure 9.5B), thus maximizing light absorption.

Under strong light, the chloroplasts move to the cell surfaces that are parallel to the incident light (the "avoidance" response; see Figure 9.5C), thus minimizing light absorption. Recent studies have shown that mesophyll cells of the *phot1* mutant have a normal avoidance response and a rudimentary accumulation response. Cells from the *phot2* mutant show a normal accumulation response but lack the avoidance response. Cells from the *phot1/phot2* double mutant lack both the avoidance and accumulation responses (Sakai et al. 2001). These results indicate that *phot2* plays a key role in the avoidance response, and that both *phot1* and *phot2* contribute to the accumulation response.

The Carotenoid Zeaxanthin Mediates Blue-Light Photoreception in Guard Cells

The carotenoid zeaxanthin has been implicated as a photoreceptor in blue light–stimulated stomatal opening. Recall from Chapters 7 and 9 that zeaxanthin is one of the three components of the xanthophyll cycle of chloroplasts, which protects photosynthetic pigments from excess excitation energy. In guard cells, however, the changes in zeaxanthin content as a function of incident radiation are distinctly different from the changes in mesophyll cells (Figure 18.18).

In sun plants such as *Vicia faba*, zeaxanthin accumulation in the mesophyll begins at about 200 μmol m^{-2} s^{-1}, and there is no detectable zeaxanthin in the early morning or late afternoon. In contrast, the zeaxanthin content in guard cells closely follows incident solar radiation at the leaf surface throughout the day, and it is nearly linearly proportional to incident photon fluxes in the early morning and late afternoon. Several key characteristics of the guard cell chloroplast strongly indicate that the primary function of the guard cell chloroplast is sensory transduction and not carbon fixation (Zeiger et al. 2002).

Compelling evidence indicates that zeaxanthin is a blue-light photoreceptor in guard cells:

- The absorption spectrum of zeaxanthin (Figure 18.19) closely matches the action spectrum for blue light–stimulated stomatal opening (see Figure 18.11).
- In daily courses of stomatal opening in intact leaves grown in a greenhouse, incident radiation, zeaxanthin content of guard cells, and stomatal apertures are closely related (see Figure 18.18).
- The blue-light sensitivity of guard cells increases as a function of their zeaxanthin concentration. Experimentally, zeaxanthin concentration in guard cells can be varied with increasing fluence rates of red light. When guard cells from epidermal peels illuminated with increasing fluence rates of red light are exposed to blue light, the resulting blue light–stimulated stomatal opening is linearly related to the fluence rate of background red-light irradiation (see the wild-type treatment in Figure 18.20) and to

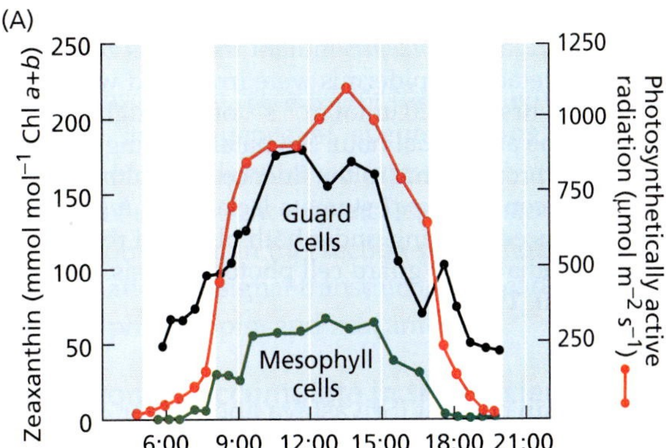

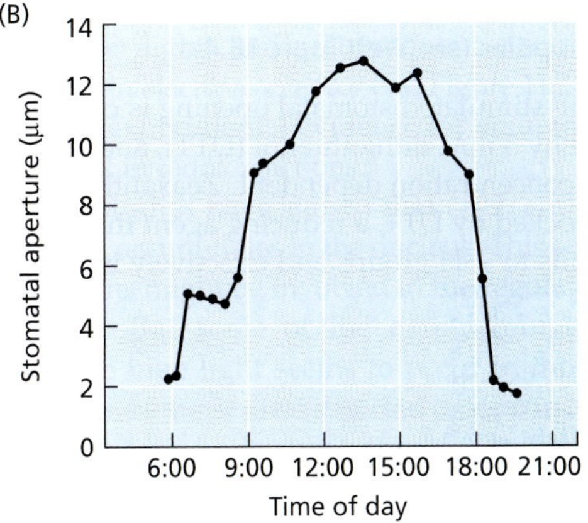

FIGURE 18.18 The zeaxanthin content of guard cells closely tracks photosynthetic active radiation and stomatal apertures. (A) Daily course of photosynthetic active radiation reaching the leaf surface, and of zeaxanthin content of guard cells and mesophyll cells of *Vicia faba* leaves grown in a greenhouse. The white areas within the graph highlight the contrasting sensitivity of the xanthophyll cycle in mesophyll and guard cell chloroplasts under the low irradiances prevailing early and late in the day. (B) Stomatal apertures in the same leaves used to measure guard cell zeaxanthin content. (After Srivastava and Zeiger 1995a.)

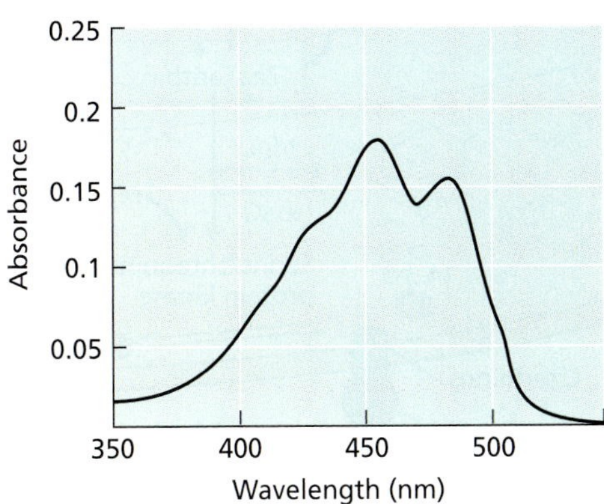

FIGURE 18.19 The absorption spectrum of zeaxanthin in ethanol.

dramatically increased our understanding of the subject. The identification of cryptochromes, phototropin, and zeaxanthin as putative blue-light photoreceptors in plant cells has stimulated great interest in this aspect of plant photobiology. Current and future work is addressing important open questions, such as the detailed sequence of the sensory transduction cascades and the precise localization and composition of the pigment proteins involved. Ongoing research on the subject virtually ensures rapid further progress.

SUMMARY

Plants utilize light as a source of energy and as a signal that provides information about their environment. A large family of blue-light responses is used to sense light quantity and direction. These blue-light signals are transduced into electrical, metabolic, and genetic processes that allow plants to alter growth, development, and function in order to acclimate to changing environmental conditions. Blue-light responses include phototropism, stomatal movements, inhibition of stem elongation, gene activation, pigment biosynthesis, tracking of the sun by leaves, and chloroplast movements within cells.

Specific blue-light responses can be distinguished from other responses that have some sensitivity to blue light by a characteristic "three-finger" action spectrum in the 400 to 500 nm region.

The physiology of blue-light responses varies broadly. In phototropism, stems grow toward unilateral light sources by asymmetric growth on their shaded side. In the inhibition of stem elongation, perception of blue light depolarizes the membrane potential of elongating cells, and the rate of elongation rapidly decreases. In gene activation, blue light stimulates transcription and translation, leading to the accumulation of gene products that are required for the morphogenetic response to light.

Blue light–stimulated stomatal movements are driven by blue light–dependent changes in the osmoregulation of guard cells. Blue light stimulates an H^+-ATPase at the guard cell plasma membrane, and the resulting pumping of protons across the membrane generates an electrochemical-potential gradient that provides a driving force for ion uptake. Blue light also stimulates starch degradation and malate biosynthesis. Solute accumulation within the guard cells leads to stomatal opening. Guard cells also utilize sucrose as a major osmotically active solute, and light quality can change the activity of different osmoregulatory pathways that modulate stomatal movements.

Cry1 and *cry2* are two *Arabidopsis* genes involved in blue light–dependent inhibition of stem elongation, cotyledon expansion, anthocyanin synthesis, the control of flowering, and the setting of circadian rhythms. It has been proposed that CRY1 and CRY2 are apoproteins of flavin-containing pigment proteins that mediate blue-light photoreception.

The cry1 and cry2 gene products have sequence similarity to photolyase but no photolyase activity. The cry1 protein, and to a lesser extent cry2, accumulates in the nucleus and might be involved in gene expression. The cry1 protein also regulates anion channel activity at the plasma membrane.

The protein phototropin has a major role in the regulation of phototropism. The C-terminal half of phototropin is a serine/threonine kinase, and the N-terminal half has two flavin-binding domains. In vitro, phototropin binds the flavin FMN and autophosphorylates in response to blue light. Mutants called *phot1* and *phot2* are defective in phototropism and in chloroplast movements. The *phot1/phot2* double mutant lacks blue light–stimulated stomatal opening.

The chloroplastic carotenoid zeaxanthin has been implicated in blue-light photoreception in guard cells. Blue light–stimulated stomatal opening is blocked if zeaxanthin accumulation in guard cells is prevented by genetic or biochemical means. Manipulation of zeaxanthin content in guard cells makes it possible to regulate their response to blue light. The signal transduction cascade for the blue-light response of guard cells comprises blue-light perception in the guard cell chloroplast, transduction of the blue-light signal across the chloroplast envelope, activation of the H^+-ATPase, turgor buildup, and stomatal opening.

Web Material

Web Topics

18.1 Guard Cell Osmoregulation and a Blue Light–Activated Metabolic Switch

Blue light controls major osmoregulatory pathways in guard cells and unicellular algae.

18.2 Historical Notes on the Research of Blue-Light Photoreceptors

Carotenoids and flavins have been the main candidates for blue-light photoreceptors.

18.3 Comparing Flavins and Carotenoids

Flavin and carotenoid photoreceptors have contrasting functional properties.

18.4 The Coleoptile Chloroplast

Both the coleoptile and the guard cell chloroplasts specialize in sensory transduction.

Web Essays

18.1 Guard Cell Photosynthesis

Photosynthesis in the guard cell chloroplast shows unique regulatory features.

18.2 Phototropins

Phototropins regulate several light responses in plants.

18.3 The Sensory Transduction of the Inhibition of Stem Elongation by Blue Light

The regulation of stem elongation rates by blue light has critical importance for plant development.

18.4 The Blue/Green Reversibility of the Blue-Light Response of Stomata

The blue/green reversal of stomatal movements is a remarkable photobiological response.

18.5 Zeaxanthin and CO_2 Sensing in Guard Cells

The functional relationship between Calvin cycle activity and zeaxanthin content of guard cells couples blue light and CO_2 sensing during stomatal movements.

Chapter References

Ahmad, M., and Cashmore, A. R. (1993) *HY4* gene of *A. thaliana* encodes a protein with characteristics of a blue light photoreceptor. *Nature* 366: 162–166.

Ahmad, M., Jarillo, J. A., Smirnova, O., and Cashmore, A. R. (1998) Cryptochrome blue light photoreceptors of *Arabidopsis* implicated in phototropism. *Nature* 392: 720–723.

Amodeo, G., Srivastava, A., and Zeiger, E. (1992) Vanadate inhibits blue light–stimulated swelling of *Vicia* guard cell protoplasts. *Plant Physiol.* 100: 1567–1570.

Assmann, S. M. (1988) Enhancement of the stomatal response to blue light by red light, reduced intercellular concentrations of carbon dioxide and low vapor pressure differences. *Plant Physiol.* 87: 226–231.

Assmann, S. M., Simoncini, L., and Schroeder, J. I. (1985) Blue light activates electrogenic ion pumping in guard cell protoplasts of *Vicia faba*. *Nature* 318: 285–287.

Briggs, W. R., and Christie, J. M. (2002) Phototropins 1 and 2: Versatile plant blue-light receptors. *Trends Plant Sci.* 7: 204–210.

Cerda-Olmedo, E., and Lipson, E. D. (1987) *Phycomyces*. Cold Spring Harbor Laboratory, Cold Spring Harbor, NY.

Cosgrove, D. J. (1994) Photomodulation of growth. In *Photomorphogenesis in Plants*, 2nd ed., R. E. Kendrick and G. H. M. Kronenberg, eds., Kluwer, Dordrecht, Netherlands, pp. 631–658.

Dietrich, P., Sanders, D., and Hedrich, R. (2001) The role of ion channels in light-dependent stomatal opening. *J. Exp. Bot.* 52: 1959–1967.

Emi, T., Kinoshita, T., and Shimazaki, K. (2001) Specific binding of vf14-3-3a isoform to the plasma membrane H+-ATPase in response to blue light and fusicoccin in guard cells of broad bean. *Plant Physiol.* 125: 1115–1125.

Firn, R. D. (1994) Phototropism. In *Photomorphogenesis in Plants*, 2nd ed., R. E. Kendrick and G. H. M. Kronenberg, eds., Kluwer, Dordrecht, Netherlands, pp. 659–681.

Frechilla, S., Talbott, L. D., Bogomolni, R. A., and Zeiger, E. (2000) Reversal of blue light-stimulated stomatal opening by green light. *Plant Cell Physiol.* 41: 171–176.

Frechilla, S., Zhu, J., Talbott, L. D., and Zeiger, E. (1999) Stomata from npq1, a zeaxanthin-less *Arabidopsis* mutant, lack a specific response to blue light. *Plant Cell Physiol.* 40: 949–954.

Horwitz, B. A. (1994) Properties and transduction chains of the UV and blue light photoreceptors. In *Photomorphogenesis in Plants*, 2nd ed., R. E. Kendrick and G. H. M. Kronenberg, eds., Kluwer, Dordrecht, Netherlands, pp. 327–350.

Iino, M., Ogawa, T., and Zeiger, E. (1985) Kinetic properties of the blue light response of stomata. *Proc. Natl. Acad. Sci. USA* 82: 8019–8023.

Karlsson, P. E. (1986) Blue light regulation of stomata in wheat seedlings. II. Action spectrum and search for action dichroism. *Physiol. Plant.* 66: 207–210.

Kinoshita, T., and Shimazaki, K. (1999) Blue light activates the plasma membrane H^+-ATPase by phosphorylation of the C-terminus in stomatal guard cells. *EMBO J.* 18: 5548–5558.

Kinoshita, T., and Shimazaki, K. (2001) Analysis of the phosphorylation level in guard-cell plasma membrane H^+-ATPase in response to fusicoccin. *Plant Cell Physiol.* 42: 424–432.

Kinoshita, T., Doi, M., Suetsugu, N., Kagawa, T., Wada, M., and Shimazaki, K. (2001) phot1 and phot2 mediate blue light regulation of stomatal opening. *Nature* 414: 656–660.

Lin, C. (2000) Plant blue-light receptors. *Trends Plant Sci.* 5: 337–342.

Matters, G. L., and Beale, S. I. (1995) Blue-light-regulated expression of genes for two early steps of chlorophyll biosynthesis in *Chlamydomonas reinhardtii*. *Plant Physiol.* 109: 471–479.

Niyogi, K. K., Grossman, A. R., and Björkman, O. (1998) *Arabidopsis* mutants define a central role for the xanthophyll cycle in the regulation of photosynthetic energy conversion. *Plant Cell* 10: 1121–1134.

Outlaw, W. H., Jr. (1983) Current concepts on the role of potassium in stomatal movements. *Physiol. Plant.* 59: 302–311.

Parks, B. M., Cho, M. H., and Spalding, E. P. (1998) Two genetically separable phases of growth inhibition induced by blue light in Arabidopsis seedlings. *Plant Physiol.* 118: 609–615.

Parks, B. M., Folta, K. M., and Spalding, E. P. (2001) Photocontrol of stem growth. *Curr. Opin. Plant Biol.* 4: 436–440.

Sakai, T., Kagawa, T., Kasahara, M., Swartz, T. E., Christie, J. M., Briggs, W. R., Wada, M., and Okada, K. (2001) Arabidopsis nph1 and npl1: Blue light receptors that mediate both phototropism and chloroplast relocation. *Proc. Natl. Acad. Sci. USA* 98: 6969–6974.

Schroeder, J. I., Allen, G. J., Hugouvieux, V., Kwak, J. M., and Waner, D. (2001) Guard cell signal transduction. *Annu. Rev. Plant Physiol. Plant Mol. Biol.* 52: 627–658.

Schwartz, A., and Zeiger, E. (1984) Metabolic energy for stomatal opening. Roles of photophosphorylation and oxidative phosphorylation. *Planta* 161: 129–136.

Senger, H. (1984) *Blue Light Effects in Biological Systems*. Springer, Berlin.

Serrano, E. E., Zeiger, E., and Hagiwara, S. (1988) Red light stimulates an electrogenic proton pump in *Vicia* guard cell protoplasts. *Proc. Natl. Acad. Sci. USA* 85: 436–440.

Shimazaki, K., Iino, M., and Zeiger, E. (1986) Blue light–dependent proton extrusion by guard cell protoplasts of *Vicia faba*. *Nature* 319: 324–326.

Spalding, E. P., and Cosgrove, D. J. (1989) Large membrane depolarization precedes rapid blue-light induced growth inhibition in cucumber. *Planta* 178: 407–410.

Srivastava, A., and Zeiger, E. (1995a) Guard cell zeaxanthin tracks photosynthetic active radiation and stomatal apertures in *Vicia faba* leaves. *Plant Cell Environ.* 18: 813–817.

Srivastava, A., and Zeiger, E. (1995b) The inhibitor of zeaxanthin formation, dithiothreitol, inhibits blue-light-stimulated stomatal opening in *Vicia faba*. *Planta* 196: 445–449.

Swartz, T. E., Corchnoy, S. B., Christie, J. M., Lewis, J. W., Szundi, I., Briggs, W. R., and Bogomolni, R. A. (2001) The photocycle of a flavin-binding domain of the blue light photoreceptor phototropin. *J. Biol. Chem.* 276: 36493–36500.

Talbott, L. D., and Zeiger, E. (1998) The role of sucrose in guard cell osmoregulation. *J. Exp. Bot.* 49: 329–337.

Talbott, L. D., Srivastava, A., and Zeiger, E. (1996) Stomata from growth-chamber-grown *Vicia faba* have an enhanced sensitivity to CO_2. *Plant Cell Environ.* 19: 1188–1194.

Tallman, G., Zhu, J., Mawson, B. T., Amodeo, G., Nouhi, Z., Levy, K., and Zeiger, E. (1997) Induction of CAM in *Mesembryanthemum crystallinum* abolishes the stomatal response to blue light and light-dependent zeaxanthin formation in guard cell chloroplasts. *Plant Cell Physiol.* 38: 236–242.

Thimann, K. V., and Curry, G. M. (1960) Phototropism and phototaxis. In *Comparative Biochemistry*, Vol. 1, M. Florkin and H. S. Mason, eds., Academic Press, New York, pp. 243–306.

Tsunoyama, Y., Morikawa, K., Shiina, T., and Toyoshima, Y. (2002) Blue light specific and differential expression of a plastid sigma factor, Sig5 in *Arabidopsis thaliana*. *FEBS Lett.* 516: 225–228.

Vogelmann, T. C. (1994) Light within the plant. In *Photomorphogenesis in Plants*, 2nd ed., R. E. Kendrick and G. H. M. Kronenberg, eds., Kluwer, Dordrecht, Netherlands, pp. 491–533.

Vogelmann, T. C., and Haupt, W. (1985) The blue light gradient in unilaterally irradiated maize coleoptiles: Measurements with a fiber optic probe. *Photochem. Photobiol.* 41: 569–576.

Yamamoto, H. Y. (1979) Biochemistry of the violaxanthin cycle in higher plants. *Pure Appl. Chem.* 51: 639–648.

Zeiger, E., and Hepler, P. K. (1977) Light and stomatal function: Blue light stimulates swelling of guard cell protoplasts. *Science* 196: 887–889.

Zeiger, E., Talbott, L. D., Frechilla, S., Srivastava, A., and Zhu, J. X. (2002) The guard cell chloroplast: A perspective for the twenty-first century. *New Phytol.* 153: 415–424.

Chapter

19

Auxin: The Growth Hormone

THE FORM AND FUNCTION of multicellular organism would not be possible without efficient communication among cells, tissues, and organs. In higher plants, regulation and coordination of metabolism, growth, and morphogenesis often depend on chemical signals from one part of the plant to another. This idea originated in the nineteenth century with the German botanist Julius von Sachs (1832–1897).

Sachs proposed that chemical messengers are responsible for the formation and growth of different plant organs. He also suggested that external factors such as gravity could affect the distribution of these substances within a plant. Although Sachs did not know the identity of these chemical messengers, his ideas led to their eventual discovery.

Many of our current concepts about intercellular communication in plants have been derived from similar studies in animals. In animals the chemical messengers that mediate intercellular communication are called **hormones**. Hormones interact with specific cellular proteins called *receptors*.

Most animal hormones are synthesized and secreted in one part of the body and are transferred to specific target sites in another part of the body via the bloodstream. Animal hormones fall into four general categories: proteins, small peptides, amino acid derivatives, and steroids.

Plants also produce signaling molecules, called *hormones*, that have profound effects on development at vanishingly low concentrations. Until quite recently, plant development was thought to be regulated by only five types of hormones: auxins, gibberellins, cytokinins, ethylene, and abscisic acid. However, there is now compelling evidence for the existence of plant steroid hormones, the brassinosteroids, that have a wide range of morphological effects on plant development. (Brassinosteroids as plant hormones are discussed in **Web Essay 19.1**.)

A variety of other signaling molecules that play roles in resistance to pathogens and defense against herbivores have also been identified, including jasmonic acid, salicylic acid, and the polypeptide systemin (see Chapter 13). Thus the number and types of hormones and hormonelike signaling agents in plants keep expanding.

The first plant hormone we will consider is auxin. Auxin deserves pride of place in any discussion of plant hormones because it was the first growth hormone to be discovered in plants, and much of the early physiological work on the mechanism of plant cell expansion was carried out in relation to auxin action.

Moreover, both auxin and cytokinin differ from the other plant hormones and signaling agents in one important respect: They are required for viability. Thus far, no mutants lacking either auxin or cytokinin have been found, suggesting that mutations that eliminate them are lethal. Whereas the other plant hormones seem to act as on/off switches that regulate specific developmental processes, auxin and cytokinin appear to be required at some level more or less continuously.

We begin our discussion of auxins with a brief history of their discovery, followed by a description of their chemical structures and the methods used to detect auxins in plant tissues. A look at the pathways of auxin biosynthesis and the polar nature of auxin transport follows. We will then review the various developmental processes controlled by auxin, such as stem elongation, apical dominance, root initiation, fruit development, and oriented, or *tropic*, growth. Finally, we will examine what is currently known about the mechanism of auxin-induced growth at the cellular and molecular levels.

THE EMERGENCE OF THE AUXIN CONCEPT

During the latter part of the nineteenth century, Charles Darwin and his son Francis studied plant growth phenomena involving tropisms. One of their interests was the bending of plants toward light. This phenomenon, which is caused by differential growth, is called **phototropism**. In some experiments the Darwins used seedlings of canary grass (*Phalaris canariensis*), in which, as in many other grasses, the youngest leaves are sheathed in a protective organ called the **coleoptile** (Figure 19.1).

Coleoptiles are very sensitive to light, especially to blue light (see Chapter 18). If illuminated on one side with a short pulse of dim blue light, they will bend (grow) toward the source of the light pulse within an hour. The Darwins found that the tip of the coleoptile perceived the light, for if they covered the tip with foil, the coleoptile would not bend. But the region of the coleoptile that is responsible for the bending toward the light, called the **growth zone**, is several millimeters below the tip.

Thus they concluded that some sort of signal is produced in the tip, travels to the growth zone, and causes the shaded side to grow faster than the illuminated side. The results of their experiments were published in 1881 in a remarkable book entitled *The Power of Movement in Plants*.

There followed a long period of experimentation by many investigators on the nature of the growth stimulus in coleoptiles. This research culminated in the demonstration in 1926 by Frits Went of the presence of a growth-promoting chemical in the tip of oat (*Avena sativa*) coleoptiles. It was known that if the tip of a coleoptile was removed, coleoptile growth ceased. Previous workers had attempted to isolate and identify the growth-promoting chemical by grinding up coleoptile tips and testing the activity of the extracts. This approach failed because grinding up the tissue released into the extract inhibitory substances that normally were compartmentalized in the cell.

Went's major breakthrough was to avoid grinding by allowing the material to diffuse out of excised coleoptile tips directly into gelatin blocks. If placed asymmetrically on top of a decapitated coleoptile, these blocks could be tested for their ability to cause bending in the absence of a unilateral light source (see Figure 19.1). Because the substance promoted the elongation of the coleoptile sections (Figure 19.2), it was eventually named **auxin** from the Greek *auxein*, meaning "to increase" or "to grow."

BIOSYNTHESIS AND METABOLISM OF AUXIN

Went's studies with agar blocks demonstrated unequivocally that the growth-promoting "influence" diffusing from the coleoptile tip was a chemical substance. The fact that it was produced at one location and transported in minute amounts to its site of action qualified it as an authentic plant hormone.

In the years that followed, the chemical identity of the "growth substance" was determined, and because of its potential agricultural uses, many related chemical analogs were tested. This testing led to generalizations about the chemical requirements for auxin activity. In parallel with these studies, the agar block diffusion technique was being applied to the problem of auxin transport. Technological advances, especially the use of isotopes as tracers, enabled plant biochemists to unravel the pathways of auxin biosynthesis and breakdown.

Our discussion begins with the chemical nature of auxin and continues with a description of its biosynthesis, transport, and metabolism. Increasingly powerful analytical methods and the application of molecular biological approaches have recently allowed scientists to identify auxin precursors and to study auxin turnover and distribution within the plant.

The Principal Auxin in Higher Plants Is Indole-3-Acetic Acid

In the mid-1930s it was determined that auxin is **indole-3-acetic acid (IAA)**. Several other auxins in higher plants were discovered later (Figure 19.3), but IAA is by far the most abundant and physiologically relevant. Because the structure of IAA is relatively simple, academic and industrial laboratories were quickly able to synthesize a wide

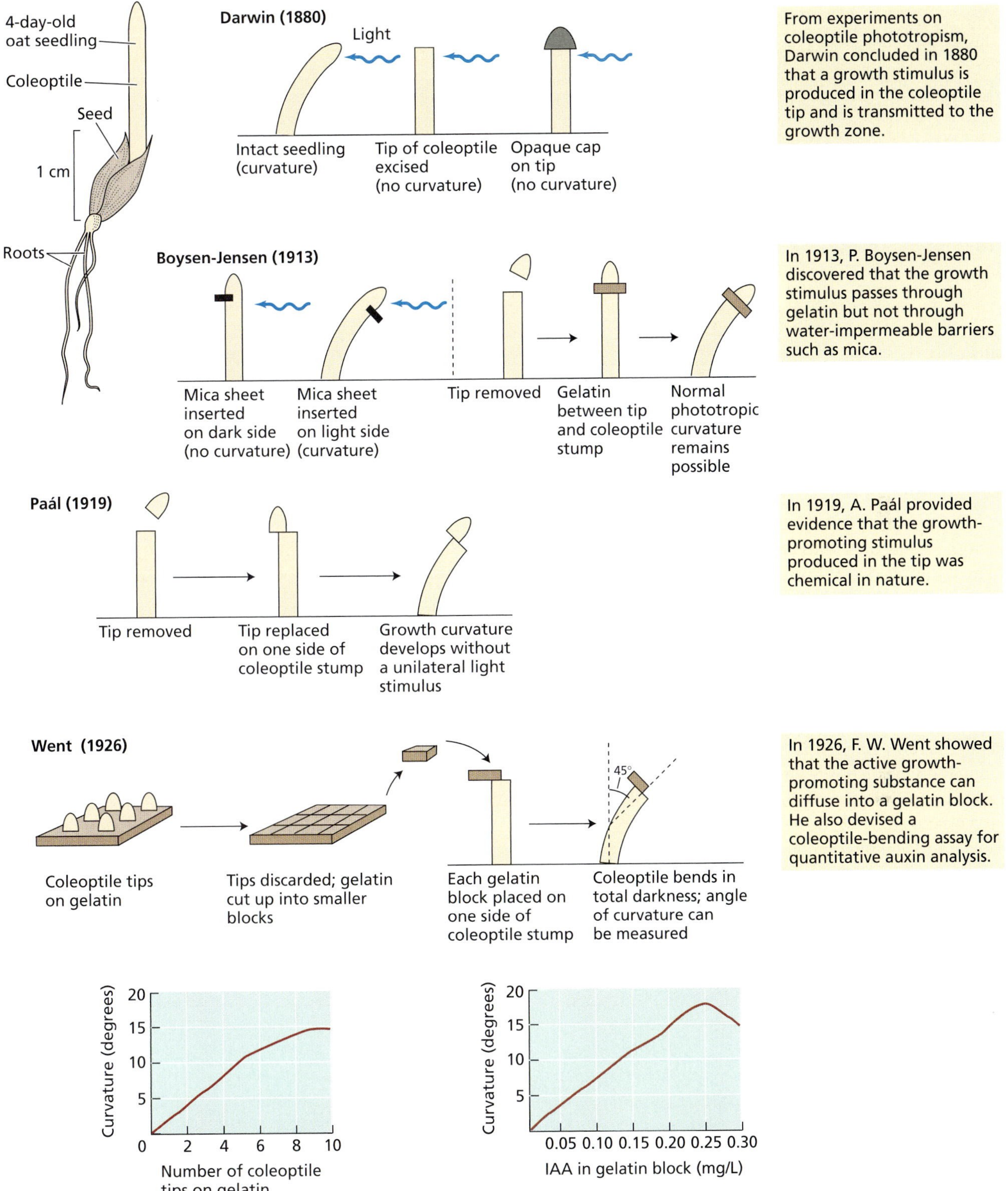

FIGURE 19.1 Summary of early experiments in auxin research.

(A)

(B)

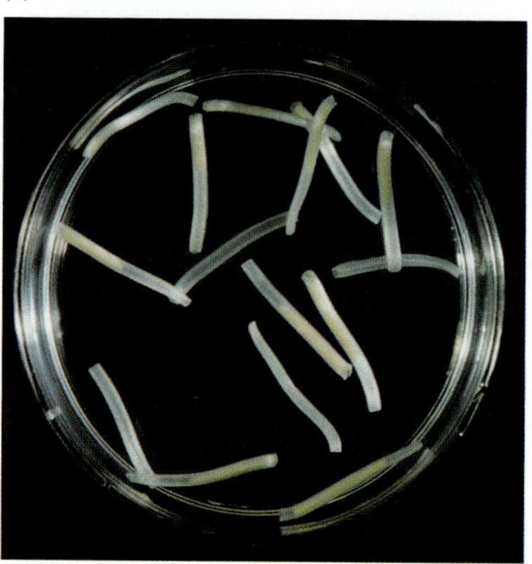

FIGURE 19.2 Auxin stimulates the elongation of oat coleoptile sections. These coleoptile sections were incubated for 18 hours in either water (A) or auxin (B). The yellow tissue inside the translucent coleoptile is the primary leaves. (Photos © M. B. Wilkins.)

array of molecules with auxin activity. Some of these are used as herbicides in horticulture and agriculture (Figure 19.4) (for additional synthetic auxins, see **Web Topic 19.1**).

An early definition of auxins included all natural and synthetic chemical substances that stimulate elongation in coleoptiles and stem sections. However, auxins affect many developmental processes besides cell elongation. Thus auxins can be defined as compounds with biological activities similar to those of IAA, including the ability to promote cell elongation in coleoptile and stem sections, cell division in callus cultures in the presence of cytokinins, formation of adventitious roots on detached leaves and stems, and other developmental phenomena associated with IAA action.

Although they are chemically diverse, a common feature of all active auxins is a molecular distance of about 0.5 nm between a fractional positive charge on the aromatic ring and a negatively charged carboxyl group (see **Web Topic 19.2**).

Auxins in Biological Samples Can Be Quantified

Depending on the information that a researcher needs, the amounts and/or identity of auxins in biological samples can be determined by bioassay, mass spectrometry, or enzyme-linked immunosorbent assay, which is abbreviated as ELISA (see **Web Topic 19.3**).

A **bioassay** is a measurement of the effect of a known or suspected biologically active substance on living material. In his pioneering work more than 60 years ago, Went used *Avena sativa* (oat) coleoptiles in a technique called the ***Avena* coleoptile curvature test** (see Figure 19.1). The coleoptile curved because the increase in auxin on one side stimulated cell elongation, and the decrease in auxin on the other side (due to the absence of the coleoptile tip) caused a decrease in the growth rate—a phenomenon called **differential growth**.

Went found that he could estimate the amount of auxin in a sample by measuring the resulting coleoptile curva-

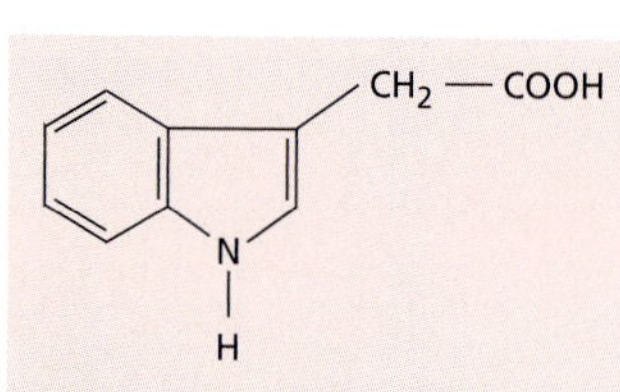

Indole-3-acetic acid (IAA)

4-Chloroindole-3-acetic acid (4-Cl-IAA)

Indole-3-butyric acid (IBA)

FIGURE 19.3 Structure of three natural auxins. Indole-3-acetic acid (IAA) occurs in all plants, but other related compounds in plants have auxin activity. Peas, for example, contain 4-chloroindole-3-acetic acid. Mustards and corn contain indole-3-butyric acid (IBA).

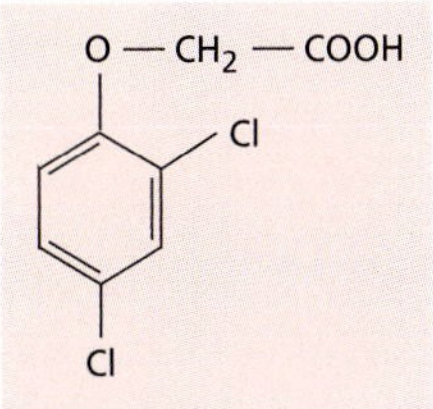

2,4-Dichlorophenoxyacetic acid (2,4-D)

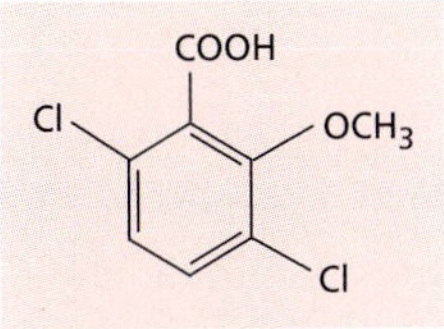

2-Methoxy-3, 6-dichlorobenzoic acid (dicamba)

FIGURE 19.4 Structures of two synthetic auxins. Most synthetic auxins are used as herbicides in horticulture and agriculture.

ture. Auxin bioassays are still used today to detect the presence of auxin activity in a sample. The *Avena* coleoptile curvature assay is a sensitive measure of auxin activity (it is effective for IAA concentrations of about 0.02 to 0.2 mg L^{-1}). Another bioassay measures auxin-induced changes in the straight growth of *Avena* coleoptiles floating in solution (see Figure 19.2). Both of these bioassays can establish the presence of an auxin in a sample, but they cannot be used for precise quantification or identification of the specific compound.

Mass spectrometry is the method of choice when information about both the chemical structure and the amount of IAA is needed. This method is used in conjunction with separation protocols involving gas chromatography. It allows the precise quantification and identification of auxins, and can detect as little as 10^{-12} g (1 picogram, or pg) of IAA, which is well within the range of auxin found in a single pea stem section or a corn kernel. These sophisticated techniques have enabled researchers to accurately analyze auxin precursors, auxin turnover, and auxin distribution within the plant.

IAA Is Synthesized in Meristems, Young Leaves, and Developing Fruits and Seeds

IAA biosynthesis is associated with rapidly dividing and rapidly growing tissues, especially in shoots. Although virtually all plant tissues appear to be capable of producing low levels of IAA, shoot apical meristems, young leaves, and developing fruits and seeds are the primary sites of IAA synthesis (Ljung et al. in press).

In very young leaf primordia of *Arabidopsis*, auxin is synthesized at the tip. During leaf development there is a gradual shift in the site of auxin production basipetally along the margins, and later, in the central region of the lamina. The basipetal shift in auxin production correlates closely with, and is probably causally related to, the basipetal maturation sequence of leaf development and vascular differentiation (Aloni 2001).

By fusing the *GUS* (β-glucuronidase) reporter gene to a promoter containing an auxin response element, and transforming *Arabidopsis* leaves with this construct in a Ti plasmid using *Agrobacterium*, it is possible to visualize the distribution of free auxin in young, developing leaves. Wherever free auxin is produced, *GUS* expression occurs—and can be detected histochemically. By use of this technique, it has recently been demonstrated that auxin is produced by a cluster of cells located at sites where hydathodes will develop (Figure 19.5).

Hydathodes are glandlike modifications of the ground and vascular tissues, typically at the margins of leaves, that allow the release of liquid water (guttation fluid) through pores in the epidermis in the presence of root pressure (see Chapter 4). As shown in Figure 19.5, during early stages of hydathode differentiation a center of high auxin synthesis is evident as a concentrated dark blue GUS stain (arrow) in the lobes of serrated leaves of *Arabidopsis* (Aloni et al. 2002). A diffuse trail of GUS activity leads down to differentiating vessel elements in a developing vascular strand. This remarkable micrograph captures the process of auxin-regulated vascular differentiation in the very act!

We will return to the topic of the control of vascular differentiation later in the chapter.

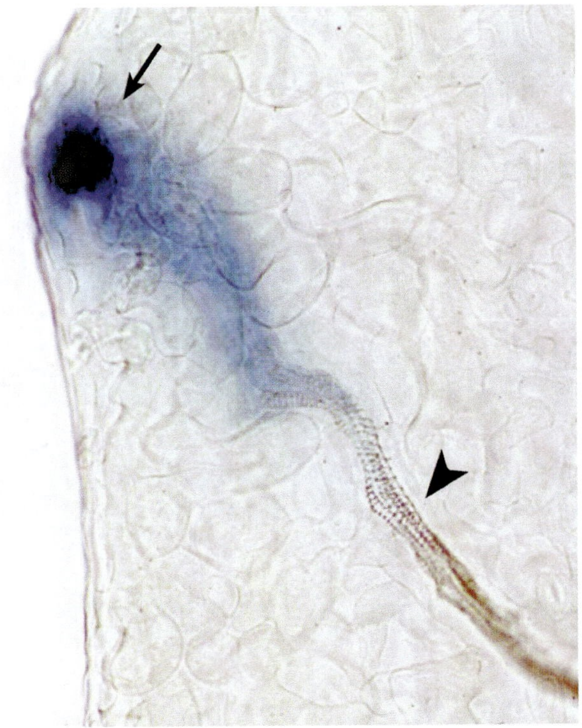

FIGURE 19.5 Detection of sites of auxin synthesis and transport in a young leaf primordium of *DR5 Arabidopsis* by means of a *GUS* reporter gene with an auxin-sensitive promoter. During the early stages of hydathode differentiation, a center of auxin synthesis is evident as a concentrated dark blue *GUS* stain (arrow) in the lobes of the serrated leaf margin. A gradient of diluted GUS activity extends from the margin toward a differentiating vascular strand (arrowhead), which functions as a sink for the auxin flow originating in the lobe. (Courtesy of R. Aloni and C. I. Ullrich.)

Multiple Pathways Exist for the Biosynthesis of IAA

IAA is structurally related to the amino acid tryptophan, and early studies on auxin biosynthesis focused on tryptophan as the probable precursor. However, the incorporation of exogenous labeled tryptophan (e.g., [^{3}H]tryptophan) into IAA by plant tissues has proved difficult to demonstrate. Nevertheless, an enormous body of evidence has now accumulated showing that plants convert tryptophan to IAA by several pathways, which are described in the paragraphs that follow.

The IPA pathway. The **indole-3-pyruvic acid (IPA)** pathway (see Figure 19.6C), is probably the most common of the tryptophan-dependent pathways. It involves a deamination reaction to form IPA, followed by a decarboxylation reaction to form indole-3-acetaldehyde (IAld). Indole-3-acetaldehyde is then oxidized to IAA by a specific dehydrogenase.

The TAM pathway. The **tryptamine (TAM) pathway** (see Figure 19.6D) is similar to the IPA pathway, except that the order of the deamination and decarboxylation reactions is reversed, and different enzymes are involved. Species that do not utilize the IPA pathway possess the TAM pathway. In at least one case (tomato), there is evidence for both the IPA and the TAM pathways (Nonhebel et al. 1993).

The IAN pathway. In the **indole-3-acetonitrile (IAN)** pathway (see Figure 19.6B), tryptophan is first converted to indole-3-acetaldoxime and then to indole-3-acetonitrile. The enzyme that converts IAN to IAA is called *nitrilase*. The IAN pathway may be important in only three plant families: the Brassicaceae (mustard family), Poaceae (grass

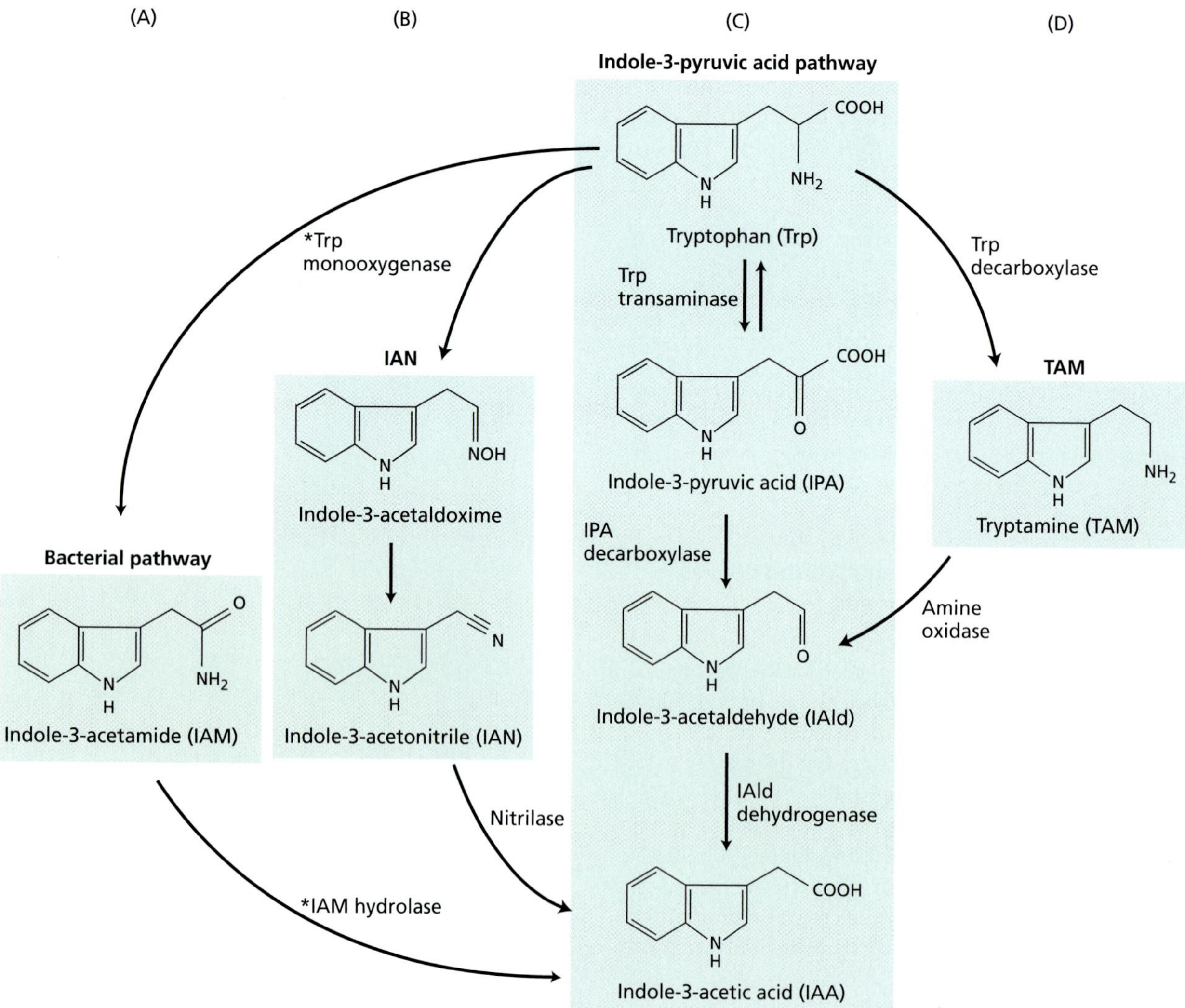

FIGURE 19.6 Tryptophan-dependent pathways of IAA biosynthesis in plants and bacteria. The enzymes that are present only in bacteria are marked with an asterisk. (After Bartel 1997.)

family), and Musaceae (banana family). Nevertheless, nitrilase-like genes or activities have recently been identified in the Cucurbitaceae (squash family), Solanaceae (tobacco family), Fabaceae (legumes), and Rosaceae (rose family).

Four genes (*NIT1* through *NIT4*) that encode nitrilase enzymes have now been cloned from *Arabidopsis*. When *NIT2* was expressed in transgenic tobacco, the resultant plants acquired the ability to respond to IAN as an auxin by hydrolyzing it to IAA (Schmidt et al. 1996).

Another tryptophan-dependent biosynthetic pathway—one that uses **indole-3-acetamide (IAM)** as an intermediate (see Figure19.6A)—is used by various pathogenic bacteria, such as *Pseudomonas savastanoi* and *Agrobacterium tumefaciens*. This pathway involves the two enzymes tryptophan monooxygenase and IAM hydrolase. The auxins produced by these bacteria often elicit morphological changes in their plant hosts.

In addition to the tryptophan-dependent pathways, recent genetic studies have provided evidence that plants can synthesize IAA via one or more tryptophan-independent pathways. The existence of multiple pathways for IAA biosynthesis makes it nearly impossible for plants to run out of auxin and is probably a reflection of the essential role of this hormone in plant development.

IAA Is Also Synthesized from Indole or from Indole-3-Glycerol Phosphate

Although a tryptophan-independent pathway of IAA biosynthesis had long been suspected because of the low levels of conversion of radiolabeled tryptophan to IAA, not until genetic approaches were available could the existence of such pathways be confirmed and defined. Perhaps the most striking of these studies in maize involves the *orange pericarp* (*orp*) mutant (Figure 19.7), in which both subunits of the enzyme tryptophan synthase are inactive (Figure 19.8). The *orp* mutant is a true tryptophan auxotroph, requiring exogenous tryptophan to survive. However, neither the *orp* seedlings nor the wild-type seedlings can convert tryptophan to IAA, even when the mutant seedlings are given enough tryptophan to reverse the lethal effects of the mutation.

FIGURE 19.7 The orange pericarp (*orp*) mutant of maize is missing both subunits of tryptophan synthase. As a result, the pericarps surrounding each kernel accumulate glycosides of anthranilic acid and indole. The orange color is due to excess indole. (Courtesy of Jerry D. Cohen.)

Despite the block in tryptophan biosynthesis, the *orp* mutant contains amounts of IAA 50-fold higher than those of a wild-type plant (Wright et al. 1991). Signficantly, when *orp* seedlings were fed [^{15}N]anthranilate (see Figure 19.8), the label subsequently appeared in IAA, but not in tryptophan. These results provided the best experimental evidence for a tryptophan-independent pathway of IAA biosynthesis.

Further studies established that the branch point for IAA biosynthesis is either indole or its precursor, indole-3-glycerol phosphate (see Figure 19.8). IAN and IPA are possible intermediates, but the immediate precursor of IAA in the tryptophan-independent pathway has not yet been identified.

The discovery of the tryptophan-independent pathway has drastically altered our view of IAA biosynthesis, but the relative importance of the two pathways (tryptophan-dependent versus tryptophan-independent) is poorly understood. In several plants it has been found that the type of IAA biosynthesis pathway varies between different tissues, and between different times of development. For example, during embryogenesis in carrot, the tryptophan-dependent pathway is important very early in development, whereas the tryptophan-independent pathway takes over soon after the root–shoot axis is established. (For more evidence of the tryptophan-independent biosynthesis of IAA, see **Web Topic 19.4**.)

Most IAA in the Plant Is in a Covalently Bound Form

Although free IAA is the biologically active form of the hormone, the vast majority of auxin in plants is found in a covalently bound state. These conjugated, or "bound," auxins have been identified in all higher plants and are considered hormonally inactive.

IAA has been found to be conjugated to both high- and low-molecular-weight compounds.

- Low-molecular-weight conjugated auxins include esters of IAA with glucose or *myo*-inositol and amide conjugates such as IAA-*N*-aspartate (Figure 19.9).
- High-molecular-weight IAA conjugates include IAA-glucan (7–50 glucose units per IAA) and IAA-glycoproteins found in cereal seeds.

The compound to which IAA is conjugated and the extent of the conjugation depend on the specific conjugating enzymes. The best-studied reaction is the conjugation of IAA to glucose in *Zea mays*.

The highest concentrations of free auxin in the living plant are in the apical meristems of shoots and in young leaves because these are the primary sites of auxin synthe-

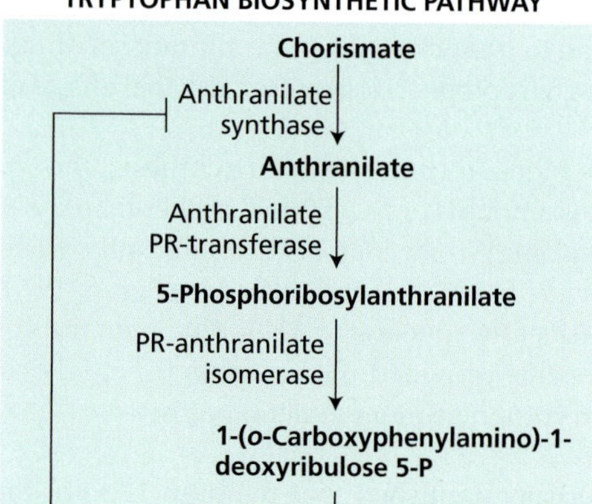
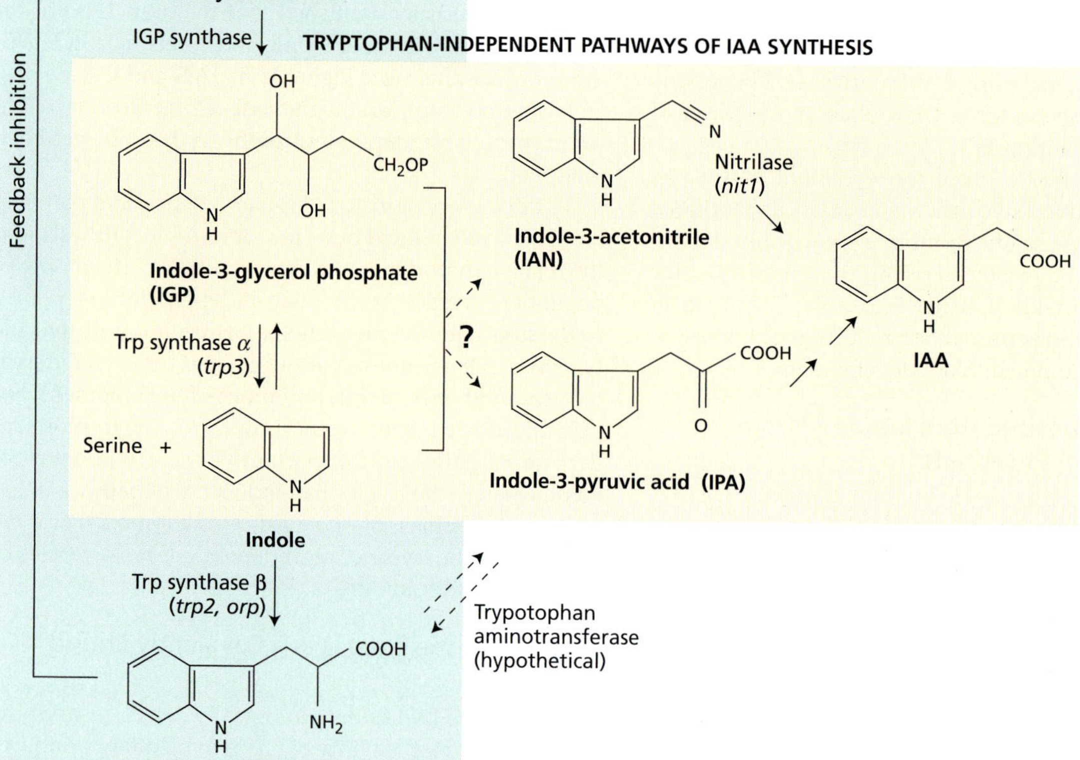

FIGURE 19.8 Tryptophan-independent pathways of IAA biosynthesis in plants. The tryptophan (Trp) biosynthetic pathway is shown on the left. Mutants discussed in **Web Topic 19.4** are indicated in parentheses. The branch-point precursor for tryptophan-independent biosynthesis is uncertain (indole-3-glycerol phosphate or indole), and IAN and IPA are two possible intermediates. PR, phosphoribosyl. (After Bartel 1997.)

sis. However, auxins are widely distributed in the plant. Metabolism of conjugated auxin may be a major contributing factor in the regulation of the levels of free auxin. For example, during the germination of seeds of *Zea mays*, IAA-*myo*-inositol is translocated from the endosperm to the coleoptile via the phloem. At least a portion of the free IAA produced in coleoptile tips of *Zea mays* is believed to be derived from the hydrolysis of IAA-*myo*-inositol.

In addition, environmental stimuli such as light and gravity have been shown to influence both the rate of auxin conjugation (removal of free auxin) and the rate of release of free auxin (hydrolysis of conjugated auxin). The formation of conjugated auxins may serve other functions as well, including storage and protection against oxidative degradation.

IAA Is Degraded by Multiple Pathways

Like IAA biosynthesis, the enzymatic breakdown (oxidation) of IAA may involve more than one pathway. For some time it has been thought that peroxidative enzymes are chiefly responsible for IAA oxidation, primarily because these enzymes are ubiquitous in higher plants and their ability to degrade IAA can be demonstrated in vitro (Figure 19.10A). However, the physiological significance of the peroxidase pathway is unclear. For example, no change in the IAA levels of transgenic plants was observed with either a tenfold increase in peroxidase expression or a tenfold repression of peroxidase activity (Normanly et al. 1995).

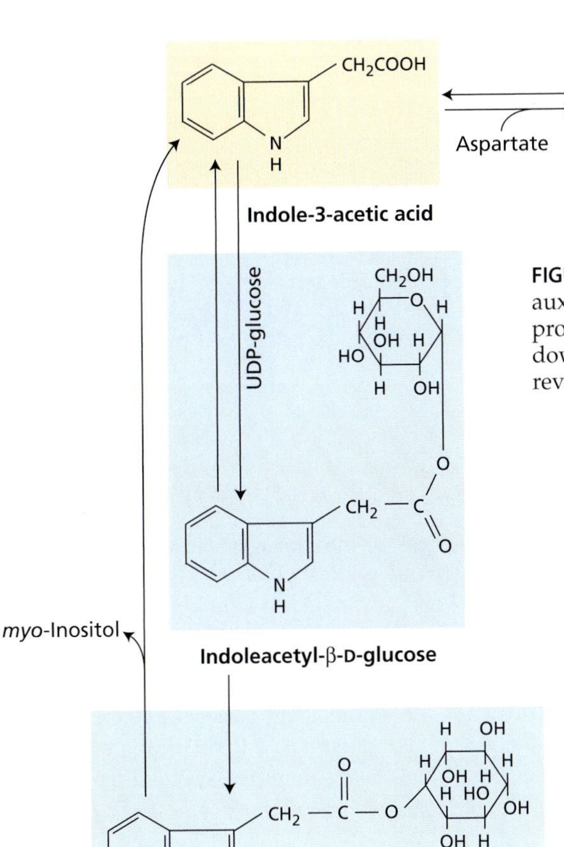

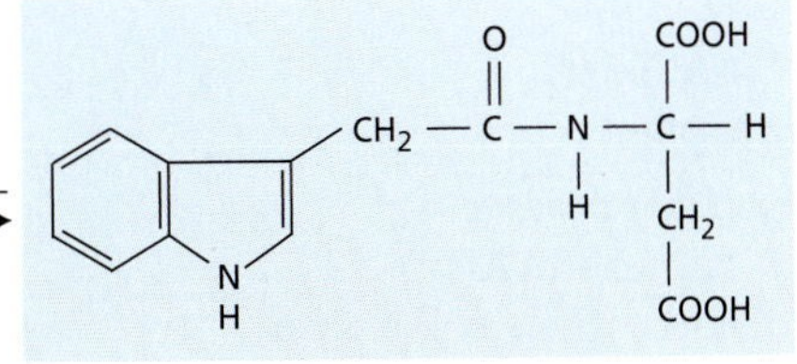

Indoleacetyl-2-*O*-*myo*-inositol

FIGURE 19.9 Structures and proposed metabolic pathways of bound auxins. The diagram shows structures of various IAA conjugates and proposed metabolic pathways involved in their synthesis and breakdown. Single arrows indicate irreversible pathways; double arrows, reversible.

On the basis of isotopic labeling and metabolite identification, two other oxidative pathways are more likely to be involved in the controlled degradation of IAA (see Figure 19.10B). The end product of this pathway is oxindole-3-acetic acid (OxIAA), a naturally occurring compound in the endosperm and shoot tissues of *Zea mays*. In one pathway, IAA is oxidized without decarboxylation to OxIAA. In another pathway, the IAA-aspartate conjugate is oxidized first to the intermediate dioxindole-3-acetylaspartate, and then to OxIAA.

In vitro, IAA can be oxidized nonenzymatically when exposed to high-intensity light, and its photodestruction in vitro can be promoted by plant pigments such as riboflavin. Although the products of auxin photooxidation have been isolated from plants, the role, if any, of the photooxidation pathway in vivo is presumed to be minor.

(A) **Decarboxylation: A minor pathway**

Indole-3-acetic acid — Peroxidase, CO_2 → 3-Methyleneoxindole

(B) **Nondecarboxylation pathways**

A; B Conjugation → Indole-3-acetylaspartate → Dioxindole-3-acetylaspartate → Oxindole-3-acetic acid (OxIAA)

FIGURE 19.10 Biodegradation of IAA. (A) The peroxidase route (decarboxylation pathway) plays a relatively minor role. (B) The two nondecarboxylation routes of IAA oxidative degradation, A and B, are the most common metabolic pathways.

Two Subcellular Pools of IAA Exist: The Cytosol and the Chloroplasts

The distribution of IAA in the cell appears to be regulated largely by pH. Because IAA^- does not cross membranes unaided, whereas IAAH readily diffuses across membranes, auxin tends to accumulate in the more alkaline compartments of the cell.

The distribution of IAA and its metabolites has been studied in tobacco cells. About one-third of the IAA is found in the chloroplast, and the remainder is located in the cytosol. IAA conjugates are located exclusively in the cytosol. IAA in the cytosol is metabolized either by conjugation or by nondecarboxylative catabolism (see Figure 19.10). The IAA in the chloroplast is protected from these processes, but it is regulated by the amount of IAA in the cytosol, with which it is in equilibrium (Sitbon et al. 1993).

The factors that regulate the steady-state concentration of free auxin in plant cells are diagrammatically summarized in **Web Topic 19.5**.

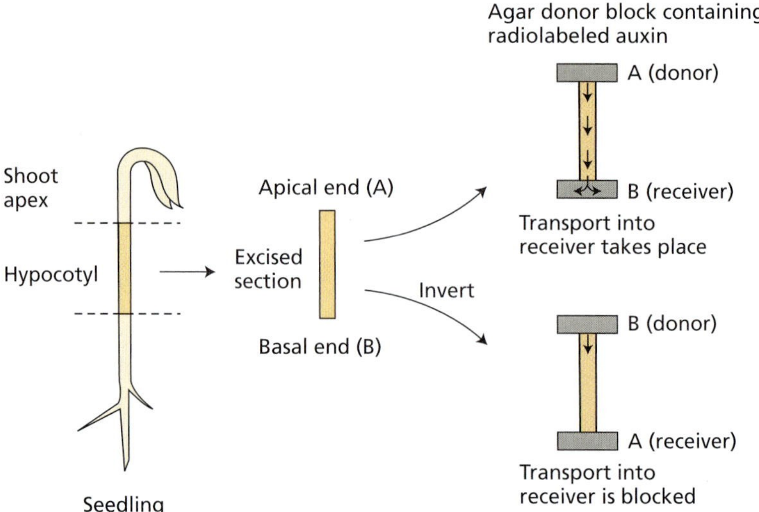

FIGURE 19.11 The standard method for measuring polar auxin transport. The polarity of transport is independent of orientation with respect to gravity.

AUXIN TRANSPORT

The main axes of shoots and roots, along with their branches, exhibit apex–base structural polarity, and this structural polarity has its origin in the polarity of auxin transport. Soon after Went developed the coleoptile curvature test for auxin, it was discovered that IAA moves mainly from the apical to the basal end (*basipetally*) in excised oat coleoptile sections. This type of unidirectional transport is termed **polar transport**. Auxin is the only plant growth hormone known to be transported polarly.

Because the shoot apex serves as the primary source of auxin for the entire plant, polar transport has long been believed to be the principal cause of an auxin gradient extending from the shoot tip to the root tip. The longitudinal gradient of auxin from the shoot to the root affects various developmental processes, including stem elongation, apical dominance, wound healing, and leaf senescence.

Recently it has been recognized that a significant amount of auxin transport also occurs in the phloem, and that the phloem is probably the principal route by which auxin is transported *acropetally* (i.e., toward the tip) in the root. Thus, more than one pathway is responsible for the distribution of auxin in the plant.

Polar Transport Requires Energy and Is Gravity Independent

To study polar transport, researchers have employed the *donor–receiver agar block method* (Figure 19.11): An agar block containing radioisotope-labeled auxin (donor block) is placed on one end of a tissue segment, and a receiver block is placed on the other end. The movement of auxin through the tissue into the receiver block can be determined over time by measurement of the radioactivity in the receiver block.

From a multitude of such studies, the general properties of polar IAA transport have emerged. Tissues differ in the degree of polarity of IAA transport. In coleoptiles, vegetative stems, and leaf petioles, basipetal transport predominates. Polar transport is not affected by the orientation of the tissue (at least over short periods of time), so it is independent of gravity.

A demonstration of the lack of effect of gravity on basipetal auxin transport is shown in Figure 19.12. When stem cuttings, in this case grape hardwood, are placed in a moist chamber, adventitious roots form at the basal ends of the cuttings, and shoots form at the apical ends, even when the cuttings are inverted. Roots form at the base because root differentiation is stimulated by auxin accumulation due to basipetal transport. Shoots tend to form at the apical end where the auxin concentration is lowest.

Polar transport proceeds in a cell-to-cell fashion, rather than via the symplast. That is, auxin exits the cell through the plasma membrane, diffuses across the compound middle lamella, and enters the cell below through its plasma membrane. The loss of auxin from cells is termed *auxin efflux*; the entry of auxin into cells is called *auxin uptake* or *influx*. The overall process requires metabolic energy, as evidenced by the sensitivity of polar transport to O_2 deprivation and metabolic inhibitors.

The velocity of polar auxin transport is 5 to 20 cm h^{-1}—faster than the rate of diffusion (see **Web Topic 3.2**), but slower than phloem translocation rates (see Chapter 10). Polar transport is also specific for active auxins, both nat-

FIGURE 19.12 Adventitious roots grow from the basal ends, and shoots grow from the apical ends, of grape hardwood cuttings, whether they are maintained in the inverted (the two cuttings on the left) or upright orientation (the cuttings on the right). The roots always form at the basal ends because polar auxin transport is independent of gravity. (From Hartmann and Kester 1983.)

ural and synthetic. Neither inactive auxin analogs nor auxin metabolites are transported polarly, suggesting that polar transport involves specific protein carriers on the plasma membrane that can recognize the hormone and its active analogs.

The major site of basipetal polar auxin transport in stems and leaves is the vascular parenchyma tissue. Coleoptiles appear to be the exception in that basipetal polar transport occurs mainly in the nonvascular tissues. Acropetal polar transport in the root is specifically associated with the xylem parenchyma of the stele (Palme and Gälweiler 1999). However, as we shall see later in the chapter, most of the auxin that reaches the root tip is translocated via the phloem.

A small amount of basipetal auxin transport from the root tip has also been demonstrated. In maize roots, for example, radiolabeled IAA applied to the root tip is transported basipetally about 2 to 8 mm (Young and Evans 1996). Basipetal auxin transport in the root occurs in the epidermal and cortical tissues, and as we shall see, it plays a central role in gravitropism.

A Chemiosmotic Model Has Been Proposed to Explain Polar Transport

The discovery of the chemiosmotic mechanism of solute transport in the late 1960s (see Chapter 6) led to the application of this model to polar auxin transport. According to the now generally accepted **chemiosmotic model** for polar auxin transport, auxin uptake is driven by the proton motive force ($\Delta E + \Delta pH$) across the plasma membrane, while auxin efflux is driven by the membrane potential, ΔE. (Proton motive force is described in more detail in **Web Topic 6.3** and Chapter 7.)

A crucial feature of the polar transport model is that the auxin efflux carriers are localized at the basal ends of the conducting cells (Figure 19.13). The evidence for each step in this model is considered separately in the discussion that follows.

Auxin influx. The first step in polar transport is auxin influx. According to the model, auxin can enter plant cells from any direction by either of two mechanisms:

1. Passive diffusion of the protonated (IAAH) form across the phospholipid bilayer
2. Secondary active transport of the dissociated (IAA^-) form via a $2H^+$–IAA^- symporter

The dual pathway of auxin uptake arises because the passive permeability of the membrane to auxin depends strongly on the apoplastic pH.

The undissociated form of indole-3-acetic acid, in which the carboxyl group is protonated, is lipophilic and readily diffuses across lipid bilayer membranes. In contrast, the dissociated form of auxin is negatively charged and therefore does not cross membranes unaided. Because the plasma membrane H^+-ATPase normally maintains the cell wall solution at about pH 5, about half of the auxin ($pK_a = 4.75$) in the apoplast will be in the undissociated form and will diffuse passively across the plasma membrane down a concentration gradient. Experimental support for pH-dependent, passive auxin uptake was first provided by the demonstration that IAA uptake by plant cells increases as the extracellular pH is lowered from a neutral to a more acidic value.

A carrier-mediated, secondary active uptake mechanism was shown to be saturable and specific for active auxins (Lomax 1986). In experiments in which the ΔpH and ΔE values of isolated membrane vesicles from zucchini (*Cucurbita pepo*) hypocotyls were manipulated artificially, the uptake of radiolabeled auxin was shown to be stimulated in the presence of a pH gradient, as in passive uptake, but also when the inside of the vesicle was negatively charged relative to the outside.

These and other experiments suggested that an H^+–IAA^- symporter cotransports two protons along with the auxin anion. This secondary active transport of auxin allows for greater auxin accumulation than simple diffusion does because it is driven across the membrane by the proton motive force.

A permease-type auxin uptake carrier, AUX1, related to bacterial amino acid carriers, has been identified in *Arabidopsis* roots (Bennett et al. 1996). The roots of *aux1* mutants are agravitropic, suggesting that auxin influx is a limiting factor for gravitropism in roots. As predicted by the chemiosmotic model, AUX1 appears to be uniformly distributed around cells in the polar transport pathway (Marchant et al. 1999). Thus in general, the polarity of auxin transport is governed by the efflux step rather than the influx step.

Plasma membrane

Permease H+-cotransport

Apex

Base

Cell wall

Cytosol

Vacuole

1. IAA enters the cell either passively in the undissociated form (IAAH) or by secondary active cotransport in the anionic form (IAA^-).

2. The cell wall is maintained at an acidic pH by the activity of the plasma membrane H^+-ATPase.

3. In the cytosol, which has a neutral pH, the anionic form (IAA^-) predominates.

4. The anions exit the cell via auxin anion efflux carriers that are concentrated at the basal ends of each cell in the longitudinal pathway.

FIGURE 19.13 The chemiosmotic model for polar auxin transport. Shown here is one cell in a column of auxin-transporting cells. (From Jacobs and Gilbert 1983.)

Auxin efflux. Once IAA enters the cytosol, which has a pH of approximately 7.2, nearly all of it will dissociate to the anionic form. Because the membrane is less permeable to IAA^- than to IAAH, IAA^- will tend to accumulate in the cytosol. However, much of the auxin that enters the cell escapes via an *auxin anion efflux carrier*. According to the chemiosmotic model, transport of IAA^- out of the cell is driven by the inside negative membrane potential.

As noted earlier, the central feature of the chemiosmotic model for polar transport is that IAA^- efflux takes place preferentially at the basal end of each cell. The repetition of auxin uptake at the apical end of the cell and preferential release from the base of each cell in the pathway gives rise to the total polar transport effect. A family of putative auxin efflux carriers known as **PIN proteins** (named after the pin-shaped inflorescences formed by the *pin1* mutant of *Arabidopsis*; Figure 19.14A) are localized precisely as the model would predict—that is, at the basal ends of the conducting cells (see Figure 19.14B).

(A)

(B)

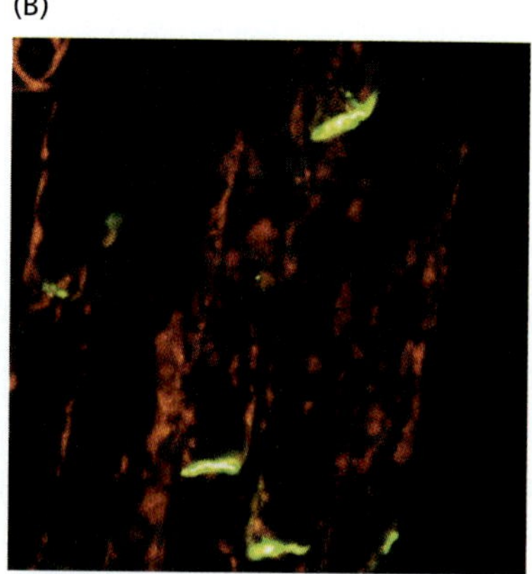

FIGURE 19.14 The *pin1* mutant of *Arabidopsis* (A) and localization of the PIN1 protein at the basal ends of conducting cells by immunofluorescence microscopy (B). (Courtesy of L. Gälweiler and K. Palme.)

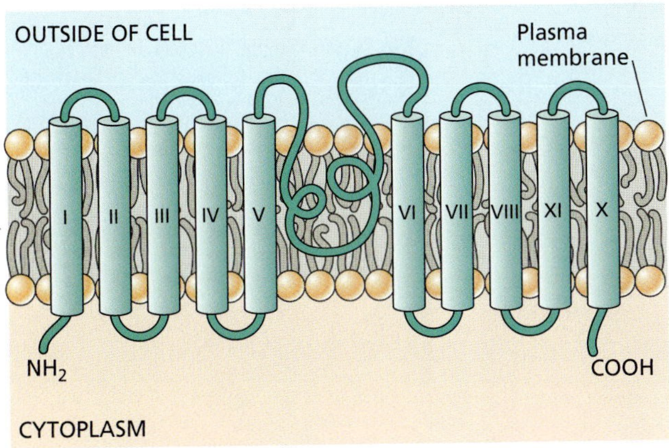

FIGURE 19.15 The topology of the PIN1 protein with ten transmembrane segments and a large hydrophilic loop in the middle. (After Palme and Gälweiler 1999.)

PIN proteins have 10 to 12 transmembrane regions characteristic of a major superfamily of bacterial and eukaryotic transporters, which include drug resistance proteins and sugar transporters (Figure 19.15). Despite topological similarities to other transporters, recent studies suggest that PIN may require other proteins for activity, and may be part of a larger protein complex.

Inhibitors of Auxin Transport Block Auxin Efflux

Several compounds have been synthesized that can act as **auxin transport inhibitors** (**ATIs**), including NPA (1-*N*-naphthylphthalamic acid) and TIBA (2,3,5-triiodobenzoic acid) (Figure 19.16). These inhibitors block polar transport by preventing auxin efflux. We can demonstrate this phenomenon by incorporating NPA or TIBA into either the donor or the receiver block in an auxin transport experiment. Both compounds inhibit auxin efflux into the receiver block, but they do not affect auxin uptake from the donor block.

Some ATIs, such as TIBA, that have weak auxin activity and are transported polarly, may inhibit polar transport in part by competing with auxin for its binding site on the efflux carrier. Others, such as NPA, are not transported polarly and are believed to interfere with auxin transport by binding to proteins associated in a complex with the efflux carrier. Such NPA-binding proteins are also found at the basal ends of the conducting cells, consistent with the localization of PIN proteins (Jacobs and Gilbert 1983).

Recently another class of ATIs has been identified that inhibits the AUX1 uptake carrier (Parry et al. 2001). For example, 1-naphthoxyacetic acid (1-NOA) (see Figure 19.16) blocks auxin uptake into cells, and when applied to *Arabidopsis* plants it causes root agravitropism similar to that of the *aux1* mutant. Like the *aux1* mutation, neither 1-NOA nor any of the other AUX1-specific inhibitors block polar auxin transport.

PIN Proteins Are Rapidly Cycled to and from the Plasma Membrane

The basal localization of the auxin efflux carriers involves targeted vesicle secretion to the basal ends of the conducting cells. Recently it has been demonstrated that PIN proteins, although stable, do not remain on the plasma membrane permanently, but are rapidly cycled to an unidentified endosomal compartment via endocytotic vesicles, and then recycled back to the plasma membrane (Geldner et al. 2001).

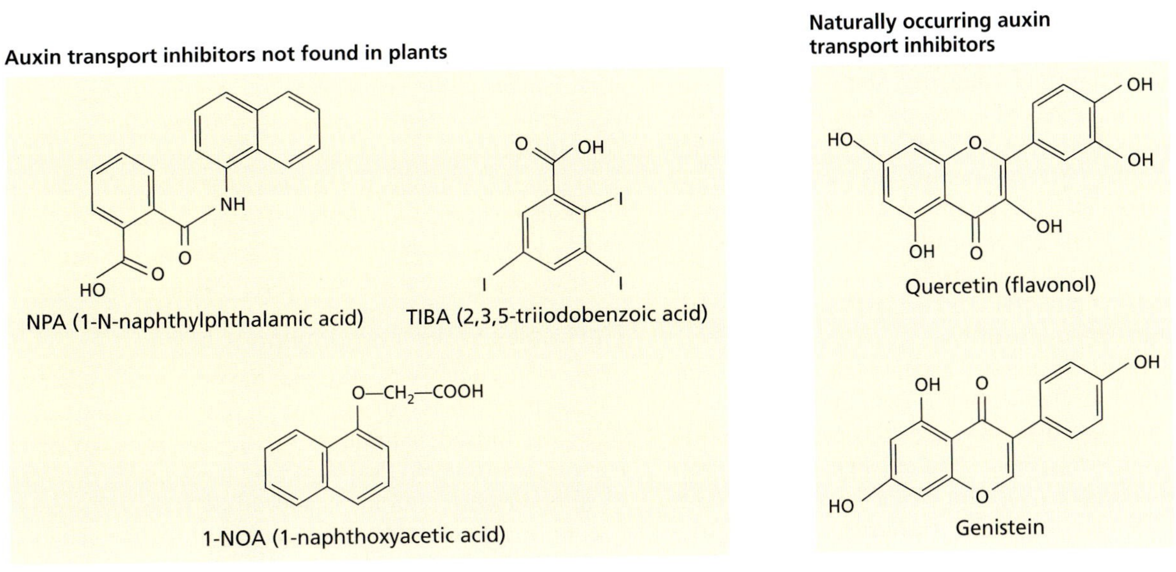

Auxin transport inhibitors not found in plants

NPA (1-N-naphthylphthalamic acid)

TIBA (2,3,5-triiodobenzoic acid)

1-NOA (1-naphthoxyacetic acid)

Naturally occurring auxin transport inhibitors

Quercetin (flavonol)

Genistein

FIGURE 19.16 Structures of auxin transport inhibitors.

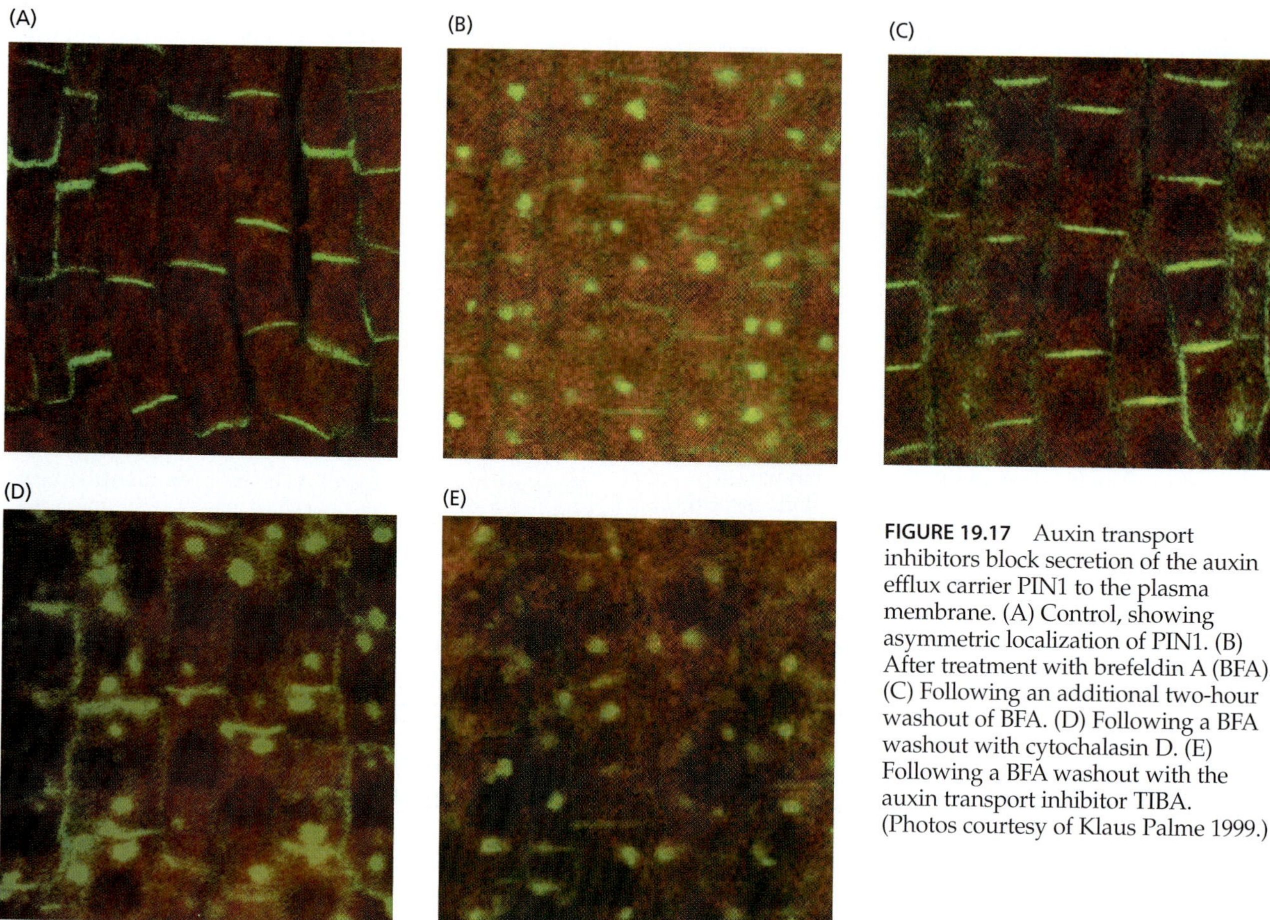

FIGURE 19.17 Auxin transport inhibitors block secretion of the auxin efflux carrier PIN1 to the plasma membrane. (A) Control, showing asymmetric localization of PIN1. (B) After treatment with brefeldin A (BFA). (C) Following an additional two-hour washout of BFA. (D) Following a BFA washout with cytochalasin D. (E) Following a BFA washout with the auxin transport inhibitor TIBA. (Photos courtesy of Klaus Palme 1999.)

Prior to treatment, the PIN1 protein is localized at the basal ends (top) of root cortical parenchyma cells (Figure 19.17A). Treatment of *Arabidopsis* seedlings with brefeldin A (BFA), which causes Golgi vesicles and other endosomal compartments to aggregate near the nucleus, causes PIN to accumulate in these abnormal intracellular compartments (see Figure 19.17B). When the BFA is washed out with buffer, the normal localization on the plasma membrane at the base of the cell is restored (see Figure 19.17C). But when cytochalasin D, an inhibitor of actin polymerization, is included in the buffer washout solution, normal relocalization of PIN to the plasma membrane is prevented (see Figure 19.17D). These results indicate that PIN is rapidly cycled between the plasma membrane at the base of the cell and an unidentified endosomal compartment by an actin-dependent mechanism.

Although they bind different targets, both TIBA and NPA interfere with vesicle traffic to and from the plasma membrane. The best way to demonstrate this phenomenon is to include TIBA in the washout solution after BFA treatment. Under these conditions, TIBA prevents the normal relocalization of PIN on the plasma membrane following the washout treatment (see Figure 19.17E) (Geldner et al. 2001).

The effects of TIBA and NPA on cycling are not specific for PIN proteins, and it has been proposed that ATIs may actually represent general inhibitors of membrane cycling (Geldner et al. 2001). On the other hand, neither TIBA nor NPA alone causes PIN delocalization, even though they block auxin efflux. Therefore, TIBA and NPA must also be able to directly inhibit the transport activity of PIN complexes on the plasma membrane—by binding either to PIN (as TIBA does) or to one or more regulatory proteins (as NPA does).

A simplified model of the effects of TIBA and NPA on PIN cycling and auxin efflux is shown in Figure 19.18. A more complete model that incorporates many of the recent findings is presented in **Web Essay 19.2**.

Flavonoids Serve as Endogenous ATIs

There is mounting evidence that flavonoids (see Chapter 13) can function as endogenous regulators of polar auxin transport. Indeed, naturally occurring aglycone flavonoid compounds (flavonoids without attached sugars) are able to compete with NPA for its binding site on membranes (Jacobs and Rubery 1988) and are typically localized on the plasma membrane at the basal ends of cells where the

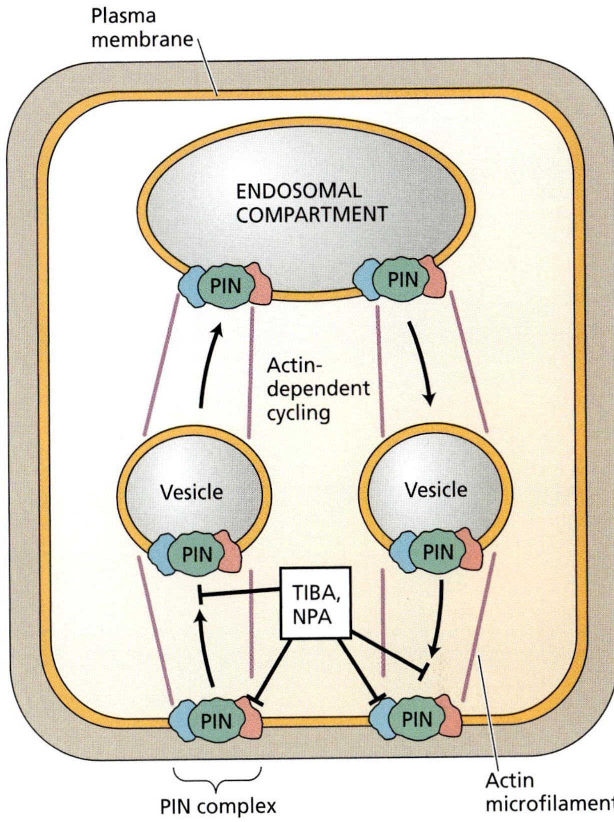

FIGURE 19.18 Actin-dependent PIN cycling between the plasma membrane and an endosomal compartment. Auxin transport inhibitors TIBA and NPA both interfere with relocalization of PIN1 proteins to basal plasma membranes after BFA washout (see Figure 19.17). This suggests that both of these auxin transport inhibitors interfere with PIN1 cycling.

efflux carrier is concentrated (Peer et al. 2001). In addition, recent studies have shown that the cells of flavonoid-deficient *Arabidopsis* mutants are less able to accumulate auxin than wild-type cells, and the mutant seedlings that lack flavonoid have altered auxin distribution profiles (Murphy et al. 1999; Brown et al. 2001).

Many of the flavonoids that displace NPA from its binding site on membranes are also inhibitors of protein kinases and protein phosphatases (Bernasconi 1996). An *Arabidopsis* mutant designated *rcn1* (*r*oots *c*url in *N*PA *1*) was identified on the basis of an enhanced sensitivity to NPA. The *RCN1* gene is closely related to the regulatory subunit of protein phosphatase 2A, a serine/threonine phosphatase (Garbers et al. 1996).

Protein phosphatases are known to play important roles in enzyme regulation, gene expression, and signal transduction by removing regulatory phosphate groups from proteins (see Chapter 14 on the web site). This finding suggests that a signal transduction pathway involving protein kinases and protein phosphatases may be involved in signaling between NPA-binding proteins and the auxin efflux carrier.

Auxin Is Also Transported Nonpolarly in the Phloem

Most of the IAA that is synthesized in mature leaves appears to be transported to the rest of the plant nonpolarly via the phloem. Auxin, along with other components of phloem sap, can move from these leaves up or down the plant at velocities much higher than those of polar transport (see Chapter 10). Auxin translocation in the phloem is largely passive, not requiring energy directly.

Although the overall importance of the phloem pathway versus the polar transport system for the long-distance movement of IAA in plants is still unresolved, the evidence suggests that long-distance auxin transport in the phloem is important for controlling such processes as cambial cell divisions, callose accumulation or removal from sieve tube elements, and branch root formation. Indeed, the phloem appears to represent the principal pathway for long-distance auxin translocation to the root (Aloni 1995; Swarup et al. 2001).

Polar transport and phloem transport are not independent of each other. Recent studies with radiolabeled IAA suggest that in pea, auxin can be transferred from the nonpolar phloem pathway to the polar transport pathway. This transfer takes place mainly in the immature tissues of the shoot apex.

A second example of transfer of auxin from the nonpolar phloem pathway to a polar transport system has recently been documented in *Arabidopsis*. It was shown that the AUX1 permease is asymmetrically localized on the plasma membrane at the upper end of root protophloem cells (i.e., the end distal from the tip) (Figure 19.19).

It has been proposed that the asymmetrically oriented AUX1 permease promotes the acropetal movement of auxin from the phloem to the root apex (Swarup et al. 2001). This type of polar auxin transport based on the asymmetric localization of AUX1 differs from the polar transport that occurs in the shoot and basal region of the root, which is based on the asymmetric distribution of the PIN complex.

Note in Figure 19.19B that AUX1 is also strongly expressed in a cluster of cells in the columella of the root cap, as well as in lateral root cap cells that overlay the cells of the distal elongation zone of the root. These cells form a minor, but physiologically important, basipetal pathway whereby auxin reaching the columella is redirected backward toward the outer tissues of the elongation zone. The importance of this pathway will become apparent when we examine the mechanism of root gravitropism.

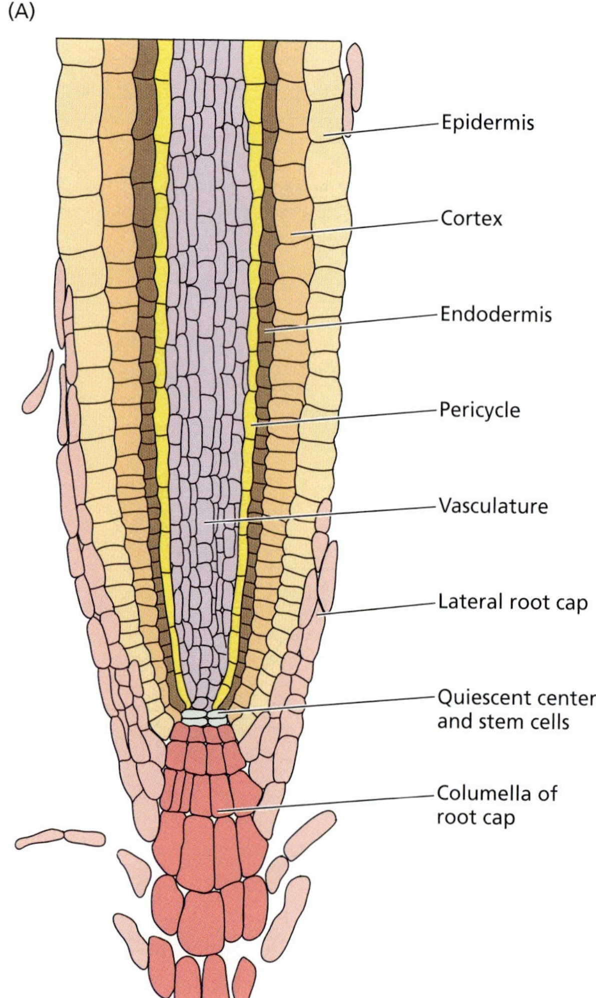

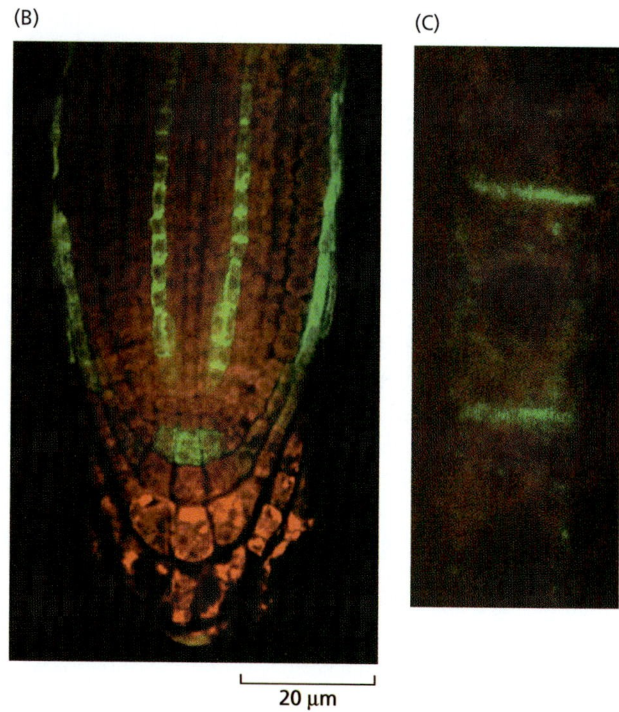

FIGURE 19.19 The auxin permease AUX1 is specifically expressed in a subset of columella, lateral root cap, and stellar tissues. (A) Diagram of tissues in the *Arabidopsis* root tip. (B) Immunolocalization of AUX1 in protophloem cells of the stele, a central cluster of cells in the columella, and lateral root cap cells. (C) Asymmetric localization of AUX1 in a file of protophloem cells. Scale bar is 2 μm in C. (From Swarup et al. 2001.)

PHYSIOLOGICAL EFFECTS OF AUXIN: CELL ELONGATION

Auxin was discovered as the hormone involved in the bending of coleoptiles toward light. The coleoptile bends because of the unequal rates of cell elongation on its shaded versus its illuminated side (see Figure 19.1). The ability of auxin to regulate the rate of cell elongation has long fascinated plant scientists. In this section we will review the physiology of auxin-induced cell elongation, some aspects of which were discussed in Chapter 15.

Auxins Promote Growth in Stems and Coleoptiles, While Inhibiting Growth in Roots

As we have seen, auxin is synthesized in the shoot apex and transported basipetally to the tissues below. The steady supply of auxin arriving at the subapical region of the stem or coleoptile is required for the continued elongation of these cells. Because the level of endogenous auxin in the elongation region of a normal healthy plant is nearly optimal for growth, spraying the plant with exogenous auxin causes only a modest and short-lived stimulation in growth, and may even be inhibitory in the case of dark-grown seedlings, which are more sensitive to supraoptimal auxin concentrations than light-grown plants are.

However, when the endogenous source of auxin is removed by excision of sections containing the elongation zones, the growth rate rapidly decreases to a low basal rate. Such excised sections will often respond dramatically to exogenous auxin by rapidly increasing their growth rate back to the level in the intact plant.

In long-term experiments, treatment of excised sections of coleoptiles (see Figure 19.2) or dicot stems with auxin stimulates the rate of elongation of the section for up to 20 hours (Figure 19.20). The optimal auxin concentration for elongation growth is typically 10^{-6} to 10^{-5} *M* (Figure 19.21).

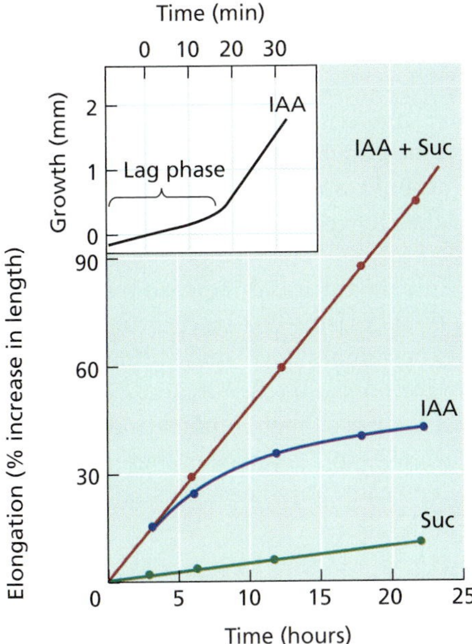

FIGURE 19.20 Time course for auxin-induced growth of *Avena* (oat) coleoptile sections. Growth is plotted as the percent increase in length. Auxin was added at time zero. When sucrose (Suc) is included in the medium, the response can continue for as long as 20 hours. Sucrose prolongs the growth response to auxin mainly by providing osmotically active solute that can be taken up for the maintenance of turgor pressure during cell elongation. KCl can substitute for sucrose. The inset shows a short-term time course plotted with an electronic position-sensing transducer. In this graph, growth is plotted as the absolute length in millimeters versus time. The curve shows a lag time of about 15 minutes for auxin-stimulated growth to begin. (From Cleland 1995.)

The inhibition beyond the optimal concentration is generally attributed to auxin-induced ethylene biosynthesis. As we will see in Chapter 22, the gaseous hormone ethylene inhibits stem elongation in many species.

Auxin control of root elongation growth has been more difficult to demonstrate, perhaps because auxin induces the production of ethylene, a root growth inhibitor. However, even if ethylene biosynthesis is specifically blocked, low concentrations (10^{-10} to 10^{-9} *M*) of auxin promote the growth of intact roots, whereas higher concentrations (10^{-6} *M*) inhibit growth. Thus, roots may require a minimum concentration of auxin to grow, but root growth is strongly inhibited by auxin concentrations that promote elongation in stems and coleoptiles.

The Outer Tissues of Dicot Stems Are the Targets of Auxin Action

Dicot stems are composed of many types of tissues and cells, only some of which may limit the growth rate. This point is illustrated by a simple experiment. When stem sections from growing regions of an etiolated dicot stem, such as pea, are split lengthwise and incubated in buffer, the two halves bend outward.

This result indicates that, in the absence of auxin the central tissues, including the pith, vascular tissues, and inner cortex, elongate at a faster rate than the outer tissues, consisting of the outer cortex and epidermis. Thus the outer tissues must be limiting the extension rate of the stem in the absence of auxin. However, when the split sections are incubated in buffer plus auxin, the two halves now curve inward, demonstrating that the outer tissues of dicot stems are the primary targets of auxin action during cell elongation.

The observation that the outer cell layers are the targets of auxin seems to conflict with the localization of polar transport in the parenchyma cells of the vascular bundles. However, auxin can move laterally from the vascular tissues of dicot stems to the outer tissues of the elongation zone. In coleoptiles, on the other hand, all of the nonvascular tissues (epidermis plus mesophyll) are capable of transporting auxin, as well as responding to it.

The Minimum Lag Time for Auxin-Induced Growth Is Ten Minutes

When a stem or coleoptile section is excised and inserted into a sensitive growth-measuring device, the growth response to auxin can be monitored at very high resolution. Without auxin in the medium, the growth rate declines rapidly. Addition of auxin markedly stimulates the growth rate after a lag period of only 10 to 12 minutes (see the inset in Figure 19.20).

Both *Avena* (oat) coleoptiles and *Glycine max* (soybean) hypocotyls (dicot stem) reach a maximum growth rate after

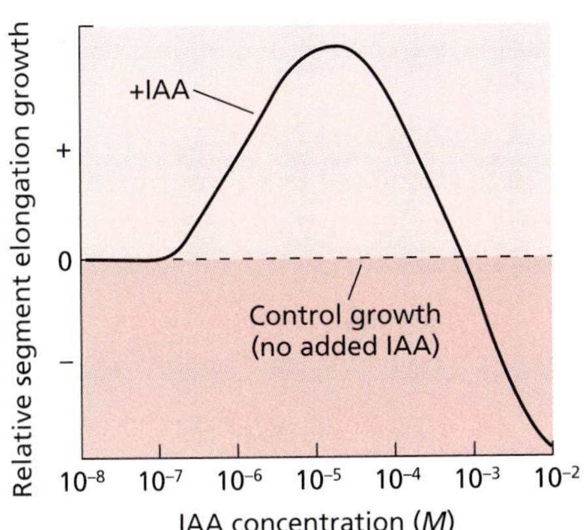

FIGURE 19.21 Typical dose–response curve for IAA-induced growth in pea stem or oat coleoptile sections. Elongation growth of excised sections of coleoptiles or young stems is plotted versus increasing concentrations of exogenous IAA. At higher concentrations (above 10^{-5} M), IAA becomes less and less effective; above about 10^{-4} M it becomes inhibitory, as shown by the fact that the curve falls below the dashed line, which represents growth in the absence of added IAA.

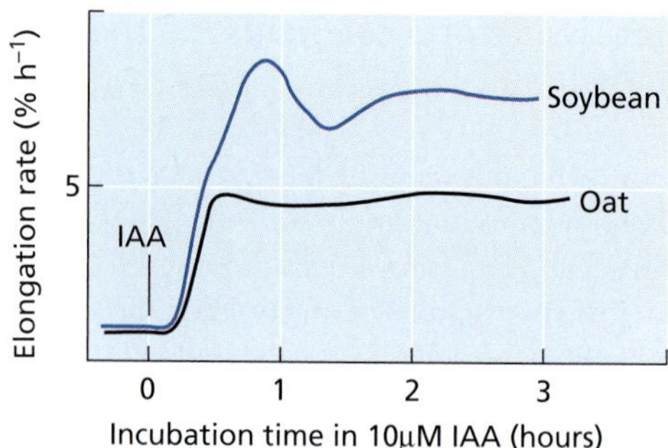

FIGURE 19.22 Comparison of the growth kinetics of oat coleoptile and soybean hypocotyl sections, incubated with 10 μM IAA and 2% sucrose. Growth is plotted as the rate at each time point, rather than the rate of the absolute length. The growth rate of the soybean hypocotyl oscillates after 1 hour, whereas that of the oat coleoptile is constant. (After Cleland 1995.)

30 to 60 minutes of auxin treatment (Figure 19.22). This maximum represents a five- to tenfold increase over the basal rate. Oat coleoptile sections can maintain this maximum rate for up to 18 hours in the presence of osmotically active solutes such as sucrose or KCl.

As might be expected, the stimulation of growth by auxin requires energy, and metabolic inhibitors inhibit the response within minutes. Auxin-induced growth is also sensitive to inhibitors of protein synthesis such as cycloheximide, suggesting that proteins with high turnover rates are involved. Inhibitors of RNA synthesis also inhibit auxin-induced growth, after a slightly longer delay (Cleland 1995).

Although the length of the lag time for auxin-stimulated growth can be increased by lowering of the temperature or by the use of suboptimal auxin concentrations, the lag time cannot be shortened by raising of the temperature, by the use of supraoptimal auxin concentrations, or by abrasion of the waxy cuticle to allow auxin to penetrate the tissue more rapidly. Thus the minimum lag time of 10 minutes is not determined by the time required for auxin to reach its site of action. Rather, the lag time reflects the time needed for the biochemical machinery of the cell to bring about the increase in the growth rate.

Auxin Rapidly Increases the Extensibility of the Cell Wall

How does auxin cause a five- to tenfold increase in the growth rate in only 10 minutes? To understand the mechanism, we must first review the process of cell enlargement in plants (see Chapter 15). Plant cells expand in three steps:

1. Osmotic uptake of water across the plasma membrane is driven by the gradient in water potential ($\Delta\Psi_w$).
2. Turgor pressure builds up because of the rigidity of the cell wall.
3. Biochemical wall loosening occurs, allowing the cell to expand in response to turgor pressure.

The effects of these parameters on the growth rate are encapsulated in the growth rate equation:

$$GR = m\,(\Psi_p - Y)$$

where GR is the growth rate, Ψ_p is the turgor pressure, Y is the yield threshold, and m is the coefficient (*wall extensibility*) that relates the growth rate to the difference between Ψ_p and Y.

In principle, auxin could increase the growth rate by increasing m, increasing Ψ_p, or decreasing Y. Although extensive experiments have shown that auxin does not increase turgor pressure when it stimulates growth, conflicting results have been obtained regarding auxin-induced decreases in Y. However, there is general agreement that auxin causes an increase in the wall extensibility parameter, m.

Auxin-Induced Proton Extrusion Acidifies the Cell Wall and Increases Cell Extension

According to the widely accepted **acid growth hypothesis**, hydrogen ions act as the intermediate between auxin and cell wall loosening. The source of the hydrogen ions is the plasma membrane H^+-ATPase, whose activity is thought to increase in response to auxin. The acid growth hypothesis allows five main predictions:

1. Acid buffers alone should promote short-term growth, provided the cuticle has been abraded to allow the protons access to the cell wall.
2. Auxin should increase the rate of proton extrusion (wall acidification), and the kinetics of proton extrusion should closely match those of auxin-induced growth.
3. Neutral buffers should inhibit auxin-induced growth.
4. Compounds (other than auxin) that promote proton extrusion should stimulate growth.
5. Cell walls should contain a "wall loosening factor" with an acidic pH optimum.

All five of these predictions have been confirmed. Acidic buffers cause a rapid and immediate increase in the growth rate, provided the cuticle has been abraded. Auxin stimulates proton extrusion into the cell wall after 10 to 15 minutes of lag time, consistent with the growth kinetics (Figure 19.23).

Auxin-induced growth has also been shown to be inhibited by neutral buffers, as long as the cuticle has been abraided. **Fusicoccin**, a fungal phytotoxin, stimulates both rapid proton extrusion and transient growth in stem and coleoptile sections (see Web Topic 19.6). And finally, wall-loosening proteins called **expansins** have been identified in the cell walls of a wide range of plant species (see Chapter 15). At acidic pH values, expansins loosen cell walls by weakening the hydrogen bonds between the polysaccharide components of the wall.

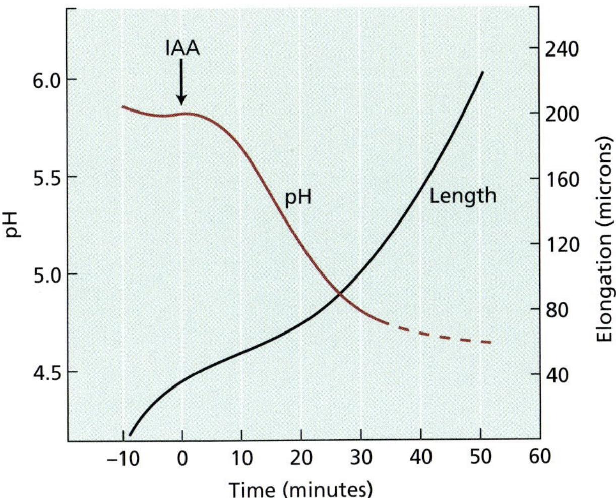

FIGURE 19.23 Kinetics of auxin-induced elongation and cell wall acidification in maize coleoptiles. The pH of the cell wall was measured with a pH microelectrode. Note the similar lag times (10 to 15 minutes) for both cell wall acidification and the increase in the rate of elongation. (From Jacobs and Ray 1976.)

Auxin-Induced Proton Extrusion May Involve Both Activation and Synthesis

In theory, auxin could increase the rate of proton extrusion by two possible mechanisms:

1. Activation of preexisting plasma membrane H^+-ATPases
2. Synthesis of new H^+-ATPases on the plasma membrane

H^+-ATPase activation. When auxin was added directly to isolated plasma membrane vesicles from tobacco cells, a small stimulation (about 20%) of the ATP-driven proton-pumping activity was observed, suggesting that auxin directly activates the H^+-ATPase. A greater stimulation (about 40%) was observed if the living cells were treated with IAA just before the membranes were isolated, suggesting that a cellular factor is also required (Peltier and Rossignol 1996).

Although an auxin receptor has not yet been unequivocally identified (as discussed later in the chapter), various auxin-binding proteins (ABPs) have been isolated and appear to be able to activate the plasma membrane H^+-ATPase in the presence of auxin (Steffens et al. 2001).

Recently an ABP from rice, ABP_{57}, was shown to bind directly to plasma membrane H^+-ATPases and stimulate proton extrusion—but only in the presence of IAA (Kim et al. 2001). When IAA is absent, the activity of the H^+-ATPase is repressed by the C-terminal domain of the enzyme, which can block the catalytic site. ABP_{57} (with bound IAA) interacts with the H^+-ATPase, activating the enzyme. A second auxin-binding site interferes with the action of the first, possibly explaining the bell-shaped curve of auxin action. This hypothetical model for the action of ABP_{57} is shown in Figure 19.24.

H^+-ATPase synthesis. The ability of protein synthesis inhibitors, such as cycloheximide, to rapidly inhibit auxin-induced proton extrusion and growth suggests that auxin might also stimulate proton pumping by increasing the synthesis of the H^+-ATPase. An increase in the amount of plasma membrane ATPase in corn coleoptiles was detected immunologically after only 5 minutes of auxin treatment, and a doubling of the H^+-ATPase was observed after 40 minutes of treatment. A threefold stimulation by auxin of an mRNA for the H^+-ATPase was demonstrated specifically in the nonvascular tissues of the coleoptiles.

In summary, the question of activation versus synthesis is still unresolved, and it is possible that auxin stimulates proton extrusion by both activation and stimulation of synthesis of the H^+-ATPase. Figure 19.25 summarizes

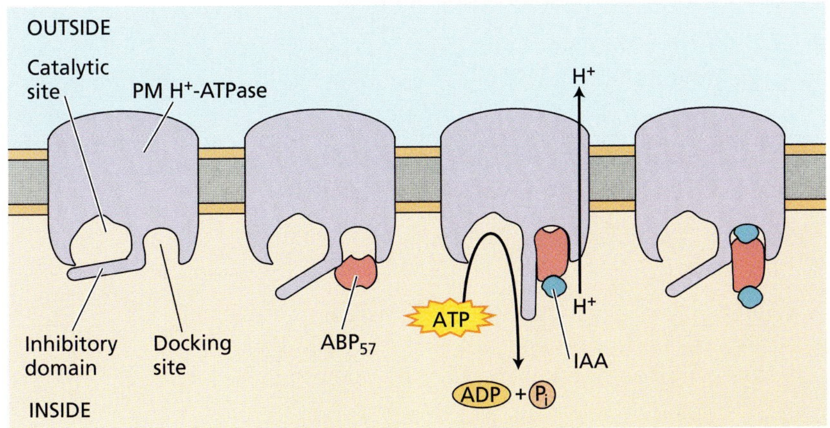

FIGURE 19.24 Model for the activation of the plasma membrane (PM) H^+-ATPase by ABP_{57} and auxin.

Activation hypothesis: Auxin binds to an auxin-binding protein (ABP1) located either on the cell surface or in the cytosol. ABP1-IAA then interacts directly with plasma membrane H^+-ATPase to stimulate proton pumping (step 1). Second messengers, such as calcium or intracellular pH, could also be involved.

Synthesis hypothesis: IAA-induced second messengers activate the expression of genes (step 2) that encode the plasma membrane H^+-ATPase (step 3). The protein is synthesized on the rough endoplasmic reticulum (step 4) and targeted via the secretory pathway to the plasma membrane (steps 5 and 6). The increase in proton extrusion results from an increase in the number of proton pumps on the membrane.

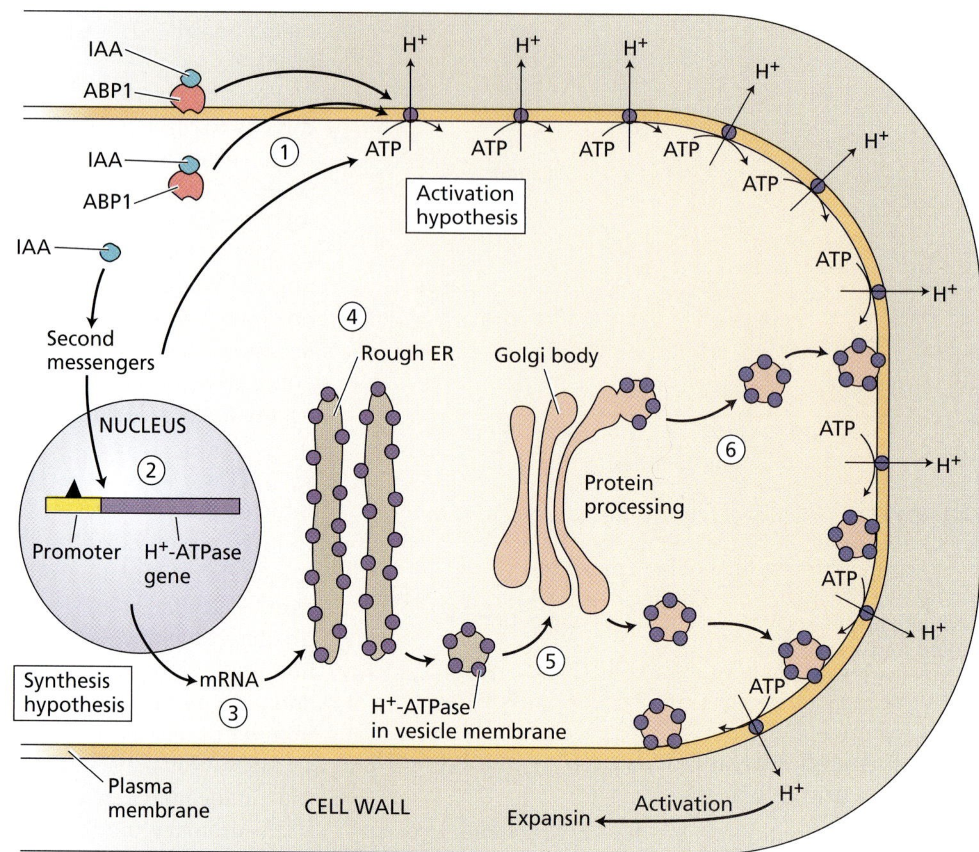

FIGURE 19.25 Current models for IAA-induced H^+ extrusion. In many plants, both of these mechanisms may operate. Regardless of how H^+ pumping is increased, acid-induced wall loosening is thought to be mediated by expansins.

the proposed mechanisms of auxin-induced cell wall loosening via proton extrusion.

PHYSIOLOGICAL EFFECTS OF AUXIN: PHOTOTROPISM AND GRAVITROPISM

Three main guidance systems control the orientation of plant growth:

1. **Phototropism**, or growth with respect to light, is expressed in all shoots and some roots; it ensures that leaves will receive optimal sunlight for photosynthesis.
2. **Gravitropism**, growth in response to gravity, enables roots to grow downward into the soil and shoots to grow upward away from the soil, which is especially critical during the early stages of germination.
3. **Thigmotropism**, or growth with respect to touch, enables roots to grow around rocks and is responsible for the ability of the shoots of climbing plants to wrap around other structures for support.

In this section we will examine the evidence that bending in response to light or gravity results from the lateral redistribution of auxin. We will also consider the cellular mechanisms involved in generating lateral auxin gradients during bending growth. Less is known about the mechanism of thigmotropism, although it, too, probably involves auxin gradients.

Phototropism Is Mediated by the Lateral Redistribution of Auxin

As we saw earlier, Charles and Francis Darwin provided the first clue concerning the mechanism of phototropism by demonstrating that the sites of perception and differential growth (bending) are separate: Light is perceived at the tip, but bending occurs below the tip. The Darwins proposed that some "influence" that was transported from the tip to the growing region brought about the observed asymmetric growth response. This influence was later shown to be indole-3-acetic acid—auxin.

When a shoot is growing vertically, auxin is transported polarly from the growing tip to the elongation zone. The polarity of auxin transport from tip to base is developmentally determined and is independent of orientation with respect to gravity. However, auxin can also be transported laterally, and this lateral movement of auxin lies at the heart of a model for tropisms originally proposed separately by the Russian plant physiologist, Nicolai Cholodny and Frits Went from the Netherlands in the 1920s.

According to the Cholodny–Went model of phototropism, the tips of grass coleoptiles have three specialized functions:

1. The production of auxin
2. The perception of a unilateral light stimulus
3. The lateral transport of IAA in response to the phototropic stimulus

Thus, in response to a directional light stimulus, the auxin produced at the tip, instead of being transported basipetally, is transported laterally toward the shaded side.

The precise sites of auxin production, light perception, and lateral transport have been difficult to define. In maize coleoptiles, auxin is produced in the upper 1 to 2 mm of the tip. The zones of photosensing and lateral transport extend farther, within the upper 5 mm of the tip. The response is also strongly dependent on the light fluence (see **Web Topic 19.7**).

Two flavoproteins, *phototropins* 1 and 2, are the photoreceptors for the blue-light signaling pathway (see **Web Essay 19.3**) that induces phototropic bending in *Arabidopsis* hypocotyls and oat coleoptiles under both high- and low-fluence conditions (Briggs et al. 2001).

Phototropins are autophosphorylating protein kinases whose activity is stimulated by blue light. The action spectrum for **blue-light** activation of the kinase activity closely matches the action spectrum for phototropism, including the multiple peaks in the blue region. Phototropin 1 displays a lateral gradient in phosphorylation during exposure to low-fluence unilateral blue light.

According to the current hypothesis, the gradient in phototropin phosphorylation induces the movement of auxin to the shaded side of the coleoptile (see **Web Topic 19.7**). Once the auxin reaches the shaded side of the tip, it is transported basipetally to the elongation zone, where it stimulates cell elongation. The acceleration of growth on the shaded side and the slowing of growth on the illuminated side (differential growth) give rise to the curvature toward light (Figure 19.26).

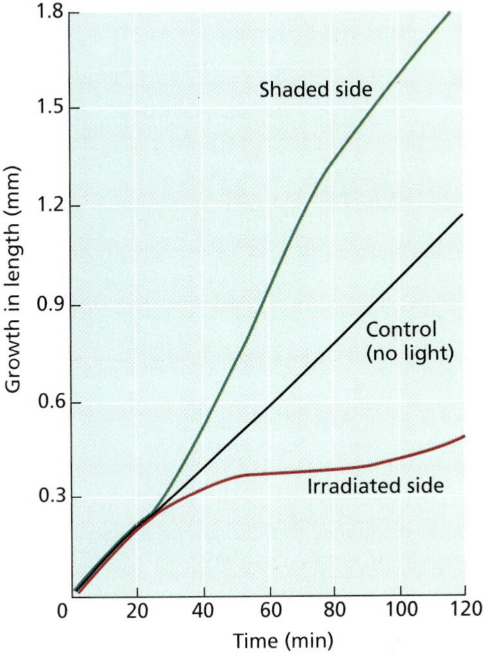

FIGURE 19.26 Time course of growth on the illuminated and shaded sides of a coleoptile responding to a 30-second pulse of unidirectional blue light. Control coleoptiles were not given a light treatment. (After Iino and Briggs 1984.)

Direct tests of the Cholodny–Went model using the agar block/coleoptile curvature bioassay have supported the model's prediction that auxin in coleoptile tips is transported laterally in response to unilateral light (Figure 19.27). The total amount of auxin diffusing out of the tip (here expressed as the angle of curvature) is the same in the presence of unilateral light as in darkness (compare Figure 19.27A and B). This result indicates that light does not cause the photodestruction of auxin on the illuminated side, as had been proposed by some investigators.

Consistent with both the Cholodny–Went hypothesis and the acid growth hypothesis, the apoplastic pH on the shaded side of a phototropically bending stem or coleoptile is more acidic than the side facing the light (Mulkey et al. 1981).

Gravitropism Also Involves Lateral Redistribution of Auxin

When dark-grown *Avena* seedlings are oriented horizontally, the coleoptiles bend upward in response to gravity. According to the Cholodny–Went model, auxin in a horizontally oriented coleoptile tip is transported laterally to the lower side, causing the lower side of the coleoptile to grow faster than the upper side. Early experimental evidence indicated that the tip of the coleoptile can perceive gravity and redistribute auxin to the lower side. For example, if coleoptile tips are oriented horizontally, a greater amount of auxin diffuses from the lower half than the upper half (Figure 19.28).

Tissues below the tip are able to respond to gravity as well. For example, when vertically oriented maize coleoptiles are decapitated by removal of the upper 2 mm of the tip and oriented horizontally, gravitropic bending occurs at a slow rate for several hours even without the tip. Application of IAA to the cut surface restores the rate of bending

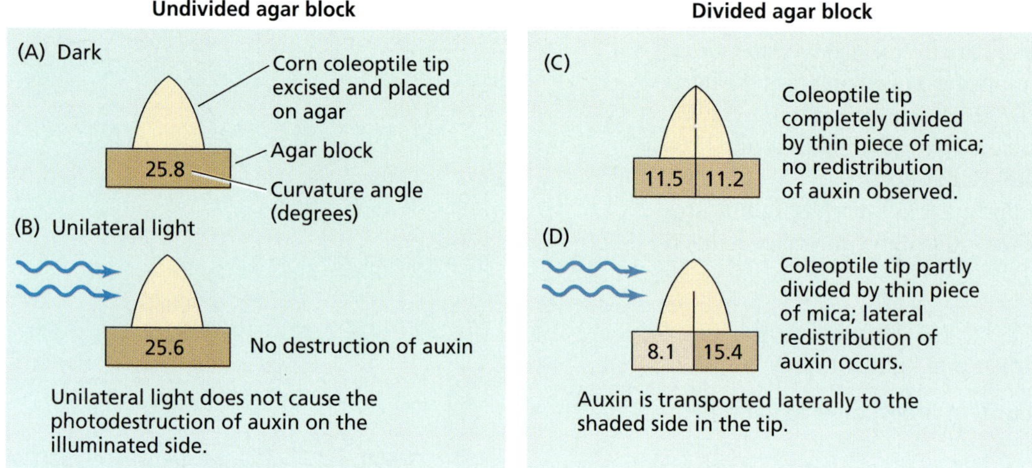

FIGURE 19.27 Evidence that the lateral redistribution of auxin is stimulated by unidirectional light in corn coleoptiles.

to normal levels. This finding indicates that both the perception of the gravitational stimulus and the lateral redistribution of auxin can occur in the tissues below the tip, although the tip is still required for auxin production.

Lateral redistribution of auxin in shoot apical meristems is more difficult to demonstrate than in coleoptiles because of the presence of leaves. In recent years, molecular markers have been widely used as reporter genes to detect lateral auxin gradients in horizontally placed stems and roots.

In soybean hypocotyls, gravitropism leads to a rapid asymmetry in the accumulation of a group of auxin-stimulated mRNAs called **SAURs** (small *a*uxin *u*p-regulated *R*NAs) (McClure and Guilfoyle 1989). In vertical seedlings, SAUR gene expression is symmetrically distributed. Within 20 minutes after the seedling is oriented horizontally, SAURs begin to accumulate on the lower half of the hypocotyl. Under these conditions, gravitropic bending first becomes evident after 45 minutes, well after the induction of the SAURs (see **Web Topic 19.8**). The existence of a lateral gradient in SAUR gene expression is indirect evidence for the existence of a lateral gradient in auxin detectable within 20 minutes of the gravitropic stimulus.

As will be discussed later in the chapter, the *GH3* gene family is also up-regulated within 5 minutes of auxin treat-

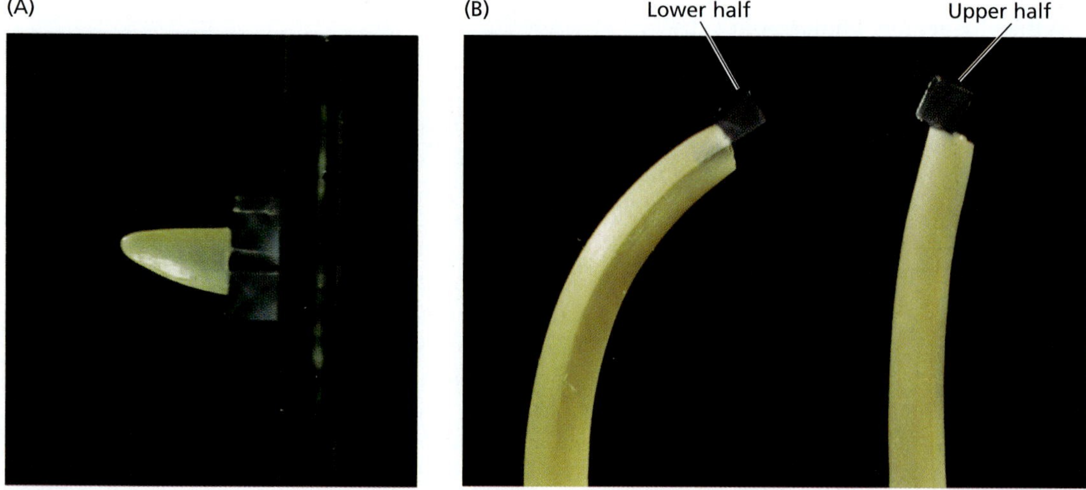

FIGURE 19.28 Auxin is transported to the lower side of a horizontally oriented oat coleoptile tip. (A) Auxin from the upper and lower halves of a horizontal tip is allowed to diffuse into two agar blocks. (B) The agar block from the lower half (left) induces greater curvature in a decapitated coleoptile than the agar block from the upper half (right). (Photo © M. B. Wilkins.)

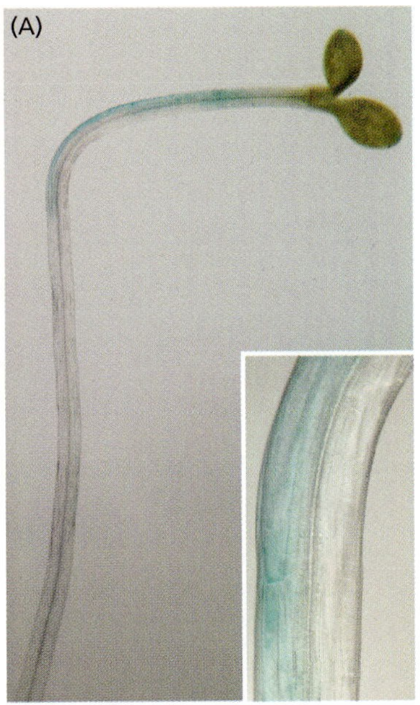

FIGURE 19.29 Lateral auxin gradients are formed in *Arabidopsis* hypocotyls during the differential growth responses to light (A). Treatment with the auxin efflux inhibitor NPA blocks auxin redistribution and bending (B). The plants were transformed with the *DR5::GUS* reporter gene. Auxin accumulation on the shaded side of the hypocotyls is indicated by the blue staining shown in the insets (A). A similar redistribution of auxin occurs in gravitropism. (Photos courtesy of Klaus Palme.)

ment and has been used as a molecular marker for the presence of auxin. By fusing an artificial promoter sequence based on the *GH3* promoter to the *GUS* reporter gene, it is possible to visualize the lateral gradient in auxin concentration that occurs during both photo- and gravitropism (Figure 19.29).

Statoliths Serve as Gravity Sensors in Shoots and Roots

Unlike unilateral light, gravity does not form a gradient between the upper and lower sides of an organ. All parts of the plant experience the gravitational stimulus equally. How do plant cells detect gravity? The only way that gravity can be sensed is through the motion of a falling or sedimenting body.

Obvious candidates for intracellular gravity sensors in plants are the large, dense amyloplasts that are present in many plant cells. These specialized amyloplasts are of sufficiently high density relative to the cytosol that they readily sediment to the bottom of the cell (Figure 19.30). Amyloplasts that function as gravity sensors are called **statoliths**, and the specialized gravity-sensing cells in which they occur are called **statocytes**. Whether the statocyte is able to detect the downward motion of the statolith as it passes through the cytoskeleton or whether the stimulus is perceived only when the statolith comes to rest at the bottom of the cell has not yet been resolved.

Shoots and Coleoptiles. In shoots and coleoptiles, gravity is perceived in the **starch sheath**, a layer of cells that surrounds the vascular tissues of the shoot. The starch sheath is continuous with the endodermis of the root, but unlike the endodermis it contains amyloplasts. *Arabidopsis* mutants lacking amyloplasts in the starch sheath display agravitropic shoot growth but normal gravitropic root growth (Fujihira et al. 2000).

As noted in Chapter 16, the *scarecrow* (*scr*) mutant of *Arabidopsis* is missing both the endodermis and the starch sheath. As a result, the hypocotyl and inflorescence of the *scr* mutant are agravitropic, while the root exhibits a normal gravitropic response. On the basis of the phenotypes of these two mutants, we can conclude the following:

- The starch sheath is required for gravitropism in shoots.
- The root endodermis, which does not contain statoliths, is not required for gravitropism in roots.

Roots. The site of gravity perception in primary roots is the root cap. Large, graviresponsive amyloplasts are located in the statocytes (see Figure 19.30A and B) in the central cylinder, or **columella**, of the root cap. Removal of the root cap from otherwise intact roots abolishes root gravitropism without inhibiting growth.

Precisely how the statocytes sense their falling statoliths is still poorly understood. According to one hypothesis, contact or pressure resulting from the amyloplast resting on the endoplasmic reticulum on the lower side of the cell triggers the response (see Figure 19.30C). The endoplasmic reticulum of columella cells is structurally unique, consisting of five to seven rough-ER sheets attached to a central nodal rod in a whorl, like petals on a flower. This specialized "nodal ER" differs from the more tubular cortical ER cisternae and may be involved in the gravity response (Zheng and Staehelin 2001).

The **starch–statolith hypothesis** of gravity perception in roots is supported by several lines of evidence. Amyloplasts are the only organelles that consistently sediment in the columella cells of different plant species, and the rate of sedimentation correlates closely with the time required to perceive the gravitational stimulus. The gravitropic responses of starch-deficient mutants are generally much slower than those of wild-type plants. Nevertheless, starchless mutants exhibit some gravitropism, suggesting that although starch is required for a normal gravitropic response, starch-independent gravity perception mechanisms may also exist.

Other organelles, such as nuclei, may be dense enough to act as statoliths. It may not even be necessary for a statolith to come to rest at the bottom of the cell. The cytoskeletal network may be able to detect a partial vertical displacement of an organelle.

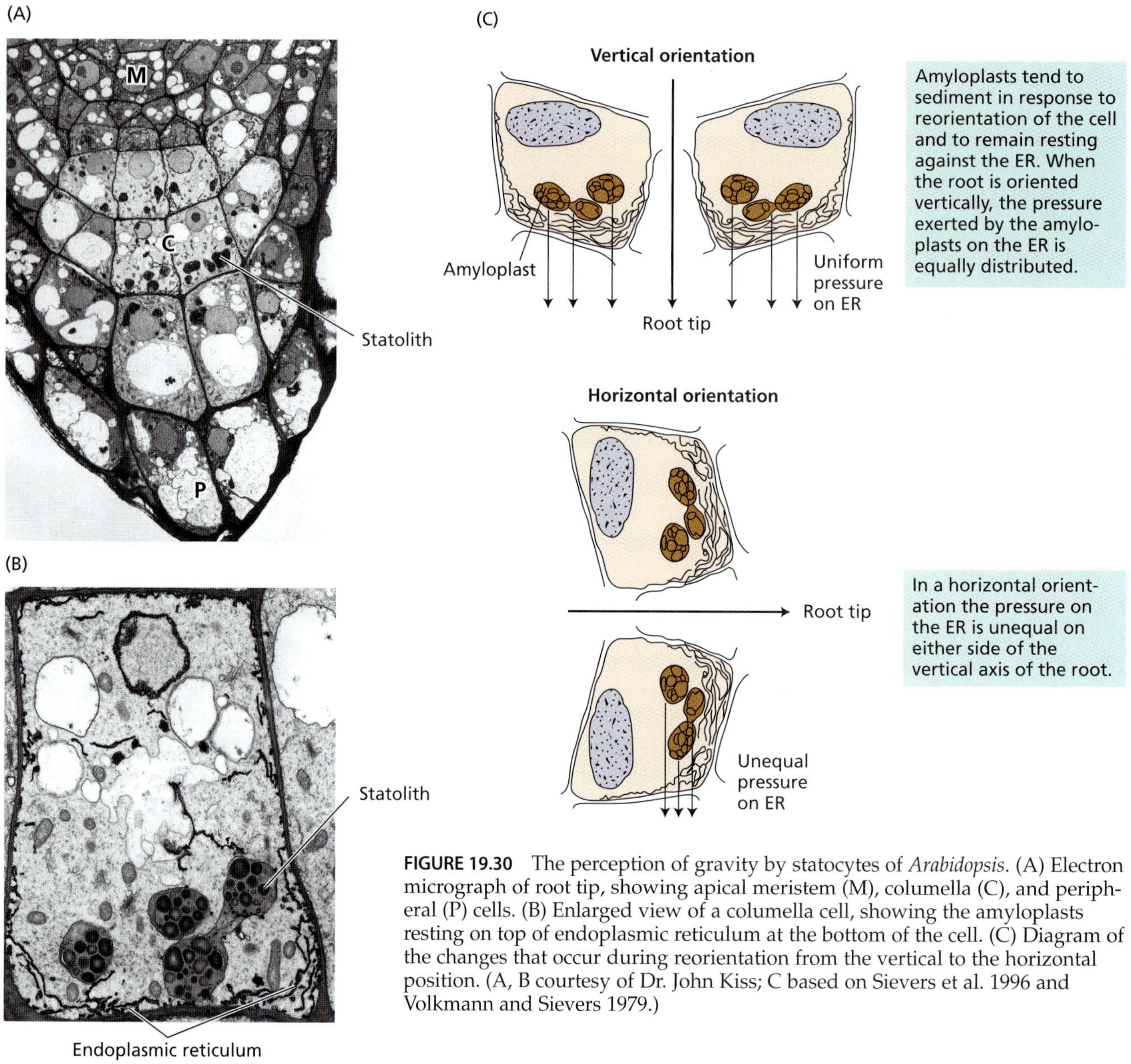

FIGURE 19.30 The perception of gravity by statocytes of *Arabidopsis*. (A) Electron micrograph of root tip, showing apical meristem (M), columella (C), and peripheral (P) cells. (B) Enlarged view of a columella cell, showing the amyloplasts resting on top of endoplasmic reticulum at the bottom of the cell. (C) Diagram of the changes that occur during reorientation from the vertical to the horizontal position. (A, B courtesy of Dr. John Kiss; C based on Sievers et al. 1996 and Volkmann and Sievers 1979.)

Recently Andrew Staehelin and colleagues proposed a new model for gravitropism, called the **tensegrity model** (Yoder et al. 2001). *Tensegrity* is an architectural term—a contraction of *tensional integrity*—coined by the innovative architect R. Buckminster Fuller. In essence, *tensegrity* refers to structural integrity created by interactive tension between the structural components. In this case the structural components consist of the meshwork of actin microfilaments that form part of the cytoskeleton of the central columella cells of the root cap. The actin network is assumed to be anchored to stretch-activated receptors on the plasma membrane. Stretch receptors in animal cells are typically mechanosensitive ion channels, and stretch-activated calcium channels have been demonstrated in plants.

According to the tensegrity model, sedimentation of the statoliths through the cytosol locally disrupts the actin meshwork, changing the distribution of tension transmitted to calcium channels on the plasma membrane, thus altering their activities. Yoder and colleagues (2001) have further proposed that the nodal ER, which is also connected to channels via actin microfilaments, may protect the cytoskeleton from being disrupted by the statoliths in specific regions, thus providing a signal for the directionality of the stimulus.

Gravity perception without statoliths? An alternative mechanism of gravity perception that does not involve statoliths has been proposed for the giant-celled freshwater alga *Chara*. See **Web Topic 19.8** for details.

Auxin Is Redistributed Laterally in the Root Cap

In addition to functioning to protect the sensitive cells of the apical meristem as the tip penetrates the soil, the root cap is the site of gravity perception. Because the cap is some distance away from the elongation zone where bending occurs, a chemical messenger is presumed to be involved in communication between the cap and the elongation zone. Microsurgery experiments in which half of the

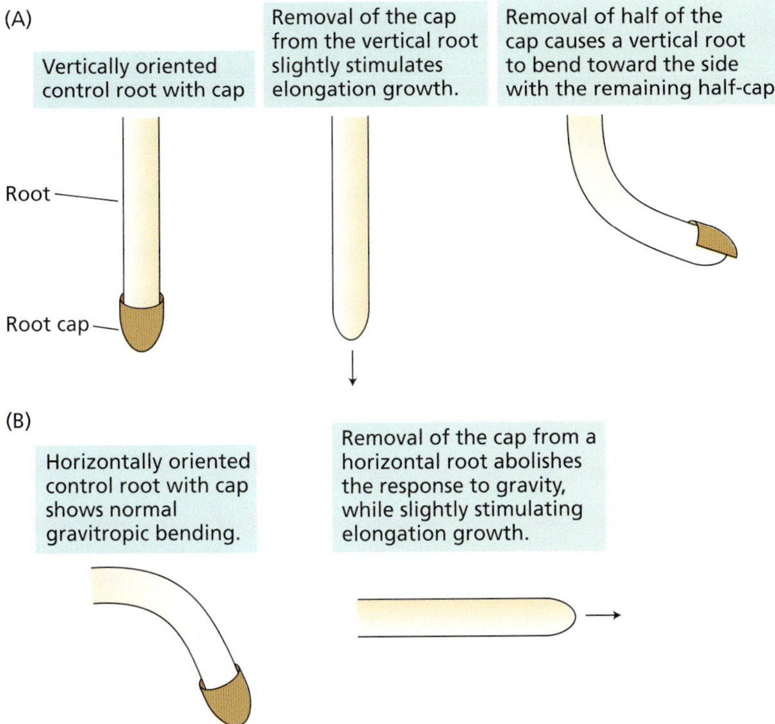

FIGURE 19.31 Microsurgery experiments demonstrating that the root cap produces an inhibitor that regulates root gravitropism. (After Shaw and Wilkins 1973.)

cap was removed showed that the cap produces a root growth inhibitor (Figure 19.31). This finding suggests that the cap supplies an inhibitor to the lower side of the root during gravitropic bending.

Although root caps contain small amounts of IAA and abscisic acid (ABA) (see Chapter 23), IAA is more inhibitory to root growth than ABA when applied directly to the elongation zone, suggesting that IAA is the root cap inhibitor. Consistent with this conclusion, ABA-deficient *Arabidopsis* mutants have normal root gravitropism, whereas the roots of mutants defective in auxin transport, such as *aux1* and *agr1*, are agravitropic (Palme and Gälweiler 1999). The *agr* mutant lacks an auxin efflux carrier related to the PIN proteins (Chen et al. 1998; Müller et al. 1998; Utsuno et al. 1998). The AGR1 protein is localized at the basal (distal) end of cortical cells near the root tip in *Arabidopsis*.

How do we reconcile the fact that the shoot apical meristem is the primary source of auxin to the root with the role of the root cap as the source of the inhibitory auxin during gravitropism? As discussed earlier in the chapter, auxin from the shoot is translocated from the stele to the root tip via protophloem cells. Asymmetrically localized AUX1 permeases on the protophloem parenchyma cells direct the acropetal transport of auxin from the phloem to a cluster of cells in the columella of the cap. Auxin is then transported radially to the lateral root cap cells, where AUX1 is also strongly expressed (see Figure 19.19).

The lateral root cap cells overlay the distal elongation zone (DEZ) of the root—the first region that responds to gravity. The auxin from the cap is taken up by the cortical parenchyma of the DEZ and transported basipetally through the elongation zone of the root. This basipetal transport, which is limited to the elongation zone, is facilitated by auxin anion carriers related to the PIN family (called AGR1), which are localized at the basal ends of the cortical parenchyma cells.

The basipetally transported auxin accumulates in the elongation zone and does not pass beyond this region. Flavonoids capable of inhibiting auxin efflux are synthesized in this region of the root and probably promote auxin retention by these cells (Figure 19.32) (Murphy et al. 2000).

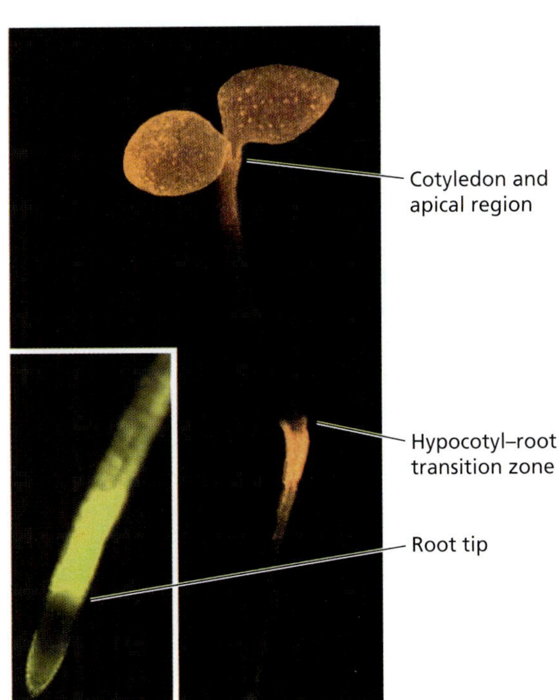

FIGURE 19.32 Flavonoid localization in a 6-day-old *Arabidopsis* seedling. The staining procedure used causes the flavonoids to fluoresce. Flavonoids are concentrated in three regions: the cotyledon and apical region, the hypocotyl–root transition zone, and the root tip area (inset). In the root tip, flavonoids are localized specifically in the elongation zone and the cap, the tissues involved in basipetal auxin transport. (From Murphy et al. 2000.)

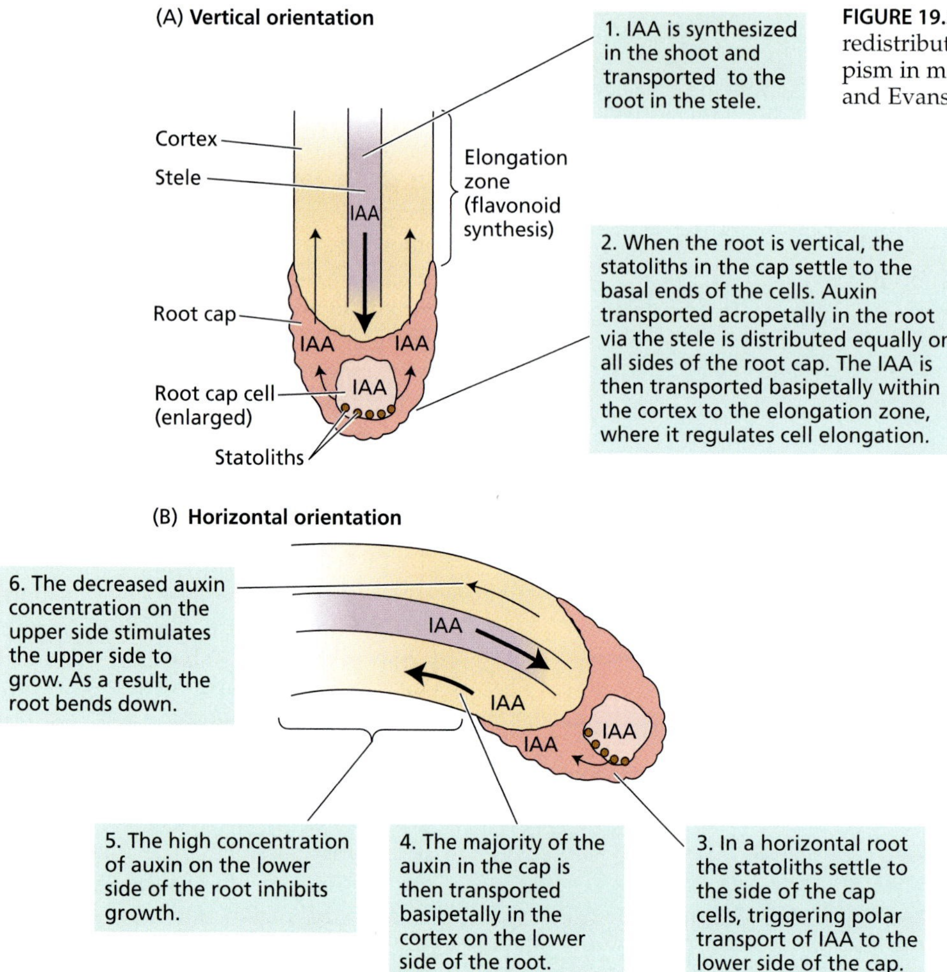

FIGURE 19.33 Proposed model for the redistribution of auxin during gravitropism in maize roots. (After Hasenstein and Evans 1988.)

According to the model, basipetal auxin transport in a vertically oriented root is equal on all sides (Figure 19.33A). When the root is oriented horizontally, however, the cap redirects the bulk of the auxin to the lower side, thus inhibiting the growth of that lower side (see Figure 19.33B). Consistent with this idea, the transport of [^{3}H]IAA across a horizontally oriented root cap is polar, with a preferential downward movement (Young et al. 1990).

PIN3 Is Relocated Laterally to the Lower Side of Root Columella Cells

Recently the mechanism of lateral auxin redistribution in the root cap has new been elucidated (Friml et al. 2002). One of the members of the PIN protein family of auxin efflux carriers, PIN3, is not only required for both photo- and gravitropism in *Arabidopsis*, but it has been shown to be relocalized to the lower side of the columella cells during root gravitropism (Figure 19.34).

As noted previously, PIN proteins are constantly being cycled between the plasma membrane and intracellular secretory compartments. This cycling allows some PIN proteins to be targeted to specific sides of the cell in response to a directional stimulus. In a vertically oriented root, PIN3 is uniformly distributed around the columella cell (see Figure 19.34A). But when the root is placed on its side, PIN3 is preferentially targeted to the lower side of the cell (see Figure 19.34B). As a result, auxin is transported polarly to the lower half of the cap.

Gravity Sensing May Involve Calcium and pH as Second Messengers

A variety of experiments have suggested that calcium–calmodulin is required for root gravitropism in maize. Some of these experiments involve EGTA (ethylene glycol-bis(β-aminoethyl ether)-N,N,N′,N′-tetraacetic acid), a compound that can chelate (form a complex with) calcium ions, thus preventing calcium uptake by cells. EGTA inhibits both root gravitropism and the asymmetric distribution of auxin in response to gravity (Young and Evans 1994).

Placing a block of agar that contains calcium ions on the side of the cap of a vertically oriented corn root induces the root to grow toward the side with the agar block (Figure 19.35). As in the case of [$^+$H]IAA, $^{45}Ca^{2+}$ is polarly transported to the lower half of the cap of a root stimulated by

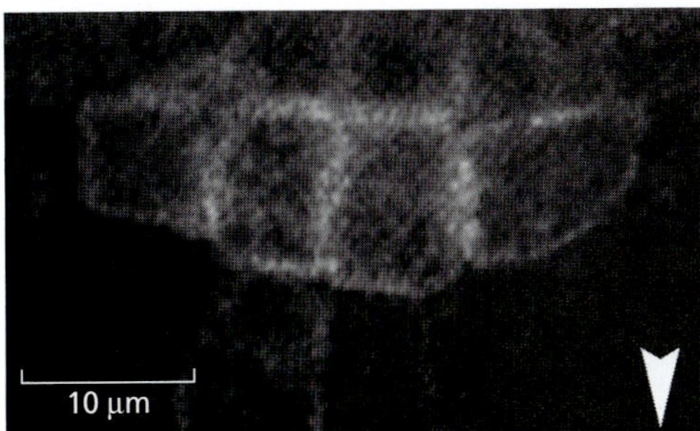

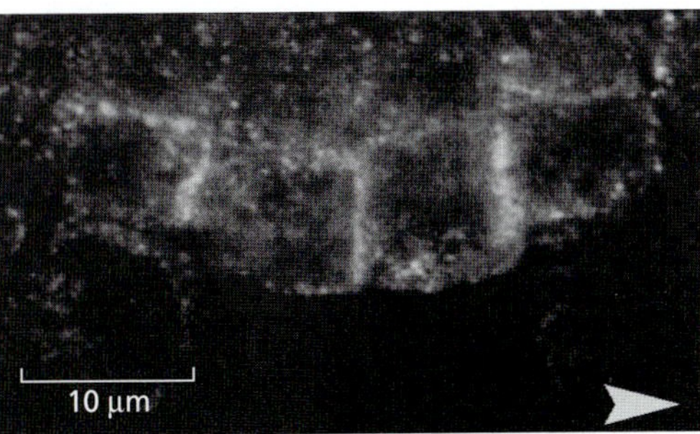

FIGURE 19.34 Relocalization of the auxin efflux carrier PIN3 during root gravitropism in *Arabidopsis*. (A) In a vertically oriented root, PIN3 is uniformly distributed around the columella cells. (B) After being oriented horizontally for 10 minutes, PIN3 has been relocalized to the lower side of the columella cells. The photo in (B) has been reorientated so that the lower side is on the right. (The direction of gravity is indicated by the arrows.) (From Friml et al. 2002, courtesy of Klaus Palme.)

gravity. However, thus far no changes in the distribution of intracellular calcium have been detected in columella cells in response to a gravitational stimulus.

Recent evidence suggests that a change in intracellular pH is the earliest detectable change in columella cells responding to gravity. Fasano et al. (2001) used pH-sensitive dyes to monitor both intracellular and extracellular pH in *Arabidopsis* roots after they were placed in a horizontal position. Within 2 minutes of gravistimulation, the cytoplasmic pH of the columella cells of the root cap increased from 7.2 to 7.6, and the apoplastic pH declined from 5.5 to 4.5. These changes preceded any detectable tropic curvature by about 10 minutes.

The alkalinization of the cytosol combined with the acidification of the apoplast suggests that an activation of the plasma membrane H^+-ATPase is one of the initial events that mediate root gravity perception or signal transduction.

FIGURE 19.35 A corn root bending toward an agar block containing calcium placed on the cap. (Courtesy of Michael L. Evans.)

DEVELOPMENTAL EFFECTS OF AUXIN

Although originally discovered in relation to growth, auxin influences nearly every stage of a plant's life cycle from germination to senescence. Because the effect that auxin produces depends on the identity of the target tissue, the response of a tissue to auxin is governed by its developmentally determined genetic program and is further influenced by the presence or absence of other signaling molecules. As we will see in this and subsequent chapters, interaction between two or more hormones is a recurring theme in plant development.

In this section we will examine some additional developmental processes regulated by auxin, including apical dominance, leaf abscission, lateral-root formation, and vascular differentiation. Throughout this discussion we assume that the primary mechanism of auxin action is comparable in all cases, involving similar receptors and signal transduction pathways. The current state of our knowledge of auxin signaling pathways will be considered at the end of the chapter.

Auxin Regulates Apical Dominance

In most higher plants, the growing apical bud inhibits the growth of lateral (axillary) buds—a phenomenon called

FIGURE 19.36 Auxin suppresses the growth of axillary buds in bean (*Phaseolus vulgaris*) plants. (A) The axillary buds are suppressed in the intact plant because of apical dominance. (B) Removal of the terminal bud releases the axillary buds from apical dominance (arrows). (C) Applying IAA in lanolin paste (contained in the gelatin capsule) to the cut surface prevents the outgrowth of the axillary buds. (Photos ©M. B. Wilkins.)

apical dominance. Removal of the shoot apex (decapitation) usually results in the growth of one or more of the lateral buds. Not long after the discovery of auxin, it was found that IAA could substitute for the apical bud in maintaining the inhibition of lateral buds of bean (*Phaseolus vulgaris*) plants. This classic experiment is illustrated in Figure 19.36.

This result was soon confirmed for numerous other plant species, leading to the hypothesis that the outgrowth of the axillary bud is inhibited by auxin that is transported basipetally from the apical bud. In support of this idea, a ring of the auxin transport inhibitor TIBA in lanolin paste (as a carrier) placed below the shoot apex released the axillary buds from inhibition.

How does auxin from the shoot apex inhibit the growth of lateral buds? Kenneth V. Thimann and Folke Skoog originally proposed that auxin from the shoot apex inhibits the growth of the axillary bud directly—the so-called *direct-inhibition model*. According to the model, the optimal auxin concentration for bud growth is low, much lower than the auxin concentration normally found in the stem. The level of auxin normally present in the stem was thought to inhibit the growth of lateral buds.

If the direct-inhibition model of apical dominance is correct, the concentration of auxin in the axillary bud should decrease following decapitation of the shoot apex. However, the reverse appears to be true. This was demonstrated with transgenic plants that contained the reporter genes for bacterial luciferase (*LUXA* and *LUXB*) under the control of an auxin-responsive promoter (Langridge et al. 1989). These reporter genes allowed researchers to study the level of auxin in different tissues by monitoring the amount of light emitted by the luciferase-catalyzed reaction.

When these transgenic plants were decapitated, the expression of the *LUX* genes increased in and around the axillary buds within 12 hours. This experiment indicated that after decapitation, the auxin content of the axillary buds *increased* rather than decreased.

Direct physical measurements of auxin levels in buds have also shown an increase in the auxin of the axillary buds after decapitation. The IAA concentration in the axillary bud of *Phaseolus vulgaris* (kidney bean) increased fivefold within 4 hours after decapitation (Gocal et al. 1991). These and other similar results make it unlikely that auxin from the shoot apex inhibits the axillary bud directly.

Other hormones, such as cytokinins and ABA, may be involved. Direct application of cytokinins to axillary buds stimulates bud growth in many species, overriding the inhibitory effect of the shoot apex. Auxin makes the shoot apex a sink for cytokinin synthesized in the root, and this may be one of the factors involved in apical dominance (see **Web Topic 19.10**).

Finally, ABA has been found in dormant lateral buds in intact plants. When the shoot apex is removed, the ABA levels in the lateral buds decrease. High levels of IAA in the shoot may help keep ABA levels high in the lateral buds. Removing the apex removes a major source of IAA, which may allow the levels of bud growth inhibitor to fall (see **Web Topic 19.11**).

Auxin Promotes the Formation of Lateral and Adventitious Roots

Although elongation of the primary root is inhibited by auxin concentrations greater than 10^{-8} *M*, initiation of lateral (branch) roots and adventitious roots is stimulated by high auxin levels. Lateral roots are commonly found above the elongation and root hair zone and originate from small groups of cells in the pericycle (see Chapter 16). Auxin stimulates these pericycle cells to divide. The dividing cells gradually form into a root apex, and the lateral root grows through the root cortex and epidermis.

Adventitious roots (roots originating from non-root tissue) can arise in a variety of tissue locations from clusters of mature cells that renew their cell division activity. These dividing cells develop into a root apical meristem in a manner somewhat analogous to the formation of lateral roots. In horticulture, the stimulatory effect of auxin on the formation of adventitious roots has been very useful for the vegetative propagation of plants by cuttings.

A series of *Arabidopsis* mutants, named *alf* (*a*berrant *l*ateral root *f*ormation), have provided some insights into the role of auxin in the initiation of lateral roots. The *alf1* mutant exhibits extreme proliferation of adventitious and lateral roots, coupled with a 17-fold increase in endogenous auxin (Figure 19.37).

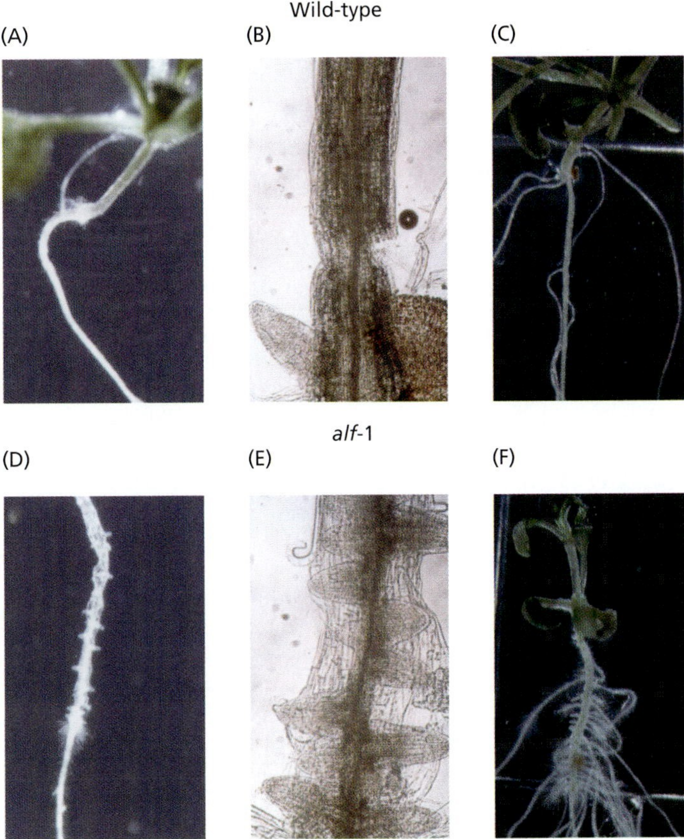

FIGURE 19.37 Root morphology of *Arabidopsis* (A–C) wild-type and *alf1* seedlings (D–F) on hormone-free medium. Note the proliferation of root primordia growing from the pericycle in the *alf1* seedlings (D and E). (From Celenza et al. 1995, courtesy of J. Celenza.)

Another mutant, *alf4*, has the opposite phenotype: It is completely devoid of lateral roots. Microscopic analysis of *alf4* roots indicates that lateral-root primordia are absent. The *alf4* phenotype cannot be reversed by application of exogenous IAA.

Yet another mutant, *alf3*, is defective in the development of lateral-root primordia into mature lateral roots. The primary root is covered with arrested lateral-root primordia that grow until they protrude through the epidermal cell layer and then stop growing. The arrested growth can be alleviated by application of exogenous IAA.

On the basis of the phenotypes of the *alf* mutants, a model in which IAA is required for at least two steps in the formation of lateral roots has been proposed (Figure 19.38) (Celenza et al. 1995):

1. IAA transported acropetally (toward the tip) in the stele is required to initiate cell division in the pericycle.
2. IAA is required to promote cell division and maintain cell viability in the developing lateral root.

Auxin Delays the Onset of Leaf Abscission

The shedding of leaves, flowers, and fruits from the living plant is known as **abscission**. These parts abscise in a region called the **abscission zone**, which is located near the base of the petiole of leaves. In most plants, leaf abscission is preceded by the differentiation of a distinct layer of cells, the **abscission layer**, within the abscission zone. During leaf senescence, the walls of the cells in the abscission layer are digested, which causes them to become soft and weak. The leaf eventually breaks off at the abscission layer as a result of stress on the weakened cell walls.

Auxin levels are high in young leaves, progressively decrease in maturing leaves, and are relatively low in senescing leaves when the abscission process begins. The role of auxin in leaf abscission can be readily demonstrated by excision of the blade from a mature leaf, leaving the petiole intact on the stem. Whereas removal of the leaf blade accelerates the formation of the abscission layer in the petiole, application of IAA in lanolin paste to the cut surface of the petiole prevents the formation of the abscission layer. (Lanolin paste alone does not prevent abscission.)

These results suggest the following:

- Auxin transported from the blade normally prevents abscission.
- Abscission is triggered during leaf senescence, when auxin is no longer being produced.

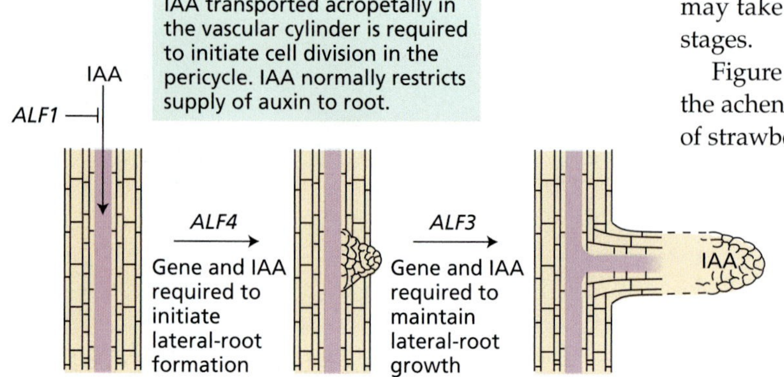

FIGURE 19.38 A model for the formation of lateral roots, based on the *alf* mutants of *Arabidopsis*. (After Celenza et al. 1995.)

However, as will be discussed in Chapter 22, ethylene also plays a crucial role as a positive regulator of abscission.

Auxin Transport Regulates Floral Bud Development

Treating *Arabidopsis* plants with the auxin transport inhibitor NPA causes abnormal floral development, suggesting that polar auxin transport in the inflorescence meristem is required for normal floral development. In *Arabidopsis*, the "pin-formed" mutant *pin1*, which lacks an auxin efflux carrier in shoot tissues, has abnormal flowers similar to those of NPA-treated plants (see Figure 19.14A). Apparently the developing floral meristem depends on auxin being transported to it from subapical tissues. In the absence of the efflux carriers, the meristem is starved for auxin, and normal phyllotaxis and floral development are disrupted (Kuhlemeier and Reinhardt 2001).

Auxin Promotes Fruit Development

Much evidence suggests that auxin is involved in the regulation of fruit development. Auxin is produced in pollen and in the endosperm and the embryo of developing seeds, and the initial stimulus for fruit growth may result from pollination.

Successful pollination initiates ovule growth, which is known as **fruit set**. After fertilization, fruit growth may depend on auxin produced in developing seeds. The endosperm may contribute auxin during the initial stage of fruit growth, and the developing embryo may take over as the main auxin source during the later stages.

Figure 19.39 shows the influence of auxin produced by the achenes of strawberry on the growth of the receptacle of strawberry.

FIGURE 19.39 (A) The strawberry "fruit" is actually a swollen receptacle whose growth is regulated by auxin produced by the "seeds," which are actually achenes–the true fruits. (B) When the achenes are removed, the receptacle fails to develop normally. (C) Spraying the achene-less receptacle with IAA restores normal growth and development. (After A. Galston 1994.)

Auxin Induces Vascular Differentiation

New vascular tissues differentiate directly below developing buds and young growing leaves (see Figure 19.5), and removal of the young leaves prevents vascular differentiation (Aloni 1995). The ability of an apical bud to stimulate vascular differentiation can be demonstrated in tissue culture. When the apical bud is grafted onto a clump of undifferentiated cells, or *callus*, xylem and phloem differentiate beneath the graft.

The relative amounts of xylem and phloem formed are regulated by the auxin concentration: High auxin concentrations induce the differentiation of xylem and phloem, but only phloem differentiates at low auxin concentrations. Similarly, experiments on stem tissues have shown that low auxin concentrations induce phloem differentiation, whereas higher IAA levels induce xylem (Aloni 1995).

The regeneration of vascular tissue following wounding is also controlled by auxin produced by the young leaf directly above the wound site (Figure 19.40). Removal of the leaf prevents the regeneration of vascular tissue, and applied auxin can substitute for the leaf in stimulating regeneration.

Vascular differentiation is polar and occurs from leaves to roots. In woody perennials, auxin produced by growing buds in the spring stimulates activation of the cambium in

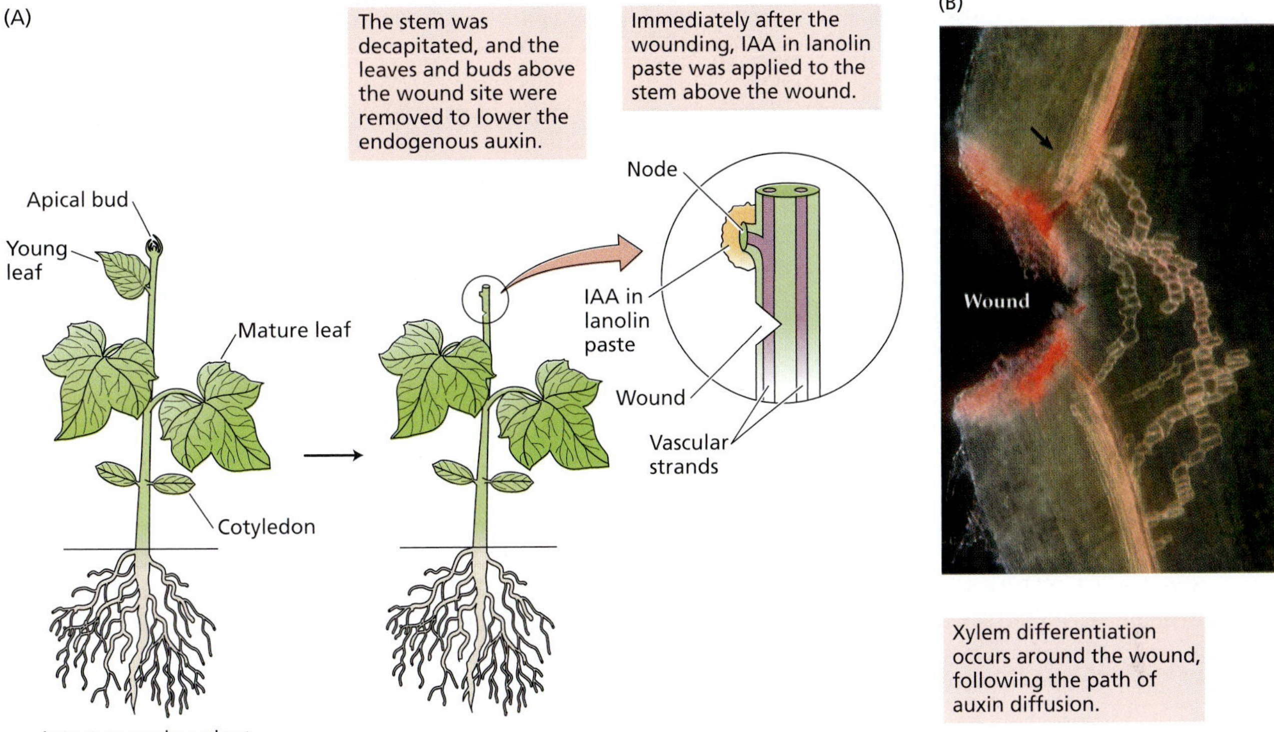

FIGURE 19.40 IAA-induced xylem regeneration around the wound in cucumber (*Cucumis sativus*) stem tissue. (A) Method for carrying out the wound regeneration experiment. (B) Fluorescence micrograph showing regenerating vascular tissue around the wound. (B courtesy of R. Aloni.)

a basipetal direction. The new round of secondary growth begins at the smallest twigs and progresses downward toward the root tip.

Further evidence for the role of auxin in vascular differentiation comes from studies in which the auxin concentration is manipulated by the transformation of plants with auxin biosynthesis genes through use of the Ti plasmid of *Agrobacterium*. When an auxin biosynthesis gene was overexpressed in petunia plants, the number of xylem tracheary elements increased. In contrast, when the level of free IAA in tobacco plants was decreased by transformation with a gene coding for an enzyme that conjugated IAA to the amino acid lysine, the number of vessel elements decreased and their sizes increased (Romano et al. 1991). Thus the level of free auxin appears to regulate the number of tracheary elements, as well as their size.

In *Zinnia elegans* mesophyll cell cultures, auxin is required for tracheary cell differentiation, but cytokinins also participate, perhaps by increasing the sensitivity of the cells to auxin. Whereas auxin is produced in the shoot and transported downward to the root, cytokinins are produced by the root tips and transported upward into the shoot. Both hormones are probably involved in the regulation of cambium activation and vascular differentiation (see Chapter 21).

Synthetic Auxins Have a Variety of Commercial Uses

Auxins have been used commercially in agriculture and horticulture for more than 50 years. The early commercial uses included prevention of fruit and leaf drop, promotion of flowering in pineapple, induction of parthenocarpic fruit, thinning of fruit, and rooting of cuttings for plant propagation. Rooting is enhanced if the excised leaf or stem cutting is dipped in an auxin solution, which increases the initiation of adventitious roots at the cut end. This is the basis of commercial rooting compounds, which consist mainly of a synthetic auxin mixed with talcum powder.

In some plant species, seedless fruits may be produced naturally, or they may be induced by treatment of the unpollinated flowers with auxin. The production of such seedless fruits is called **parthenocarpy**. In stimulating the formation of parthenocarpic fruits, auxin may act primarily to induce fruit set, which in turn may trigger the endogenous production of auxin by certain fruit tissues to complete the developmental process.

Ethylene is also involved in fruit development, and some of the effects of auxin on fruiting may result from the promotion of ethylene synthesis. The control of ethylene in the commercial handling of fruit is discussed in Chapter 22.

In addition to these applications, today auxins are widely used as herbicides. The chemicals 2,4-D and dicamba (see Figure 19.4) are probably the most widely used synthetic auxins. Synthetic auxins are very effective because they are not metabolized by the plant as quickly as IAA is. Because maize and other monocotyledons can rapidly inactivate synthetic auxins by conjugation, these auxins are used by farmers for the control of dicot weeds, also called *broad-leaved weeds*, in commercial cereal fields, and by home gardeners for the control of weeds such as dandelions and daisies in lawns.

AUXIN SIGNAL TRANSDUCTION PATHWAYS

The ultimate goal of research on the molecular mechanism of hormone action is to reconstruct each step in the signal transduction pathway, from receptor binding to the physiological response. In this last section of the chapter, we will examine candidates for the auxin receptor and then discuss the various signaling pathways that have been implicated in auxin action. Finally we will turn our attention to auxin-regulated gene expression.

ABP1 Functions as an Auxin Receptor

In addition to its possible direct role in plasma membrane H^+-ATPase activation (discussed earlier), the auxin-binding protein ABP1 appears to function as an auxin receptor in other signal transduction pathways. ABP1 homologs have been identified in a variety of monocot and dicot species (Venis and Napier 1997). Knockouts of the *ABP1* gene in *Arabidopsis* are lethal, and less severe mutations result in altered development (Chen et al. 2001). Recent studies indicate that, despite being localized primarily on the endoplasmic reticulum (ER), a small amount of ABP1 is secreted to the plasma membrane outer surface where it interacts with auxin to cause protoplast swelling and H^+-pumping (Venis et al. 1996; Steffens et al. 2001).

However, it is unlikely that ABP1 mediates all auxin response pathways because expression of a number of auxin-responsive genes is not affected when protoplasts are incubated with anti-ABP1 antibodies. It is also unclear what role the ABP1 in the ER plays in auxin-responsive signal transduction. Finally, it remains to be determined whether ABP_{57}, the soluble and unrelated ABP from rice that activates the H^+-ATPase (see Figure 19.24), is involved in a signal transduction pathway.

Calcium and Intracellular pH Are Possible Signaling Intermediates

Calcium plays an important role in signal transduction in animals and is thought to be involved in the action of certain plant hormones as well. The role of calcium in auxin action seems very complex and, at this point in time, very uncertain. Nevertheless, some experimental evidence shows that auxin increases the level of free calcium in the cell.

Changes in cytoplasmic pH can also serve as a second messenger in animals and plants. In plants, auxin induces a decrease in cytosolic pH of about 0.2 units within 4 minutes of application. The cause of this pH drop is not known. Since the cytosolic pH is normally around 7.4, and the pH optimum of the plasma membrane H^+-ATPase is 6.5, a decrease in the cytosolic pH of 0.2 units could cause a marked increase in the activity of the plasma membrane H^+-ATPase. The decrease in cytosolic pH might also account for the auxin-induced increase in free intracellular calcium, by promoting the dissociation of bound forms.

MAP kinases (see Chapter 14 on the web site) that play a role in signal transduction by phosphorylating proteins in a cascade that ultimately activates transcription factors have also been implicated in auxin responses. When tobacco cells are deprived of auxin, they arrest at the end of either the G_1 or the G_2 phase and cease dividing; if auxin is added back into the culture medium, the cell cycle resumes (Koens et al. 1995). (For a description of the cell cycle, see Chapter 1.) Auxin appears to exert its effect on the cell cycle primarily by stimulating the synthesis of the major cyclin-dependent protein kinase (CDK): Cdc2 (*c*ell *d*ivision *c*ycle 2) (see Chapter 14 on the web site).

Auxin-Induced Genes Fall into Two Classes: Early and Late

One of the important functions of the signal transduction pathway(s) initiated when auxin binds to its receptor is the activation of a select group of transcription factors. The activated transcription factors enter the nucleus and promote the expression of specific genes. Genes whose expression is stimulated by the activation of preexisting transcription factors are called **primary response genes** or **early genes**.

This definition implies that all of the proteins required for auxin-induced expression of the early genes are present in the cell at the time of exposure to the hormone; thus, early-gene expression cannot be blocked by inhibitors of protein synthesis such as cycloheximide. As a consequence, the time required for the expression of the early genes can be quite short, ranging from a few minutes to several hours (Abel and Theologis 1996).

In general, primary response genes have three main functions: (1) Some of the early genes encode proteins that regulate the transcription of **secondary response genes**, or **late genes**, that are required for the long-term responses to the hormone. Because late genes require de novo protein synthesis, their expression can be blocked by protein synthesis inhibitors. (2) Other early genes are involved in intercellular communication, or cell-to-cell signaling. (3) Another group of early genes is involved in adaptation to stress.

Five major classes of early auxin-responsive genes have been identified:

- Genes involved in auxin-regulated growth and development:
    1. The *AUX/IAA* gene family
    2. The *SAUR* gene family
    3. The *GH3* gene family
- Stress response genes:
    1. Genes encoding glutathione S-transferases
    2. Genes encoding 1-aminocyclopropane-1-carboxylic acid (ACC) synthase, the key enzyme in the ethylene biosynthetic pathway (see Chapter 22)

Early genes for growth and development. Members of the *AUX/IAA* gene family encode short-lived transcription factors that function as repressors or activators of the expression of late auxin-inducible genes. The expression of most of the *AUX/IAA* family of genes is stimulated by auxin within 5 to 60 minutes of hormone addition All the genes encode small hydrophilic polypeptides that have putative DNA-binding motifs similar to those of bacterial repressors. They also have short half-lives (about 7 minutes), indicating that they are turning over rapidly.

The *SAUR* gene family was mentioned earlier in the chapter in relation to tropisms. Auxin stimulates the expression of *SAUR* genes within 2 to 5 minutes of treatment, and the response is insensitive to cycloheximide. The five *SAUR* genes of soybean are clustered together, contain no introns, and encode highly similar polypeptides of unknown function. Because of the rapidity of the response, expression of *SAUR* genes has proven to be a convenient probe for the lateral transport of auxin during photo- and gravitropism.

GH3 early-gene family members, identified in both soybean and *Arabidopsis*, are stimulated by auxin within 5 minutes. Mutations in *Arabidopsis GH3*-like genes result in dwarfism (Nakazawa et al. 2001) and appear to function in light-regulated auxin responses (Hsieh et al. 2000). Because *GH3* expression is a good reflection of the presence of endogenous auxin, a synthetic *GH3*-based **reporter** gene known as *DR5* is widely used in auxin bioassays (see Figure 19.5 and **Web Topic 19.12**) (Ulmasov et al. 1997).

Early genes for stress adaptations. As mentioned earlier in the chapter, auxin is involved in stress responses, such as wounding. Several genes encoding glutathione-S-transferases (GSTs), a class of proteins stimulated by various stress conditions, are induced by elevated auxin concentrations. Likewise, ACC synthase, which is also induced by stress and is the rate-limiting step in ethylene biosynthesis (see Chapter 22), is induced by high levels of auxin.

To be induced, the promoters of the early auxin genes must contain response elements that bind to the transcription factors that become activated in the presence of auxin. A limited number of these response elements appear to be arranged combinatorily within the promoters of a variety of auxin-induced genes.

Auxin-Responsive Domains Are Composite Structures

A conserved **auxin response element** (**AuxRE**) within the promoters of the early auxin genes, like *GH3*, is usually combined with other response elements to form **auxin response domains** (**AuxRDs**). For example, the *GH3* gene promoter of soybean is composed of three independently acting AuxRDs (each containing multiple AuxREs) that contribute incrementally to the strong auxin inducibility of the promoter.

Early Auxin Genes Are Regulated by Auxin Response Factors

As noted previously, early auxin genes are by definition insensitive to protein synthesis inhibitors such as cycloheximide. Instead of being inhibited, the expression of many of the early auxin genes has been found to be stimulated by cycloheximide.

Cycloheximide stimulation of gene expression is accomplished both by transcriptional activation and by mRNA stabilization. Transcriptional activation of a gene by inhibitors of protein synthesis usually indicates that the gene is being repressed by a short-lived repressor protein or by a regulatory pathway that involves a protein with a high turnover rate.

A family of **auxin response factors** (**ARFs**) function as transcriptional activators by binding to the auxin response element TGTCTC, which is present in the promoters of *GH3* and other early auxin response genes. Mutations in ARF genes result in severe developmental defects. To bind the AuxRE stably, ARFs must form dimers. It has been proposed that ARF dimers promote transcription by binding to two AuxREs arranged in a palindrome (Ulmasov et al. 1997).

Recent studies also indicate that proteins encoded by the *AUX/IAA* gene family (itself one of the early auxin response gene families) can inhibit the transcription of early auxin response genes by forming inactive heterodimers with ARFs. These inactive heterodimers may act to inhibit ARF–AuxRE binding, thereby blocking either gene activation or repression. AUX/IAA proteins may thus function as ARF inhibitors.

It is now believed that auxin induces the transcription of the early response genes by promoting the proteolytic degradation of the inhibitory AUX/IAA proteins so that active ARF dimers can form. The precise mechanism by

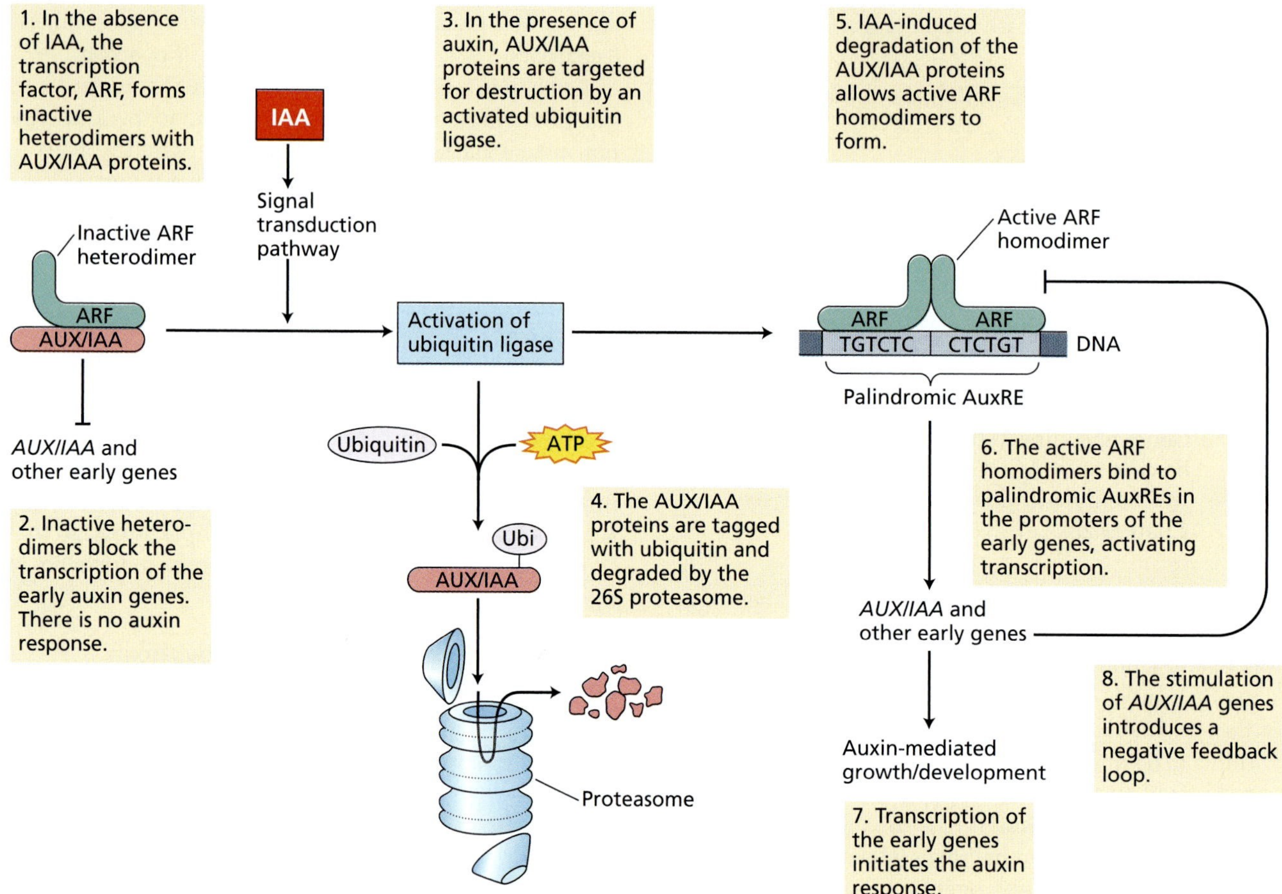

FIGURE 19.41 A model for auxin regulation of transcriptional activation of early response genes by auxin. (After Gray et al. 2001.)

which auxin causes AUX/IAA turnover is unknown, although it is known to involve ubiquitination by a ubiquitin ligase and proteolysis by the massive 26S proteasome complex (see Chapter 14 on the web site) (Gray et al. 2001; Zenser et al. 2001). Note that a negative feedback loop is introduced into the pathway by virtue of the fact that one of the gene families turned on by auxin is *AUX/IAA*, which inhibits the response.

A model for auxin regulation of the early response genes based on the findings described here is shown in Figure 19.41.

SUMMARY

Auxin was the first hormone to be discovered in plants and is one of an expanding list of chemical signaling agents that regulate plant development. The most common naturally occurring form of auxin is indole-3-acetic acid (IAA). One of the most important roles of auxin in higher plants is the regulation of elongation growth in young stems and coleoptiles. Low levels of auxin are also required for root elongation, although at higher concentrations auxin acts as a root growth inhibitor.

Accurate measurement of the amount of auxin in plant tissues is critical for understanding the role of this hormone in plant physiology. Early coleoptile-based bioassays have been replaced by more accurate techniques, including physicochemical methods and immunoassay.

Regulation of growth in plants may depend in part on the amount of free auxin present in plant cells, tissues, and organs. There are two main pools of auxin in the cell: the cytosol and the chloroplasts. Levels of free auxin can be modulated by several factors, including the synthesis and breakdown of conjugated IAA, IAA metabolism, compartmentation, and polar auxin transport. Several pathways have been implicated in IAA biosynthesis, including tryptophan-dependent and tryptophan-independent pathways. Several degradative pathways for IAA have also been identified.

IAA is synthesized primarily in the apical bud and is transported polarly to the root. Polar transport is thought to occur mainly in the parenchyma cells associated with the vascular tissue. Polar auxin transport can be divided into two main processes: IAA influx and IAA efflux. In accord with the chemiosmotic model for polar transport, there are two modes of IAA influx: by a pH-dependent passive transport of the undissociated form, or by an active H^+

cotransport mechanism driven by the plasma membrane H^+-ATPase.

Auxin efflux is thought to occur preferentially at the basal ends of the transporting cells via anion efflux carriers and to be driven by the membrane potential generated by the plasma membrane H^+-ATPase. Auxin transport inhibitors (ATIs) can interrupt auxin transport directly by competing with auxin for the efflux channel pore or by binding to regulatory or structural proteins associated with the efflux channel. Auxin can be transported nonpolarly in the phloem.

Auxin-induced cell elongation begins after a lag time of about 10 minutes. Auxin promotes elongation growth primarily by increasing cell wall extensibility. Auxin-induced wall loosening requires continuous metabolic input and is mimicked in part by treatment with acidic buffers.

According to the acid growth hypothesis, one of the important actions of auxin is to induce cells to transport protons into the cell wall by stimulating the plasma membrane H^+-ATPase. Two mechanisms have been proposed for auxin-induced proton extrusion: direct activation of the proton pump and enhanced synthesis of the plasma membrane H^+-ATPase. The ability of protons to cause cell wall loosening is mediated by a class of proteins called expansins. Expansins loosen the cell wall by breaking hydrogen bonds between the polysaccharide components of the wall. In addition to proton extrusion, long-term auxin-induced growth involves the uptake of solutes and the synthesis and deposition of polysaccharides and proteins needed to maintain the acid-induced wall-loosening capacity.

Promotion of growth in stems and coleoptiles and inhibition of growth in roots are the best-studied physiological effects of auxins. Auxin-promoted differential growth in these organs is responsible for the responses to directional stimuli (i.e., light, gravity) called tropisms. According to the Cholodny–Went model, auxin is transported laterally to the shaded side during phototropism and to the lower side during gravitropism. Statoliths (starch-filled amyloplasts) in the statocytes are involved in the normal percepton of gravity, but they are not absolutely required.

In addition to its roles in growth and tropisms, auxin plays central regulatory roles in apical dominance, lateral-root initiation, leaf abscission, vascular differentiation, floral bud formation, and fruit development. Commercial applications of auxins include rooting compounds and herbicides.

The auxin-binding soluble protein ABP1 is a strong candidate for the auxin receptor. ABP1 is located primarily in the ER lumen. Studies of the signal transduction pathways involved in auxin action have implicated other signaling intermediates such as Ca^{2+}, intracellular pH, and kinases in auxin-induced cell division.

Auxin-induced genes fall into two categories: early and late. Induction of early genes by auxin does not require protein synthesis and is insensitive to protein synthesis inhibitors. The early genes fall into three functional classes: expression of the late genes (secondary response genes), stress adaptation, and intercellular signaling. The auxin response domains of the promoters of the auxin early genes have a composite structure in which an auxin-inducible response element is combined with a constitutive response element. Auxin-induced genes may be negatively regulated by repressor proteins that are degraded via a ubiquitin activation pathway.

Web Material

Web Topics

19.1 Additional Synthetic Auxins

Biologically active synthetic auxins have suprisingly diverse structures.

19.2 The Structural Requirements for Auxin Activity

Comparisons of a wide variety of compounds that possess auxin activity have revealed common features at the molecular level that are essential for biological activity.

19.3 Auxin Measurement by Radioimmunoassy

Radioimmunoassay (RIA) allows the measurement of physiological levels (10^{-9} g = 1 ng) of IAA in plant tissues.

19.4 Evidence for the Tryptophan-Independent Biosynthesis of IAA

Additional experimental evidence for the tryptophan-independent biosynthesis of IAA is provided.

19.5 The Multiple Factors That Regulate Steady-State IAA Levels

The steady-state level of free IAA in the cytosol is determined by several interconnected processes, including synthesis, degradation, conjugation, compartmentation and transport.

19.6 The Mechanism of Fusicoccin Activation of the Plasma Membrane H^+-ATPase

Fusicoccin, a phytotoxin produced by the fungus *Fusicoccum amygdale*, causes membrane hyperpolarization and proton extrusion in nearly all plant tissues, and acts as a "super-auxin" in elongation assays.

19.7 The Fluence Response of Phototropism

The effect of light dose on phototropism is described and a model explaining the phenomenon is presented.

19.8 Differential *SAUR* Gene Expression during Gravitropism

SAUR gene expression is used to detect the lateral auxin gradient during gravitropism.

19.9 Gravity Perception without Statoliths in *Chara*

The giant-celled freshwater alga, *Chara*, bends in response to gravity without any apparent statoliths.

19.10 The Role of Cytokinins in Apical Dominance

In Douglas fir *Pseudotsuga menziesii*, there is a correlation between cytokinin levels and axillary bud growth.

19.11 The Role of ABA in Apical Dominance

In Quackgrass (*Elytrigia repens*) axillary bud growth is correlated with a reduction in ABA.

19.12 The Facilitation of IAA Measurements by *GH3*-Based Reporter Constructs

Because *GH3* expression is a good reflection of the presence of endogenous auxin, a *GH3*-based **reporter** gene, known as *DR5*, is widely used in auxin bioassays.

19.13 The Effect of Auxin on Ubiquitin-Mediated Degradation of AUX/IAA Proteins

A model for auxin-regulated degradation of AUX/IAA proteins is discussed.

Web Essays

19.1 Brassinosteroids: A New Class of Plant Steroid Hormones

Brassinosteroids have been implicated in a wide range of developmental phenomena in plants, including stem elongation, inhibition of root growth, and ethylene biosynthesis.

19.2 Exploring the Cellular Basis of Polar Auxin Transport.

Experimental evidence indicates that the polar transport of the plant hormone auxin is regulated at the cellular level. This implies that proteins involved in auxin transport must be asymmetrically distributed on the plasma membrane. How those transport proteins get to their destination is the focus of ongoing research.

19.3 Phototropism: From Photoperception to Auxin-Dependent Changes in Gene Expression

How photoperception by phototropins is coupled to auxin signaling is the subject of this essay.

Chapter References

Abel, S., Ballas, N., Wong, L-M., and Theologis, A. (1996) DNA elements responsive to auxin. *Bioessays* 18: 647–654.

Aloni, R. (2001) Foliar and axial aspects of vascular differentiation: Hypotheses and evidence. *J. Plant Growth Regul.* 20: 22–34.

Aloni, R. (1995) The induction of vascular tissue by auxin and cytokinin. In *Plant Hormones and Their Role in Plant Growth Development*, 2nd ed., P. J. Davies, ed., Kluwer, Dordrecht, Netherlands, pp. 531–546.

Aloni, R., Schwalm, K., Langhans, M., and Ullrich, C. I. (2002) Gradual shifts in sites and levels of auxin synthesis during leaf-primordium development and their role in vascular differentiation and leaf morphogenesis in *Arabidopsis*. Manuscript submitted for publication.

Bartel, B. (1997) Auxin biosynthesis. *Annu. Rev. Plant Physiol. Plant Mol. Biol.* 48: 51–66.

Bennett, M. J., Marchand, A., Green, H. G., May, S. T., Ward, S. P., Millner, P. A., Walker, A. R., Schultz, B., and Feldmann, K. A. (1996) *Arabidopsis* AUX1 gene: A permease-like regulator of root gravitropism. *Science* 273: 948–950.

Bernasconi, P. (1996) Effect of synthetic and natural protein tyrosine kinase inhibitors on auxin efflux in zucchini (*Cucurbita pepo*) hypocotyls. *Physiol. Plant.* 96: 205–210.

Briggs, W. R., Beck, C. F., Cashmore, A. R., Christie, J. M., Hughes, J., Jarillo, J. A., Kagawa, T., Kanegae, H., Liscum, E., Nagatani, A., Okada, K., Salomon, M., Ruediger, W., Sakai, T., Takano, M., Wada, M., and Watson, J. C. (2001) The phototropin family of photoreceptors. *Plant Cell* 13: 993–997.

Brown, D. E., Rashotte, A. M, Murphy, A. S., Normanly, J., Tague, B.W., Peer W. A., Taiz, L., and Muday, G. K. (2001) Flavonoids act as negative regulators of auxin transport *in vivo* in *Arabidopsis*. *Plant Physiol.* 126: 524–535.

Celenza, J. L., Grisafi, P. L., and Fink, G. R. (1995) A pathway for lateral root formation in *Arabidopsis thaliana*. *Genes Dev.* 9: 2131–2142.

Chen, J. G., Ullah, H., Young, J. C., Sussman, M. R., and Jones, A. M. (2001) ABP1 is required for organized cell elongation and division in *Arabidopsis* embryogenesis. *Genes Dev.* 15: 902–911.

Chen, R., Hilson, P., Sedbrook, J., Rosen, E., Caspar, T., and Masson, P. H. (1998) The *Arabidopsis thaliana* AGRAVITROPIC 1 gene encodes a component of the polar-auxin-transport efflux carrier. *Proc. Natl. Acad. Sci. USA* 95: 15112–15117.

Cleland, R. E. (1995) Auxin and cell elongation. In *Plant Hormones and Their Role in Plant Growth and Development*, 2nd ed., P. J. Davies, ed., Kluwer, Dordrecht, Netherlands, pp. 214–227.

Fasano, J. M., Swanson, S. J., Blancaflor, E. B., Dowd, P. E., Kao, T. H., and Gilroy, S. (2001) Changes in root cap pH are required for the gravity response of the *Arabidopsis* root. *Plant Cell* 13: 907–921.

Friml, J., Wlšniewska, J., Benková, E., Mendgen, K., and Palme, K. (2002) Lateral relocation of auxin efflux regulator PIN3 mediates tropism in *Arabidopsis*. *Nature* 415: 806–809.

Fujihira, K., Kurata, T., Watahiki, M. K., Karahara, I., and Yamamoto, K. T. (2000) An agravitropic mutant of Arabidopsis, *endodermal-amyloplast less 1*, that lacks amyloplasts in hypocotyl endodermal cell layer. *Plant Cell Physiol.* 41: 1193–1199.

Galston, A. (1994) *Life Processes of Plants*. Scientific American Library, New York.

Garbers, C., DeLong, A., Deruere, J., Bernasconi, P., and Soll, D. (1996) A mutation in protein phosphatase 2A regulatory subunit affects auxin transport in *Arabidopsis*. *EMBO J.* 15: 2115–2124.

Geldner, N., Friml, J., Stierhof, Y. D., Jurgens, G., and Palme, K. (2001) Auxin transport inhibitors block PIN1 cycling and vesicle trafficking. *Nature*. 413: 425–428.

Gocal, G. F. W., Pharis, R. P., Yeung, E. C., and Pearce, D. (1991) Changes after decapitation in concentrations of IAA and abscisic acid in the larger axillary bud of *Phaseolus vulgaris* L. cultivar Tender Green. *Plant Physiol.* 95: 344–350.

Gray, W. M., Kepinski, S., Rouse, D., Leyser, O., and Estelle, M. (2001) Auxin regulates the SCFTIR$_1$–dependent degradation of AUX/IAA proteins. *Nature* 414: 271–276.

Hartmann, H. T., and Kester, D. E. (1983) *Plant Propagation: Principles and Practices*, 4th edition. Prentice-Hall, Inc., N.J.

Hasenstein, K. H., and Evans, M. L. (1988) Effects of cations on hormone transport in primary roots of *Zea mays*. *Plant Physiol.* 86: 890–894.

Hsieh, H. L., Okamoto, H, Wang, M. L., Ang, L. H., Matsui, M., Goodman, H., Deng, XW. (2000) *FIN219*, an Auxin-regulated gene, defines a link between phytochrome A and the downstream regulator COP1 in light control of *Arabidopsis* development. *Genes Dev.* 14: 1958–1970.

Iino, M., and Briggs, W. R. (1984) Growth distribution during first positive phototropic curvature of maize coleoptiles. *Plant Cell Environ.* 7: 97–104.

Jacobs, M., and Gilbert, S. F. (1983) Basal localization of the presumptive auxin carrier in pea stem cells. *Science* 220: 1297–1300.

Jacobs, M., and Rubery, P. H. (1988) Naturally occurring auxin transport regulators. *Science* 241: 346–349.

Jacobs, M, and Ray, P. M. (1976) Rapid auxin-induced decrease in the free space pH and its relationship to auxin-induced growth in maize and pea. *Plant Physiol.* 58: 203–209.

Kim, Y.-S., Min, J.-K., Kim, D., and Jung, J. (2001) A soluble auxin-binding protein, ABP57. *J. Biol. Chem.* 276: 10730–10736.

Koens, K. B., Nicoloso, F. T., Harteveld, M., Libbenga, K. R., and Kijne, J. W. (1995) Auxin starvation results in G2-arrest in suspension-cultured tobacco cells. *J. Plant Physiol.* 147: 391–396.

Kuhlemeier, C. and Reinhardt, D. (2001) Auxin and Phyllotaxis. *Trends in Plant Science*. 6: 187–189.

Langridge, W. H. R., Fitzgerald, K. J., Koncz, C., Schell, J., and Szalay, A. A. (1989) Dual promoter of *Agrobacterium tumefaciens* mannopine synthase genes is regulated by plant growth hormones. *Proc. Natl. Acad. Sci. USA* 86: 3219–3223.

Ljung, K., Bhalerao, R. P., and Sandberg, G. (2001) Sites and homeostatic control of auxin biosynthesis in *Arabidopsis* during vegetative growth. *Plant J.* 29: 465–474.

Lomax, T. L. (1986) Active auxin uptake by specific plasma membrane carriers. In *Plant Growth Substances*, M. Bopp, ed., Springer, Berlin, pp. 209–213.

Marchant, A., Kargul, J., May, S. T., Muller, P., Delbarre, A., Perrot-Rechenmann, C., and Bennet, M. J. (1999) AUX1 regulates root gravitropism in *Arabidopsis* by facilitating auxin uptake within root apical tissues. *EMBO J.* 18: 2066–2073.

McClure, B. A., and Guilfoyle, T. (1989) Rapid redistribution of auxin-regulated RNAs during gravitropism. *Science* 243: 91–93.

Mulkey, T. I., Kuzmanoff, K. M., and Evans, M. L. (1981) Correlations between proton-efflux and growth patterns during geotropism and phototropism in maize and sunflower. *Planta* 152: 239–241.

Müller, A., Guan, C., Gälweiler, L., Taenzler, P., Huijser, P., Marchant, A., Parry, G., Bennett, M., Wisman, E., and Palme, K. (1998) AtPIN2 defines a locus of *Arabidopsis* for root gravitropism control. *EMBO J.* 17: 6903–6911.

Murphy, A. S., Peer W. A., and Taiz, L. (2000) Regulation of auxin transport by aminopeptidases and endogenous flavonoids. *Planta* 211: 315–324.

Nakazawa, M., Yabe, N., Ishikawa, T., Yamamoto, Y. Y. Yoshizumi, T., Hasunuma, K., Matsui, M. (2001) *DFL1*, an auxin-responsive *GH3* gene homologue, negatively regulates shoot cell elongation and lateral root formation, and positively regulates the light response of hypocotyls length. *Plant J.* 25: 213–221.

Nonhebel, H. M., Cooney, T. P., and Simpson, R. (1993) The route, control and compartmentation of auxin synthesis. *Aust J. Plant Physiol.* 20: 527–539.

Normanly, J. P., Slovin, J., and Cohen, J. (1995) Rethinking auxin biosynthesis and metabolism. *Plant Physiol.* 107: 323–329.

Palme, K., and Gälweiler, L. (1999) PIN-pointing the molecular basis of auxin transport. *Curr. Opin. Plant Biol.* 2: 375–381.

Parry, G., Delbarre, A., Marchant, A., Swarup, R., Napier, R., Perrot-Rechenmann, C., Bennett, M. J. (2001) Novel auxin transport inhibitors phenocopy the auxin influx carrier mutation aux1. *Plant J.* 25: 399–406.

Peer, W. A., Brown, D., Taiz, L., Muday, G. K., and Murphy, A. S. (2001) Flavonol accumulation patterns correlate with developmental phenotypes of transparent testa mutants of *Arabidopsis thaliana*. *Plant Physiol.* 126: 536–548.

Peltier, J.-B., and Rossignol, M. (1996) Auxin-induced differential sensitivity of the H^+-ATPase in plasma membrane subfractions from tobacco cells. *Biochem. Biophys. Res. Commun.* 219: 492–496.

Romano, C. P., Hein, M. B., and Klee, H. J. (1991) Inactivation of auxin in tobacco transformed with the indoleacetic acid-lysine synthetase gene of *Pseudomonas savastanoi*. *Genes Dev.* 5: 438–446.

Schmidt, R. C., Müller, A., Hain, R., Bartling, D., and Weiler, E. W. (1996) Transgenic tobacco plants expressing the *Arabidopsis thaliana* nitrilase II enzyme. *Plant J.* 9: 683–691.

Shaw, S., and Wilkins, M. B. (1973) The source and lateral transport of growth inhibitors in geotropically stimulated roots of *Zea mays* and *Pisum sativum*. *Planta* 109: 11–26.

Sievers, A., Buchen, B., and Hodick, D. (1996) Gravity sensing in tip-growing cells. *Trends Plant Sci.* 1: 273–279.

Sitbon, F., Edlund, A., Gardestrom, P., Olsson, O., and Sandberg, G. (1993) Compartmentation of indole-3-acetic acid metabolism in protoplasts isolated from leaves of wild-type and IAA-overproducing transgenic tobacco plants. *Planta* 191: 274–279.

Steffens, B., Feckler, C., Palme, K., Christian, M., Bottger, M., and Luethen, H. (2001) The auxin signal for protoplast swelling is perceived by extracellular ABP1. *Plant J.* 27: 591–599.

Swarup, R., Friml, J., Marchant, A., Ljung, K., Sandberg, G., Palme, K., and Bennett, M. (2001) Localization of the auxin permease AUX1 suggests two functionally distinct hormone transport pathways operate in the *Arabidopsis* root apex. *Genes Dev.* 15: 2648–2653.

Ulmasov, T., Murfett, J., Hagen, G., and Guilfoyle, T. J. (1997) Aux/IAA proteins repress expression of reporter genes containing natural and highly active synthetic auxin response elements. *Plant Cell.* 9: 1963–1971.

Utsuno, K., Shikanai, T., Yamada, Y., and Hashimoto, T. (1998) AGR, an AGRAVITROPIC locus of *Arabidopsis thaliana*, encodes a novel membrane protein family member. *Plant Cell Physiol.* 39: 1111–1118.

Venis, M. A., and Napier, R. M. (1997) Auxin perception and signal transduction. In *Signal Transduction in Plants*, P. Aducci, ed., Birkhäuser, Basel, Switzerland, pp. 45–63.

Venis, M. A., Napier, R. M., Oliver, S. (1996) Molecular analysis of auxin-specific signal transduction. *Plant Growth Regulation.* 18: 1–6.

Volkmann, D., and Sievers, A. (1979) Graviperception in multicellular organs. In *Encyclopedia of Plant Physiology*, New Series, Vol. 7, W. Haupt and M. E. Feinleib, eds., Springer, Berlin, pp. 573–600.

Wright, A. D., Sampson, M. B., Neuffer, M. G., Michalczuk, L. P., Slovin, J., and Cohen, J. (1991) Indole-3-acetic acid biosynthesis in the mutant maize orange pericarp, a tryptophan auxotroph. *Science* 254: 998–1000.

Yoder, T. L., Zheng, H.-Q., Todd, P., and Staehelin, L. A. (2001) Amyloplast sedimentation dynamics in maize columella cells support a new model for the gravity-sensing apparatus of roots. *Plant Physiol.* 125: 1045–1060.

Young, L. M., and Evans, M. L. (1994) Calcium-dependent asymmetric movement of 3H-indole-3-acetic acid across gravistimulated isolated root caps of maize. *Plant Growth Regul.* 14: 235–242.

Young, L. M., and Evans, M. L. (1996) Patterns of auxin and abscisic acid movement in the tips of gravistimulated primary roots of maize. *Plant Growth Regul.* 20: 253–258.

Young, L. M., Evans, M. L., and Hertel, R. (1990) Correlations between gravitropic curvature and auxin movement across gravistimulated roots of *Zea mays*. *Plant Physiol.* 92: 792–796.

Zenser, N., Ellsmore, A., Leasure, C., and Callis, J. (2001) Auxin modulates the degradation rate of Aux/IAA proteins. *Proc. Natl. Acad. Sci. USA* 98: 11795–11800.

Zheng, H. Q., and Staehelin, L. A. (2001) Nodal endoplasmic reticulum, a specialized form of endoplasmic reticulum found in gravity-sensing root tip columella cells. *Plant Physiol.* 125: 252–265.

Chapter

20 Gibberellins: Regulators of Plant Height

FOR NEARLY 30 YEARS after the discovery of auxin in 1927, and more than 20 years after its structural elucidation as indole-3-acetic acid, Western plant scientists tried to ascribe the regulation of all developmental phenomena in plants to auxin. However, as we will see in this and subsequent chapters, plant growth and development are regulated by several different types of hormones acting individually and in concert.

In the 1950s the second group of hormones, the gibberellins (GAs), was characterized. The gibberellins are a large group of related compounds (more than 125 are known) that, unlike the auxins, are defined by their chemical structure rather than by their biological activity. Gibberellins are most often associated with the promotion of stem growth, and the application of gibberellin to intact plants can induce large increases in plant height. As we will see, however, gibberellins play important roles in a variety of physiological phenomena.

The biosynthesis of gibberellins is under strict genetic, developmental, and environmental control, and numerous gibberellin-deficient mutants have been isolated. Mendel's tall/dwarf alleles in peas are a famous example. Such mutants have been useful in elucidating the complex pathways of gibberellin biosynthesis.

We begin this chapter by describing the discovery, chemical structure, and role of gibberellins in regulating various physiological processes, including seed germination, mobilization of endosperm storage reserves, shoot growth, flowering, floral development, and fruit set. We then examine biosynthesis of the gibberellins, as well as identification of the active form of the hormone.

In recent years, the application of molecular genetic approaches has led to considerable progress in our understanding of the mechanism of gibberellin action at the molecular level. These advances will be discussed at the end of the chapter.

THE DISCOVERY OF THE GIBBERELLINS

Although gibberellins did not become known to American and British scientists until the 1950s, they had been discovered much earlier by Japanese scientists. Rice farmers in Asia had long known of a disease that makes the rice plants grow tall but eliminates seed production. In Japan this disease was called the "foolish seedling," or *bakanae*, disease.

Plant pathologists investigating the disease found that the tallness of these plants was induced by a chemical secreted by a fungus that had infected the tall plants. This chemical was isolated from filtrates of the cultured fungus and called *gibberellin* after *Gibberella fujikuroi*, the name of the fungus.

In the 1930s Japanese scientists succeeded in obtaining impure crystals of two fungal growth-active compounds, which they termed *gibberellin A* and *B*, but because of communication barriers and World War II, the information did not reach the West. Not until the mid-1950s did two groups—one at the Imperial Chemical Industries (ICI) research station at Welyn in Britain, the other at the U.S. Department of Agriculture (USDA) in Peoria, Illinois—succeed in elucidating the structure of the material that they had purified from fungal culture filtrates, which they named *gibberellic acid*:

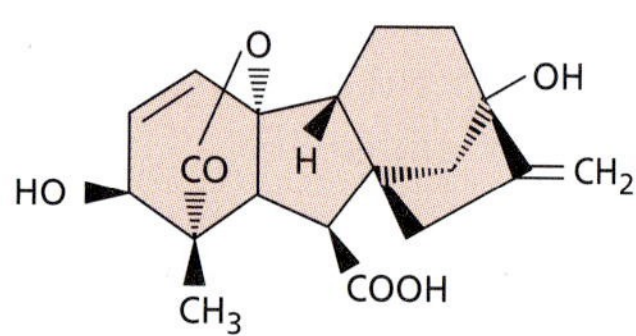

Gibberellic acid (GA_3)

At about the same time scientists at Tokyo University isolated three gibberellins from the original gibberellin A and named them gibberellin A_1, gibberellin A_2, and gibberellin A_3. Gibberellin A_3 and gibberellic acid proved to be identical.

It became evident that an entire family of gibberellins exists and that in each fungal culture different gibberellins predominate, though gibberellic acid is always a principal component. As we will see, the structural feature that all gibberellins have in common, and that defines them as a family of molecules, is that they are derived from the *ent*-kaurene ring structure:

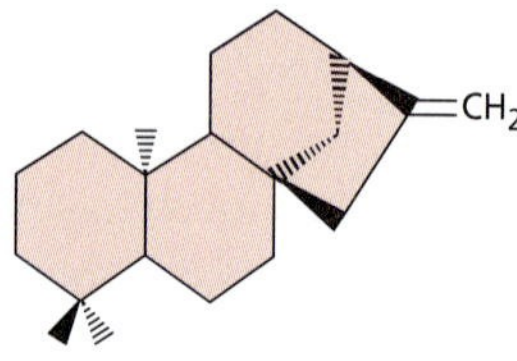

***ent*-Kaurene**

As gibberellic acid became available, physiologists began testing it on a wide variety of plants. Spectacular responses were obtained in the elongation growth of dwarf and rosette plants, particularly in genetically dwarf peas (*Pisum sativum*), dwarf maize (*Zea mays*), and many rosette plants.

In contrast, plants that were genetically very tall showed no further response to applied gibberellins. More recently, experiments with dwarf peas and dwarf corn have confirmed that the natural elongation growth of plants is regulated by gibberellins, as we will describe later.

Because applications of gibberellins could increase the height of dwarf plants, it was natural to ask whether plants contain their own gibberellins. Shortly after the discovery of the growth effects of gibberellic acid, gibberellin-like substances were isolated from several species of plants.[1] *Gibberellin-like substance* refers to a compound or an extract that has gibberellin-like biological activity, but whose chemical structure has not yet been defined. Such a response indicates, but does not prove, that the tested substance is a gibberellin.

In 1958 a gibberellin (gibberellin A_1) was conclusively identified from a higher plant (runner bean seeds, *Phaseolus coccineus*):

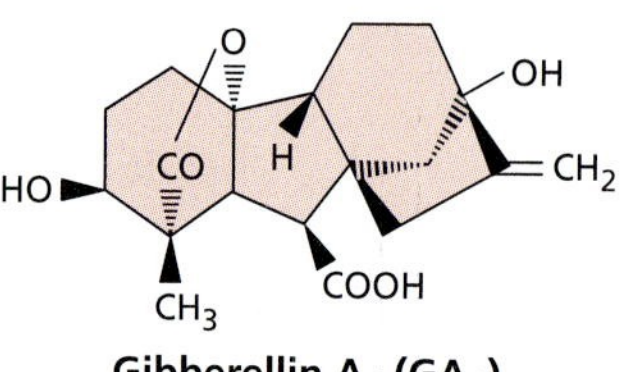

Gibberellin A_1 (GA_1)

Because the concentration of gibberellins in immature seeds far exceeds that in vegetative tissue, immature seeds were the tissue of choice for gibberellin extraction. However, because the concentration of gibberellins in plants is very low (usually 1–10 parts per billion for the active gibberellin in vegetative tissue and up to 1 part per million of total gibberellins in seeds), chemists had to use truckloads of seeds.

As more and more gibberellins from fungal and plant sources were characterized, they were numbered as gibberellin A_X (or GA_X), where *X* is a number, in the order of their discovery. This scheme was adopted for all gibberellins in 1968. However, the number of a gibberellin is simply a cataloging convenience, designed to prevent chaos in the naming of the gibberellins. The system implies no close chemical similarity or metabolic relationship between gibberellins with adjacent numbers.

All gibberellins are based on the *ent*-gibberellane skeleton:

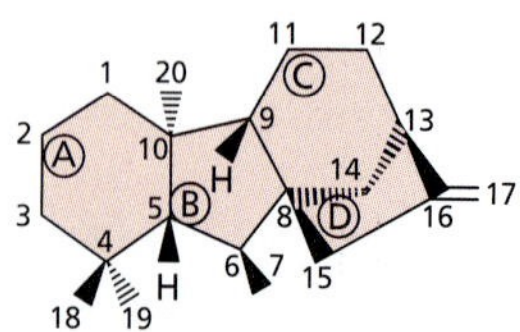

***ent*-Gibberellane structure**

[1] Phinney (1983) provides a wonderful personal account of the history of gibberellin discoveries.

Some gibberellins have the full complement of 20 carbons (C_{20}-GAs):

GA_{12} (a C_{20}-gibberellin)

Others have only 19 (C_{19}-GAs), having lost one carbon to metabolism.

There are other variations in the basic structure, especially the oxidation state of carbon 20 (in C_{20}-GAs) and the number and position of hydroxyl groups on the molecule (see **Web Topic 20.1**). Despite the plethora of gibberellins present in plants, genetic analyses have demonstrated that only a few are biologically active as hormones. All the others serve as precursors or represent inactivated forms.

EFFECTS OF GIBBERELLIN ON GROWTH AND DEVELOPMENT

Though they were originally discovered as the cause of a disease of rice that stimulated internode elongation, endogenous gibberellins influence a wide variety of developmental processes. In addition to stem elongation, gibberellins control various aspects of seed germination, including the loss of dormancy and the mobilization of endosperm reserves. In reproductive development, gibberellin can affect the transition from the juvenile to the mature stage, as well as floral initiation, sex determination, and fruit set. In this section we will review some of these gibberellin-regulated phenomena.

Gibberellins Stimulate Stem Growth in Dwarf and Rosette Plants

Applied gibberellin promotes internodal elongation in a wide range of species. However, the most dramatic stimulations are seen in dwarf and rosette species, as well as members of the grass family. Exogenous GA_3 causes such extreme stem elongation in dwarf plants that they resemble the tallest varieties of the same species (Figure 20.1). Accompanying this effect are a decrease in stem thickness, a decrease in leaf size, and a pale green color of the leaves.

Some plants assume a rosette form in short days and undergo shoot elongation and flowering only in long days (see Chapter 24). Gibberellin application results in *bolting* (stem growth) in plants kept in short days (Figure 20.2), and normal bolting is regulated by endogenous gibberellin.

In addition, as noted earlier, many long-day rosette plants have a cold requirement for stem elongation and flowering, and this requirement is overcome by applied gibberellin.

GA also promotes internodal elongation in members of the grass family. The target of gibberellin action is the **intercalary meristem**—a meristem near the base of the internode that produces derivatives above and below. Deepwater rice is a particularly striking example. We will examine the effects of gibberellin on the growth of deepwater rice in the section on the mechanism of gibberellin-induced stem elongation later in the chapter.

Although stem growth may be dramatically enhanced by GAs, gibberellins have little direct effect on root growth. However, the root growth of extreme dwarfs is less than that of wild-type plants, and gibberellin application to the shoot enhances both shoot and root growth. Whether the effect of gibberellin on root growth is direct or indirect is currently unresolved.

Gibberellins Regulate the Transition from Juvenile to Adult Phases

Many woody perennials do not flower until they reach a certain stage of maturity; up to that stage they are said to

FIGURE 20.1 The effect of exogenous GA_1 on normal and dwarf (*d1*) corn. Gibberellin stimulates dramatic stem elongation in the dwarf mutant but has little or no effect on the tall wild-type plant. (Courtesy of B. Phinney.)

FIGURE 20.2 Cabbage, a long-day plant, remains as a rosette in short days, but it can be induced to bolt and flower by applications of gibberellin. In the case illustrated, giant flowering stalks were produced. (© Sylvan Wittwer/Visuals Unlimited.)

be juvenile (see Chapter 24). The juvenile and mature stages often have different leaf forms, as in English ivy (*Hedera helix*) (see Figure 24.9). Applied gibberellins can regulate this juvenility in both directions, depending on the species. Thus, in English ivy GA_3 can cause a reversion from a mature to a juvenile state, and many juvenile conifers can be induced to enter the reproductive phase by applications of nonpolar gibberellins such as $GA_4 + GA_7$. (The latter example is one instance in which GA_3 is not effective.)

Gibberellins Influence Floral Initiation and Sex Determination

As already noted, gibberellin can substitute for the long-day or cold requirement for flowering in many plants, especially rosette species (see Chapter 24). Gibberellin is thus a component of the flowering stimulus in some plants, but apparently not in others.

In plants where flowers are unisexual rather than hermaphroditic, floral sex determination is genetically regulated. However, it is also influenced by environmental factors, such as photoperiod and nutritional status, and these environmental effects may be mediated by gibberellin. In maize, for example, the staminate flowers (male) are restricted to the tassel, and the pistillate flowers (female) are contained in the ear. Exposure to short days and cool nights increases the endogenous gibberellin levels in the tassels 100-fold and simultaneously causes feminization of the tassel flowers. Application of exogenous gibberellic acid to the tassels can also induce pistillate flowers.

For studies on genetic regulation, a large collection of maize mutants that have altered patterns of sex determination have been isolated. Mutations in genes that affect either gibberellin biosynthesis or gibberellin signal transduction result in a failure to suppress stamen development in the flowers of the ear (Figure 20.3). Thus the primary role of gibberellin in sex determination in maize seems to be to suppress stamen development (Irish 1996).

In dicots such as cucumber, hemp, and spinach, gibberellin seems to have the opposite effect. In these species, application of gibberellin promotes the formation of staminate flowers, and inhibitors of gibberellin biosynthesis promote the formation of pistillate flowers.

Gibberellins Promote Fruit Set

Applications of gibberellins can cause *fruit set* (the initiation of fruit growth following pollination) and growth of some fruits, in cases where auxin may have no effect. For example, stimulation of fruit set by gibberellin has been observed in apple (*Malus sylvestris*).

Gibberellins Promote Seed Germination

Seed germination may require gibberellins for one of several possible steps: the activation of vegetative growth of

FIGURE 20.3 Anthers develop in the ears of a gibberellin-deficient dwarf mutant of corn (*Zea mays*). (Bottom) Unfertilized ear of the dwarf mutant *an1*, showing conspicuous anthers. (Top) Ear from a plant that has been treated with gibberellin. (Courtesy of M. G. Neuffer.)

the embryo, the weakening of a growth-constraining endosperm layer surrounding the embryo, and the mobilization of stored food reserves of the endosperm. Some seeds, particularly those of wild plants, require light or cold to induce germination. In such seeds this dormancy (see Chapter 23) can often be overcome by application of gibberellin. Since changes in gibberellin levels are often, but not always, seen in response to chilling of seeds, gibberellins may represent a natural regulator of one or more of the processes involved in germination.

Gibberellin application also stimulates the production of numerous hydrolases, notably α-amylase, by the aleurone layers of germinating cereal grains. This aspect of gibberellin action has led to its use in the brewing industry in the production of malt (discussed in the next section). Because this is the principal system in which gibberellin signal transduction pathways have been analyzed, it will be treated in detail later in the chapter.

FIGURE 20.4 Gibberellin induces growth in Thompson's seedless grapes. The bunch on the left is an untreated control. The bunch on the right was sprayed with gibberellin during fruit development. (© Sylvan Wittwer/Visuals Unlimited.)

Gibberellins Have Commercial Applications

The major uses of gibberellins (GA_3, unless noted otherwise), applied as a spray or dip, are to manage fruit crops, to malt barley, and to increase sugar yield in sugarcane. In some crops a reduction in height is desirable, and this can be accomplished by the use of gibberellin synthesis inhibitors (see **Web Topic 20.1**).

Fruit production. A major use of gibberellins is to increase the stalk length of seedless grapes. Because of the shortness of the individual fruit stalks, bunches of seedless grapes are too compact and the growth of the berries is restricted. Gibberellin stimulates the stalks to grow longer, thereby allowing the grapes to grow larger by alleviating compaction, and it promotes elongation of the fruit (Figure 20.4).

A mixture of benzyladenine (a cytokinin; see Chapter 21) and $GA_4 + GA_7$ can cause apple fruit to elongate and is used to improve the shape of Delicious-type apples under certain conditions. Although this treatment does not affect yield or taste, it is considered commercially desirable.

In citrus fruits, gibberellins delay senescence, allowing the fruits to be left on the tree longer to extend the market period.

Malting of barley. Malting is the first step in the brewing process. During malting, barley seeds (*Hordeum vulgare*) are allowed to germinate at temperatures that maximize the production of hydrolytic enzymes by the aleurone layer. Gibberellin is sometimes used to speed up the malting process. The germinated seeds are then dried and pulverized to produce "malt," consisting mainly of a mixture of amylolytic (starch-degrading) enzymes and partly digested starch.

During the subsequent "mashing" step, water is added and the amylases in the malt convert the residual starch, as well as added starch, to the disaccharide maltose, which is converted to glucose by the enzyme maltase. The resulting "wort" is then boiled to stop the reaction. In the final step, yeast converts the glucose in the wort to ethanol by fermentation.

Increasing sugarcane yields. Sugarcane (*Saccharum officinarum*) is one of relatively few plants that store their carbohydrate as sugar (sucrose) instead of starch (the other important sugar-storing crop is sugar beet). Originally from New Guinea, sugarcane is a giant perennial grass that can grow from 4 to 6 m tall. The sucrose is stored in the central vacuoles of the internode parenchyma cells. Spraying the crop with gibberellin can increase the yield of raw cane by up to 20 tons per acre, and the sugar yield by 2 tons per acre. This increase is a result of the stimulation of internode elongation during the winter season.

Uses in plant breeding. The long juvenility period in conifers can be detrimental to a breeding program by preventing the reproduction of desirable trees for many years. Spraying with $GA_4 + GA_7$ can considerably reduce the time to seed production by inducing cones to form on very young trees. In addition, the promotion of male flowers in cucurbits, and the stimulation of bolting in biennial rosette crops such as beet (*Beta vulgaris*) and cabbage (*Brassica oleracea*), are beneficial effects of gibberellins that are occasionally used commercially in seed production.

Gibberellin biosynthesis inhibitors. Bigger is not always better. Thus, gibberellin biosynthesis inhibitors are used commercially to prevent elongation growth in some plants. In floral crops, short, stocky plants such as lilies, chrysanthemums, and poinsettias are desirable, and restrictions on elongation growth can be achieved by applications of gibberellin synthesis inhibitors such as ancymidol (known commercially as A-Rest) or paclobutrazol (known as Bonzi).

Tallness is also a disadvantage for cereal crops grown in cool, damp climates, as occur in Europe, where lodging can be a problem. *Lodging*—the bending of stems to the ground caused by the weight of water collecting on the ripened heads—makes it difficult to harvest the grain with a combine harvester. Shorter internodes reduce the tendency of the plants to lodge, increasing the yield of the crop. Even genetically dwarf wheats grown in Europe are sprayed with gibberellin biosynthesis inhibitors to further reduce stem length and lodging.

Yet another application of gibberellin biosynthesis inhibitors is the restriction of growth in roadside shrub plantings.

BIOSYNTHESIS AND METABOLISM OF GIBBERELLIN

Gibberellins constitute a large family of diterpene acids and are synthesized by a branch of the **terpenoid pathway**, which was described in Chapter 13. The elucidation of the gibberellin biosynthetic pathway would not have been possible without the development of sensitive methods of detection. As noted earlier, plants contain a bewildering array of gibberellins, many of which are *biologically inactive*. In this section we will discuss the biosynthesis of GAs, as well as other factors that regulate the steady-state levels of the biologically active form of the hormone in different plant tissues.

Gibberellins Are Measured via Highly Sensitive Physical Techniques

Systems of measurement using a biological response, called *bioassays*, were originally important for detecting gibberellin-like activity in partly purified extracts and for assessing the biological activity of known gibberellins (Figure 20.5). The use of bioassays, however, has declined with the development of highly sensitive physical techniques that allow precise identification and quantification of specific gibberellins from small amounts of tissue.

High-performance liquid chromatography (HPLC) of plant extracts, followed by the highly sensitive and selective analytical method of gas chromatography combined with mass spectrometry (GC-MS), has now become the method of choice. With the availability of published mass spectra, researchers can now identify gibberellins without possessing pure standards. The availability of heavy-isotope-labeled standards of common gibberellins, which can themselves be separately detected on a mass spectrometer, allows the accurate measurement of levels in plant tissues by mass spectrometry with these heavy-isotope-labeled gibberellins as internal standards for quantification (see **Web Topic 20.2**).

Gibberellins Are Synthesized via the Terpenoid Pathway in Three Stages

Gibberellins are tetracyclic diterpenoids made up of four isoprenoid units. Terpenoids are compounds made up of five-carbon (isoprene) building blocks:

$$-CH_2-\overset{\overset{\large CH_3}{|}}{C}=CH-CH_2-$$

joined head to tail. Researchers have determined the entire gibberellin biosynthetic pathway in seed and vegetative tissues of several species by feeding various radioactive precursors and intermediates and examining the production of the other compounds of the pathway (Kobayashi et al. 1996).

The gibberellin biosynthetic pathway can be divided into three stages, each residing in a different cellular compartment (Figure 20.6) (Hedden and Phillips 2000).

FIGURE 20.5 Gibberellin causes elongation of the leaf sheath of rice seedlings, and this response is used in the dwarf rice leaf sheath bioassay. Here 4-day-old seedlings were treated with different amounts of GA and allowed to grow for another 5 days. (Courtesy of P. Davies.)

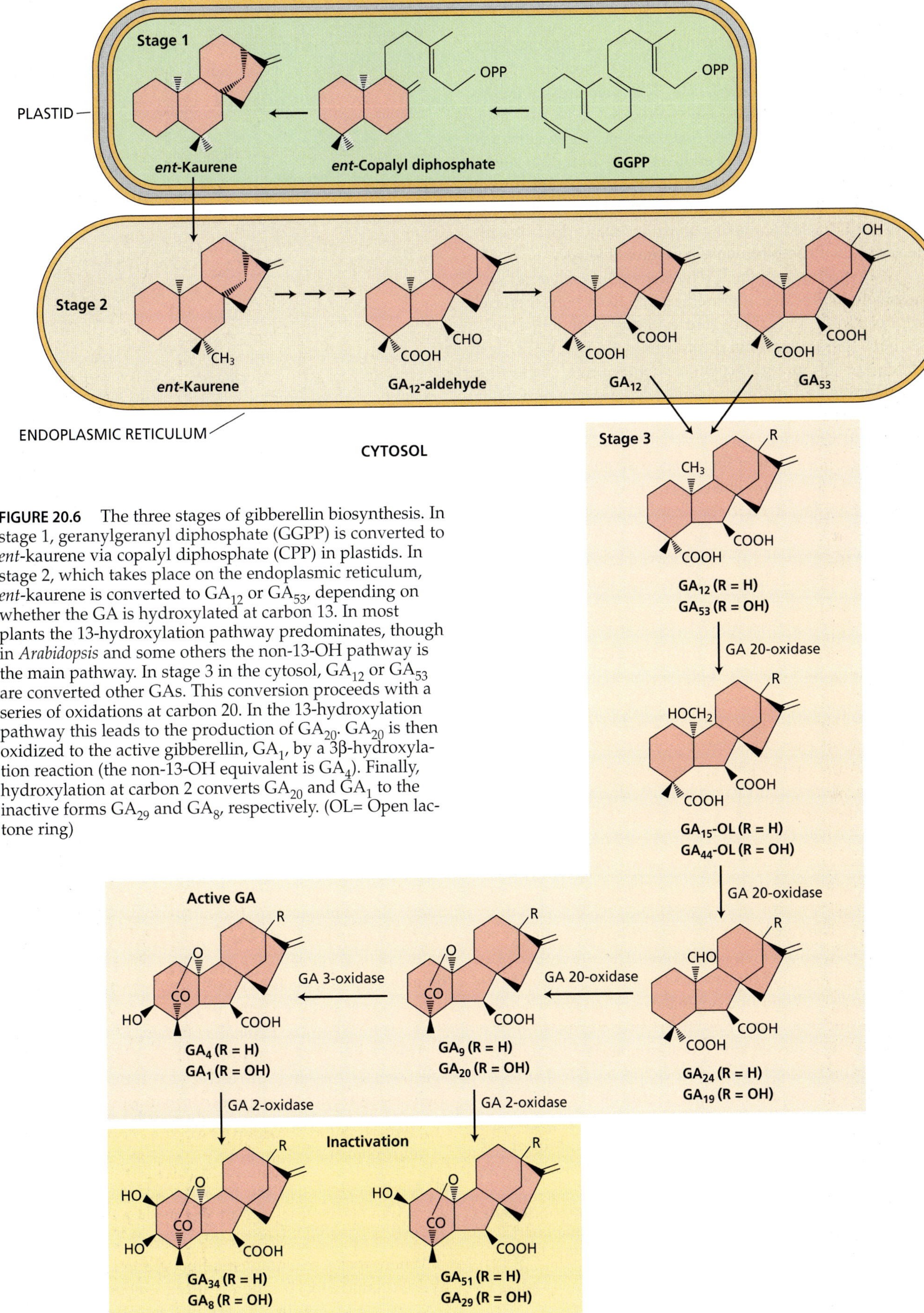

FIGURE 20.6 The three stages of gibberellin biosynthesis. In stage 1, geranylgeranyl diphosphate (GGPP) is converted to *ent*-kaurene via copalyl diphosphate (CPP) in plastids. In stage 2, which takes place on the endoplasmic reticulum, *ent*-kaurene is converted to GA_{12} or GA_{53}, depending on whether the GA is hydroxylated at carbon 13. In most plants the 13-hydroxylation pathway predominates, though in *Arabidopsis* and some others the non-13-OH pathway is the main pathway. In stage 3 in the cytosol, GA_{12} or GA_{53} are converted other GAs. This conversion proceeds with a series of oxidations at carbon 20. In the 13-hydroxylation pathway this leads to the production of GA_{20}. GA_{20} is then oxidized to the active gibberellin, GA_1, by a 3β-hydroxylation reaction (the non-13-OH equivalent is GA_4). Finally, hydroxylation at carbon 2 converts GA_{20} and GA_1 to the inactive forms GA_{29} and GA_8, respectively. (OL= Open lactone ring)

Stage 1: Production of terpenoid precursors and ent-kaurene in plastids. The basic biological isoprene unit is isopentenyl diphosphate (IPP).[2] IPP used in gibberellin biosynthesis in green tissues is synthesized in plastids from glyceraldehyde-3-phosphate and pyruvate (Lichtenthaler et al. 1997). However, in the endosperm of pumpkin seeds, which are very rich in gibberellin, IPP is formed in the cytosol from mevalonic acid, which is itself derived from acetyl-CoA. Thus the IPP used to make gibberellins may arise from different cellular compartments in different tissues.

Once synthesized, the IPP isoprene units are added successively to produce intermediates of 10 carbons (geranyl diphosphate), 15 carbons (farnesyl diphosphate), and 20 carbons (geranylgeranyl diphosphate, GGPP). GGPP is a precursor of many terpenoid compounds, including carotenoids and many essential oils, and it is only after GGPP that the pathway becomes specific for gibberellins.

The cyclization reactions that convert GGPP to *ent*-kaurene represent the first step that is specific for the gibberellins (Figure 20.6). The two enzymes that catalyze the reactions are localized in the proplastids of meristematic shoot tissues, and they are not present in mature chloroplasts (Aach et al. 1997). Thus, leaves lose their ability to synthesize gibberellins from IPP once their chloroplasts mature.

Compounds such as AMO-1618, Cycocel, and Phosphon D are specific inhibitors of the first stage of gibberellin biosynthesis, and they are used as growth height reducers.

Stage 2: Oxidation reactions on the ER form GA_{12} and GA_{53}. In the second stage of gibberellin biosynthesis, a methyl group on *ent*-kaurene is oxidized to a carboxylic acid, followed by contraction of the B ring from a six- to a five-carbon ring to give GA_{12}-aldehyde. GA_{12}-aldehyde is then oxidized to $\mathbf{GA_{12}}$, the first gibberellin in the pathway in all plants and thus the precursor of all the other gibberellins (see Figure 20.6).

Many gibberellins in plants are also hydroxylated on carbon 13. The hydroxylation of carbon 13 occurs next, forming GA_{53} from GA_{12}. All the enzymes involved are monooxygenases that utilize cytochrome P450 in their reactions. These P450 monooxygenases are localized on the endoplasmic reticulum. Kaurene is transported from the plastid to the endoplasmic reticulum, and is oxidized *en route* to kaurenoic acid by kaurene oxidase, which is associated with the plastid envelope (Helliwell et al. 2001).

Further conversions to GA_{12} take place on the endoplasmic reticulum. Paclobutrazol and other inhibitors of P450 monooxygenases specifically inhibit this stage of gibberellin biosynthesis before GA_{12}-aldehyde, and they are also growth retardants.

[2] As noted in Chapter 13, IPP is the abbreviation for isopentenyl *pyro*phosphate, an earlier name for this compound. Similarly, the other pyrophosphorylated intermediates in the pathway are now referred to as *di*phosphates, but they continue to be abbreviated as if they were called *pyro*phosphates.

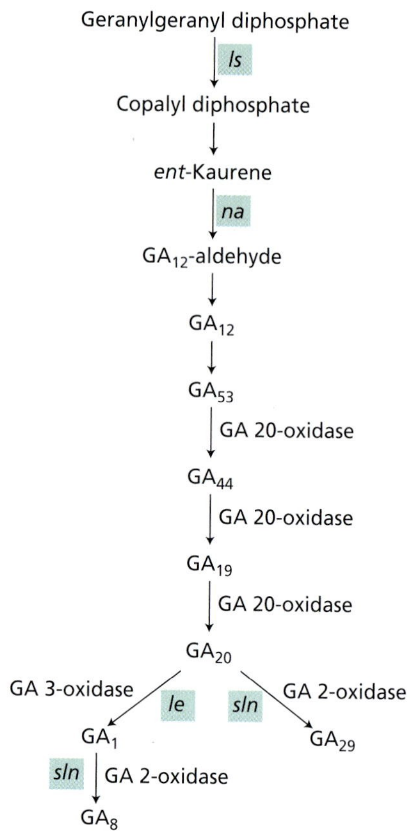

FIGURE 20.7 A portion of the gibberellin biosynthetic pathway showing the abbreviations and location of the mutant genes that block the pathway in pea and the enzymes involved in the metabolic steps after GA_{53}.

Stage 3: Formation in the cytosol of all other gibberellins from GA_{12} or GA_{53}. All subsequent steps in the pathway (see Figure 20.6) are carried out by a group of soluble dioxygenases in the cytosol. These enzymes require 2-oxoglutarate and molecular oxygen as cosubstrates, and they use Fe^{2+} and ascorbate as cofactors.

The specific steps in the modification of GA_{12} vary from species to species, and between organs of the same species. Two basic chemical changes occur in most plants:

1. Hydroxylation at carbon 13 (on the endoplasmic reticulum) or carbon 3, or both.

2. A successive oxidation at carbon 20 ($CH_2 \rightarrow CH_2OH \rightarrow CHO$). The final step of this oxidation is the loss of carbon 20 as CO_2 (see Figure 20.6).

When these reactions involve gibberellins initially hydroxylated at C-13, the resulting gibberellin is GA_{20}. GA_{20} is then converted to the biologically active form,

GA_1, by hydroxylation of carbon 3. (Because this is in the beta configuration [drawn as if the bond to the hydroxyl group were toward the viewer], it is referred to as 3β-hydroxylation.)

Finally, GA_1 is inactivated by its conversion to GA_8 by a hydroxylation on carbon 2. This hydroxylation can also remove GA_{20} from the biosynthetic pathway by converting it to GA_{29}.

Inhibitors of the third stage of the gibberellin biosynthetic pathway interfere with enzymes that utilize 2-oxoglutarate as cosubstrates. Among these, the compound prohexadione (BX-112), is especially useful because it specifically inhibits GA 3-oxidase, the enzyme that converts inactive GA_{20} to growth-active GA_1.

The Enzymes and Genes of the Gibberellin Biosynthetic Pathway Have Been Characterized

The enzymes of the gibberellin biosynthetic pathway are now known, and the genes for many of these enzymes have been isolated and characterized (see Figure 20.7). Most notable from a regulatory standpoint are two biosynthetic enzymes—GA 20-oxidase (GA20ox)[3] and GA 3-oxidase (GA3ox)—and an enzyme involved in gibberellin metabolism, GA 2-oxidase (GA2ox):

- **GA 20-oxidase** catalyzes all the reactions involving the successive oxidation steps of carbon 20 between GA_{53} and GA_{20}, including the removal of C-20 as CO_2.
- **GA 3-oxidase** functions as a 3β-hydroxylase, adding a hydroxyl group to C-3 to form the active gibberellin, GA_1. (The evidence demonstrating that GA_1 is the active gibberellin will be discussed shortly.)
- **GA 2-oxidase** *inactivates* GA_1 by catalyzing the addition of a hydroxyl group to C-2.

The transcription of the genes for the two gibberellin biosynthetic enzymes, as well as for GA 2-oxidase, is highly regulated. All three of these genes have sequences in common with each other and with other enzymes utilizing 2-oxoglutarate and Fe^{2+} as cofactors. The common sequences represent the binding sites for 2-oxoglutarate and Fe^{2+}.

Gibberellins May Be Covalently Linked to Sugars

Although active gibberellins are free, a variety of gibberellin glycosides are formed by a covalent linkage between gibberellin and a sugar. These gibberellin conjugates are particularly prevalent in some seeds. The conjugating sugar is usually glucose, and it may be attached to the gibberellin via a carboxyl group forming a gibberellin glycoside, or via a hydroxyl group forming a gibberellin glycosyl ether.

When gibberellins are applied to a plant, a certain proportion usually becomes glycosylated. Glycosylation therefore represents another form of inactivation. In some cases, applied glucosides are metabolized back to free GAs, so glucosides may also be a storage form of gibberellins (Schneider and Schmidt 1990).

GA_1 Is the Biologically Active Gibberellin Controlling Stem Growth

Knowledge of biosynthetic pathways for gibberellins reveals where and how dwarf mutations act (Figure 20.7). Although it had long been assumed that gibberellins were natural growth regulators because gibberellin application caused dwarf plants to grow tall, direct evidence was initially lacking. In the early 1980s it was demonstrated that tall stems do contain more bioactive gibberellin than dwarf stems have, and that the level of the endogenous bioactive gibberellin mediates the genetic control of tallness (Reid and Howell 1995).

The gibberellins of tall pea plants containing the homozygous *Le* allele (wild type) were compared with dwarf plants having the same genetic makeup, except containing the *le* allele (mutant). *Le* and *le* are the two alleles of the gene that regulates tallness in peas, the genetic trait first investigated by Gregor Mendel in his pioneering study in 1866. We now know that tall peas contain much more bioactive GA_1 than dwarf peas have (Ingram et al. 1983).

As we have seen, the precursor of GA_1 in higher plants is GA_{20} (GA_1 is 3β-OH GA_{20}). If GA_{20} is applied to homozygous dwarf (*le*) pea plants, they fail to respond, although they do respond to applied GA_1. The implication is that the *Le* gene enables the plants to convert GA_{20} to GA_1. Metabolic studies using both stable and radioactive isotopes demonstrated conclusively that the *Le* gene encodes an enzyme that 3β-hydroxylates GA_{20} to produce GA_1 (Figure 20.8).

Mendel's *Le* gene was isolated, and the recessive *le* allele was shown to have a single base change leading to a defective enzyme only one-twentieth as active as the wild-type

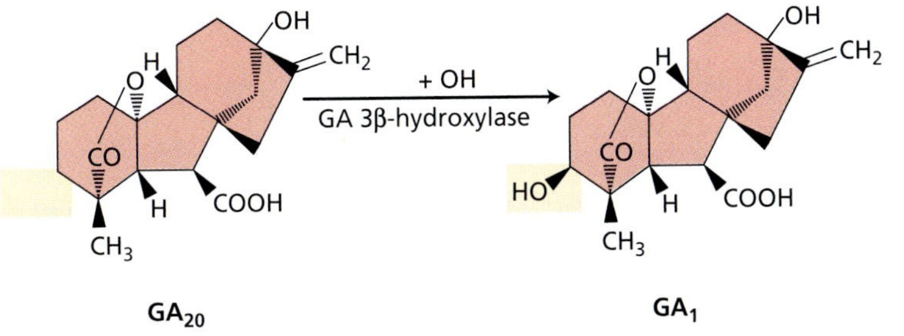

OH, CH_2, O, H, CO, H, COOH, CH_3 — + OH, GA 3β-hydroxylase → OH, CH_2, O, H, HO, CO, H, COOH, CH_3

GA_{20} GA_1

FIGURE 20.8 Conversion of GA_{20} to GA_1 by GA 3β-hydroxylase, which adds a hydroxyl group (OH) to carbon 3 of GA_{20}.

[3] *GA 20-oxidase* means an enzyme that oxidizes at carbon 20; it is not the same as GA_{20}, which is gibberellin 20 in the GA numbering scheme.

enzyme, so much less GA_1 is produced and the plants are dwarf (Lester et al. 1997).

Endogenous GA_1 Levels Are Correlated with Tallness

Although the shoots of gibberellin-deficient *le* dwarf peas are much shorter than those of normal plants (internodes of 3 cm in mature dwarf plants versus 15 cm in mature normal plants), the mutation is "leaky" (i.e., the mutated gene produces a partially active enzyme) and some endogenous GA_1 remains to cause growth. Different *le* alleles give rise to peas differing in their height, and the height of the plant has been correlated with the amount of endogenous GA_1 (Figure 20.9).

There is also an extreme dwarf mutant of pea that has even fewer gibberellins. This dwarf has the allele *na* (the wild-type allele is *Na*), which completely blocks gibberellin biosynthesis between *ent*-kaurene and GA_{12}-aldehyde (Reid and Howell 1995). As a result, homozygous (*nana*) mutants, which are almost completely free of gibberellins, achieve a stature of only about 1 cm at maturity (Figure 20.10).

However, *nana* plants may still possess an active GA 3β-hydroxylase encoded by *Le*, and thus can convert GA_{20} to GA_1. If a *nana naLe* shoot is grafted onto a dwarf *le* plant, the resulting plant is tall because the *nana* shoot tip can convert the GA_{20} from the dwarf into GA_1.

Such observations have led to the conclusion that GA_1 is the biologically active gibberellin that regulates tallness in peas (Ingram et al. 1986; Davies 1995). The same result has been obtained for maize, a monocot, in parallel studies using genotypes that have blocks in the gibberellin biosynthetic pathway. Thus the control of stem elongation by GA_1 appears to be universal.

Although GA_1 appears to be the primary active gibberellin in stem growth for most species, a few other gib-

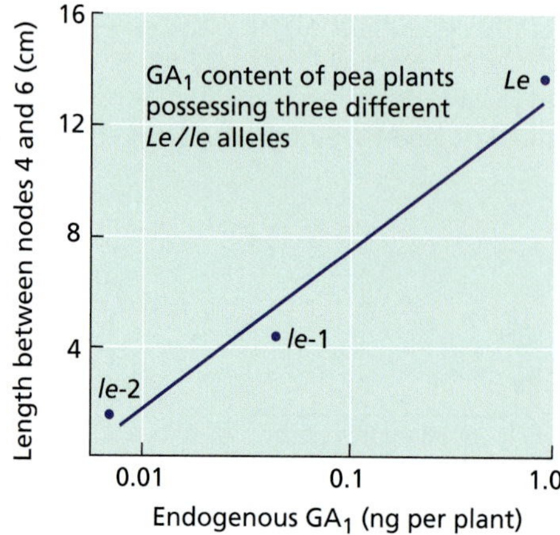

FIGURE 20.9 Stem elongation corresponds closely to the level of GA_1. Here the GA_1 content in peas with three different alleles at the *Le* locus is plotted against the internode elongation in plants with those alleles. The allele *le*-2 is a more intense dwarfing allele of *Le* than is the regular *le*-1 allele. There is a close correlation between the GA level and internode elongation. (After Ross et al. 1989.)

FIGURE 20.10 Phenotypes and genotypes of peas that differ in the gibberellin content of their vegetative tissue. (All alleles are homozygous.) (After Davies 1995.)

berellins have biological activity in other species or tissues. For example, GA_3, which differs from GA_1 only in having one double bond, is relatively rare in higher plants but is able to substitute for GA_1 in most bioassays:

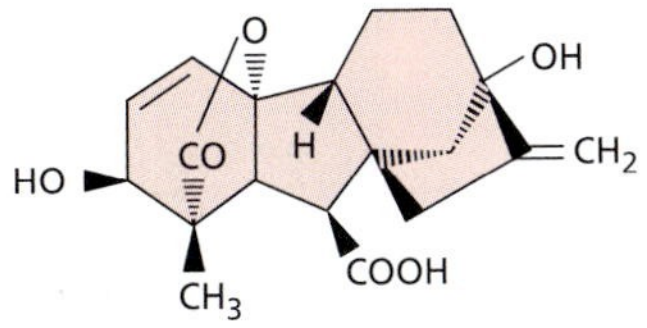

Gibberellic acid (GA_3)

GA_4, which lacks an OH group at C-13, is present in both *Arabidopsis* and members of the squash family (Cucurbitaceae). It is as active as GA_1, or even more active, in some bioassays, indicating that GA_4 is a bioactive gibberellin in the species where it occurs (Xu et al. 1997). The structure of GA_4 looks like this:

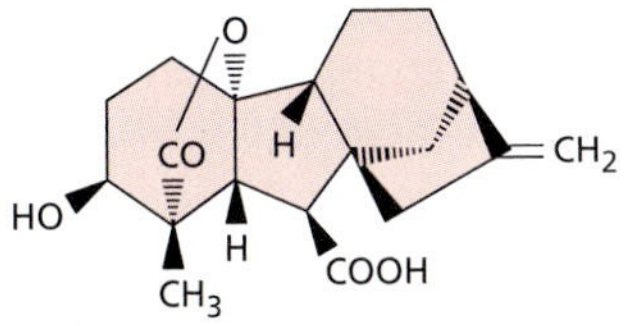

Gibberellin A_4 (GA_4)

Gibberellins Are Biosynthesized in Apical Tissues

The highest levels of gibberellins are found in immature seeds and developing fruits. However, because the gibberellin level normally decreases to zero in mature seeds, there is no evidence that seedlings obtain any active gibberellins from their seeds.

Work with pea seedlings indicates that the gibberellin biosynthetic enzymes and GA3ox are specifically localized in young, actively growing buds, leaves, and upper internodes (Elliott et al. 2001). In *Arabidopsis*, GA20ox is expressed primarily in the apical bud and young leaves, which thus appear to be the principal sites of gibberellin synthesis (Figure 20.11).

The gibberellins that are synthesized in the shoot can be transported to the rest of the plant via the phloem. Intermediates of gibberellin biosynthesis may also be translocated in the phloem. Indeed, the initial steps of gibberellin biosynthesis may occur in one tissue, and metabolism to active gibberellins in another.

Gibberellins also have been identified in root exudates and root extracts, suggesting that roots can also synthesize gibberellins and transport them to the shoot via the xylem.

Gibberellin Regulates Its Own Metabolism

Endogenous gibberellin regulates its own metabolism by either switching on or inhibiting the transcription of the genes that encode enzymes of gibberellin biosynthesis and degradation (feedback and feed-forward regulation, respectively). In this way the level of active gibberellins is kept within a narrow range, provided that precursors are available and the enzymes of gibberellin biosynthesis and degradation are functional.

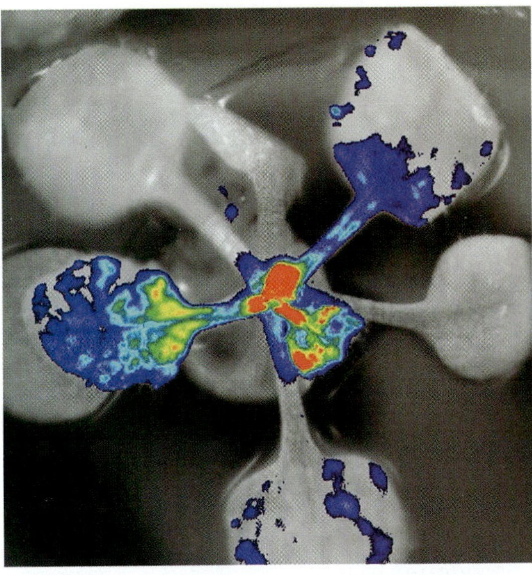

FIGURE 20.11 Gibberellin is synthesized mainly in the shoot apex and in young developing leaves. This false color image shows light emitted by transgenic *Arabidopsis* plants expressing the firefly luciferase coding sequence coupled to the GA20ox gene promoter. The emitted light was recorded by a CCD camera after the rosette was sprayed with the substrate luciferin. The image was then color-coded for intensity and superimposed on a photograph of the same plant. The red and yellow regions correspond to the highest light intensity. (Courtesy of Jeremy P. Coles, Andrew L. Phillips, and Peter Hedden, IACR-Long Ashton Research Station.)

For example, the application of gibberellin causes a down-regulation of the biosynthetic genes—GA20ox and GA3ox—and an elevation in transcription of the degradative gene—GA2ox (Hedden and Phillips 2000; Elliott et al. 2001). A mutation in the GA 2-oxidase gene, which prevents GA_1 from being degraded, is functionally equivalent to applying exogenous gibberellin to the plant, and produces the same effect on the biosynthetic gene transcription.

Conversely, a mutation that lowers the level of active gibberellin, such as GA_1, in the plant stimulates the transcription of the biosynthetic genes—GA20ox and GA3ox—and down-regulates the degradative enzyme—GA2ox. In peas this is particularly evident in very dwarf plants, such as those with a mutation in the *LS* gene (CPP synthase) or even more severely dwarf *na* plants (defective GA_{12}-aldehyde synthase) (Figure 20.12).

Environmental Conditions Can Alter the Transcription of Gibberellin Biosynthesis Genes

Gibberellins play an important role in mediating the effects of environmental stimuli on plant development. Environmental factors such as photoperiod and temperature can alter the levels of active gibberellins by affecting gene transcription for specific steps in the biosynthetic pathway (Yamaguchi and Kamiya 2000).

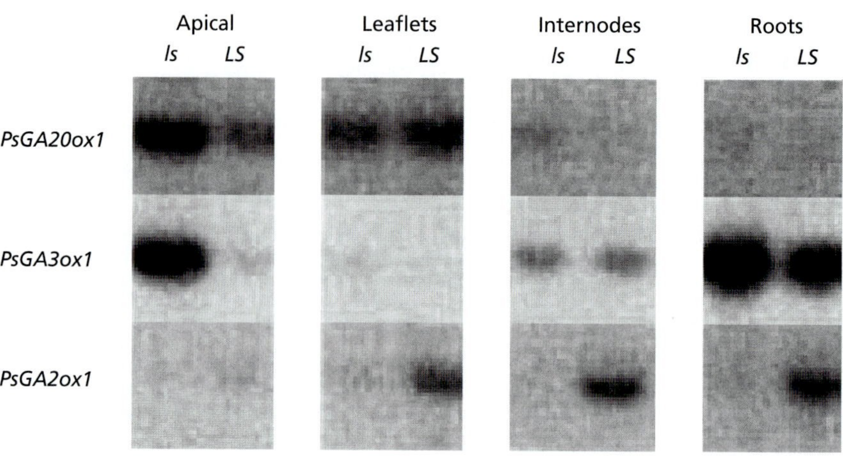

FIGURE 20.12 Northern blots of the mRNA for the enzymes of gibberellin biosynthesis in different tissues of peas. The more intense the band, the more mRNA was present. The plants designated *LS* are tall wild-type plants. Those designated *ls* are very dwarf mutants due to a defective copalyl diphosphate synthase that creates a block in the GA biosynthesis pathway. The differences in the spot intensity show that a low level of GA_1 in the mutant *ls* plants causes the upregulation of GA_1 biosynthesis by GA20ox and GA3ox, and a repression of GA_1 breakdown by GA2ox. (From Elliott et al. 2001.)

Light regulation of GA_1 biosynthesis. The presence of light has many profound effects. Some seeds germinate only in the light, and in such cases gibberellin application can stimulate germination in darkness. The promotion of germination by light has been shown to be due to increases in GA_1 levels resulting from a light-induced increase in the transcription of the gene for GA3ox, which converts GA_{20} to GA_1 (Toyomasu et al. 1998). This effect shows red/far-red photoreversibility and is mediated by phytochrome (see Chapter 17).

When a seedling becomes exposed to light as it emerges from the soil, it changes its form (see Chapter 17)—a process referred to as de-etiolation. One of the most striking changes is a decrease in the rate of stem elongation such that the stem in the light is shorter than the one in the dark. Initially it was assumed that the light-grown plants would contain less GA_1 than dark-grown plants. However, light-grown plants turned out to contain *more* GA_1 than dark-grown plants—indicating that de-etiolation is a complex process involving changes in the level of GA_1, as well as changes in the responsiveness of the plant to GA_1.

In peas, for example, the level of GA_1 initially falls within 4 hours of exposure to light because of an increase in transcription of the gene for GA2ox, leading to an increase in GA_1 breakdown (Figure 20.13A). The level of GA_1 remains low for a day but then increases, so that by

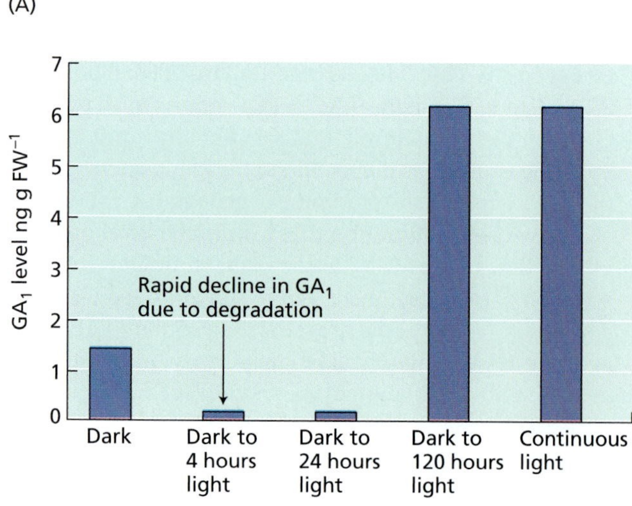

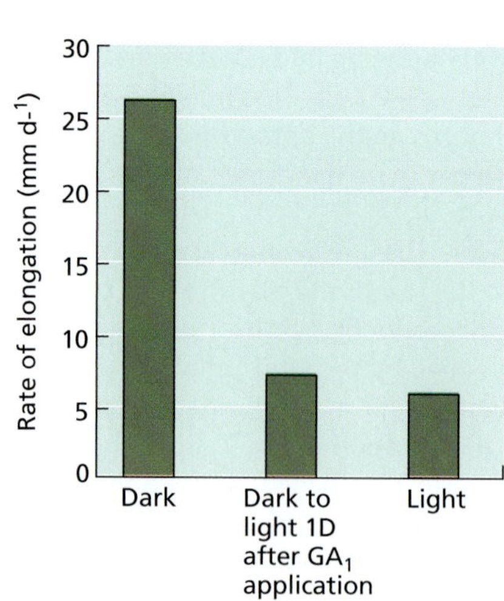

FIGURE 20.13 When a plant grows in the light, the rate of extension slows down through regulation by changes in hormone levels and sensitivity. (A) When dark-grown pea seedlings are transferred to light, GA_1 level drops rapidly because of metabolism of GA_1, but then increases to a higher level, similar to that of light-grown plants, over the next 4 days. (B) To investigate the GA_1 response in various light regimes, 10 μm of GA_1 was applied to the internode of GA-deficient *na* plants in darkness, 1 day after the start of the light, or 6 days of continuous light, and growth in the next 24 hours was measured. The results show that the gibberellin sensitivity of pea seedlings falls rapidly upon transfer from darkness to light, so the elongation rate of plants in the light is lower than in the dark, even though their total GA_1 content is higher. (After O'Neill et al. 2000.)

FIGURE 20.14 Spinach plants undergo stem and petiole elongation only in long days, remaining in a rosette form in short days. Treatment with the GA biosynthesis inhibitor AMO-1618 prevents stem and petiole elongation and maintains the rosette growth habit even under long days. Gibberellic acid can reverse the inhibitory effect of AMO-1618 on stem and petiole elongation. As shown in Figure 20.16, long days cause changes in the gibberellin content of the plant. (Courtesy of J. A. D. Zeevaart.)

5 days there is a fivefold increase in the GA_1 content of the stems, even though the stem elongation rate is lower (Figure 20.13B) (O'Neill et al. 2000). The reason that growth slows down despite the increase in GA_1 level is that the plants are now severalfold *less sensitive* to the GA_1 present.

As will be discussed later in the chapter, sensitivity to active gibberellin is governed by components of the gibberellin signal transduction pathway.

Photoperiod regulation of GA_1 biosynthesis. When plants that require long days to flower (see Chapter 24) are shifted from short days to long days, gibberellin metabolism is altered. In spinach (*Spinacia oleracea*), in short days, when the plants maintain a rosette form (Figure 20.14), the level of gibberellins hydroxylated at carbon 13 is relatively low. In response to increasing day length, the shoots of spinach plants begin to elongate after about 14 long days.

The levels of all the gibberellins of the carbon 13–hydroxylated gibberellin pathway ($GA_{53} \rightarrow GA_{44} \rightarrow GA_{19} \rightarrow GA_{20} \rightarrow GA_1 \rightarrow GA_8$) start to increase after about 4 days (Figure 20.15). Although the level of GA_{20} increases 16-fold during the first 12 days, it is the fivefold increase in GA_1 that induces stem growth (Zeevaart et al. 1993).

The dependence of stem growth on GA_1 has been shown through the use of different inhibitors of gibberellin synthesis and metabolism. The inhibitors AMO-1618 and BX-112 both prevent internode elongation (bolting). The effect of AMO-1618, which blocks gibberellin biosynthesis prior to GA_{12}-aldehyde, can be overcome by applications of GA_{20} (Figure 20.16A). However, the effect of another inhibitor, BX-112, which blocks the production of GA_1 from GA_{20}, can be overcome only by GA_1 (Figure 20.16B). This result demonstrates that the rise in GA_1 is the crucial factor in regulating spinach stem growth.

The level of GA 20-oxidase mRNA in spinach tissues, which occurs in the highest amount in shoot tips and elongating stems (see Figure 20.11), is increased under long-day conditions (Wu et al. 1996). The fact that GA 20-oxidase is the enzyme that converts GA_{53} to GA_{20} (see Figure 20.7) explains why the concentration of GA_{20} was found to be higher in spinach under long-day conditions (Zeevaart et al. 1993).

Photoperiod control of tuber formation. Potato tuberization is another process regulated by photoperiod (Figure 20.17). Tubers form on wild potatoes only in short days (although the requirement for short days has been bred out of many cultivated varieties), and this tuberization can be blocked by applications of gibberellin. The transcription of GA20ox was found to fluctuate during the light–dark cycle, leading to lower levels of GA_1 in short days. Potato plants overexpressing the GA20ox gene showed delayed tuberization, whereas trans-

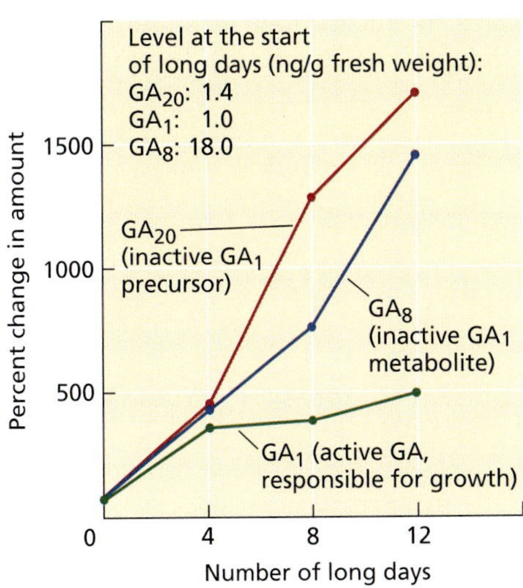

FIGURE 20.15 The fivefold increase in GA_1 is what causes growth in spinach exposed to an increasing number of long days but before stem elongation starts at about 14 days. (After Davies 1995; redrawn from data in Zeevaart et al. 1993.)

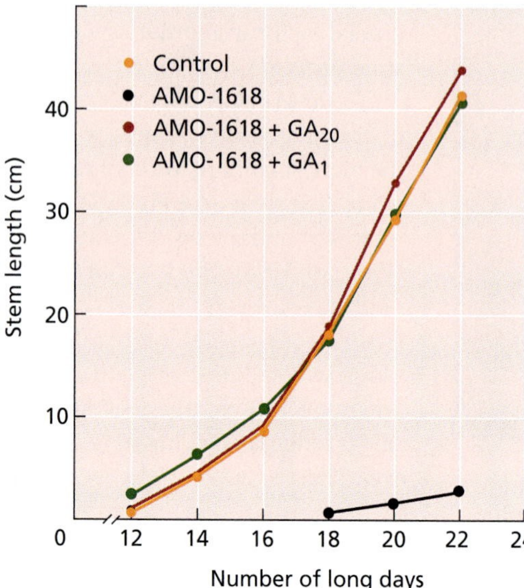

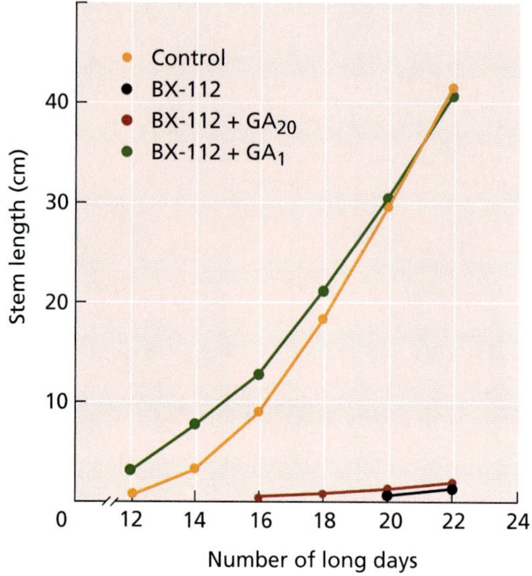

FIGURE 20.16 The use of specific growth retardants (GA biosynthesis inhibitors) and the reversal of the effects of the growth retardants by different GAs can show which steps in GA biosynthesis are regulated by environmental change, in this case the effect of long days on stem growth in spinach. The control lacks inhibitors or added GA. (After Zeevaart et al. 1993.)

formation with the antisense gene for GA20ox promoted tuberization, demonstrating the importance of the transcription of this gene in the regulation of potato tuberization (Carrera et al. 2000).

In general, de-etiolation, light-dependent seed germination, and the photoperiodic control of stem growth in rosette plants and tuberization in potato are all mediated by phytochromes (see Chapter 17). There is mounting evidence that many phytochrome effects are in part due to modulation of the levels of gibberellins through changes in the transcription of the genes for gibberellin biosynthesis and degradation.

Temperature effects. Cold temperatures are required for the germination of certain seeds (stratification) and for flowering in certain species (vernalization) (see Chapter

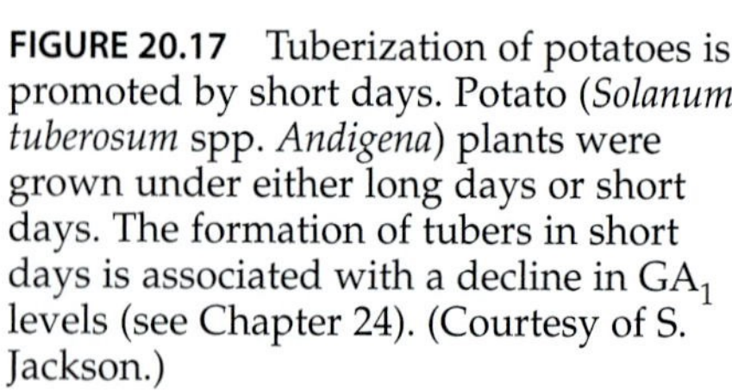

FIGURE 20.17 Tuberization of potatoes is promoted by short days. Potato (*Solanum tuberosum* spp. *Andigena*) plants were grown under either long days or short days. The formation of tubers in short days is associated with a decline in GA_1 levels (see Chapter 24). (Courtesy of S. Jackson.)

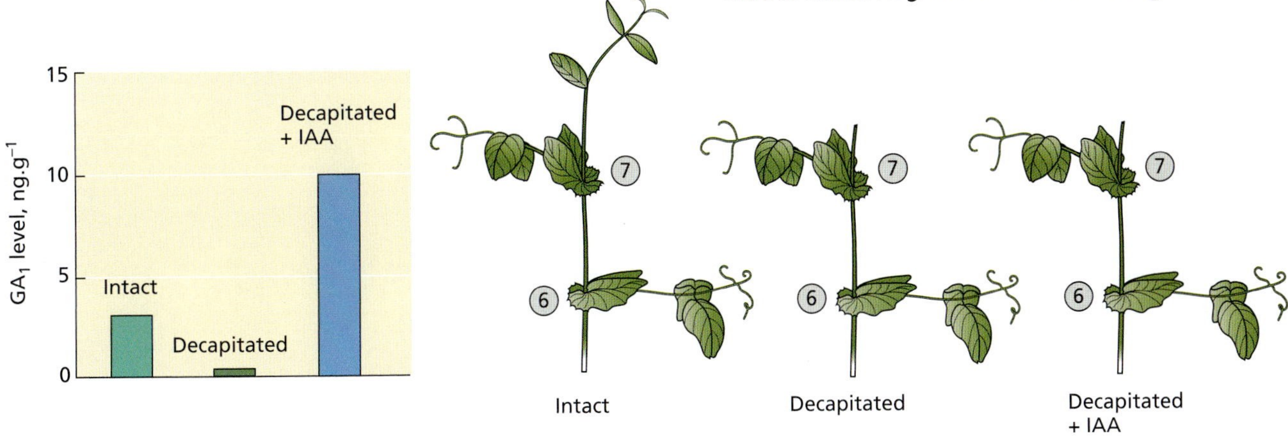

FIGURE 20.18 Decapitation reduces, and IAA (auxin) restores, endogenous GA_1 content in pea plants. Numbers refer to the leaf node. (From Ross et al. 2000.)

24). For example, a prolonged cold treatment is required for both the stem elongation and the flowering of *Thlaspi arvense* (field pennycress), and gibberellins can substitute for the cold treatment.

In the absence of the cold treatment, *ent*-kaurenoic acid accumulates to high levels in the shoot tip, which is also the site of perception of the cold stimulus. After cold treatment and a return to high temperatures, the *ent*-kaurenoic acid is converted to GA_9, the most active gibberellin for stimulating the flowering response. These results are consistent with a cold-induced increase in the activity of *ent*-kaurenoic acid hydroxylase in the shoot tip (Hazebroek and Metzger 1990).

Auxin Promotes Gibberellin Biosynthesis

Although we often discuss the action of hormones as if they act singly, the net growth and development of the plant are the results of many combined signals. In addition, hormones can influence each other's biosynthesis so that the effects produced by one hormone may in fact be mediated by others.

For example, it has long been known that auxin induces ethylene biosynthesis. It is now evident that gibberellin can induce auxin biosynthesis and that auxin can induce gibberellin biosynthesis. If pea plants are decapitated, leading to a cessation in stem elongation, not only is the level of auxin lowered because its source has been removed, but the level of GA_1 in the upper stem drops sharply. This change can be shown to be an auxin effect because replacing the bud with a supply of auxin restores the GA_1 level (Figure 20.18).

The presence of auxin has been shown to promote the transcription of *GA3ox* and to repress the transcription of *GA2ox* (Figure 20.19). In the absence of auxin the reverse occurs. Thus the apical bud promotes growth not only through the direct biosynthesis of auxin, but also through the auxin-induced biosynthesis of GA_1 (Figure 20.20) (Ross et al. 2000; Ross and O'Neill 2001).

Figure 20.21 summarizes some of the factors that modulate the active gibberellin level through regulation of the transcription of the genes for gibberellin biosynthesis or metabolism.

Dwarfness Can Now Be Genetically Engineered

The characterization of the gibberellin biosynthesis and metabolism genes—*GA20ox*, *GA3ox*, and *GA2ox*—has

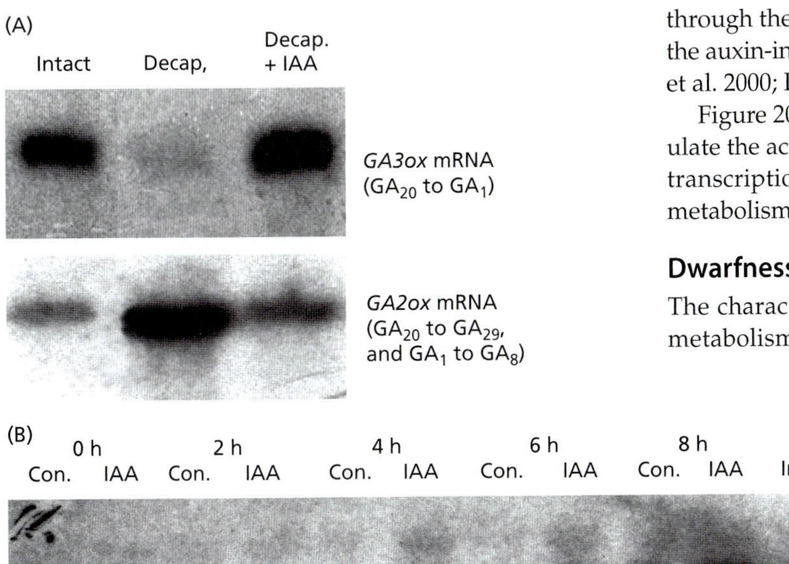

FIGURE 20.19 (A) IAA up-regulates the transcription of GA 3β-hydroxylase (forming GA_1), and down-regulates that of GA 2-oxidase, which destroys GA_1. (B) The increase in GA 3β-hydroxylase in response to IAA can be seen by 2 hours. Con., control. (From Ross et al. 2000.)

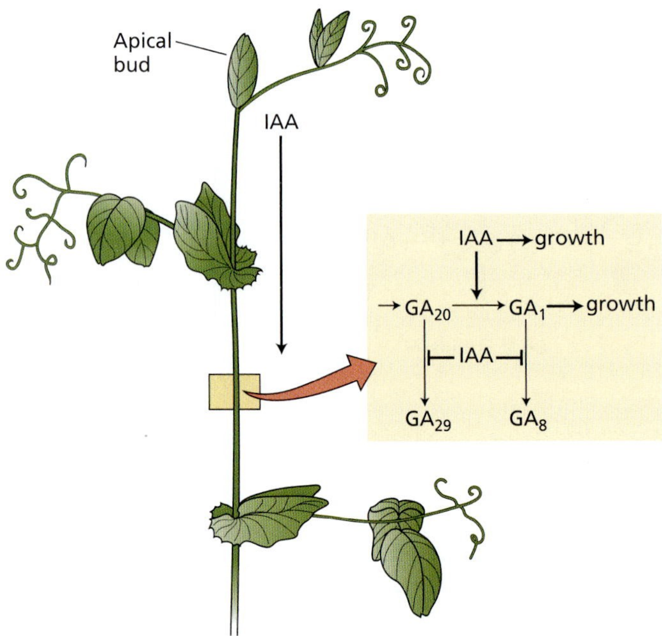

FIGURE 20.20 IAA (from the apical bud) promotes and is required for GA_1 biosynthesis in subtending internodes. IAA also inhibits GA_1 breakdown. (From Ross and O'Neill 2001.)

FIGURE 20.22 Genetically engineered dwarf wheat plants. The untransformed wheat is shown on the extreme left. The three plants on the right were transformed with a gibberellin 2-oxidase cDNA from bean under the control of a constitutive promoter, so that the endogenous active GA_1 was degraded. The varying degrees of dwarfing reflects varying degrees of overexpression of the foreign gene. (Photo from Hedden and Phillips 2000, courtesy of Andy Phillips.)

enabled genetic engineers to modify the transcription of these genes to alter the gibberellin level in plants, and thus affect their height (Hedden and Phillips 2000). The desired effect is usually to increase dwarfness because plants grown in dense crop communities, such as cereals, often grow too tall and thus are prone to lodging. In addition, because gibberellin regulates bolting, one can prevent bolting by inhibiting the rise in gibberellin. An example of the latter is the inhibition of bolting in sugar beet.

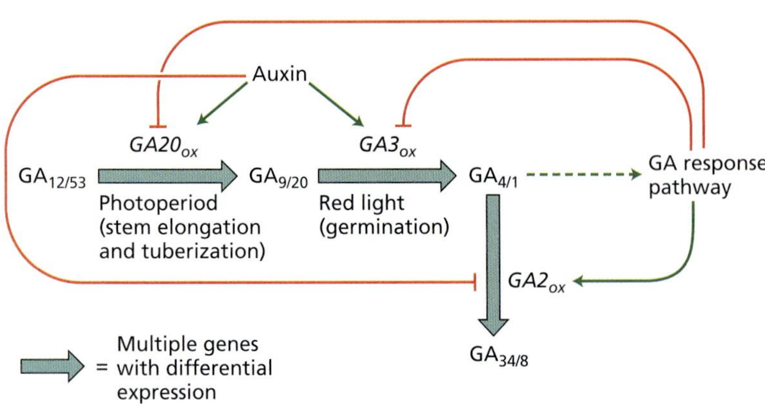

FIGURE 20.21 The pathway of gibberellin biosynthesis showing the identities of the genes for the metabolic enzymes and the way that their transcription is regulated by feedback, environment, and other endogenous hormones.

Sugar beet is a biennial, forming a swollen storage root in the first season and a flower and seed stalk in the second. To extend the growing season and obtain bigger beets, farmers sow the beets as early as possible in the spring, but sowing too early leads to bolting in the first year, with the result that no storage roots form. A reduction in the capacity to make gibberellin inhibits bolting, allowing earlier sowing of the seeds and thus the growth of larger beets.

Reductions in GA_1 levels have recently been achieved in such crops as sugar beet and wheat, either by the transformation of plants with antisense constructs of the *GA20ox* or *GA3ox* genes, which encode the enzymes leading to the synthesis of GA_1, or by overexpressing the gene responsible for GA_1 metabolism: *GA2ox*. Either approach results in dwarfing in wheat (Figure 20.22) or an inhibition of bolting in rosette plants such as beet.

The inhibition of seed production in such transgenic plants can be overcome by sprays of gibberellin solution, provided that the reduction in gibberellin has been achieved by blocking the genes for GA20ox or GA3ox, the gibberellin biosynthetic enzymes. A similar strategy has recently been applied to turf grass, keeping the grass short with no seedheads, so that mowing can be virtually eliminated—a boon for homeowners!

PHYSIOLOGICAL MECHANISMS OF GIBBERELLIN-INDUCED GROWTH

As we have seen, the growth-promoting effects of gibberellin are most evident in dwarf and rosette plants. When dwarf plants are treated with gibberellin, they resemble the tallest varieties of the same species (see Figure 20.1). Other examples of gibberellin action include the elongation of hypocotyls and of grass internodes.

A particularly striking example of internode elongation is found in deep-water rice (*Oryza sativa*). In general, rice plants are adapted to conditions of partial submergence. To enable the upper foliage of the plant to stay above water, the internodes elongate as the water level rises. Deep-water rice has the greatest potential for rapid internode elongation. Under field conditions, growth rates of up to 25 cm per day have been measured.

The initial signal is the reduced partial pressure of O_2 resulting from submergence, which induces ethylene biosynthesis (see Chapter 22). The ethylene trapped in the submerged tissues, in turn, reduces the level of abscisic acid (see Chapter 23), which acts as an antagonist of gibberellin. The end result is that the tissue becomes more responsive to its endogenous gibberellin (Kende et al. 1998). Because inhibitors of gibberellin biosynthesis block the stimulatory effect of both submergence and ethylene on growth, and exogenous gibberellin can stimulate growth in the absence of submergence, gibberellin appears to be the hormone directly responsible for growth stimulation.

GA-stimulated growth in deep-water rice can be studied in an excised stem system (Figure 20.23). The addition of gibberellin causes a marked increase in the growth rate after a lag period of about 40 minutes. Cell elongation accounts for about 90% of the length increase during the first 2 hours of gibberellin treatment.

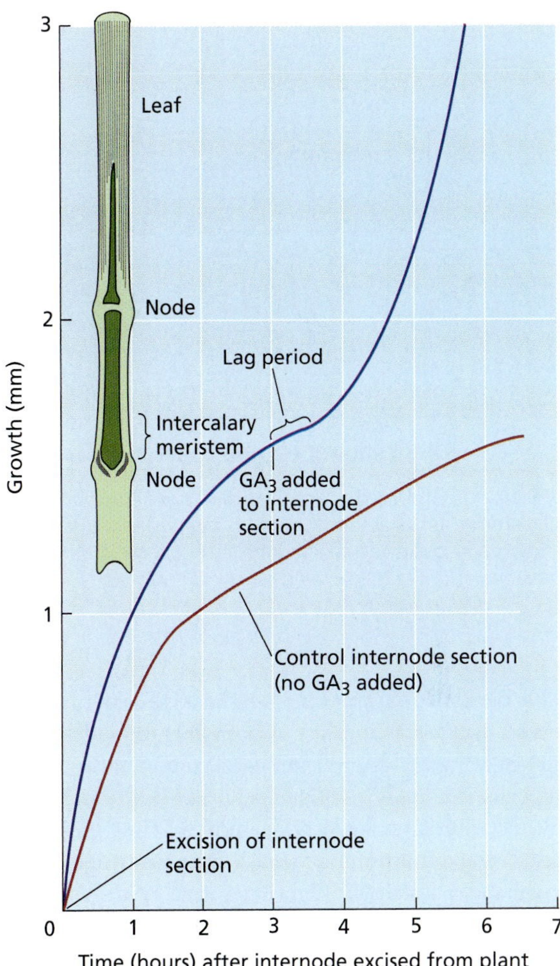

FIGURE 20.23 Continuous recording of the growth of the upper internode of deep-water rice in the presence or absence of exogenous GA_3. The control internode elongates at a constant rate after an initial growth burst during the first 2 hours after excision of the section. Addition of GA after 3 hours induced a sharp increase in the growth rate after a 40-minute lag period (upper curve). The difference in the initial growth rates of the two treatments is not significant here, but reflects slight variation in experimental materials. The inset shows the internode section of the rice stem used in the experiment. The intercalary meristem just above the node responds to GA. (After Sauter and Kende 1992.)

Gibberellins Stimulate Cell Elongation and Cell Division

The effect of gibberellins applied to intact dwarf plants is so dramatic that it would seem to be a simple task to determine how they act. Unfortunately, this is not the case because, as we have seen with auxin, so much about plant cell growth is not understood. However, we do know some characteristics of gibberellin-induced stem elongation.

Gibberellin increases both cell elongation and cell division, as evidenced by increases in cell length and cell number in response to applications of gibberellin. For example, internodes of tall peas have more cells than those of dwarf peas, and the cells are longer. Mitosis increases markedly in the subapical region of the meristem of rosette long-day plants after treatment with gibberellin (Figure 20.24). The dramatic stimulation of internode elongation in deep-water rice is due in part to increased cell division activity in the intercalary meristem. Moreover, only the cells of the intercalary meristem whose division is increased by gibberellin exhibit gibberellin-stimulated cell elongation.

Because gibberellin-induced cell elongation appears to precede gibberellin-induced cell division, we begin our discussion with the role of gibberellin in regulating cell elongation.

Gibberellins Enhance Cell Wall Extensibility without Acidification

As discussed in Chapter 15, the elongation rate can be influenced by both cell wall extensibility and the osmotically driven rate of water uptake. Gibberellin has no effect

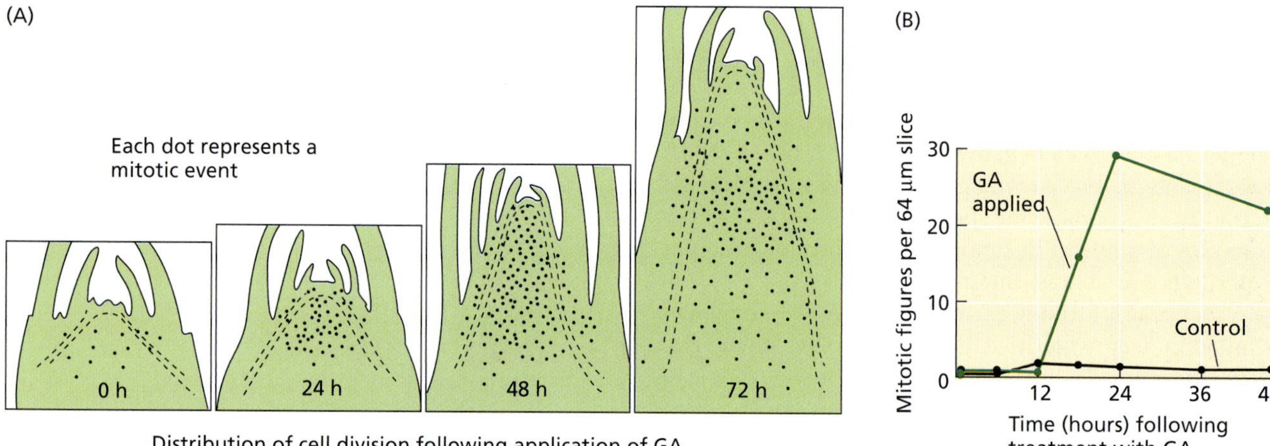

FIGURE 20.24 Gibberellin applications to rosette plants induce stem internode elongation in part by increasing cell division. (A) Longitudinal sections through the axis of *Samolus parviflorus* (brookweed) show an increase in cell division after application of GA. (Each dot represents one mitotic figure in a section 64 μm thick.) (B) The number of such mitotic figures with and without GA in stem apices of *Hyoscyamus niger* (black henbane). (After Sachs 1965.)

on the osmotic parameters but has consistently been observed to cause an increase in both the mechanical extensibility of cell walls and the stress relaxation of the walls of living cells. An analysis of pea genotypes differing in gibberellin content or sensitivity showed that gibberellin decreases the minimum force that will cause wall extension (the wall yield threshold) (Behringer et al. 1990). Thus, both gibberellin and auxin seem to exert their effects by modifying cell wall properties.

In the case of auxin, cell wall loosening appears to be mediated in part by cell wall acidification (see Chapter 19). However, this does not appear to be the mechanism of gibberellin action. In no case has a gibberellin-stimulated increase in proton extrusion been demonstrated. On the other hand, gibberellin is never present in tissues in the complete absence of auxin, and the effects of gibberellin on growth may depend on auxin-induced wall acidification.

The typical lag time before gibberellin-stimulated growth begins is longer than for auxin; as noted already, in deepwater rice it is about 40 minutes (see Figure 20.23), and in peas it is 2 to 3 hours (Yang et al. 1996). These longer lag times point to a growth-promoting mechanism distinct from that of auxin. Consistent with the existence of a separate gibberellin-specific wall-loosening mechanism, the growth responses to applied gibberellin and auxin are additive.

Various suggestions have been made regarding the mechanism of gibberellin-stimulated stem elongation, and all have some experimental support, but as yet none provide a clear-cut answer. For example, there is evidence that the enzyme xyloglucan endotransglycosylase (XET) is involved in gibberellin-promoted wall extension. The function of XET may be to facilitate the penetration of expansins into the cell wall. (Recall that expansins are cell wall proteins that cause wall loosening in acidic conditions by weakening hydrogen bonds between wall polysaccharides [see Chapter 15].) Both expansins and XET may be required for gibberellin-stimulated cell elongation (see **Web Topic 20.3**).

Gibberellins Regulate the Transcription of Cell Cycle Kinases in Intercalary Meristems

As noted earlier, the growth rate of the internodes of deepwater rice dramatically increases in response to submergence, and part of this response is due to increased cell divisions in the intercalary meristem. To study the effect of gibberellin on the cell cycle, researchers isolated nuclei from the intercalary meristem and quantified the amount of DNA per nucleus (Figure 20.25) (Sauter and Kende 1992).

In submergence-induced plants, gibberellin activates the cell division cycle first at the transition from G_1 to S phase, leading to an increase in mitotic activity. To do this, gibberellin induces the expression of the genes for several **cyclin-dependent protein kinases (CDKs)**, which are involved in regulation of the cell cycle (see Chapter 1). The transcription of these genes—first those regulating the transition from G_1 to S phase, followed by those regulating the transition from G_2 to M phase—is induced in the intercalary meristem by gibberellin. The result is a gibberellin-induced increase in the progression from the G_1 to the S phase through to mitosis and cell division (see **Web Topic 20.4**) (Fabian et al. 2000).

Gibberellin Response Mutants Have Defects in Signal Transduction

Single-gene mutants impaired in their response to gibberellin provide valuable tools for identifying genes that encode possible gibberellin receptors or components of signal transduction pathways. In screenings for such mutants,

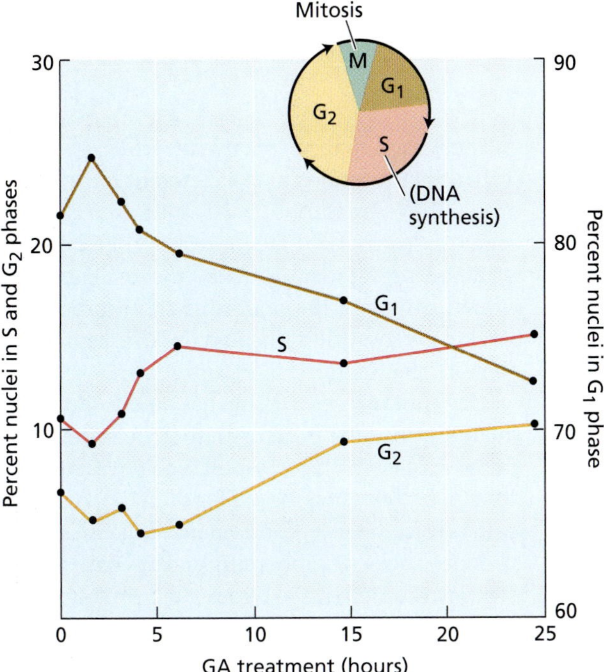

FIGURE 20.25 Changes in the cell cycle status of nuclei from the intercalary meristems of deep-water rice internodes treated with GA_3. Note that the scale for the G_1 nuclei is on the right side of the graph. (After Sauter and Kende 1992.)

three main classes of mutations affecting plant height have been selected:

1. Gibberellin-insensitive dwarfs
2. Gibberellin-deficient mutants in which the gibberellin deficiency has been overcome by a second "suppressor" mutation, so the plants look closer to normal
3. Mutants with a constitutive gibberellin response ("slender" mutants)

All three types of gibberellin response mutants have been generated in *Arabidopsis*, but equivalent mutations have also been found in several other species; in fact, some have been in agricultural use for many years.

The three types of mutant screens have sometimes identified genes encoding the same signal transduction components, even though the phenotypes being selected are completely different. This is possible because mutations at different sites in the same protein can produce vastly different phenotypes, depending on whether the mutation is in a regulatory domain or in an activity, or functional, domain. Some examples of the different phenotypes that can result from changes at different sites in the same protein are described in the sections that follow.

Functional domain (repression). The principal gibberellin signal transduction components that have been identified so far are *repressors of gibberellin signaling*; that is, they repress what we regard as gibberellin-induced tall growth and make the plant dwarf. The repressor proteins are negated or turned off by gibberellin so that the default-type growth—namely, tall—is allowed to proceed. The loss of function resulting from a mutation in the functional domain of such a *negative regulator* results in the mutant appearing as if it has been treated with gibberellin; that is, it has a tall phenotype. Thus a loss-of-function mutation of a negative regulator is like a double negative in English grammar: It translates into a positive.

Because the effects of these loss-of-function mutations are pleiotropic—that is, they also affect developmental processes other than stem elongation—the steps in the pathway involved in the growth response are probably common to all gibberellin responses.

Regulatory domain. If a mutation in the gene for the same negative regulator causes a change in the *regulatory domain* (i.e., that part of the protein that receives a signal from the gibberellin receptor indicating the presence of gibberellin), the protein is unable to receive the signal, and it retains its growth-repressing activity. The phenotype of such a mutant will be that of a gibberellin-insensitive dwarf. Thus, different mutations in the same gene can give opposite phenotypes (tall versus dwarf), depending on whether the mutation is located in the repression domain or the regulatory domain.

The regulatory domain mutations that confer loss of gibberellin sensitivity result in the synthesis of a constitutively active form of the repressor than cannot be turned off by gibberellin. The more of this type of mutant repressor that is present in the cell, the more dwarf the plant will be. Hence, such regulatory domain mutations are semidominant.

In contrast, mutations in the repression domain inactivate the negative regulator (i.e., they act as "knockout" alleles) so that it no longer represses growth; such mutations are recessive because in a heterozygote half the proteins will still be able to repress growth in the absence of gibberellin. *All* of the negative regulators have to be nonfunctional for the plant to grow tall without gibberellin.

With this as background, we now examine specific examples of mutations in the genes that encode proteins in the gibberellin signal transduction pathway.

Different Genetic Screens Have Identified the Related Repressors GAI and RGA

Several gibberellin-insensitive dwarf mutants have been isolated from various species. The first to be isolated in *Arabidopsis* was the *gai-1* mutant (Figure 20.26) (Sun 2000). The *gai-1* mutants resemble gibberellin-deficient mutants, except that they do not respond to exogenous gibberellin.

Another mutant was obtained by screening for a second mutation in a gibberellin-deficient *Arabidopsis* mutant that restores, or partially restores, wild-type growth. The origi-

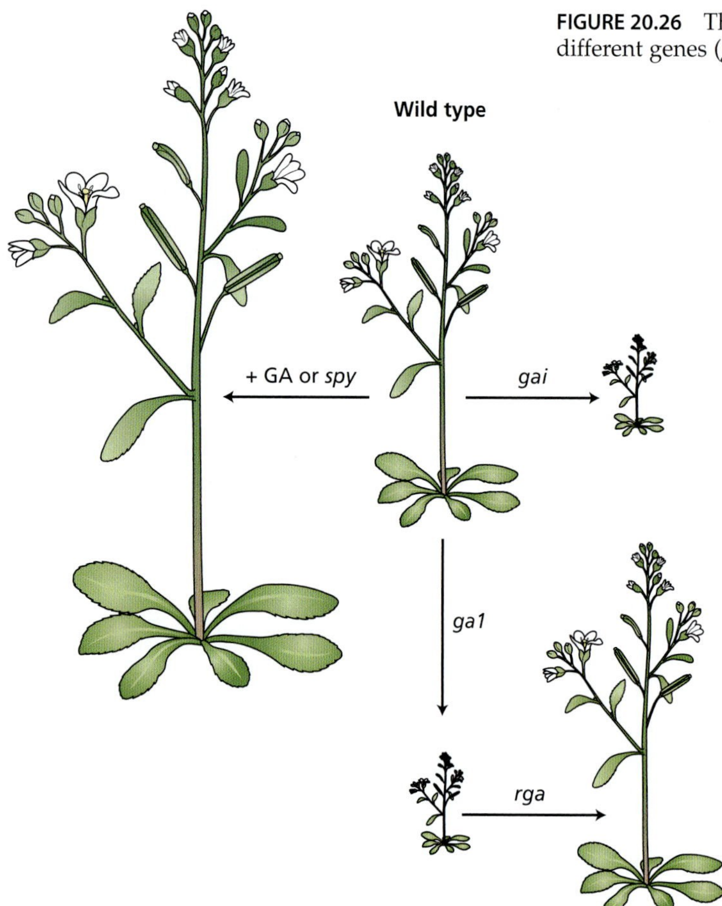

FIGURE 20.26 The effects of gibberellin treatment and mutations in three different genes (*gai, ga1*, and *rga*) on the phenotype of *Arabidopsis*.

nal gibberellin-deficient mutant was *ga1-3,* and the second mutation that partially "rescued" the phenotype (i.e., restored normal growth) was called *rga* (for repressor of *ga1-3*).[4] The *rga* mutation is a recessive mutation that, when present in double copy, gives a plant of intermediate height (see Figure 20.26).

Despite the contrasting phenotypes of the mutants, the wild-type *GAI* and *RGA* genes turned out to be closely related, with a very high (82%) sequence identity. The *gai-1* mutation is semidominant, as are similar gibberellin-insensitive dwarf mutations in other species.

Genetic analyses have indicated that both the GAI and RGA proteins normally act as repressors of gibberellin responses. Gibberellin acts indirectly through an unidentified signaling intermediate, which is thought to bind to the regulatory domains of the GAI and RGA proteins (Figure 20.27). The repressor is no longer able to inhibit growth, and the resulting plant is tall.

The reason that *gai-1* is dwarf, while *rga* is tall, is that the mutations are in different parts of the protein. Whereas the *gai-1* mutation (which negates sensitivity of the repressor to gibberellin) is in the regulatory domain, the *rga* mutation (which prevents the action of the repressor in blocking growth) is located in the repression domain, as illustrated in Figure 20.28.

The mutant *gai-1* gene has been shown to encode a mutant protein with a deletion of 17 amino acids, which corresponds to the regulatory domain of the repressor (Dill et al. 2001). A similar mutation in the receptor domain of the *RGA* gene also produces a gibberellin-insensitive dwarf, demonstrating that the two related proteins have overlapping functions. Because of this deletion in the *gai-1* mutant, the action of the repressor cannot be alleviated by gibberellin, and growth is constitutively inhibited.

Gibberellins Cause the Degradation of RGA Transcriptional Repressors

The *Arabidopsis* wild-type *GAI* and *RGA* genes are members of a large gene family encoding tran-

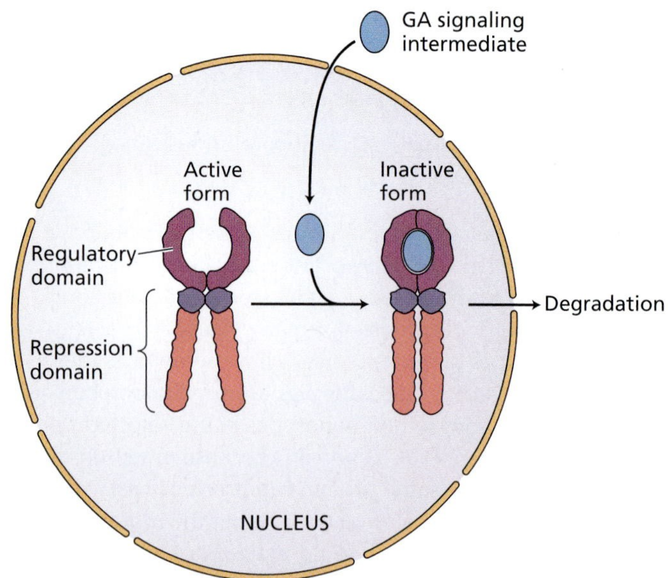

FIGURE 20.27 Two main functional domains of GAI and RGA: the regulatory domain and the repression domain. The repression domain is active in the absence of gibberellin. A gibberellin-induced signaling intermediate binds to the regulatory domain, targeting it for destruction. Note that the protein forms homodimers.

[4] Be careful not to confuse *gai* (gibberellin insensitive) and *ga1* (gibberellin-deficient #1), which can look alike in print.

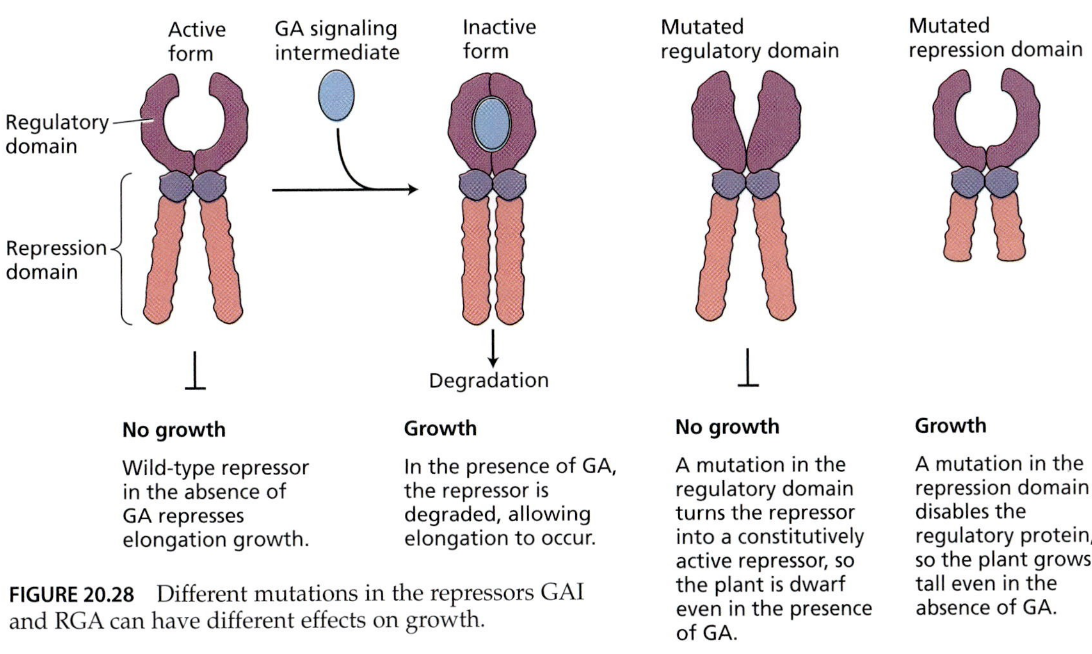

FIGURE 20.28 Different mutations in the repressors GAI and RGA can have different effects on growth.

scriptional repressors that have highly conserved regions with nuclear localization signals. To demonstrate the nuclear localization and repressor nature of the RGA product, the *RGA* promoter was fused to the gene for a green fluorescent protein whose product can be visualized under the microscope. The green color could be seen in cell nuclei.

When the plants were treated with gibberellin, there was no green color, showing that the RGA protein was not present following gibberellin treatment. However, when the gibberellin content was severely lowered by treatment with the gibberellin biosynthesis inhibitor paclobutrazol, the nuclei acquired a very intense green fluorescence, demonstrating both the presence and nuclear localization of the RGA protein only when gibberellin was absent or low (Figure 20.29) (Silverstone et al. 2001).

Both GAI and RGA also have a conserved region at the amino terminus of the protein referred to as DELLA, after

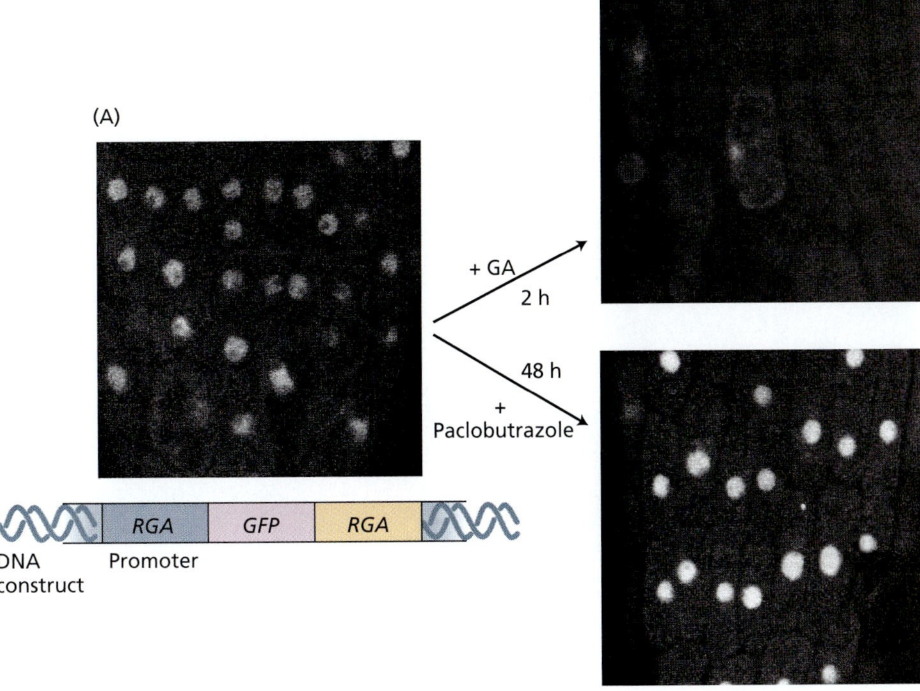

FIGURE 20.29 The RGA protein is found in the cell nucleus, consistent with its identity as a transcription factor, and its level is affected by the level of GA. (A) Plant cells were transformed with the gene for RGA fused to the gene for green fluorescent protein (GFP), allowing detection of RGA in the nucleus by fluorescence microscopy. (B) Effect of GA on RGA. A 2-hour pretreatment with gibberellin causes the loss of RGA from the cell (top). When the gibberellin biosynthesis is inhibited in the presence of paclobutrazole, the RGA content in the nucleus increases (bottom). (From Silverstone et al. 2001.)

the code letters for the amino acids in that sequence. This region is involved in the gibberellin response because it is the location of the mutation in *gai-1* that renders it nonresponsive to gibberellin. It turns out that the RGA protein is synthesized all the time; in the presence of gibberellin this protein is targeted for destruction, and the DELLA region is required for this response (Dill et al. 2001).

It is likely that gibberellin also brings about the turnover of GAI. *RGA* and *GAI* have partially redundant functions in maintaining the repressed state of the gibberellin signaling pathway. However, *RGA* appears to play a more dominant role than *GAI* because in a gibberellin-deficient mutant, a second mutation in the repression domain of *gai* (*gai-t6*) does not restore growth, whereas a comparable mutation in *rga* does. On the other hand, the existence of repression domain mutations in both of these genes allows for complete expression of many characteristics induced by GA, including plant height, in the absence of gibberellin (see Figure 20.26) (Dill and Sun 2001; King et al. 2001).

DELLA Repressors Have Been Identified in Crop Plants

Functional DELLA repressors have been found in several crop plants that have dwarfing mutations, analogous to *gai-1*, in the genes encoding these proteins. Most notable are the *rht* (*r*educed *h*eigh*t*) mutations of wheat that have been in use in agriculture for 30 years. These alleles encode gibberellin response modulators that lack gibberellin responsiveness, leading to dwarfness (Peng et al. 1999; Silverstone and Sun 2000).

Cereal dwarfs such as these are very important as the foundations of the green revolution that enabled large increases in yield to be obtained. Normal cereals grow too tall when close together in a field, especially with high levels of fertilizer. The result is that plants fall down (lodge), and the yield decreases concomitantly. The use of these stiff-strawed dwarf varieties that resist lodging enables high yields.

The Negative Regulator SPINDLY Is an Enzyme That Alters Protein Activity

"Slender mutants" resemble wild-type plants that have been treated with gibberellin repeatedly. They exhibit elongated internodes, parthenocarpic (seed-free) fruit growth (in dicots), and poor pollen production. Slender mutants are rare compared to dwarf mutants.

One possible explanation of the slender phenotype could be simply that the mutants have higher-than-normal levels of endogenous gibberellins. For example, in the *sln* mutation of peas, a gibberellin deactivation step is blocked in the seed. As a result, the mature seed, which in the wild type contains little or no GA, has abnormally high levels of GA_{20}. The GA_{20} from the seed is then taken up by the germinating seedling and converted to the bioactive GA_1, giving rise to the slender phenotype. However, once the seedling runs out of GA_{20} from the seed, its phenotype returns to normal (Reid and Howell 1995).

If, on the other hand, the slender phenotype is *not* due to an overproduction of endogenous gibberellin, the mutant is considered to be a **constitutive response mutant** (Sun 2000). The best characterized of such mutants are the ultratall mutants: *la cry*^{s} in pea, (representing mutations at two loci: *La* and *Cry*^{s}) (see Figure 20.10); *procera* (*pro*) in tomato; *slender* (*sln*) in barley; and *spindly* (*spy*) in *Arabidopsis* (Figure 20.30). All of these mutations are recessive and appear to be loss-of-function mutations in negative regulators of the gibberellin response pathway, as in the case of the DELLA regulators.

SPINDLY (*SPY*) in *Arabidopsis* and related genes in other species are similar in sequence to genes that encode glucosamine transferases in animals (Thornton et al. 1999). These enzymes modify target proteins by the glycosylation of serine or threonine residues. Glycosylation can modify protein activity either directly or indirectly by interfering with or blocking sites of phosphorylation by protein kinases. The target protein for spindly proteins has not yet been identified.

FIGURE 20.30 The *Arabidopsis spy* mutation causes the negation of a growth repressor, so the plants look as if they were treated with gibberellin. From left to right: wild type, *ga1* (GA-deficient), *ga1/spy* double mutant, and *spy*. (Courtesy of N. Olszewski.)

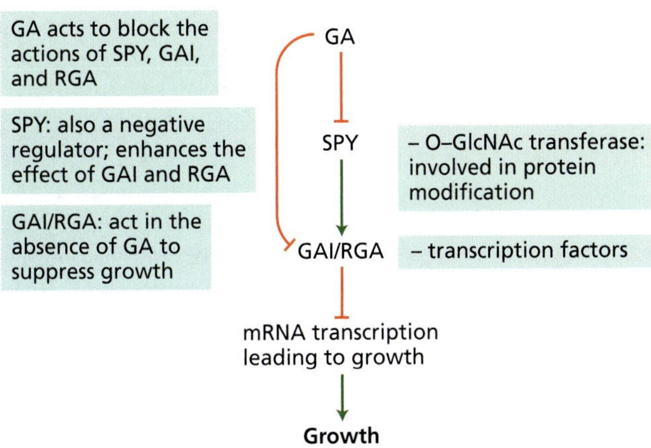

FIGURE 20.31 Interactions between gibberellin and the genes *SPY*, *GAI*, and *RGA* in the regulation of stem elongation.

SPY Acts Upstream of *GAI* and *RGA* in the Gibberellin Signal Transduction Chain

On the basis of the evidence presented in the preceding sections and other studies on the expression of *SPY*, *GAI*, and *RGA* (Sun 2000; Dill et al. 2001), we can begin to sketch out the following elements of the gibberellin signal transduction chain (Figures 20.31 and 20.32):

- Two or more transcriptional regulators encoded by *GAI* and *RGA* act as inhibitors of the transcription of genes that directly or indirectly promote growth.
- SPY appears to be a signal transduction intermediate acting upstream of GAI and RGA that, itself, turns on or enhances the transcription or action of *GAI* and *RGA*, or another negative regulator.
- In the presence of gibberellin, *SPY*, *GAI*, and *RGA* are all negated or turned off.

GA-deficient plant cell: No growth

GA receptor
CYTOPLASM
SPY
GAI
RGA
Transcription of GA-induced genes
NUCLEUS
Plasma membrane

In a GA-deficient cell in a GA biosynthesis mutant, or a wild-type cell without the GA signal, the transmembrane GA receptor is inactive in the absence of GA signal. In this situation, SPY is an active O-GlcNAc transferase that catalyzes the addition of a signal GlcNAc residue (from UDP-GlcNAc) via an O linkage to specific serine and/or threonine residues of target proteins, possibly RGA and GAI. Active RGA and GAI function as repressors of transcription, and they indirectly or directly inhibit the expression of GA-induced genes.

Plant cell with GA: Growth

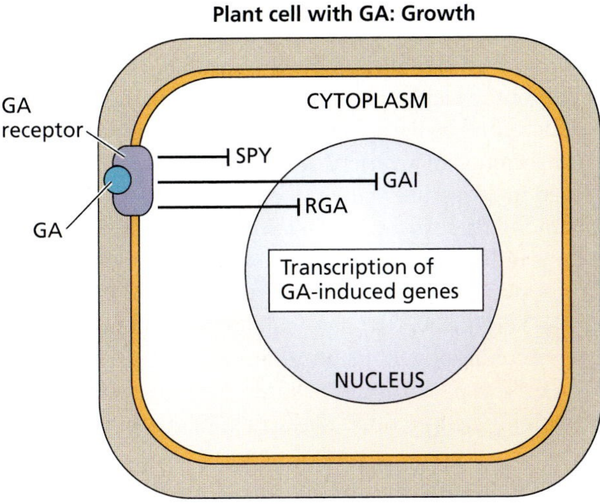

In the presence of GA the GA receptor is activated by binding of bioactive GA. The GA signal inhibits RGA and GAI repressors both directly and by deactivating SPY. In the absence of repression by RGA and GAI, GA-induced genes are transcribed.

FIGURE 20.32 Proposed roles of the active SPY, GAI, and RGA proteins in the GA signaling pathway within a plant cell.

- The RGA protein is degraded, and it is likely that GAI is similarly destroyed.

Whether gibberellin negates *GAI* and *RGA* through *SPY*, or independently, or both, is currently under investigation. However, the basic message in this case and in the cases of other plant hormones, such as ethylene (see Chapter 22) and the photoreceptor phytochrome (see Chapter 17), is that the default developmental program is for the induced type of growth to occur, but the default pathway is prevented by the presence of various negative regulators. Rather than directly promoting an effect, the arrival of the developmental signal—in this case gibberellin—negates the growth repressor, enabling the default condition.

GIBBERELLIN SIGNAL TRANSDUCTION: CEREAL ALEURONE LAYERS

Genetic analyses of gibberellin-regulated growth, such as the studies described in the previous section, have identified some of the genes and their gene products, but not the biochemical pathways involved in gibberellin signal transduction. The biochemical and molecular mechanisms, which are probably common to all gibberellin responses, have been studied most extensively in relation to the gibberellin-stimulated synthesis and secretion of α-amylase in cereal aleurone layers (Jacobsen et al. 1995).

In this section we will describe how such studies have shed light on the location of the gibberellin receptor, the transcriptional regulation of the genes for α-amylase and other proteins, and the possible signal transduction pathways involved in the control of α-amylase synthesis and secretion by gibberellin.

Gibberellin from the Embryo Induces α-Amylase Production by Aleurone Layers

Cereal grains (*caryopses*; singular *caryopsis*) can be divided into three parts: the diploid embryo, the triploid endosperm, and the fused testa–pericarp (seed coat–fruit wall). The embryo part consists of the plant embryo proper, along with its specialized absorptive organ, the *scutellum* (plural *scutella*), which functions in absorbing the solubilized food reserves from the endosperm and transmitting them to the growing embryo. The endosperm is composed of two tissues: the centrally located starchy endosperm and the aleurone layer (Figure 20.33A).

The starchy endosperm, typically nonliving at maturity, consists of thin-walled cells filled with starch grains. The aleurone layer surrounds the starchy endosperm and is cytologically and biochemically distinct from it. Aleurone cells are enclosed in thick primary cell walls and contain large numbers of protein-storing vacuoles called *protein bodies* (Figures 20.33B–D), enclosed by a single membrane. The protein bodies also contain phytin, a mixed cation salt (mainly Mg^{2+} and K^{+}) of *myo*-inositolhexaphosphoric acid (phytic acid).

During germination and early seedling growth, the stored food reserves of the endosperm—chiefly starch and protein—are broken down by a variety of hydrolytic enzymes, and the solubilized sugars, amino acids, and other products are transported to the growing embryo. The two enzymes responsible for starch degradation are α- and β-amylase. α-Amylase hydrolyzes starch chains internally to produce oligosaccharides consisting of α-1,4-linked glucose residues. β-Amylase degrades these oligosaccharides from the ends to produce maltose, a disaccharide. Maltase then converts maltose to glucose.

α-Amylase is secreted into the starchy endosperm of cereal seeds by both the scutellum and the aleurone layer (see Figure 20.33A). The sole function of the aleurone layer of the seeds of graminaceous monocots (e.g., barley, wheat, rice, rye, and oats) appears to be the synthesis and release of hydrolytic enzymes. After completing this function, aleurone cells undergo programmed cell death.

Experiments carried out in the 1960s confirmed Gottlieb Haberlandt's original observation of 1890 that the secretion of starch-degrading enzymes by barley aleurone layers depends on the presence of the embryo. When the embryo was removed (i.e., the seed was de-embryonated), no starch was degraded. However, when the de-embryonated "half-seed" was incubated in close proximity to the excised embryo, starch was digested, demonstrating that the embryo produced a diffusible substance that triggered α-amylase release by the aleurone layer.

It was soon discovered that gibberellic acid (GA_3) could substitute for the embryo in stimulating starch degradation. When de-embryonated half-seeds were incubated in buffered solutions containing gibberellic acid, secretion of α-amylase into the medium was greatly stimulated after an 8-hour lag period (relative to the control half-seeds incubated in the absence of gibberellic acid).

The significance of the gibberellin effect became clear when it was shown that the embryo synthesizes and releases gibberellins (chiefly GA_1) into the endosperm during germination. Thus the cereal embryo efficiently regulates the mobilization of its own food reserves through the secretion of gibberellins, which stimulate the digestive function of the aleurone layer (see Figure 20.33A).

Gibberellin has been found to promote the production and/or secretion of a variety of hydrolytic enzymes that are involved in the solubilization of endosperm reserves; principal among these is α-amylase. Since the 1960s, investigators have utilized isolated aleurone layers, or even aleurone cell protoplasts (see Figure 20.33C and D), rather than half-seeds (see Figure 20.33B). The isolated aleurone layer, consisting of a homogeneous population of target cells, provides a unique opportunity to study the molecular aspects of gibberellin action in the absence of nonresponding cell types.

In the following discussion of gibberellin-induced α-amylase production we focus on three questions:

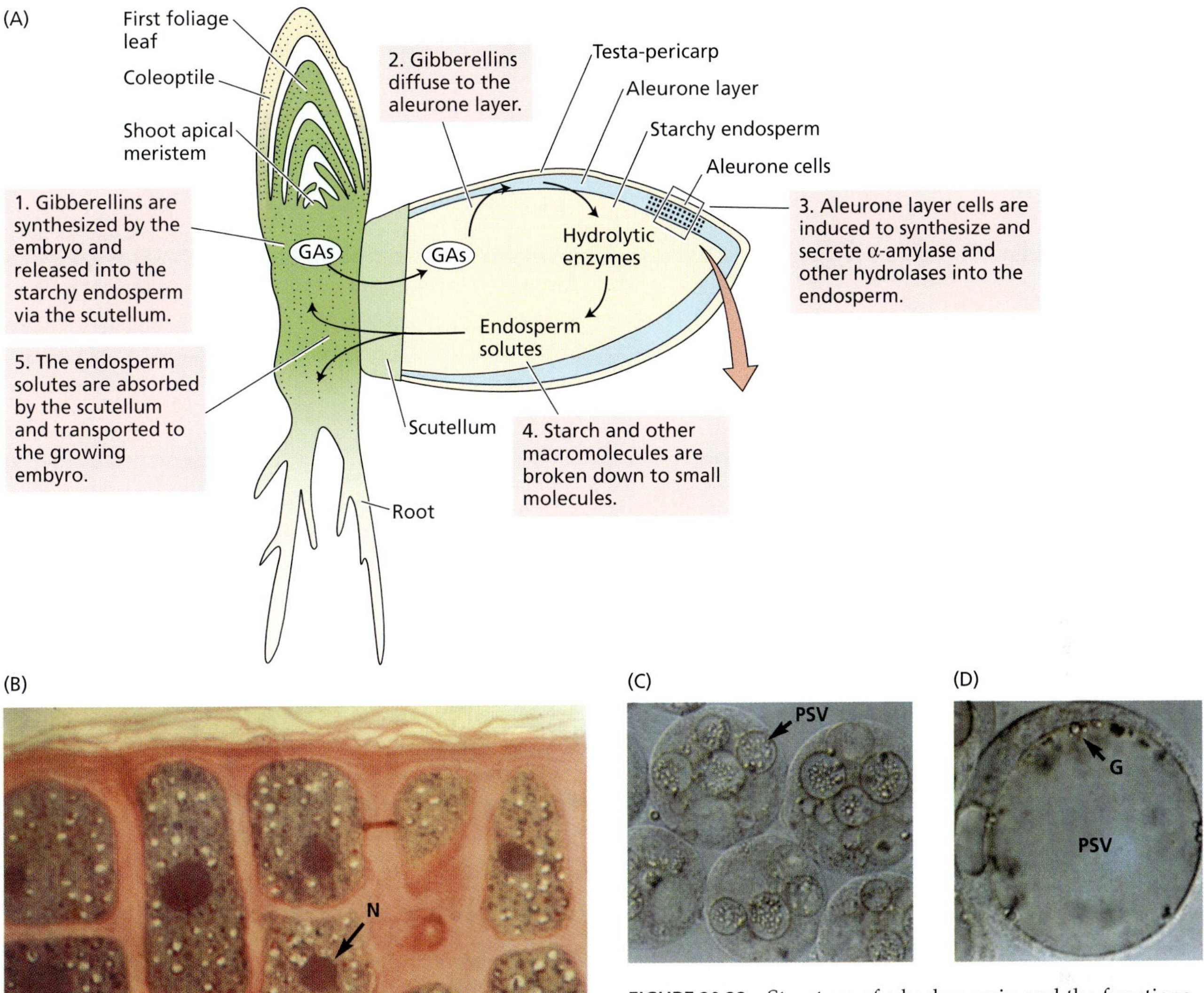

FIGURE 20.33 Structure of a barley grain and the functions of various tissues during germination (A). Microscope photos of the barley aleurone layer (B) and barley aleurone protoplasts at an early (C) and late stage (D) of amylase production. Protein storage vesicles (PSV) can be seen in each cell. G = phytin globoid; N = nucleus. (Photos from Bethke et al. 1997, courtesy of P. Bethke.)

1. How does gibberellin regulate the increase in a-amylase?
2. Where is the gibberellin receptor located in the cell?
3. What signal transduction pathways operate between the gibberellin receptor and a-amylase production?

Gibberellic Acid Enhances the Transcription of α-Amylase mRNA

Before molecular biological approaches were developed, there was already physiological and biochemical evidence that gibberellic acid might enhance α-amylase production at the level of gene transcription (Jacobsen et al. 1995). The two main lines of evidence were as follows:

1. GA_3-stimulated α-amylase production was shown to be blocked by inhibitors of transcription and translation.
2. Heavy-isotope- and radioactive-isotope-labeling studies demonstrated that the stimulation of α-amylase activity by gibberellin involved de novo synthesis of the enzyme from amino acids, rather than activation of preexisting enzyme.

Definitive molecular evidence now shows that gibberellin acts primarily by inducing the expression of the

(A) **Enzyme synthesis**

Synthesis of α-amylase by isolated barley aleurone layers is evident after 6–8 hours of treatment with GA_3 (10^{-6} *M*).

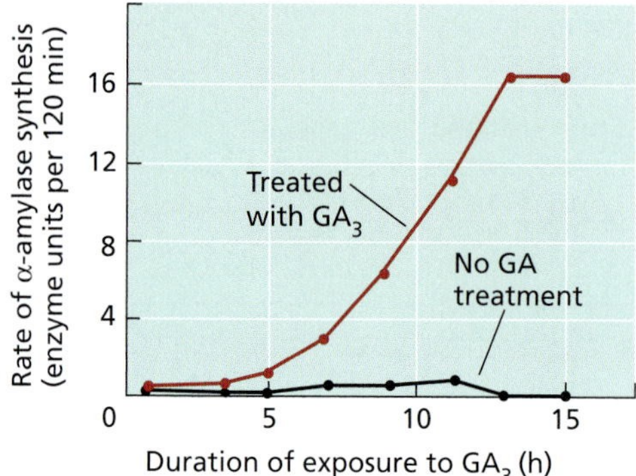

(B) **mRNA synthesis**

A gibberellin-induced increase in translatable α-amylase mRNA precedes the release of the α-amylase from the aleurone cells by several hours.

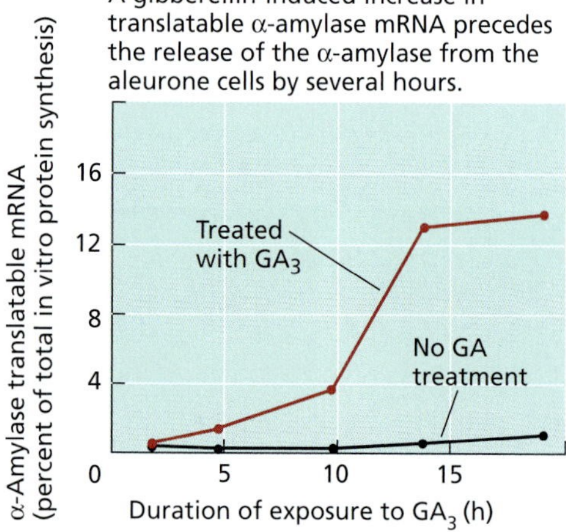

FIGURE 20.34 Gibberellin effects on enzyme synthesis and mRNA synthesis. The α-amylase mRNA in this case was measured by the in vitro production of α-amylase as a percentage of the protein produced by the translation of the bulk mRNA. (From Higgins et al. 1976.)

gene for α-amylase. It has been shown that GA_3 enhances the level of translatable mRNA for α-amylase in aleurone layers (Figure 20.34). Furthermore, by using isolated nuclei, investigators also demonstrated that there was an enhanced transcription of the α-amylase gene rather than a decrease in mRNA turnover (see **Web Topic 20.5**).

The purification of α-amylase mRNA, which is produced in relatively large amounts in aleurone cells, enabled the isolation of genomic clones containing both the structural gene for α-amylase and its upstream promoter sequences. These promoter sequences were then fused to the reporter gene that encodes the enzyme β-glucuronidase (GUS), which yields a blue color in the presence of an artificial substrate when the gene is expressed. The regulation of transcription by gibberellin was proved when such chimeric genes containing α-amylase promoters that were fused to reporter genes were introduced into aleurone protoplasts and the production of the blue color was shown to be stimulated by gibberellin (Jacobsen et al. 1995).

The partial deletion of known sequences of bases from α-amylase promoters from several cereals indicates that the sequences conferring gibberellin responsiveness, termed *gibberellin response elements*, are located 200 to 300 base pairs upstream of the transcription start site (see **Web Topic 20.6**).

A GA-MYB Transcription Factor Regulates α-Amylase Gene Expression

The stimulation of α-amylase gene expression by gibberellin is mediated by a specific transcription factor that binds to the promoter of the α-amylase gene (Lovegrove and Hooley 2000). To demonstrate such DNA-binding proteins in rice, a technique called a *mobility shift assay* was used (see **Web Topic 20.7**). This assay detects the increase in size that occurs when the α-amylase promoter binds to a protein isolated from gibberellin-treated aleurone cells (Ou-Lee et al. 1988). The mobility shift assay also allowed identification of the regulatory DNA sequences (**gibberellin response elements**) in the promoter that are involved in binding the protein.

Identical gibberellin response elements were found to occur in all cereal α-amylase promoters, and their presence was shown to be essential for the induction of α-amylase gene transcription by gibberellin. These studies demonstrated that gibberellin increases either the level or the activity of a transcription factor protein that switches on the production of α-amylase mRNA by binding to an upstream regulatory element in the α-amylase gene promoter.

The sequence of the gibberellin response element in the α-amylase gene promoter turned out to be similar to that of the binding sites for MYB transcription factors that are known to regulate growth and development in phytochrome responses (see Chapter 14 on the Web site and Chapter 17) (Jacobsen et al. 1995). This knowledge enabled the isolation of mRNA for a MYB transcription factor, named GA-MYB, associated with the gibberellin induction of α-amylase gene expression.

The synthesis of *GA-MYB* mRNA in aleurone cells increases within 3 hours of gibberellin application, several hours before the increase in α-amylase mRNA (Gubler et al. 1995) (Figure 20.35). The inhibitor of translation, cycloheximide, has no effect on the production of *MYB* mRNA, indicating that *GA-MYB* is a *primary response gene*, or *early gene*. In contrast, the α-amylase gene is a *secondary response gene*, or *late gene*, as indicated by the fact that its transcription is blocked by cycloheximide.

How does gibberellin cause the *MYB* gene to be expressed? Because protein synthesis is not involved, gibberellin may bring about the activation of one or more *preexisting* transcription factors. The activation of transcription factors is typically mediated by protein phosphorylation events occurring at the end of a signal transduction pathway. We will now examine what is known about the signaling pathways involved in gibberellin-induced α-amylase production up to the point of GA-MYB production.

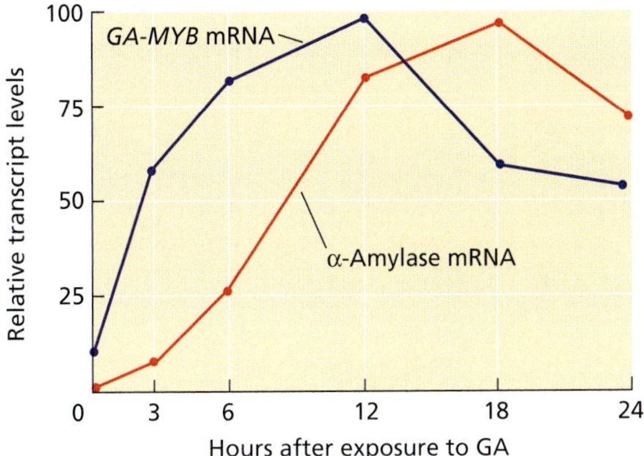

FIGURE 20.35 Time course for the induction of *GA-MYB* and α-amylase mRNA by gibberellic acid. The production of *GA-MYB* mRNA precedes α-amylase mRNA by about 5 hours. This result is consistent with the role of *GA-MYB* as an early *GA* response gene that regulates the transcription of the gene for α-amylase. In the absence of GA, the levels of both *GA-MYB* and α-amylase mRNAs are negligible. (After Gubler et al. 1995.)

Gibberellin Receptors May Interact with G-Proteins on the Plasma Membrane

A cell surface localization of the gibberellin receptor is suggested from the fact that gibberellin that has been bound to microbeads that are unable to cross the plasma membrane is still active in inducing α-amylase production in aleurone protoplasts (Hooley et al. 1991). In addition, microinjection of GA_3 into aleurone protoplasts had no effect, but when the protoplasts were immersed in GA_3 solution, they produced α-amylase (Gilroy and Jones 1994). These results suggest that gibberellin acts on the outer face of the plasma membrane.

Two gibberellin-binding plasma membrane proteins have been isolated through the use of purified plasma membrane and a radioactively labeled gibberellin that was chemically modified to permanently attach to protein to which it was weakly bound. Because the presence of excess gibberellin reduces binding, and these proteins from a semidwarf, gibberellin-insensitive sweet pea bind gibberellin less strongly, they may represent the gibberellin receptors (Lovegrove et al. 1998).

In animal cells, heterotrimeric GTP-binding proteins (G-proteins) in the cell membrane are often involved as first steps in a pathway between a hormone receptor and subsequent cytosolic signals. Evidence has been obtained that G-proteins are also involved in the early gibberellin signaling events in aleurone cells (Jones et al. 1998).

Treatment of oat aleurone protoplasts with a peptide called Mas7, which stimulates GTP/GDP exchange by G-proteins, was found to induce α-amylase gene expression and to stimulate α-amylase secretion, suggesting that such a GTP/GDP exchange on the cell membrane is a reaction en route to the induction of α-amylase biosynthesis by gibberellin. In addition, gibberellin-induced α-amylase gene expression and secretion were inhibited by a guanine nucleotide analog that binds to the α subunit of heterotrimeric G-proteins and inhibits GTP/GDP exchange, further supporting the preceding conclusion.

Recent genetic studies have provided further support for the role of G-proteins as intermediates in the gibberellin signal transduction pathway. The rice dwarf mutant *dwarf 1* (*d1*) has a defective gene encoding the α subunit. Besides being dwarf, the aleurone layers of the *d1* mutant synthesize less α-amylase in response to gibberellin than wild-type aleurone layers do. This reduction in α-amylase production by the *d1* mutant demonstrates that G-proteins are one of the components of the gibberellin signal transduction pathway involved in both the growth response and the production of α-amylase. However, the difference in α-amylase production between the mutant and the wild type goes away with increasing gibberellin concentration, suggesting that gibberellin can also stimulate α-amylase production by a G-protein-independent pathway (Ashikari et al. 1999; Ueguchi-Tanaka et al. 2000).

Cyclic GMP, Ca^{2+}, and Protein Kinases Are Possible Signaling Intermediates

In animal cells, G-proteins can activate the enzyme guanylyl cyclase, the enzyme that synthesizes cGMP from GTP, leading to an increase in cGMP concetration. Cyclic GMP, in turn, can regulate ion channels, Ca^{2+} levels, protein kinase activity, and gene transcription (see Chapter 14 on the Web site). Gibberellin has been reported to cause a transient rise in cGMP levels in barley aleurone layers, suggesting a possible role for cGMP in α-amylase production (Figure 20.36) (see **Web Topic 20.8**) (Pensen et al. 1996).

Calcium and the calcium-binding protein calmodulin act as second messengers for many hormonal responses in

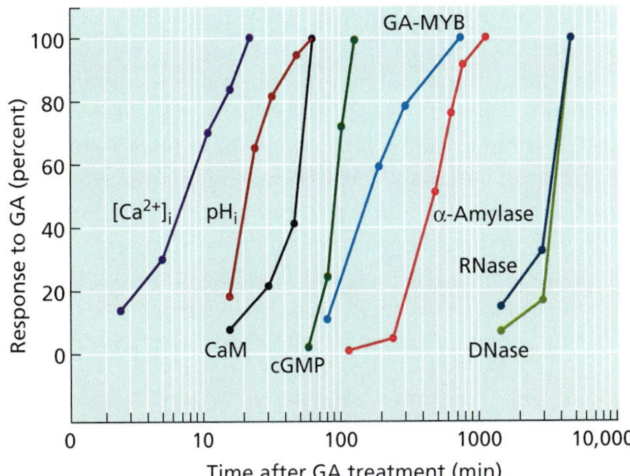

FIGURE 20.36 Following the addition of GA to barley aleurone protoplasts, a multiple signal transduction pathway is initiated. The timing of some of these events is shown. (From Bethke et al. 1997.)

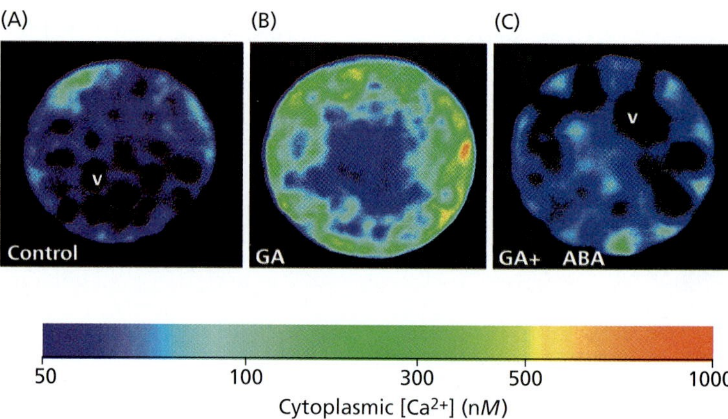

FIGURE 20.37 Increase in the calcium in barley aleurone protoplasts following GA addition can be seen from this false color image. The level of calcium corresponding to the colors is in the lower scale. (A) Untreated protoplast. (B) GA-treated protoplast. (C) Protoplast treated with both abscisic acid (AB) and GA. Abscisic acid opposes the effects of GA in aleurone cells. (From Ritchie and Gilroy 1998b.)

animal cells (see Chapter 14 on the Web site), and they have been implicated in various plant responses to environmental and hormonal stimuli. The earliest event in aleurone protoplasts after the application of gibberellin is a rise in the cytoplasmic calcium concentration that occurs well before the onset of α-amylase synthesis (see Figures 20.36 and 20.37) (Bethke et al. 1997). Without calcium, α-amylase secretion does not occur, though in barley aleurone protoplasts its synthesis goes ahead normally, so we have to conclude that, in barley, calcium is not on the signaling pathway to α-amylase gene transcription, though it does play a role in enzyme secretion.

Protein phosphorylation by protein kinases is another component in many signaling pathways, and gibberellin appears to be no exception. The injection of a protein kinase substrate into barley aleurone protoplasts to compete with endogenous protein phosphorylation inhibited α-amylase secretion, suggesting the involvement of protein phosphorylation in the α-amylase secretion pathway (Ritchie and Gilroy 1998a). This did not affect the gibberellin-stimulated increase in calcium, indicating that the protein kinase step is downstream of the calcium signaling event.

In conclusion, gibberellin signal transduction in aleurone cells seems to involve G-proteins as well as cyclic GMP, leading to production of the transcription factor GA-MYB, which induces α-amylase gene transcription. α-Amylase secretion has similar initial components but also involves an increase in cytoplasmic calcium and protein phosphorylation. The detailed signaling pathways remain to be worked out. A model of the known biochemical components of the gibberellin signal transduction pathways in aleurone cells is illustrated in Figure 20.38.

The Gibberellin Signal Transduction Pathway Is Similar for Stem Growth and α-Amylase Production

It is widely believed that gibberellin initially acts through a common pathway or pathways in all of its effects on development. As we have seen, the genetic approaches applied to the study of gibberellin-stimulated growth led to the identification of the *SPY/GAI/RGA* negative regulatory pathway. The proteins SPY, GAI, and RGA act as repressors of gibberellin responses. Gibberellin deactivates these repressors.

Because the aleurone layers of gibberellin-insensitive dwarf wheat are also insensitive to GA, the same signal transduction pathways that regulate growth appear to regulate gibberellin-induced α-amylase production. Indeed a *SPY*-type gene associated with α-amylase production has been isolated from barley (*HvSPY*), and its expression is able to inhibit gibberellin-induced α-amylase synthesis, while GA-MYB-type factors are also implicated in the gibberellin transduction chain regulating stem growth.

Rice with the *dwarf 1* mutation also produces little α-amylase in response to gibberellin. As noted earlier, the mutation causing *dwarf 1* is known to be in the α subunit of the G-protein complex, providing evidence that the action of gibberellin in both stem elongation and the production of α-amylase are regulated by plasma membrane heterotrimeric G-proteins.

As the entire elongation growth and α-amylase signaling pathways are worked out, it will be interesting to see how much they have in common and where they diverge.

SUMMARY

Gibberellins are a family of compounds defined by their structure. They now number over 125, some of which are found only in the fungus *Gibberella fujikuroi*. Gibberellins induce dramatic internode elongation in certain types of plants, such as dwarf and rosette species and grasses.

FIGURE 20.38 Composite model for the induction of α-amylase synthesis in barley aleurone layers by gibberellin. A calcium-independent pathway induces α-amylase gene transcription; a calcium-dependent pathway is involved in α-amylase secretion. (The SPY negative regulator was omitted for clarity.) ▶

1. GA_1 from the embryo first binds to a cell surface receptor.

2. The cell surface GA receptor complex interacts with a heterotrimeric G-protein, initiating two separate signal transduction chains.

3. A calcium-independent pathway, involving cGMP, results in the activation of a signaling intermediate.

4. The activated signaling intermediate binds to DELLA repressor proteins in the nucleus.

5. The DELLA repressors are degraded when bound to the GA signal.

6. The inactivation of the DELLA repressors allows the expression of the *MYB* gene, as well as other genes, to proceed through transcription, processing, and translation.

7. The newly synthesized MYB protein then enters the nucleus and binds to the promoter genes for α-amylase and other hydrolytic enzymes.

8. Transcription of α-amylase and other hydrolytic genes is activated.

9. α-Amylase and other hydrolases are synthesized on the rough ER.

10. Proteins are secreted via the Golgi.

11. The secretory pathway requires GA stimulation via a calcium–calmodulin-dependent signal transduction pathway.

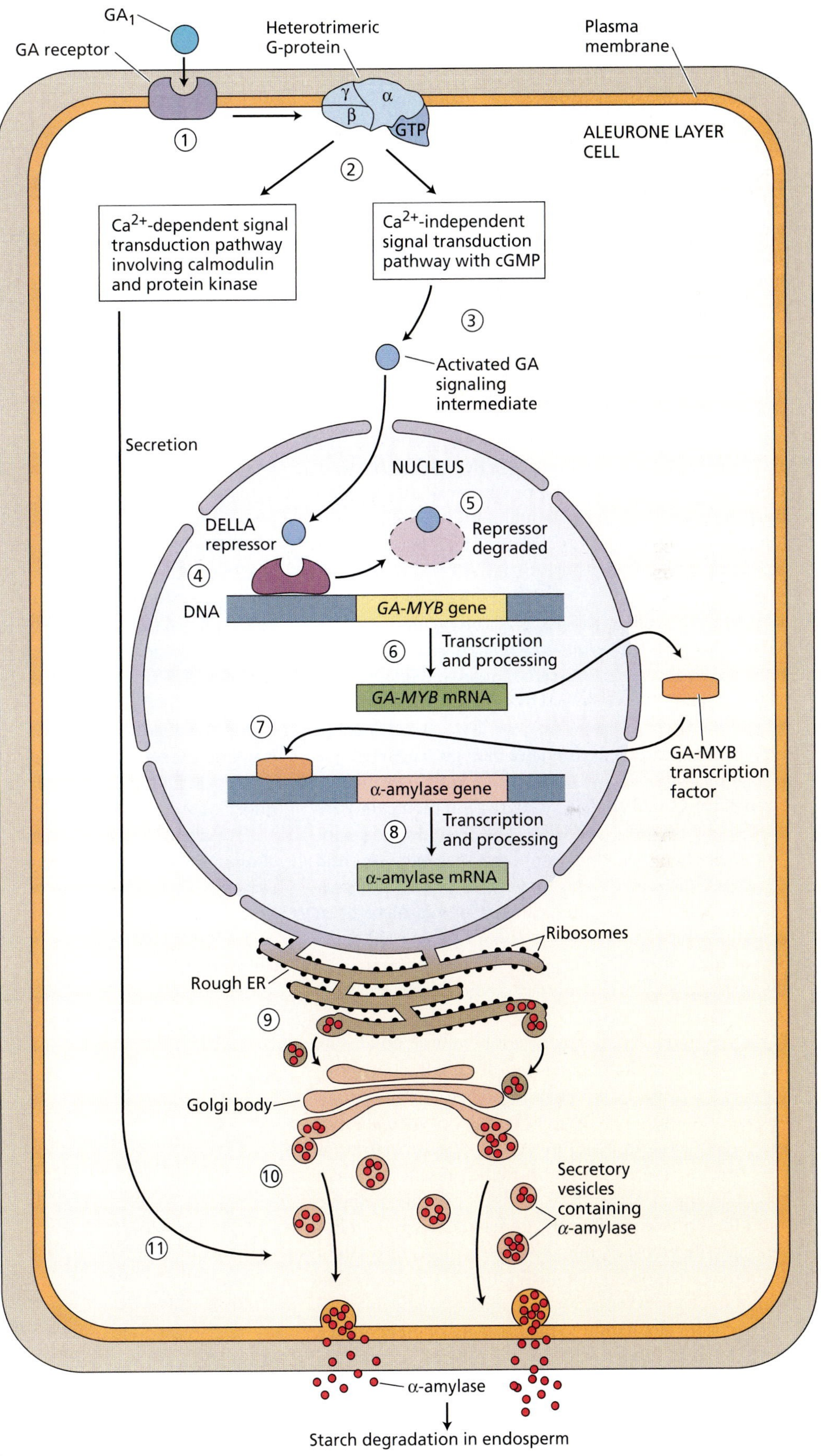

Other physiological effects of gibberellin include changes in juvenility and flower sexuality, and the promotion of fruit set, fruit growth, and seed germination. Gibberellins have several commercial applications, mainly in enhancement of the size of seedless grapes and in the malting of barley. Gibberellin synthesis inhibitors are used as dwarfing agents.

Gibberellins are identified and quantified by gas chromatography combined with mass spectrometry, following separation by high-performance liquid chromatography. Bioassays may be used to give an initial idea of the gibberellins present in a sample. Only certain GAs, notably GA_1 and GA_4, are responsible for the effects in plants; the others are precursors or metabolites.

Gibberellins are terpenoid compounds, made up of isoprene units. The first compound in the isoprenoid pathway committed to gibberellin biosynthesis is *ent*-kaurene. The biosynthesis up to *ent*-kaurene occurs in plastids. *ent*-Kaurene is converted to GA_{12}—the precursor of all the other gibberellins—on the plastid envelope and then on the endoplasmic reticulum via cytochrome P450 monooxygenases. Commonly a hydroxylation at C-13 also takes place to give GA_{53}.

GA_{53} or GA_{12}, each of which has 20 carbon atoms, is converted to other gibberellins by sequential oxidation of carbon 20, followed by the loss of this carbon to give 19-carbon gibberellins. This process is followed by hydroxylation at carbon 3 to give the growth-active GA_1 or GA_4. A subsequent hydroxylation at carbon 2 eliminates biological activity.

The steps after GA_{53} or GA_{12} occur in the cytoplasm. The genes for GA 20-oxidase (*GA20ox*), which catalyzes the steps between GA_{53} and GA_{20}, GA 3β-hydroxylase (or GA 3-oxidase; *GA3ox*), which converts GA_{20} into GA_1, and GA 2-oxidase (*GA2ox*), which converts active GA_1 to inactive GA_8, have been isolated. Dwarf plants have been genetically engineered by the use of antisense *GA20ox* or *GA3ox*, or overexpression of *GA2ox*. Gibberellins may also be glycosylated to give either an inactivated form or a storage form.

The endogenous level of active gibberellin regulates its own synthesis by switching on or inhibiting the transcription of the genes for the enzymes of gibberellin biosynthesis and degradation. Environmental factors such as photoperiod (e.g., leading to bolting and potato tuberization) and temperature (vernalization), and the presence of auxin from the stem apex, also modulate gibberellin biosynthesis through the transcription of the genes for the gibberellin biosynthetic enzymes. Light regulates both GA_1 biosynthesis through regulation of the transcription of the gibberellin degradation gene and also causes a decrease in the responsiveness of stem elongation to the presence of gibberellin.

The most pronounced effect of applied gibberellins is stem elongation in dwarf and rosette plants. Gibberellins stimulate stem growth by promoting both cell elongation and cell division. The activity of some wall enzymes has been correlated with gibberellin-induced growth and cell wall loosening. Gibberellin-stimulated cell divisions in deep-water rice are regulated at the transition between DNA replication and cell division.

Three types of gibberellin response mutants have been useful in the identification of genes involved in the gibberellin signaling pathway involved in stem growth: (1) gibberellin-insensitive dwarfs (e.g., *gai-1*), (2) gibberellin deficiency reversion mutants (e.g., *rga*), and (3) constitutive gibberellin responders (slender mutants) (e.g., *spy*).

GAI and RGA are related nuclear transcription factors that repress growth. In the presence of gibberellin they are degraded. The mutant *gai-1*, and the related wheat dwarfing gene mutant *rht*, have lost the ability to respond to gibberellin. *SPY* encodes a glycosyl transferase that is a member of a signal transduction chain prior to GAI/RGA. When a mutation interferes with the repressor function of any of these, the plants grow tall.

Gibberellin induces transcription of the gene for α-amylase biosynthesis in cereal grain aleurone cells. This process is mediated by the transcription of a specific transcription factor, GA-MYB, which binds to the upstream region of the α-amylase gene, thus switching it on. The gibberellin receptor is located on the surface of aleurone cells. G-proteins and cyclic GMP have been implicated as members of the signal transduction chain on the way to GA-MYB. Calcium is not on the route to α-amylase gene transcription, though it does play a role in α-amylase secretion via protein phosphorylation.

The gibberellin signal transduction pathway is probably similar for stem elongation and α-amylase production. Dwarf wheat and rice also have impaired α-amylase gene transcription. Gibberellin acts by deactivating repressors, such as SPY, GAI, and RGA en route to both an increase in cell elongation and the production of α-amylase.

Web Material

Web Topics

20.1 Structures of Some Important Gibberellins, Their Precursors and Derivatives, and Inhibitors of Gibberellin Biosynthesis

The chemical structures of various gibberellins and inhibitors of gibberellin biosynthesis are presented.

20.2 Gibberellin Detection

Gibberellin quantitation is now routine thanks to sensitive modern physical methods of detection.

20.3 Gibberellin-Induced Stem Elongation

Various mechanisms of gibberellin-induced cell wall loosening are discussed.

20.4 CDKs and Gibberellin-Induced Cell Division

Additional information on the mechanism of gibberellin regulation of the cell cycle is given.

20.5 Gibberellin-Induction of α-amylase mRNA

Evidence is provided for gibberellin-induced transcription of α-amylase mRNA.

20.6 Promoter Elements and Gibberellin Responsiveness

Gibberellin response elements mediate the effects of gibberellin on α-amylase transcription.

20.7 Regulation of α-amylase Gene Expression by Transcription Factors

Experiments identifying MYB transcription factors as mediators of gibberellin-induced gene transcription are described.

20.8 Gibberellin Signal Transduction

Various signaling intermediates have been implicated in gibberellin responsiveness.

Chapter References

Aach, H., Bode, H., Robinson, D.G., and Graebe, J. E. (1997) *ent*-Kaurene synthase is located in proplastids of meristematic shoot tissues. *Planta* 202: 211–219.

Ashikari, M., Wu, J., Yano, M., Sasaki, T., and Yoshimura, A. (1999) Rice gibberellin-insensitive dwarf mutant gene *Dwarf 1* encodes the α-subunit of GTP-binding protein. *Proc. Natl. Acad. Sci. USA* 96: 10284–10289.

Behringer, F. J., Cosgrove, D. J., Reid, J. B., and Davies, P. J. (1990) Physical basis for altered stem elongation rates in internode length mutants of *Pisum*. *Plant Physiol.* 94: 166–173.

Bethke, P. C., Schuurink, R., and Jones, R. L. (1997) Hormonal signalling in cereal aleurone. *J. Exp. Bot.* 48: 1337–1356.

Campbell, N. A., Reece, J. B., and Mitchell, L. G. (1999) *Biology*, 5th ed. Benjamin Cummings, Menlo Park, CA.

Carrera, E., Bou, J., Garcia-Martinez, J. L., and Prat, S. (2000) Changes in GA 20-oxidase gene expression strongly affect stem length, tuber induction and tuber yield of potato plants. *Plant J.* 22: 247–256.

Davies, P. J. (1995) The plant hormones: Their nature, occurrence, and functions. In *Plant Hormones: Physiology, Biochemistry and Molecular Biology*, P. J. Davies, ed., Kluwer, Dordrecht, Netherlands, pp. 1–12.

Dill, A., and Sun, T. P. (2001) Synergistic derepression of gibberellin signaling by removing RGA and GAI function in *Arabidopsis thaliana*. *Genetics* 159: 777–785.

Dill, A., Jung, H. S., and Sun, T. P. (2001) The DELLA motif is essential for gibberellin-induced degradation of RGA. *Proc. Natl. Acad. Sci. USA* 98: 14162–14167.

Elliott, R. C., Ross, J. J., Smith, J. J., and Lester, D. R. (2001) Feed-forward regulation of gibberellin deactivation in pea. *J. Plant Growth Regul.* 20: 87–94.

Fabian, T., Lorbiecke, R., Umeda, M., and Sauter, M. (2000) The cell cycle genes *cycA1;1* and *cdc2Os-3* are coordinately regulated by gibberellin in planta. *Planta* 211: 376–383.

Gilroy, S., and Jones, R. L. (1994) Perception of gibberellin and abscisic acid at the external face of the plasma membrane of barley (*Hordeum vulgare* L.) aleurone protoplasts. *Plant Physiol.* 104: 1185–1192.

Gubler, F., Kalla, R., Roberts, J. K., and Jacobsen, J. V. (1995) Gibberellin-regulated expression of a *myb* gene in barley aleurone cells: Evidence of myb transactivation of a high-pl alpha-amylase gene promoter. *Plant Cell* 7: 1879–1891.

Hazebroek, J. P., and Metzger, J. D. (1990) Thermoinductive regulation of gibberellin metabolism in *Thlaspi arvense* L. I. Metabolism of [2H]-ent-Kaurenoic acid and [14C]gibberellin A_{12}-aldehyde. *Plant Physiol.* 94: 157–165.

Hedden, P., and Kamiya, Y. (1997) Gibberellin biosynthesis: Enzymes, genes and their regulation. *Annu. Rev. Plant Physiol. Plant Mol. Biol.* 48: 431–460.

Hedden, P., and Phillips, A. L. (2000) Gibberellin metabolism: New insights revealed by the genes. *Trends Plant Sci.* 5: 523–530.

Helliwell, C. A., Sullivan, J. A., Mould, R. M., Gray, J. C., Peacock, W. J., and Dennis, E. S. (2001) A plastid envelope location of *Arabidopsis ent*-kaurene oxidase links the plastid and endoplasmic reticulum steps of the gibberellin biosynthesis pathway. *Plant J.* 28: 201–208.

Higgins, T. J. V., Zwar, J. A., and Jacobsen, J. V. (1976) Gibberellic acid enhances the level of translatable mRNA for α-amylase in barley aleurone layers. *Nature* 260: 166–169.

Hooley, R., Beale, M. H., and Smith, S. J. (1991) Gibberellin perception at the plasma membrane of *Avena fatua* aleurone protoplasts. *Planta* 183: 274–280.

Ingram, T. J., Reid, J. B., and Macmillan, J. (1986) The quantitative relationship between gibberellin A_1 and internode growth in *Pisum sativum* L. *Planta* 168: 414–420.

Ingram, T. J., Reid, J. B., Potts, W. C., and Murfet, I. C. (1983) Internode length in *Pisum*. IV The effect of the *Le* gene on gibberellin metabolism. *Physiol. Plant.* 59: 607–616.

Irish, E. E. (1996) Regulation of sex determination in maize. *Bioessays* 18: 363–369.

Jacobsen, J. V., Gubler, F., and Chandler, P. M. (1995) Gibberellin action in germinated cereal grains. In *Plant Hormones: Physiology, Biochemistry and Molecular Biology*, P. J. Davies, ed., Kluwer, Dordrecht, Netherlands, pp. 246–271.

Jones, H. D., Smith, S. J., Desikan, R., Plakidou, D. S., Lovegrove, A., and Hooley, R. (1998) Heterotrimeric G proteins are implicated in gibberellin induction of α-amylase gene expression in wild oat aleurone. *Plant Cell* 10: 245–253.

Kende, H., van-der, K. E., and Cho, H. T. (1998) Deepwater rice: A model plant to study stem elongation. *Plant Physiol.* 118: 1105–1110.

King, K. E., Moritz, T., and Harberd, N. P. (2001) Gibberellins are not required for normal stem growth in *Arabidopsis thaliana* in the absence of GAI and RGA. *Genetics* 159: 767–776.

Kobayashi, M., Spray, C. R., Phinney, B. O., Gaskin, P., and MacMillan, J. (1996) Gibberellin metabolism in maize: The stepwise conversion of gibberellin A_{12}-aldehyde to gibberellin A_{20}. *Plant Physiol.* 110: 413–418.

Lester, D. R., Ross, J. J., Davies, P. J., and Reid, J. B. (1997) Mendel's stem length gene (*Le*) encodes a gibberellin 3β-hydroxylase. *Plant Cell* 9: 1435–1443.

Lichtenthaler, H. K., Rohmer, M., and Schwender, J. (1997) Two independent biochemical pathways for isopentenyl diphosphate and isoprenoid biosynthesis in higher plants. *Physiol. Plant.* 101: 643–652.

Lovegrove, A., and Hooley, R. (2000) Gibberellin and abscisic acid signalling in aleurone. *Trends Plant Sci.* 5: 102–110.

Lovegrove, A., Barratt, D. H. P., Beale, M. H., and Hooley, R. (1998) Gibberellin-photoaffinity labelling of two polypeptides in plant plasma membranes. *Plant J.* 15: 311–320.

O'Neill, D. P., Ross, J. J., and Reid, J. B. (2000) Changes in gibberellin A_1 levels and response during de-etiolation of pea seedlings. *Plant Physiol.* 124: 805–812.

Ou-Lee, T. M., Turgeon, R., and Wu, R. (1988) Interaction of a gibberellin-induced factor with the upstream region of an α-amylase gene in rice aleurone tissue. *Proc. Natl. Acad. Sci. USA* 85: 6366–6369.

Peng, J., Richards, D. E., Hartley, N. M., Murphy, G. P., Flintham, J. E., Beales, J., Fish, L. J., Pelica, F., Sudhakar, D., Christou, P., Snape, J. W., Gale, M. D., and Harberd, N. P. (1999) 'Green revolution' genes encode mutant gibberellin response modulators. *Nature* 400: 256–261.

Pensen, S. P., Schuurink, R. C., Fath, A., Gubler, F., Jacobsen, J. V., and Jones, R. L. (1996) cGMP is required for gibberellic acid-induced gene expression in barley aleurone. *Plant Cell* 8: 2325–2333.

Phinney, B. O. (1983) The history of gibberellins. In *The Biochemistry and Physiology of Gibberellins*, A. Crozier (ed.), Praeger, New York, pp. 15–52.

Reid, J. B., and Howell, S. H. (1995) Hormone mutants and plant development. In *Plant Hormones: Physiology, Biochemistry and Molecular Biology*, P. J. Davies, ed., Kluwer, Dordrecht, Netherlands, pp. 448–485.

Ritchie, S., and Gilroy, S. (1998a) Calcium-dependent protein phosphorylation may mediate the gibberellic acid response in barley aleurone. *Plant Physiol.* 116: 765–776.

Ritchie, S., and Gilroy, S. (1998b) Tansley Review No. 100: Gibberellins: Regulating genes and germination. *New Phytol.* 140: 363–383.

Ross, J., and O'Neill, D. (2001) New interactions between classical plant hormones. *Trends Plant Sci.* 6: 2–4.

Ross, J. J., O'Neill, D. P., Smith, J. J., Kerckhoffs, L. H. J., and Elliott, R. C. (2000) Evidence that auxin promotes gibberellin A_1 biosynthesis in pea. *Plant J.* 21: 547–552.

Ross, J. J., Reid, J. B., Gaskin, P. and Macmillan, J. (1989) Internode length in *Pisum*. Estimation of GA_1 levels in genotypes *Le, le* and *led*. *Physiol. Plant.* 76: 173–176.

Sachs, R. M. (1965) Stem elongation. *Annu. Rev. Plant. Physiol.* 16: 73–96.

Sauter, M., and Kende, H. (1992) Gibberellin-induced growth and regulation of the cell division cycle in deepwater rice. *Planta* 188: 362–368.

Schneider, G., and Schmidt, J. (1990) Conjugation of gibberellins in *Zea mays* L. In *Plant Growth Substances, 1988*, R. P. Pharis and S. B. Rood eds., Springer, Heidelberg, Germany, pp. 300–306.

Silverstone, A. L., and Sun, T. P. (2000) Gibberellins and the green revolution. *Trends Plant Sci.* 5: 1–2.

Silverstone, A. L., Jung, H. S., Dill, A., Kawaide, H., Kamiya, Y., and Sun, T. P. (2001) Repressing a repressor: Gibberellin-induced rapid reduction of the RGA protein in *Arabidopsis. Plant Cell* 13: 1555–1565.

Sun, T. P. (2000) Gibberellin signal transduction. *Curr. Opin. Plant Biol.* 3: 374–380.

Thornton, T. M., Swain, S. M., and Olszewski, N. E. (1999) Gibberellin signal transduction presents . . . the SPY who O-GlcNAc'd me. *Trends Plant Sci.* 4: 424–428.

Toyomasu, T., Kawaide, H., Mitsuhashi, W., Inoue, Y., and Kamiya, Y. (1998) Phytochrome regulates gibberellin biosynthesis during germination of photoblastic lettuce seeds. *Plant Physiol.* 118: 1517–1523.

Ueguchi-Tanaka, M., Fujisawa, Y., Kobayashi, M., Ashikari, M., Iwasaki, Y., Kitano, H., and Matsuoka, M. (2000) Rice dwarf mutant *d1*, which is defective in the alpha subunit of the heterotrimeric G protein, affects gibberellin signal transduction. *Proc. Natl. Acad. Sci. USA* 97: 11638–11643.

Wu, K., Li, L., Gage, D. A., and Zeevaart, J. A. D. (1996) Molecular cloning and photoperiod-regulated expression of gibberellin 20-oxidase from the long-day plant spinach. *Plant Physiol.* 110: 547–554.

Xu, Y. L., Gage, D. A., and Zeevaart, J. A. D. (1997) Gibberellins and stem growth in *Arabidopsis thaliana. Plant Physiol.* 114: 1471–1476.

Yamaguchi, S., and Kamiya, Y. (2000) Gibberellin biosynthesis: Its regulation by endogenous and environmental signals. *Plant Cell Physiol.* 41: 251–257.

Yang, T., Davies, P. J., and Reid, J. B. (1996) Genetic dissection of the relative roles of auxin and gibberellin in the regulation of stem elongation in intact light-grown peas. *Plant Physiol.* 110: 1029–1034.

Zeevaart, J. A. D., Gage, D. A., and Talon, M. (1993) Gibberellin A_1 is required for stem elongation in spinach. *Proc. Natl. Acad. Sci. USA* 90: 7401–7405.

Chapter

21

Cytokinins: Regulators of Cell Division

THE CYTOKININS WERE DISCOVERED in the search for factors that stimulate plant cells to divide (i.e., undergo cytokinesis). Since their discovery, cytokinins have been shown to have effects on many other physiological and developmental processes, including leaf senescence, nutrient mobilization, apical dominance, the formation and activity of shoot apical meristems, floral development, the breaking of bud dormancy, and seed germination. Cytokinins also appear to mediate many aspects of light-regulated development, including chloroplast differentiation, the development of autotrophic metabolism, and leaf and cotyledon expansion.

Although cytokinins regulate many cellular processes, the control of cell division is central in plant growth and development and is considered diagnostic for this class of plant growth regulators. For these reasons we will preface our discussion of cytokinin function with a brief consideration of the roles of cell division in normal development, wounding, gall formation, and tissue culture.

Later in the chapter we will examine the regulation of plant cell proliferation by cytokinins. Then we will turn to cytokinin functions not directly related to cell division: chloroplast differentiation, the prevention of leaf senescence, and nutrient mobilization. Finally, we will consider the molecular mechanisms underlying cytokinin perception and signaling.

CELL DIVISION AND PLANT DEVELOPMENT

Plant cells form as the result of cell divisions in a primary or secondary meristem. Newly formed plant cells typically enlarge and differentiate, but once they have assumed their function—whether transport, photosynthesis, support, storage, or protection—usually they do not divide again during the life of the plant. In this respect they appear to be similar to animal cells, which are considered to be terminally differentiated.

However, this similarity to the behavior of animal cells is only superficial. Almost every type of plant cell that retains its nucleus at maturity has been shown to be capable of dividing. This property comes into play during such processes as wound healing and leaf abscission.

Differentiated Plant Cells Can Resume Division

Under some circumstances, mature, differentiated plant cells may resume cell division in the intact plant. In many species, mature cells of the cortex and/or phloem resume division to form secondary meristems, such as the vascular cambium or the cork cambium. The abscission zone at the base of a leaf petiole is a region where mature parenchyma cells begin to divide again after a period of mitotic inactivity, forming a layer of cells with relatively weak cell walls where abscission can occur (see Chapter 22).

Wounding of plant tissues induces cell divisions at the wound site. Even highly specialized cells, such as phloem fibers and guard cells, may be stimulated by wounding to divide at least once. Wound-induced mitotic activity typically is self-limiting; after a few divisions the derivative cells stop dividing and redifferentiate. However, when the soil-dwelling bacterium *Agrobacterium tumefaciens* invades a wound, it can cause the neoplastic (tumor-forming) disease known as **crown gall**. This phenomenon is dramatic natural evidence of the mitotic potential of mature plant cells.

Without *Agrobacterium* infection, the wound-induced cell division would subside after a few days and some of the new cells would differentiate as a protective layer of cork cells or vascular tissue. However, *Agrobacterium* changes the character of the cells that divide in response to the wound, making them tumorlike. They do not stop dividing; rather they continue to divide throughout the life of the plant to produce an unorganized mass of tumorlike tissue called a **gall** (Figure 21.1). We will have more to say about this important disease later in this chapter.

FIGURE 21.1 Tumor that formed on a tomato stem infected with the crown gall bacterium, *Agrobacterium tumefaciens*. Two months before this photo was taken the stem was wounded and inoculated with a virulent strain of the crown gall bacterium. (From Aloni et al. 1998, courtesy of R. Aloni.)

Diffusible Factors May Control Cell Division

The considerations addressed in the previous section suggest that mature plant cells stop dividing because they no longer receive a particular signal, possibly a hormone, that is necessary for the initiation of cell division. The idea that cell division may be initiated by a diffusible factor originated with the Austrian plant physiologist G. Haberlandt, who, in about 1913, demonstrated that vascular tissue contains a water-soluble substance or substances that will stimulate the division of wounded potato tuber tissue. The effort to determine the nature of this factor (or factors) led to the discovery of the cytokinins in the 1950s.

Plant Tissues and Organs Can Be Cultured

Biologists have long been intrigued by the possibility of growing organs, tissues, and cells in culture on a simple nutrient medium, in the same way that microorganisms can be cultured in test tubes or on petri dishes. In the 1930s, Philip White demonstrated that tomato roots can be grown indefinitely in a simple nutrient medium containing only sucrose, mineral salts, and a few vitamins, with no added hormones (White 1934).

In contrast to roots, isolated stem tissues exhibit very little growth in culture without added hormones in the medium. Even if auxin is added, only limited growth may occur, and usually this growth is not sustained. Frequently this auxin-induced growth is due to cell enlargement only. The shoots of most plants cannot grow on a simple medium lacking hormones, even if the cultured stem tissue contains apical or lateral meristems, until adventitious roots form. Once the stem tissue has rooted, shoot growth resumes, but now as an integrated, whole plant.

These observations indicate that there is a difference in the regulation of cell division in root and shoot meristems. They also suggest that some root-derived factor(s) may regulate growth in the shoot.

Crown gall stem tissue is an exception to these generalizations. After a gall has formed on a plant, heating the plant to 42°C will kill the bacterium that induced gall formation. The plant will survive the heat treatment, and its gall tissue will continue to grow as a bacteria-free tumor (Braun 1958).

Tissues removed from these bacteria-free tumors grow on simple, chemically defined culture media that would not support the proliferation of normal stem tissue of the same species. However, these stem-derived tissues are not organized. Instead they grow as a mass of disorganized, relatively undifferentiated cells called **callus tissue**.

Callus tissue sometimes forms naturally in response to wounding, or in graft unions where stems of two different plants are joined. Crown gall tumors are a specific type of callus, whether they are growing attached to the plant or in culture. The finding that crown gall callus tissue can be cultured demonstrated that cells derived from stem tissues are capable of proliferating in culture and that contact with the bacteria may cause the stem cells to produce cell division–stimulating factors.

THE DISCOVERY, IDENTIFICATION, AND PROPERTIES OF CYTOKININS

A great many substances were tested in an effort to initiate and sustain the proliferation of normal stem tissues in culture. Materials ranging from yeast extract to tomato juice were found to have a positive effect, at least with some tissues. However, culture growth was stimulated most dramatically when the liquid endosperm of coconut, also known as coconut milk, was added to the culture medium.

Philip White's nutrient medium, supplemented with an auxin and 10 to 20% coconut milk, will support the continued cell division of mature, differentiated cells from a wide variety of tissues and species, leading to the formation of callus tissue (Caplin and Steward 1948). This finding indicated that coconut milk contains a substance or substances that stimulate mature cells to enter and remain in the cell division cycle.

Eventually coconut milk was shown to contain the cytokinin *zeatin*, but this finding was not obtained until several years after the discovery of the cytokinins (Letham 1974). The first cytokinin to be discovered was the synthetic analog kinetin.

Kinetin Was Discovered as a Breakdown Product of DNA

In the 1940s and 1950s, Folke Skoog and coworkers at the University of Wisconsin tested many substances for their ability to initiate and sustain the proliferation of cultured tobacco pith tissue. They had observed that the nucleic acid base adenine had a slight promotive effect, so they tested the possibility that nucleic acids would stimulate division in this tissue. Surprisingly, autoclaved herring sperm DNA had a powerful cell division–promoting effect.

After much work, a small molecule was identified from the autoclaved DNA and named **kinetin**. It was shown to be an adenine (or aminopurine) derivative, 6-furfurylaminopurine (Miller et al. 1955):

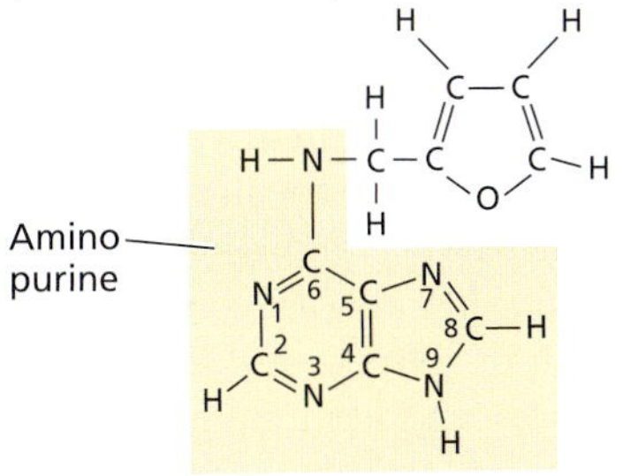

Amino purine

Kinetin

In the presence of an auxin, kinetin would stimulate tobacco pith parenchyma tissue to proliferate in culture. No kinetin-induced cell division occurs without auxin in the culture medium. (For more details, see **Web Topic 21.1**.)

Kinetin is not a naturally occurring plant growth regulator, and it does not occur as a base in the DNA of any species. It is a by-product of the heat-induced degradation of DNA, in which the deoxyribose sugar of adenosine is converted to a furfuryl ring and shifted from the 9 position to the 6 position on the adenine ring.

The discovery of kinetin was important because it demonstrated that cell division could be induced by a simple chemical substance. Of greater importance, the discovery of kinetin suggested that naturally occurring molecules with structures similar to that of kinetin regulate cell division activity within the plant. This hypothesis proved to be correct.

Zeatin Is the Most Abundant Natural Cytokinin

Several years after the discovery of kinetin, extracts of the immature endosperm of corn (*Zea mays*) were found to contain a substance that has the same biological effect as kinetin. This substance stimulates mature plant cells to divide when added to a culture medium along with an auxin. Letham (1973) isolated the molecule responsible for this activity and identified it as *trans*-6-(4-hydroxy-3-methylbut-2-enylamino)purine, which he called **zeatin**:

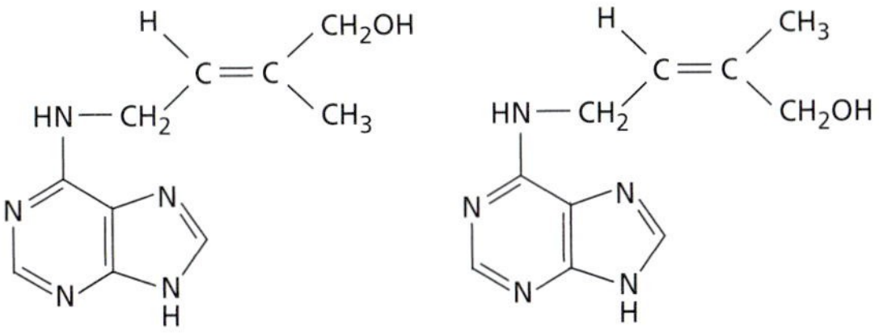

trans-Zeatin *cis*-Zeatin
6-(4-Hydroxy-3-methylbut-2-enylamino)purine

The molecular structure of zeatin is similar to that of kinetin. Both molecules are adenine or aminopurine derivatives. Although they have different side chains, in both cases the side chain is attached to the 6 nitrogen of the aminopurine. Because the side chain of zeatin has a double bond, it can exist in either the *cis* or the *trans* configuration.

In higher plants, zeatin occurs in both the *cis* and the *trans* configurations, and these forms can be interconverted by an enzyme known as *zeatin isomerase*. Although the *trans* form of zeatin is much more active in biological assays, the *cis* form may also play important roles, as suggested by the fact that it has been found in high levels in a number of plant species and particular tissues. A gene encoding a glucosyl transferase enzyme specific to *cis*-zeatin has recently been cloned, which further supports a biological role for this isoform of zeatin.

Since its discovery in immature maize endosperm, zeatin has been found in many plants and in some bacteria. It is the most prevalent cytokinin in higher plants, but other substituted aminopurines that are active as cytokinins have been isolated from many plant and bacterial species. These aminopurines differ from zeatin in the nature of the side chain attached to the 6 nitrogen or in the attachment of a side chain to carbon 2:

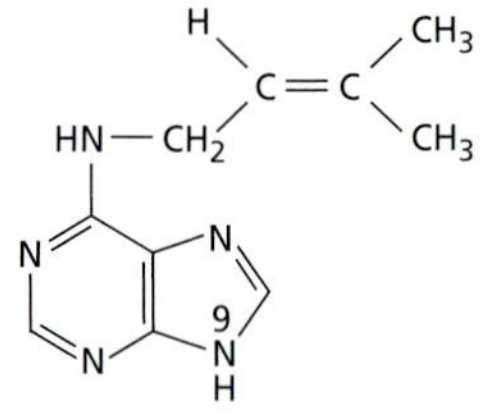

N^6-(Δ^2-Isopentenyl)-adenine (iP)

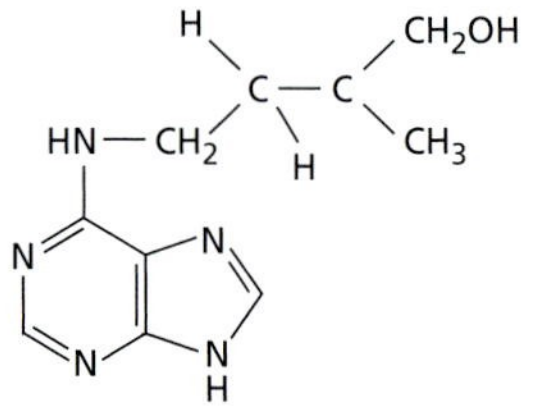

Dihydrozeatin (DZ)

In addition, these cytokinins can be present in the plant as a **riboside** (in which a ribose sugar is attached to the 9 nitrogen of the purine ring), a **ribotide** (in which the ribose sugar moiety contains a phosphate group), or a **glycoside** (in which a sugar molecule is attached to the 3, 7, or 9 nitrogen of the purine ring, or to the oxygen of the zeatin or dihydrozeatin side chain) (see **Web Topic 21.2**).

Some Synthetic Compounds Can Mimic or Antagonize Cytokinin Action

Cytokinins are defined as compounds that have biological activities similar to those of *trans*-zeatin. These activities include the ability to do the following:

- Induce cell division in callus cells in the presence of an auxin
- Promote bud or root formation from callus cultures when in the appropriate molar ratios to auxin
- Delay senescence of leaves
- Promote expansion of dicot cotyledons

Many chemical compounds have been synthesized and tested for cytokinin activity. Analysis of these compounds provides insight into the structural requirements for activity. Nearly all compounds active as cytokinins are N^6-substituted aminopurines, such as benzyladenine (BA):

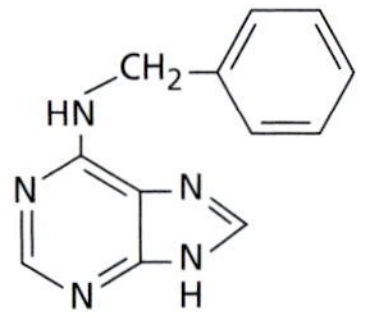

Benzyladenine (benzylaminopurine) (BA)

and all the naturally occurring cytokinins are aminopurine derivatives. There are also synthetic cytokinin compounds that have not been identified in plants, most notable of which are the diphenylurea-type cytokinins, such as thidiazuron, which is used commercially as a defoliant and an herbicide:

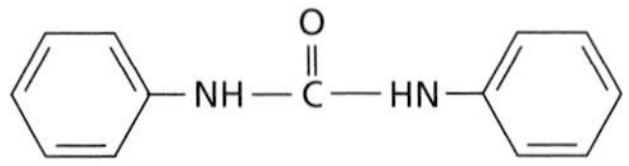

***N,N'*-Diphenylurea** (nonamino purine with weak activity)

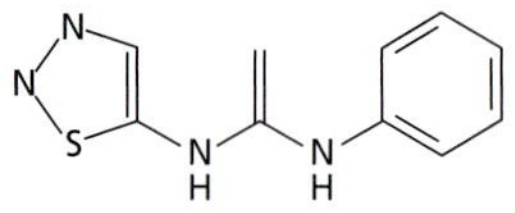

Thidiazuron

In the course of determining the structural requirements for cytokinin activity, investigators found that some molecules act as *cytokinin antagonists*:

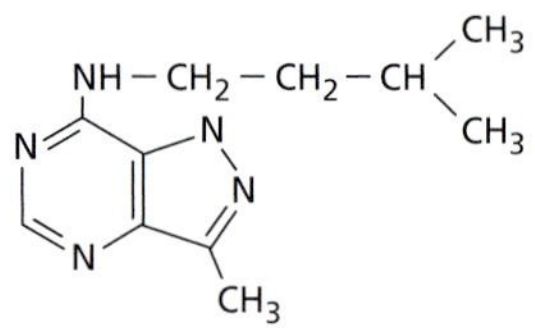

3-Methyl-7-(3-methylbutylamino)pyrazolo[4,3-D]pyrimidine

These molecules are able to block the action of cytokinins, and their effects may be overcome by the addition of more cytokinin. Naturally occurring molecules with cytokinin activity may be detected and identified by a combination of physical methods and bioassays (see **Web Topic 21.3**).

Cytokinins Occur in Both Free and Bound Forms

Hormonal cytokinins are present as free molecules (not covalently attached to any macromolecule) in plants and certain bacteria. Free cytokinins have been found in a wide spectrum of angiosperms and probably are universal in this group of plants. They have also been found in algae, diatoms, mosses, ferns, and conifers.

The regulatory role of cytokinins has been demonstrated only in angiosperms, conifers, and mosses, but they may function to regulate the growth, development, and metabolism of all plants. Usually zeatin is the most abundant naturally occurring free cytokinin, but *dihydrozeatin* (*DZ*) and *isopentenyl adenine* (*iP*) also are commonly found in higher plants and bacteria. Numerous derivatives of these three cytokinins have been identified in plant extracts (see the structures illustrated in Figure 21.6).

Transfer RNA (tRNA) contains not only the four nucleotides used to construct all other forms of RNA, but also some unusual nucleotides in which the base has been modified. Some of these "hypermodified" bases act as cytokinins when the tRNA is hydrolyzed and tested in one of the cytokinin bioassays. Some plant tRNAs contain *cis*-

zeatin as a hypermodified base. However, cytokinins are not confined to plant tRNAs. They are part of certain tRNAs from all organisms, from bacteria to humans. (For details, see **Web Topic 21.4**.)

The Hormonally Active Cytokinin Is the Free Base

It has been difficult to determine which species of cytokinin represents the active form of the hormone, but the recent identification of the cytokinin receptor CRE1 has allowed this question to be addressed. The relevant experiments have shown that the free-base form of *trans*-zeatin, but not its riboside or ribotide derivatives, binds directly to CRE1, indicating that the free base is the active form (Yamada et al. 2001).

Although the free-base form of *trans*-zeatin is thought to be the hormonally active cytokinin, some other compounds have cytokinin activity, either because they are readily converted to zeatin, dihydrozeatin, or isopentenyl adenine, or because they release these compounds from other molecules, such as cytokinin glucosides. For example, tobacco cells in culture do not grow unless cytokinin ribosides supplied in the culture medium are converted to the free base.

In another example, excised radish cotyledons grow when they are cultured in a solution containing the cytokinin base benzyladenine (BA, an N^6-substituted aminopurine cytokinin). The cultured cotyledons readily take up the hormone and convert it to various BA glucosides, BA ribonucleoside, and BA ribonucleotide. When the cotyledons are transferred back to a medium lacking a cytokinin, their growth rate declines, as do the concentrations of BA, BA ribonucleoside, and BA ribonucleotide in the tissues. However, the level of the BA glucosides remains constant. This finding suggests that the glucosides cannot be the active form of the hormone.

Ribosylzeatin (zeatin riboside)

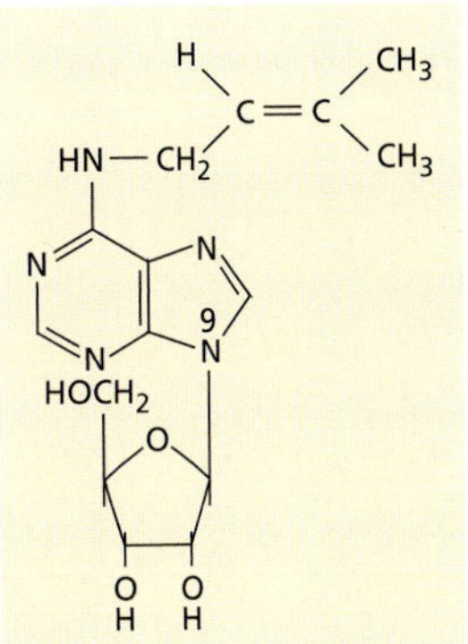

N^6-(Δ^2-Isopentenyl)adenosine ([9R]iP)

FIGURE 21.2 Structures of ribosylzeatin and N6-(Δ2-isopentenyl)adenosine ([9R]iP).

Some Plant Pathogenic Bacteria, Insects, and Nematodes Secrete Free Cytokinins

Some bacteria and fungi are intimately associated with higher plants. Many of these microorganisms produce and secrete substantial amounts of cytokinins and/or cause the plant cells to synthesize plant hormones, including cytokinins (Akiyoshi et al. 1987). The cytokinins produced by microorganisms include *trans*-zeatin, [9R]iP, *cis*-zeatin, and their ribosides (Figure 21.2). Infection of plant tissues with these microorganisms can induce the tissues to divide and, in some cases, to form special structures, such as mycorrhizae, in which the microorganism can reside in a mutualistic relationship with the plant.

In addition to the crown gall bacterium, *Agrobacterium tumefaciens*, other pathogenic bacteria may stimulate plant cells to divide. For example, *Corynebacterium fascians* is a major cause of the growth abnormality known as **witches'-broom** (Figure 21.3). The shoots of plants infected by *C. fascians* resemble an old-fashioned straw broom because the lateral buds, which normally remain dormant, are stimulated by the bacterial cytokinin to grow (Hamilton and Lowe 1972).

FIGURE 21.3 Witches' broom on balsam fir (*Abies balsamea*). (Photo © Gregory K. Scott/Photo Researchers, Inc.)

Infection with a close relative of the crown gall organism, *Agrobacterium rhizogenes*, causes masses of roots instead of callus tissue to develop from the site of infection. *A. rhizogenes* is able to modify cytokinin metabolism in infected plant tissues through a mechanism that will be described later in this chapter.

Certain insects secrete cytokinins, which may play a role in the formation of galls utilized by these insects as feeding sites. Root-knot nematodes also produce cytokinins, which may be involved in manipulating host development to produce the giant cells from which the nematode feeds (Elzen 1983).

BIOSYNTHESIS, METABOLISM, AND TRANSPORT OF CYTOKININS

The side chains of naturally occurring cytokinins are chemically related to rubber, carotenoid pigments, the plant hormones gibberellin and abscisic acid, and some of the plant defense compounds known as phytoalexins. All of these compounds are constructed, at least in part, from isoprene units (see Chapter 13).

Isoprene is similar in structure to the side chains of zeatin and iP (see the structures illustrated in Figure 21.6). These cytokinin side chains are synthesized from an isoprene derivative. Large molecules of rubber and the carotenoids are constructed by the polymerization of many isoprene units; cytokinins contain just one of these units. The precursor(s) for the formation of these isoprene structures are either mevalonic acid or pyruvate plus 3-phosphoglycerate, depending on which pathway is involved (see Chapter 13). These precursors are converted to the biological isoprene unit dimethylallyl diphosphate (DMAPP).

Crown Gall Cells Have Acquired a Gene for Cytokinin Synthesis

Bacteria-free tissues from crown gall tumors proliferate in culture without the addition of any hormones to the culture medium. Crown gall tissues contain substantial amounts of both auxin and free cytokinins. Furthermore, when radioactively labeled adenine is fed to periwinkle (*Vinca rosea*) crown gall tissues, it is incorporated into both zeatin and zeatin riboside, demonstrating that gall tissues contain the cytokinin biosynthetic pathway. Control stem tissue, which has not been transformed by *Agrobacterium*, does not incorporate labeled adenine into cytokinins.

During infection by *Agrobacterium tumefaciens*, plant cells incorporate bacterial DNA into their chromosomes. The virulent strains of *Agrobacterium* contain a large plasmid known as the **Ti plasmid**. Plasmids are circular pieces of extrachromosomal DNA that are not essential for the life of the bacterium. However, plasmids frequently contain genes that enhance the ability of the bacterium to survive in special environments.

A small portion of the Ti plasmid, known as the **T-DNA**, is incorporated into the nuclear DNA of the host plant cell (Figure 21.4) (Chilton et al. 1977). T-DNA carries genes necessary for the biosynthesis of *trans*-zeatin and auxin, as well as a member of a class of unusual nitrogen-containing compounds called *opines* (Figure 21.5). Opines are not synthesized by plants except after crown gall transformation.

The T-DNA gene involved in cytokinin biosynthesis—known as the *ipt*[1] gene—encodes an ***isopentenyl transferase* (IPT)** enzyme that transfers the isopentenyl group from DMAPP to AMP (adenosine monophosphate) to form isopentenyl adenine ribotide (Figure 21.6) (Akiyoshi et al. 1984; Barry et al. 1984). The *ipt* gene has been called the *tmr* locus because, when *inactivated* by mutation, it results in "rooty" tumors. Isopentenyl adenine ribotide can be converted to the active cytokinins isopentenyl adenine, *trans*-zeatin, and dihydrozeatin by endogenous enzymes in plant cells. This conversion route is similar to the pathway for cytokinin synthesis that has been postulated for normal tissue (see Figure 21.6).

The T-DNA also contains two genes encoding enzymes that convert tryptophan to the auxin indole-3-acetic acid (IAA). This pathway of auxin biosynthesis differs from the one in nontransformed cells and involves indoleacetamide as an intermediate (see Figure 19.6). The *ipt* gene and the two auxin biosynthetic genes of T-DNA are **phyto-oncogenes**, since they can induce tumors in plants (see **Web Topic 21.5**).

Because their promoters are plant eukaryotic promoters, none of the T-DNA genes are expressed in the bacterium; rather they are transcribed after they are inserted into the plant chromosomes. Transcription of the genes leads to synthesis of the enzymes they encode, resulting in the production of zeatin, auxin, and an opine. The bacterium can utilize the opine as a nitrogen source, but cells of higher plants cannot. Thus, by transforming the plant cells, the bacterium provides itself with an expanding environment (the gall tissue) in which the host cells are directed to produce a substance (the opine) that only the bacterium can utilize for its nutrition (Bomhoff et al. 1976).

An important difference between the control of cytokinin biosynthesis in crown gall tissues and in normal tissues is that the T-DNA genes for cytokinin synthesis are expressed in all infected cells, even those in which the native plant genes for biosynthesis of the hormone are normally repressed.

IPT Catalyzes the First Step in Cytokinin Biosynthesis

The first committed step in cytokinin biosynthesis is the transfer of the isopentenyl group of dimethylallyl diphos-

[1] Bacterial genes, unlike plant genes, are written in lowercase italics.

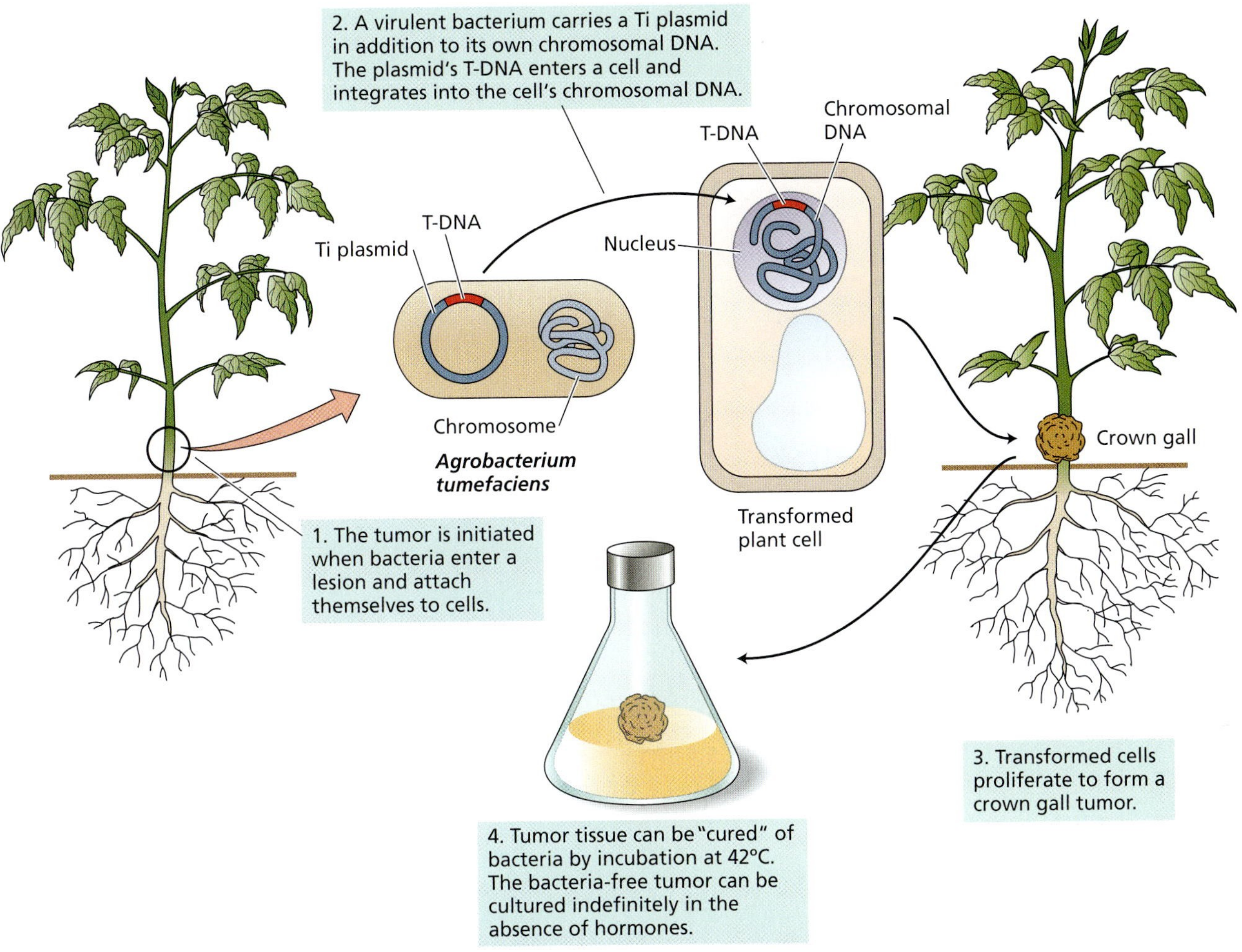

FIGURE 21.4 Tumor induction by *Agrobacterium tumefaciens*. (After Chilton 1983.)

phate (DMAPP) to an adenosine moiety. An enzyme that catalyzes such an activity was first identified in the cellular slime mold *Dictyostelium discoideum*, and subsequently the *ipt* gene from *Agrobacterium* was found to encode such an enzyme. In both cases, DMAPP and AMP are converted to isopentenyladenosine-5′-monophosphate (iPMP).

As noted earlier, cytokinins are also present in the tRNAs of most cells, including plant and animal cells. The tRNA cytokinins are synthesized by modification of specific adenine residues within the fully transcribed tRNA. As with the free cytokinins, isopentenyl groups are transferred to the adenine molecules from DMAPP by an enzyme call tRNA-IPT. The genes for tRNA-IPT have been cloned from many species.

NH_2(NH=)C—NH—$(CH_2)_3$—CH—COOH
|
NH
|
CH_3—CH—COOH

Octopine

NH_2(NH=)C—NH—$(CH_2)_3$—CH—COOH
|
NH
|
COOH—$(CH_2)_2$—CH—COOH

Nopaline

FIGURE 21.5 The two major opines, octopine and nopaline, are found only in crown gall tumors. The genes required for their synthesis are present in the T-DNA from *Agrobacterium tumefaciens*. The bacterium, but not the plant, can utilize the opines as a nitrogen source.

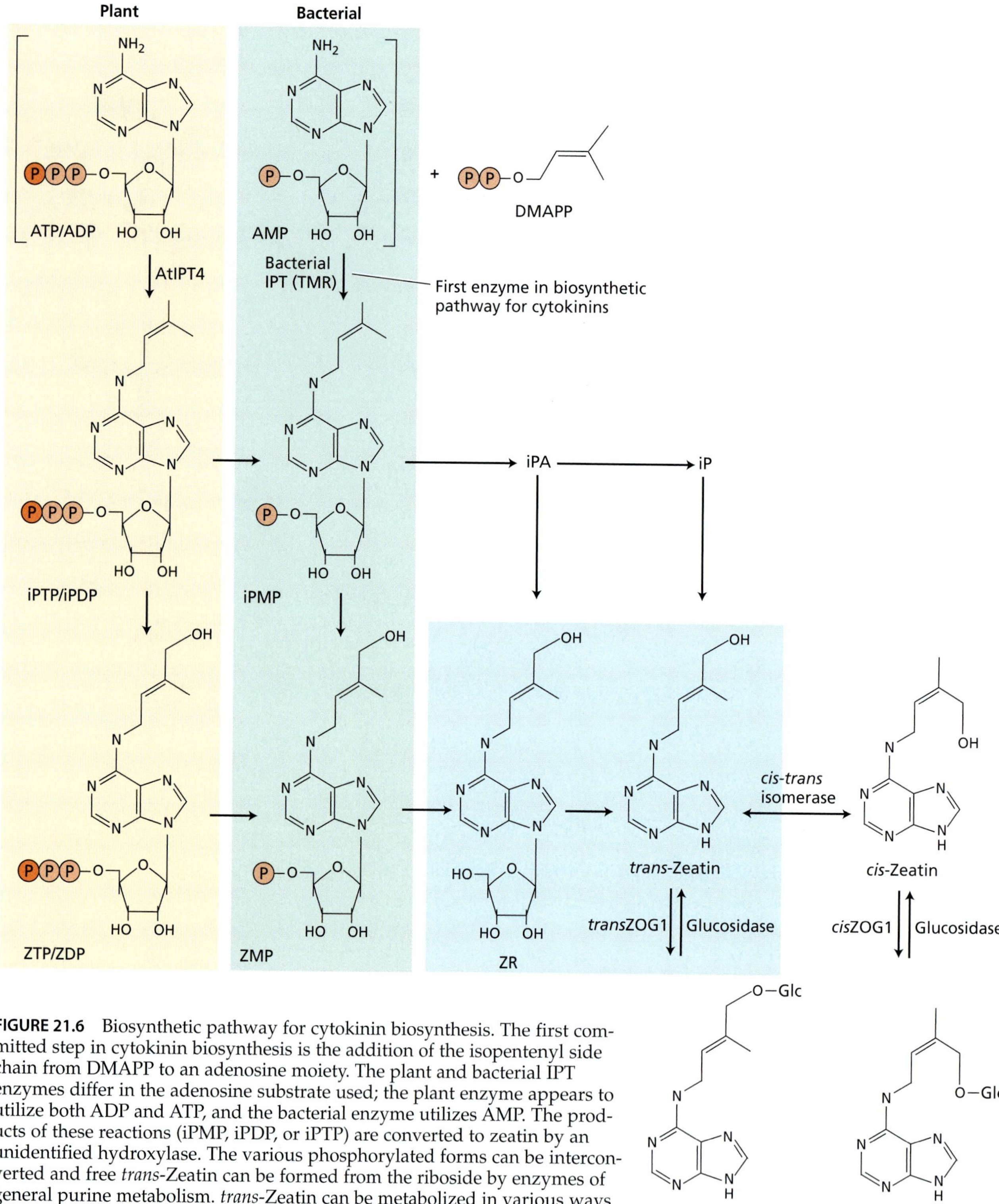

FIGURE 21.6 Biosynthetic pathway for cytokinin biosynthesis. The first committed step in cytokinin biosynthesis is the addition of the isopentenyl side chain from DMAPP to an adenosine moiety. The plant and bacterial IPT enzymes differ in the adenosine substrate used; the plant enzyme appears to utilize both ADP and ATP, and the bacterial enzyme utilizes AMP. The products of these reactions (iPMP, iPDP, or iPTP) are converted to zeatin by an unidentified hydroxylase. The various phosphorylated forms can be interconverted and free *trans*-Zeatin can be formed from the riboside by enzymes of general purine metabolism. *trans*-Zeatin can be metabolized in various ways as shown, and these reactions are catalyzed by the indicated enzymes.

The possibility that free cytokinins are derived from tRNA has been explored extensively. Although the tRNA-bound cytokinins can act as hormonal signals for plant cells if the tRNA is degraded and fed back to the cells, it is unlikely that any significant amount of the free hormonal cytokinin in plants is derived from the turnover of tRNA.

An enzyme with IPT activity was identified from crude extracts of various plant tissues, but researchers were unable to purify the protein to homogeneity. Recently, plant *IPT* genes were cloned after the *Arabidopsis* genome was analyzed for potential *ipt*-like sequences (Kakimoto 2001; Takei et al. 2001). Nine different *IPT* genes were identified in *Arabidopsis*—many more than are present in animal genomes, which generally contain only one or two such genes used in tRNA modification.

Phylogenetic analysis revealed that one of the *Arabidopsis IPT* genes resembles bacterial tRNA-*ipt*, another resembles eukaryotic tRNA-*IPT*, and the other seven form a distinct group or clade together with other plant sequences (see **Web Topic 21.6**). The grouping of the seven *Arabidopsis IPT* genes in this unique plant clade provided a clue that these genes may encode the cytokinin biosynthetic enzyme.

The proteins encoded by these genes were expressed in *E. coli* and analyzed. It was found that, with the exception of the gene most closely related to the animal tRNA-*IPT* genes, these genes encoded proteins capable of synthesizing free cytokinins. Unlike their bacterial counterparts, however, the *Arabidopsis* enzymes that have been analyzed utilize ATP and ADP preferentially over AMP (see Figure 21.6).

Cytokinins from the Root Are Transported to the Shoot via the Xylem

Root apical meristems are major sites of synthesis of the free cytokinins in whole plants. The cytokinins synthesized in roots appear to move through the xylem into the shoot, along with the water and minerals taken up by the roots. This pathway of cytokinin movement has been inferred from the analysis of xylem exudate.

When the shoot is cut from a rooted plant near the soil line, the xylem sap may continue to flow from the cut stump for some time. This xylem exudate contains cytokinins. If the soil covering the roots is kept moist, the flow of xylem exudate can continue for several days. Because the cytokinin content of the exudate does not diminish, the cytokinins found in it are likely to be synthesized by the roots. In addition, environmental factors that interfere with root function, such as water stress, reduce the cytokinin content of the xylem exudate (Itai and Vaadia 1971). Conversely, resupply of nitrate to nitrogen-starved maize roots results in an elevation of the concentration of cytokinins in the xylem sap (Samuelson 1992), which has been correlated to an induction of cytokinin-regulated gene expression in the shoots (Takei et al. 2001).

Although the presence of cytokinin in the xylem is well established, recent grafting experiments have cast doubt on the presumed role of this root-derived cytokinin in shoot development. Tobacco transformed with an inducible *ipt* gene from *Agrobacterium* displayed increased lateral bud outgrowth and delayed senescence.

To assess the role of cytokinin derived from the root, the tobacco root stock engineered to overproduce cytokinin was grafted to a wild-type shoot. Surprisingly, no phenotypic consequences were observed in the shoot, even though an increased concentration of cytokinin was measured in the transpiration stream (Faiss et al. 1997). Thus the excess cytokinin in the roots had no effect on the grafted shoot.

Roots are not the only parts of the plant capable of synthesizing cytokinins. For example, young maize embryos synthesize cytokinins, as do young developing leaves, young fruits, and possibly many other tissues. Clearly, further studies will be needed to resolve the roles of cytokinins transported from the root versus cytokinins synthesized in the shoot.

A Signal from the Shoot Regulates the Transport of Zeatin Ribosides from the Root

The cytokinins in the xylem exudate are mainly in the form of zeatin ribosides. Once they reach the leaves, some of these nucleosides are converted to the free-base form or to glucosides (Noodén and Letham 1993). Cytokinin glucosides may accumulate to high levels in seeds and in leaves, and substantial amounts may be present even in senescing leaves. Although the glucosides are active as cytokinins in bioassays, often they lack hormonal activity after they form within cells, possibly because they are compartmentalized in such a way that they are unavailable. Compartmentation may explain the conflicting observations that cytokinins are transported readily by the xylem but that radioactive cytokinins applied to leaves in intact plants do not appear to move from the site of application.

Evidence from grafting experiments with mutants suggests that the transport of zeatin riboside from the root to the shoot is regulated by signals from the shoot. The *rms4* mutant of pea (*Pisum sativum* L.) is characterized by a 40-fold decrease in the concentration of zeatin riboside in the xylem sap of the roots. However, grafting a wild-type shoot onto an *rms4* mutant root increased the zeatin riboside levels in the xylem exudate to wild-type levels. Conversely, grafting an *rms4* mutant shoot onto a wild-type root lowered the concentration of zeatin riboside in the xylem exudate to mutant levels (Beveridge et al. 1997).

These results suggest that a signal from the shoot can regulate cytokinin transport from the root. The identity of this signal has not yet been determined.

Cytokinins Are Rapidly Metabolized by Plant Tissues

Free cytokinins are readily converted to their respective nucleoside and nucleotide forms. Such interconversions likely involve enzymes common to purine metabolism.

Many plant tissues contain the enzyme **cytokinin oxidase**, which cleaves the side chain from zeatin (both *cis* and *trans*), zeatin riboside, iP, and their *N*-glucosides, but not their *O*-glucoside derivatives (Figure 21.7). However, dihydrozeatin and its conjugates are resistant to cleavage. Cytokinin oxidase irreversibly inactivates cytokinins, and it could be important in regulating or limiting cytokinin effects. The activity of the enzyme is induced by high cytokinin concentrations, due at least in part to an elevation of the RNA levels for a subset of the genes.

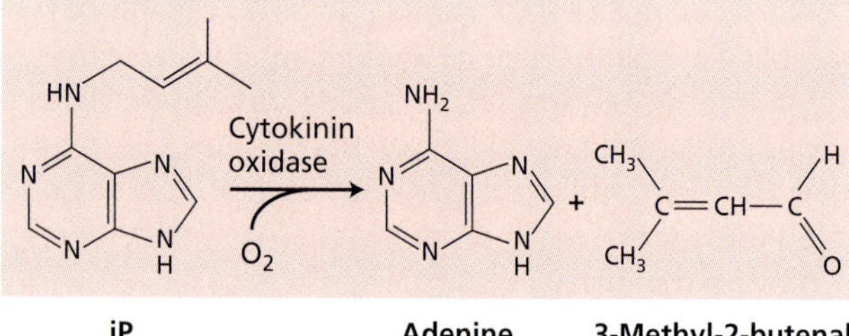

FIGURE 21.7 Cytokinin oxidase irreversibly degrades some cytokinins.

A gene encoding cytokinin oxidase was first identified in maize (Houba-Herin et al. 1999; Morris et al. 1999). In *Arabidopsis*, cytokinin oxidase is encoded by a multigene family whose members show distinct patterns of expression. Interestingly, several of the genes contain putative secretory signals, suggesting that at least some of these enzymes may be extracellular.

Cytokinin levels can also be regulated by conjugation of the hormone at various positions. The nitrogens at the 3, 7, and 9 positions of the adenine ring of cytokinins can be conjugated to glucose residues. Alanine can also be conjugated to the nitrogen at the 9 positon, forming lupinic acid. These modifications are generally irreversible, and such conjugated forms of cytokinin are inactive in bioassays, with the exception of the N^3-glucosides.

The hydroxyl group of the side chain of cytokinins is also the target for conjugation to glucose residues, or in some cases xylose residues, yielding *O*-glucoside and *O*-xyloside cytokinins. *O*-glucosides are resistant to cleavage by cytokinin oxidases, which may explain why these derivatives have higher biological activity in some assays than their corresponding free bases have.

Enzymes that catalyze the conjugation of either glucose or xylose to zeatin have been purified, and their respective genes have been cloned (Martin et al. 1999). These enzymes have stringent substrate specificities for the sugar donor and the cytokinin bases. Only free *trans*-zeatin and dihydrozeatin bases are efficient substrates; the corresponding nucleosides are not substrates, nor is *cis*-zeatin. The specificity of these enzymes suggests that the conjugation to the side chain is precisely regulated.

The conjugations at the side chain can be removed by glucosidase enzymes to yield free cytokinins, which, as discussed earlier, are the active forms. Thus, cytokinin glucosides may be a storage form, or metabolically inactive state, of these compounds. A gene encoding a glucosidase that can release cytokinins from sugar conjugates has been cloned from maize, and its expression could play an important role in the germination of maize seeds (Brzobohaty et al. 1993).

Dormant seeds often have high levels of cytokinin glucosides but very low levels of hormonally active free cytokinins. Levels of free cytokinins increase rapidly, however, as germination is initiated, and this increase in free cytokinins is accompanied by a corresponding decrease in cytokinin glucosides.

THE BIOLOGICAL ROLES OF CYTOKININS

Although discovered as a cell division factor, cytokinins can stimulate or inhibit a variety of physiological, metabolic, biochemical, and developmental processes when they are applied to higher plants, and it is increasingly clear that endogenous cytokinins play an important role in the regulation of these events in the intact plant.

In this section we will survey some of the diverse effects of cytokinin on plant growth and development, including a discussion of its role in regulating cell division. The discovery of the tumor-inducing Ti plasmid in the plant-pathogenic bacterium *Agrobacterium tumefaciens* provided plant scientists with a powerful new tool for introducing foreign genes into plants, and for studying the role of cytokinin in development. In addition to its role in cell proliferation, cytokinin affects many other processes, including differentiation, apical dominance, and senescence.

Cytokinins Regulate Cell Division in Shoots and Roots

As discussed earlier, cytokinins are generally required for cell division of plant cells in vitro. Several lines of evidence suggest that cytokinins also play key roles in the regulation of cell division in vivo.

Much of the cell division in an adult plant occurs in the meristems (see Chapter 16). Localized expression of the *ipt* gene of *Agrobacterium* in somatic sectors of tobacco leaves causes the formation of ectopic (abnormally located) meristems, indicating that elevated levels of cytokinin are sufficient to initiate cell divisions in these leaves (Estruch et al. 1991). Elevation of endogenous cytokinin levels in transgenic *Arabidopsis* results in overexpression of the KNOTTED homeobox transcription factor homologs *KNAT1* and *STM*—genes that are important in the regulation of meristem function (see Chapter 16) (Rupp et al. 1999). Interestingly, overexpression of *KNAT1* also appears to elevate cytokinin levels in transgenic tobacco, suggesting an interdependent relationship between *KNAT* and the level of cytokinins.

Overexpression of several of the *Arabidopsis* cytokinin oxidase genes in tobacco results in a reduction of endogenous cytokinin levels and a consequent strong retardation of shoot development due to a reduction in the rate of cell proliferation in the shoot apical meristem (Figures 21.8 and 21.9) (Werner et al. 2001). This finding strongly supports the notion that endogenous cytokinins regulate cell division in vivo.

Surprisingly, the same overexpression of cytokinin oxidase in tobacco led to an *enhancement* of root growth (Figure 21.10), primarily by increasing the size of the root api-

FIGURE 21.8 Tobacco plants overexpressing the gene for cytokinin oxidase. The plant on the left is wild type. The two plants on the right are overexpressing two different constructs of the *Arabidopsis* gene for cytokinin oxidase: *AtCKX1* and *AtCKX2*. Shoot growth is strongly inhibited in the transgenic plants. (From Werner et al. 2001.)

cal meristem (Figure 21.11). Since the root is a major source of cytokinin, this result may indicate that cytokinins play opposite roles in regulating cell proliferation in root and shoot meristems.

An additional line of evidence linking cytokinin to the regulation of cell division in vivo came from analyses of mutations in the cytokinin receptor (which will be discussed later in the chapter). Mutations in the cytokinin receptor disrupt the development of the root vasculature. Known as *cre1*, these mutants have no phloem in their roots; the root vascular system is composed almost entirely of xylem (see Chapters 4 and 10).

Further analysis revealed that this defect was due to an insufficient number of vasculature stem cells. That is, at the time of differentiation of the phloem and xylem, the pool of stem cells is abnormally small in *cre1* mutants; all the cells become committed to a xylem fate, and no stem cells remain to specify phloem. These results indicate that cytokinin plays a key role in regulating proliferation of the vasculature stem cells of the root.

Cytokinins Regulate Specific Components of the Cell Cycle

Cytokinins regulate cell division by affecting the controls that govern the passage of the cell through the cell division cycle. Zeatin levels were found to peak in synchronized culture tobacco cells at the end of S phase, mitosis, and G_1 phase.

Cytokinins were discovered in relation to their ability to stimulate cell division in tissues supplied with an optimal level of auxin. Evidence suggests that both auxin and cytokinins participate in regulation of the cell cycle and that they do so by controlling the activity of cyclin-dependent kinases. As discussed in Chapter 1, *cyclin-dependent protein kinases* (*CDKs*), in concert with their regulatory subunits, the *cyclins*, are enzymes that regulate the eukaryotic cell cycle.

The expression of the gene that encodes the major CDK, Cdc2 (cell *d*ivision *c*ycle 2), is regulated by auxin (see Chapter 19). In pea root tissues, *CDC2* mRNA was induced within 10 minutes after treatment with auxin, and high levels of CDK are induced in tobacco pith when it is cultured on medium containing auxin (John et al. 1993). However, the CDK induced by auxin is enzymatically inactive, and

(A)

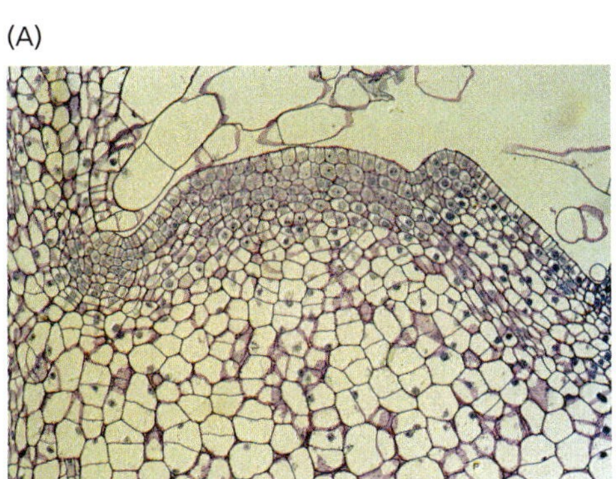

(B)

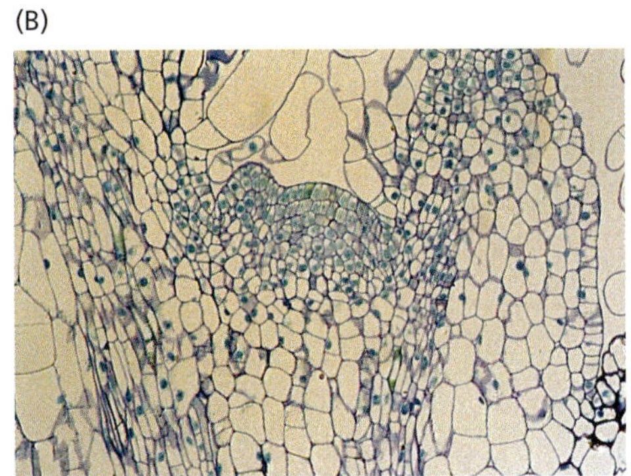

FIGURE 21.9 Cytokinin is required for normal growth of the shoot apical meristem. (A) Longitudinal section through the shoot apical meristem of a wild-type tobacco plant. (B) Longitudinal section through the shoot apical meristem of a transgenic tobacco overexpressing the gene that encodes cytokinin oxidase (*AtCKX1*). Note the reduction in the size of the apical meristem in the cytokinin-deficient plant. (From Werner et al. 2001.)

FIGURE 21.10 Cytokinin suppresses the growth of roots. The cytokinin-deficient *AtCKX1* roots (right) are larger than those of the wild-type tobacco plant (left). (From Werner et al. 2001.)

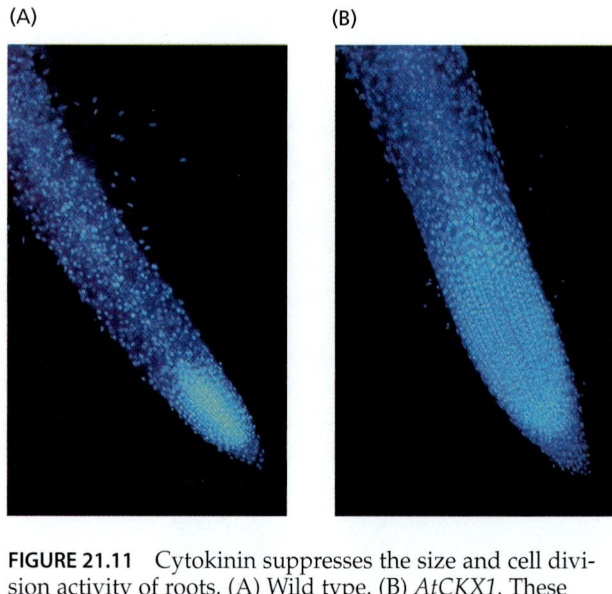

FIGURE 21.11 Cytokinin suppresses the size and cell division activity of roots. (A) Wild type. (B) *AtCKX1*. These roots were stained with the fluorescent dye, 4′, 6-diamidino-2-phenylindole, which stains the nucleus. (From Werner et al. 2001.)

high levels of CDK alone are not sufficient to permit cells to divide.

Cytokinin has been linked to the activation of a Cdc25-like phosphatase, whose role is to remove an inhibitory phosphate group from the Cdc2 kinase (Zhang et al. 1996). This action of cytokinin provides one potential link between cytokinin and auxin in regulating the cell cycle.

Recently, a second major input for cytokinin in regulating the cell cycle has emerged. Cytokinins elevate the expression of the *CYCD3* gene, which encodes a *D-type cyclin* (Soni et al. 1995; Riou-Khamlichi et al. 1999). In animal cells, D-type cyclins are regulated by a wide variety of growth factors and play a key role in regulating the passage through the restriction point of the cell cycle in G_1. D-type cyclins are thus key players in the regulation of cell proliferation.

In *Arabidopsis*, *CYCD3* is expressed in proliferating tissues such as shoot meristems and young leaf primordia. In a crucial experiment, it was found that overexpression of *CYCD3* can bypass the cytokinin requirement for cell proliferation in culture (Figure 21.12) (Riou-Khamlichi et al. 1999). These and other results suggest that a major mechanism for cytokinin's ability to stimulate cell division is its increase of *CYCD3* function.

The Auxin: Cytokinin Ratio Regulates Morphogenesis in Cultured Tissues

Shortly after the discovery of kinetin, it was observed that the differentiation of cultured callus tissue derived from tobacco pith segments into either roots or shoots depends on the ratio of auxin to cytokinin in the culture medium. Whereas high auxin:cytokinin ratios stimulated the formation of roots, low auxin:cytokinin ratios led to the formation of shoots. At intermediate levels the tissue grew as an undifferentiated callus (Figure 21.13) (Skoog and Miller 1965).

The effect of auxin: cytokinin ratios on morphogenesis can also be seen in crown gall tumors by mutation of the T-DNA of the *Agrobacterium* Ti plasmid (Garfinkel et al. 1981). Mutating the *ipt* gene (the *tmr* locus) of the Ti plasmid blocks zeatin biosynthesis in the infected cells. The resulting high auxin:cytokinin ratio in the tumor cells causes the proliferation of roots instead of undifferentiated callus tissue. In contrast, mutating either of the genes for auxin biosynthesis (*tms* locus) low-

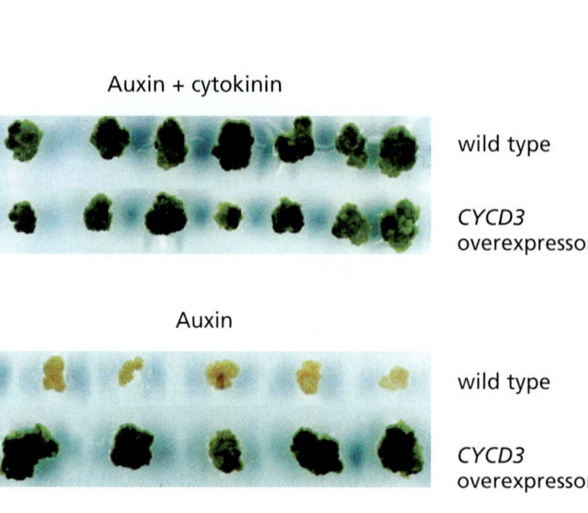

FIGURE 21.12 *CYCD3*-expressing callus cells can divide in the absence of cytokinin. Leaf explants from transgenic Arabidopsis plants expressing *CYCD3* under a cauliflower mosaic virus 35S promoter were induced to form calluses through culturing in the presence of auxin plus cytokinin or auxin alone. The wild-type control calluses required cytokinin to grow. The *CYCD3*-expressing calluses grew well on medium containing auxin alone. The photographs were taken after 29 days. (From Riou-Khamlichi et al. 1999.)

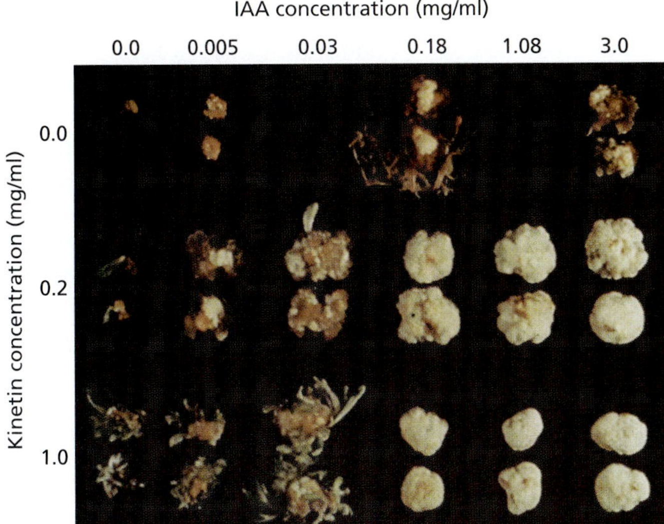

FIGURE 21.13 The regulation of growth and organ formation in cultured tobacco callus at different concentrations of auxin and kinetin. At low auxin and high kinetin concentrations (lower left) buds developed. At high auxin and low kinetin concentrations (upper right) roots developed. At intermediate or high concentrations of both hormones (middle and lower right) undifferentiated callus developed. (Courtesy of Donald Armstrong.)

ers the auxin:cytokinin ratio and stimulates the proliferation of shoots (Figure 21.14) (Akiyoshi et al. 1983). These partially differentiated tumors are known as teratomas.

Cytokinins Modify Apical Dominance and Promote Lateral Bud Growth

One of the primary determinants of plant form is the degree of apical dominance (see Chapter 19). Plants with strong apical dominance, such as maize, have a single growing axis with few lateral branches. In contrast, many lateral buds initiate growth in shrubby plants.

Although apical dominance may be determined primarily by auxin, physiological studies indicate that cytokinins play a role in initiating the growth of lateral buds. For example, direct applications of cytokinins to the axillary buds of many species stimulate cell division activity and growth of the buds.

The phenotypes of cytokinin-overproducing mutants are consistent with this result. Wild-type tobacco shows strong apical dominance during vegetative development, and the lateral buds of cytokinin overproducers grow vigorously, developing into shoots that compete with the main shoot. Consequently, cytokinin-overproducing plants tend to be bushy.

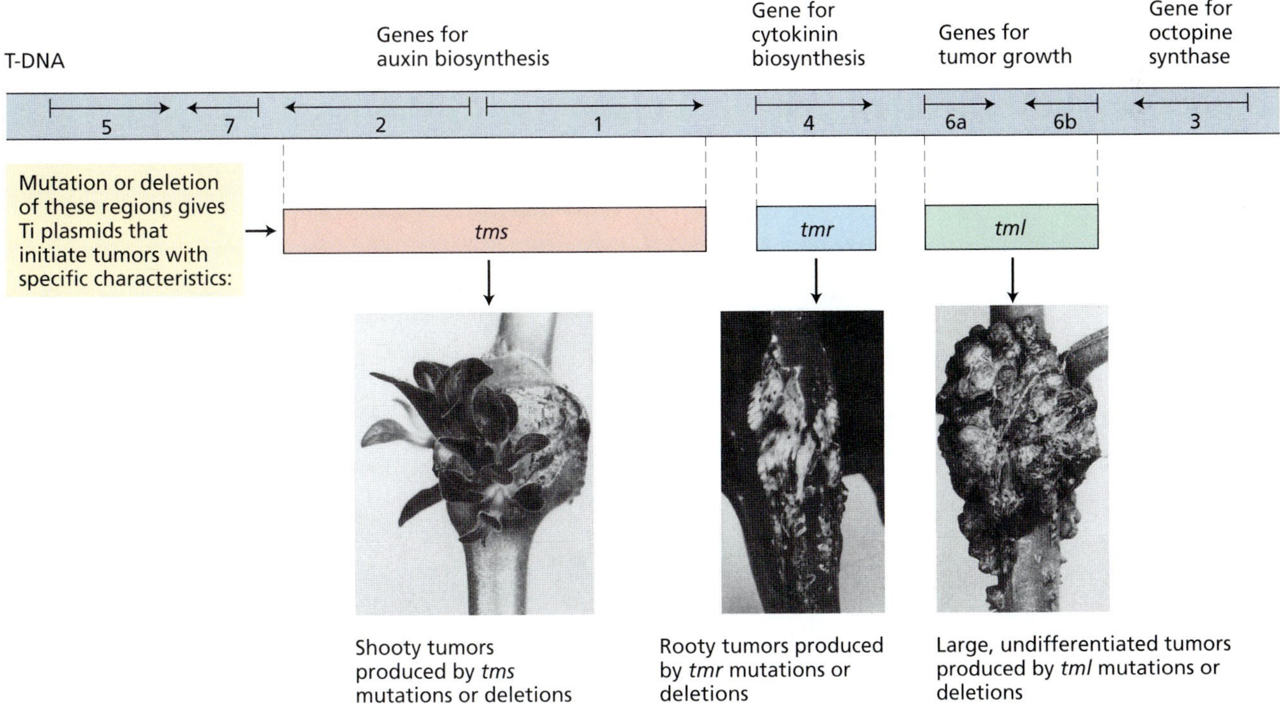

FIGURE 21.14 Map of the T-DNA from an *Agrobacterium* Ti plasmid, showing the effects of T-DNA mutations on crown gall tumor morphology. Genes 1 and 2 encode the two enzymes involved in auxin biosynthesis; gene 4 encodes a cytokinin biosynthesis enzyme. Mutations in these genes produce the phenotypes illustrated. (From Morris 1986, courtesy of R. Morris.)

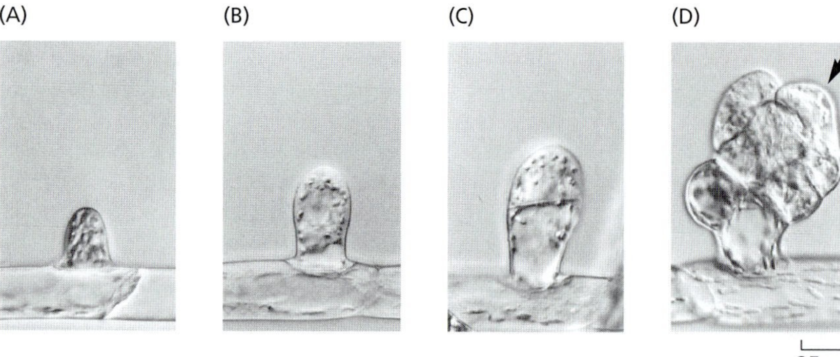

FIGURE 21.15 Bud formation in the moss *Funaria* begins with the formation of a protuberance at the apical ends of certain cells in the protonema filament. A–D show various stages of bud development. Once formed, the bud goes on to produce the leafy gametophyte stage of the moss. (Courtesy of K. S. Schumaker.)

Cytokinins Induce Bud Formation in a Moss

Thus far we have restricted our discussion of plant hormones to the angiosperms. However, many plant hormones are present and developmentally active in representative species throughout the plant kingdom. The moss *Funaria hygrometrica* is a well-studied example. The germination of moss spores gives rise to a filament of cells called a *protonema* (plural *protonemata*). The protonema elongates and undergoes cell divisions at the tip, and it forms branches some distance back from the tip (**see Web Essay 21.1**).

The transition from filamentous growth to leafy growth begins with the formation of a swelling or protuberance near the apical ends of specific cells (Figure 21.15). An asymmetric cell division follows, creating the **initial cell**. The initial cell then divides mitotically to produce the **bud**, the structure that gives rise to the leafy gametophyte. During normal growth, buds and branches are regularly initiated, usually beginning at the third cell from the tip of the filament.

Light, especially red light, is required for bud formation in *Funaria*. In the dark, buds fail to develop, but cytokinin added to the medium can substitute for the light requirement. Cytokinin not only stimulates normal bud development; it also increases the total number of buds (Figure 21.16). Even very low levels of cytokinin (picomolar, or 10^{-12} *M*) can stimulate the first step in bud formation: the swelling at the apical end of the specific protonemal cell.

Cytokinin Overproduction Has Been Implicated in Genetic Tumors

Many species in the genus *Nicotiana* can be crossed to generate interspecific hybrids. More than 300 such interspecific hybrids have been produced; 90% of these hybrids are normal, exhibiting phenotypic characteristics intermediate between those of both parents. The plant used for cigarette tobacco, *Nicotiana tabacum*, for example, is an interspecific hybrid. However, about 10% of these interspecific crosses result in progeny that tend to form spontaneous tumors called **genetic tumors** (Figure 21.17) (Smith 1988).

Genetic tumors are similar morphologically to those induced by *Agrobacterium tumefaciens*, discussed at the beginning of this chapter, but genetic tumors form spontaneously in the absence of any external inducing agent. The tumors are composed of masses of rapidly proliferating cells in regions of the plant that ordinarily would contain few dividing cells. Furthermore, the cells divide without differentiating into the cell types normally associated with the tissues giving rise to the tumor.

Nicotiana hybrids that produce genetic tumors have abnormally high levels of both auxin and cytokinins. Typically, the cytokinin

(A)

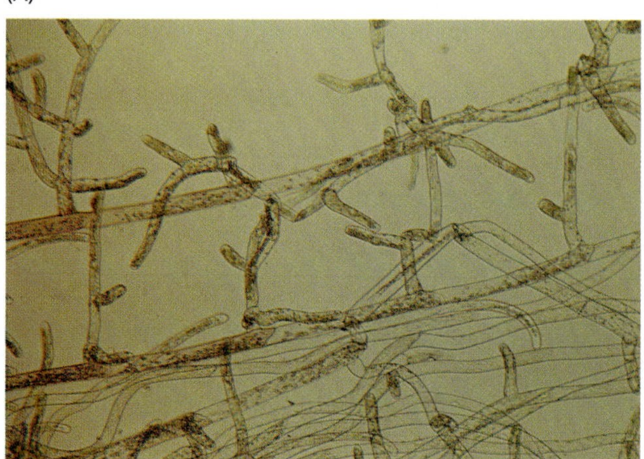

(B)

FIGURE 21.16 Cytokinin stimulates bud development in *Funaria*. (A) Control protonemal filaments. (B) Protonemal filaments treated with benzyladenine. (Courtesy of H. Kende.)

FIGURE 21.17 Expression of genetic tumors in the hybrid *Nicotiana langsdorffii* × *N. glauca*. (From Smith 1988.)

levels in tumor-prone hybrids are five to six times higher than those found in either parent.

Cytokinins Delay Leaf Senescence

Leaves detached from the plant slowly lose chlorophyll, RNA, lipids, and protein, even if they are kept moist and provided with minerals. This programmed aging process leading to death is termed **senescence** (see Chapters 16 and 23). Leaf senescence is more rapid in the dark than in the light. Treating isolated leaves of many species with cytokinins will delay their senescence.

Although applied cytokinins do not prevent senescence completely, their effects can be dramatic, particularly when the cytokinin is sprayed directly on the intact plant. If only one leaf is treated, it remains green after other leaves of similar developmental age have yellowed and dropped off the plant. Even a small spot on a leaf will remain green if treated with a cytokinin, after the surrounding tissues on the same leaf begin to senesce.

Unlike young leaves, mature leaves produce little if any cytokinin. Mature leaves may depend on root-derived cytokinins to postpone their senescence. Senescence is initiated in soybean leaves by seed maturation—a phenomenon known as *monocarpic senescence*—and can be delayed by seed removal. Although the seedpods control the onset of senescence, they do so by controlling the delivery of root-derived cytokinins to the leaves.

The cytokinins involved in delaying senescence are primarily zeatin riboside and dihydrozeatin riboside, which may be transported into the leaves from the roots through the xylem, along with the transpiration stream (Noodén et al. 1990).

To test the role of cytokinin in regulating the onset of leaf senescence, tobacco plants were transformed with a chimeric gene in which a senescence-specific promoter was used to drive the expression of the *ipt* gene (Gan and Amasino 1995). The transformed plants had wild-type levels of cytokinins and developed normally, up to the onset of leaf senescence.

As the leaves aged, however, the senescence-specific promoter was activated, triggering the expression of the *ipt* gene within leaf cells just as senescence would have been initiated. The resulting elevated cytokinin levels not only blocked senescence, but also limited further expression of the *ipt* gene, preventing cytokinin overproduction (Figure 21.18). This result suggests that cytokinins are a natural regulator of leaf senescence.

FIGURE 21.18 Leaf senescence is retarded in a transgenic tobacco plant containing a cytokinin biosynthesis gene, *ipt*. The *ipt* gene is expressed in response to signals that induce senescence. (From Gan and Amasino 1995, courtesy of R. Amasino.)

Cytokinins Promote Movement of Nutrients

Cytokinins influence the movement of nutrients into leaves from other parts of the plant, a phenomenon known as *cytokinin-induced nutrient mobilization*. This process is revealed when nutrients (sugars, amino acids, and so on) radiolabeled with ^{14}C or ^{3}H are fed to plants after one leaf or part of a leaf is treated with a cytokinin. Later the whole plant is subjected to autoradiography to reveal the pattern of movement and the sites at which the labeled nutrients accumulate.

Experiments of this nature have demonstrated that nutrients are preferentially transported to, and accumulated in, the cytokinin-treated tissues. It has been postulated that the hormone causes nutrient mobilization by creating a new source–sink relationship. As discussed in Chapter 10, nutrients translocated in the phloem move from a site of production or storage (the source) to a site of utilization (the sink). The metabolism of the treated area may be stimulated by the hormone so that nutrients move toward it. However, it is not necessary for the nutrient itself to be metabolized in the sink cells because even nonmetabolizable substrate analogs are mobilized by cytokinins (Figure 21.19).

Cytokinins Promote Chloroplast Development

Although seeds can germinate in the dark, the morphology of dark-grown seedlings is very different from that of light-grown seedlings (see Chapter 17): Dark-grown seedlings are said to be **etiolated**. The hypocotyl and internodes of etiolated seedlings are more elongated, cotyledons and leaves do not expand, and chloroplasts do not mature. Instead of maturing as chloroplasts, the proplastids of dark-grown seedlings develop into **etioplasts**, which do not synthesize chlorophyll or most of the enzymes and structural proteins required for the formation of the chloroplast thylakoid system and photosynthesis machinery. When seedlings germinate in the light, chloroplasts mature directly from the proplastids present in the embryo, but etioplasts also can mature into chloroplasts when etiolated seedlings are illuminated.

If the etiolated leaves are treated with cytokinin before being illuminated, they form chloroplasts with more extensive grana, and chlorophyll and photosynthetic enzymes are synthesized at a greater rate upon illumination (Figure 21.20). These results suggest that cytokinins—along with other factors, such as light, nutrition, and development—regulate the synthesis of photosynthetic pigments and proteins. The ability of exogenous cytokinin to enhance de-etiolation of dark-grown seedlings is mimicked by certain mutations that lead to cytokinin overproduction. (For more on how cytokinins promote light-mediated development, see **Web Topic 21.7**.)

Cytokinins Promote Cell Expansion in Leaves and Cotyledons

The promotion of cell enlargement by cytokinins is most clearly demonstrated in the cotyledons of dicots with leafy cotyledons, such as mustard, cucumber, and sunflower. The cotyledons of these species expand as a result of cell enlargement during seedling growth. Cytokinin treatment promotes additional cell expansion, with no increase in the dry weight of the treated cotyledons.

Leafy cotyledons expand to a much greater extent when the seedlings are grown in the light than in the dark, and cytokinins promote cotyledon growth in both light- and dark-grown seedlings (Figure 21.21). As with auxin-

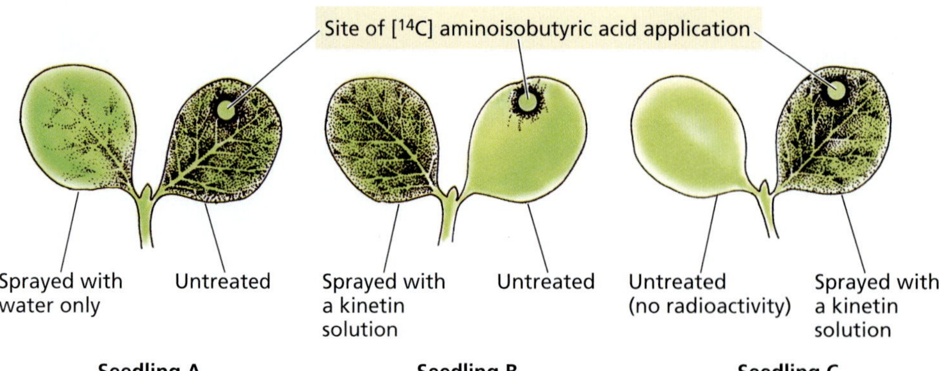

FIGURE 21.19 The effect of cytokinin on the movement of an amino acid in cucumber seedlings. A radioactively labeled amino acid that cannot be metabolized, such as aminoisobutyric acid, was applied as a discrete spot on the right cotyledon of each of these seedlings. (Drawn from data obtained by K. Mothes.)

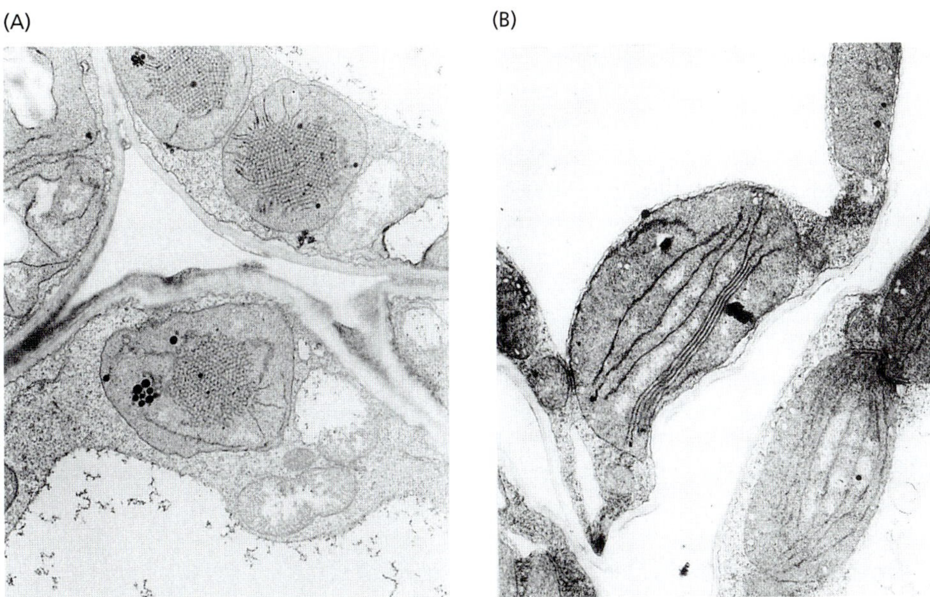

FIGURE 21.20 Cytokinin influence on the development of wild-type *Arabidopsis* seedlings grown in darkness. (A) Plastids develop as etioplasts in the untreated, dark grown control. (B) Cytokinin treatment resulted in thylakoid formation in the plastids of dark-grown seedlings. (From Chory et al. 1994, courtesy of J. Chory, © American Society of Plant Biologists, reprinted with permission.)

induced growth, cytokinin-stimulated expansion of radish cotyledons is associated with an increase in the mechanical extensibility of the cell walls. However, cytokinin-induced wall loosening is not accompanied by proton extrusion. Neither auxin nor gibberellin promotes cell expansion in cotyledons.

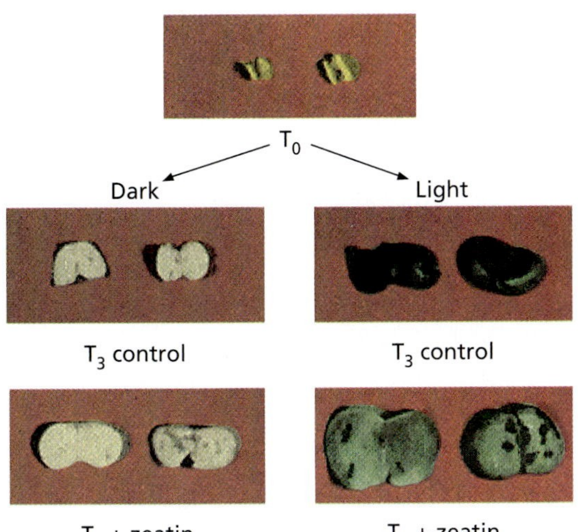

FIGURE 21.21 The effect of cytokinin on the expansion of radish cotyledons. The experiment described here shows that the effects of light and cytokinin are additive. T_0 represents germinating radish seedlings before the experiment began. The detached cotyledons were incubated for 3 days (T_3) in either darkness or light with or without 2.5 m*M* zeatin. In both the light and the dark, zeatin-treated cotyledons expanded more than in the control. (From Huff and Ross 1975.)

Cytokinins Regulate Growth of Stems and Roots

Although endogenous cytokinins are clearly required for normal cell proliferation in the apical meristem, and therefore normal shoot growth (see Figure 21.9), applied cytokinins typically inhibit the process of cell elongation in both stems and roots. For example, exogenous cytokinin inhibits hypocotyl elongation at concentrations that promote leaf and cotyledon expansion in the dark-grown seedlings.

In related experiments, internode and root elongation are both inhibited in transgenic plants expressing the *ipt* gene and in cytokinin-overproducing mutants. It is likely that the inhibition of hypocotyl and internode elongation induced by excess cytokinin is due to the production of ethylene, and this inhibition thus may represent another example of the interdependence of hormonal regulatory pathways (Cary et al. 1995; Vogel et al. 1998).

On the other hand, other experiments suggest that endogenous cytokinins at normal physiological concentrations inhibit root growth. For example, a weak allele of a cytokinin receptor mutant and a loss-of-function allele of a cytokinin signaling element both have longer roots than the wild type (Inoue et al. 2001; Sakai et al. 2001). As previously noted, transgenic tobacco engineered to overexpress cytokinin oxidase (and thus to have lower levels of cytokinin) also has longer roots than its wild-type counterpart (see Figure 21.10) (Werner et al. 2001). These results indicate that endogenous cytokinins may negatively regulate root elongation.

Cytokinin-Regulated Processes Are Revealed in Plants That Overproduce Cytokinin

The *ipt* gene from the *Agrobacterium* Ti plasmid has been introduced into many species of plants, resulting in cytokinin overproduction. These transgenic plants exhibit an array of developmental abnormalities that tell us a great deal about the biological role of cytokinins.

As discussed earlier, plant tissues transformed by *Agrobacterium* carrying a wild-type Ti plasmid proliferate as tumors as a result of the overproduction of both auxin and cytokinin. And as mentioned already, if all of the other genes in the T-DNA are deleted and plant tissues are transformed with T-DNA containing only a selective antibiotic resistance marker gene and the *ipt* gene, shoots proliferate instead of callus.

The shoot teratomas formed by *ipt*-transformed tissues are difficult to root, and when roots are formed, they tend to be stunted in their growth. As a result, it is difficult to obtain plants from shoots expressing the *ipt* gene under the control of its own promoter because the promoter is a constitutive promoter and the gene is continuously expressed.

To circumvent this problem, a variety of promoters whose expression can be regulated have been used to drive the expression of the *ipt* gene in the transformed tissues. For example, several studies have employed a heat shock promoter, which is induced in response to elevated temperature, to drive inducible expression of the *ipt* gene in transgenic tobacco and *Arabidopsis*. In these plants, heat induction substantially increased the level of zeatin, zeatin riboside and ribotide, and *N*-conjugated zeatin.

These cytokinin-overproducing plants exhibit several characteristics that point to roles played by cytokinin in plant physiology and development:

- The shoot apical meristems of cytokinin-overproducing plants produce more leaves.
- The leaves have higher chlorophyll levels and are much greener.
- Adventitious shoots may form from unwounded leaf veins and petioles.
- Leaf senescence is retarded.
- Apical dominance is greatly reduced.
- The more extreme cytokinin-overproducing plants are stunted, with greatly shortened internodes.
- Rooting of stem cuttings is reduced, as is the root growth rate.

Some of the consequences of cytokinin overproduction could be highly beneficial for agriculture if synthesis of the hormone can be controlled. Because leaf senescence is delayed in the cytokinin-overproducing plants, it should be possible to extend their photosynthetic productivity (which we'll discuss shortly).

In addition, cytokinin production could be linked to damage caused by predators. For example, tobacco plants transformed with an *ipt* gene under the control of the promoter from a wound-inducible protease inhibitor II gene were more resistant to insect damage. The tobacco hornworm consumed up to 70% fewer tobacco leaves in plants that expressed the *ipt* gene driven by the protease inhibitor promoter (Smigocki et al. 1993).

CELLULAR AND MOLECULAR MODES OF CYTOKININ ACTION

The diversity of the effects of cytokinin on plant growth and development is consistent with the involvement of signal transduction pathways with branches leading to specific responses. Although our knowledge of how cytokinin works at the cellular and molecular levels is still quite fragmentary, significant progress has been achieved. In this section we will discuss the nature of the cytokinin receptor and various cytokinin-regulated genes, as well as a model for cytokinin signaling based on current information.

A Cytokinin Receptor Related to Bacterial Two-Component Receptors Has Been Identified

The first clue to the nature of the cytokinin receptor came from the discovery of the *CKI1* gene. *CKI1* was identified in a screen for genes that, when overexpressed, conferred cytokinin-independent growth on *Arabidopsis* cells in culture. As discussed already, plant cells generally require cytokinin in order to divide in culture. However, a cell line that overexpresses *CKI1* is capable of growing in culture in the absence of added cytokinin.

CKI1 encodes a protein similar in sequence to bacterial two-component sensor histidine kinases, which are ubiquitous receptors in prokaryotes (see Chapter 14 on the web site and Chapter 17). Bacterial two-component regulatory systems mediate a range of responses to environmental stimuli, such as osmoregulation and chemotaxis. Typically these systems are composed of two functional elements: a *sensor histidine kinase*, to which a signal binds, and a downstream *response regulator*, whose activity is regulated via phosphorylation by the sensor histidine kinase. The sensor histidine kinase is usually a membrane-bound protein that contains two distinct domains, called the input and histidine kinase, or "transmitter," domains (Figure 21.22).

Detection of a signal by the input domain alters the activity of the histidine kinase domain. Active sensor kinases are dimers that transphosphorylate a conserved histidine residue. This phosphate is then transferred to a conserved aspartate residue in the receiver domain of a cognate response regulator (see Figure 21.22), and this phosphorylation alters the activity of the kinases. Most response regulators also contain *output* domains that act as transcription factors.

The phenotype resulting from *CKI1* overexpression, combined with its similarity to bacterial receptors, suggested that the CKI1 and/or similar histidine kinases are cytokinin receptors. Support for this model came from identification of the *CRE1* gene (Inoue et al. 2001).

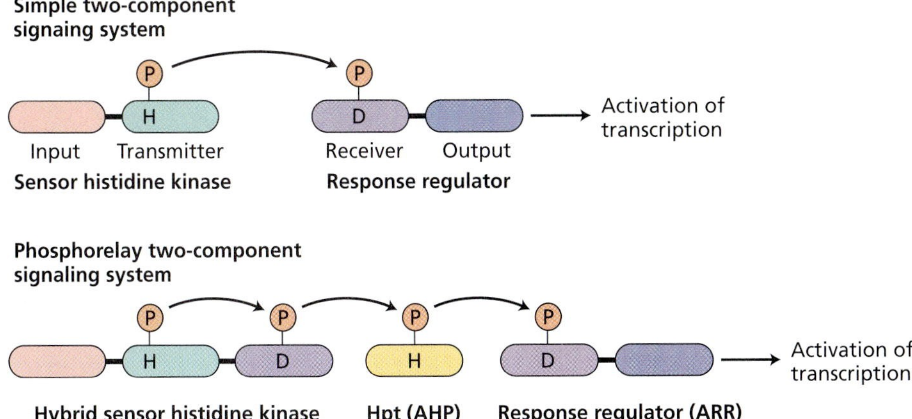

FIGURE 21.22 Simple versus phosphorelay types of two-component signaling systems. (A) In simple two-component systems, the input domain is the site where the signal is sensed. This regulates the activity of the histidine kinase domain, which when activated autophosphorylates on a conserved His residue. The phosphate is then transferred to an Asp residue that resides within the receiver domain of a response regulator. Phosphorylation of this Asp regulates the activity of the output domain of the response regulator, which in many cases is a transcription factor. (B) In the phosphorelay-type two-component signaling system, an extra set of phosphotransfers is mediated by a histidine phosphotransfer protein (Hpt), called AHP in *Arabidopsis*. The *Arabidopsis* response regulators are called ARRs. H = histidine, D = aspartate.

Like *CKI1*, *CRE1* encodes a protein similar to bacterial histidine kinases. Loss-of-function *cre1* mutations were identified in a genetic screen for mutants that failed to develop shoots from undifferentiated tissue culture cells in response to cytokinin. This is essentially the opposite screen from the one just described, from which the *CKI1* gene was identified by a gain-of-function (ability to divide in the absence of cytokinin) mutation. The *cre1* mutants are also resistant to the inhibition of root elongation observed in response to cytokinin.

Convincing evidence that *CRE1* encodes a cytokinin receptor came from analysis of the expression of the protein in yeast. Yeast cells also contain a sensor histidine kinase, and deletion of the gene that encodes this kinase—*SLN1*—is lethal. Expression of *CRE1* in *SLN1*-deficient yeast can restore viability, *but only if cytokinins are present in the medium*. Thus the activity of CRE1 (i.e., its ability to replace SLN1) is dependent on cytokinin, which, coupled with the cytokinin-insensitive phenotype of the *cre1* mutants in *Arabidopsis*, unequivocally demonstrates that CRE1 is a cytokinin receptor. It remains to be determined if CKI1 is also a cytokinin receptor.

Two other genes in the *Arabidopsis* genome (*AHK2* and *AHK3*) are closely related to *CRE1*, suggesting that, like the ethylene receptors (see Chapter 22), the cytokinin receptors are encoded by a multigene family. Indeed, it has been demonstrated that cytokinins bind to the predicted extracellular domains of CRE1, AHK2, and AHK3 with high affinity, confirming that they are indeed cytokinin receptors (Yamada et al. 2001). This raises the possibility that these genes are at least partially genetically redundant (as are the ethylene receptors), which may explain the relatively mild phenotypes that result from loss-of-function *cre1* mutations.

Cytokinins Cause a Rapid Increase in the Expression of Response Regulator Genes

One of the primary effects of cytokinin is to alter the expression of various genes. The first set of genes to be upregulated in response to cytokinin are the ***ARR*** (*Arabidopsis response regulator*) genes. These genes are homologous to the receiver domain of bacterial two-component response regulators, the downstream target of sensor histidine kinases (see the previous section).

In *Arabidopsis*, response regulators are encoded by a multigene family. They fall into two basic classes: the **type-A *ARR*** genes, which are made up solely of a receiver domain, and the **type-B *ARR*** genes, which contain a transcription factor domain in addition to the receiver domain (Figure 21.23). The rate of transcription of the type-A gene is increased within 10 minutes in response to applied cytokinin (Figure 21.24) (D'Agostino et al. 2000). This rapid induction is specific for cytokinin and does not require new protein synthesis. Both of these features are hallmarks of primary response genes (discussed in Chapters 17 and 19).

The rapid induction of the type-A genes, coupled with their similarity to signaling elements predicted to act downstream of sensor histidine kinases, suggests that these elements act downstream of the CRE1 cytokinin receptor family to mediate the primary cytokinin response. Interestingly, one of these type-A genes, *ARR5*, is expressed primarily in the apical meristems of both shoots and roots (Figure 21.25), consistent with a role in regulating cell proliferation, a key aspect of cytokinin action.

ARR7
ARR5
ARR6
ARR15
ARR4
ARR3
ARR16
ARR19
ARR8
ARR17
ARR13
ARR14
ARR11
ARR1
ARR2
ARR10
ARR12
Type A ARRs
Type B ARRs
D
COOH
Receiver domain
Output domain (transcription factor)

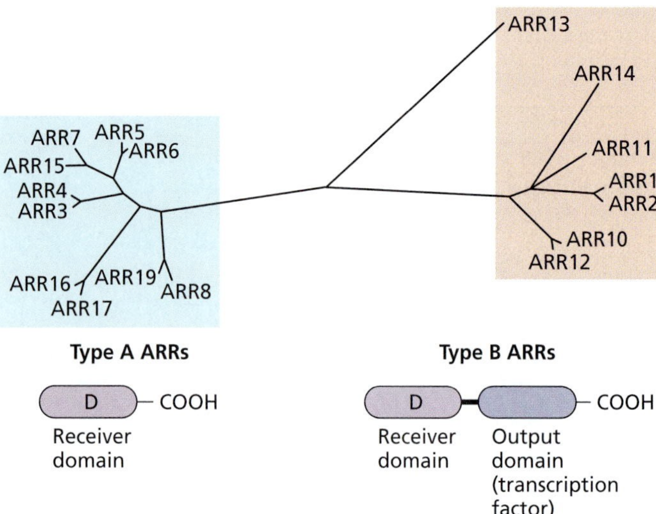

FIGURE 21.23 Phylogenetic tree of *Arabidopsis* response regulators. The top part of the figure shows a phylogenetic tree that represents the degree of relatedness of the receiver domains present in the *Arabidopsis* genome. The closer two proteins are on the tree, the more similar are their amino acid sequences. Note that these proteins fall into two distinct groups, or clades, called the type-A ARRs (blue) and the type-B ARRs (red). These differences in sequence are also reflected in a distinct domain structure, as depicted below the tree. The type-A ARRs consist solely of a receiver domain, but the type-A proteins also contain a fused output domain at the carboxy-terminus.

The expression of a wide variety of other genes is altered in response to cytokinin, but generally with slower kinetics than the type-A genes. These include the gene that encodes nitrate reductase, light-regulated genes such as *LHCB* and *SSU*, and defense-related genes such as *PR1*, as well as genes that encode an extensin (cell wall protein rich in hydroxyproline), rRNAs, cytochrome P450s, and peroxidase. Cytokinin elevates the expression of these genes both by increasing the rate of transcription (as in the case of the type-A *ARRs*) and/or by a stabilization of the RNA transcript (e.g., the β-expansin gene).

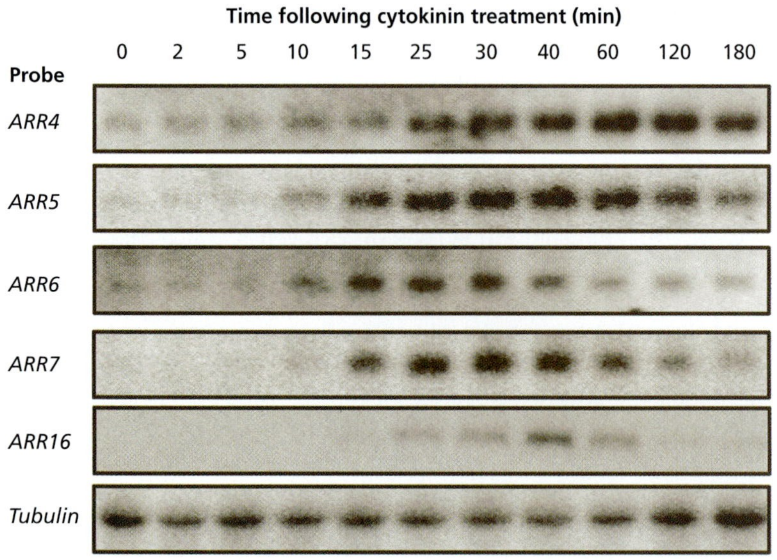

Time following cytokinin treatment (min)
0 2 5 10 15 25 30 40 60 120 180
Probe
ARR4
ARR5
ARR6
ARR7
ARR16
Tubulin

FIGURE 21.24 Induction of type-A *ARR* genes in response to cytokinin. RNA from *Arabidopsis* seedlings treated for the indicated time with cytokinin was isolated and analyzed by Northern blotting. Each row shows the result of probing the Northern blot with an individual type-A gene, and each lane contains RNA derived from *Arabidopsis* seedlings treated for the indicated time with cytokinin. The darker the band, the higher the level of *ARR* mRNA in that sample. (From D'Agostino et al. 2000.)

Histidine Phosphotransferases May Mediate the Cytokinin Signaling Cascade

From the preceding discussions we have seen that cytokinin binds to the CRE1 receptors to initiate a response that culminates in the elevation of transcription of the type-A *ARRs*. The type-A ARR proteins, in turn, may regulate the expression of numerous other genes, as well as the activities of various target proteins that ultimately alter cellular function. How is the signal propagated from CRE1 (which is at the plasma membrane) to the nucleus to alter type-A *ARR* transcription?

One set of genes that are likely to be involved in this signaling cascade encode the **AHP** (*Arabidopsis histidine phosphotransfer*) proteins. In two-component systems that involve a sensor kinase fused to a receiver domain (the structure of most eukaryotic sensor histidine kinases, including those of the CRE1 family), there is an additional set of phosphotransfers that are mediated by a **histidine phosphotransfer protein** (**Hpt**).

Phosphate is first transferred from ATP to a histidine within the histidine kinase domain, and then transferred to an aspartate residue on the fused receiver. From the aspartate residue the phosphate group is then transferred to a histidine on the Hpt protein and then finally to an aspartate on the receiver domain of the response regulator (see Figure 21.22). This phosphorylation of the receiver domain of the response regulator alters its activity. Thus, Hpt proteins are predicted to mediate the phosphotransfer between sensor kinases and response regulators.

In *Arabidopsis* there are 5 Hpt genes, called *AHPs*. The AHP proteins have been shown to physically associate with receiver

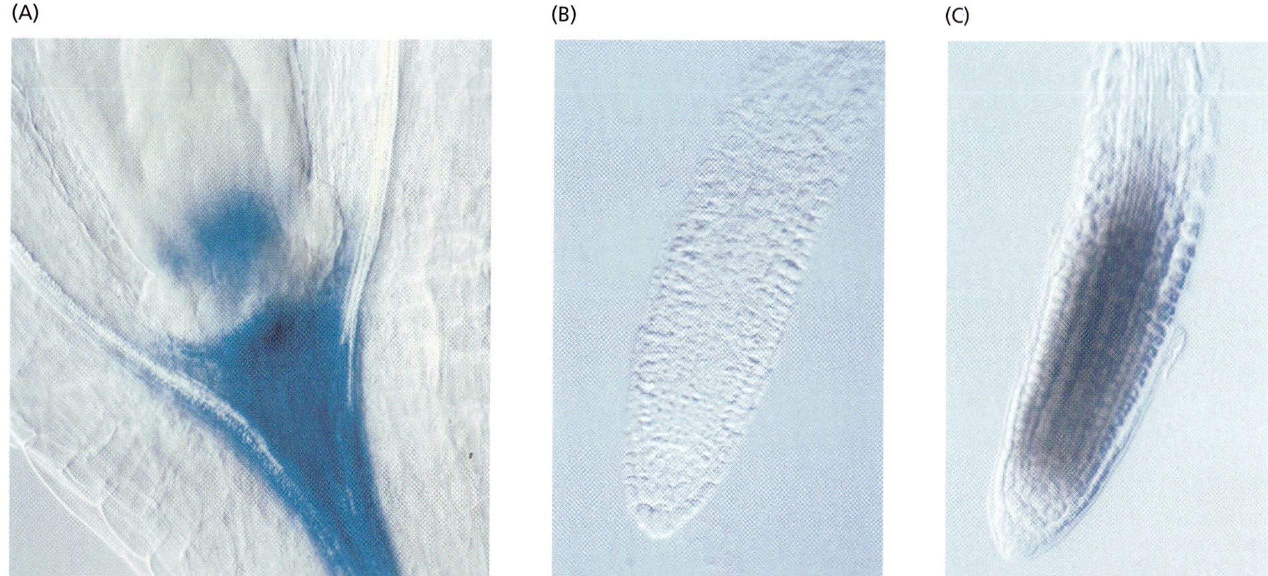

FIGURE 21.25 Expression of *ARR5*. The pattern of *ARR5* expression was examined by fusion of the promoter to a *GUS* reporter gene (A) or by whole-mount in situ hybridization (B and C). For the latter, the tissue is hybridized with labeled single-stranded *ARR5* RNA in either the sense orientation (B) or the antisense (C). The sense RNA is a negative control and reveals background, nonspecific staining. The antisense probe specifically hybridizes with the *ARR5* mRNA present in the tissue, thereby revealing its spatial distribution. With both methods, *ARR5* expression is observed primarily in the apical meristems. (From D'Agostino et al. 2000.)

domains from several *Arabidopsis* histidine kinases, including CRE1, and a subset of the AHPs have been demonstrated to transiently translocate from the cytoplasm to the nucleus in response to cytokinin (Figure 21.26) (Hwang and Sheen 2001). This finding suggests that the AHPs are the immediate downstream targets of the activated CRE receptors, and that these proteins transduce the cytokinin signal into the nucleus.

Cytokinin-Induced Phosphorylation Activates Transcription Factors

The question now becomes, How do the activated AHPs, once in the nucleus, act to regulate gene transcription? Genetic studies in intact *Arabidopsis* plants and overexpression studies in isolated *Arabidopsis* protoplasts using a cytokinin responsive reporter have provided a likely answer (Hwang and Sheen 2001; Sakai et al. 2001).

Disruption of *ARR1*, one of the type-B *ARR* genes, reduces the induction of the type-A *ARR* genes in response

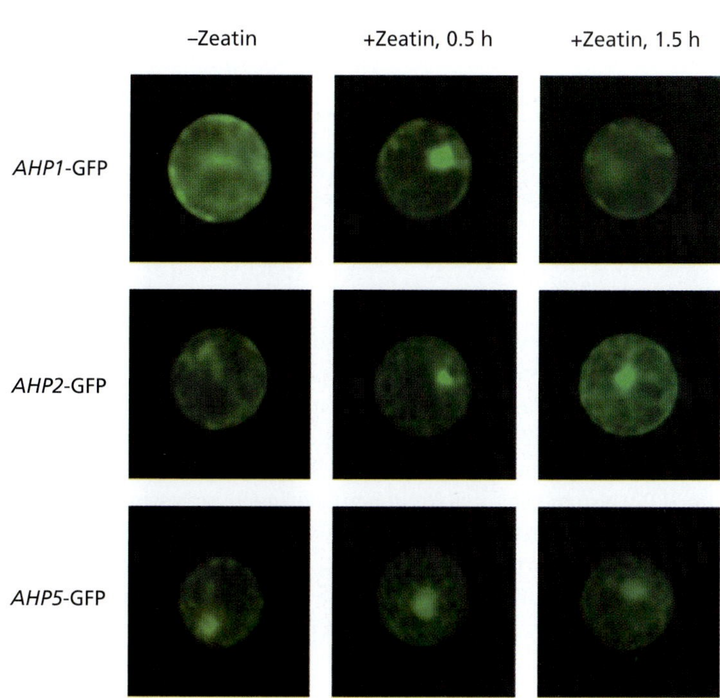

FIGURE 21.26 Cytokinin induces the transient movement of some AHP proteins into the nucleus. *Arabidopsis* protoplasts expressing various *AHP* genes fused to green fluorescent protein (GFP) as a reporter were treated with zeatin and monitored for 1.5 hours. *AHP1*-GFP and *AHP2*-GFP show nuclear localization after 30 minutes, but this localization is transient in the case of *AHP1*-GFP. Zeatin did not seem to affect the distribution of *AHP5*-GFP. (From Hwang and Sheen 2001.)

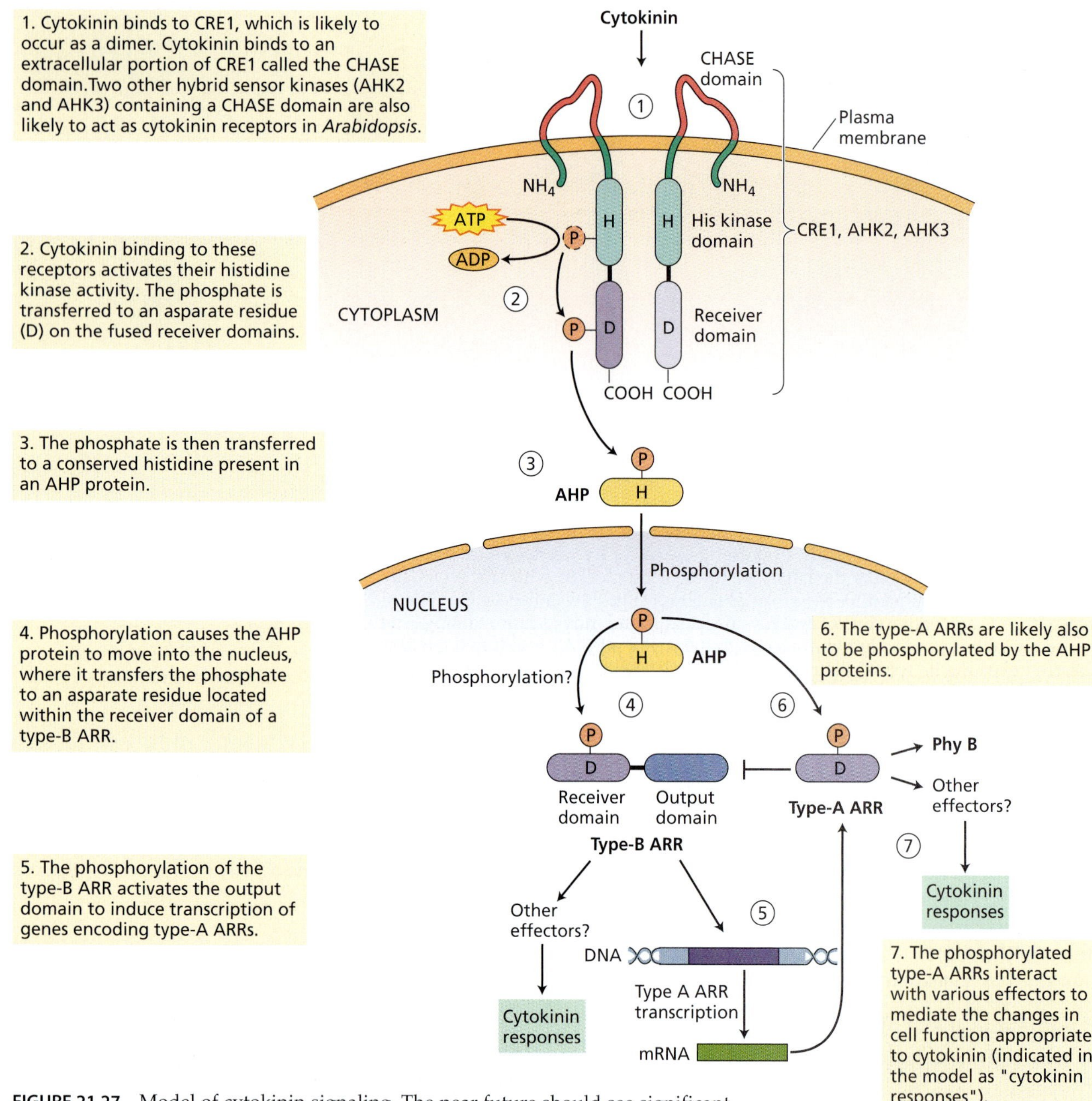

FIGURE 21.27 Model of cytokinin signaling. The near future should see significant refinement of this model, the tools are now in hand to analyze the interactions among these elements.

to cytokinin. Conversely, an increase in *ARR*1 function increases the response of the type-A genes to cytokinin. This suggests that ARR1, which is a transcription factor, directly regulates transcription of the type-A *ARR*s, and that by analogy other members of the type-B *ARR* family (see Figure 21.23) also mediate cytokinin-regulated gene expression.

This conclusion is supported by the findings that type-B ARRs operate as transcriptional activators and that there are multiple binding sites for ARR1, a type-B ARR, in the 5′ DNA regulatory sequences of the type-A ARR genes.

A model of cytokinin signaling is presented in Figure 21.27. Cytokinin binds to the CRE1 receptor and initiates a phosphorylation cascade that results in the phosphorylation and activation of a subset of the type-B ARR proteins. Activation of the type-B proteins (transcription factors) leads to the transcriptional activation of the type-A genes. The type-A ARR proteins are likely also phosphorylated in response to cytokinin, and perhaps together with the type-B proteins, they interact with various targets to mediate the changes in cellular function, such as an activation of the cell cycle. Type-A ARRs are also able to inhibit their own expression by an unknown mechanism, providing a negative feedback loop (see Figure 21.27). Much work needs to be done to confirm and refine this model, but we are beginning to glimpse for the first time the molecular basis for cytokinin action in plants.

SUMMARY

Mature plant cells generally do not divide in the intact plant, but they can be stimulated to divide by wounding, by infection with certain bacteria, and by plant hormones, including cytokinins. Cytokinins are N^6-substituted aminopurines that will initiate cell proliferation in many plant cells when they are cultured on a medium that also contains an auxin. The principal cytokinin of higher plants—zeatin, or *trans*-6-(4-hydroxy-3-methylbut-2-enylamino)purine—is also present in plants as a riboside or ribotide and as glycosides. These forms are generally also active as cytokinins in bioassays through their enzymatic conversion to the free zeatin base by plant tissue.

The first committed step in cytokinin biosynthesis—the transfer of the isopentenyl group from DMAPP to the 6 nitrogen of adenosine tri- and diphosphate—is catalyzed by isopentenyl transferase (IPT). The product of this reaction is readily converted to zeatin and other cytokinins. Cytokinins are synthesized in roots, in developing embryos, young leaves, fruits, and crown gall tissues. Cytokinins are also synthesized by plant-associated bacteria, insects, and nematodes.

Cytokinin oxidases degrade cytokinin irreversibly and may play a role in regulation of the levels of this hormone. Conjugation of both the side chain and the adenosine moiety to sugars (mostly glucose) also may play a role in the regulation of cytokinin levels and may target subpools of the hormone for distinct roles, such as transport. Cytokinins are also interconverted among the free base and the nucleoside and nucleotide forms.

Crown galls originate from plant tissues that have been infected with *Agrobacterium tumefaciens*. The bacterium injects a specific region of its Ti plasmid called T-DNA into wounded plant cells, and the T-DNA is incorporated into the host nuclear genome. The T-DNA contains a gene for cytokinin biosynthesis, as well as genes for auxin biosynthesis. These phyto-oncogenes are expressed in the plant cells, leading to hormone synthesis and unregulated proliferation of the cells to form the gall.

Cytokinins are most abundant in the young, rapidly dividing cells of the shoot and root apical meristems. They do not appear to be actively transported through living plant tissues. Instead, they are transported passively into the shoot from the root through the xylem, along with water and minerals. At least in pea, however, the shoot can regulate the flow of cytokinin from the root.

Cytokinins participate in the regulation of many plant processes, including cell division, morphogenesis of shoots and roots, chloroplast maturation, cell enlargement, and senescence. Both cytokinin and auxin regulate the plant cell cycle and are needed for cell division. The roles of cytokinins have been elucidated from application of exogenous cytokinins, the phenotype of transgenic plants designed to overexpress cytokinins as a result of introduction of the bacterial *ipt* gene, and recently from transgenic plants that have a reduced endogenous cytokinin content as a result of overexpression of cytokinin oxidase.

In addition to cell division, the ratio of auxin to cytokinin determines the differentiation of cultured plant tissues into either roots or buds: High ratios promote roots; low ratios, buds. Cytokinins also have been implicated in the release of axillary buds from apical dominance. In the moss *Funaria*, cytokinins greatly increase the number of "buds," the structures that give rise to the leafy gametophyte stage of development.

The mechanism of action of cytokinin is just beginning to emerge. A cytokinin receptor has been identified in *Arabidopsis*. This transmembrane protein is related to the bacterial two-component sensor histidine kinases. Cytokinins increase the abundance of several specific mRNAs. Some of these are primary response genes that are similar to bacterial two-component response regulators. The signal transduction mechanism from CRE1 to transcriptional activation of the type-A *ARR*s involves other homologs of two-component elements.

Web Material

Web Topics

21.1 Cultured Cells Can Acquire the Ability to Synthesize Cytokinins

The phenomenon of habituation is described, whereby callus tissues become cytokinin independent.

21.2 Structures of Some Naturally Occurring Cytokinins

The structures of various naturally occurring cytokinins are presented.

21.3 Various Methods Are Used to Detect and Identify Cytokinins

Cytokinins can be qualified using immunological and sensitive physical methods.

21.4 Cytokinins Are Also Present in Some tRNAs in Animal and Plant Cells

Modified adenosines near the 3′ end of the anticodons of some tRNAs have cytokinin activity.

21.5 The Ti Plasmid and Plant Genetic Engineering

Applications of the Ti plasmid of *Agrobacterium* in bioengineering are described.

21.6 Phylogenetic Tree of *IPT* Genes

Arabidopsis contains nine different *IPT* genes, several of which form a distinct clade with other plant sequences.

21.7 Cytokinin Can Promote Light-Mediated Development

Cytokinins can mimic the effect of the *det* mutation on chloroplast development and de-etiolation.

Web Essay

21.1 Cytokinin-Induced Form and Structure in Moss

The effects of cytokinins on the development of moss protonema are described.

Chapter References

Akiyoshi, D. E., Klee, H., Amasino, R. M., Nester, E. W., and Gordon, M. P. (1984) T-DNA of *Agrobacterium tumefaciens* encodes an enzyme of cytokinin biosynthesis. *Proc. Natl. Acad. Sci. USA* 81: 5994–5998.

Akiyoshi, D. E., Morris, R. O., Hinz, R., Mischke, B. S., Kosuge, T., Garfinkel, D. J., Gordon, M. P., and Nester, E. W. (1983) Cytokinin/auxin balance in crown gall tumors is regulated by specific loci in the T-DNA. *Proc. Natl. Acad. Sci. USA* 80: 407–411.

Akiyoshi, D. E., Regier, D. A., and Gordon, M. P. (1987) Cytokinin production by *Agrobacterium* and *Pseudomonas* spp. *J. Bacteriol.* 169: 4242–4248.

Aloni, R., Wolf, A., Feigenbaum, P. Avni, A., and Klee, H. J. (1998) The Never ripe mutant provides evidence that tumor-induced ethylene controls the morphogenesis of *Agrobacterium tumefaciens*-induced crown galls in tomato stems. *Plant Physiol.* 117: 841–849.

Barry, G. F., Rogers, R. G., Fraley, R. T., and Brand, L. (1984) Identification of cloned biosynthesis gene. *Proc. Natl. Acad. Sci. USA* 81: 4776–4780.

Beveridge, C. A., Murfet, I. C., Kerhoas, L., Sotta, B., Miginiac, E., and Rameau, C. (1997) The shoot controls zeatin riboside export from pea roots. Evidence from the branching mutant *rms4*. *Plant J.* 11: 339–345.

Bomhoff, G., Klapwijk, P. M., Kester, H. C. M., and Schilperoort, R. A. (1976) Octopine and nopaline synthesis and breakdown genetically controlled by plasmid of *Agrobacterium tumefaciens*. *Mol. Gen. Genet.* 145: 177–181.

Braun, A. C. (1958) A physiological basis for autonomous growth of the crown-gall tumor cell. *Proc. Natl. Acad. Sci. USA* 44: 344–349.

Brzobohaty, B., Moore, I., Kristoffersen, P., Bako, L., Campos, N., Schell, J., and Palme, K. (1993) Release of active cytokinin by a β-glucosidase localized to the maize root meristem. *Science* 262: 1051–1054.

Caplin, S. M., and Steward, F. C. (1948) Effect of coconut milk on the growth of the explants from carrot root. *Science* 108: 655–657.

Cary, A. J., Liu, W., and Howell, S. H. (1995) Cytokinin action is coupled to ethylene in its effects on the inhibition of root and hypocotyl elongation in *Arabidopsis thaliana* seedlings. *Plant Physiol.* 107: 1075–1082.

Chilton, M.-D. (1983) A vector for introducing new genes into plants. *Sci. Am.* 248(00): 50–59.

Chilton, M.-D., Drummond, M. H., Merlo, D. J., Sciaky, D., Montoya, A. L., Gordon, M. P., and Nester, E. W. (1977) Stable incorporation of plasmid DNA into higher plant cells: The molecular basis of crown gall tumorigenesis. *Cell* 11: 263–271.

Chory, J., Reinecke, D., Sim, S., Washburn, T., and Brenner, M. (1994) A role for cytokinins in de-etiolation in *Arabidopsis*. *Det* mutants have an altered response to cytokinins. *Plant Physiol.* 104: 339–347.

D'Agostino, I. B., Deruère, J., and Kieber, J. J. (2000) Characterization of the response of the *Arabidopsis ARR* gene family to cytokinin. *Plant Physiol.* 124: 1706–1717.

Elzen, G. W. (1983) Cytokinins and insect galls. *Comp. Biochem. Physiol.* 76A(1): 17–19.

Estruch, J. J., Chriqui, D., Grossmann, K., Schell, J., and Spena, A. (1991) The plant oncogene *RolC* is responsible for the release of cytokinins from glucoside conjugates. *EMBO J.* 10: 2889–2895.

Faiss, M., Zalubìová, J., Strnad, M., and Schmülling, T. (1997) Conditional transgenic expression of the *ipt* gene indicates a function for cytokinins in paracrine signaling in whole tobacco plants. *Plant J.* 12: 410–415.

Gan, S., and Amasino, R. M. (1995) Inhibition of leaf senescence by autoregulated production of cytokinin. *Science* 270: 1986–1988.

Garfinkel, D. J., Simpson, R. B., Ream, L. W., White, F. F., Gordon, M. P., and Nester, E. W. (1981) Genetic analysis of crown gall: Fine structure map of the T-DNA by site-directed mutagenesis. *Cell* 27: 143–153.

Hamilton, J. L., and Lowe, R. H. (1972) False broomrape: A physiological disorder caused by growth-regulator imbalance. *Plant Physiol.* 50: 303–304.

Houba-Herin, N., Pethe, C., d'Alayer, J., and Laloue M. (1999) Cytokinin oxidase from *Zea mays*: Purification, cDNA cloning and expression in moss protoplasts. *Plant J.* 17: 615–626.

Huff, A. K., and Ross, C. W. (1975) Promotion of radish cotyledon enlargement and reducing sugar content by zeatin and red light. *Plant Physiol.* 56: 429–433.

Hwang, I., and Sheen, J. (2001). Two-component circuitry in *Arabidopsis* signal transduction. *Nature* 413: 383–389.

Inoue, T., Higuchi, M., Hashimoto, Y., Seki, M., Kobayashi, M., Kato, T., Tabata, S., Shinozaki, K., and Kakimoto, T. (2001) Identification of CRE1 as a cytokinin receptor from *Arabidopsis*. *Nature* 409: 1060–1063.

Itai, C., and Vaadia, Y. (1971) Cytokinin activity in water-stressed shoots. *Plant Physiol.* 47: 87–90.

John, P. C. L., Zhang, K., Don, C., Diederich, L., and Wightman, F. (1993) P34-cdc2 related proteins in control of cell cycle progression, the switch between division and differentiation in tissue development, and stimulation of division by auxin and cytokinin. *Aust. J. Plant Physiol.* 20: 503–526.

Kakimoto, T. (2001) Identification of plant cytokinin biosynthetic enzymes as dimethylallyl diphosphate: ATP/ADP isopentenyltransferases. *Plant Cell. Physiol.* 42: 677–685.

Letham, D. S. (1973) Cytokinins from *Zea mays*. *Phytochemistry* 12: 2445–2455.

Letham, D. S. (1974) Regulators of cell division in plant tissues XX. The cytokinins of coconut milk. *Physiol. Plant.* 32: 66–70.

Martin R. C., Mok M. C., and Mok D. W. S. (1999) Isolation of a cytokinin gene, ZOG1, encoding zeatin O-glucosyltransferase from *Phaseolus lunatus*. *Proc. Natl. Acad. Sci. USA* 96: 284–289.

Miller, C. O., Skoog, F., Von Saltza, M. H., and Strong, F. (1955) Kinetin, a cell division factor from deoxyribonucleic acid. *J. Am. Chem. Soc.* 77: 1392–1393.

Morris, R. O. (1986) Genes specifying auxin and cytokinin biosynthesis in phytopathogens. *Annu. Rev. Plant Physiol.* 37: 509–538.

Morris, R., Bilyeu, K., Laskey, J., and Cheikh, N. (1999) Isolation of a gene encoding a glycosylated cytokinin oxidase from maize. *Biochem. Biophys. Res. Commun.* 225: 328–333.

Noodén, L. D., and Letham, D. S. (1993) Cytokinin metabolism and signaling in the soybean plant. *Aust. J. Plant Physiol.* 20: 639–653.

Noodén, L. D., Singh, S., and Letham, D. S. (1990) Correlation of xylem sap cytokinin levels with monocarpic senescence in soybean. *Plant Physiol.* 93: 33–39.

Riou-Khamlichi, C., Huntley, R., Jacqmard, A., and Murray, J. A. (1999) Cytokinin activation of *Arabidopsis* cell division through a D-type cyclin. *Science* 283: 1541–1544.

Rupp, H.-M., Frank, M., Werner, T., Strnad, M., and Schmülling, T. (1999) Increased steady state mRNA levels of the STM and KNATI homeobox genes in cytokinin overproducing *Arabidopsis thaliana* indicate a role for cytokinins in the shoot apical meristem. *Plant J.* 18: 557–563.

Sakai, H., Honma, T., Aoyama, T., Sato, S., Kato, T., Tabata, S., and Oka, A. (2001) *Arabidopsis ARR*1 is a transcription factor for genes immediately responsive to cytokinins. *Science*. 294: 1519–1521.

Samuelson, M. E., Eliasson, L., Larsson, C. M. (1992) Nitrate-regulated growth and cytokinin responses in seminal roots of barley. *Plant Physiol.* 98: 309–315.

Skoog, F., and Miller, C. O. (1965) Chemical regulation of growth and organ formation in plant tissues cultured *in vitro*. In *Molecular and Cellular Aspects of Development*, E. Bell, ed., Harper and Row, New York, pp. 481–494.

Smigocki, A., Neal, J. W., Jr., McCanna, I., and Douglass, L. (1993) Cytokinin-mediated insect resistance in *Nicotiana* plants transformed with the *ipt* gene. *Plant Mol. Biol.* 23: 325–335.

Smith, H. H. (1988) The inheritance of genetic tumors in *Nicotiana* hybrids. *J. Hered.* 79: 277–284.

Soni, R., Carmichael, J. P., Shah, Z. H., and Murray, J. A. H. (1995) A family of cyclin D homologs from plants differentially controlled by growth regulators and containing the conserved retinoblastoma protein interaction motif. *Plant Cell* 7: 85–103.

Takei, K., Sakakibara, H., and Sugiyama, T. (2001) Identification of genes encoding adenylate isopentyltransferase, a cytokinin biosynthetic enzyme, in *Arabidopsis thaliana*. *J. Biol. Chem.* 276: 26405–26410.

Takei, K., Sakakibara, H., Taniguchi, M., and Sugiyama, T. (2001) Nitrogen-dependent accumulation of cytokinins in roots and the translocation to leaf: Implication of cytokinin species that induces gene expression of maize response regulator. *Plant Cell Physiol.* 42: 85–93.

Vogel, J. P., Woeste, K., Theologis, A., and Kieber, J. J. (1998) Recessive and dominant mutations in the ethylene biosynthetic gene AC55 of *Arabidopsis* confer cytokinin-insensitivity and ethylene overproduction respectively. *Proc. Natl. Acad. Sci. USA* 95: 4766–4771.

Werner, T., Motyka, V., Strnad, M., and Schmülling, T. (2001) Regulation of plant growth by cytokinin. *Proc. Natl. Acad. Sci. USA* 98: 10487–10492.

White, P. R. (1934) Potentially unlimited growth of excised tomato root tips in a liquid medium. *Plant Physiol.* 9: 585–600.

Yamada, H., Suzuki, T., Terada, K., Takei, K., Ishikawa, K., Miwa, K., Yamashino, T., and Mizuno, T. (2001). The *Arabidopsis* AHK4 histidine kinase is a cytokinin-binding receptor that transduces cytokinin signals across the membrane. *Plant Cell Physiol.* 42: 1017–1023.

Zhang, K., Letham, D. S., and John, P. C. L. (1996) Cytokinin controls the cell cycle at mitosis by stimulating the tyrosine dephosphorylation and activation of p34cdc2-like H1 histone kinase. *Planta* 200: 2–12.

Chapter

22 Ethylene: The Gaseous Hormone

DURING THE NINETEENTH CENTURY, when coal gas was used for street illumination, it was observed that trees in the vicinity of streetlamps defoliated more extensively than other trees. Eventually it became apparent that coal gas and air pollutants affect plant growth and development, and ethylene was identified as the active component of coal gas.

In 1901, Dimitry Neljubov, a graduate student at the Botanical Institute of St. Petersburg in Russia, observed that dark-grown pea seedlings growing in the laboratory exhibited symptoms that were later termed the *triple response*: reduced stem elongation, increased lateral growth (swelling), and abnormal, horizontal growth. When the plants were allowed to grow in fresh air, they regained their normal morphology and rate of growth. Neljubov identified ethylene, which was present in the laboratory air from coal gas, as the molecule causing the response.

The first indication that ethylene is a natural product of plant tissues was published by H. H. Cousins in 1910. Cousins reported that "emanations" from oranges stored in a chamber caused the premature ripening of bananas when these gases were passed through a chamber containing the fruit. However, given that oranges synthesize relatively little ethylene compared to other fruits, such as apples, it is likely that the oranges used by Cousins were infected with the fungus *Penicillium*, which produces copious amounts of ethylene. In 1934, R. Gane and others identified ethylene chemically as a natural product of plant metabolism, and because of its dramatic effects on the plant it was classified as a hormone.

For 25 years ethylene was not recognized as an important plant hormone, mainly because many physiologists believed that the effects of ethylene were due to auxin, the first plant hormone to be discovered (see Chapter 19). Auxin was thought to be the main plant hormone, and ethylene was considered to play only an insignificant and indirect physiological role. Work on ethylene was also hampered by the lack of chemical techniques for its quantification. However, after gas chromatography was introduced in ethylene research in 1959, the importance of ethylene was rediscovered and its physiological significance as a plant growth regulator was recognized (Burg and Thimann 1959).

In this chapter we will describe the discovery of the ethylene biosynthetic pathway and outline some of the important effects of ethylene on plant growth and development. At the end of the chapter we will consider how ethylene acts at the cellular and molecular levels.

STRUCTURE, BIOSYNTHESIS, AND MEASUREMENT OF ETHYLENE

Ethylene can be produced by almost all parts of higher plants, although the rate of production depends on the type of tissue and the stage of development. In general, meristematic regions and nodal regions are the most active in ethylene biosynthesis. However, ethylene production also increases during leaf abscission and flower senescence, as well as during fruit ripening. Any type of wounding can induce ethylene biosynthesis, as can physiological stresses such as flooding, chilling, disease, and temperature or drought stress.

The amino acid methionine is the precursor of ethylene, and ACC (1-aminocyclopropane-1-carboxylic acid) serves as an intermediate in the conversion of methionine to ethylene. As we will see, the complete pathway is a cycle, taking its place among the many metabolic cycles that operate in plant cells.

The Properties of Ethylene Are Deceptively Simple

Ethylene is the simplest known olefin (its molecular weight is 28), and it is lighter than air under physiological conditions:

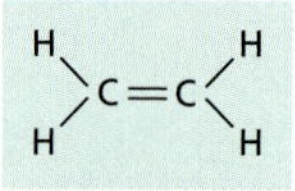

$$H_2C=CH_2$$

Ethylene

It is flammable and readily undergoes oxidation. Ethylene can be oxidized to ethylene oxide:

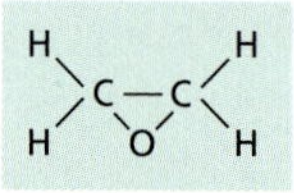

$$H_2C{-}CH_2 \text{ (epoxide, bridged by O)}$$

Ethylene oxide

and ethylene oxide can be hydrolyzed to ethylene glycol:

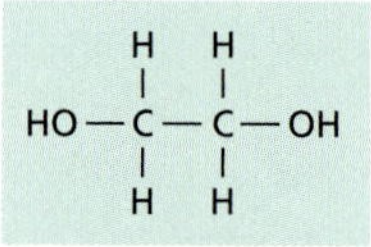

$$HO{-}CH_2{-}CH_2{-}OH$$

Ethylene glycol

In most plant tissues, ethylene can be completely oxidized to CO_2, in the following reaction:

Complete oxidation of ethylene

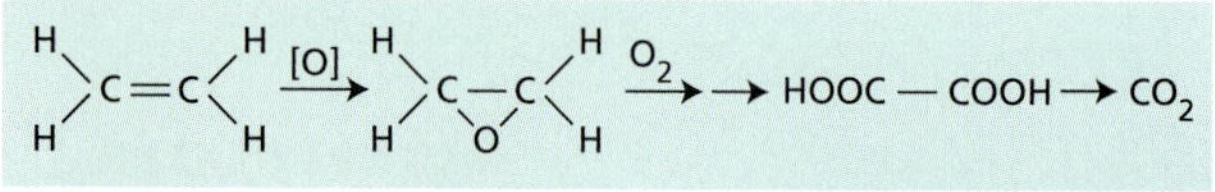

$$H_2C=CH_2 \xrightarrow{[O]} H_2C{-}O{-}CH_2 \xrightarrow{O_2} HOOC-COOH \rightarrow CO_2$$

Ethylene **Ethylene oxide** **Oxalic acid** **Carbon dioxide**

Ethylene is released easily from the tissue and diffuses in the gas phase through the intercellular spaces and outside the tissue. At an ethylene concentration of 1 μL L^{-1} in the gas phase at 25°C, the concentration of ethylene in water is 4.4×10^{-9} *M*. Because they are easier to measure, gas phase concentrations are normally given for ethylene.

Because ethylene gas is easily lost from the tissue and may affect other tissues or organs, ethylene-trapping systems are used during the storage of fruits, vegetables, and flowers. Potassium permanganate ($KMnO_4$) is an effective absorbent of ethylene and can reduce the concentration of ethylene in apple storage areas from 250 to 10 μL L^{-1}, markedly extending the storage life of the fruit.

Bacteria, Fungi, and Plant Organs Produce Ethylene

Even away from cities and industrial air pollutants, the environment is seldom free of ethylene because of its production by plants and microorganisms. The production of ethylene in plants is highest in senescing tissues and ripening fruits (>1.0 nL g-fresh-weight^{-1} h^{-1}), but all organs of higher plants can synthesize ethylene. Ethylene is biologically active at very low concentrations—less than 1 part per million (1 μL L^{-1}). The internal ethylene concentration in a ripe apple has been reported to be as high as 2500 μL L^{-1}.

Young developing leaves produce more ethylene than do fully expanded leaves. In bean (*Phaseolus vulgaris*), young leaves produce 0.4 nL g^{-1} h^{-1}, compared with 0.04 nL g^{-1} h^{-1} for older leaves. With few exceptions, nonsenescent tissues that are wounded or mechanically perturbed will temporarily increase their ethylene production severalfold within 30 minutes. Ethylene levels later return to normal.

Gymnosperms and lower plants, including ferns, mosses, liverworts, and certain cyanobacteria, all have shown the ability to produce ethylene. Ethylene production by fungi and bacteria contributes significantly to the ethylene content of soil. Certain strains of the common enteric bacterium *Escherichia coli* and of yeast (a fungus) produce large amounts of ethylene from methionine.

There is no evidence that healthy mammalian tissues produce ethylene, nor does ethylene appear to be a metabolic product of invertebrates. However, recently it was found that both a marine sponge and cultured mammalian

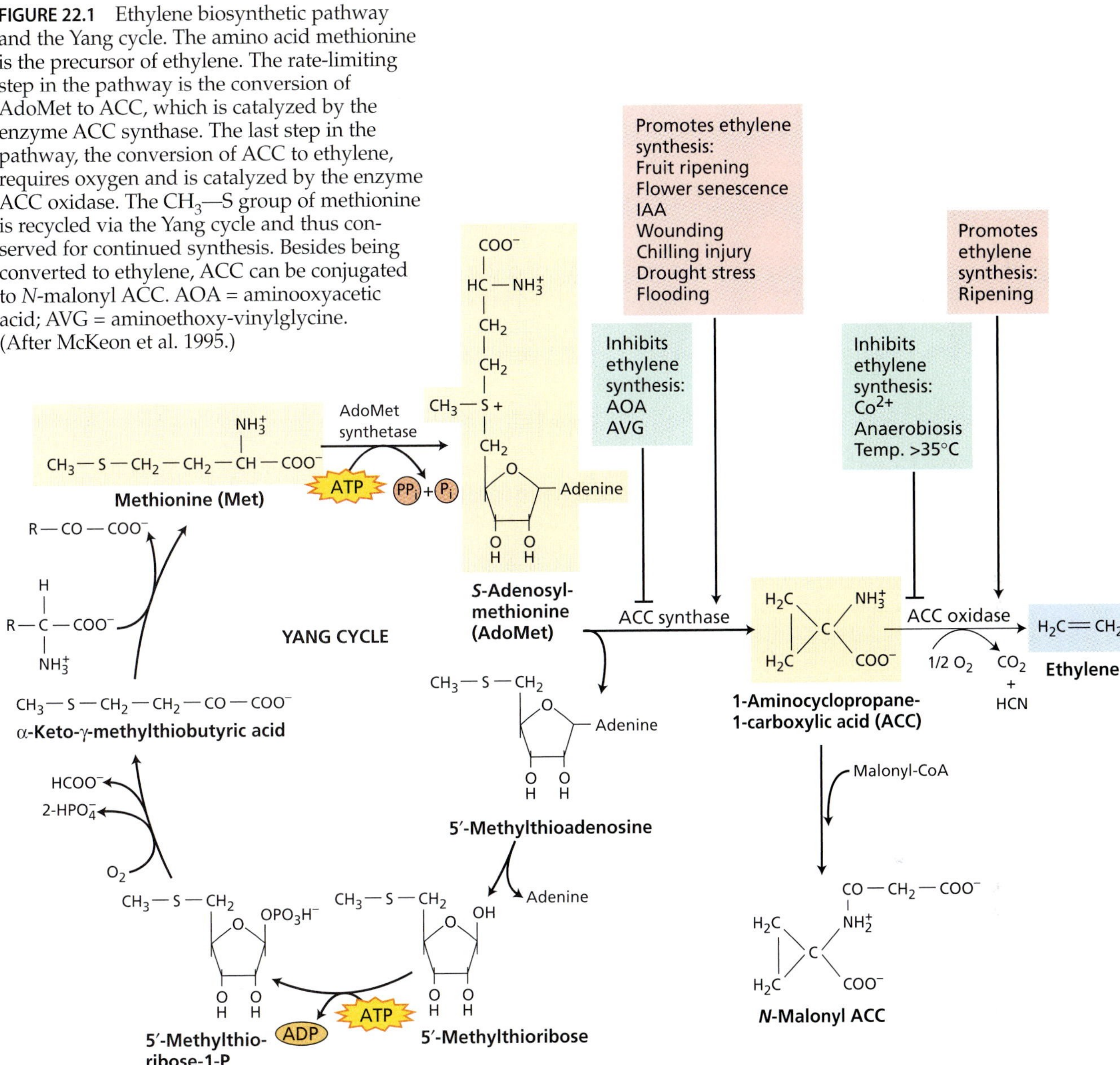

FIGURE 22.1 Ethylene biosynthetic pathway and the Yang cycle. The amino acid methionine is the precursor of ethylene. The rate-limiting step in the pathway is the conversion of AdoMet to ACC, which is catalyzed by the enzyme ACC synthase. The last step in the pathway, the conversion of ACC to ethylene, requires oxygen and is catalyzed by the enzyme ACC oxidase. The CH_3—S group of methionine is recycled via the Yang cycle and thus conserved for continued synthesis. Besides being converted to ethylene, ACC can be conjugated to *N*-malonyl ACC. AOA = aminooxyacetic acid; AVG = aminoethoxy-vinylglycine. (After McKeon et al. 1995.)

cells can respond to ethylene, raising the possibility that this gaseous molecule acts as a signaling molecule in animal cells (Perovic et al. 2001).

Regulated Biosynthesis Determines the Physiological Activity of Ethylene

In vivo experiments showed that plant tissues convert l-[^{14}C]methionine to [^{14}C]ethylene, and that the ethylene is derived from carbons 3 and 4 of methionine (Figure 22.1). The CH_3—S group of methionine is recycled via the Yang cycle. Without this recycling, the amount of reduced sulfur present would limit the available methionine and the synthesis of ethylene. *S*-adenosylmethionine (AdoMet), which is synthesized from methionine and ATP, is an intermediate in the ethylene biosynthetic pathway, and the immediate precursor of ethylene is **1-aminocyclopropane-1-carboxylic acid** (**ACC**) (see Figure 22.1).

The role of ACC became evident in experiments in which plants were treated with [^{14}C]methionine. Under anaerobic conditions, ethylene was not produced from the [^{14}C]methionine, and labeled ACC accumulated in the tissue. On exposure to oxygen, however, ethylene production surged. The labeled ACC was rapidly converted to ethylene in the presence of oxygen by various plant tissues, suggesting that ACC is the immediate precursor of ethylene in higher plants and that oxygen is required for the conversion.

In general, when ACC is supplied exogenously to plant tissues, ethylene production increases substantially. This observation indicates that the synthesis of ACC is usually the biosynthetic step that limits ethylene production in plant tissues.

ACC synthase, the enzyme that catalyzes the conversion of AdoMet to ACC (see Figure 22.1), has been characterized in many types of tissues of various plants. ACC synthase is an unstable, cytosolic enzyme. Its level is regulated by environmental and internal factors, such as wounding, drought stress, flooding, and auxin. Because ACC synthase is present in such low amounts in plant tissues (0.0001% of the total protein of ripe tomato) and is very unstable, it is difficult to purify the enzyme for biochemical analysis (see **Web Topic 22.1**).

ACC synthase is encoded by members of a divergent multigene family that are differentially regulated by various inducers of ethylene biosynthesis. In tomato, for example, there are at least nine ACC synthase genes, different subsets of which are induced by auxin, wounding, and/or fruit ripening.

ACC oxidase catalyzes the last step in ethylene biosynthesis: the conversion of ACC to ethylene (see Figure 22.1). In tissues that show high rates of ethylene production, such as ripening fruit, ACC oxidase activity can be the rate-limiting step in ethylene biosynthesis. The gene that encodes ACC oxidase has been cloned (see **Web Topic 22.2**). Like ACC synthase, ACC oxidase is encoded by a multigene family that is differentially regulated. For example, in ripening tomato fruits and senescing petunia flowers, the mRNA levels of a subset of ACC oxidase genes are highly elevated.

The deduced amino acid sequences of ACC oxidases revealed that these enzymes belong to the Fe^{2+}/ascorbate oxidase superfamily. This similarity suggested that ACC oxidase might require Fe^{2+} and ascorbate for activity—a requirement that has been confirmed by biochemical analysis of the protein. The low abundance of ACC oxidase and its requirement for cofactors presumably explain why the purification of this enzyme eluded researchers for so many years.

Catabolism. Researchers have studied the catabolism of ethylene by supplying $^{14}C_2H_4$ to plant tissues and tracing the radioactive compounds produced. Carbon dioxide, ethylene oxide, ethylene glycol, and the glucose conjugate of ethylene glycol have been identified as metabolic breakdown products. However, because certain cyclic olefin compounds, such as 1,4-cyclohexadiene, have been shown to block ethylene breakdown without inhibiting ethylene action, ethylene catabolism does not appear to play a significant role in regulating the level of the hormone (Raskin and Beyer 1989).

Conjugation. Not all the ACC found in the tissue is converted to ethylene. ACC can also be converted to a conjugated form, *N*-malonyl ACC (see Figure 22.1), which does not appear to break down and accumulates in the tissue. A second conjugated form of ACC, 1-(γ-L-glutamylamino) cyclopropane-1-carboxylic acid (GACC), has also been identified. The conjugation of ACC may play an important role in the control of ethylene biosynthesis, in a manner analogous to the conjugation of auxin and cytokinin.

Environmental Stresses and Auxins Promote Ethylene Biosynthesis

Ethylene biosynthesis is stimulated by several factors, including developmental state, environmental conditions, other plant hormones, and physical and chemical injury. Ethylene biosynthesis also varies in a circadian manner, peaking during the day and reaching a minimum at night.

Fruit ripening. As fruits mature, the rate of ACC and ethylene biosynthesis increases. Enzyme activities for both ACC oxidase (Figure 22.2) and ACC synthase increase, as do the mRNA levels for subsets of the genes encoding each enzyme. However, application of ACC to unripe fruits only slightly enhances ethylene production, indicating that an increase in the activity of ACC oxidase is the rate-limiting step in ripening (McKeon et al. 1995).

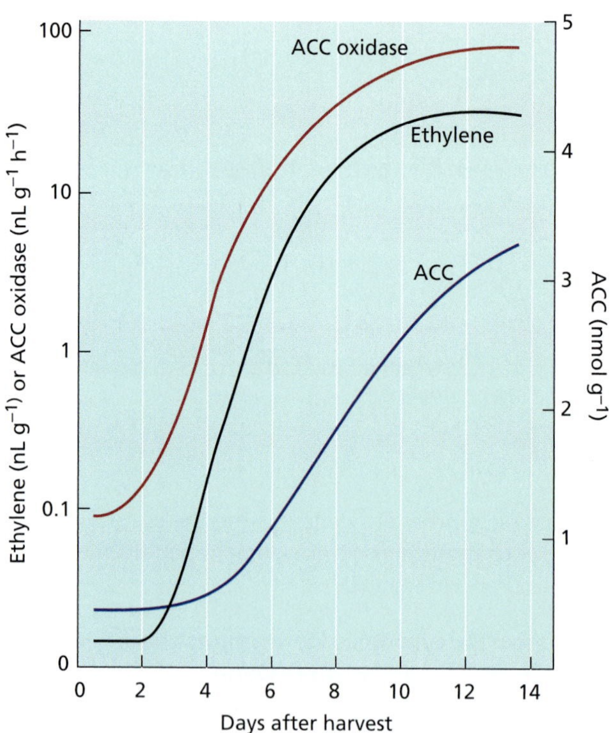

FIGURE 22.2 Changes in ethylene and ACC content and ACC oxidase activity during fruit ripening. Changes in the ACC oxidase activity and ethylene and ACC concentrations of Golden Delicious apples. The data are plotted as a function of days after harvest. Increases in ethylene and ACC concentrations and in ACC oxidase activity are closely correlated with ripening. (From Yang 1987.)

Stress-induced ethylene production. Ethylene biosynthesis is increased by stress conditions such as drought, flooding, chilling, exposure to ozone, or mechanical wounding. In all these cases ethylene is produced by the usual biosynthetic pathway, and the increased ethylene production has been shown to result at least in part from an increase in transcription of ACC synthase mRNA. This "stress ethylene" is involved in the onset of stress responses such as abscission, senescence, wound healing, and increased disease resistance (see Chapter 25).

Auxin-induced ethylene production. In some instances, auxins and ethylene can cause similar plant responses, such as induction of flowering in pineapple and inhibition of stem elongation. These responses might be due to the ability of auxins to promote ethylene synthesis by enhancing ACC synthase activity. These observations suggest that some responses previously attributed to auxin (indole-3-acetic acid, or IAA) are in fact mediated by the ethylene produced in response to auxin.

Inhibitors of protein synthesis block both ACC and IAA-induced ethylene synthesis, indicating that the synthesis of new ACC synthase protein caused by auxins brings about the marked increase in ethylene production. Several ACC synthase genes have been identified whose transcription is elevated following application of exogenous IAA, suggesting that increased transcription is at least partly responsible for the increased ethylene production observed in response to auxin (Nakagawa et al. 1991; Liang et al. 1992).

Posttranscriptional regulation of ethylene production. Ethylene production can also be regulated posttranscriptionally. Cytokinins also promote ethylene biosynthesis in some plant tissues. For example, in etiolated *Arabidopsis* seedlings, application of exogenous cytokinins causes a rise in ethylene production, resulting in the triple-response phenotype (see Figure 22.5A).

Molecular genetic studies in *Arabidopsis* have shown that cytokinins elevate ethylene biosynthesis by increasing the stability and/or activity of one isoform of ACC synthase (Vogel et al. 1998). The carboxy-terminal domain of this ACC synthase isoform appears to be the target for this posttranscriptional regulation. Consistent with this, the carboxy-terminal domain of an ACC synthase isoform from tomato has been shown to be the target for a calcium-dependent phosphorylation (Tatsuki and Mori 2001).

Ethylene Production and Action Can Be Inhibited

Inhibitors of hormone synthesis or action are valuable for the study of the biosynthetic pathways and physiological roles of hormones. Inhibitors are particularly helpful when it is difficult to distinguish between different hormones that have identical effects in plant tissue or when a hormone affects the synthesis or the action of another hormone.

For example, ethylene mimics high concentrations of auxins by inhibiting stem growth and causing epinasty (a downward curvature of leaves). Use of specific inhibitors of ethylene biosynthesis and action made it possible to discriminate between the actions of auxin and ethylene. Studies using inhibitors showed that ethylene is the primary effector of epinasty and that auxin acts indirectly by causing a substantial increase in ethylene production.

Inhibitors of ethylene synthesis. **Aminoethoxy-vinylglycine (AVG)** and **aminooxyacetic acid (AOA)** block the conversion of AdoMet to ACC (see Figure 22.1). AVG and AOA are known to inhibit enzymes that use the cofactor pyridoxal phosphate. The cobalt ion (Co^{2+}) is also an inhibitor of the ethylene biosynthetic pathway, blocking the conversion of ACC to ethylene by ACC oxidase, the last step in ethylene biosynthesis.

Inhibitors of ethylene action. Most of the effects of ethylene can be antagonized by specific ethylene inhibitors. Silver ions (Ag^+) applied as silver nitrate ($AgNO_3$) or as silver thiosulfate ($Ag(S_2O_3)_2^{3-}$) are potent inhibitors of ethylene action. Silver is very specific; the inhibition it causes cannot be induced by any other metal ion.

Carbon dioxide at high concentrations (in the range of 5 to 10%) also inhibits many effects of ethylene, such as the induction of fruit ripening, although CO_2 is less efficient than Ag^+. This effect of CO_2 has often been exploited in the storage of fruits, whose ripening is delayed at elevated CO_2 concentrations. The high concentrations of CO_2 required for inhibition make it unlikely that CO_2 acts as an ethylene antagonist under natural conditions.

The volatile compound *trans*-cyclooctene, but not its isomer *cis*-cyclooctene, is a strong competitive inhibitor of ethylene binding (Sisler et al. 1990); *trans*-cyclooctene is thought to act by competing with ethylene for binding to the receptor. A novel inhibitor, **1-methylcyclopropene (MCP)**, was recently found that binds almost irreversibly to the ethylene receptor (Figure 22.3) (Sisler and Serek 1997). MCP shows tremendous promise in commercial applications.

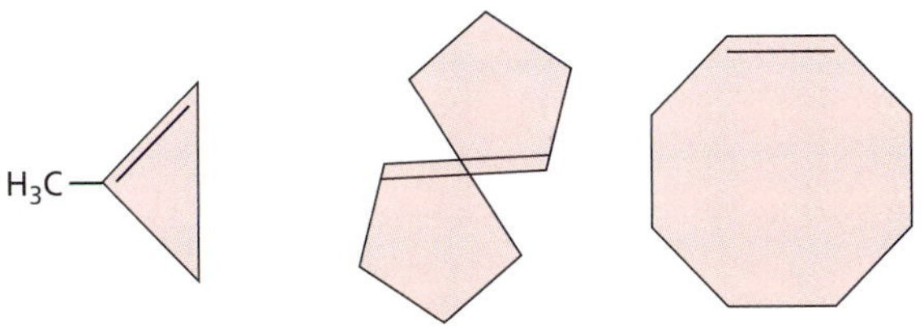

FIGURE 22.3 Inhibitors that block ethylene binding to its receptor. Only the *trans* form of cyclooctene is active.

Ethylene Can Be Measured by Gas Chromatography

Historically, bioassays based on the seedling triple response were used to measure ethylene levels, but they have been replaced by **gas chromatography**. As little as 5 parts per billion (ppb) (5 pL per liter)[1] of ethylene can be detected, and the analysis time is only 1 to 5 minutes.

Usually the ethylene produced by a plant tissue is allowed to accumulate in a sealed vial, and a sample is withdrawn with a syringe. The sample is injected into a gas chromatograph column in which the different gases are separated and detected by a flame ionization detector. Quantification of ethylene by this method is very accurate. Recently a novel method to measure ethylene was developed that uses a laser-driven photoacoustic detector that can detect as little as 50 parts per trillion (50 ppt = 0.05 pL L^{-1}) ethylene (Voesenek et al. 1997).

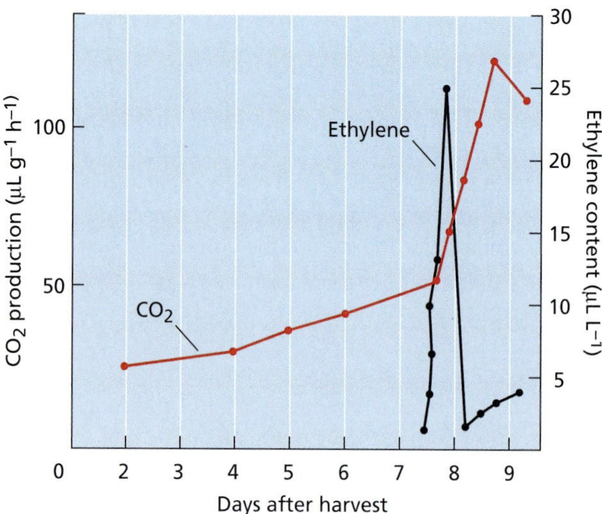

FIGURE 22.4 Ethylene production and respiration. In banana, ripening is characterized by a climacteric rise in respiration rate, as evidenced by the increased CO_2 production. A climacteric rise in ethylene production precedes the increase in CO_2 production, suggesting that ethylene is the hormone that triggers the ripening process. (From Burg and Burg 1965.)

DEVELOPMENTAL AND PHYSIOLOGICAL EFFECTS OF ETHYLENE

As we have seen, ethylene was discovered in connection with its effects on seedling growth and fruit ripening. It has since been shown to regulate a wide range of responses in plants, including seed germination, cell expansion, cell differentiation, flowering, senescence, and abscission. In this section we will consider the phenotypic effects of ethylene in more detail.

Ethylene Promotes the Ripening of Some Fruits

In everyday usage, the term *fruit ripening* refers to the changes in fruit that make it ready to eat. Such changes typically include softening due to the enzymatic breakdown of the cell walls, starch hydrolysis, sugar accumulation, and the disappearance of organic acids and phenolic compounds, including tannins. From the perspective of the plant, fruit ripening means that the seeds are ready for dispersal.

For seeds whose dispersal depends on animal ingestion, *ripeness* and *edibility* are synonymous. Brightly colored anthocyanins and carotenoids often accumulate in the epidermis of such fruits, enhancing their visibility. However, for seeds that rely on mechanical or other means for dispersal, *fruit ripening* may mean drying followed by splitting. Because of their importance in agriculture, the vast majority of studies on fruit ripening have focused on edible fruits.

Ethylene has long been recognized as the hormone that accelerates the ripening of edible fruits. Exposure of such fruits to ethylene hastens the processes associated with ripening, and a dramatic increase in ethylene production accompanies the initiation of ripening. However, surveys of a wide range of fruits have shown that not all of them respond to ethylene.

All fruits that ripen in response to ethylene exhibit a characteristic respiratory rise before the ripening phase called a **climacteric**.[2] Such fruits also show a spike of ethylene production immediately before the respiratory rise (Figure 22.4). Inasmuch as treatment with ethylene induces the fruit to produce additional ethylene, its action can be described as **autocatalytic**. Apples, bananas, avocados, and tomatoes are examples of climacteric fruits.

In contrast, fruits such as citrus fruits and grapes do not exhibit the respiration and ethylene production rise and are called **nonclimacteric** fruits. Other examples of climacteric and nonclimacteric fruits are given in Table 22.1.

When unripe climacteric fruits are treated with ethylene, the onset of the climacteric rise is hastened. When nonclimacteric fruits are treated in the same way, the magnitude of the respiratory rise increases as a function of the ethylene concentration, but the treatment does not trigger production of endogenous ethylene and does not accelerate ripening. Elucidation of the role of ethylene in the ripening of climacteric fruits has resulted in many practical applications aimed at either uniform ripening or the delay of ripening.

Although the effects of exogenous ethylene on fruit ripening are straightforward and clear, establishing a causal relation between the level of endogenous ethylene and fruit ripening is more difficult. Inhibitors of ethylene biosynthe-

[1] pL = picoliter = 10^{-12} L.

[2] The term *climacteric* can be used either as a noun, as in "most fruits exhibit a climacteric during ripening" or as an adjective, as in "a climacteric rise in respiration." The term *nonclimacteric*, however, is used only as an adjective.

TABLE 22.1
Climacteric and nonclimacteric fruits

Climacteric	Nonclimacteric
Apple	Bell pepper
Avocado	Cherry
Banana	Citrus
Cantaloupe	Grape
Cherimoya	Pineapple
Fig	Snap bean
Mango	Strawberry
Olive	Watermelon
Peach	
Pear	
Persimmon	
Plum	
Tomato	

sis (such as AVG) or of ethylene action (such as CO_2, MCP, or Ag^+) have been shown to delay or even prevent ripening. However, the definitive demonstration that ethylene is required for fruit ripening was provided by experiments in which ethylene biosynthesis was blocked by expression of an antisense version of either ACC synthase or ACC oxidase in transgenic tomatoes (see **Web Topic 22.3**). Elimination of ethylene biosynthesis in these transgenic tomatoes completely blocked fruit ripening, and ripening was restored by application of exogenous ethylene (Oeller et al. 1991).

Further demonstration of the requirement for ethylene in fruit ripening came from the analysis of the *never-ripe* mutation in tomato. As the name implies, this mutation completely blocks the ripening of tomato fruit. Molecular analysis revealed that *never-ripe* was due to a mutation in an ethylene receptor that rendered it unable to bind ethylene (Lanahan et al. 1994). These experiments provided unequivocal proof of the role of ethylene in fruit ripening, and they opened the door to the manipulation of fruit ripening through biotechnology.

In tomatoes several genes have been identified that are highly regulated during ripening (Gray et al. 1994). During tomato fruit ripening, the fruit softens as the result of cell wall hydrolysis and changes from green to red as a consequence of chlorophyll loss and the synthesis of the carotenoid pigment lycopene. At the same time, aroma and flavor components are produced.

Analysis of mRNA from tomato fruits from wild-type and transgenic tomato plants genetically engineered to lack ethylene has revealed that gene expression during ripening is regulated by at least two independent pathways:

1. *An ethylene-dependent pathway* includes genes involved in lycopene and aroma biosynthesis, respiratory metabolism, and ACC synthase.
2. *A developmental, ethylene-independent pathway* includes genes encoding ACC oxidase and chlorophyllase.

Thus, not all of the processes associated with ripening in tomato are ethylene dependent.

Leaf Epinasty Results When ACC from the Root Is Transported to the Shoot

The downward curvature of leaves that occurs when the upper (adaxial) side of the petiole grows faster than the lower (abaxial) side is termed **epinasty** (Figure 22.5B). Ethylene and high concentrations of auxin induce epinasty, and it has now been established that auxin acts indirectly by inducing ethylene production. As will be discussed later in the chapter, a variety of stress conditions, such as salt stress or pathogen infection, increase ethylene production and also induce epinasty. There is no known physiological function for the response.

In tomato and other dicots, flooding (waterlogging) or anaerobic conditions around the roots enhances the synthesis of ethylene in the shoot, leading to the epinastic response. Because these environmental stresses are sensed by the roots and the response is displayed by the shoots, a signal from the roots must be transported to the shoots. This signal is ACC, the immediate precursor of ethylene. ACC levels were found to be significantly higher in the xylem sap after flooding of tomato roots for 1 to 2 days (Figure 22.6) (Bradford and Yang 1980).

Because water fills the air spaces in waterlogged soil and O_2 diffuses slowly through water, the concentration of oxygen around flooded roots decreases dramatically. The elevated production of ethylene appears to be caused by the accumulation of ACC in the roots under anaerobic conditions, since the conversion of ACC to ethylene requires oxygen (see Figure 22.1). The ACC accumulated in the anaerobic roots is then transported to shoots via the transpiration stream, where it is readily converted to ethylene.

Ethylene Induces Lateral Cell Expansion

At concentrations above 0.1 $\mu L\ L^{-1}$, ethylene changes the growth pattern of seedlings by reducing the rate of elongation and increasing lateral expansion, leading to swelling of the region below the hook. These effects of ethylene are common to growing shoots of most dicots, forming part of the **triple response**. In *Arabidopsis*, the triple response consists of inhibition and swelling of the hypocotyl, inhibition of root elongation, and exaggeration of the apical hook (Figure 22.7).

As discussed in Chapter 15, the directionality of plant cell expansion is determined by the orientation of the cellulose microfibrils in the cell wall. Transverse microfibrils reinforce the cell wall in the lateral direction, so that turgor pressure is channeled into cell elongation. The orientation of the microfibrils in turn is determined by the orientation of the cortical array of microtubules in the cortical (peripheral) cytoplasm. In typical elongating plant cells, the cortical microtubules are arranged transversely, giving rise to transversely arranged cellulose microfibrils.

increases, equalizing the growth rates on both sides. The kinematic aspects of hook growth (i.e., maintenance of the hook shape over time) were discussed in Chapter 16.

Red light induces hook opening, and far-red light reverses the effect of red, indicating that phytochrome is the photoreceptor involved in this process (see Chapter 17). A close interaction between phytochrome and ethylene controls hook opening. As long as ethylene is produced by the hook tissue in the dark, elongation of the cells on the inner side is inhibited. Red light inhibits ethylene formation, promoting growth on the inner side, thereby causing the hook to open.

The auxin-insensitive mutation *axr1* and treatment of wild-type seedlings with NPA (*1*-N-naphthylphthalamic acid), an inhibitor of polar auxin transport, both block the formation of the apical hook in *Arabidopsis*. These and other results indicate a role for auxin in maintaining hook structure. The more rapid growth rate of the outer tissues relative to the inner tissues could reflect an ethylene-dependent auxin gradient, analogous to the lateral auxin gradient that develops during phototropic curvature (see Chapter 19).

A gene required for formation of the apical hook, *HOOKLESS1* (so called because mutations in this gene result in seedlings lacking an apical hook), was identified in *Arabidopsis* (Lehman et al. 1996). Disruption of this gene severely alters the pattern of expression of auxin-responsive genes. When the gene is overexpressed in *Arabidopsis*, it causes constitutive hook formation even in the light. *HOOKLESS1* encodes a putative *N*-acetyltransferase that is hypothesized to regulate—by an unknown mechanism—differential auxin distribution in the apical hook induced by ethylene.

Ethylene Breaks Seed and Bud Dormancy in Some Species

Seeds that fail to germinate under normal conditions (water, oxygen, temperature suitable for growth) are said to be dormant (see Chapter 23). Ethylene has the ability to break dormancy and initiate germination in certain seeds, such as cereals. In addition to its effect on dormancy, ethylene increases the rate of seed germination of several species. In peanuts (*Arachis hypogaea*), ethylene production and seed germination are closely correlated. Ethylene can also break bud dormancy, and ethylene treatment is sometimes used to promote bud sprouting in potato and other tubers.

Ethylene Promotes the Elongation Growth of Submerged Aquatic Species

Although usually thought of as an inhibitor of stem elongation, ethylene is able to promote stem and petiole elongation in various submerged or partially submerged aquatic plants. These include the dicots *Ranunculus sceleratus*, *Nymphoides peltata*, and *Callitriche platycarpa*, and the fern *Regnellidium diphyllum*. Another agriculturally important example is the cereal deepwater rice (see Chapter 20).

In these species, submergence induces rapid internode or petiole elongation, which allows the leaves or upper parts of the shoot to remain above water. Treatment with ethylene mimics the effects of submergence.

Growth is stimulated in the submerged plants because ethylene builds up in the tissues. In the absence of O_2, ethylene synthesis is diminished, but the loss of ethylene by diffusion is retarded under water. Sufficient oxygen for growth and ethylene synthesis in the underwater parts is usually provided by aerenchyma tissue.

As we saw in Chapter 20, in deepwater rice it has been shown that ethylene stimulates internode elongation by increasing the amount of, and the sensitivity to, gibberellin in the cells of the intercalary meristem. The increased sensitivity to GA (gibberellic acid) in these cells in response to ethylene is brought about by a decrease in the level of abscisic acid (ABA), a potent antagonist of GA.

Ethylene Induces the Formation of Roots and Root Hairs

Ethylene is capable of inducing adventitious root formation in leaves, stems, flower stems, and even other roots. Ethylene has also been shown to act as a positive regulator of root hair formation in several species (see Figure 22.5D). This relationship has been best studied in *Arabidopsis*, in which root hairs normally are located in the epidermal cells that overlie a junction between the underlying cortical cells (Dolan et al. 1994).

In ethylene-treated roots, extra hairs form in abnormal locations in the epidermis; that is, cells not overlying a cortical cell junction differentiate into hair cells (Tanimoto et al. 1995). Seedlings grown in the presence of ethylene inhibitors (such as Ag^+), as well as ethylene-insensitive mutants, display a reduction in root hair formation in response to ethylene. These observations suggest that ethylene acts as a positive regulator in the differentiation of root hairs.

Ethylene Induces Flowering in the Pineapple Family

Although ethylene inhibits flowering in many species, it induces flowering in pineapple and its relatives, and it is used commercially in pineapple for synchronization of fruit set. Flowering of other species, such as mango, is also initiated by ethylene. On plants that have separate male and female flowers (monoecious species), ethylene may change the sex of developing flowers (see Chapter 24). The promotion of female flower formation in cucumber is one example of this effect.

Ethylene Enhances the Rate of Leaf Senescence

As described in Chapter 16, senescence is a genetically programmed developmental process that affects all tissues of the plant. Several lines of physiological evidence support roles for ethylene and cytokinins in the control of leaf senescence:

- Exogenous applications of ethylene or ACC (the precursor of ethylene) accelerate leaf senescence, and treatment with exogenous cytokinins delays leaf senescence (see Chapter 21).
- Enhanced ethylene production is associated with chlorophyll loss and color fading, which are characteristic features of leaf and flower senescence (see Figure 22.5C); an inverse correlation has been found between cytokinin levels in leaves and the onset of senescence.
- Inhibitors of ethylene synthesis (e.g., AVG or Co^{2+}) and action (e.g., Ag^+ or CO_2) retard leaf senescence.

Taken together, the physiological studies suggest that senescence is regulated by the balance of ethylene and cytokinin. In addition, abscisic acid (ABA) has been implicated in the control of leaf senescence. The role of ABA in senescence will be discussed in Chapter 23.

Senescence in ethylene mutants. Direct evidence for the involvement of ethylene in the regulation of leaf senescence has come from molecular genetic studies on *Arabidopsis*. As will be discussed later in the chapter, several mutants affecting the response to ethylene have been identified. The specific bioassay employed was the triple-response assay in which ethylene significantly inhibits seedling hypocotyl elongation and promotes lateral expansion.

Ethylene-insensitive mutants, such as *etr1* (*et*hylene-*r*esistant *1*) and *ein2* (*e*thylene-*in*sensitive *2*), were identified by their failure to respond to ethylene (as will be described later in the chapter). The *etr1* mutant is unable to perceive the ethylene signal because of a mutation in the gene that codes for the ethylene receptor protein; the *ein2* mutant is blocked at a later step in the signal transduction pathway.

Consistent with a role for ethylene in leaf senescence, both *etr1* and *ein2* were found to be affected not only during the early stages of germination, but throughout the life cycle, including senescence (Zacarias and Reid 1990; Hensel et al. 1993; Grbič and Bleecker 1995). The ethylene mutants retained their chlorophyll and other chloroplast components for a longer period of time compared to the wild type. However, because the total life spans of these mutants were increased by only 30% over that of the wild type, ethylene appears to increase the *rate* of senescence, rather than acting as a developmental switch that initiates the senescence process.

Use of genetic engineering to probe senescence. Another very useful genetic approach that offers direct evidence for the function of specific gene(s) is based on transgenic plants. Through genetic engineering technology, the roles of both ethylene and cytokinins in the regulation of leaf senescence have been confirmed.

One way to suppress the expression of a gene is to transform the plant with antisense DNA, which consists of the gene of interest in the reverse orientation with respect to the promoter. When the antisense gene is transcribed, the resulting antisense mRNA is complementary to the sense mRNA and will hybridize to it. Because double-stranded RNA is rapidly degraded in the cell, the effect of the antisense gene is to deplete the cell of the sense mRNA.

Transgenic plants expressing antisense versions of genes that encode enzymes involved in the ethylene biosynthetic pathway, such as ACC synthase and ACC oxidase, can synthesize ethylene only at very low levels. Consistent with a role for ethylene in senescence, such antisense mutants have been shown to exhibit delayed leaf senescence, as well as fruit ripening, in tomato (see **Web Topic 22.1**).

The Role of Ethylene in Defense Responses Is Complex

Pathogen infection and disease will occur only if the interactions between host and pathogen are genetically compatible. However, ethylene production generally increases in response to pathogen attack in both compatible (i.e., pathogenic) and noncompatible (nonpathogenic) interactions.

The discovery of ethylene-insensitive mutants has allowed the role of ethylene in the response to various pathogens to be assessed. The emerging picture is that the involvement of ethylene in pathogenesis is complex and depends on the particular host–pathogen interaction. For example, blocking the ethylene response does not affect the resistance response to *Pseudomonas* bacteria in *Arabidopsis* or to tobacco mosaic virus in tobacco. In compatible interactions of these pathogens and hosts, however, elimination of ethylene responsiveness prevents the development of disease symptoms, even though the growth of the pathogen appears to be unaffected.

On the other hand, ethylene, in combination with jasmonic acid (see Chapter 13), is required for the activation of several plant defense genes. In addition, ethylene-insensitive tobacco and *Arabidopsis* mutants become susceptible to several necrotrophic (cell-killing) soil fungal pathogens that are normally not plant pathogens. Thus, ethylene appears to be involved in the resistance response to some pathogens, but not others.

Ethylene Biosynthesis in the Abscission Zone Is Regulated by Auxin

The shedding of leaves, fruits, flowers, and other plant organs is termed **abscission** (see **Web Topic 22.4**). Abscission takes place in specific layers of cells, called **abscission layers**, which become morphologically and biochemically differentiated during organ development. Weakening of the cell walls at the abscission layer depends on cell wall–degrading enzymes such as cellulase and polygalacturonase (Figure 22.9).

used extensively to increase the longevity of cut carnations and several other flowers. The potent inhibitor AVG retards fruit ripening and flower fading, but its commercial use has not yet been approved by regulatory agencies. The strong,

$$Cl—CH_2—CH_2—\underset{O^-}{\overset{O}{\overset{\|}{\underset{|}{P}}}}—OH + OH^- \longrightarrow CH_2{=}CH_2 + H_2PO_4^- + Cl^-$$

2-Chloroethylphosphonic acid (ethephon) **Ethylene**

offensive odor of *trans*-cyclooctene precludes its use in agriculture. Currently, 1-methylcyclopropene (MCP) is being developed for use in a variety of postharvest applications.

The near future may see a variety of agriculturally important species that have been genetically modified to manipulate the biosynthesis of ethylene or its perception. The inhibition of ripening in tomato by expression of an antisense version of ACC synthase and ACC oxidase has already been mentioned. Another example of this technology is in petunia, in which ethylene biosynthesis has been blocked by transformation of an antisense version of ACC oxidase. Senescence and petal wilting of cut flowers are delayed for weeks in these transgenic plants.

CELLULAR AND MOLECULAR MODES OF ETHYLENE ACTION

Despite the broad range of ethylene's effects on development, the primary steps in ethylene action are assumed to be similar in all cases: They all involve binding to a receptor, followed by activation of one or more signal transduction pathways (see Chapter 14 on the web site) leading to the cellular response. Ultimately, ethylene exerts its effects primarily by altering the pattern of gene expression. In recent years, remarkable progress has been made in our understanding of ethylene perception, as the result of molecular genetic studies of *Arabidopsis thaliana.*

One key to the elucidation of ethylene signaling components has been the use of the triple-response morphology of etiolated *Arabidopsis* seedlings to isolate mutants affected in their response to ethylene (see Figure 22.7) (Guzman and Ecker 1990). Two classes of mutants have been identified by experiments in which mutagenized *Arabidopsis* seeds were grown on an agar medium in the presence or absence of ethylene for 3 days in the dark:

1. Mutants that fail to respond to exogenous ethylene (ethylene-resistant or ethylene-insensitive mutants)
2. Mutants that display the response even in the absence of ethylene (constitutive mutants)

Ethylene-insensitive mutants are identified as tall seedlings extending above the lawn of short, triple-responding seedlings when grown in the presence of ethylene. Conversely, constitutive ethylene response mutants are identified as seedlings displaying the triple response in the absence of exogenous ethylene.

Ethylene Receptors Are Related to Bacterial Two-Component System Histidine Kinases

The first ethylene-insensitive mutant isolated was *etr1* (*e*thylene-*r*esistant *1*) (Figure 22.12). The *etr1* mutant was identified in a screen for mutations that block the response of *Arabidopsis* seedlings to ethylene. The amino acid sequence of the carboxy-terminal half of *ETR1* is similar to bacterial two-component histidine kinases—receptors used by bacteria to perceive various environmental cues, such as chemo-sensory stimuli, phosphate availability, and osmolarity.

Bacterial two-component systems consist of a sensor histidine kinase and a response regulator, which often acts as a transcription factor (see Chapter 14 on the web site). ETR1 was the first example of a eukaryotic histidine kinase,

FIGURE 22.12 Screen for the *etr1* mutant of *Arabidopsis.* Seedlings were grown for 3 days in the dark in ethylene. Note that all but one of the seedlings are exhibiting the triple response: exaggeration in curvature of the apical hook, inhibition and radial swelling of the hypocotyl, and horizontal growth. The *etr1* mutant is completely insensitive to the hormone and grows like an untreated seedling. (Photograph by K. Stepnitz of the MSU/DOE Plant Research Laboratory.)

but others have since been found in yeast, mammals, and plants. Both phytochrome (see Chapter 17) and the cytokinin receptor (see Chapter 21) also share sequence similarity to bacterial two-component histidine kinases.

The similarity to bacterial receptors and the ethylene insensitivity of the *etr1* mutants suggested that ETR1 might be an ethylene receptor. Consistent with this hypothesis, *ETR1* expression in yeast conferred the ability to bind radiolabeled ethylene with an affinity that closely parallels the dose-response curve of *Arabidopsis* seedlings to ethylene (see **Web Topic 22.5**).

The *Arabidopsis* genome encodes four additional proteins similar to ETR1 that also function as ethylene receptors: ETR2, ERS1 (*E*TR1-*r*elated *s*equence *1*), ERS2, and EIN4 (Figure 22.13). Like ETR1, these receptors have been shown to bind ethylene, and missense mutations in the genes that encode these proteins, analogous to the original *etr1* mutation, prevent ethylene binding to the receptor while allowing the receptor to function normally as a regulator of the ethylene response pathway in the absence of ethylene.

All of these proteins share at least two domains:

1. The amino-terminal domain spans the membrane at least three times and contains the ethylene-binding site. Ethylene can readily access this site because of its hydrophobicity.
2. The middle portion of the ethylene receptors contains a histidine kinase catalytic domain.

A subset of the ethylene receptors also have a carboxy-terminal domain that is similar to bacterial two-component receiver domains. In other two-component systems, binding of ligand regulates the activity of the histidine kinase domain, which autophosphorylates a conserved histidine residue. The phosphate is then transferred to an aspartic acid residue located within the fused receiver domain.Although histidine kinase activity has been demonstrated for one of the ethylene receptors—ETR1—several others are missing critical amino acids, making it unlikely that they possess histidine kinase activity. Thus the biochemical mechanism of these ethylene receptors is not known.

Recent studies indicate that ETR1 is located on the *endoplasmic reticulum*, rather than on the plasma membrane as originally assumed. Such an intracellular location for the ethylene receptor is consistent with the hydrophobic nature of ethylene, which enables it to pass freely through the plasma membrane into the cell. In this respect ethylene is similar to the hydrophobic signaling molecules of animals, such as steroids and the gas nitric oxide, which also bind to intracellular receptors.

High-Affinity Binding of Ethylene to Its Receptor Requires a Copper Cofactor

Even prior to the identification of its receptor, scientists had predicted that ethylene would bind to its receptor via a transition metal cofactor, most likely copper or zinc. This prediction was based on the high affinity of olefins, such as ethylene, for these transition metals. Recent genetic and biochemical studies have borne out these predictions.

Analysis of the ETR1 ethylene receptor expressed in yeast demonstrated that a copper ion was coordinated to the protein and that this copper was necessary for high-affinity ethylene binding (Rodriguez et al. 1999). Silver ion could substitute for copper to yield high-affinity binding, which indicates that silver blocks the action of ethylene not by interfering with ethylene binding, but by preventing the changes in the protein that normally occur when ethylene binds to the receptor.

Evidence that copper binding is required for ethylene receptor function in vivo came from identification of the *RAN1* gene in *Arabidopsis* (Hirayama et al. 1999). Strong *ran1* mutations block the formation of functional ethylene receptors (Woeste and Kieber 2000). Cloning of *RAN1* revealed that it encodes a protein similar to a yeast protein required for the transfer of a copper ion cofactor to an iron transport protein. In an analogous manner, RAN1 is likely to be involved in the addition of a copper ion cofactor necessary for the function of the ethylene receptors.

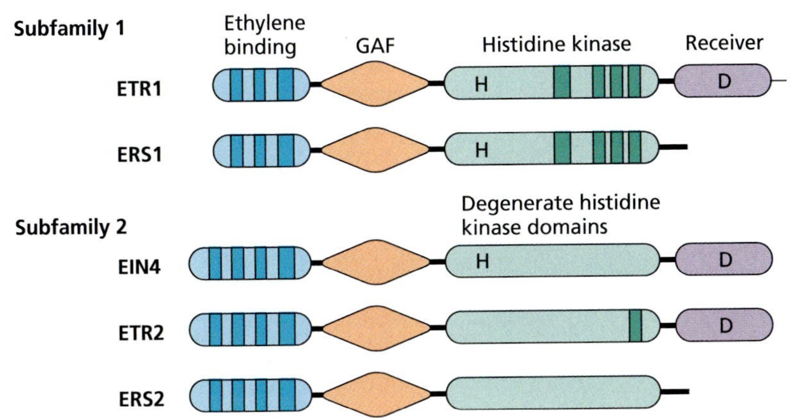

FIGURE 22.13 Schematic diagram of five ethylene receptor proteins and their functional domains. The GAF domain is a conserved cGMP-binding domain found in a diverse group of proteins. Note that EIN4, ETR2, and ERS2 have degenerate histidine kinase domains.

(A)

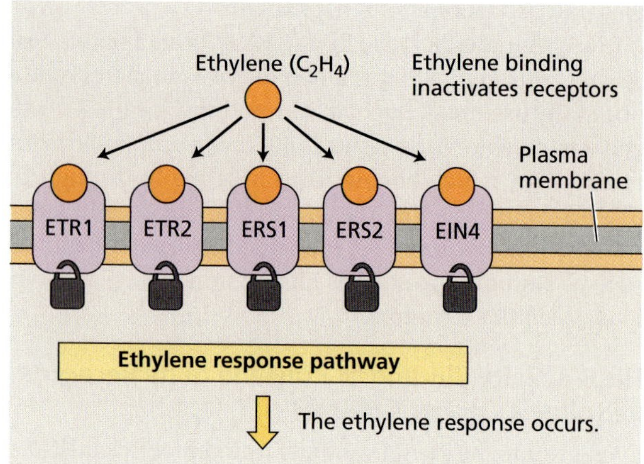

(B)

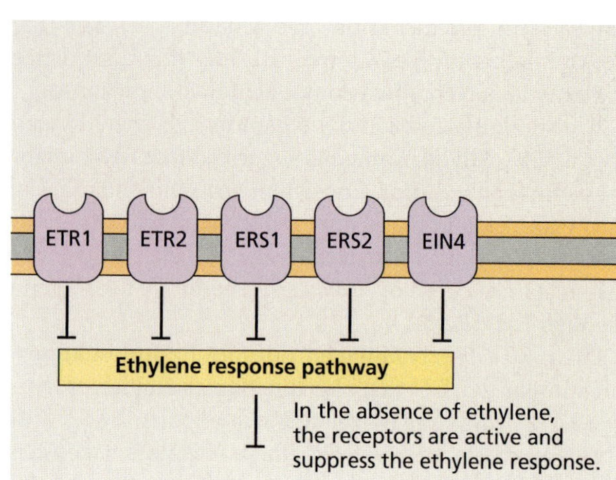

(C)

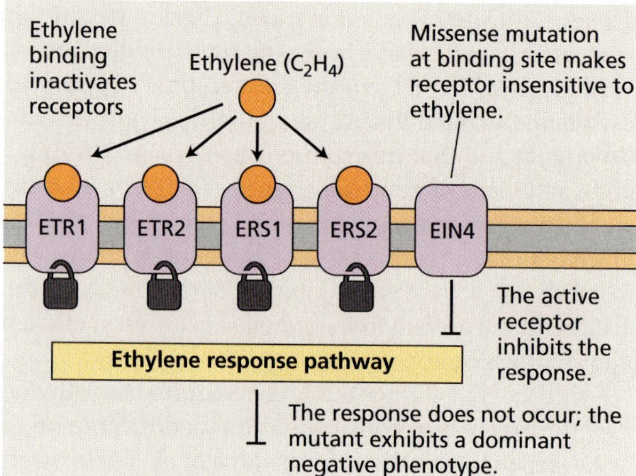

(D)

Disruptions in the regulatory domains of multiple ethylene receptors (at least three)

ETR1 ETR2 ERS1 ERS2 EIN4

Disrupted receptors are inactive in the presence or absence of ethylene.

Ethylene response pathway

The ethylene response occurs.

FIGURE 22.14 Model for ethylene receptor action based on the phenotype of receptor mutants. (A) In the wild type, ethylene binding inactivates the receptors, allowing the response to occur. (B) In the absence of ethylene the receptors act as negative regulators of the response pathway. (C) A missense mutation that interferes with ethylene binding to its receptor, but leaves the regulatory site active, results in a dominant negative phenotype. (D) Disruption mutations in the regulatory sites result in a constitutive ethylene response.

Unbound Ethylene Receptors Are Negative Regulators of the Response Pathway

In *Arabidopsis*, tomato, and probably most other plant species, the ethylene receptors are encoded by multigene families. Targeted disruption (complete inactivation) of the five *Arabidopsis* ethylene receptors (ETR1, ETR2, ERS1, ERS2, and EIN4) has revealed that they are functionally redundant (Hua and Meyerowitz 1998). That is, disruption of any single gene encoding one of these proteins has no effect, but a plant with disruptions in all five receptor genes exhibits a constitutive ethylene response phenotype (Figure 22.14D).

The observation that ethylene responses, such as the triple response, become constitutive when the receptors are disrupted indicates that the receptors are normally "on" (i.e., in the active state) in the *absence* of ethylene, and that the function of the receptor *minus* its ligand (ethylene), is to *shut off* the signaling pathway that leads to the response (Figure 22.14B). Binding of ethylene turns off the receptors, thus allowing the response pathway to proceed (Figure 22.14A).

This somewhat counterintuitive model for ethylene receptors as negative regulators of a signaling pathway is unlike the mechanism of most animal receptors, which, after binding their ligands, serve as positive regulators of their respective signal transduction pathways.

In contrast to the disrupted receptors, receptors with missense mutations at the ethylene binding site (as occurs in the original *etr1* mutant) are unable to bind ethylene, but are still active as negative regulators of the ethylene response pathway. Such missense mutations result in a plant that expresses a subset of receptors that can no longer be turned off by ethylene, and thus confer a *dominant ethylene-insensitive phenotype* (Figure 22.14C). Even though the normal receptors can all be turned off by ethylene, the mutant receptors continue to signal the cell to suppress ethylene responses whether ethylene is present or not.

A Serine/Threonine Protein Kinase Is Also Involved in Ethylene Signaling

The recessive *ctr1* (*c*onstitutive *t*riple *r*esponse *1* = triple response in the absence of ethylene) mutation was identified in screens for mutations that constitutively activated ethylene responses (Figure 22.15). The fact that the mutation caused an *activation* of the ethylene response suggests that the wild-type protein also acts as a *negative regulator* of the response pathway (Kieber et al. 1993), similar to the ethylene receptors.

CTR1 appears to be related to RAF-1, a MAPKKK serine/threonine protein kinase (*m*itogen-*a*ctivated *p*rotein *k*inase *k*inase *k*inase) that is involved in the transduction of various external regulatory signals and developmental signaling pathways in organisms ranging from yeast to humans (see Chapter 14 on the web site). In animal cells, the final product in the MAP kinase cascade is a phosphorylated transcription factor that regulates gene expression in the nucleus.

EIN2 Encodes a Transmembrane Protein

The *ein2* (*e*thylene-*in*sensitive *2*) mutation blocks all ethylene responses in both seedling and adult *Arabidopsis* plants. The *EIN2* gene encodes a protein containing 12 membrane-spanning domains that is most similar to the N-RAMP (*n*atural *r*esistance–*a*ssociated *m*acrophage *p*rotein) family of cation transporters in animals (Alonso et al. 1999), suggesting that it may act as a channel or pore. To date, however, researchers have failed to demonstrate a transport activity for this protein, and the intracellular location of the protein is not known.

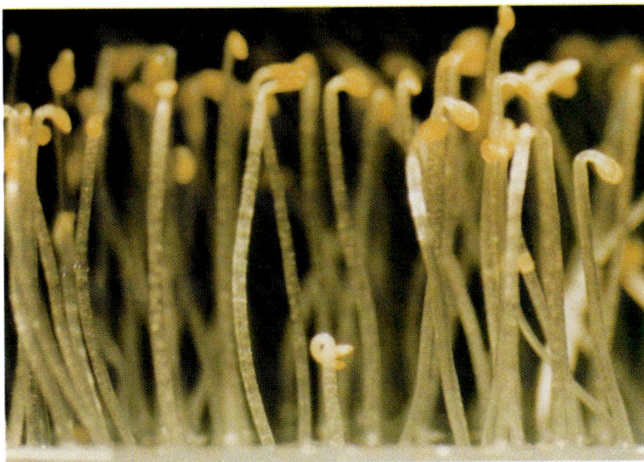

FIGURE 22.15 Screen for *Arabidopsis* mutants that constitutively display the triple response. Seedlings were grown for 3 days in the dark in air. A single *ctr1* mutant seedling is evident among the taller, wild-type seedlings. (Courtesy of J. Kieber.)

Interestingly, mutations in the *EIN2* gene have also been identified in genetic screens for resistance to other hormones, such as jasmonic acid and ABA, suggesting that EIN2 may be a common intermediate in the signal transduction pathways of various hormones and other chemical signals.

Ethylene Regulates Gene Expression

One of the primary effects of ethylene signaling is an alteration in the expression of various target genes. Ethylene affects the mRNA transcript levels of numerous genes, including the genes that encode cellulase, as well as ripening-related genes and ethylene biosynthesis genes. Regulatory sequences called **ethylene response elements,** or **EREs,** have been identified from the ethylene-regulated genes.

Key components mediating ethylene's effects on gene expression are the EIN3 family of transcription factors (Chao et al. 1997). There are at least four *EIN3*-like genes in *Arabidopsis*, and homologs have been identified in both tomato and tobacco. In response to an ethylene signal, homodimers of EIN3 or its paralogs (closely related proteins), bind to the promoter of a gene called *ERF1* (*e*thylene *r*esponse *f*actor *1*) and activate its transcription (Solano et al. 1998).

ERF1 encodes a protein that belongs to the **ERE-binding protein** (**EREBP**) family of transcription factors, which were first identified in tobacco as proteins that bind to ERE sequences (Ohme-Takagi and Shinshi 1995). Several EREBPs are rapidly up-regulated in response to ethylene. The EREBP genes exist in *Arabidopsis* as a very large gene family, but only a few of the genes are inducible by ethylene.

Genetic Epistasis Reveals the Order of the Ethylene Signaling Components

The order of action of the genes *ETR1*, *EIN2*, *EIN3*, and *CTR1* has been determined by the analysis of how the mutations interact with each other (i.e., their epistatic order). Two mutants with opposite phenotypes are crossed, and a line harboring both mutations (the double mutant) is identified in the F_2 generation. In the case of the ethylene response mutants, researchers constructed a line doubly mutant for *ctr1*, a constitutive ethylene response mutant, and one of the ethylene-insensitive mutations.

The phenotype that the double mutant displays reveals which of the mutations is epistatic to the other. For example, if an *etr1/ctr1* double mutant displays a *ctr1* mutant phenotype, the *ctr1* mutation is said to be epistatic to *etr1*. From this it can be inferred that CTR1 acts downstream of

ETR1 (Avery and Wasserman 1992). In this way, the order of action of *ETR1*, *EIN2*, and *EIN3* were determined relative to *CTR1*.

The ETR1 protein has been shown to interact physically with the predicted downstream protein, CTR1, suggesting that the ethylene receptors may directly regulate the kinase activity of CTR1 (Clark et al. 1998). The model in Figure 22.16 summarizes these and other data. Genes that are similar to several of these *Arabidopsis* signaling genes have been found in other species (see **Web Topic 22.6**).

This model is still incomplete because other ethylene response mutations have been identified that act in this pathway. In addition, we are only beginning to understand the biochemical properties of these proteins and how they interact. However, we are beginning to glimpse the outline of the molecular basis for the perception and transduction of this hormonal signal.

SUMMARY

Ethylene is formed in most organs of higher plants. Senescing tissues and ripening fruits produce more ethylene than do young or mature tissues. The precursor of ethylene in vivo is the amino acid methionine, which is converted to AdoMet (*S*-adenosylmethionine), ACC (1-aminocyclopropane-1-carboxylic acid), and ethylene. The rate-limiting step of this pathway is the conversion of AdoMet to ACC, which is catalyzed by ACC synthase. ACC synthase is encoded by members of a multigene family that are differentially regulated in various plant tissues and in response to various inducers of ethylene biosynthesis.

Ethylene biosynthesis is triggered by various developmental processes, by auxins, and by environmental stresses. In all these cases the level of activity and of mRNA of ACC synthase increases. The physiological effects of ethylene can

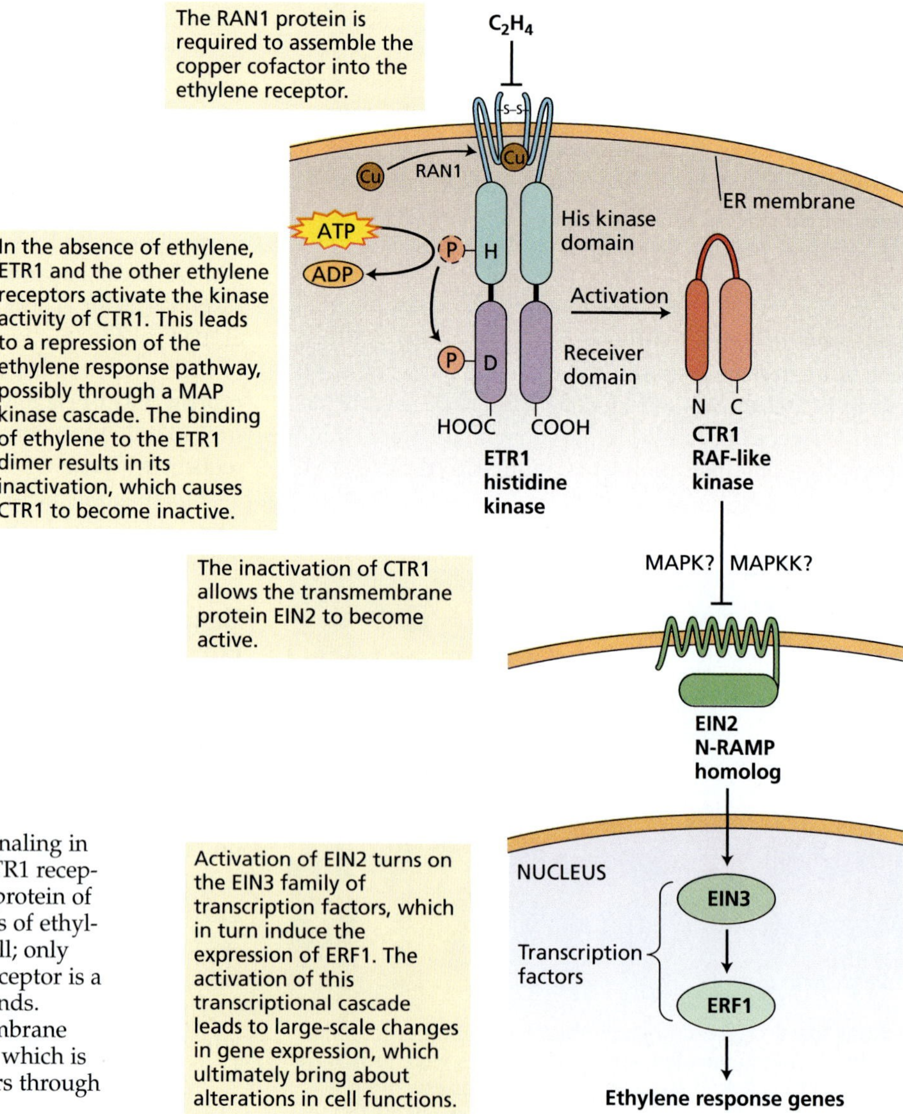

FIGURE 22.16 Model of ethylene signaling in *Arabidopsis*. Ethylene binds to the ETR1 receptor, which is an integral membrane protein of the ER membrane. Multiple isoforms of ethylene receptors may be present in a cell; only ETR1 is shown for simplicity. The receptor is a dimer, held together by disulfide bonds. Ethylene binds within the trans-membrane domain, through a copper co-factor, which is assembled into the ethylene receptors through the RAN1 protein.

be blocked by biosynthesis inhibitors or by antagonists. AVG (aminoethoxy-vinylglycine) and AOA (aminooxy-acetic acid) inhibit the synthesis of ethylene; carbon dioxide, silver ions, *trans*-cyclooctene, and MCP inhibit ethylene action. Ethylene can be detected and measured by gas chromatography.

Ethylene regulates fruit ripening and other processes associated with leaf and flower senescence, leaf and fruit abscission, root hair development, seedling growth, and hook opening. Ethylene also regulates the expression of various genes, including ripening-related genes and pathogenesis-related genes.

The ethylene receptor is encoded by a family of genes that encode proteins similar to bacterial two-component histidine kinases. Ethylene binds to these receptors in a transmembrane domain through a copper cofactor. Downstream signal transduction components include CTR1, a member of the RAF family of protein kinases; and EIN2, a channel-like transmembrane protein. The pathway activates a cascade of transcription factors, including the EIN3 and EREBP families, which then modulate gene expression.

Web Material

Web Topics

22.1 Cloning of ACC Synthase

A brief description of the cloning of the gene for ACC synthase using antibodies raised against the partially purified protein.

22.2 Cloning of the ACC Oxidase Gene

The ACC oxidase gene was cloned by a circuitous route using antisense DNA.

22.3 ACC Synthase Gene Expression and Biotechnology

A discussion of the use of the ACC synthase gene in biotechnology.

22.4 Abscission and the Dawn of Agriculture

A short essay on the domestication of modern cereals based on artificial selection for non-shattering rachises.

22.5 Ethylene Binding to ETR1 and Seedling Response to Ethylene

Ethylene-binding to its receptor ETR1 was first demonstrated by expressing the gene in yeast.

22.6 Conservation of Ethylene Signaling Components in Other Plant Species

The evidence suggests that ethylene signaling is similar in all plant species.

Chapter References

Abeles, F. B., Morgan, P. W., and Saltveit, M. E., Jr. (1992) *Ethylene in Plant Biology*, 2nd ed. Academic Press, San Diego.

Alonso, J. M., Hirayama, T., Roman, G., Nourizadeh, S., and Ecker, J. R. (1999) EIN2, a bifunctional transducer of ethylene and stress responses in *Arabidopsis*. *Science* 284: 2148–2152.

Avery, L., and Wasserman, S. (1992) Ordering gene function: The interpretation of epistasis in regulatory hierarchies. *Trends Genet.* 8: 312–316.

Bradford, K. J., and Yang, S. F. (1980) Xylem transport of 1-aminocyclopropane-1-carboxylic acid, an ethylene precursor, in waterlogged tomato plants. *Plant Physiol.* 65: 322–326.

Burg, S. P., and Burg, E. A. (1965) Relationship between ethylene production and ripening in bananas. *Bot. Gaz.* 126: 200–204.

Burg, S. P., and Thimann, K. V. (1959) The physiology of ethylene formation in apples. *Proc. Natl. Acad. Sci. USA* 45: 335–344.

Chao, Q., Rothenberg, M., Solano, R., Roman, G., Terzaghi, W., and Ecker, J. R. (1997) Activation of the ethylene gas response pathway in Arabidopsis by the nuclear protein ETHYLENE-INSENSITIVE3 and related proteins. *Cell* 89: 1133–1144.

Clark, K. L., Larsen, P. B., Wang, X., and Chang, C. (1998) Association of the *Arabidopsis* CTR1 Raf-like kinase with the ETR1 and ERS ethylene receptors. *Proc. Natl. Acad. Sci. USA* 95: 5401–5406.

Dolan, L., Duckett, C. M., Grierson, C., Linstead, P., Schneider, K., Lawson, E., Dean, C., Poethig, S., and Roberts, K. (1994) Clonal relationships and cell patterning in the root epidermis of *Arabidopsis*. *Development* 120: 2465–2474.

Gray, J. E., Picton, S., Giovannoni, J. J., and Grierson, D. (1994) The use of transgenic and naturally occurring mutants to understand and manipulate tomato fruit ripening. *Plant Cell Environ.* 17: 557–571.

Grbić, V., and Bleecker, A. B. (1995) Ethylene regulates the timing of leaf senescence in *Arabidopsis*. *Plant J.* 8: 595–602.

Guzman, P., and Ecker, J. R. (1990) Exploiting the triple response of *Arabidopsis* to identify ethylene-related mutants. *Plant Cell* 2: 513–523.

Hensel, L. L., Grbić, V., Baumgarten, D. A., and Bleecker, A. B. (1993) Developmental and age-related processes that influence the longevity and senescence of photosynthetic tissues in *Arabidopsis*. *Plant Cell* 5: 553–564.

Hirayama, T., Kieber, J. J., Hirayama, N., Kogan, M., Guzman, P., Nourizadeh, S., Alonso, J. M., Dailey, W. P., Dancis, A., and Ecker, J. R. (1999) *RESPONSIVE-TO-ANTAGONIST1*, a Menkes/Wilson disease-related copper transporter, is required for ethylene signaling in Arabidopsis. *Cell* 97: 383–393.

Hoffman, N. E., and Yang, S. F. (1980) Changes of 1-aminocyclopropane-1-carboxylic acid content in ripening fruits in relation to their ethylene production rates. *J. Amer. Soc. Hort. Sci.* 105: 492–495.

Hua, J., and Meyerowitz, E. M. (1998) Ethylene responses are negatively regulated by a receptor gene family in *Arabidopsis thaliana*. *Cell* 94: 261–271.

Kieber, J. J., Rothenburg, M., Roman, G., Feldmann, K. A., and Ecker, J. R. (1993) CTR1, a negative regulator of the ethylene response pathway in *Arabidopsis*, encodes a member of the Raf family of protein kinases. *Cell* 72: 427–441.

Lanahan, M., Yen, H.-C., Giovannoni, J., and Klee, H. (1994) The *Never-ripe* mutation blocks ethylene perception in tomato. *Plant Cell* 6: 427–441.

Lehman, A., Black, R., and Ecker, J. R. (1996) *Hookless1*, an ethylene response gene, is required for differential cell elongation in the *Arabidopsis* hook. *Cell* 85: 183–194.

Liang, X., Abel, S., Keller, J., Shen, N., and Theologis, A. (1992) The 1-aminocyclopropane-1-carboxylate synthase gene family of *Arabidopsis thaliana*. *Proc. Natl. Acad. Sci. USA* 89: 11046–11050.

McKeon, T. A., Fernández-Maculet, J. C., and Yang, S. F. (1995) Biosynthesis and metabolism of ethylene. In *Plant Hormones: Physiology, Biochemistry and Molecular Biology*, 2nd ed., P. J. Davies, ed., Kluwer, Dordrecht, Netherlands, pp. 118–139.

Morgan, P. W. (1984) Is ethylene the natural regulator of abscission? In *Ethylene: Biochemical, Physiological and Applied Aspects*, Y. Fuchs and E. Chalutz, eds., Martinus Nijhoff, The Hague, Netherlands, pp. 231–240.

Nakagawa, J. H., Mori, H., Yamazaki, K., and Imaseki, H. (1991) Cloning of the complementary DNA for auxin-induced 1-aminocyclopropane-1-carboxylate synthase and differential expression of the gene by auxin and wounding. *Plant Cell Physiol.* 32: 1153–1163.

Oeller, P., Min-Wong, L., Taylor, L., Pike, D., and Theologis, A. (1991) Reversible inhibition of tomato fruit senescence by antisense RNA. *Science* 254: 437–439.

Ohme-Takagi, M., and Shinshi, H. (1995) Ethylene-inducible DNA binding proteins that interact with an ethylene-responsive element. *Plant Cell* 7: 173–182.

Olson, D. C., White, J. A., Edelman, L., Harkins, R. N., and Kende, H. (1991) Differential expression of two genes for 1-aminocyclopropane-1-carboxylate synthase in tomato fruits. *Proc. Natl. Acad. Sci. USA* 88: 5340–5344.

Perovic, S., Seack, J., Gamulin, V., Müller, W. E. G., and Schröder, H. C. (2001) Modulation of intracellular calcium and proliferative activity of invertebrate and vertebrate cells by ethylene. *BMC Cell Biol.* 2: 7.

Raskin, I., and Beyer, E. M., Jr. (1989) Role of ethylene metabolism in *Amaranthus retroflexus*. *Plant Physiol.* 90: 1–5.

Reid, M. S. (1995) Ethylene in plant growth, development and senescence. In *Plant Hormones: Physiology, Biochemistry and Molecular Biology*, 2nd ed., P. J. Davies, ed., Kluwer, Dordrecht, Netherlands, pp. 486–508.

Rodriguez, F. I., Esch, J. J., Hall, A. E., Binder, B. M., Schaller, E. G., and Bleecker, A. B. (1999) A copper cofactor for the ethylene receptor ETR1 from *Arabidopsis*. *Science* 283: 396–398.

Sexton, R., Burdon, J. N., Reid, J. S. G., Durbin, M. L., and Lewis, L. N. (1984) Cell wall breakdown and abscission. In *Structure, Function, and Biosynthesis of Plant Cell Walls*, W. M. Dugger and S. Bartnicki-Garcia, eds., American Society of Plant Physiologists, Rockville, MD, pp. 383–406.

Sisler, E. C., and Serek, M. (1997) Inhibitors of ethylene responses in plants at the receptor level: Recent developments. *Physiol. Plant.* 100: 577–582.

Sisler, E., Blankenship, S., and Guest, M. (1990) Competition of cyclooctenes and cyclooctadienes for ethylene binding and activity in plants. *Plant Growth Regul.* 9: 157–164.

Solano, R., Stepanova, A., Chao, Q., and Ecker, J. R. (1998) Nuclear events in ethylene signaling: A transcriptional cascade mediated by ETHYLENE-INSENSITIVE3 and ETHYLENE-RESPONSE-FACTOR1. *Gene Dev.* 12: 3703–3714.

Tanimoto, M., Roberts, K., and Dolan, L. (1995) Ethylene is a positive regulator of root hair development in *Arabidopsis thaliana*. *Plant J.* 8: 943–948.

Tatsuki, M., and Mori, H. (2001) Phosphorylation of tomato 1-aminocyclopropane-1-carboxylic acid synthase, LE-ACS2, at the C-terminal region. *J. Biol. Chem.* 276: 28051–28057.

Voesenek, L. A. C. J., Banga, M., Rijnders, J. H. G. M., Visser, E. J. W., Harren, F. J. M., Brailsford, R. W., Jackson, M. B., and Blom, C. W. P. M. (1997) Laser-driven photoacoustic spectroscopy: What we can do with it in flooding research. *Ann. Bot.* 79: 57–65.

Vogel, J. P., Schuerman, P., Woeste, K., Brandstatter, I. Kieber, J. J. (1998) Isolation and characterization of *Arapidopsis* mutants defective in the induction of ethylene biosynthesis by cytokinin. *Genetics* 149: 417–427.

Woeste, K., and Kieber, J. J. (2000) A strong loss-of-function allele of *RAN1* results in constitutive activation of ethylene responses as well as a rosette-lethal phenotype. *Plant Cell* 12: 443–455.

Yang, S. F. (1987) The role of ethylene and ethylene synthesis in fruit ripening. In *Plant Senescence: Its Biochemistry and Physiology*, W. W. Thomson, E. A. Nothnagel, and R. C. Huffaker, eds., American Society of Plant Physiologists, Rockville, MD, pp. 156–166.

Yuan, M., Shaw, P. J., Warn, R. M., and Lloyd, C. W. (1994) Dynamic reorientation of cortical microtubules, from transverse to longitudinal, in living plant cells. *Proc. Natl. Acad. Sci. USA* 91: 6050–6053.

Zacarias, L., and Reid, M. S. (1990) Role of growth regulators in the senescence of *Arabidopsis thaliana* leaves. *Physiol. Plant.* 80: 549–554.

Chapter 23

Abscisic Acid: A Seed Maturation and Antistress Signal

THE EXTENT AND TIMING OF PLANT GROWTH are controlled by the coordinated actions of positive and negative regulators. Some of the most obvious examples of regulated nongrowth are seed and bud dormancy, adaptive features that delay growth until environmental conditions are favorable. For many years, plant physiologists suspected that the phenomena of seed and bud dormancy were caused by inhibitory compounds, and they attempted to extract and isolate such compounds from a variety of plant tissues, especially dormant buds.

Early experiments used paper chromatography for the separation of plant extracts, as well as bioassays based on oat coleoptile growth. These early experiments led to the identification of a group of growth-inhibiting compounds, including a substance known as *dormin* purified from sycamore leaves collected in early autumn, when the trees were entering dormancy. Upon discovery that dormin was chemically identical to a substance that promotes the abscission of cotton fruits, *abscisin II*, the compound was renamed **abscisic acid** (**ABA**) (see Figure 23.1), to reflect its supposed involvement in the abscission process.

It is now known that ethylene is the hormone that triggers abscission and that ABA-induced abscission of cotton fruits is due to ABA's ability to stimulate ethylene production. As will be discussed in this chapter, ABA is now recognized as an important plant hormone in its own right. It inhibits growth and stomatal opening, particularly when the plant is under environmental stress. Another important function is its regulation of seed maturation and dormancy. In retrospect, *dormin* would have been a more appropriate name for this hormone, but the name *abscisic acid* is firmly entrenched in the literature.

OCCURRENCE, CHEMICAL STRUCTURE, AND MEASUREMENT OF ABA

Abscisic acid has been found to be a ubiquitous plant hormone in vascular plants. It has been detected in mosses but appears to be absent in

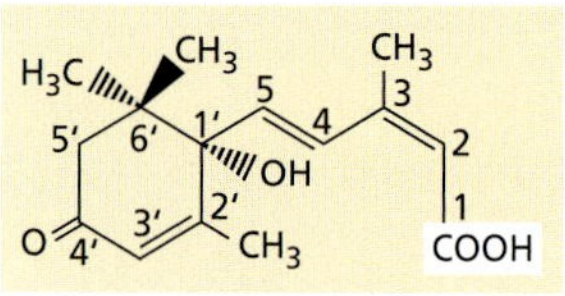

(*S*)-*cis*-ABA (naturally occurring active form)

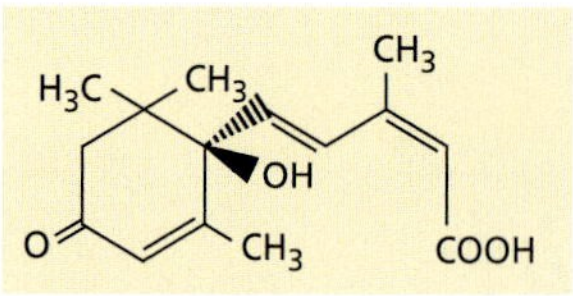

(*R*)-*cis*-ABA (inactive in stomatal closure)

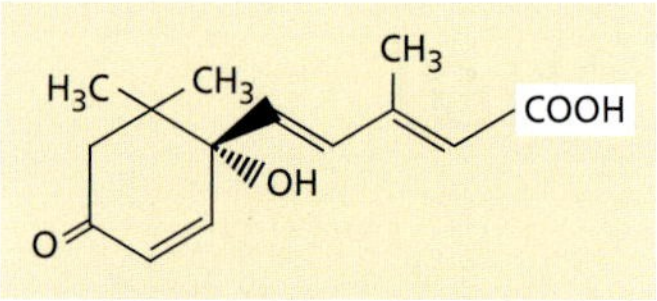

(*S*)-2-*trans*-ABA (inactive, but interconvertible with active [*cis*] form)

FIGURE 23.1 The chemical structures of the *S* (counterclockwise array) and *R* (clockwise array) forms of *cis*-ABA, and the (*S*)-2-*trans* form of ABA. The numbers in the diagram of (*S*)-*cis*-ABA identify the carbon atoms.

liverworts (see **Web Topic 23.1**). Several genera of fungi make ABA as a secondary metabolite (Milborrow 2001). Within the plant, ABA has been detected in every major organ or living tissue from the root cap to the apical bud. ABA is synthesized in almost all cells that contain chloroplasts or amyloplasts.

The Chemical Structure of ABA Determines Its Physiological Activity

ABA is a 15-carbon compound that resembles the terminal portion of some carotenoid molecules (Figure 23.1). The orientation of the carboxyl group at carbon 2 determines the *cis* and *trans* isomers of ABA. Nearly all the naturally occurring ABA is in the *cis* form, and by convention the name *abscisic acid* refers to that isomer.

ABA also has an asymmetric carbon atom at position 1′ in the ring, resulting in the *S* and *R* (or + and –, respectively) enantiomers. The *S* enantiomer is the natural form; commercially available synthetic ABA is a mixture of approximately equal amounts of the *S* and *R* forms. The *S* enantiomer is the only one that is active in fast responses to ABA, such as stomatal closure. In long-term responses, such as seed maturation, both enantiomers are active. In contrast to the *cis* and *trans* isomers, the *S* and *R* forms cannot be interconverted in the plant tissue.

Studies of the structural requirements for biological activity of ABA have shown that almost any change in the molecule results in loss of activity (see **Web Topic 23.2**).

ABA Is Assayed by Biological, Physical, and Chemical Methods

A variety of bioassays have been used for ABA, including inhibition of coleoptile growth, germination, or GA-induced α-amylase synthesis. Alternatively, promotion of stomatal closure and gene expression are examples of rapid inductive responses (see **Web Topic 23.3**).

Physical methods of detection are much more reliable than bioassays because of their specificity and suitability for quantitative analysis. The most widely used techniques are those based on gas chromatography or high-performance liquid chromatography (HPLC). Gas chromatography allows detection of as little as 10^{-13} g ABA, but it requires several preliminary purification steps, including thin-layer chromatography. Immunoassays are also highly sensitive and specific.

BIOSYNTHESIS, METABOLISM, AND TRANSPORT OF ABA

As with the other hormones, the response to ABA depends on its concentration within the tissue and on the sensitivity of the tissue to the hormone. The processes of biosynthesis, catabolism, compartmentation, and transport all contribute to the concentration of active hormone in the tissue at any given stage of development. The complete biosynthetic pathway of ABA was elucidated with the aid of ABA-deficient mutants blocked at specific steps in the pathway.

ABA Is Synthesized from a Carotenoid Intermediate

ABA biosynthesis takes place in chloroplasts and other plastids via the pathway depicted in Figure 23.2. Several ABA-deficient mutants have been identified with lesions at specific steps of the pathway. These mutants exhibit abnormal phenotypes that can be corrected by the application of exogenous ABA. For example, *flacca* (*flc*) and *sitiens* (*sit*) are "wilty mutants" of tomato in which the tendency of the leaves to wilt (due to an inability to close their stomata) can be prevented by the application of exogenous ABA. The *aba* mutants of *Arabidopsis* also exhibit a wilty phenotype. These and other mutants have been useful in elucidating the details of the pathway (Milborrow 2001).

The pathway begins with isopentenyl diphosphate (IPP), the biological isoprene unit, and leads to the synthesis of the C_{40} xanthophyll (i.e., oxygenated carotenoid) **violaxanthin** (see Figure 23.2). Synthesis of violaxanthin is catalyzed by zeaxanthin epoxidase (ZEP), the enzyme encoded by the *ABA1* locus of *Arabidopsis*. This discovery provided conclusive evidence that ABA synthesis occurs via the "indirect" or carotenoid pathway, rather than as a small molecule. Maize mutants (*vp*) that are blocked at other steps in the carotenoid pathway also have reduced levels of ABA and exhibit **vivipary**—the precocious germination of seeds in the fruit while still attached to the plant (Figure 23.3). Vivipary is a feature of many ABA-deficient seeds.

Violaxanthin is converted to the C_{40} compound **9′-*cis*-neoxanthin**, which is then cleaved to form the C_{15} com-

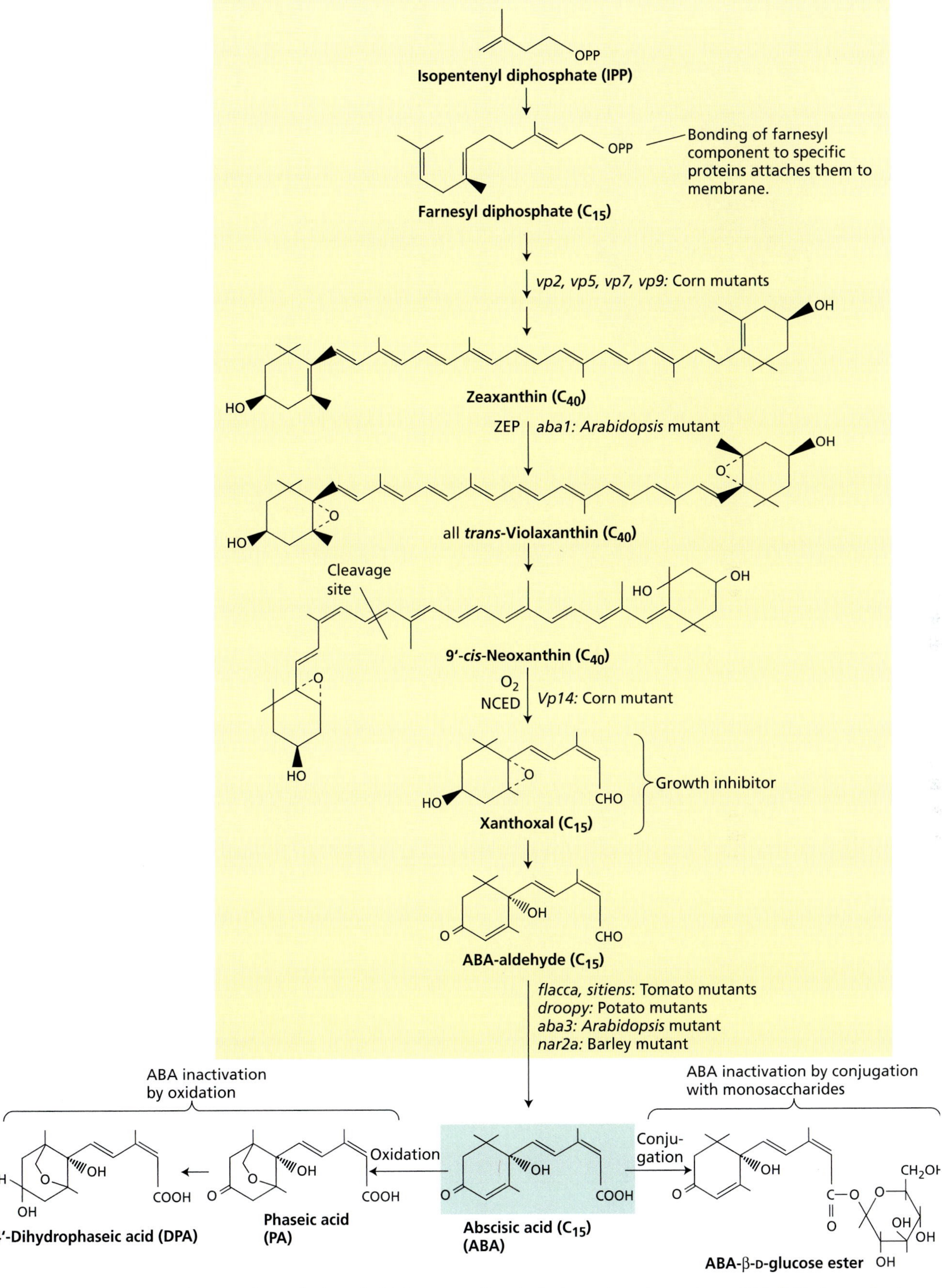

FIGURE 23.2 ABA biosynthesis and metabolism. In higher plants, ABA is synthesized via the terpenoid pathway (see Chapter 13). Some ABA-deficient mutants that have been helpful in elucidating the pathway are shown at the steps at which they are blocked. The pathways for ABA catabolism include conjugation to form ABA-β-D-glucosyl ester or oxidation to form phaseic acid and then dihydrophaseic acid. ZEP = zeaxanthin epoxidase; NCED = 9-cis-epoxycarotenoids dioxygenase.

FIGURE 23.3 Precocious germination in the ABA-deficient *vp14* mutant of maize. The VP14 protein catalyzes the cleavage of 9-*cis*-epoxycarotenoids to form xanthoxal, a precursor of ABA. (Courtesy of Bao Cai Tan and Don McCarty.)

pound **xanthoxal**, previously called *xanthoxin*, a neutral growth inhibitor that has physiological properties similar to those of ABA. The cleavage is catalyzed by **9-*cis*-epoxycarotenoid dioxygenase** (**NCED**), so named because it can cleave both 9-*cis*-violaxanthin and 9′-*cis*-neoxanthin.

Synthesis of NCED is rapidly induced by water stress, suggesting that the reaction it catalyzes is a key regulatory step for ABA synthesis. The enzyme is localized on the thylakoids, where the carotenoid substrate is located. Finally, xanthoxal is converted to ABA via oxidative steps involving the intermediate(s) **ABA-aldehyde** and/or possibly xanthoxic acid. This final step is catalyzed by a family of aldehyde oxidases that all require a molybdenum cofactor; the *aba3* mutants of *Arabidopsis* lack a functional molybdenum cofactor and are therefore unable to synthesize ABA.

ABA Concentrations in Tissues Are Highly Variable

ABA biosynthesis and concentrations can fluctuate dramatically in specific tissues during development or in response to changing environmental conditions. In developing seeds, for example, ABA levels can increase 100-fold within a few days and then decline to vanishingly low levels as maturation proceeds. Under conditions of water stress, ABA in the leaves can increase 50-fold within 4 to 8 hours. Upon rewatering, the ABA level declines to normal in the same amount of time.

Biosynthesis is not the only factor that regulates ABA concentrations in the tissue. As with other plant hormones, the concentration of free ABA in the cytosol is also regulated by degradation, compartmentation, conjugation, and transport. For example, cytosolic ABA increases during water stress as a result of synthesis in the leaf, redistribution within the mesophyll cell, import from the roots, and recirculation from other leaves. The concentration of ABA declines after rewatering because of degradation and export from the leaf, as well as a decrease in the rate of synthesis.

ABA Can Be Inactivated by Oxidation or Conjugation

A major cause of the inactivation of free ABA is oxidation, yielding the unstable intermediate 6-hydroxymethyl ABA, which is rapidly converted to **phaseic acid** (**PA**) and **dihydrophaseic acid** (**DPA**) (see Figure 23.2). PA is usually inactive, or it exhibits greatly reduced activity, in bioassays. However, PA can induce stomatal closure in some species, and it is as active as ABA in inhibiting gibberellic acid–induced α-amylase production in barley aleurone layers. These effects suggest that PA may be able to bind to ABA receptors. In contrast to PA, DPA has no detectable activity in any of the bioassays tested.

Free ABA is also inactivated by covalent conjugation to another molecule, such as a monosaccharide. A common example of an ABA conjugate is **ABA-β-D-glucosyl ester** (**ABA-GE**). Conjugation not only renders ABA inactive as a hormone; it also alters its polarity and cellular distribution. Whereas free ABA is localized in the cytosol, ABA-GE accumulates in vacuoles and thus could theoretically serve as a storage form of the hormone.

Esterase enzymes in plant cells could release free ABA from the conjugated form. However, there is no evidence that ABA-GE hydrolysis contributes to the rapid increase in ABA in the leaf during water stress. When plants were subjected to a series of stress and rewatering cycles, the ABA-GE concentration increased steadily, suggesting that the conjugated form is not broken down during water stress.

ABA Is Translocated in Vascular Tissue

ABA is transported by both the xylem and the phloem, but it is normally much more abundant in the phloem sap. When radioactive ABA is applied to a leaf, it is transported both up the stem and down toward the roots. Most of the radioactive ABA is found in the roots within 24 hours. Destruction of the phloem by a stem girdle prevents ABA accumulation in the roots, indicating that the hormone is transported in the phloem sap.

ABA synthesized in the roots can also be transported to the shoot via the xylem. Whereas the concentration of ABA in the xylem sap of well-watered sunflower plants is between 1.0 and 15.0 n*M*, the ABA concentration in water-stressed sunflower plants increases to as much as 3000 n*M* (3.0 μ*M*) (Schurr et al. 1992). The magnitude of the stress-induced change in xylem ABA content varies widely among species, and it has been suggested that ABA also is transported in a conjugated form, then released by hydrolysis in leaves. However, the postulated hydrolases have yet to be identified.

As water stress begins, some of the ABA carried by the xylem stream is synthesized in roots that are in direct contact with the drying soil. Because this transport can occur before the low water potential of the soil causes any measurable change in the water status of the leaves, ABA is believed to be a root signal that helps reduce the transpiration rate by closing stomata in leaves (Davies and Zhang 1991).

Although a concentration of 3.0 μM ABA in the apoplast is sufficient to close stomata, not all of the ABA in the xylem stream reaches the guard cells. Much of the ABA in the transpiration stream is taken up and metabolized by the mesophyll cells. During the early stages of water stress, however, the pH of the xylem sap becomes more alkaline, increasing from about pH 6.3 to about pH 7.2 (Wilkinson and Davies 1997).

The major control of ABA distribution among plant cell compartments follows the "anion trap" concept: The dissociated (anion) form of this weak acid accumulates in alkaline compartments and may be redistributed according to the steepness of the pH gradients across membranes. In addition to partitioning according to the relative pH of compartments, specific uptake carriers contribute to maintaining a low apoplastic ABA concentration in unstressed plants.

Stress-induced alkalinization of the apoplast favors formation of the dissociated form of abscisic acid, ABA^-, which does not readily cross membranes. Hence, less ABA enters the mesophyll cells, and more reaches the guard cells via the transpiration stream (Figure 23.4). Note that ABA is redistributed in the leaf in this way without any increase in the total ABA level. This increase in xylem sap pH may function as a root signal that promotes early closure of the stomata.

DEVELOPMENTAL AND PHYSIOLOGICAL EFFECTS OF ABA

Abscisic acid plays primary regulatory roles in the initiation and maintenance of seed and bud dormancy and in the plant's response to stress, particularly water stress. In addition, ABA influences many other aspects of plant development by interacting, usually as an antagonist, with auxin, cytokinin, gibberellin, ethylene, and brassinosteroids. In this section we will explore the diverse physiological effects of ABA, beginning with its role in seed development.

ABA Levels in Seeds Peak during Embryogenesis

Seed development can be divided into three phases of approximately equal duration:

1. During the first phase, which is characterized by cell divisions and tissue differentiation, the zygote undergoes embryogenesis and the endosperm tissue proliferates.

2. During the second phase, cell divisions cease and storage compounds accumulate.

3. In the final phase, the embryo becomes tolerant to desiccation, and the seed dehydrates, losing up to 90% of its water. As a consequence of dehydration, metabolism comes to a halt and the seed enters a **quiescent** ("resting") state. In contrast to dormant seeds, quiescent seeds will germinate upon rehydration.

The latter two phases result in the production of viable seeds with adequate resources to support germination and

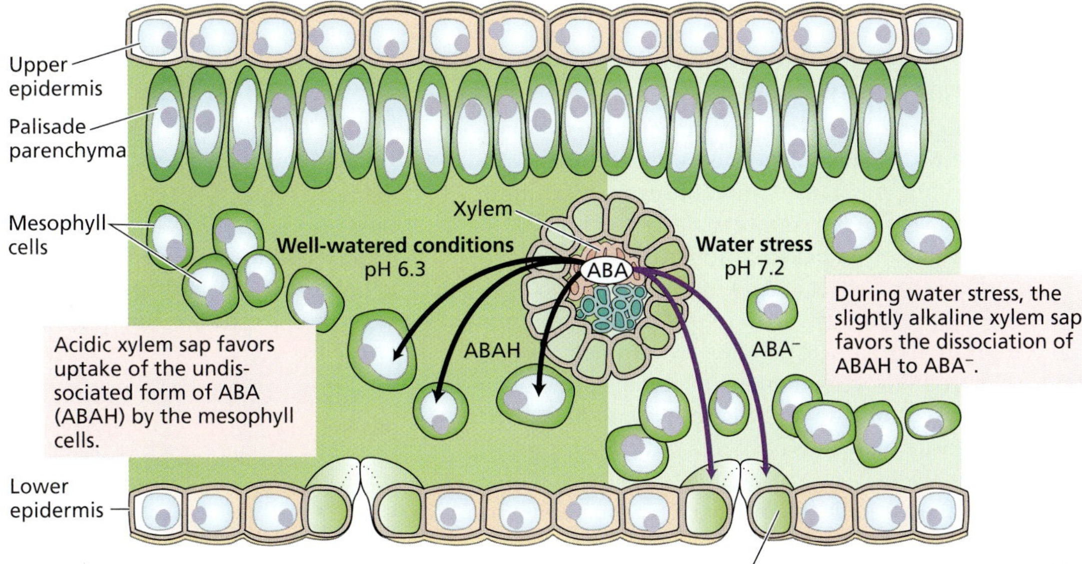

FIGURE 23.4 Redistribution of ABA in the leaf resulting from alkalinization of the xylem sap during water stress.

the capacity to wait weeks to years before resuming growth. Typically, the ABA content of seeds is very low early in embryogenesis, reaches a maximum at about the halfway point, and then gradually falls to low levels as the seed reaches maturity. Thus there is a broad peak of ABA accumulation in the seed corresponding to mid- to late embryogenesis.

The hormonal balance of seeds is complicated by the fact that not all the tissues have the same genotype. The seed coat is derived from maternal tissues (see **Web Topic 1.2**); the zygote and endosperm are derived from both parents. Genetic studies with ABA-deficient mutants of *Arabidopsis* have shown that the zygotic genotype controls ABA synthesis in the embryo and endosperm and is essential to dormancy induction, whereas the maternal genotype controls the major, early peak of ABA accumulation and helps suppress vivipary in midembryogenesis (Raz et al. 2001).

ABA Promotes Desiccation Tolerance in the Embryo

An important function of ABA in the developing seed is to promote the acquisition of desiccation tolerance. As will be described in Chapter 25 (on stress physiology), desiccation can severely damage membranes and other cellular constituents. During the mid- to late stages of seed development, specific mRNAs accumulate in embryos at the time of high levels of endogenous ABA. These mRNAs encode so-called **late-embryogenesis-abundant** (**LEA**) proteins thought to be involved in desiccation tolerance. Synthesis of many LEA proteins, or related family members, can be induced by ABA treatment of either young embryos or vegetative tissues. Thus the synthesis of most LEA proteins is under ABA control (see **Web Topic 23.4**).

ABA Promotes the Accumulation of Seed Storage Protein during Embryogenesis

Storage compounds accumulate during mid- to late embryogenesis. Because ABA levels are still high, ABA could be affecting the translocation of sugars and amino acids, the synthesis of the reserve materials, or both.

Studies in mutants impaired in both ABA synthesis and response showed no effect of ABA on sugar translocation. In contrast, ABA has been shown to affect the amounts and composition of storage proteins. For example, exogenous ABA promotes accumulation of storage proteins in cultured embryos of many species, and some ABA-deficient or -insensitive mutants have reduced storage protein accumulation. However, storage protein synthesis is also reduced in other seed developmental mutants with normal ABA levels and responses, indicating that ABA is only one of several signals controlling the expression of storage protein genes during embryogenesis.

ABA not only regulates the accumulation of storage proteins during embryogenesis; it can also maintain the mature embryo in a dormant state until the environmental conditions are optimal for growth. Seed dormancy is an important factor in the adaptation of plants to unfavorable environments. As we will discuss in the next few sections, plants have evolved a variety of mechanisms, some of them involving ABA, that enable them to maintain their seeds in a dormant state.

Seed Dormancy May Be Imposed by the Coat or the Embryo

During seed maturation, the embryo enters a quiescent phase in response to desiccation. Seed germination can be defined as the resumption of growth of the embryo of the mature seed; it depends on the same environmental conditions as vegetative growth does. Water and oxygen must be available, the temperature must be suitable, and there must be no inhibitory substances present.

In many cases a viable (living) seed will not germinate even if all the necessary environmental conditions for growth are satisfied. This phenomenon is termed **seed dormancy**. Seed dormancy introduces a temporal delay in the germination process that provides additional time for seed dispersal over greater geographic distances. It also maximizes seedling survival by preventing germination under unfavorable conditions. Two types of seed dormancy have been recognized: coat-imposed dormancy and embryo dormancy.

Coat-imposed dormancy. Dormancy imposed on the embryo by the seed coat and other enclosing tissues, such as endosperm, pericarp, or extrafloral organs, is known as **coat-imposed dormancy**. The embryos of such seeds will germinate readily in the presence of water and oxygen once the seed coat and other surrounding tissues have been either removed or damaged. There are five basic mechanisms of coat-imposed dormancy (Bewley and Black 1994):

1. *Prevention of water uptake.*
2. *Mechanical constraint.* The first visible sign of germination is typically the radicle breaking through the seed coat. In some cases, however, the seed coat may be too rigid for the radicle to penetrate. For the seeds to germinate, the endosperm cell walls must be weakened by the production of cell wall–degrading enzymes.
3. *Interference with gas exchange.* Lowered permeability of seed coats to oxygen suggests that the seed coat inhibits germination by limiting the oxygen supply to the embryo.
4. *Retention of inhibitors.* The seed coat may prevent the escape of inhibitors from the seed.
5. *Inhibitor production.* Seed coats and pericarps may contain relatively high concentrations of growth inhibitors, including ABA, that can suppress germination of the embryo.

Embryo dormancy. The second type of seed dormancy is **embryo dormancy**, a dormancy that is intrinsic to the embryo and is not due to any influence of the seed coat or other surrounding tissues. In some cases, embryo dormancy can be relieved by amputation of the cotyledons. Species in which the cotyledons exert an inhibitory effect include European hazel (*Corylus avellana*) and European ash (*Fraxinus excelsior*).

A fascinating demonstration of the cotyledon's ability to inhibit growth is found in species (e.g., peach) in which the isolated dormant embryos germinate but grow extremely slowly to form a dwarf plant. If the cotyledons are removed at an early stage of development, however, the plant abruptly shifts to normal growth.

Embryo dormancy is thought to be due to the presence of inhibitors, especially ABA, as well as the absence of growth promoters, such as GA (gibberellic acid). The loss of embryo dormancy is often associated with a sharp drop in the ratio of ABA to GA.

Primary versus secondary seed dormancy. Different types of seed dormancy also can be distinguished on the basis of the timing of dormancy onset rather than the cause of dormancy:

- Seeds that are released from the plant in a dormant state are said to exhibit **primary dormancy**.
- Seeds that are released from the plant in a nondormant state, but that become dormant if the conditions for germination are unfavorable, exhibit **secondary dormancy**. For example, seeds of *Avena sativa* (oat) can become dormant in the presence of temperatures higher than the maximum for germination, whereas seeds of *Phacelia dubia* (small-flower scorpionweed) become dormant at temperatures below the minimum for germination. The mechanisms of secondary dormancy are poorly understood.

Environmental Factors Control the Release from Seed Dormancy

Various external factors release the seed from embryo dormancy, and dormant seeds typically respond to more than one of three factors:

1. *Afterripening*. Many seeds lose their dormancy when their moisture content is reduced to a certain level by drying—a phenomenon known as **afterripening**.
2. *Chilling*. Low temperature, or **chilling**, can release seeds from dormancy. Many seeds require a period of cold (0–10°C) while in a fully hydrated (imbibed) state in order to germinate.
3. *Light*. Many seeds have a light requirement for germination, which may involve only a brief exposure, as in the case of lettuce, an intermittent treatment (e.g., succulents of the genus *Kalanchoe*), or even a specific photoperiod involving short or long days.

For further information on environmental factors affecting seed dormancy, see **Web Topic 23.5**. For a discussion of seed longevity, see **Web Topic 23.6**.

Seed Dormancy Is Controlled by the Ratio of ABA to GA

Mature seeds may be either dormant or nondormant, depending on the species. Nondormant seeds, such as pea, will germinate readily if provided with water only. Dormant seeds, on the other hand, fail to germinate in the presence of water, and instead require some additional treatment or condition. As we have seen, dormancy may arise from the rigidity or impermeability of the seed coat (coat-imposed dormancy) or from the persistence of the state of arrested development of the embryo. Examples of the latter include seeds that require afterripening, chilling, or light to germinate.

ABA mutants have been extremely useful in demonstrating the role of ABA in seed dormancy. Dormancy of *Arabidopsis* seeds can be overcome with a period of afterripening and/or cold treatment. ABA-deficient (*aba*) mutants of *Arabidopsis* have been shown to be nondormant at maturity. When reciprocal crosses between *aba* and wild-type plants were carried out, the seeds exhibited dormancy only when the embryo itself produced the ABA. Neither maternal nor exogenously applied ABA was effective in inducing dormancy in an *aba* embryo.

On the other hand, maternally derived ABA constitutes the major peak present in seeds and is required for other aspects of seed development—for example, helping suppress vivipary in midembryogenesis. Thus the two sources of ABA function in different developmental pathways. Dormancy is also greatly reduced in seeds from the ABA-insensitive mutants *abi1* (*ABA-insensitive1*), *abi2*, and *abi3*, even though these seeds contain higher ABA concentrations than those of the wild type throughout development, possibly reflecting feedback regulation of ABA metabolism. ABA-deficient tomato mutants seem to function in the same way, indicating that the phenomenon is probably a general one. However, other mutants with reduced dormancy, but normal ABA levels and sensitivity, point to additional regulators of dormancy.

Although the role of ABA in initiating and maintaining seed dormancy is well established, other hormones contribute to the overall effect. For example, in most plants the peak of ABA production in the seed coincides with a decline in the levels of IAA and GA.

An elegant demonstration of the importance of the ratio of ABA to GA in seeds was provided by the genetic screen that led to isolation of the first ABA-deficient mutants of *Arabidopsis* (Koornneef et al. 1982). Seeds of a GA-deficient mutant that could not germinate in the absence of exogenous GA were mutagenized and then grown in the greenhouse. The seeds produced by these mutagenized plants were then screened for **revertants**—that is, seeds that had regained their ability to germinate.

Revertants were isolated, and they turned out to be mutants of abscisic acid synthesis. The revertants germinated because dormancy had not been induced, so subsequent synthesis of GA was no longer required to overcome it. This study elegantly illustrates the general principle that the balance of plant hormones is often more critical than are their absolute concentrations in regulating development. However, ABA and GA exert their effects on seed dormancy at different times, so their antagonistic effects on dormancy do not necessarily reflect a direct interaction.

Recent genetic screens for suppressors of ABA insensitivity have identified additional antagonistic interactions between ABA and ethylene or brassinosteroid effects on germination. In addition, many new alleles of ABA-deficient or *ABA-insensitive4* (*abi4*) mutants have been identified in screens for altered sensitivity to sugar. These studies show that a complex regulatory web integrates hormonal and nutrient signaling.

ABA Inhibits Precocious Germination and Vivipary

When immature embryos are removed from their seeds and placed in culture midway through development before the onset of dormancy, they germinate precociously—that is, without passing through the normal quiescent and/or dormant stage of development. ABA added to the culture medium inhibits precocious germination. This result, in combination with the fact that the level of endogenous ABA is high during mid- to late seed development, suggests that ABA is the natural constraint that keeps developing embryos in their embryogenic state.

Further evidence for the role of ABA in preventing precocious germination has been provided by genetic studies of vivipary. The tendency toward vivipary, also known as *preharvest sprouting*, is a varietal characteristic in grain crops that is favored by wet weather. In maize, several viviparous (*vp*) mutants have been selected in which the embryos germinate directly on the cob while still attached to the plant. Several of these mutants are ABA deficient (*vp2*, *vp5*, *vp7*, and *vp14*) (see Figure 23.3); one is ABA insensitive (*vp1*). Vivipary in the ABA-deficient mutants can be partially prevented by treatment with exogenous ABA. Vivipary in maize also requires synthesis of GA early in embryogenesis as a positive signal; double mutants deficient in both GA and ABA do not exhibit vivipary (White et al. 2000).

In contrast to the maize mutants, single-gene mutants of *Arabidopsis* (*aba1*, *aba3*, *abi1*, and *abi3*) fail to exhibit vivipary, although they are nondormant. The lack of vivipary might reflect a lack of moisture because such seeds will germinate within the fruits under conditions of high relative humidity. However, other *Arabidopsis* mutants with a normal ABA response and only moderately reduced ABA levels (e.g., *fusca3*, which belongs to a class of mutants[1] defective in regulating the transition from embryogenesis to germination) exhibit some vivipary even at low humidities. Furthermore, double mutants combining either defects in ABA biosynthesis or ABA response with the *fusca3* mutation have a high frequency of vivipary (Nambara et al. 2000), suggesting that redundant control mechanisms suppress vivipary in *Arabidopsis*.

[1] Named after the Latin term for the reddish brown color of the embryos.

ABA Accumulates in Dormant Buds

In woody species, dormancy is an important adaptive feature in cold climates. When a tree is exposed to very low temperatures in winter, it protects its meristems with bud scales and temporarily stops bud growth. This response to low temperatures requires a sensory mechanism that detects the environmental changes (sensory signals), and a control system that transduces the sensory signals and triggers the developmental processes leading to bud dormancy.

ABA was originally suggested as the dormancy-inducing hormone because it accumulates in dormant buds and decreases after the tissue is exposed to low temperatures. However, later studies showed that the ABA content of buds does not always correlate with the degree of dormancy. As we saw in the case of seed dormancy, this apparent discrepancy could reflect interactions between ABA and other hormones as part of a process in which bud dormancy and growth are regulated by the balance between bud growth inhibitors, such as ABA, and growth-inducing substances, such as cytokinins and gibberellins.

Although much progress has been achieved in elucidating the role of ABA in seed dormancy by the use of ABA-deficient mutants, progress on the role of ABA in bud dormancy, which applies mainly to woody perennials, has lagged because of the lack of a convenient genetic system. This discrepancy illustrates the tremendous contribution that genetics and molecular biology have made to plant physiology, and it underscores the need for extending such approaches to woody species.

Analyses of traits such as dormancy are complicated by the fact that they are often controlled by the combined action of several genes, resulting in a gradation of phenotypes referred to as *quantitative traits*. Recent genetic mapping studies suggest that homologs of *ABI1* may regulate bud dormancy in poplar trees. For a description of such studies, see **Web Topic 23.7**.

ABA Inhibits GA-Induced Enzyme Production

ABA inhibits the synthesis of hydrolytic enzymes that are essential for the breakdown of storage reserves in seeds. For example, GA stimulates the aleurone layer of cereal grains to produce α-amylase and other hydrolytic enzymes that break down stored resources in the endosperm during germination (see Chapter 20). ABA inhibits this GA-dependent enzyme synthesis by inhibiting the transcription of α-amylase mRNA. ABA exerts this inhibitory effect via at least two mechanisms:

1. VP1, a protein originally identified as an activator of ABA-induced gene expression, acts as a transcriptional repressor of some GA-regulated genes (Hoecker et al. 1995).

2. ABA represses the GA-induced expression of GA-MYB, a transcription factor that mediates the GA induction of α-amylase expression (Gomez-Cadenas et al. 2001).

ABA Closes Stomata in Response to Water Stress

Elucidation of the roles of ABA in freezing, salt, and water stress (see Chapter 25) led to the characterization of ABA as a stress hormone. As noted earlier, ABA concentrations in leaves can increase up to 50 times under drought conditions—the most dramatic change in concentration reported for any hormone in response to an environmental signal. Redistribution or biosynthesis of ABA is very effective in causing stomatal closure, and its accumulation in stressed leaves plays an important role in the reduction of water loss by transpiration under water stress conditions (Figure 23.5).

Stomatal closing can also be caused by ABA synthesized in the roots and exported to the shoot. Mutants that lack the ability to produce ABA exhibit permanent wilting and are called *wilty* mutants because of their inability to close their stomata. Application of exogenous ABA to such mutants causes stomatal closure and a restoration of turgor pressure.

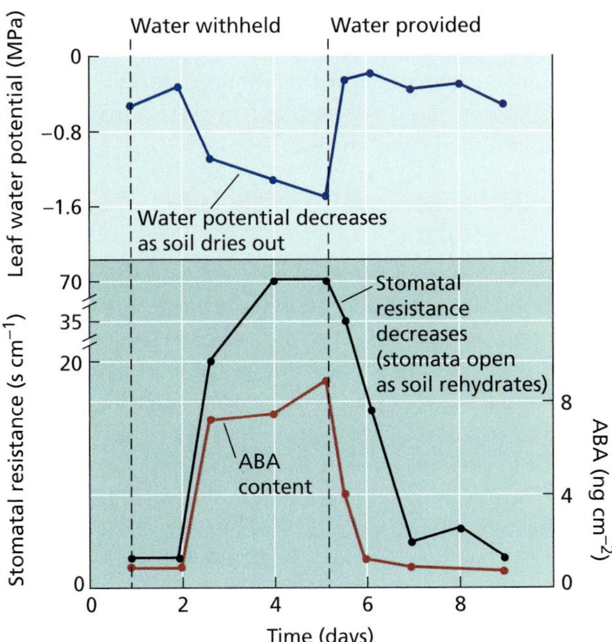

FIGURE 23.5 Changes in water potential, stomatal resistance (the inverse of stomatal conductance), and ABA content in maize in response to water stress. As the soil dried out, the water potential of the leaf decreased, and the ABA content and stomatal resistance increased. The process was reversed by rewatering. (After Beardsell and Cohen 1975.)

ABA Promotes Root Growth and Inhibits Shoot Growth at Low Water Potentials

ABA has different effects on the growth of roots and shoots, and the effects are strongly dependent on the water status of the plant. Figure 23.6 compares the growth of shoots and roots of maize seedlings grown under either abundant water conditions (high water potential) or dehydrating conditions (low water potential). Two types of seedlings were used: (1) wild-type seedlings with normal ABA levels and (2) an ABA-deficient, viviparous mutant.

When the water supply is ample (high water potential), shoot growth is greater in the wild-type plant (normal endogenous ABA levels) than in the ABA-deficient mutant. The reduced shoot growth in the ABA-deficient mutant could be due in part to excessive water loss from the leaves. In maize and tomato, however, the stunted shoot growth of ABA-deficient plants at high water potentials seems to be due to the overproduction of ethylene, which is normally inhibited by endogenous ABA (Sharp et al. 2000). This finding suggests that endogenous ABA promotes shoot growth in well-watered plants by suppressing ethylene production.

When water is limiting (i.e., at low water potentials), the opposite occurs: Shoot growth is greater in the ABA-deficient mutant than in the wild type. Thus, endogenous ABA acts as a signal to reduce shoot growth only under water stress conditions.

Now let's examine how ABA affects roots. When water is abundant, root growth is slightly greater in the wild type (normal endogenous ABA) than in the ABA-deficient mutant, similar to growth in shoots. Therefore, at high water potentials (when the total ABA levels are low), endogenous ABA exerts a slight positive effect on the growth of both roots and shoots.

Under dehydrating conditions, however, the growth of the roots is much higher in the wild type than in the ABA-deficient mutant, although growth is still inhibited relative to root growth of either genotype when water is abundant. In this case, endogenous ABA promotes root growth, apparently by inhibiting ethylene production during water stress (Spollen et al. 2000).

To summarize, under dehydrating conditons, when ABA levels are high, the endogenous hormone exerts a strong positive effect on root growth by suppressing ethylene production, and a slight negative effect on shoot growth. The overall effect is a dramatic increase in the root:shoot ratio at low water potentials (see Figure 23.6C), which, along with the effect of ABA on stomatal closure, helps the plant cope with water stress. For another example of the role of ABA in the response to dehydration, see **Web Essay 1**.

ABA Promotes Leaf Senescence Independently of Ethylene

Abscisic acid was originally isolated as an abscission-causing factor. However, it has since become evident that ABA stimulates abscission of organs in only a few species and

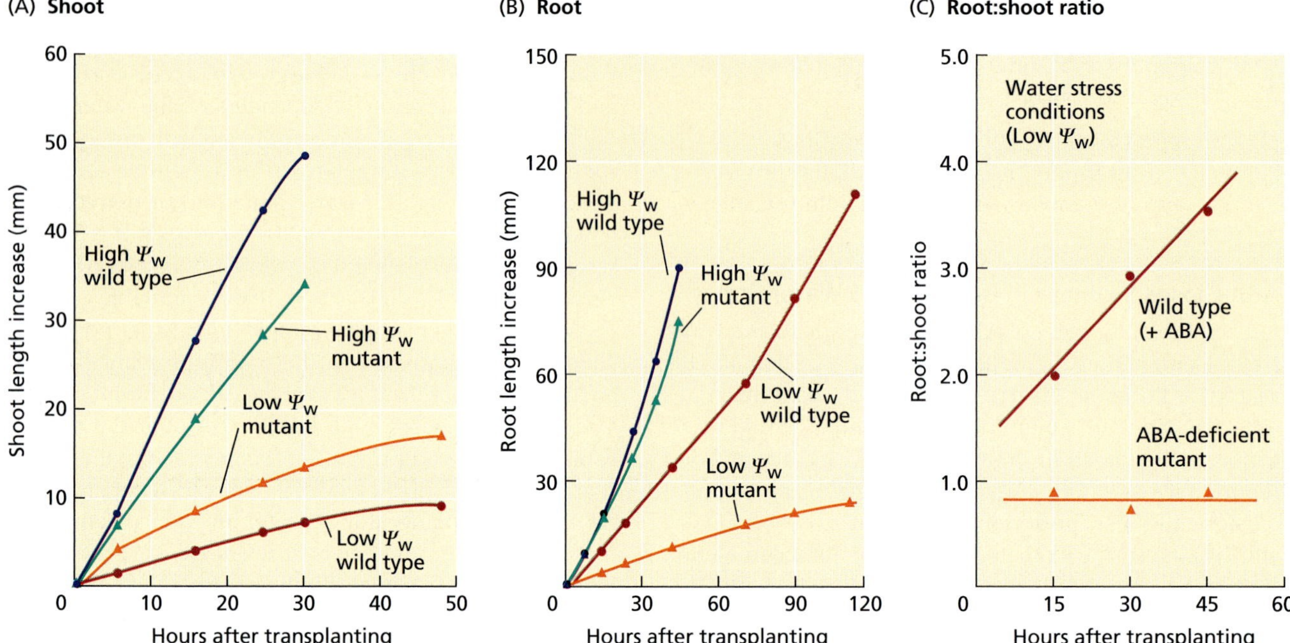

FIGURE 23.6 Comparison of the growth of the shoots (A) and roots (B) of normal versus ABA-deficient (viviparous) maize plants growing in vermiculite maintained either at high water potential (–0.03 MPa) or at low water potential (–0.3 Mpa in A and –1.6 MPa in B). Water stress (low water potential) depresses the growth of both shoots and roots compared to the controls. (C) Note that under water stress conditions (low Ψ_w), the ratio of root growth to shoot growth is much higher when ABA is present (i.e., in the wild type) than when it is absent (in the mutant). (From Saab et al. 1990.)

that the primary hormone causing abscission is ethylene. On the other hand, ABA is clearly involved in leaf senescence, and through its promotion of senescence it might indirectly increase ethylene formation and stimulate abscission. (For more discussion on the relationship between ABA and ethylene, see **Web Topic 23.8**.)

Leaf senescence has been studied extensively, and the anatomical, physiological, and biochemical changes that take place during this process were described in Chapter 16. Leaf segments senesce faster in darkness than in light, and they turn yellow as a result of chlorophyll breakdown. In addition, the breakdown of proteins and nucleic acids is increased by the stimulation of several hydrolases. ABA greatly accelerates the senescence of both leaf segments and attached leaves.

CELLULAR AND MOLECULAR MODES OF ABA ACTION

ABA is involved in short-term physiological effects (e.g., stomatal closure), as well as long-term developmental processes (e.g., seed maturation). Rapid physiological responses frequently involve alterations in the fluxes of ions across membranes and may involve some gene regulation as well, and long-term processes inevitably involve major changes in the pattern of gene expression.

Signal transduction pathways, which amplify the primary signal generated when the hormone binds to its receptor, are required for both the short-term and the long-term effects of ABA. Genetic studies have shown that many conserved signaling components regulate both short- and long-term responses, indicating that they share common signaling mechanisms. In this section we will describe what is known about the mechanism of ABA action at the cellular and molecular levels.

ABA Is Perceived Both Extracellularly and Intracellularly

Although ABA has been shown to interact directly with phospholipids, it is widely assumed that the ABA receptor is a protein. To date, however, the protein receptor for ABA has not been identified. Experiments have been performed to determine whether the hormone must enter the cell to be effective, or whether it can act externally by binding to a receptor located on the outer surface of the plasma membrane. The results so far suggest multiple sites of perception.

Some experiments point to a receptor on the outer surface of the cell. For example, microinjected ABA fails to alter stomatal opening in the spiderwort *Commelina*, or to inhibit GA-induced α-amylase synthesis in barley aleurone protoplasts (Anderson et al. 1994; Gilroy and Jones 1994). Furthermore, impermeant ABA–protein conjugates have been shown to activate both ion channel activity and gene expression (Schultz and Quatrano 1997; Jeannette et al. 1999).

Other experiments, however, support an intracellular location for the ABA receptor:

(A)

UV

Photolyzable caged ABA

(S)-cis-ABA

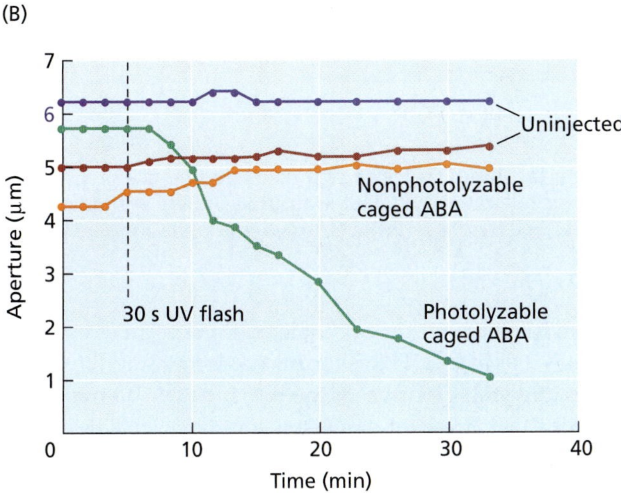

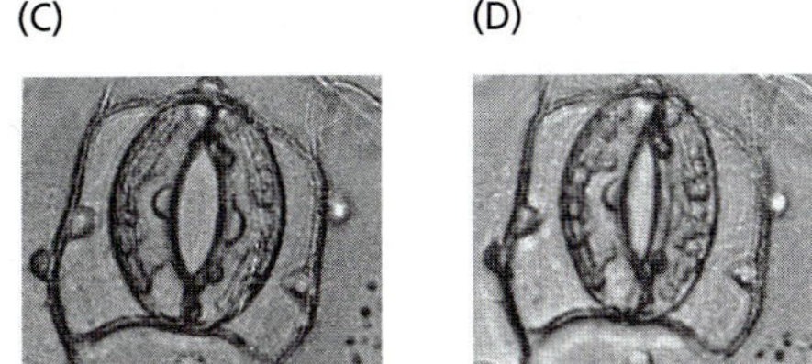

FIGURE 23.7 Stomatal closure induced by UV photolysis of caged ABA in the guard cell cytoplasm. Single guard cells in stomatal complexes of *Commelina* were microinjected with caged ABA. (A) Photolysis reaction induced by UV irradiation. (B) The stomatal apertures recorded before and after a 30-second exposure of the cells to UV. (C, D) Light micrographs of the same stomatal complex in which the right-hand guard cell was loaded with the photolyzable cages ABA 10 minutes before UV photolysis (C) and 30 minutes after photolysis (D). (A and B from Allan et al. 1994; C and D courtesy of A. Allan, from Allan et al. 1994; © American Society of Plant Biologists, reprinted with permission.)

- Extracellular application of ABA was nearly twice as effective at inhibiting stomatal opening at pH 6.15, when it is fully protonated and readily taken up by guard cells, versus at pH 8, when it is largely dissociated to the anionic form that does not readily cross membranes (Anderson et al. 1994).
- ABA supplied directly and continuously to the cytosol via a patch pipette inhibited K^+_{in} channels, which are required for stomatal opening (Schwartz et al. 1994).
- Microinjection of an inactive "caged" form of ABA into guard cells of *Commelina* resulted in stomatal closure after the stomata were treated briefly with UV irradiation to activate the hormone—that is, release it from its molecular cage (Figure 23.7) (Allan et al. 1994). Control guard cells injected with a nonphotolyzable form of the caged ABA did not close after UV irradiation.

Taken together, these results indicate that extracellular perception of ABA can prevent stomatal opening and regulate gene expression, and intracellular ABA can both induce stomatal closure and inhibit the K^+_{in} current required for opening. Thus there appear to be both extracellular and intracellular ABA receptors. However, they have yet to be identified or localized.

ABA Increases Cytosolic Ca^{2+}, Raises Cytosolic pH, and Depolarizes the Membrane

As discussed in Chapter 18, stomatal closure is driven by a reduction in guard cell turgor pressure caused by a massive long-term efflux of K^+ and anions from the cell. During the subsequent shrinkage of the cell due to water loss, the surface area of the plasma membrane may contract by as much as 50%. Where does the extra membrane go? The answer seems to be that it is taken up as small vesicles by endocytosis—a process that also involves reorganization of the actin cytoskeleton. However, the first changes detected after exposure of guard cells to ABA are transient membrane depolarization caused by the net influx of positive charge, and transient increases in the cytosolic calcium concentration (Figure 23.8).

ABA stimulates elevations in the concentration of cytosolic Ca^{2+} by inducing both influx through plasma membrane channels and release of calcium into the cytosol from internal compartments, such as the central vacuole (Schroeder et al. 2001). Stimulation of influx occurs via a pathway that uses **reactive oxygen species** (**ROS**), such as hydrogen peroxide (H_2O_2) or superoxide ($O_2^{\bullet -}$), as secondary messengers leading to plasma membrane channel activation (Pei et al. 2000).

Calcium release from intracellular stores can be induced by a variety of second messengers, including inositol 1,4,5-trisphosphate (IP_3), cyclic ADP-ribose (cADPR), and self-

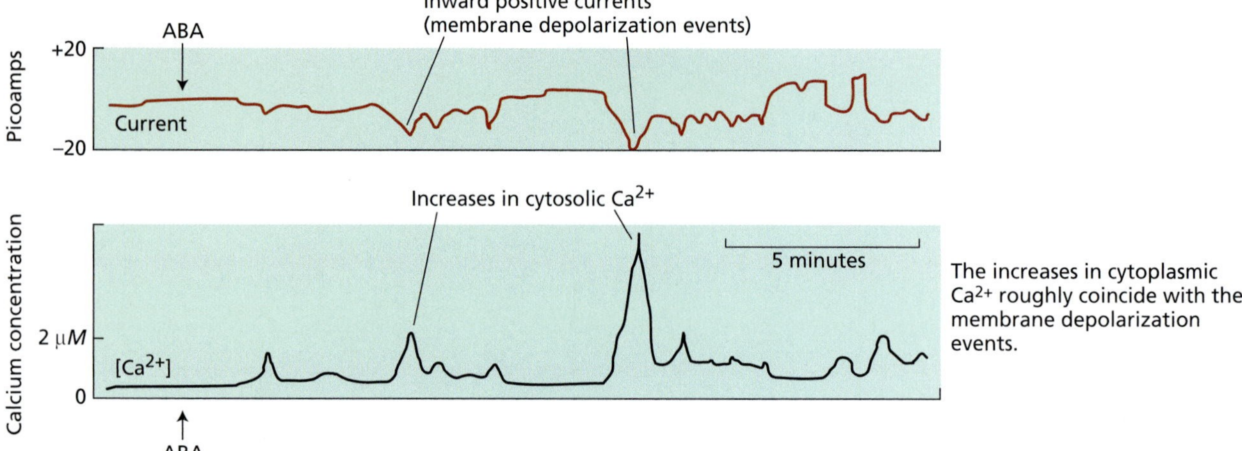

FIGURE 23.8 Simultaneous measurements of ABA-induced inward positive currents and ABA-induced increases in cytosolic Ca^{2+} concentrations in a guard cell of *Vicia faba* (broad bean). The current was measured by the patch clamp technique; calcium was measured by use of a fluorescent indicator dye. ABA was added to the system at the arrow in each case. (From Schroeder and Hagiwara 1990.)

amplifying (calcium-induced) Ca^{2+} release. Recent studies have shown that ABA stimulates **nitric oxide** (**NO**) synthesis in guard cells, which induces stomatal closure in a cADPR-dependent manner, indicating that NO is an even earlier secondary messenger in this response pathway (Neill et al. 2002) (for background on NO, see Chapter 14 on the Web site).

The combination of calcium influx and the release of calcium from internal stores raises the cytosolic calcium concentration from 50 to 350 n*M* to as high as 1100 n*M* (1.1 m*M*) (Figure 23.9) (Mansfield and McAinsh, in Davies 1995). This increase is sufficient to cause stomatal closure, as demonstrated by the following experiment.

As in the experiment described earlier, calcium was microinjected into guard cells in a caged form that could be hydrolyzed by a pulse of UV light. This method allowed the investigators to control both the concentration of free calcium and the time of release to the cytosol. At cytosolic concentrations of 600 n*M* or more, release of calcium from its cage triggered stomatal closure (Gilroy et al. 1990). This level of intracellular calcium is well within the concentration range observed after ABA treatment.

In the preceding studies, intracellular free calcium was measured by the use of microinjected calcium-sensitive ratiometric fluorescent dyes[2], such as fura-2 or indo-1. However, microinjections of fluorescent dyes into single plant cells are difficult and often result in cell death. Success rates of viable injections into *Arabidopsis* guard cells can be less than 3%. In contrast, transgenic plants expressing the gene for the calcium indicator protein **yellow cameleon** make it possible to monitor several fluorescing cells in parallel, without the need for invasive injections (Allen et al. 1999b) (see **Web Topic 23.9**). Such studies have demonstrated that the cytosolic Ca^{2+} concentration oscillates with distinct periodicities, depending on the signals received (Figure 23.10).

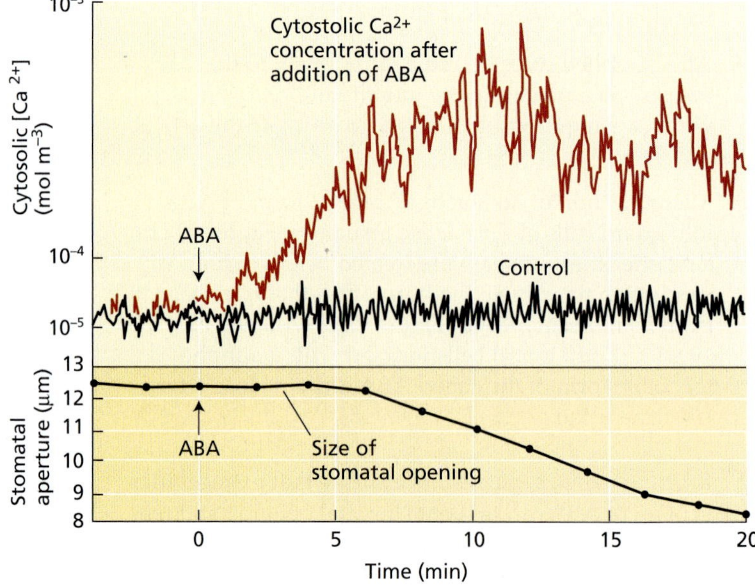

FIGURE 23.9 Time course of the ABA-induced increase in guard cell cytosolic Ca^{2+} concentration (upper panel) and ABA-induced stomatal aperture (lower panel). (From Mansfield and McAinsh 1995.)

[2] Ratiometric fluorescent dyes undergo a shift in their excitation and emission spectra when they bind calcium. On the basis of property, one can determine the intracellular concentrations of both forms of the dye (with and without bound calcium) by exciting them with the appropriate two wavelengths. The ratio of the two emissions provides a measure of the calcium concentration that is independent of dye concentration.

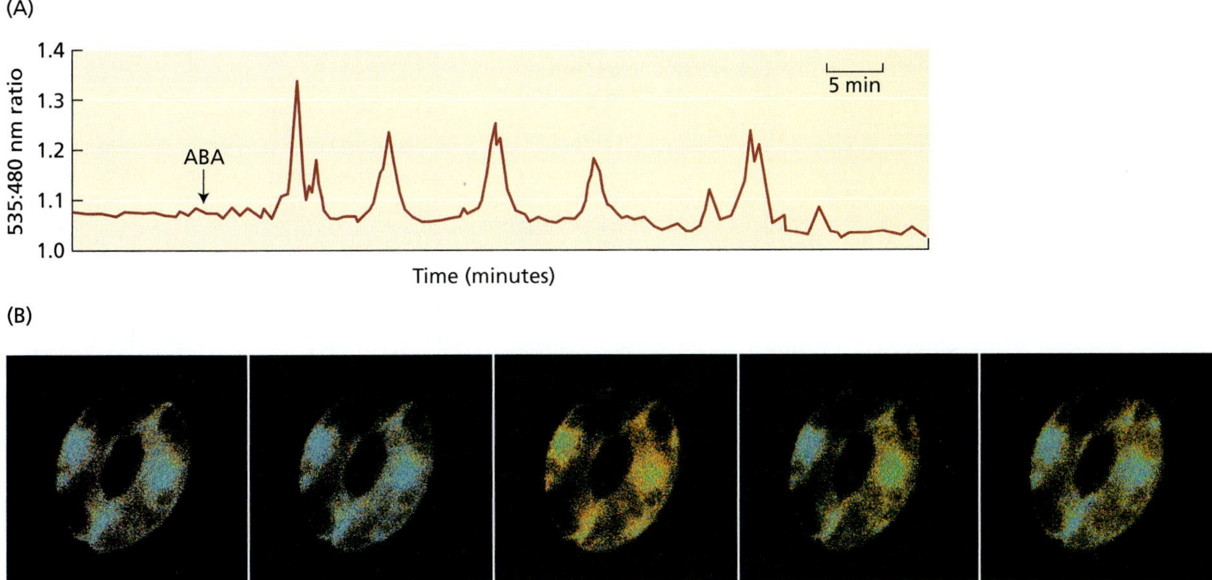

FIGURE 23.10 ABA-induced calcium oscillations in *Arabidopsis* guard cells expressing yellow cameleon, a calcium indicator protein dye. (A) Oscillations elicited by ABA are indicated by increases in the ratio of flourescence emission at 535 and 480 nm. (B) Pseudo colored images of fluorescence in *Arabidopsis* guard cells, where blue, green, yellow and red represent increasing cytosolic calcium concentration. (From Schroeder et al. 2001.)

These results support the hypothesis that an increase in cytosolic calcium, partly derived from intracellular stores, is responsible for ABA-induced stomatal closure. However, the growth hormone auxin can induce stomatal opening, and this auxin-induced stomatal opening, like ABA-induced stomatal closure, is accompanied by *increases* in cytosolic calcium. This finding suggests that the detailed characteristics of the location and periodicity of Ca^{2+} oscillations (the "Ca^{2+} signature"), rather than the overall concentration of cytosolic calcium, determine the cellular response.

In addition to increasing the cytosolic calcium concentration, ABA causes an alkalinization of the cytosol from about pH 7.67 to pH 7.94. The increase in cytosolic pH has been shown to activate the K^+ efflux channels on the plasma membrane apparently by increasing the number of channels available for activation (see Chapter 6).

ABA Activation of Slow Anion Channels Causes Long-Term Membrane Depolarization

The rapid, transient depolarizations induced by ABA are insufficient to open the K^+ efflux channels, which require long-term membrane depolarization in order to open. However, long-term depolarizations in response to ABA have been demonstrated. According to a widely accepted model, long-term membrane depolarization is triggered by two factors: (1) an ABA-induced transient depolarization of the plasma membrane, coupled with (2) an increase in cytosolic calcium. Both of these conditions are required to open calcium-activated slow (S-type) anion channels on the plasma membrane (Schroeder and Hagiwara 1990) (see Chapter 6). ABA has been shown to activate slow anion channels in guard cells (Grabov et al. 1997; Pei et al. 1997).

The prolonged opening of these slow anion channels permits large quantities of Cl^- and malate^{2-} ions to escape from the cell, moving down their electrochemical gradients. (The inside of the cell is negatively charged, thus pushing Cl^- and malate^{2-} out of the cell, and the outside has lower Cl^- and malate^{2-} concentrations than the interior.) The outward flow of negatively charged Cl^- and malate^{2-} ions generated in this way strongly depolarizes the membrane, triggering the voltage-gated K^+ efflux channels to open.

In support of this model, inhibitors that block slow anion channels, such as 5-nitro-2,3-phenylpropylaminobenzoic acid (NPPB), also block ABA-induced stomatal closing. Inhibitors of the rapid (R-type) anion channels, such as 4,4′-diisothiocyanatostilbene-2,2′-disulfonic acid (DIDS), have no effect on ABA-induced stomatal closing (Schwartz et al. 1995).

Another factor that can contribute to membrane depolarization is inhibition of the plasma membrane H^+-ATPase. ABA inhibits blue light–stimulated proton pumping by guard cell protoplasts (Figure 23.11), consistent with the model that the depolarization of the plasma membrane by ABA is partially caused by a decrease in the activity of the plasma membrane H^+-ATPase. However, ABA does not inhibit the proton pump directly.

In *Vicia faba* (broad bean), at least, the plasma membrane H^+-ATPase of the leaves is strongly inhibited by calcium. A calcium concentration of 0.3 μM blocks 50% of the activity of H^+-ATPase, and 1 μM calcium blocks the enzyme completely (Kinoshita et al. 1995). It appears that two fac-

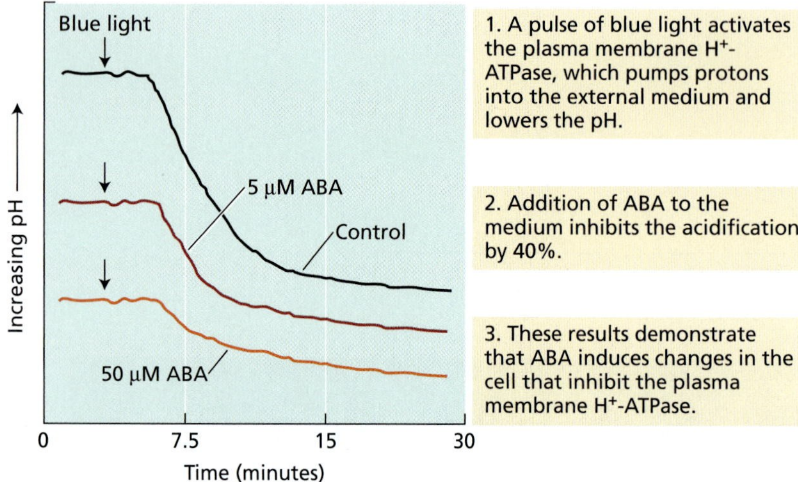

FIGURE 23.11 ABA inhibition of blue light–stimulated proton pumping by guard cell protoplasts. A suspension of guard cell protoplasts was incubated under red-light irradiation, and the pH of the suspension medium was monitored with a pH electrode. The starting pH was the same in all cases (the curves are displaced for ease of viewing). (After Shimazaki et al. 1986.)

tors contribute to ABA inhibition of the plasma membrane proton pump: an increase in the cytosolic Ca^{2+} concentration, and alkalinization of the cytosol.

In addition to causing stomatal closure, ABA prevents light-induced stomatal opening. In this case ABA acts by inhibiting the inward K^+ channels, which are open when the membrane is hyperpolarized by the proton pump (see Chapters 6 and 18). Inhibition of the inward K^+ channels is mediated by the ABA-induced increase in cytosolic calcium concentration. Thus calcium and pH affect guard cell plasma membrane channels in two ways:

1. They prevent stomatal opening by inhibiting inward K^+ channels and plasma membrane proton pumps.
2. They promote stomatal closing by activating outward anion channels, thus leading to activation of K^+ efflux channels.

ABA Stimulates Phospholipid Metabolism

As discussed previously, much evidence supports a role for calcium both in the promotion of stomatal closing and in the inhibition of stomatal opening. According to the classic calcium-dependent signal transduction pathway of animal cells, IP_3 is released, along with diacylglycerol (DAG), when phospholipase C is activated by a G-protein in the plasma membrane (see Chapter 14 on the web site). Does ABA use the same pathway when it induces stomatal closure?

In agreement with this model, ABA has been shown to stimulate phosphoinositide metabolism in *Vicia faba* (broad bean) guard cells. To detect the effect of ABA on IP_3 release, it was necessary to include Li^+ in the incubation medium as an inhibitor of inositol phosphatase, which rapidly removes phosphate groups from IP_3. Under these conditions, a 90% ABA-induced increase in the level of IP_3 was measured within 10 seconds of hormone treatment (Lee et al. 1996). Recent studies in *Arabidopsis* using antisense DNA to block expression of an ABA-induced phospholipase C have shown that this enzyme is required for ABA effects on germination, growth, and gene expression (Sanchez and Chua 2001).

Heterotrimeric G-proteins may mediate the effects of ABA on stomatal movements. For example, in *Vicia faba* most studies have shown that G-protein activators, such as GTPγS, can inhibit the activity of the inward K^+ channels. Consistent with the inhibitor results, ABA failed to inhibit inward K^+ channels or light-induced stomatal opening in an *Arabidopsis* mutant with a defective Gα subunit (Wang et al. 2001). However, ABA still promoted stomatal closure in this mutant, indicating that inhibition of opening and promotion of closing take two distinct paths to the same end point—that is, closed stomata.

Other potential second messengers mediating the ABA response, such as phosphatidic acid and *myo*-inositol-hexaphosphate (IP_6) have been identified, but the relationship of these compounds to IP_3 and Ca^{2+} signaling is not yet known.

All of these experiments indicate that stomatal guard cells respond to multiple signals, possibly involving multiple receptors and overlapping signal transduction pathways.

Protein Kinases and Phosphatases Participate in ABA Action

Nearly all biological signaling systems involve protein phosphorylation and dephosphorylation reactions at some step in the pathway. Thus we can expect that signal transduction in guard cells, with their multiple sensory inputs, involves protein kinases and phosphatases. Artificially raising the ATP concentration inside guard cells by allowing the cytoplasm to equilibrate with the solution inside a patch pipette (see Chapter 6) strongly activates the slow anion channels.

This activation of the slow anion channels by ATP is abolished by the inclusion of protein kinase inhibitors in the patch pipette solution (Schmidt et al. 1995). Protein kinase inhibitors also block ABA-induced stomatal closing. In contrast, lowering the concentration of ATP in the cytosol inactivates the slow anion channels. Additional experiments confirm that this inactivation is due to the presence of protein phosphatases, which remove phosphate groups that are covalently attached to proteins. In view of these results, it appears that protein phosphorylation and dephosphorylation play important roles in the ABA signal transduction pathway in guard cells.

There is now direct evidence for an ABA-activated protein kinase (AAPK) in *Vicia faba* guard cells (Li and Assmann 1996; Mori and Muto 1997). AAPK activity appears to be required for ABA activation of S-type anion currents and stomatal closing. This enzyme is an autophosphorylating protein kinase that either forms part of a Ca^{2+}-independent signal transduction pathway for ABA, or acts farther downstream of calcium-induced signaling events. (The presence of both Ca^{2+}-dependent and Ca^{2+}-independent pathways for ABA action will be discussed shortly.) In addition, two Ca^{2+}-dependent protein kinases, as well as MAP kinases, have been implicated in the ABA regulation of stomatal aperture.

The analysis of ABA-insensitive mutants has begun to help in the identification of genes coding for components of the signal transduction pathway. The *Arabidopsis abi1-1* and *abi2-1* mutations result in insensitivity to ABA in both seeds and adult plants. These *abi* mutants display phenotypes consistent with a defect in ABA signaling, including reduced seed dormancy, a tendency to wilt (due to improper regulation of stomatal aperture), and decreased expression of various ABA-inducible genes.

The defects in stomatal response include the ABA insensitivity of S-type anion channels—both inward and outward K^+ channels—and actin reorganization. Although nonresponsive to ABA, the mutant stomata will close when exposed to high external concentrations of Ca^{2+}, suggesting that they are defective in their ability to initiate Ca^{2+} signaling. Consistent with this finding, ABA does not induce Ca^{2+} oscillations in these mutants (Allen et al. 1999a).

ABI Protein Phosphatases Are Negative Regulators of the ABA Response

The *Arabidopsis ABI1* and *ABI2* genes have been cloned and identified as encoding two closely related serine/threonine protein phosphatases. This finding suggests that ABI1 and ABI2 regulate the activity of target proteins by dephosphorylating specific serine or threonine residues, but none of their substrates have been definitively identified.

Because the *abi1-1* and *abi2-1* mutations result in decreased response to ABA, it was initially assumed that the wild-type genes *promote* the ABA response. However, the original mutations turned out to be dominant rather than recessive, and recent studies have shown that they act as "dominant negatives"; that is, one defective copy of the gene is sufficient to disrupt the ABA response by poisoning the activity of the functional gene products from the remaining wild-type allele.

Subsequently, recessive mutants of *ABI1* were obtained that exhibited a simple loss of *ABI1* activity. These recessive mutants of *ABI1* actually showed increased ABA sensitivity (Gosti et al. 1999). Furthermore, overproducing the wild-type gene products or their homologs (closely related proteins) by reintroducing the gene into plants, under control of a highly expressed promoter, confers *reduced* ABA sensitivity (Sheen 1998). Thus the wild-type function of these protein phosphatases is to inhibit the ABA response.

ABA Signaling Also Involves Ca^{2+}-Independent Pathways

Although an ABA-induced increase in cytosolic calcium concentration is a key feature of the current model for ABA-induced guard cell closure, ABA is able to induce stomatal closure even in guard cells that show no increase in cytosolic calcium (Allan et al. 1994). In other words, ABA seems to be able to act via one or more calcium-independent pathways.

In addition to calcium, ABA can utilize cytosolic pH as a signaling intermediate. As previously discussed, a rise in cytosolic pH can lead to the activation of outward K^+ channels, and one effect of the *abi1* mutation is to render these K^+ channels insensitive to pH.

Such redundancy in the signal transduction pathways explains how guard cells are able to integrate a wide range of hormonal and environmental stimuli that affect stomatal aperture, and such redundancy is probably not unique to guard cells.

A simplified general model for ABA action in stomatal guard cells is shown in Figure 23.12. For clarity, only the cell surface receptors are shown.

ABA Regulation of Gene Expression Is Mediated by Transcription Factors

Downstream of the early ABA signal transduction processes already discussed, ABA causes changes in gene expression. ABA has been shown to regulate the expression of numerous genes during seed maturation and under certain stress conditions, such as heat shock, adaptation to low temperatures, and salt tolerance (Rock 2000). The ABA- and stress-induced genes are presumed to contribute to adaptive aspects of induced tolerance (see Chapter 25). They include genes encoding proteases, chaperonins, proteins similar to LEA proteins, enzymes of sugar or other compatible solute[3] metabolism, ion and water channel proteins, enzymes that detoxify active oxygen species, and regulatory proteins such as transcription factors and protein kinases.

In a few cases, stimulation of transcription by ABA has been demonstrated directly. Gene activation by ABA is mediated by transcription factors. Four main classes of regulatory sequences conferring ABA inducibility have been identified, and proteins that bind to these sequences have

[3] An organic compound that can serve as a nontoxic, osmotically active solute in the cytosol; such compounds usually accumulate during water or salt stress (see Chapter 25).

1. ABA binds to its receptors.
2. ABA-binding induces the formation of reactive oxygen species, which activate plasma membrane Ca^{2+} channels.
3. ABA increases the levels of cyclic ADP-ribose and IP_3, which activate additional calcium channels on the tonoplast.
4. The influx of calcium initiates intracellular calcium oscillations and promotes the further release of calcium from vacuoles.
5. The rise in intracellular calcium blocks K^+_{in} channels.
6. The rise in intracellular calcium promotes the opening if Cl^-_{out} (anion) channels on the plasma membrane, causing membrane depolarization.
7. The plasma membrane proton pump is inhibited by the ABA-induced increase in cytostolic calcium and a rise in intracellular pH, further depolarizing the membrane.
8. Membrane depolarization activates K^+_{out} channels.
9. K^+ and anions to be released across the plasma membrane are first released from vacuoles into the cytosol.

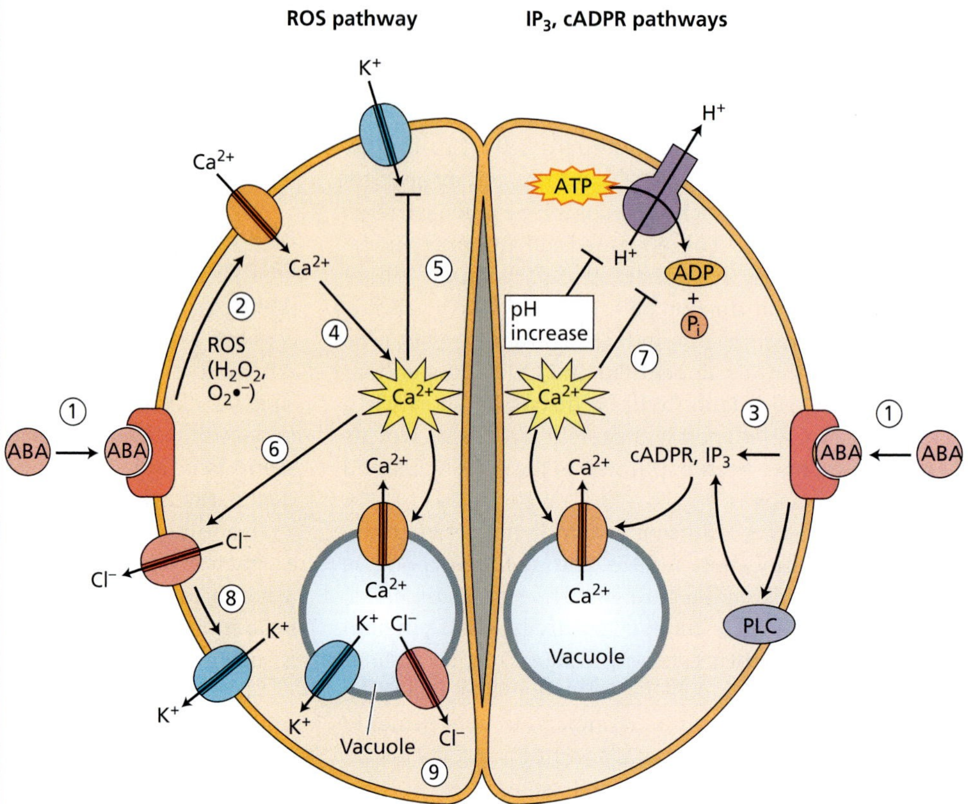

FIGURE 23.12 Simplified model for ABA signaling in stomatal guard cells. The net effect is the loss of potassium and its anion (Cl^- or malate^{2-}) from the cell. (R = receptor; ROS = reactive oxygen species; cADPR = cyclic ADP-ribose; G-protein = GTP-binding protein; PLC = phospholipiase C.)

been characterized (see **Web Topic 23.10**). Under stress conditions, induction of gene expression may be ABA dependent or ABA independent, and additional transcription factors have been identified that specifically mediate responses to cold, drought, or salt (see Chapter 25).

A few DNA elements have been identified that are involved in transcriptional repression by ABA. The best-characterized of these are the gibberellin response elements (GAREs) that mediate the gibberellin-inducible, ABA-repressible expression of the barley α-amylase gene (see Chapter 20).

Four transcription factors involved in ABA gene activation in maturing seeds have been identified by genetic means; mutations in the genes encoding any of these proteins reduce seed ABA responsiveness. The maize *VP1* (*VIVIPAROUS-1*) and *Arabidopsis ABI3* (*ABA-INSENSITIVE3*) genes encode highly similar proteins, and the *ABI4* and *ABI5* genes encode members of two other transcription factor families. *VP1/ABI3*, and *ABI4* are members of gene families found only in plants. In contrast, *ABI5* is a member of the basic leucine zipper (bZIP) family, whose members are present in all eukaryotes (Finkelstein and Lynch 2000).

Additional members of the *ABI5* subfamily have been identified by nongenetic means and are also correlated with ABA-, embryonic-, drought-, or salt stress–induced gene expression. Characterization of *vp1*, *abi4*, and *abi5* mutants has shown that each of these genes can either activate or repress transcription, depending on the target gene. Because the promoter of any given gene contains binding sites for a variety of regulators, it is likely that these transcription factors act in complexes made up of varying combinations of regulators, whose composition is determined by the combination of available regulators and binding sites.

To date, the protein ABI3/VP1 has been shown to interact physically with a variety of proteins, including ABI5 and its rice homolog (TRAB1). ABI5 also forms homodimers and heterodimers with other bZIP family members. There is additional evidence for indirect interactions that may be mediated by 14-3-3 proteins, a class of acidic proteins that dimerize and facilitate protein–protein interactions in a variety of signaling, transport, and enzymatic functions (**see**

Web Topic 23.11). These studies demonstrate the capacity for specific binding among a variety of transcription factors predicted to interact as components of regulatory complexes involved in ABA-induced gene expression.

Other Negative Regulators of the ABA Response Have Been Identified

As described already, negative regulators of the ABA response (protein phosphatases) have been identified by isolation of dominant negative mutants such as *abi1* and *abi2* that result in ABA-insensitive phenotypes (analogous to the dominant negative effects of the ethylene receptor mutant *etr1*; see Chapter 22).

Other negative regulators have been identified through isolation of mutants exhibiting enhanced responses to ABA. Mutants showing increased sensitivity to ABA during germination include *era* (*e*nhanced *r*esponse to *A*BA) and *abh* (*ABA h*ypersensitive) (Cutler et al. 1996; Hugouvieux et al. in press). The *era* and *abh* mutants both confer ABA hypersensitivity in both stomatal closing and germination, making these mutants resistant to wilting and mildly drought tolerant.

Farnesyl transferase. The *ERA1* gene was cloned, and its protein product was identified as a subunit of the enzyme farnesyl transferase. Farnesyl transferases catalyze attachment of the isoprenoid intermediate farnesyl diphosphate (see Chapter 13) to proteins that contain a specific signal sequence of amino acids. Many proteins that have been shown to participate in signal transduction are farnesylated. Farnesylated proteins are anchored to the membrane via hydrophobic interactions between the farnesyl group and the membrane lipids (see Figure 1.6). The identification of ERA1 as part of farnesyl transferase suggests that a protein that normally suppresses the ABA response requires farnesylation and is possibly anchored to the membrane.

mRNA processing. *ABH1* encodes an mRNA 5′ cap–binding protein that may be involved in mRNA processing of negative regulators of ABA signaling. (Recall that eukaryotic messenger RNAs have a "cap" consisting of methylated guanosine at the 5′ end.) Comparison of transcript accumulation in wild-type and *abh1* plants showed a small number of misexpressed genes in the mutant, including some encoding possible signaling molecules.

Ethylene insensitivity. *ERA3* was found to be allelic to a previously identified ethylene signaling locus, *ETHYLENE-INSENSITIVE 2* (*EIN2*) (Ghassemian et al. 2000) (see Chapter 22). In addition to displaying defects in ABA and ethylene responses, mutations in this gene result in defects in the responses to auxin, jasmonic acid, and stress. This gene encodes a membrane-bound protein that appears to represent a point of "cross-talk"—i.e., a common signaling intermediate—mediating the responses to many different signals.

IP_3 catabolism. Other screens have identified ABA signaling mutants on the basis of incorrect expression of reporter genes controlled by ABA-responsive promoters. Although the defects in some of these mutants are limited to gene expression, others affect plant growth responses. One such mutant, termed *fiery* (*fry*) to reflect the intensity of light emission by its ABA/stress-responsive luciferase reporter, is also hypersensitive to ABA and stress inhibition of germination and growth. The *FIERY* gene encodes an enzyme required for IP_3 catabolism (Xiong et al. 2001). The mutant phenotype demonstrates that the ability to attenuate, as well as induce, stress signaling is important for successful induction of stress tolerance.

Similar to the signaling mechanisms documented for other plant hormones, ABA signaling involves the coordinated action of positive and negative regulators affecting processes as diverse as transcription, RNA processing, protein phosphorylation or farnesylation, and metabolism of secondary messengers. As the signaling components are identified, and often are found to function in responses to multiple signals, the next challenge is to determine how they can lead to ABA-specific responses.

SUMMARY

Abscisic acid plays major roles in seed and bud dormancy, as well as responses to water stress. ABA is a 15-carbon terpenoid compound derived from the terminal portion of carotenoids. ABA in tissues can be measured by bioassays based on growth, germination, or stomatal closure. Gas chromatography, HPLC, and immunoassays are the most reliable and accurate methods available for measuring ABA levels.

ABA is produced by cleavage of a 40-carbon carotenoid precursor that is synthesized from isopentenyl diphosphate via the plastid terpenoid pathway. ABA is inactivated by both oxidative degradation and conjugation.

ABA is synthesized in almost all cells that contain plastids and is transported via both the xylem and the phloem. The level of ABA fluctuates dramatically in response to developmental and environmental changes. During seed maturation, ABA levels peak in mid- to late embryogenesis.

ABA is required for the development of desiccation tolerance in the developing embryo, the synthesis of storage proteins, and the acquisition of dormancy. Seed dormancy and germination are controlled by the ratio of ABA to GA, and ABA-deficient embryos may exhibit precocious germination and vivipary. ABA is also antagonized by ethylene and brassinosteroid promotion of germination. Although less is known about the role of ABA in buds, ABA is one of the inhibitors that accumulates in dormant buds.

During water stress, the ABA level of the leaf can increase 50-fold. In addition to closing stomata, ABA increases the hydraulic conductivity of the root and increases the root:shoot ratio at low water potentials. ABA and an alkalinization of the xylem sap are thought to be two chemical signals that the root sends to the shoot as the soil dries. The increased pH of the xylem sap may allow more of the ABA of the leaf to be translocated to the stomata via the transpiration stream.

ABA exerts both short-term and long-term control over plant development. The long-term effects are mediated by ABA-induced gene expression. ABA stimulates the synthesis of many classes of proteins during seed development and during water stress, including the LEA family, proteases and chaperonins, ion and water channels, and enzymes catalyzing compatible solute metabolism or detoxification of active oxygen species. These proteins may protect membranes and other proteins from desiccation damage, or they may aid in recovery from the deleterious effects of stress. ABA response elements and several transcription factors that bind to them have been identified. ABA also suppresses GA-induced gene expression—for example, the synthesis of GA-MYB and α-amylase by barley aleurone layers.

There is evidence for both extracellular and intracellular ABA receptors in guard cells. ABA closes stomata by causing long-term depolarization of the guard cell plasma membrane. Depolarization is believed to be caused by an increase in cytosolic Ca^{2+}, as well as alkalinization of the cytosol. The increase in cytosolic calcium is due to a combination of calcium uptake and release of calcium from internal stores. This calcium increase leads to the opening of slow anion channels, which results in membrane depolarization. IP_3, IP_6, cADPR, PA, and reactive oxygen species all function as secondary messengers in ABA-treated guard cells, and G-proteins participate in the response. Outward K^+ channels open in response to membrane depolarization and to the rise in pH, bringing about massive K^+ efflux.

In general, the ABA response appears to be regulated by more than one signal transduction pathway, even within a single cell type. This redundancy is consistent with the ability of plant cells to respond to multiple sensory inputs. There is genetic evidence for cross-talk between ABA signaling and the signaling of all other major classes of phytohormones, as well as sugars.

Web Material

Web Topics

23.1 The Structure of Lunularic Acid from Liverworts

Although inactive in higher plants, lunularic acid appears to have a function similar to ABA in liverworts.

23.2 Structural Requirements for Biological Activity of Abscisic Acid

To be active as a hormone, ABA requires certain functional groups

23.3 The Bioassay of ABA

Several ABA-responding tissues have been used to detect and measure ABA.

23.4 Proteins Required for Desiccation Tolerance

ABA induces the synthesis of proteins that protect cells from damage due to desiccation.

23.5 Types of Seed Dormancy and the Roles of Environmental Factors

This discussion expands on the various types of seed dormancy and describes how environmental factors affect seed dormancy.

23.6 The Longevity of Seeds

Under certain conditions, seeds can remain dormant for hundreds of years.

23.7 Genetic Mapping of Dormancy: Quantitative Trait Locus (QTL) Scoring of Vegetative Dormancy Combined with a Candidate Gene Approach

A genetic method for determining the number and chromosomal locations of genes affecting a quantitative trait affected by many unlinked genes is described.

23.8 ABA-Induced Senescence and Ethylene

Hormone-insensitive mutants have made it possible to distinguish the effects of ethylene from those of ABA on senescence.

23.9 Yellow Cameleon: A Noninvasive Tool for Measuring Intracellular Calcium

The features of the yellow cameleon protein that enable it to act as a reporter for calcium concentration are described.

23.10 Promoter Elements That Regulate ABA Induction of Gene Expression

A table of the different ABA response elements is presented.

23.11 The Two-Hybrid System

The GAL4 transcription factor can be used to detect protein-protein interactions in yeast.

Web Essay

23.1 Heterophylly in Aquatic Plants

Abscisic acid induces aerial-type leaf morphology in many aquatic plants.

Chapter References

Allan, A. C., Fricker, M. D., Ward, J. L., Beale, M. H., and Trewavas, A. J. (1994) Two transduction pathways mediate rapid effects of abscisic acid in *Commelina* guard cells. *Plant Cell* 6: 1319–1328.

Allen, G. J., Kuchitsu, K., Chu, S. P., Murata, Y., and Schroeder, J. I. (1999a) *Arabidopsis* abi1-1 and abi2-1 phosphatase mutations reduce abscisic acid-induced cytoplasmic calcium rises in guard cells. *Plant Cell* 11: 1785–1798.

Allen, G. J., Kwak, J. M., Chu, S. P., Llopis, J., Tsien, R. Y., Harper, J. F., and Schroeder, J. I. (1999b) Cameleon calcium indicator reports cytoplasmic calcium dynamics in *Arabidopsis* guard cells. *Plant J.* 19: 735–747.

Anderson, B. E., Ward, J. M., and Schroeder, J. I. (1994) Evidence for an extracellular reception site for abscisic acid in *Commelina* guard cells. *Plant Physiol.* 104: 1177–1183.

Beardsell, M. F., and Cohen, D. (1975) Relationships between leaf water status, abscisic acid levels, and stomatal resistance in maize and sorghum. *Plant Physiol.* 56: 207–212.

Bewley, J. D., and Black, M. (1994) *Seeds: Physiology of Development and Germination*, 2nd ed. Plenum, New York.

Cutler, S., Ghassemian, M., Bonetta, D., Cooney, S., and McCourt, P. (1996) A protein farnesyl transferase involved in abscisic acid signal transduction in *Arabidopsis*. *Science* 273: 1239–1241.

Davies, P. J., ed. (1995) *Plant Hormones: Physiology, Biochemistry and Molecular Biology*, 2nd ed. Kluwer, Dordrecht, Netherlands.

Davies, W. J., and Zhang, J. (1991) Root signals and the regulation of growth and development of plants in drying soil. *Annu. Rev. Plant Physiol. Plant Mol. Biol.* 42: 55–76.

Finkelstein, R. R., and Lynch, T. J. (2000) The *Arabidopsis* abscisic acid response gene ABI5 encodes a basic leucine zipper transcription factor. *Plant Cell* 12: 599–609.

Finkelstein, R. R., Wang, M. L., Lynch, T. J., Rao, S., and Goodman, H. M. (1998) The *Arabidopsis* abscisic acid response locus ABI4 encodes an APETALA2 domain protein. *Plant Cell* 10: 1043–1054.

Ghassemian, M., Nambara, E., Cutler, S., Kawaide, H., Kamiya, Y., and McCourt, P. (2000) Regulation of abscisic acid signaling by the ethylene response pathway in *Arabidopsis*. *Plant Cell* 12: 1117–1126.

Gilroy, S., and Jones, R. L. (1994) Perception of gibberellin and abscisic acid at the external face of the plasma membrane of barley (*Hordeum vulgare* L.) aleurone protoplasts. *Plant Physiol.* 104: 1185–1192.

Gilroy, S., Read, N. D., and Trewavas, A. J. (1990) Elevation of cytoplasmic calcium by caged calcium or caged inositol trisphosphate initiates stomatal closure. *Nature* 343: 769–771.

Gomez-Cadenas, A., Zentella, R., Walker-Simmons, M. K., and Ho, T.-H. D. (2001) Gibberellin/abscisic acid antagonism in barley aleurone cells: Site of action of the protein kinase PKABA1 in relation to gibberellin signaling molecules. *Plant Cell* 13: 667–679.

Gosti, F., Beaudoin, N., Serizet, C., Webb, A. A. R., Vartanian, N., and Giraudat, J. (1999) ABI1 protein phosphatase 2C is a negative regulator of abscisic acid signaling. *Plant Cell* 11: 1897–1909.

Grabov, A., Leung, J., Giraudat, J., and Blatt, M. (1997) Alteration of anion channel kinetics in wild-type and *abi1-1* transgenic *Nicotiana benthamiana* guard cells by abscisic acid. *Plant J.* 12: 203–213.

Hoecker, U., Vasil, I. K., and McCarty, D. R. (1995) Integrated control of seed maturation and germination programs by activator and repressor functions of Viviparous-1 of maize. *Genes Dev.* 9: 2459–2469.

Hugouvieux, V., Kwak, J. M., and Schroeder, J. I. (In press) A mRNA cap binding protein, ABH1, modulates early abscisic acid signal transduction in *Arabidopsis*. *Cell.*

Jeannette, E., Rona, J.-P., Bardat, F., Cornel, D., Sotta, B., and Miginiac, E. (1999) Induction of *RAB18* gene expression and activation of K^+ outward rectifying channels depend on an extracellular perception of ABA in *Arabidopsis thaliana* suspension cells. *Plant J.* 18: 13–22.

Kinoshita, T., Nishimura, M., and Shimazaki, K.-I. (1995) Cytosolic concentration of Ca^{2+} regulates the plasma membrane H^+-ATPase in guard cells of fava bean. *Plant Cell* 7: 1333–1342.

Koornneef, M., Jorna, M. L., Brinkhorst-van der Swan, D. L. C., and Karssen, C. M. (1982) The isolation of abscisic acid (ABA) deficient mutants by selection of induced revertants in non-germinating gibberellin sensitive lines of *Arabidopsis thaliana* L. *Heynh. Theor. Appl. Genet.* 61: 385–393.

Lee, Y., Choi, Y. B., Suh, S., Lee, J., Assmann, S. M., Joe, C. O., Kelleher, J. F., and Crain, R. C. (1996) Abscisic acid-induced phosphoinositide turnover in guard cell protoplasts of *Vicia faba*. *Plant Physiol.* 110: 987–996.

Li, J., and Assmann, S. M. (1996) An abscisic acid-activated and calcium-independent protein kinase from guard cells of fava bean. *Plant Cell* 8: 2359–2368.

Mansfield, T. A., and McAinsh, M. R. (1995) Hormones as regulators of water balance. In *Plant Hormones: Physiology, Biochemistry and Molecular Biology*, 2nd ed., P. J. Davies, ed., Kluwer, Dordrecht, Netherlands, pp. 598–616.

Milborrow, B. V. (2001) The pathway of biosynthesis of abscisic acid in vascular plants: A review of the present state of knowledge of ABA biosynthesis. *J. Exp. Bot.* 52: 1145–1164.

Mori, I. C., and Muto, S. (1997) Abscisic acid activates a 48-kilodalton protein kinase in guard cell protoplasts. *Plant Physiol.* 113: 833–839.

Nambara, E., Hayama, R., Tsuchiya, Y., Nishimura, M., Kawaide, H., Kamiya, Y., and Naito, S. (2000) The role of *abi3* and *fus3* loci in *Arabidopsis thaliana* on phase transition from late embryo development to germination. *Dev. Biol.* 220: 412–423.

Neill, S. J., Desikan, R., Clarke, A., and Hancock, J. T. (2002) Nitric oxide is a novel component of abscisic acid signaling in stomatal guard cells. *Plant Physiol.* 128: 13–16.

Pei, Z.-M., Kuchitsu, K., Ward, J. M., Schwarz, M., and Schroeder, J. I. (1997) Differential abscisic acid regulation of guard cell slow anion channels in *Arabidopsis* wild-type and abi1 and abi2 mutants. *Plant Cell* 9: 409–423.

Pei, Z. M., Murata, Y., Benning, G., Thomine, S., Klusener, B., Allen, G. J., Grill, E., and Schroeder, J. I. (2000) Calcium channels activated by hydrogen peroxide mediate abscisic acid signalling in guard cells. *Nature* 406: 731–734.

Raz, V., Bergervoet, J. H. W., and Koornneef, M. (2001) Sequential steps for developmental arrest in *Arabidopsis* seeds. *Development* 128: 243–252.

Rock, C. D. (2000) Pathways to abscisic acid-regulated gene expression. *New Phytol.* 148: 357–396.

Saab, I. N., Sharp, R. E., Pritchard, J., and Voetberg, G. S. (1990) Increased endogenous abscisic acid maintains primary root growth and inhibits shoot growth of maize seedlings at low water potentials. *Plant Physiol.* 93: 1329–1336.

Sanchez, J.-P., and Chua, N.-H. (2001) *Arabidopsis* PLC1 is required for secondary responses to abscisic acid signals. *Plant Cell* 13: 1143–1154.

Schmidt, C., Schelle, I., Liao, Y.-J., and Schroeder, J. I. (1995) Strong regulation of slow anion channels and abscisic acid signaling in guard cells by phosphorylation and dephosphorylation events. *Proc. Natl. Acad. Sci. USA* 92: 9535–9539.

Schroeder, J. I., and Hagiwara, S. (1990) Repetitive increases in cytosolic Ca^{2+} of guard cells by abscisic acid activation of nonselective Ca^{2+} permeable channels. *Proc. Natl. Acad. Sci. USA* 87: 9305–9309.

Schroeder, J. I., Allen, G. J., Hugouvieux, V., Kwak, J. M., and Waner, D. (2001) Guard cell signal transduction. *Annu. Rev. Plant Phys. Plant Mol. Biol.* 52: 627–658.

Schultz, T. F., and Quatrano, R. S. (1997) Evidence for surface perception of abscisic acid by rice suspension cells as assayed by Em gene expression. *Plant Sci.* 130: 63–71.

Schurr, U., Gollan, T., and Schulze, E.-D. (1992) Stomatal response to drying soil in relation to changes in the xylem sap composition of *Helianthus annuus*. II. Stomatal sensitivity to abscisic acid imported from the xylem sap. *Plant Cell Environ.* 15: 561–567.

Schwartz, A., Ilan, N., Schwartz, M., Scheaffer, J., Assmann, S. M., and Schroeder, J. I. (1995) Anion-channel blockers inhibit S-type anion channels and abscisic acid responses in guard cells. *Plant Physiol.* 109: 651–658.

Schwartz, A., Wu, W.-H., Tucker, E. B., and Assmann, S. M. (1994) Inhibition of inward K^+ channels and stomatal response by abscisic acid: An intracellular locus of phytohormone action. *Proc. Natl. Acad. Sci. USA* 91: 4019–4023.

Sharp, R. E., LeNoble, M. E., Else, M. A., Thorne, E. T., and Gherardi, F. (2000) Endogenous ABA maintains shoot growth in tomato independently of effects on plant water balance: Evidence for an interaction with ethylene. *J. Exp. Bot.* 51: 1575–1584.

Sheen, J. (1998) Mutational analysis of protein phosphatase 2C involved in abscisic acid signal transduction in higher plants. *Proc. Natl. Acad. Sci. USA* 95: 975–980.

Shimazaki, K., Iino, M., and Zeiger, E. (1986) Blue light–dependent proton extrusion by guard cell protoplasts of *Vicia faba*. *Nature* 319: 324–326.

Spollen, W. G., LeNoble, M. E., Samuels, T. D., Bernstein, N., and Sharp, R. E. (2000) Abscisic acid accumulation maintains maize primary root elongation at low water potentials by restricting ethylene production. *Plant Physiol.* 122: 967–976.

Wang, X.-Q., Ullah, H., Jones, A. M., and Assmann, S. M. (2001) G protein regulation of ion channels and abscisic acid signaling in *Arabidopsis* guard cells. *Science* 292: 2070–2072.

White, C. N., Proebsting, W. M., Hedden, P., and Rivin, C.J. (2000) Gibberellins and seed development in maize. I. Evidence that gibberellin/abscisic acid balance governs germination versus maturation pathways. *Plant Physiol.* 122: 1081–1088.

Wilkinson, S., and Davies, W. J. (1997) Xylem sap pH increase: A drought signal received at the apoplastic face of the guard cell that involves the suppression of saturable abscisic acid uptake by the epidermal symplast. *Plant Physiol.* 113: 559–573.

Xiong, L., Lee, H., Ishitani, M., Zhang, C., and Zhu, J.-K. (2001) *FIERY1* encoding an inositol polyphosphate 1-phosphatase is a negative regulator of abscisic acid and stress signaling in *Arabidopsis*. *Genes Dev.* 15: 1971–1984.

Chapter

24

The Control of Flowering

MOST PEOPLE LOOK FORWARD to the spring season and the profusion of flowers it brings. Many vacationers carefully time their travels to coincide with specific blooming seasons: *Citrus* along Blossom Trail in southern California, tulips in Holland. In Washington, D.C., and throughout Japan, the cherry blossoms are received with spirited ceremonies. As spring progresses into summer, summer into fall, and fall into winter, wildflowers bloom at their appointed times.

Although the strong correlation between flowering and seasons is common knowledge, the phenomenon poses fundamental questions that will be addressed in this chapter:

- How do plants keep track of the seasons of the year and the time of day?
- Which environmental signals control flowering, and how are those signals perceived?
- How are environmental signals transduced to bring about the developmental changes associated with flowering?

In Chapter 16 we discussed the role of the root and shoot apical meristems in vegetative growth and development. The transition to flowering involves major changes in the pattern of morphogenesis and cell differentiation at the shoot apical meristem. Ultimately this process leads to the production of the floral organs—sepals, petals, stamens, and carpels (see Figure 1.2.A in **Web Topic 1.2**).

Specialized cells in the anther undergo meiosis to produce four haploid microspores that develop into pollen grains. Similarly, a cell within the ovule divides meiotically to produce four haploid megaspores, one of which survives and undergoes three mitotic divisions to produce the cells of the embryo sac (see Figure 1.2.B in **Web Topic 1.2**). The embryo sac represents the mature female gametophyte. The pollen grain, with its germinating pollen tube, is the mature male gametophyte generation. The two gametophytic structures produce the gametes (egg and sperm cells), which fuse to form the diploid zygote, the first stage of the new sporophyte generation.

Clearly, flowers represent a complex array of functionally specialized structures that differ substantially from the vegetative plant body in form and cell types. The transition to flowering therefore entails radical changes in cell fate within the shoot apical meristem. In the first part of this chapter we will discuss these changes, which are manifested as *floral development*. Recently genes have been identified that play crucial roles in the formation of the floral organs. Such studies have shed new light on the genetic control of plant reproductive development.

The events occurring in the shoot apex that specifically commit the apical meristem to produce *flowers* are collectively referred to as **floral evocation**. In the second part of this chapter we will discuss the events leading to floral evocation. The developmental signals that bring about floral evocation include endogenous factors, such as *circadian rhythms*, *phase change*, and *hormones*, and external factors, such as day length (*photoperiod*) and temperature (*vernalization*). In the case of photoperiodism, transmissible signals from the leaves, collectively referred to as the **floral stimulus**, are translocated to the shoot apical meristem. The interactions of these endogenous and external factors enable plants to synchronize their reproductive development with the environment.

FLORAL MERISTEMS AND FLORAL ORGAN DEVELOPMENT

Floral meristems usually can be distinguished from vegetative meristems, even in the early stages of reproductive development, by their larger size. The transition from vegetative to reproductive development is marked by an increase in the frequency of cell divisions within the central zone of the shoot apical meristem. In the vegetative meristem, the cells of the central zone complete their division cycles slowly. As reproductive development commences, the increase in the size of the meristem is largely a result of the increased division rate of these central cells. Recently, genetic and molecular studies have identified a network of genes that control floral morphogenesis in *Arabidopsis*, snapdragon (*Antirrhinum*), and other species.

In this section we will focus on floral development in *Arabidopsis*, which has been studied extensively (Figure 24.1). First we will outline the basic morphological changes that occur during the transition from the vegetative to the reproductive phase. Next we will consider the arrangement of the floral organs in four whorls on the meristem, and the types of genes that govern the normal pattern of floral development. According to the widely accepted ABC model (which is described in Figure 24.6), the specific locations of floral organs in the flower are regulated by the overlapping expression of three types of floral organ identity genes.

The Characteristics of Shoot Meristems in *Arabidopsis* Change with Development

During the vegetative phase of growth, the *Arabidopsis* vegetative apical meristem produces phytomeres with very short internodes, resulting in a basal rosette of leaves (see Figure 24.1A). (Recall from Chapter 16 that a phytomere consists of a leaf, the node to which the leaf is attached, the axillary bud, and the internode below the node.)

As plants initiate reproductive development, the vegetative meristem is transformed into an indeterminate **primary inflorescence meristem** that produces floral meristems on its flanks (Figure 24.2). The lateral buds of the

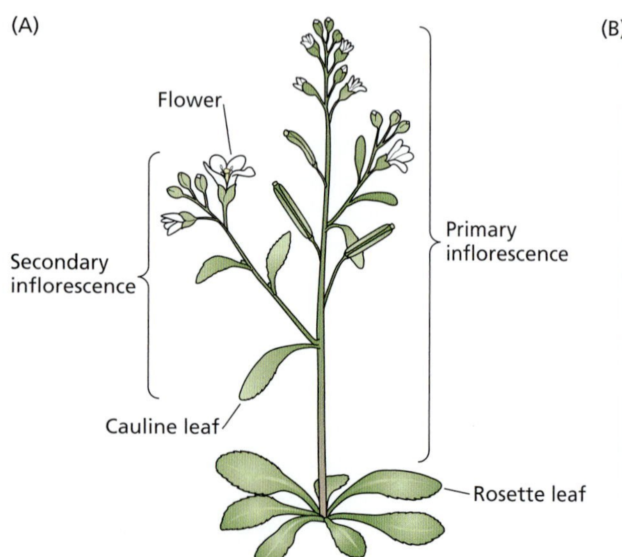

FIGURE 24.1 (A) The shoot apical meristem in *Arabidopsis thaliana* generates different organs at different stages of development. Early in development the shoot apical meristem forms a rosette of basal leaves. When the plant makes the transition to flowering, the shoot apical meristem is transformed into a primary inflorescence meristem that ultimately produces an elongated stem bearing flowers. Leaf primordia initiated prior to the floral transition become cauline leaves, and secondary inflorescences develop in the axils of the cauline leaves. (B) Photograph of an *Arabidopsis* plant. (Photo courtesy of Richard Amasino.)

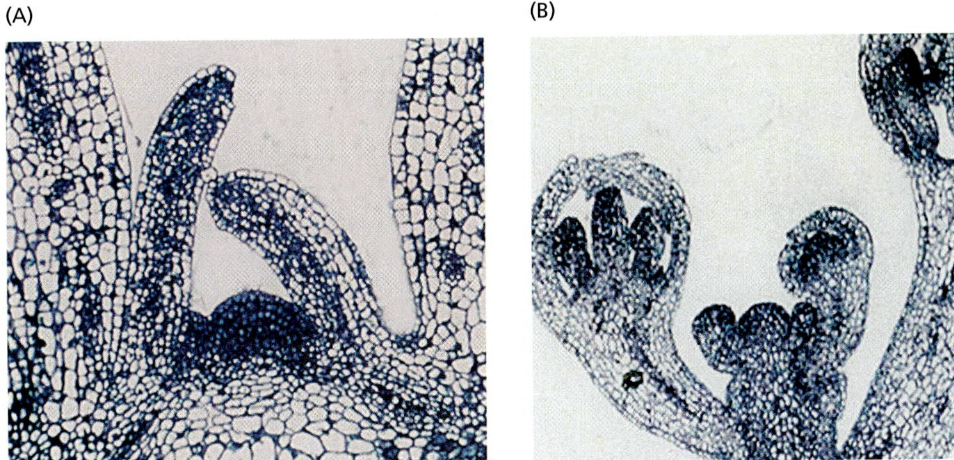

FIGURE 24.2 Longitudinal sections through a vegetative (A) and a reproductive (B) shoot apical region of *Arabidopsis*. (Photos courtesy of V. Grbic´ and M. Nelson.)

cauline leaves (inflorescence leaves) develop into **secondary inflorescence meristems**, and their activity repeats the pattern of development of the primary inflorescence meristem, as shown in Figure 24.1A.

The Four Different Types of Floral Organs Are Initiated as Separate Whorls

Floral meristems initiate four different types of floral organs: sepals, petals, stamens, and carpels (Coen and Carpenter 1993). These sets of organs are initiated in concentric rings, called **whorls**, around the flanks of the meristem (Figure 24.3). The initiation of the innermost organs, the carpels, consumes all of the meristematic cells in the apical dome, and only the floral organ primordia are present as the floral bud develops. In the wild-type *Arabidopsis* flower, the whorls are arranged as follows:

- The first (outermost) whorl consists of four sepals, which are green at maturity.

The second whorl is composed of four petals, which are white at maturity.

- The third whorl contains six stamens, two of which are shorter than the other four.
- The fourth whorl is a single complex organ, the gynoecium or pistil, which is composed of an ovary with two fused carpels, each containing numerous ovules, and a short style capped with a stigma (Figure 24.4).

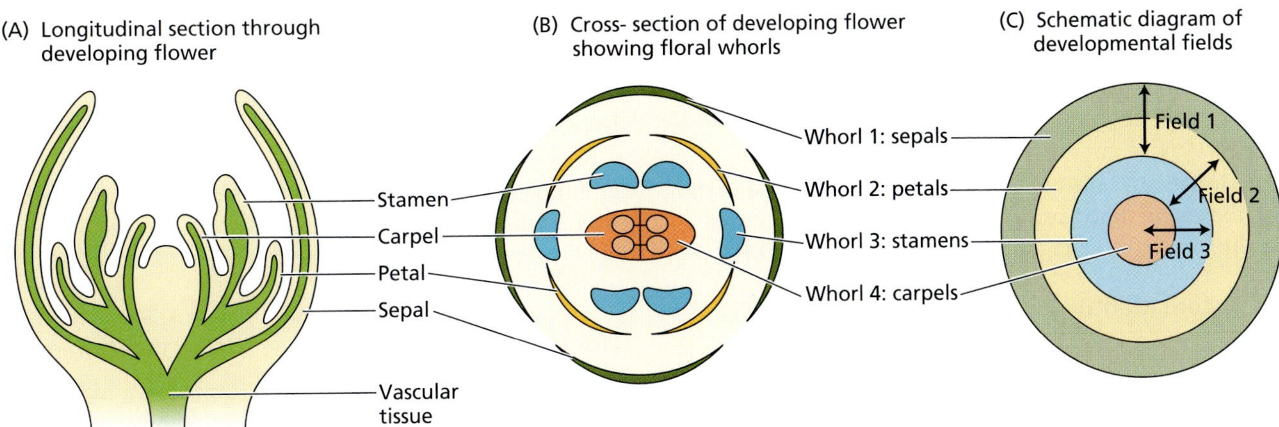

FIGURE 24.3 The floral organs are initiated sequentially by the floral meristem of *Arabidopsis*. (A and B) The floral organs are produced as successive whorls (concentric circles), starting with the sepals and progressing inward. (C) According to the combinatorial model, the functions of each whorl are determined by overlapping developmental fields. These fields correspond to the expression patterns of specific floral organ identity genes. (From Bewley et al. 2000.)

(A)

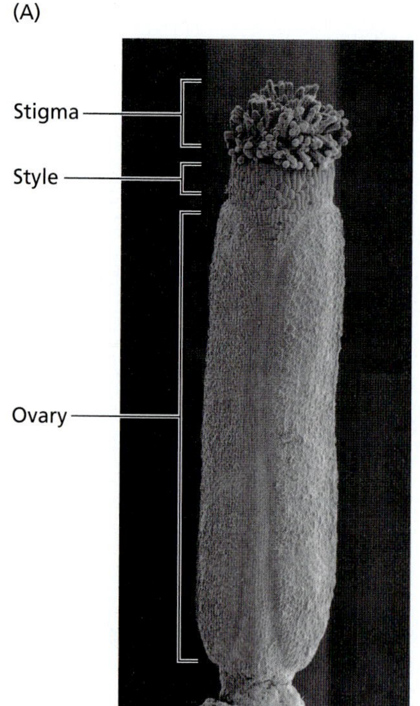

(B)

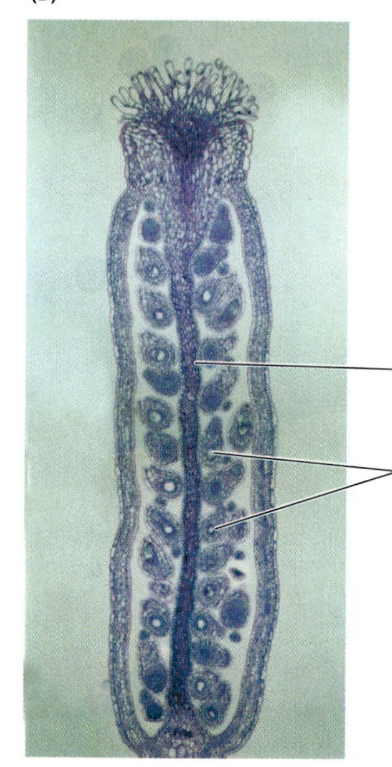

FIGURE 24.4 The *Arabidopsis* pistil consists of two fused carpels, each containing many ovules. (A) Scanning electron micrograph of a pistil, showing the stigma, a short style, and the ovary. (B) Longitudinal section through the pistil, showing the many ovules. (From Gasser and Robinson-Beers 1993, courtesy of C. S. Gasser, © American Society of Plant Biologists, reprinted with permission.)

Three Types of Genes Regulate Floral Development

Mutations have identified three classes of genes that regulate floral development: floral organ identity genes, cadastral genes, and meristem identity genes.

1. **Floral organ identity genes** directly control floral identity. The proteins encoded by these genes are transcription factors that likely control the expression of other genes whose products are involved in the formation and/or function of *floral* organs.
2. **Cadastral genes** act as spatial regulators of the floral organ identity genes by setting boundaries for their expression. (The word *cadastre* refers to a map or survey showing property boundaries for taxation purposes.)
3. **Meristem identity genes** are necessary for the initial induction of the organ identity genes. These genes are the positive regulators of floral organ identity.

Meristem Identity Genes Regulate Meristem Function

Meristem identity genes must be active for the primordia formed at the flanks of the apical meristem to become floral meristems. (Recall that an apical meristem that is forming floral meristems on its flanks is known as an inflorescence meristem.) For example, mutants of *Antirrhinum* (snapdragon) that have a defect in the meristem identity gene *FLORICAULA* develop an inflorescence that does not produce flowers. Instead of causing floral meristems to form in the axils of the bracts, the mutant *floricaula* gene results in the development of additional inflorescence meristems at the bract axils. The wild-type *floricaula* (*FLO*) gene controls the determination step in which floral meristem identity is established.

In *Arabidopsis*, *AGAMOUS-LIKE 20*[1] (*AGL20*), *APETALA1* (*AP1*), and *LEAFY* (*LFY*) are all critical genes in the genetic pathway that must be activated to establish floral meristem identity. *LFY* is the *Arabidopsis* version of the snapdragon *FLO* gene. *AGL20* plays a central role in floral evocation by integrating signals from several different pathways involving both environmental and internal cues (Borner et al. 2000). *AGL20* thus appears to serve as a master switch initiating floral development.

Once activated, *AGL20* triggers the expression of *LFY*, and *LFY* turns on the expression of *AP1* (Simon et al. 1996). In *Arabidopsis*, *LFY* and *AP1* are involved in a positive feedback loop; that is, *AP1* expression also stimulates the expression of *LFY*.

Homeotic Mutations Led to the Identification of Floral Organ Identity Genes

The genes that determine floral organ identity were discovered as **floral homeotic mutants** (see Chapter 14 on the

[1] Also known as *SUPPRESSOR OF OVEREXPRESSION OF CONSTANS 1* (*SOC1*).

web site). As discussed in Chapter 14, mutations in the fruit fly, *Drosophila*, led to the identification of a set of homeotic genes encoding transcription factors that determine the locations at which specific structures develop. Such genes act as major developmental switches that activate the entire genetic program for a particular structure. The expression of homeotic genes thus gives organs their identity.

As we have seen already in this chapter, dicot flowers consist of successive whorls of organs that form as a result of the activity of floral meristems: sepals, petals, stamens, and carpels. These organs are produced when and where they are because of the orderly, patterned expression and interactions of a small group of homeotic genes that specify floral organ identity.

The floral organ identity genes were identified through homeotic mutations that altered floral organ identity so that some of the floral organs appeared in the wrong place. For example, *Arabidopsis* plants with mutations in the *APETALA2* (*AP2*) gene produce flowers with carpels where sepals should be, and stamens where petals normally appear.

The homeotic genes that have been cloned so far encode transcription factors—proteins that control the expression of other genes. Most plant homeotic genes belong to a class of related sequences known as **MADS box genes**, whereas animal homeotic genes contain sequences called homeoboxes (see Chapter 14 on the web site).

Many of the genes that determine floral organ identity are MADS box genes, including the *DEFICIENS* gene of snapdragon and the *AGAMOUS*, *PISTILLATA1*, and *APETALA3* genes of *Arabidopsis*. The MADS box genes share a characteristic, conserved nucleotide sequence known as a *MADS box*, which encodes a protein structure known as the *MADS domain*. The MADS domain enables these transcription factors to bind to DNA that has a specific nucleotide sequence.

Not all genes containing the MADS box domain are homeotic genes. For example, *AGL20* is a MADS box gene, but it functions as a meristem identity gene.

Three Types of Homeotic Genes Control Floral Organ Identity

Five different genes are known to specify floral organ identity in *Arabidopsis*: *APETALA1* (*AP1*), *APETALA2* (*AP2*), *APETALA3* (*AP3*), *PISTILLATA* (*PI*), and *AGAMOUS* (*AG*) (Bowman et al. 1989; Weigel and Meyerowitz 1994). The organ identity genes initially were identified through mutations that dramatically alter the structure and thus the identity of the floral organs produced in two adjacent whorls (Figure 24.5). For example, plants with the *ap2* mutation lack sepals and petals (see Figure 24.5B). Plants bearing *ap3* or *pi* mutations produce sepals instead of petals in the second whorl, and carpels instead of stamens in the third whorl (see Figure 24.5C). And plants homozygous for the *ag* mutation lack both stamens and carpels (see Figure 24.5D).

Because mutations in these genes change floral organ identity without affecting the initiation of flowers, they are homeotic genes. These homeotic genes fall into three classes—types A, B, and C—defining three different kinds of activities (Figure 24.6):

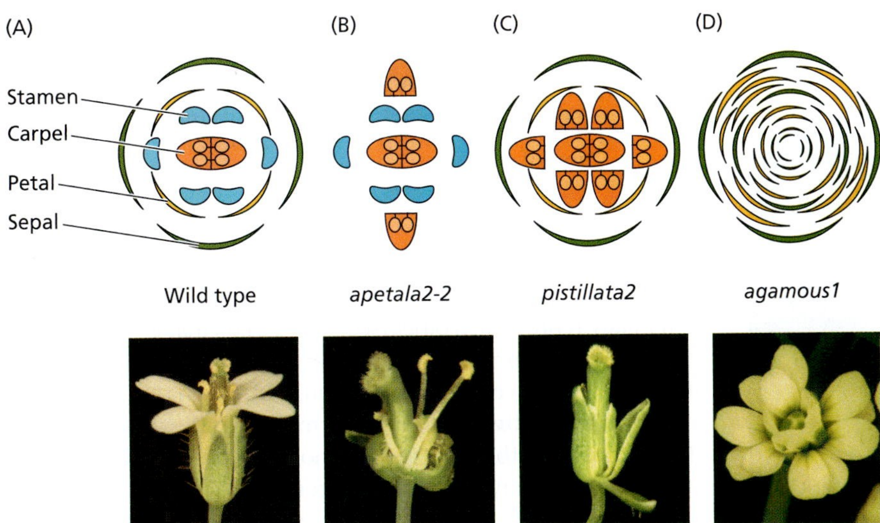

FIGURE 24.5 Mutations in the floral organ identity genes dramatically alter the structure of the flower. (A) Wild type; (B) *apetala2-2* mutants lack sepals and petals; (C) *pistillata2* mutants lack petals and stamens; (D) *agamous1* mutants lack both stamens and carpels. (From Bewley et al. 2000.)

tion is the promotion of flowering—at subsequent higher temperatures—brought about by exposure to cold. Other signals, such as total light radiation and water availability, can also be important external cues.

The evolution of both internal (autonomous) and external (environment-sensing) control systems enables plants to carefully regulate flowering at the optimal time for reproductive success. For example, in many populations of a particular species, flowering is synchronized. This synchrony favors crossbreeding and allows seeds to be produced in favorable environments, particularly with respect to water and temperature.

THE SHOOT APEX AND PHASE CHANGES

All multicellular organisms pass through a series of more or less defined developmental stages, each with its characteristic features. In humans, infancy, childhood, adolescence, and adulthood represent four general stages of development, and puberty is the dividing line between the nonreproductive and the reproductive phases. Higher plants likewise pass through developmental stages, but whereas in animals these changes take place throughout the entire organism, in higher plants they occur in a single, dynamic region, the **shoot apical meristem**.

Shoot Apical Meristems Have Three Developmental Phases

During postembryonic development, the shoot apical meristem passes through three more or less well-defined developmental stages in sequence:

1. The juvenile phase
2. The adult vegetative phase
3. The adult reproductive phase

The transition from one phase to another is called **phase change**.

The primary distinction between the juvenile and the adult vegetative phases is that the latter has the ability to form reproductive structures: flowers in angiosperms, cones in gymnosperms. However, actual expression of the reproductive competence of the adult phase (i.e., flowering) often depends on specific environmental and developmental signals. Thus the absence of flowering itself is not a reliable indicator of juvenility.

The transition from juvenile to adult is frequently accompanied by changes in vegetative characteristics, such as leaf morphology, phyllotaxy (the arrangement of leaves on the stem), thorniness, rooting capacity, and leaf retention in deciduous plants (Figure 24.9; see also **Web Topic 24.1**). Such changes are most evident in woody perennials, but they are apparent in many herbaceous species as well. Unlike the abrupt transition from the adult vegetative phase to the reproductive phase, the transition from juvenile to vegetative adult is usually gradual, involving intermediate forms.

FIGURE 24.9 Juvenile and adult forms of ivy (*Hedera helix*). The juvenile form has lobed palmate leaves arranged alternately, a climbing growth habit, and no flowers. The adult form (projecting out to the right) has entire ovate leaves arranged in spirals, an upright growth habit, and flowers. (Photo by L. Taiz.)

Sometimes the transition can be observed in a single leaf. A dramatic example of this is the progressive transformation of juvenile leaves of the leguminous tree *Acacia heterophylla* into phyllodes, a phenomenon noted by Goethe. Whereas the juvenile pinnately compound leaves consist of rachis (stalk) and leaflets, adult phyllodes are specialized structures representing flattened petioles (Figure 24.10).

Intermediate structures also form during the transition from aquatic to aerial leaf types of aquatic plants such as *Hippuris vulgaris* (common marestail). As in the case of *A. heterophylla*, these intermediate forms possess distinct regions with different developmental patterns. To account for intermediate forms during the transition from juvenile to adult in maize (see **Web Topic 24.2**), a **combinatorial model** has been proposed (Figure 24.11). According to this model, shoot development can be described as a series of independently regulated, *overlapping* programs (juvenile, adult, and reproductive) that modulate the expression of a common set of developmental processes.

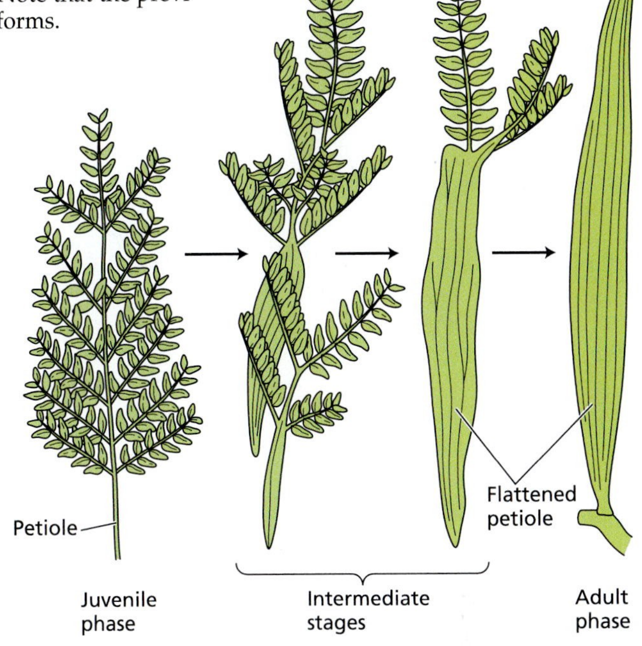

FIGURE 24.10 Leaves of *Acacia heterophylla*, showing transitions from pinnately compound leaves (juvenile phase) to phyllodes (adult phase). Note that the previous phase is retained at the top of the leaf in the intermediate forms.

In the transition from juvenile to adult leaves, the intermediate forms indicate that different regions of the same leaf can express different developmental programs. Thus the cells at the tip of the leaf remain committed to the juvenile program, while the cells at the base of the leaf become committed to the adult program. The developmental fates of the two sets of cells in the same leaf are quite different.

Juvenile Tissues Are Produced First and Are Located at the Base of the Shoot

The sequence in time of the three developmental phases results in a spatial gradient of juvenility along the shoot axis. Because growth in height is restricted to the apical meristem, the juvenile tissues and organs, which form first, are located at the base of the shoot. In rapidly flowering herbaceous species, the juvenile phase may last only a few days, and few juvenile structures are produced. In contrast, woody species have a more prolonged juvenile phase, in some cases lasting 30 to 40 years (Table 24.1). In these cases the juvenile structures can account for a significant portion of the mature plant.

Once the meristem has switched over to the adult phase, only adult vegetative structures are produced, culminating in floral evocation. The adult and reproductive phases are therefore located in the upper and peripheral regions of the shoot.

Attainment of a sufficiently large size appears to be more important than the plant's chronological age in determining the transition to the adult phase. Conditions that retard growth, such as mineral deficiencies, low light, water stress, defoliation, and low temperature tend to prolong the juvenile phase or even cause **rejuvenation** (reversion to juvenility) of adult shoots. In contrast, conditions that promote vigorous growth accelerate the transition to the adult phase. When growth is accelerated, exposure to the correct flower-inducing treatment can result in flowering.

Although plant size seems to be the most important factor, it is not always clear which specific component associated with size is critical. In some *Nicotiana* species, it appears that plants must produce a certain number of leaves to transmit a sufficient amount of the floral stimulus to the apex.

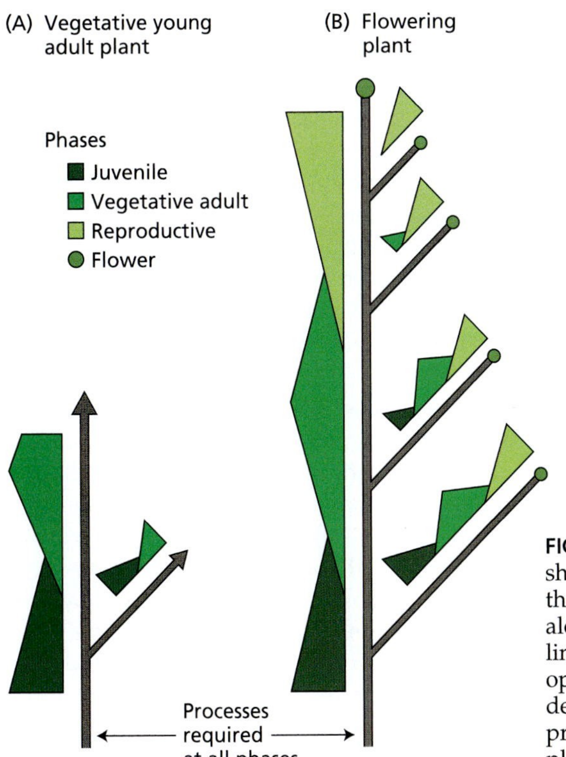

FIGURE 24.11 Schematic representation of the combinatorial model of shoot development in maize. Overlapping gradients of expression of the juvenile, vegetative adult, and reproductive phases are indicated along the length of the main axis and branches. The continuous black line represents processes that are required during all phases of development. Each of the three phases may be regulated by separated developmental programs, with intermediate phases arising when the programs overlap. (A) Vegetative young adult plant. (B) Flowering plant. (After Poethig 1990.)

TABLE 24.1
Length of juvenile period in some woody plant species

Species	Length of juvenile period
Rose (*Rosa* [hybrid tea])	20–30 days
Grape (*Vitis* spp.)	1 year
Apple (*Malus* spp.)	4–8 years
Citrus spp.	5–8 years
English ivy (*Hedera helix*)	5–10 years
Redwood (*Sequoia sempervirens*)	5–15 years
Sycamore maple (*Acer pseudoplatanus*)	15–20 years
English oak (*Quercus robur*)	25–30 years
European beech (*Fagus sylvatica*)	30–40 years

Source: Clark 1983.

Once the adult phase has been attained, it is relatively stable, and it is maintained during vegetative propagation or grafting. For example, in mature plants of English ivy (*Hedera helix*), cuttings taken from the basal region develop into juvenile plants, while those from the tip develop into adult plants. When scions were taken from the base of the flowering tree silver birch (*Betula verrucosa*) and grafted onto seedling rootstocks, there were no flowers on the grafts within the first 2 years. In contrast, the grafts flowered freely when scions were taken from the top of the flowering tree.

In some species, the juvenile meristem appears to be capable of flowering but does not receive sufficient floral stimulus until the plant becomes large enough. In mango (*Mangifera indica*), for example, juvenile seedlings can be induced to flower when grafted to a mature tree. In many other woody species, however, grafting to an adult flowering plant does not induce flowering.

Phase Changes Can Be Influenced by Nutrients, Gibberellins, and Other Chemical Signals

The transition at the shoot apex from the juvenile to the adult phase can be affected by transmissible factors from the rest of the plant. In many plants, exposure to low-light conditions prolongs juvenility or causes reversion to juvenility. A major consequence of the low-light regime is a reduction in the supply of carbohydrates to the apex; thus carbohydrate supply, especially sucrose, may play a role in the transition between juvenility and maturity. Carbohydrate supply as a source of energy and raw material can affect the size of the apex. For example, in the florist's chrysanthemum (*Chrysanthemum morifolium*), flower primordia are not initiated until a minimum apex size has been reached.

The apex receives a variety of hormonal and other factors from the rest of the plant in addition to carbohydrates and other nutrients. Experimental evidence shows that the application of gibberellins causes reproductive structures to form in young, juvenile plants of several conifer families. The involvement of *endogenous* GAs in the control of reproduction is also indicated by the fact that other treatments that accelerate cone production in pines (e.g., root removal, water stress, and nitrogen starvation) often also result in a buildup of GAs in the plant.

On the other hand, although gibberellins promote the attainment of reproductive maturity in conifers and many herbaceous angiosperms as well, GA_3 causes rejuvenation in *Hedera* and in several other woody angiosperms. The role of gibberellins in the control of phase change is thus complex, varies among species, and probably involves interactions with other factors.

Competence and Determination Are Two Stages in Floral Evocation

The term *juvenility* has different meanings for herbaceous and woody species. Whereas juvenile herbaceous meristems flower readily when grafted onto flowering adult plants (see **Web Topic 24.3**), juvenile woody meristems generally do not. What is the difference between the two?

Extensive studies in tobacco have demonstrated that floral evocation requires the apical bud to pass through two developmental stages (Figure 24.12) (McDaniel et al. 1992). One stage is the acquisition of competence. A bud is said to be **competent** if it is able to flower when given the appropriate developmental signal.

For example, if a vegetative shoot (scion) is grafted onto a flowering stock and the scion flowers immediately, it is demonstrably capable of responding to the level of floral stimulus present in the stock and is therefore competent. Failure of the scion to flower would indicate that the shoot apical meristem has not yet attained competence. Thus the juvenile meristems of herbaceous plants are competent to flower, but those of woody species are not.

The next stage that a competent vegetative bud goes through is determination. A bud is said to be **determined** if it progresses to the next developmental stage (flowering) even after being removed from its normal context. Thus a florally determined bud will produce flowers even if it is grafted onto a vegetative plant that is not producing any floral stimulus.

In a day-neutral tobacco, for example, plants typically flower after producing about 41 leaves or nodes. In an experiment to measure the floral determination of the axillary buds, flowering tobacco plants were decapitated just above the thirty-fourth leaf (from the bottom). Released from apical dominance, the axillary bud of the thirty-fourth leaf grew out, and after producing 7 more leaves (for a total of 41), it flowered (Figure 24.13A) (McDaniel 1996). However, if the thirty-fourth bud was excised from the plant and either rooted or grafted onto a stock without leaves near the base, it produced a complete set of leaves (41) before flowering. This result shows that the thirty-fourth bud was not yet florally determined.

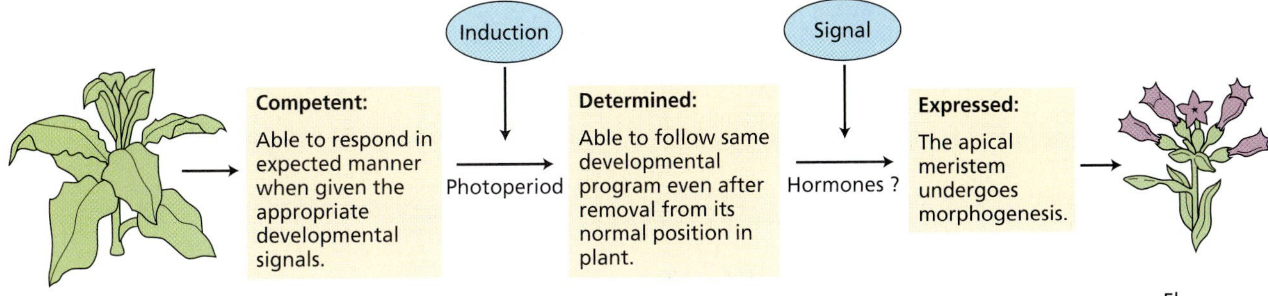

FIGURE 24.12 A simplified model for floral evocation at the shoot apex in which the cells of the vegetative meristem acquire new developmental fates. To initiate floral development, the cells of the meristem must first become competent. A competent vegetative meristem is one that can respond to a floral stimulus (induction) by becoming florally determined (committed to producing a flower). The determined state is usually expressed, but this may require an additional signal. (After McDaniel et al. 1992.)

In another experiment, the donor plant was decapitated above the thirty-seventh leaf. This time the thirty-seventh axillary bud flowered after producing four leaves *in all three situations* (see Figure 24.13B). This result demonstrates that the terminal bud became florally determined after initiating 37 leaves.

Extensive grafting of shoot tips among tobacco varieties has established that the number of nodes a meristem produces before flowering is a function of two factors: (1) the strength of the floral stimulus from the leaves and (2) the competence of the meristem to respond to the signal (McDaniel et al. 1996).

In some cases the **expression** of flowering may be delayed or arrested even after the apex becomes determined, unless it receives a second developmental signal that stimulates expression (see Figure 24.12). For example, intact *Lolium temulentum* (darnel ryegrass) plants become committed to flowering after a single exposure to a long day. If the *Lolium* shoot apical meristem is excised 28 hours after the beginning of the long day and cultured in vitro, it will produce normal inflorescences in culture, but only if the hormone gibberellic acid (GA) is present in the medium. Because apices cultured from plants grown exclusively in short days never flower, even in the presence of

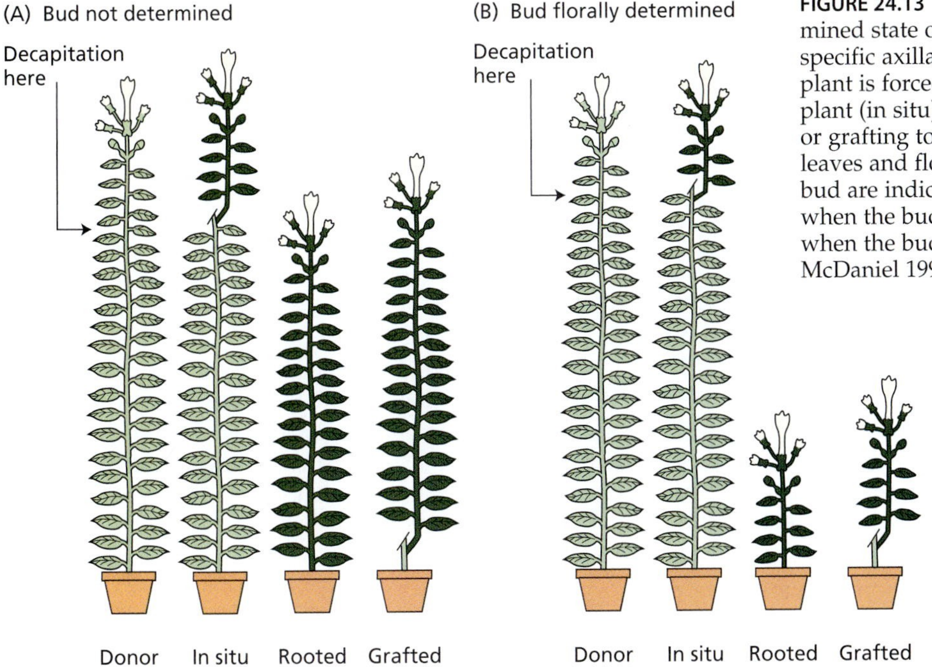

FIGURE 24.13 Demonstration of the determined state of axillary buds in tobacco. A specific axillary bud of a flowering donor plant is forced to grow, either directly on the plant (in situ) by decapitation, or by rooting or grafting to the base of the plant. The new leaves and flowers produced by the axillary bud are indicated by shading. (A) Result when the bud is not determined. (B) Result when the bud is florally determined. (After McDaniel 1996.)

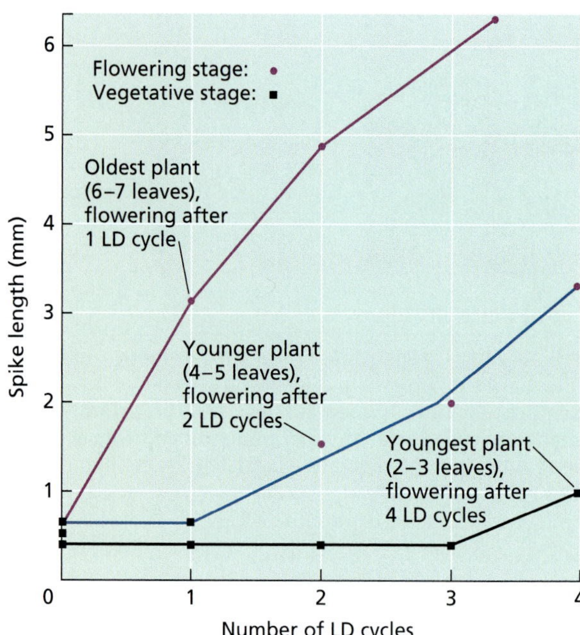

FIGURE 24.14 Effect of plant age on the number of long-day (LD) inductive cycles required for flowering in the long-day plant *Lolium temulentum* (darnel ryegrass). An inductive long-day cycle consisted of 8 hours of sunlight followed by 16 hours of low-intensity incandescent light. The older the plant is, the fewer photoinductive cycles are needed to produce flowering.

GA, we can conclude that long days are required for determination in *Lolium*, whereas GA is required for *expression* of the determined state.

In general, once a meristem has become competent, it exhibits an increasing tendency to flower with age (leaf number). For example, in plants controlled by day length, the number of short-day or long-day cycles necessary to achieve flowering is often fewer in older plants (Figure 24.14). As will be discussed later in the chapter, this increasing tendency to flower with age has its physiological basis in the greater capacity of the leaves to produce a floral stimulus.

Before discussing how plants perceive day length, however, we will lay the foundation by examining how organisms measure time in general. This topic is known as **chronobiology**, or the study of **biological clocks**. The best-understood biological clock is the circadian rhythm.

CIRCADIAN RHYTHMS: THE CLOCK WITHIN

Organisms are normally subjected to daily cycles of light and darkness, and both plants and animals often exhibit rhythmic behavior in association with these changes. Examples of such rhythms include leaf and petal movements (day and night positions), stomatal opening and closing, growth and sporulation patterns in fungi (e.g., *Pilobolus* and *Neurospora*), time of day of pupal emergence (the fruit fly *Drosophila*), and activity cycles in rodents, as well as metabolic processes such as photosynthetic capacity and respiration rate.

When organisms are transferred from daily light–dark cycles to continuous darkness (or continuous dim light), many of these rhythms continue to be expressed, at least for several days. Under such uniform conditions the period of the rhythm is then close to 24 hours, and consequently the term circadian rhythm is applied (see Chapter 17). Because they continue in a constant light or dark environment, these circadian rhythms cannot be direct responses to the presence or absence of light but must be based on an internal pacemaker, often called an endogenous oscillator. A molecular model for a plant endogenous oscillator was described in Chapter 17.

The endogenous oscillator is coupled to a variety of physiological processes, such as leaf movement or photosynthesis, and it maintains the rhythm. For this reason the endogenous oscillator can be considered the clock mechanism, and the physiological functions that are being regulated, such as leaf movements or photosynthesis, are sometimes referred to as the hands of the clock.

Circadian Rhythms Exhibit Characteristic Features

Circadian rhythms arise from cyclic phenomena that are defined by three parameters:

1. **Period**, the time between comparable points in the repeating cycle. Typically the period is measured as the time between consecutive maxima (peaks) or minima (troughs) (Figure 24.15A).
2. **Phase**[2], any point in the cycle that is recognizable by its relationship to the rest of the cycle. The most obvious phase points are the peak and trough positions.
3. **Amplitude**, usually considered to be the distance between peak and trough. The amplitude of a biological rhythm can often vary while the period remains unchanged (as, for example, in Figure 24.15C).

In constant light or darkness, rhythms depart from an exact 24-hour period. The rhythms then drift in relation to solar time, either gaining or losing time depending on whether the period is shorter or longer than 24 hours. Under natural conditions, the endogenous oscillator is

[2] The term *phase* should not be confused with the term *phase change* in meristem development, discussed earlier.

(A)

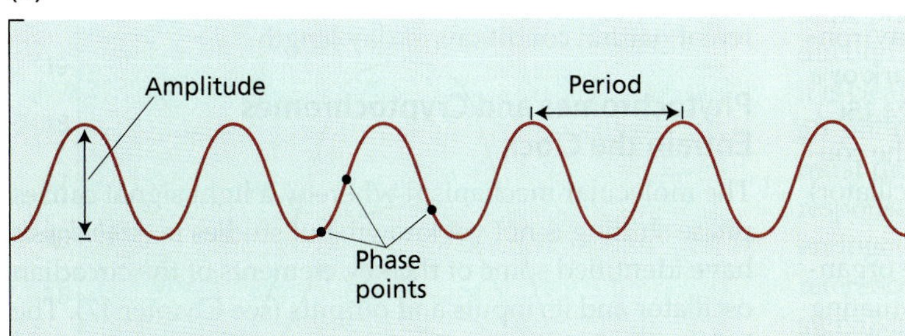

A typical circadian rhythm. The **period** is the time between comparable points in the repeating cycle; the **phase** is any point in the repeating cycle recognizable by its relationship with the rest of the cycle; the **amplitude** is the distance between peak and trough.

(B)

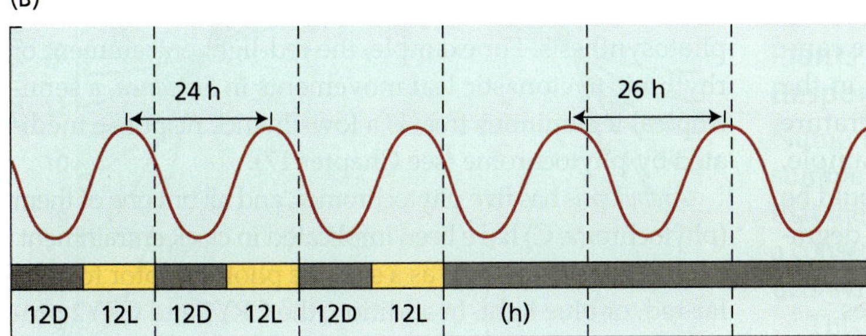

A circadian rhythm entrained to a 24 h light–dark (L–D) cycle and its reversion to the free-running period (26 h in this example) following transfer to continuous darkness.

(C)

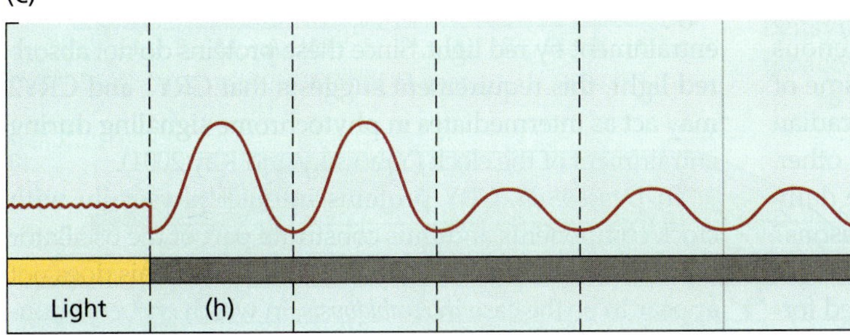

Suspension of a circadian rhythm in continuous bright light and the release or restarting of the rhythm following transfer to darkness.

(D)

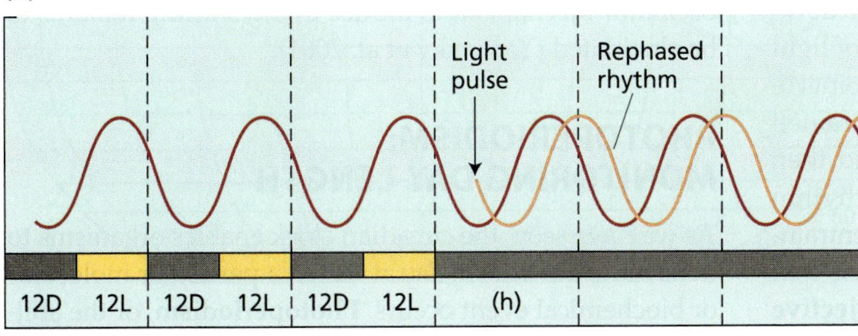

Typical phase-shifting response to a light pulse given shortly after transfer to darkness. The rhythm is rephased (delayed) without its period being changed.

FIGURE 24.15 Some characteristics of circadian rhythms.

entrained (synchronized) to a true 24-hour period by environmental signals, the most important of which are the light-to-dark transition at dusk and the dark-to-light transition at dawn (see Figure 24.15B).

Such environmental signals are termed **zeitgebers** (German for "time givers"). When such signals are removed—for example, by transfer to continuous darkness—the rhythm is said to be **free-running,** and it reverts to the circadian period that is characteristic of the particular organism (see Figure 24.15B).

Although the rhythms are generated internally, they normally require an environmental signal, such as exposure to light or a change in temperature, to initiate their expression. In addition, many rhythms damp out (i.e., the

(*Xanthium strumarium*) or 10 hours in soybean (*Glycine max*). The duration of darkness was also shown to be important in LDPs (see Figure 24.19). These plants were found to flower in short days, provided that the accompanying night length was also short; however, a regime of long days followed by long nights was ineffective.

Night Breaks Can Cancel the Effect of the Dark Period

A feature that underscores the importance of the dark period is that it can be made ineffective by interruption with a short exposure to light, called a **night break** (see Figure 24.19A). In contrast, interrupting a long day with a brief dark period does not cancel the effect of the long day (see Figure 24.19B). Night-break treatments of only a few minutes are effective in *preventing* flowering in many SDPs, including *Xanthium* and *Pharbitis*, but much longer exposures are often required to *promote* flowering in LDPs.

In addition, the effect of a night break varies greatly according to the time when it is given. For both LDPs and SDPs, a night break was found to be most effective when given near the middle of a dark period of 16 hours (Figure 24.20).

The discovery of the night-break effect, and its time dependence, had several important consequences. It established the central role of the dark period and provided a valuable probe for studying photoperiodic timekeeping. Because only small amounts of light are needed, it became possible to study the action and identity of the photoreceptor without the interfering effects of photosynthesis and other nonphotoperiodic phenomena. This discovery has also led to the development of commercial methods for regulating the time of flowering in horticultural species, such as *Kalanchoe*, chrysanthemum, and poinsettia (*Euphorbia pulcherrima*).

The Circadian Clock Is Involved in Photoperiodic Timekeeping

The decisive effect of night length on flowering indicates that measuring the passage of time in darkness is central to photoperiodic timekeeping. Most of the available evidence favors a mechanism based on a circadian rhythm (Bünning 1960). According to the **clock hypothesis**, photoperiodic timekeeping depends on an endogenous circadian oscillator of the type involved in the daily rhythms described in Chapter 17 in relation to phytochrome. The central oscillator is coupled to various physiological processes that involve gene expression, including flowering in photoperiodic species.

Measurements of the effect of a night break on flowering can be used to investigate the role of circadian rhythms in photoperiodic timekeeping. For example, when soybean

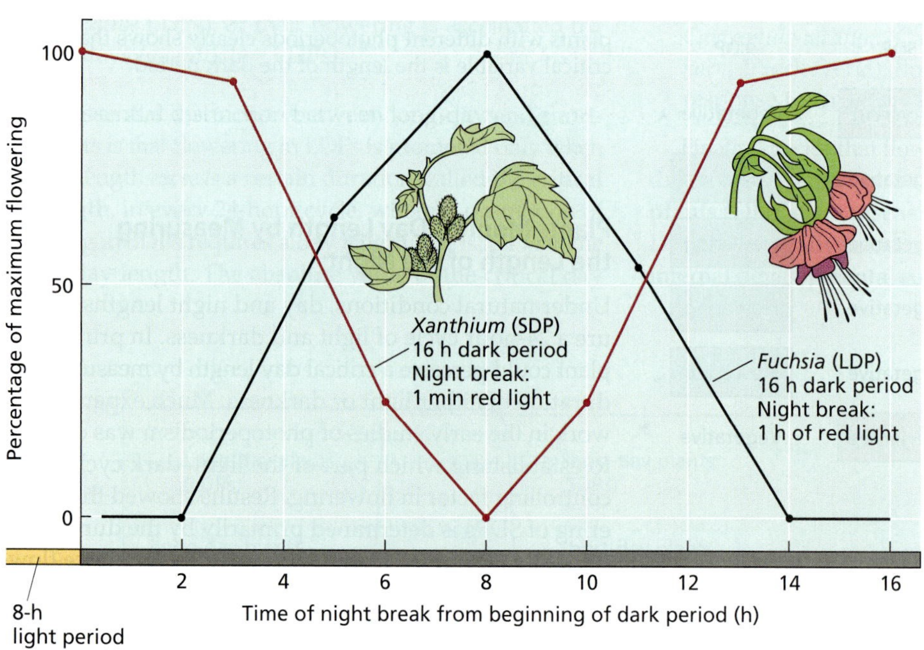

FIGURE 24.20 The time when a night break is given determines the flowering response. When given during a long dark period, a night break promotes flowering in LDPs and inhibits flowering in SDPs. In both cases, the greatest effect on flowering occurs when the night break is given near the middle of the 16-hour dark period. The LDP *Fuchsia* was given a 1-hour exposure to red light in a 16-hour dark period. *Xanthium* was exposed to red light for 1 minute in a 16-hour dark period. (Data for *Fuchsia* from Vince-Prue 1975; data for *Xanthium* from Salisbury 1963 and Papenfuss and Salisbury 1967.)

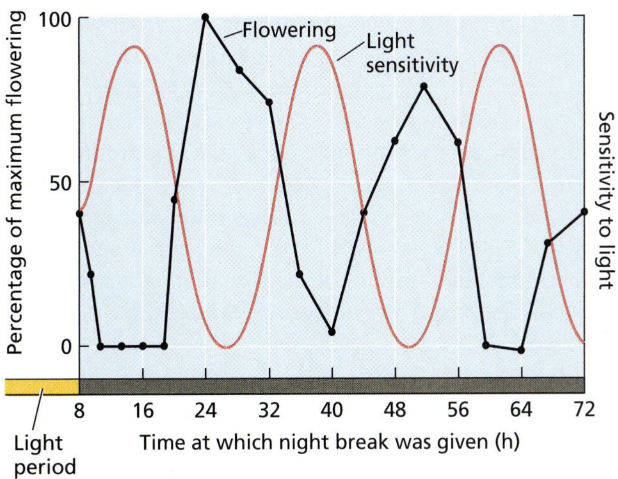

FIGURE 24.21 Rhythmic flowering in response to night breaks. In this experiment, the SDP soybean (*Glycine max*) received cycles of an 8-hour light period followed by a 64-hour dark period. A 4-hour night break was given at various times during the long inductive dark period. The flowering response, plotted as the percentage of the maximum, was then plotted for each night break given. Note that a night break given at 26 hours induced maximum flowering, while no flowering was obtained when the night break was given at 40 hours. Moreover, this experiment demonstrates that the sensitivity to the effect of the night break shows a circadian rhythm. These data support a model in which flowering in SDPs is induced only when dawn (or a night break) occurs after the completion of the light-sensitive phase. In LDPs the light break must coincide with the light-sensitive phase for flowering to occur. (Data from Coulter and Hamner 1964.)

plants, which are SDPs, are transferred from an 8-hour light period to an extended 64-hour dark period, the flowering response to night breaks shows a circadian rhythm (Figure 24.21).

This type of experiment provides strong support for the clock hypothesis. If this SDP were simply measuring the length of night by the accumulation of a particular intermediate in the dark, any dark period greater than the critical night length should cause flowering. Yet long dark periods are not inductive for flowering if the light break is given at a time that does not properly coincide with a certain phase of the endogenous circadian oscillator. This finding demonstrates that flowering in SDPs requires both a dark period of sufficient duration and a dawn signal at an appropriate time in the circadian cycle (see Figure 24.15).

Further evidence for the role of a circadian oscillator in photoperiod measurement is the observation that the photoperiodic response can be phase-shifted by light treatments (see **Web Topic 24.4**).

The Coincidence Model Is Based on Oscillating Phases of Light Sensitivity

The involvement of a circadian oscillator in photoperiodism poses an important question: How does an oscillation with a 24-hour period measure a critical duration of darkness of, say, 8 to 9 hours, as in the SDP *Xanthium*? Erwin Bünning proposed in 1936 that the control of flowering by photoperiodism is achieved by an oscillation of phases with different sensitivities to light. This proposal has evolved into a **coincidence model** (Bünning 1960), in which the circadian oscillator controls the timing of light-sensitive and light-insensitive phases.

The ability of light either to promote or to inhibit flowering depends on the phase in which the light is given. When a light signal is administered during the light-sensitive phase of the rhythm, the effect is either to *promote* flowering in LDPs or to *prevent* flowering in SDPs. As shown in Figure 24.21, the phases of sensitivity and insensitivity to light continue to oscillate in darkness in SDPs. Flowering in SDPs is induced only when exposure to light from a night break or from dawn occurs after completion of the light-sensitive phase of the rhythm. In other words, *flowering is induced when the light exposure is coincident with the appropriate phase of the rhythm.* This continued oscillation of sensitive and insensitive phases in the absence of dawn and dusk light signals is characteristic of a variety of processes controlled by the circadian oscillator.

The Leaf Is the Site of Perception of the Photoperiodic Stimulus

The photoperiodic stimulus in both LDPs and SDPs is perceived by the leaves. For example, treatment of a single leaf of the SDP *Xanthium* with short photoperiods is sufficient to cause the formation of flowers, even when the rest of the plant is exposed to long days. Thus, in response to photoperiod the leaf transmits a signal that regulates the transition to flowering at the shoot apex. The photoperiod-regulated processes that occur in the leaves resulting in the transmission of a floral stimulus to the shoot apex are referred to collectively as **photoperiodic induction**.

Photoperiodic induction can take place in a leaf that has been separated from the plant. For example, in the SDP *Perilla crispa*, an excised leaf exposed to short days can cause flowering when subsequently grafted to a noninduced plant maintained in long days (Zeevaart and Boyer 1987). This result indicates that photoperiodic induction depends on events that take place exclusively in the leaf.

Grafting experiments, which have contributed greatly to our understanding of the floral stimulus, will be discussed in more detail later in the chapter.

The Floral Stimulus Is Transported via the Phloem

Once produced, the flowering stimulus appears to be transported to the meristem via the phloem, and it appears to be chemical rather than physical in nature. Treatments that block phloem transport, such as girdling or localized heat-killing (see Chapter 10), prevent movement of the floral signal.

It is possible to measure rates of movement of the flowering stimulus by removing a leaf at different times after induction, and comparing the time it takes for the signal to reach two buds located at different distances from the induced leaf. The rationale for this type of measurement is that a threshold amount of the signaling compound has reached the bud when flowering takes place, despite the removal of the leaf.

Studies using this method have shown that the rate of transport of the flowering signal is comparable to, or somewhat slower than, the rate of translocation of sugars in the phloem (see Chapter 10). For example, export of the floral stimulus from adult leaves of the SDP *Chenopodium* is complete within 22.5 hours from the beginning of the long night period. In the LDP *Sinapis*, movement of the floral stimulus out of the leaf is complete by as early as 16 hours after the start of the long-day treatment. These rates are consistent with a floral stimulus that moves in the phloem (Zeevaart 1976).

Because the floral stimulus is translocated along with sugars in the phloem, it is subject to source–sink relations. An induced leaf positioned close to the shoot apex is more likely to cause flowering than an induced leaf at the base of a stem, which normally feeds the roots. Similarly, noninduced leaves positioned between the induced leaf and the apical bud will tend to inhibit flowering by serving as the preferred source leaves for the bud, thus preventing the floral stimulus from the more distal induced leaf from reaching its target. This inhibition also explains why a minimum amount of photosynthesis is required by the induced leaf to drive translocation.

Phytochrome Is the Primary Photoreceptor in Photoperiodism

Night-break experiments are well suited for studying the nature of the photoreceptors involved in the reception of light signals during the photoperiodic response. The inhibition of flowering in SDPs by night breaks was one of the first physiological processes shown to be under the control of phytochrome (Figure 24.22).

In many SDPs, a night break becomes effective only when the supplied dose of light is sufficient to saturate the photoconversion of Pr (phytochrome that absorbs red light) to Pfr (phytochrome that absorbs far-red light) (see Chapter 17). A subsequent exposure to far-red light, which photoconverts the pigment back to the physiologically inactive Pr form, restores the flowering response.

In some LDPs, red and far-red reversibility has also been demonstrated. In these plants, a night break of red light promoted flowering, and a subsequent exposure to far-red light prevented this response.

Action spectra for the inhibition and restoration of the flowering response in SDPs are shown in Figure 24.23. A peak at 660 nm, the absorption maximum of Pr (see Chapter 17), is obtained when dark-grown *Pharbitis* seedlings are

FIGURE 24.22 Phytochrome control of flowering by red (R) and far-red (FR) light. A flash of red light during the dark period induces flowering in an LDP, and the effect is reversed by a flash of far-red light. This response indicates the involvement of phytochrome. In SDPs, a flash of red light prevents flowering, and the effect is reversed by a flash of far-red light.

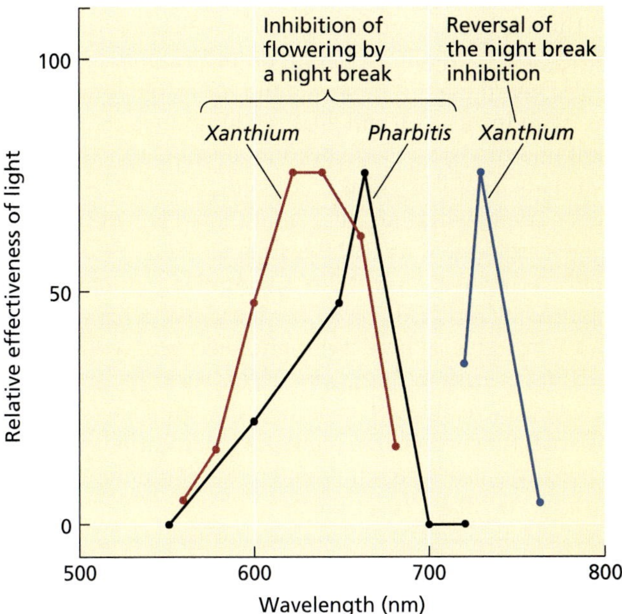

FIGURE 24.23 Action spectra for the control of flowering by night breaks implicates phytochrome. Flowering in SDPs is inhibited by a short light treatment (night break) given in an otherwise inductive period. In the SDP *Xanthium strumarium*, red-light night breaks of 620 to 640 nm are the most effective. Reversal of the red-light effect is maximal at 725 nm. In the dark-grown SDP *Pharbitis nil*, which is devoid of chlorophyll and its interference with light absorption, night breaks of 660 nm are the most effective. This 660 nm maximum coincides with the absorption maximum of phytochrome. (Data for *Xanthium* from Hendricks and Siegelman 1967; data for *Pharbitis* from Saji et al. 1983.)

used to avoid interference from chlorophyll. In contrast, the spectra for *Xanthium* provide an example of the response in green plants, in which the presence of chlorophyll can cause some discrepancy between the action spectrum and the absorption spectrum of Pr. These action spectra and the reversibility between red light and far-red light confirm the role of phytochrome as the photoreceptor that is involved in photoperiod measurement in SDPs.

In LDPs the role of phytochrome is more complex, and a blue-light photoreceptor (which will be discussed shortly) also plays a role in controlling flowering.

Far-Red Light Modifies Flowering in Some LDPs

Circadian rhythms have also been found in LDPs. A circadian rhythm in the promotion of flowering by far-red light has been observed in barley (*Hordeum vulgare*) and *Arabidopsis* (Deitzer 1984), as well as in darnel ryegrass (*Lolium temulentum*) (Figure 24.24). The response is proportional to the irradiance and duration of far-red light and is therefore a high-irradiance response (HIR). Like other HIRs, PHYA is the phytochrome that mediates the response to far-red light (see Chapter 17). In both cases, when the plant is exposed to far-red light for 4 to 6 hours, flowering is promoted compared with plants maintained under continuous white or red light—a response mediated by PHYB. The rhythm continues to run in the light.

In SDPs, on the other hand, a characteristic feature of the timing mechanism is that the rhythm of the response to far-red light damps out after a few hours in continuous light and is restarted upon transfer to darkness.

The response to far-red light is not the only rhythmic feature in LDPs. Although relatively insensitive to a night break of only a few minutes, many LDPs can be induced to flower with a longer night break, usually of at least 1 hour. A circadian oscillation in the flowering response to such a long night break has been observed in LDPs, showing that a rhythm of responsiveness to light continues to run in darkness.

Thus, circadian rhythms that modify the flowering response in LDPs have been shown to run both in the light (promotion by far-red light) and in the dark (promotion by red or white light). However, we do not yet know how the circadian rhythm is coupled to the photoperiodic response.

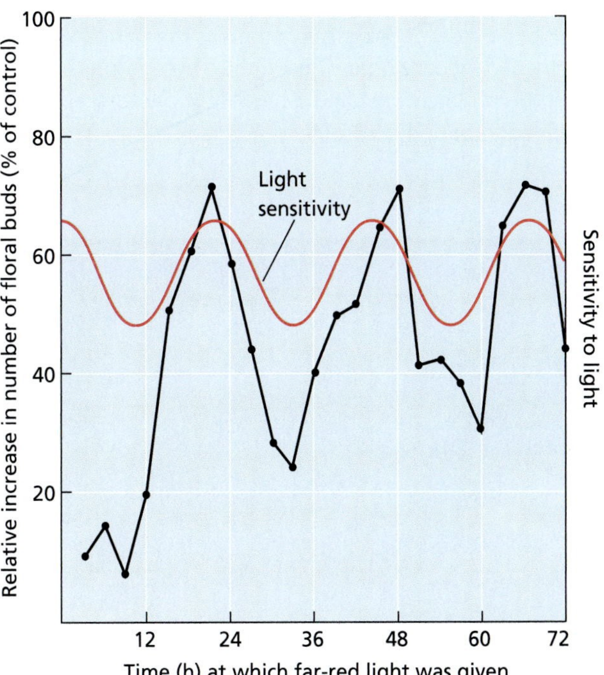

FIGURE 24.24 Effect of far-red light on floral induction in *Arabidopsis*. Four hours of far-red light was added at the indicated times during a continuous 72-hour daylight period. Data points in the graph are plotted at the centers of the 6-hour treatments. The data show a circadian rhythm of sensitivity to the far-red promotion of flowering (red line). This supports a model in which flowering in LDPs is promoted when the light treatment (in this case far-red light) coincides with the peak of light sensitivity. (After Deitzer 1984.)

A Blue-Light Photoreceptor Also Regulates Flowering

In some LDPs, such as *Arabidopsis*, blue light can promote flowering, suggesting the possible participation of a blue-light photoreceptor in the control of flowering. The role of blue light in flowering and its relationship to circadian rhythms have been investigated by use of the luciferase reporter gene construct mentioned in **Web Topic 24.6**. In continuous white light, the cyclic luminescence has a period of 24.7 hours, but in constant darkness the period lengthens to 30 to 36 hours. Either red or blue light, given individually, shortens the period to 25 hours.

To distinguish between the effects of phytochrome and a blue-light photoreceptor, researchers transformed phytochrome-deficient *hy1* mutants, which are defective in chromophore synthesis and are therefore deficient in *all* phytochromes (see Chapter 17), with the luciferase construct to determine the effect of the mutation on the period length (Millar et al. 1995).

Under continuous white light, the *hy1* plants had a period similar to that of the wild type, indicating that little or no phytochrome is required for white light to affect the period. Furthermore, under continuous red light, which would be perceived only by PHYB (see Chapter 17), the period of *hy1* was significantly lengthened (i.e., it became more like constant darkness), whereas the period was not lengthened by continuous blue light. These results indicate that both phytochrome and a blue-light photoreceptor are involved in period control.

The role of blue light in regulating both circadian rhythmicity and flowering is also supported by studies with an *Arabidopsis* flowering-time mutant: *elf3* (*e*arly *f*lowering *3*) (see **Web Topics 24.5 and 24.6**). Confirmation that a blue-light photoreceptor is involved in sensing inductive photoperiods in *Arabidopsis* was recently provided by experiments demonstrating that mutations in one of the cryptochrome genes, *CRY2* (see Chapter 18), caused a delay in flowering and an inability to perceive inductive photoperiods (Guo et al. 1998). As discussed in Chapter 18, *CRY1* encodes a blue-light photoreceptor controlling seedling growth in *Arabidopsis*. Thus, various *CRY* family members have, through evolution, become specialized for different functions in the plant. As noted earlier, the CRY protein has also been implicated in the entrainment of the circadian oscillator (see Chapter 17).

VERNALIZATION: PROMOTING FLOWERING WITH COLD

Vernalization is the process whereby flowering is promoted by a cold treatment given to a fully hydrated seed (i.e., a seed that has imbibed water) or to a growing plant. Dry seeds do not respond to the cold treatment. Without the cold treatment, plants that require vernalization show delayed flowering or remain vegetative. In many cases these plants grow as rosettes with no elongation of the stem (Figure 24.25).

In this section we will examine some of the characteristics of the cold requirement for flowering, including the

FIGURE 24.25 Vernalization induces flowering in the winter-annual types of *Arabidopsis thaliana*. The plant on the left is a winter-annual type that has not been exposed to cold. The plant on the right is a genetically identical winter-annual type that was exposed to 40 days of temperatures slightly above freezing (4°C) as a seedling. It flowered 3 weeks after the end of the cold treatment with about 9 leaves on the primary stem. (Courtesy of Colleen Bizzell.)

range and duration of the inductive temperatures, the sites of perception, the relationship to photoperiodism, and a possible molecular mechanism.

Vernalization Results in Competence to Flower at the Shoot Apical Meristem

Plants differ considerably in the age at which they become sensitive to vernalization. Winter annuals, such as the winter forms of cereals (which are sown in the fall and flower in the following summer), respond to low temperature very early in their life cycle. They can be vernalized before germination if the seeds have imbibed water and become metabolically active. Other plants, including most biennials (which grow as rosettes during the first season after sowing and flower in the following summer), must reach a minimal size before they become sensitive to low temperature for vernalization.

The effective temperature range for vernalization is from just below freezing to about 10°C, with a broad optimum usually between about 1 and 7°C (Lang 1965). The effect of cold increases with the duration of the cold treatment until the response is saturated. The response usually requires several weeks of exposure to low temperature, but the precise duration varies widely with species and variety.

Vernalization can be lost as a result of exposure to devernalizing conditions, such as high temperature (Figure 24.26), but the longer the exposure to low temperature, the more permanent the vernalization effect.

Vernalization appears to take place primarily in the shoot apical meristem. Localized cooling causes flowering when only the stem apex is chilled, and this effect appears to be largely independent of the temperature experienced by the rest of the plant. Excised shoot tips have been successfully vernalized, and where seed vernalization is possible, fragments of embryos consisting essentially of the shoot tip are sensitive to low temperature.

In developmental terms, vernalization results in the acquisition of competence of the meristem to undergo the floral transition. Yet, as discussed earlier in the chapter, competence to flower does not guarantee that flowering will occur. A vernalization requirement is often linked with a requirement for a particular photoperiod (Lang 1965). The most common combination is a requirement for cold treatment *followed* by a requirement for long days—a combination that leads to flowering in early summer at high latitudes (see **Web Topic 24.7**). Unless devernalized, the vernalized meristem can remain competent to flower for as long as 300 days in the absence of the inductive photoperiod.

Vernalization May Involve Epigenetic Changes in Gene Expression

It is important to note that for vernalization to occur, active metabolism is required during the cold treatment. Sources of energy (sugars) and oxygen are required, and temperatures below freezing at which metabolic activity is suppressed are not effective for vernalization. Furthermore, cell division and DNA replication also appear to be required.

One model for how vernalization affects competence is that there are stable changes in the pattern of gene expression in the meristem after cold treatment. Changes in gene expression that are stable even after the signal that induced the change (in this case cold) is removed are known as **epigenetic regulation**. Epigenetic changes of gene expression in many organisms, from yeast to mammals, often require cell division and DNA replication, as is the case for vernalization.

The involvement of epigenetic regulation in the vernalization process has been confirmed in the LDP *Arabidopsis*. In winter-annual ecotypes of *Arabidopsis* that require both vernalization and long days to flower, a gene that acts as a repressor of flowering has been identified: ***FLOWERING LOCUS C*** **(*FLC*)**. *FLC* is highly expressed in nonvernalized shoot apical meristems (Michaels and Amasino 2000). After vernalization, this gene is epigenetically switched off by an unknown mechanism for the remainder of the plant's life cycle, permitting flowering in response to long days to occur (Figure 24.27). In the next generation, however, the gene is switched on again, restoring the requirement for

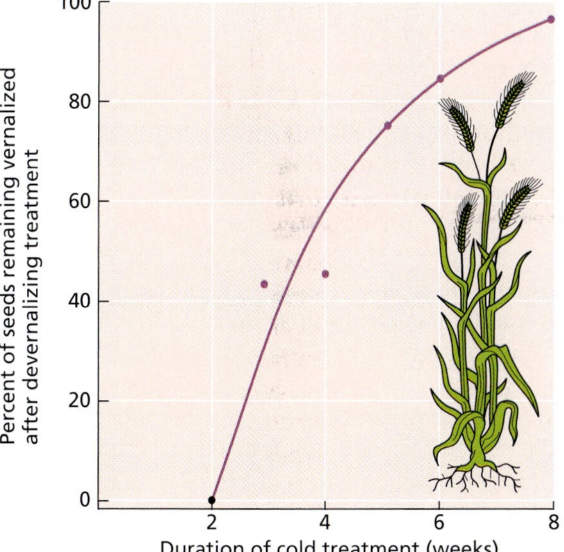

FIGURE 24.26 The duration of exposure to low temperature increases the stability of the vernalization effect. The longer that winter rye (*Secale cereale*) is exposed to a cold treatment, the greater the number of plants that remain vernalized when the cold treatment is followed by a devernalizing treatment. In this experiment, seeds of rye that had imbibed water were exposed to 5°C for different lengths of time, then immediately given a devernalizing treatment of 3 days at 35°C. (Data from Purvis and Gregory 1952.)

FIGURE 24.27 (Left) Vernalization blocks the expression of the gene *FLOWERING LOCUS C* (*FLC*) in cold-requiring winter annual ecotypes of *Arabidopsis*. (Right) A winter annual with an *FLC* mutation exhibits early flowering without cold treatment. (Photo courtesy of R. Amasino.)

cold. Thus in *Arabidopsis*, the state of expression of the *FLC* gene represents a major determinant of meristem competence (Michaels and Amasino 2000).

BIOCHEMICAL SIGNALING INVOLVED IN FLOWERING

In the preceding sections we examined the influence of environmental conditions (such as temperature and day length) versus that of autonomous factors (such as age) on flowering. Although floral evocation occurs at the apical meristems of the shoots, some of the events that result in floral evocation are triggered by biochemical signals arriving at the apex from other parts of the plant, especially from the leaves. Mutants have been isolated that are deficient in the floral stimulus (see **Web Topic 24.6**).

In this section we will consider the nature of the biochemical signals arriving from the leaves and other parts of the plant in response to photoperiodic stimuli. Such signals may serve either as activators or as inhibitors of flowering. After years of investigation, no single substance has been identified as the universal floral stimulus, although certain hormones, such as gibberellins and ethylene, can induce flowering in some species. Hence, most current models of the floral stimulus are based on multiple factors.

Grafting Studies Have Provided Evidence for a Transmissible Floral Stimulus

The production in photoperiodically induced leaves of a biochemical signal that is transported to a distant target tissue (the shoot apex) where it stimulates a response (flowering) satisfies an important criterion for a hormonal effect. In the 1930s, Mikhail Chailakhyan, working in Russia, postulated the existence of a universal flowering hormone, which he named **florigen**.

The evidence in support of florigen comes mainly from early grafting experiments in which noninduced receptor plants were stimulated to flower by being grafted onto a leaf or shoot from photoperiodically induced donor plants. For example, in the SDP *Perilla crispa*, a member of the mint family, grafting a leaf from a plant grown under inductive short days onto a plant grown under noninductive long days causes the latter to flower (Figure 24.28). Moreover, the floral stimulus seems to be the same in plants with different photoperiodic requirements. Thus, grafting an induced leaf from the LDP *Nicotiana sylvestris*, grown under long days, onto the SDP Maryland Mammoth tobacco caused the latter to flower under noninductive (long day) conditions.

The leaves of DNPs have also been shown to produce a graft-transmissible floral stimulus (Table 24.2). For example, grafting a single leaf of a day-neutral variety of soy-

FIGURE 24.28 Demonstration by grafting of a leaf-generated floral stimulus in the SDP *Perilla*. (Left) Grafting an induced leaf from a plant grown under short days onto a noninduced shoot causes the axillary shoots to produce flowers. The donor leaf has been trimmed to facilitate grafting, and the upper leaves have been removed from the stock to promote phloem translocation from the scion to the receptor shoots. (Right) Grafting a noninduced leaf from a plant grown under LDs results in the formation of vegetative branches only. (Photo courtesy of J. A. D. Zeevaart.)

bean, Agate, onto the short-day variety, Biloxi, caused flowering in Biloxi even when the latter was maintained in noninductive long days. Similarly, a leaf from a day-neutral variety of tobacco (*Nicotiana tabacum*, cv. Trapezond) grafted onto the LDP *Nicotiana sylvestris* induced the latter to flower under noninductive short days.

In a few cases, flowering has been induced by grafts between different genera. The SDP *Xanthium strumarium* flowered under long-day conditions when shoots of flowering *Calendula officinalis* were grafted onto a vegetative *Xanthium* stock. Similarly, grafting a shoot from the LDP *Petunia hybrida* onto a stock of the cold-requiring biennial *Hyoscyamus niger* (henbane) caused the latter to flower under long days, even though it was nonvernalized (Figure 24.29).

In *Perilla* (see Figure 24.28), the movement of the floral stimulus from a donor leaf to the stock across the graft union

FIGURE 24.29 Successful transfer of the floral stimulus between different genera: The scion (right branch) is the LDP *Petunia hybrida*, and the stock is nonvernalized *Hyoscyamus niger* (henbane). The graft combination was maintained under LDs. (Photo courtesy of J. A. D. Zeevaart.)

TABLE 24.2
Transmissible factors regulate flowering.

Donor plants maintained under flower-inducing conditions	Photoperiod type[a,b]	Vegetative receptor plant induced to flower	Photoperiod type[a,b]
Helianthus annus	DNP in LD	*H. tuberosus*	SDP in LD
Nicotiana tabacum Delcrest	DNP in SD	*N. sylvestris*	LDP in SD
Nicotiana sylvestris	LDP in LD	*N. tabacum* Maryland Mammoth	SDP in LD
Nicotiana tabacum Maryland Mammoth	SDP in SD	*N. sylvestris*	LDP in SD

Note: The successful transfer of a flowering induction signal by grafting between plants of different photoperiodic response groups shows the existence of a transmissible floral hormone that is effective.
[a]LDPs = Long-day plants; SDPs = Short- day plants; DNPs = Day-neutral plants.
[b]LD, long days; SD, short days.

correlated closely with the translocation of ^{14}C-labeled assimilates from the donor, and this movement was dependent on the establishment of vascular continuity across the graft union (Zeevaart 1976). These results confirmed earlier girdling studies showing that the floral stimulus is translocated along with photoassimilates in the phloem.

Indirect Induction Implies That the Floral Stimulus Is Self-Propagating

In at least three cases—*Xanthium* (SDP), *Bryophyllum* (SLDP), and *Silene* (LDP)—the induced state appears to be self-propagating (Zeevaart 1976). That is, young leaves that develop on the receptor plant after it has been induced to flower by a donor leaf can themselves be used as donor leaves in subsequent grafting experiments, even though these leaves have never been subjected to an inductive photoperiod. This phenomenon is called *indirect induction*.

It is characteristic of indirect induction that the strength of the floral stimulus from the donor leaf remains constant even after serial grafting of new donors to several plants has taken place (Figure 24.30A). This suggests that the induced state is in some way propagated throughout the

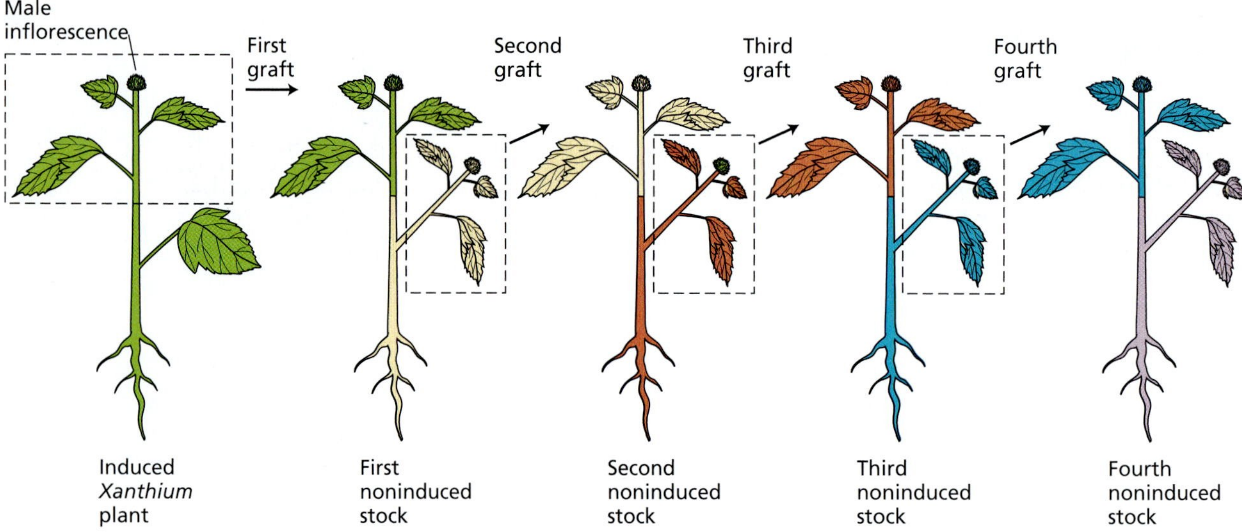

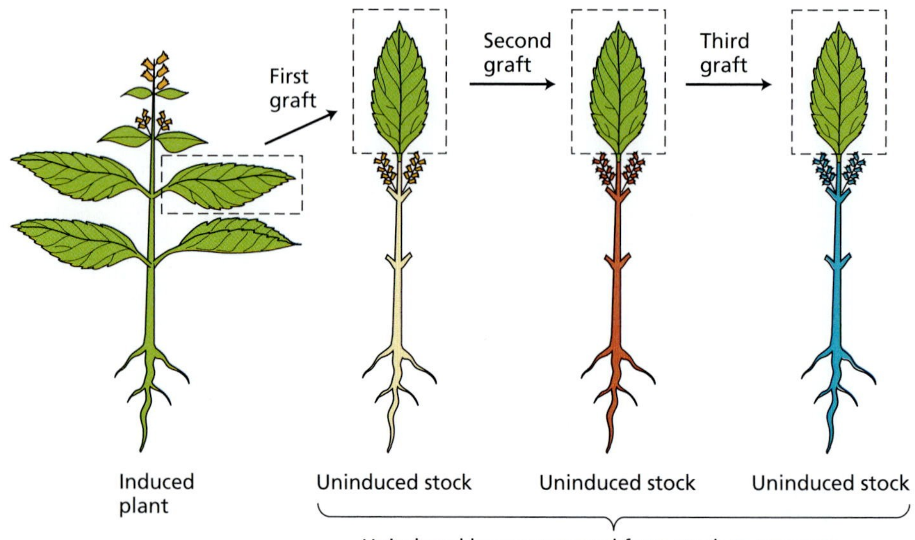

FIGURE 24.30 Different types of leaf induction in *Xanthium* and *Perilla*. (A) *Xanthium* exhibits indirect induction. Noninduced leaves from a plant induced to flower are capable of inducing other plants to flower even though they have never received an inductive photoperiod. This suggests that the floral stimulus is self-propagating. (B) In *Perilla*, only the leaf given the inductive photoperiod is capable of serving as a donor for the floral stimulus. In *Perilla* as well as *Xanthium*, one leaf can continue to induce flowering in successive grafting experiments (Lang 1965).

plant. Although this feature of the floral stimulus has sometimes been described as viruslike, it is unlikely that the floral stimulus can replicate itself like a virus. Rather, the floral stimulus is likely to be a molecule that induces its own production in a positive feedback loop. In *Xanthium* (cocklebur), removal of all buds from the shoot blocks indirect induction, indicating that meristematic tissue, or perhaps auxin, is required for propagation of the induced state.

On the other hand, indirect induction does not occur in the SDP *Perilla*. In *Perilla*, only the leaf actually given an inductive photoperiod is capable of transmitting the floral stimulus in a grafting experiment (see Figure 24.30B). Thus the floral stimulus of *Perilla* is not self-propagating as it is in *Xanthium*, *Bryophyllum*, and *Silene*. Either the mechanism for a positive feedback loop is absent in *Perilla* leaves, or translocation of the floral stimulus is restricted to the meristem so that it never enters the leaves.

Unlike *Xanthium*, which requires the presence of a bud for stable induction, *Perilla* leaves can be stably induced even when detached from the plant. Once induced, *Perilla* leaves cannot be uninduced, and the same leaf can continue to serve as a donor of the floral stimulus in successive grafting experiments without any reduction in potency (Zeevaart 1976).

FIGURE 24.31 Graft transmission of an inhibitor of flowering. Non-induced rosettes from the LDP *Nicotiana sylvestris* were grafted onto the day-neutral tobacco (*Nicotiana tabacum*, cv. Trapezond). Flowering of the day-neutral plant was suppressed under short days (left branch of plant on right), but not under long days (left branch of plant on left). Arrowheads indicate graft unions. (From Lang et al. 1977.)

Evidence for Antiflorigen Has Been Found in Some LDPs

Grafting studies have implicated transmissible inhibitors in flowering regulation as well. Such inhibitors have been called **antiflorigen**, but (like florigen) antiflorigen may consist of multiple compounds. For example, grafting an uninduced leafy shoot from the LDP *Nicotiana sylvestris* onto the day-neutral tobacco cultivar Trapezond suppressed flowering in the day-neutral plant under short days but not long-day conditions (Figure 24.31). On the other hand, when an uninduced donor from the SDP Maryland Mammoth was grafted onto Trapezond, it had no effect on flowering in either short-day or long-day conditions. This and similar results suggest that the leaves of LDPs, but not SDPs, produce flowering inhibitors under noninductive conditions.

Similar studies in peas have led to the identification of several genetic loci that regulate steps in the biosynthetic pathways of both floral activators and floral inhibitors (see **Web Topic 24.5**).

Attempts to Isolate Transmissible Floral Regulators Have Been Unsuccessful

The many attempts to isolate and characterize the floral stimulus have been largely unsuccessful. The most common approach has been to make extracts from induced leaf tissue and test for their ability to elicit flowering in noninduced plants. In other experiments, investigators have extracted and analyzed phloem sap from induced plants. In some studies, extracts from one of these sources have induced flowering in test plants, but these results have not been consistently reproduced. Most of these extractions have focused on small molecules.

Recent studies using fluorescent tracers have shown that in *Arabidopsis* there is actually a *decrease* in the movement of small molecules from the leaf-to-shoot apex via the symplast at the time of floral induction (Gisel et al. 2002). The lack of tracer movement from the leaf to the shoot apex may indicate either a reduction in overall symplastic transport to the shoot apex, or a change in the selectivity of the plasmodesmata during floral induction. There is increasing evidence that macromolecular traffic between cells via plasmodesmata plays essential roles in normal meristem development and function (see Chapter 16). Particles as large as viruses can move from cell to cell via plasmodesmata, and throughout the plant via the phloem. Phloem translocation of small RNAs has recently been implicated in the spread of a viral resistance mechanism throughout plants (Hamilton and Baulcombe 1999). It is therefore possible that the floral stimulus is a macromolecule, such as RNA or protein, that is translocated via the phloem from the leaf to the apical meristem, where it functions as a regulator of gene expression (Crawford and Zambryski 1999).

However, thus far attempts to identify such a signal have been unsuccessful.

Efforts to isolate a specific, graft-transmissible inhibitor of flowering have also been unsuccessful. Thus, despite unequivocal data from grafting experiments showing that transmissible factors regulate flowering (see Table 24.2) (Zeevaart 1976), the substances involved remain elusive.

Gibberellins and Ethylene Can Induce Flowering in Some Plants

Among the naturally occurring growth hormones, gibberellins (GAs) (see Chapter 20) can have a strong influence on flowering (see **Web Topic 24.8**). Recent studies suggest that gibberellin promotes flowering in *Arabidopsis* by activating expression of the *LFY* gene (Blazquez and Weigel 2000). Exogenous gibberellin can evoke flowering when applied either to rosette LDPs like *Arabidopsis*, or to dual–day length plants such as *Bryophyllum*, when grown under short days (Lang 1965; Zeevaart 1985).

In addition, application of GAs can evoke flowering in a few SDPs in noninductive conditions, and in cold-requiring plants that have not been vernalized. As previously discussed, cone formation can also be promoted in juvenile plants of several gymnosperm families by addition of GAs. Thus, in some plants exogenous GAs can bypass the endogenous trigger of age in autonomous flowering, as well as the primary environmental signals of day length and temperature.

As discussed in Chapter 20, plants contain many GA-like compounds. Most of these compounds are either precursors to, or inactive metabolites of, the active forms of GA. In some situations different GAs have markedly different effects on flowering and stem elongation, such as in the long-day plant *Lolium temulentum* (darnel ryegrass) (see **Web Topic 24.9**).

These observations suggest that the regulation of flowering may be associated with specific GAs, but they do not prove that GA is the hypothetical flowering hormone. In fact, a certain level of GA is likely to be required for flowering in many species, but other pathways to flowering are necessary as well. For example, a mutation in GA biosynthesis renders the quantitative LDP *Arabidopsis thaliana* unable to flower in noninductive short days but has little effect on flowering in long days, demonstrating that endogenous GA is required for flowering in specific situations (Wilson et al. 1992).

Considerable attention has been given to the effects of day length on GA metabolism in the plant (see Chapter 20). For example, in the long-day plant spinach (*Spinacia oleracea*), the levels of gibberellins are relatively low in short days, and the plants maintain a rosette form. After the plants are transferred to long days, the levels of all the gibberellins of the 13-hydroxylated pathway ($GA_{53} \rightarrow GA_{44} \rightarrow GA_{19} \rightarrow GA_{20} \rightarrow GA_1$; see Chapter 20) increase. However, the fivefold increase in the physiologically active gibberellin, GA_1, is what causes the marked stem elongation that accompanies flowering.

In addition to GAs, other growth hormones can either inhibit or promote flowering. One commercially important example is the striking promotion of flowering in pineapple (*Ananas comosus*) by ethylene and ethylene-releasing compounds—a response that appears to be restricted to members of the pineapple family (Bromeliaceae). Thus, as discussed next, the floral stimulus may be composed of many components, and these components may differ in different groups of plants.

The Transition to Flowering Involves Multiple Factors and Pathways

It is clear that the transition to flowering involves a complex system of interacting factors that include, among others, carbohydrates, gibberellins, cytokinins, and, in the bromeliads, ethylene (see **Web Topic 24.10**). Leaf-generated transmissible signals are required for determination of the shoot apex in both autonomously regulated and photoperiodic species. Determining whether these transmissible signals consist of single or multiple components is a major challenge for the future.

Recent genetic studies have established that there are four genetically distinct developmental pathways that control flowering in the LDP *Arabidopsis* (Blazquez 2000). Figure 24.32 shows a simplified version of the four pathways:

1. The *photoperiodic pathway* involves phytochromes and cryptochromes. (Note that PHYA and PHYB have contrasting effects on flowering; see **Web Topic 24.11**.) The interaction of these photoreceptors with a circadian clock initiates a pathway that eventually results in the expression of the gene *CONSTANS* (*CO*), which encodes a zinc-finger transcription factor that promotes flowering. *CO* acts through other genes to increase the expression of the floral meristem identity gene *LEAFY* (*LFY*).

2. In the dual *autonomous/vernalization pathway*, flowering occurs either in response to internal signals—the production of a fixed number of leaves—or to low temperatures. In the autonomous pathway of *Arabidopsis*, all of the genes associated with the pathway are expressed in the meristem. The autonomous pathway acts by reducing the expression of the flowering repressor gene *FLOWERING LOCUS C* (*FLC*), an inhibitor of *LFY* (Michaels and Amasino 2000). Vernalization also represses *FLC*, but perhaps by a different mechanism (an epigenetic switch). Because the *FLC* gene is a common target, the autonomous and vernalization pathways are grouped together.

3. The *carbohydrate*, or *sucrose*, *pathway* reflects the metabolic state of the plant. Sucrose stimulates flowering in *Arabidopsis* by increasing *LFY* expression, although the genetic pathway is unknown.

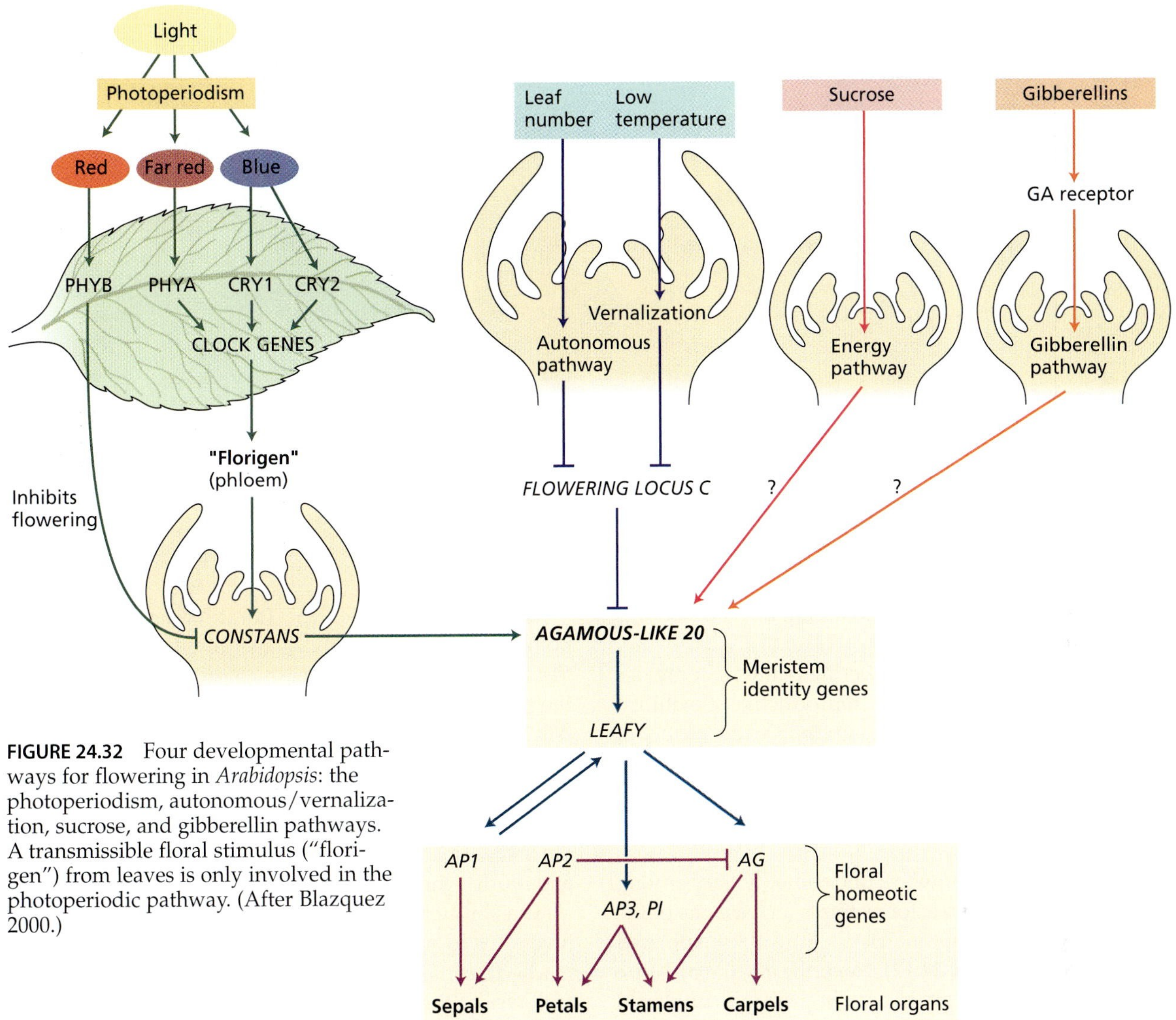

FIGURE 24.32 Four developmental pathways for flowering in *Arabidopsis*: the photoperiodism, autonomous/vernalization, sucrose, and gibberellin pathways. A transmissible floral stimulus ("florigen") from leaves is only involved in the photoperiodic pathway. (After Blazquez 2000.)

4. The *gibberellin pathway* is required for early flowering and for flowering under noninductive short days.

All four pathways converge by increasing the expression of the key floral meristem identity gene *AGAMOUS-LIKE 20* (*AGL20*). The role of *AGL20,* a MADS box–containing transcription factor, is to integrate the signals coming from all four pathways into a unitary output. Obviously the strongest output signal occurs when all four pathways are activated.

Figure 24.33 shows the level of *AGL20* gene expression in the shoot apical meristem of an *Arabidopsis* plant after shifting from noninductive short days (8-hour day length) to inductive long days (16-hour day length). Note that an increase in *AGL20* expression can be detected as early as 18 hours after the beginning of the long-day treatment (Borner et al. 2000). Thus it takes only 10 hours beyond an 8-hour short day for the meristem to begin responding to the floral stimulus from the leaves. This timing is consistent with previous measurements of the rates of export of the floral stimulus from induced leaves (discussed earlier in the chapter).

Although many pathways feed into *AGL20*, there must be some redundancy in the system because flowering is only delayed, but not completely blocked, in *agl20* mutants. Thus, one or two other genes must be able to take over the role of *AGL20* when it is mutated.

Once turned on by *AGL20*, *LFY* activates the floral homeotic genes—*APETALA1* (*AP1*), *APETALA3* (*AP3*), *PISTILLATA* (*PI*), and *AGAMOUS* (*AG*)—that are required for floral organ development. *APETALA2* (*AP2*) is expressed in both vegetative and floral meristems and is therefore not affected by *LFY*. However, as discussed earlier in the chapter, *AP2* exerts a negative effect on *AG* expression (see Figure 24.6).

Besides serving as a floral homeotic gene, *AP1* functions as a meristem identity gene in *Arabidopsis* because it is involved in a positive feedback loop with *LFY*. Conse-

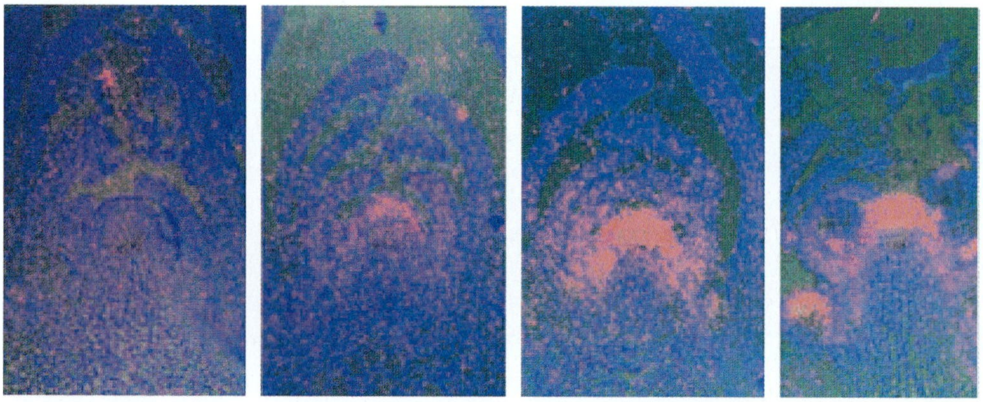

FIGURE 24.33 Increase in expression of the gene *AGAMOUS-LIKE 20* (*AGL20*) during floral evocation in the shoot apical meristem of *Arabidopsis*. The times after shifting the plants from SDs to LDs are indicated. (From Borner et al. 2000.)

quently, once the transition to flowering has reached this stage, flowering is irreversible.

The existence of multiple flowering pathways provides angiosperms with maximum reproductive flexibility, enabling them to produce seeds under a wide variety of conditions. Redundancy within the pathways ensures that reproduction, the most crucial of all physiological functions, will be relatively insensitive to mutations and evolutionarily robust.

The details of the pathways undoubtedly vary among different species. In maize, for example, at least one of the genes involved in the autonomous pathway is expressed in leaves (see **Web Topic 24.12**). Nevertheless, the presence of multiple flowering pathways is probably universal among angiosperms.

SUMMARY

Flower formation occurs at the shoot apical meristem and is a complex morphological event. The rosette plant *Arabidopsis* has been an important model for studies on floral development. The four floral organs (sepals, petals, stamens, and carpels) are initiated as successive whorls. Three classes of genes regulate floral development. The first class contains positive regulators of the floral meristem identity. *APETALA1* (*AP1*) and *LEAFY* (*LFY*) are the most important *Arabidopsis* floral meristem identity genes.

Meristem identity genes are positive regulators of another class of genes that determine floral organ identity. There are five known floral organ identity genes in *Arabidopsis*: *APETALA1* (*AP1*), *APETALA2* (*AP2*), *APETALA3* (*AP3*), *PISTILLATA* (*PI*), and *AGAMOUS* (*AG*). Cadastral genes make up the third group. Cadastral genes act as spatial regulators of the floral organ identity genes by setting boundaries for their expression.

The genes that control floral organ identity are homeotic. Most homeotic genes in plants contain the MADS box. Mutations in these genes alter the identity of the floral organs produced in two adjacent whorls. The ABC model seeks to explain how the floral homeotic genes control organ identity through the unique combinations of their products. Type A genes control organ identity in the first and second whorls. Type B activity controls organ determination in the second and third whorls. The third and fourth whorls are controlled by type C activity.

The ability to flower (i.e., to make the transition from juvenility to maturity) is attained when the plant has reached a certain age or size. In some plants, the transition to flowering then occurs independently of the environment (autonomously). Other plants require exposure to appropriate environmental conditions. The most common environmental inputs for flowering are day length and temperature.

The response to day length—photoperiodism—promotes flowering at a particular time of year, and several different categories of responses are known. The photoperiodic signal is perceived by the leaf. Exposure to low temperature—vernalization—is required for flowering in some plants, and this requirement is often coupled with a day length requirement. Vernalization occurs at the shoot apical meristem. Photoperiodism and vernalization interact in several ways.

Daily rhythms—circadian rhythms—can locate an event at a particular time of day. Timekeeping in these rhythms is based on an endogenous circadian oscillator. Keeping the rhythm on local time depends on the phase response of the rhythm to environmental signals. The most important signals are dawn and dusk.

Short-day plants flower when a critical duration of darkness is exceeded. Long-day plants flower when the length of the dark period is less than a critical value. Light given at certain times in a dark period that is longer than the critical value—a night break—prevents the effect of the dark period. Light also acts on the circadian oscillator to entrain the photoperiodic rhythm, an effect that is important for timekeeping in the dark. The photoperiodic mechanism shows some variation in short-day and long-day responses, but both appear to involve phytochrome and a circadian oscillator.

When photoperiod-responsive plants are induced to flower by exposure to appropriate day lengths, leaves send a chemical signal to the apex to bring about flowering. This transmissible signal is able to cause flowering in plants of different photoperiodic response groups. In noninductive day lengths, a transmissible inhibitor of flowering may be produced by the leaves of LDPs.

Although physiological experiments, especially grafting, indicate the existence of a transmissible floral stimulus and, in some cases, flowering inhibitors, the chemical identity of these factors is not known. Plant growth hormones, especially the gibberellins, can modify flowering in many plants.

The transition to flowering is regulated by multiple signals and multiple pathways. In *Arabidopsis*, flowering is controlled by four pathways: the photoperiodic, autonomous/vernalization, sucrose, and GA pathways. All of these pathways converge to regulate the meristem identity genes *AGAMOUS-LIKE 20* (*AGL20*) and *LEAFY* (*LFY*). *AGL20* and *LFY*, in turn, regulate the floral homeotic genes to produce the floral organs. The existence of multiple pathways for flowering provides angiosperms with the flexibility to reproduce under a variety of environmental conditions, thus increasing their evolutionary fitness.

Web Material

Web Topics

24.1 Contrasting the Characteristics of Juvenile and Adult Phases of English Ivy (*Hedera helix*) and Maize (*Zea mays*)

A table of juvenile vs. adult morphological characteristics is presented.

24.2 Regulation of Juvenility by the *TEOPOD* (*TP*) Genes in Maize

The genetic control of juvenility in maize is discussed.

24.3 Flowering of Juvenile Meristems Grafted to Adult Plants

The competence of juvenile meristems to flower can be tested in grafting experiments.

24.4 Characteristics of the Phase-Shifting Response in Circadian Rhythms

Petal movements in *Kalenchoe* have been used to study circadian rhythms.

24.5 Genes That Control Flowering Time

A discussion of genes that control different apects of flowering time is presented.

24.6 Support for the Role of Blue-Light Regulation of Circadian Rhythms

The role of ELF3 in mediating the effects of blue light on flowering time is discussed.

24.7 Regulation of Flowering in Canterbury Bell by Both Photoperiod and Vernalization

Short days acting on the leaf can substitute for vernalization at the shoot apex in Canterbury Bell.

24.8 Examples of Floral Induction by Gibberellins in Plants with Different Environmental Requirements for Flowering

A table of the effects of gibberellins on plants with different photoperiodic requirements.

24.9 The Different Effects of Two Different Gibberellins on Flowering (Spike Length) and Elongation (Stem Length)

GA_1 and GA_{32} have different effects on flowering in *Lolium*.

24.10 The Influence of Cytokinins and Polyamines on Flowering

Other growth regulators beside gibberellins may participate in the flowering response.

24.11 The Contrasting Effects of Phytochromes A and B on Flowering

A brief discussion of the effects of phyA and phyB on flowering in *Arabidopsis* and other species.

24.12 A Gene That Regulates the Floral Stimulus in Maize

The INDETERMINATE 1 gene of maize regulates the transition to flowering and is expressed in young leaves.

Chapter References

Bewley, J. D., Hempel, F. D., McCormick, S., and Zambryski, P. (2000) Reproductive Development. In: *Biochemistry and Molecular Biology of Plants*, B. B. Buchanan, W. Gruissem, and R. L. Jones (eds.), American Society of Plant Biologists, Rockville, MD.

Blazquez, M. A. (2000) Flower development pathways. *J. Cell Sci.* 113: 3547–3548.

Blazquez, M. A., and Weigel, D. (2000) Integration of floral inductive signals in *Arabidopsis*. *Nature* 404: 889–892.

Borner, R., Kampmann, G., Chandler, J., Gleissner, R., Wisman, E., Apel, K., and Melzer, S. (2000) A *MADS* domain gene involved in the transition to flowering in *Arabidopsis*. *Plant J.* 24: 591–599.

Bowman, J. L., Smyth, D. R., and Meyerowitz, E. M. (1989) Genes directing flower development in *Arabidopsis*. *Plant Cell* 1: 37–52.

Bünning, E. (1960) Biological clocks. *Cold Spring Harbor Symp. Quant. Biol.* 15: 1–9.

Clark, J. R. (1983) Age-related changes in trees. *J. Arboriculture* 9: 201–205.

Coen, E. S., and Carpenter, R. (1993) The metamorphosis of flowers. *Plant Cell* 5: 1175–1181.

Coulter, M. W., and Hamner, K. C. (1964) Photoperiodic flowering response of Biloxi soybean in 72 hour cycles. *Plant Physiol.* 39: 848–856.

Crawford, K., and Zambryski, P. (1999) Phylem transport: Are you chaperoned? *Curr. Biol.* 9: R281–R285.

Deitzer, G. (1984) Photoperiodic induction in long-day plants. In *Light and the Flowering Process,* D. Vince-Prue, B. Thomas, and K. E. Cockshull eds., Academic Press, New York, pp. 51–63.

Devlin, P. F., and Kay, S. A. (2000) Cryptochromes are required for phytochrome signaling to the circadian clock but not for rhythmicity. *Plant Cell* 12: 2499–2509.

Gasser, C. S., and Robinson-Beers, K. (1993) Pistil development. *Plant Cell* 5: 1231–1239.

Gisel, A., Hempel, F. D., Barella, S., and Zambryski, P. (2002) Leaf-to-shoot apex movement of symplastic tracer is restricted coincident with flowering *Arabidopsis*. *Proc. Nat'l Acad. Sci. USA* 99: 1713–1717.

Guo, H., Yang, H., Mockler, T. C., and Lin, C. (1998) Regulation of flowering time by *Arabidopsis* photoreceptors. *Science* 279: 1360–1363.

Hamilton, A. J., and Baulcombe, D. C. (1999) A species of small antisense RNA in posttranscriptional gene silencing in plants. *Science* 286: 950–952.

Hendricks, S. B., and Siegelman, H. W. (1967) Phytochrome and photoperiodism in plants. *Comp. Biochem.* 27: 211–235.

Lang, A. (1965) Physiology of flower initiation. In *Encyclopedia of Plant Physiology* (Old Series, Vol. 15), W. Ruhland, ed., Springer, Berlin, pp. 1380–1535.

Lang, A., Chailakhyan, M. K., and Frolova, I. A. (1977) Promotion and inhibition of flower formation in a dayneutral plant in grafts with a short-day plant and a long-day plant. *Proc. Natl. Acad. Sci. USA* 74: 2412–2416.

McDaniel, C. N. (1996) Developmental physiology of floral initiation in *Nicotiana tabacum* L. *J. Exp. Bot.* 47: 465–475.

McDaniel, C. N., Hartnett, L. K., and Sangrey, K. A. (1996) Regulation of node number in day-neutral *Nicotiana tabacum*: A factor in plant size. *Plant J.* 9: 56–61.

McDaniel, C. N., Singer, S. R., and Smith, S. M. E. (1992) Developmental states associated with the floral transition. *Dev. Biol.* 153: 59–69.

Michaels, S. D., and Amasino, R. M. 2000. Memories of winter: Vernalization and the competence to flower. *Plant Cell Environ.* 23: 1145–1154.

Millar, A. J., Carre, I. A., Strayer, C. A., Chua, N.-H., and Kay, S. A. (1995) Circadian clock mutants in *Arabidopsis* identified by luciferase imaging. *Science* 267: 1161–1163.

Papenfuss, H. D., and Salisbury, F. B. (1967) Aspects of clock resetting in flowering of *Xanthium*. *Plant Physiol.* 42: 1562–1568.

Poethig, R. S. (1990) Phase change and the regulation of shoot morphogenesis in plants. *Science* 250: 923–930.

Purvis, O. N., and Gregory, F. G. (1952) Studies in vernalization of cereals. XII. The reversibility by high temperature of the vernalized condition in Petkus winter rye. *Ann. Bot.* 1: 569–592.

Reid, J. B., Murfet, I. C., Singer, S. R., Weller, J. L., and Taylor, S.A. (1996) Physiological genetics of flowering in *Pisum*. *Sem. Cell Dev. Biol.* 7: 455–463.

Saji, H., Vince-Prue, D., and Furuya, M. (1983) Studies on the photoreceptors for the promotion and inhibition of flowering in dark-grown seedlings of *Pharbitis nil* Choisy. *Plant Cell Physiol.* 67: 1183–1189.

Salisbury, F. B. (1963) Biological timing and hormone synthesis in flowering of *Xanthium*. *Planta* 49: 518–524.

Simon, R., Igeno, M. I., and Coupland, G. (1996) Activation of floral meristem identity genes in *Arabidopsis*. *Nature* 384: 59–62.

Vince-Prue, D. (1975) *Photoperiodism in Plants.* McGraw-Hill, London.

Weigel, D., and Meyerowitz, E. M. (1994) The ABCs of floral homeotic genes. *Cell* 78: 203–209.

Wilson, R. A., Heckman, J. W., and Sommerville, C. R. (1992) Gibberellin is required for flowering in *Arabidopsis thaliana* under short days. *Plant Physiol.* 100: 403–408.

Yanovsky, M. J., and Kay. S. A. (2001) Signaling networks in the plant circadian rhythm. *Curr. Opinion in Plant Biol* 4: 429–435.

Yanovsky, M. J., Mazzella, M. A., Whitelam, G. C., and Casal, J. J. (2001) Resetting the circadian clock by phytochromes and cryptochromes in *Arabidopsis*. *J. Biol. Rhythms* 16: 523–530.

Zeevaart, J. A. D. (1976) Physiology of flower formation. *Ann. Rev. Plant Physiol.* 27: 321–348.

Zeevaart, J. A. D. (1985) *Bryophyllum*. In *Handbook of Flowering*, Vol. II, A. H. Halevy, ed., CRC Press, Boca Raton, FL, pp. 89–100.

Zeevaart, J. A. D. (1986) Perilla. In *Handbook of Flowering*, Vol. 5, A. H. Halevy, ed., CRC Press, Boca Raton, FL, pp. 239–252.

Zeevaart, J. A. D., and Boyer, G. L. (1987) Photoperiodic induction and the floral stimulus in *Perilla*. In *Manipulation of Flowering*, J. G. Atherton, ed., Butterworths, London, pp. 269–277.

Chapter

25

Stress Physiology

IN BOTH NATURAL AND AGRICULTURAL CONDITIONS, plants are frequently exposed to environmental stresses. Some environmental factors, such as air temperature, can become stressful in just a few minutes; others, such as soil water content, may take days to weeks, and factors such as soil mineral deficiencies can take months to become stressful. It has been estimated that because of stress resulting from climatic and soil conditions (abiotic factors) that are suboptimal, the yield of field-grown crops in the United States is only 22% of the genetic potential yield (Boyer 1982).

In addition, stress plays a major role in determining how soil and climate limit the distribution of plant species. Thus, understanding the physiological processes that underlie stress injury and the adaptation and acclimation mechanisms of plants to environmental stress is of immense importance to both agriculture and the environment.

The concept of plant stress is often used imprecisely, and stress terminology can be confusing, so it is useful to start our discussion with some definitions. **Stress** is usually defined as an external factor that exerts a disadvantageous influence on the plant. This chapter will concern itself with environmental or abiotic factors that produce stress in plants, although biotic factors such as weeds, pathogens, and insect predation can also produce stress. In most cases, stress is measured in relation to plant survival, crop yield, growth (biomass accumulation), or the primary assimilation processes (CO_2 and mineral uptake), which are related to overall growth.

The concept of stress is intimately associated with that of **stress tolerance**, which is the plant's fitness to cope with an unfavorable environment. In the literature the term *stress resistance* is often used interchangeably with *stress tolerance*, although the latter term is preferred. Note that an environment that is stressful for one plant may not be stressful for another. For example, pea (*Pisum sativum*) and soybean (*Glycine max*) grow best at about 20°C and 30°C, respectively. As temperature increases, the pea shows signs of heat stress much sooner than the soybean. Thus the soybean has greater heat stress tolerance.

If tolerance increases as a result of exposure to prior stress, the plant is said to be **acclimated** (or hardened). Acclimation can be distinguished from **adaptation**, which usually refers to a *genetically* determined level of resistance acquired by a process of selection over many generations. Unfortunately, the term *adaptation* is sometimes used in the literature to indicate acclimation. And to add to the complexity, we will see later that gene expression plays an important role in acclimation.

Adaptation and acclimation to environmental stresses result from integrated events occurring at all levels of organization, from the anatomical and morphological level to the cellular, biochemical, and molecular level. For example, the wilting of leaves in response to water deficit reduces both water loss from the leaf and exposure to incident light, thereby reducing heat stress on leaves.

Cellular responses to stress include changes in the cell cycle and cell division, changes in the endomembrane system and vacuolization of cells, and changes in cell wall architecture, all leading to enhanced stress tolerance of cells. At the biochemical level, plants alter metabolism in various ways to accommodate environmental stresses, including producing osmoregulatory compounds such as proline and glycine betaine. The molecular events linking the perception of a stress signal with the genomic responses leading to tolerance have been intensively investigated in recent years.

In this chapter we will examine these principles, and the ways in which plants adapt and acclimate to water deficit, salinity, chilling and freezing, heat, and oxygen deficiency in the root biosphere. Air pollution, an important source of plant stress, is discussed in **Web Essay 25.1**. Although it is convenient to examine each of these stress factors separately, most are interrelated, and a common set of cellular, biochemical, and molecular responses accompanies many of the individual acclimation and adaptation processes.

For example, water deficit is often associated with salinity in the root biosphere and with heat stress in the leaves (resulting from decreased evaporative cooling due to low transpiration), and chilling and freezing lead to reductions in water activity and osmotic stress. We will also see that plants often display cross-tolerance—that is, tolerance to one stress induced by acclimation to another. This behavior implies that mechanisms of resistance to several stresses share many common features.

WATER DEFICIT AND DROUGHT RESISTANCE

In this section we will examine some drought resistance mechanisms, which are divided into several types. First we can distinguish between **desiccation postponement** (the ability to maintain tissue hydration) and **desiccation tolerance** (the ability to function while dehydrated), which are sometimes referred to as drought tolerance at high and low water potentials, respectively. The older literature often uses the term *drought avoidance* (instead of *drought tolerance*), but this term is a misnomer because drought is a meteorological condition that is tolerated by all plants that survive it and avoided by none. A third category, **drought escape**, comprises plants that complete their life cycles during the wet season, before the onset of drought. These are the only true "drought avoiders."

Among the desiccation postponers are water savers and water spenders. *Water savers* use water conservatively, preserving some in the soil for use late in their life cycle; *water spenders* aggressively consume water, often using prodigious quantities. The mesquite tree (*Prosopis* sp.) is an example of a water spender. This deeply rooted species has ravaged semiarid rangelands in the southwestern United States, and because of its prodigious water use, it has prevented the reestablishment of grasses that have agronomic value.

Drought Resistance Strategies Vary with Climatic or Soil Conditions

The water-limited productivity of plants (Table 25.1) depends on the total amount of water available and on the water-use efficiency of the plant (see Chapters 4 and 9). A plant that is capable of acquiring more water or that has higher water-use efficiency will resist drought better. Some plants possess adaptations, such as the C_4 and CAM modes of photosynthesis that allow them to exploit more arid environments. In addition, plants possess acclimation mechanisms that are activated in response to water stress.

Water deficit can be defined as any water content of a tissue or cell that is below the highest water content exhibited at the most hydrated state. When water deficit develops slowly enough to allow changes in developmental processes, water stress has several effects on growth, one of which is a limitation in leaf expansion. Leaf area is important because photosynthesis is usually proportional to it. However, rapid leaf expansion can adversely affect water availability.

TABLE 25.1
Yields of corn and soybean crops in the United States

	Crop yield (percentage of 10-year average)		
Year	Corn	Soybean	
1979	104	106	
1980	87	88	Severe drought
1981	104	100	
1982	108	104	
1983	77	87	Severe drought
1984	101	93	
1985	112	113	
1986	113	110	
1987	114	111	
1988	80	89	Severe drought

Source: U.S. Department of Agriculture 1989.

If precipitation occurs only during winter and spring, and summers are dry, accelerated early growth can lead to large leaf areas, rapid water depletion, and too little residual soil moisture for the plant to complete its life cycle. In this situation, only plants that have some water available for reproduction late in the season or that complete the life cycle quickly, before the onset of drought (exhibiting drought escape), will produce seeds for the next generation. Either strategy will allow some reproductive success.

The situation is different if summer rainfall is significant but erratic. In this case, a plant with large leaf area, or one capable of developing large leaf area very quickly, is better suited to take advantage of occasional wet summers. One acclimation strategy in these conditions is a capacity for both vegetative growth and flowering over an extended period. Such plants are said to be *indeterminate* in their growth habit, in contrast to *determinate* plants, which develop preset numbers of leaves and flower over only very short periods.

In the discussions that follow, we will examine several acclimation strategies, including inhibited leaf expansion, leaf abscission, enhanced root growth, and stomatal closure.

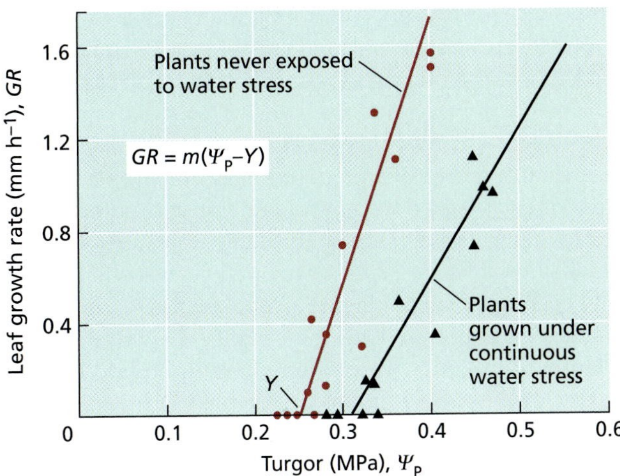

FIGURE 25.1 Dependence of leaf expansion on leaf turgor. Sunflower (*Helianthus annuus*) plants were grown either with ample water or with limited soil water to produce mild water stress. After rewatering, plants of both treatment groups were stressed by the withholding of water, and leaf growth rates (*GR*) and turgor (Ψ_p) were periodically measured. Both decreased extensibility (*m*) and increased threshold turgor for growth (*Y*) limit the leaf's capacity to grow after exposure to stress. (After Matthews et al. 1984.)

Decreased Leaf Area Is an Early Adaptive Response to Water Deficit

Typically, as the water content of the plant decreases, its cells shrink and the cell walls relax (see Chapter 3). This decrease in cell volume results in lower turgor pressure and the subsequent concentration of solutes in the cells. The plasma membrane becomes thicker and more compressed because it covers a smaller area than before. Because turgor reduction is the earliest significant biophysical effect of water stress, turgor-dependent activities such as leaf expansion and root elongation are the most sensitive to water deficits (Figure 25.1).

Cell expansion is a turgor-driven process and is extremely sensitive to water deficit. Cell expansion is described by the relationship

$$GR = m(\Psi_p - Y) \tag{25.1}$$

where *GR* is growth rate, Ψ_p is turgor, *Y* is the yield threshold (the pressure below which the cell wall resists plastic, or nonreversible, deformation), and *m* is the wall extensibility (the responsiveness of the wall to pressure).

This equation shows that a decrease in turgor causes a decrease in growth rate. Note also that besides showing that growth slows down when stress reduces Ψ_p, Equation 25.1 shows that Ψ_p need decrease only to the value of *Y*, not to zero, to eliminate expansion. In normal conditions, *Y* is usually only 0.1 to 0.2 MPa less than Ψ_p, so small decreases in water content and turgor can slow down or fully stop growth.

Water stress not only decreases turgor, but also decreases *m* and increases *Y*. Wall extensibility (*m*) is normally greatest when the cell wall solution is slightly acidic. In part, stress decreases *m* because cell wall pH typically rises during stress. The effects of stress on *Y* are not well understood, but presumably they involve complex structural changes of the cell wall (see Chapter 15) that may not be readily reversed after relief of stress. Water-deficient plants tend to become rehydrated at night, and as a result substantial leaf growth occurs at that time. Nonetheless, because of changes in *m* and *Y*, the growth rate is still lower than that of unstressed plants having the same turgor (see Figure 25.1).

Because leaf expansion depends mostly on cell expansion, the principles that underlie the two processes are similar. Inhibition of cell expansion results in a slowing of leaf expansion early in the development of water deficits. The smaller leaf area transpires less water, effectively conserving a limited water supply in the soil over a longer period. Reduction in leaf area can thus be considered a first line of defense against drought.

In indeterminate plants, water stress limits not only leaf size, but also leaf number, because it decreases both the number and the growth rate of branches. Stem growth has been studied less than leaf expansion, but stem growth is probably affected by the same forces that limit leaf growth during stress.

Keep in mind, too, that cell and leaf expansion also depend on biochemical and molecular factors beyond those that control water flux. Much evidence supports the view that plants change their growth rates in response to stress by coordinately controlling many other important processes such as cell wall and membrane biosynthesis, cell division, and protein synthesis (Burssens et al. 2000).

Water Deficit Stimulates Leaf Abscission

The total leaf area of a plant (number of leaves × surface area of each leaf) does not remain constant after all the leaves have matured. If plants become water stressed after a substantial leaf area has developed, leaves will senesce and eventually fall off (Figure 25.2). Such a leaf area adjustment is an important long-term change that improves the plant's fitness in a water-limited environment. Indeed, many drought-deciduous, desert plants drop all their leaves during a drought and sprout new ones after a rain. This cycle can occur two or more times in a single season. Abscission during water stress results largely from enhanced synthesis of and responsiveness to the endogenous plant hormone ethylene (see Chapter 22).

Water Deficit Enhances Root Extension into Deeper, Moist Soil

Mild water deficits also affect the development of the root system. Root-to-shoot biomass ratio appears to be governed by a functional balance between water uptake by the root and photosynthesis by the shoot (see Figure 23.6). Simply stated, *a shoot will grow until it is so large that water uptake by the roots becomes limiting to further growth*; conversely, *roots will grow until their demand for photosynthate from the shoot equals the supply*. This functional balance is shifted if the water supply decreases.

As discussed already, leaf expansion is affected very early when water uptake is curtailed, but photosynthetic activity is much less affected. Inhibition of leaf expansion reduces the consumption of carbon and energy, and a greater proportion of the plant's assimilates can be distributed to the root system, where they can support further root growth. At the same time, the root apices in dry soil lose turgor.

All these factors lead to a preferential root growth into the soil zones that remain moist. As water deficits progress, the upper layers of the soil usually dry first. Thus, plants commonly show a mainly shallow root system when all soil layers are wetted, and a loss of shallow roots and proliferation of deep roots as water in top layers of the soil is depleted. Deeper root growth into wet soil can be considered a second line of defense against drought.

Enhanced root growth into moist soil zones during stress requires allocation of assimilates to the growing root tips. During water deficit, assimilates are directed to the fruits and away from the roots (see Chapter 10). For this reason the enhanced water uptake resulting from root growth is less pronounced in reproductive plants than in vegetative plants. Competition for assimilates between roots and fruits is one explanation for the fact that plants are generally more sensitive to water stress during reproduction.

FIGURE 25.2 The leaves of young cotton (*Gossypium hirsutum*) plants abscise in response to water stress. The plants at left were watered throughout the experiment; those in the middle and at right were subjected to moderate stress and severe stress, respectively, before being watered again. Only a tuft of leaves at the top of the stem is left on the severely stressed plants. (Courtesy of B. L. McMichael.)

Stomata Close during Water Deficit in Response to Abscisic Acid

The preceding sections focused on changes in plant development during slow, long-term dehydration. When the onset of stress is more rapid or the plant has reached its full leaf area before initiation of stress, other responses protect the plant against immediate desiccation. Under these conditions, stomata closure reduces evaporation from the existing leaf area. Thus, stomatal closure can be considered a third line of defense against drought.

Uptake and loss of water in guard cells changes their turgor and modulates stomatal opening and closing (see Chapters 4 and 18). Because guard cells are located in the leaf epidermis, they can lose turgor as a result of a direct loss of water by evaporation to the atmosphere. The decrease in turgor causes stomatal closure by **hydropassive closure**. This closing mechanism is likely to operate in air of low humidity, when direct water loss from the guard cells is too rapid to be balanced by water movement into the guard cells from adjacent epidermal cells.

A second mechanism, called **hydroactive closure**, closes the stomata when the whole leaf or the roots are dehydrated and depends on metabolic processes in the guard cells. A reduction in the solute content of the guard cells results in water loss and decreased turgor, causing the stomata to close; thus the hydraulic mechanism of hydroactive closure is a reversal of the mechanism of stomatal opening. However, the control of hydroactive closure differs in subtle but important ways from stomatal opening.

Solute loss from guard cells can be triggered by a decrease in the water content of the leaf, and abscisic acid (ABA) (see Chapter 23) plays an important role in this

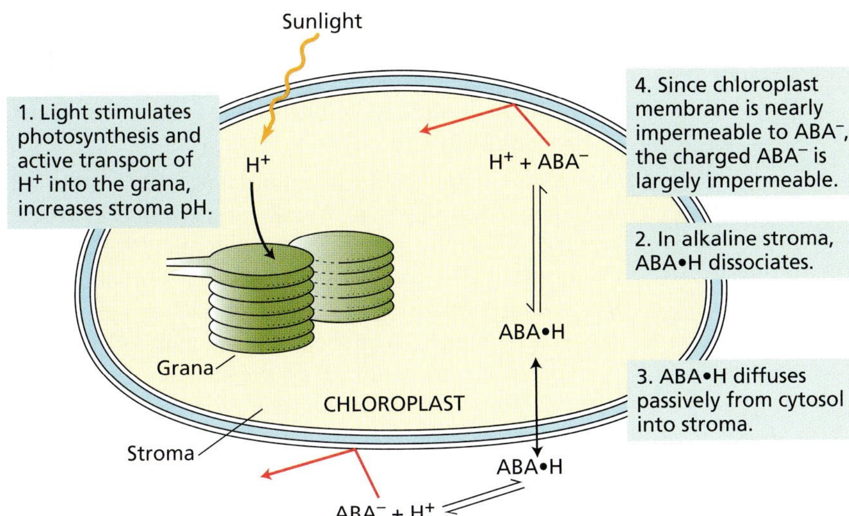

FIGURE 25.3 Accumulation of ABA by chloroplasts in the light. Light stimulates proton uptake into the grana, making the stroma more alkaline. The increased alkalinity causes the weak acid ABA•H to dissociate into H^+ and the ABA^- anion. The concentration of ABA•H in the stroma is lowered below the concentration in the cytosol, and the concentration difference drives the passive diffusion of ABA•H across the chloroplast membrane. At the same time, the concentration of ABA^- in the stroma increases, but the chloroplast membrane is almost impermeable to the anion (red arrows), which thus remains trapped. This process continues until the ABA•H concentrations in the stroma and the cytosol are equal. But as long as the stroma remains more alkaline, the total ABA concentration (ABA•H + ABA^-) in the stroma greatly exceeds the concentration in the cytosol.

process. Abscisic acid is synthesized continuously at a low rate in mesophyll cells and tends to accumulate in the chloroplasts. When the mesophyll becomes mildly dehydrated, two things happen:

1. Some of the ABA stored in the chloroplasts is released to the apoplast (the cell wall space) of the mesophyll cell (Hartung et al. 1998). The redistribution of ABA depends on pH gradients within the leaf, on the weak-acid properties of the ABA molecule, and on the permeability properties of cell membranes (Figure 25.3). The redistribution of ABA makes it possible for the transpiration stream to carry some of the ABA to the guard cells.

2. ABA is synthesized at a higher rate, and more ABA accumulates in the leaf apoplast. The higher ABA concentrations resulting from the higher rates of ABA synthesis appear to enhance or prolong the initial closing effect of the stored ABA. The mechanism of ABA-induced stomatal closure is discussed in Chapter 23.

Stomatal responses to leaf dehydration can vary widely both within and across species. The stomata of some dehydration-postponing species, such as cowpea (*Vigna unguiculata*) and cassava (*Manihot esculenta*), are unusually responsive to decreasing water availability, and stomatal conductance and transpiration decrease so much that leaf water potential (Ψ_w; see Chapters 3 and 4) may remain nearly constant during drought.

Chemical signals from the root system may affect the stomatal responses to water stress (Davies et al. 2002). Stomatal conductance is often much more closely related to soil water status than to leaf water status, and the only plant part that can be directly affected by soil water status is the root system. In fact, dehydrating only part of the root system may cause stomatal closure even if the well-watered portion of the root system still delivers ample water to the shoots.

When corn (*Zea mays*) plants were grown with roots trained into two separate pots and water was withheld from only one of the pots, the stomata closed partially, and the leaf water potential increased, just as in the dehydration postponers already described. These results show that stomata can respond to conditions sensed in the roots. Besides ABA (Sauter et al. 2001), other signals, such as pH and inorganic ion redistribution, appear to play a role in long-distance signaling between the roots and the shoots (Davies et al. 2002).

Water Deficit Limits Photosynthesis within the Chloroplast

The photosynthetic rate of the leaf (expressed per unit leaf area) is seldom as responsive to mild water stress as leaf expansion is (Figure 25.4) because photosynthesis is much less sensitive to turgor than is leaf expansion. However, mild water stress does usually affect both leaf photosynthesis and stomatal conductance. As stomata close during early stages of water stress, water-use efficiency (see Chapters 4 and 9) may increase (i.e., more CO_2 may be taken up per unit of water transpired) because stomatal closure inhibits transpiration more than it decreases intercellular CO_2 concentrations.

As stress becomes severe, however, the dehydration of mesophyll cells inhibits photosynthesis, mesophyll metabolism is impaired, and water-use efficiency usually decreases. Results from many studies have shown that the relative effect of water stress on stomatal conductance is significantly larger than that on photosynthesis. The response of photosynthesis and stomatal conductance to water stress can be partitioned by exposure of stressed

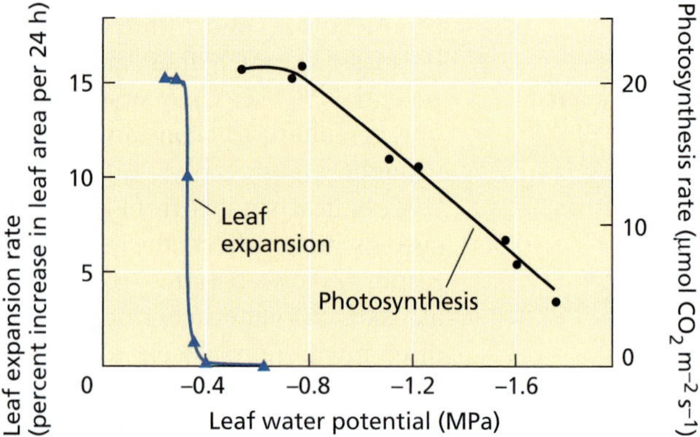

FIGURE 25.4 Effects of water stress on photosynthesis and leaf expansion of sunflower (*Helianthus annuus*). This species is typical of many plants in which leaf expansion is very sensitive to water stress, and it is completely inhibited under mild stress levels that hardly affect photosynthetic rates. (After Boyer 1970.)

leaves to air containing high concentrations of CO_2. Any effect of the stress on stomatal conductance is eliminated by the high CO_2 supply, and differences between photosynthetic rates of stressed and unstressed plants can be directly attributed to damage from the water stress to photosynthesis.

Does water stress directly affect translocation? Water stress decreases both photosynthesis and the consumption of assimilates in the expanding leaves. As a consequence, water stress indirectly decreases the amount of photosynthate exported from leaves. Because phloem transport depends on turgor (see Chapter 10), decreased water potential in the phloem during stress may inhibit the movement of assimilates. However, experiments have shown that translocation is unaffected until late in the stress period, when other processes, such as photosynthesis, have already been strongly inhibited (Figure 25.5).

This relative insensitivity of translocation to stress allows plants to mobilize and use reserves where they are needed (e.g., in seed growth), even when stress is extremely severe. The ability to continue translocating assimilates is a key factor in almost all aspects of plant resistance to drought.

Osmotic Adjustment of Cells Helps Maintain Plant Water Balance

As the soil dries, its matric potential (see **Web Topic 3.3**) becomes more negative. Plants can continue to absorb water only as long as their water potential (Ψ_w) is lower (more negative) than that of the soil water. Osmotic adjustment, or accumulation of solutes by cells, is a process by which water potential can be decreased without an accompanying decrease in turgor or decrease in cell volume. Recall Equation 3.6 from Chapter 3: $\Psi_w = \Psi_s + \Psi_p$. The change in cell water potential results simply from changes in solute potential (Ψ_s), the osmotic component of Ψ_w.

Osmotic adjustment is a net increase in solute content per cell that is independent of the volume changes that result from loss of water. The decrease in Ψ_s is typically limited to about 0.2 to 0.8 MPa, except in plants adapted to extremely dry conditions. Most of the adjustment can usually be accounted for by increases in concentration of a variety of common solutes, including sugars, organic acids, amino acids, and inorganic ions (especially K^+).

Cytosolic enzymes of plant cells can be severely inhibited by high concentrations of ions. The accumulation of ions during osmotic adjustment appears to be restricted to the vacuoles, where the ions are kept out of contact with enzymes in the cytosol or subcellular organelles. Because of this compartmentation of ions, other solutes must accumulate in the cytoplasm to maintain water potential equilibrium within the cell.

These other solutes, called **compatible solutes** (or compatible osmolytes), are organic compounds that do not interfere with enzyme functions. Commonly accumulated compatible solutes include the amino acid proline, sugar alcohols (e.g., sorbitol and mannitol), and a quaternary amine called glycine betaine. Synthesis of compatible solutes helps plants adjust to increased salinity in the rooting zone, as discussed later in this chapter.

Osmotic adjustment develops slowly in response to tissue dehydration. Over a time course of several days, other changes (such as growth or photosynthesis) are also taking place. Thus it can be argued that osmotic adjustment is not an independent and direct response to water deficit, but a result of another factor, such as decreased growth rate.

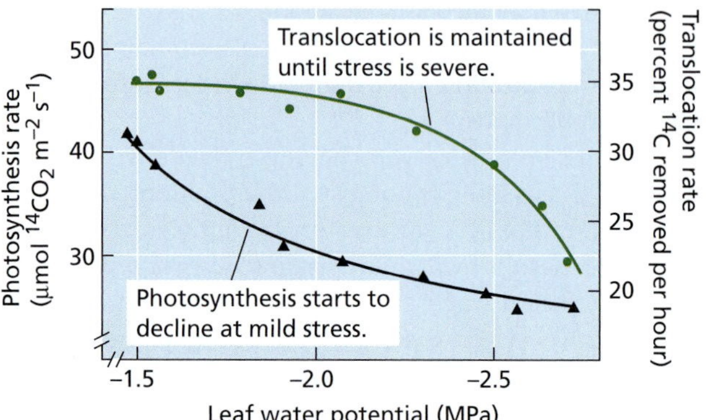

FIGURE 25.5 Relative effects of water stress on photosynthesis and translocation in sorghum (*Sorghum bicolor*). Plants were exposed to $^{14}CO_2$ for a short time interval. The radioactivity fixed in the leaf was taken as a measure of photosynthesis, and the loss of radioactivity after removal of the $^{14}CO_2$ source was taken as a measure of the rate of assimilate translocation. Photosynthesis was affected by mild stress, whereas, translocation was unaffected until stress was severe. (After Sung and Krieg 1979.)

Nonetheless, leaves that are capable of osmotic adjustment clearly can maintain turgor at lower water potentials than nonadjusted leaves. Maintaining turgor enables the continuation of cell elongation and facilitates higher stomatal conductances at lower water potentials. This suggests that osmotic adjustment is an acclimation that enhances dehydration tolerance.

How much extra water can be acquired by the plant because of osmotic adjustment in the leaf cells? Most of the extractable soil water is held in spaces (filled with water and air) from which it is readily removed by roots (see Chapter 4). As the soil dries, this water is used first, leaving behind the small amount of water that is held more tightly in small pores.

Osmotic adjustment enables the plant to extract more of this tightly held water, but the increase in total available water is small. Thus the cost of osmotic adjustment in the leaf is offset by rapidly diminishing returns in terms of water availability to the plant, as can be seen by a comparison of the water relations of adjusting and nonadjusting species (Figure 25.6). These results show that osmotic adjustment promotes dehydration tolerance but does not have a major effect on productivity (McCree and Richardson 1987).

Osmotic adjustment also occurs in roots, although the process in roots has not been studied so extensively as in leaves. The absolute magnitude of the adjustment is less in roots than in leaves, but as a percentage of the original tissue solute potential (Ψ_s), it can be larger in roots than in leaves. As with leaves, these changes may in many cases increase water extraction from the previously explored soil only slightly. However, osmotic adjustment can occur in the root meristems, enhancing turgor and maintaining root growth. This is an important component of the changes in root growth patterns as water is depleted from the soil.

Does osmotic adjustment increase plant productivity? Researchers have engineered the accumulation of osmoprotective solutes by conventional plant breeding, by physiological methods (inducing adjustment with controlled water deficits), and through the use of transgenic plants expressing genes for solute synthesis and accumulation. However, the engineered plants grow more slowly, and they are only slightly more tolerant to osmotic stresses. Thus the use of osmotic adjustment to improve agricultural performance is yet to be perfected.

Water Deficit Increases Resistance to Liquid-Phase Water Flow

When a soil dries, its resistance to the flow of water increases very sharply, particularly near the *permanent wilting point*. Recall from Chapter 4 that at the permanent wilting point (usually about –1.5 MPa), plants cannot regain turgor pressure even if all transpiration stops (for more details on the relationship between soil hydraulic conductivity and soil water potential, see **Figure 4.2.A in Web Topic 4.2**). Because of the very large soil resistance to water flow, water delivery to the roots at the permanent wilting point is too slow to allow the overnight rehydration of plants that have wilted during the day.

Rehydration is further hindered by the resistance within the plant, which has been found to be larger than the resistance within the soil over a wide range of water deficits (Blizzard and Boyer 1980). Several factors may contribute to the increased plant resistance to water flow during drying. As plant cells lose water, they shrink. When roots shrink, the root surface can move away from the soil particles that hold the water, and the delicate root hairs may be damaged. In addition, as root extension slows during soil drying, the outer layer of the root cortex (the hypodermis) often becomes more extensively covered with suberin,

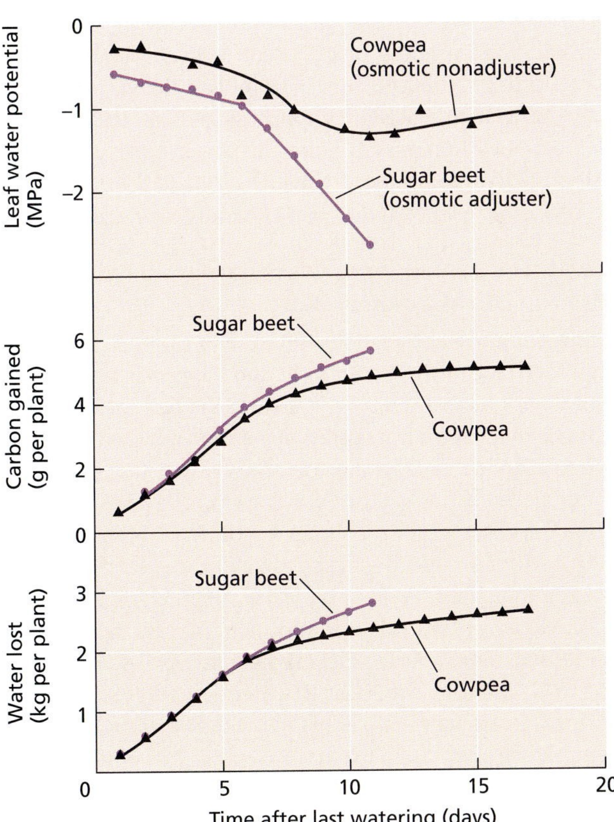

FIGURE 25.6 Water loss and carbon gain by sugar beet (*Beta vulgaris*), an osmotically adjusting species, and cowpea (*Vigna unguiculata*), a nonadjusting species that conserves water during stress by stomatal closure. Plants were grown in pots and subjected to water stress. On any given day after the last watering, the sugar beet leaves maintained a lower water potential than the cowpea leaves, but photosynthesis and transpiration during stress were only slightly greater in the sugar beet. The major difference between the two plants was the leaf water potential. These results show that osmotic adjustment promotes dehydration tolerance but does not have a major effect on productivity. (After McCree and Richardson 1987.)

a water-impermeable lipid (see Figure 4.4), increasing the resistance to water flow.

Another important factor that increases resistance to water flow is *cavitation*, or the breakage of water columns under tension within the xylem. As we saw in Chapter 4, transpiration from leaves "pulls" water through the plant by creating a tension on the water column. The cohesive forces that are required to support large tensions are present only in very narrow columns in which the water adheres to the walls.

Cavitation begins in most plants at moderate water potentials (–1 to –2 MPa), and the largest vessels cavitate first. For example, in trees such as oak (*Quercus*), the large-diameter vessels that are laid down in the spring function as a low-resistance pathway early in the growing season, when ample water is available. As the soil dries out during the summer, these large vessels cease functioning, leaving the small-diameter vessels produced during the stress period to carry the transpiration stream. This shift has long-lasting consequences: Even if water becomes available, the original low-resistance pathway remains nonfunctional, reducing the efficiency of water flow.

Water Deficit Increases Wax Deposition on the Leaf Surface

A common developmental response to water stress is the production of a thicker cuticle that reduces water loss from the epidermis (cuticular transpiration). Although waxes are deposited in response to water deficit both on the surface and within the cuticle inner layer, the inner layer may be more important in controlling the rate of water loss in ways that are more complex than by just increasing the amount of wax present (Jenks et al. in press).

A thicker cuticle also decreases CO_2 permeability, but leaf photosynthesis remains unaffected because the epidermal cells underneath the cuticle are nonphotosynthetic. Cuticular transpiration, however, accounts for only 5 to 10% of the total leaf transpiration, so it becomes significant only if stress is extremely severe or if the cuticle has been damaged (e.g., by wind-driven sand).

Water Deficit Alters Energy Dissipation from Leaves

Recall from Chapter 9 that evaporative heat loss lowers leaf temperature. This cooling effect can be remarkable: In Death Valley, California—one of the hottest places in the world—leaf temperatures of plants with access to ample water were measured to be 8°C below air temperatures. In warm, dry climates, an experienced farmer can decide whether plants need water simply by touching the leaves because a rapidly transpiring leaf is distinctly cool to the touch. When water stress limits transpiration, the leaf heats up unless another process offsets the lack of cooling. Because of these effects of transpiration on leaf temperature, water stress and heat stress are closely interrelated (see the discussion of heat stress later in this chapter).

Maintaining a leaf temperature that is much lower than the air temperature requires evaporation of vast quantities of water. This is why adaptations that cool leaves by means other than evaporation (e.g., changes in leaf size and leaf orientation) are very effective in conserving water. When transpiration decreases and leaf temperature becomes warmer than the air temperature, some of the extra energy in the leaf is dissipated as sensible heat loss (see Chapter 9). Many arid-zone plants have very small leaves, which minimize the resistance of the boundary layer to the transfer of heat from the leaf to the air (see Figure 9.14).

Because of their low boundary layer resistance, small leaves tend to remain close to air temperature even when transpiration is greatly slowed. In contrast, large leaves have higher boundary layer resistance and dissipate less thermal energy (per unit leaf area) by direct transfer of heat to the air.

In larger leaves, leaf movement can provide additional protection against heating during water stress. Leaves that orient themselves away from the sun are called *paraheliotropic*; leaves that gain energy by orienting themselves normal (perpendicular) to the sunlight are referred to as *diaheliotropic* (see Chapter 9). Figure 25.7 shows the strong effect of water stress on leaf position in soybean. Other factors that can alter the interception of radiation include wilting, which changes the angle of the leaf, and leaf rolling in grasses, which minimizes the profile of tissue exposed to the sun.

Absorption of energy can also be decreased by hairs on the leaf surface or by layers of reflective wax outside the cuticle. Leaves of some plants have a gray-white appearance because densely packed hairs reflect a large amount of light. This hairiness, or **pubescence**, keeps leaves cooler by reflecting radiation, but it also reflects the visible wavelengths that are active in photosynthesis and thus it decreases carbon assimilation. Because of this problem, attempts to breed pubescence into crops to improve their water-use efficiency have been generally unsuccessful.

Osmotic Stress Induces Crassulacean Acid Metabolism in Some Plants

Crassulacean acid metabolism (CAM) is a plant adaptation in which stomata open at night and close during the day (see Chapters 8 and 9). The leaf-to-air vapor pressure difference that drives transpiration is much reduced at night, when both leaf and air are cool. As a result, the water-use efficiencies of CAM plants are among the highest measured. A CAM plant may gain 1 g of dry matter for only 125 g of water used—a ratio that is three to five times greater than the ratio for a typical C_3 plant (see Chapter 4).

CAM is very prevalent in succulent plants such as cacti. Some succulent species display facultative CAM, switching to CAM when subjected to water deficits or saline conditions (see Chapter 8). This switch in metabolism is a remarkable adaptation to stress, involving accumulation of the enzymes phosphoenolpyruvate (PEP) carboxylase (Figure 25.8), pyruvate–orthophosphate dikinase, and NADP malic enzyme, among others.

(A) Well-watered

(B) Mild water stress

(C) Severe water stress

FIGURE 25.7 Orientation of leaflets of field-grown soybean (*Glycine max*) plants in the normal, unstressed, position (A); during mild water stress (B); and during severe water stress (C). The large leaf movements induced by mild stress are quite different from wilting, which occurs during severe stress. Note that during mild stress (B), the terminal leaflet has been raised, whereas the two lateral leaflets have been lowered; each is almost vertical. (Courtesy of D. M. Oosterhuis.)

As discussed in Chapters 8 and 9, CAM metabolism involves many structural, physiological, and biochemical features, including changes in carboxylation and decarboxylation patterns, transport of large quantities of malate into and out of the vacuoles, and reversal of the periodicity of stomatal movements. Thus, CAM induction is a remarkable adaptation to water deficit that occurs at many levels of organization.

Osmotic Stress Changes Gene Expression

As noted earlier, the accumulation of compatible solutes in response to osmotic stress requires the activation of the metabolic pathways that biosynthesize these solutes. Several genes coding for enzymes associated with osmotic adjustment are turned on (up-regulated) by osmotic stress and/or salinity, and cold stress. These genes encode enzymes such as the following (Buchanan et al. 2000):

- Δ'^{1}-Pyrroline-5-carboxylate synthase, a key enzyme in the proline biosynthetic pathway
- Betaine aldehyde dehydrogenase, an enzyme involved in glycine betaine accumulation
- *myo*-Inositol 6-*O*-methyltransferase, a rate-limiting enzyme in the accumulation of the cyclic sugar alcohol called pinitol

Several other genes that encode well-known enzymes are induced by osmotic stress. The expression of glyceraldehyde-3-phosphate dehydrogenase increases during osmotic stress, perhaps to allow an increase of carbon flow into organic solutes for osmotic adjustment. Enzymes involved in lignin biosynthesis are also controlled by osmotic stress.

Reduction in the activities of key enzymes also takes place. The accumulation of the sugar alcohol mannitol in response to osmotic stress appears not to be brought about by the up-regulation of genes producing enzymes involved in mannitol biosynthesis, but rather by the down-regulation of genes associated with sucrose production and mannitol degradation. In this way mannitol accumulation is enhanced during episodes of osmotic stress.

Other genes regulated by osmotic stress encode proteins associated with membrane transport, including ATPases

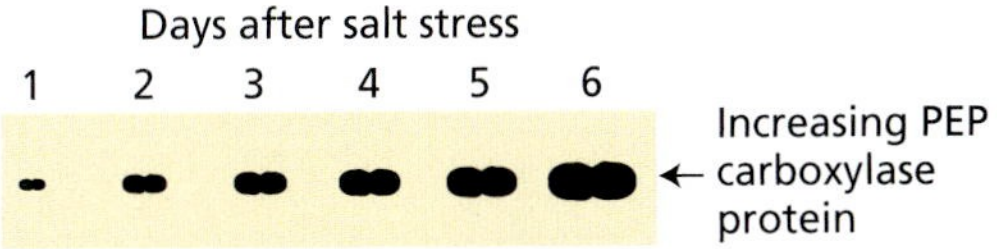

FIGURE 25.8 Increases in the content of phosphoenolpyruvate (PEP) carboxylase in ice plant, *Mesembryanthemum crystallinum*, during the salt-induced shift from C_3 metabolism to CAM. Salt stress was induced by the addition of 500 mM NaCl to the irrigation water. The PEP carboxylase protein was revealed in the gels by the use of antibodies and a stain. (After Bohnert et al. 1989.)

Table 25.2
The five groups of late embryogenesis abundant (LEA) proteins found in plants

Group (family name)[a]	Protein(s) in the group	Structural characteristics and motifs	Functional information/ proposed function
Group 1 (D-19 family)	Cotton D-19 Wheat Em (early methionine-labeled protein) Sunflower Ha ds10 Barley B19	Conformation is predominantly random coil with some predicted short α helices Charged amino acids and glycine are abundant	Contains more water of hydration than typical globular proteins Overexpression confers water deficit tolerance on yeast cells
Group 2 (D-11 family) (also referred to as dehydrins)	Maize DHN1, M3, RAB17 Cotton D-11 *Arabidopsis* pRABAT1, ERD10, ERD14 *Craterostigma* pcC 27-04, pcC 6-19 Tomato pLE4, TAS14 Barley B8, B9, B17 Rice pRAB16A Carrot pcEP40	Variable structure includes α helix–forming lysine-rich regions The consensus sequence for group 2 dehydrins is EKKGIMDKIKELPG The number of times this consensus repeats per protein varies Often contains a poly(serine) region Often contains regions of variable length rich in polar residues and either Gly or Ala., and Pro	Often localized to the cytoplasm or nucleus More acidic members of the family are associated with the plasma membrane May act to stabilize macromolecules at low water potential
Group 3 (D-7 family)	Barley HVA1 (ABA-induced) Cotton D-7 Wheat pMA2005, pMA1949 *Craterostigma* pcC3-06	Eleven amino-acid consensus sequence motif TAQAAKEKAXE is repeated in the protein Contains apparent amphipathic α helices Dimeric protein	Transgenic plants expressing HVA1 demonstrate enhanced water deficit stress tolerance D-7 is an abundant protein in cotton embryos (estimated concentration 0.25 m*M*) Each putative dimer of D-7 may bind as many as ten inorganic phosphates and their counterions
Group 4 (D-95 family)	Soybean D-95 *Craterostigma* pcC27-45	Slightly hydrophobic N-terminal region is predicted to form amphipathic α helices	In tomato, a gene encoding a similar protein is expressed in response to nematode feeding
Group 5 (D-113 family)	Tomato LE25 Sunflower Hads11 Cotton D-113	Family members share sequence homology at the conserved N terminus N-terminal region is predicted to form α helices C-terminal domain is predicted to be a random coil of variable length and sequence Ala, Gly, and Thr are abundant in the sequence	Binds to membranes and/or proteins to maintain structure during stress Possibly functions in ion sequestration to protect cytosolic metabolism When LE25 is expressed in yeast, it confers salt and freezing tolerance D-113 is abundant in cottonseeds (up to 0.3 m*M*)

[a]The protein family names are derived from the cotton seed proteins that are most similar to the family.
Source: After Bray et al. 2000.

(Niu et al. 1995) and the water channel proteins, *aquaporins* (see Chapter 3) (Maggio and Joly 1995). Several protease genes are also induced by stress, and these enzymes may degrade (remove and recycle) other proteins that are denatured by stress episodes. The protein *ubiquitin* tags proteins that are targeted for proteolytic degradation. Synthesis of the mRNA for ubiquitin increases in *Arabidopsis* upon desiccation stress. In addition, some *heat shock proteins* are osmotically induced and may protect or renature proteins inactivated by desiccation.

The sensitivity of cell expansion to osmotic stress (see Figure 25.1) has stimulated studies of various genes that encode proteins involved in the structural composition and integrity of cell walls. Genes coding for enzymes such as S-adenosylmethionine synthase and peroxidases, which may be involved in lignin biosynthesis, have been shown to be controlled by stress.

A large group of genes that are regulated by osmotic stress was discovered by examination of naturally desiccating embryos during seed maturation. These genes code for so-called **LEA proteins** (named for *l*ate *e*mbryogenesis *a*bundant), and they are suspected to play a role in cellular membrane protection. Although the function of LEA proteins is not well understood (Table 25.2), they accumulate in vegetative tissues during episodes of osmotic stress. The proteins encoded by these genes are typically hydrophilic and strongly bind water. Their protective role might be associated with an ability to retain water and to prevent crystallization of important cellular proteins and other molecules during desiccation. They might also contribute to membrane stabilization.

More recently, microarray techniques have been used to examine the expression of whole genomes of some plants in response to stress. Such studies have revealed that large numbers of genes display changes in expression after plants are exposed to stress. Stress-controlled genes reflect up to 10% of the total number of rice genes examined (Kawasaki et al. 2001)

Osmotic stress typically leads to the accumulation of ABA (see Chapter 23), so it is not surprising that products of ABA-responsive genes accumulate during osmotic stresses. Studies of ABA-insensitive and ABA-deficient mutants have shown that numerous genes that are induced by osmotic stress are in fact induced by the ABA accumulated during the stress episode. However, not all genes that are up-regulated by osmotic stresses are ABA regulated. As discussed in the next section, other mechanisms for regulating gene expression of osmotic stress–regulated genes have been uncovered.

Stress-Responsive Genes Are Regulated by ABA-Dependent and ABA-Independent Processes

Gene transcription is controlled through the interaction of regulatory proteins (transcription factors) with specific regulatory sequences in the promoters of the genes they regulate (Chapter 14 on the web site discusses these processes in detail). Different genes that are induced by the same signal (desiccation or salinity, for example) are controlled by a signaling pathway leading to the activation of these specific transcription factors.

Studies of the promoters of several stress-induced genes have led to the identification of specific regulatory sequences for genes involved in different stresses. For example, the *RD29* gene contains DNA sequences that can be activated by osmotic stress, by cold, and by ABA (Yamaguchi-Shinozaki and Shinozaki 1994; Stockinger et al. 1997).

The promoters of ABA-regulated genes contain a six-nucleotide sequence element referred to as the **ABA response element** (**ABRE**), which probably binds transcriptional factors involved in ABA-regulated gene activation (see Chapter 23). The promoters of these genes, which are regulated by osmotic stress in an ABA-dependent manner, contain an alternative nine-nucleotide regulatory sequence element, the **dehydration response element** (**DRE**) which is recognized by an alternative set of proteins regulating transcription. Thus the genes that are regulated by osmotic stresses appear to be regulated either by signal transduction pathways mediated by the action of ABA (**ABA-dependent genes**), or by an **ABA-independent**, osmotic stress–responsive signal transduction pathway.

At least two signaling pathways have been implicated in the regulation of gene expression in an ABA-independent manner (Figure 25.9). Transacting *transcription factors* (called DREB1 and DREB2) that bind to the DRE elements in the promoters of osmotic stress–responsive genes are apparently activated by an ABA-independent signaling cascade. Other ABA-independent, osmotic stress–respon-

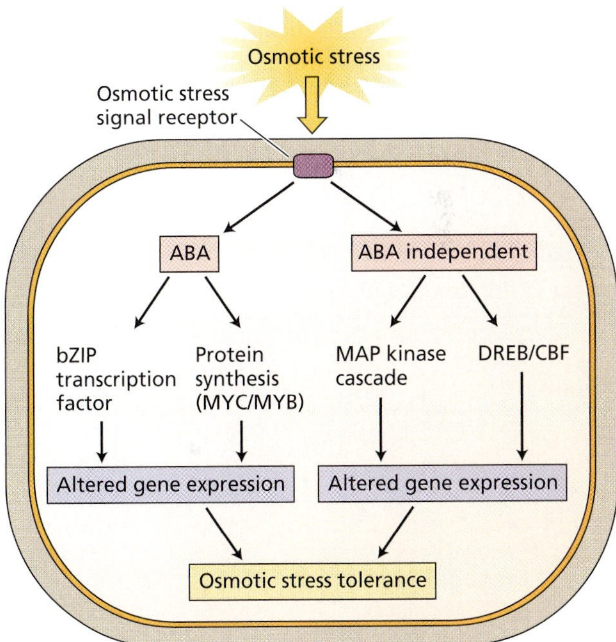

FIGURE 25.9 Signal transduction pathways for osmotic stress in plant cells. Osmotic stress is perceived by an as yet unknown receptor in the plasma membrane activating ABA-independent and an ABA-dependent signal transduction pathways. Protein synthesis participates in one of the ABA-dependent pathways involving MYC/MYB. The bZIP ABA-dependent pathway involves recognition of ABA-responsive elements in gene promoters. Two ABA-independent pathways, one involving the MAP kinase signaling cascade and the other involving DREBP/CBF-related transcription factors have also been demonstrated. (After Shinozaki and Yamaguchi-Shinozaki, 2000.)

sive genes appear to be directly controlled by the so-called MAP kinase signaling cascade of protein kinases (discussed in detail in Chapter 14 on the web site). Other changes in gene expression appear to be mediated via other mechanisms not involving DREBs.

This complexity and "cross-talk" found in signaling cascades, exemplified here by both ABA-dependent and ABA-independent pathways, is typical of eukaryotic signaling. Such complexity reflects the wealth of interaction between gene expression and the physiological processes mediating adaptation to osmotic stress.

HEAT STRESS AND HEAT SHOCK

Most tissues of higher plants are unable to survive extended exposure to temperatures above 45°C. Nongrowing cells or dehydrated tissues (e.g., seeds and pollen) can survive much higher temperatures than hydrated, vegetative, growing cells (Table 25.3). Actively growing tissues rarely survive temperatures above 45°C, but dry seeds can endure 120°C, and pollen grains of some species can endure 70°C. In general, only single-celled eukaryotes can complete their life cycle at temperatures above 50°C, and only prokaryotes can divide and grow above 60°C.

Periodic brief exposure to sublethal heat stresses often induces tolerance to otherwise lethal temperatures, a phenomenon referred to as **induced thermotolerance**. The mechanisms mediating induced thermotolerance will be discussed later in the chapter. As mentioned earlier, water and temperature stress are interrelated; shoots of most C_3 and C_4 plants with access to abundant water supply are maintained below 45°C by evaporative cooling; if water becomes limiting, evaporative cooling decreases and tissue temperatures increase. Emerging seedlings in moist soil may constitute an exception to this general rule. These seedlings may be exposed to greater heat stress than those in drier soils because wet, bare soil is typically darker and absorbs more solar radiation than drier soil.

TABLE 25.3
Heat-killing temperatures for plants

Plant	Heat-killing temperature (C°)	Time of exposure
Nicotiana rustica (wild tobacco)	49–51	10 min
Cucurbita pepo (squash)	49–51	10 min
Zea mays (corn)	49–51	10 min
Brassica napus (rape)	49–51	10 min
Citrus aurantium (sour orange)	50.5	15–30 min
Opuntia (cactus)	>65	—
Sempervivum arachnoideum (succulent)	57–61	—
Potato leaves	42.5	1 hour
Pine and spruce seedlings	54–55	5 min
Medicago seeds (alfalfa)	120	30 min
Grape (ripe fruit)	63	—
Tomato fruit	45	—
Red pine pollen	70	1 hour
Various mosses		
Hydrated	42–51	—
Dehydrated	85–110	—

Source: After Table 11.2 in Levitt 1980.

High Leaf Temperature and Water Deficit Lead to Heat Stress

Many CAM, succulent higher plants, such as *Opuntia* and *Sempervivum*, are adapted to high temperatures and can tolerate tissue temperatures of 60 to 65°C under conditions of intense solar radiation in summer (see Table 25.3). Because CAM plants keep their stomata closed during the day, they cannot cool by transpiration. Instead, they dissipate the heat from incident solar radiation by re-emission of long-wave (infrared) radiation and loss of heat by conduction and convection (see Chapter 9).

On the other hand, typical, nonirrigated C_3 and C_4 plants rely on transpirational cooling to lower leaf temperature. In these plants, leaf temperature can readily rise 4 to 5°C above ambient air temperature in bright sunlight near midday, when soil water deficit causes partial stomatal closure or when high relative humidity reduces the potential for evaporative cooling. The physiological consequences of these increases in tissue temperature are discussed in the next section.

Increases in leaf temperature during the day can be pronounced in plants from arid and semiarid regions experiencing drought and high irradiance from sunshine. Heat stress is also a potential danger in greenhouses, where low air speed and high humidity decrease the rate of leaf cooling. A moderate degree of heat stress slows growth of the whole plant. Some irrigated crops, such as cotton, use transpirational cooling to dissipate heat. In irrigated cotton, enhanced transpirational cooling is associated with higher agronomic yields (see **Web Topic 25.1**).

At High Temperatures, Photosynthesis Is Inhibited before Respiration

Both photosynthesis and respiration are inhibited at high temperatures, but as temperature increases, photosynthetic rates drop before respiratory rates (Figure 25.10A and B). The temperature at which the amount of CO_2 fixed by photosynthesis, equals the amount of CO_2 released by respiration, in a given time interval is called the **temperature compensation point**.

At temperatures above the temperature compensation point, photosynthesis cannot replace the carbon used as a substrate for respiration. As a result, carbohydrate reserves decline, and fruits and vegetables lose sweetness. This imbalance between photosynthesis and respiration is one of the main reasons for the deleterious effects of high temperatures.

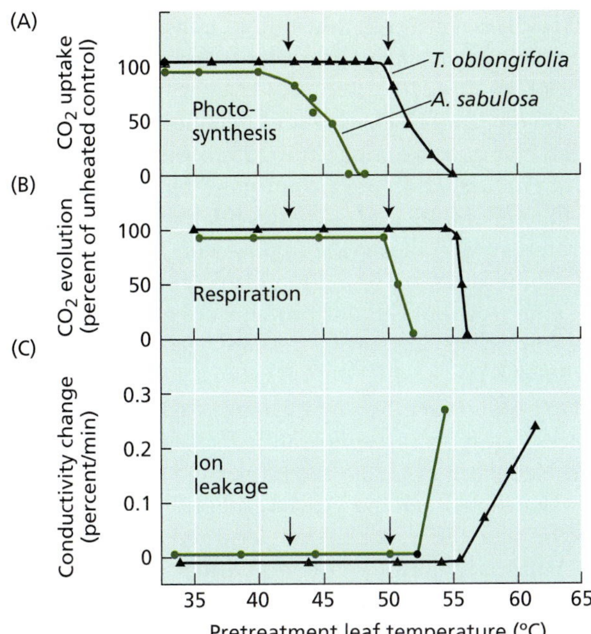

FIGURE 25.10 Response of frosted orache (*Atriplex sabulosa*) and Arizona honeysweet (*Tidestromia oblongifolia*) to heat stress. Photosynthesis (A) and respiration (B) were measured on attached leaves, and ion leakage (C) was measured in leaf slices submerged in water. At the beginning of the experiment, control rates were measured at a noninjurious 30°C. Attached leaves were then exposed to the indicated temperatures for 15 minutes and returned to the initial control conditions before the rates were recorded. Arrows indicate the temperature thresholds for inhibition of photosynthesis in each of the two species. Photosynthesis, respiration, and membrane permeability were all more sensitive to heat damage in *A. sabulosa* than in *T. oblongifolia*. In both species, however, photosynthesis was more sensitive to heat stress than either of the other two processes, and photosynthesis was completely inhibited at temperatures that were noninjurious to respiration. (From Björkman et al. 1980.)

In the same plant the temperature compensation point is usually lower for shade leaves than for sun leaves that are exposed to light (and heat). Enhanced respiration rates relative to photosynthesis at high temperatures are more detrimental in C_3 plants than in C_4 or CAM plants because the rates of both dark respiration and photorespiration are increased in C_3 plants at higher temperatures (see Chapter 8).

Plants Adapted to Cool Temperatures Acclimate Poorly to High Temperatures

The extent to which plants that are genetically adapted to a given temperature range can acclimate to a contrasting temperature range is illustrated by a comparison of the responses of two C_4 species: *Atriplex sabulosa* (frosted orache, family Chenopodiaceae) and *Tidestromia oblongifolia* (Arizona honeysweet, family Amaranthaceae).

A. sabulosa is native to the cool climate of coastal northern California, and *T. oblongifolia* is native to the very hot climate of Death Valley, California, where it grows in a temperature regime that is lethal for most plant species. When these species were grown in a controlled environment and their growth rates were recorded as a function of temperature, *T. oblongifolia* barely grew at 16°C, while *A. sabulosa* was at 75% of its maximum growth rate. By contrast, the growth rate of *A. sabulosa* began to decline between 25 and 30°C, and growth ceased at 45°C, the temperature at which *T. oblongifolia* growth showed a maximum (see Figure 25.10A) (Björkman et al. 1980). Clearly, neither species could acclimate to the temperature range of the other.

High Temperature Reduces Membrane Stability

The stability of various cellular membranes is important during high-temperature stress, just as it is during chilling and freezing. Excessive fluidity of membrane lipids at high temperatures is correlated with loss of physiological function. In oleander (*Nerium oleander*), acclimation to high temperatures is associated with a greater degree of saturation of fatty acids in membrane lipids, which makes the membranes less fluid (Raison et al. 1982).

At high temperatures there is a decrease in the strength of hydrogen bonds and electrostatic interactions between polar groups of proteins within the aqueous phase of the membrane. High temperatures thus modify membrane composition and structure and can cause leakage of ions (see Figure 25.10C). Membrane disruption also causes the inhibition of processes such as photosynthesis and respiration that depend on the activity of membrane-associated electron carriers and enzymes.

Photosynthesis is especially sensitive to high temperature (see Chapter 9). In their study of *Atriplex* and *Tidestromia*, O. Björkman and colleagues (1980) found that electron transport in photosystem II was more sensitive to high temperature in the cold-adapted *A. sabulosa* than in the heat-adapted *T. oblongifolia*. In these plants the enzymes ribulose-1,5-bisphosphate carboxylase, NADP:glyceraldehyde-3-phosphate dehydrogenase, and phosphoenolpyruvate carboxylase were less stable at high temperatures in *A. sabulosa* than in *T. oblongifolia*.

However, the temperatures at which these enzymes began to denature and lose activity were distinctly higher than the temperatures at which photosynthesis began to decline. These results suggest that early stages of heat injury to photosynthesis are more directly related to changes in membrane properties and to uncoupling of the energy transfer mechanisms in chloroplasts than to a general denaturation of proteins.

Several Adaptations Protect Leaves against Excessive Heating

In environments with intense solar radiation and high temperatures, plants avoid excessive heating of their leaves by decreasing their absorption of solar radiation. This adaptation is important in warm, sunny environments in which a transpiring leaf is near its upper limit of temperature tolerance. In these conditions, any further warming arising from decreased evaporation of water or increased energy absorption can damage the leaf.

Both drought resistance and heat resistance depend on the same adaptations: reflective leaf hairs and leaf waxes; leaf rolling and vertical leaf orientation; and growth of small, highly dissected leaves to minimize the boundary layer thickness and thus maximize convective and conductive heat loss (see Chapters 4 and 9). Some desert shrubs—for example, white brittlebush (*Encelia farinosa*, family Compositae)—have dimorphic leaves to avoid excessive heating: Green, nearly hairless leaves found in the winter are replaced by white, pubescent leaves in the summer.

At Higher Temperatures, Plants Produce Heat Shock Proteins

In response to sudden, 5 to 10°C rises in temperature, plants produce a unique set of proteins referred to as **heat shock proteins** (**HSPs**). Most HSPs function to help cells withstand heat stress by acting as molecular chaperones. Heat stress causes many cell proteins that function as enzymes or structural components to become unfolded or misfolded, thereby leading to loss of proper enzyme structure and activity.

Such misfolded proteins often aggregate and precipitate, creating serious problems within the cell. HSPs act as molecular chaperones and serve to attain a proper folding of misfolded, aggregated proteins and to prevent misfolding of proteins. This facilitates proper cell functioning at elevated, stressful temperatures.

Heat shock proteins were discovered in the fruit fly (*Drosophila melanogaster*) and have since been identified in other animals, and in humans, as well as in plants, fungi, and microorganisms. For example, when soybean seedlings are suddenly shifted from 25 to 40°C (just below the lethal temperature), synthesis of the set of mRNAs and proteins commonly found in the cell is suppressed, while transcription and translation of a set of 30 to 50 other proteins (HSPs) is enhanced. New mRNA transcripts for HSPs can be detected 3 to 5 minutes after heat shock (Sachs and Ho 1986).

Although plant HSPs were first identified in response to sudden changes in temperature (25 to 40°C) that rarely occur in nature, HSPs are also induced by more gradual rises in temperature that are representative of the natural environment, and they occur in plants under field conditions. Some HSPs are found in normal, unstressed cells, and some essential cellular proteins are homologous to HSPs but do not increase in response to thermal stress (Vierling 1991).

Plants and most other organisms make HSPs of different sizes in response to temperature increases (Table 25.4). The molecular masses of the HSPs range from 15 to 104 kDa (kilodaltons), and they can be grouped into five classes based on size. Different HSPs are localized to the nucleus, mitochondria, chloroplasts, endoplasmic reticulum, and cytosol. Members of the HSP60, HSP70, HSP90, and HSP100 groups act as molecular chaperones, involving ATP-dependent stabilization and folding of proteins, and the assembly of oligomeric proteins. Some HSPs assist in polypeptide transport across membranes into cellular compartments. HSP90s are associated with hormone receptors in animal cells and may be required for their activation, but there is no comparable information for plants.

Low-molecular-weight (15–30 kDa) HSPs are more abundant in higher plants than in other organisms. Whereas plants contain five to six classes of low-molecular-weight HSPs, other eukaryotes show only one class (Buchanan et al. 2000). The different classes of 15–30 kDa molecular-weight HSPs (smHSPs) in plants are distributed in the cytosol, chloroplasts, ER and mitochondria. The function of these small HSPs is not understood.

Cells that have been induced to synthesize HSPs show improved thermal tolerance and can tolerate exposure to temperatures that are otherwise lethal. Some of the HSPs are not unique to high-temperature stress. They are also induced by widely different environmental stresses or conditions, including water deficit, ABA treatment, wounding, low temperature, and salinity. Thus, cells previously

TABLE 25.4
The five classes of heat shock proteins found in plants

HSP class	Size (kDa)	Examples (Arabidopsis / prokaryotic)	Cellular location
HSP100	100–114	AtHSP101 / ClpB, ClpA/C	Cytosol, mitochondria, chloroplasts
HSP90	80–94	AtHSP90 / HtpG	Cytosol, endoplasmic reticulum
HSP70	69–71	AtHSP70 / DnaK	Cytosol/nucleus, mitochondria, chloroplasts
HSP60	57–60	AtTCP-1 / GroEL, GroES	Mitochondria, chloroplasts
smHSP	15–30	Various AtHSP22, AtHSP20, AtHSP18.2, AtHSP17.6 / IBPA/B	Cytosol, mitochondria, chloroplasts, endoplasmic reticulum

Source: After Boston et al. 1996.

exposed to one stress may gain cross-protection against another stress. Such is the case with tomato fruits, in which heat shock (48 hours at 38°C) has been observed to promote HSP accumulation and to protect cells for 21 days from chilling at 2°C.

A Transcription Factor Mediates HSP Accumulation in Response to Heat Shock

All cells seem to contain molecular chaperones that are constitutively expressed and function like HSPs. These chaperones are called **heat shock cognate proteins**. However, when cells are subjected to a stressful, but nonlethal heat episode, the synthesis of HSPs dramatically increases while the continuing translation of other proteins is dramatically lowered or ceases. This heat shock response appears to be mediated by a specific transcription factor (**HSF**) that acts on the transcription of HSP mRNAs.

In the absence of heat stress, HSF exists as monomers that are incapable of binding to DNA and directing transcription (Figure 25.11). Stress causes HSF monomers to associate into trimers that are then able to bind to specific sequence elements in DNA referred to as heat shock elements (HSEs). Once bound to the HSE, the trimeric HSF is phosphorylated and promotes the transcription of HSP mRNAs. HSP70 subsequently binds to HSF, leading to the dissociation of the HSF/HSE complex, and the HSF is subsequently recycled to the monomeric HSF form. Thus, by the action of HSF, HSPs accumulate until they become abundant enough to bind to HSF, leading to the cessation of HSP mRNA production.

HSPs Mediate Thermotolerance

Conditions that induce thermal tolerance in plants closely match those that induce the accumulation of HSPs, but that correlation alone does not prove that HSPs play an essential role in acclimation to heat stress. More conclusive experiments show that expression of an activated HSF induces constitutive synthesis of HSPs and increases the thermotolerance of *Arabidopsis*. Studies with *Arabidopsis* plants containing an antisense DNA sequence that reduces HSP70 synthesis showed that the high-temperature extreme at which the plants could survive was reduced by 2°C compared with controls, although the mutant plants grew normally at optimum temperatures (Lee and Schoeffl 1996).

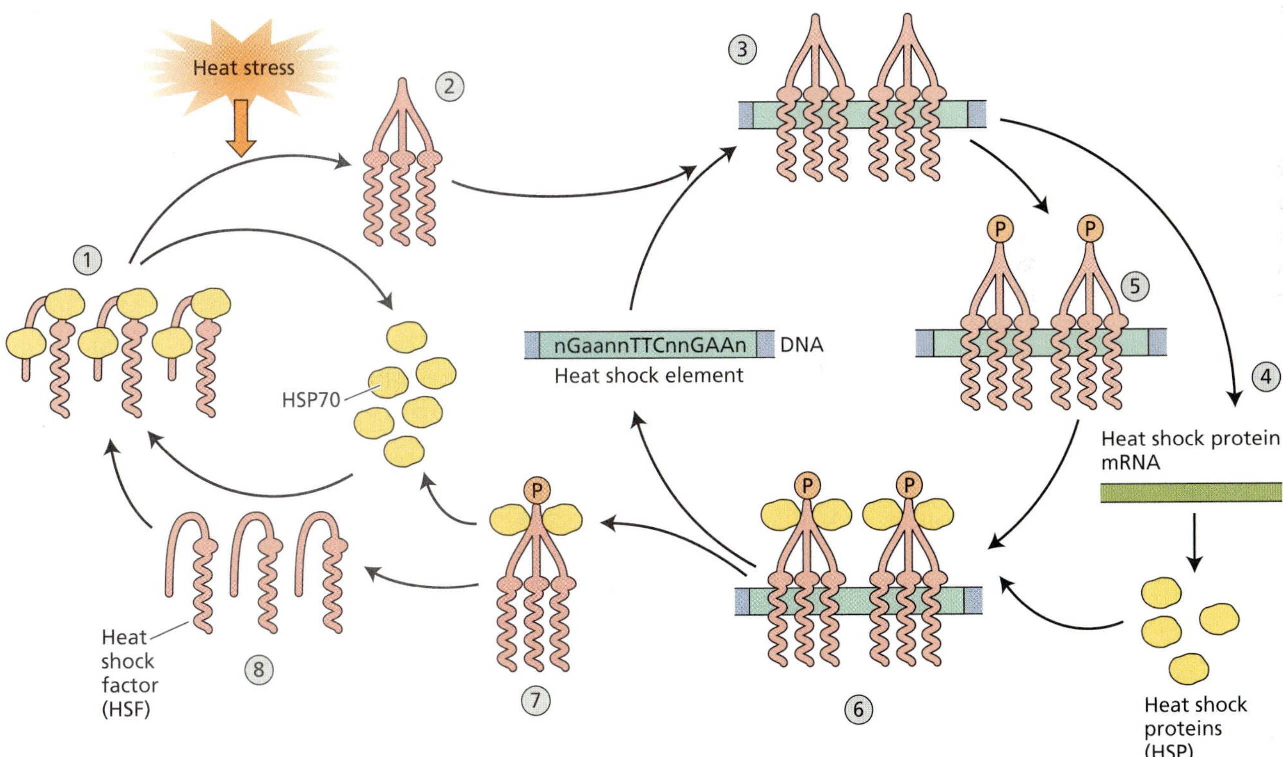

FIGURE 25.11 The heat shock factor (HSF) cycle activates the synthesis of heat shock protein mRNAs. In nonstressed cells, HSF normally exists in a monomeric state (1) associated with HSP70 proteins. Upon the onset of an episode of heat stress, HSP70 dissociates from HSF which subsequently trimerizes (2). Active trimers bind to heat shock elements (HSE) in the promoter of heat shock protein (HSP) genes (3), and activate the transcription of HSP mRNAs leading to the translation of HSPs among which are HSP70 (4). The HSF trimers associated with the HSE are phosphorylated (5) facilitating the binding of HSP70 to the phosphorylated trimers (6). The HSP70 trimer complex (7) dissociates from the HSE and disassembles and dephosphorylates into HSF monomers (8), which subsequently bind HSP reforming the resting HSP70/HSF complex. (After Bray et al. 2000.)

Presumably failure to synthesize the entire range of HSPs that are usually induced in the plant would lead to a much more dramatic loss of thermotolerance. Other studies with both *Arabidopsis* mutants (Hong and Vierling 2000) and transgenic plants (Queitsch et al. 2000) demonstrate that at least HSP101 is a critical component of both induced and constitutive thermotolerance in plants.

Adaptation to Heat Stress Is Mediated by Cytosolic Calcium

Enzymes participating in metabolic pathways can have different temperature responses, and such differential thermostability may affect specific steps in metabolism before HSPs can restore activity by their molecular chaperone capacity. Heat stress can therefore cause changes in metabolism leading to the accumulation of some metabolites and the reduction of others. Such changes can dramatically alter the function of metabolic pathways and lead to imbalances that can be difficult to correct.

In addition, heat stress can alter the rate of metabolic reactions that consume or produce protons, and it can affect the activity of proton-pumping ATPases that pump protons from the cytosol into the apoplast or vacuoles (see Chapter 6). This might lead to an acidification of the cytosol, which could cause additional metabolic perturbations during stress. Cells can have metabolic acclimation mechanisms that ameliorate these effects of heat stress on metabolism.

One of the metabolic acclimations to heat stress is the accumulation of the nonprotein amino acid γ-aminobutyric acid (GABA). During episodes of heat stress, GABA accumulates to levels six- to tenfold higher than in unstressed plants. GABA is synthesized from the amino acid L-glutamate, in a single reaction catalyzed by the enzyme glutamate decarboxylase (GAD). GAD is one of several enzymes whose activity is modulated by the calcium-activated, regulatory protein *calmodulin* (for details on the mode of action of calmodulin, see Chapter 14 on the web site).

Calcium-activated calmodulin activates GAD (Figure 25.12) and increases the biosynthesis rate of GABA (Snedden et al. 1995). In transgenic plants expressing the calcium-sensing protein aequorin, it has been shown that

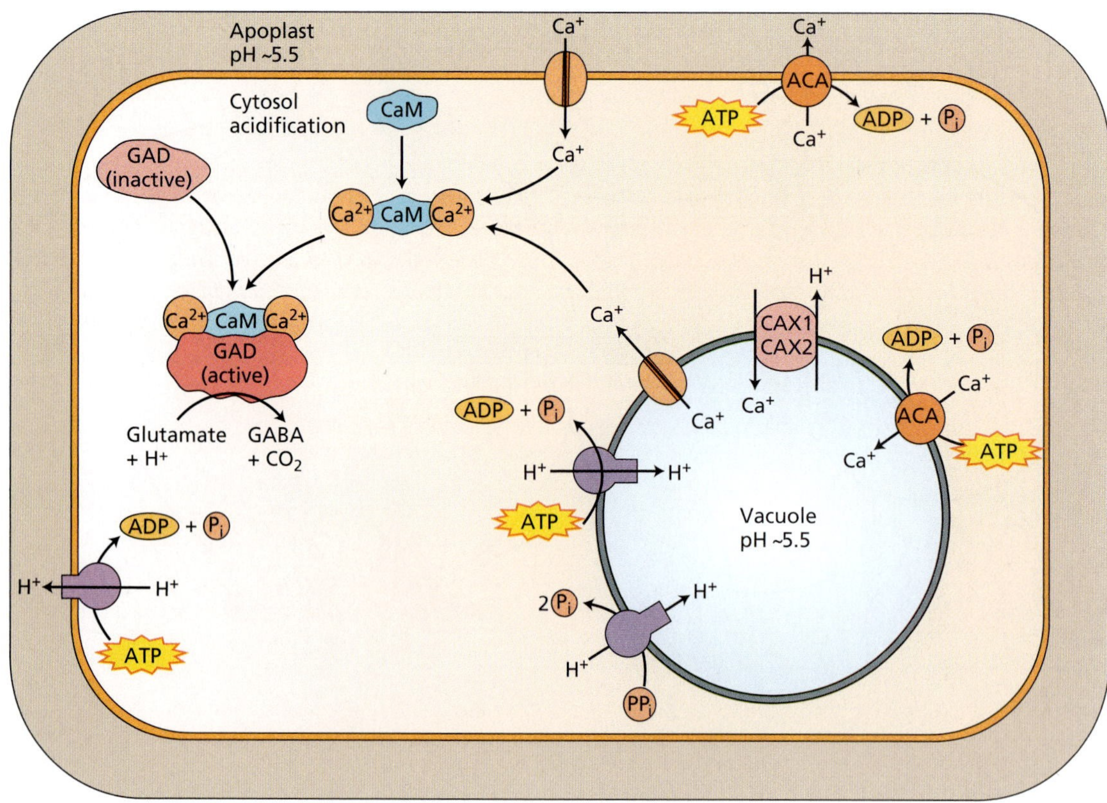

FIGURE 25.12 Heat stress causes a reduction in cytosolic pH from the normal slightly alkaline value, probably by inhibiting proton-pumping ATPases and pyrophosphatases that pump protons across the plasma membrane or into the vacuole. Additionally, heat stress effects a change in calcium homeostasis inside the cell by affecting the influx of calcium into the cytosol through either plasma membrane or vacuolar calcium channels, or by action on efflux ATPases or proton cotransporters. This increase in cytosolic calcium leads to the activation of calmodulin (CaM), which binds to glutamate decarboxylase (GAD) converting it from the inactive to the active form. Glutamate conversion to γ-aminobutyric acid (GABA) is then accomplished consuming protons in the process and mediating an increase in cytosolic pH. CAX1 and CAX2 are transport proteins, ACA: Ca^{2+} ATPase.

high-temperature stress increases cytosolic levels of calcium, and that these increases lead to the calmodulin-mediated activation of GAD and the high-temperature induced accumulation of GABA.

Although GABA is an important signaling molecule in mammalian brain tissue, there is no evidence that it functions as a signaling molecule in plants. Possible functions of GABA in heat stress resistance are under investigation.

CHILLING AND FREEZING

Chilling temperatures are too low for normal growth but not low enough for ice to form. Typically, tropical and subtropical species are susceptible to chilling injury. Among crops, maize, *Phaseolus* bean, rice, tomato, cucumber, sweet potato, and cotton are chilling sensitive. *Passiflora, Coleus,* and *Gloxinia* are examples of susceptible ornamentals.

When plants growing at relatively warm temperatures (25 to 35°C) are cooled to 10 to 15°C, **chilling injury** occurs: Growth is slowed, discoloration or lesions appear on leaves, and the foliage looks soggy, as if soaked in water for a long time. If roots are chilled, the plants may wilt.

Species that are generally sensitive to chilling can show appreciable variation in their response to chilling temperatures. Genetic adaptation to the colder temperatures associated with high altitude improves chilling resistance (Figure 25.13). In addition, resistance often increases if plants are first hardened (acclimated) by exposure to cool, but noninjurious, temperatures. Chilling damage thus can be minimized if exposure is slow and gradual. Sudden exposure to temperatures near 0°C, called *cold shock*, greatly increases the chances of injury.

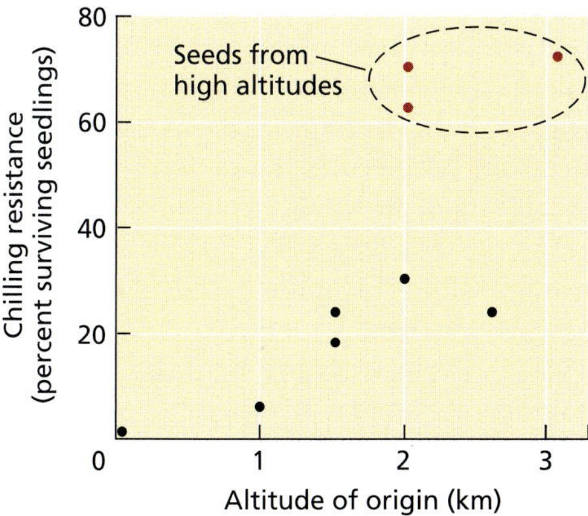

FIGURE 25.13 Survival at low temperature of seedlings of different populations of tomato collected from different altitudes in South America. Seed was collected from wild tomato (*Lycopersicon hirsutum*) and grown in the same greenhouse at 18 to 25°C. All seedlings were then chilled for 7 days at 0°C and then kept for 7 days in a warm growth room, after which the number of survivors was counted. Seedlings from seed collected from high altitudes showed greater resistance to chilling (cold shock) than those from seed collected from lower altitudes. (From Patterson et al. 1978.)

Freezing injury, on the other hand, occurs at temperatures below the freezing point of water. Full induction of tolerance to freezing, as with chilling, requires a period of acclimation at cold temperatures.

In the discussion that follows we will examine how chilling injury alters membrane properties, how ice crystals damage cells and tissues, and how ABA, gene expression, and protein synthesis mediate acclimation to freezing.

Membrane Properties Change in Response to Chilling Injury

Leaves from plants injured by chilling show inhibition of photosynthesis, slower carbohydrate translocation, lower respiration rates, inhibition of protein synthesis, and increased degradation of existing proteins. All of these responses appear to depend on a common primary mechanism involving loss of membrane function during chilling.

For instance, solutes leak from the leaves of chilling-sensitive *Passiflora maliformis* (conch apple) floated on water at 0°C, but not from those of chilling-resistant *Passiflora caerulea* (passionflower). Loss of solutes to the water reflects damage to the plasma membrane and possibly also to the tonoplast. In turn, inhibition of photosynthesis and of respiration reflects injury to chloroplast and mitochondrial membranes.

Why are membranes affected by chilling? Plant membranes consist of a lipid bilayer interspersed with proteins and sterols (see Chapters 1 and 11). The physical properties of the lipids greatly influence the activities of the integral membrane proteins, including H^+-ATPases, carriers, and channel-forming proteins that regulate the transport of ions and other solutes (see Chapter 6), as well as the transport of enzymes on which metabolism depends.

In chilling-sensitive plants, the lipids in the bilayer have a high percentage of saturated fatty acid chains, and membranes with this composition tend to solidify into a semicrystalline state at a temperature well above 0°C. Keep in mind that saturated fatty acids that have no double bonds and lipids containing *trans*-monounsaturated fatty acids solidify at higher temperatures than do membranes composed of lipids that contain unsaturated fatty acids.

As the membranes become less fluid, their protein components can no longer function normally. The result is inhibition of H^+-ATPase activity, of solute transport into and out of cells, of energy transduction (see Chapters 7 and 11), and of enzyme-dependent metabolism. In addition, chilling-sensitive leaves exposed to high photon fluxes and chilling temperatures are photoinhibited (see Chapter 7), causing acute damage to the photosynthetic machinery.

Membrane lipids from chilling-resistant plants often have a greater proportion of unsaturated fatty acids than those from chilling-sensitive plants (Table 25.5), and during acclimation to cool temperatures the activity of desaturase enzymes increases and the proportion of unsaturated lipids rises (Williams et al. 1988; Palta et al. 1993). This modification lowers the temperature at which the mem-

TABLE 25.5
Fatty acid composition of mitochondria isolated from chilling-resistant and chilling-sensitive species

	Percent weight of total fatty acid content					
	Chilling-resistant species			Chilling-sensitive species		
Major fatty acids[a]	Cauliflower bud	Turnip root	Pea shoot	Bean shoot	Sweet potato	Maize shoot
Palmitic (16:0)	21.3	19.0	12.8	24.0	24.9	28.3
Stearic (18:0)	1.9	1.1	2.9	2.2	2.6	1.6
Oleic (18:0)	7.0	12.2	3.1	3.8	0.6	4.6
Linoleic (18:2)	16.4	20.6	61.9	43.6	50.8	54.6
Linolenic (18:3)	49.4	44.9	13.2	24.3	10.6	6.8
Ratio of unsaturated to saturated fatty acids	3.2	3.9	3.8	2.8	1.7	2.1

[a] Shown in parentheses are the number of carbon atoms in the fatty acid chain and the number of double bonds.
Source: After Lyons et al. 1964.

brane lipids begin a gradual phase change from fluid to semicrystalline and allows membranes to remain fluid at lower temperatures. Thus, desaturation of fatty acids provides some protection against damage from chilling.

The importance of membrane lipids to tolerance of low temperatures has been demonstrated by work with mutant and transgenic plants in which the activity of particular enzymes led to a specific change in membrane lipid composition independent of acclimation to low temperature. For example, *Arabidopsis* was transformed with a gene from *Escherichia coli* that raised the proportion of high-melting-point (saturated) membrane lipids. This gene greatly increased the chilling sensitivity of the transformed plants.

Similarly, the *fab1* mutants of *Arabidopsis* have increased levels of saturated fatty acids, particularly 16:0 (see Table 25.5, and Tables 11.3 and 11.4). During a period of 3 to 4 weeks at chilling temperatures, photosynthesis and growth were gradually inhibited, and exposure to chilling temperature eventually destroyed the chloroplasts of this mutant. At nonchilling temperatures, the mutant grew as well as wild-type controls did (Wu et al. 1997). (For additional transformation examples, see **Web Topic 25.2**.)

Ice Crystal Formation and Protoplast Dehydration Kill Cells

The ability to tolerate freezing temperatures under natural conditions varies greatly among tissues. Seeds, other partly dehydrated tissues, and fungal spores can be kept indefinitely at temperatures near absolute zero (0 K, or –273°C), indicating that these very low temperatures are not intrinsically harmful.

Fully hydrated, vegetative cells can also retain viability if they are cooled very quickly to avoid the formation of large, slow-growing ice crystals that would puncture and destroy subcellular structures. Ice crystals that form during very rapid freezing are too small to cause mechanical damage. Conversely, rapid warming of frozen tissue is required to prevent the growth of small ice crystals into crystals of a damaging size, or to prevent loss of water vapor by sublimation, both of which take place at intermediate temperatures (–100 to –10°C).

Under natural conditions, cooling of intact, multicellular plant organs is never fast enough to limit crystal formation in fully hydrated cells to only small, harmless ice crystals. Ice usually forms first within the intercellular spaces, and in the xylem vessels, along which the ice can quickly propagate. This ice formation is not lethal to hardy plants, and the tissue recovers fully if warmed. However, when plants are exposed to freezing temperatures for an extended period, the growth of extracellular ice crystals results in the movement of liquid water from the protoplast to the extracellular ice, causing excessive dehydration (for a detailed description of this process, see **Web Topic 25.3**).

During rapid freezing, the protoplast, including the vacuole, supercools; that is, the cellular water remains liquid even at temperatures several degrees below its theoretical freezing point. Several hundred molecules are needed for an ice crystal to begin forming. The process whereby these hundreds of water molecules start to form a stable ice crystal is called **ice nucleation**, and it strongly depends on the properties of the involved surfaces. Some large polysaccharides and proteins facilitate ice crystal formation and are called ice nucleators.

Some ice nucleation proteins made by bacteria appear to facilitate ice nucleation by aligning water molecules along repeated amino acid domains within the protein. In plant cells, ice crystals begin to grow from endogenous ice nucleators, and the resulting, relatively large intracellular ice crystals cause extensive damage to the cell and are usually lethal.

Limitation of Ice Formation Contributes to Freezing Tolerance

Several specialized plant proteins may help limit the growth of ice crystals by a noncolligative mechanism—that is, an effect that does not depend on the lowering of the freezing point of water by the presence of solutes. These *antifreeze proteins* are induced by cold temperatures, and they bind to the surfaces of ice crystals to prevent or slow further crystal growth.

In rye leaves, antifreeze proteins are localized in the epidermal cells and cells surrounding the intercellular spaces, where they can inhibit the growth of extracellular ice. Plants and animals may use similar mechanisms to limit ice crystals: A cold-inducible gene identified in *Arabidopsis* has DNA homology to a gene that encodes the antifreeze protein in fishes such as winter flounder. Antifreeze proteins are discussed in more detail later in the chapter.

Sugars and some of the cold-induced proteins are suspected to have cryoprotective (*cryo-* = "cold") effects; they stabilize proteins and membranes during dehydration induced by low temperature. In winter wheat, the greater the sucrose concentration, the greater the freezing tolerance. Sucrose predominates among the soluble sugars associated with freezing tolerance that function in a colligative fashion, but in some species raffinose, fructans, sorbitol, or mannitol serves the same function.

During cold acclimation of winter cereals, soluble sugars accumulate in the cell walls, where they may help restrict the growth of ice. A cryoprotective glycoprotein has been isolated from leaves of cold-acclimated cabbage (*Brassica oleracea*). In vitro, the protein protects thylakoids isolated from nonacclimated spinach (*Spinacia oleracea*) against damage from freezing and thawing.

Some Woody Plants Can Acclimate to Very Low Temperatures

When in a dormant state, some woody plants are extremely resistant to low temperatures. Resistance is determined in part by previous acclimation to cold, but genetics plays an important role in determining the degree of tolerance to low temperatures. Native species of *Prunus* (cherry, plum, and other pit fruits) from northern cooler climates in North America are hardier after acclimation than those from milder climates. When the species were tested together in the laboratory, those with a northern geographic distribution showed greater ability to avoid intracellular ice formation, underscoring distinct genetic differences (Burke and Stushnoff 1979).

Under natural conditions, woody species acclimate to cold in two distinct stages (Weiser 1970):

1. In the first stage, hardening is induced in the early autumn by exposure to short days and nonfreezing chilling temperatures, both of which combine to stop growth. A diffusible factor that promotes acclimation (probably ABA) moves in the phloem from leaves to overwintering stems and may be responsible for the changes. During this period, woody species also withdraw water from the xylem vessels, thereby preventing the stem from splitting in response to the expansion of water during later freezing. Cells in this first stage of acclimation can survive temperatures well below 0°C, but they are not fully hardened.
2. In the second stage, direct exposure to freezing is the stimulus; no known translocatable factor can confer the hardening resulting from exposure to freezing. When fully hardened, the cells can tolerate exposure to temperatures of –50 to –100°C.

Resistance to Freezing Temperatures Involves Supercooling and Slow Dehydration

In many species of the hardwood forests of southeastern Canada and the eastern United States, acclimation to freezing involves the suppression of ice crystal formation at temperatures far below the theoretical freezing point (see **Web Topic 25.3** for details). This *deep supercooling* is seen in species such as oak, elm, maple, beech, ash, walnut, hickory, rose, rhododendron, apple, pear, peach, and plum (Burke and Stushnoff 1979). Deep supercooling also takes place in the stem and leaf tissue of tree species such as Engelmann spruce (*Picea engelmannii*) and subalpine fir (*Abies lasiocarpa*) growing in the Rocky Mountains of Colorado.

Resistance to freezing is quickly weakened once growth has resumed in the spring (Becwar et al. 1981). Stem tissues of subalpine fir, which undergo deep supercooling and remain viable to below –35°C in May, lose their ability to suppress ice formation in June and can be killed at –10°C.

Cells can supercool only to about –40°C, at which temperature ice forms spontaneously. Spontaneous ice formation sets the *low-temperature limit* at which many alpine and subarctic species that undergo deep supercooling can survive. It also explains why the altitude of the timberline in mountain ranges is at or near the –40°C minimum isotherm.

The cell protoplast suppresses ice nucleation when undergoing deep supercooling. In addition, the cell wall acts as a barrier both to the growth of ice from the intercellular spaces into the wall, and to the loss of liquid water from the protoplast to the extracellular ice, which is driven by a steep vapor pressure gradient (Wisniewski and Arora 1993).

Many flower buds (e.g., grape, blueberry, peach, azalea, and flowering dogwood) survive the winter by deep supercooling, and serious economic losses, particularly of peach, can result from the decline in freezing tolerance of the flower buds in the spring. The cells then no longer supercool, and ice crystals that form extracellularly in the bud scales draw water from the apical meristem, killing the floral apex by dehydration.

The floral buds of apple and pear, the vegetative buds of all temperate fruit trees, and the living cells in their bark do not supercool, but they resist dehydration during extracellular ice formation. Resistance to cellular dehydration is highly developed in woody species that are subject to average annual temperature minima below –40°C, particularly species found in northern Canada, Alaska, northern Europe, and Asia.

Ice formation starts at –3 to –5°C in the intercellular spaces, where the crystals continue to grow, fed by the gradual withdrawal of water from the protoplast, which remains unfrozen. Resistance to freezing temperatures depends on the capacity of the extracellular spaces to accommodate the volume of growing ice crystals and on the ability of the protoplast to withstand dehydration.

This restriction of ice crystal formation to extracellular spaces, accompanied by gradual protoplast dehydration, may explain why some woody species that are resistant to freezing are also resistant to water deficit during the growing season. For example, species of willow (*Salix*), white birch (*Betula papyrifera*), quaking aspen (*Populus tremuloides*), pin cherry (*Prunus pensylvanica*), chokecherry (*Prunus virginiana*), and lodgepole pine (*Pinus contorta*) tolerate very low temperatures by limiting the formation of ice crystals to the extracellular spaces. However, acquisition of resistance depends on slow cooling and gradual extracellular ice formation and protoplast dehydration. Sudden exposure to very cold temperatures before full acclimation causes intracellular freezing and cell death.

Some Bacteria That Live on Leaf Surfaces Increase Frost Damage

When leaves are cooled to temperatures in the –3 to –5°C range, the formation of ice crystals on the surface (frost) is accelerated by certain bacteria that naturally inhabit the leaf surface, such as *Pseudomonas syringae* and *Erwinia herbicola*, which act as ice nucleators. When artificially inoculated with cultures of these bacteria, leaves of frost-sensitive species freeze at warmer temperatures than leaves that are bacteria free (Lindow et al. 1982). The surface ice quickly spreads to the intercellular spaces within the leaf, leading to cellular dehydration.

Bacterial strains can be genetically modified so that they lose their ice-nucleating characteristics. Such strains have been used commercially in foliar sprays of valuable frost-sensitive crops like strawberry to compete with native bacterial strains and thus minimize the number of potential ice nucleation points.

ABA and Protein Synthesis Are Involved in Acclimation to Freezing

In seedlings of alfalfa (*Medicago sativa* L.), tolerance to freezing at –10°C is greatly improved by previous exposure to cold (4°C) or by treatment with exogenous ABA without exposure to cold. These treatments cause changes in the pattern of newly synthesized proteins that can be resolved on two-dimensional gels. Some of the changes are unique to the particular treatment (cold or ABA), but some of the newly synthesized proteins induced by cold appear to be the same as those induced by ABA (see Chapter 23) or by mild water deficit.

Protein synthesis is necessary for the development of freezing tolerance, and several distinct proteins accumulate during acclimation to cold, as a result of changes in gene expression (Guy 1999). Isolation of the genes for these proteins reveals that several of the proteins that are induced by low temperature share homology with the RAB/LEA/DHN (responsive to ABA, late embryo abundant, and dehydrin, respectively) protein family. As described earlier in the section on gene regulation by osmotic stress, these proteins accumulate in tissues exposed to different stresses, such as osmotic stress. Their functions are under investigation.

ABA appears to have a role in inducing freezing tolerance. Winter wheat, rye, spinach, and *Arabidopsis thaliana* are all cold-tolerant species, and when they are hardened by water shortages, their freezing tolerance also increases. This tolerance to freezing is increased at nonacclimating temperatures by mild water deficit, or at low temperatures, either of which increases endogenous ABA concentrations in leaves.

Plants develop freezing tolerance at nonacclimating temperatures when treated with exogenous ABA. Many of the genes or proteins expressed at low temperatures or under water deficit are also inducible by ABA under nonacclimating conditions. All these findings support a role of ABA in tolerance to freezing.

Mutants of *Arabidopsis* that are insensitive to ABA (*abi1*) or are ABA deficient (*aba1*) are unable to undergo low-temperature acclimation to freezing. Only in *aba1*, however, does exposure to ABA restore the ability to develop freezing tolerance (Mantyla et al. 1995). On the other hand, not all the genes induced by low temperature are ABA dependent, and it is not yet clear whether expression of ABA-induced genes is critical for the full development of freezing tolerance. For instance, research on the tolerance of rye crowns to freezing has found that the lethal temperature for 50% of the crowns (LT_{50}) is –2 to –5°C for controls grown at 25°C, –8°C for ABA-treated crowns, and –28°C after acclimation at 2°C.

Clearly exogenous ABA cannot confer the same freezing acclimation that exposure to low temperatures does. Cell cultures of bromegrass (*Bromus inermis*) show a more dramatic induction of freezing tolerance when treated with ABA: Whereas controls grown at 25°C could survive to –9°C, 7 days of exposure to ABA improved the freezing tolerance to –40°C (Gusta et al. 1996).

Typically, a minimum of several days of exposure to cool temperatures is required for freezing resistance to be induced fully. Potato requires 15 days of exposure to cold. On the other hand, when rewarmed, plants lose their freezing tolerance rapidly, and they can become susceptible to freezing once again in 24 hours.

The need for cool temperatures to induce acclimation to chilling or freezing, and the rapid loss of acclimation upon warming, explain the susceptibility of plants in the southern United States (and similar climatic zones with highly variable winters) to extremes of temperature in the winter months, when air temperature can drop from 20 to 25°C to below 0°C in a few hours.

Numerous Genes Are Induced during Cold Acclimation

Expression of certain genes and synthesis of specific proteins are common to both heat and cold stress, but some aspects of cold-inducible gene expression differ from that produced by heat stress (Thomashow 2001). Whereas during cold episodes the synthesis of "housekeeping" proteins (proteins made in the absence of stress) is not substantially down-regulated, during heat stress housekeeping-protein synthesis is essentially shut down.

On the other hand, the synthesis of several heat shock proteins that can act as molecular chaperones is up-regulated under cold stress in the same way that it is during heat stress. This suggests that protein destabilization accompanies both heat and cold stress and that mechanisms for stabilizing protein structure during both heat and cold episodes are important for survival.

Another important class of proteins whose expression is up-regulated by cold stress is the **antifreeze proteins**. Antifreeze proteins were first discovered in fishes that live in water under the polar ice caps. As discussed earlier, these proteins have the ability to inhibit ice crystal growth in a noncolligative manner, thus preventing freeze damage at intermediate freezing temperatures. Antifreeze proteins confer to aqueous solutions the property of *thermal hysteresis* (transition from liquid to solid is promoted at a lower temperature than is transition from solid to liquid), and thus they are sometimes referred to as **thermal hysteresis proteins** (**THPs**).

Several types of cold-induced, antifreeze proteins have been discovered in cold-acclimated winter-hardy monocots. When the specific genes coding for these proteins were cloned and sequenced, it was found that all antifreeze proteins belong to a class of proteins such as endochitinases and endoglucanases, which are induced upon infection of different pathogens. These proteins, called **pathogenesis-related** (**PR**) **proteins** are thought to protect plants against pathogens. It thus appears that at least in monocots, the dual role of these proteins as antifreeze and pathogenesis-related proteins might protect plant cells against both cold stress and pathogen attack.

Another group of proteins found to be associated with osmotic stress (see the discussion earlier in this chapter) are also up-regulated during cold stress. This group includes proteins involved in the synthesis of *osmolytes*, proteins for membrane stabilization, and the LEA proteins. Because the formation of extracellular ice crystals generates significant osmotic stresses inside cells, coping with freezing stress also requires the means to cope with osmotic stress.

A Transcription Factor Regulates Cold-Induced Gene Expression

More than 100 genes are up-regulated by cold stress. Because cold stress is clearly related to ABA responses and to osmotic stress, not all the genes up-regulated by cold stress necessarily need to be associated with cold tolerance, but most likely many of them are. Many cold stress–induced genes are activated by transcriptional activators called **C-repeat binding factors** (**CBF1**, **CBF2**, **CBF3**; also called DREB1b, DREB1c, and DREB1a, respectively) (Shinozaki and Yamaguchi-Shinozaki 2000).

CBF/DREB1-type transcription factors bind to **CRT/DRE elements** (C-repeat/dehydration-responsive, ABA-independent sequence elements) in gene promoter sequences, which were discussed earlier in the chapter. CBF/DREB1 is involved in the coordinate transcriptional response of numerous cold and osmotic stress–regulated genes, all of which contain the CRT/DRE elements in their promoters. CBF1/DREB1b is unique in that it is specifically induced by cold stress and not by osmotic or salinity stress, whereas the DRE-binding elements of the DREB2 type (discussed earlier in the section on osmotic stresses) are induced only by osmotic and salinity stresses and not by cold.

The expression of CBF1/DREB1b is controlled by a separate transcription factor, called ICE (*i*nducer of CBF *e*xpression). ICE transcription factors do not appear to be induced by cold, and it is presumed that ICE or an associated protein is posttranscriptionally activated, permitting activation of CBF1/DRE1b, but the precise signaling pathway(s) of cold perception, calcium signaling, and the activation of ICE are presently under investigation.

Transgenic plants constitutively expressing CBF1 have more cold–up-regulated gene transcripts than wild-type plants have, suggesting that numerous cold–up-regulated proteins that may be involved in cold acclimation are being produced in the absence of cold in these CBF1 transgenic plants. In addition, CBF1 tansgenic plants are more cold tolerant than control plants.

SALINITY STRESS

Under natural conditions, terrestrial higher plants encounter high concentrations of salts close to the seashore and in estuaries where seawater and freshwater mix or replace each other with the tides. Far inland, natural salt seepage from geologic marine deposits can wash into adjoining areas, rendering them unusable for agriculture. However, a much more extensive problem in agriculture is the accumulation of salts from irrigation water.

Evaporation and transpiration remove pure water (as vapor) from the soil, and this water loss concentrates solutes in the soil. When irrigation water contains a high concentration of solutes and when there is no opportunity to flush out accumulated salts to a drainage system, salts can quickly reach levels that are injurious to salt-sensitive species. It is estimated that about one-third of the irrigated land on Earth is affected by salt.

In this section we discuss how plant function is affected by water and soil salinity, and we examine the processes that assist plants in avoiding salinity stress.

TABLE 25.6
Properties of seawater and of good quality irrigation water

Property	Seawater	Irrigation water
Concentration of ions (m*M*)		
Na^+	457	<2.0
K^+	9.7	<1.0
Ca^{2+}	10	0.5–2.5
Mg^{2+}	56	0.25–1.0
Cl^-	536	<2.0
SO_4^{2-}	28	0.25–2.5
HCO_3^-	2.3	<1.5
Osmotic potential (MPa)	–2.4	–0.039
Total dissolved salts (mg L^{-1} or ppm)	32,000	500

Salt Accumulation in Soils Impairs Plant Function and Soil Structure

In discussing the effects of salts in the soil, we distinguish between high concentrations of Na^+, referred to as **sodicity**, and high concentrations of total salts, referred to as **salinity**. The two concepts are often related, but in some areas Ca^{2+}, Mg^{2+}, and $SO^2{}_4{}^-$, as well as NaCl, can contribute substantially to salinity. The high Na^+ concentration of a sodic soil can not only injure plants directly but also degrade the soil structure, decreasing porosity and water permeability. A sodic clay soil known as caliche is so hard and impermeable that dynamite is sometimes required to dig through it!

In the field, the salinity of soil water or irrigation water is measured in terms of its electrical conductivity or in terms of osmotic potential. Pure water is a very poor conductor of electric current; the conductivity of a water sample is due to the ions dissolved in it. The higher the salt concentration in water, the greater its electrical conductivity and the lower its osmotic potential (higher osmotic pressure) (Table 25.6).

The quality of irrigation water in semiarid and arid regions is often poor. In the United States the salt content of the headwaters of the Colorado River is only 50 mg L^{-1}, but about 2000 km downstream, in southern California, the salt content of the same river reaches about 900 mg L^{-1}, enough to preclude growth of some salt-sensitive crops, such as maize. Water from some wells used for irrigation in Texas may contain as much as 2000 to 3000 mg salt L^{-1}. An annual application of irrigation water totaling 1 m from such wells would add 20 to 30 tons of salts per hectare (8–12 tons per acre) to the soil. These levels of salts are damaging to all but the most resistant crops.

Salinity Depresses Growth and Photosynthesis in Sensitive Species

Plants can be divided into two broad groups on the basis of their response to high concentrations of salts. **Halophytes** are native to saline soils and complete their life cycles in that environment. **Glycophytes** (literally "sweet plants"), or nonhalophytes, are not able to resist salts to the same degree as halophytes. Usually there is a threshold concentration of salt above which glycophytes begin to show signs of growth inhibition, leaf discoloration, and loss of dry weight.

Among crops, maize, onion, citrus, pecan, lettuce, and bean are highly sensitive to salt; cotton and barley are moderately tolerant; and sugar beet and date palms are highly tolerant (Greenway and Munns 1980). Some species that are highly tolerant of salt, such as *Suaeda maritima* (a salt marsh plant) and *Atriplex nummularia* (a saltbush), show growth stimulation at Cl^- concentrations many times greater than the lethal level for sensitive species (Figure 25.14).

Salt Injury Involves Both Osmotic Effects and Specific Ion Effects

Dissolved solutes in the rooting zone generate a low (more negative) osmotic potential that lowers the soil water potential. The general water balance of plants is thus affected because leaves need to develop an even lower water potential to maintain a "downhill" gradient of water potential between the soil and the leaves (see Chapter 4). This effect of dissolved solutes is similar to that of a soil water deficit (as discussed earlier in this chapter), and most plants respond to excessive levels of soil salinity in the same way as described earlier for water deficit.

A major difference between the low-water-potential environments caused by salinity versus soil desiccation is the total amount of water available. During soil desiccation a finite amount of water can be obtained from the soil profile by the plant, causing ever decreasing water potentials. In most saline environments a large (essentially unlimited) amount of water at a constant, low water potential is available.

Of particular importance here is the fact that most plants can adjust osmotically when growing in saline soils. Such adjustment prevents loss of turgor (which would slow cell growth; see Figure 25.1) while generating a lower water potential, but these plants often continue to grow more slowly after this adjustment for an unknown reason that curiously is not related to insufficient turgor (Bressan et al. 1990)

In addition to the plant responses to low water potential, specific ion **toxicity effects** also occur when injurious concentrations of ions—particularly Na^+, Cl^-, or SO_4^{2-}—accumulate in cells. Under nonsaline conditions, the cytosol of higher-plant cells contains 100 to 200 m*M* K^+ and 1 to 10 m*M* Na^+, an ionic environment in which many enzymes function optimally. An abnormally high ratio of Na^+ to K^+ and high concentrations of total salts inactivate enzymes and inhibit protein synthesis. At a high concentration, Na^+ can displace Ca^{2+} from the plasma membrane of cotton root hairs, resulting in a change in plasma membrane permeability that can be detected as leakage of K^+ from the cells (Cramer et al. 1985).

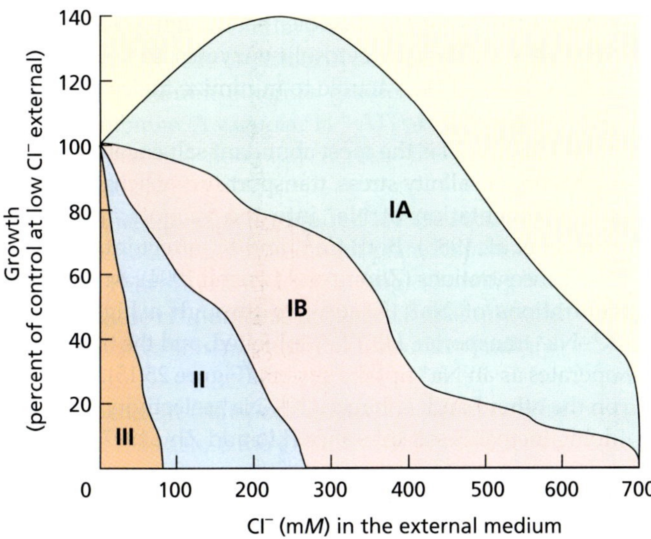

Group IA (halophytes) includes sea blite (*Suaeda maritima*) and salt bush (*Atriplex nummularia*). These species show growth stimulation with Cl^- levels below 400 n*M*.

Group IB (halophytes) includes Townsend's cordgrass (*Spartina x townsendii*) and sugar beet (*Beta vulgaris*).These plants tolerate salt, but their growth is retarded.

Group II (halophytes and nonhalophytes) includes salt-tolerant halophytic grasses that lack salt glands, such as *Festuca rubra* subsp. red fescue (*littoralis*) and *Puccinellia peisonis*, and nonhalophytes, such as cotton (*Gossypium* spp.) and barley (*Hordeum vulgare*). All are inhibited by high salt concentrations. Within this group, tomato (*Lycopersicon esculentum*) is intermediate, and common bean (*Phaseolus vulgaris*) and soybean (*Glycine max*) are sensitive.

The species in **Group III (very salt-sensitive nonhalophytes)** are severely inhibited or killed by low salt concentrations. Included are many fruit trees, such as citrus, avocado, and stone fruit.

FIGURE 25.14 The growth of different species subjected to salinity relative to that of unsalinized controls. The curves dividing the regions are based on data for different species. Plants were grown for 1 to 6 months. (From Greenway and Munns 1980.)

Photosynthesis is inhibited when high concentrations of Na^+ and/or Cl^- accumulate in chloroplasts. Since photosynthetic electron transport appears relatively insensitive to salts, either carbon metabolism or photophosphorylation may be affected. Enzymes extracted from salt-tolerant species are just as sensitive to the presence of NaCl as enzymes from salt-sensitive glycophytes are. Hence the resistance of halophytes to salts is not a consequence of salt-resistant metabolism. Instead, other mechanisms come into play to avoid salt injury, as discussed in the following section.

Plants Use Different Strategies to Avoid Salt Injury

Plants minimize salt injury by excluding salt from meristems, particularly in the shoot, and from leaves that are actively expanding and photosynthesizing. In plants that are salt sensitive, resistance to moderate levels of salinity in the soil depends in part on the ability of the roots to prevent potentially harmful ions from reaching the shoots.

Recall from Chapter 4 that the Casparian strip imposes a restriction to the movements of ions into the xylem. To bypass the Casparian strips, ions need to move from the apoplast to the symplastic pathway across cell membranes. This transition offers salt-resistant plants a mechanism to partially exclude harmful ions.

Sodium ions enter roots passively (by moving down an electrochemical-potential gradient; see Chapter 6), so root cells must use energy to extrude Na^+ actively back to the outside solution. By contrast, Cl^- is excluded by negative electric potential across the cell membrane, and the low permeability of root plasma membranes to this ion. Movement of Na^+ into leaves is further minimized by absorption of Na^+ from the transpiration stream (xylem sap) during its movement from roots to shoots and leaves.

Some salt-resistant plants, such as salt cedar (*Tamarix* sp.) and salt bush (*Atriplex* sp.), do not exclude ions at the root, but instead have salt glands at the surface of the leaves. The ions are transported to these glands, where the salt crystallizes and is no longer harmful. In general, halophytes have a greater capacity than glycophytes for ion accumulation in shoot cells.

Although some plants, such as mangroves, grow in saline environments with abundant water supplies, the ability to acquire that water requires that they make osmotic adjustments to obtain water from the low-water-potential external environment. As discussed earlier in relation to water deficit, plant cells can adjust their water potential (Ψ_w) in response to osmotic stress by lowering their solute potential (Ψ_s). Two intracellular processes contribute to the decrease in Ψ_s: the accumulation of ions in the vacuole and the synthesis of compatible solutes in the cytosol.

As mentioned earlier in the chapter, compatible solutes include glycine betaine, proline, sorbitol, mannitol, pinitol, and sucrose. Specific plant families tend to use one or two of these compounds in preference to others. The amount of carbon used for the synthesis of these organic solutes can be rather large (about 10% of the plant weight). In natural vegetation this diversion of carbon to adjust water potential does not affect survival, but in agricultural crops it can reduce growth and therefore total biomass and harvestable yields.

Many halophytes exhibit a growth optimum at moderate levels of salinity, and this optimum is correlated with the capacity to accumulate ions in the vacuole, where they can contribute to the cell osmotic potential without damaging the salt-sensitive enzymes. To a lesser extent, this process also occurs in more salt-sensitive glycophytes, but the adjustment may be slower.

Buchanan, B. B., Gruissem, W., and Jones, R. eds. (2000) *Biochemistry & Molecular Biology of Plants*. American Society of Plant Physiologists, Rockville, MD.

Burke, M. J., and Stushnoff, C. (1979) Frost hardiness: A discussion of possible molecular causes of injury with particular reference to deep supercooling of water. In *Stress Physiology in Crop Plants*, H. Mussell and R. C. Staples, eds., Wiley, New York, pp. 197–225.

Burssens, S., Himanen, K., van de Cotte, B., Beeckman, T., Van Montagu, M., Inze, D., and Verbruggen, N. (2000) Expression of cell cycle regulatory genes and morphological alterations in response to salt stress in *Arabidopsis thaliana*. *Planta* 211: 632–640.

Cramer, G. R., Läuchli, A., and Polito, V. S. (1985) Displacement of Ca^{2+} by Na^+ from the plasmalemma of root cells. A primary response to salt stress? *Plant Physiol.* 79: 207–211.

Davies, W. J., Wilkinson, S., and Loveys, B. (2002) Stomatal control by chemical signaling and the exploitation of this mechanism to increase water-use efficiency in agriculture. *New Phytol.* 153: 449–460.

Drew, M. C. (1997) Oxygen deficiency and root metabolism: Injury and acclimation under hypoxia and anoxia. *Annu. Rev. Plant Physiol. Plant Mol. Biol.* 48: 223–250.

Drew, M. C., He, C. J., and Morgan P. W. (2000) Programmed cell death and aerenchyma formation in roots. *Trends in Plant Science* 5: 123–127.

Greenway, H., and Munns, R. (1980) Mechanisms of salt tolerance in nonhalophytes. *Annu. Rev. Plant Physiol. Plant Mol. Biol.* 31: 149–190.

Gusta, L. V., Wilen, R. W., and Fu, P. (1996) Low-temperature stress tolerance: The role of abscisic acid, sugars, and heat-stable proteins. *Hort. Sci.* 31: 39–46.

Guy, C. L. (1999) Molecular responses of plants to cold shock and cold acclimation. *J. Mol. Microbiol. Biotechnol.* 1: 231–242.

Hartung, W., Wilkinson, S., and Davies, W. J. (1998) Factors that regulate abscisic acid concentrations at the primary site of action at the guard cell. *J. Exp. Bot.* 49: 361–367.

Hasegawa, P. M., Bressan, R. A., Zhu, J. K., and Bohnert, H. J. (2000) Plant cellular and molecular responses to high salinity. *Annu. Rev. Plant Physiol. Plant Mol. Biol.* 51: 463–499.

Hong, S. W., and Vierling, E. (2000) Mutants of *Arabidopsis thaliana* defective in the acquisition of tolerance to high temperature stress. *Proc. Natl. Acad. Sci. USA* 97: 4392–4397.

Jenks, M. A., Eigenbrode, S., and Lemeiux, B. (In press) Cuticular waxes of *Arabidopsis*. In *The Arabidopsis Book*, C. Somerville and E. Meyerowitz, eds., American Society of Plant Physiologists, Rockville, MD.

Kawasaki, S., Brochert, C., Deyholos, M., Wang, H., Brazille, S., Kawai, K., Galbraith, D. W., and Bohnert, H. J. (2001) Gene expression profiles during the initial phase of salt stress in rice. *Plant Cell* 13: 889–906.

Lee, J. H., and Schoeffl, F. (1996) An Hsp70 antisense gene affects the expression of HSP70/HSC70, the regulation of HSF, and the acquisition of thermotolerance in transgenic Arabidopsis thaliana. *Mol. Gen. Genet.* 252: 11–19.

Levitt, J. (1980) *Responses of plants to environmental stresses*, Vol. 1, 2nd ed. Academic Press, New York.

Lindow, S. E., Arny, D. C., and Upper, C. D. (1982) Bacterial ice nucleation: A factor in frost injury to plants. *Plant Physiol.* 70: 1084–1089.

Liu, J. P., and Zhu, J. K. (1997) An Arabidopsis mutant that requires increased calcium for potassium nutrition and salt tolerance. *Proc. Natl. Acad. Sci. USA* 94: 14960–14964.

Lyons, J. M., Wheaton, T. A., and Pratt, H. K. (1964) Relationship between the physical nature of mitochondrial membranes and chilling sensitivity in plants. *Plant Physiol.* 39: 262–268.

Maggio, A., and Joly, R. J. (1995) Effects of mercuric chloride on the hydraulic conductivity of tomato root systems: Evidence for a channel-mediated water pathway. *Plant Physiol.* 109: 331–335.

Mantyla, E., Lang, V., and Palva, E. T. (1995) Role of abscisic acid in drought-induced freezing tolerance, cold acclimation, and accumulation of LTI78 and RAB18 proteins in Arabidopsis thaliana. *Plant Physiol.* 107: 141–148.

Matthews, M. A., Van Volkenburgh, E., and Boyer, J. S. (1984) Acclimation of leaf growth to low water potentials in sunflower. *Plant Cell Environ.* 7: 199–206.

McCree, K. J., and Richardson, S. G. (1987) Stomatal closure vs. osmotic adjustment: A comparison of stress responses. *Crop Sci.* 27: 539–543.

Niu, X., Bressan, R. A., Hasegawa, P. M., and Pardo, J. M. (1995) Ion homeostasis in NaCl stress environments. *Plant Physiol.* 109: 735–742.

Palta, J. P., Whitaker, B. D., and Weiss, L. S. (1993) Plasma membrane lipids associated with genetic variability in freezing tolerance and cold acclimation of Solanum species. *Plant Physiol.* 103: 793–803.

Patterson, B. D., Paull, R., and Smillie, R. M. (1978) Chilling resistance in Lycopersicon hirsutum Humb. & Bonpl., a wild tomato with a wide altitudinal distribution. *Aust. J. Plant Physiol.* 5: 609–617.

Queitsch, C., Hong, S. W., Vierling, E., and Lindquist, S. (2000) Heat shock protein 101 plays a crucial role in thermotolerance in *Arabidopsis*. *Plant Cell* 12: 479–492.

Quintero, F. J., Blatt, M. R., and Pardo, J. M. (2000) Functional conservation between yeast and plant endosomal Na^+/H^+ antiporters. *FEBS Lett.* 471: 224–228.

Raison, J. K., Pike, C. S., and Berry, J. A. (1982) Growth temperature-induced alterations in the thermotropic properties of Nerium oleander membrane lipids. *Plant Physiol.* 70: 215–218.

Roberts, J. K. M., Hooks, M. A., Miaullis, A. P., Edwards, S., and Webster, C. (1992) Contribution of malate and amino acid metabolism to cytoplasmic pH regulation in hypoxic maize root tips studied using nuclear magnetic resonance spectroscopy. *Plant Physiol.* 98: 480–487.

Sachs, M. M., and Ho, D. T. H. (1986) Alteration of gene expression during environmental stress in plants. *Annu. Rev. Plant Physiol. Plant Mol. Biol.* 37: 363–376.

Sachs, M. M., Subbaiah, C. G., and Saab, I. N. (1996) Anaerobic gene expression and flooding tolerance in maize. *J. Exp. Bot.* 47: 1–15.

Sauter, A., Davies W. J., and Hartung W. (2001) The long distance abscisic acid signal in the droughted plant: The fate of the hormone on its way from the root to the shoot. *J. Exp. Bot.* 52: 1–7.

Shi, H., Ishitani, M., Kim, C., and Zhu, J. K. (2000) The *Arabidopsis thaliana* salt tolerance gene SOS1 encodes a putative Na^+/H^+ antiporter. *Proc. Natl. Acad. Sci. USA* 97: 6896–6901.

Shinozaki, K., and Yamaguchi-Shinozaki, K. (2000) Molecular responses to dehydration and low temperature: Differences and cross-talk between two stress signaling pathways. *Curr. Opinion in Plant Biol.* 3: 217–223.

Snedden, W. A., Arazi, T., Fromm, H., and Shelp, B. J. (1995) Calcium/calmodulin activation of soybean glutamate decarboxylase. *Plant Physiol.* 108: 543–549.

Stockinger, E. J., Gilmour, S. J., and Thomashow, M. F. (1997) *Arabidopsis thaliana* CBF1 encodes an AP2 domain-containing transcriptional activator that binds to the C-repeat-DRE, a cis-acting DNA regulatory element that stimulates transcription in response to low temperature and water deficit. *Proc. Natl. Acad. Sci. USA* 94: 1035–1040.

Sung, F. J. M., and Krieg, D. R. (1979) Relative sensitivity of photosynthetic assimilation and translocation of ^{14}carbon to water stress. *Plant Physiol.* 64: 852–856.

Thomashow, M. (2001) So what's new in the field of plant cold acclimation? Lots! *Plant Physiol.* 125: 89–93.

U. S. Department of Agriculture (1989) *Agricultural Statistics*, U. S. Government Printing Office, Washington DC.

Vierling, E. (1991) The roles of heat shock proteins in plants. *Annu. Rev. Plant Physiol. Plant Mol. Biol.* 42: 579–620.

Weiser, C. J. (1970) Cold resistance and injury in woody plants. *Science* 169: 1269–1278.

Williams, J. P., Khan, M. U., Mitchell, K., and Johnson, G. (1988) The effect of temperature on the level and biosynthesis of unsaturated fatty acids in diacylglycerols of *Brassica napus* leaves. *Plant Physiol.* 87: 904–910.

Wisniewski, M., and Arora, R. (1993) Adaptation and response of fruit trees to freezing temperatures. In *Cytology, Histology and Histochemistry of Fruit Tree Diseases*, A. Biggs, ed., CRC Press, Boca Raton, FL, pp. 299–320.

Wu, J., Lightner, J., Warwick, N., and Browse, J. (1997) Low-temperature damage and subsequent recovery of fab1 mutant *Arabidopsis* exposed to 2°C. *Plant Physiol.* 113: 347–356.

Zhang, J., and Zhang, X. (1994) Can early wilting of old leaves account for much of the ABA accumulation in flooded pea plants? *J. Exp. Bot.* 45: 1335–1342.

Zhong, H., and Läuchli, A. (1994) Spacial distribution of solutes, K, Na, Ca and their deposition rates in the growth zone of primary cotton roots: Effects of NaCl and $CaCl_2$. *Planta* 194: 34–41.

Cell plate Wall-like structure that separates newly divided cells. Formed by the phragmoplast and later becomes the cell wall.

Cell wall matrix Plant cell wall material consisting of hemicelluloses and pectins plus a small amount of structural protein.

Cellobiose A 1→4-linked β-D-glucose disaccharide that makes up cellulose.

Cellulose Linear chains of 1→4-linked β-D-glucose. The repeating unit is cellobiose.

Cellulose microfibril Thin, ribbon-like structure of indeterminate length and variable width composed of 1→4-linked β-D-glucan chains tightly packed in crystalline arrays alternating with less organized amorphous regions. Provides structural integrity to the cell walls of plants and determines the directionality of cell expansion.

Cellulose synthase Enzyme that catalyzes the synthesis of individual 1→4-linked β-D-glucans that make up the cellulose microfibril.

Central zone A central cluster of relatively large, highly vacuolate, slow-dividing cells in shoot apical meristems, comparable to the quiescent center of root meristems.

Cereal grains Seeds of grasses consisting of the diploid embryo, the triploid endosperm, and the fused seed coat–fruit wall.

CF_0F_1 ATPase A multi-protein complex associated with the thylakoid membrane that couples the passage of protons across the membrane to the synthesis of ATP from ADP and phosphate. Similar to F_0F_1ATP synthase in oxidative phosphorylation but much less sensitive to oligomycin.

Channels Transmembrane proteins that function as selective pores for passive transport of ions or water across the membrane.

Chelates Substances such as EGTA that form a complex with divalent cations eliminating their biological activity.

Chelator A carbon compound that can form a noncovalent complex with certain cations facilitating their uptake (e.g., malic acid, citric acid).

Chemical potential The free energy associated with a substance that is available to perform work.

Chemical potential of water *See* water potential.

Chemical-potential gradient A change in the free energy per mole of a substance, measured over a given distance. A substance moving spontaneously does so down its chemical potential gradient.

Chemiosmotic mechanism The mechanism whereby the electrochemical gradient of protons established across the thylakoid membrane by the electron transport process between photosystems II and I is used to drive energy requiring cellular processes such as ATP synthesis. It also operates in mitochondrial respiration and at the cell plasma membrane.

Chilling injury Changes that occur when plants growing at 25 to 35°C are cooled to 10 to 15°C. Includes slowed growth, leaf discoloration and/or lesions. Contrast with freezing injury.

Chilling-sensitive plants Plants that experience a sharp reduction in growth rate at temperatures between 0 and 12°C.

2-chloroethylphosphonic acid *See* ethephon.

Chlorophyll A group of light absorbing green pigments active in photosynthesis.

Chlorophyll *a*/*b* antenna proteins *See* light harvesting complex proteins

Chlorophyllase An enzyme that removes the phytol from chlorophyll as part of the chlorophyll breakdown process.

Chlorophyte Unicellular photosynthetic eukaryote whose chloroplasts contain chlorophyll *a* and *b* (green algae).

Chloroplast The organelle that is the site of photosynthesis in eukaryotic photosynthetic organisms.

Chlorosis The yellowing of older, lower plant leaves characteristic of prolonged nitrogen deficiency.

Cholodny–Went model Early mechanism proposed for tropisms involving stimulation of the bending of the plant axis by lateral transport of auxin in response to a stimulus, such as light, gravity, or touch. The original model has been supported and expanded by recent experimental evidence.

Chromophore A light-absorbing pigment molecule that is usually bound to a protein (an apoprotein).

Chronic photoinhibition Photoinhibition of photosynthentic activity in which both quantum efficiency and the maximum rate of photosynthesis are decreased. Occurs under excess light.

Cinnamic acid A phenylpropanoid derived from the amino acid phenylalanine that is a key intermediate in the biosynthesis of many phenolic compounds.

Circadian rhythm A biological activity that shows a cycle of high-activity and low-activity independent of external stimuli, with a regular periodicity of about 24 hours (L. *circa diem*: about a day).

Citric acid cycle (Krebs cycle, tricarboxylic acid cyle) A cycle of reactions localized in the mitochondrial matrix that catalize the oxidation of pyruvate to CO_2, ATP, and NADH are generated in the oxidation process.

***CKI1* gene** Gene whose overexpression confers cytokinin-independent growth on *Arabidopsis* cells in culture. Encodes a protein similar to bacterial histidine kinases functioning in signal transduction.

Climacteric Marked rise in respiration at the onset of ripening that occurs in all fruits that ripen in response to ethylene, and in the senescence process of detached leaves and flowers.

Clock hypothesis Currently accepted hypothesis of how plants measure night length. Proposes that photoperiodic timekeeping depends on the endogenous oscillator of circadian rhythms. *See* hourglass hypothesis.

CoA *See* coenzyme A.

Coconut milk The liquid endosperm of coconut seeds that contains cytokinins and other nutritional factors. Stimulates the growth of normal stem tissues when added to liquid culture media.

Coenzyme A A coenzyme with an —SH group that serves as an acyl group carrier for many enzymatic reactions.

Cohesion–tension theory A model for sap ascent in the xylem up the stem of the plant, stating that evaporation of water from the leaves at the top of the stem causes a tension (negative hydrostatic pressure) that pulls water up the long water columns in the xylem.

Colchicine A drug that destroys microtubules and blocks cell division.

Coleoptile A modified ensheathing leaf that covers and protects the young primary leaves of a grass seedling as it grows through the soil. Unilateral light perception, especially blue light, by the tip results in asymmetric growth and bending due to unequal auxin distribution in the lighted and shaded sides.

Collenchyma A specialized parenchyma with irregularly thickened, pectin-rich, primary cell walls that function in support in growing parts of a stem or leaf.

Colligative properties Properties of solutions that depend on the number of dissolved particles and not on their chemical characteristics.

Columella root cap The central region of the root cap that contains the statocytes—cells containing large, dense amyloplasts that function in gravity perception during root gravitropism.

Columella stem cells Root cap stem cells that divide to generate a sector of the root cap, the columella.

Combinatorial model Proposal that during the transition from juvenile to adult shoot in maize, a series of independently regulated, overlapping programs (juvenile, adult, and reproductive) modulate the expression of a common set of developmental processes.

Companion cell In angiosperms, a metabolically active cell that is connected to its sieve element by large, branched plasmodesmata and takes over many of the metabolic activities of the sieve element. In source leaves, it functions in the transport of photosynthate into the sieve elements.

Compatible solutes (compatible osmolytes) Organic compounds that are accumulated in the cytosol during osmotic adjustment. Compatible solutes do not inhibit cytosolic enzymes as do high concentrations of ions. Examples of compatible solutes include proline, sorbitol, mannitol, glycine betaine.

Competence The capacity of a particular cell or group of cells to respond in the expected manner when given the appropriate developmental signal.

Complex I A protein complex in the mitochondrial electron transport chain that oxidizes NADH and reduces ubiquinone.

Complex II A protein complex in the mitochondrial electron transport chain that oxidizes succinate and reduces ubiquinone.

Complex III A protein complex in the mitochondrial electron transport chain that oxidizes reduced ubiquinone (ubiquinol) and reduces cytochrome *c*.

Complex IV A protein complex in the mitochondrial electron transport chain that oxidizes reduced cytochrome *c* and reduces O_2 to H_2O.

Complex V *See* F_oF_1-ATP synthase.

Compound (or **mixed**) **fertilizers** Contain two or more mineral nutrients; numbers such as 10–14–10 refer to the effective percentages of nitrogen, phosphorus, and potassium.

Condensed tannins Tannins that are polymers of flavonoid units. Require use of strong acid for hydrolysis.

Constitutive Constantly present or expressed, whether there is demand or not. Refers to the ongoing synthesis of a particular protein. Contrast with inducible.

Constitutive ethylene response mutants Mutants that show the ethylene triple response in the absence of exogenous ethylene. *See ctr* mutant.

COP *See* critical oxygen pressure.

Cork cambium A layer of lateral meristem that develops within mature cells of the cortex and the secondary phloem. Produces the secondary protective layer, the periderm.

Corpus The internal cytohistological zones of the shoot apical meristem: central zone, peripheral zone, rib meristem.

Cortical cytoplasm The outer region or layer of cytoplasm adjacent to the plasma membrane.

Cortical–endodermal stem cells A ring of stem cells that surround the quiescent center and generate the cortical and endodermal layers in roots.

Cotransport The simultaneous transport of two solutes by the same carrier. Usually one solute is moving down its chemical-potential gradient, while the other is moving against its chemical-potential gradient. *See* symport and antiport.

Cotyledons The one or more seed leaves contained in the seed of seed plants. In some seeds they are storage organs supporting early nonphotosynthetic growth of the seedling. In other seeds, they absorb and transmit to the seedling resources stored in the endosperm. *See also* monocot and dicot.

Coumarins A group of phenylpropanoid compounds including the phototoxic furanocoumarins, and other substances responsible for the odor of fresh hay.

Coupled reactions *See* coupling.

Coupling A process by which a chemical reaction releasing free energy is linked to a reaction requiring free energy.

Crassulacean acid metabolism (CAM) A biochemical process for concentrating CO_2 at the carboxylation site of rubisco. Found in the family Crassulaceae (*Crassula*, *Kalanchoe*, *Sedum*) and numerous other families of angiosperms. In CAM, CO_2 uptake and fixation take place at night, and decarboxylation and reduction of the internally released CO_2 occur during the day.

***CRE1* gene** *Arabidopsis* gene that encodes a cytokinin receptor protein, similar to bacterial two-component histidine kinases.

Cristae Folds in the inner mitochondrial membrane that project

the root moves through the soil. Site for the perception of gravity and signaling for the gravitropic response.

Root cap–epidermal stem cells Generate the epidermis of the root cap by anticlinal cell divisions and generate the lateral root cap by periclinal divisions followed by anticlinal divisions.

Root cap stem cells Meristematic cells that give rise to the root cap.

Root hairs Microscopic extensions of root epidermal cells that greatly increase the surface area of the root, thus providing greater capacity for absorption of soil ions and, to a lesser extent, soil water.

Root pressure A positive hydrostatic pressure in the xylem of roots.

Rotenone Specific inhibitor of complex I.

R-type channels A type of gated channel for anions that opens or closes very rapidly in response a voltage change.

Rubisco The acronym for the chloroplast enzyme *ribu*lose *bis*phosphate *c*arboxylase/*o*xygenase. In a carboxylase reaction, rubisco uses atmospheric CO_2 and ribulose-1,5-bisphosphate to form two molecules of 3-phosphoglycerate. It also functions as an oxygenase that incorporates O_2 to ribulose-1,5-bisphosphate to yield one molecule of 3-phosphoglycerate and another of 2-2-phosphoglycolate. The competition between CO_2 and O_2 for ribulose-1,5-bisphosphate limits net CO_2 fixation.

Rubisco activase An enzyme that facilitates the dissociation of sugar bisphosphates-rubisco complexes and, in so doing, activates rubisco.

RUBs A family of small, ubiquitin-related proteins. Proteins linked to RUB are usually activated rather than degraded.

***SAG* genes** Senescence-associated genes whose expression is induced during senescence.

Salicyl hydroxamic acid Specific inhibitor of the alternative oxidase.

Salicylic acid A benzoic acid derivative believed to be an endogenous signal for *SAR*.

Salinity Refers to high concentrations of total salts in the soil. Contrast with sodicity.

Salinization The accumulation of mineral ions, particularly sodium chloride and sodium sulfate, in soil often due to irrigation.

Salt stress The adverse effects of excess minerals on plants

Salt-tolerant plants Plants that can survive or even thrive in high-salt soils. *See also* halophytes.

Sap Fluid content of the xylem, sieve elements of the phloem and the cell vacuole.

Saponins Toxic glycosides of steroids and triterpenes with detergent properties. They may interfere with sterol uptake from the digestive system or disrupt cell membranes.

SAR *See* systemic acquired resistance.

Saturation Refers to a condition under which an increase in a stimulus does not elicit a further increase in a response. A maximum state; not capable of further increase, movement, or inclusion.

***SAUR* genes** In soybean, a group of primary response genes stimulated by auxin within 2 to 5 minutes of treatment.

Sclereid A type of nonelongated sclerenchyma cell commonly found in hard structures such as seed coats.

Sclerenchyma Plant tissue composed of cells, often dead at maturity, with thick, lignified secondary cell walls. It functions in support of nongrowing regions of the plant.

***SCR* gene** *Arabidopsis SCARECROW* gene that controls tissue organization and cell differentiation in the embryo, hypocotyl, primary roots, and secondary roots.

***scr* mutant** *Arabidopsis* mutant in which hypocotyl and inflorescence are agravitropic and lack both endodermis and starch sheath.

Scutellum The single cotyledon of the grass embryo, specialized for nutrient absorption from the endosperm.

Second messenger Intracellular molecule (e. g., cyclic AMP, cyclic GMP, calcium, IP_3, or diacylglycerol) whose production has been elicited by a systemic hormone (the primary messenger) binding to a receptor (often on the plasma membrane). Diffuses intracellularly to the target enzymes or intracellular receptor to produce and amplify the response.

Secondary active transport Active transport that uses energy stored in the proton motive force or other ion gradient, and operates by symport or antiport.

Secondary meristems Meristems that are formed after seed germination and include axillary meristems and lateral meristems. Their activity may be suppressed by active primary meristem.

Secondary metabolites (secondary products) Plant compounds that have no direct role in plant growth and development, but function as defenses against herbivores and microbial infection by microbial pathogens, attractants for pollinators and seed-dispersing animals, and as agents of plant–plant competition.

Secondary products (natural products) *See* secondary metabolites.

Secondary response genes ("late genes") Genes whose expression requires protein synthesis and follows that of primary response genes.

Secondary transport Active transport driven by the proton gradient established by the proton pump.

Secondary wall Cell wall synthesized by nongrowing cells. Often multilayered and containing lignin, it differs in composition and structure from the primary wall. Forms during cell differentiation after cell expansion ceases.

Seed Develops from the ovule after fertilization of the egg, consisting of protective layers enclosing embryo of seed plants. May contain nutritive endosperm tissue separate from the embryo.

Seed coat The outer layer of the seed, derived from the integument of the ovule.

Seed plants Plants in which the embryo is protected and nourished within a seed. The gymnosperms and angiosperms.

Selectively permeability (of a membrane) Membrane property that allows diffusion of some molecules across the membrane to a different extent than other molecules.

Selectivity filter The domain of a channel protein that determines its specificity of transport.

Self-assembly Tendency for large molecules under appropriate conditions to aggregate spontaneously into organized structures.

Senescence An active, genetically controlled, developmental process in which cellular structures and macromolecules are broken down and translocated away from the senescing organ (typically leaves) to actively growing regions that serve as nutrient sinks. Initiated by environmental cues, regulated by hormones.

Sense RNA RNA capable of translation into a functional protein. *See* antisense RNA.

Sensible heat loss Loss of heat from leaf surfaces to the air circulating around the leaf, under conditions in which leaf surface temperature is higher than that of the air.

Sepals Green, leaf-like structures that form the outermost part of a flower. In bud, they enclose and protect other flower parts. *See also* petals, stamens, and carpels.

Sesquiterpene lactones Bitter, antiherbivore, 15-carbon terpenes found in members of the composite family, such as sunflower and sagebrush.

Sesquiterpenes Terpenes having fifteen carbons, three five-carbon isoprene units.

Sex determination The process whereby unisexual flowers are produced by the early selective abortion of either the stamen or the pistil primordia. Genetically regulated, but also influenced by photoperiod and nutritional status. Mediated by GA.

Shade avoidance response A response to shading; includes lengthening of the stem.

Shikimic acid pathway Reactions that convert simple carbohydrate precursors to the aromatic amino acids—phenylalanine, tyrosine, and tryptophan.

Shoot apex Consists of the shoot apical meristem plus the most recently formed leaf primordia (organs derived from the apical meristem).

Shoot apical meristem Meristem at the tip of a shoot. Consists of tunica layers, central zone, peripheral zone, rib meristem.

Short-day plant (SDP) A plant that flowers only in short days (qualitative SDP) or whose flowering is accelerated by short days (quantitative SDP).

Short-distance transport Transport over a distance of only two or three cell diameters. Involved in phloem loading, when sugars move from the mesophyll to the vicinity of the smallest veins of the source leaf, and in phloem unloading, when sugars move from the veins to the sink cells.

Sieve cells The relatively unspecialized sieve elements of gymnosperms. Contrast with sieve tube elements.

Sieve effect The penetration of photosynthetically active light through several layers of cells due to the gaps between chloroplasts permitting the passage of light.

Sieve element loading The movement of sugars into the sieve elements and companion cells of source leaves, where they become more concentrated than in the mesophyll.

Sieve element unloading The process by which imported sugars leave the sieve elements of sinks.

Sieve element–companion cell complex A functional unit consisting of a sieve element and its companion cell.

Sieve elements Cells of the phloem that conduct sugars and other organic materials throughout the plant. Refers to both sieve tube elements (angiosperms) and sieve cells (gymnosperms).

Sieve plates Sieve areas found in angiosperm sieve-tube elements; they have larger pores (sieve-plate pores) than other sieve areas and are generally found in end walls of sieve tube elements.

Sieve tube elements The highly differentiated sieve elements typical of the angiosperms. Contrast with sieve cells.

Sieve tube Tube formed by the joining together of individual sieve tube elements at their end walls.

Signal transduction A sequence of processes by which an extracellular signal (typically light, a hormone or neurotransmitter) interacts with a receptor at the cell surface, causing a change in the level of a second messenger and ultimately a change in cell functioning.

Signal transduction cascade *See* cascade, signal transduction.

Singlet oxygen ($^1O_2$*) An extremely reactive and damaging form of oxygen formed by reaction of excited chlorophyll with molecular oxygen. Causes damage to cellular components especially lipids.

Sink Any organ that imports photosynthate, including nonphotosynthetic organs and organs that do not produce enough photosynthetic products to support their own growth or storage needs, e.g., roots, tubers, developing fruits, and immature leaves. Contrast with source.

Sink activity The rate of uptake of photosynthate per unit weight of sink tissue.

Sink size The total weight of the sink.

Sink strength The ability of a sink organ to mobilize assimilates toward itself. Depends on two factors: sink size and sink activity.

***sln* mutant** In peas, a mutant having abnormally high levels of GA_{20} in the seed due to the impairment of a hydroxylation step in GA deactivation. In barley, an ultra-tall mutant resulting from a recessive allele causing negation of a negative signal transduction factor.

Sodicity Refers to the high concentration of Na^+ in the soil. Contrast with salinity.

Soil hydraulic conductivity A measure of the ease with which water moves through the soil.

About the Book

Editor: Andrew D. Sinauer

Project Editor: Kathaleen Emerson

Editorial Assistant: Sydney Carroll

Production Manager: Christopher Small

Electronic Book Production: Joan Gemme

Illustration Program: Elizabeth Morales

Copy Editor: Stephanie Hiebert

Indexer: Grant Hackett

Photo Researcher: David McIntyre

Book Design: Joan Gemme

Cover Design: Jefferson Johnson

Book Manufacturer: Courier Companies, Inc.